A Collection of
Technical Papers

AIAA
Guidance, Navigation, and Control Conference

Part 1

New Orleans, LA
August 11- 13, 1997

ISBN
1-56347-234-1

Printed in the U.S.A.

Recent Events in Guidance, Navigation and Control
by
Michael Polites, Darin Groll, and Johnny Evers

In Weapons and Missiles

Subscale hardware tests of Lockheed Martin's control system for an AirBorne Laser (ABL) indicate that the system can be precisely targeted to deliver enough energy and power to kill a thin-skinned, liquid-fuel ballistic missile from several hundred kilometers range. In the Boeing, Lockheed Martin, and TRW ABL concept, main telescope optics guide imaging, beacon, and kill laser beams onto the target. A multibeam 1.06 micron laser illuminates the target for imaging by a high-resolution infrared sensor. A beacon laser aimed onto the target nose cone provides a measurement of optical distortion caused by the atmosphere and heated laser optics. Corrections are calculated from this measurement process and applied to the 1.3 micron chemical oxygen-iodine kill laser at a rate of several kilohertz.

THAAD (Theater High Altitude Area Defense) failed the last four attempted intercepts of simulated ballistic missile targets. No single subsystem has been faulted; the latest failure was likely due to a failure in the divert and attitude control system which had functioned properly in previous tests. The ground-based radar and other subsystems operated correctly on the last test flight.

The Arrow 2 anti-missile missile, a $1.6 billion Israel Aircraft Industries project which is partly funded by the United States, had two successful test intercepts. The latest resulted in a direct impact with the target, although the radar fuse failed a few seconds before intercept, which prevented warhead detonation until impact. Arrow 2 has a range of 2000-3000 km and limited transportability over secondary roads.

Raytheon's Patriot PAC2 successfully intercepted Scud missile targets on two different occasions. The latest was over the Kwajalein Missile Range and used the Guidance Enhanced Missile version of the interceptor.

A prototype Standard Missile 2 Block 4A destroyed a Lance target at the White Sands Missile Range. The missile velocity reached Mach 6 and had a closing velocity in excess of Mach 8 during the 54000 ft altitude intercept. The Block 4A missile relies on the Airborne Early Warning Ground Environment Integrated System (AEGIS) radar for approach guidance and for range and range-rate measurements in the terminal phase of the intercept. An on-board Hughes infrared seeker provides angle and angle-rate measurements during the terminal phase. The infrared seeker is covered until after second stage rocket ignition when the missile performs a pitch-to-ditch maneuver to discard the cover. The existing Block 4 RF side-looking fuse will be paired with a new forward-looking fuse, which will be designed during the engineering and manufacturing development phase of the program.

The Ballistic Missile Defense Organization's (BMDO's) Midcourse Space Experiment (MSX) satellite successfully tracked a pair of suborbital rockets over their entire trajectories. Prior to this test, the Army Space and Strategic Defense Command had tested the satellite's tracking capability during the MSX Dedicated Target mission. Here, the satellite assisted in tracking a target complex of 25 deployed objects plus the post-boost vehicle over the Kwajalein Missile Range. A GPS/INS guided Strategic Target System (STARS) booster provided targets.

The Fly-Away Sensor Package (FASP) supports BMDO's Theater Missile Defense Critical Measurements Program. Ejected from the side of a missile, the FASP flies in formation with the booster and payload during reentry. Imagery from infrared and visible-light cameras allows researchers to characterize target and countermeasure behavior in and out of the atmosphere. This data will be used to validate performance models, algorithms, and simulations of high-altitude targets. Playing the video in reverse gives a good approximation of what an interceptor would see as it closes on a target. The cone-shaped FASP contains a small beacon system and thrusters to ensure it remains pointed on the targets during flight.

The Standoff Land Attack Missile Expanded Response (SLAM ER) successfully completed its first flight, having been launched from an F/A-18C aircraft. SLAM ER is a planar winged version of the SLAM with a new GPS/INS, upgraded data link, adaptive terrain following, stop motion aimpoint update, and a terminal Kalman filter. The flight controls and terminal guidance were also redesigned for the bank-to-turn airframe.

The Hughes design submission for the British FMRAAM (Future Medium-Range Air-to-Air Missile) is based on the latest AMRAAM (Advanced Medium-Range Air-to-Air Missile) seeker and warhead, powered by a liquid fuel ramjet developed by Aerospatiale. The Hughes missile will be compatible with all aircraft which carry AMRAAM and, according to the Hughes program manager, will be marketed to all 14 nations currently approved to receive AMRAAM. Hughes' candidate is competing against the METEOR, which is offered by a European consortium consisting of British Aerospace, Matra, GEC-Marconi, Daimler-Benz, Alenia and Saab.

Hughes captured Aim-9X competition with a design based on the Boxoffice 2 airframe, coupled with an infrared seeker developed for Britain's Advanced Short Range Air-to-Air Missile (ASRAAM), and employing the Sidewinder Mk-36 motor and warhead. The seeker has a two-axis gimbaled telescope which optically feeds the seeker array attached to the inner airframe skin. The design uses an innovative low-cost Stirling cryogenic cooler to cool the array. The Stirling cryocooler provides unrestricted active time in flight compared to limits of roughly 1-2 hours with stored gas cooling and also eliminates the orifice clogging problem associated with Sidewinder class seekers. Raytheon's design candidate had a high-resolution "rotate-to-view" seeker which yielded asymmetrical off-boresight capability greater than 90 degrees. Otherwise, performance of the two candidates was very similar. The Pentagon gave Hughes the competition award because its design entry was assessed as being more affordable.

Joint Direct Attack Munition (JDAM), a joint Air Force, Navy, and Marine Corps production program, completed over 50 drops this year. The GPS/INS guided bomb achieved an accuracy of 10.4 meters CEP (circular error probable) in tests from F/A-18, F-17, and B-2 aircraft. McDonnell Douglas Aerospace is the prime contractor on the program.

The Wright Laboratory Armament Directorate's Miniaturized Munition Technology Demonstration (MMTD) program is transitioning its 250lb penetrator concept to a new Air Force System Program Office for procurement. The 18-month technology development project demonstrated that a differential Global-Navigation-System/Inertial-Navigation-System (GPS/INS) guidance scheme could guide a 6-inch diameter, 12-foot long airframe to an average of 2.1 meters miss distance over 5 test flights. The last test involved a simultaneous drop of two bombs, one inert and one with a 50-lb live warhead, against two different target aircraft bunkers. Each weapon pierced its target and the live warhead detonated precisely at the target's center, destroying the aircraft and rendering the shelter unserviceable. McDonnell Douglas Aerospace repackaged their Joint Direct Attack Munitions (JDAM) guidance system, updated to take advantage of differential correction, for the MMTD project.

USAF Phillips Lab is using Draper Labs' ultra high-g event recorder, known as AUGER, in tests designed to collect information on high velocity impacts. The data will be used to develop new smart fuses for penetrators targeting high-value deeply buried underground structures.

Lockheed Martin has been awarded the potentially \$1-billion Wind-Corrected Munitions Dispenser (WCMD) program. The tail-kit guidance units, which are bolted to standard tactical munitions dispensers in the field, compensate for wind encountered during a drop, launch errors, and ballistic errors from high altitude launches. The reported 85-ft accuracy is more than adequate for dispensing submunitions over a large area.

The Army's Enhanced Fiber Optic Guided Missile (EFOGM) completed its first unguided test flight at Redstone Arsenal. The test demonstrated that the 240-micron fiber-optic cable dispensed properly and that communication was maintained throughout the flight. Raytheon Electronics Systems, the EFOGM prime contractor, is preparing for a series of 1997-1998 guided test flights at White Sands Missile Range to demonstrate the imaging infrared seeker, warhead, and solid rocket sustainer motor.

The Naval Surface Warfare Center conducted the first successful test of a GPS/INS guided projectile fired from a fielded gun system. The rocket-assisted 5-inch projectile, fired from an Mk-45 gun, reached an apogee of approximately 8 miles over a 19-mile downrange flight. The navigation hardware, a Draper Labs' micromechanical Inertial Measurement Unit (IMU) integrated with a Rockwell Collins GPS receiver, survived the 6500-g gun launch, acquired the GPS satellite constellation in "P" code within 15 seconds, and developed a navigation solution of the projectile's trajectory.

In Aircraft

The McDonnell Douglas F/A-18 E/F Super Hornet made its first carrier landing and catapult during Initial Sea Trials (IST). Pilots reported excellent flying qualities. IST was conducted with control laws that were unchanged from the first flight.

The upgraded Royal Air Force Tornado completed its maiden flight at British Aerospace's Wharton facility. Upgrades include a Forward Looking InfraRed (FLIR) system, night-vision goggles, new head-up and head-down displays, and integrated GPS.

F-22 rollout ceremonies were held at Lockheed Martin's plant in Marietta, GA. The aircraft has thrust vectoring for superior maneuverability, a stealthy radar cross section, and supercruise capability.

Boeing and Lockheed Martin were selected as finalists in the Joint Strike Fighter competition. The main difference between the two designs is the way thrust vectoring is provided.

NASA-Langley and NASA-Dryden demonstrated the first closed-loop control of an aircraft using forebody controls in the NASA High-Alpha Research Vehicle (HARV). Flight tests confirmed that forebody controls can significantly improve the maneuverability of fighter aircraft by providing yaw control to coordinate roll maneuvers. Above 25-degrees angle-of-attack, the actuated forebody strakes were roughly equivalent to the HARV's yaw thrust vectoring system.

Barron Associates, with Lockheed Martin, Wright Labs, and CALSPAN, demonstrated fully adaptive control technologies in the Self-Designing Controller Program using the Variable Stability In-Flight Test Aircraft (VISTA). On-line parameter identification and gain calculation algorithms allowed the flight control system to continuously adjust itself to meet desired flying qualities, even with control surface failures. The flight test program culminated on the fifth flight with a successful landing of the aircraft with the left horizontal tail inoperative.

The F-15 ACTIVE (Advanced Control Technology for Integrated Vehicles) aircraft demonstrated engine thrust vector operation at the highest speed in aviation history, Mach 1.97 at 45,000 ft. altitude. The jet engine nozzles were vectored to allowable limits in pitch and yaw at that flight condition. ACTIVE is a joint NASA/ USAF technology development program with contractors McDonnell Douglas Aerospace and Pratt & Whitney.

United Kingdom's Defense Evaluation and Research Agency (DERA) successfully test flew its VAAC fly-by-wire flight control system through the entire flight envelope of a modified Harrier. This is the first demonstration of a full-authority all-axis digital fly-by-wire Vertical/Short TakeOff and Landing (VSTOL) aircraft.

The tailless X-36 remotely piloted aircraft produced by McDonnell Douglas Aerospace and NASA-Ames successfully completed its first flight at NASA-Dryden. The subscale prototype flew over 5 minutes and reached an altitude of about 4,900 feet. The X-36 has no vertical tails, which makes it more stealthy and potentially more agile than current fighter aircraft. The design reduces its weight and drag and explores new flight control technologies, such as split ailerons and thrust vectoring. A pilot in a ground-station cockpit flies the X-36 by remote control. Plans call for 24 additional flights at NASA-Dryden over the next six months.

The Pentagon's newest Unmanned Aerial Vehicle (UAV), the Alliant Techsystems Outrider, successfully completed its first flight, a 2-mile-long figure-8 course for 8.5 minutes at speeds of up to 112 knots. The UAV includes an autopilot and an automatic takeoff and landing system.

The Darkstar UAV received permission from the Pentagon to fly again this summer. The flight control laws have been modified from those flown last year.

A quarter-scale prototype of the Centurion solar-powered drone was flight tested by AeroVironment. The craft, which is an unswept high-aspect-ratio flying wing, was found to be stable and controllable with enough authority to recover from turbulence upsets near the ground. A reflex airfoil provides pitch stability. The pilot uses symmetrical thrust to climb and differential thrust to steer. The vehicle has no active yaw stabilization.

The USAF's and NASA's 100-inch, 80-lb subscale model of an aircraft designed to fly at Mach 5.5, made its maiden 34-second flight, reaching an altitude of 150 ft and making a 180-degree turn. This is the first time that a waverider concept vehicle has taken off under its own power.

Auburn University researchers flew the first remotely piloted helicopter that uses a solid state adaptive main rotor system for flight control. The system uses directionally attached piezoelectric adaptive sheets that essentially twist when a small electric current is applied. The movement of these sheets changes the pitch of two servopaddles attached as short, secondary blades to the rotor shaft. As they rotate, the lightweight graphite-epoxy servopaddles exert aerodynamic pressures that tilt the rotor for the desired direction of flight. During adaptive control flight, the 23.5-inch-diameter, 2-mainblade minirotor demonstrated full pitch and roll control.

The RAH-66 Comanche Prototype No. 1 helicopter flew its first "wheels-up" flight at the Sikorsky Experimental Flight Test Center. The helicopter's digital fly-by-wire flight controls and handling qualities are now being tested.

The German federal aviation agency granted Lufthansa CityLine an operating license for a head-up guidance system which allows pilots to land its Canadair Regional Jets when visibility along the runway is down to 200 meters and at a decision height of 50 ft. Lufthansa CityLine is the first airline in Germany to be cleared for manual Category 3A instrument landings.

NASA-Ames completed flight tests of a low altitude collision avoidance system for helicopters. The system, designed to detect obstacles in and near a helicopter's flight path, uses a nose-mounted Honeywell 35-GHz millimeter wave radar sensor. The collision avoidance works well at night and under certain bad-visibility weather conditions, such as rain and fog. The system is suitable for approach and landing but is insufficient for general low-level flight operations.

The final version of software for the Traffic Alert and Collision Avoidance System-Two (TCAS-2) is undergoing flight test. The first version of TCAS-2, which went into operational service in 1990, gave two merging TCAS-2-equipped aircraft the ability to negotiate an escape maneuver and advise the pilots to climb, descend, or maintain altitude to avoid passing too close. The TCAS-1 only warned of the presence of another transponder-equipped aircraft, but could not negotiate for avoidance maneuvers.

The McDonnell Douglas joint-USAF/Navy fourth-generation ACES 2 ejection seat is undergoing rocket sled testing at Holloman AFB. Testing shows that pintle nozzles provide improved seat steering and attitude control during initial stages of ejection. Pintle nozzles were developed by Aerojet, who is under contract to McDonnell Douglas.

Fibersense Technology Corp. (FTC) is developing a low-cost medium-performance fiber-optic gyro (FOG) that uses a mass-produced commercial part. This FOG will use an integrated opto-electronics module developed by AT&T's Lucent Technologies for fiber-optic communications systems. FTC expects to supply it with drift rates of 1-10 deg/hr, for a few hundred dollars in production quantities. The USAF is considering replacing the spinning-wheel rate gyros in many of their aircraft autopilots with smaller, more-reliable FOGs like these. In fact, FTC's unit is a leading candidate in the replacement program.

Nonlinear Bifurcation Analysis has been applied successfully by DERA in the UK to identify the full range of equilibrium conditions and associated stability of a military fast-jet aircraft. A range of modes have been identified including flight, various deep stall and spin modes, wing rock, and autorotation. The theoretical results have been verified in piloted flight simulation.

Nonlinear Dynamic Inversion (NDI) has been successfully used by DERA as a fast-prototyping tool for developing the functional design of control laws for a wide range of air vehicles. The method is used routinely in both off-line analysis and piloted simulation to support work on the control of fixed wing and VSTOL aircraft and weapons.

In Space

NASA-Johnson and NASA-Goddard completed the second servicing mission of the Hubble Space Telescope with replacement of two primary components in the pointing control subsystem. One of the three fine guidance sensors and one of the four reaction wheels were successfully replaced. Post-servicing mission pointing and jitter analysis showed nominal performance.

SatCon Technology Corporation was awarded a $2 million contract to develop and deliver for demonstration a composite-flywheel-based integrated-power-and-attitude control system for low earth orbiting satellites. The contract, funded through the USAF Phillips Lab, will be conducted over the next two years. Completion of it will be a significant step toward production of a space-qualified flywheel energy storage and attitude control system. These systems are projected to provide lower cost and longer operational life for commercial satellites by combining the functions currently performed by separate battery and momentum wheel systems, thereby eliminating the need for batteries.

NASA-Lewis developed small electric propulsion thrusters called Pulsed Plasma Thrusters (PPTs) under a contract to Primex Aerospace. This technology is currently being prepared for use in pitch axis control of NASA-Goddard's EO-1 spacecraft. Flight hardware delivery from Primex is set for January 1998. PPTs are also being considered for the attitude control system functions on JPL's DS-3 mission.

Sira Ltd. of England and Sodern of France are entering the commercial satellite market with a new way of stabilizing spacecraft. They are developing star trackers that would be used in conjunction with GPS receivers for a low-cost, highly-reliable approach to spacecraft attitude control. With this approach, mechanical gyroscopes or other rate sensors are not needed.

A NASA-Goddard/CTA-Space-Systems GPS experiment on the REX-II spacecraft achieved three historic firsts. This was the first demonstration of spacecraft attitude control using exclusively GPS as an attitude sensor. This was the first long-term, real-time on-board GPS attitude determination experiment. This was the first GPS experiment to provide spacecraft Orbital Element data downlink. The GPS experiment used two Trimble TANS Vector receivers with specialized on-orbit software developed by NASA-Goddard and Stanford University. The algorithms embedded in these receivers support real-time on-orbit attitude determination at a 1 Hz update rate using differential carrier phase signals obtained from four patch antennas mounted on the anti-nadir side of the vehicle. In addition to the attitude data, Standard Positioning Service (SPS) accuracy orbit knowledge and Keplerian Orbital elements are produced using this receiver. This information can be used autonomously in an on-board orbit propagator to maintain on-board orbit knowledge or can be downlinked to the ground for science post processing.

American GNC Corporation (AGNC) developed a fully-coupled kinematic GPS/IMU algorithm utilizing GPS carrier phase observables to determine very accurate time, position and velocity information for high dynamic environments. The design accounts for on-the fly phase ambiguity resolution, carrier cycle slip detection and failure identification and isolation. AGNC also developed a multiple micromechanical sensor IMU deriving higher performance from low-cost, miniature gyros and accelerometers. The design integrates electronic filters, error compensation, optimal estimation and failure detection and identification.

Honeywell was awarded a $7.2 million NASA-Johnson contract to develop and qualify a Space Integrated GPS/INS, called SIGI. SIGI will greatly enhance space navigation by providing NASA with a GPS position, velocity and attitude determination capability integrated with an advanced inertial system. Following qualification, the system will ultimately be applied to a variety of space vehicles, including the Space Shuttle, the International Space Station (ISS), and the Crew Return Vehicle (CRV) for ISS. The system will allow NASA space voyagers the choice of three independent navigation solutions: INS, GPS and blended GPS/INS.

NASA-Marshall and Mayflower Communications Company developed a 19-state Kalman filter for relative GPS navigation that can estimate position and velocity to 0.7 meters (1-sigma) and 0.02 meters/second (1-sigma), respectively. This performance requires two GPS receivers with at least six continuous tracking channels, raw pseudorange and carrier phase output, and a receiver output rate of 1 Hz. The filter has been tested extensively at NASA-Marshall's GPS Lab and Flight Robotics Lab and will be used in an automated rendezvous and capture flight demonstration that is scheduled for the fall of 1997.

NASA-Johnson is designing the CRV flight control system using design tools such as Matlab. Once the design is complete, it will be autocoded and put into super blocks using Matrix_X. The flight control system will then be combined with the remaining Guidance, Navigation and Control (GN&C) software and tested at NASA-Johnson's Rapid Development Lab.

NASA-Marshall is developing the X-33 Guidance and Control (G&C) flight software algorithms for the ascent, transition, and entry phases of flight. Lockheed Martin is responsible for the software in the terminal area energy management and approach and landing phases. To meet the aggressive X-33 schedule, NASA-Marshall built a simulation testbed for the development of the flight software called MAVERIC (Marshall Aerospace Vehicle Representation In C). This testbed allows both 3 and 6 degree of freedom simulations for separate development of the guidance and control software. This high fidelity simulation features: numerous atmosphere and wind models including the GRAM 95 model, Monte Carlo dispersion analysis capability, a Mission Manager that sequences flight phases and abort conditions, and the prototype G&C flight software.

The space shuttle Atlantis performed a series of test maneuvers during reentry to help determine if future missions can use more aerodynamic control and less thruster propellant on rentry. If so, this could save several hundred pounds of launch weight for more upbound space station cargo. At about Mach 16 and 200,000 ft, the orbiter's flight control system automatically began a series of five pitch maneuvers that rotated the vehicle up to 20 deg at 5 deg/sec using only elevon control with no attitude thruster firings. These test maneuvers, interspersed between four normal thruster-controlled reentry maneuvers to dissipate energy, continued down to Mach 3.

Microcosm developed a low-cost GN&C system for launch vehicles that exploits recent developments in GPS and computer technology. The system was designed by choosing low cost, commercial, off-the-shelf components that together would provide the lowest life-cycle cost with adequate performance. The navigation technique chosen was to use an integrated GPS/INS system, which combines an inexpensive GPS receiver with inexpensive inertial sensors together with a Kalman filter in a single package. The complementary features of the two technologies provide good navigation and attitude determination for an accelerating vehicle at low cost, with high data rates, and with resistance to dropouts.

Alignment of the inertial platform is not a problem since determination of the absolute position is not dependent upon the initial orientation.

The Jet Propulsion Laboratory (JPL)/Lockheed Martin Mars Global Surveyor (MGS) spacecraft was launched on a McDonnell Douglas Delta booster. Liftoff occurred at the moment the rotation of the Earth moved the Delta launch pad into the proper geometric plane for the booster to perform an unusual two-burn Earth escape maneuver. That maneuver placed the spacecraft on a trajectory curving 435 million miles around the Sun to reach Mars when the planet is 157.6 million miles from Earth. Commanding is being done both from JPL and Lockheed Martin in Denver, CO. This is the first time in the history of a deep-space planetary mission that JPL has shared flight control with an outside facility. This new shared system enables the contractor to use the same engineers and test hardware to both build and control the spacecraft, avoiding costly duplication.

JPL's Mars Pathfinder spacecraft was launched and is enroute to the red planet. It represents a new breed of interplanetary spacecraft, having been designed to place a science payload on the surface of Mars with three years of development at 1/15 the cost of Viking. Landing will involve a parachute, a radar altimeter triggering solid rocket motors for braking, and airbags to cushion the impact. Once landed, the spacecraft will deploy various science instruments, including a rover.

JPL developed and demonstrated, in clear weather, a variable-buoyancy research balloon for maintaining an approximately constant altitude over land and sea. This dual-balloon system uses for lift a 16-ft-high balloon that is typically filled with helium. Under it is suspended a radiosonde, strobelight and 10-ft-long bag or balloon filled with a phase-change fluid such as butane, ethyl chloride or Refrigerant 114. The type and amount of phase-change fluid is chosen according to prevailing conditions and the altitude desired. The fluid condenses above the desired altitude and evaporates below it, causing the balloon system to "bob" around the desired altitude. JPL is studying the possible use of this system for exploring the atmospheres of Venus and other planets.

Two differential-GPS-guided bombs hit their separate targets in a simultaneous test drop at Eglin AFB test range.

The VISTA aircraft was used to demonstrate adaptive control technologies.

The Space Shuttle Atlantis performed test maneuvers to determine if more aerodynamic control can be used on reentry.

AIAA Guidance, Navigation, and Control Conference and Exhibit

New Orleans, LA / August 11–13, 1997

TABLE OF CONTENTS

Session 16-GNC-8 Hypersonic Trajectories/Launch Vehicle

Session 17-GNC-9 Neural Networks I

Session 18-GNC-10 Practical Techniques for Air Traffic Management

NA - Not Available

WD - Withdrawn

Session 19-GNC-11 GN & C Components

Session 20-GNC-12 Robotics

Session 21-GNC-13 Guidance, Navigation, and Control I

NA - Not Available   WD - Withdrawn

NA - Not Available WD - Withdrawn

NA - Not Available WD - Withdrawn

NA - Not Available WD - Withdrawn

Session 46-GNC-24 Fuzzy Logic

Session 47-GNC-25 Guidance, Navigation, and Control II

Session 48-GNC-26 Optimal Flight Control

Session 49-GNC-27 Nonlinear Flight Control

Session 50-GNC-28 Handling Qualities and Other Applications

Session 51-GNC-29 Trajectory and Guidance Analysis

Session 58-GNC-30 Analysis of Air Traffic Management Models

NA - Not Available

WD - Withdrawn

Session 59-GNC-31 Fault Tolerance and Parameter Identification

Session 60-GNC-32 Flexible Structure II

Session 61-GNC-33 Spacecraft Attitude Control

Session 62-GNC-34 Neural Networks II

Session 63-GNC-35 Missile Autopilot

Session 64-GNC-36 New and Emerging Technologies and Challenges in Aerospace Guidance, Navigation, and Control I

Session 65-GNC-37 Flight Control Theory

AIAA-97-3450

PERIODICALLY PRECONDITIONED CONJUGATE GRADIENT-RESTORATION ALGORITHM FOR OPTIMAL CONTROL: THE HYBRID APPPROACH

James R. Cloutier[1]
Navigation and Control Branch
U.S. Air Force Armament Directorate
Eglin AFB, FL 32542-6810

Richard A. Hull[2]
Sverdrup, Inc.
P. O. Box 1935
Eglin AFB, FL 32542-0935

Abstract

In this paper, a hybrid approach (HA) is used in the development of a periodically preconditioned conjugate gradient-restoration algorithm (PPCGRA). The hybrid approach takes advantage of the best features of the direct and indirect approaches to optimal control. The original infinite-dimensional problem is transcribed into a mathematical programming problem using high-order integration methods, a characteristic of the direct approach, yet conjugate gradient searches are conducted only over the control variables and control parameters, a characteristic of the indirect approach.

The hybrid approach has the following attributes. Ill-conditioned problems can be preconditioned and periodically reconditioned from knowledge of the Hessian. The dimension of the gradient search space is reduced by eliminating the state variables from the search. The solution of the linearized discrete two-point boundary-value problem is obtained by solving a linear algebraic equation instead of having to perform integration sweeps.

The particular class of control problems that is studied not only has differential constraints and terminal constraints, but also algebraic constraints which must be satisfied over the entire interval of integration. The development of a PPCGRA for this class of problems is important since numerous control problems arise in this form or can be brought to this form through the use of various transformations and/or the use of slack variables.

[1]Associate Fellow AIAA
[2]Member AIAA

1. Introduction

The theory of scaling for function minimization is well-established and can be found in many books on numerical optimization, as in [1] and [2], to name only two. Depending on the problem and the nature of its ill-conditioning, scaling can be used to balance derivatives, balance constraints, or modify the range of the objective function. It can take the form of pure scaling, in which a diagonal matrix is used to transform the axial scales, or the form of generalized scaling, in which a positive definite matrix is used to alter both the scale and the direction of the axes. The primary uses for scaling are to reduce the effects of roundoff errors in numerical computation and to alter the shape of performance index surfaces to enhance the convergence rates of certain algorithms, particularly gradient based algorithms.

With specific regard to conjugate gradient algorithms, it is known that convergence can be accelerated by the introduction of a scaling transformation $y = M^{1/2}x$, where M is a symmetric positive-definite matrix. In the transformed variables, the Hessian is given by $M^{-1/2}\mathcal{H}(x)M^{-1/2\,T}$ and the steepest descent direction by $-M^{-1}f_x(x)$ where $\mathcal{H}(x)$ and $f_x(x)$ are the Hessian and gradient, respectively, in the x-frame. The method resulting from this transformation is called the preconditioned conjugate gradient algorithm (PCGA) and the matrix M is called the preconditioner [3]. Golub and O'Leary in [4] give an excellent history of conjugate gradient algorithms and provide an annotated bibliography which includes numerous papers on preconditioning. Most of that work and more recent work on preconditioning [5]-[14] address the problem of solving large sparse linear systems and involve the concept of partial preconditioning. In partial preconditioning, incomplete factorization of the Hessian is used to retain, in the factors, the sparseness or bandedness that exists in the original matrix.

In [15], the concept of preconditioning for mathematical programming problems was extended to the nonlinear constrained optimization problem, and both partial and complete preconditioning was addressed. A theory and efficient synthesis procedure was developed concerning the optimal preconditioner for both positive definite and indefinite Hessians. It was shown that the optimal preconditioner is a matrix, denoted by $\mathcal{H}_+$, which is the positive definite reflection of the Hessian $\mathcal{H}$. It was further shown how $\mathcal{H}_+$ could be computed from a singular value decomposition of $\mathcal{H}$. In the transformed space, the optimal preconditioner produces a transformed Hessian with a condition number of one, which, if $\mathcal{H}$ is positive definite, translates to local circular surface contours in the neighborhood of the point at which the preconditioner was computed. These results were then used to construct a PPCGRA which exhibited significant convergence rate enhancements over its partially-preconditioned and non-preconditioned counterparts for ill-conditioned nonlinear constrained mathematical programming problems.

Until recently, in contrast to the rich literature on scaling for mathematical programming problems, scaling theory for optimal control problems was scant and largely *ad hoc*. Common practice was to balance the constraints or the states or the controls using, typically, a constant diagonal scaling matrix. In [16], this void in the literature was filled by the development of a preconditioning theory for the standard optimal control problem. The preconditioning theory revealed that the states and the controls must not only be balanced over their vector components, but also over the time or spatial parameter. This work was then extended in [17] to the class of optimal control problems having not only differential and terminal constraints but also algebraic constraints which have to be satisfied over the entire interval of integration.

The preconditioning theory presented in [16] and [17] was then used in conjunction with the direct approach (DA) to optimal control to construct a PPCGRA [18]. Being based on the direct approach, the algorithm in [18] performs gradient searches over all of the discrete variables, including the discretized state variables. In [19], the theory associated with using a hybrid approach to optimal control problems having both differential and algebraic constraints was developed. The hybrid approach not only allows for the preconditioning of ill-conditioned problems through knowledge of the Hessian, but also reduces the dimension of the gradient search space by eliminating the discrete state variables from the search. In this paper, the hybrid approach is utilized to develop a PPCGRA whose gradient search space consists of only the discrete control variables and control parameters. The PPCGRA:HA is then compared to the PPCGRA:DA both in terms of algorithm structure and numerical performance.

2. The Control Problem with Algebraic Constraints

Consider the optimal control problem of minimizing the functional

$$I = \int_0^1 f(x(t), u(t), \pi, t)dt + g(x(1), \pi) \quad (1)$$

with respect to the state $x(t)$, control $u(t)$, and parameter π subject to the differential constraint

$$\dot{x} = \phi(x(t), u(t), \pi, t), \quad 0 \leq t \leq 1 \quad (2)$$

the algebraic constraint

$$S(x(t), u(t), \pi, t) = 0, \quad 0 \leq t \leq 1 \quad (3)$$

and the boundary conditions

$$x(0) = x_0^* \qquad \psi(x(1), \pi) = 0. \quad (4)$$

In Eqs. (1)-(4), f and g are scalar functions, ϕ is a n-vector function, S is a k-vector function, and ψ is a q-vector function; the state variable x is an n-vector, the control variable u is an m-vector, and the parameter π is a p-vector.

It is should be noted here that the above formulation is a general structure which can accomodate optimal control problems with initial and final times other than zero and one, respectively. These latter control problems can be brought to the form (1)-(4) through normalization of the actual time and by regarding the final time, if it is free, as one of the components of π.

From the calculus of variations, it is known that problem (1)-(4) is one of the Bolza type. It can be recast as that of minimizing the augmented functional

$$J = \int_0^1 \left[f + \lambda^T(t)(\dot{x} - \phi) + \rho^T(t)S\right] dt + \left(g + \mu^T\psi\right)_1 \quad (5)$$

subject to (2)-(4). In the above expression, $\lambda(t)$ is an $n \times 1$ time-varying Lagrange multiplier, $\rho(t)$ is a $k \times 1$ time-varying Lagrange multiplier, and μ is a $q \times 1$ constant Lagrange multiplier. Defining the Hamiltonian to be

$$H(x(t), u(t), \pi, \lambda(t), t) = f(x(t), u(t), \pi, t) - \lambda^T(t)\phi(x(t), u(t), \pi, t) + \rho^T(t)S(x(t), u(t), \pi, t) \quad (6)$$

the standard techniques of variational control theory lead to the following necessary conditions for optimality:

$$\dot{\lambda} = H_x \quad (7)$$

$$H_u = 0 \quad (8)$$

$$\int_0^1 H_\pi dt + (G_\pi)_1 = 0 \quad (9)$$

$$(\lambda + G_x)_1 = 0 \quad (10)$$

where the augmented Bolza function, G, is given by

$$G(x(1), \pi) = g(x(1), \pi) + \mu^T \psi(x(1), \pi). \quad (11)$$

3. Direct Transcription via High-Order Integration

Now consider the high-order approximation of problem (1)-(4) which is given as follows:

$$\text{Minimize} \quad F(z) = \sum_{i=1}^{N} \beta_i f(x_i, u_i, \pi, t_i)\Delta t_i + g(x_N, \pi) \quad (12)$$

with respect to z subject to

$$\Phi(z) = \begin{bmatrix} x_1 - x_0^* \\ x_2 - x_1 - \Gamma(\phi_1)\Delta t_1 \\ x_3 - x_2 - \Gamma(\phi_2)\Delta t_2 \\ \vdots \\ x_N - x_{N-1} - \Gamma(\phi_{N-1})\Delta t_{N-1} \\ \psi(x_N, \pi) \\ S(x_1, u_1, \pi, t_1) \\ S(x_2, u_2, \pi, t_2) \\ \vdots \\ S(x_N, u_N, \pi, t_N) \end{bmatrix} = 0 \quad (13)$$

where

$$z = \left[x_1^T \; x_2^T \; \ldots \; x_N^T \; u_1^T \; u_2^T \; \ldots \; u_N^T \; \pi^T\right]^T \quad (14)$$

$$\Gamma(\phi_i) \triangleq \Gamma(\phi; x_i, u_i, u_{i+1}, \pi, t_i; \Delta t), \quad i = 1, \ldots, N-1 \quad (15)$$

and

$$t_1 = 0 \; ; \; t_N = 1$$

$$t_{i+1} = t_i + \Delta t_i, \quad i = 1, \ldots, N-1 \quad (16)$$

$$\Delta t_i = \Delta t_{i+1} = \Delta t, \quad i = 1, \ldots, N-1 \quad (17)$$

$$x_i = x(t_i) \qquad u_i = u(t_i). \quad (18)$$

In the above, the β_i weightings in Eq. (12) represent a general high-order quadrature scheme while $\Gamma(\phi; x_i, u_i, u_{i+1}, \pi, t_i; \Delta t)$ in Eq. (13) is an increment function which generically represents a high-order integration method such as fourth-order Runge-Kutta. High-order integrators typically require a value of the control at the midpoint of the incremental interval of integration. We assume here that the estimate of that value is supplied by linearly interpolating between the values of the control at the endpoints of the interval. Hence, Γ is a function of both u_i and u_{i+1}. If a more sophisticated interpolation scheme were employed, then Γ would be a function of several of the surrounding u_i's.

The augmented function J for the problem (12)-(13) is given by

$$J(z, \lambda) = F(z) + \lambda^T \Phi(z) \quad (19)$$

where

$$\lambda = \begin{bmatrix} \lambda_1 \\ \lambda_2 \\ \vdots \\ \lambda_N \\ \lambda_{N+1} \\ \lambda_{N+2} \\ \vdots \\ \lambda_{2N+1} \end{bmatrix} = \begin{bmatrix} \lambda(t_1) \\ \lambda(t_2) \\ \vdots \\ \lambda(t_N) \\ \mu \\ \beta_1 \rho_1 \Delta t_1 \\ \vdots \\ \beta_N \rho_N \Delta t_N \end{bmatrix}. \quad (20)$$

Here, λ is an $[(n \times N) + q + (k \times N)] \times 1$ Lagrange multiplier, β_i is the i-th quadrature weight, $\lambda_i = \lambda(t_i)$, and $\rho_i = \rho(t_i)$, where $\lambda(t)$ is the time-varying Lagrange multiplier associated with the dynamics of $x(t)$ in Eq. (2) and $\rho(t)$ is the time-varying Lagrange multiplier associated with the algebraic constraint $S(t)$ defined in Eq. (3). Note that by defining $\lambda_{N+1+i} = \beta_i \rho_i \Delta t_i, \quad i = 1, \ldots, N$, the discrete time function $S(x_i, u_i, \pi, t_i)$ is being brought under the numerical quadrature sum in the same way the continuous time function $S(x(t), u(t), \pi, t)$ was brought under the integral in Eq. (5). For $i = 2, \ldots, N$, λ_i is associated with the constraint

$$-\Gamma(\phi; x_{i-1}, u_{i-1}, u_i, \pi, t_{i-1}; \Delta t_{i-1})\Delta t_{i-1} + x_i - x_{i-1} = 0.$$

Each of these constraints involve two different times, t_{i-1} and t_i, and while we have set $\lambda_i = \lambda(t_i)$, for limit purposes as $\Delta t \to 0$, λ_i can be time tagged with any $t \in [t_{i-1}, t_i]$. Finally, note that λ_1 is associated with the initial condition constraint $x_1 - x_0^* = 0$ while λ_{N+1} is associated with the terminal constraint $\psi(x_N, \pi) = 0$.

The augmented function J can be rewritten as

$$J(z, \lambda) = \sum_{i=1}^{N} \lambda_i (x_i - x_{i-1})$$

$$
\begin{aligned}
&+ \sum_{i=1}^{N} H(x_i, u_i, u_{i+1}, \pi, t_i)\Delta t_i \\
&+ G(x_N, \pi)
\end{aligned} \tag{21}
$$

where

$$G(x_N, \pi) = g(x_N, \pi) + \lambda_{N+1}^T \psi(x_N, \pi) \tag{22}$$

is the augmented Bolza function for $g(x_N, \pi)$ and $H(x_i, u_i, u_{i+1}, \pi, t_i)$ is the discrete Hamiltonian given by

$$
\begin{aligned}
H(x_i, u_i, u_{i+1}, \pi, t_i) &= \beta_i \left[f(x_i, u_i, \pi, t_i) \right. \\
&+ \left. \rho_i^T S(x_i, u_i, \pi, t_i) \right] \\
&- \lambda_{i+1}^T \Gamma(\phi; x_i, u_i, u_{i+1}, \pi, t_i; \Delta t) \\
& \qquad i = 1, \ldots, N
\end{aligned} \tag{23}
$$

with $\Gamma(\phi; x_N, u_N, u_{N+1}, \pi, t_N) \triangleq 0$ for notational convenience. Note that the discrete Hamiltonian $H(x_i, u_i, u_{i+1}, \pi, t_i)$ is not the sampled continuous Hamiltonian evaluated at the discrete points x_i, u_i, π, t_i; instead, it is the discrete resultant of the continuous Hamiltonian being operated on by the high-order integration methods.

If $J_z(z, \lambda)$ denotes the gradient of the augmented function, then the optimal solution (z, λ) must satisfy simultaneously the equations

$$
\begin{aligned}
J_{z_1}(z, \lambda) &= 0 && (24) \\
J_{z_2}(z, \lambda) &= 0 && (25) \\
\Phi(z) &= 0 && (26)
\end{aligned}
$$

where

$$
\begin{aligned}
z_1 &= \left[x_1^T \; x_2^T \; \ldots \; x_N^T \right]^T && (27) \\
z_2 &= \left[u_1^T \; u_2^T \; \ldots \; u_N^T \; \pi^T \right]^T && (28) \\
z &= \left[z_1^T \; z_2^T \right]^T . && (29)
\end{aligned}
$$

Eqs.(24) and (25) can be expanded using Eq. (21) to yield

$$
J_{z_1}(z, \lambda) = \begin{bmatrix} \lambda_1 - \lambda_2 + H_{x_1}(\varsigma_1)\Delta t \\ \lambda_2 - \lambda_3 + H_{x_2}(\varsigma_2)\Delta t \\ \vdots \\ \lambda_{N-1} - \lambda_N + H_{x_{N-1}}(\varsigma_{N-1})\Delta t \\ \lambda_N + H_{x_N}(\varsigma_N)\Delta t + G_{x_N}(x_N, \pi) \end{bmatrix} \tag{30}
$$

$$
J_{z_2}(z, \lambda) = \begin{bmatrix} \left(H_{u_1}(\varsigma_0) + H_{u_1}(\varsigma_1) \right) \Delta t \\ \left(H_{u_2}(\varsigma_1) + H_{u_2}(\varsigma_2) \right) \Delta t \\ \vdots \\ \left(H_{u_N}(\varsigma_{N-1}) + H_{u_N}(\varsigma_N) \right) \Delta t \\ \sum_{i=1}^{N} H_\pi(\varsigma_i)\Delta t + G_\pi(x_N, \pi) \end{bmatrix} \tag{31}
$$

where, for compactness

$$H(\varsigma_i) = H(x_i, u_i, u_{i+1}, \pi, t_i) \tag{32}$$

In deriving Eqs. (30) and (31), the following substitutions were made. First, since $f_{u_{i+1}}(x_i, u_i, \pi, t_i) = 0$, and $S_{u_{i+1}}(x_i, u_i, \pi, t_i) = 0$,

$$\beta_i \left[f_{u_{i+1}}(x_i, u_i, \pi, t_i) + S_{u_{i+1}}(x_i, u_i, \pi, t_i)\rho_i \right] = 0 \tag{33}$$

which implies that

$$
\begin{aligned}
&\Gamma_{u_{i+1}}(\phi; x_i, u_i, u_{i+1}, \pi, t_i; \Delta t)\lambda_{i+1} \\
&\quad = H_{u_{i+1}}(x_i, u_i, u_{i+1}, \pi, t_i)
\end{aligned} \tag{34}
$$

Second, since $\Gamma(\phi; x_N, u_N, u_{N+1}, \pi, t_N; \Delta t) = 0$ by definition,

$$
\begin{aligned}
&\beta_N \left[f_{x_N}(x_N, u_N, \pi, t_N) + S_{x_N}(x_N, u_N, \pi, t_N)\rho_N \right] \\
&\quad = H_{x_N}(x_N, u_N, u_{N+1}, \pi, t_N) && (35) \\
&\beta_N \left[f_{u_N}(x_N, u_N, \pi, t_N) + S_{u_N}(x_N, u_N, \pi, t_N)\rho_N \right] \\
&\quad = H_{u_N}(x_N, u_N, u_{N+1}, \pi, t_N) && (36) \\
&\beta_N \left[f_\pi(x_N, u_N, \pi, t_N) + S_\pi(x_N, u_N, \pi, t_N)\rho_N \right] \\
&\quad = H_\pi(x_N, u_N, u_{N+1}, \pi, t_N) && (37)
\end{aligned}
$$

Additionally, $H_{u_1}(x_0, u_0, u_1, \pi, t_0) \triangleq 0$ for notational convenience noting that x_0, u_0, and t_0 are undefined.

The Hessian of the transcribed problem (12)-(13) is given by

$$\mathcal{H}(z, \lambda) = J_{zz}(z, \lambda) = F_{zz}(z) + \sum_{j=1}^{n \times N + q + k \times N} \Phi_{j_{zz}}(z)\lambda_j \tag{38}$$

where $\Phi_j(z)$ and λ_j are the j-th scalar components of $\Phi(z)$ and λ, respectively. Expanding Eq. (38) by taking the partial with respect to z of Eqs. (30) and (31) yields the Hessian matrix given in Figure 1.

The matrix subblocks corresponding to the cross-partials of H with respect to x and u and the second partials of H with respect to u are bidiagonal and tridiagonal, respectively, due to the fact that Γ is a function of both u_i and u_{i+1}; for the same reason, the cross-partials of H with respect to u and π and the diagonal elements of the second partial of H with respect to u consist of the sum of two terms.

4. Preconditioning Theory

Preconditioning is the execution of a scaling transformation $z' = M^{1/2}z$, where M is a symmetric positive definite matrix. In the case of a nonlinear constrained optimization problem, in the transformed variables the Hessian is given by $M^{-1/2}\mathcal{H}M^{-1/2\,T}$ and the steepest descent direction

$$\mathcal{H}(z,\lambda) = J_{zz}(z,\lambda) = \begin{bmatrix} J_{z1z1} & J_{z1z2} \\ J_{z2z1} & J_{z2z2} \end{bmatrix}$$

$$J_{z1z1} = \begin{bmatrix} H_{x_1x_1}\Delta t & & & 0 \\ & \ddots & & \vdots \\ & & H_{x_{N-1}x_{N-1}}\Delta t & 0 \\ 0 & \cdots & 0 & H_{x_Nx_N}\Delta t + G_{x_Nx_N} \end{bmatrix}$$

$$J_{z1z2} = \begin{bmatrix} H_{x_1u_1}\Delta t & H_{x_1u_2}\Delta t & & & H_{x_1\pi}\Delta t \\ & \ddots & \ddots & & \vdots \\ & & H_{x_{N-1}u_{N-1}}\Delta t & H_{x_{N-1}u_N}\Delta t & H_{x_{N-1}\pi}\Delta t \\ 0 & \cdots & 0 & H_{x_Nu_N}\Delta t & H_{x_N\pi}\Delta t + G_{x_N\pi} \end{bmatrix}$$

$$J_{z2z1} = \begin{bmatrix} H_{u_1x_1}\Delta t & H_{u_2x_1}\Delta t & & & H_{\pi x_1}\Delta t \\ & \ddots & \ddots & & \vdots \\ & & H_{u_{N-1}x_{N-1}}\Delta t & H_{u_Nx_{N-1}}\Delta t & H_{\pi x_{N-1}}\Delta t \\ 0 & \cdots & 0 & H_{u_Nx_N}\Delta t & H_{\pi x_N}\Delta t + G_{\pi x_N} \end{bmatrix}^T$$

$$J_{z2z2} = \mathcal{H}_{LR}(z,\lambda) =$$

$$\begin{bmatrix} \sum_{i=0}^{1} H_{u_1u_1}(\varsigma_i)\Delta t & H_{u_1u_2}\Delta t & & & \sum_{i=0}^{1} H_{u_1\pi}(\varsigma_i)\Delta t \\ H_{u_2u_1}\Delta t & \sum_{i=1}^{2} H_{u_2u_2}(\varsigma_i)\Delta t & H_{u_2u_3}\Delta t & & \sum_{i=1}^{2} H_{u_2\pi}(\varsigma_i)\Delta t \\ \ddots & \ddots & \ddots & & \vdots \\ & H_{u_{N-1}u_{N-2}}\Delta t & \sum_{i=N-2}^{N-1} H_{u_{N-1}u_{N-1}}(\varsigma_i)\Delta t & H_{u_{N-1}u_N}\Delta t & \vdots \\ & & H_{u_Nu_{N-1}}\Delta t & \sum_{i=N-1}^{N} H_{u_Nu_N}(\varsigma_i)\Delta t & \sum_{i=N-1}^{N} H_{u_N\pi}(\varsigma_i)\Delta t \\ \sum_{i=0}^{1} H_{\pi u_1}(\varsigma_i)\Delta t & \sum_{i=1}^{2} H_{\pi u_2}(\varsigma_i)\Delta t & \cdots & \sum_{i=N-1}^{N} H_{\pi u_N}(\varsigma_i)\Delta t & \sum_{i=1}^{N} H_{\pi\pi}(\varsigma_i)\Delta t + G_{\pi\pi} \end{bmatrix}$$

Figure 1: The Finite Dimensional Hessian

by $-M^{-1}J_z(z,\lambda)$, where $\mathcal{H}$ and J_z, are the original Hessian and the augmented gradient, respectively, evaluated at a particular point (z_0, λ_0).

The matrix M above is called the preconditioner. The optimal preconditioner is that M which results in the transformed Hessian having a condition number of one; if $\mathcal{H}$ is positive definite, this is equivalent to the existence of circular surface contours in a neighborhood about the point (z_0, λ_0) in the general nonlinear case, and global circular surface contours in the quadratic-linear case.

It was shown in [15] that the optimal preconditioner, $\mathcal{H}_+$, is the positive definite reflection of the Hessian matrix $\mathcal{H}(z_0, \lambda_0)$, i.e., that matrix whose eigenvectors are identical to those of the Hessian and whose eigenvalues are the moduli of the latter's eigenvalues. It was also shown how $\mathcal{H}_+$ could be efficiently obtained from a singular value decomposition of $\mathcal{H}(z_0, \lambda_0)$ as detailed below.

For the hybrid approach, the constrained relationship between the discrete state variables, the discrete control variables and the control parameters is utilized to reduce the size of the search space (see Section 5). This results in a search space that consists of only the discrete control variables and the control parameters. In this case, only $\mathcal{H}_{LR}(z,\lambda)$, the lower right mega-subblock of the Hessian (see Figure 1) involving the second partials and cross partials of the discrete Hamiltonian with respect to $u_i, i = 1, \ldots, N$ and π, needs to be used in preconditioning, and perhaps periodically reconditioning, the problem. To construct the optimal preconditioner for the hybrid approach, $\mathcal{H}_{LR}(z,\lambda)$ is first decomposed into its singular value decomposition:

$$\mathcal{H}_{LR}(z,\lambda) = USV^T \tag{39}$$

The optimal mega-subblock preconditioner is then

given by

$$\mathcal{H}_{LR+} = USU^T = VSV^T \tag{40}$$

Finally, the optimal preconditioner is given by

$$\mathcal{H}_{+} = \begin{bmatrix} I & \vdots & 0 \\ \dots & \vdots & \dots \\ 0 & \vdots & \mathcal{H}_{LR_+} \end{bmatrix} \tag{41}$$

Note that the coordinate axes corresponding to the discrete state components $x_i, i = 1, \dots, N$ of z are not, and do not need to be, transformed.

5. Periodically Preconditioned Conjugate Gradient-Restoration Algorithm Using the Hybrid Approach (PPCGRA:HA)

The PPCGRA:HA consists of a sequence of conjugate gradient-restoration cycles. The conjugate gradient portion of each cycle is based upon a conjugate gradient algorithm that is derived for the special case of a quadratic function subject to linear constraints. This portion of the cycle involves a single step and is designed to decrease the value of the function while satisfying the constraints to first order. The restoration portion of each cycle involves one or more iterations and is designed to restore the norm of the constraint function to within a predetermined tolerance about zero.

The conjugate gradient-restoration sequence is reinitialized with a simple gradient step every $n^* - q^*$ or less cyles, where n^* is the number of optimizing variables and q^* is the number of constraints. Here $n^* = (n+m)N + p$ and $q^* = nN + q + kN$. At the beginning of each simple gradient step, the preconditioning matrix $\mathcal{H}_{LR_+}$ is computed via Eq. (40), for the purpose of accelerating the convergence rate of the algorithm.

The non-preconditioned version of the algorithm can be obtained from the periodically preconditioned algorithm by setting $\mathcal{H}_+$ equal to the identity matrix I in Eq. (41), i.e., setting $\mathcal{H}_{LR_+} = I$. It should be noted that whenever $\mathcal{H}_{LR}$ is singular, $\mathcal{H}_{LR_+}$ is made positive definite by adding the last nonzero singular value to the zero singular values in Eq. (40). This enables the algorithm to simply step through the singularities.

Conjugate Gradient Phase

Let $z = [z_1^T \;\; z_2^T]^T$ (see Eqs. (27)-(29)) denote a nominal point satisfying the constraint equation (13) to within a predetermined accuracy. Let $\Delta z = [\Delta z_1^T \;\; \Delta z_2^T]^T$ be a displacement from z which leads to the varied point $\tilde{z} = [\tilde{z}_1^T \;\; \tilde{z}_2^T]^T$. Let $p = [p_1^T \;\; p_2^T]^T$ be the present search direction, $\hat{p} = [\hat{p}_1^T \;\; \hat{p}_2^T]^T$ the previous search direction,$(\hat{z}, \hat{\lambda})$ the solution values at the beginning of the previous search, γ the directional coefficient, and α the conjugate gradient stepsize. Finally, let A be the variation of the discrete state variables normalized with respect to the gradient stepsize, i.e., $A = \Delta z_1/\alpha$. The conjugate gradient phase is represented by the following equations:

$$\begin{bmatrix} \Phi_{z_1}^T(z) & -\Phi_{z_2}^T(z)\mathcal{H}_{LR_+}^{-1}\Phi_{z_2}(z) \\ 0 & \Phi_{z_1}(z) \end{bmatrix} \begin{bmatrix} A \\ \lambda \end{bmatrix} = \begin{bmatrix} \Phi_{z_2}^T(z)\mathcal{H}_{LR_+}^{-1}F_{z_2}(z) \\ -F_{z_1}(z) \end{bmatrix} \tag{42}$$

$$\begin{aligned} Q(z,\lambda) &= J_{z_2}^T(z,\lambda)\mathcal{H}_{LR_+}^{-1}J_{z_2}(z,\lambda) && (43) \\ \gamma &= Q(z,\lambda)/Q(\hat{z},\hat{\lambda}) && (44) \\ p_1 &= A && (45) \\ p_2 &= -J_{z_2}(z,\lambda) + \gamma\hat{p}_2 && (46) \\ \begin{bmatrix} \Delta z_1 \\ \Delta z_2 \end{bmatrix} &= \alpha\mathcal{H}_+^{-1}p = \alpha \begin{bmatrix} p_1 \\ \mathcal{H}_{LR_+}^{-1}p_2 \end{bmatrix} && (47) \\ \tilde{z} &= z + \Delta z && (48) \end{aligned}$$

It can be shown that the values of A and λ that satisfy Eq. (42) (a) ensure satisfaction of the necessary condition (24) and minimize the norm of the error in the necessary condition (25), (b) yield a projected gradient that satisfies the constraints to first order, and (c) yield a conjugate search direction that approximately satisfies the constraints to first order provided that the gradient of the constraint function is not changing too rapidly. For the first iteration, or any simple gradient step, Eq. (44) is bypassed and γ is set to zero.

After determining the search direction, the optimum stepsize α is computed via a one-dimensional minimizing search on the augmented function

$$J(\tilde{z},\lambda) = J(z + \alpha\mathcal{H}_+^{-1}p, \lambda) = J(\alpha) \tag{49}$$

If $J_\alpha(0) < 0$, the search for the optimium gradient stepsize is initiated. If $J_\alpha(0) \geq 0$ and $\gamma \neq 0$, the conjugate gradient-restoration sequence is reinitialized with $\gamma = 0$. If the optimum stepsize causes the norm of the constraint function to exceed some huge predetermined bound, the stepsize is bisected until the bound is satisfied and the conjugate gradient sequence is reinitialized in the next cycle with $\gamma = 0$.

The simultaneous conditions

$$J_\alpha(0) \geq 0, \qquad \gamma = 0 \tag{50}$$

represent the nonconvergence of the algorithm due to roundoff errors.

It is important to note that Eq. (42) represents the solution of the linearized two-point boundary-value problem given by

$$\Phi_{z_1}^T(z)A + \Phi_{z_2}^T(z)\begin{bmatrix} B \\ C \end{bmatrix} = 0 \tag{51}$$

$$J_{z_1}(z,\lambda) = 0 \tag{52}$$

$$\begin{bmatrix} B \\ C \end{bmatrix} = -\mathcal{H}_{LR+}^{-1} J_{z_2}(z,\lambda) \tag{53}$$

where

$$\begin{bmatrix} B \\ C \end{bmatrix} = \Delta z_2/\alpha \tag{54}$$

is the variation in the discrete control variables and control parameters normalized with respect to the gradient stepsize. The first N components of Eq. (51) represent the integration of the linearized and discretized dynamics equation (2). The first $N-1$ components of Eq. (52) represent the integration of the discretized necessary condition $\dot{\lambda} = H_x$ as given by Equations (24) and (30).

If the preconditioner is set to the identity matrix, the above two-point boundary-value problem has the same structure as those given in [20] and [21]. The gradient algorithms in [20] and [21] were derived using the indirect approach to optimal control and the linearized two-point boundary-value problems were solved using integration sweeps in conjunction with the method of particular solutions. Similarly, using the hybrid approach in the absence of preconditioning, it was shown in [19] that integration sweeps could be employed in solving Eqs. (51)-(53) and moreover that the integration methods were already embedded in the equations. A significant computational advantage of the hybrid approach, however, is that integration sweeps can be bypassed and the linearized two-point boundary-value problem can be solved directly via Eq. (42). Additionally, the presence of preconditioning renders the integration sweep approach impossible since the matrix multiplier $\mathcal{H}_{LR+}^{-1}$ in Eq. (53) makes each component B_i of the normalized variation in the discrete controls a function of all the discrete state and control variables which are referenced to the discrete times $t_i, i = 1, \ldots, N$.

Restoration Phase

A restoration phase involving one or more iterations is conducted after each conjugate gradient step. The restoration phase is designed to restore the norm of the constraint function to within a predetermined tolerance about zero. Modified Newton's method is applied to the constraint function $\Phi(z)$ with an auxillary minimum distance criterion enforced for solution uniqueness.

Let z be nominal point and let Δz be a displacement from z which leads to the varied point $\tilde{z}$. The restoration algorithm is defined by the solution of the problem of minimizing

$$\frac{1}{2}\Delta z^T \mathcal{H}_+ \Delta z \tag{55}$$

subject to

$$\beta\Phi(z) + \Phi_z^T(z)\Delta z = 0 \tag{56}$$

where the constraint (56) is the underdetermined modified Newton equation. Given that p is the search direction, β is the restoration stepsize, and $\beta\sigma$ is the Lagrange multiplier for the constraint (56), the restoration phase is represented by the following equations:

$$\Phi_z^T(z)\mathcal{H}_+^{-1}\Phi_z(z)\sigma - \Phi(z) = 0 \tag{57}$$

$$p = -\Phi_z(z)\sigma \tag{58}$$

$$\Delta z = \beta\mathcal{H}_+^{-1}p \tag{59}$$

$$\tilde{z} = z + \Delta z \tag{60}$$

For each iteration, the stepsize β is initially set to one and is bisected as required to ensure a monotonic descent property on the norm of the constraint function. The restorative iterations are terminated when the norm of the constraint function drops below a predetermined small value.

6. Comparisons Between the Algorithms

The periodically preconditioned conjugate gradient-restoration algorithm based on the direct approach (PPCGRA:DA) performs gradient searches over all of the discrete variables, including the discrete state variables. In this case, the linearized two-point boundary-value problem is given by

$$\Phi_z^T(z)\begin{bmatrix} A \\ B \\ C \end{bmatrix} = 0 \tag{61}$$

$$\begin{bmatrix} A \\ B \\ C \end{bmatrix} = -\mathcal{H}_+^{-1} J_z(z,\lambda) \tag{62}$$

where the preconditioner $\mathcal{H}_+$ is computed from the singular value decomposition of the entire Hessian $\mathcal{H}(z,\lambda)$ via

$$\mathcal{H}(z,\lambda) = USV^T \tag{63}$$
$$\mathcal{H}_+ = USU^T = VSV^T \tag{64}$$

Thus, the conjugate gradient phase of the PPCGRA:DA is given by

$$\Phi_z^T(z)\mathcal{H}_+^{-1}\Phi_z(z)\lambda + \Phi_z^T(z)\mathcal{H}_+^{-1}F_z(z) = 0 \tag{65}$$
$$Q(z,\lambda) = J_z^T(z,\lambda)\mathcal{H}_+^{-1}J_z(z,\lambda) \tag{66}$$
$$\gamma = Q(z,\lambda)/Q(\hat{z},\hat{\lambda}) \tag{67}$$
$$p = -J_z(z,\lambda) + \gamma\hat{p} \tag{68}$$
$$\Delta z = \alpha\mathcal{H}_+^{-1}p \tag{69}$$
$$\tilde{z} = z + \Delta z \tag{70}$$

where Eq. (65) has been obtained by substituting Eq. (62) into Eq. (61). The restoration phase of the PPCGRA:DA has the same structure as that of the PPCGRA:HA but with the preconditioner being based on the singular value decomposition of the entire Hessian matrix $\mathcal{H}(z,\lambda)$.

Algorithmic Comparisons

Comparing the conjugate gradient phases of the PPCGRA:HA and PPCGRA:DA, two potential advantages of the PPCGRA:HA over the PPCGRA:DA can be seen. First, the PPCGRA:HA only performs gradient searches only over the discrete controls and control parameters while ensuring that $J_{z_1}(z,\lambda) = 0$, the necessary condition corresponding to $\dot{\lambda} = H_x$, is enforced at the beginning of every step. This reduction in the search space can be substantial. For example, if $N = 100$, $n = 5$, $m = 2$, and $p = 1$, the dimension of the search space of the PPCGRA:HA would be 201 as compared to 701 for the PPCGRA:DA. Second, construction of the preconditioner for the PPCGRA:HA involves performing a singular value decomposition only on $\mathcal{H}_{LR}$ instead of the entire Hessian matrix $\mathcal{H}$. In the example above, the PPCGRA:HA would have to decompose a 201×201 matrix as compared to a 701×701 for the PPCGRA:DA.

Numerical Results

The PPCGRA:HA and the PPCGRA:DA were applied to several optimal control problems and compared in terms of convergence rate and approximate number of computer operations. Three examples will be presented to illustrate the performance of these algorithms with and without preconditioning.

The algorithms were implemented in MATLAB, and the one dimensional line searchs over β in equation (59) and α in equation (69) were conducted using a golden section search with parabolic interpolation. The tolerance for convergence of the one dimensional line search was set at 10^{-6}. Similarily, the tolerance for termination of the restoration loop was 10^{-6}, and the least positive singular value allowed in computing $\mathcal{H}_+^{-1}$ or $\mathcal{H}_{LR+}^{-1}$ was 10^{-12}. Convergence of the algorithm was determined whenever:

$$\|J_z(z,\lambda)\| \leq 10^{-6}$$

For each example the functions $f(x(t),u(t),\pi,t)$, $g(x(1),\pi)$, $\phi(x(t),u(t),\pi,t)$, $S(x(t),u(t),\pi,t)$, and $\psi(x(1),\pi)$ were defined in continuous form as matrices in MATLAB .m files. The first and second partial derivatives of these functions with respect to the variables x, u, and π were likewise defined. MATLAB functions were developed to transcribe the continuous problem to a discrete problem by numerical evaluation of the function, constraint and the necessary partial derivatives of each for any given number of time points. The integration method used for the quadrature scheme in equation (12) was a modified composite method using Simpson's rule, in which midpoints of the function are added using linear interpolation. A fourth-order Runga-Kutta method was used for the increment function $\Gamma(\phi_i)$ in the constraint equation (13), which represents integration of the system differential equations between time points.

Example 1 - The first example represents a second order nonlinear multiple input system with quadratic performance index. The objective is to minimize:

$$J = \int_0^1 [x_1^2 + x_2^2 + 2(u_1^2 + u_2^2)]dt$$

subject to the differential constraints:

$$\dot{x} = \phi = \begin{bmatrix} x_1 - x_1^3 + x_2 + u_1 \\ x_1 + x_1^2 x_2 - x_2 + u_2 \end{bmatrix}$$

with initial conditions:

$$x(0) = [1 \quad 2]^T$$

Figure 2 presents a plot of the optimum state trajectories and control histories obtained numerically using the preconditioned hybrid approach algorithm with 20 time increments. These results agree with the those obtained using the direct approach. The convergence rates of the two algorithms are presented in the following table, where # Steps represents the number of conjugate gradient-restoration

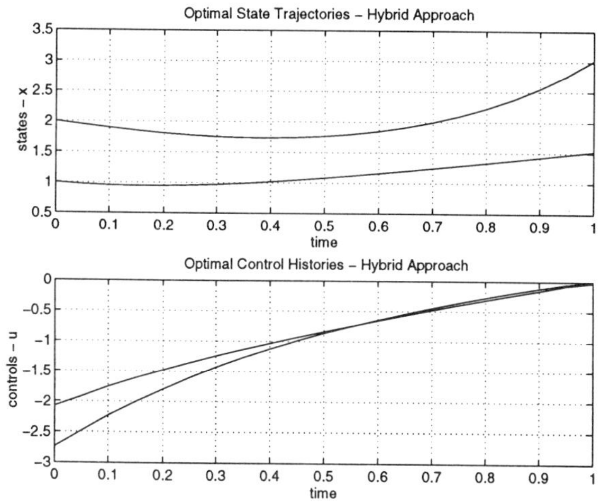

Figure 2: Example 1

steps taken to converge, and Mega Flops represents the approximate number of floating point operations required, in millions.

Example 1 Convergence Rates

Precon-ditioned	# Steps		Mega Flops	
	Direct	Hybrid	Direct	Hybrid
No	14	27	160.9	503.2
Yes	6	14	97.2	294.3

Example 2 - This example presents a second order linear system minimum time problem with an inequality constraint upon the control. This problem has been presented in previous work utilizing the direct approach [18]. The objective is to minimize final time, t_f:

$$J = t_f$$

subject to:

$$\dot{x} = \phi = \begin{bmatrix} x_2 \\ u \end{bmatrix}$$

with initial conditions:

$$x(0) = [0 \quad 0]^T$$

terminal constraints:

$$x(t_f) = [1 \quad 0]^T$$

and the control contraint:

$$|u| \leq 1.$$

Final time, which is free in this problem, can be represented as a parameter ($\pi = t_f$), but in order to ensure a positive solution, π^2 must be used in the performance index, and the state equations must then be time scaled by π^2. The inequality constaint on u is transformed into an equality constraint using a slack variable u_2 and a Valentine transformation. These transformations put the original system into the requisite form of equations (1)-(4) and result in the following equivalent optimization problem with nonlinear constraints. Minimize:

$$J = \pi^2$$

subject to:

$$\dot{x} = \phi = \begin{bmatrix} \pi^2 x_2 \\ \pi^2 u_1 \end{bmatrix}$$

$$S = u_1^2 + u_2^2 - 1 = 0$$

with terminal constraint:

$$\psi = \begin{bmatrix} x_1 - 1 \\ x_2 \end{bmatrix} = 0$$

and initial conditions:

$$x(0) = [0 \quad 0]^T.$$

Constrained minimum time problems such as this are notoriously difficult to solve due to the discontinuous nature of the solution. Figure 3 presents the optimal solution obtained numerically using the hybrid approach with 20 time increments. It is obvious that the control switching resolution is dependent upon the number of time points used in discretizing the problem. Again, these results agree with the those obtained using the direct approach and the convergence rates of the two algorithms are presented in Table 2.

Example 2 Convergence Rates

Precon-ditioned	# Steps		# Mega Flops	
	Direct	Hybrid	Direct	Hybrid
No	26	27	1011.8	1284.4
Yes	11	12	501.9	656.1

Example 3 - The final example is a second order linear system with an ill-conditioned quadratic performance index. This example demonstrates the importance of preconditioning for both the direct approach and the hybrid approach. The objective is to minimize:

$$J = \int_0^1 [x_1^2 + x_2^2 + .005u^2]dt + g(x, \pi)$$

subject to:

$$\dot{x} = \phi = \begin{bmatrix} x_2 \\ -x_2 + u \end{bmatrix}$$

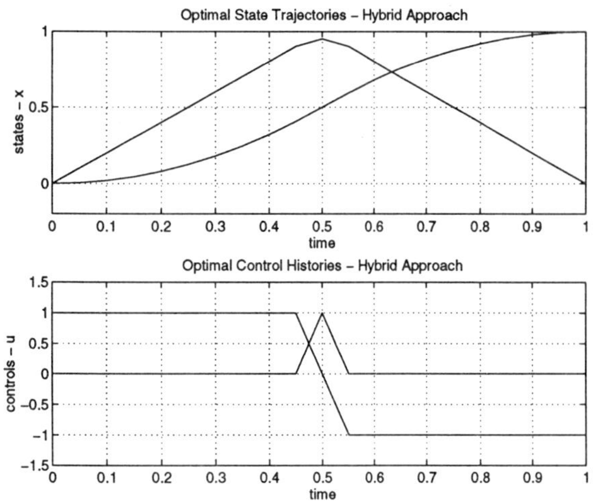

Figure 3: Example 2

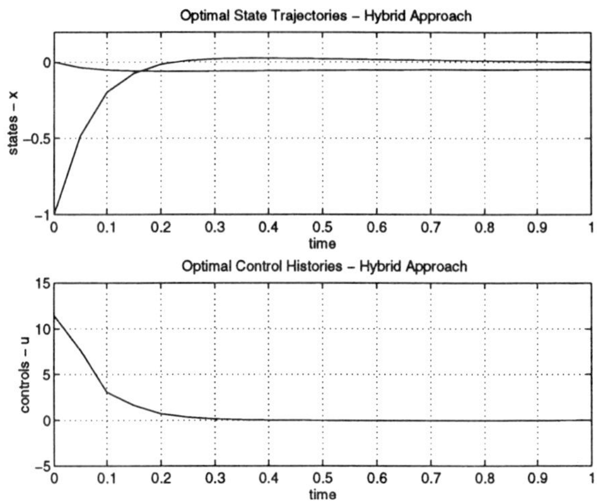

Figure 4: Example 3

with initial conditions:

$$x(0) = [0 \quad -1]^T$$

Figure 4 presents a plot of the optimum state trajectories and control histories obtained using the preconditioned hybrid approach algorithm with 20 time increments. These results also agree with the those obtained using the direct approach, and the convergence rates of the two algorithms are presented in the following table.

Example 3 Convergence Rates

Precon-	# Steps		# Mega Flops	
ditioned	Direct	Hybrid	Direct	Hybrid
No	341	363	2019.5	6236.0
Yes	1	44	15.7	724.3

For linear quadratic problems it is expected that the conjugate gradient-restoration algorithm will converge in $n^* - q^*$ $(= 20)$ steps or less without preconditioning, and that it will converge in one step with preconditioning. Due to the ill-conditioning of the performance index in this example, numerical round-off errors result in the poor performance observed for both approaches without preconditioning. For the direct approach, preconditioning reduces the number of steps required to converge from 341 to 1. One step convergence is not obtained for the hybrid approach using preconditioning, since the solution for the Lagrange multiplier is not optimal at each step in the hybrid approach. In terms of computer floating point operations, preconditioning provides nearly an order of magnitude improvement for the hybrid approach, and two orders of magnitude improvement for the direct approach.

7. Summary

A hybrid approach has been used to develop a periodically preconditioned conjugate gradient-restoration algorithm for optimal control problems. The algorithm conducts gradient searches only over the discrete control variables and control parameters. The linearized two-point boundary-value problem is solved via a linear algebraic equation, eliminating the need to perform integration sweeps. Preconditioning is made possible through knowledge of the Hessian which is available as a result of direct transcription of the original infinite-dimensional problem.

Initially it was hypothesized that the combination of a reduced gradient search space, and an efficient means of solving the linearized two-point boundary-value problem would result in significant convergence rate improvements for the hybrid approach over the direct approach. Numerical results did not support this conjecture, and in fact the hybrid algorithm required slightly more iterations and increased computational time to converge. Furthermore, preconditioning did not produce one step convergence in the linear quadratic problem using the hybrid approach, as it did using the direct approach. Preconditioning however, was shown to be a powerful enhancement for the convergence rate using either approach.

These results indicate that the direct approach is preferable to the hybrid approach in the examples studied. The hybrid approach may yet provide a useful tool in the numerical solution of ill-conditioned constrained optimal control problems of large dimension, where preconditioning is required, and the size of the Hessian becomes the dominant computational concern. In this regard however, future studies must also compare the hybrid approach using preconditioning to the direct approach using suboptimal preconditioning.

References

[1] Luenberger, D.G., *Introduction to Linear and Nonlinear Programming*, Addison-Wesley Publishing Company, Reading, Massachusetts, 1973.

[2] Gill, P.E., Murray, W., and Wright, M.H., *Practical Optimization*, Academic Press, New York, New York, 1981.

[3] Dennis, J.E., Jr., and Schnable, R.B., "A view of Unconstrained Optimization," Report No. CU-CS-376-87, Computer Science Department, University of Colorado, 1987.

[4] Golub, G.H., and O'Leary, D.P., "Some History of the Conjugate Gradient and Lanczos Algorithms: 1948-1976," SIAM Review, Vol 31, No. 1, pp. 50-102, 1989.

[5] Hoffman, K., and Kunze, R., *Linear Algebra*, Prentice-Hall, Englewood Cliffs, New Jersey, 1971.

[6] Nour-Omid, B., and Parlett, B.N., "A Comparison of Lanczos with Conjugate Gradient Using Element Preconditioning," Report No. PAM-363, Center for Pure and Applied Mathematics, University of California, Berkeley, California, 1987.

[7] Nour-Omid, B., and Parlett, B.N., "Element Preconditioning Using Splitting Techniques," SIAM Journal of Scientific and Statistical Computing, Vol. 6, pp. 761-770, 1985.

[8] Winget, J.M., and Hughes, J.T.R., "Solution Algorithms for Nonlinear Transient Heat Conduction Analysis Employing Element-by-Element Iterative Strategies," Computer Methods in Applied Mathematics and Engineering, Vol. 52, pp. 711-815, 1985.

[9] Gary, J., McCormick, S., and Sweet, R., "Successive Overrelaxation, Multigrad, and Preconditioned Conjugate Gradient Algorithms for Solving a Diffusion Problem on a Vector Computer," Applied Mathematics and Computation, Vol. 13, pp. 285-309.

[10] Wong, Y.S., "Solving Large Elliptic Difference Equations on Cyber 205," Parallel Computing, Vol 6, PP. 195-205, 1988.

[11] Ashby, S.F., "Polynomial Preconditioning for Conjugate Gradient Methods," Report No. UIUCDCS-R-87-1355, Department of Computer Science, University of Illinois, Urbana, Illinois, 1987.

[12] Chronopoulos, A.T., and Gear, C.W., "Implementation of Preconditioned s-Step Conjugate Gradient Methods on a Multiprocessor System with Memory Hierarchy," Report No. UIUCDCS-R-87-1347, Department of Computer Science, University of Illinois, Urbana, Illinois, 1987.

[13] Poole, E.L., "Multi-Color Incomplete Choleski-Conjugate Gradient Methods for Vector Computers," PhD Thesis, University of Virginia, Charlottesville, Virginia, 1986.

[14] Kershaw, D.S., "The Incomplete Cholski-Conjugate Gradient Method for Solution of Systems of Linear Equations," Journal of Computational Physics, Vol. 26, pp.3-65, 1978.

[15] Cloutier, J.R., and Wilson R.F., "Periodically Preconditioned Conjugate Gradient-Restoration Algorithm," Journal of Optimization Theory and Applications, Vol. 70, No. 1, July 1991.

[16] Cloutier, J.R., and D'Souza C.N.,"A Preconditioning Theory for the Standard Optimal Control Problem," Proceedings of the AIAA Guidance, Navigation, and Control Conference, Hilton Head, SC, August 1992.

[17] Cloutier, J.R., and D'Souza C.N.,"A Preconditioning Theory for Optimal Control Problems with Nondifferential Constraints," Proceedings of the AIAA Guidance, Navigation, and Control Conference, Monterey, CA, August 1993.

[18] Mracek, C. P., Cloutier, J.R., and D'Souza C.N.,"A Periodically Preconditioned Conjugate Gradient-Restoration Algorithm: The Direct Approach," Proceedings of the AIAA Guidance, Navigation, and Control Conference, San Diego, CA, July 1996.

[19] Cloutier, J.R., and D'Souza C.N.,"A Hybrid Approach to Optimal Control Problems with Nondifferential Constraints," Proceedings of the American Control Conference, Seattle, WA, June 1995.

[20] Miele, A., Damoulakis, J. N., Cloutier, J. R., and Tietze, J. L., "Sequential Gradient-Restoration Algorithm for Optimal Control Problems with Nondifferential Constraints," Journal of Optimization Theory and Applications, Vol. 13, No. 2, pp. 218-255, 1974.

[21] Miele, A. and Cloutier, J. R., "Sequential Conjugate Gradient-Restoration Algorithm for Optimal Control with Nondifferential Constraints," *Applied Nonlinear Analysis*, Edited by V. Lakshmikantham, J. Eisenfeld, and A. R. Mitchell, Academic Press, New York, New York, pp. 89-93, 1979.

AIAA-97-3451

MINIMAL-PARAMETER ATTITUDE MATRIX ESTIMATION FROM VECTOR OBSERVATIONS

Yaakov Oshman* and F. Landis Markley†
Guidance, Navigation and Control Branch, Code 712
NASA Goddard Space Flight Center, Greenbelt, Maryland 20771

ABSTRACT

A computationally efficient, sequential method is presented for attitude matrix estimation using gyro and vector measurements. The method is based on a recently introduced, minimal-parameter third-order method for solving the orthogonal matrix differential equation. A nonlinear algorithm is developed, which uses a minimal set of three parameters to estimate the nine-parameter direction-cosine matrix. An orthogonalization procedure, incorporated into the measurement processing stage, enhances the accuracy and stability of the resulting algorithm, yet retains reasonable simplicity. The performance of the estimator is demonstrated via a numerical simulation study.

I Introduction

USING a sequence of vector measurements for attitude determination has been intensively investigated over the last three decades. First proposed in 1965 by Wahba, the problem is to estimate the attitude of a spacecraft based on a sequence of noisy vector observations, resolved in the body-fixed coordinate system and in a reference system. The use of attitude determination from vector observations was reported in Ref. 1.

Inertial reference systems typically utilize vector measurements in combination with strap-down gyros to estimate both the spacecraft attitude and the gyro drift rate biases. Several approaches have been proposed for the design of such systems, differing mainly in their choice of attitude representation method.

*National Research Council Research Associate, currently on sabbatical, Department of Aerospace Engineering, Technion—Israel Institute of Technology, Haifa 32000, Israel. Senior Member AIAA.

†Staff Engineer. Associate Fellow AIAA.

The quaternion was used in Refs. 2–4, in the framework of extended Kalman filtering (EKF) algorithms. Vector observations were also used to estimate both the quaternion and the attitude rate of the spacecraft, in a gyroless attitude determination and control setting.[5] The main advantage of using the quaternion representation is that it is not singular for any rotation. Moreover, its kinematic equation is linear and the computation of the associated attitude matrix involves only algebraic expressions. However, the quaternion representation is not minimal since it is four-dimensional. This leads to a normalization constraint that has to be addressed in filtering algorithms, and increases the associated computation load.

Euler angles were used in Refs. 6,7. Although this parametrization is minimal, its use imposes a large computational burden due to the transcendental expressions involved in the computation of the attitude matrix. Moreover, Euler angles are singular, as are all three-dimensional (3-D) attitude representations.

The Rodrigues parameters (also known as the Gibbs vector) were utilized in Ref. 8. Being a minimal set, they render the resulting estimator computationally efficient. However, the Rodrigues parameters are singular for 180 degree rotations. The modified Rodrigues parameters (MRP), which allow rotations up to 360 degrees, were recently used in Ref. 9.

The direction-cosine matrix (DCM), a natural attitude representation, was used in a gyro-star tracker setting by several researchers.[10] Since it is inherently nonsingular, it requires no special singularity-handling procedures. Moreover, its kinematic equation is linear, as is its associated vector measurement equation, which greatly facilitate the filter implementation. Although the advantages of directly parameterizing the attitude using the DCM are clear, the main disadvantage of this approach is computational, as it requires the estimation of a nine-dimensional parameter vector.

In this paper we propose to sequentially estimate the attitude matrix using a minimal-dimension filter, thus alleviating the computational burden normally associated with DCM identification. It is assumed, as usual, that the body-referenced angular velocity is measured by an orthogonal triad of rate gyros. The approach taken to this end is motivated by the idea of finding a minimal-parameter solution to the orthogonal matrix differential equation in $\mathbb{R}^n$, first introduced by Bar-Itzhack and Markley.[11] In a recent work targeted at solving a problem first raised in Ref. 12, they presented a minimal-parameter solution to the orthogonal matrix differential equation

$$\dot{V}(t) = W(t)V(t), \qquad V(t_0) = V_0 \tag{1}$$

where $V(t) \in \mathbb{R}^{n,n}$, $V_0 V_0^T = I$, $W(t) = -W^T(t)$ $\forall t \geq t_0$ and the raised dot indicates the temporal derivative. Exploiting the properties of V and W, Bar-Itzhack and Markley introduced a novel minimal parametrization of the orthogonal matrix V. Based on extended Rodrigues parameters (ERP), this parametrization enabled solving Eq. (1) using only $n(n-1)/2$ parameters, as opposed to n^2 integrations implied by a straightforward solution of Eq. (1).

Motivated by Ref. 11, Ronen and Oshman[13] have recently introduced a third-order method for the solution of the orthogonal matrix differential equation in $\mathbb{R}^n$. The method is based on a third-order, minimal parametrization of the orthogonal matrix V using the $n(n-1)/2$ off-diagonal terms of the skew-symmetric matrix $A(t, t_0) \triangleq \int_{t_0}^{t} W(\tau)\, d\tau$. For the 3-D case, these parameters, hereafter called integrated rate parameters (IRP), are the angles resulting from time-integration of the body-frame components of the spacecraft angular velocity.

The work presented herein utilizes the minimal, three-dimensional IRP vector to estimate the nine-parameter attitude matrix. Building upon this state vector, the resulting three-dimensional filter possesses an extremely simple time-propagation procedure, which is at the heart of its computational efficiency. The DCM orthogonality constraint is also addressed, contributing to the filter's accuracy and numerical stability, yet keeping its structure reasonably simple.

In the following section we briefly review the IRP minimal-parameter, third-order method. In Sec. III we develop the attitude estimator. Special attention is given to the analysis of the orthogonalization step. Sec. VI presents a numerical study, consisting of Monte Carlo simulations of the new algorithm. Concluding remarks are offered in the last section.

II Third-Order, Minimal Attitude Parameters

For completeness, this section briefly reviews the minimal-parameter problem and the IRP method. Then, the method is adapted to the 3-D attitude kinematic equation.

The Minimal-Parameter Problem

The minimal-parameter problem is the following.[11]

Given the matrix differential equation Eq. (1), find: 1) a set of $m = n(n-1)/2$ parameters which unambiguously define $V(t)$, 2) the differential equation satisfied by these parameters, 3) the transformation which maps these parameters into the matrix V, and 4) a simple and efficient method to solve the parameters' differential equation and to compute $V(t)$.

The recently introduced, third-order, minimal-parameter method[13] for the solution of Eq. (1) is summarized in the following.

Let the skew-symmetric matrix $A(t, t_0)$ be defined as

$$A(t, t_0) \triangleq \int_{t_0}^{t} W(\tau)\, d\tau \tag{2}$$

Then, a third-order approximation of the solution $V(t)$ in terms of the entries of the matrix $A(t, t_0)$ is given by the matrix $\tilde{V}(t, t_0)$, defined as

$$\tilde{V}(t, t_0) \triangleq \Big\{ I + A(t, t_0) + \frac{A^2(t, t_0)}{2!} + \frac{A^3(t, t_0)}{3!} + \frac{(t - t_0)}{3!} \Big[A(t, t_0) W_0 - W_0 A(t, t_0) \Big] \Big\} V_0 \tag{3}$$

where $W_0 = W(t_0)$. Moreover, $\tilde{V}$ is a third-order approximation of an orthogonal matrix, in the sense that

$$\tilde{V}(t, t_0) \tilde{V}^T(t, t_0) = I + \mathcal{O}\big((t - t_0)^4\big) \tag{4}$$

where $\mathcal{O}(x)$ denotes a function of x that has the property that $\mathcal{O}(x)/x$ is bounded as $x \to 0$.

Referring now to the minimal-parameter problem, the new parameters, which define the third-order solution of Eq. (1), are the $n(n-1)/2$ off-diagonal terms of $A(t, t_0)$. For the 3-D case, these parameters have a simple geometric interpretation: they are the angles resulting from a temporal-integration of the three components of the angular velocity vector

$$\omega(t) \triangleq \Big[\omega_1(t) \quad \omega_2(t) \quad \omega_3(t) \Big]^T \tag{5}$$

where ω_i is the angular velocity component along the i-axis of the initial coordinate system, and $i = 1, 2, 3$ for x, y, z, respectively.

The differential equation satisfied by these parameters is

$$\dot{A}(t, t_0) = W(t), \qquad A(t_0, t_0) = 0 \tag{6}$$

which can be easily solved using any quadrature scheme. As demonstrated in Ref. 13, this minimal-parameter method is both computationally efficient and accurate.

The Attitude Matrix Kinematic Equation

In the 3-D case, the orthogonal matrix referred to is the attitude matrix, or the direction cosine matrix (DCM), denoted by $D(t)$. The well-known differential equation satisfied by this matrix is

$$\dot{D}(t) = \Omega(t)D(t), \qquad D(t_0) = D_0 \tag{7}$$

where

$$\Omega(t) = -\big[\omega(t)\times\big] \tag{8}$$

and $\big[\omega(t)\times\big]$ is the *cross product matrix* corresponding to $\omega(t)$, which is defined according to

$$\big[a\times\big] \triangleq \begin{bmatrix} 0 & -a_3 & a_2 \\ a_3 & 0 & -a_1 \\ -a_2 & a_1 & 0 \end{bmatrix} \qquad \forall a \in \mathbb{R}^3 \tag{9}$$

In this case, the matrix $A(t, t_0)$ takes the form

$$A(t, t_0) \triangleq -\big[\theta(t)\times\big] \tag{10}$$

where the parameter vector $\theta(t)$ is defined as

$$\theta(t) \triangleq \big[\theta_1(t) \quad \theta_2(t) \quad \theta_3(t)\big]^T \tag{11}$$

and

$$\theta_i(t) \triangleq \int_{t_0}^{t} \omega(\tau)\, d\tau, \qquad i = 1, 2, 3 \tag{12}$$

III The Attitude Estimator

In this section we develop the attitude estimation algorithm. The development of the algorithm relies on the choice of the parameter vector θ, defined in Eqs. (11) and (12), to be the estimator's state vector.

Let the sampling period be denoted by $T \triangleq t_{k+1} - t_k$. Using the notation $\theta(k) \triangleq \theta(t_k)$, the state vector at time t_k is

$$\theta(k) = \big[\theta_1(k) \quad \theta_2(k) \quad \theta_3(k)\big]^T \tag{13}$$

and Eq. (12) implies

$$\theta_i(k) = \int_{t_0}^{t_k} \omega_i(\tau)\, d\tau, \qquad i = 1, 2, 3 \tag{14}$$

where $\omega(t)$ is the spacecraft angular velocity vector, defined in Eq. (5). From Eq. (14) we have

$$\theta(k+1) = \theta(k) + \int_{t_k}^{t_{k+1}} \omega(\tau)\, d\tau \tag{15}$$

Defining $A(k+1, k)$ to be the discrete-time analog of $A(t, t_0)$, i.e.,

$$A(k+1, k) \triangleq -\big[\big(\theta(k+1) - \theta(k)\big)\times\big] \tag{16}$$

Eq. (3) is rewritten as

$$\begin{aligned} D(k+1) = \Big\{ & I + A(k+1,k) + \frac{1}{2}A^2(k+1,k) \\ & + \frac{1}{6}A^3(k+1,k) + \frac{1}{6}T\Big[A(k+1,k)\Omega(k) \\ & - \Omega(k)A(k+1,k)\Big]\Big\} D(k) \end{aligned} \tag{17}$$

In practice we only have access to the *measured* angular velocity, denoted by $\hat{\omega}(t)$, which satisfies

$$\hat{\omega}(t) = \omega(t) + \delta\omega(t) \tag{18}$$

Here $\delta\omega(t)$ is the rate-gyro (RG) measurement noise. For simplicity, this noise is assumed to be a zero-mean, white, Gaussian process, denoted as

$$\delta\omega(t) \sim \mathcal{WN}\big(0, Q(t)\big) \tag{19}$$

where its intensity, $Q(t)$ (the power spectral density matrix), is defined by

$$E\big\{\delta\omega(t)\delta\omega^T(s)\big\} = Q(t)\delta(t-s) \tag{20}$$

(The incorporation of more elaborate gyro noise models is straightforward).

The estimation algorithm comprises three sub-algorithm: 1) filtering, 2) DCM orthogonalization, and 3) prediction. These three procedures are developed in the ensuing.

Filtering

Let the minimum mean-squared estimate (MMSE) of $\theta(j)$ based on measurements up to and including t_k be denoted by $\hat{\theta}(j|k)$. Assume that at t_{k+1} we have on hand the predicted parameter vector $\hat{\theta}(k+1|k)$ and its corresponding prediction error covariance matrix, $P(k+1|k) \triangleq E\{\tilde{\theta}(k+1|k)\tilde{\theta}^T(k+1|k)\}$, where the estimation error is defined as

$$\tilde{\theta}(j|k) \triangleq \theta(j) - \hat{\theta}(j|k) \tag{21}$$

The purpose of the filtering scheme, to be developed in the sequel, is to compute the a posteriori estimate and the corresponding error covariance matrix by way of incorporating the new vector measurements acquired at t_{k+1}.

As the first step in developing the measurement update algorithm, we next derive the observation equation, relating the acquired vector measurements to the state.

The Observation Equation

Let $\mathcal{S}_u$ and $\mathcal{S}_v$ denote the reference Cartesian coordinate system and the body-fixed Cartesian coordinate system, respectively. The new pair of corresponding noisy vector measurements consists of the *unit* vectors $u(k+1)$ and $v(k+1)$, which represent the *measured* values of the same vector $r(k+1)$, resolved in $\mathcal{S}_u$ and in $\mathcal{S}_v$, respectively. The direction-cosine matrix $D(k+1)$ transforms the *true* vector representation u_0 in $\mathcal{S}_u$ into its corresponding *true* representation v_0 in $\mathcal{S}_v$ according to

$$v_0(k+1) = D(k+1)u_0(k+1) \tag{22}$$

Assuming no constraint on the measurement noise direction, the body-frame measured unit vector, $v(k+1)$, is related to the true vector according to

$$v(k+1) = \frac{v_0(k+1) + n'_v(k+1)}{\|v_0(k+1) + n'_v(k+1)\|} \tag{23}$$

where the sensor measurement noise n'_v is a white, Gaussian noise sequence with $n'_v(k+1) \sim \mathcal{N}\big(0, R'_v(k+1)\big)$. Since both $v_0(k+1)$ and $v(k+1)$ are unit vectors, it follows from Eq. (23) that

$$v(k+1) = v_0(k+1) + \mathcal{P}^{\perp}_{v_0}(k+1)n'_v(k+1) + \mathcal{O}\big(\|n'_v(k+1)\|^2\big) \tag{24}$$

where the idempotent matrix

$$\mathcal{P}^{\perp}_{v_0}(k+1) \triangleq I - v_0(k+1)v_0^T(k+1) \tag{25}$$

represents the orthogonal projector onto the orthogonal complement of $\text{span}\{v_0(k+1)\}$. Defining, therefore, the *effective* measurement noise associated with the measurement $v(k+1)$ as

$$n_v(k+1) \triangleq \mathcal{P}^{\perp}_{v_0}(k+1)n'_v(k+1) \tag{26}$$

yields the following measurement model

$$v(k+1) = v_0(k+1) + n_v(k+1) \tag{27}$$

where the effective measurement noise is, to a good approximation, a white, Gaussian sequence with $n_v(k+1) \sim \mathcal{N}\big(0, R_v(k+1)\big)$ and the measurement noise covariance matrix is

$$R_v(k+1) \triangleq \mathcal{P}^{\perp}_{v_0}(k+1)R'_v(k+1)\mathcal{P}^{\perp}_{v_0}(k+1) \tag{28}$$

Remark 1. This measurement model is similar to the model derived in Ref. 4 for *complete vector sensors.*

Remark 2. If, in the particular sensor used, n'_v is constrained to be orthogonal to v_0, then Eq. (28) reduces to

$$R_v(k+1) = R'_v(k+1) \tag{29}$$

To account for non-ideal effects (e.g., star catalog errors), it is assumed in this work that the true unit vector and the measured unit vector are related according to

$$u(k+1) = u_0(k+1) + n_u(k+1) \tag{30}$$

where $n_u \perp u_0$ is a white, Gaussian measurement noise, that is uncorrelated with n_v and satisfies

$$n_u(k) \sim \mathcal{N}\big(0, R_u(k)\big) \tag{31}$$

with $R_u(k)$ being a known covariance matrix.

Since it is desired to relate the information contained in the measurements to the state vector at the corresponding time point, Eq. (22) is rewritten as

$$v_0(k+1) = D\big[\theta(k+1) - \theta(k), D(k)\big]u_0(k+1) \tag{32}$$

where the notation $D\big[\theta(k+1) - \theta(k), D(k)\big]$ reflects the fact that the attitude at time t_{k+1} is related to the attitude at time t_k via the IRP vector difference $\theta(k+1) - \theta(k)$ [see Eqs. (16) and (17)].

To exploit the information contained in the new vector measurements, the nonlinear measurement equation (32) is linearized about a *nominal* state, consisting of the most recent state estimate. Assuming that immediately after the previous measurement update (at t_k) linearization has been carried out about the *a posteriori* state estimate, the resulting nominal state at the current measurement update is the predicted estimate, $\hat{\theta}(k+1|k)$. Therefore, the predicted parameters are assumed to be related to the true ones according to

$$\theta(k+1) = \hat{\theta}(k+1|k) + \delta\theta(k+1) \tag{33}$$

where $\delta\theta(k+1)$ is the perturbation of the parameter vector about the nominal (i.e., predicted) state. Using now the most recent estimates for $D(k)$ and $\theta(k)$, namely $\hat{D}(k|k)$ and $\hat{\theta}(k|k)$, respectively, in Eq. (32), it follows from Eqs. (27), (30) and (33) that

$$v(k+1) - n_v(k+1) = D\big[\hat{\theta}(k+1|k) + \delta\theta(k+1) - \hat{\theta}(k|k), \hat{D}(k|k)\big]\big(u(k+1) - n_u(k+1)\big) \quad (34)$$

However, as shown in the sequel, the a posteriori state estimate is zeroed after each measurement update (due to full reset control of the state). Hence, hereinafter, the reset value of the state estimate, $\hat{\theta}^c(k|k) = 0$, is used. This allows us to rewrite Eq. (34) as

$$v(k+1) - n_v(k+1) = D\big[\hat{\theta}(k+1|k) + \delta\theta(k+1), \hat{D}(k|k)\big]\big(u(k+1) - n_u(k+1)\big) \quad (35)$$

Now expand D about the nominal parameter vector using a first-order Taylor series expansion, i.e.,

$$D\big[\hat{\theta}(k+1|k) + \delta\theta(k+1), \hat{D}(k|k)\big] = D\big[\hat{\theta}(k+1|k), \hat{D}(k|k)\big] + \sum_{i=1}^{3} \frac{\partial}{\partial\theta_i} D\big[\theta(k+1), \hat{D}(k|k)\big]\big|_{\hat{\theta}(k+1|k)} \delta\theta_i(k+1) \quad (36)$$

where $(\bullet)|_{\hat{\theta}(k+1|k)}$ denotes 'evaluated at $\hat{\theta}(k+1|k)$'. Using Eq. (17), the sensitivity matrices appearing in Eq. (36) are computed as

$$\frac{\partial}{\partial\theta_i} D\big[\theta(k+1), \hat{D}(k|k)\big] = G_i(k+1)\hat{D}(k|k), \quad i = 1,2,3 \quad (37)$$

where the matrices $\{G_i\}_{i=1}^3$ appear in Table 1. In Table 1, the components of θ are evaluated at t_{k+1}, and the components of ω are evaluated at t_k. Notice that in a typical application, it can be assumed that the parameters $\{\theta_i\}_{i=1}^3$ are small, such that the second-order quantities $\{\theta_i\theta_j\}_{i,j=1}^3$ are negligible. Using this small-angle approximation results in much simpler forms for $G_i(k+1)$.

Using now Eq. (36) in Eq. (35) and neglecting second-order terms yields

$$v(k+1) - D\big[\hat{\theta}(k+1|k), \hat{D}(k|k)\big]u(k+1) = \sum_{i=1}^{3} \frac{\partial}{\partial\theta_i} D\big[\theta(k+1), \hat{D}(k|k)\big]\big|_{\hat{\theta}(k+1|k)} \times \delta\theta_i(k+1)u(k+1) - D\big[\hat{\theta}(k+1|k), \hat{D}(k|k)\big]n_u(k+1) + n_v(k+1) \quad (38)$$

Now let the (observation) matrix $H(k+1) \in \mathbb{R}^{3,3}$ be

$$H(k+1) \triangleq \Big[h_1(k+1) \,\Big|\, h_2(k+1) \,\Big|\, h_3(k+1)\Big] \quad (39)$$

where its columns are

$$h_i(k+1) = \frac{\partial}{\partial\theta_i} D\big[\theta(k+1), \hat{D}(k|k)\big]\big|_{\hat{\theta}(k+1|k)} u(k+1), \quad i = 1,2,3 \quad (40)$$

Also, define the *effective* measurement $y(k+1)$ to be

$$y(k+1) \triangleq v(k+1) - D\big[\hat{\theta}(k+1|k), \hat{D}(k|k)\big]u(k+1) \quad (41)$$

and the *effective* measurement noise to be

$$n(k+1) \triangleq n_v(k+1) - D\big[\hat{\theta}(k+1|k), \hat{D}(k|k)\big]n_u(k+1) \quad (42)$$

Then, using these definitions in Eq. (38) yields the following measurement equation:

$$y(k+1) = H(k+1)\delta\theta(k+1) + n(k+1) \quad (43)$$

The measurement noise is a white, Gaussian sequence with

$$n(k+1) \sim \mathcal{N}(0, R(k+1)) \quad (44)$$

where

$$R(k+1) \triangleq R_v(k+1) + D\big[\hat{\theta}(k+1|k), \hat{D}(k|k)\big] \times R_u(k+1)D^T\big[\hat{\theta}(k+1|k), \hat{D}(k|k)\big] \quad (45)$$

Having the linearized measurement equation (43) and the statistical characterization of the measurement noise (44) on hand, we can now derive the MMSE estimator for the parameter vector.

State and Covariance Update

Using Eqs. (21) and (33), we have

$$\delta\theta(k+1) = \theta(k+1) - \hat{\theta}(k+1|k) = \tilde{\theta}(k+1|k) \quad (46)$$

Since $\hat{\theta}(k+1|k)$ is an unbiased, MMSE predictor, we have

$$E\{\delta\theta(k+1)\} = E\{\tilde{\theta}(k+1|k)\} = 0 \quad (47)$$

and

$$\operatorname{cov}\{\delta\theta(k+1)\} = \operatorname{cov}\{\tilde{\theta}(k+1|k)\} = P(k+1|k) \quad (48)$$

Table 1 Sensitivity matrices

$$G_1 = \begin{bmatrix} 0 & \frac{1}{2}\theta_2 - \frac{1}{3}\theta_1\theta_3 - \frac{1}{6}T\omega_2 & \frac{1}{2}\theta_3 + \frac{1}{3}\theta_1\theta_2 - \frac{1}{6}T\omega_3 \\ \frac{1}{2}\theta_2 + \frac{1}{3}\theta_1\theta_3 + \frac{1}{6}T\omega_2 & -\theta_1 & 1 - \frac{1}{6}(\theta_2^2 + \theta_3^2) - \frac{1}{2}\theta_1^2 \\ \frac{1}{2}\theta_3 - \frac{1}{3}\theta_1\theta_2 + \frac{1}{6}T\omega_3 & -1 + \frac{1}{6}(\theta_2^2 + \theta_3^2) + \frac{1}{2}\theta_1^2 & -\theta_1 \end{bmatrix}$$

$$G_2 = \begin{bmatrix} -\theta_2 & \frac{1}{2}\theta_1 - \frac{1}{3}\theta_2\theta_3 + \frac{1}{6}T\omega_1 & -1 + \frac{1}{6}(\theta_1^2 + \theta_3^2) + \frac{1}{2}\theta_2^2 \\ \frac{1}{2}\theta_1 + \frac{1}{3}\theta_2\theta_3 - \frac{1}{6}T\omega_1 & 0 & \frac{1}{2}\theta_3 - \frac{1}{3}\theta_1\theta_2 - \frac{1}{6}T\omega_3 \\ 1 - \frac{1}{6}(\theta_1^2 + \theta_3^2) - \frac{1}{2}\theta_2^2 & \frac{1}{2}\theta_3 + \frac{1}{3}\theta_1\theta_2 + \frac{1}{6}T\omega_3 & -\theta_2 \end{bmatrix}$$

$$G_3 = \begin{bmatrix} -\theta_3 & 1 - \frac{1}{6}(\theta_1^2 + \theta_2^2) - \frac{1}{2}\theta_3^2 & \frac{1}{2}\theta_1 + \frac{1}{3}\theta_2\theta_3 + \frac{1}{6}T\omega_1 \\ -1 + \frac{1}{6}(\theta_1^2 + \theta_2^2) + \frac{1}{2}\theta_3^2 & -\theta_3 & \frac{1}{2}\theta_2 - \frac{1}{3}\theta_1\theta_3 + \frac{1}{6}T\omega_2 \\ \frac{1}{2}\theta_1 - \frac{1}{3}\theta_2\theta_3 - \frac{1}{6}T\omega_1 & \frac{1}{2}\theta_2 + \frac{1}{3}\theta_1\theta_3 - \frac{1}{6}T\omega_2 & 0 \end{bmatrix}$$

Hence

$$\delta\theta(k+1) \sim \mathcal{N}\big(0, P(k+1|k)\big) \tag{49}$$

Using the linearized measurement equation (43) and the statistical properties of the measurement and prediction errors, Eqs. (44) and (49), respectively, the MMSE estimator of $\delta\theta(k+1)$ is

$$\widehat{\delta\theta}(k+1|k+1) = K(k+1)y(k+1) \tag{50}$$

where $K(k+1)$, the estimator gain matrix, is computed as

$$K(k+1) = P(k+1|k)H^T(k+1) \times \big[H(k+1)P(k+1|k)H^T(k+1) + R(k+1)\big]^{-1} \tag{51}$$

Also, from Eq. (46) we have

$$\widehat{\delta\theta}(k+1|k+1) = \hat{\theta}(k+1|k+1) - \hat{\theta}(k+1|k) \tag{52}$$

Using Eq. (52) in Eq. (50) finally yields the state measurement update equation

$$\hat{\theta}(k+1|k+1) = \hat{\theta}(k+1|k) + K(k+1)y(k+1) \tag{53}$$

To derive the covariance update equation, we subtract $\theta(k+1)$ from both sides of Eq. (53) and use Eqs. (43) and (46) to obtain

$$\tilde{\theta}(k+1|k+1) = \big[I - K(k+1)H(k+1)\big]\tilde{\theta}(k+1|k) - K(k+1)n(k+1) \tag{54}$$

from which the familiar, Joseph-form, covariance update equation results

$$P(k+1|k+1) = \big[I - K(k+1)H(k+1)\big]P(k+1|k) \times \big[I - K(k+1)H(k+1)\big]^T + K(k+1)R(k+1)K^T(k+1) \tag{55}$$

where $P(k+1|k+1) \triangleq E\{\tilde{\theta}(k+1|k+1)\tilde{\theta}^T(k+1|k+1)\}$ is the filtering error covariance matrix.

Attitude Matrix Update

To compute the measurement-updated attitude matrix at time t_{k+1}, we use the most recent estimate of the parameter vector $\hat{\theta}(k+1|k+1)$, the estimated attitude matrix corresponding to time t_k, and the measured angular velocity matrix, defined as

$$\hat{\Omega}(k) \triangleq -\big[\hat{\omega}(k)\times\big] \tag{56}$$

in Eq. (17). This yields

$$\hat{D}(k+1|k+1) = \Big\{I + \hat{A}(k+1,k) + \frac{1}{2}\hat{A}^2(k+1,k) + \frac{1}{6}\hat{A}^3(k+1,k) + \frac{1}{6}T\Big[\hat{A}(k+1,k)\hat{\Omega}(k) - \hat{\Omega}(k)\hat{A}(k+1,k)\Big]\Big\}\hat{D}^*(k|k) \tag{57}$$

where the *a posteriori* estimate of $A(k+1,k)$ is defined as

$$\hat{A}(k+1,k) \triangleq -\big[\hat{\theta}(k+1|k+1)\times\big] \tag{58}$$

and $\hat{D}^*(k|k)$ is the a posteriori, *orthogonalized* estimate of the attitude matrix at time t_k, to be discussed in the sequel.

Estimate Reset

As shown in Eq. (57), the a posteriori attitude matrix, $\hat{D}(k+1|k+1)$, is computed based on the a posteriori estimate, $\hat{\theta}(k+1|k+1)$. This estimate of the attitude matrix is then used in consecutive prediction and filtering steps, which, in turn, implies a full *reset control* of the parameter vector

$$\theta^c(k+1) = \theta(k+1) - \hat{\theta}(k+1|k+1) \tag{59}$$

where $\theta^c(k+1)$ is the *reset* state vector at t_{k+1}, and a corresponding reset of the state estimate

$$\hat{\theta}^c(k+1|k+1) = 0 \tag{60}$$

which is then used in the ensuing time propagation step.

Attitude Matrix Orthogonalization

Although the true attitude matrix is orthogonal, the filtered DCM will not be orthogonal, due to 1) numerical implementation errors, 2) the approximations involved in the development of the nonlinear filtering algorithm and 3) the third-order formula used to compute the attitude from the estimated parameters. To improve the algorithm's accuracy and to enhance its stability, an additional orthogonalization stage is introduced into the estimator, immediately following the measurement update stage. In the orthogonalization stage, the filtered attitude matrix is orthogonalized, that is, the orthogonal matrix closest to the filtered attitude matrix is found. This orthogonal matrix is then propagated to the next measurement update point.

In the sequel, the Euclidean norm (2-norm) will be used for vectors, and the Frobenius norm (F-norm) will be used for matrices.

Given the *a posteriori* attitude matrix $\hat{D}(k+1|k+1)$, the matrix orthogonalization problem is the following.

Find

$$\hat{D}^*(k+1|k+1) \triangleq \arg\min_{D\in\mathbb{R}^{3,3}} \|\hat{D}(k+1|k+1) - D\| \tag{61a}$$

subject to

$$D^T D = I \tag{61b}$$

Being a special case of the orthogonal Procrustes problem, the matrix orthogonalization problem can be easily solved using the singular value decomposition (SVD). Thus, if

$$\hat{D}(k+1|k+1) = U(k+1)\Sigma(k+1)V^T(k+1) \tag{62}$$

is the SVD of the matrix $\hat{D}(k+1|k+1)$ where $U(k+1)$ and $V(k+1)$ are the left and right singular vector matrices, respectively, and $\Sigma(k+1)$ is the singular value matrix, then

$$\hat{D}^*(k+1|k+1) = U(k+1)V^T(k+1) \tag{63}$$

The excessive computational burden associated with the SVD might render its use prohibitive in certain applications, e.g., in *real-time* attitude determination and control. In such cases, an alternative orthogonalization scheme, introduced by Bar-Itzhack and Meyer,[14] can be used. According to this scheme, the orthogonalized matrix $\hat{D}^*(k+1|k+1)$ can be computed iteratively using the recursion:

$$X_{j+1} = \frac{3}{2}X_j - \frac{1}{2}X_j X_j^T X_j \tag{64a}$$

$$X_0 = \hat{D}(k+1|k+1) \tag{64b}$$

where $X_j \xrightarrow{j\to\infty} \hat{D}^*(k+1|k+1)$. This scheme was shown in Ref. 14 to be globally convergent, and to possess a quadratic convergence rate.

Noting the fast convergence rate of the recursive orthogonalization method shown above, this scheme is incorporated into our estimation algorithm using just a *single* step of the recursion (64). Thus, an improved (nearly orthogonal), a posteriori estimate for the attitude matrix, is computed as

$$\hat{D}^*(k+1|k+1) = N(k+1)\hat{D}(k+1|k+1) \tag{65}$$

where the linear transformation that maps the a posteriori attitude matrix into its orthogonal version is defined by

$$N(k+1) \triangleq \frac{3}{2}I - \frac{1}{2}\hat{D}(k+1|k+1)\hat{D}^T(k+1|k+1) \tag{66}$$

DCM Orthogonalization: Analysis

The introduction of the external orthogonalization step into the estimator may, conceivably, affect its performance and statistical characteristics, thus calling for appropriate adjustments in the algorithm in order to preserve its theoretical properties. In the remainder of this section, the possible effects of the orthogonalization procedure (65) on the a posteriori state estimate and error covariance matrix are analyzed. In fact, as the next theorem shows, to first-order accuracy the orthogonalization procedure does not affect the estimator.

Theorem 1. *Let $\hat{D}^*(k+1|k+1)$ denote the orthogonalized version of $\hat{D}(k+1|k+1)$, computed in Eq. (65). Then*

$$\hat{D}^*(k+1|k+1) = \left[I + \eta[\hat{\theta}(k+1|k+1)]\right]\hat{D}(k+1|k+1) \tag{67}$$

where $\eta[\hat{\theta}(k+1|k+1)]$ is a matrix-valued function that satisfies

$$\|\eta[\hat{\theta}(k+1|k+1)]\| \sim \mathcal{O}\left(\|\hat{\theta}(k+1|k+1)\|^2\right) \tag{68}$$

Proof. Rewrite Eq. (57) as

$$\hat{D}(k+1|k+1) = \Phi\big[\hat{A}(k+1,k)\big]\hat{D}^*(k|k) \quad (69)$$

which is an implied definition of the matrix-valued function $\Phi(\bullet)$. Using Eq. (69) in Eq. (66) and noting the orthogonality of $\hat{D}^*(k|k)$ yields

$$N(k+1) = \frac{3}{2}I - \frac{1}{2}\Phi\big[\hat{A}(k+1,k)\big]\Phi^T\big[\hat{A}(k+1,k)\big] \quad (70)$$

Using now the definition of $\hat{A}(k+1,k)$, Eq. (58) and the properties of the cross product matrix, we have

$$\begin{aligned}\Phi\big[\hat{A}(k+1,k)\big] &= I - \big[\hat{\theta}(k+1|k+1)\times\big] \\ &- \frac{1}{2}\|\hat{\theta}(k+1|k+1)\|^2 I \\ &+ \frac{1}{2}\hat{\theta}(k+1|k+1)\hat{\theta}^T(k+1|k+1) \\ &+ \frac{1}{6}\|\hat{\theta}(k+1|k+1)\|^2\big[\hat{\theta}(k+1|k+1)\times\big] \\ &+ \frac{1}{6}\big[\big(\hat{\theta}(k+1|k+1)\times\hat{\omega}(k)T\big)\times\big] \end{aligned} \quad (71)$$

Now let $\{\mu_i[\hat{\theta}(k+1|k+1)]\}_{i=1}^3$ be the following matrix-valued functions

$$\begin{aligned}\mu_1[\hat{\theta}(k+1|k+1)] &\triangleq \frac{1}{2}\Big[-\|\hat{\theta}(k+1|k+1)\|^2 I \\ &+ \hat{\theta}(k+1|k+1)\hat{\theta}^T(k+1|k+1)\Big] \end{aligned} \quad (72a)$$

$$\mu_2[\hat{\theta}(k+1|k+1)] \triangleq \frac{1}{6}\|\hat{\theta}(k+1|k+1)\|^2\big[\hat{\theta}(k+1|k+1)\times\big] \quad (72b)$$

$$\mu_3[\hat{\theta}(k+1|k+1)] \triangleq \frac{1}{6}\big[\big(\hat{\theta}(k+1|k+1)\times\hat{\omega}(k)T\big)\times\big] \quad (72c)$$

It is easy to show that

$$\|\mu_1[\hat{\theta}(k+1|k+1)]\| \le \sqrt{\frac{3}{2}}\|\hat{\theta}(k+1|k+1)\|^2 \quad (73)$$

$$\|\mu_2[\hat{\theta}(k+1|k+1)]\| = \frac{1}{3\sqrt{2}}\|\hat{\theta}(k+1|k+1)\|^3 \quad (74)$$

Consider now the vector product $\hat{\theta}(k+1|k+1) \times \hat{\omega}(k)T$, appearing in Eq. (72c). As will be shown in the sequel [see the state prediction equation (79)], regarding $\hat{\omega}(t)$ as approximately constant over the small sampling interval $[t_k, t_{k+1}]$ yields

$$\hat{\omega}(k)T \approx \hat{\theta}(k+1|k) \quad (75)$$

Hence, using Eq. (52) we have

$$\begin{aligned}&\|\hat{\theta}(k+1|k+1)\times\hat{\omega}(k)T\| \\ &= \|\hat{\theta}(k+1|k+1)\times\widehat{\delta\theta}(k+1|k+1)\| \\ &\le \|\hat{\theta}(k+1|k+1)\|^2 \end{aligned} \quad (76)$$

whence

$$\|\mu_3[\hat{\theta}(k+1|k+1)]\| \le \frac{1}{3\sqrt{2}}\|\hat{\theta}(k+1|k+1)\|^2 \quad (77)$$

Using Eqs. (73), (74) and (77) in Eqs. (71) and (70), and noting Eq. (65), completes the proof. □

Now, from Eq. (67) we conclude that, after the initial transient period, the effects of the orthogonalization procedure on the filtered DCM are only of second-order in $\hat{\theta}(k+1|k+1)$. Hence, to first-order accuracy, no changes in $\hat{\theta}(k+1|k+1)$ are necessary as a result of the orthogonalization, and, consequently, no changes are required in the a posteriori covariance matrix.

Prediction

In the prediction step, the reset a posteriori state estimate at time t_k and its corresponding error covariance matrix, $\hat{\theta}^c(k|k)$ and $P(k|k)$, respectively, are propagated to time t_{k+1}.

Noting Eqs. (15) and (59) we have, after the state reset at time t_k,

$$\theta(k+1) = \tilde{\theta}(k|k) + \int_{t_k}^{t_{k+1}} \omega(\tau)\,d\tau \quad (78)$$

In practice, we only have access to the measured value of the angular velocity. Thus, using the measured velocity in Eq. (78) yields the following state prediction equation

$$\hat{\theta}(k+1|k) = \int_{t_k}^{t_{k+1}} \hat{\omega}(\tau)\,d\tau \quad (79)$$

Subtracting Eq. (79) from Eq. (78) and noting Eq. (18), the corresponding prediction error equation is

$$\tilde{\theta}(k+1|k) = \tilde{\theta}(k|k) - \int_{t_k}^{t_{k+1}} \delta\omega(\tau)\,d\tau \quad (80)$$

Noting that the two terms in the RHS of Eq. (80) are uncorrelated, the following, trivially simple covariance propagation equation results

$$P(k+1|k) = P(k|k) + \int_{t_k}^{t_{k+1}} Q(\tau)\,d\tau \quad (81)$$

Attitude Matrix Prediction

To predict the attitude matrix at t_{k+1} we use the most recent estimate of the parameter vector,

$\hat{\theta}(k+1|k)$, the orthogonalized estimate of the attitude matrix corresponding to t_k, and the measured angular velocity matrix, in Eq. (17). This yields

$$\hat{D}(k+1|k) = \Big\{ I + \bar{A}(k+1,k) + \frac{1}{2}\bar{A}^2(k+1,k) + \frac{1}{6}\bar{A}^3(k+1,k) + \frac{1}{6}T\big[\bar{A}(k+1,k)\hat{\Omega}(k) - \hat{\Omega}(k)\bar{A}(k+1,k)\big]\Big\}\hat{D}^*(k|k) \tag{82}$$

where the *a priori* estimate of $A(k+1,k)$ is defined as

$$\bar{A}(k+1,k) \triangleq -\big[\hat{\theta}(k+1|k)\times\big] \tag{83}$$

IV Numerical Study

To demonstrate the performance of the new attitude estimation algorithm, a numerical Monte Carlo simulation study was performed. In this study, the estimated attitude matrix was compared to the true matrix using the following metrics.

The *convergence index*, used to measure the convergence of the estimated attitude to the true attitude, was defined as the following ensemble average over M Monte Carlo runs

$$J_c(k+1) \triangleq \frac{1}{M}\sum_{n=1}^{M}\|\hat{D}_n^*(k+1|k+1) - D_n(k+1)\| \tag{84}$$

The *orthogonality index*, used to measure the degree of orthogonality of the estimated attitude, was defined as

$$J_o(k+1) \triangleq \frac{1}{M}\sum_{n=1}^{M}\|\hat{D}_n^{*T}(k+1|k+1)\hat{D}_n^*(k+1|k+1) - I\| \tag{85}$$

In each case tested, these metrics were evaluated using $M = 100$ Monte Carlo runs.

Two cases were investigated. In both, the standard deviation of the gyro noise power spectral density was 0.01 deg/h$^{1/2}$. *Both* the body-frame and the reference frame vector measurements were contaminated by zero-mean, white, Gaussian noise sequences, orthogonal to the true directions, with

$$R_v = \sigma_v^2 I, \qquad R_u = \sigma_u^2 I \tag{86}$$

The noise equivalent angles were set to $\sigma_u = \sigma_v = 100$ arc-seconds (these values are very conservative, compared to the current technology state-of-the-art[1]).

In all runs, the initial attitude estimate was set to the identity matrix (thus assuming that $\mathcal{S}_u$ and $\mathcal{S}_v$ coincide at t_0) while the true attitude corresponded to Euler angles of 30 deg, 20 deg and 10 deg in roll, pitch and yaw, respectively. This constitutes a conservative assumption, as we can always use the first vector measurements to obtain a rough initial estimate of the attitude using some point-estimation scheme.[10,15] However, in this work it was found that there was no need to use such an initialization scheme, as all simulation runs starting at the identity matrix successfully converged, thus demonstrating the robustness of the algorithm.

Case I

In this case (taken from Ref. 8), the angular velocity of $\mathcal{S}_v$ relative to $\mathcal{S}_u$ was

$$\omega(t) = \begin{bmatrix}2 & 3 & 6\end{bmatrix}^T \sin(0.2t) \quad \text{deg/s} \tag{87}$$

The filter was run at a rate of 20 Hz, i.e., the sampling interval was $T = 0.05$ s, while the measurement processing rate was a slow 1 Hz.

The performance indices obtained from a 100-run Monte Carlo simulation are presented in Figs. 1 and 2. The estimator performed satisfactorily, converging in all runs to an accurate and orthogonal DCM estimate after a relatively short transient period.

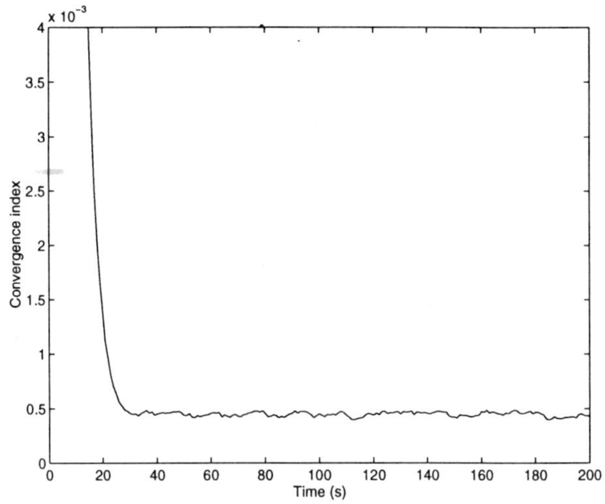

Fig. 1 Case I: Convergence index.

Figs. 3 and 4 show the true roll angle and the roll estimation error, respectively, for a typical run. A

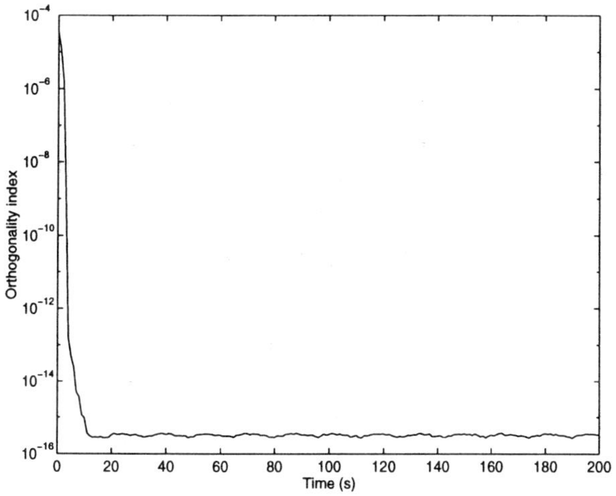

Fig. 2 Case I: Orthogonality index.

3-2-1 Euler angle sequence was assumed. Similar results were obtained for the pitch and yaw angles. The steady-state Euler angle estimation errors in a typical run were lower than 0.015 degrees (1-σ).

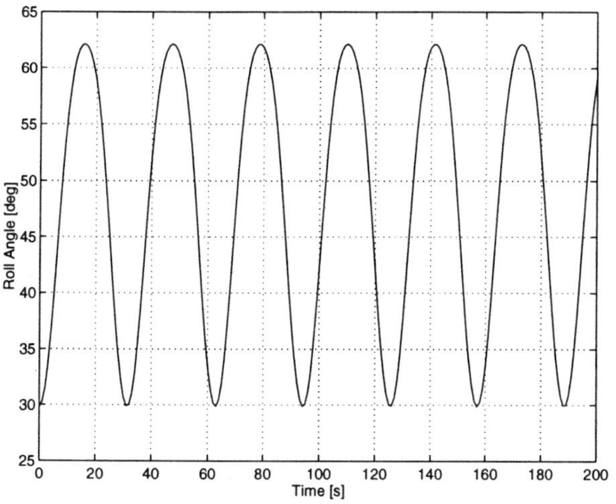

Fig. 3 Case I: True Roll angle.

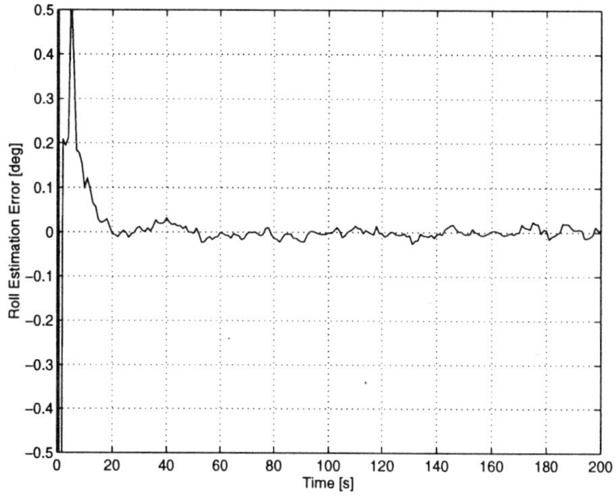

Fig. 4 Case I: Roll angle estimation error.

larger values than those obtained in the previous example. The orthogonality index values could be further decreased, if needed, by using a smaller discretization interval (raising the filter time update rate).

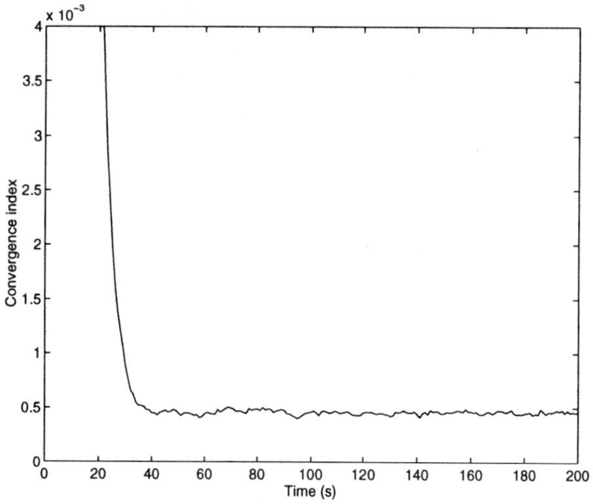

Fig. 5 Case II: Convergence index.

Case II

In this case (used also in Refs. 3, 10), the nominal angular velocity was a constant vector whose components in body axes were 0.628 rad/s along each axis. For this faster dynamics, the filter was run at a rate of 40 Hz, while the measurement processing rate was a slow 1 Hz, as before.

The convergence and orthogonality indices for this case are shown in Figs. 5 and 6, respectively. Again, it can be observed that the filter performed satisfactorily, even though the orthogonality index exhibits

V Conclusions

A nonlinear estimator has been presented, that uses vector measurements and gyro readings to estimate the attitude matrix. The algorithm is based on a recently introduced, third-order minimal parametrization of the attitude matrix, which facilitates the use of a three-dimensional filter to estimate the nine-parameter attitude matrix.

The simple kinematics obeyed by the parame-

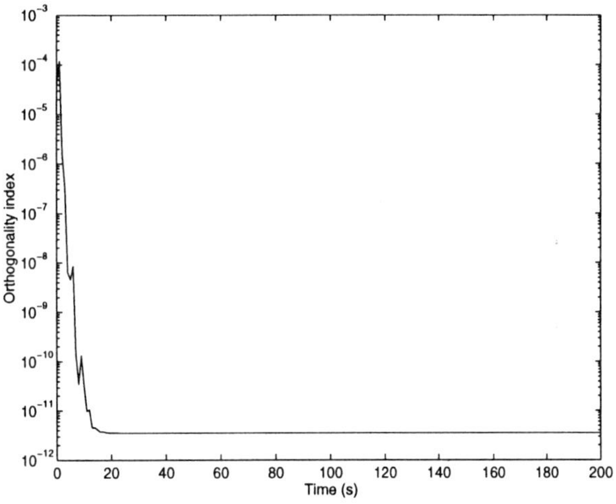

Fig. 6 Case II: Orthogonality index.

ter vector chosen is inherited by the estimator's time propagation equations, rendering the algorithm computationally efficient. DCM orthogonality is dealt with by incorporating a simple orthogonalization procedure following each measurement update.

Simulation studies have demonstrated the performance of the proposed algorithm. Assuming conservative gyro and vector measurement errors, the new algorithm has been shown to be both accurate and robust with respect to initialization errors.

Acknowledgement

This work was performed while the first author held a National Research Council - NASA Goddard Space Flight Center Research Associateship.

REFERENCES

[1] P. Kudva and A. Throckmorton. Attitude Determination Studies for the Earth Observation System AM1 (EOS-AM1) Mission. *Journal of Guidance, Control, and Dynamics*, 19(6):1326–1331, Nov.–Dec. 1996.

[2] E. J. Lefferts, F. L. Markley, and M. D. Shuster. Kalman Filtering for Spacecraft Attitude Estimation. *Journal of Guidance, Control, and Dynamics*, 5(5):417–429, Sept.-Oct. 1982.

[3] I. Y. Bar-Itzhack and Y. Oshman. Attitude Determination from Vector Observations: Quaternion Estimation. *IEEE Transactions on Aerospace and Electronic Systems*, AES-21(1):128–136, January 1985.

[4] M.D. Shuster. Kalman Filtering of Spacecraft Attitude and the Quest Model. *The Journal of the Astronautical Sciences*, 38(3):377–393, Jul.-Sept. 1990.

[5] E. Gai, K. Daly, J. Harrison, and L. Lemons. Star-Sensor-Based Satellite Attitude/Attitude Rate Estimator. *Journal of Guidance, Control, and Dynamics*, 8(5):560–565, Sept.-Oct. 1985.

[6] J. L. Farrell. Attitude Determination by Kalman Filtering. *Automatica*, 6:419–430, 1970.

[7] I. Y. Bar-Itzhack and M. Idan. Recursive Attitude Determination from Vector Observations: Euler Angle Estimation. *Journal of Guidance, Control, and Dynamics*, 10(2):152–157, Mar.–Apr. 1987.

[8] M. Idan. Estimation of Rodrigues Parameters From Vector Observations. *IEEE Transactions on Aerospace and Electronic Systems*, AES-32(2):578–586, April 1996.

[9] J. L. Crassidis and F. L. Markley. Attitude Estimation Using Modified Rodrigues Parameters. In *Proceedings of Flight Mechanics/Estimation Theory Symposium*, pages 71–83, NASA Goddard Space Flight Center, May 14–16 1996. NASA.

[10] I. Y. Bar-Itzhack and J. Reiner. Recursive Attitude Determination from Vector Measurements: DCM Identification. *Journal of Guidance, Control, and Dynamics*, 7(1):51–56, Jan.–Feb. 1984.

[11] I. Y. Bar-Itzhack and F. L. Markley. Minimal Parameter Solution of the Orthogonal Matrix Differential Equation. *IEEE Transactions on Automatic Control*, AC-35(3):314–317, March 1990.

[12] Y. Oshman and I. Y. Bar-Itzhack. Eigenfactor Solution of the Matrix Riccati Equation—A Continuous Square Root Algorithm. *IEEE Transactions on Automatic Control*, AC-30(10):971–978, October 1985.

[13] M. Ronen and Y. Oshman. A Third-Order, Minimal-Parameter Solution of The Orthogonal Matrix Differential Equation. To appear, *Journal of Guidance, Control and Dynamics.*

[14] I. Y. Bar-Itzhack and J. Meyer. On the Convergence of Iterative Orthogonalization Processes. *IEEE Transactions on Aerospace and Electronic Systems*, AES-12:146–151, March 1976.

[15] M. D. Shuster and S. D. Oh. Three Axis Attitude Determination from Vector Obsevations. *Journal of Guidance, Control, and Dynamics*, 4:70–77, Jan.-Feb. 1981.

AIAA-97-3452

ATTITUDE DETERMINATION USING GLOBAL POSITIONING SYSTEM SIGNALS

John L. Crassidis
Assistant Professor
Department of Mechanical Engineering
Catholic University of America
Washington, D.C. 20064

F. Landis Markley
Staff Engineer
Guidance, Navigation, and Control Branch
NASA-Goddard Space Flight Center
Greenbelt, MD 20771

Abstract

In this paper, a novel technique for finding a point-by-point (deterministic) attitude solution of a vehicle using Global Positioning System phase difference measurements is presented. This technique transforms a general cost function into a more numerically efficient form by determining three-dimensional vectors in either the body or reference coordinate system. Covariance relationships for the new algorithm, as well as methods which minimize the general cost function, are also derived. The equivalence of the general cost function and transformed cost function is shown for the case of orthogonal baselines or sightlines. Simulation results are shown which demonstrate the usefulness of the new algorithm and covariance expressions.

Introduction

The utilization of phase difference measurements from Global Positioning System (GPS) receivers provides a novel approach for three-axis attitude determination and/or estimation. These measurements have been successfully used to determine the attitude of both aircraft[1] and spacecraft.[2,3] Recently, much attention has been placed on spacecraft-based applications. One of the first space-based GPS experiments for attitude determination was flown on the RADCAL (RADar CALibration) spacecraft.[4] To obtain maximum GPS visibility, and to reduce signal interference due to multipath reflection, GPS patch antennas were placed on the top surface of the spacecraft bus. Although the antenna baselines were short for attitude determination, accuracies between 0.5 to 1.0 degrees (root-mean-square) were achieved.

In this paper, the problem of finding the attitude from GPS phase difference measurements using deterministic approaches is addressed. Error sources, such as integer sign ambiguity,[5] are not investigated. These errors are assumed to be accounted for before the attitude determination problem is solved. The most common GPS attitude determination scheme minimizes a cost function constituting the sum weighted two-norm residuals between the estimated and determined phase difference quantities. However, as of this writing, the optimal attitude solution which minimizes this general cost function can only be found using iterative techniques, such as gradient search methods. A suboptimal solution involves transforming the general cost function into a form which can be minimized without iterative intense methods. One such technique, developed by Cohen,[1] transforms the general cost function into a form identical to Wahba's problem.[6] Therefore, fast algorithms such as QUEST[7] and FOAM[8] can then be used to determine the attitude. Cohen showed that the solution based on Wahba's problem is almost an order of magnitude faster than a conventional nonlinear least-squares algorithm.

Cohen's approach involves a two step process. The first step involves finding a weighting matrix, using a Singular Value Decomposition (SVD), which transforms the baseline configuration to an equivalent orthonormal basis. At least three non-collinear baselines must exist to perform this transformation. If this is not the case, the transformation can still be accomplished as long as three non-collinear sightlines exist. However, a SVD must be performed for each new sightline, which can be computationally expensive, whereas the baseline transformation has to be done only once. The second step involves finding the attitude using the fast algorithms such as QUEST or FOAM. Since the weighting matrix transforms the baseline configuration to an equivalent orthonormal basis, suboptimal attitude solutions may arise if the baseline configuration does not already form an orthonormal basis.[1] An example of this scenario is when three baselines are coplanar. In order to determine the optimal attitude, iterative techniques which minimize the general cost function must be used. The method presented in this paper also is suboptimal for the case where the baseline (or sightline) configuration does not form an orthonormal basis. However, it does not require a SVD of a 3×3 dimensional matrix to perform the orthonormal transformation.

Bar-Itzhack et. al.[9] show another analytical conversion of the basic GPS scalar difference measurements into unit vectors to be used in Wahba's problem. This is accomplished by expressing the angle determined by one of the baselines, which describes a cone around the baseline vector, and likewise for the second baseline, into a three-dimensional vector resolved in a reference coordinate system. Attitude solutions are provided for baselines which constitute Cartesian and non-Cartesian coordinate systems; however, these solutions shown in Ref. [9] involve only two baseline vectors. This paper generalizes these results to multiple baseline vectors. Also, covariance relations are shown for the new approach, as well as for techniques which minimize the general cost function directly. This allows users to quantify any additional errors produced by transforming the general cost function into Wahba's form.

The organization of this paper proceeds as follows. First, the concept of the GPS phase difference measurement is introduced. Then, the general cost function used for GPS-based attitude determination is reviewed. Next, Cohen's method for transforming the general cost function into Wahba's problem is shown. Also, system observability using two baselines is discussed. Then, a general technique for transforming the general cost function is developed. Also, the equivalence of the general and transformed (Wahba) cost functions for orthogonal baselines and/or sightlines is discussed. Next, a covariance analysis is performed on the new algorithm, and on algorithms which minimize the general cost function directly. Finally, results are shown for a simulated vehicle with near-orthogonal baselines, non-orthogonal baselines, and baselines which are nearly collinear.

Background

In this section, a brief background of the GPS phase difference measurement is shown. The GPS constellation of spacecraft was developed for accurate navigation information of land-based, air, and spacecraft user systems. Spacecraft applications initially involved obtaining accurate orbit information and accurate time-tagging of spacecraft operations. However, attitude determination of vehicles, such as spacecraft or aircraft, has gained much attention. The main measurement used for attitude determination is the phase difference of the GPS signal received from two antennas separated by a baseline. The principle of the wavefront angle and wavelength, which are used to develop a phase difference, is illustrated in Figure 1.

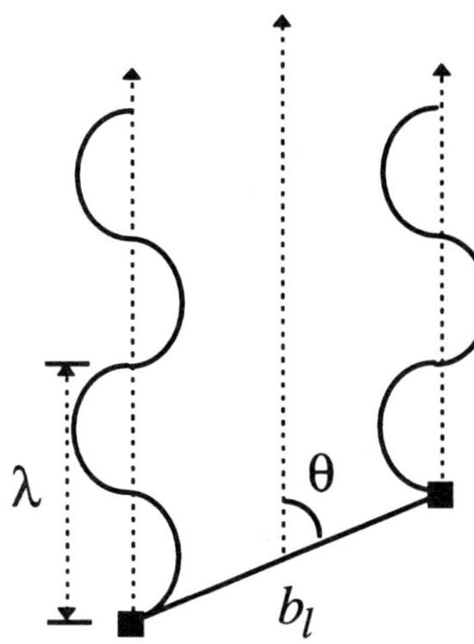

Fig. 1 GPS Wavelength and Wavefront Angle

The phase difference measurement is obtained by

$$b_l \cos\theta = \lambda\left(n + \Delta\phi^0/2\pi\right) \tag{1}$$

where b_l is the baseline length, θ is the angle between the baseline and the line of sight to the GPS spacecraft, n is the number of integer wavelengths between two receivers, $\Delta\phi^0$ is the actual phase difference measurement, and λ is the wavelength of the GPS signal. The two GPS frequency carriers are L1 at 1575.42 MHz and L2 at 1227.6 MHz. Then, assuming no integer offset, we define a normalized phase difference measurement $\Delta\phi$ used for attitude determination by

$$\Delta\phi \equiv \frac{\lambda\,\Delta\phi^0}{2\pi\, b_l} = \underline{b}^T A\, \underline{s} \tag{2}$$

where $\underline{s} \in R^3$ is the normalized line of sight vector to the GPS spacecraft in an inertial frame, $\underline{b} \in R^3$ is the normalized baseline vector, which is the relative position vector from one antenna to another, and the attitude matrix A is in $SO(3)$, which is a Lie group of orthogonal matrices with determinant 1 (i.e., $A^T A = I$ and $\det A = 1$).

Cohen's Method

In this section, Cohen's method[1] for determining the attitude of a vehicle using Equation (2) is reviewed. The general cost function to be minimized is given by

$$J(A) = \frac{1}{2}\sum_{i=1}^{m}\sum_{j=1}^{n} w_{ij}\left(\Delta\phi_{ij} - \underline{b}_i^T A\,\underline{s}_j\right)^2 \tag{3}$$

where m represents the number of baselines, and n represents the number of observed GPS spacecraft. The parameters, w_{ij}, serve to weight each individual phase measurement. The phase measurement can contain noise, which is modeled by

$$\Delta\phi_{ij} = \Delta\phi_{ij}^{\text{true}} + v_{ij} \tag{4}$$

where v_{ij} is a zero-mean stationary Gaussian process with covariance given by σ_{ij}^2. The maximum likelihood estimate for w_{ij} is given by $1/\sigma_{ij}^2$. If the weights w_{ij} factor into a satellite-dependent and a baseline-dependent factor, i.e. $w_{ij} = w_{bi}w_{sj}$, then the cost function in Equation (3) may be re-written as[1]

$$J(A) = \left\| W_B^{1/2}\left(\Delta\Phi - B^T A\,S\right)W_S^{1/2}\right\|_F^2 \tag{5}$$

where $\| \; \|_F^2$ denotes the Frobenius norm, and

$$\Delta\Phi = \begin{bmatrix} \Delta\phi_{11} & \Delta\phi_{12} & \cdots & \Delta\phi_{1n} \\ \Delta\phi_{21} & & & \\ & & \ddots & \vdots \\ \Delta\phi_{m1} & & \cdots & \Delta\phi_{mn} \end{bmatrix} \tag{6a}$$

$$B = \left[\underline{b}_1 \mid \underline{b}_2 \mid \cdots \mid \underline{b}_m\right] \tag{6b}$$

$$S = \left[\underline{s}_1 \mid \underline{s}_2 \mid \cdots \mid \underline{s}_n\right] \tag{6c}$$

The weighting matrices W_B and W_S are applicable to the rows (baselines) and columns (spacecraft) of $\Delta\Phi$, respectively.

If the quaternion[10] representation is used for the attitude matrix, then Equation (5) leads to a quartic dependence in the quaternions. In Wahba's problem, this dependence cancels out of the cost function. In order to cancel this dependence in Equation (5), Cohen chooses the following weighting matrix for W_B

$$W_B = V_B \Sigma_B^{-2} V_B^T \tag{7}$$

where V_B and Σ_B are given from a SVD of B, i.e. $B = U_B \Sigma_B V_B^T$. From Equation (7), the matrix B must be full rank, which means that at least three non-collinear baselines must be used. However, if this is not true a solution can still be found as long as three non-collinear sightlines exist. This can be accomplished by performing a SVD of S, and choosing W_S as in the same form in Equation (7). However, a SVD must be performed for each sightline. This is more computationally expensive than using Equation (7), which may be done once for constant baselines. It is also not obvious that Equation (7) is consistent with Equation (3). Substituting Equation (7) into the general cost function in Equation (5) leads to Wahba's problem, which maximizes

$$J'(A) = \text{trace}\left(A\,S\,W_S \Delta\Phi^T W_B B^T\right) \equiv \text{trace}\left(A\,G^T\right) \tag{8}$$

However, by constraining W_B or W_S, the solution using the transformed cost function in Equation (8) is suboptimal for non-orthogonal baselines or sightlines. The concept of the suboptimal solution will be discussed in detail later.

Attitude Determination from Vectorized Measurements

In this section, a new method for attitude determination from GPS phase measurements is developed. This new method extends the method shown in Ref. [9], which converts the phase measurements into vector measurements. The general method for the vectorized measurements is based on an algorithm given by Shuster.[11] Also, a covariance analysis is performed for the new method, and for methods which minimize the general cost function in Equation (3) directly.

The vectorized measurement problem involves determining the sightline vector in the body frame, denoted by $\underline{\tilde{s}}_j \equiv A\,\underline{s}_j$, or the baseline in an inertial frame, denoted by $\underline{\bar{b}}_i \equiv A^T \underline{b}_i$. For the sightline case, the following cost function is minimized

$$J_j\left(\underline{\tilde{s}}_j\right) = \frac{1}{2}\sum_{i=1}^{m}\frac{1}{\sigma_{ij}^2}\left(\Delta\phi_{ij} - \underline{b}_i^T \underline{\tilde{s}}_j\right)^2 \quad \text{for } j = 1, 2, \ldots, n \tag{9}$$

The minimization of Equation (9) is straightforward and leads to[11]

$$\underline{\tilde{s}}_j = M_j^{-1}\underline{y}_j \tag{10}$$

where

$$M_j = \sum_{i=1}^{m} \frac{1}{\sigma_{ij}^2} \underline{b}_i \underline{b}_i^T \quad \text{for } j = 1,2,\ldots,n \tag{11a}$$

$$\underline{y}_j = \sum_{i=1}^{m} \frac{1}{\sigma_{ij}^2} \Delta\phi_{ij} \underline{b}_i \quad \text{for } j = 1,2,\ldots,n \tag{11b}$$

The error covariance of $\underline{\tilde{s}}_j$ is given by

$$P_j = M_j^{-1} \tag{12}$$

If the sightline in the body is required to be normalized, then the cost function in Equation (9) must be minimized subject to the constraint $\underline{\tilde{s}}^T \underline{\tilde{s}} = 1$. However, Shuster[11] showed that the error introduced by ignoring this constraint is on the order of $m^{-1}\text{trace}\left[\left|\underline{\tilde{s}}_j\right|^{-2}\left(I - \underline{\tilde{s}}_j \underline{\tilde{s}}_j^T\right) P_j \left(I - \underline{\tilde{s}}_j \underline{\tilde{s}}_j^T\right)\right]$, which is usually negligible. The solution of Wahba's problem as shown below will determine the optimal attitude (in the least-squares sense) which results in a normalized body vector. Therefore, the normalization constraint may be ignored. Also, the normalized error covariance is singular, as shown in Ref. [11]. This singularity is avoided by using the covariance given by Equation (12). For a discussion of singularity issues for measurement covariances see Ref. [12].

From Equation (10) it is seen that at least three non-collinear baselines are required to determine the sightlines in the body frame. This is analogous to the problem posed by Cohen.[1] However, if only two non-collinear baselines exist, a solution is again possible as long as three non-collinear sightlines exist. This approach determines the baselines in the inertial frame, using the following cost function

$$J_i(\underline{\bar{b}}_i) = \frac{1}{2} \sum_{j=1}^{n} \frac{1}{\sigma_{ij}^2} \left(\Delta\phi_{ij} - \underline{\bar{b}}_i^T \underline{s}_j\right)^2 \quad \text{for } i = 1,2,\ldots,m \tag{13}$$

The minimization of Equation (13) is again straightforward and leads to

$$\underline{\bar{b}}_i = N_i^{-1} \underline{z}_i \tag{14}$$

where

$$N_i = \sum_{j=1}^{n} \frac{1}{\sigma_{ij}^2} \underline{s}_j \underline{s}_j^T \quad \text{for } i = 1,2,\ldots,m \tag{15a}$$

$$\underline{z}_i = \sum_{j=1}^{n} \frac{1}{\sigma_{ij}^2} \Delta\phi_{ij} \underline{s}_j \quad \text{for } i = 1,2,\ldots,m \tag{15b}$$

The error covariance of $\underline{\bar{b}}_i$ is given by

$$Q_i = N_i^{-1} \tag{16}$$

The case with two non-collinear baselines and two non-collinear sightlines can also be solved for either the baseline inertial case or sightline body case. Solving for the latter case yields

$$\underline{\tilde{s}}_j = a_{1j} \underline{b}_1 + a_{2j} \underline{b}_2 + a_{3j} \left(\underline{b}_1 \times \underline{b}_2\right) \quad \text{for } j = 1,2 \tag{17}$$

where

$$a_{1j} = \left|\underline{b}_1 \times \underline{b}_2\right|^{-2} \left[\Delta\phi_{1j} - \Delta\phi_{2j} \left(\underline{b}_1 \cdot \underline{b}_2\right)\right] \text{for } j = 1,2 \tag{18a}$$

$$a_{2j} = \left|\underline{b}_1 \times \underline{b}_2\right|^{-2} \left[\Delta\phi_{2j} - \Delta\phi_{1j} \left(\underline{b}_1 \cdot \underline{b}_2\right)\right] \text{for } j = 1,2 \tag{18b}$$

$$\begin{aligned} a_{3j} = \pm \left|\underline{b}_1 \times \underline{b}_2\right|^{-2} \Big\{ & f_j \left|\underline{b}_1 \times \underline{b}_2\right|^2 - \Delta\phi_{1j}^2 \\ & -2\Delta\phi_{1j}\Delta\phi_{2j}\left(\underline{b}_1 \cdot \underline{b}_2\right) + \Delta\phi_{2j}^2 \Big\}^2 \quad \text{for } j = 1,2 \end{aligned} \tag{18c}$$

where $f_j = \left|\underline{\tilde{s}}_j\right|^2$. Equation (18c) involves knowledge of $\left|\underline{\tilde{s}}_j\right|^2$. However, this quantity can be assumed to be 1 with reasonable accuracy. Also, from Equation (18c), there are two possible solutions for the body sightlines. However, this sign ambiguity can usually be resolved from the geometry of vehicle to the GPS spacecraft. The error covariance is given by[11]

$$P_j = T L_j T^T \tag{19}$$

where

$$T = \left[\underline{b}_1 \ \vdots \ \underline{b}_2 \ \vdots \ \underline{b}_1 \times \underline{b}_2\right] \tag{20a}$$

$$L_j = \begin{bmatrix} D_j & \underline{l}_j \\ \underline{l}_j^T & d_j \end{bmatrix} \tag{20b}$$

and

$$D_j = U P_{\phi_j} U \tag{21a}$$

$$U = \left|\underline{b}_1 \times \underline{b}_2\right|^{-2} \begin{bmatrix} 1 & -\underline{b}_1 \cdot \underline{b}_2 \\ -\underline{b}_1 \cdot \underline{b}_2 & 1 \end{bmatrix} \tag{21b}$$

$$P_{\phi_j} = \begin{bmatrix} \sigma_{1j}^2 & 0 \\ 0 & \sigma_{2j}^2 \end{bmatrix} \tag{21c}$$

$$\underline{l}_j = \mp \left|\underline{b}_1 \times \underline{b}_2\right|^{-1} \left[1 - \underline{\psi}_j^T D_j \underline{\psi}_j\right]^{-1/2} U P_{\phi_j} U \underline{\psi}_j \tag{21d}$$

$$d_j = \left|\underline{b}_1 \times \underline{b}_2\right|^{-2} \left[1 - \underline{\psi}_j^T D_j \underline{\psi}_j\right]^{-1} \underline{\psi}_j^T U P_{\phi_j} U \underline{\psi}_j \tag{21e}$$

$$\underline{\psi}_j \equiv \begin{bmatrix} \Delta\phi_{1j} \\ \Delta\phi_{2j} \end{bmatrix} \tag{21f}$$

The covariance in Equation (19) is singular. However, this does not affect the determination of the attitude error covariance, as will be shown. Also, the method can be trivially modified to determine the baselines in inertial space.

Attitude Determination

The attitude determination problem using body sightlines is very similar to that using inertial baselines, so we may consider only the former case. The attitude is determined by using the following cost function

$$J(A) = \frac{1}{2}\sum_{j=1}^{n}\left(\underline{\tilde{s}}_j - A\underline{s}_j\right)^T M_j\left(\underline{\tilde{s}}_j - A\underline{s}_j\right) \tag{22}$$

This cost function is not identical to Wahba's problem since the quartic dependence in the quaternion does not cancel, unless the baselines form an orthonormal basis so that M_j is given by a scalar times the identity matrix. The cost function in Equation (22) is in fact equivalent to the general cost function in Equation (3). This is shown by substituting Equation (10) and (11) into (22) and expanding terms, giving

$$J(A) = \frac{1}{2}\sum_{j=1}^{n}\Big(\underline{y}_j^T M_j^{-1}\underline{y}_j - 2\underline{y}_j^T A\underline{s}_j + \underline{s}_j^T A^T M_j A\underline{s}_j\Big) \tag{23}$$

Expanding Equation (23) now yields

$$J(A) = \frac{1}{2}\sum_{j=1}^{n}\left(\underline{y}_j^T M_j^{-1}\underline{y}_j - \sum_{i=1}^{m}\frac{1}{\sigma_{ij}^2}\Delta\phi_{ij}^2\right) + \frac{1}{2}\sum_{i=1}^{m}\sum_{j=1}^{n}\frac{1}{\sigma_{ij}^2}\left(\Delta\phi_{ij} - \underline{b}_i^T A\underline{s}_j\right)^2 \tag{24}$$

Since the first term in Equation (24) is independent of attitude, it is clear that this cost function is equivalent to the general cost function in Equation (3). In order to reduce the cost function in Equation (22) into a form corresponding to Wahba's problem the condition that M_j is given by a scalar times the identity matrix must be valid. Therefore, if the baselines do not form an orthonormal basis, then the attitude solution is suboptimal.

Attitude Covariance

Wahba posed the three-axis determination problem in terms of finding the proper orthogonal attitude matrix that minimizes

$$J(A) = \frac{1}{2}\sum_{j=1}^{n} a_j\left|\underline{\tilde{s}}_j - A\underline{s}_j\right|^2 \tag{25}$$

Several efficient algorithms have been developed to solve this problem (e.g., QUEST[7] and FOAM[8]). Another solution for the attitude matrix is given by performing a SVD of the following matrix

$$F = \sum_{j=1}^{n} a_j \underline{\tilde{s}}_j \underline{s}_j^T = U\Sigma V^T \tag{26}$$

The optimal solution for the attitude matrix is given by[13]

$$A_{\text{opt}} = U_+ V_+^T \tag{27}$$

where

$$U_+ = U\left[\text{diag}(1, 1, \det U)\right] \tag{28a}$$

$$V_+ = V\left[\text{diag}(1, 1, \det V)\right] \tag{28b}$$

The covariance of the estimation error angle vector in the body frame is given by[13]

$$E\{\delta\underline{\alpha}\,\delta\underline{\alpha}^T\} = P_{\text{body}} = \left(I - F A_{opt}^T\right)^{-1} \times \sum_{i=1}^{m}\sum_{j=1}^{n} a_i a_j \left[\underline{\tilde{s}}_j \times\right] E\{\underline{e}_i \underline{e}_j^T\}\left[\underline{\tilde{s}}_j \times\right]^T \left(I - F A_{opt}^T\right)^{-1} \tag{29}$$

where $[\underline{\tilde{s}}_j \times]$ represents the cross product matrix (see Ref. [13]), $\delta\underline{\alpha}$ is the small error angle, and

$$\underline{e}_k \equiv A_{opt}\delta\underline{s}_k - \delta\underline{\tilde{s}}_k, \quad \text{for any } k \tag{30}$$

The terms $\delta\underline{s}$ and $\delta\underline{\tilde{s}}$ represent variations in the inertial and body sightlines, respectively. The expectation in Equation (29) can be written as

$$E\{\underline{e}_i \underline{e}_j^T\} = A_{opt} E\{\delta\underline{s}_i \delta\underline{s}_j^T\} A_{opt}^T + E\{\delta\underline{\tilde{s}}_i \delta\underline{\tilde{s}}_j^T\} \tag{31}$$

Assuming that the only errors are in the effective phase measurements reduces Equation (29) to

$$P_{\text{body}} = \left(I - F A_{opt}^T\right)^{-1}\sum_{j=1}^{n} a_j^2 \left[\underline{\tilde{s}}_j \times\right] P_j \left[\underline{\tilde{s}}_j \times\right]^T \times \left(I - F A_{opt}^T\right)^{-1} \tag{32}$$

Now using the approximation of

$$\tilde{\underline{s}}_j \approx A_{\text{opt}} \underline{s}_j \tag{33}$$

yields

$$\begin{aligned} I - F A_{opt}^T &\approx \sum_{j=1}^{n} a_j \left(I - \tilde{\underline{s}}_j \tilde{\underline{s}}_j^T \right) \\ &= \sum_{j=1}^{n} a_j [\tilde{\underline{s}}_j \times][\tilde{\underline{s}}_j \times]^T \equiv X \end{aligned} \tag{34}$$

and thus the error angle covariance is given by

$$P_{\text{body}} \approx X^{-1} \left\{ \sum_{j=1}^{n} a_j^2 [\tilde{\underline{s}}_j \times] P_j [\tilde{\underline{s}}_j \times]^T \right\} X^{-1} \tag{35}$$

Note that if the covariances P_j are multiples of the identity, $P_j = \sigma_j^2 I$, and then setting $a_j = \sigma_j^{-2}$ yields

$$P_{\text{body}} \approx \left[\sum_{j=1}^{n} \sigma_j^{-2} [\tilde{\underline{s}}_j \times][\tilde{\underline{s}}_j \times]^T \right]^{-1} = X^{-1} \tag{36}$$

Therefore, in this case the covariance in Equation (36) would be identical to the covariance given by QUEST.[7] The best suboptimal weighting factor a_j in Equation (35) can be found by minimizing the trace of P_{body}. However, this is extremely complex. If Equation (36) is still a good approximation, then a_j can be chosen to minimize some matrix norm of the following

$$J(a_j) = \| a_j P_j - I \| \tag{37}$$

An alternative to Equation (37) is to minimize the following cost function for some matrix norm

$$J(a_j) = \| a_j I - P_j^{-1} \| \tag{38}$$

For example, minimizing Equation (38) with a Frobenius norm results in

$$a_j = \frac{1}{3} \text{trace}\left(P_j^{-1}\right) = \frac{1}{3} \text{trace}\left(M_j\right) \tag{39}$$

Once a proper weight is determined, then Wahba's problem in Equation (25) can be solved. The covariance of the attitude errors is given by Equation (35).

Transforming the general cost function in Equation (3) results in a suboptimal solution. In order to quantify the errors introduced by the suboptimal solution, the error attitude covariance for the general cost function is derived. This is accomplished by using results from maximum likelihood estimation.[14] The Fisher information matrix for a parameter vector $\underline{x}$ is given by

$$F_{xx} = E\left\{ \frac{\partial}{\partial \underline{x} \partial \underline{x}^T} J(\underline{x}) \right\}_{\underline{x}_{\text{true}}} \tag{40}$$

where $J(\underline{x})$ is the negative log likelihood function, which is the loss function in this case. If the measurements are Gaussian and linear in the parameter vector, then the error covariance is given by

$$P_{xx} = F_{xx}^{-1} \tag{41}$$

Since the cost function in Equation (22) is equivalent to the full cost function in Equation (3), Equation (22) can be used to determine the covariance of the optimal solution. First, the attitude matrix is approximated by

$$\begin{aligned} A &= e^{-[\delta \underline{\alpha} \times]} A_{\text{true}} \\ &\approx \left(I - [\delta \underline{\alpha} \times] + \frac{1}{2} [\delta \underline{\alpha} \times]^2 \right) A_{\text{true}} \end{aligned} \tag{42}$$

Equations (42) and (33) are next substituted into Equations (22) and (40) to determine the Fisher information matrix. First-order terms vanish in the partials, and third-order terms become zero since $E\{\delta \underline{\alpha}\} = \underline{0}$. Also, assuming that the quartic terms are negligible leads to the following simple form for the optimal covariance

$$P_{\text{opt}} \approx \left[\sum_{j=1}^{n} [\tilde{\underline{s}}_j \times] P_j^{-1} [\tilde{\underline{s}}_j \times]^T \right]^{-1} \tag{43}$$

Note that the optimal covariance in Equation (43) reduces to the covariance in Equation (36) if the condition $P_j = \sigma_j^2 I$ is true. The errors introduced when using a suboptimal solution can now be compared to the expected performance of minimizing the general cost function in Equation (3). Also, for the case of two baselines and two sightlines, the optimal covariance can be derived by using Equation (3) in the Fisher information matrix, which leads to

$$P_{\text{opt}} \approx \left[\sum_{i=1}^{2} \sum_{j=1}^{2} \frac{1}{\sigma_{ij}^2} \left(\tilde{\underline{s}}_j \times \underline{b}_i \right) \left(\tilde{\underline{s}}_j \times \underline{b}_i \right)^T \right]^{-1} \tag{44}$$

The covariance analysis can be easily extended to the case where the baselines in inertial space are determined. The body covariance for the transformed cost function in this case becomes

$$P_{\text{body}} \approx B_a \sum_{i=1}^{m} a_i^2 [\underline{b}_i \times] A Q_i A^T [\underline{b}_i \times]^T B_a \quad (45a)$$

$$B_a = \left[\sum_{i=1}^{m} a_i [\underline{b}_i \times]^2 \right]^{-1} \quad (45b)$$

The error covariance for the optimal solution is given by

$$P_{\text{opt}} \approx \left[\sum_{i=1}^{m} [\underline{b}_i \times] A Q_i^{-1} A^T [\underline{b}_i \times]^T \right]^{-1} \quad (46)$$

Simulation Results

In this section, simulation results are shown using the new algorithm and covariance expressions. Three case are presented. The first case involves three baselines which are nearly orthogonal. The second involves three baselines which do not constitute an orthogonal set. The third case involves three baselines, where the first two baselines are far from constituting an orthogonal set (i.e., nearly collinear). Although the third case would most likely never be used in a practical application, it provides a radical test comparison between the optimal and suboptimal solutions. It is assumed that the vehicle is always in the view of two GPS spacecraft with constant and normalized sightlines given by

$$\underline{s}_1 = \frac{1}{\sqrt{3}} [1 \quad 1 \quad 1]^T, \qquad \underline{s}_2 = \frac{1}{\sqrt{2}} [0 \quad 1 \quad 1]^T \quad (47)$$

The three normalized baseline cases are given by

Case 1

$$\underline{b}_1 = \frac{1}{\sqrt{1.09}} [1 \quad 0.3 \quad 0]^T, \quad \underline{b}_2 = [0 \quad 1 \quad 0]^T, \quad \underline{b}_3 = [0 \quad 0 \quad 1]^T \quad (48a)$$

Case 2

$$\underline{b}_1 = \frac{1}{\sqrt{2}} [1 \quad 1 \quad 0]^T, \quad \underline{b}_2 = [0 \quad 1 \quad 0]^T, \quad \underline{b}_3 = [0 \quad 0 \quad 1]^T \quad (48b)$$

Case 3

$$\underline{b}_1 = \frac{1}{\sqrt{1.02}} [0.1 \quad 1 \quad 0.1]^T, \quad \underline{b}_2 = [0 \quad 1 \quad 0]^T, \quad \underline{b}_3 = [0 \quad 0 \quad 1]^T \quad (48c)$$

The noise for each phase difference measurement is assumed to have a normalized standard deviation of $\sigma = 0.001$ (corresponding to an attitude error of about 0.5 degrees). Also, the attitude of the vehicle is assumed to be Earth-pointing with a rotation of 236 deg/hr about the vehicle's y-axis (negative orbit-normal), while holding the remaining axis rotations to zero. The spacecraft z-axis is defined to be pointed nadir, and the x-axis completes the triad.

If the baselines do not constitute an orthogonal set, the solution of the transformed cost function to Wahba's form is suboptimal. However, the covariance analysis shown in this paper can be used to assess the errors introduced from the transformation. All simulation results presented in the figures use a_j given by Equation (39). Figure 2 shows the attitude errors and three-sigma bounds by solving Wahba's form for Case 1. This shows the excellent agreement between theory and simulated measurement processes.

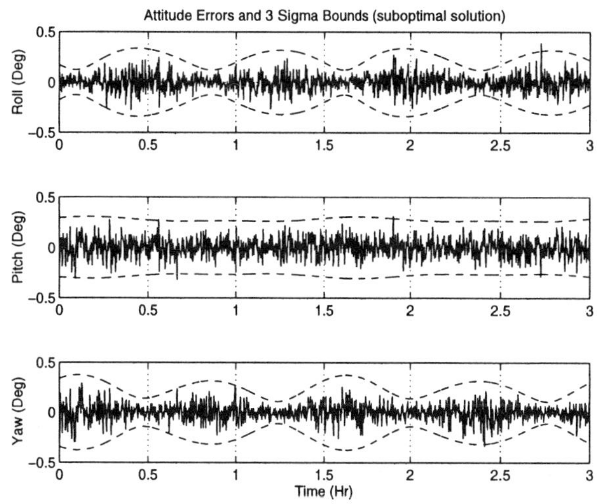

Fig. 2 Attitude Errors and Bounds for Case 1

In order to quantify the error introduced by using the suboptimal solution, the following error factor is used

$$f = \frac{1}{m_{\text{tot}}} \sum_{k=1}^{m_{\text{tot}}} \frac{\text{trace}\left\{\text{diag}\left[P_{\text{body}}^{1/2}\right]\right\}}{\text{trace}\left\{\text{diag}\left[P_{\text{opt}}^{1/2}\right]\right\}} \quad (49)$$

where m_{tot} represents the total number of measurements used in the simulation. A plot of the error factor at each time is shown in Figure 3. Equation (49) represents the average of the curve shown in Figure 3. Clearly, the suboptimal solution is adequate, with a maximum error of about 3%.

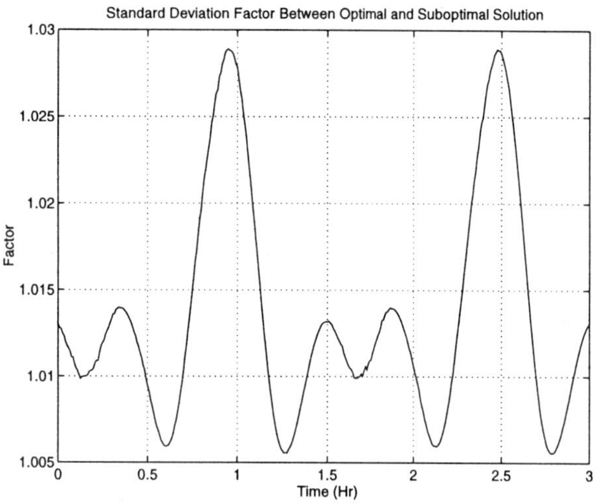

Fig. 3 Error Factor for Case 1

A plot of the standard deviation errors for the suboptimal and optimal solutions for Case 2 is shown in Figure 4. The optimal standard deviation error is always lower than the suboptimal solution.

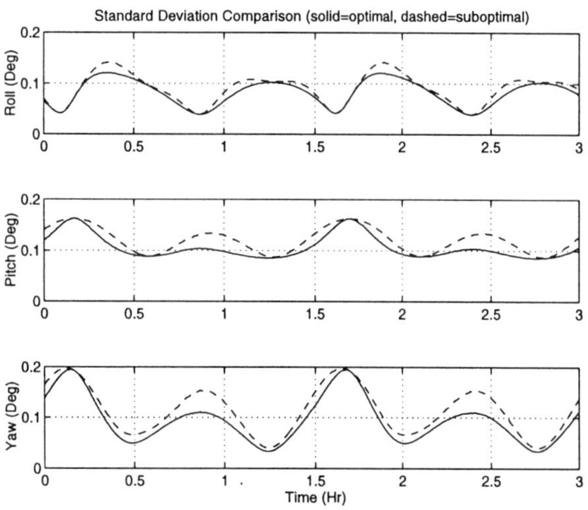

Fig. 4 Standard Deviation Comparison for Case 2

A plot of the error factor at each time is shown in Figure 5. For this case, the suboptimal solution can produce large errors, with a maximum error of about 35%. This is due to the non-orthogonal baselines, and due to the attitude of the vehicle.

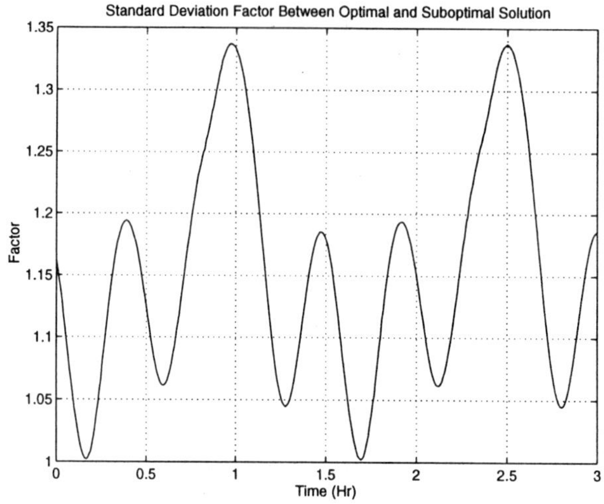

Fig. 5 Error Factor for Case 2

Results using various values of a_j for all three case are shown in Table 1 (the subscripts max and min denote eigenvalues). The performance factor f represents the average error. Clearly, various choices for the weighting factors a_j do not affect system performance. Also, Case 3 where the baselines are nearly collinear results in a substantial degradation in performance when using the suboptimal solution as compared to the optimal solution. Therefore, the covariance analysis is extremely helpful for determining whether or not the suboptimal and/or the optimal solution meets required performance specifications.

Table 1 Weighting Factor *f* Performance Comparisons

	Case 1	Case 2	Case 3
$a_j = \dfrac{1}{(P_j)_{\max}}$	1.014	1.151	58.13
$a_j = \dfrac{2}{(P_j)_{\max} + (P_j)_{\min}}$	1.014	1.151	58.16
$a_j = \dfrac{(P_j^{-1})_{\max} + (P_j^{-1})_{\min}}{2}$	1.014	1.151	58.29
$a_j = \dfrac{1}{3}\mathrm{trace}(P_j^{-1})$	1.014	1.151	58.13

Conclusions

The problem of determining the attitude of a vehicle using GPS phase measurements was addressed in this paper. A general method which transforms the general GPS cost function into a Wahba cost function was presented. Covariance equations for both the new method, and methods which solve the general cost function were developed. It was shown that the transformation produces suboptimal attitude solutions for non-orthogonal baselines and sightlines. The equivalence of both covariance equations for orthogonal baselines and/or sightlines was also shown. Simulation results indicate that the new method is adequate for nearly orthogonal baselines or sightlines, but can produce large errors for nearly collinear baselines or sightlines, as compared to methods which minimize the general cost function directly. This paper provides a means of accessing various performance criteria, such as computational efficiency versus attitude accuracy, for the particular application.

Acknowledgments

The first author's work is supported by a National Research Council Postdoctoral Fellowship tenured at NASA-Goddard Space Flight Center. The author greatly appreciates this support.

References

[1] Cohen, C.E., "Attitude Determination Using GPS," Ph.D. Dissertation, Stanford University, Dec. 1992.

[2] Melvin, P.J., and Hope, A.S., "Satellite Attitude Determination with GPS," *Advances in the Astronautical Sciences*, Vol. 85, Part 1, AAS #93-556, pp. 59-78.

[3] Melvin, P.J., Ward, L.M., and Axelrad, P., "The Analysis of GPS Attitude Data from a Slowly Rotating, Symmetrical Gravity Gradient Satellite," *Advances in the Astronautical Sciences*, Vol. 89, Part 1, AAS #95-113, pp. 539-558.

[4] Lightsey, E.G., Cohen, C.E., Feess, W.A., and Parkinson, B.W., "Analysis of Spacecraft Attitude Measurements Using Onboard GPS," *Advances in the Astronautical Sciences*, Vol. 86, AAS #94-063, pp. 521-532.

[5] Cohen, C.E., and Parkinson, B.W., "Integer Ambiguity Resolution of the GPS Carrier for Spacecraft Attitude Determination," *Advances in the Astronautical Sciences*, Vol. 78, AAS #92-015, pp. 107-118.

[6] Wahba, G., "A Least Squares Estimate of Spacecraft Attitude," *SIAM Review*, Vol. 7, No. 3, July 1965, pg. 409.

[7] Shuster, M.D., and Oh, S.D., "Attitude Determination from Vector Observations," *Journal of Guidance and Control*, Vol. 4, No. 1, Jan.-Feb. 1981, pp. 70-77.

[8] Markley, F.L., "Attitude Determination from Vector Observations: A Fast Optimal Matrix Algorithm," *The Journal of the Astronautical Sciences*, Vol. 41, No. 2, April-June 1993, pp. 261-280.

[9] Bar-Itzhack, I.Y., Montgomery, P.Y., and Garrick, J.C., "Algorithms for Attitude Determination Using GPS," *Proceedings of the 37th Israel Annual Conference on Aerospace Sciences*, Tel-Aviv and Haifa, Israel, Feb. 1997, pp. 233-244.

[10] Shuster, M.D., "A Survey of Attitude Representations," *The Journal of the Astronautical Sciences*, Vol. 41, No. 4, Oct.-Dec. 1993, pp. 439-517.

[11] Shuster, M.D., "Efficient Algorithms for Spin-Axis Attitude Estimation," *The Journal of the Astronautical Sciences*, Vol. 31, No. 2, April-June 1983, pp. 237-249.

[12] Shuster, M.D., "Kalman Filtering of Spacecraft Attitude and the QUEST Model," *The Journal of the Astronautical Sciences*, Vol. 38, No. 3, July-Sept. 1990, pp. 377-393.

[13] Markley, F.L., "Attitude Determination Using Vector Observations and the Singular Value Decomposition," *The Journal of the Astronautical Sciences*, Vol. 36, No. 3, July-Sept. 1988, pp. 245-258.

[14] Shuster, M.D., "Maximum Likelihood Estimation of Spacecraft Attitude," *The Journal of the Astronautical Sciences*, Vol. 37, No. 1, Jan.-March 1989, pp. 79-88.

MODEL-GROUP SWITCHING ALGORITHM FOR MULTIPLE-MODEL ESTIMATION WITH VARIABLE STRUCTURE

X. Rong Li, Xiaorong Zhi and Youmin Zhang*
University of New Orleans, New Orleans, LA 70148
Phone: 504-280-7416, FAX: 504-280-3950, E-mail: xrlee@uno.edu

Abstract

A general multiple-model estimator with variable structure (VSMM), called model-group switching (MGS) algorithm, is presented. It assumes that the total set of models can be covered by a number of model groups, each representing a cluster of closely related system behavior patterns or structures, and a particular group is running at any given time determined by a hard decision. This algorithm is the first VSMM estimator that is generally applicable to a large class of problem with hybrid (continuous and discrete) uncertainties and easily implementable. The algorithm is implemented for a problem of tracking a maneuvering target using a time-varying set (group) of models. Simulation results show that the proposed MGS algorithm provides a substantial reduction in computation while having virtually identical performance with the fixed-structure Interacting Multiple-Model (IMM) estimator.

1 Introduction

Most existing multiple-model (MM) estimators have a fixed structure (FSMM) in the sense that they use a fixed set of models at all times. They have found great success in solving many state estimation problems compounded with structural or parametric uncertainty in many areas over the last three decades, in particular target tracking (see [8, 2, 3] and the long lists of references therein). Existing MM estimation techniques with a fixed structure have, however, arrived at such a stage that great improvement can no longer be expected within their fixed structure, especially for problems involving many structural modes.

The limitation of MM estimation with a fixed structure has been more or less perceived for some time. Ad hoc remedies were proposed for particular applications, such as the moving-bank MM estimators in [15]. A serious attempt was initiated in [11] and continued in [10, 12] to lay down a theoretical foundation for MM estimation without the limitation of the fixed structure; that is, with a variable structure (VSMM). Theoretical results fundamental to the MM estimation with a variable structure were also presented in [12, 10].

As manifested by the great impact of the monumental success of the Interacting Multiple-Model (IMM) estimator [4] on the application of MM estimation techniques, it is fair to say that no matter how promising VSMM estimation may appear, its ultimate success relies on the development of good practical VSMM algorithms that can be readily implemented and are general enough to be applicable to a large class of hybrid estimation problems (i.e., those with both continuous and discrete uncertainties) [8]. If succeeded, such development will be a new milestone in the history of MM estimation. The primary value of theoretical work on the variable structure is to facilitate this development. Development and design of VSMM algorithms have received increasing attention recently due to their great practical significance. Toward this end, three practical VSMM schemes were outlined in [12, 11] that are general enough to be applicable to many problems. Designs of several ad hoc VSMM algorithms were reported in [5, 15, 14, 16, 6, 7] for the particular applications considered.

This paper presents a VSMM estimator, called model-group switching (MGS) algorithm, that is generally applicable to most hybrid estimation problems and easily implementable. It belongs to one of the three VSMM schemes outlined in [12, 11] — the digraph switching scheme. Its development relies heavily on the theoretical results of [9].

2 Variable Structure MM Estimation

Consider the following one of the simplest stochastic hybrid systems, known as the (time-varying) "jump linear systems,"

$$x_{k+1} = F(s_{k+1}, k)x_k + G(s_{k+1}, k)w(s_{k+1}, k) \quad (1)$$
$$z_k = H(s_k, k)x_k + v(s_k, k) \quad (2)$$

where x is the (conventional) *base state* vector; z is the noisy measurement vector; s_k is the *modal state* or *system mode* (system mode index) at time k, which denotes the mode in effect during the sampling period ending at t_k; w and v are the mode-dependent process and measurement noise sequences, respectively. The system mode sequence $\langle s_k \rangle$ is assumed to be a first-order homogeneous Markov chain with the following transition

*Research supported in part by NSF Grant ECS-9496319, ONR Grant N00014-97-1-0570 and LEQSF Grant (1996–99)-RD-A-32.

probabilities

$$P\{m_{k+1}^j | m_k^i\} = \pi_{ij} \qquad \forall m_i, m_j \in \mathbf{S} \tag{3}$$

where the event that mode m_j is in effect at time k is denoted as $m_k^j \triangleq \{s_k = m_j\}$ and $\mathbf{S}$ is the set of all possible system modes.

In the MM approach, it is assumed that the nonlinear system (1)–(2) at any given time can be adequately approximated by one or (a combination of) more models in a set M of the following conventional linear models:

$$x_{k+1} = F_k^j x_k + G_k^j w_k^j \qquad \forall m_j \in M \tag{4}$$

$$z_k = H_k^j x_k + v_k^j \qquad \forall m_j \in M \tag{5}$$

and a jump between the system modes can be modeled by switching from one model to another, governed by the Markov law (3). Here subscript j denotes quantities pertaining to model m_j.

The state estimate and its associated covariance matrix are calculated in an MM estimator using the minimum mean square error (MMSE) criterion as follows:

$$\hat{x}_{k|k} = \sum_j \hat{x}_{k|k}^j P\{H_k^j | z^k\} \tag{6}$$

$$P_{k|k} = \sum_j [P_{k|k}^j + (\hat{x}_{k|k} - \hat{x}_{k|k}^j)(\hat{x}_{k|k} - \hat{x}_{k|k}^j)'] P\{H_k^j | z^k\} \tag{7}$$

where $\hat{x}_{k|k}^j$ is the optimal estimate at time k under the hypothesis $H_k^j \triangleq$ {model sequence j through time k matches the sequence of the true system modes through time k}, and $P_{k|k}^j$ is the associated covariance; $z^k = \langle z_i \rangle_{i \le k}$ is the measurement sequence through time k. The summations in (6)–(7) of a full-hypothesis-tree (FHT) MM estimator are over all possible hypotheses (mode sequences).

Most existing MM algorithms have a fixed structure (FSMM) in the sense that the model set M in (4)–(5) is assumed time invariant, even though the models themselves may be time varying or adaptive. A variable-structure MM (VSMM) estimator is one with a time-varying set M of models.

The FSMM approach has certain inherent limitations, which stem from its fundamental assumption that the system mode at any time can be represented (with a sufficient accuracy) by one of a fixed set of models that can be determined before measurements are received in real time, and its inability to incorporate certain types of a priori information. The FSMM estimators perform reasonably well for problems that can be handled with a small set of models. When they are applied to solve real-world problems, however, it is often the case that use of only a few models is not good enough. The computational complexity increases considerably as the number of models increases. More importantly, as shown in [10, 12], the use of more models does not necessarily improve the performance; in fact, the performance will deteriorate if too many models are used due to the excessive "competition" from the "unnecessary" (excess) models. Thus one may face a dilemma: more models have to be used to improve the accuracy, but the use of too many models will degrade the performance, let alone the increase in computation. To find a way out of this dilemma, [12, 11] proposed the use of a variable structure in MM estimation. Not only can a VSMM estimator take advantage of the real-time system mode information contained in the measurement sequence, but it also is able to incorporate certain a priori information of the system mode that would be difficult or impossible for an FSMM estimator to include. An example of such a priori information is that the system mode is unlikely (but not impossible) to be in a certain subset of the total mode set for the particular application under consideration. Another example is the a priori knowledge of the magnitude of jumps in the system mode.

As shown in [12], a key difference between the optimal VSMM and FSMM estimators is that the former is a probabilistically weighted sum of all estimators based on *admissible mode-set* sequences, rather than of all estimators based on *possible mode* sequences as in the latter. In other words, the former is based on partitioning the *range* of the mapping defined by the state dependency of system mode set; whereas the FSMM estimator is based on partitioning the *domain* of the mapping. Although the probabilistically weighted sum of all estimators based on the admissible *mode*-set sequences, as required by the optimal VSMM estimator, is computationally infeasible, a practical VSMM estimator may take advantage of its suggested two-level hierarchical structure: multiple *model-set* sequences at the higher level and multiple *model* sequences at the lower level. For many applications, the higher level with multiple model-set sequences should be replaced, due to computational constraints, with a single model-set sequence (most likely along with a proper merging of model-set sequences) obtained by model-set adaptation. This is the recursive adaptive model-set (RAMS) approach.

The RAMS approach has two functional components: model-sequence conditioned estimation and model-set adaptation. [9] deals with both components in a general setting. What were presented there include a fairly satisfactory solution to the problem of the model-sequence conditioned estimation and important theoretical results for the model-set adaptation. In fact, a major component of the algorithm proposed in this paper is an application of some relevant theoretical results given there.

Foreseeing the difficulty in obtaining general solution to model-set adaptation, [12] proposed and outlined the following three practical VSMM schemes: active digraph (active model-set), digraph switching (model-group switching), and adaptive grid schemes. The simple and efficient *model-set switching scheme* is the topic of this paper. The algorithms presented in [14, 16] also belong to this class. The study of other VSMM schemes will be reported in near future.

3 The Proposed MGS Estimator

The model set used for an MM estimator can be made adaptive by switching among a number of predetermined model groups according to certain rules. This is the basic idea of the *model-group switching* (MGS) scheme. In this scheme, a predetermined cover or partition of the total model set is designed first. This cover or partition consists of model groups representing certain collections (clusters) of closely related system modes. Since a model "group" implies that its member models are (closely) connected in some sense, "model-group switching" is a more proper name for this scheme than "model-set (or model-subset) switching." A partition consists of disjoint groups of models whose union is the total set (i.e., any model in the total set is in one and only one group). A cover has its model groups collectively exhaustive but not necessarily disjoint (some models might belong to more than one group).

Two types of model-group switching are of major interest: hard switching and soft switching. They both have the following two-level hierarchical structure: a higher level consisting of model groups and a lower level consisting of models within a group.

The soft model-group switching scheme is based on the assumption that each of the model groups at any time has a certain probability of having a member model matching (closely) the system mode currently in effect and the overall estimate is the probabilistically weighted sum of the MM estimators based on these model groups. Assuming the sequence of the model group in effect is a Markov or semi-Markov process, it is possible to determine/design the transition probability matrix for this process and to apply the FSMM estimator to it. This can be referred to as the *soft switching* of the model group, just like the soft model switching in a decision-free MM estimator [8]. In this scheme, the FSMM approach with soft decision is applied in two levels: model (lower) level and model group (higher) level. Since probabilistic weights for the model groups are used in this scheme, the model groups have to be disjoint and thus a partition (instead of a cover) has to be used. In the strict sense of variable structure, this scheme is not really one with model-group switching, or not even a VSMM algorithm since it is nothing but a probabilistically weighted sum of a fixed number of *fixed structure* estimators, each based on a model group. This can also be understood from one of the remarks on the optimal VSMM estimator given in [12], which states that an algorithm based on a *partition of the total set* requires implicitly the use of a fixed structure. In other words, this soft switching algorithm is a degenerated VSMM (or disguised FS) scheme. A proper name for it seems "multiple model-group algorithm" rather than "model-group soft switching algorithm." Note also that in order to use soft switching, all model groups have to be run at all times and thus its computational complexity is comparable with that of the corresponding FSMM estimator based on the total model set.

3.1 Hard Switching of Model Groups

In the *hard switching* of model groups, the active model group is made adaptive by switching among a number of predetermined model groups according to a set of certain "hard" rules (i.e., by a hard decision). As such, only one model group has to be run at any time and thus it may provide a substantial saving in computation over the FSMM estimator based on the total model set. The adjacent model groups should have common models since they are valuable for the design of an adaptation logic and the initialization of newly activated filters. It is thus clear that a hard switching MGS algorithm is in general based on a cover rather than a partition of the total model set.

Following the idea proposed in [9], model-group switching is completed in two stages: activation and termination. The advantage of such a two-stage switching is clear: It can reduce the chance of false switching dramatically at the cost of an insignificant increase in computation, rather than a significant increase in the delay of correct switching, as would be for a single-stage switching algorithm. The use of model-group activation and termination, rather than a single-stage switching, is one of the key ideas of the MGS algorithm. It will become clear that part of the superiority of the proposed MGS algorithm comes from this decomposition.

General speaking, one cycle of an MGS algorithm consists of the following conceptual steps:

- S1. Model-group adaptation: Decide whether a candidate model group should be activated. If it is activated, check if it should be maintained for the next time cycle. The model group used in the previous cycle should be retained for the current cycle but it may be terminated for the next cycle by a logic.
- S2. Initialization of newly activated models/filters: If a candidate model group is activated, then assign proper initial probabilities to the newly activated models and determine the initial estimates and error covariances of the filters based on these models.
- S3. MM estimation: Make MM estimate for the current cycle based on all the models in the selected group(s).

Step 3 is identical to a cycle of a recursive FSMM estimator. Steps 1 and 2 are unique to the MGS algorithm

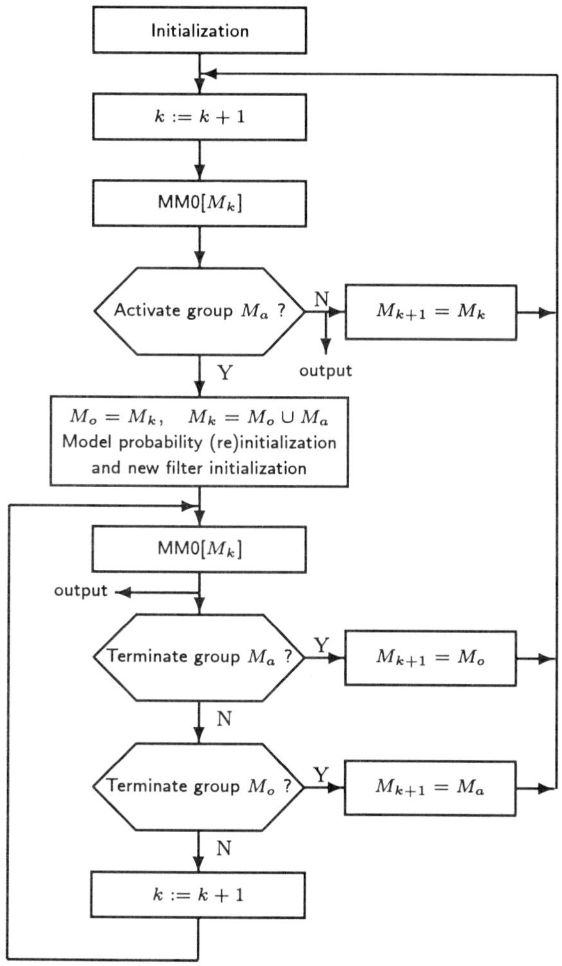

Figure 1: Conceptual flowchart of the MGS algorithm.

and thus are investigated in the sequel. Steps 2 and 3 are actually integrated in our MGS algorithm by using the VSIMM recursion of [9].

Fig. 1 gives the conceptual flowchart of the model-group switching algorithm proposed in this paper.

In the flowchart, MM0[M_k] stands for one cycle of an MM estimator of zero memory depth based on model group M_k in effect at time k. An MM estimator is of a zero memory depth if it calculates all its quantities in the current time cycle using only the quantities obtained in the most recent cycle as well as a priori information. Notable examples of the zero-memory-depth MM estimators include the IMM and GPB1 (the first-order generalized pseudo-Bayesian) algorithms. Even though MM0 is used in the flowchart, the algorithm actually works for other MM estimators as well (with minor changes).

3.2 Model-Group Adaptation

With the decomposition of model-group switching into activation and termination, the decision for model-group adaptation includes

- Decision for candidate model-group activation.
- Decision for the termination of the newly activated candidate model group M_a.
- Decision for the termination of the model group M_o currently in effect.

3.2.1 Model-group activation

In general, the decision (logic) for model-group adaptation should consists of a set of rules based on both a priori and a posteriori information about the current system mode in effect. The a priori mode information is incorporated mostly in the topology of the total model set (more rigorously, digraph) and the corresponding transition probability matrix, even though they are also possible to be made time varying or adaptive. Use of heuristic rules or tricks can also be justified as utilizing the a priori information, which is definitely problem dependent. The a posteriori mode information comes only from the measurement sequence and is most compactly summarized (but not in full) in the sequences of model probabilities and likelihoods (since a mode may actually be different from every model). As such, it can be expected that a more or less general logic should rely primarily on the model probabilities and/or likelihoods, based on the model-set topology and possibly with the transition probabilities taken into account.

Although some useful theoretical results were presented in [9], the actual candidate group activation logic should be problem dependent. How to obtain it is a design task. Its design relies on the topology of the total model set, the particular cover used, the physical meaning of the models, desired performance versus computational constraints, etc. This design should be integrated with the selection of the thresholds for the model-group termination.

3.2.2 Model-Group Termination

Based on Theorems 3 and 4 of [9], the following tests for the termination of a candidate or a current model group are proposed, which are generally applicable and have some optimality properties, as shown in [9].

Candidate model-group termination:

The candidate model group M_a is terminated for the next time ($k+1$) while competing with the current model group M_o if and only if *either* of the following two conditions (criteria) is satisfied

$$\text{probability condition:} \qquad \frac{\mu_k^{M_a}}{\mu_k^{M_o}} < t_1^{\mu} \tag{8}$$

$$\text{likelihood condition:} \qquad \prod_{n=N}^{k} \frac{L_n^{M_a}}{L_n^{M_o}} < t_1^{L} \tag{9}$$

where N is the time that the candidate model group is activated. In the above, the model-group probability

μ^{M_l} is the sum of all probabilities of the models in group M_l calculated based on $M_a \cup M_o$, given by, for $l = a$ or o,

$$\mu_k^{M_l} \triangleq P\{M_k^l | M_k^{a+o}, z^k\} = \sum_{m_n \in M_l} P\{m_k^n | M_k^{a+o}, z^k\} \tag{10}$$

where M_k^l (or M_k^{a+o}) is the event that one of the models in M_l (or $M_a \cup M_o$) matches the system mode at k:

$$M_k^l = \{s_k \in M_l\} \tag{11}$$

$$M_k^{a+o} = \{s_k \in (M_a \cup M_o)\} \tag{12}$$

The model-group likelihood L^{M_l} is the probabilistically weighted sum of all likelihoods of the models in group M_l, given by, for $l = a$ or o,

$$\begin{aligned} L_k^{M_l} &\triangleq p[z_k | M_k^l, z^{k-1}] \\ &= \sum_{m_n \in M_l} p[z_k | m_k^n, M_k^l, z^{k-1}] P\{m_k^n | M_k^l, z^{k-1}\} \\ &= \sum_{m_n \in M_l} p[z_k | m_k^n, z^{k-1}] \frac{P\{m_k^n, M_k^l | M_k^{a+o}, z^{k-1}\}}{P\{M_k^l | M_k^{a+o}, z^{k-1}\}} \\ &= \sum_{m_n \in M_l} L_k^n \hat{\mu}_{k|k-1}^{n|M_l} \end{aligned} \tag{13}$$

where L^n is the likelihood of model m_n, defined by

$$L_k^n = p[z_k | m_k^n, z^{k-1}] \tag{14}$$

$\hat{\mu}_{k|k-1}^{n|M_l}$ is the predicted probability of model m_n at time k based on group M_l and z^{k-1}:

$$\hat{\mu}_{k|k-1}^{n|M_l} \triangleq P\{m_k^n | M_k^l, z^{k-1}\} = \frac{1}{\hat{\mu}_{k|k-1}^{M_l}} \hat{\mu}_{k|k-1}^n \tag{15}$$

$\hat{\mu}_{k|k-1}^n$ is the predicted probability of model m_n at time k based on the total model set $\mathbf{M}$:

$$\hat{\mu}_{k|k-1}^n \triangleq P\{m_k^n | z^{k-1}\} = \sum_{m_m \in M_{k-1}} \pi_{mn} \mu_{k-1}^m \tag{16}$$

and $\hat{\mu}_{k|k-1}^{M_l}$ is a normalization factor for $\hat{\mu}_{k|k-1}^n$ within M_l, given explicitly by

$$\hat{\mu}_{k|k-1}^{M_l} \triangleq P\{M_k^l | \mathbf{M}_k, z^{k-1}\} = \sum_{m_n \in M_l} \hat{\mu}_{k|k-1}^n \tag{17}$$

Note that if the IMM algorithm is used as the MM0 cycle, then $\hat{\mu}_{k|k-1}^n$ has been obtained in the IMM cycle.

The thresholds t_1^μ and t_1^L are design parameters.

Current model-group termination:

The current model-group M_o is terminated for the next time $(k+1)$ while competing with the candidate model-group M_a if and only if *both* of the following two conditions (criteria) are satisfied

$$\text{probability condition:} \qquad \frac{\mu_k^{M_a}}{\mu_k^{M_o}} > t_2^\mu \tag{18}$$

$$\text{likelihood condition:} \qquad \prod_{n=N}^{k} \frac{L_n^{M_a}}{L_n^{M_o}} > t_2^L \tag{19}$$

The thresholds t_2^μ and t_2^L are design parameters.

Based on Theorems 3 and 4 of [9], the two separate tasks of termination of the candidate and current model groups can be combined by using the Sequential Model-Set Likelihood Ratio Test and the Sequential Mode-Set Probability Ratio Test, which are optimal in the sense of most efficient use of the information in the observations with guaranteed expected type I and type II error (false alarm and miss) probabilities.

The rationale for the use of "*and*" logic (for the probability and likelihood) in the current group termination and "*or*" logic in the candidate group lies in the fact that the model-group probabilities and likelihoods are calculated more accurately for the current group than for the candidate group in the MGS algorithm since the filters in the candidate group are newly initialized.

3.3 Initialization of a Model Group

Initialization of the newly activated filters consists of

- Assign the probabilities, at one time cycle before, to the models on which these filters are based, and
- Determine the state estimates and the associated error covariances for these filters, at one time cycle before.

How these tasks could be done systematically and reasonably is a major issue related to the ad hoc techniques of [5, 15, 14, 16, 6, 7].

The concept of state-dependent system mode set, established in [12], is a powerful one, particularly useful for filter initialization. It implies that in plain terms, given the current system mode, the set of possible system modes at the next time is a subset of the total mode set, which is determined by the Markovian mode transition law (i.e., the graph-theoretic adjacence of the mode sets). As applied to filter initialization, the assignment of the initial probability to a model should account only for the probabilities of those models from which the former model is allowed to be switched; and the initial state estimate of a filter based on a model should be determined only from the estimates (and the probabilities) of those filters based on the models from which the former model is allowed to be switched.

Specifically, if m_n is a *newly* activated model at time k, then its probability of being true (more precisely, matching to the system mode in effect) at $k-1$ is (has to be) zero:

$$\mu_{k-1}^n = P\{m_{k-1}^n | M^{k-1}, z^{k-1}\} = 0, \qquad \forall m_n \in M_n \tag{20}$$

where M_n is the set of new and only new models; $M^k = \langle M_i \rangle_{i \le k}$ is the sequence of the model sets through time k; and the expected probability of m_n being true at k without information in the observation at k (i.e., the predicted mode probability) is given by

$$\begin{aligned}\hat{\mu}^n_{k|k-1} &= P\{m^n_k|M^{k-1}, z^{k-1}\} \\ &= \sum_{m_j \in E^n} P\{m^n_k|m^j_{k-1}\}P\{m^j_{k-1}|M^{k-1}, z^{k-1}\} \\ &= \sum_{m_j \in E^n} \pi_{jn}\mu^j_{k-1} \end{aligned} \tag{21}$$

where E^n is the set of models in M_{k-1} that are allowed to switch to m_n:

$$E^n = \{m_l : m_l \in M_{k-1}, \pi_{ln} \neq 0\} \tag{22}$$

The initial probabilities and the filter estimates (and associated covariances) of all the models in the current group should be kept unchanged in initialization.

The initialization of every filter based on a newly activated model in the candidate group should be done as follows. In general, when calculating $E[x_k|m^n_k]$, only the previous estimates in the set $\{\hat{x}^l_{k-1|k-1}, m_l \in E^n\}$ should be (and need be) used, where E^n was given by (22). Specifically, the initial estimate for the filter based on model m_n at time k can be obtained by, similarly to the mixing step in the IMM estimator,

$$\begin{aligned}\hat{x}^{0n}_{k-1|k-1} &= E[x_{k-1}|m^n_k, M^{k-1}] \\ &= E\left[E[x_{k-1}|m_{k-1}, m^n_k, M^{k-2}]|m^n_k, M^{k-1}\right] \\ &= \sum_{m_l \in E^n} E[x_{k-1}|m^l_{k-1}, M^{k-2}]P\{m^l_{k-1}|m^n_k, M^{k-1}\} \\ &= \sum_{m_l \in E^n} \hat{x}^l_{k-1|k-1}\mu^{l|n}_{k-1|k-1} \end{aligned} \tag{23}$$

where conditioning on z^{k-1} has been dropped for simplicity; the mixing weight is given by

$$\mu^{l|n}_{k-1|k-1} = P\{m^l_{k-1}|m^n_k, M^{k-1}\} = \frac{\pi_{ln}\mu^l_{k-1}}{\sum_{m_i \in E^n} \pi_{in}\mu^i_{k-1}} \tag{24}$$

The associated error covariance can be calculated accordingly. In simpler words, if model m_n can only be switched from model m_l in M_{k-1}, then $\hat{x}^l_{k-1|k-1}$ and $P^l_{k-1|k-1}$ should clearly be used as the initial estimate and covariance for the filter based on m_n at time k. If m_n can be switched from several models in M_{k-1}, then the probabilistically weighted sum of the estimates at $k-1$ from the filters based on these models should be used as the initial estimate for the filter based on m_n at k and the initial covariance can be determined accordingly.

If the VSIMM recursion of [9] is used for the MM0 cycle, then the initialization discussed above can actually be omitted because it is incorporated by the VSIMM recursion automatically. Due to space limitation, the reader is referred to [9] for details.

3.4 Fusion of Estimates Based on Current and Newly Activated Models

In general, the current and newly activated group may have models in common. When this is indeed the case, the MGS algorithm as outlined in Fig. 1 has some waste in computation because the estimates based on these common models as calculated in the second MM0[M_k] should not differ from those calculated in the first MM0[M_k]. But how to combine the estimates from two model sets to yield the overall estimate? The optimal fusion of [9] provides a satisfactory answer. Following this result, the computational waste can be eliminated as follows:

- Replace the second MM0[M_k] in Fig. 1 by MM0[M_n], where $M_n = M_a - M_o$ is the set of new and only new models.
- Use the optimal fusion theorem of [9] to obtain the overall estimate.

3.5 Initialization of MGS Algorithm

How the MGS algorithm should be initialized depends on what a priori information is available concerning the initial system mode. If a priori information indicates that the initial system mode is likely to be in a certain subset of the total mode set, then the MGS algorithm should start from the corresponding model group. In general, one of the following three procedures may be used for initialization if no a priori information for the initial system mode is available:

- Run a set of the MGS algorithms using all model groups for a few cycles and then choose the one with the highest group likelihood as the initial model group.
- Run an FSMM estimator using all models for a few cycles and then choose the model group which has the highest group probability as the initial one.
- Choose as the initial one the "nominal" model group that represents the nominal system mode(s) with a mechanism to switch to other nonnominal groups. A good example of such nominal model groups is M_1 of (27).

3.6 The MGS Algorithm

Summarizing the results of this section and assuming that an IMM cycle is used as the MM0 estimation cycle, one cycle of the proposed MGS algorithm is listed in Table 1, where VSIMM cycle was given in [9].

4 Design of MGS Algorithm

The design of an MGS algorithm consists of the following two major elements:

- Design of the model-set cover (i.e., model groups).

Table 1: One cycle of the MGS algorithm.

S1. Increase the time counter k by 1. Run the VSIMM$[M_k, M_{k-1}]$ cycle.

S2. Check if a candidate model-group is activated. If no model group is activated, then output $\hat{x}_{k|k}$, $P_{k|k}$, and $\{\mu_k^i\}_{m_i \in M_k}$ obtained from the VSIMM$[M_k, M_{k-1}]$ cycle. Let $M_{k+1} = M_k$ and go to Step 1.

S3. If a model group M_a is activated, then let $N = k$, $M_o = M_k$ and

- Run the VSIMM$[M_k^n, M_{k-1}]$ cycle, where $M_n = M_a - M_o$ is the set of new and only new models.
- Calculate the estimates, error covariances, and mode probabilities for the union set $M = M_n \cup M_o = M_a \cup M_o$:

$$\mu_k^i = \frac{L_k^i \hat{\mu}_{k|k-1}^i}{\sum_{m_i \in M} L_k^i \hat{\mu}_{k|k-1}^i}, \qquad \forall m_i \in M$$

$$\hat{x}_{k|k} = \sum_{m_i \in M} \hat{x}_{k|k}^i \mu_k^i$$

$$P_{k|k} = \sum_{m_i \in M} \{P_{k|k}^i + [\hat{x}_{k|k}^i - \hat{x}_{k|k}][\hat{x}_{k|k}^i - \hat{x}_{k|k}]'\}\mu_k^i$$

  where the likelihoods $\{L_k^i\}$ and predicted probabilities $\{\hat{\mu}_{k|k-1}^i\}$ were obtained in the VSIMM$[M_k, M_{k-1}]$ cycle.
- Output $\hat{x}_{k|k}$, $P_{k|k}$, and $\{\mu_k^i\}_{m_i \in M}$.

S4. For model groups $M_l = M_o, M_a$, compute

$$\mu_k^{M_l} = \sum_{m_i \in M_l} \mu_k^i$$

$$\hat{\mu}_{k|k-1}^{M_l} = \sum_{m_i \in M_l} \hat{\mu}_{k|k-1}^i$$

$$L_k^{M_l} = \frac{1}{\hat{\mu}_{k|k-1}^{M_l}} \sum_{m_i \in M_l} L_k^i \hat{\mu}_{k|k-1}^i$$

If

$$\frac{\mu_k^{M_a}}{\mu_k^{M_o}} < t_1^\mu \qquad \text{or} \qquad \prod_{n=N}^{k} \frac{L_n^{M_a}}{L_n^{M_o}} < t_1^L$$

then terminate group M_a: Let $M_{k+1} = M_o$, $M_k = M_o \cup M_a$, and go to Step 1.

If

$$\frac{\mu_k^{M_a}}{\mu_k^{M_o}} > t_2^\mu \qquad \text{and} \qquad \prod_{n=N}^{k} \frac{L_n^{M_a}}{L_n^{M_o}} > t_2^L$$

then terminate group M_o: Let $M_{k+1} = M_a$, $M_k = M_o \cup M_a$, and go to Step 1.

S5. Increase the time counter k by 1. Let $M_k = M_{k-1} = M = M_o \cup M_a$. Run the VSIMM$[M_k, M_{k-1}]$ cycle and output $\hat{x}_{k|k}$, $P_{k|k}$, and $\{\mu_k^i\}_{m_i \in M}$. Go to Step 4.

- Design of candidate model-group activation logic and selection of the thresholds for the termination of model groups.

In addition, the following design elements should also be completed, which are common to most MM estimators:

- Select the model transition probability matrix.
- Tune the process noise covariance matrix of each filter.

These design elements should be evaluated under carefully determined test scenarios.

In our design, the following problem, adopted from [1], is considered. To track a maneuvering target in a planar motion that may have a piecewise constant acceleration with a (coordinate-combined) maximum value of 40m/s^2 in any direction, [1] compared the performance of an IMM estimator with that of a Viterbi algorithm based on the following set of 12 time-invariant models, characterized by the expected acceleration vector $\mathbf{a}$,

$$\begin{array}{ll} m_1 : \mathbf{a} = [0,\ 0]' & m_2 : \mathbf{a} = [20,\ 0]' \\ m_3 : \mathbf{a} = [0,\ 20]' & m_4 : \mathbf{a} = [-20,\ 0]' \\ m_5 : \mathbf{a} = [0,\ -20]' & m_6 : \mathbf{a} = [20,\ 20]' \\ m_7 : \mathbf{a} = [-20,\ 20]' & m_8 : \mathbf{a} = [-20,\ -20]' \\ m_9 : \mathbf{a} = [20,\ -20]' & m_{10} : \mathbf{a} = [40,\ 0]' \\ m_{11} : \mathbf{a} = [0,\ 40]' & m_{12} : \mathbf{a} = [-40,\ 0]' \end{array} \tag{25}$$

These models were obtained by a quantization of the possible system mode set (space). To cover the assumed acceleration range, the following model

$$m_{13} : \mathbf{a} = [0,\ -40]' \tag{26}$$

is (and should be) added to the model set in our work.

The following cover of the total model set were used in our design (see [13] for details):

$$M_1 = \{m_1, m_2, m_3, m_4, m_5\} \tag{27}$$
$$M_2^2 = \{m_2, m_{10}, m_6, m_1, m_9\} \tag{28}$$
$$M_3^2 = \{m_3, m_{11}, m_7, m_1, m_6\} \tag{29}$$
$$M_4^2 = \{m_4, m_{12}, m_8, m_1, m_7\} \tag{30}$$
$$M_5^2 = \{m_5, m_{13}, m_9, m_1, m_8\} \tag{31}$$

Based on the considerations presented in [13] and the general discussion of Subsection 3.2.1 after a brief trial-and-error adjustment, the following simple activation logic was adopted in our design:

- Activate group M_j while $M_k = M_1$, where $j \neq 1$, if *both* of the following two conditions are satisfied

$$\text{likelihood condition:} \quad L_k^j = \max_{m_l \in M_k} L_k^l \tag{32}$$

$$\text{probability condition:} \quad \mu_k^j > t_0 \tag{33}$$

  where μ^j and L^j are the probability and likelihood of model m_j, respectively. For example, if model m_2 has the highest likelihood and its probability exceeds the threshold t_0 while group M_1 is in effect, then group M_2 is activated.

- Activate group M_1 while $M_k = M_j$, where $j \neq 1$, if *both* of the following two conditions are satisfied

$$\text{likelihood condition:} \quad L_k^1 = \max_{m_l \in M_k} L_k^l \quad (34)$$

$$\text{probability condition:} \quad \mu_k^1 > t_1 \quad (35)$$

- Activate group M_j while $M_k = M_i$, where $i, j \neq 1$, if both of the following two conditions are satisfied for some $m_n \in M_j$, $m_n \neq m_1$,

$$\text{likelihood condition:} \quad L_k^n = \max_{m_l \in M_i} L_k^l \quad (36)$$

$$\text{probability condition:} \quad \mu_k^n > t_2 \quad (37)$$

- Direct switching between M_2 and M_4, or M_3 and M_5 is not allowed. Such transition may go through M_1 as an intermediate group.

In our design, $t_0 = t_1 = t_2 = 0.3$ were used for simplicity.

In our design, the following termination thresholds were used for simplicity:

$$t_1^\mu = t_1^L = 0.95, \qquad t_2^\mu = 1.1, \qquad t_2^L = 1 \quad (38)$$

The MGS algorithm was initialized by the model group M_1 with even model probabilities. To have a fair comparison, the fixed-structure IMM estimator was initialized with the same initial conditions as those of the MGS algorithm, including

$$\mu_0^i = 1/5, \qquad i = 1, \ldots, 5 \quad (39)$$

$$\mu_0^j = 0, \qquad j = 6, \ldots, 13 \quad (40)$$

5 Design of Test Scenarios

It should be emphasized that a comparison between an FSMM algorithm and a VSMM algorithm depends to a some degree on the scenarios used for the evaluation.

Both deterministic and random scenarios were designed and used for testing and comparing the performance and computational requirements of the proposed MGS algorithm with the fixed-structure (FS) IMM estimator using all 13 models.

For the random scenario, it is assumed that the acceleration vector $\mathbf{a}(t) = a(t)\angle\theta(t)$ is a semi-Markov process; specifically, it is a 2-dimensional process that would be Markov except that the sojourn time τ for each of its state is random. The amplitude a and angle θ are both random. In simple terms, it implies that the acceleration process undergoes sudden jumps from a state with a magnitude a and phase θ to another one after staying in it for a random period of time. The following models for τ, a and θ are proposed:

- Denoting by τ_i the sojourn time for the state $a = a_i$, its conditional probability density is

$$p(\tau_i|a_i) = \frac{1}{c}\mathcal{N}(\tau_i; \bar{\tau}_a, \sigma_{\tau_a}^2), \qquad \tau_i > 0 \quad (41)$$

where $\mathcal{N}(x; \bar{x}, \sigma^2)$ stands for Gaussian (normal) probability density function of x with mean $\bar{x}$ and variance σ^2; and $\bar{\tau}_a$ and σ_{τ_a} are two constants, depending on a_i.

- Denoting by a_{i+1} the magnitude of acceleration over the period $(t_i, t_{i+1}]$, its conditional probability densities are

$$p(a_{i+1}|a_i) = P_0\delta(a_{i+1}) + P_M\delta(a_{i+1} - a_M) + (1 - P_0 - P_M)\frac{1}{a_M}1[a_{i+1}; (0, a_M)]$$

where δ is the delta function; $1(x; R)$ is the indicator function, defined by

$$1(x; R) = \begin{cases} 1 & x \in R \\ 0 & x \notin R \end{cases} \quad (42)$$

- Denoting by θ_{i+1} the angle of acceleration over $(t_i, t_{i+1}]$, its conditional probability density is

$$p(\theta_{i+1}|a_i = 0) = \frac{1}{2\pi}1(\theta_{i+1}; (-\pi, \pi]) \quad (43)$$

$$p(\theta_{i+1}|a_i > 0) = \mathcal{N}(\theta_{i+1}; \theta_i, \sigma_\theta^2) \quad (44)$$

The following parameters were used in our design:

$$\bar{\tau}_a = \bar{\tau}_M + \frac{a_M - a}{a_M}(\bar{\tau}_0 - \bar{\tau}_M), \quad \bar{\tau}_M = 10 \quad \bar{\tau}_0 = 30$$

$$P_M = 0.1, \quad a_M = 37, \quad \sigma_\theta = \pi/12, \quad \sigma_{\tau_a} = \frac{1}{12}\bar{\tau}_a$$

$$P_0 = \begin{cases} 0.6 & a_i \neq a_M \\ 0.8 & a_i = a_M \end{cases}$$

The random sojourn time was rounded to its nearest integer and the initial acceleration a_1 was set to zero.

With such a random scenario, it is difficult, if not virtually impossible, to design a VSMM algorithm with subtle tricks that are effective only for certain scenarios. Of course, it cannot prevent intentional cheat.

A drawback of such a random scenario is that the peak estimation errors and the steady-state errors are no longer separable. For this reason, several deterministic scenarios were also used so that the peak and steady-state errors can be seen well. Due to space limitation, however, only some results for the random scenario and the following deterministic scenario were included in this paper: The system mode sequence, characterized by the sequence of accelerations $\mathbf{a}_k$, is specified by

$$\mathbf{a}_k = \begin{cases} [0,0]' & 1 \le k \le 30 \\ [18,22]' & 31 \le k \le 45 \\ [2,37]' & 46 \le k \le 55 \\ [0,0]' & 56 \le k \le 80 \\ [25,2]' & 81 \le k \le 98 \\ [-2,19]' & 99 \le k \le 119 \\ [0,-1]' & 120 \le k \le 139 \\ [38,-1]' & 140 \le k \le 150 \\ [0,0]' & 151 \le k \le 160 \end{cases} \quad (45)$$

Note that both the random and deterministic scenarios designed here have more frequent mode transitions than in most practical situations. This is less favorable to a VSMM algorithm, including the MGS algorithm, than an FSMM estimator. This will be explained later. Scenario (45) with so frequent mode transitions is designed so that many different mode transitions are incorporated into a single sequence of modes.

6 Simulation Results and Discussions

Fig. 2 shows, for the deterministic scenario, the RMS position errors, velocity errors, and their corresponding standard deviations, over 500 Monte Carlo runs, of the MGS estimator and the fixed-structure (FS) IMM estimator that uses all 13 models. Note that both algorithms have their RMS errors significantly higher than their corresponding standard deviations during the mode transition transients.

Fig. 3 shows some model probabilities of the MGS estimator and the FS-IMM estimator for the deterministic scenario over 500 Monte Carlo runs.

Fig. 4 shows, for the random scenario, the RMS position errors, velocity errors, and their corresponding standard deviations, over 500 Monte Carlo runs, of the MGS estimator and the FS-IMM estimator. Again, note that both algorithms have their RMS errors significantly higher than their corresponding standard deviations during the mode transition transients.

Table 2 lists the computational complexity in FLOPs (per cycle) and the quality of maneuver (mode) status reports, in terms of the percentages of correct mode identification (ID), incorrect mode ID, and no mode ID (indecisive), of the MGS and FS-IMM estimators. A correct ID is obtained if the model that is closest to the system mode in effect at the given time has a probability higher than 0.5. An incorrect ID is obtained if the model with a probability over 0.5 is not the one closest to the system mode in effect at the given time. It is indecisive (NID) if no model has a probability above 0.5. The CUP time used for the two algorithms could not be reported because according to MATLAB's technical support, there is a bug in the MATLAB "cputime" function (it may report negative time). Also given in the table are the average distances between the system mode in effect at a given time and the models used in the MGS or IMM algorithm at that time, given by

$$\text{distance} = \sum_{k=1}^{k_{\max}} \sum_{m_i \in M_k} \|s_k - m_i\| P\{s_k = m_i | s_k \in M_k, z^k\} \tag{46}$$

where $\|s_k - m_i\|$ is simply the distance in the acceleration space.

From our simulation results, including those given in Figs. 2–4 and Table 2, it is clear that the MGS algorithm has computational complexity only about one third of that of the FS-IMM estimator while they have almost identical performance. On the other hand, if the same computational constraints are imposed on both the MGS and FS-IMM algorithms, then the MGS algorithm will have performance that is substantially superior to that of the FS-IMM algorithm.

Table 2: Percentages of correct (CID), incorrect (IID), and no (NID) mode identification, relative flop ratio, and average modal distance of MGS and IMM algorithms.

	flops	CID (%)	IID (%)	NID	distance
Deterministic scenario:					
MGS	1	75.97	20.72	3.31	8.0465
IMM	2.81	76.24	20.15	3.61	8.0484
Random scenario:					
MGS	1	75.32	20.19	4.49	7.2474
IMM	2.73	75.24	19.59	5.17	7.3085

Our design is fairly general (not quite specific to the problem considered), except for the candidate activation logic, because our purpose is to compare the general MGS algorithm with the IMM algorithm. Better results can be obtained for the MGS algorithm by fine tuning the design according to the specifics of the problem.

It should be emphasized that a comparison of the computational complexity between a VSMM algorithm and an FSMM algorithm depends to a large degree on the test scenarios used since the VSMM algorithm operates fewer filters at any given time than the FSMM algorithm at the cost of computing group adaptation logic and filter initialization (which are independent of how sophisticated the model-based filters are): A scenario with more frequent mode transitions, as those used here, will increase the computation of the VSMM algorithm significantly while that of an FSMM algorithm will remain unchanged. On the other hand, the more sophisticated each model-based filter is, the more superior the MGS algorithm is to the IMM algorithm. In view of these facts, the test scenarios considered here are in favor of the FSMM estimator rather than the MGS algorithm since 1) second-order (nearly constant velocity) Kalman filters were used, which are among the simplest; 2) the system mode undergoes frequent jumps in our test scenarios. As such, it can be reasonably concluded that *the superiority of the proposed MGS algorithm will be more substantial when tested with more realistic scenarios with more complex model-based filters.*

In fact, comparisons between the MGS and FS-IMM algorithms were also conducted for several other deterministic scenarios in which the mode transitions occurred less frequently and/or with a smaller jump in magnitude (i.e., with a smaller distance in the mode space). The results show better improvement of the MGS algorithm over IMM algorithm. Due to space limitation, these results are not presented here.

It should be pointed out that the actual mode identification quality of the algorithms is more or less better than those listed in Table 2, especially for the random scenario, because the system mode may locate somewhere which has an almost equal distance to more than one model.

It was found that the use of second-order (CV) models with specified expected accelerations led to significantly better results than third-order (CA) models for both MGS and FS-IMM algorithms. Also, poor results would be obtained if the accelerations of a model were taken as part of the state. This is due to the fact that the accelerations of all models in the group would be driven by the measurement sequence from their specified (expected) values towards the true system mode and thus the model set (group) would no longer be able to provide a coverage wide enough to handle effectively sudden jumps in the system mode that may occur later.

7 Conclusions

A multiple-model estimator, called model-group switching (MGS) algorithm, has been presented in this paper. It is the first MM estimator with variable structure that is general enough to be applicable to a large class of state problems compounded with structural and/or parametric uncertainties or changes and is easily implementable. The algorithm runs at any given time a time-varying set (group) of closely related models of the total model set. The determination of which model group to use is done by a hard decision, which consists of two-stage switching logics (hypothesis tests) for activation and termination of model groups. The model-group termination is done by sequential model probability/likelihood ratio test, based on the theoretical results presented in [9]. It has been demonstrated that the design of the activation logic is fairly simple and easy. The general VSIMM recursion of [9] is adopted in the MGS algorithm, which initializes the newly activated filters and assigns the initial probabilities of the corresponding models near optimally. A problem of tracking a maneuvering target using the proposed MGS algorithm was considered. Simulation results in terms of e.g., RMS errors, mode probabilities, percentages of correct, incorrect mode identification and no identification, have been presented for both a fairly complicated deterministic scenario and a more realistic and random scenario in which the acceleration is assumed a semi-Markov process. They demonstrate substantial superiority of the proposed MGS algorithm, in terms of the performance and computational complexity, to the fixed-structure IMM estimator using all models. The random test scenario and the performance indices proposed can also be used for the evaluation of many other algorithms.

References

[1] A. Averbuch, S. Itzikowitz, and T. Kapon, "Radar Target Tracking — Viterbi versus IMM," *IEEE Trans. Aerospace and Electronic Systems*, vol. AES-27, pp. 550–563, May 1991.

[2] Y. Bar-Shalom and X. R. Li, *Estimation and Tracking: Principles, Techniques, and Software.* Boston, MA: Artech House, 1993.

[3] Y. Bar-Shalom and X. R. Li, *Multitarget-Multisensor Tracking: Principles and Techniques.* Storrs, CT: YBS Publishing, 1995.

[4] H. A. P. Blom and Y. Bar-Shalom, "The Interacting Multiple Model Algorithm for Systems with Markovian Switching Coefficients," *IEEE Trans. Automatic Control*, vol. AC-33, pp. 780–783, Aug. 1988.

[5] M. Gauvrit, "Bayesian Adaptive Filter for Tracking with Measurements of Uncertain Origin," *Automatica*, vol. 20, pp. 217–224, Mar. 1984.

[6] V. P. Jilkov and D. S. Angelova, "Performance Evaluation and Comparison of Variable Structure Multiple-Model Algorithms for Tracking Maneuvering Radar Targets," in *Proc. 26th European Microwave Conf.*, (Prague, Czech), Sept. 1996.

[7] V. P. Jilkov, D. S. Angelova, and T. A. Semerdjiev, "Mode-Set Adaptive IMM for Maneuvering Target Tracking," *IEEE Trans. Aerospace and Electronic Systems* (submitted), 1997.

[8] X. R. Li, "Hybrid Estimation Techniques," in *Control and Dynamic Systems: Advances in Theory and Applications,* vol. 76, (C. T. Leondes, ed.), pp. 213–287, New York: Academic Press, 1996.

[9] X. R. Li, "Multiple-Model Estimation with Variable Structure: II — Recursive Adaptive Model-Set Approach," submitted for journal and conference publication, 1997.

[10] X. R. Li, "Multiple-Model Estimation with Variable Structure: Some Theoretical Considerations," in *Proc. 33rd IEEE Conf. Decision and Control,* (Orlando, FL), pp. 1199–1204, Dec. 1994.

[11] X. R. Li and Y. Bar-Shalom, "Mode-Set Adaptation in Multiple-Model Estimators for Hybrid Systems," in *Proc. 1992 American Control Conf.*, (Chicago, IL), pp. 1794–1799, June 1992.

[12] X. R. Li and Y. Bar-Shalom, "Multiple-Model Estimation with Variable Structure," *IEEE Trans. Automatic Control*, vol. AC-41, pp. 478–493, 1996.

[13] X. R. Li, X. R. Zhi, and Y. M. Zhang, "Multiple-Model Estimation with Variable Structure: III — Model-Group Switching Algorithm," *Proc. 36th IEEE Conf. Decision and Control*, 1997.

[14] H. Lin and D. P. Atherton, "An Investigation of the SFIMM Algorithm for Tracking Manoeuvring Targets," in *Proc. 32nd IEEE Conf. Decision and Control*, (San Antonio, TX), pp. 930–935, Dec. 1993.

[15] P. S. Maybeck and K. P. Hentz, "Investigation of Moving-Bank Multiple Model Adaptive Algorithms," *AIAA J. Guidance, Control, and Dynamics*, vol. 10, pp. 90–96, Jan.-Feb. 1987.

[16] A. Munir and D. P. Atherton, "Adaptive Interacting Multiple Model Algorithm for Tracking a Manoeuvring Target," *IEE Proc.-Radar, Sonar, and Navigation*, vol. 142, pp. 11–17, Feb. 1995.

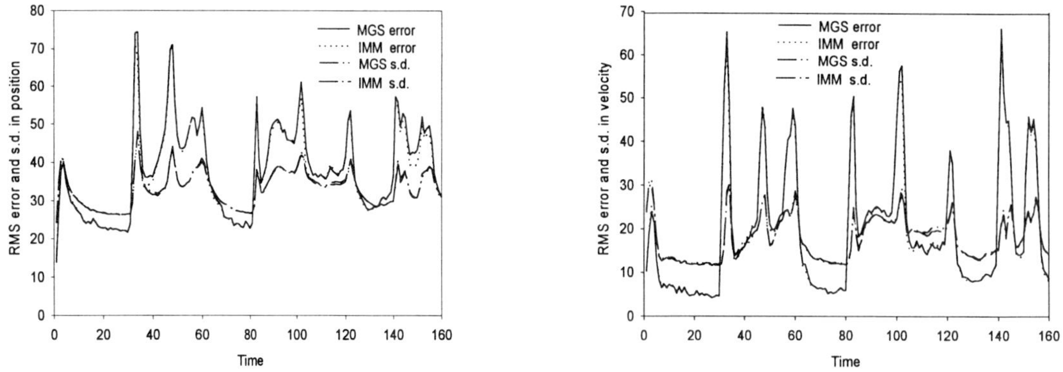

Figure 2: RMS position and velocity errors and deviations for the deterministic scenario.

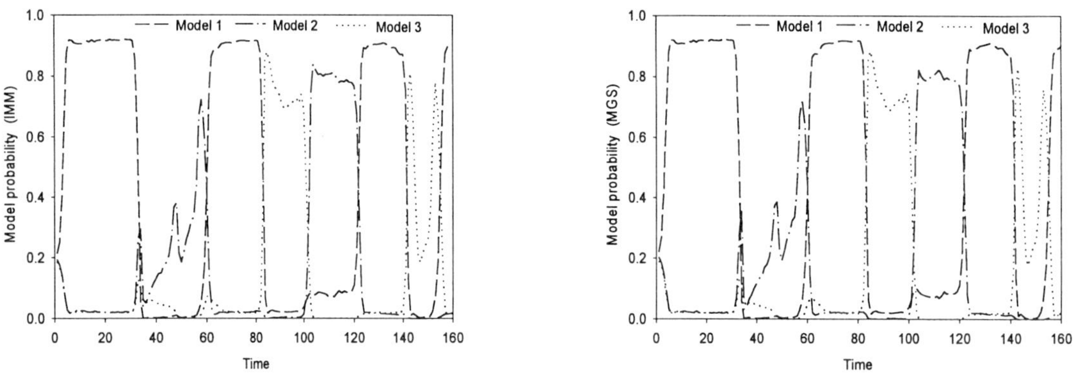

Figure 3: Average model probabilities for the deterministic scenario (models 1, 2, and 3).

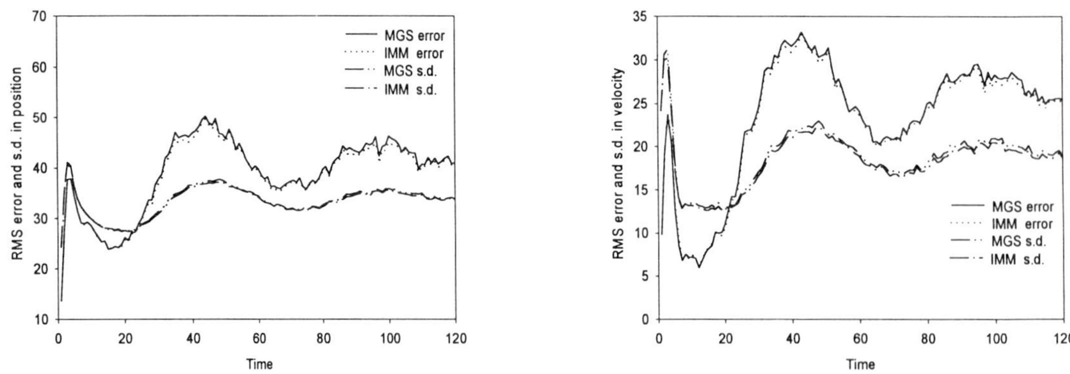

Figure 4: RMS position and velocity errors and deviations for the random scenario.

AIAA-97-3454

H_∞ ROBUST POLE PLACEMENT OF SINGLE INPUT UNCERTAIN SYSTEMS FOR CONTROL OF AIRCRAFT

Shyh-Pyng Shue*, Peng Shi†

ABSTRACT

H_∞ robust pole placement using weight selection for single-input controllable uncertain systems is presented in this paper. The system is considered with two kinds of robust requirements. One is robust control with respect to system uncertainties, and the other is with respect to system disturbance. The H_∞ robust criteria based on the selection of the weighting matrix is used to show that the closed loop poles can be asymptotically stable under all uncertainties which are parameter-bounded. The purpose of using the weighting matrix instead of the performance output matrix is to ensure that the uncertainties on the performance output can be ignored. It will be shown that the weighting coefficients can be determined by linear algebraic equations. Therefore, this weighting matrix can be used for the algebraic Riccati equation to shift all closed loop poles to a stable area. An aircraft model is used to demonstrate the proposed method.

NOMENCLATURE

$A, \hat{A}, \tilde{A}, A_{0.5}, A_{0.6}, A_{0.7}$	state matrices
ΔA	state matrix uncertainty
A_c	closed loop state matrix
a_i	fixed state coefficients
$B, \hat{B}, \tilde{B}, B_{0.5}, B_{0.6}, B_{0.7}$	control matrices
ΔB	control matrix uncertainty
b_i	uncertainties of the state matrix
C_i	output matrix
c_i	bounds of coefficients for the Routh criteria
D_i	control matrix for the output
e_0	uncertainty of the control coefficient
$\hat{F}(s)$	transfer function
d_i	coefficients of the characteristic polynomial for prespecified eigenvalues
f_i	coefficients from computing of Routh criteria
G	disturbance matrix
g_i	coefficients from computing of Routh criteria
i	positive integer
n	dimension of the state
M_∞	Mach number
$P, \hat{P}, \overline{P}, \tilde{P}$	solutions of ARE, Lyapunov matrices
$p_{i,j}, P_{i,j}$	coefficients of the solution of ARE
$P_c, \hat{P}_c$	controllability matrices
Q	weighting matrix
q_i, $Q_{i,i}$	diagonal coefficients of the weighting matrix
q	pitching rate
r_1, r_2	state uncertainties for the first example
s	Laplace operator
s_1	control uncertainty for the first example
$T, T_{0.5}, T_{0.6}, T_{0.7}$	transformation matrices
u	control input
$u(i)$	discrete step function
U_0	steady state airspeed
V_x	horizontal velocity
V_z	vertical velocity
w	system disturbance
$X, \hat{x}, \tilde{x}$	state vector
x	state variables, $[V_x \ \theta \ V_z \ q]^T$
x_0	initial condition of the system response
y	measurement output
z	performance output

Greek Symbol

α	angle of attack (A. O. A)
α_i	lower bounds of the state uncertainties
β	$\sqrt{1-M_\infty^2}$
β_i	upper bounds of the state uncertainties
δ_0	lower bound of the control uncertainty
δ_e	control input (elevator)
ε_0	upper bound of the control uncertainty
γ_0	steady state climb angle
γ	the prescribed value for the system disturbance
η	bounded constant for selection of γ
κ	constant value = $(1-\gamma^{-2})$
θ	pitching angle
λ_i	eigenvalues of the closed loop system

*. Department of Aerospace Engineering, Wichita State University, Wichita, KS, 67260-0044, USA. student member of AIAA. Fax: (1-316) 978-3307, Email: shue@twsuvm.uc.twsu.edu

†. Centre of Industrial and Applied Mathematics, School of Mathematics, The University of South Australia, The Levels, SA 5095, Australia. Fax:(61-8)83025785, Email:matsp@zarniwoop.levels.unisa.edu.au

I. INTRODUCTION

Robust control of an uncertain system has had considerable attention in the recent years. Usually, the uncertainties of the system can be presumed by several forms. The most usual one is to assume that the uncertainties are in the state matrix. The paper presented in [1] has shown that this kind of uncertain system can be stabilized by the 'matching condition'. The 'matching condition' assumes that the uncertainties on the state matrix satisfy the form of the control matrix multiplying a constant matrix. Applying the H_∞ method to the uncertain control system results in the Hamiltonian system satisfying an algebraic Riccati equation. Through adjusting the algebraic Riccati equation (ARE), the robust controller can be obtained based on this ARE solution. Reference [2] shows that the necessity of the matching condition to robustly control the uncertain system.

The second kind of uncertain systems assumes that the uncertainties of the system are not only in the state matrix, but also in the control matrix, as well as the output matrix. This kind of uncertain system is the most general one. Applying the 'matching condition' [3-4], the above uncertain system can also be robustly controlled.

References [5-7] showed while 'matching condition' is no longer available, the system can still be controlled. However, the paper [7] also mentioned that not all of these types of systems can be robustly controlled.

Robust pole placement [8-10] is a method recently developed for robustly stabilizing the uncertain system. The purpose of this method is to shift the poles of closed loop to a stable area for any changes of system uncertainties. In fact, the robust pole placement method can also shift the closed loop poles to the desired region. For example, Reference [9] presented a method in which the closed loop poles are moved to the vertical strip, and the method [10] is presented to shift the closed loop poles to a specified disk.

The aircraft model is known to be very nonlinear, because of the changes of the aerodynamic derivatives. Considering the certain ranges of airspeed, and certain altitude, the model of aircraft can be seen as a certain system with some bounded uncertainties. Therefore, if the method of control of the uncertain system applies for the aircraft system, a certain control gain can be used to control all changes of the uncertainties due to the aerodynamic derivatives.

In this paper, single input controllable uncertain systems will be used to show that the matching condition is not necessary to be used on the control of uncertain systems. It will be shown that the general uncertain model can be controlled via selecting a suitable weighting matrix for the ARE. An uncertain aircraft model is used to demonstrate the proposed method. In addition, the formulas developed in this paper can be used based on the change of the prescribed constant, γ.

II. PROBLEM STATEMENT

Consider a parameter uncertain single input controllable system as

$$\frac{d^n x}{dt^n} = (a_{n-1}+b_{n-1})\frac{d^{n-1}x}{dt^{n-1}} + (a_{n-2}+b_{n-2})\frac{d^{n-2}x}{dt^{n-2}} + \cdots$$

$$+(a_1+b_1)\frac{dx}{dt} + (a_0+b_0)x + (1+e_0)u + w, \quad (1)$$

where x and its time derivatives are the state variables, u is the single control input, and w is the system disturbance. Note that $(a_{n-1}, a_{n-2}, \cdots, a_1, a_0)$ are known and fixed state coefficients, $(b_{n-1}, b_{n-2}, \cdots, b_1, b_0)$ are the parameter uncertainties, and e_0 is the control coefficient uncertainty. The state parameter uncertainties have the following bounded conditions,

$$\begin{array}{ll} b_{n-1} \in [\alpha_{n-1}, \beta_{n-1}] & b_{n-2} \in [\alpha_{n-2}, \beta_{n-2}] \\ \vdots & \vdots \\ b_1 \in [\alpha_1, \beta_1] & \\ b_0 \in [\alpha_0, \beta_0], & \end{array} \quad (2)$$

where $\beta_i \geq \alpha_i$ for $i = 0, 1, \cdots, n-1$, are known and fixed values. Furthermore, the control coefficient uncertainty is bounded by the following condition,

$$e_0 \in [\delta_0, \varepsilon_0] \quad (3)$$

where $\delta_0 \leq \varepsilon_0$.

The above system (1) can be rewritten as the following state space form,

$$\dot{X} = (A + \Delta A)X + (B + \Delta B)u + Gw, \quad (4)$$

where $X = \begin{bmatrix} x & dx/dt & \cdots & d^{n-1}x/dt^{n-1} \end{bmatrix}^T$

$$A = \begin{bmatrix} 0 & 1 & 0 & 0 \\ \vdots & \ddots & \ddots & \vdots \\ 0 & \cdots & 0 & 1 \\ -a_0 & -a_1 & \cdots & -a_{n-1} \end{bmatrix} \quad (6)$$

$$\Delta A = \begin{bmatrix} 0 & 0 & 0 & 0 \\ \vdots & \vdots & \vdots & \vdots \\ 0 & 0 & 0 & 0 \\ -b_0 & -b_1 & \cdots & -b_{n-1} \end{bmatrix} \quad (7)$$

$$B^T = G^T = \begin{bmatrix} 0 & \cdots & 0 & 1 \end{bmatrix} \quad (8)$$

and

$$\Delta B^T = \begin{bmatrix} 0 & \cdots & 0 & e_0 \end{bmatrix}. \quad (9)$$

Assume that the system has the performance output as

$$z = (C_1 + \Delta C)X + Du, \quad (10)$$

where ΔC is the uncertain matrix of the performance output. It will be shown that the output matrix and its uncertainties can be ignored by the weighting selection. As long as $(1+e_0)$ is not zero for any $e_0 \in [\delta_0, \varepsilon_0]$, the uncertainties, ΔA, will not affect the controllability of the system. This is because the controllability matrix will be non-singular for any changes in uncertainties, (ΔA and ΔB).

The objective is to find an H_∞ state feedback controller, $u = u(X)$, such that the uncertain system will be asymptotically stable under disturbance attenuation. Note that this problem is known as robustness of the parameter uncertainties and robustness of the system disturbance. The approach here is a little different from the standard H_∞ method. Instead of finding a suitable algebraic Riccati inequality (ARI) for the uncertain system, the weighting matrix method is introduced such that the ARI can be changed to algebraic Riccati equation. As long as the weighting matrix is satisfied by some certain constraints, then this weighting matrix can be used to the ARE whose solution will be used for the designing of robust control.

III. METHOD OF ANALYSIS

3.1 PRELIMINARY:

Consider a general linear system

$$\dot{\hat{x}} = \hat{A}\hat{x} + \hat{B}\hat{u} + \hat{G}\hat{w} \quad (11)$$

with its measurement and performance outputs as

$$\hat{y} = \hat{C}_2\hat{x} + \hat{D}_{21}\hat{w} \quad (12)$$

$$\hat{z} = \hat{C}_1\hat{x} + \hat{D}_{12}\hat{u}. \quad (13)$$

The objective of the H_∞ technique is to find a state feedback controller, such as the following H_∞ performance

$$\left\|\hat{F}(s)\right\|^2 \le \hat{\gamma}^2 \quad (14)$$

where $\|\cdot\|^2$ is the Euclidean norm, and $\hat{\gamma}$ is the prescribed constant. Note that the transfer function is denoted as

$$\hat{F}(s) = \begin{bmatrix} \hat{A} & \hat{G} & \hat{B} \\ \hat{C}_1 & 0 & \hat{D}_{12} \\ \hat{C}_2 & \hat{D}_{21} & 0 \end{bmatrix}. \quad (15)$$

The solution, $\hat{P}$, of the H_∞ state feedback case can be shown to satisfy the following algebraic Riccati inequality,

$$\hat{A}^T\hat{P} + \hat{P}\hat{A} + \hat{C}_1^T\hat{C}_1 - \hat{P}(\hat{B}\hat{B}^T - \hat{\gamma}^{-2}\hat{G}\hat{G}^T)\hat{P} \le 0, \quad (16)$$

Therefore, the state feedback controller can be obtained as

$$\hat{u} = -\hat{B}^T\hat{P}\hat{x} \quad (17)$$

The exogenous input will be

$$\hat{w} = \gamma^{-2}\hat{G}^T\hat{P}\hat{x}. \quad (18)$$

In this paper, we will concentrate on the state feedback case. The procedure resulting in the above ARI is referred to Reference [11]. Some assumptions can also be found in this reference.

3.2 WEIGHTING MATRIX:

Extended the preliminary results to the single input uncertain system (4), the following similar ARI can be obtained,

$$\begin{aligned}&(A+\Delta A)^T P + P(A+\Delta A) + (C_1+\Delta C)^T(C_1+\Delta C)\\ &-P[(B+\Delta B)(B+\Delta B)^T - \gamma^{-2}GG^T]P \le 0\end{aligned} \quad (19)$$

where γ is the disturbance attenuation to be achieved. The above inequality can be changed to an equation by adding a negative semi-definite weighting matrix on the right hand side of (19) as follows,

$$\begin{aligned}&(A+\Delta A)^T P + P(A+\Delta A) + (C_1+\Delta C)^T(C_1+\Delta C)\\ &-P[(B+\Delta B)(B+\Delta B)^T - \gamma^{-2}GG^T]P = -Q\end{aligned} \quad (20)$$

where Q is a positive semi-definite matrix. Note that if the weighting matrix has the following property,

$$Q \ge (C+\Delta C)^T(C+\Delta C) \quad (20.a)$$

then following lemma is always true for any system uncertainties.

Lemma 3.1: Instead of solving the ARI (19), the solution of the ARI can be changed to the following ARI which will provide the solution for the H_∞ controller such that the uncertain system will be robust,

$$\begin{aligned}&(A+\Delta A)^T P + P(A+\Delta A) + Q -\\ &P[(B+\Delta B)(B+\Delta B)^T - \gamma^{-2}GG^T]P \le 0\end{aligned} \quad (21)$$

Proof of *Lemma* 3.1: Since the third term of the left hand side of (20), $[C_1+\Delta C]^T[C_1+\Delta C]$, is positive semi-definite, exchanging this term with the weighting matrix results in (21). This simplification shows that the uncertainty inside the performance output matrix can be replaced by some suitable weighting matrix, Q.

If the weighting matrix is computed and selected from the worst case of the system uncertainties, then equation (20.a) will be true for any uncertainties in the system.

Furthermore, the H_∞ controller of the uncertain system can be shown to be

$$u(X) = -(B+\Delta B)^T PX, \quad (22)$$

and its exogenous input is

$$w = \gamma^2 G^T PX. \quad (23)$$

Therefore, the closed loop state matrix can be obtained as

$$A_c = (A+\Delta A) - \left[(B+\Delta B)(B+\Delta B)^T - \gamma^{-2}GG^T\right]P. \quad (24)$$

The objective of this paper is to find the formulations of the weighting matrix, Q, based on the different prescribed values, γ, such that the closed loop state matrix, A_c, is Hurwitz. Note that the formulations of the weighting selection are computed from the ARE. There may be more than one positive semi-definite matrix, which will satisfy the above requirement. However, the proposed method will be shown to result in the linear formulations for the weighting matrix, Q. The advantages of using this weighting matrix will be analyzed on the discussion portion.

3.3 METHODOLOGIES:

Assume that the solution, P, of the ARI (20) is

$$P = \begin{bmatrix} p_{0,0} & p_{0,1} & \cdots & p_{0,n-1} \\ p_{1,0} & p_{1,1} & \ddots & p_{1,n-1} \\ \vdots & \ddots & \ddots & \vdots \\ p_{n-1,0} & p_{n-1,1} & \cdots & p_{n-1,n-1} \end{bmatrix} > 0, \quad (25)$$

where $p_{i,j} = p_{j,i}$ for $0 \le i,j \le n-1$ and $i \ne j$. The uncertain system (1) is said to be robustly controlled by the H_∞ controller, $u(X)$, if and only if the closed loop state matrix (24), A_c, has all closed loop poles located on the left hand side of the complex plane for any bounded parameter uncertainties, ΔA, ΔB, and ΔC. The uncertainties of the performance output, ΔC, can be recovered from the selection of the weighting matrix. This has been proven on *Lemma* 3.1. The uncertainties of the control coefficient, ΔB, contributed to the ARI are negative. Therefore, the worst case of the control coefficient will be used to explore the solution of the ARI, *P*. This implies that $B+\Delta B=\begin{bmatrix}0 & \cdots & 0 & 1+\delta_0\end{bmatrix}^T$. As long as the solution, *P*, of the ARI is computed from the worst case, then this solution can be used for any uncertainties of the control coefficient.

Furthermore, the closed loop state matrix contributed from the H_∞ controller can be found as

$$(B+\Delta B)(B+\Delta B)^T P = \begin{bmatrix} 0 & 0 & \cdots & 0 \\ \vdots & \vdots & \vdots & \vdots \\ 0 & 0 & \cdots & 0 \\ (1+\delta_0)^2 p_{n-1,0} & (1+\delta_0)^2 p_{n-1,1} & \cdots & (1+\delta_0)^2 p_{n-1,n-1} \end{bmatrix} \tag{26}$$

Likewise, the closed loop state matrix from the exogenous input is

$$\gamma^{-2}GG^T P = \begin{bmatrix} 0 & 0 & \cdots & 0 \\ \vdots & \vdots & \vdots & \vdots \\ 0 & 0 & \cdots & 0 \\ \gamma^{-2} p_{n-1,0} & \gamma^{-2} p_{n-1,1} & \cdots & \gamma^{-2} p_{n-1,n-1} \end{bmatrix}. \tag{27}$$

Substituting the above two equations (26) and (27) into (24) results in the following closed loop state matrix

$$A_c = A+\Delta A-\left[(B+\Delta B)(B+\Delta B)^T-\gamma^{-2}GG^T\right]P =$$

$$\begin{bmatrix} 0 & 1 & \cdots & 0 \\ \vdots & \vdots & \ddots & \vdots \\ 0 & \cdots & 0 & 1 \\ -a_0-b_0-\eta p_{n-1,0} & -a_1-b_1-\eta p_{n-1,1} & \cdots & -a_{n-1}-b_{n-1}-\eta p_{n-1,n-1} \end{bmatrix} \tag{28}$$

where

$$\eta=\begin{cases}(1+\delta_0)^2-\gamma^{-2} & \text{if } \delta_0\le 0 \\ 1-\gamma^{-2} & \text{if } \delta_0>0\end{cases}.$$

Since the solution of the ARI has to be positive definite, therefore, the constraint of the prescribed value can be obtained from the above closed loop state matrix as

$$\eta=(1+\delta_0)^2-\gamma^{-2}>0, \tag{29}$$

assuming $\delta_0\le 0$. This constraint gives the prescribed value, γ, the following condition,

$$\gamma>\left|1+\delta_0\right|^{-1}. \tag{30}$$

3.4 ROUTH STABILITY CRITERION:

In this subsection, Routh stability criterion is used to determine the necessarily robust formulations for the *n*-coefficients of the ARI, $(p_{n-1,0}, p_{n-1,1}, \cdots, p_{n-1,n-1})$. Since the formulation of the uncertain closed loop state matrix has been shown in (28), the characteristic polynomial of the uncertain closed loop state matrix can be easily found as follows,

$$\begin{aligned} s^n &+\left[a_{n-1}+b_{n-1}+\eta p_{n-1,n-1}\right]s^{n-1}+\cdots \\ &+\left[a_{n-i}+b_{n-i}+\eta p_{n-1,n-i}\right]s^{n-i}+\cdots \\ &+\left[a_1+b_1+\eta p_{n-1,1}\right]s+\left[a_0+b_0+\eta p_{n-1,0}\right]=0, \end{aligned} \tag{31}$$

where *s* is the Laplace operator. Assume that the above polynomial is

$$c_n s^n+c_{n-1}s^{n-1}+\cdots+c_1 s+c_0=0, \tag{32}$$

where $c_n=1$. Applying Routh stability criterion for the above characteristic polynomial, the following *n* constraints can be obtained first,

$$\begin{aligned} c_{n-1} &= \left[a_{n-1}+b_{n-1}+\eta p_{n-1,n-1}\right]\ge 0, \\ &\vdots \\ c_{n-i} &= \left[a_{n-i}+b_{n-i}+\eta p_{n-1,n-i}\right]\ge 0, \\ &\vdots \\ c_0 &= \left[a_0+b_0+\eta p_{n-1,0}\right]\ge 0 \end{aligned} \tag{33}$$

These constraints must be satisfied by any changes of the parameter-bounded uncertainties in (2). Furthermore, the following Routh criteria procedures also need to be satisfied,

s^n	c_n	c_{n-2}	c_{n-4}	••••
s^{n-1}	c_{n-1}	c_{n-3}	c_{n-5}	••••
•	f_1	f_2	f_3	••••
•	g_1	g_2	g_3	••••
•				
•	•••••	•••••	•••••	••••

where

$$f_1=\frac{c_{n-1}c_{n-2}-c_n c_{n-3}}{c_{n-1}},\; f_2=\frac{c_{n-1}c_{n-4}-c_n c_{n-5}}{c_{n-1}} \quad \text{etc.}$$

$$g_1=\frac{f_1 c_{n-3}-c_{n-1}f_2}{f_1},\; g_2=\frac{f_1 c_{n-5}-c_{n-1}f_3}{f_1} \quad \text{etc.}$$

Note that all coefficients for the Routh stability test have to be positive; that means

$$f_1\ge 0, f_2\ge 0, \cdots \text{ etc, and } g_1\ge 0, g_2\ge 0, \cdots \text{ etc}.$$

The above procedures results in these *n*-coefficients of the ARI, $(p_{n-1,0}, p_{n-1,1}, \cdots, p_{n-1,n-1})$, to be known and fixed constants. In fact, as long as the *n*-conditions shown in (33) are satisfied, the controlled system will be robust for any changes of uncertainties, ΔA, ΔB, and ΔC, and prescribed values, $\gamma>\left|1+\delta_0\right|^{-1}$. Since the necessary formulations of the *n*-coefficients of the ARI, $(p_{n-1,0}, p_{n-1,1}, \cdots, p_{n-1,n-1})$ have been found, the algebraic Riccati Equation will be employed to find the corresponding weighting matrix such that the H_∞ controller without uncertainty can be used for the uncertain system.

3.5 ALGEBRAIC RICCATI EQUATION:

In this subsection, Since the effect of the uncertainties of the system cannot be added into the calculation of the ARI. The following algebraic Riccati equation will be used for the system to find the corresponding weighting matrix,

$$A^T P + PA + Q - P[BB^T - \gamma^{-2}GG^T]P = 0. \quad (34)$$

The bounded coefficients, $(p_{n-1,0}, p_{n-1,1}, \cdots, p_{n-1,n-1})$, are the only coefficients from the solution of the ARE used to the closed loop state matrix. These coefficients have been determined by the worst case for uncertainties, $(\Delta A, \Delta B)$. Therefore, these coefficients will be used to determine the formulations for the limit value of the weighting matrix. Note that this method is very conservative. The minimum value of the weighting matrix is the limit of the ARE for the prescribed value, γ. However, as long as the weighting matrix is selected larger than the above weighting matrix, the controlled system will be robust.

The algebraic Riccati equation has $[n(n+1)/2]$ coefficients needed to be determined. Through the previous procedure, the n-coefficients of the ARE, $(p_{n-1,0}, p_{n-1,1}, \cdots, p_{n-1,n-1})$, have been found. In order to yield the unique solution for the ARE, the following assumption on the weighting matrix is made,

$$Q = \begin{bmatrix} q_1 & 0 & \cdots & 0 \\ 0 & q_2 & \ddots & \vdots \\ \vdots & \ddots & \ddots & 0 \\ 0 & \cdots & 0 & q_n \end{bmatrix}. \quad (35)$$

Note that the above assumption resulting in unknown coefficients of the ARI is still $n(n+1)/2$. However, the formulations of n-coefficients on the diagonal weighting matrix, and the rest of coefficients of the ARE must be determined. These formulations can be found based on the following procedures. Note that the equality sign is used to develop all formulations.

It is known that the ARE is a nonlinear matrix equation. The nonlinear portion of the ARE is from the terms of control and disturbance matrix combination, which are $P\left[BB^T - \gamma^{-2}GG^T\right]P$. Expanding these terms, the following equation can be obtained,

$$P\left[(B+\Delta B)(B+\Delta B)^T - \gamma^{-2}GG^T\right]P =$$

$$\begin{bmatrix} p_{n-1,0}^{\;2} & \cdots & p_{n-1,0}\cdot p_{n-1,n-2} & \kappa p_{n-1,0}\cdot p_{n-1,n-1} \\ \vdots & \vdots & \ddots & \vdots \\ p_{n-1,0}\cdot p_{n-1,n-2} & \ddots & \ddots & \kappa p_{n-1,n-2}\cdot p_{n-1,n-1} \\ \kappa p_{n-1,0}\cdot p_{n-1,n-1} & \cdots & \kappa p_{n-1,n-2}\cdot p_{n-1,n-1} & \kappa p_{n-1,n-1}\cdot p_{n-1,n-1} \end{bmatrix} \quad (36)$$

where $\kappa = 1 - \gamma^{-2} > 0$. All coefficients of the right hand side of (36) are known. The difficulty of solving the nonlinear portion of the ARE is now removed. Expanding the ARE (34), the rest coefficients of P can be obtained from the off-diagonal terms of the ARI expanding matrix. These coefficients can be formulated as the following general forms,

$$\begin{aligned} P_{i,j} &= P_{i,n-1}\cdot a_{j+1} + P_{n-1,j+1}\cdot a_i \\ &+\kappa\cdot(P_{i,n-1}P_{n-1,j+1}) - P_{i-1,j+1}\cdot u(i) \end{aligned} \quad (37)$$

where $0 \le i, j < n-1$, for $i \ne j$, and

$$\begin{aligned} P_{i,i} &= P_{i,n-1}\cdot a_{i+1} + P_{n-1,i+1}\cdot a_i \\ &+\kappa\cdot(P_{i,n-1}P_{n-1,i+1}) - P_{i+1,i-1}\cdot u(i) \end{aligned} \quad (38)$$

where $0 \le i < n-1$. Note that $u(i)$ is a discrete step function, which is defined as

$$u(i) = \begin{cases} 0 & \text{when } i = 0 \\ 1 & \text{when } i > 0 \end{cases}.$$

The diagonal terms of the ARE expanding matrix determine the coefficients of the weighting matrix. These coefficients have the following general formulations,

$$q_i = \kappa \cdot P_{i,n-1}^{\;2} + 2\cdot P_{i,n-1}\cdot(a_i + b_i) - 2\cdot P_{i,i-1}\cdot u(i). \quad (39)$$

Note that formulations for all coefficients, the ARI solution and the weighting matrix, are linear. The necessary computational time to find these coefficients is very little. Following the above procedures, the limit of the weighting matrix can be determined. As long as the weighting matrix is selected larger than the above constraints (39), the solution of the ARE will provide the necessary Lyapunov matrix to the control input. This control input will always allow the closed loop system to be asymptotically stable for any uncertainties, $(\Delta A, \Delta B, \Delta C)$. Furthermore, the robustness of closed loop poles can be designed based on the desired zone. This procedure will be presented in the discussion portion.

IV. EXAMPLE

Consider a second order system from [12] as follows,

$$\ddot{x} + (-1 - r_2)\dot{x} + (-1 - r_1)x = (1 + s_1)u + w$$

where (r_1, r_2, s_1) are the parameter uncertainties. These uncertainties have the following constraints,

$$r_1 \in [-1/2,\ 1],\ r_2 \in [-1/2,\ 1],$$
$$\text{and } s_1 \in [-1/4,\ 1/2].$$

Note that w is the system disturbance here. This assumption is different from Reference [12]. The above system can be written as the following state space form,

$$A = \begin{bmatrix} 0 & 1 \\ 1 & 1 \end{bmatrix},\ B = G = \begin{bmatrix} 0 \\ 1 \end{bmatrix}.$$

Note that from equation (29), $\gamma \ge |1 - 1/4|^{-1} = 4/3$. In this case, the prescribed value for $\gamma = 2$ is chosen to demonstrate the proposed method. Therefore,

$$\eta = (1 - 1/4)^2 - \gamma^{-2} = 5/16 \text{ and } \kappa = (1 - \gamma^{-2}) = 3/4.$$

Assume the solution of ARE is

$$P = \begin{bmatrix} p_{0,0} & p_{0,1} \\ p_{1,0} & p_{1,1} \end{bmatrix}.$$

Therefore, the closed loop state matrix becomes,

$$A_c = \begin{bmatrix} 0 & 1 \\ (1+r_1-\eta p_{1,0}) & (1+r_2-\eta p_{1,1}) \end{bmatrix}.$$

The characteristic polynomial of the closed loop state matrix is

$$s^2-(1+r_2-\eta p_{1,1})s-(1+r_1-\eta p_{1,0})=0.$$

From Routh stability criteria, the following constraints can be obtained,

$$p_{1,1}\geq 32/5, \text{ and } p_{1,0}\geq 32/5.$$

Applying the formulation for the rest of ARE solution, the following result is obtained,

$$p_{0,0}\geq p_{0,1}\cdot 1+p_{1,2}\cdot 1+\kappa\cdot(p_{0,0}\cdot p_{0,1})=17.92.$$

Therefore, the diagonal weighting coefficients can be determined as

$$q_0\geq \kappa p_{1,1}^2+2p_{0,1} \text{ and } q_1\geq \kappa p_{1,1}^2+2p_{1,1}-2p_{0,1}.$$

One can verify that the above two coefficients are positive for any admissible uncertainties, r_1 and r_2, Therefore, as long as the weighting matrix, Q, is selected to be larger than the following value,

$$Q\geq\begin{bmatrix} 17.92 & 0 \\ 0 & 5.12 \end{bmatrix},$$

the solution of the ARE will provide the robust controller for the uncertain systems under bounded parameters, for any prescribed values, $\gamma\geq 2$.

In order to avoid the high gain control, the prescribed value is necessarily chosen to be large. However, the smaller the prescribed value, the more stability of the system to the disturbance. These two conditions are contradicted. These two values are necessary to be designed based on the saturation of the control input. In this case, the following weighting matrix is selected

$$Q=\begin{bmatrix} 18 & 0 \\ 0 & 5.12 \end{bmatrix}.$$

Using this weighting matrix based on $\gamma=2$, one can find the following solution,

$$P=\begin{bmatrix} 17.9705 & 6.4105 \\ 6.4105 & 6.4028 \end{bmatrix}$$

for the following ARE,

$$A^TP+PA+Q-P(BB^T-\frac{1}{\gamma^2}GG^T)P=0.$$

One can verify that the solution of the above ARE, P, makes the following closed loop state matrix be stable for any admissible uncertainties, (r_1 and r_2) and s_1,

$$A_c=A+\Delta A-[(B+\Delta B)(B+\Delta B)^T-\gamma^{-2}GG^T]P.$$

V. DISCUSSIONS

In this section, the advantages of the proposed method will be discussed. Combination of the current method with the transformation matrix technique will be examined minutely. Detectability of the uncertain system and contribution of the weighting matrix will be analyzed. Furthermore, the robust pole placement to some certain zones based on the weight selection will be prospected.

5.1 TRANSFORMATION AND DETECTABILITY

From Reference [16], it is shown that single input controllable systems can be transformed into second canonical forms using transformation matrix. Combining this technique with the current proposed method results in any single input uncertain systems can be robustly controlled, as long as a suitable weighting matrix can be determined. The transformation technique is the method of transferring a single input controllable system to a second canonical system, which is described as follows.

Consider a single-input controllable system

$$\dot{\hat{x}}=\hat{A}\hat{x}+\hat{B}\hat{u},\ \hat{x}(0)=x_0 \tag{40}$$

where $\hat{x}$, $\hat{A}$, and $\hat{B}$ have appropriate dimensions. Since the system is controllable, the controllability matrix, $\hat{P}_c$, is invertible and can be determined by the state matrix and control matrix,

$$\hat{P}_c=\begin{bmatrix}\hat{B} & \hat{A}\hat{B} & \hat{A}^2\hat{B} & \cdots & \hat{A}^{n-1}\hat{B}\end{bmatrix}. \tag{41}$$

If the following second canonical form is desired,

$$\dot{\tilde{x}}=\tilde{A}\tilde{x}+\tilde{B}\tilde{u},\ \tilde{x}(0)=T^{-1}\hat{x}_0, \tag{42}$$

then the transformation matrix, T, can be used to achieve this goal. The transformation matrix operator is defined as

$$\left.\begin{matrix}\hat{x}\\ \hat{A}\\ \hat{B}\end{matrix}\right\}\Rightarrow\hat{x}=T\tilde{x}\Rightarrow\left\{\begin{matrix}\tilde{x}=T^{-1}\hat{x}\\ \tilde{A}=T^{-1}\hat{A}T.\\ \tilde{B}=T^{-1}\hat{B}\end{matrix}\right. \tag{43}$$

The following procedure is used to find the transformation matrix, T.

Step 1. Compute the characteristic polynomial of $\hat{A}$,

$$\det(sI-\hat{A})=s^n+\hat{a}_{n-1}s^{n-1}+\cdots+\hat{a}_1s+\hat{a}_0. \tag{44}$$

Step 2 Determine the state matrix of the second canonical form as

$$\tilde{A}=\begin{bmatrix} 0 & 1 & \cdots & 0 \\ 0 & 0 & \ddots & \vdots \\ \vdots & \ddots & \ddots & 1 \\ -\hat{a}_0 & -\hat{a}_1 & \cdots & -\hat{a}_{n-1}\end{bmatrix} \tag{45}$$

and the control matrix as

$$\tilde{B}=\begin{bmatrix}0 & \cdots & 0 & 1\end{bmatrix}^T. \tag{46}$$

Step 3. Calculate the second canonical controllability matrix,

$$\tilde{P}_c=\begin{bmatrix}\tilde{B} & \tilde{A}\tilde{B} & \tilde{A}^2\tilde{B} & \cdots & \tilde{A}^{n-1}\tilde{B}\end{bmatrix}. \tag{47}$$

Step 4. Determine the transformation matrix, T, by the two controllability matrices,

$$T=\hat{P}_c\tilde{P}_c^{-1}. \tag{48}$$

Since the transformation matrix has been found, the back transformation technique can be applied to transform the second canonical form system back to the original system. For example, if $\overline{P}$ is the solution of the ARE for the second canonical form system, then the solution of the original system, $\hat{P}$, will be

$$\hat{P} = T^{-T}\bar{P}T^{-1}. \tag{49}$$

The proof of the above equation is to simply substitute the above equation into the original ARE. Comparing the difference of two AREs results in the above equation. Therefore, the transformation matrix with the current proposed method allows any single input controllable uncertain systems to be resolved. An aircraft example is used to illustrate the above description in the next section.

Furthermore, if the original system has the unstable pole and zero cancellation due to the unobservability of the original system, then the ARI will have no positive definite solution for the system. This means that the H_∞ technique cannot apply for this kind of situation. However, the selection of the weighting matrix can recover the unobservability of the original system. Therefore, the system can still be robustly controlled. Similarly, for an uncertain system with uncertainties on output matrix, the ranges of the uncertainties may contain a mode such that the system is unobservable. This kind of problems can also be resolved by weighing selection for the ARE.

5.2 ROBUST POLE PLACEMENT TO SPECIFIED ZONES

In general, the robust control problem is to have all closed loop poles on the left hand side of the complex plane under any kinds of admissible uncertainties. This situation is shown in Figure 1. In this figure, the dash-line area is the desired region for the closed loop poles. As long as the closed loop poles are stable for any parameter uncertainties, the controlled system will be called robustness.

However, using the proposed method, it is possible for an uncertain system not only to be robustly stable but also to have a good performance. These kind of problems are known as robust pole placement of an uncertain system to specified regions. For instance, Figure 2 shows the desired region for a robust criterion. The necessary formulations on the current paper to have the proposed method reaching the above performance can be modified from the Routh stability criterion. Note that all closed loop poles can be selected to be stable through the Routh stability criterion. Assume that the closed loop poles will be selected to satisfy the robust region shown in Figure 2. The solid lines of the stable region are constructed by the following eigenvalues, $\lambda = (\lambda_1, \lambda_2, \cdots, \lambda_n)$. The characteristic polynomial of these eigenvalues can be found as

$$\prod_{i=1}^{n}(s-\lambda_i) = s^n + d_{n-1}s^{n-1} + \cdots + d_1 s + d_0. \tag{50}$$

Simply subtracting the coefficients of the above equation based on powers of the Laplace operator into equation (31) results in

$$\begin{aligned} &s^n + \left[a_{n-1} + b_{n-1} - d_{n-1} + \eta p_{n-1,n-1}\right]s^{n-1} + \cdots \\ &+\left[a_{n-i} + b_{n-i} - d_{n-i} + \eta p_{n-1,n-i}\right]s^{n-i} + \cdots \\ &+\left[a_1 + b_1 - d_1 + \eta p_{n-1,1}\right]s + \left[a_0 + b_0 - d_0 + \eta p_{n-1,0}\right] = 0. \end{aligned} \tag{51}$$

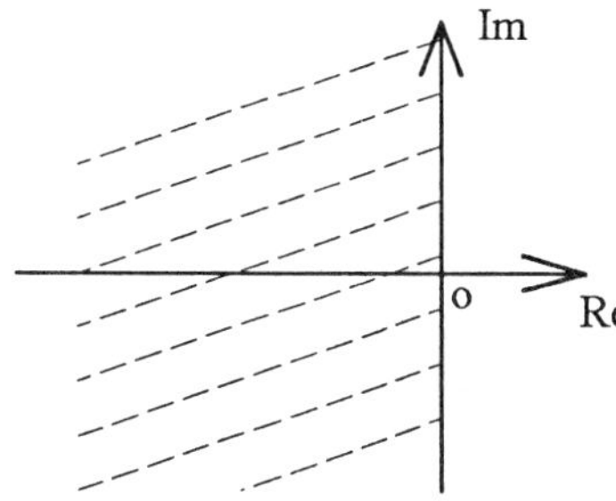

Figure 1: Robustly stable region in general.

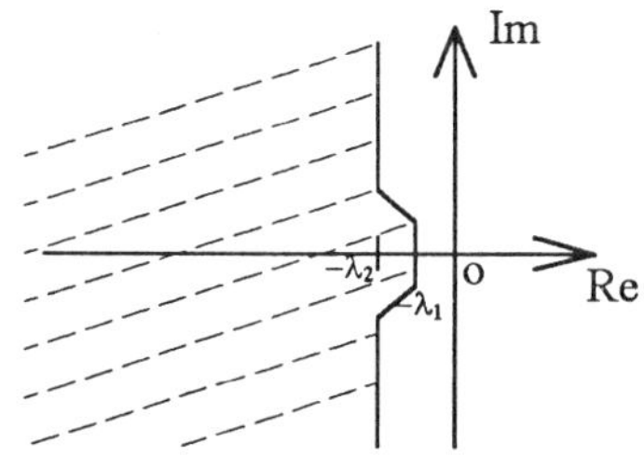

Figure 2: Required robustly stable region.

As long as the coefficients of the above characteristic polynomial satisfy the Routh stability criterion for all n-coefficients of the ARE solution, $(p_{n-1,0}, p_{n-1,1}, \cdots, p_{n-1,n-1})$, the formulations of the weighting matrix computed based on these n-coefficients used for the ARE (34) will provide the sufficient control gain such that all closed loop poles are within the desired region. In fact, these n-coefficients of the ARE solution can be found to satisfy the following constraints

$$\begin{aligned} &p_{n-1,n-1} \geq \max\left[(-a_{n-1} - b_{n-1} + d_{n-1})/\eta\right], \\ &\quad\vdots \qquad\qquad \vdots \qquad\qquad \vdots \\ &p_{n-1,n-i} = p_{n-i,n-1} \geq \max\left[(-a_{n-i} - b_{n-i} + d_{n-i})/\eta\right], \\ &\quad\vdots \qquad\qquad \vdots \qquad\qquad \vdots \\ &p_{n-1,0} = p_{0,n-1} = \max\left[(-a_0 - b_0 + d_0)/\eta\right] \end{aligned} \tag{52}$$

where max represents the maximum value of the denoted equation for any changes of the uncertainties, b_i. An aircraft model in the next section will be used to illustrate the above procedure.

VI. ALGORITHM AND APPLICATION

In this section, an algorithm is presented. The combination of the weight selection and transformation matrix is used to provide a flow chart for control of aircraft in longitudinal motion. This flow chart is shown to be able to control the uncertain aircraft system with any aerodynamic derivative changes between Mach = $0.5 - 0.7$.

6.1 ALGORITHM:

In this subsection, an algorithm will be presented based on the aircraft system on the next sub-section. This will allow readers to easily follow the significant contribution of the proposed technique.

Step 1: Introduce longitudinal motion of the aircraft model, which is single-input controllable systems for a range of airspeed, $M_\infty = 0.5 - 0.7$.

Step 2: Use the transformation matrix technique to transform the above systems to second canonical forms.

Step 3: Select the model of $M_\infty = 0.6$ as the certain aircraft model.

Step 4: The differences between the above certain model and the rest of the models of aircraft systems are assumed to be the uncertainties, b_i, of the aircraft system for $M_\infty = 0.5 - 0.7$. Therefore, the bounded parameters, α_i and β_i, can be determined.

Step 5: Determine the prescribed value, γ, based on the above uncertainties. Note that this prescribed value has to satisfy the constraint in (30).

Step 6: Assume that the closed loop poles of uncertain aircraft systems will be shifted to a specified zone for the purpose of better performance. The bounded eigenvalues of this zone can be pre-selected. Therefore, the formulations from the previous section can be used to determine the four coefficients of the ARE solution.

Step 7: Apply the formulations for the rest of the ARE solution to find the desired Lyapunov matrix, P.

Step 8: Use the formulations for the weighting coefficients to determine the minimum value of the weighting coefficients.

Step 9: Select the positive weighting matrix for the ARE based on the above minimum weighting coefficient.

Step 10: Use the above positive weighting matrix to find the new solution of the ARE.

Step 11: Back transform the new ARE solution, P, into the original system using the back transformation technique. Therefore, this original Lyapunov matrix, $\hat{P}$, can be used for the robust controller.

Note that after the illustration of robust control for the uncertain aircraft system, comparisons based on the simulation results are made for different Mach numbers. It will be shown that the certain system will be the interpolated case of the responses for any different Mach numbers. Furthermore, the simulation results will be not only robustness, but also better performance for any different admissible uncertainties.

6.2 ILLUSTRATION:

The airplane model of longitudinal motion can be found in Reference [12]. Applying the aerodynamic derivative coefficients [13 and 15] of a medium size commercial airplane to the above equation for $M_\infty = 0.5$ to 0.7 results in the following equations,

$$\dot{x} = A_{0.5}x + B_{0.5}\delta_e$$
$$\dot{x} = A_{0.6}x + B_{0.6}\delta_e$$
$$\dot{x} = A_{0.6}x + B_{0.6}\delta_e$$

where

$$A_{0.5} = \begin{bmatrix} -0.0139 & -32.1694 & 0.0322 & 0 \\ 0 & 0 & 0 & 1.0000 \\ -0.1457 & -1.4024 & -0.5374 & 509.8912 \\ 0.0 & 0.0 & -0.0036 & -0.4194 \end{bmatrix}, \quad B_{0.5} = \begin{bmatrix} 0 \\ 0 \\ -23.5326 \\ -2.1764 \end{bmatrix}$$

$$A_{0.6} = \begin{bmatrix} -0.0123 & -32.1694 & 0.0258 & 0 \\ 0 & 0 & 0 & 1.0000 \\ -0.1333 & -1.4023 & -0.6724 & 611.6848 \\ 0.0 & 0.0 & -0.0044 & -0.5185 \end{bmatrix}, \quad B_{0.6} = \begin{bmatrix} 0 \\ 0 \\ -33.8853 \\ -3.0474 \end{bmatrix}$$

$$A_{0.7} = \begin{bmatrix} -0.0117 & -32.1694 & 0.0209 & 0 \\ 0 & 0 & 0 & 1.0000 \\ -0.1320 & -1.4024 & -0.8276 & 713.3459 \\ 0.0 & 0.0 & -0.0055 & -0.6284 \end{bmatrix}, \quad B_{0.5} = \begin{bmatrix} 0 \\ 0 \\ -46.1189 \\ -4.0300 \end{bmatrix}.$$

Since the performance output uncertainties will not affect the calculation of ARI, these equations are ignored here.

The following matrices from the transformation matrix technique (48),

$$T_{0.5} = 10^3 \begin{bmatrix} 0.0350 & 0.0340 & -0.00076 & 0 \\ -0.000025 & -0.0011 & -0.00218 & 0 \\ -0.01016 & -0.01254 & -1.1199 & -0.0235 \\ 0 & -0.000025 & -0.0011 & -0.00218 \end{bmatrix},$$

$$T_{0.6} = 10^3 \begin{bmatrix} 0.0612 & 0.0496 & -0.0009 & 0 \\ -0.00003 & -0.0019 & -0.0030 & 0 \\ -0.0130 & -0.0188 & -1.8820 & -0.0339 \\ 0 & -0.00003 & -0.0019 & -0.003 \end{bmatrix},$$

$$T_{0.7} = 10^3 \begin{bmatrix} 0.0993 & 0.0691 & -0.00096 & 0 \\ -0.000047 & -0.0031 & -0.00403 & 0 \\ -0.0171 & -0.0284 & -2.9043 & -0.0461 \\ 0 & -0.000047 & -0.0031 & -0.00403 \end{bmatrix},$$

can be used to transform the above single input system to the second canonical form,

$$\tilde{A}_{0.5} = \begin{bmatrix} 0 & 1 & 0 & 0 \\ 0 & 0 & 1 & 0 \\ 0 & 0 & 0 & 1 \\ -0.0166 & -0.0254 & -2.0554 & -0.9707 \end{bmatrix}$$

$$\tilde{A}_{0.6} = \begin{bmatrix} 0 & 1 & 0 & 0 \\ 0 & 0 & 1 & 0 \\ 0 & 0 & 0 & 1 \\ -0.0190 & -0.0332 & -3.0818 & -1.2032 \end{bmatrix}$$

$$\tilde{A}_{0.7} = \begin{bmatrix} 0 & 1 & 0 & 0 \\ 0 & 0 & 1 & 0 \\ 0 & 0 & 0 & 1 \\ -0.0231 & -0.0458 & -4.4335 & -1.4677 \end{bmatrix}$$

$$\tilde{B}_{0.5} = \tilde{B}_{0.6} = \tilde{B}_{0.7} = \begin{bmatrix} 0 & 0 & 0 & 1 \end{bmatrix}^T.$$

Assume that the model of $M_\infty = 0.6$ is desired, and the rest of models are bounded systems which will be controlled simultaneously. Furthermore, the system disturbance is assumed due to the small perturbation from the highest time derivative. The above assumptions result in the aircraft model satisfying the system described in (4), which is

$$\dot{X} = (\tilde{A}_{0.6} + \Delta A)X + \tilde{B}_{0.6}\delta_e + Gw$$

where

$$\tilde{A}_{0.6} = \begin{bmatrix} 0 & 1 & 0 & 0 \\ 0 & 0 & 1 & 0 \\ 0 & 0 & 0 & 1 \\ -a_0 & -a_1 & -a_2 & -a_3 \end{bmatrix}, \quad \tilde{B}_{0.6} = G = \begin{bmatrix} 0 \\ 0 \\ 0 \\ 1 \end{bmatrix},$$

and $a_0 = 0.0190$, $a_1 = 0.0332$, $a_2 = 3.0818$, $a_3 = 1.2032$. Note that the uncertainty, ΔA, satisfies the bounded conditions as shown in (2) which are

$$b_0 = [-0.0023,\ 0.0041],\ b_1 = [-0.0078,\ 0.0126],$$
$$b_2 = [-1.0265,\ 1.3517],\ b_3 = [-0.2325,\ 0.2445].$$

Note that the above upper and lower bounds are computed from the differences between the $(M_\infty = 0.6)$ system and the $(M_\infty = 0.5$ and $0.7)$ systems, respectively.

The objective is to shift the closed loop poles to a specified zone which is under the left hand side of the following eigenvalues,

$$\prod_{i=1}^{4}(s-\lambda_i) = s^4 + c_3 s^3 + c_2 s^2 + c_1 s + c_0,$$
$$= s^4 + 0.22 s^3 + 0.0141 s^2 + 0.00022 s + 10^{-6}$$

where $\lambda_1 = \lambda_2 = -0.01$, and $\lambda_3 = \lambda_4 = -0.1$.

In order to avoid the high gain for the control input of the aircraft system, the prescribed value is assumed to be $\gamma = 5.0$. Note that the uncertainties from the control coefficients have been shifted to the output matrix. The uncertainties on the output matrix can be recovered via selection of the weighting matrix. This procedure has been introduced in Section III. Furthermore, the following two values can be determined,

$$\eta = \kappa = 1 - \gamma^{-2} = 24/25.$$

Applying the formulations with coefficients for the above specified zone, the following four coefficients can be obtained,

$$P_{3,3} \geq (-a_3 - b_3 + d_3)/\kappa = -0.7819$$
$$P_{3,2} = P_{2,3} \geq (-a_2 - b_2 + d_2)/\kappa = -2.1263$$
$$P_{3,1} = P_{1,3} \geq (-a_1 - b_1 + d_1)/\kappa = -0.0262$$
$$P_{3,0} = P_{0,3} \geq (-a_0 - b_0 + d_0)/\kappa = -0.0173.$$

Since that the above four coefficients are determined, the formulation from (37) can be employed to find the rest of the Lyapunov coefficients,

$$P_{0,1} = P_{1,0} \geq P_{3,0}a_2 + P_{3,2}a_0 + P_{3,2}P_{3,0} \cdot k = -0.0583$$
$$P_{0,2} = P_{0,2} \geq P_{3,0}a_3 + P_{3,3}a_1 + P_{3,3}P_{3,0} \cdot k = -0.0227$$
$$P_{1,2} = P_{2,1} \geq P_{3,1}a_3 + P_{3,3}a_1 + P_{3,3}P_{3,1} \cdot k - P_{3,0} = -0.0205.$$

Furthermore, from (38), the following coefficients are obtained,

$$P_{0,0} \geq P_{3,0}a_1 + P_{3,1}a_0 + P_{3,1}P_{3,0} \cdot k = -0.0006$$
$$P_{1,1} \geq P_{3,1}a_2 + P_{3,2}a_1 + P_{3,2}P_{3,1} \cdot k = -0.0751$$
$$P_{2,2} \geq P_{3,2}a_3 + P_{3,3}a_2 + P_{3,2}P_{3,3} \cdot k = -3.3457.$$

Therefore, the following coefficients of the weighting matrix can be determined through the formulation (39)

$$Q_{0,0} \geq P_{0,3}^2 \cdot \kappa + 2P_{0,3}a_0 = -0.0004$$
$$Q_{1,1} \geq P_{1,3}^2 \cdot \kappa + 2P_{1,3}a_1 - 2P_{0,1} = 0.1156$$
$$Q_{2,2} \geq P_{2,3}^2 \cdot \kappa + 2P_{2,3}a_2 - 2P_{1,2} = -8.7244$$
$$Q_{3,3} \geq P_{3,3}^2 \cdot \kappa + 2P_{3,3}a_3 - 2P_{2,3} = 2.9581.$$

One can verify that using the above weighting matrix for ARE (34) will result in the same Lyapunov matrix, P. Both solutions, P and Q, are invertible and unique. However, In order to generate the robust controller for the uncertain system, the weighting matrix must be selected to be larger than the above weighting matrix. Therefore, the following weighting matrix is employed to find the solution of ARE for the H_∞ controller,

$$Q = \begin{bmatrix} 0.01 & 0 & 0 & 0 \\ 0 & 0.2 & 0 & 0 \\ 0 & 0 & 0.01 & 0 \\ 0 & 0 & 0 & 3.0 \end{bmatrix}.$$

Note that the above corresponding weighting matrix is selected to be positive definite. While the corresponding coefficients of the weighting matrix are negative, then the constant value, 0.01, is selected. While the corresponding coefficients of the weighting matrix are positive, then the bigger values are used for the corresponding weighting matrix. Using this positive weighting matrix for the ARE results in the following solution of the ARE,

$$P = \begin{bmatrix} 0.0929 & 0.3209 & 0.2202 & 0.0842 \\ 0.3209 & 3.1147 & 2.0736 & 0.9025 \\ 0.2202 & 2.0736 & 4.2148 & 0.6155 \\ 0.0842 & 0.9025 & 0.6155 & 1.1917 \end{bmatrix}.$$

Note that the above matrix is positive definite.

Now, the back transformation technique will be used to transform the above Lyapunov matrix back into the original system. Therefore, the Lyapunov matrix for the original system can be obtained as

$$\hat{P}_{0.6} = T_{0.6}^{-T} P T_{0.6}^{-1} =$$

$$\begin{bmatrix} 2.34\cdot10^{-5} & -0.0019 & 1.9\cdot10^{-6} & -4.5\cdot10^{-4} \\ -0.0019 & 0.7358 & -0.0008 & 0.1513 \\ 1.9\cdot10^{-6} & -0.0008 & 1.8\cdot10^{-6} & -2.78\cdot10^{-4} \\ -4.5\cdot10^{-4} & 0.1513 & -2.78\cdot10^{-4} & 0.1343 \end{bmatrix}.$$

Substituting the above matrix into the formulation of the robust controller results in the following controller

$$u = 4.5\cdot10^{-4}V_x - 0.1513\theta + 2.78\cdot10^{-4}V_z - 0.1343q.$$

Note that the above controller will be used to control the aircraft for any changes of the aerodynamic derivatives due to various airspeeds between $M_\infty = 0.5$ to $M_\infty = 0.7$. The simulation results show the potential of the proposal technique.

6.3 COMPARISON

The Lyapunov matrix, $\hat{P}_{0.6}$, obtained by the corresponding weighting matrix, will be used for the H_∞ controllers to stabilize airplane systems generated by aerodynamic derivative coefficients for Mach number between $0.5-0.7$. The following initial condition is used to see differences between the controlled system for various airspeeds,

$$x_0 = [20 ft/\sec \quad 0 \quad 5 ft/\sec \quad 0].$$

Time histories of the initial response based on the above initial condition are shown in Figure 3, 4, 5, and 6.

Figures 3 and 4 are the slow mode of the system. These two figures show that the model of Mach = 0.6 is interpolated between Mach =0.5 and 0.7. The robust controller from the solution of the ARE for the H_∞ controller gradually changes the slow mode of the controlled system. The differences of the responses for the admissible uncertainties are very small for the slow mode. Figures 5 and 6 are the state responses for the fast mode of the controlled system. The fast responses of the system with the higher Mach number will have the larger amplitude. However, the transition time of this kind of response is fast. The differences of the fast states between the various Mach numbers are also very limited. In fact, one can verify that the interpolated states of the initial response by the robust controller will be always located on the cases between Mach =0.5 and Mach = 0.7. This implies that the robust controller indeed recovers the uncertainties of the system due to the changes of aerodynamic derivatives.

All four state responses show that the robust controller provides not only reasonable amplitudes to the system, but also better performances to the system. Furthermore, the robust controller reduces the differences of the responses due to the various airspeeds. This means that all state responses generated by the robust controller perform very closely.

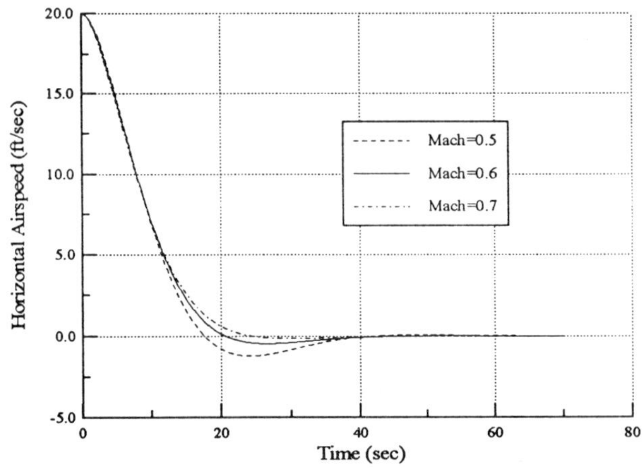

Figure 3: Time response of the horizontal airspeed for different Mach numbers

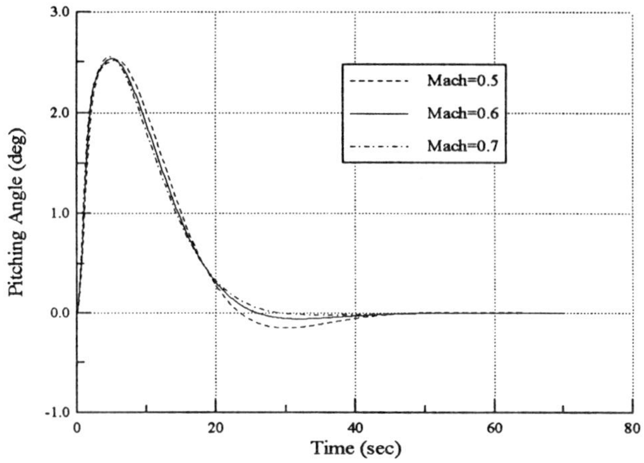

Figure 4: Time response of the pitching angle for different Mach numbers

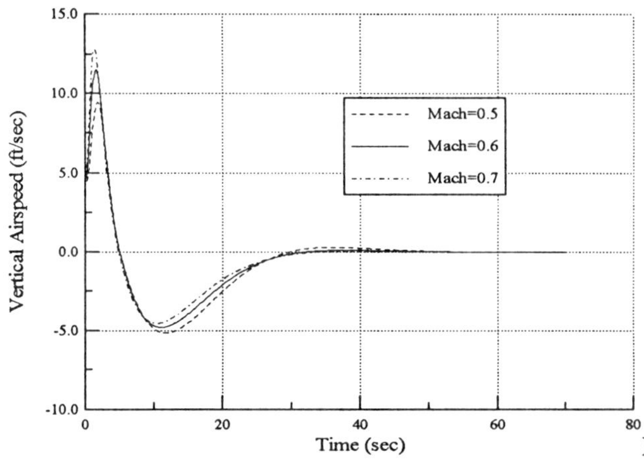

Figure 5: Time response of the vertical airspeed for different Mach numbers

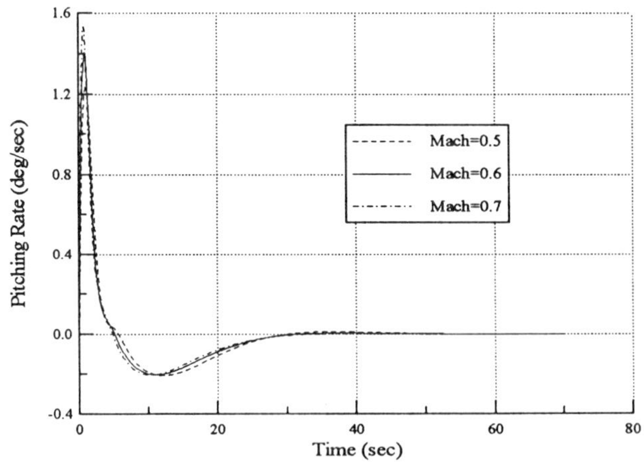

Figure 6: Time response of the pitching rate for different Mach numbers

Finally, the robust controllers used for the above three Mach numbers are shown in Figure 7. This figure shows that the robust control gains used to achieve the above performances are not necessarily high. Again, the interpolated model is the case for $M_\infty = 0.6$. The differences between these three modes are also very small.

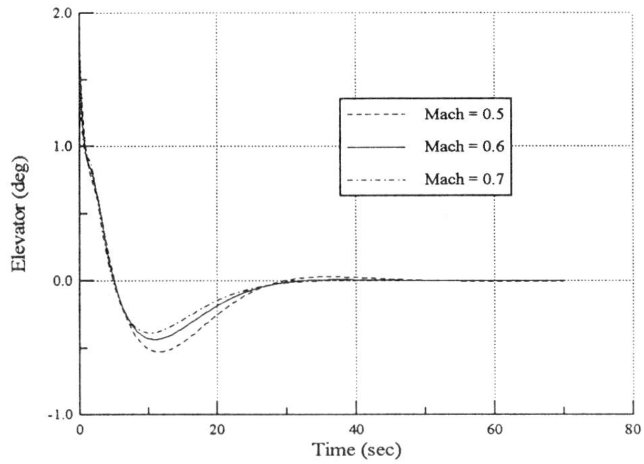

Figure 7: Time response of the state feedback control input for different Mach numbers

VII. CONCLUDING REMARK

An algorithm is presented for H_∞ control of single input parameter uncertain systems based on weight selection. The H_∞ control problem has been shown to be governed by an ARI. Through changing the weighting matrix instead of the performance output matrix, the system uncertainties on the performance output can be neglected. Applying Routh criteria for the parameter uncertain system, n-coefficient limited values of the ARI solution can be determined. These coefficients can be used to find the bounded coefficients for the weighting matrix. Any weighting matrices larger than the one selected based on the Routh criteria method can be used for the ARE. The solution of the ARE will always provide a sufficient large stability region for any uncertainties in the system. Combination of the introduced method and the transformation technique is used to illustrate the robust control of an uncertain aircraft model due to changes of aerodynamic derivatives. It was shown that the parameter-bounded uncertain airplane can be robustly controlled by the proposed method. Furthermore, the employed technique also provides a very good performance for the system under uncertainties and disturbances.

REFERENCES

1. Barmish, B. R., Corless, M. J., and Letmann, G., "A New Class of Stabilizing Controllers for Uncertain Dynamical Systems," *SIAM Journal of Control and Optimization*, Vol. 21, No. 2, 1983, pp. 246-255.
2. Petersen, I. R., "Structural Stabilization of Uncertain Systems: Necessity of Matching Condition," *SIAM Journal of Control and Optimization*, Vol. 23, pp. 286-296, 1985.
3. Petersen, I. R., "Stabilization of an Uncertain Linear System in Which Uncertain parameters Enter Into the Output Matrix," *SIAM Journal of Control and Optimization*, Vol. 26, No. 6, 1988, pp. 1257-1264.
4. Hopp, T. H., and Schmitendorf, W. E., "Design of a Linear Controller for Robust Tracking and Model Following," *ASME Journal of Dynamic Systems, Measurement, and Control*, Vol. 112, pp. 552-558, 1990.
5. Qu, Z., and Dorsey, J., "Robust Control of Generalized Dynamic Systems Without the Matching Conditions," *ASME Journal of Dynamic Systems, Measurement, and Control*, Vol. 113, pp. 582-589, 1991.
6. Stalford, H. L., "Robust Control of Uncertain Systems in the absence of Matching Conditions: Scalar Input," *Proceeding of IEEE Conference on Decision and Control*, Los Angles, CA, Dec. 1987, pp.1298-1303.
7. Chen, Y. H., and Leitmann, G., "Robustness of Uncertain Dynamical Systems in the Absence of Matching Conditions," *International Journal of Control*, Vol. 45, No. 5, 1987, pp. 1527-1542.
8. Wang, Y. W., and Bernstein, D. S., "H_2 / H_∞ Optimal Control Synthesis with an α-Shifted Pole Placement," *Proceeding of the 31st Conference on Decision and Control*, Tucson, AZ, Dec. 1992, pp. 3711-3716.
9. Wang, S.-G., Shieh, L. S., and Sunkel, J. W., "Robust Optimal Pole Placement in a Vertical Strip and Disturbance Rejection," *Proceedings of the 32nd Conference on Decision and Control*, San Antonio, Texas, Dec. 1993, pp. 1134-1139.
10. Garcia, G. and Bernussou, J., "Pole Assignment for Uncertain Systems in a Specified Disk by State Feedback," *IEEE Transaction on Automatic Control*, Vol. 40, No. 1, Jan. 1995, pp. 184-190.
11. Doyle, J. C., Glover, K., Khargonekar, P. P., and Francis, B. A., "State Space Solutions to Standard H_2 / H_∞ Control Problems," *IEEE Transaction on Automatic Control*, Vol. 34, No. 8, Aug. 1989, pp. 831-847.
12. Leitmann, G., "Guaranteed Asymptotic Stability for Some Linear Systems with Bounded Uncertainties," *ASME Journal of Dynamic Systems, Measurement, and Control*, Sep. 1979, Vol. 101, pp. 212 - 216.
13. Roskam, J., *Airplane Flight Dynamics and Automatic Flight Controls, Part I*, Roskam Aviation and Engineering Corporation, Route 4, Box 274, Ottawa, KS 66067, 1979.
14. Mclean, D., *Automatic Flight Control Systems*, Prentice Hall, Anglewood Cliffs, NJ 07632, 1990.
15. Nelson, R., *Flight Stability and Automatic Control*, McGraw Hill, 1989.
16. Brogan, W. L., *Modern Control Theory*, Third Edition, Prentice Hall, 1991, pp. 319-325.

AIAA-97-3455

A COMPLEXITY THEORETIC PERSPECTIVE ON ROBUSTNESS ANALYSIS

M. Mesbahi *
Jet Propulsion Laboratory
California Institute of Technology
Pasadena, Ca 91109-8099

Abstract

Let δ_Σ be a measure of the relative stability of a stable dynamical system Σ defined over the n-dimensional Euclidean space. Let $\tau_{\mathcal{A}(\Sigma)}$ be a measure of the computational efficiency of a particular algorithm $\mathcal{A}$ which *verifies* the stability property of Σ by computing a certificate of stability P. We demonstrate the existence of a particular measure δ_Σ and an algorithm $\mathcal{A}$ such that, $\delta_\Sigma \, \tau_{\mathcal{A}(\Sigma)} = O(n)$. In addition, we show that δ_Σ determines the size of the certificate P. These results provide the foundation for an algorithmic theory of stability and robustness.

1 Introduction

1.1

The purpose of the paper is to present the steps that have been taken towards an algorithm theory of stability and robustness.[10,11,12] The algorithmic theory has its roots in the robustness analysis of dynamical systems on one hand, and computational complexity of the optimization problems on the other.[15] The results in the present paper substantiate, an further motivate, our perspective on the deep relationship between these two areas of research.

*Technical Staff, Guidance and Control Analysis Group, Automation and Control Section; Email: *mesbahi@hafez.jpl.nasa.gov*; Tel: +1-818-354-8626.

In control and system theory we use computation to study the stability issues of the dynamical systems, but in order to come up with the computational methods, we often use stability results (for example to guarantee reasonable convergence behavior). In effect we have been led to investigate the behavior of one dynamical system by examining the stationary point of the other.

Consider for example the stability analysis of the matrix difference equation,

$$x_{k+1} = A x_k, \quad A \in R^{n\times n}. \tag{1.1}$$

In order to study the stability properties of (1.1), we study the solution of the Lyapunov inequalities,

$$X > 0 \qquad AXA' - X < 0. \tag{1.2}$$

Let us call the solution to (1.2), if it exists, $\bar{X}$, and refer to it as the *certificate* of stability. In order to obtain a feasible point for (1.2), let us employ a variant of the interior point methods (ipms),[7,13] not because ipms are the best methods for solving inequalities of the form (1.2), but due to their general nature which makes them applicable to much larger classes of inequality systems.

Given (1.2) we consider the optimization problem,

$$\inf\ t \tag{1.3}$$

$$AXA' - X < tI, \tag{1.4}$$

$$X > 0. \tag{1.5}$$

It should be clear that for a large enough value t_0 for t, the above optimization problem is at least *feasible*. Our algorithm can be constructed based on the following idea: starting from a feasible pair (t_0, P_0), at every step reduce the value of t without leaving the feasible region (by appropriate centering, etc.), and stop the algorithm when you reach a negative value for t. This idea parallels that of a variant of the ipms called the barrier method; by carefully choosing the desired reduction in the value of t and an appropriate way of staying away from the boundary of the feasible region, we can in fact prove very nice theoretical complexity bounds for checking the stability of the system (1.1): given that A is stable this method is guaranteed to output the certificate of stability $\bar{X}$ (approximately) in a reasonable time.

We notice that they way we go about constructing an algorithm for solving (1.2) is independent of the actual instance of the problem, that is, its "flowchart" does not depend on a particular matrix A. On the other hand, as long as this flowchart involves any kind of branching (statements of the form: if $x > 0$ then start all over again), the running time of the algorithm depends on the particular instance of the problem– *the running time of the algorithm depends on the properties of the particular input matrix A.* Algorithmic Stability Theory is concerned with the computational efficiency of the algorithms which are used to study systems and control problems, expressed in terms of the fundamental quantities in control theory, for example stability margins. What is rather surprising is that this correspondence can be established in a very concrete and elegant framework. For example, we show that not only the robustness measures find their way in the complexity analysis of a particular class of algorithms for solving the LMIs,[4] but also the size of the certificate is a function of these quantities. This phenomena transcends beyond the stability analysis in terms of the Lyapunov inequalities; in fact these parameters appear in the computational efficiency estimates for checking the positive real and bounded realness of transfer matrices, putting the corresponding results at the center stage of the modern robust control theory.

The organization of the paper is as follows. In the next section we initially consider the Lyapunov equation, and demonstrate that the product of the running time of the conjugate gradient method and a particular variant of the interior point methods (ipms) on one hand, and the corresponding robustness measure for linear systems on the other, are inversely proportional. We then to proceed to show the importance that the robustness measures play in defining the solution set of the Lyapunov inequality. The result which extends the above observations to the more general problem of checking the positive realness of a transfer matrix are then presented. The paper is concluded with a brief afterthought on the implications of the results. Some of the proofs are omitted for brevity; the interested reader is referred to the two manuscripts.[10,12]

A few words on the notation before we present our main results. For two symmetric matrices $A = A'$ and $B = B'$, $A > B$ ($A \geq B$) indicates that $A - B$ is positive definite (positive semi-definite, respectively). $\tau_{\mathcal{A}(\Sigma)}$ designates the running time of the algorithm $\mathcal{A}$ which verifies the stability properties of the dynamical system Σ. The notation $f(n) = O(g(n))$ indicates that there exist positive constants c and m such that $0 \leq f(n) \leq cg(n)$ for all $n \geq m$; herm$\{A\}$ denotes the hermitian part of the matrix A, $\frac{1}{2}(A+A')$. For a given transfer matrix $H(jw)$, $H^*(jw)$ denotes its conjugate transpose.

2 Conjugate Gradients and Stability Analysis

2.1

We now provide some basic facts regarding the numerical algorithm which is at the cornerstone of the main theorem of this section, namely the conjugate gradient method (cgm).

The cgm is devised to solve a given system of linear equation

$$Ax = b, \tag{2.1}$$

where $A \in R^{n \times n}$ is a symmetric positive definite matrix, and $b \in R^n$. The cgm proceeds to solve the linear system (2.1) for its solution x^*, by considering it as the minimization problem,

$$\min_x \frac{1}{2}x'Ax - b'x$$

Given the pair (A, b) and the initial point x_0, the method produces the sequence $\{x_i\}_{i \geq 1}$, according the following rules:[8]

1. $k = 0; d_0 := -g_0 = b - Ax_0$
2. $\alpha_k = -g_k'd_k/d_k'Ad_k = g_k'g_k/d_k'Ag_k$
3. $x_{k+1} = x_k + \alpha_k d_k$
4. $\beta_k = g_{k+1}'Ad_k/d_k'Ad_k = g_{k+1}'g_{k+1}/g_k'g_k$

5. $d_{k+1} = -g_{k+1} + \beta_k d_k$

6. Go to 2,

where $g_k := Ax_k - b$. The cgm can be viewed as an optimal process, where "optimality" is with respect to the following interesting property: The point x_i generated by the conjugate gradient method satisfies

$$\min_{p \in \mathcal{P}_i} \|p(A)e_0\| = \|x^* - x_i\|_A \qquad (2.2)$$

where $\|y\|_A := (y'Ay)^{1/2}$ for all $y \in R^n$, $\mathcal{P}_i$ is the set of polynomials p of degree at most i, such that $p(0) = 1$, and $e_0 = x^* - x_0$.

A direct consequence of the optimality relation (2.2) is that if A has $k \leq n$ distinct eigenvalues, then the cgm terminates in k steps, since the minimal polynomial ψ of A (the minimum degree polynomial such that $\psi(A) = 0$) would be of degree k. In particular, in view of the Cayley-Hamilton theorem the cgm terminates in at most n steps.[5]

Let us order the eigenvalues of the matrix A as $0 \leq \lambda_1 \leq \lambda_2 \leq \ldots \leq \lambda_n$. Another consequence of the optimality property (2.2) (via the Chebyshev iteration and polynomials) is that,[2,14]

$$\frac{\|x_i - x^*\|_A}{\|x_0 - x^*\|_A} \leq \min_{p \in \mathcal{P}_i} \|p(A)\| \leq \min_{p \in \mathcal{P}_i} \max_{\lambda_i} |p(\lambda_i)|$$

$$\leq \min_{p \in \mathcal{P}_i} \max_{\lambda_1 \leq \lambda \leq \lambda_n} |p(\lambda)| = \frac{1}{T_i(\frac{\lambda_n + \lambda_1}{\lambda_n - \lambda_1})} = 2\frac{\sigma^i}{1 + \sigma^{2i}} \qquad (2.3)$$

where T_i is the Chebyshev polynomial of the first kind,

$$T_i(z) = 1/2[(z + \sqrt{z^2 - 1})^i + (z - \sqrt{z^2 - 1})^i]$$

and

$$\sigma = \frac{1 - \sqrt{\lambda_1/\lambda_n}}{1 + \sqrt{\lambda_1/\lambda_n}}$$

The second equality from the right in (2.3) follows from an important property of the Chebyshev polynomials.

Define $\rho := \sqrt{\lambda_n/\lambda_1}$ to be the relative condition number of $\sqrt{A}$. Then it can be shown that,

$$\|x_i - x^*\|_A \leq 2(\frac{1 - \rho^{-1}}{1 + \rho^{-1}})^i \|x^* - x_0\|_A$$

Thus, in order to guarantee that for $0 < \epsilon < 1$, $\|x^* - x_i\|_A / \|x^* - x_0\|_A \leq \epsilon$, one must have

$$(\frac{1 - \rho^{-1}}{1 + \rho^{-1}})^i \leq \frac{\epsilon}{2}$$

and thereby,

$$i = O(\log_2 \frac{2}{\epsilon}\rho)$$

Hence if $i = \Omega(\log \frac{2}{\epsilon}\rho)$, then $\|x_i - x^*\|_A \leq \epsilon \|x_0 - x^*\|_A$.

For the case that the matrix A is not positive definite, and $A'A$ is not singular, one can consider solving $A'Ax = A'b$, and the above statements still hold with A being replaced by $A'A$.

2.2

We now consider the efficiency of the conjugate gradient method in connection with solving the linear system arising from the Lyapunov equation.

The problem considered by Lyapunov is as follows: Given a matrix $A \in R^{n \times n}$, determine whether $x = 0$ is the globally asymptotically stable equilibrium point of Σ_1:

$$\Sigma_1 : \quad \dot{x} = Ax. \qquad (2.4)$$

In particular, one is interested to know whether the trajectories of Σ_1 goes back to the origin, if Σ_1 is *disturbed* by any non-zero initial condition. This property on the other hand, is equivalent to A being Hurwitz, i.e., all eigenvalues of A have negative real parts.[5]

As it is well known Lyapunov proved that the origin is the globally asymptotically stable equilibrium point of Σ_1 if and only if, given a matrix $Q > 0$, there exists a matrix $P > 0$ such that,[9]

$$A'P + PA = -Q \qquad (2.5)$$

Suppose that the matrix A is in fact Hurwitz, and consider solving (2.5) using Kronecker products, as first proposed by Bellman.[3] Our goal is to demonstrate that the *running time of the conjugate gradient method for approximately solving the Lyapunov equation conveys an estimate of the relative stability of A*.

Let us rewrite (2.5) as

$$B(\mathbf{vec}\ P) = -\mathbf{vec}\ Q \qquad (2.6)$$

where

$$B := I \otimes A' + A' \otimes I \in R^{n^2 \times n^2}.$$

Each eigenvalue of B is the sum of a pair of eigenvalues of A.

In order to use the cgm, we consider solving the system of linear equation

$$B'B(\mathbf{vec}\ P) = -B'(\mathbf{vec}\ Q).$$

Since the origin is not in the spectrum of $B'B$ (A is assumed to be Hurwitz), starting with an initial matrix P_0, if

$$i = \Omega(\varrho \log \frac{2}{\epsilon})$$

then the cgm produces the matrix P_i such that,

$$||P^* - P_i||_{B'B} \leq \epsilon||P^* - P_0||_{B'B}$$

where,

$$\begin{aligned} \varrho &:= \lambda_{\max}(\sqrt{B'B})/\lambda_{\min}(\sqrt{B'B}) \\ &= \sigma_{\max}(B)/\sigma_{\min}(B) = ||B||/\sigma_{\min}(B) \end{aligned} \quad (2.7)$$

Since $\sigma_{\min}(B) \geq \lambda_1(\text{herm}(-B)) = |\lambda|_{\min}(A + A')$,

$$\varrho \leq ||B||/|\lambda|_{\min}(A + A').$$

Let us define

$$\delta_{\Sigma_1} := |\lambda|_{\min}(A + A')/||B||$$

Hence if $i = \Omega(1/\delta_{\Sigma_1})$, then $||P^* - P_i||_{B'B} \leq \epsilon||P^* - P_0||_{B'B}$. Thereby, in order to obtain an ϵ-approximation of the *certificate* of stability P^* (in the $B'B$ norm), the termination time of the cgm $\tau_{\text{cgm}(\Sigma_1)}$, is at least of the order of the magnitude of $1/\delta_{\Sigma_1}$. Denote by $\tau_{\text{cgm}(\Sigma_1)}$ the time it takes the cgm to approximately solve the Lyapunov equation associated with Σ_1. One thus has

$$\tau_{\text{cgm}(\Sigma_1)}\,\delta_{\Sigma_1} = O(1).$$

A closer look at the quantity δ_{Σ_1} reveals that it can be viewed as a *robustness measure* for the system Σ_1, since,

$$\delta_{\Sigma_1} = |\lambda|_{\min}(A + A')/||B|| = |\bar{\Delta}|/||B||$$

where

$$\bar{\Delta} := \{\inf \Delta \in R : A + \Delta I \text{ is not Hurwitz}\}. \quad (2.8)$$

Consequently we have proved the following theorem.

Theorem 2.1 *Given the system Σ_1, there is an algorithm $\mathcal{A}$ and a stability robustness measure δ_{Σ_1} such that,*

$$\delta_{\Sigma_1}\tau_{\mathcal{A}(\Sigma_1)} = O(1).$$

We now state a result which demonstrate the role that robustness measures play in providing a bound for the certificate of stability.

Given a matrix $A \in R^{n\times n}$ such that (1.2) is feasible, define

$$\alpha := \inf_{\Delta} \frac{||\Delta||}{||A||},$$

such that $A + \Delta$ does not lead to a feasible system (1.2). Let

$$\delta_A := \frac{1}{\log(\sqrt{n} + \frac{1}{\alpha})}. \quad (2.9)$$

For obvious reasons, let us call δ_A the robustness parameter of the matrix A. One then has the following correspondence between the size of the certificate of stability and robustness parameter δ_A.

Theorem 2.2 *The certificate of stability for a given matrix A, $\bar{X}$, satisfies,*

$$||\bar{X}|| = O(\frac{n}{\delta_A}), \quad (2.10)$$

provided that $\delta_A > 0$.

3 Interior Point Methods and Stability Analysis

3.1

This section is devoted to the results pertaining to the relationship between the computational efficiency of the interior point methods and the stability properties of dynamical systems. We initially state and prove this connection in terms of the Lyapunov inequalities, and then proceed to state the similar results for the positive real systems.

3.1.1 Stability

In order to establish the stability of the origin for the system Σ_1 defined by the matrix $A \in R^{n\times n}$,

$$\Sigma_1 : \quad \dot{x} = Ax \quad (3.11)$$

one can check the feasibility of the following system of linear matrix inequalities (LMIs):

$$\mathcal{L}_1: \quad A'P + PA < 0 \quad (3.12)$$

$$P > 0 \quad (3.13)$$

Let us for the moment forget that these matrix inequalities can somehow be solved via a system of linear equations (Section 2). We approach the problem of finding a feasible point of the set defined by (3.12)–(3.13) via the interior point methods. This will provide us with an opportunity to, rather informally, review the interior point methods (ipms), as well as presenting the idea which shall be generalized in the subsequent section. The material on the ipms which follow have been presented in a more general setting.[13,15]

In order to find a feasible point of (3.12)–(3.13), one can consider instead the following optimization problem,

$$\begin{aligned} \mathcal{L}_2: \quad & \inf \quad t & (3.14)\\ \text{s.t.} \quad & A'P + PA < t(A'\bar{P} + \bar{P}A + I) & (3.15)\\ & P > 0 & (3.16)\\ & t \geq 0 & (3.17) \end{aligned}$$

where the matrix $\bar{P}$ is chosen to be positive definite, e.g., $\bar{P} = I$. One might wonder why $\bar{P}$ is introduced in (3.15). The reason is that by doing so, a feasible point of $\mathcal{L}_2$ is readily available: $(P_0, t_0) = (\bar{P}, 1)$. Our task is now to initiate the algorithm from $(\bar{P}, 1)$ with the objective value of 1, and try to somehow reduce the objective value to zero (which would be the case if and only if Σ_1 is stable), *without* leaving the feasible region of $\mathcal{L}_2$. This is exactly what an interior point method does (more specifically, we have in mind the barrier method).

Few comments, and a reformulation of $\mathcal{L}_2$ precedes our description of the barrier method. Suppose that we were to solve the following optimization problem:

$$\begin{aligned} \mathcal{L}_3: \quad & \inf \quad t & (3.18)\\ \text{s. t.} \quad & A'P + PA < tI & (3.19)\\ & P > 0 & (3.20)\\ & \|P\| < 1 & (3.21)\\ & -1 < t < 2 & (3.22) \end{aligned}$$

Let $t_{\inf}$ and $t_{\sup}$ denote the value of the infimum and the supremum of the objective functional on the respective region (e.g., $t_{\inf} = 0$ in $\mathcal{L}_2$ if Σ_1 is stable). The value of $t_{\inf}$ in $\mathcal{L}_3$ clearly is a measure of relative stability; intuitively, the more negative one can choose t, the more "stable" Σ_1 is. The lower bound for t and the norm constraint on P are chosen for normalization purposes; otherwise the problem would be unbounded, if feasible. The choice of the upper bound for t would be justifies shortly.

It is not clear however how a feasible point for $\mathcal{L}_3$ can be chosen to initiate the algorithm from. We thus consider instead a combination of $\mathcal{L}_2$ and $\mathcal{L}_3$:

$$\begin{aligned} \mathcal{L}: \quad & \inf \quad t & (3.23)\\ \text{s. t.} \quad & A'P + PA < t(A'\bar{P} + \bar{P}A + I) & (3.24)\\ & P > 0 & (3.25)\\ & \|P\| < 1 & (3.26)\\ & -1 < t < 2 & (3.27) \end{aligned}$$

with $\bar{P} > 0$ and $\|\bar{P}\| < 1$. The initial point $(\bar{P}, 1)$ is now readily available as an initial point. Again, the value of $t_{\inf}$ for $\mathcal{L}$ somehow conveys information regarding the relative stability of Σ_1, an observation which shall be made more precise shortly.

Let us denote the feasible region of $\mathcal{L}$ by $\mathcal{F}_{\mathcal{L}}$. Note that $\mathcal{F}_{\mathcal{L}} \subseteq SR_+^{n \times n} \times R$ and that it is an open and convex set. It turns out that associated with the set $\mathcal{F}_{\mathcal{L}}$, there is a functional b : interior $\mathcal{F}_{\mathcal{L}} \rightarrow R$, which acts as a "self-concordant barrier." The term "self concordant" refers to certain properties of the gradient and the Hessian of the functional b evaluated at points in $\mathcal{F}_{\mathcal{L}}$; for the purpose of our presentation, we shall bypass the exact definition and direct the readers' attention to the references given above for the ipm theory.

There are two important points however that need to be emphasized regarding the functional b. First, if $\{x_k\}_{k \geq 1} \in \mathcal{F}_{\mathcal{L}}$ is a sequence that approaches the boundary of $\mathcal{F}_{\mathcal{L}}$, $b(x_k) \rightarrow \infty$ as $k \rightarrow \infty$. Second, there is a parameter K associated with b which determines the computational efficiency of the interior point method for minimizing (or maximizing) a linear functional over $\mathcal{F}_{\mathcal{L}}$, the so-called self-concordant parameter. For brevity, we shall simply write down the self-concordant barrier for $\mathcal{F}_{\mathcal{L}}$ and its associated self-concordant parameter K. Subsequently, we provide a description of the algorithm for solving $\mathcal{L}$ using b, and its efficiency in terms of the self-concordant parameter of b.

Let $b : \mathcal{F}_{\mathcal{L}} \rightarrow R$ be defined as:

$$b(P, t) = -12 \log \det P$$

$$-12\log\det(-A'P - PA + t(A'\bar{P} + \bar{P}A + I)) - 12\log(1 - ||P||^2) - 12\log(t+1) - 12\log(2-t). \quad (3.28)$$

Note that indeed when $(P_k, t_k) \to$ boundary of $\mathcal{F}_{\mathcal{L}}$, $b(P_k, t_k) \to \infty$. The associated self-concordant parameter for the functional b (3.28) turns out to be,

$$K := \sqrt{12n + 12n + 48} = O(\sqrt{n}). \quad (3.29)$$

We are now ready to describe the interior point method for solving $\mathcal{L}$. Starting from the initial point $(P_0, t_0) = (\bar{P}, 1)$, and parameter $\mu = \mu_0$, (1) Let $k = 0$, (2) solve the unconstraint optimization problem

$$\min \Phi(P_k, t_k, \mu_k)$$

where $\Phi(P_k, t_k, \mu_k) := \mu_k t + b(P_k, t_k)$, (3) let $\mu_{k+1} = (1 + \frac{1}{6K})\mu_k$, (4) let $k = k+1$; Go to 2. Of course Step 2 cannot be solved exactly; a lot of research has been devoted to come up with a stopping criterion for this step which is sufficient for proving nice theoretical efficiency of the complete algorithm. One such criterion, rather interestingly, is to take "one" Newton step (with a "nice" initial point), and then increase μ_k accordingly (the so-called short step method). Intuitively, as $k \to \infty$, $\mu_k \to \infty$, and the sequence of minimal values of $\Phi(P_k, t_k, \mu_k)$ will approach t_{inf}.

The resulting complexity bound stated below is the upshot of the interior point approach.

Theorem 3.1 [15] *For solving $\mathcal{L}$, starting with $(\bar{P}, 1)$ using*

$$O(K\log(K + \frac{1}{\epsilon})) \quad (3.30)$$

iterations, the above barrier method computes $(P^*, t^*) \in \mathcal{F}_{\mathcal{L}}$, and t^* is known to satisfy

$$\frac{t^* - t_{\text{inf}}}{t_{\text{sup}} - t_{\text{inf}}} \le \epsilon \quad (3.31)$$

i.e., after $O(K\log(K + \frac{1}{\epsilon}))$, a (relative) ϵ-optimal point is found by the barrier method.

Theorem 3.1 has few implications which are not difficult to realize.[15]

1. Starting with $(\bar{P}, 1)$, given an $\epsilon > 0$, the barrier method produces the pair (P_i, t_i), known to satisfy

$$(P_i, t_i) \in \mathcal{F}_{\mathcal{L}}, \quad t_i - t_{\text{inf}} \le \epsilon, \quad (3.32)$$

if,

$$i = \Omega(K\log(K + \frac{t_{\text{sup}} - t_{\text{inf}}}{\epsilon})) \quad (3.33)$$

2. If $t_{\text{inf}} < \alpha < t_{\text{sup}}$, after

$$O(K\log(K + \frac{t_{\text{sup}} - t_{\text{inf}}}{\min\{t_{\text{sup}} - \alpha, \alpha - t_{\text{inf}}\}})) \quad (3.34)$$

iterations, for which the last pair is (P, t), is guaranteed to satisfy $(P, t) \in \mathcal{F}_{\mathcal{L}}$ and $t = \alpha$.

The second ramification above will now be employed to shed light on the efficiency of the interior point method for determining the stability of Σ_1.

Consider solving $\mathcal{L}$ using the interior point method described above. To check the stability of Σ_1 it is necessary and sufficient to stop the algorithm after the i-th iteration, when $t_i = 0$. According to (2) this is guaranteed after

$$\tau_{\Sigma_1} = O(K\log(K + \frac{t_{\text{sup}} - t_{\text{inf}}}{\min\{t_{\text{sup}}, -t_{\text{inf}}\}})) \quad (3.35)$$

i.e., τ_{Σ_1} is the termination time of the barrier method for checking the stability of Σ_1.

Let $\bar{P} = \frac{1}{2}I$ and start the interior point method described earlier from $(\bar{P}, 1)$. Thus $t_{\text{sup}} \ge 1$ and trivially $t_{\text{sup}} < 2$. Note that $t_{\text{inf}} \le 0$, since if $t_{\text{inf}} > 0$ and the pair (t_{inf}, P^*) is the solution to $\mathcal{L}$, then for $0 < \epsilon < 1$, $(\epsilon t_{\text{inf}}, \epsilon P^*)$ is also a solution to $\mathcal{L}$, which is a contradiction.

Referring to (3.35), we observe that τ_{Σ_1} is essentially a function of K which is itself a function of n only, and a combination of t_{sup} and t_{inf}.

Given a Hurwitz matrix A, let $\alpha := \inf_\Delta \frac{||\Delta||}{||A||}$ such that $A + \Delta$ is not Hurwitz. Define

$$\delta_{\Sigma_1} := \frac{1}{\log(\sqrt{n} + \frac{1}{\alpha})} \quad (3.36)$$

Since $-1 < t_{\text{inf}} \le 0$, for small $\epsilon > 0$,

$$A'P + PA < (t_{\text{inf}} - \epsilon)(A'\bar{P} + \bar{P}A + I) \quad (3.37)$$
$$P > 0 \quad (3.38)$$
$$||P|| < 1 \quad (3.39)$$

is inconsistent. Thus,

$$\alpha||A|| \le |t_{\text{inf}} - \epsilon|||A + I|| \quad (3.40)$$

$$\leq \quad (|t_{\inf}| + \epsilon)(\|A\| + 1) \quad (3.41)$$

$$\Rightarrow \frac{\alpha\|A\|}{\|A\|+1} \leq |t_{\inf}| + \epsilon \quad (3.42)$$

$$\Rightarrow t_{\inf} \leq \frac{-\|A\|}{\|A\|+1}\alpha \quad (3.43)$$

since $t_{\inf} \leq 0$.

Consequently, since $0 \leq -t_{\inf} \leq t_{\sup} < 2$,

$$\begin{aligned} \tau_{\Sigma_1} &= O(K \log(K + \frac{3}{-t_{\inf}})) \\ &= O(K \log(K + \frac{1}{\alpha})) \end{aligned}$$

Thereby we have established the following theorem.

Theorem 3.2

$$\tau_{\Sigma_1}\delta_{\Sigma_1} = O(K) = O(\sqrt{n}). \quad (3.44)$$

Theorem 3.2 establishes a natural, but very interesting relationship between robustness properties of Σ_1 and the efficiency of ipms for determining whether Σ_1 is stable. More specifically, given that (3.44) holds, fixing n and using the ipm for the solution of the Lyapunov equation, information pertaining to the relative stability of the corresponding system is somehow revealed! This observation has consequences which go far beyond the stability analysis of Σ_1.

3.1.2 Positive Realness

Consider the linear time invariant system Σ_2:

$$\Sigma_2 : \quad \dot{x} = A\,x + B\,u \quad (3.45)$$

$$y = C\,x + D\,u \quad (3.46)$$

such that the quadruple (A, B, C, D) is the minimal realization of the transfer matrix

$$H(s) = C(sI - A)^{-1}B + D$$

in which case we write $H \sim (A, B, C, D)$. We shall assume that the pairs (A, B) and (A, C) are respectively, controllable and observable. The matrix A is also assumed to be Hurwitz. For further reference let $A \in R^{n\times n}$, $B \in R^{n\times m}$, $C \in R^{m\times n}$ and $D \in R^{m\times n}$ and without loss of generality assume that $m < n$. Given an initial condition x_0 and a control function η which maps y to u, the equations (3.45)–(3.46) define a trajectory for the feedback system Σ_2.

Given that $H \sim (A, B, C, D)$ and assuming that A is Hurwitz, the transfer matrix H is called (generalized) strongly positive real (GSPR) if there exists $\epsilon > 0$ such that

$$H(jw) + H^*(jw) > \epsilon I \quad \forall w$$

where $H^*(jw)$ denotes the conjugate transpose of the transfer matrix $H(jw)$. The (generalized) Positive Real (GPR) Lemma states that $H(s)$ is GSPR and stable if and only if the following system of linear matrix inequalities is feasible,[1,16]

$$\text{herm}\{\begin{pmatrix} P & 0 \\ 0 & -I \end{pmatrix}\begin{pmatrix} A & B \\ C & D \end{pmatrix}\} < 0, \quad (3.47)$$

$$P > 0. \quad (3.48)$$

Let us define two robustness measures for a GSPR system. Denote by $E := \begin{pmatrix} A & B \\ C & D \end{pmatrix}$. Now let,

$$\alpha := \inf_{\Delta} \|\Delta\|/\|E\| \quad (3.49)$$

such that there *does not* exist a matrix P that satisfies the following set of linear matrix inequalities,

$$\text{herm}\{\begin{pmatrix} P & 0 \\ 0 & -I \end{pmatrix}(E + \Delta)\} < 0, \quad (3.50)$$

$$P > 0, \quad (3.51)$$

and let

$$\delta_H := 1/\log(\sqrt{n+m} + \frac{1}{\alpha}). \quad (3.52)$$

The quantity $\delta^1_{\Sigma_2}$ is a measure of the relative perturbation that a stable GSPR system Σ_2 can tolerate, and remain stable and GSPR. The perturbation Δ, can for example be the result of the uncertainty in the modeling of the plant, or due to the finite accuracy of the computer arithmetic (which for example, is used to check the GSPR property of the system).

The *robustness measure* for a GPR systems just introduced is related in an interesting way to the *computational efficiency* of the barrier method (when applied to solve the system of LMIs resulting from the GPR Lemma).

Theorem 3.3 [10] *Given the system Σ_2, there is an algorithm $\mathcal{A}$ such that for the robustness measures $\delta^1_{\Sigma_2}$ and $\delta^2_{\Sigma_2}$,*

$$\delta_H \tau_{\mathcal{A}(\Sigma_2)} = O(\sqrt{n+m})$$

Similar to the case of the Lyapunov inequality, the robustness parameter effects, in a very precise manner, the size of the certificate of positive realness.

Theorem 3.4 *The certificate of positive realness for a GSPR system, $\bar{P}$, satisfies,*

$$||\bar{P}|| = O(\frac{n+m}{\delta_H}), \tag{3.53}$$

provided that $\delta_H > 0$.

The precise relationship which were presented above between the computational efficiency of certain numerical algorithms on one hand, and the robustness measures for dynamical systems on the other, can in principle be used to give an *algorithmic definition* of the relative stability. In particular, one can *define* robustness in terms of purely algorithmic consideration to the effect that,

> the degree of the relative stability of a stable system is the inverse of the time it takes to check its stability via the interior point methods.

The objection to this approach would be that the stability properties of a system should in principle be coordinate free, and thus, should not depend on a particular algorithm. Nevertheless, since at the present time, we are far from obtaining *optimal* algorithms for solving stability problems (e.g., linear matrix inequalities), a *machine independent theory of stability* is far from being realized. Moreover, in order to check the stability of a dynamical system, an algorithm *has* to be introduced (on some particular model of computation), and thereby one can argue that, stability properties can be viewed with the running time of that algorithm as our frame of reference. The contribution of the paper is thus to demonstrate that the above approach can indeed be adopted for certain important problems in system analysis.

4 Conclusion

The relationship between the parameters δ_A and δ_H and the complexity estimates of the algorithms which are used in system analysis provide a very natural and intuitive way to understand the interaction between systems and optimization algorithms. At a deeper level however, it is indeed fascinating to see that the connection between efficiency and robustness can be stated in such a concise manner, as the results of this paper indicate.

Acknowledgment

The research of the author was performed at the Jet Propulsion Laboratory, California Institute of Technology, under contract with the National Aeronautics and Space Administration.

References

[1] B. D. O. Anderson. A system theory criterion for positive real matrices. *SIAM Journal of Control and Optimization*, (6)2:171–192, 1967.

[2] O. Axelsson. *Iterative Solution Methods*. Cambridge University Press, Cambridge, 1994.

[3] R. Bellman. Kronecker products and the second method of Lyapunov. *Mathematische Nachrichten*, 17–19, 1959.

[4] S. P. Boyd and L. EL Ghaoui and E. Feron and V. Balakrishnan. Linear Matrix Inequalities in System and Control Theory, SIAM, Philadelphia, 1994.

[5] C-T. Chen. *Linear System Theory and Design*. Holt, Rinehart and Winston, Inc., New York, 1984.

[6] Z. Gajic and M. T. J. Qureshi. Lyapunov Matrix Equation in System Stability and Control, Academic Press, CA, 1995.

[7] N. Karmarkar. A new polynomial-time algorithm for linear programming. *Combinatorica*, 4:373–395, 1984.

[8] D. G. Luenberger. *Linear and Nonlinear Programming*. Addison-Wesley Publishing Company, Reading, Massachusetts, second edition, 1984.

[9] M. Lyapunov. General stability problem of motion. *Annals of Mathematics*, 1949; photoreproduction of the French translation of 1907.

[10] M. Mesbahi and G. P. Papavssilopoulos. LMIs, Interior Point Methods, Complexity Theory, and Robustness Analysis, Proceedings of the IEEE Conference on Decision and Control, Dec. 1996, Kobe, Japan.

[11] M. Mesbahi. More Steps Toward an Algorithmic Theory of Stability, Submitted to the IEEE Conference on Decision and Control, Dec. 1997, San Diego, CA.

[12] M. Mesbahi and G. P. Papavassilopoulos. Towards an Algorithmic Theory of Stability and Robustness. Submitted to IEEE Transactions on Automatic Control, July 1996.

[13] Y. Nesterov and A. Nemirovskii. *Interior-Point Polynomial Algorithms in Convex Programming.* SIAM, Philadelphia, 1994.

[14] J. Renegar. Condition numbers, the barrier method and the conjugate gradient method. Technical report, School of Operations Research and Industrial Engineering, Cornell University, Itacha, New York, 1994.

[15] J. Renegar. Linear programming, complexity theory and elementary functional analysis. Technical report, School of Operations Research and Industrial Engineering, Cornell University, Itacha, New York, 1994.

[16] M. G. Safonov. Course notes: Advances in robust control, Spring 1996. Department of Electrical Engineering-Systems, University of Southern California, Los Angeles, California.

AIAA-97-3456

MODEL REDUCTION WITH A FINITE HORIZON H_∞ CRITERION

M. Bala Subrahmanyam*
Naval Air Warfare Center, Patuxent River, Maryland 20670

Saroj K. Biswas†
Temple University, Philadelphia, Pennsylvania 19122

ABSTRACT

An important problem in flight control and flying qualities is the approximation of a complex high order system by a low order model. In this paper, for a given reduced order model, we define the correlation measure between the plant and the model outputs to be the minimum of the ratio of weighted signal energy to weighted error energy. We give a criterion for the evaluation of the correlation measure in terms of minimization of a parameter occurring in a two-point boundary value problem. Once the correlation measure for a given reduced order model can be evaluated, a nonlinear programming algorithm can be used to select a model which maximizes the correlation between the plant and model outputs. The correlation index used can be regarded as an extension of the H_∞ performance criterion to the finite-interval time-varying case. However, the usual H_∞ problem seeks an optimal controller, whereas our problem is to select the reduced order model matrices which give the best correlation index. We also give an expression for the variation of the correlation owing to parameter variations. The utilization of the theory is demonstrated by means of an example.

I. Introduction

Model reduction is an important problem in the case of airplanes with significant aeroservoelastic dynamics. The original model in such cases is of high order and thus, the resulting controller will have a complex structure, especially if it uses full state feedback. Also for highly augmented aircraft with flight and propulsive controls, it is useful to develop low order models to analyze flying qualities.

If the aim is to design a low order controller for a high order plant, there are at least three broad approaches to achieve this. A general account of these three approaches is given in Ref. 1. The so called direct design methods assume a stabilizing controller of fixed degree and seek to find the controller that maximizes a quadratic performance index [2,3].

Another approach is to get a high order controller by some design technique, such as LQG or H_∞, and then to approximate the high order controller by a low order one which possesses certain desirable properties. The third approach is to approximate the high order plant by a low order one. Then a low order controller is designed and used to control the original plant. In this paper we concentrate on this approach and consider the problem of approximating the original plant by a low order model in an optimal sense.

There are several model reduction techniques based on balanced realization of the state space system of a transfer function. In these methods we can have an estimate of the infinity norm of the error between the reduced transfer function and the plant transfer function. Ref. 4 initiated the work on balanced realizations, Ref. 5 extended the work for frequency weighting, and Ref. 6 gives an alternate computational method.

In this paper we present a model reduction technique based on an H_∞ time domain criterion. It has the advantage of comparing directly the output of the reduced order model with that of the plant utilizing a minimax criterion. Preliminary results in this direction were reported in Refs. 7 and 8. We also give criteria that need to be maximized at the optimal reduced order model. At the optimal reduced order model, the correlation measure will be maximized. We also give expressions for the derivatives of the correlation measure with respect to the reduced order model parameters. Also a new numerical method for the evaluation of the correlation measure will be presented. The new method will be utilized in the computation of the optimal reduced order model in a number of cases.

We now state the main problem. For the sake of generality, we pose it for time-varying systems. Let

---

* Electronics Engineer, Code 4326, Flight Control, Associate Fellow AIAA.

† Associate Professor, Department of Electrical Engineering.

the plant be described by

$$\dot{x}_p = A_p(t)x_p + B_p(t)u, \qquad x_p(t_0) = 0, \tag{1}$$
$$y_p = C_p(t)x_p + D_p(t)u, \tag{2}$$

where $x_p(t)$, $u(t)$, and $y_p(t)$ denote the plant state vector, the control vector, and the plant output vector respectively. Let the reduced order model which approximates the plant be chosen to be

$$\dot{x}_m = A_m(t)x_m + B_m(t)u, \qquad x_m(t_0) = 0, \tag{3}$$
$$y_m = C_m(t)x_m + D_m(t)u, \tag{4}$$

where $x_m(t)$ and $y_m(t)$ denote respectively the state vector and the output vector of the reduced order model.

For given $A_m(t), B_m(t), C_m(t)$, and $D_m(t)$, let u be chosen such that the correlation index given by

$$\frac{\int_{t_0}^{T} \frac{1}{2}u^*(t)R(t)u(t)\,dt}{\int_{t_0}^{T} \frac{1}{2}(y_p - y_m)^* Q(t)(y_p - y_m)\,dt} \tag{5}$$

is minimized. The superscript * denotes matrix or vector transpose. Let this minimum value be denoted by ρ. Thus u represents the worst input and ρ gives a measure of the worst-case correlation between the plant output and the model output. The problem is to choose $A_m(t), B_m(t), C_m(t)$, and $D_m(t)$ such that ρ is maximized.

Since Eq. (5) represents the ratio of weighted signal energy to weighted error energy, the above problem may be regarded as a modified H_∞ problem except for a few differences. We consider time-varying systems and in our case the interval of control is finite. There are extensions of the H_∞ results to the finite-horizon time-varying case[9]. However, our approach is different and is based on considering the inherent two-point boundary value problem. Also, the general aim of H_∞ problems is the design of an optimal controller, whereas in this paper we are interested in the selection of reduced model matrices. It is necessary in our case to use nonlinear programming algorithms in order to select the model matrices which maximize ρ.

In the case of time-invariant systems, a nonlinear programming algorithm can be used to find at least a local maximum of ρ. For the time-varying case, the matrices $A_m(t), B_m(t), C_m(t)$, and $D_m(t)$ need to be expressed in terms of basis functions and a nonlinear programming algorithm needs to be used to maximize ρ with respect to the coefficients of the basis functions.

We do not require the plant and the model to be open loop stable. This is significant since many of the modern aircraft have open loop unstable poles. We show in Section IV by means of examples that the method is indeed applicable to such cases. There is yet another advantage of our method. One of the criticisms in the approach of getting a low order model from a high order plant is that satisfactory approximation of the plant requires some knowledge in advance of the controller[1]. Since we maximize the correlation between the plant and model outputs for the worst possible input, the correlation in the case of any other controller is bound to be better. Thus, our method furnishes a satisfactory approximation without requiring an *a priori* knowledge of the controller.

We now give a summary of the results of the paper. In Section II, conditions that characterize the worst input are derived for a given model. A two-point boundary value problem needs to be solved for the least positive ρ to obtain the worst-case correlation between the outputs of the plant and the model. We present an algorithm for the evaluation of ρ. A nonlinear programming algorithm can then be used to find the model matrices which maximize ρ.

In Section III we derive an expression for the variation of correlation between the plant and model outputs as a functional of the variations in system parameters. Making use of this result, we derive expressions for the derivatives of ρ with respect to the elements of A_m, B_m and C_m. These expressions are useful in case the nonlinear programming algorithm makes use of derivatives. From the expressions for the derivatives, conditions that need to be satisfied at the optimal reduced order model can be derived.

In Section IV a simple example is worked out. Finally, certain conclusions are given in Section V.

II. Computation of ρ for a given reduced order model

Assume that the matrices $A_m(t), B_m(t), C_m(t)$, and $D_m(t)$ are given certain values which are not necessarily the optimal values. In this section, we characterize ρ as the minimum positive value for which a certain two-point boundary value problem has a nontrivial solution. Also, we derive a computationally useful characterization of ρ.

Let

$$x = (\,x_p^* \quad x_m^*\,)^*, \tag{6}$$
$$y = y_p - y_m, \tag{7}$$
$$A(t) = \begin{pmatrix} A_p & 0 \\ 0 & A_m \end{pmatrix}, \tag{8}$$
$$B(t) = \begin{pmatrix} B_p \\ B_m \end{pmatrix}, \tag{9}$$
$$C(t) = (\,C_p \quad -C_m\,), \tag{10}$$

and

$$D(t) = D_p - D_m. \tag{11}$$

Assuming that $D_m(t) = D_p(t)$, we can write Eqs. (1) – (4) as

$$\dot{x} = A(t)x + B(t)u, \qquad x(t_0) = 0, \tag{12}$$

$$y = C(t)x. \tag{13}$$

The performance index given by Eq. (5) can be put in the form

$$J(u) = \frac{\int_{t_0}^{T} \frac{1}{2}u^*(t)R(t)u(t)\,dt}{\int_{t_0}^{T} \frac{1}{2}x^*(t)W(t)x(t)\,dt}, \tag{14}$$

where

$$W(t) = C^*(t)Q(t)C(t). \tag{15}$$

The problem is to characterize $u(t)$ that minimizes the index in Eq. (14).

Let $\rho = \inf_u J(u)$. We assume that for all t

$$\begin{aligned} R(t) &> 0, \\ Q(t) &\geq 0. \end{aligned} \tag{16}$$

Eq. (16) guarantees that the numerator and denominator of Eq. (14) are nonnegative for any u. The necessary conditions that characterize the worst input can be stated as follows.

Theorem 1. *Consider the system given by Eqs. (12)-(15). If $u(t)$ minimizes the correlation index given by Eq. (14), then there exists an adjoint vector $\psi(t)$, not identically zero, such that*

$$\frac{d\psi}{dt} = -A^*\psi - \rho W x, \qquad \psi(T) = 0, \tag{17}$$

with

$$u(t) = R^{-1}B^*\psi. \tag{18}$$

Proof. For a proof, see Theorem 4.1, Chapter 3, of Ref. 8.

Let

$$\hat{A} = A, \tag{19}$$

$$\hat{B} = BR^{-1}B^*, \tag{20}$$

and

$$\hat{C} = -\rho W. \tag{21}$$

Thus, we have a two-point boundary value problem given by

$$\begin{pmatrix} \dot{x} \\ \dot{\psi} \end{pmatrix} = \begin{pmatrix} \hat{A} & \hat{B} \\ \hat{C} & -\hat{A}^* \end{pmatrix} \begin{pmatrix} x \\ \psi \end{pmatrix} \tag{22}$$

with

$$x(t_0) = 0, \qquad \psi(T) = 0. \tag{23}$$

The following theorem follows from Theorem 4.2, Chapter 3, of Ref. 8.

Theorem 2. *Let (x, ψ) satisfy the boundary value problem given by Eqs. (22) and (23) for the least positive ρ such that $\int_{t_0}^{T} \frac{1}{2}x^*Wx\,dt > 0$, where $u = R^{-1}B^*\psi$. Then ρ is the minimum value of the index given by Eq. (14) and u is the worst input.*

In order to find the least positive value ρ which yields a nontrivial solution for Eqs. (22) and (23), a computational technique is given in Chapter 3 of Ref. 8, which utilizes the transition matrix associated with Eq. (22). However, the transition matrix of a Hamiltonian matrix is very ill-conditioned, especially as the time t becomes large, and the above mentioned numerical method becomes highly unstable. In this paper, we use a different technique developed by the second author which we found to be quite stable numerically for any final time T. The theory behind the technique is given below.

Consider Eqs. (22) and (23) for the time-invariant case. Let

$$\begin{pmatrix} x \\ \psi \end{pmatrix} = \begin{pmatrix} w_{11} & w_{12} \\ w_{21} & w_{22} \end{pmatrix} \begin{pmatrix} p \\ q \end{pmatrix} \tag{24}$$

so that

$$\begin{pmatrix} \dot{p} \\ \dot{q} \end{pmatrix} = \begin{pmatrix} \Lambda & 0 \\ 0 & -\Lambda \end{pmatrix} \begin{pmatrix} p \\ q \end{pmatrix} \tag{25}$$

In Eq. (25), Λ could be in Jordan canonical form if it has real roots. If it has complex roots, we found that the numerical method is more stable if the block associated with a complex pair contains only real elements.

From Eq. (23), we have

$$\begin{aligned} w_{11}p(0) + w_{12}q(0) &= 0 \\ w_{21}p(T) + w_{22}q(T) &= 0. \end{aligned} \tag{26}$$

From (25)

$$\begin{aligned} p(T) &= \exp(\Lambda T)p(0) \\ q(0) &= \exp(\Lambda T)q(T). \end{aligned} \tag{27}$$

From (26), we get

$$\begin{pmatrix} w_{11} & w_{12}e^{\Lambda T} \\ w_{21}e^{\Lambda T} & w_{22} \end{pmatrix} \begin{pmatrix} p(0) \\ q(T) \end{pmatrix} = 0. \tag{28}$$

Eq. (28) has a nontrivial solution for $p(0)$ and $q(T)$ if and only if

$$\det \begin{pmatrix} w_{11} & w_{12}e^{\Lambda T} \\ w_{21}e^{\Lambda T} & w_{22} \end{pmatrix} = 0. \tag{29}$$

Thus, we can characterize ρ as the least positive value for which Eq. (29) holds.

If the determinant in Eq. (29) varied smoothly with ρ, we could have picked the first positive value at which the determinant changes sign. It so happens that the determinant in Eq. (29) changes sign arbitrarily under the implementation with MATLAB and hence is not a smooth function of ρ. However, if we take the absolute value of the determinant, it is a smooth function of ρ except at the point where the determinant becomes zero. This does not provide a viable method with a computer since it is much easier to find the zero of a function when it changes sign at the zero point.

III. Derivatives and Optimality Conditions

In this section we derive an expression for the variation in the correlation measure ρ in terms of variations in the system matrices. Making use of this expression, for the time-invariant case, we can derive expressions for the derivative of ρ with respect to the elements of the reduced order matrices. We can further derive the conditions that need to be satisfied at the optimal reduced order model.

Consider Eqs. (1) – (10). The system equations are given by

$$\dot{x} = A(t)x + B(t)u, \qquad x(t_0) = 0, \tag{30}$$

$$y = C(t)x. \tag{31}$$

We can write Eq. (5) as

$$\frac{\int_{t_0}^{T} \frac{1}{2} u^*(t)R(t)u(t)\, dt}{\int_{t_0}^{T} \frac{1}{2} x^*(t)C^*(t)Q(t)C(t)x(t)\, dt.} \tag{32}$$

For given $A_m(t), B_m(t)$, and $C_m(t)$, let ρ be the minimum of the correlation index in Eq. (32) over $u(t)$. Let the elemental variations in $A_m(t), B_m(t)$, and $C_m(t)$ be denoted by $\delta A_m(t), \delta B_m(t)$, and $\delta C_m(t)$ respectively. Let $\delta A(t), \delta B(t)$, and $\delta C(t)$ be the variations in the matrices $A(t), B(t)$, and $C(t)$ corresponding to the elemental variations $\delta A_m(t), \delta B_m(t)$, and $\delta C_m(t)$. Notice that

$$\delta A(t) = \begin{pmatrix} 0 & 0 \\ 0 & \delta A_m(t) \end{pmatrix}, \tag{33}$$

$$\delta B(t) = \begin{pmatrix} 0 \\ \delta B_m(t) \end{pmatrix}, \tag{34}$$

and

$$\delta C(t) = \begin{pmatrix} 0 & \delta C_m(t) \end{pmatrix}. \tag{35}$$

Let $\delta\rho$ denote the variation in ρ caused by $\delta A, \delta B$, and δC. We now derive an expression for $\delta\rho$ in terms of $\delta A(t), \delta B(t)$, and $\delta C(t)$. For given $A_m(t), B_m(t)$, and $C_m(t)$, let u minimize the index in Eq. (32). In the following, we suppress the dependence of the matrices on t for simplicity of notation. From Eqs. (19)–(23), we get the following boundary value problem which needs to be satisfied by the corresponding pair (x, ψ).

$$\dot{x} = Ax + BR^{-1}B^*\psi, \tag{36}$$

$$\dot{\psi} = -\rho C^*QCx - A^*\psi, \tag{37}$$

$$x(t_0) = \psi(T) = 0. \tag{38}$$

Let x_1 and ψ_1 represent the variations in x and ψ due to $\delta A, \delta B$, and δC. From Eqs. (36)–(38), we have the following equations satisfied by x_1 and ψ_1.

$$\dot{x}_1 = Ax_1 + \delta A\, x + BR^{-1}B^*\psi_1 + (BR^{-1}\delta B^* + \delta B\, R^{-1}B^*)\psi, \tag{39}$$

$$\dot{\psi}_1 = -\delta\rho C^*QCx - \rho(\delta C^*QC + C^*Q\delta C)x - \rho C^*QCx_1 - A^*\psi_1 - \delta A^*\psi, \tag{40}$$

$$x_1(t_0) = \psi_1(T) = 0. \tag{41}$$

Theorem 3. *Consider Eqs. (36)–(41). Let*

$$\Omega = -\int_{t_0}^{T} \psi^*\delta A\, x\, dt - \int_{t_0}^{T} \psi^*B^*R^{-1}\delta B\, \psi\, dt - \rho\int_{t_0}^{T} x^*C^*Q\delta C\, x\, dt. \tag{42}$$

Then the variation in ρ is given by

$$\delta\rho = \frac{\Omega}{\int_{t_0}^{T} \frac{1}{2} x^*C^*QCx\, dt}. \tag{43}$$

Proof. From Eq. (40), we get

$$\int_{t_0}^{T} x^*\dot{\psi}_1\, dt = -\delta\rho\int_{t_0}^{T} x^*C^*QCx\, dt - \rho\int_{t_0}^{T} x^*(\delta C^*QC + C^*Q\delta C)x\, dt - \rho\int_{t_0}^{T} x^*C^*QCx_1\, dt - \int_{t_0}^{T} x^*A^*\psi_1\, dt - \int_{t_0}^{T} x^*\delta A^*\psi\, dt. \tag{44}$$

Also, by an integration by parts and by Eqs. (36), (38), and (41),

$$\int_{t_0}^{T} x\dot{\psi}_1\, dt = -\int_{t_0}^{T} x^*A\psi_1\, dt - \int_{t_0}^{T} \psi^*BR^{-1}B^*\psi_1\, dt. \tag{45}$$

Since

$$\int_{t_0}^{T} x^*(\delta C^* QC + C^* Q\delta C)x\, dt = 2\int_{t_0}^{T} x^* C^* Q\delta C\ x\, dt, \tag{46}$$

from Eqs. (44) and (45), we get

$$\begin{aligned} \delta\rho \int_{t_0}^{T} x^* C^* QCx\, dt + 2\rho \int_{t_0}^{T} x^* C^* Q\delta C\ x\, dt \\ +\rho \int_{t_0}^{T} x^* C^* QCx_1\, dt \\ + \int_{t_0}^{T} x^* \delta A^* \psi\, dt = \int_{t_0}^{T} \psi^* BR^{-1}B^* \psi_1\, dt. \end{aligned} \tag{47}$$

From Eq. (37),

$$\rho \int_{t_0}^{T} x^* C^* QCx_1\, dt = -\int_{t_0}^{T} (\dot\psi + A^*\psi)^* x_1\, dt. \tag{48}$$

Integrating the first term of the integrand from the right side of Eq. (48) by parts, and using Eqs. (38) and (41), we get

$$\begin{aligned} \rho \int_{t_0}^{T} x^* C^* QCx_1\, dt = \int_{t_0}^{T} \psi^* \delta A\ x\, dt \\ + \int_{t_0}^{T} \psi^* BR^{-1}B^* \psi_1\, dt \\ + \int_{t_0}^{T} \psi^*(BR^{-1}\delta B^* + \delta B\ R^{-1}B^*)\psi\, dt. \end{aligned} \tag{49}$$

Incorporating Eq. (49) in Eq. (47) and using the fact that

$$\int_{t_0}^{T} \psi^*(BR^{-1}\delta B^* + \delta B\ R^{-1}B^*)\psi\, dt = 2\int_{t_0}^{T} \psi^* BR^{-1}\delta B\ \psi\, dt, \tag{50}$$

we get Eq. (43). □

Using Eq. (43), the variation in the correlation measure ρ due to parameter variations can be computed for any given $A_m(t), B_m(t)$, and $C_m(t)$. Incorporating Eqs. (33)–(35) in Eq. (43), we get

$$\delta\rho = \frac{\left[\begin{array}{c} -\int_{t_0}^{T} \psi_m^* \delta A_m\ x_m\, dt - \int_{t_0}^{T} \psi^* BR^{-1}\delta B_m^*\ \psi_m\, dt \\ +\rho \int_{t_0}^{T} x^* C^* Q\delta C_m\ x_m\, dt \end{array}\right]}{\int_{t_0}^{T} \frac{1}{2} x^* C^* QCx\, dt}. \tag{51}$$

For any matrix M, let M^{ij} denote the element in the i-th row and j-th column. For the time-invariant case, from Eq. (51), we get

$$\frac{\partial\rho}{\partial A_m^{ij}} = \frac{-\int_{t_0}^{T} \psi_m x_m^*\, dt}{\int_{t_0}^{T} \frac{1}{2} x^* C^* QCx\, dt}, \tag{52}$$

$$\frac{\partial\rho}{\partial B_m^{ij}} = \frac{-\int_{t_0}^{T} \psi_m \psi^* BR^{-1}\, dt}{\int_{t_0}^{T} \frac{1}{2} x^* C^* QCx\, dt}, \tag{53}$$

$$\frac{\partial\rho}{\partial C_m^{ij}} = \frac{\rho\int_{t_0}^{T} QCxx_m^*\, dt}{\int_{t_0}^{T} \frac{1}{2} x^* C^* QCx\, dt}, \tag{54}$$

Thus, at the optimal A_m, B_m, and C_m, we have

$$\int_{t_0}^{T} \psi_m x_m^*\, dt = 0, \tag{55}$$

$$\int_{t_0}^{T} \psi_m \psi^* BR^{-1}\, dt = 0, \tag{56}$$

$$\int_{t_0}^{T} QCxx_m^*\, dt = 0. \tag{57}$$

We can reduce the computation of various integrals in the numerators of Eqs. (52)–(54) to solving an appropriate algebraic Lyapunov equation. Gradient computation requires the evaluation of

$$\int_{t_0}^{T} \begin{pmatrix} x \\ \psi \end{pmatrix} \begin{pmatrix} x^* & \psi^* \end{pmatrix} dt, \tag{58}$$

out of which one picks appropriate partitions as needed according to the numerators of Eqs. (52)–(54).

We have

$$\begin{pmatrix} \dot x \\ \dot\psi \end{pmatrix} = M \begin{pmatrix} x \\ \psi \end{pmatrix}, \tag{59}$$

where M is defined by Eq. (22). Denote

$$X = \begin{pmatrix} x \\ \psi \end{pmatrix}. \tag{60}$$

Now consider

$$\frac{d}{dt}(XX^*) = \dot X X^* + X\dot X^*. \tag{61}$$

Substituting Eq. (59) into the above and integrating over t_0 to T, it is easy to verify that

$$XX^*|_{t_0}^{T} = MP + PM^* \tag{62}$$

where

$$P = \int_{t_0}^{T} XX^*\, dt = \int_{t_0}^{T} \begin{pmatrix} x \\ \psi \end{pmatrix} \begin{pmatrix} x^* & \psi^* \end{pmatrix} dt. \tag{63}$$

From Eqs. (23) and (62), we get

$$MP + PM^* = S \tag{64}$$

where

$$S = \begin{pmatrix} x(T)x^*(T) & 0 \\ 0 & -\psi(t_0)\psi^*(t_0) \end{pmatrix}. \quad (65)$$

Thus the problem of computing the integral in Eq. (58) is reduced to solving the algebraic equation (64).

IV. Numerical Example

We will demonstrate the reduction of a second order system to a first order one. The system is of the form

$$\dot{x}_p = A_p x_p + B_p u, \qquad x_p(0) = 0, \quad (66)$$

$$y_p = C_p x_p, \quad (67)$$

where

$$A_p = \begin{pmatrix} 0 & 1 \\ -10 & -11 \end{pmatrix}.$$

The other matrices are $B_p = \begin{pmatrix} 0 \\ 1 \end{pmatrix}$ and $C_p = \begin{pmatrix} 1 & 0 \end{pmatrix}$.

Our first order model equations are

$$\dot{x}_m = a_m x_m + b_m u, \quad x_m(0) = 0, \quad y_m = c_m x_m. \quad (68)$$

The weighting matrices in Eq. (5) are $R = .0001, Q = 1$, and the final time T is taken to be 1 second. Our optimization technique is the fminu algorithm in the Optimization Toolbox of MATLAB. We also gave the gradients as input at each step.

The value of ρ was computed using the simple techniques listed in Chapter 3 of Ref. 8. In the first instance, the initial values were $a_m = -0.2, b_m = 0.9$, and $c_m = 0.2$. The value of ρ increased from an initial value of 0.0232 to a maximum of 3.365877. The final reduced order model is given by $a_m = -0.254207, b_m = 0.899641$, and $c_m = 0.0776696$. The final value of the gradient of ρ with respect to a_m, b_m, and c_m were $0.0122 \times 1.0e-5, -0.0167 \times 1.0e-5$, and $-0.8754 \times 1.0e-5$, respectively.

In the second instance, the intial values were $a_m = -1, b_m = 1$, and $c_m = 0.5$ and the final values are given by $a_m = -0.254206, b_m = 0.5912354$, and $c_m = 0.11818422$. In this case, the value of λ increased from 0.003 to a final maximum of 3.365877. The final value of the gradient of ρ with respect to a_m, b_m, and c_m were $-0.023 \times 1.0e-3, -0.136 \times 1.0e-3$, and $-0.681 \times 1.0e-3$, respectively. Note that there is no unique reduced order model in this case. In both instances the value of $b_m c_m$ is the same.

IV. Conclusions

In this paper we presented a technique for reduced order modeling using a finite horizon H_∞ optimality criterion. A characterization for the determination of the correlation between plant and model outputs is given. We need to do further work in utilizing the algorithms that are developed in this paper for computing the correlation coefficient and the gradients of the correlation coefficient with respect to reduced order model parameters.

References

[1]Anderson, B. D. and Liu, L. Y., "Controller Reduction : Concepts and Approaches," *I.E.E.E. Transactions on Automatic Control,* Vol. 34, 1989, pp. 802 – 812.

[2]Ly, U. -L., "A Design Algorithm for Robust Low Order Controller," Ph. D. dissertation, Department of Aeronautics and Astronautics, Stanford University, 1982.

[3]Bernstein, D. S. and Hyland, D. C., "The Optimal Projection Equations for Fixed-Order Dynamic Compensation," *I.E.E.E. Transactions on Automatic Control,* Vol. 29, 1984, pp. 1034 – 1037.

[4]Moore, B. C., "Principal Component Analysis in Linear Systems: Controllability, Observability, and Model Reduction," *I.E.E.E. Transactions on Automatic Control,* Vol. 26, 1981, pp. 17 – 31.

[5]Enns, D. F., "Model Reduction with Balanced Realizations: An Error Bound and a Frequency Weighted Generalization," 1984.

[6]Safonov, M. G. and Chiang, R. Y., "A Schur Method for Balanced Model Reduction," *Proceedings of the I.E.E.E. Conference on Decision and Control,* 1987.

[7]Subrahmanyam, M. B. and Steinberg, M., "Model Reduction with a Finite-Interval H_∞ Criterion," *Proceedings of the AIAA Guidance, Navigation and Control Conference*, Portland, OR, 1990, pp. 1419-1427.

[8]Subrahmanyam, M. B., Optimal Control with a Worst-Case Performance Criterion and Applications, *Lecture Notes in Control and Information Sciences,* Vol. 145, Springer-Verlag, Heidelberg, New York, 1990.

[9]Subrahmanyam, M. B., Finite Horizon H_∞ and Related Control Problems, Birkhäuser, Boston, 1996.

AIAA-97-3457

On H_∞ Controller Design for Singularly Perturbed Uncertain Systems

Peng Shi,* Shyh-Pyng Shue,† Ramesh K. Agarwal†

Abstract.[1] In this paper, the problem of robust stability and robust disturbance attenuation is investigated for a class of singularly perturbed linear systems with norm-bounded parameter uncertainties in both state and output equations. Based on the slow and fast subsystems, a composite linear controller is designed such that both robust stability and a prescribed H_∞ performance for the full-order system are achieved, irrespective of the uncertainties. It has been shown that, by using Riccati equation approach, the above problem can be solved if an H_∞ control problem for a related singularly perturbed linear system without parameter uncertainty can be solved. Thus, the solution to the latter is independent of the singular perturbation ε when ε tends to zero.

1 Introduction

In the past three decades, a great deal of interest has been devoted to the singularly perturbed control systems (see, for example [3, 8, 9, 10, 13, 14, 20, 21, 23, 24, 29] and the references therein). An important approach adopted to deal with these systems is based on the so called reduced technique [19, 22]. The composite design based on separate designs for slow and fast subsystems has been systematically reviewed in [29]. The issue on controller designing for singularly perturbed systems in H_∞ sense has been addressed by [17, 30, 34]. Recently, the H_∞-optimal control of singularly perturbed linear systems, under either perfect state measurements or imperfect state measurements, for both finite and infinite horizons has been studied by [23, 24] via a differential game theoretic approach. The asymptotic expansions for game theoretic Riccati equations has been tackled by [5] and it has been shown how they may be used in singularly perturbed H_∞ control. More recently, [6, 7] also considered a construction of high-order approximations to a controller that guarantees a desired performance level on the basis of the exact decomposition of the full-order Riccati equations to the reduced-order slow and fast equations. A singularly perturbed zero-sum dynamic game with full information has been considered in [12] and it has been demonstrated that when the singular perturbations parameter tends to zero, similar to singularly perturbed differential game [11], the upper(lower) value function of the dynamic game has a limit which coincides with a viscosity solution of a Hamilton-Jacobi-Isaacs type equation.

On the other hand, robust control of uncertain dynamical systems has been a subject of recurring interest over the past ten years. Considerable attention has been given to both the problems of robust stabilization and robust performance and a number of results covering these issues have been reported in the literature; see e.g. [1, 18, 25, 26, 31, 32, 33, 36, 35] and the references therein. However, to the best of authors' knowledge, the design of robust controllers for singularly perturbed linear continuous-time systems with parameter uncertainty has not been fully investigated.

In this paper, we consider the problem of robust disturbance attenuation with stability for a class of uncertain singularly perturbed linear continuous-time systems. The class of uncertain systems is described by a state space model with slow and fast dynamics and time-varying norm-bounded parameter uncertainty in both the state and output equations. Attention is focused on the design of a composite linear controllers based on the slow and fast problems. We study the problem of robust H_∞ control in which both robust stability and a prescribed disturbance attenuation for the full order system are required to be achieved irrespective of the uncertainties. Our results show that the above problem can be recast into H_∞ syntheses for

*Research of this author was supported in part by the Australian Research Council under Grant A49532206. Centre for Industrial and Applicable Mathematics, School of Mathematics, The University of South Australia, The Levels SA 5095, Australia. Fax: (+61 8) 83025785. Email: matsp@zarniwoop.levels.unisa.edu.au

†Department of Aerospace Engineering, Wichita State University, Wichita, KS 67260-0044, USA. Fax: (+1 316) 9783307. Email: shue@twsuvm.uc.twsu.edu; agarwal@ae.twsu.edu. S. P. Shue is a student member of AIAA. R. K. Agarwal is a fellow of AIAA.

a related singularly perturbed linear system without parameter uncertainty. Therefore, the existing results on singularly perturbed H_∞ control such as those in [5, 7, 23, 24] can be applied to obtain a solution to the problem of robust H_∞ control of uncertain singularly perturbed systems. Particularly, the solution above is independent of singular perturbation if the latter is sufficiently small.

Notation. The notations used in this paper are fairly standard. The superscript "T" denotes matrix transposition and the notation $X \geq Y$(respectively, $X > Y$) where X and Y are symmetric matrices, means that $X - Y$ is positive semi-definite (respectively, positive definite). I_n denotes the $n \times n$ identity matrix. $\rho(X)$ is the spectral radius of the matrix X. $L_2[0, \infty)$ stands for the space of square integrable vector functions over the interval $[0, \infty)$. $\|\cdot\|$ will refer to the Euclidean vector norm whereas $\|\cdot\|_2$ denotes the $L_2[0, \infty)$-norm over $[0, \infty)$.

2 Problem Formulation and Preliminaries

Consider the uncertain systems with slow and fast dynamics described in the standard "singularly perturbed" form by

$$(\Sigma) : \dot{x}_1(t) = [A_{11} + \Delta A_{11}(t)]x_1(t) + A_{12}x_2(t) + B_{11}w(t) + [B_{12} + \Delta B_{12}(t)]u(t), \quad x_1(0) = 0 \tag{2.1}$$

$$\varepsilon\dot{x}_2(t) = [A_{21} + \Delta A_{21}(t)]x_1(t) + A_{22}x_2(t) + B_{21}w(t) + [B_{22} + \Delta B_{22}(t)]u(t), \quad x_2(0) = 0 \tag{2.2}$$

$$z(t) = C_{11}x_1(t) + C_{12}x_2(t) + D_{12}u(t) \tag{2.3}$$

$$y(t) = [C_{21} + \Delta C_{21}(t)]x_1(t) + C_{22}x_2(t) + D_{21}w(t) \tag{2.4}$$

where $x_1 \in \Re^{n_1}$ is the slow dynamics state, $x_2 \in \Re^{n_2}$ is the fast dynamics state, $u \in \Re^q$ is the control input, $w \in \Re^p$ is the disturbance input, which belongs to $L_2[0, \infty)$, $y \in \Re^m$ is the measurement, $z \in \Re^r$ is the controlled output, $\varepsilon > 0$ is a small parameter, A_{11}, A_{12}, A_{21}, A_{22}, B_{11}, B_{12}, B_{21}, B_{22}, C_{11}, C_{21}, D_{12} and D_{21} are known constant matrices of appropriate dimensions that describe the nominal system, $\Delta A_{11}, \Delta A_{21}, \Delta B_{12}$, ΔB_{22} and ΔC_{21} are real time-varying matrices representing norm-bounded parameter uncertainties.

The admissible parameter uncertainties are of the form

$$\begin{bmatrix} \Delta A_{11}(t) & \Delta B_{12}(t) \\ \Delta A_{21}(t) & \Delta B_{22}(t) \end{bmatrix} = \begin{bmatrix} H_1 \\ H_2 \end{bmatrix} F(t)[E_1 \ \ E_2] \tag{2.5}$$

$$\Delta C_{21}(t) = H_3F(t)E_1 \tag{2.6}$$

where E_1, E_2, H_1, H_2 and H_3 are known constant matrices with appropriate dimensions. $F(t)$ is an unknown time-varying matrix satisfying

$$F^T(t)F(t) \leq I, \ \forall t \in [0, \infty) \tag{2.7}$$

with the elements of $F(t)$ being Lebesgue measurable.

Remark 2.1 *The matrices $F(t)$ is allowed to be state-dependent, i.e. $F(t) = F[t, x_1(t), x_2(t)]$ as long as (2.7) is satisfied along all possible state trajectories. The matrix $F(t)$ contains the uncertain parameters in the dynamics state and control input matrices of the system (Σ). The matrices E_1, E_2, H_1, H_2 and H_3 specify how the uncertain parameter in $F(t)$ affects the nominal matrices of the system (Σ). Observe that the unit overbound for $F(t)$ does not cause any loss of generality. Indeed, $F(t)$ can be always normalized, in the sense of (2.7), by appropriate choosing the matrices E_1, E_2, H_1, H_2 and H_3.* □

Remark 2.2 *The system (2.1)-(2.4) is a singularly perturbed system consisting of parametric uncertainties. The smallness of ε allows the characterization of the variable $x_1(t)$ as slow and the variable $x_2(t)$ as fast since $x_2(t)$ may change their value with the velocity of the order $O(\varepsilon^{-1})$ [10]. It should be noted that the models of the form (2.1)-(2.4) can be used to represent many important physical systems, such as power networks, transportation, aerospace, water resources. The parameter uncertainty structure in (2.5)-(2.6) has been widely used in the problems of robust stabilization and robust control of uncertain linear systems (see, e.g. [16, 18, 26, 33, 32]) and can represent parameter uncertainty in many physical cases, e.g. systems satisfying "matching conditions" [1]. Also, many practical systems possess parameter uncertainties which can be either exactly modelled, or overbounded by (2.7). Furthermore, any possible parameter uncertainty in B_{11}, B_{21} and D_{21} are assumed to be absorbed in $w(t)$.* □

Define

$$x(t) = [x_1^T(t) \ \ x_2^T(t)]^T, \ \ A(\varepsilon) = \begin{bmatrix} A_{11} & A_{12} \\ \frac{1}{\varepsilon}A_{21} & \frac{1}{\varepsilon}A_{22} \end{bmatrix} \tag{2.8}$$

$$B_1(\varepsilon) = \begin{bmatrix} B_{11} \\ \frac{1}{\varepsilon}B_{21} \end{bmatrix}, \ B_2(\varepsilon) = \begin{bmatrix} B_{12} \\ \frac{1}{\varepsilon}B_{22} \end{bmatrix} \tag{2.9}$$

$$C_1 = [C_{11} \ \ C_{12}], \ C_2 = [C_{21} \ \ C_{22}] \tag{2.10}$$

$$\Delta A(t, \varepsilon) = \begin{bmatrix} \Delta A_{11}(t) & 0 \\ \frac{1}{\varepsilon}\Delta A_{21}(t) & 0 \end{bmatrix} \tag{2.11}$$

$$\Delta B_2(t, \varepsilon) = \begin{bmatrix} \Delta B_{12}(t) \\ \frac{1}{\varepsilon}\Delta B_{22}(t) \end{bmatrix} \tag{2.12}$$

$$\Delta C_2(t) = [\Delta C_{21}(t) \ \ 0] \tag{2.13}$$

Then (2.1)-(2.4) can be rewritten as

$$(\Sigma_1): \quad \dot{x}(t) = [A(\varepsilon) + \Delta A(t,\varepsilon)]x(t) + B_1(\varepsilon)w(t)$$
$$+[B_2(\varepsilon) + \Delta B_2(t,\varepsilon)]u(t), \ x(0) = 0 \quad (2.14)$$
$$z(t) = C_1 x(t) + D_{12}u(t) \quad (2.15)$$
$$y(t) = [C_2 + \Delta C_2(t)]x(t) + D_{21}w(t) \quad (2.16)$$

Notice that in (2.14)-(2.16), the parameter uncertainties $\Delta A(t,\varepsilon)$, $\Delta B_2(t,\varepsilon)$ and $\Delta C_2(t)$ have the same structure as that in (2.5)-(2.6). More precisely,

$$\Delta A(t,\varepsilon) = \begin{bmatrix} \Delta A_{11}(t) & 0 \\ \frac{1}{\varepsilon}\Delta A_{21}(t) & 0 \end{bmatrix} = \begin{bmatrix} H_1 \\ \frac{1}{\varepsilon}H_2 \end{bmatrix} F(t)[E_1 \ \ 0] \quad (2.17)$$

$$\Delta B_2(t,\varepsilon) = \begin{bmatrix} \Delta B_{12}(t) \\ \frac{1}{\varepsilon}\Delta B_{22}(t) \end{bmatrix} = \begin{bmatrix} H_1 \\ \frac{1}{\varepsilon}H_2 \end{bmatrix} F(t)E_2 \quad (2.18)$$

$$\Delta C_2(t) = [\Delta C_{21}(t) \ \ 0] = H_3 F(t)[E_1 \ \ 0] \quad (2.19)$$

We assume that the system (2.14)-(2.16) is *stable*. To be more precise, the following definition is introduced.

Consider the system

$$(\Sigma_0): \ \dot{x}(t) = A(\varepsilon)x(t) + B(\varepsilon)w(t), \ x(0) = 0 \quad (2.20)$$
$$z(t) = C(\varepsilon)x(t) + D(\varepsilon)w(t) \quad (2.21)$$

Denote the transfer function from w to z by

$$G(s,\varepsilon) = C(\varepsilon)(sI - A(\varepsilon))^{-1}B(\varepsilon) + D(\varepsilon).$$

where $A(\varepsilon), B(\varepsilon), C(\varepsilon)$ and $D(\varepsilon)$ are constant matrices for any fixed $\varepsilon > 0$.

Definition 2.1 *Given a scalar $\gamma > 0$, system (Σ_0) is said to be stable with H_∞ disturbance attenuation γ if the matrix $A(\varepsilon)$ is stable and $||G(s,\varepsilon)||_\infty < \gamma$.*

In the above, $||G(s,\varepsilon)||_\infty$ stands for the infinity norm of $G(s,\varepsilon)$ which is also given by

$$||G(s,\varepsilon)||_\infty = \sup_{0 \neq w \in L_2[0,\infty)} \frac{||z||_2}{||w||_2}$$

We shall make the following assumptions for the system (2.14)-(2.16), which are typical in H_∞ control problem.

Assumption 2.1

(A.1) $(A(\varepsilon) \ \ B_2(\varepsilon))$ *is stabilizable and*
$\begin{bmatrix} A(\varepsilon) - j\omega I & B_2(\varepsilon) \\ C_1 & D_{12} \end{bmatrix}$ *has full row rank,*
$\forall \varepsilon > 0, \ \omega \geq 0;$

(A.2) $(C_2 \ \ A(\varepsilon))$ *is detectable and*
$\begin{bmatrix} A(\varepsilon) - j\omega I & B_1(\varepsilon) \\ C_2 & D_{21} \end{bmatrix}$ *has full column rank,*
$\forall \varepsilon > 0, \ \omega \geq 0;$

(A.3) $\begin{bmatrix} D_{12} \\ E_2 \end{bmatrix}$ *is of full column rank;*

(A.4) $[D_{21} \ \ H_3]$ *is of full row rank.*

Remark 2.3 *It should be noted that in Assumption 2.1, (A.1) and (A.2) is sufficient for the existence of a stabilizing controller for the nominal system of (Σ_1); while (A.3) and (A.4) means that the robust control problem is "nonsingular". Observe that if the parameter uncertainties in the control and output matrices disappear, i.e. $E_2 = 0$ and $H_3 = 0$, (A.3) and (A.4) in Assumption 2.1 will reduce to $D_{12}^T D_{12} > 0$ and $D_{21}D_{21}^T > 0$, which are standard assumptions in the nonsingular H_∞ control problem for the nominal system of (Σ_1) [15, 27].* □

It is well known that when the parameter uncertainty in (2.14)-(2.16) does not exist, i.e. $\Delta A(t,\varepsilon) \equiv 0, \Delta B_2(t,\varepsilon) \equiv 0$ and $\Delta C_2(t) \equiv 0$, we have the standard H_∞ output feedback control result.

Theorem 2.1 *([4]) Consider the system (2.14)-(2.16) (with $\Delta A(t,\varepsilon) \equiv 0$, $\Delta B_2(t,\varepsilon) \equiv 0$ and $\Delta C_2(t) \equiv 0$) satisfying Assumption 2.1 and let $\gamma > 0$ and $\varepsilon > 0$ be given. Then this system is stabilizable with disturbance attenuation γ via linear dynamic output feedback if and only if the following conditions are satisfied:*

(a) There exists a solution $X(\varepsilon) \geq 0$ to the following algebraic Riccati equation(ARE):

$$A^T(\varepsilon)X(\varepsilon) + X(\varepsilon)A(\varepsilon) + \frac{1}{\gamma^2}X(\varepsilon)B_1(\varepsilon)B_1^T(\varepsilon)X(\varepsilon)$$
$$-(X(\varepsilon)B_2(\varepsilon) + C_1^T D_{12})(D_{12}^T D_{12})^{-1}$$
$$\cdot(B_2^T(\varepsilon)X(\varepsilon) + D_{12}^T C_1) + C_1^T C_1 = 0 \quad (2.22)$$

and $A(\varepsilon) + B_1(\varepsilon)F_1(\varepsilon) + B_2(\varepsilon)F_2(\varepsilon)$ is stable[2], where

$$F_1(\varepsilon) = \gamma^{-2}B_1^T(\varepsilon)X(\varepsilon)$$
$$F_2(\varepsilon) = -(D_{12}^T D_{12})^{-1}(B_2^T(\varepsilon)X(\varepsilon) + D_{12}^T C_1)$$

(b) There exists a solution $Y(\varepsilon) \geq 0$ to the following algebraic Riccati equation(ARE):

$$A(\varepsilon)Y(\varepsilon) + Y(\varepsilon)A^T(\varepsilon) + \frac{1}{\gamma^2}Y(\varepsilon)C_1^T C_1 Y(\varepsilon)$$
$$-(Y(\varepsilon)C_2^T + B_1(\varepsilon)D_{21}^T)(D_{21}D_{21}^T)^{-1} \quad (2.23)$$
$$\cdot(C_2 Y(\varepsilon) + D_{21}B_1^T(\varepsilon)) + B_1(\varepsilon)B_1^T(\varepsilon) = 0 \quad (2.24)$$

[2] $X(\varepsilon)$ is called as a stabilizing solution of (2.22).

and $A(\varepsilon) + G_1(\varepsilon)C_1 + G_2(\varepsilon)C_2$ *is stable*[3], *where*

$$G_1(\varepsilon) = \gamma^{-2}Y(\varepsilon)C_1^T$$
$$G_2(\varepsilon) = -(Y(\varepsilon)C_2^T + B_1(\varepsilon)D_{21}^T)(D_{21}D_{21}^T)^{-1}$$

(c) $I - \gamma^{-2}Y(\varepsilon)X(\varepsilon) > 0$.

Moreover, if conditions (a)-(c) are satisfied, then, a suitable stabilizing dynamic output feedback controller is given by

$$\dot{\theta}(t) = A_0(\varepsilon)\theta(t) + L(\varepsilon)y(t) \quad (2.25)$$
$$u(t) = K(\varepsilon)\theta(t) \quad (2.26)$$

where

$$\begin{aligned} K(\varepsilon) &= -(D_{12}^T D_{12})^{-1}(B_2^T(\varepsilon)X(\varepsilon) + D_{12}^T C_1) \\ L(\varepsilon) &= (I - \gamma^{-2}Y(\varepsilon)X(\varepsilon))^{-1}(Y(\varepsilon)C_2^T \\ &\quad + B_1(\varepsilon)D_{21}^T)(D_{21}D_{21}^T)^{-1} \\ A_0(\varepsilon) &= A(\varepsilon) + B_2(\varepsilon)K(\varepsilon) + \gamma^{-2}B_1(\varepsilon)B_1^T(\varepsilon)X(\varepsilon) \\ &\quad - L(\varepsilon)(C_2 + \gamma^{-2}D_{21}B_1^T(\varepsilon)X(\varepsilon)). \end{aligned}$$

In this paper, we are concerned with the design of a robust feedback control law $u = \mathcal{G}y$ for system (2.14)-(2.16). We will investigate the design of a linear strictly proper dynamic output feedback controller ($\mathcal{G}$) for (2.14)-(2.16) that reduces z uniformly for any w in the sense that given a scalar $\gamma > 0$ and for any $\varepsilon \in (0, \varepsilon_0]$, the closed-loop system of (2.14)-(2.16) with the controller ($\mathcal{G}$) is robustly stable and satisfies

$$\|z\|_2 < \gamma\|w\|_2 \quad (2.27)$$

for any non-zero $w \in L_2[0, \infty)$ and for all admissible uncertainties. In this situation, the closed-loop system of (2.14)-(2.16) with ($\mathcal{G}$) is said to have a *robust disturbance attenuation* γ *over the horizon* $[0, \infty)$.

In the above, '*robustly stable*' means that the closed-loop system is globally uniformly asymptotically stable about the origin for all admissible parameter uncertainties.

We also need the following assumption for system (2.1)-(2.4):

Assumption 2.2 A_{22} is invertible.

Note that Assumption 2.2 is a standard requirement for singularly perturbed linear control system, see, e.g. [2, 23, 24].

Associated with system (2.1)-(2.4), the reduced model (slow dynamics) is obtained as follows [23, 24].

[3] $Y(\varepsilon)$ is called as a stabilizing solution of (2.24).

Letting $\varepsilon = 0$ and solving $x_2(t)$ in (2.2) (to be denoted by $\bar{x}_2(t)$), we have

$$\begin{aligned} \bar{x}_2(t) = -A_{22}^{-1}\,[(A_{21} + \Delta A_{21}(t))x_s(t) + B_{21}w_s(t) \\ +(B_{22} + \Delta B_{22}(t))u_s(t)] \end{aligned} \quad (2.28)$$

Now, substituting (2.28) into (2.1) and (2.3)-(2.4), we get the slow subsystem of (2.1)-(2.4)

$$\begin{aligned} \dot{x}_s(t) &= [\tilde{A} + \Delta\tilde{A}(t)]x_s(t) + \tilde{B}_1 w_s(t) \\ &\quad + [\tilde{B}_2 + \Delta\tilde{B}_2(t)]u_s(t), \; x_s(0) = 0 \quad (2.29) \\ z_s(t) &= [\tilde{C}_1 + \Delta\tilde{C}_1(t)]x_s(t) + \tilde{D}_{11}w_s(t) \\ &\quad + [\tilde{D}_{12} + \Delta\tilde{D}_{12}(t)]u_s(t) \quad (2.30) \\ y_s(t) &= [\tilde{C}_2 + \Delta\tilde{C}_2(t)]x_s(t) + \tilde{D}_{21}w_s(t) \\ &\quad + [\tilde{D}_{22} + \Delta\tilde{D}_{22}(t)]u_s(t) \quad (2.31) \end{aligned}$$

where

$$\begin{aligned} & x_1 =: x_s, \; u =: u_s, \; w =: w_s, \; z =: z_s, \; y =: y_s \\ & \tilde{A} = A_{11} - A_{12}A_{22}^{-1}A_{21} \\ & \tilde{B}_1 = B_{11} - A_{12}A_{22}^{-1}B_{21}, \quad \tilde{B}_2 = B_{12} - A_{12}A_{22}^{-1}B_{22} \\ & \tilde{C}_1 = C_{11} - C_{12}A_{22}^{-1}A_{21}, \quad \tilde{C}_2 = C_{21} - C_{22}A_{22}^{-1}A_{21} \\ & \tilde{D}_{11} = -C_{12}A_{22}^{-1}B_{21}, \quad \tilde{D}_{12} = D_{12} - C_{12}A_{22}^{-1}B_{22} \\ & \tilde{D}_{21} = D_{21} - C_{22}A_{22}^{-1}B_{21}, \quad \tilde{D}_{22} = -C_{22}A_{22}^{-1}B_{22} \\ & \Delta\tilde{A}(t) = \Delta A_{11}(t) - A_{12}A_{22}^{-1}\Delta A_{21}(t) \\ & \qquad = (H_1 - A_{12}A_{22}^{-1}H_2)F(t)E_1 \\ & \Delta\tilde{B}_2(t) = \Delta B_{12}(t) - A_{12}A_{22}^{-1}\Delta B_{22}(t) \\ & \qquad = (H_1 - A_{12}A_{22}^{-1}H_2)F(t)E_2 \\ & \Delta\tilde{C}_1(t) = -C_{12}A_{22}^{-1}\Delta A_{21}(t) \\ & \qquad = -C_{12}A_{22}^{-1}H_2F(t)E_1 \\ & \Delta\tilde{C}_2(t) = \Delta C_{21}(t) - C_{22}A_{22}^{-1}\Delta A_{21}(t) \\ & \qquad = (H_3 - C_{22}A_{22}^{-1}H_2)F(t)E_1 \\ & \Delta\tilde{D}_{12}(t) = -C_{12}A_{22}^{-1}\Delta B_{22}(t) \\ & \qquad = -C_{12}A_{22}^{-1}H_2F(t)E_2 \\ & \Delta\tilde{D}_{22}(t) = -C_{22}A_{22}^{-1}\Delta B_{22}(t) \\ & \qquad = -C_{22}A_{22}^{-1}H_2F(t)E_2 \end{aligned}$$

For system (2.29)-(2.31), we introduce the complementary structure assumption:

Assumption 2.3

(a) $\begin{bmatrix} \tilde{D}_{12} \\ E_2 \end{bmatrix}$ *is of full column rank;*

(b) $[\tilde{D}_{21} \;\; H_3 - C_{22}A_{22}^{-1}H_2]$ *is of full row rank.*

Note that by Assumption 2.1, Assumption 2.3 is automatically satisfied if $D_{12}^T C_{12} = 0$ and $B_{21}D_{21}^T = 0$.

When there is no parameter uncertainty in (2.29)-(2.31), i.e. $\Delta\tilde{A}(t) \equiv 0$, $\Delta\tilde{B}_2(t) \equiv 0$, $\Delta\tilde{C}_1(t) \equiv 0$, $\Delta\tilde{C}_2(t) \equiv 0$, $\Delta\tilde{D}_{12}(t) \equiv 0$ and $\Delta\tilde{D}_{22}(t) \equiv 0$, by the result of [28], the algebraic Riccati equations associated with system (2.29)-(2.31) and γ are as follows:

$$\begin{aligned}\tilde{A}^T X_s + X_s\tilde{A} - (X_s\tilde{B} + \tilde{C}_1^T\tilde{D}_1)\tilde{R}^{-1}(\tilde{B}^T X_s + \tilde{D}_1^T\tilde{C}_1)\\ +\tilde{C}_1^T\tilde{C}_1 = 0 \qquad (2.32)\end{aligned}$$

$$\begin{aligned}Y_s\tilde{A}^T + \tilde{A}Y_s - (Y_s\tilde{C}^T + \tilde{B}_1\tilde{D}_2^T)\tilde{S}^{-1}(\tilde{C}Y_s + \tilde{D}_2\tilde{B}_1^T)\\ +\tilde{B}_1\tilde{B}_1^T = 0 \qquad (2.33)\end{aligned}$$

where

$$\tilde{B} = [\tilde{B}_1 \;\; \tilde{B}_2],\; \tilde{C} = \begin{bmatrix}\tilde{C}_1\\ \tilde{C}_2\end{bmatrix},\; \tilde{D}_1 = [\tilde{D}_{11} \;\; \tilde{D}_{12}]$$

$$\tilde{D}_2 = \begin{bmatrix}\tilde{D}_{11}\\ \tilde{D}_{21}\end{bmatrix},\; \tilde{R} = \tilde{D}_1^T\tilde{D}_1 - \begin{bmatrix}\gamma^2 I_p & 0\\ 0 & 0\end{bmatrix}$$

$$\tilde{S} = \tilde{D}_2\tilde{D}_2^T - \begin{bmatrix}\gamma^2 I_r & 0\\ 0 & 0\end{bmatrix}$$

and p and r are the dimensions of $w(t)$ and $z(t)$, respectively.

In order to ensure equations (2.32)-(2.33) are well defined, we introduce the following assumption:

Assumption 2.4 $\tilde{R}$ and $\tilde{S}$ are invertible.

Note that Assumption 2.4 will become superfluous if $C_{12} = 0$.

In connection with the fast dynamics associated with (2.1)-(2.4), letting

$$\tau =: \frac{t}{\varepsilon},\;\; x_f =: x_2 - \bar{x}_2,\;\; w_f =: w - w_s$$
$$u_f =: u - u_s,\; z_f =: z - z_s,\; y_f = y - y_s$$

where $\bar{x}_2$ is as in (2.28), we may define the fast subsystem of (2.1)-(2.4) (see, i.e. [2, 21, 24])

$$\begin{aligned}\dot{x}_f(\tau) &= A_{22}x_f(\tau) + B_{21}w_f(\tau)\\ &\quad +[B_{22} + \Delta B_{22}(\tau)]u_f(\tau),\; x_f(0) = 0 &(2.34)\\ z_f(\tau) &= C_{12}x_f(\tau) + D_{12}u_f(\tau) &(2.35)\\ y_f(\tau) &= C_{22}x_f(\tau) + D_{21}w_f(\tau) &(2.36)\end{aligned}$$

If the parameter uncertainty does not exist in (2.34)-(2.36), i.e. $\Delta B_{22}(\tau) \equiv 0$, similar to (2.32)-(2.33), we have the following algebraic equations associated with system (2.34)-(2.36) and γ (see e.g. [4]):

$$\begin{aligned}A_{22}^T X_f + X_f A_{22} - (X_f\hat{B}_2 + C_{12}^T\hat{D}_1)R^{-1}\\ \cdot(\hat{B}_2^T X_f + \hat{D}_1^T C_{12}) + C_{12}^T C_{12} = 0 \qquad (2.37)\end{aligned}$$

$$\begin{aligned}A_{22}Y_f + Y_f A_{22}^T - (Y_f\hat{C}_2^T + B_{21}\hat{D}_2^T)S^{-1}\\ \cdot(\hat{C}_2 Y_f + \hat{D}_2 B_{21}^T) + B_{21}B_{21}^T = 0 \qquad (2.38)\end{aligned}$$

where

$$\hat{B}_2 = [B_{21} \;\; B_{22}],\; \hat{C}_2 = \begin{bmatrix}C_{12}\\ C_{22}\end{bmatrix}$$

$$\hat{D}_1 = [0 \;\; D_{12}],\; \hat{D}_2 = \begin{bmatrix}0\\ D_{21}\end{bmatrix}$$

$$R = \begin{bmatrix}-\gamma^2 I_p & 0\\ 0 & D_{12}^T D_{12}\end{bmatrix}$$

$$S = \begin{bmatrix}-\gamma^2 I_r & 0\\ 0 & D_{21}D_{21}^T\end{bmatrix}.$$

It should be remarked that the solutions $X(\varepsilon)$ and $Y(\varepsilon)$ of (2.22) and (2.24), respectively, are functions of the singular perturbation ε. But, the solutions X_s, Y_s, X_f and Y_f of (2.32), (2.33), (2.37) and (2.38), respectively, are independent of the singular perturbation ε. The following theorem will establish the relations between $X(\varepsilon)$ and X_s with X_f.

Theorem 2.2 *([5]) Assume that Assumptions 2.1-2.4 are satisfied. If the equations (2.32) and (2.37) have stabilizing solutions $X_s \geq 0$ and $X_f \geq 0$, respectively, then there exists $\hat{\varepsilon} > 0$ such that, for any $\varepsilon \in (0,\hat{\varepsilon})$, equation (2.22) has a stabilizing solution $X(\varepsilon) \geq 0$. Moreover,*

$$X(\varepsilon) = \begin{bmatrix}X_{11}(\varepsilon) & X_{12}(\varepsilon)\\ X_{12}^T(\varepsilon) & X_{22}(\varepsilon)\end{bmatrix}$$

has the asymptotic structure

$$X_{11}(\varepsilon) = X_s + \varepsilon\hat{X}_{11}(\varepsilon),\quad X_{12}(\varepsilon) = \varepsilon\tilde{X}_{12} + \varepsilon^2\hat{X}_{12}(\varepsilon)$$
$$X_{22}(\varepsilon) = \varepsilon X_f + \varepsilon^2\hat{X}_{22}(\varepsilon)$$
$$\tilde{X}_{12} = -\left(A_{21}^T X_f + X_s A_{12} + C_{11}^T C_{12} - \hat{F}_1^T R\hat{F}_2\right)A_{22}^{-1}$$
$$\hat{F}_1 = \left(I_{p+q} + \hat{F}_2 A_{22}^{-1}\hat{B}_2\right)\tilde{F} + \hat{F}_2 A_{22}^{-1}A_{21}$$
$$\tilde{F} = -\tilde{R}^{-1}(\tilde{B}^T X_s + \tilde{D}_1^T\tilde{C}_1)$$
$$\hat{F}_2 = -R^{-1}(\hat{B}_2^T X_f + \hat{D}_1^T C_{12})$$

$$\|\hat{X}_{ij}(\varepsilon)\| \leq c < \infty,\; \forall\varepsilon \in (0,\hat{\varepsilon}),\; i,j = 1,2,$$

where q is the dimension of $u(t)$.

Furthermore, the feedback gain is given as follows:

$$F(\varepsilon) = \begin{bmatrix}F_1(\varepsilon)\\ F_2(\varepsilon)\end{bmatrix}$$

with

$$F_1(\varepsilon) = \frac{1}{\gamma^2}B_1^T(\varepsilon)X(\varepsilon)$$

$$F_2(\varepsilon) = -(D_{12}^T D_{12})^{-1}(B_2^T(\varepsilon)X(\varepsilon) + D_{12}^T C_1)$$

having the asymptotic structure

$$F(\varepsilon) = [\hat{F}_1 \;\; \hat{F}_2] + \varepsilon\bar{F}(\varepsilon)$$
$$\|\bar{F}(\varepsilon)\| \leq c < \infty,\; \forall\varepsilon \in (0,\hat{\varepsilon}).$$

The relations between $Y(\varepsilon)$ and Y_s with Y_f are given in the next theorem.

Theorem 2.3 *([5]) Assume that Assumptions 2.1-2.4 are satisfied. If the equations (2.33) and (2.38) have stabilizing solutions $Y_s \geq 0$ and $Y_f \geq 0$, respectively, then there exists $\hat{\varepsilon} > 0$ such that, for any $\varepsilon \in (0, \hat{\varepsilon})$, equation (2.24) has a stabilizing solution $Y(\varepsilon) \geq 0$. Moreover,*

$$Y(\varepsilon) = \begin{bmatrix} Y_{11}(\varepsilon) & Y_{12}(\varepsilon) \\ Y_{12}^T(\varepsilon) & Y_{22}(\varepsilon) \end{bmatrix}$$

has the asymptotic structure

$$\begin{aligned} Y_{11}(\varepsilon) &= Y_s + \varepsilon \hat{Y}_{11}(\varepsilon), \quad Y_{12}(\varepsilon) = \tilde{Y}_{12} + \varepsilon \hat{Y}_{12}(\varepsilon) \\ Y_{22}(\varepsilon) &= \frac{1}{\varepsilon}(Y_f + \varepsilon \hat{Y}_{22}(\varepsilon)) \\ \tilde{Y}_{12} &= -\left(A_{12}Y_f + Y_s A_{21}^T + B_{11}B_{21}^T - \hat{K}_1 S \hat{K}_2^T\right) A_{22}^{-T} \\ \hat{K}_1 &= \tilde{K}\left(I_{m+r} + \hat{C}_2 A_{22}^{-1}\hat{K}_2\right) + A_{12}A_{22}^{-1}\hat{K}_2 \\ \tilde{K} &= -(Y_s\tilde{C}^T + \tilde{B}_1\tilde{D}_2^T)\tilde{S}^{-1} \\ \hat{K}_2 &= -(Y_f\hat{C}_2^T + B_{21}\hat{D}_2^T)S^{-1} \\ &\|\hat{Y}_{ij}(\varepsilon)\| \leq c < \infty, \ \forall \varepsilon \in (0, \hat{\varepsilon}), \ i, j = 1, 2, \end{aligned}$$

where m is the dimension of $y(t)$.

Furthermore,

$$K(\varepsilon) = \begin{bmatrix} K_1(\varepsilon) \\ K_2(\varepsilon) \end{bmatrix}$$

with

$$\begin{aligned} K_1(\varepsilon) &= \frac{1}{\gamma^2} Y(\varepsilon) C_1^T \\ K_2(\varepsilon) &= -(Y(\varepsilon)C_2^T + B_1(\varepsilon)D_{21}^T)(D_{21}D_{21}^T)^{-1} \end{aligned}$$

having the asymptotic structure

$$\begin{aligned} K(\varepsilon) &= [\hat{K}_1 \ \ \hat{K}_2] + \varepsilon \bar{K}(\varepsilon), \\ \|\bar{K}(\varepsilon)\| &\leq c < \infty, \ \forall \varepsilon \in (0, \hat{\varepsilon}). \end{aligned}$$

Remark 2.4 *The results of Theorems 2.2 and 2.3 have shown that the existence of solutions $X(\varepsilon)$ and $Y(\varepsilon)$ to (2.22) and (2.24), for $\varepsilon > 0$ small enough, can be guaranteed by the existence of solutions X_s, Y_s, X_f and Y_f to (2.32), (2.33), (2.37) and (2.38), respectively. Note that the latter solutions are independent of singular parameter ε.*

3 Main Results

In this section, we provide a solution to the robust H_∞ control problem for the system (2.1)-(2.4). A controller is designed to achieve a robust H_∞ performance in the presence of both singular perturbation and parameter uncertainty.

Before proceeding to formulate our results, let us recall the following lemma which will be used in the proof of our main results.

Lemma 3.1 *([16]) Consider the system*

$$\begin{aligned} \dot{x}(t) &= [A + \Delta A(t)]x(t) + [B_w + \Delta B_w(t)]w(t) \\ &\quad + [B_u + \Delta B_u(t)]u(t) \\ z(t) &= [C_z + \Delta C_z(t)]x(t) + [D_{zw} + \Delta D_{zw}(t)]w(t) \\ &\quad + [D_{zu} + \Delta D_{zu}(t)]u(t) \\ y(t) &= [C_y + \Delta C_y(t)]x(t) + [D_{yw} + \Delta D_{yw}(t)]w(t) \\ &\quad + [D_{yu} + \Delta D_{yu}(t)]u(t) \end{aligned}$$

where $u(t)$ is the control input, $w(t)$ is the disturbance input, $y(t)$ is the measured output and $z(t)$ is the controlled output, with uncertainties

$$\begin{bmatrix} \Delta A(t) & \Delta B_w(t) & \Delta B_u(t) \\ \Delta C_z(t) & \Delta D_{zw}(t) & \Delta D_{zu}(t) \\ \Delta C_y(t) & \Delta D_{yw}(t) & \Delta D_{yu}(t) \end{bmatrix} = \begin{bmatrix} H_x \\ H_z \\ H_y \end{bmatrix} F(t)[E_x \ \ E_w \ \ E_u], \quad F^T(t)F(t) \leq I.$$

This system is stabilizable with disturbance attenuation $\gamma > 0$ by a strictly proper linear output feedback control if and only if there exists a $\lambda > 0$ such that the uncertainty-free system

$$\begin{aligned} \dot{x}(t) &= Ax(t) + [B_w \ \ \gamma\lambda H_x] \begin{bmatrix} w(t) \\ \hat{w}(t) \end{bmatrix} + B_u u(t) \\ \begin{bmatrix} z(t) \\ \hat{z}(t) \end{bmatrix} &= \begin{bmatrix} C_z \\ \frac{1}{\lambda}E_x \end{bmatrix} x(t) \\ &\quad + \begin{bmatrix} D_{zw} & \gamma\lambda H_z \\ \frac{1}{\lambda}E_w & 0 \end{bmatrix} \begin{bmatrix} w(t) \\ \hat{w}(t) \end{bmatrix} + \begin{bmatrix} D_{zu} \\ \frac{1}{\lambda}E_u \end{bmatrix} u(t) \\ y(t) &= C_y(t) + [D_{yw} \ \ \gamma\lambda H_y] \begin{bmatrix} w(t) \\ \hat{w}(t) \end{bmatrix} + D_{yu}u(t) \end{aligned}$$

with $u(t)$ the control input, $[w^T(t) \ \ \hat{w}^T(t)]^T$ the disturbance input, $y(t)$ the measured output and $[z^T(t) \ \ \hat{z}^T(t)]^T$ the controlled output, is stabilizable with disturbance attenuation γ via an output feedback control. □

Remark 3.1 *It has been established in Lemma 3.1 that the problem of output feedback H_∞ control of uncertainty systems can be converted to a standard H_∞ control problem for systems without parameter uncertainty. Note that the latter can be solved by the results in [4, 28].* □

We now present a solution to the problem of robust H_∞ control of system (2.14)-(2.16). To this end, associated with system (2.14)-(2.16), we introduce the following auxiliary system:

$$\dot{x}(t) = A(\varepsilon)x(t) + [B_1(\varepsilon) \quad \gamma\lambda\bar{H}_1(\varepsilon)] \begin{bmatrix} w(t) \\ \hat{w}(t) \end{bmatrix} + B_2(\varepsilon)u(t), \; x(0) = 0 \tag{3.1}$$

$$\begin{bmatrix} z(t) \\ \hat{z}(t) \end{bmatrix} = \begin{bmatrix} C_1 \\ \frac{1}{\lambda}\bar{E}_1 \end{bmatrix} x(t) + \begin{bmatrix} D_{12} \\ \frac{1}{\lambda}\bar{E}_2 \end{bmatrix} u(t) \tag{3.2}$$

$$y(t) = C_2 x(t) + [D_{21} \quad \gamma\lambda\bar{H}_2] \begin{bmatrix} w(t) \\ \hat{w}(t) \end{bmatrix} \tag{3.3}$$

where $x(t) = [x_1^T(t) \; x_2^T(t)]^T$ is the state, $[w^T(t) \; \hat{w}^T(t)]^T$ is the exogenous input, $y(t)$ is the measured output, $[z^T(t) \; \hat{z}^T(t)]^T$ is the controlled output, $\gamma > 0$ is the desired H_∞ performance for system (2.14)-(2.16) we wish to achieve, $A(\varepsilon), B_1(\varepsilon), B_2(\varepsilon), C_1, C_2, D_{12}$ and D_{21} are the same as in (2.8)-(2.10) and (2.3)-(2.4), respectively, and $\bar{H}_1(\varepsilon), \bar{H}_2, \bar{E}_1$ and $\bar{E}_2$ are defined as follows:

$$\bar{H}_1(\varepsilon) = \begin{bmatrix} H_1 \\ \frac{1}{\varepsilon}H_2 \end{bmatrix}, \quad \bar{H}_2 = H_3$$
$$\bar{E}_1 = [E_1 \quad 0], \quad \bar{E}_2 = E_2$$

with H_1, H_2, H_3, E_1 and E_2 as given in (2.5)-(2.6), $\varepsilon > 0$ is a small parameter and $\lambda > 0$ is a scaling parameter.

Define

$$\bar{B}_1(\varepsilon) = [B_1(\varepsilon) \quad \gamma\lambda\bar{H}_1(\varepsilon)], \; \bar{C}_1 = \begin{bmatrix} C_1 \\ \frac{1}{\lambda}\bar{E}_1 \end{bmatrix}$$
$$\bar{D}_{12} = \begin{bmatrix} D_{12} \\ \frac{1}{\lambda}\bar{E}_2 \end{bmatrix}, \; \bar{D}_{21} = [D_{21} \quad \gamma\lambda\bar{H}_2] \tag{3.4}$$

Then, the game-theoretic Riccati equations associated with (3.1)-(3.3) and the disturbance attenuation level γ (see, for example [4, 28]) are

$$A^T(\varepsilon)X(\varepsilon) + X(\varepsilon)A(\varepsilon) + \frac{1}{\gamma^2}X(\varepsilon)\bar{B}_1(\varepsilon)\bar{B}_1^T(\varepsilon)X(\varepsilon) - (X(\varepsilon)B_2(\varepsilon) + \bar{C}_1^T\bar{D}_{12})(\bar{D}_{12}^T\bar{D}_{12})^{-1} \cdot (B_2^T(\varepsilon)X(\varepsilon) + \bar{D}_{12}^T\bar{C}_1) + \bar{C}_1^T\bar{C}_1 = 0 \tag{3.5}$$

$$A(\varepsilon)Y(\varepsilon) + Y(\varepsilon)A^T(\varepsilon) + \frac{1}{\gamma^2}Y(\varepsilon)\bar{C}_1^T\bar{C}_1Y(\varepsilon) - (Y(\varepsilon)C_2^T + \bar{B}_1(\varepsilon)\bar{D}_{21}^T)(\bar{D}_{21}\bar{D}_{21}^T)^{-1} \cdot (C_2Y(\varepsilon) + \bar{D}_{21}\bar{B}_1^T(\varepsilon)) + \bar{B}_1(\varepsilon)\bar{B}_1^T(\varepsilon) = 0 \tag{3.6}$$

Similar to (2.29)-(2.31), the reduced model (slow dynamics) associated with (3.1)-(3.3) is given as follows:

$$\dot{x}_s(t) = A^s x_s(t) + B_1^s w_s(t) + B_2^s u_s(t), \; x_s(0) = 0 \tag{3.7}$$

$$z_s(t) = C_1^s x_s(t) + D_{11}^s w_s(t) + D_{12}^s u_s(t) \tag{3.8}$$

$$y_s(t) = C_2^s x_s(t) + D_{21}^s w_s(t) + D_{22}^s u_s(t) \tag{3.9}$$

where

$$x_1 =: x_s, \; u =: u_s, \; [w^T \quad \hat{w}^T]^T =: w_s$$
$$[z^T \quad \hat{z}^T]^T =: z_s, \; y =: y_s$$
$$A^s = A_{11} - A_{12}A_{22}^{-1}A_{21} \tag{3.10}$$
$$B_1^s = \bar{B}_{11} - A_{12}A_{22}^{-1}\bar{B}_{21},$$
$$B_2^s = B_{12} - A_{12}A_{22}^{-1}B_{22} \tag{3.11}$$
$$C_1^s = \begin{bmatrix} C_{11} - C_{12}A_{22}^{-1}A_{21} \\ \frac{1}{\lambda}E_1 \end{bmatrix},$$
$$C_2^s = C_{21} - C_{22}A_{22}^{-1}A_{21} \tag{3.12}$$
$$D_{11}^s = -C_{12}A_{22}^{-1}\bar{B}_{21},$$
$$D_{12}^s = \begin{bmatrix} D_{12} - C_{12}A_{22}^{-1}B_{22} \\ \frac{1}{\lambda}E_2 \end{bmatrix} \tag{3.13}$$
$$D_{21}^s = \bar{D}_{21} - C_{22}A_{22}^{-1}\bar{B}_{21}, \; D_{22}^s = -C_{22}A_{22}^{-1}B_{22} \tag{3.14}$$
$$\bar{B}_{11} = [B_{11} \quad \gamma\lambda H_1], \; \bar{B}_{21} = [B_{21} \quad \gamma\lambda H_2] \tag{3.15}$$

and $\bar{D}_{21}$ is as in (3.4).

The game-theoretic Riccati equations associated with (3.7)-(3.9) and the disturbance attenuation level γ (see, for example [4, 28]) are

$$A^{sT}X_s + X_sA^s - (X_sB^s + C_1^{sT}D_1^s)(R^s)^{-1} \cdot (B^{sT}X_s + D_1^{sT}C_1^s) + C_1^{sT}C_1^s = 0 \tag{3.16}$$

$$Y_sA^{sT} + AY_s - (Y_sC^{sT} + B_1^sD_2^{sT})(S^s)^{-1} \cdot (C^sY_s + D_2^sB_1^{sT}) + B_1^sB_1^{sT} = 0 \tag{3.17}$$

where

$$B^s = [B_1^s \quad B_2^s], \; C^s = \begin{bmatrix} C_1^s \\ C_2^s \end{bmatrix},$$
$$D_1^s = [D_{11}^s \quad D_{12}^s], \; D_2^s = \begin{bmatrix} D_{11}^s \\ D_{21}^s \end{bmatrix} \tag{3.18}$$
$$R^s = D_1^{sT}D_1^s - \begin{bmatrix} \gamma^2 I_{s1} & 0 \\ 0 & 0 \end{bmatrix},$$
$$S^s = D_2^sD_2^{sT} - \begin{bmatrix} \gamma^2 I_{s2} & 0 \\ 0 & 0 \end{bmatrix} \tag{3.19}$$

and $s1$ and $s2$ are the dimensions of $w_s(t)$ and $z_s(t)$, respectively.

Associated with system (3.1)-(3.3), the fast dynamics is as follows:

$$\dot{x}_f(\tau) = A^f x_s(\tau) + B_1^f w_f(\tau) + B_2^f u_f(\tau), \; x_f(0) = 0 \tag{3.20}$$

$$z_f(\tau) = C_1^f x_f(\tau) + D_{12}^f u_f(\tau) \tag{3.21}$$

$$y_f(\tau) = C_2^f x_f(\tau) + D_{21}^f w_f(\tau) \tag{3.22}$$

where

$$\tau =: \frac{t}{\varepsilon}, \; x_f =: x_2 - \bar{x}_2, \; u_f =: u - u_s$$

$$w_f =: [w^T \ \ \hat{w}^T]^T - w_s,$$
$$z_f =: [z^T \ \ \hat{z}^T]^T - z_s, \ y_f =: y - y_s$$
$$A^f = A_{22}, \ B_1^f = \bar{B}_{21},$$
$$B_2^f = B_{22}, \ C_1^f = \begin{bmatrix} C_{12} \\ 0 \end{bmatrix} \tag{3.23}$$
$$C_2^f = C_{22}, \ D_{12}^f = \begin{bmatrix} D_{12} \\ \frac{1}{\lambda}E_2 \end{bmatrix}, \ D_{21}^f = \bar{D}_{21} \tag{3.24}$$

$\bar{x}_2$ is the solution for x_2 in (3.2) setting $\varepsilon = 0$, and $\bar{B}_{21}$ and $\bar{D}_{21}$ are given in (3.15) and (3.4), respectively.

Parallel to (3.16)-(3.17), we have the following game-theoretic Riccati equations associated with (3.20)-(3.22) and the disturbance attenuation level γ

$$A^{f^T} X_f + X_f A^f - (X_f B^f + C_1^{f^T} D_1^f)(R^f)^{-1} \cdot (B^{f^T} X_f + D_1^{f^T} C_1^f) + C_1^{f^T} C_1^f = 0 \tag{3.25}$$
$$A^f Y_f + Y_f A^{f^T} - (Y_f C^{f^T} + B_1^f D_2^{f^T})(S^f)^{-1} \cdot (C^f Y_f + D_2^f B_1^{f^T}) + B_1^f B_1^{f^T} = 0 \tag{3.26}$$

where

$$B^f = [B_1^f \ \ B_2^f], \ C^f = \begin{bmatrix} C_1^f \\ C_2^f \end{bmatrix},$$
$$D_1^f = [0 \ \ D_{12}^f], \ D_2^f = \begin{bmatrix} 0 \\ D_{21}^f \end{bmatrix} \tag{3.27}$$
$$R^f = \begin{bmatrix} -\gamma^2 I_{s1} & 0 \\ 0 & D_{12}^{f^T} D_{12}^f \end{bmatrix},$$
$$S^f = \begin{bmatrix} -\gamma^2 I_{s2} & 0 \\ 0 & D_{21}^f D_{21}^{f^T} \end{bmatrix}. \tag{3.28}$$

Now we are in a position to establish our main result in this paper.

Theorem 3.1 *Consider the system (2.14)-(2.16) satisfying Assumptions 2.1-2.4 and let $\mathcal{G}$ denote a controller of the form (2.25)-(2.26). Then, given a scalar $\gamma > 0$, the closed-loop system of (2.14)-(2.16) with $\mathcal{G}$ is robustly stable and has a robust disturbance attenuation γ over the horizon $[0, \infty)$ for some $\varepsilon_0 > 0$, $\varepsilon \in (0, \varepsilon_0)$ if there exists a positive scalar λ such that the Riccati equations (3.16)-(3.17) and (3.25)-(3.26) have solutions $X_s \geq 0, Y_s \geq 0$ and $X_f \geq 0, Y_f \geq 0$, respectively, with $\rho(X_s Y_s) < \gamma^2$ and $\rho(X_f Y_f) < \gamma^2$.*

Proof. By Theorem 2.2, the existence of nonnegative solutions X_s and X_f to (3.16) and (3.25) implies that there exists an $\hat{\varepsilon}_1 > 0$ such that for all $\varepsilon \in (0, \hat{\varepsilon}_1)$, the Riccati equation (3.5) has a stabilizing solution $X(\varepsilon) \geq 0$. Similarly, using Theorem 2.3, the existence of nonnegative solutions Y_s and Y_f to (3.17) and (3.26) implies that there exists an $\hat{\varepsilon}_2 > 0$ such that for all $\varepsilon \in (0, \hat{\varepsilon}_2)$, the Riccati equation (3.6) has a stabilizing solution $Y(\varepsilon) \geq 0$.

From Corollary 2.3 in [5], the product $X(\varepsilon)Y(\varepsilon)$ has the asymptotic structure

$$X(\varepsilon)Y(\varepsilon) = \begin{bmatrix} X_s Y_s & X_s Y_{12}^* + X_{12}^* Y_f \\ 0 & X_f Y_f \end{bmatrix} + \varepsilon Z(\varepsilon),$$
$$\|Z(\varepsilon)\| \leq c < \infty, \ \forall \varepsilon \in (0, \varepsilon_0) \tag{3.29}$$

where X_{12}^* and Y_{12}^* will be defined in Corollaries 3.1 and 3.2, respectively, and $\varepsilon_0 = \min\{\hat{\varepsilon}_1, \hat{\varepsilon}_2\}$.

Since $\rho(X_s Y_s) < \gamma^2$ and $\rho(X_f Y_f) < \gamma^2$, then (3.29) implies $\rho(X(\varepsilon)Y(\varepsilon)) < \gamma^2$, $\forall \varepsilon \in (0, \varepsilon_0)$. Hence, by Theorem 2.1, the system (3.1)-(3.3) is stabilizable with disturbance attenuation γ via a linear dynamic output feedback controller.

Finally, the desired result can be obtained by applying Lemma 3.1 to the system (2.14)-(2.16) and the system (3.1)-(3.3). ∇∇∇

Remark 3.2 *It has been shown in Theorem 3.1 that the existence of nonnegative solutions X_s, Y_s and X_f, Y_f to (3.16)-(3.17) and (3.25)-(3.26), respectively, ensures the existence of stabilizing solutions $X(\varepsilon), Y(\varepsilon)$ to (3.5)-(3.6) for $\varepsilon \in (0, \varepsilon_0)$, which, together with the spectral radius conditions of $X_s Y_s$ and $X_f Y_f$, guarantees that the uncertain system (2.14)-(2.16) is robustly stabilizable and has a robust H_∞ performance γ. Note that the equations (3.16)-(3.17) and (3.25)-(3.26) are independent of the singular parameter ε.*

Let us partition the solutions $X(\varepsilon)$ and $Y(\varepsilon)$ of (3.5)-(3.6) obtained in Theorem 3.1 as

$$X(\varepsilon) = \begin{bmatrix} X_{11}(\varepsilon) & X_{12}(\varepsilon) \\ X_{12}^T(\varepsilon) & X_{22}(\varepsilon) \end{bmatrix}$$
$$Y(\varepsilon) = \begin{bmatrix} Y_{11}(\varepsilon) & Y_{12}(\varepsilon) \\ Y_{12}^T(\varepsilon) & Y_{22}(\varepsilon) \end{bmatrix}$$

Then, from Theorems 2.2 and 2.3, we have the following corollaries.

Corollary 3.1 *$X(\varepsilon)$ has the following asymptotic structures:*

$$X_{11}(\varepsilon) = X_s + \varepsilon \bar{X}_{11}(\varepsilon), \ X_{12}(\varepsilon) = \varepsilon X_{12}^* + \varepsilon^2 \bar{X}_{12}(\varepsilon)$$
$$X_{22}(\varepsilon) = \varepsilon X_f + \varepsilon^2 \bar{X}_{22}(\varepsilon)$$
$$X_{12}^* = -\left(A_{21}^T X_f + X_s A_{12} + \bar{C}_1^T C_1^f - \bar{F}_1^T R^f \bar{F}_2\right) A_{22}^{-1}$$
$$\bar{F}_1 = \left(I_{s1+q} + \bar{F}_2 A_{22}^{-1} B^f\right) \bar{F} + \bar{F}_2 A_{22}^{-1} A_{21}$$
$$\bar{F} = -R^{s^{-1}}(B^{s^T} X_s + D_1^{s^T} C_1^s)$$
$$\bar{F}_2 = -R^{f^{-1}}(B^{f^T} X_f + D_1^{f^T} C_1^f)$$
$$\|\bar{X}_{ij}(\varepsilon)\| \leq c < \infty, \ \forall \varepsilon \in (0, \min\{\hat{\varepsilon}_1, \hat{\varepsilon}_2\}), \ i,j = 1,2.$$

Furthermore, the feedback gain is given as follows:

$$F(\varepsilon)=\begin{bmatrix} F_1(\varepsilon) \\ F_2(\varepsilon) \end{bmatrix}$$

with

$$F_1(\varepsilon)=\frac{1}{\gamma^2}\bar{B}_1(\varepsilon)^T(\varepsilon)X(\varepsilon)$$
$$F_2(\varepsilon)=-(\bar{D}_{12}^T\bar{D}_{12})^{-1}(B_2^T(\varepsilon)X(\varepsilon)+\bar{D}_{12}^T\bar{C}_1)$$

having the asymptotic structure

$$F(\varepsilon)=[\bar{F}_1 \quad \bar{F}_2]+\varepsilon F^*(\varepsilon),$$
$$\|F^*(\varepsilon)\|\le c<\infty,\ \forall\varepsilon\in(0,\min\{\hat{\varepsilon}_1,\hat{\varepsilon}_2\}),$$

where $A_{12}, A_{21}, A_{22}, B_2(\varepsilon), \bar{C}_1, \bar{D}_{12}, B^s, C_1^s, D_1^s, R^s, B^f, D_1^f$ *and* R^f *are as in (2.1)-(2.2), (2.14), (3.4), (3.18)-(3.19), (3.11)-(3.13), (3.27)-(3.28) and (3.23)-(3.24), respectively.* □

Remark 3.3 *From the asymptotic structure of* $X(\varepsilon)$*, it can be easily seen that the "leading element" of* $X(\varepsilon)$ *is* $X_{11}(\varepsilon)$*, or* X_s*, which shows that* $X(\varepsilon)$ *can be arbitrarily "approximated" by* X_s *when* $\varepsilon>0$ *small enough.*

Corollary 3.2 $Y(\varepsilon)$ *has the following asymptotic structures:*

$$Y_{11}(\varepsilon)=Y_s+\varepsilon\bar{Y}_{11}(\varepsilon),\quad Y_{12}(\varepsilon)=Y_{12}^*+\varepsilon\bar{Y}_{12}(\varepsilon),$$
$$Y_{22}(\varepsilon)=\frac{1}{\varepsilon}(Y_f+\varepsilon\bar{Y}_{22}(\varepsilon))$$
$$Y_{12}^*=-\left(A_{12}Y_f+Y_sA_{21}^T+\bar{B}_{11}\bar{B}_{21}^T-\bar{K}_1S^f\bar{K}_2^T\right)A_{22}^{-1}$$
$$\bar{K}_1=\bar{K}\left(I_{s2+m}+C^fA_{22}^{-1}\bar{K}_2\right)+A_{12}A_{22}^{-1}\bar{K}_2$$
$$\bar{K}=(Y_sC^{sT}+B_1^sD_2^{sT})S^{s-1}$$
$$\bar{K}_2=-(Y_fC^{fT}+\bar{B}_{21}D_2^{fT})S^{f-1}$$
$$\|\bar{Y}_{ij}(\varepsilon)\|\le c<\infty,\ \forall\varepsilon\in(0,\min\{\hat{\varepsilon}_1,\hat{\varepsilon}_2\}),\ i,j=1,2.$$

Furthermore,

$$K(\varepsilon)=\begin{bmatrix} K_1(\varepsilon) \\ K_2(\varepsilon) \end{bmatrix}$$

with

$$K_1(\varepsilon)=\frac{1}{\gamma^2}Y(\varepsilon)\bar{C}_1^T$$
$$K_2(\varepsilon)=-(Y(\varepsilon)C_2^T+\bar{B}_1(\varepsilon)\bar{D}_{21}^T)(\bar{D}_{21}\bar{D}_{21}^T)^{-1}$$

having the asymptotic structure

$$K(\varepsilon)=[\bar{K}_1 \quad \bar{K}_2]+\varepsilon K^*(\varepsilon)$$
$$\|K^*(\varepsilon)\|\le c<\infty,\ \forall\varepsilon\in(0,\min\{\hat{\varepsilon}_1,\hat{\varepsilon}_2\}),$$

where $A_{12}, A_{21}, A_{22}, C_2, \bar{B}(\varepsilon), \bar{C}_1, \bar{D}_{21}, \bar{B}_{11}, \bar{B}_{21}, B_1^s, C^s, D_2^s, C^f, D_2^f$ *and* S^f *are as in (2.1)-(2.2), (2.16), (3.4), (3.11), (3.15), (3.18)-(3.19) and (3.27)-(3.28), respectively.* □

If the state information is available for feedback, then we have the following result.

Theorem 3.2 *Consider the system (2.14)-(2.16) satisfying Assumptions 2.1-2.4. Then, given a scalar* $\gamma>0$ *and for any* $\varepsilon>0$ *small enough, the closed-loop system of (2.14)-(2.16) with the controller*

$$u(t)=F_{21}^0x_1(t)+F_{22}^0x_2(t)$$

is robustly stable and has a robust disturbance attenuation γ *over the horizon* $[0,\infty)$ *if there exists a positive scalar* λ *such that the Riccati equations (3.16) and (3.25) have solutions* $X_s\ge 0$ *and* $X_f\ge 0$*, respectively, where*

$$\bar{F}_1=\begin{bmatrix} F_{11}^0 \\ F_{21}^0 \end{bmatrix},\quad \bar{F}_2=\begin{bmatrix} F_{12}^0 \\ F_{22}^0 \end{bmatrix},$$

with $\bar{F}_1$ *and* $\bar{F}_2$ *as in Corollary 3.1.*

Proof. It can be established along the same line as that of Theorem 2.5 in [5]. ∇∇∇

Remark 3.4 *It should be noted that the controller,* $u(t)$*, in Theorem 3.2 does not depend upon* ε*, which robustly stabilizes the uncertain system (2.14)-(2.16) and achieves the disturbance attenuation* γ*.* □

4 Conclusion

The problem of robust stability with robust disturbance attenuation for a class of singularly perturbed uncertain linear systems has been studied in this paper. The uncertainties we considered consist of norm-bounded parameter uncertainty in both the state and output equations. We decomposite the whole dynamics into slow and fast dynamics, and show that the solutions to slow and fast dynamics can be used to construct a solution to the problem of robust H_∞ control for the full order singularly perturbed uncertain system. It has been shown that the controller for the above problem may be designed by solving H_∞ syntheses for a related system without parameter uncertainties. Furthermore, the composite controller is independent of the singular perturbation ε, when ε small enough.

Acknowledgment

The authors would like to thank V. Dragan for a number of fruitful discussions and valuable suggestions during this work which helped in improving the quality of the paper.

References

[1] Barmish, B. R., (1985). Necessary and sufficient conditions for quadratic stabilizability of an uncertain system. *J. Optimiz. Theory Appl.*, **46**, 399–408.

[2] Chow, J. H. and P. V. Kokotovic, (1976). A decomposition of near-optimum regulators for systems with slow and fast modes. *IEEE Trans. Automat. Control*, **21**(6), 701–705.

[3] Corless, M., G. Leitmann and E. P. Ryan, (1990). Control of uncertain systems with neglected dynamics. in A.S.I. Zinober(Ed.), *Deterministic Control of Uncertain Systems,* IEE Press, London, 252–268.

[4] Doyle, J. C., K. Glover, P. P. Khargonekar and B. A. Francis, (1989). State space solutions to the standard H^2 and H^∞ control problems. *IEEE Trans. Automat. Control*, **34**(8), 831–847.

[5] Dragan, V., (1993). Asymptotic expansions for game-theoretic Riccati equations and stabilization with disturbance attenuation for singularly perturbed systems. *System & Control Letters*, **20**(6), 455–463.

[6] Fridman, E., (1995). Exact decomposition of linear singularly perturbed H^∞-optimal control problem. *Kybernetika*, **31**(6), 589–597.

[7] Fridman, E., (1996). Near-optimal H^∞ control of linear singularly perturbed systems. *IEEE Trans. Automat. Control*, **41**(2), 236–240.

[8] Gaitsgory, V., (1992). Suboptimization of singularly perturbed control systems. *SIAM J. Contr. & Optimiz.*, **30**(5), 1228–1249.

[9] Gaitsgory, V., (1993). *Control of systems with slow and fast motions.* Nauka, Moscow. (in Russian).

[10] Gaitsgory, V., (1993). Suboptimal control of singularly perturbed systems and periodic optimization. *IEEE Trans. Automat. Control*, **38**(6), 888–903.

[11] Gaitsgory, V., (1996). Limit Hamilton-Jacobi-Isaace equations for singularly perturbed zero-sum differential games. *Int. J. Mathematical Analysis and Applications*, **202**, 862–899.

[12] Gaitsgory, V. and P. Shi, (1996). Limit Hamilton-Jacobi-Isaacs equations for singularly perturbed zero-sum dynamic (discrete time) games. *Proc. of 7th Int. Symp. Dynamic Games & Applications.* , Kanagawa, Japan. to appear.

[13] Garofalo, F. and G. Leitmann, (1988). A composite controller ensuring ultimate boundedness for a class of singularly perturbed uncertain systems. *Dynamics and Stability of Systems*, **3**, 135–145.

[14] Garofalo, F. and G. Leitmann, (1990). Nonlinear composite control of a class of nominally linear singularly perturbed uncertain systems. in A.S.I. Zinober(Ed.), *Deterministic Control of Uncertain Systems,* IEE Press, London, 269–288.

[15] Glover, K. and J. C. Doyle, (1988). State-space formulate for all stabilizing controllers that satisfy an H_∞-norm bound and relations to risk sensitivity. *System & Control Letters*, **11**(3), 162–172.

[16] Gu, K., (1994). H_∞ control of systems under norm bounded uncertainties in all system matrices. *IEEE Trans. Automat. Control*, **39**(6), 1320–1322.

[17] Khalil, H. K. and F. C. Chen, (1992). H_∞ control of two-time-scale systems. *System & Control Letters*, **19**(1), 35–42.

[18] Khargonekar, P. P., I. R. Petersen and K. Zhou, (1990). Robust stabilization of uncertain linear systems : quadratic stabilizability and H_∞ control theory. *IEEE Trans. Automat. Control*, **35**(3), 356–361.

[19] Kokotovic, P. V., R. E. O'Malley and P. Sannuti, (1976). Singular perturbations and order reduction in control theory. *Automatica*, **12**(1), 123–132.

[20] Kokotovic, P. V., (1984). Applications of singular perturbations techniques to control problems. *SIAM Review*, **26**, 501–550.

[21] Kokotovic, P. V., R. E. Khalil and J. O'Reilly, (1986). *Singular perturbation methods in control: analysis and design.* Academic Press, New York.

[22] O'Malley, R. E., (1974). *Introduction to singular perturbations.* Academic Press, New York.

[23] Pan, Z. and T. Basar, (1993). H^∞-optimal control for singularly perturbed systems. Part I: Perfect state measurements. *Automatica*, **29**(2), 401–423.

[24] Pan, Z. and T. Basar, (1994). H^∞-optimal control for singularly perturbed systems. Part II: Imperfect state measurements. *IEEE Trans. Automat. Control*, **39**(2), 280–299.

[25] Petersen, I. R. and C. V. Hollot, (1986). A Riccati equation approach to the stabilization of uncertain linear systems. *Automatica*, **22**(4), 397–411.

[26] Petersen, I. R., (1987). A stabilization algorithm for a class of uncertain linear systems. *System & Control Letters*, **8**(4), 351–357.

[27] Petersen, I. R., B. D. O. Andersen and E. A. Jonckheere, (1991). A first principles solution to the non-singular H_∞ control problem. *Int. J. Robust & Nonlinear Control*, **1**(3), 171–185.

[28] Safonov, M. G., D. J. N. Limebeer and R. Y. Chiang, (1989). Simplifying the H_∞ theory via loop-shifting, matrix-pencil and descriptor concepts. *Int. J. Control*, **50**(6), 2467–2488.

[29] Saksena, V. R., J. O'Reilly and P. V. Kokotovic, (1984). Singular perturbations and time-scale methods in control theory: survey 1976-1983. *Automatica*, **20**(2), 273–293.

[30] Shahruz, S. M., (1989). H_∞-optimal compensators for singularly perturbed systems. *Proc. 28th IEEE Conf. Decision & Control.* , 2397–2398, Tampa, FL.

[31] Shi, P., L. Xie and C. E. de Souza, (1992). Output feedback control of sampled-data systems with parametric uncertainties. *Proc. 31st IEEE Conf. Decision & Control.* , 2876–2881, Tucson, AZ.

[32] Shi, P., (1997). Controller design for uncertain systems under sampled measurements. *IMA J. Mathe. Contr. Inform.* to appear.

[33] Shi, P., (1997). Robust filtering for uncertain systems with sampled measurements. *Int. J. Systems Science*, **27**(12), 1403–1415.

[34] Vian, J. L. and M. E. Sawan, (1991). H_∞ control for singularly perturbed systems. *Proc. 30th IEEE Conf. Decision & Control.* , 1072–1074, Brighton, England.

[35] Xie, L., M. Fu and C. E. de Souza, (1992). H_∞ control and quadratic stabilization of systems with parameter uncertainty via output feedback. *IEEE Trans. Automat. Control*, **37**(8), 1253–1256.

[36] Xie, L., P. Shi and C. E. de Souza, (1993). On designing controllers for a class of uncertain sampled-data nonlinear systems. *IEE Proc.-Control Theory Appl.*, **140**(2), 119–126.

AIAA-97-3458

ADAPTIVE $\mathcal{H}_\infty$-CONTROL PROBLEMS AND PARAMETER PROJECTION TECHNIQUES

Wei-Min Lu and Fred Y. Hadaegh *

Jet Propulsion Laboratory
California Institute of Technology
Pasadena, Ca 91109-8099

Abstract

Parameter projection techniques play very important roles in adaptive control. In this paper, two scaled parameter projection techniques with respect to nonsmooth convex parameter sets are developed. The two projection techniques are then used in adaptive $\mathcal{H}_\infty$-control problems to derive the adaptive control laws.

1 Introduction

New technological development in space engineering and science requires sophisticated control systems with both high performance and reliability. How to achieve the required performance and reliability against various uncertainties has been a very challenging issue for control system design in recent years. Adaptive and robust control techniques are considered as useful methods to achieve this goal. Generally speaking, adaptive control is effective in dealing with parametric uncertainty, while the robust control schemes are good at handling dynamic uncertainty. However, oftentimes, control systems have both types of uncertainty; therefore, an integrated treatment in control system design is necessary.

In this paper, we consider adaptive robust control design for an uncertain system with both parametric and dynamic uncertainties, which results in an adaptive system with $\mathcal{H}_\infty$-performance. As the $\mathcal{H}_\infty$-performance has a close relation with dissipativity, a dissipation theoretic approach is used in dealing with the adaptive $\mathcal{H}_\infty$-control problem. Dissipation theoretic technique as a generalization of Lyapunov method [13] for adaptive control yields a simple adaptive control law which can significantly reduce implementation cost. However, the methods generally do not guarantee boundedness and convergence of parameters. Though the parameter convergence is not the objective of adaptive control, the boundedness of the estimated parameter should be guaranteed in the adaptive control law. One of the methods to guarantee the boundedness is to use the parameter projection techniques.

In applications, the unknown parameters of the parameterized systems are usually in a bounded set. In this paper, it is assumed that allowable parameter sets are compact and convex, but not necessarily smooth, e.g., a cube in the parameter space. During the adaptation, one of the requirements is that the parameter-adjustment mechanism of an adaptive control system keep the adjusted parameters in the parameter sets so as not to invalidate the solvability conditions. Here, both vector and direct parameter projection techniques are used to achieve this goal. The vector projection, which was originally introduced as a gradient projection method to generate the feasible directions in constrained optimization [11], is probably the most extensively used projection technique in adaptive parameter estimation and adaptive control [4, 16, 15, 5, 7]. However, in these cases, the projections are considered only for smooth sets. The vector projection is generalized to a more general setting such that the non-smooth parameter sets are allowed. The direct parameter projection is relatively new in adaptive control (another version appeared in [2]). It will be seen that this technique is suitable for the adaptive

*Associate Fellow AIAA

control problems where integral performance specifications are involved, in particular adaptive $\mathcal{H}_\infty$-control problem. The two projections not only play a very important role in adaptive control problems, but also are of interest in their own right.

This paper is a generalization of the work reported in [9]. Other work related to the adaptive $\mathcal{H}_\infty$-control includes [2, 23, 12, 17]. In this paper, the emphasis is the development of general projection techniques for nonsmooth sets and their applications in adaptive $\mathcal{H}_\infty$-control; the existence of adaptive $\mathcal{H}_\infty$-controllers are characterized in terms of solutions of parameter-dependent Hamilton-Jacobi inequalities, and adaptive controllers are constructed from the solutions and the projection techniques. This paper is thus divided into two parts. In Section 2, the general projection techniques are developed; both direct and direct parameter projections techniques are rigorously treated with respect to compact, convex, but possibly nonsmooth parameter sets. In Section 3, an adaptive $\mathcal{H}_\infty$-control problem is stated; the solutions for the adaptive $\mathcal{H}_\infty$-control problem are derived for the case when the original storage functions are independent of the parameters. Both vector and direct parameter projection techniques are used in the derivation of adaptive control laws.

2 Parameter Projection Techniques

In this section, vector and direct parameter (scaled) projections using the techniques from nonsmooth analysis and viability theory [1, 18]are presented. Both projection techniques will play a very important role in the adaptive $\mathcal{H}_\infty$-control design. The special non-scaled projection techniques were considered in [9].

2.1 Invariance and Contingent Cone

Consider a differential equation:

$$\dot{x} = f(x, t) \tag{1}$$

where $f : \mathbf{R}^n \times \mathbf{R}^+ \to \mathbf{R}^n$ is continuous in x and measurable in t. Suppose for all $x_0 \in \mathbf{R}^n$, the differential equation has a unique solution starting from $x(0) = x_0$ defined for $t \in \mathbf{R}^+$. A set $\mathbf{X} \subset \mathbf{R}^n$ is a invariant set of (1), if for all $x_0 \in \mathbf{X}$, its solution will stay in the set $\mathbf{X}$ for all t. It is known that if $\mathbf{X}$ is an invariant set, so is its closure $\overline{\mathbf{X}}$.

Given a compact set $\mathbf{K} \subset \mathbf{R^n}$, we next examine the invariance for the differential equation (1). The invariance is characterized in terms of contingent cones [1]. The contingent cone to $\mathbf{K}$ is well defined as a set-valued map $T_\mathbf{K} : \mathbf{K} \hookrightarrow X$:

$$T_\mathbf{K}(x) := \{v | \liminf_{h \to 0^+} \frac{d_\mathbf{K}(x + hv)}{h} \le 0\}. \tag{2}$$

where $d_\mathbf{K}(x) := \inf_{z \in \mathbf{K}} \|x - z\|$. For all $x \in \mathbf{K}$, the value $T_\mathbf{K}(x)$ is a closed cone. If $\mathbf{K}$ is convex, then $T_\mathbf{K}(x)$ is a tangent cone. Here, we are concerned with only the convex sets.

Suppose $\mathbf{K}$ is convex and $0 \in Int(\mathbf{K})$. The Minkowski function of $\mathbf{K}$ is defined as (see, e.g., [11])

$$\Psi_\mathbf{K}(x) = \inf\{\lambda \in \mathbf{R}^+ : x \in \lambda\mathbf{K}\};$$

Given $r \ge 0$, we define the set $\mathbf{K}_r$ as

$$\mathbf{K}_r := \{x \in \mathbf{R}^n : \Psi_\mathbf{K}(x) \le r\}.$$

Then $\mathbf{K} = \mathbf{K}_1$. The Minkowski functions of convex sets are convex, but not differentiable in general [11]. The contingent cone to a convex set is a lower semi-continuous set-valued map on $\mathbf{K}$ with closed convex value [1]. The following lemma provides a more explicit representation of contingent cones to convex sets [18].

Lemma 2.1 *If* $\mathbf{K}$ *is convex, then*

$$T_\mathbf{K}(x) = \overline{\{y : \exists t > 0 : x + ty \in \mathbf{K}\}}.$$

Another useful notion related to a convex set is its normal cone. If $\mathbf{K} \subset \mathbf{R}^n$ is convex, then we can define its normal cone as follows:

$$N_\mathbf{K}(x) = \{z \in \mathbf{R}^n : y^T z \le 0, \ \ \forall y \in T_\mathbf{K}(x)\}$$

Therefore, $N_\mathbf{K}(x) = \{0\}$ if $x \in Int(\mathbf{K})$.

From the above characterizations, we give the following propositions.

Proposition 2.2 *Consider a linear invertible map* $Q : \mathbf{R}^n \to \mathbf{R}^n$. *Suppose* $\mathbf{K}$ *is a convex compact set in* $\mathbf{R}^n$, $x \in \mathbf{K}$; *let* $\mathbf{M} = Q\mathbf{K}$ *which is also convex compact, and* $y = Qx$. *Then the Minkowski functions, contingent cones, and normal cones satisfy* $\Psi_\mathbf{M}(y) = \Psi_\mathbf{K}(x)$, $T_\mathbf{M}(y) = QT_\mathbf{K}(x)$, *and* $N_\mathbf{M}(y) = Q^{-1}N_\mathbf{K}(x)$.

Proposition 2.3 *Given a convex compact set* $\mathbf{K} \subset \mathbf{R}^n$*, it is an invariant set of differential equation (1) if and only if for all* $x \in \mathbf{K}$*,*

$$f(x,t) \in T_{\mathbf{K}}(x) \tag{3}$$

for all $t \in \mathbf{R}^+$.

Therefore, the invariance of a set can be characterized by its contingent cone. Given a convex and compact set $\mathbf{K} \subset \mathbf{R^n}$, the solutions to differential equation (1) are not necessarily always constrained inside the set $\mathbf{K}$. However, in adaptive control problems, we usually require some parameters, which are governed by differential equations, stay inside given sets during the evolution (see Sections 5 and 6), and some properties still remain. In the following two subsections, we will introduce two projection methods to achieve this goal.

2.2 Direct Parameter Projection

Consider a convex and compact set $\mathbf{K} \subset \mathbf{R}^n$. The projection $\Pi^Q_{\mathbf{K}}(x)$ of a point $x \in \mathbf{R}^n$ onto $\mathbf{K}$ is defined as follows:

$$\Pi^Q_{\mathbf{K}}(x) = \arg\min_{z \in \mathbf{K}} \|x - z\|_Q := \sqrt{(x-z)^T Q (x-z)} \tag{4}$$

The above projection is well-defined, since $\mathbf{K} \subset \mathbf{R}^n$ is convex and compact; in addition, $\Pi^Q_{\mathbf{K}}(x)$ is continuous. We first have the following characterizations.

Proposition 2.4 *Given a convex compact set* $\mathbf{K} \subset \mathbf{R}^n$*. Take* $x \in \mathbf{R}^n$*; the following statements are equivalent:*

(i) $\xi \in \mathbf{K}$ *is such that* $\xi = \Pi^Q_{\mathbf{K}}(x)$*;*

(ii) $(x-\xi)^T Q (x_0 - \xi) \leq 0$.

(iii) $x - \xi \in Q^{-1} N_{\mathbf{K}}(\xi)$.

Proof [**(i)** $\Leftrightarrow$ **(ii)**] From the definition,

$$\xi = \Pi^Q_{\mathbf{K}}(x) = \arg\min_{z \in \mathbf{K}} \left\| Q^{1/2}(x - z) \right\|.$$

From Theorem 1 in [11, p.69], the above inequality holds if and only if

$$(Q^{1/2}x - Q^{1/2}\xi)^T (Q^{1/2}x - Q^{1/2}\xi) \leq 0,$$

which is exactly (ii).

[**(ii)** $\Leftrightarrow$ **(iii)**] From Lemma 2.1 and the definition of normal cones (see also [18, Proposition 2G]),

$$Q^{-1/2} N_{\mathbf{K}}(\xi) = N_{Q^{1/2}\mathbf{K}}(Q^{1/2}\xi)$$
$$= \{Q^{1/2}v \mid v^T Q (x_0 - \xi) \leq 0, \forall x_0 \in \mathbf{K}\}. \tag{5}$$

From Proposition 2.2,

$$N_{Q^{1/2}\mathbf{K}}(Q^{1/2}\xi) = Q^{-1/2} N_{\mathbf{K}}(\xi),$$

thus

$$Q^{-1} N_K(\xi) = \{v \mid v^T Q (x_0 - \xi) \leq 0, \forall x_0 \in \mathbf{K}\}.$$

Therefore, (ii) holds if and only if

$$x - \xi \in Q^{-1} N_K(\xi).$$

In the following, we have the following property of the direct projection.

Proposition 2.5 *Let a convex compact set* $\mathbf{K} \subset \mathbf{R}^n$*. Then for any absolutely continuous function* $x : \mathbf{R} \to \mathbf{R}^n$*, its projection:*

$$\xi(t) := \Pi^Q_{\mathbf{K}}(x(t))$$

is also absolutely continuous.

Proof As Q is invertible, it is sufficient to show that the map $Q^{1/2}\Pi^Q_{\mathbf{K}} : \mathbf{R}^n \to \mathbf{R}^n$ is Lipschitz. In fact, we will show that for all $x, y \in \mathbf{R}^n$,

$$\left\| Q^{1/2}\Pi^Q_{\mathbf{K}}(x) - Q^{1/2}\Pi^Q_{\mathbf{K}}(y) \right\| \leq \left\| Q^{1/2} \right\| \|x - y\|. \tag{6}$$

Indeed,

$$\begin{aligned}
&\left\| Q^{1/2}(x-y) \right\|^2 \\
&= \left\| \Pi^Q_{\mathbf{K}}(x) - \Pi^Q_{\mathbf{K}}(y) + (x - \Pi^Q_{\mathbf{K}}(x)) - (y - \Pi^Q_{\mathbf{K}}(y)) \right\|_Q^2 \\
&= \left\| \Pi^Q_{\mathbf{K}}(x) - \Pi^Q_{\mathbf{K}}(y) \right\|_Q^2 + \left\| (x - \Pi^Q_{\mathbf{K}}(x)) - (y - \Pi^Q_{\mathbf{K}}(y)) \right\|_Q^2 + \\
&\quad -2(x - \Pi^Q_{\mathbf{K}}(x))^T Q (\Pi^Q_{\mathbf{K}}(y) - \Pi^Q_{\mathbf{K}}(x)) - \\
&\quad 2(y - \Pi^Q_{\mathbf{K}}(y))^T Q (\Pi^Q_{\mathbf{K}}(x) - \Pi^Q_{\mathbf{K}}(y)) \\
&\geq \left\| \Pi^Q_{\mathbf{K}}(x) - \Pi^Q_{\mathbf{K}}(y) \right\|_Q^2 \\
&= \left\| Q^{1/2}\Pi^Q_{\mathbf{K}}(x) - Q^{1/2}\Pi^Q_{\mathbf{K}}(y) \right\|^2
\end{aligned}$$

where the inequality follows from the above proposition (ii), e.g.,

$$(x - \Pi^Q_{\mathbf{K}}(x))^T Q(\Pi^Q_{\mathbf{K}}(y) - \Pi^Q_{\mathbf{K}}(x)) \leq 0.$$

The direct parameter projection has the following property which is useful in the adaptive $\mathcal{H}_\infty$-control problem.

Theorem 2.6 *Given the convex and compact* $\mathbf{K}$. *Then for any absolute continuous function* $x : \mathbf{R}^+ \to \mathbf{R}^n$ *with* $x(0) \in \mathbf{K}$, *and the projection:* $\xi(t) = \Pi^Q_{\mathbf{K}}(x(t))$, *the following inequality holds:*

$$\int_0^T (\xi(t) - x^*)^T Q(\dot{\xi}(t) - \dot{x}(t))dt \leq 0 \qquad (7)$$

for all $x^* \in \mathbf{K}$ *and* $T \geq 0$.

Proof Notice that the left-hand side of (7) is well defined from the previous proposition. Thus, given $x^* \in \mathbf{K}$ and $T \geq 0$, we have

$$\begin{aligned}
&\int_0^T (\xi(t) - x^*)^T Q(\dot{\xi}(t) - \dot{x}(t))dt \\
&= (\xi(t) - x(t))^T Q(\xi(t) - x^*)|_0^T + \\
&\quad - \int_0^T (\xi(t) - x(t))^T Q d\xi(t) \\
&= (\xi(T) - x(T))^T Q(\xi(T) - x^*) + \\
&\quad - \int_0^T (\xi(t) - x(t))^T Q\dot{\xi}(t)dt
\end{aligned}$$

where the last equality holds as $\xi(0) = x(0) \in \mathbf{K}$.

As $\xi(t) = \Pi^Q(x(t))$ for $t \in [0, T]$, Proposition 2.4 (iii) gives

$$x(t) - \xi(t) \in Q^{-1} N_{\mathbf{K}}(\xi(t)) \qquad (8)$$

for all $t \in [T_0, T]$.

Also from Proposition 2.4 (ii), one has the following:

$$\begin{aligned}
&(\xi(T) - x(T))^T Q(\xi(T) - x^*) \\
&= (x(T) - \xi(T))^T Q(x^* - \xi(T)) \leq 0.
\end{aligned}$$

On the other hand, if $\xi(t) \in \mathbf{K}$ is differentiable at $t \in (T_o, T)$, there exists a positive sequence $\{h_n\}$ with $h_n \to 0$ as $n \to \infty$ such that $\xi(t + h_n) \in \mathbf{K}$; denote

$$d_n(t) := \frac{\xi(t + h_n) - \xi(t)}{h_n},$$

then $d_n(t) \to \dot{\xi}(t)$ as $n \to \infty$. Since $\xi(t) + h_n d_n(t) = \xi(t + h_n) \in \mathbf{K}$, $d_n(t) \in T_{\mathbf{K}}(\xi(t))$ by Lemma 2.1. Therefore, $\dot{\xi}(t) = \lim_{n\to\infty} d_n(t) \in T_{\mathbf{K}}(\xi(t))$ as $T_K(\xi(t))$ is a closed cone. Thus, (8) as well as the definition of normal cone implies

$$(\xi(t) - x(t))^T Q\dot{\xi}(t) = -(x(t) - \xi(t))^T Q\dot{\xi}(t) \geq 0.$$

for all $t \in [T_0, T]$. Therefore,

$$\begin{aligned}
&\int_0^T (\xi(t) - x^*)^T Q(\dot{\xi}(t) - \dot{x}(t))dt \\
&= (\xi(T) - x(T))^T Q(\xi(T) - x^*) - \int_0^T (\xi(t) - x(t))^T Q\dot{\xi}(t)dt \leq 0
\end{aligned}$$

In the above proof, we have also proved the following useful result.

Corollary 2.7 *Let* $\xi : \mathbf{R}^+ \to \mathbf{R}^n$ *be an absolutely continuous function. Give a compact and convex set* $\mathbf{K} \in \mathbf{R}^n$; *if* $\xi(t) \in \mathbf{K}$ *for all* $t \in \mathbf{R}^n$, *then*

$$\dot{\xi}(t) \in T_{\mathbf{K}}(\xi(t))$$

for almost all $t \in \mathbf{R}^n$.

2.3 Vector Projection

Vector projection technique is introduced for constrained optimization (see, e.g., [11]). It is an indirect parameter projection technique. It has been widely used in the adaptive parameter estimation and adaptive control (see e.g., [4, 16, 5]).

Consider a convex and compact set $\mathbf{K} \subset \mathbf{R}^n$ and its Minkowski function $\Psi_{\mathbf{K}}$. Given a positive definite scaling matrix Q, we first define the scaled projection of a vector $v \in \mathbf{R}^n$ at a point $x \in \mathbf{R}^n$ on the contingent cone $T_{\mathbf{K}}(x)$ as follows:

$$\pi^Q_{\mathbf{K}}(x, v) = \begin{cases} v & \text{if } x \in Int(\mathbf{K}) \text{ or } v \in T_{\mathbf{K}_r}(x); \\ 0 & \text{if } v \in Q^{-1} N_{\mathbf{K}_r}(x); \\ v^T Q w_0 w_0 & \text{otherwise.} \end{cases} \qquad (9)$$

where $r = \Psi_{\mathbf{K}}(x)$ and

$$w_0 = \arg\max\{v^T Q w | w \in T_{\mathbf{K}_r}(x), \|w\|_Q = 1\}.$$

We have the following theorem.

Theorem 2.8 *Given a convex and compact set* $\mathbf{K} \subset \mathbf{R}^n$, *the projection is defined by (9), then we have the following assertions:*

(i) $\pi^Q_{\mathbf{K}}(x,v) \in T_{\mathbf{K}}(x)$ *for all* $x \in \mathbf{K}$ *and* $v \in \mathbf{R}^n$.

(ii) $\left\|\pi^Q_{\mathbf{K}}(x,v)\right\|_Q \le \|v\|_Q$ *for all* $x \in \mathbf{K}$ *and* $v \in \mathbf{R}^n$.

(iii) $(x - x^*)^T Q(\pi^Q_{\mathbf{K}}(x,v) - v) \le 0$ *for* $x^* \in \mathbf{K}$, $x \in \mathbf{K}$, *and* $v \in \mathbf{R}^n$.

(iv) Consider the system (1) under the above projection:

$$\dot{x} = \pi^Q_{\mathbf{K}}(x, f(x,t)); \tag{10}$$

then the set $\mathbf{K}$ *is an invariant set for the projected system.*

Proof Given $x \in \mathbf{K}$ and $y \in \mathbf{R}^n$.

(i) If $x \in Int(\mathbf{K})$, then $T_{\mathbf{K}}(x) = \mathbf{R}^n$, so $\pi^Q_{\mathbf{K}}(x,v) = v \in T_{\mathbf{K}}(x)$; if $v \in T_{\mathbf{K}}(x)$, then $\pi^Q_{\mathbf{K}}(x,v) = v \in T_{\mathbf{K}}(x)$; if $v \in Q^{-1}N_{\mathbf{K}_r}(x) = \{w : v^T Q w \le 0, \forall v \in T_{\mathbf{K}_r}(x)\}$, then $\pi^Q_{\mathbf{K}}(x,v) = 0 \in T_{\mathbf{K}}(x)$. We only need to show the last case with $\Psi(x) \ge 1$. In fact, as $w_0 \in T_{\mathbf{K}_r}(x)$ and $v^T Q w_0 \ge 0$, then $\pi^Q_{\mathbf{K}}(x,v) = (v^T Q w_0) w_0 \in T_{\mathbf{K}_r}(x)$.

(ii) As the inequality is satisfied trivially in the first two cases, we only need to show the case when $\pi^Q_{\mathbf{K}}(x,v) = v^T Q w_0 w_0$. Indeed,

$$\left\|\pi^Q_{\mathbf{K}}(x,v)\right\|_Q = \left\|v^T Q w_0 w_0\right\|_Q$$

$$= \left\|v^T Q w_0\right\| \le \|v\|_Q \|w_0\|_Q = \|v\|_Q .$$

(iii) If $x \in Int(\mathbf{K})$ or $v \in T_{\mathbf{K}}(x)$, then $\pi^Q_{\mathbf{K}}(x,v) = v$, so $(x - x^*)^T Q(\pi^Q_{\mathbf{K}}(x,v) - v) = 0$. Now we consider the other cases.

Notice that, for all $x^* \in \mathbf{K}$, as $x + (x^* - x) = x \in \mathbf{K}$, then $x^* - x \in T_{\mathbf{K}}(x)$ by Lemma 2.1. So if $v \in Q^{-1}N_{\mathbf{K}}(x)$, then $Qv \in N_{\mathbf{K}}(x)$, and by the definition of normal cone,

$$(x - x^*)^T Q(\pi^Q_{\mathbf{K}}(x,v) - v) = (x^* - x)^T Q v \le 0.$$

Next, let's consider the remaining case: $\pi^Q_{\mathbf{K}}(x,v) = v^T Q w_0 w_0$. We first show the following:

$$Q(v - \pi^Q_{\mathbf{K}}(x,v)) \in N_{\mathbf{K}}(x).$$

From the definition of the projection,

$$w_0 = \arg\max\{(Q^{1/2}v)^T(Q^{1/2}w)|$$

$$Q^{1/2}w \in Q^{1/2}T_{\mathbf{K}_r}(x), \left\|Q^{1/2}w\right\| = 1\}$$

Therefore,

$$v^T Q w_0 Q^{1/2} w_0 = (Q^{1/2}v)^T (Q^{1/2}w_0)(Q^{1/2}w_0)$$

$$= \arg \min_{z \in Q^{1/2}T_{\mathbf{K}}(x)} \left\|Q^{1/2}v - z\right\|,$$

or

$$v^T Q w_0 w_0 = \arg \min_{Q^{1/2}z_0 \in Q^{1/2}T_{\mathbf{K}}(x)} \|v - z_0\|_Q .$$

By the use of Proposition 2.4, we have

$$(v - v^T Q w_0 w_0)^T Q(z - v^T Q w_0 w_0) \le 0 \tag{11}$$

for all $z \in T_{\mathbf{K}}(x)$. On the other hand, as $T_{\mathbf{K}}(x)$ is a convex cone, so for all $u \in T_{\mathbf{K}}(x)$, $u + v^T Q w_0 w_0 \in T_{\mathbf{K}}(x)$; so (11) implies

$$(v - v^T Q w_0 w_0)^T Q u$$

$$= (v - v^T Q w_0 w_0)^T Q(u + v^T Q w_0 w_0 - v^T Q w_0 w_0) \le 0.$$

Therefore, $Q(v - v^T Q w_0 w_0) \in N_{\mathbf{K}}(x)$ as claimed. Again by the definition of normal cone,

$$(x - x^*)^T Q(\pi^Q_{\mathbf{K}}(x,v) - y) = (x^* - x)^T Q(y - \pi^Q_{\mathbf{K}}(x,v)) \le 0.$$

(iv) As $\mathbf{K}$ is convex, then its Minkowski function $\Psi_{\mathbf{K}}$ is convex, so it is absolutely continuous. Now for any absolutely continuous function $x(t)$ that is a solution of (10) with $x(0) \in \mathbf{K}$, then $\eta(t) := \Psi_{\mathbf{K}}(x(t))$ is also absolutely continuous. It is sufficient to show $x(t) \in \mathbf{K}$. Indeed, if it is not true, then there exists $T > 0$, such that $x(T) \in \mathbf{R}^n \backslash \mathbf{K}$; so $\Psi_{\mathbf{K}}(x(T)) > 0$. Suppose $T_0 < T$ is such that

$$T_0 = \inf\{t \ge 0 : x(t) \in \mathbf{R}^n \backslash \mathbf{K}\}$$

Therefore, $x(T_0) \in \mathbf{K}$ as K is compact, so $\Psi_{\mathbf{K}}(x(0)) \le 1$. Let $t \in (T_0, T)$ be a point on $\mathbf{R}^+$ where both $\dot{x}(t)$ and $\frac{\partial \Psi_{\mathbf{K}}}{\partial x}(x(t))$ exist, then there exists $\epsilon(h)$ with $\lim_{h\to 0} \frac{\epsilon(h)}{h} = 0$ such that

$$x(t+h) = x(t) + h\dot{x}(t) + \epsilon(h).$$

Then

$$\dot{\eta}(t) = \lim_{h \to 0^+} \frac{\Psi_{\mathbf{K}}(x(t) + h\dot{x}(t) + \epsilon(h)) - \Psi_{\mathbf{K}}(x(t))}{h}$$

$$= \frac{\partial \Psi_{\mathbf{K}}}{\partial x}(x(t))\dot{x}(t) = \frac{\partial \Psi_{\mathbf{K}}}{\partial x}(x(t))\pi^Q_{\mathbf{K}}(x, f(x,t))$$

Notice that $\frac{\partial \Psi_{\mathbf{K}}}{\partial x}(x(t)) \in N_{\mathbf{K}_r}(x(t))$ where $r = \Psi_{\mathbf{K}}(x(t))$; by the same argument as (i), we can show $\pi^Q_{\mathbf{K}}(x, f(x,t)) \in T_{\mathbf{K}_r}(x(t))$. Thus,

$$\dot{\eta}(t) = \frac{\partial \Psi_{\mathbf{K}}}{\partial x}(x(t))\pi^Q_{\mathbf{K}}(x, f(x,t)) \le 0.$$

almost everywhere on $[T_0, T]$. Thus,

$$0 < \Psi_{\mathbf{K}}(x(T)) - \Psi_{\mathbf{K}}(x(T_0)) = \eta(T) - \eta(T_0)$$

$$= \int_0^T \dot{\eta}(t)dt \leq 0,$$

which is a contradiction. Therefore, $\mathbf{K}$ is an invariant set.

Notice that in the above theorem, (iv) is not the conclusion of (i), as we don't assume the solutions of (10) are unique for $x(0) \in \mathbf{K}$ (see Proposition 2.3). In the proof of (iv), the projection property of a vector outside $\mathbf{K}$ is used which is also true if the projection (2.8) of a vector is defined onto the exterior contingent cones instead of $T_{\mathbf{K}_r}(x)$ with $r > 1$ (see [1, Definition 5.1.1]). (iii) is a useful property for the adaptive control design.

It is also noticed that the right-hand side of (10) is not necessarily continuous, even if f is continuous. In [16], with some relaxation, the authors define a projection which is Lipschitzian and guarantees the projected system to have an invariant set larger than the parameter set.

3 Adaptive $\mathcal{H}_\infty$-Control

3.1 Adaptive $\mathcal{H}_\infty$-Control Problem and Dissipativity

In this section, we will consider the adaptive attenuation of disturbances for nonlinear systems with emphasis on the application of projection techniques. The uncertain nonlinear system $G(\theta)$ to be considered is governed by the following parameterized dynamical equation:

$$\left\{ \begin{array}{rcl} \dot{x} &=& f(x,\theta) + g_1(x,\theta)w + g_2(x,\theta)u \\ z &=& h(x) + k_1(x)w + k_2(x)u \\ y &=& h_2(x) + k_{21}(x)w + k_{22}(x)u \end{array} \right. \tag{12}$$

where θ is a r-dimensional vector of unknown constant parameters with $\theta \in \Theta \in \mathbf{R}^r$. In the adaptive control problem, we will consider the parameter-dependence in the following fashion:

$$f(x,\theta) = f_0(x) + \sum_{i=1}^{r} \theta_i f_i(x)$$

$$g_j(x,\theta) = g_{j0}(x) + \sum_{i=1}^{r} \theta_i g_{ji}(x), \quad j = 1, 2.$$

It is assumed that $f_i, g_{ji}, h, k_j \in \mathbf{C}^0$, $f_i(0) = 0, h(0) = 0$, and $R(x) := I - k_1^T(x)k_1(x) > 0$ for all $x \in \mathbf{R}^n$; x, w, u, z, and y are state, exogenous disturbance, control input, regulated output, and measured output with dimensions n, p_1, p_2, q, and $n + p_1$, respectively.

The objective of the adaptive $\mathcal{H}_\infty$-controller design is to attenuate the impact z of the exogenous disturbance w and the error induced by the initial guess of the parameter. The magnitudes of signals z, w and z are measured by their $\mathcal{L}_2$-norms. The adaptive controllers to be sought have the following form.

$$K : \left\{ \begin{array}{l} \dot{p} = \phi(p,y,u) \\ u = \kappa(p)y \end{array} \right. \tag{13}$$

where $p \in \mathbf{R}^r$ is the estimation of the real parameter θ, $\phi \in \mathbf{C}^0$, and $\dot{p} = \phi(p,y,u)$ is the parameter update law. For fixed p, $u = \kappa(p)y$ is a I/O map from y to u; it is taken as a (possibly modified) gain-scheduled controller in the sequel. An adaptive $\mathcal{H}_\infty$ control system is illustrated in Figure 1. The precise statement of the adaptive control problem is given next.

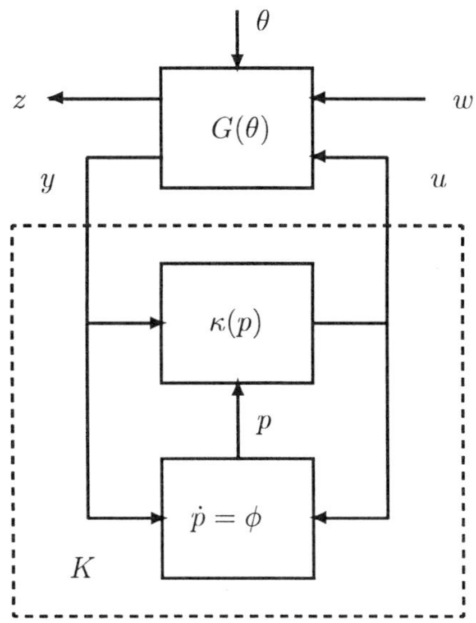

Figure 1: Adaptive $\mathcal{H}_\infty$-Control System

Definition 3.1 (Adaptive $\mathcal{H}_\infty$-Control Problem) *Suppose $\epsilon > 0$ is given. The adaptive $\mathcal{H}_\infty$-control design is to seek a controller (13) such that the resulting closed loop system with $x(0) = 0$ satisfies*

$$\int_0^T \|z(t)\|^2 dt \leq \int_0^T \|w(t)\|^2 dt + (p(0)-\theta)^T Q(p(0)-\theta) \tag{14}$$

for all $T \in \mathbf{R}^+$, $w \in \mathcal{L}_2[0,\infty)$, $p(0) \in \Theta$, *and* $\theta \in \Theta$.

The performance can be interpreted as the attenuation of exogenous disturbance and the error of the initial parameter guess. Note that in this statement the initial state is assumed to be at the origin. If the initial state is unknown, the performance can be modified accordingly.

In the following, we consider full-information feedback, in which case both x and w are available to the control input u. Moreover, we make the following assumptions to simplify the process.

Assumption 3.2 *Consider the system (12).*

[A1] The parameter set Θ *is convex and compact, and* $0 \in Int(\Theta)$.

[A2] $y = \begin{bmatrix} x \\ w \end{bmatrix}$.

[A3] $k_1(x) = 0$ *and* $k_2^T(x)\begin{bmatrix} h(x) & k_2(x) \end{bmatrix} = \begin{bmatrix} 0 & I \end{bmatrix}$ *for all* $x \in \mathbf{R}^n$.

Assumption [A2] just restates the full information problem. [A3] is a standard assumption in the $\mathcal{H}_\infty$-control problem [3, 8] in most of the derivations in the following.

Remark 3.3 *It will be seen that for the parameter-dependent system (12) with above assumptions, if the multiplier* $g_1(x,\theta)$ *of the disturbance* w *is independent of* θ*, then only the state information, in stead of full information, is needed to construct the adaptive* $\mathcal{H}_\infty$*-control law.*

It is known that if the parameter θ is known, then the $\mathcal{H}_\infty$-control problem has a state-feedback solution if there exists a non-negative function $V : \mathbf{R}^n \times \Theta \to \mathbf{R}^+$ which is positive definite with respect to x such that the following parameter dependent Hamilton-Jacobi inequality is satisfied for all $\theta \in \Theta$:

$$\frac{\partial V}{\partial x}(x,\theta)f(x,\theta) + \frac{1}{4}\frac{\partial V}{\partial x}(x,\theta)(g_1(x,\theta)g_1^T(x,\theta) + -g_2(x,\theta)g_2^T(x,\theta))\frac{\partial V^T}{\partial x}(x,\theta) + h^T(x)h(x) \leq 0 \tag{15}$$

and the parameter-dependent $\mathcal{H}_\infty$-controller is

$$u = -\frac{1}{2}g_2(x,\theta)\frac{\partial V^T}{\partial x}(x,\theta). \tag{16}$$

Moreover, the closed loop system with the above controller is dissipative with respect to the supply rate $\|w\|^2 - \|z\|^2$, and the function V is a storage function for the closed-loop system satisfying the dissipation inequality:

$$\dot{V}(x,\theta) \leq \|w\|^2 - \|z\|^2. \tag{17}$$

However, if the parameter is unknown before the system is in operation, we need to design an adaptive mechanism to estimate the parameter on-line and use the estimated parameter to adjust the necessary control action; in which case, the controller

$$u = \psi(x,w,p)$$

is used instead, where ψ is the state feedback (or its modification) defined in (16), p is an estimation of θ, and its update law has the following general form:

$$\dot{p} = \phi(p,x,w,u)$$

To guarantee the $\mathcal{H}_\infty$-performance (14) for the closed system, we need to show that the adaptive (closed-loop) system is dissipative with respect to the supply rate $\|w\|^2 - \|z\|^2$; it is enough to find a storage function $W_\theta : \mathbf{R}^n \times \mathbf{R}^r \to \mathbf{R}^+$ for each $\theta \in \Theta$, such that the following dissipation inequality is satisfied:

$$W_\theta(x(T),p(T)) - W_\theta(x(0),p(0)) \leq \int_0^T (\|w(t)\|^2 - \|z(t)\|^2)dt.$$

Its differential version is satisfied if W is differentiable:

$$\frac{\partial W_\theta}{\partial x}(x,p)(f(x,\theta) + g_1(x,\theta)w + g_2(x,\theta)\psi(x,w,p)) + \frac{\partial W_\theta}{\partial p}(x,p)\phi(p,x,w,\psi(x,w,p)) \leq \|w\|^2 - \|z\|^2. \tag{18}$$

for each $\theta \in \Theta$ and $(x,p) \in \mathbf{R}^n \times \mathbf{R}^r$. Next, we will explicitly construct storage functions such that (18) is satisfied. From the discussion in Section 2.2, suppose the Hamilton-Jacobi inequality (3.1) has a solution $V(x,\theta)$, a possible choice for the storage function is $V(x,p)$ where the unknown parameter θ is replaced by its estimation p. However, it does not reflect the parameter estimation nature for the update law. On the other hand, the parameter enters the system in an affine fashion. Therefore, a meaningful choice of the storage function of the adaptive control system is the one with an additional quadratic p-term:

$$W_\theta(x,p) = V(x,p) + (p-\theta)^TQ(p-\theta) \tag{19}$$

Note that this idea was first introduced to construct Lyapunov functions for stable adaptive systems [13], and has been used in many adaptive control problems [6, 16, 23, 7]. For the sake of simplicity, we will assume that the function $V : \mathbf{R}^n \times \Theta \to \mathbf{R}^+$ satisfying the the above Hamilton-Jacobi inequality be continuously differentiable with respect to both arguments. Detailed solutions to the adaptive $\mathcal{H}_\infty$-control problem with full information feedback in different cases when V is independent of the parameter is presented next.

3.2 Solutions to Adaptive $\mathcal{H}_\infty$-Control Problem

In this section, we mainly consider the case where there exists a positive definite function $V : \mathbf{R}^n \to \mathbf{R}^+$ which is independent of θ such that it satisfies the Hamilton-Jacobi inequality (3.1), i.e.,

$$\frac{\partial V}{\partial x}(x)f(x,\theta) + \frac{1}{4}\frac{\partial V}{\partial x}(x)(g_1(x,\theta)g_1^T(x,\theta) +$$

$$-g_2(x,\theta)g_2^T(x,\theta))\frac{\partial V^T}{\partial x}(x) + h^T(x)h(x) \leq 0. \quad (20)$$

for all $\theta \in \Theta$.

Let $W : \mathbf{R}^n \times \Theta \to \mathbf{R}^+$ be a positive definite function defined as

$$W(x,p) = V(x) + (p-\theta)^T Q(p-\theta). \quad (21)$$

where Q is the positive definite matrix defined in the definition. Take W as a storage function candidate of the adaptive $\mathcal{H}_\infty$-control system. Then

$$\dot{W}(x,p) = \dot{V}(x) + 2(p-\theta)^T Q\dot{p}$$

$$= \frac{\partial V}{\partial x}(x)(f(x,\theta)+g_1(x,\theta)w+g_2(x,\theta)u)+2(p-\theta)^T Q\dot{p}$$

$$= \frac{\partial V}{\partial x}(x)(f(x,p) + g_1(x,p)w + g_2(x,p)u)+$$

$$+\sum_{i=1}^{r}\{\frac{\partial V}{\partial x}(x)(\theta_i - p_i)(f_i(x) + g_{1i}(x)w + g_{2i}(x)u)\}$$

$$+2(p-\theta)^T Q\dot{p}$$

Notice that if $p \in \mathbf{\Theta}$, then from the assumption (3.2), then

$$\frac{\partial V}{\partial x}(x)f(x,p) \leq -(\frac{1}{4}\frac{\partial V}{\partial x}(x)(g_1(x,p)g_1^T(x,p)+$$

$$-g_2(x,p)g_2^T(x,p))\frac{\partial V^T}{\partial x}(x) + h^T(x)h(x))$$

Replace the above inequality and use the completion of square, one has

$$\dot{W}(x,p) \leq \|w(t)\|^2 - \|z(t)\|^2 +$$

$$+\left\|u(t) + \frac{1}{2}g_2^T(x,p)\frac{\partial V^T}{\partial x}(x))\right\|^2 +$$

$$-\left\|w(t) - \frac{1}{2}g_1^T(x,p)\frac{\partial V^T}{\partial x}(x)\right\|^2 + 2(p-\theta)^T Q(\dot{p}-\Phi(x,w,u)), \quad (22)$$

where $\Phi : \mathbf{R}^n \times \mathbf{R}^{p_1} \times \mathbf{R}^{p_2} \to \mathbf{R}^r$ is defined as

$$\Phi(x,w,u) = \frac{1}{2}Q^{-1}\begin{bmatrix} \frac{\partial V}{\partial x}(x)(f_1(x) + g_{11}(x)w + g_{21}(x)u) \\ \frac{\partial V}{\partial x}(x)(f_2(x) + g_{12}(x)w + g_{22}(x)u) \\ \vdots \\ \frac{\partial V}{\partial x}(x)(f_r(x) + g_{1r}(x)w + g_{2r}(x)u) \end{bmatrix}. \quad (23)$$

From (22), one has that if $p \in \Theta$ and $u = -\frac{1}{2}g_2^T(x,p)\frac{\partial V^T}{\partial x}(x)$, then

$$\dot{W}(x,p) \leq \|w(t)\|^2 - \|z(t)\|^2 + 2(p-\theta)^T Q(\dot{p}-\Phi(x,w,u)), \quad (24)$$

Now integrate both sides of (24) from 0 to T and notice $W(x(T),p(T)) \geq 0$ and $x(0) = 0$, we have

$$\int_0^T \|z(t)\|^2 dt \leq \int_0^T \|w(t)\|^2 dt + (p(0)-\theta)^T Q(p(0)-\theta) +$$

$$+2\int_0^T (p(t)-\theta)^T Q(\dot{p}(t) - \Phi(x(t),w(t),u(t)))dt$$

$$\leq \int_0^T \|w(t)\|^2 dt + (p(0)-\theta)^T Q(p(0)-\theta)$$

$$+2\int_0^T (p(t)-\theta)^T Q(\dot{p}(t) - \Phi(x(t),w(t),u(t)))dt. \quad (25)$$

Therefore, if we can find a parameter update law for p such that

$$\int_0^T (p(t)-\theta)^T Q(\dot{p}(t) - \Phi(x(t),w(t),u(t)))dt \leq 0$$

and

$$p(t) \in \Theta, \quad \forall t \in \mathbf{R}^+,$$

then the adaptive $\mathcal{H}_\infty$-control problem is solved. Fortunately, we can use the projection techniques developed in Section 3 to achieve the above requirements.

Theorem 3.4 (Adaptive $\mathcal{H}_\infty$-Control with Vector Projection) *Consider the parameter-dependent system (12). Suppose there exists a nonnegative function $V : \mathbf{R}^n \to \mathbf{R}^+$ such that for each*

$\theta \in \Theta$, (3.2) is satisfied. Then given a positive definite matrix $Q \in \mathbf{R}^{r\times r}$, the adaptive $\mathcal{H}_\infty$-control problem has a solution. And an adaptive control law is given by

$$\begin{cases} \dot{p} = \pi^Q_\Theta(p, \Phi(x, w, -\frac{1}{4}g_2^T(x,p)\frac{\partial V^T}{\partial x}(x)) \\ u = -\frac{1}{2}g_2^T(x,p)\frac{\partial V^T}{\partial x}(x) \end{cases} \quad (26)$$

where Φ is defined by (23) and π^Q_Θ is the scaled vector projection with respect to the set Θ.

Proof Consider the adaptive control law (26). From Theorem 2.8, one has that the given parameter update law:

$$\dot{p} = \pi^Q_\Theta(p, \Phi(x, w, u)),$$

insures $p(t) \in \Theta$ and

$$(p-\theta)^T Q(\dot{p} - \Phi(x,w,u))$$

$$= (p-\theta)^T Q(\pi^Q_\Theta(p, \Phi(x,w,u)) - \Phi(x,w,u)) \le 0,$$

which implies

$$\int_0^T (p(t)-\theta)^T Q(\dot{p}(t) - \Phi(x(t), w(t), u(t)))dt \le 0$$

for all $T \in \mathbf{R}^+$. Now apply the adaptive control law (26), we have the relation (3.2), which implies

$$\int_0^T \|z(t)\|^2 dt \le \int_0^T \|w(t)\|^2 dt + (p(0)-\theta)^T Q(p(0)-\theta),$$

for all $T \in \mathbf{R}^+$.

The direct parameter projection method can be also applied.

Theorem 3.5 (Adaptive $\mathcal{H}_\infty$-Control with Direct Parameter Projection) *Consider the parameter-dependent system (12). Suppose there exists a non-negative function $V : \mathbf{R}^n \to \mathbf{R}^+$ such that for each $\theta \in \Theta$, (3.2) is satisfied. Then the adaptive $\mathcal{H}_\infty$-control problem has a solution with a positive definite matrix $Q \in \mathbf{R}^{r\times r}$. And an adaptive control law is given by*

$$\begin{cases} \dot{\pi} = \Phi(x, w, -\frac{1}{4}g_2^T(x,p)\frac{\partial V^T}{\partial x}(x)) \\ p = \Pi^Q_\Theta(\pi) \\ u = -\frac{1}{2}g_2^T(x,p)\frac{\partial V^T}{\partial x}(x) \end{cases} \quad (27)$$

where Φ is defined by (23) and Π^Q_Θ is the scaled direct parameter projection with respect to the set Θ.

Proof Consider the adaptive control law (27). Suppose $p(t)$ for $t \in [0, \infty)$ is generated by the resulting update law:

$$\dot{\pi} = \Phi(x, w, u),$$

and

$$p(t) = \Pi^Q_\Theta(\pi(t));$$

then from Theorem 2.6, one has $p(t) \in \Theta$ and for all $T \in \mathbf{R}^+$,

$$\int_0^T (p(t)-\theta)^T Q(\dot{p}(t) - \Phi(x(t), w(t), u(t)))dt$$

$$= \int_0^T (p(t)-\theta)^T Q(\dot{p}(t) - \dot{\pi}(t))dt \le 0.$$

Now apply the adaptive control law (27), we have the relation (3.2), which implies

$$\int_0^T \|z(t)\|^2 dt \le \int_0^T \|w(t)\|^2 dt + (p(0)-\theta)^T Q(p(0)-\theta),$$

for all $T \in \mathbf{R}^+$.

Remark 3.6 *It is interesting to compare Theorem 3.5 with the sufficient condition result for the minimax adaptive problem in [2]. As in the above theorem, the sufficient condition in [2] for the minimax adaptive control problem to have solution is that there exists a non-negative function $V : \mathbf{R}^n \to \mathbf{R}^+$ independent of the parameter such that the Hamilton-Jacobi inequality (3.2) is satisfied; and the control action u is also obtained by the parameter projection. However, the implications of the function $V(x)$ in the two papers are different. In this paper, $V(x)$ is just the storage function of the parameterized $\mathcal{H}_\infty$-control system, but not the resulting adaptive $\mathcal{H}_\infty$-control system. In [2], $V(x)$ is the storage function of the resulting minimax adaptive control system, i.e., an upper bound of the value function. The adaptive controller in this paper is simpler than that given in [2]. However, the use and implications of parameter projection in [2] and this paper are different. The scaling matrix of parameter projection in [2] is dependent on state, while the projection in this paper is constant in this sense.*

In conclusion, it is noted that the direct parameter projection guarantees that the adaptive control law (27) is continuous, while the adaptive control law (26) using vector projection is not. The latter control law can be made continuous using the vector projection defined by (9) under some smoothness assumption.

Acknowledgement

This research was performed at the Jet Propulsion Laboratory, California Institute of Technology, under contract with the National Aeronautics and Space Administration.

References

[1] J.-P. Aubin, *Viability Theory*, Boston, MA: Birkhauser, 1991.

[2] G. Didinsky and T. Basar, "Minimax Adaptive Control of Uncertain Plants," *Proc. of 33rd IEEE CDC*, Orlando, FL, 1994.

[3] J.C. Doyle, K. Glover, P. Khargonekar, and B. Francis, "State-Space Solutions to Standard $\mathcal{H}_2$ and $\mathcal{H}_\infty$ Control Problems", *IEEE T-AC*, Vol.34, pp.831 - 847, 1989.

[4] G.C. Goodwin and D.Q. Mayne, "A Parameter Estimation Perspective of Continuous Time Model Reference Adaptive Control," *Automatica*, Vol.23(1), pp.57–70, 1987.

[5] P.A. Ioannou and J. Sun, *Robust Adaptive Control,* Upper Saddle River, NJ: PTR Prentice-Hall, 1996.

[6] I. Kanellakopoulos, P. Kokotovic, and R. Marino, "An Extended Direct Scheme for Robust Adaptive Nonlinear Control," *Automatica*, Vol.27(2), pp.247–255, 1991.

[7] M. Krstic, I. Kanellakopoulos, and P. Kokotovic, *Nonlinear and Adaptive Control Design*, 1995.

[8] W.-M. Lu and J.C. Doyle, "$\mathcal{H}_\infty$ Control of Nonlinear Systems: A Class of Controllers," *Caltech CDS Tech. Memo.*, No. CIT-CDS-93-008, also *Proc. 1993 IEEE CDC*, San Antonio, TX, 1993.

[9] W.-M. Lu and A. Packard, "Adaptive $\mathcal{H}_\infty$-Control for Nonlinear Systems: A Dissipation Theoretical Approach," *Proc. 1997 ACC*, Albuquerque, NM, June, 1997.

[10] W.-M. Lu, F. Hadaegh, and A. Packard, "Adaptive Robust Control of Nonlinear Uncertain Systems with Structured Uncertainty," *1997 IEEE CDC* (submitted), San Diego, CA.

[11] D.G. Luenberger, *Optimization by Vector Space Methods*, New York: John Wiley & Sons, Inc., 1969.

[12] Z. Pan and T. Basar, "Adaptive Controller Design for Tracking and Disturbance Attenuation in Parameter-Strict-Feedback Nonlinear Systems," *IEEE T-AC* (submitted), 1996.

[13] P.C. Parks, "Lyapunov Redesign of Model Reference Adaptive Control Systems," *IEEE T-AC*, Vol.11, pp.362–367, 1966.

[14] M.M. Polycarpou and P.A. Ioannou, "On the Existence and Uniqueness of Solutions in Adaptive Control Systems," *IEEE T-AC*, Vol.38(3), pp.474–479, 1993.

[15] J-B. Pomet and L. Praly, "Adaptive Nonlinear Regulation: Estimation from the Lyapunov Equation," *IEEE T-AC*, Vol.37(6), pp.729–740, 1992.

[16] L. Praly, G. Bastin, J.-B. Pomet, and Z.P. Jiang, "Adaptive Stabilization of Nonlinear Systems," *Foundations of Adaptive Control* (P.V. Kokotovic, ed), Berlin: Springer-Verlag, pp.347–433, 1991.

[17] S. Rangan and K. Poolla, "Adaptive $\mathcal{H}_\infty$ Control with Multiple Nonlinear Models," *Preprint*, 1996.

[18] R.T. Rockafellar, *The Theory of Subgradients and Its Applications to Problems of Optimization: Convex and Nonconvex Functions*, Berlin: Heldermann Verlag, 1981.

[19] N. Rouche, P. Habets, and M. Laloy, *Stability Theory by Liapunov's Direct Method*, New York: Springer-Verlag, 1977.

[20] D.G. Taylor, P.V. Kokotovic, R. Marino, and I. Kanellakopoulos, "Adaptive Regulation of Nonlinear Systems with Unmodeled Dynamics," *IEEE T-AC*, Vol.34(4), pp.405–412, 1989.

[21] A.J. van der Schaft, "$\mathcal{L}_2$-Gain Analysis of Nonlinear Systems and Nonlinear State Feedback $\mathcal{H}_\infty$-Control", *IEEE Trans. AC*, Vol.AC-37,pp.770–784, 1992.

[22] J.C. Willems, "Dissipative Dynamical Systems," (Parts I and II), *Arch. Rat. Mech. Anal.*, vol.45, pp.321–393, 1972.

[23] X.H. Yang, F. Wu, and A. Packard, "Adaptive Control of Full Information Problem," *Proc. 1995 ACC*, Seattle, WA, pp.3371–3372, 1995.

AIAA-97-3459

ROBUST ADAPTIVE REGULATION OF POTENTIALLY INVERSELY UNSTABLE FIRST-ORDER HYBRID SYSTEMS

M. De la Sen and S. Alonso-Quesada
Instituto de Investigación y Desarrollo de Procesos IIDP,
Departamento de Electricidad y Electrónica,
Facultad de Ciencias, Universidad del Pais Vasco,
Leioa (Bizkaia), Apdo. 644 de Bilbao, 48080-Bilbao, SPAIN,
Tfno. 34 4 4647700, Ext. 2543 (M. De la Sen), Ext. 2481 (S. Alonso-Quesada).

Abstract. This note presents a robust adaptive stabilization scheme for a class of hybrid time-invariant linear systems involving both continuous and discrete signals. The usual assumptions of inverse stability of the plant and knowledge of the high-frequency gain are not required. The design philosophy used relies on the separation of the continuous and discrete dynamics by generating two additive terms to build the control action. The estimation scheme is unified in the sense that both continuous and discrete estimated parameters are generated from the same adaptation error. The controllability of the nominally estimated plant model is maintained by using an hysteresis switching function under the controllability of the nominal plant. This allows relaxing the usual assumption of stability of the plant inverse. Also, an adaptation dead zone is used for robust stabilization by using a known overbounding function of the contribution of the unmodelled dynamics.

Keywords. Adaptive stabilization, Parameter estimation, Unmodelled dynamics.

1. Introduction

The adaptive stabilization and control of linear continuous and discrete systems has been succesfully developed in the last years[1-6]. Usually, the plant relative degree (interpreted as a delay in the discrete case) and the high-frequency gain sign are assumed to be known together with the absolute upper-bound for that gain in the discrete case. Also, upper-bounds of the numbers of the plant poles and zeros are typically assumed known while the plant is assumed to be of stable inverse. Attemps of relaxing those assumptions have been recently made. In this way, a formulation valid for synthesis of universal stabilizing controllers for systems of relative degree one or two has been formulated[5]. Also, the assumption of knowledge of the high-frequency gain sign for continuous systems has been removed[6,7]. Its discrete counterpart has been relaxed[11] including the lack of necessity of previously required knowledge of the absolute upper-bound of the high-frequency gain. The knowledge of upper-bounds of the polynomial orders can be overcome in the robustness framework through the decomposition of the plant in its nominal and unmodelled parts. For this purpose additional hypothesis are made related to the exponential stability of the unmodelled dynamics and availability of an overbounding function of its related contribution[4,12]. The assumption of stability of the inverse of the plant has been successfully removed in the discrete case and, more recently, in the continuous one[8-10]. In the continuous case, the problem is solved by using either excitation of the plant signals or a modification of the least-squares estimation by exploiting the properties of the standard least-squares covariance matrix[8,9]. This note is motivated by the obvious fact that parallel formulations for hybrid models could potentially be developed with the same guidelines. Such guidelines consist of using hysteresis switching functions to maintain the controllability of the estimated model as well as adaptation dead zones for robust stabilization. Hybrid systems are usually defined as continuous plants subject to discrete parameter estimation and/or discrete control, or systems having coupled continuous and discrete dynamics[13-15]. In this note a class of nominally controllable first-order time-invariant differential-difference plant which belongs to the class of hybrid systems[17] is adaptively stabilized while subject to unmodelled dynamics. The stabilizer design is based upon a pole-placement scheme for the continuous subsystem. An hysteresis switching function is used to maintain the controllability of the nominal estimated model in the estimation scheme. Also, an adaptation dead zone that uses a known overbounding function of the contribution of the unmodelled dynamics, assumed to be exponentially stable, is used for robust adaptive stabilization. It is not assumed that any of both (continuous and discrete) dynamics are of stable inverses. The presence of unmodelled dynamics can be interpreted in this approach as the adaptive stabilization of a plant of arbitrary order n subject to the constraint that (n-1) poles are stable and having known upper-bound of the convergence abscissa. Furthermore, the knowledge of the time-derivatives of the input and output is not required since filtered variables are used in the controller implementation. The proposed stabilizers are an alternative design to previous universal stabilizers[5,18,19] without requiring either nonlinear input gains or hypotesis on prefixed plant relative orders.

2. Robust adaptive stabilizer for a first-order hybrid plant

2.1. Problem statement for nominal known plants

Consider the differential-difference controllable nominally first-order system:

$$\dot{y}(t) + a_1 y(t) + a_2 y[k] + a_3 y[k-1] = \\ = b_0 \dot{u}(t) + b_1 u(t) + b_2 u[k] + b_3 u[k-1] + \eta(t) \quad (1)$$

for $t \in [kT, (k+1)T)$, all integer $k \geq 0$, where "(t)" and "[k]" are continuous time and discrete time arguments, namely, [k] applies for sampled signals at sampling instants kT (T being the sampling period). The term $\eta(t)$ is the contribution of the unmodelled dynamics in (1). The continuous-time argument is suppressed in the following for notational abbreviation unless confusion can arise.The filtered signals subscripted by "f" are introduced being governed by the differential equations[10]:

$$\dot{u}_f = -qu_f + u \; ; \; \dot{y}_f = -qy_f + y \; ; \\ \dot{g}_f = -qg_f + g[k] \; ; \; \dot{\eta}_f = -q\eta_f + \eta \quad (2)$$

for $t \in [kT, (k+1)T)$, all integer $k \geq 0$, where q is a prefixed positive constant, and

$$g[k] := -a_2 y[k] - a_3 y[k-1] + b_2 u[k] + b_3 u[k-1] \quad (3)$$

all integer $k \geq 0$ with a similar equation arising for $g_f[k]$ by substituting all the signals by their filtered versions obtained from (2). Eqn. (1) can be rewritten by using (2)-(3) as follows:

$$[\frac{d}{dt} + q]\,[\dot{y}_f + a_1 y_f - b_0 \dot{u}_f - b_1 u_f - g_f - \eta_f] = 0 \quad (4)$$

for all real $t \geq 0$ and, since $q > 0$,

$$\dot{y}_f + a_1 y_f = b_0 \dot{u}_f + b_1 u_f + g_f + \eta_f + \varepsilon_0 e^{-qt} \quad (5)$$

for all real $t \geq 0$ where $\varepsilon_0\, e^{-qt}$ is an exponentially decaying term depending on initial conditions. By using the change of variable $x_f := y_f - b_0 u_f$, for all real $t \geq 0$, eqn. (5) takes the form:

$$\dot{x}_f = -a_1 x_f + (b_1 - a_1 b_0) u_f + g_f + \eta_f + \varepsilon_0 e^{-qt} \quad (6)$$

where, through the same change of variable, eqns. (3) can be rewritten as:

$$g[k] := -a_2 x[k] - a_3 x[k-1] + \\ +(b_2 - a_2 b_0) u[k] + (b_3 - a_3 b_0) u[k-1] \quad (7)$$

with a similar equation arising for $g_f[k]$ by substituting all the signals by their filtered versions obtained from (2). The next relation is obtained directly from (2) and (6):

$$x_f(kT+\tau) = e^{-a_1\tau} x_f[k] + +y_f'(kT+\tau) + \\ +e^{-a_1\tau} \int_0^{\tau} e^{a_1\tau'} [(b_1 - a_1 b_0) u_f(kT+\tau') + \eta_f(kT+\tau')] d\tau' \quad (8)$$

for all $\tau \in [0, T]$, all integer $k \geq 0$, where

$$y_f'(kT+\tau) = \frac{e^{-q\tau} - e^{-a_1\tau}}{a_1 - q} g_f[k] + \frac{\varepsilon_0}{a_1 - q}\left(e^{-q\tau} - e^{-a_1\tau}\right) + \\ + \left[\frac{1 - e^{-a_1\tau}}{qa_1} + \frac{e^{-q\tau} - e^{-a_1\tau}}{q(a_1 - q)}\right] g[k] \qquad \text{if } q \neq a_1 \quad (9)$$

$$y_f'(kT+\tau) = (g_f[k] - \frac{g[k]}{a_1} + \varepsilon_0)\, e^{-a_1\tau}\, \tau + \\ + \frac{1 - e^{-a_1\tau}}{a_1^2} g[k] \qquad \text{if } \quad q = a_1 \quad (10)$$

for all $\tau \in [0, T]$, all integer $k \geq 0$. Using the inverse change of variable from x(.) to y(.), the solution can be equivalently written as:

$$y_f(kT+\tau) = e^{-a_1\tau} y_f[k] + b_0 (u_f(kT+\tau) - e^{-a_1\tau} u_f[k]) + \\ +e^{-a_1\tau} \int_0^{\tau} e^{a_1\tau'} [(b_1 - a_1 b_0)\, u_f(kT+\tau') + \eta_f(kT+\tau')]\, d\tau' + \\ +y_f'(kT+\tau) \quad (11)$$

for all $\tau \in [0, T]$, all integer $k \geq 0$. In view of (2), (5) and (11), the next controllability result holds since controllability of the filtered output implies that of the unfiltered output.

Proposition 1 (Controllability of the nominal plant). The perfectly modelled ideal system (1) (i.e., eqn. (1) particularized to $\eta \equiv 0$) is controllable iff $b_1 \neq 0$ provided that $b_0 = 0$ and iff $b_0 \neq 0$ provided that $b_1 = 0$. An equivalent condition is that controllability holds iff $b_1 - a_1 b_0 \neq 0$. ***

Note from (11) that if η_f does not involve derivatives of u_f of order higher than two and it is only dependent on previous discretized values of the input, then the above result also holds for the non-ideal situation when $\eta \neq 0$. One gets from (2) and (3),

$$g_f = e^{-q(t-kT)} g_f[k] + \frac{1}{q}(1 - e^{-q(t-kT)})\, g[k] = \theta_d^T \phi_d \quad (12)$$

for all $t \in [kT, (k+1)T)$, all integer $k \geq 0$, where

$$\theta_d^T = [b_2, b_3, -a_2, -a_3] \; ; \; \phi_d^T = [\lambda_u, \lambda_u', \lambda_y, \lambda_y'] \quad (13)$$

with

$$\lambda_v = e^{-q(t-kT)} v_f[k] + \frac{1}{q}(1 - e^{-q(t-kT)})\, v[k] \; ;$$

$$\lambda'_v = e^{-q(t-kT)} v_f[k-1] + \frac{1}{q}(1 - e^{-q(t-kT)})\, v[k-1] \quad (14)$$

with $k = k(T, t) := \max\{\text{integer } z \geq 0 : kT \leq t\}$, for the sampled signal v(t) with $\dot{v}_f = -qv_f + v[k]$. Using (13) and (14), (5) can be rewritten as

$$\dot{y}_f = \theta^T \phi + \eta_f = \theta_c^T \phi_c + \theta_d^T \phi_d + \eta_f \quad (15)$$

where

$$\theta^T := [\theta_c^T, \theta_d^T] \; ; \; \phi^T := [\phi_c^T, \phi_d^T] \quad (16.a)$$

$$\theta_c^T := [b_0, b_1, -a_1, \varepsilon_0] \; ; \; \phi_c^T := [\dot{u}_f, u_f, y_f, e^{-qt}] \quad (16.b)$$

$$\theta_d^T := [b_2, b_3, -a_2, -a_3] \; ; \; \phi_d^T := [\lambda_u, \lambda_u', \lambda_y, \lambda_y'] \quad (16.c)$$

Assume that the following control law is used

$$u_f = u_{f1} + u_{f2} \quad (17)$$

where

$$\dot{u}_{f1} = -s_1\, u_{f1} - r_0\, y_f \quad (18.a)$$

$$\dot{u}_{f2} = -n_0 u_{f2} - n_1 \lambda_u - m_1 \lambda'_u - n_2 \lambda_y - m_2 \lambda'_y \quad \text{if } b_0 \neq 0 \\ u_{f2} = -\left(n_1 \lambda_u + m_1 \lambda'_u + n_2 \lambda_y + m_2 \lambda'_u\right) \quad \text{if } b_0 = 0 \quad (18.b)$$

where s_1, r_0, n_i, m_j (i=0, 1, 2; j=1, 2) are scalar parameters. Note that the second equation in (18.b) reduces to the first one when $b_0 = 0$. In the case of known parameters, the controls u_{f1} and u_{f2} have the respective objectives of closed-loop pole-placement in the ideal case of perfectly modelled dynamics and compensation for their influence of the discrete dynamics in (1). The substitution of (16)-(18) into (15) yields the following closed-loop differential-difference equation for (5):

$$\dot{y}_f = \theta_1^T \phi_1 + \eta_f, \quad \text{all } t \geq 0 \tag{19}$$

where

$$\theta_1^T := [b_1 - b_0 s_1, b_1 - b_0 n_0, b_2 - b_0 n_1, b_3 - b_0 m_1, -(a_1 + b_0 r_0), -(a_2 + b_0 n_2), -(a_3 + b_0 m_2), \varepsilon_0] \tag{20.a}$$

$$\phi_1^T := [u_{f1}, u_{f2}, \lambda_u, \lambda'_u, y_f, \lambda_y, \lambda'_y, e^{-qt}] \tag{20.b}$$

2.2. Pole-placement and compensation of the discrete dynamics

In the case of perfectly modelled dynamics with zero initial conditions (i.e., $e_0 = \eta_f = 0$, all $t \geq 0$), the pole-placement can be achieved by u_f if it exists a function $u_{f2} \in R$ in (17)-(18) such that the following identities hold from (3) and (12)

$$b_0 \dot{u}_{f2} + b_1 u_{f2} + e^{-q(t-kT)} g_f[k] + \frac{1}{q}(1 - e^{-q(t-kT)}) g[k] =$$

$$= b_0 \dot{u}_{f2} + b_1 u_{f2} + b_2 \lambda_u + b_3 \lambda'_u - a_2 \lambda_y - a_3 \lambda'_y = 0 \qquad t \geq 0 \tag{21.a}$$

with y_f being the solution of

$$\dot{y}_f + a_1 y_f = b_0 \dot{u}_{f1} + b_1 u_{f1} \tag{21.b}$$

since $\varepsilon_0 = 0$. Note that (5) follows by summing up eqns. (21). This implies that if the additive control u_{f2} in (17) is generated from (21.a) with zero initial conditions then (21.b) is equivalent to (5) and describes fully the overall hybrid system after filtering since the discrete contribution is anihilated by such a control component. The substitution of the control law (17)-(18) into eqns. (21) yields the two relationships (a)-(b), eqns. (22) below:

(a) $$[(s + a_1)(s + s_1) + r_0(b_0 s + b_1)]\, y_f(s) = C(s)\, y_f(s) \tag{22.a}$$

which arises by taking Laplace transforms with zero initial conditions (i.e., $\varepsilon_0 = 0$) through the substitution of (18.a) into (21.b) provided that the nominal closed-loop poles are fixed to the roots of the strictly Hurwitz polynomial $C(s) = s^2 + c_1 s + c_2$, where s is the Laplace operator and s_1 and r_0 are the controller parameters that satisfy the algebraic linear system

$$\begin{bmatrix} 1 & 0 & 0 \\ a_1 & 1 & b_0 \\ 0 & a_1 & b_1 \end{bmatrix} \begin{bmatrix} 1 \\ s_1 \\ r_0 \end{bmatrix} = \begin{bmatrix} 1 \\ c_1 \\ c_2 \end{bmatrix} \tag{22.b}$$

which is always uniquely solvable in s_1 and r_0 iff system (1) is controllable (i.e., iff $b_1 \neq a_1 b_0$ from Proposition 1):

(b)

$$(b_1 - b_0 n_0) u_{f2} + (b_2 - b_0 n_1)\lambda_u + (b_3 - b_0 m_1)\lambda'_u - -(a_2 + b_0 n_2)\lambda_y - (a_3 + b_0 m_2)\lambda'_y = 0 \qquad \text{if } b_0 \neq 0$$

or

$$(b_2 - b_1 n_1)\lambda_u + (b_3 - b_1 m_1)\lambda'_u - (a_2 + b_1 n_2)\lambda_y - -(a_3 + b_1 m_2)\lambda'_y = 0 \qquad \text{if } b_0 = 0 \tag{22.c}$$

from (18.b) into (21.a). Assume that the nominal plant is controllable. Then, b_0 and b_1 cannot be simultaneously zero from Proposition 1. Thus, u_{f2} can be generated from (18.b) fulfilling the constraint (22.c) as follows. First, then define a small positive real constant ε fulfilling $\varepsilon < \left|\frac{b_1}{a_1}\right|$ if $a_1 \neq 0$, and arbitrary otherwise, and execute the next two actions:

b1) If $|b_0| \geq \varepsilon$, then choose:

$$n_0 = b_1 / b_0; n_1 = b_2 / b_0; m_1 = b_3 / b_0; \quad n_2 = -a_2 / b_0; m_2 = -a_3 / b_0 \tag{23.a}$$

b2) If $|b_0| < \varepsilon$ then $b_1 \neq 0$ (from Proposition 1 since the plant is controllable and $\varepsilon < \left|\frac{b_1}{a_1}\right|$ if $|b_1 - a_1 b_0| > |b_1| - |a_1|\varepsilon > 0$). The choices of the parameters in (18.b) are:

$$n_1 = b_2 / b_1; m_1 = b_3 / b_1; \quad n_2 = -a_2 / b_1; m_2 = -a_3 / b_1 \tag{23.b}$$

2. 3. Robust adaptive stabilization

Now, $\eta_f \neq 0$ and ε_0 (perhaps being nonzero) are unknown. Eqn. (15) is subject to the adaptation error:

$$e := \dot{y}_f - \hat{\theta}^T \phi = -\tilde{\theta}^T \phi + \eta_f, \quad \text{all } t \geq 0 \tag{24}$$

where $\tilde{\theta} = \hat{\theta} - \theta = [\tilde{\theta}_c^T, \tilde{\theta}_d^T]^T$; $\tilde{\theta}_c = \hat{\theta}_c - \theta_c$; $\tilde{\theta}_d = \hat{\theta}_d - \theta_d$ are parametrical errors of the parameter estimates $\hat{\theta} := [\hat{\theta}_c^T, \hat{\theta}_d^T]^T$; $\hat{\theta}_c$ and $\hat{\theta}_d$ being the estimates of the true parameter vectors θ, θ_c and θ_d from the estimation algorithm given below. The control law is evaluated from (17)-(18) after replacing θ by a modification of the parameter estimates given in the algorithm of section 2.4 below. The following assumption is introduced to address the stability proofs in the adaptive case[16]:

Assumption 1. An overbounding function $\overline{\eta}_f$ of the absolute value of the unmodelled dynamics contribution to (5) [or, equivalently, to (15) and (24)], $|\eta_f|$ (i.e., $\overline{\eta}_f \geq |\eta_f|$) is known for all $t \geq 0$. ***

Remark 1. Assumption 1 allows the implementation of a dead zone in the estimation algorithm for robust stabilization by using $\overline{\eta}_f$. Since the plant is controllable, an hysteresis switching function allowing the above referred to modification of the parameter estimates guarantees the controllability of the nominal estimated plant model at all time[10]. This guarantees automatically the uniform stabilizability assumption of the estimated model[16], which is a key technical point to address the mathematical proof of the closed-loop stability, established in that paper as follows: There exists a finite set $\{D_1, D_2, \ldots, D_p\}$ of not necessarily disjoint convex sets such that the parametrization $\theta \in \bigcup_{i=1}^{p} D_i$ and the corresponding model is uniformly stabilizable for all $\hat{\theta} \in \bigcup_{i=1}^{p} D_i$. ***

The next result is related to finding a practical overbounding function $\overline{\eta}_f$.

Proposition 2 (Overbounding function for the contribution of the unmodelled dynamics). Assume that η_f is the sum of a bounded term plus a term related to u_f by a strictly proper exponentially stable transfer function. Thus, $\overline{\eta}_f = \varepsilon_1 \rho + \varepsilon_2$ where $\rho(t) := \operatorname{Sup}_{0 \le \tau \le t} \left\{ \left| \alpha y_f(\tau) + \beta u_f(\tau) \right| e^{-\sigma_0 (t-\tau)} \right\}$ for some real constants $\sigma_0 \in (0,1)$, $\varepsilon_1 \ge 0$, and $\varepsilon_2 \ge 0$, and $\alpha > 0$ and $\beta > 0$.

Proof (Outline): It is addressed in for the case when $g[k] = 0$[16]. If there is a non identically zero $g[k]$ and possibly associated unmodelled terms including a term $\tilde{g}[k]$ being a sum of terms including past input and output signals at sapling instants. Thus, the overbounding function for η_f can be written as:

$$\overline{\eta}_f = \overline{\eta}_{f0} + \sum_{j=0}^{k} e^{-q(t-jT)} \left| \tilde{g}[k] \right| \le$$

$$\le \overline{\eta}_{f0} + \alpha_1 \sum_{j=0}^{k} e^{-q(t-jT)} \operatorname{Sup}_{0 \le \tau \le t} \left(\alpha \left| y_f(\tau) \right| + \beta \left| u_f(\tau) \right| \right)$$

for all $t \in [kT, (k+1)T)$, all integer $k \ge 0$; α_1 being a positive real constant; where $\overline{\eta}_{f0} := \varepsilon_1' \rho + \varepsilon_2'$ is the value of $\overline{\eta}_f$ when $\tilde{g}[k] = 0$ (i.e., there is no unmodelled discrete dynamics contribution), some real constants, $\sigma_0 \in (0,1)$, $\varepsilon_1' \ge 0$ and $\varepsilon_2' \ge 0$ ($\overline{\eta}_{f0}$ being the contribution of the continuous unmodelled dynamics to y_f). Since $q > 0$,

$$e^{-qt} e^{qjT} \le e^{q(j-k)T} = e^{-qlT} \quad (l = k - j)$$

all $t \in [kT, (k+1)T)$, all integer $j \le k \Rightarrow$

$$\sum_{j=0}^{k} e^{-qt} e^{jqT} \le \sum_{l=k}^{0} e^{-lqT} \le \frac{1}{1 - e^{-qT}} \Rightarrow$$

$$\overline{\eta}_f \le \varepsilon_2 + \varepsilon_1 \operatorname{Sup}_{0 \le \tau \le t} \left\{ \left| \alpha y_f(\tau) + \beta u_f(\tau) \right| e^{-\sigma_0 (t-\tau)} \right\}$$

and the result follows with $\varepsilon_2 = \varepsilon_2' + \varepsilon_2''$ (ε_2'' being a constant term associated with initial conditions in $\tilde{g}[k]$) and $\varepsilon_1 := \varepsilon'_1 + \dfrac{\alpha_1}{(1 - e^{-qT})}$. ***

The term ρ can be calculated by knowing the convergence abscissa of the unmodelled part of (1) and, eventually, the number of past inputs and outputs at sampling instants under the stability Assumption 1. The next adaptation mechanism is used which consists of an algorithm for estimation of the parameters of the adaptive controller and the associated adaptive control law:

2.4. Estimation Algorithm and Control Law

Step 1 [A priori Parameter Estimation]:

$$\dot{\tilde{\theta}} = \dot{\hat{\theta}} = b P \phi e \qquad (25.a)$$

$$\dot{P} = -b P \phi \phi^T P \text{ (Covarince adaptation)} \qquad (25.b)$$

with $0 < P(0) = P^T(0)$ and bounded.

$$b := \frac{g s}{1 + \gamma \phi^T P \phi} \qquad (25.c)$$

with $g : [0, \infty) \to [g_1, g_2]$; $\gamma : [0, \infty) \to [\gamma_1, \gamma_2]$ for some real constants g_i, γ_i (i=1, 2) with $0 < g_1 \le g_2 < \infty$; $0 < \gamma_1 \le \gamma_2 < \infty$.

Step 2 [Hysteresis Switching]:

$$\hat{\beta}(t^+) := \begin{cases} \hat{\beta}_j(t^-) \text{ some } j \in \{1,2,\ldots,7\} \text{ with } j \ne i \\ \quad \text{if } c(\hat{\beta}_i(t^+)) = 0 \text{ and} \\ \quad c(\hat{\beta}_j(t^+)) = c(\hat{\beta}_j(t^-)) \ge (1 + \delta_0)\, c(\hat{\beta}_i(t^-)) \\ \hat{\beta}_i(t^-) \text{ otherwise} \end{cases} \qquad (25.d)$$

provided that $\hat{\beta}(t^-) = \hat{\beta}_i(t^-)$ with $c(\hat{\beta}) := \left| \hat{b}_0' \hat{a}_1' - \hat{b}_1' \right|$ for any $t \in R_0^+ := R^+ \cup \{0\}$, some prefixed real constant $\delta_0 \in (0,1)$, where

$$\hat{\beta}_1 = [0, \ldots, 0]^T;\ \hat{\beta}_2 = b\, v;\ \hat{\beta}_3 = -\hat{\beta}_2;\ \hat{\beta}_4 = b(p_1 + p_3);$$
$$\hat{\beta}_5 = -\hat{\beta}_4;\ \hat{\beta}_6 = b(p_1 - p_3);\ \hat{\beta}_7 = -\hat{\beta}_6;$$
$$v := \hat{b}_0 p_3 + \hat{a}_1 p_1 - p_2 \qquad (25.e)$$

p_i being the i-column of the matrix P.

Step 3 [Adaptation Dead Zone]:

$$s := \begin{cases} 0 & \text{if } t \in I_1 := \left\{ t \in R_0^+ : \left| e \right| \le \mu \overline{\eta}_f \right\} \\ f(\mu \overline{\eta}_f, e) / e & \text{otherwise (i.e., for } t \in I_2 := R_0^+ - I_1) \end{cases} \qquad (25.f)$$

where the adaptation error e is defined in (24), $\mu : [0, \infty) \in [\mu_1, \mu_2]$ for some real constants μ_i (i=1, 2) with $\infty > \mu_2 \ge \mu_1 > 1$; and $f : [0, \infty) \in R$ is given by

$$f(\sigma, e) := \begin{cases} e - \sigma & \text{if } e > \sigma \\ 0 & \text{if } |e| \le \sigma \\ e + \sigma & \text{if } e < -\sigma \end{cases} \qquad (25.g)$$

Step 4 [Parameter Estimation Modification]:

$$\hat{\theta}' = \hat{\theta} + \overline{P}\, \overline{\hat{\beta}} \qquad (25.h)$$

$$P = P^T = \begin{bmatrix} P_c & P_{cd} \\ P_{cd}^T & P_d \end{bmatrix} = \begin{bmatrix} p_1^T \\ p_2^T \\ \cdots \\ p_8^T \end{bmatrix}; \qquad (25.i)$$

$$\overline{P} = \begin{bmatrix} P_c & 0 \\ 0 & 0 \end{bmatrix};\ \overline{\hat{\beta}} = [\hat{\beta}^T, 0^T]^T$$

with the four matrix blocks of P and $\overline{P}$ being of orders 4x4.

Step 5 [Computation of the Parameters of the Adaptive Controller]
Calculate $\hat{s}_1$, $\hat{r}_0$, and $\hat{n}_{(.)}$ and $\hat{m}_{(.)}$ from (22.b) and (23) by replacing $a_1 \to \hat{a}_1 \to \hat{a}_1'$, $b_0 \to \hat{b}_0 \to \hat{b}_0'$, $b_1 \to \hat{b}_1 \to \hat{b}_1'$ (the first replacements arising in (22.b) and the second ones in (23)), $a_i \to \hat{a}_i$ and $b_i \to \hat{b}_i$ (i=2, 3).

Step 6 [Control Law]:
The adaptive control law is given by (17), (18) and (23) where the control parameters are substituted by their estimates obtained by including all the above steps of this algorithm.

Remarks 2. (Comments on the Estimation Algorithm). The estimation modification (Step 4) combined with the hysteresis switching (Step 2) is used to maintain the controllability of the nominal estimated model at all time as proved in Lemma 1 below related to the convergence properties of the algorithm. If $\hat{\beta} = 0$ then no modification takes place. If $\hat{\beta}$ converges to a constant in finite time then $P\hat{\beta}$ converges, under mild conditions in Lemma 1, then $\hat{\theta}'$ converges, since $\hat{\theta}$ also converges so that the parameter estimation (First part of Step 1) as well as its modification (Step 4) asymptotically stop[10]. The use of the adaptation dead zone (Step 3) is used for robust stabilization in the presence of unmodelled dynamics[16]. Note from (15)-(16), (25.b) and (25.i) that $\hat{\theta}_d' = \hat{\theta}_d$ at all time. ***

3. Convergence and Stability Results

The next auxiliary result is given about robust stability and convergence properties of the adaptation algorithm which is independent of the control law where L_∞ and L_p denote the spaces of uniformly bounded and p-integrable functions on $(0, \infty)$.

Lemma 1 (Properties of the Parameter Estimation Algorithm). The algorithm of eqns. (25) has the next properties irrespective of the control law:

(i) $\|\hat{\theta}\| \in L_\infty$; $\|\hat{\theta}'\| \in L_\infty$; $\|\hat{\beta}\| \in L_\infty$

(ii) (a) $\bar{f} := \dfrac{|f|}{\sqrt{1+\gamma\phi^T P \phi}} \in L_\infty \cap L_2$, and

(b) $|f| \in L_\infty \cap L_2$; $\dfrac{1}{\sqrt{1+\gamma\phi^T P\phi}} \in L_\infty \cap L_2$

with (b) being valid if $\|\phi\|$ is uniformly bounded; and $b|\eta_f^2 - e^2| \in L_1 \cap L_\infty$; $b^{\frac{1}{2}}|\eta_f + e| \in L_1 \cap L_\infty$: $b^{\frac{1}{2}}|\eta_f - e| \in L_1 \cap L_\infty$ and $\|\dot{\hat{\theta}}\| \in L_\infty$.

(iii) $\|\dot{\tilde{\theta}}(t)\| = \|\dot{\hat{\theta}}(t)\| \in L_\infty \cap L_2$ and $\hat{\theta}$ converges to a finite limit as time tends to infinity.

(iv) $b^{\frac{1}{2}} P\phi \in L_\infty \cap L_2$ and P is uniformly bounded and it converges to a constant limit as time tends to infinity.

(v) The modified estimator $\hat{\theta}'$ of (25.h)-(25.i) has the same properties as $\hat{\theta}$ in (iii) with $\hat{\beta}$ in (25.d) and (25.e) switching a finite number of times.

(vi) The time-varying controller parameters $\hat{s}_1$, $\hat{r}_0$, $\hat{n}_0$, $\hat{n}_i$ and $\hat{m}_i$ (i=1, 2) are uniformly bounded.

The proof is given in Appendix A. ***

Remarks 3. Note from (25.c) and (25.d) to (25.f) that $\|\hat{\beta}\| = b = 0$ and $\hat{\theta}' = \hat{\theta}$ on $\mathbf{I_1}$. Note that $\hat{\theta} = [\hat{\theta}_c^T, \hat{\theta}_d^T]^T$ is always bounded from Lemma 1 (i) and it converges to a limit. The L_2-integrability of $\|\dot{\hat{\theta}}\|$ (Lemma 1 (iii)) or, at least its boundedness and that of the controller parameters [Lemma 1 (i) and (vi)] are used to prove the robust stability of the closed-loop system. ***

The stability of the overall system is investigated by building a time-varying system of state $[y_f, u_f]^T$ involving the adaptive control law. Eqns. (15) or, equivalently, (19) can be rewritten by using the parameter estimation with the modification of (25.h)-(25.i) as follows:

$$
\begin{aligned}
e &= \dot{y}_f - \hat{\theta}^T\phi = -\tilde{\theta}^T\phi + \eta_f = \dot{y}_f - \hat{\theta}'^T\phi + \bar{\hat{\beta}}^T \bar{P}\phi = \\
&= \dot{y}_f - \hat{\theta}'^T\phi + \hat{\beta}^T P_c \phi_c = \dot{y}_f - \hat{\theta}_1^T\phi_1 = -\tilde{\theta}_1^T\phi_1 + \eta_f
\end{aligned} \tag{26}
$$

with the regressor ϕ containing the input time-derivative in contrast to the regressor ϕ_1 from (15) and (19). Since $\dot{u}_{f1} = \dot{u}_f - \dot{u}_{f2} = -\hat{r}_0 y_f - \hat{s}_1 u_{f1}$ with $\hat{r}_0$ and $\hat{s}_1$ being calculated after replacing $\hat{\theta}'$ in (22.b), the next dynamic equivalent system can be built from (26)-(27) by taking into account the parametrization and regressor in (20) as follows:

$$\dot{z} = A z + \vartheta \tag{27}$$

where

$$z := [y_f, u_{f1}]^T;\ \vartheta := [\omega, 0]^T;\ \omega = \omega_1 + \omega_2 \tag{28.a}$$

$$A := \begin{bmatrix} -(\hat{a}_1' + \hat{b}_0'\hat{r}_0) & \hat{b}_1' - \hat{b}_0'\hat{s}_1 \\ -\hat{r}_0 & -\hat{s}_1 \end{bmatrix} \tag{28.b}$$

$$
\begin{aligned}
\omega_1 &:= e + \hat{\varepsilon}_0' e^{-qt} - \hat{\beta}^T P_c \phi_c = \\
&= -\tilde{\theta}^T\phi + \eta_f + \hat{\varepsilon}_0' e^{-qt} - \hat{\beta}^T P_c \phi_c
\end{aligned} \tag{28.c}
$$

$$
\omega_2 := \begin{cases}
(\hat{b}_1' - \hat{b}_0'\hat{n}_0)u_{f2} + (\hat{b}_2 - \hat{b}_0'\hat{n}_1)\lambda_u + (\hat{b}_3 - \hat{b}_0'\hat{m}_1)\lambda_u' - \\
-(\hat{a}_2 + \hat{b}_0'\hat{n}_2)\lambda_y - (\hat{a}_3 + \hat{b}_0'\hat{m}_2)\lambda_y' = 0 \\
\text{if } |\hat{b}_0'| \geq \varepsilon' \text{ (some arbitrary prefixed small positive} \\
\text{real time - dependent constant } \varepsilon' \text{ fulfilling} \\
\varepsilon' < \left|\dfrac{\hat{b}_1'}{\hat{a}_1'}\right| \text{ when } \hat{a}_1' \neq 0\text{), and} \\
(\hat{b}_2 - \hat{b}_1'\hat{n}_1)\lambda_u + (\hat{b}_3 - \hat{b}_1'\hat{m}_1)\lambda_u' - (\hat{a}_2 + \hat{b}_1'\hat{n}_2)\lambda_y - \\
-(\hat{a}_3 + \hat{b}_2\hat{m}_2)\lambda_y' = 0 \\
\text{if } |\hat{b}_0'| < \varepsilon' (\Rightarrow |\hat{b}_1'| > \varepsilon'|\hat{a}_1'| \text{ since } c(\hat{\beta}) > 0)
\end{cases} \tag{28.d}
$$

$\omega_2 = 0$ since from Remark 2, $\hat{\theta}'_d = \hat{\theta}_d$ and all the coefficients in the constraints of (28.d) are zero for all $t \geq 0$ from the substitution of (23) with the estimates obtained in (25). Note that the constant ε' is introduced in (28.d) to avoid division by zero since the estimates are continuous functions. The next result holds for the homogeneous system (27) (i.e., $\vartheta = 0$ in (27)):

Lemma 2 (Exponential Stability of the Homogeneous Equivalent System). The homogeneous linear time-varying system (27)-(28) is globally exponentially stable for any bounded initial conditions if the nominal plant is controllable and Assumption 1 holds. This also implies that Lemma 1 (ii-b) holds.

Proof. The homogeneous system (27) is globally exponentially stable under the next sufficient conditions[16]:

(a) The A-matrix is uniformly bounded for all $t \geq 0$ from the use of Lemma 1 (i) into (28.b) and its eigenvalues lie in the open left hand-side complex plane from (22.a) since the polynomial C(s) is Hurtwitz.

(b) $\int_{t_1}^{t_1+t} \|\dot{A}(\tau)\|^2 d\tau \leq k'_1$, all $t_1 \geq 0$ and $t > 0$ for some real constant $k'_1 > 0$ from the use of Lemma 1 (iii) into (28.b) since $\|\dot{\tilde{\theta}}\| \in L_2$. The proof has been completed. ***

The stability of the inhomogeneous system (27)-(28) is now investigated based on Lemma 2 so as to establish sufficient robust stability conditions for the closed-loop system. The solution of (27) is

$$z(t) = \psi(t,0)\, z(0) + \int_0^t \psi(t,\tau)\, \vartheta(\tau)\, d\tau \qquad (29)$$

where $\psi(t,\tau)$ is the state transition matrix corresponding to A( ). Since the homogeneous system of (27) is exponentially stable, one gets from (29) by using the estimation algorithm (25), Lemma 2, Assumption 1 and Schwartz's inequality:

$$\begin{aligned}
&\|z\| \leq k_0 + k_1 \int_0^t e^{-\sigma(t-\tau)} \|\vartheta(\tau)\| d\tau \leq \\
&\leq k_0 + k_1 \int_0^t e^{-\sigma(t-\tau)} [\mu_2(\varepsilon_2 + \varepsilon_1\, \rho(\tau)) + \\
&+ |\hat{\varepsilon}'_0(\tau)|\, e^{-q\tau} + |\beta^T(\tau) P_c(\tau) \phi_c(\tau)|]\, d\tau + \\
&+ \frac{k_1}{\sqrt{2\sigma}} \left\{ \int_0^t f^2(\tau)\, d\tau \right\}^{\frac{1}{2}} \leq k_{01} + \\
&+ \frac{k_1\, \mu_2\, \varepsilon_1\, k_2}{\sigma} \operatorname*{Sup}_{0 \leq \tau \leq t} \{\|z(\tau)\|\} + \frac{k_1}{\sqrt{2\sigma}} \left\{ \int_0^t f^2(\tau)\, d\tau \right\}^{\frac{1}{2}} + \\
&+ \beta_0\, k_1 \left[\int_0^t e^{-2\sigma(t-\tau)}\, d\tau \right]^{1/2} \left[\int_0^t \|P_c(\tau) \phi_c(\tau)\|^2 d\tau \right]^{1/2}
\end{aligned} \qquad (30)$$

for some real constants $k_0 \geq 0$, $k_1 \geq 0$, $k_2 > 0$ and $\sigma > 0$ where $\infty > \beta_0 \geq \operatorname*{Sup}_{t \in [0,\infty)} \{\|\hat{\beta}(\tau)\|\}$ [from Lemma 1 (i)] and $k_{01} \geq k_0 + k_1 \int_0^t e^{-\sigma t}\, e^{(\sigma - q)\tau} |\hat{\varepsilon}'_0(\tau)|\, d\tau$. Besides $\sup_{t \in [0,\infty)} \{\|\phi_c(\tau)\|\} \leq k_2 \sup_{t \in [0,\infty)} \{\|z(\tau)\|\} + k'_2$, all $t \geq 0$, some real constant $k'_2 \geq 0$ since $z = [y_f, u_{f1}]^T$, ϕ_c is defined in (16) and $|\dot{u}_{f1}| \leq k'_3 u_{f1} + k'_4 y_f \leq |k^T z|$ with $k^T := [k_4, k_3]$ because $\hat{\theta} \in L_\infty$ from Lemma 1 (i). Using $\bar{f} := b^{\frac{1}{2}} |f| = \frac{|f|}{\sqrt{1 + \gamma \phi^T P \phi}}$, one gets from (30):

$$\begin{aligned}
&\|z(t)\|^2 \leq k_3 + k_4 \int_0^t \bar{f}^2(\tau)\, d\tau + \\
&+ k_5 \int_0^t \bar{f}^2(\tau)\, \|z(\tau)\|^2 d\tau + k_6 k_2 k_0 \int_0^t \|z(\tau)\|^2 d\tau
\end{aligned} \qquad (31)$$

provided that $\varepsilon_1 < \sigma / (k_1 k_2 \mu_2)$. Since $\bar{f} \in L_2$ from Lemma 1 (ii) and the application of Gronwall's lemma[10] to (31) implies that $z \in L_\infty$. This leads to the following robust stability result of the closed-loop system.

Theorem 1 (Main Result on Robust Stability). Assume that the nominal system (1) is controllable, Assumption 1 holds with the overbounding function for the unmodelled dynamics satisfying $\varepsilon_1 < \sigma / (k_1 k_2 \mu_2)$ in proposition 2, σ being an available upper-bound of the convergence abscissa of the eigenvalues of the A-matrix in (28.b). Then, the closed-loop system is globally stable. Furthermore, the filtered and unfiltered input and output as well as their time-derivatives converge asymptotically to zero in the absence of unmodelled dynamics.

Outline of proof. $|y_f|$ and $|u_{f1}|$ are in L_∞ since $\|z\| \in L_\infty$ from the above developments. Assume that $|u_{f2}|$ is unbounded so that $|u_{f1}| \in L_\infty$. If $|u_f|$ is unbounded then $|y_f|$ would be unbounded what would contradict $|y_f| \in L_\infty$ so that $|u_{f2}| \in L_\infty$ and $|u_f| \in L_\infty$. Thus, $|\dot{u}_f| \in L_\infty$ and $|\dot{y}_f| \in L_\infty$ from the adaptive control law obtained from (17)-(18) since $\|\hat{\theta}\| \in L_\infty$ from Lemma 1 (i). The results imply also that $\|\phi_c\|$ and $\|\phi_d\|$ and, then, $\|\phi\|$ and $\|\phi_1\|$ are in L_∞. If $\eta_f \equiv 0$ then, $s = 1$, $f = e$ from (25.f)-(25.g). This implies that $\bar{f} = \bar{e}$ (normalized adaptation error) $\in L_\infty \cap L_2$ from Lemma 1 (ii) and $e \in L_\infty \cap L_2$ since $\|P\| \in L_\infty$ from Lemma 1 (iv) and $\|\phi\| \in L_\infty$ $\Rightarrow \operatorname*{Inf}_{t \geq 0} \left(\frac{1}{1 + \gamma \phi^T P \phi} \right) > 0$. This implies that the system (27)-(28) is globally asymptotically stable from the exponential stability for the nominal inhomogeneous system from Lemma 2. As a result, the filtered and unfiltered input and output of the original plant as well as their time-derivatives are uniformly bounded and converge asymptotically to zero. Their time-derivatives are also uniformly bounded and converge asymptotically to zero from the adaptive version of the control law (17)-(18) [Step 6 of the algorithm of section 2.4] since the parameter vectors are uniformly bounded (Lemma 1 (i)) and the input and output converge asymptotically to zero. ***

4. Simulations

In this section we present the results obtained in five simulations. The common dates considered in all experiments are: the sample time T=0.1, the known plant parameters $b_0=1$, $b_3=0$ and $a_3=0$, the unknown plant parameters $b_1=-1$, $a_1=-2$, $b_2=2$, $a_2=-1$ and $\varepsilon_0=10$, the filter parameter q=1, the functions of the algorithm $\gamma(t)=0.1$ and $\mu(t)=1.01$. The unmodelled dynamics is additive and given by $\Delta(s)=\dfrac{\eta(s)}{u(s)}=\dfrac{0.3}{s+4.5}$. In the bound of the unmodelled dynamics we have considered $\sigma_0=0.8$, $\alpha=0.01$ and $\beta=0.55$. The initial conditions in the algorithm are $P(0)=10^5\,I_5$, $\hat{b}_1(0)=1$, $-\hat{a}_1(0)=-2$, $\hat{\varepsilon}_0(0)=1$, $\hat{b}_2(0)=-1.5$ and $-\hat{a}_2(0)=-1.3$. We consider $C(s)=s^2+5s+3$ as Hurwitz polynomial for the pole-placement for the closed-loop system. Since there are some design parameters as the hysteresis width δ_0 and the scalars ε_1 and ε_2 governing the overbounding function for the unmodelled dynamics contribution, the various subsequent simulations are made under different mixed conditions as small versus large hysteresis width and small versus large constants defining the overbounding function for the unmodelled dynamics.

4.1. Simulation 1

We choose $\varepsilon_1=0.4$ and $\varepsilon_2=1$ for the unmodelled dynamics upper-bound. The hysteresis switching function is defined by $\delta_0=0.02$. At each time instant β_j is chosen such that $c(\beta_j)=\max\limits_{1\le k\le 7}\{c(\beta_k)\}$ with k integer.

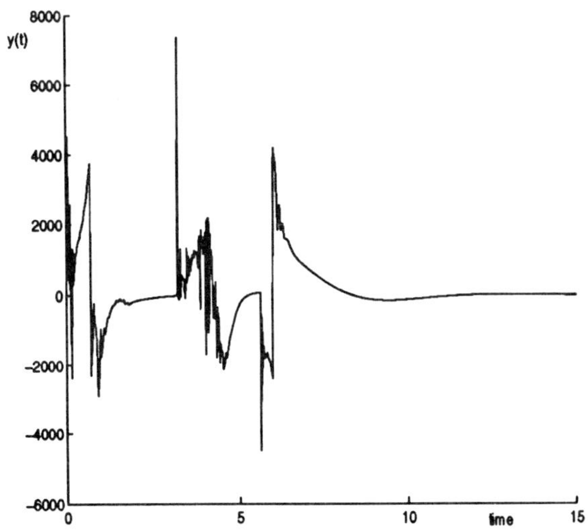

Fig 1.1: Output signal

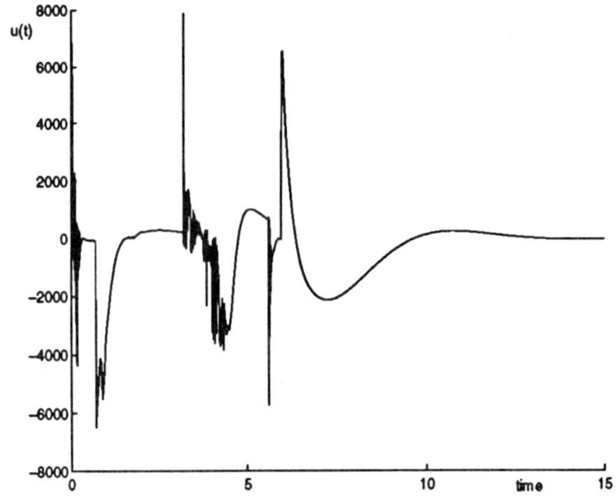

Fig 1.2: Control signal

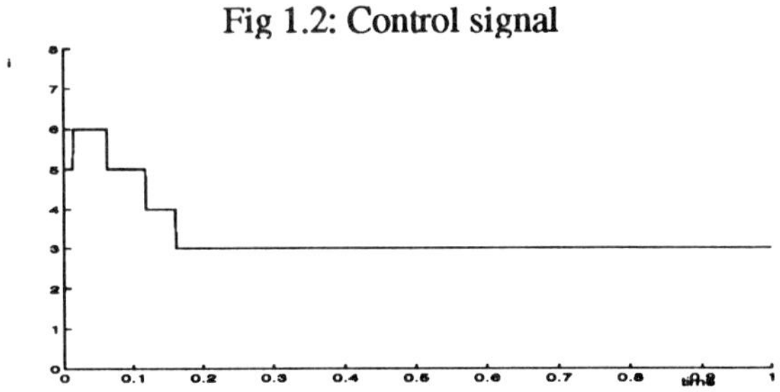

Fig 1.3: Index of the chosen beta-function

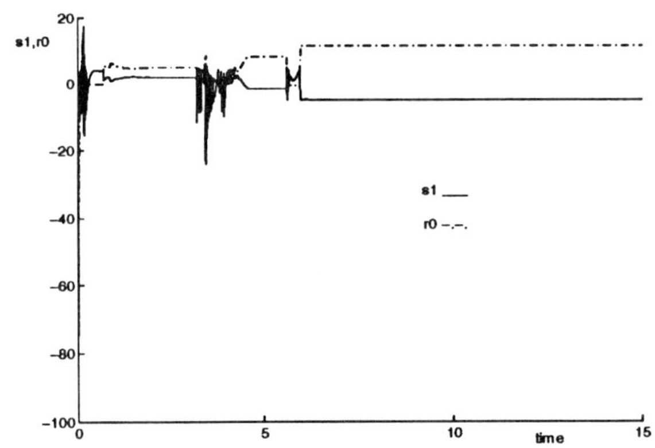

Fig 1.4: Control parameters s_1, r_0

4.2. Simulation 2

$\varepsilon_1=0.75$; $\varepsilon_2=1.5$; $\delta_0=0.02$ and β_j such that $c(\beta_j)=\max\limits_{1\le k\le 7}\{c(\beta_k)\}$.

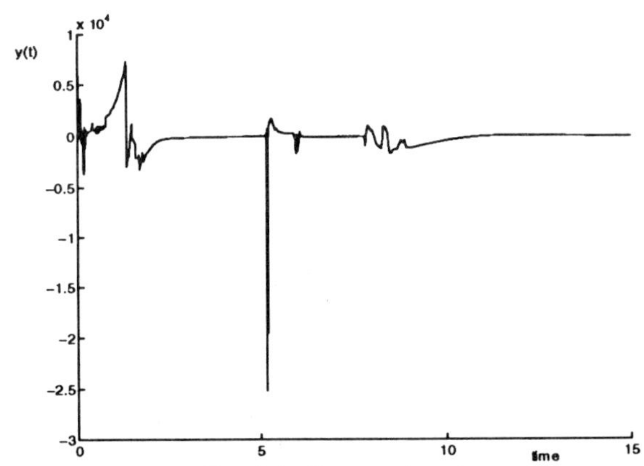

Fig 2.1: Output signal

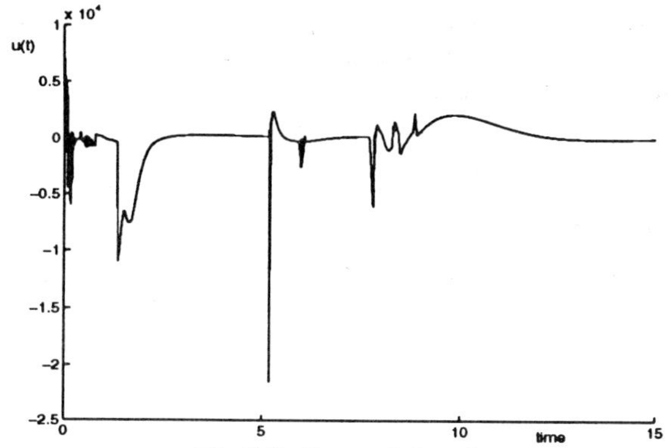

Fig 2.2: Control signal

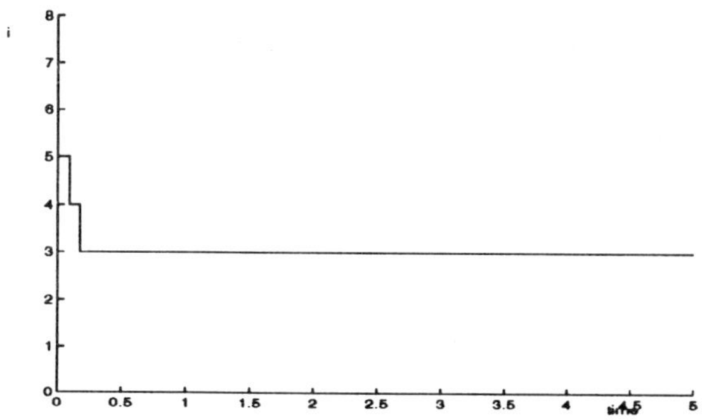

Fig 2.3: Index of the chosen beta-function

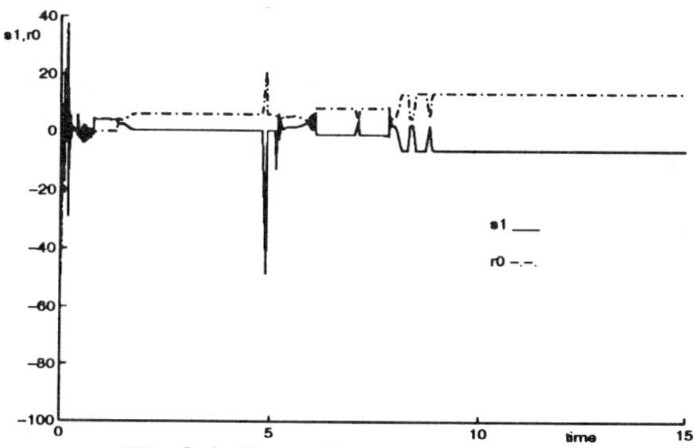

Fig 2.4: Control parameters s_1,, r_0

4.3. Simulation 3

$\varepsilon_1 = 0.75$; $\varepsilon_2 = 1.5$; $\delta_0 = 0.5$ and

β_j such that $c(\beta_j) = \max_{1 \le k \le 7} \{c(\beta_k)\}$.

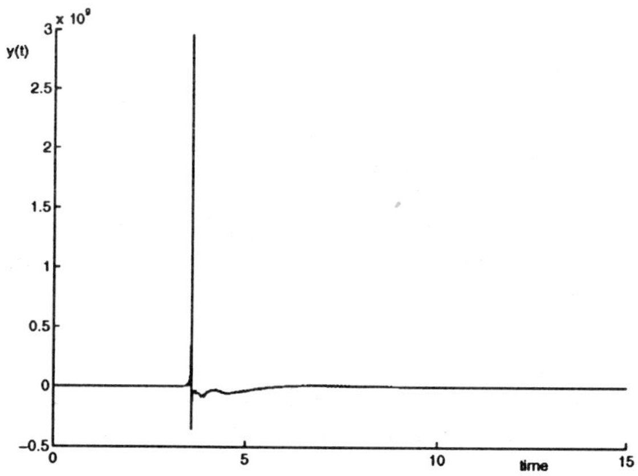

Fig 3.1: Output signal

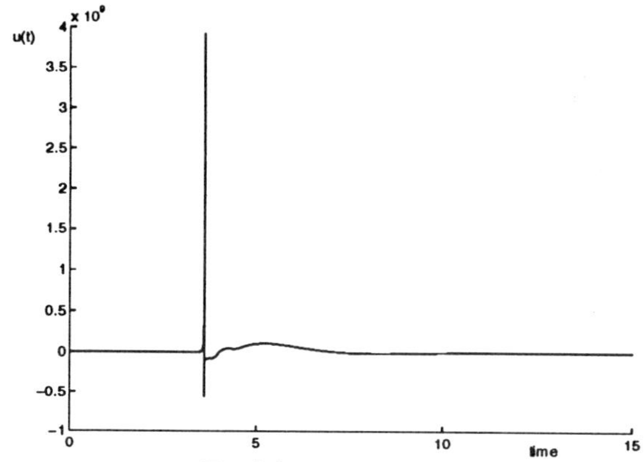

Fig 3.2: Input signal

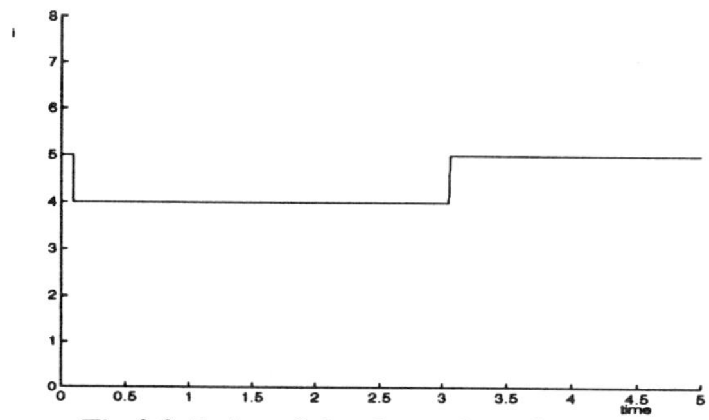

Fig 3.3: Index of the chosen beta-function

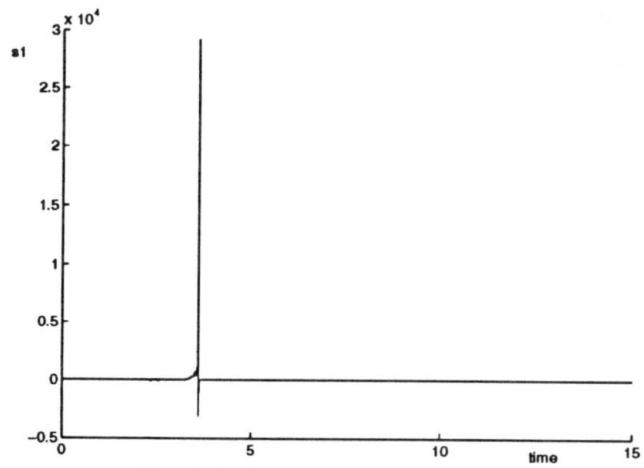

Fig 3.4: Control parameters s_1

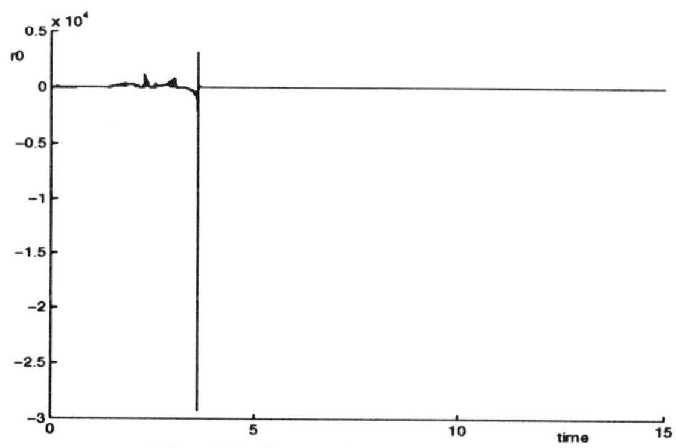

Fig 3.5: Control parameters r_0

4.4. Simulation 4

$\varepsilon_1 = 0.75$; $\varepsilon_2 = 1.5$; $\delta_0 = 0.02$

β_j such that $c(\beta_j) = \min_{1 \le k \le 7} \{c(\beta_k) \mid c(\beta_k) \ge 0.1\}$.

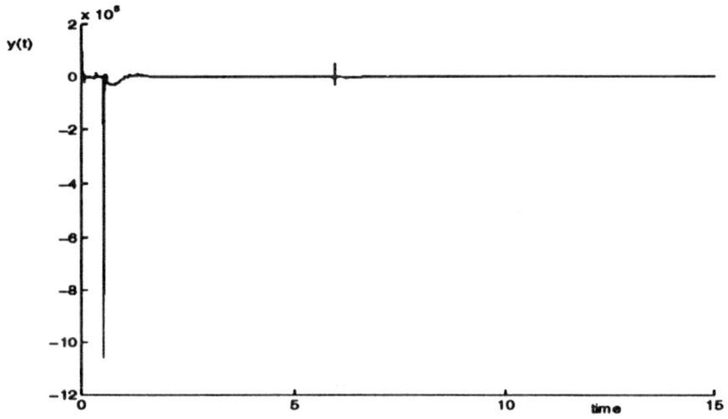

Fig 4.1: Output signal

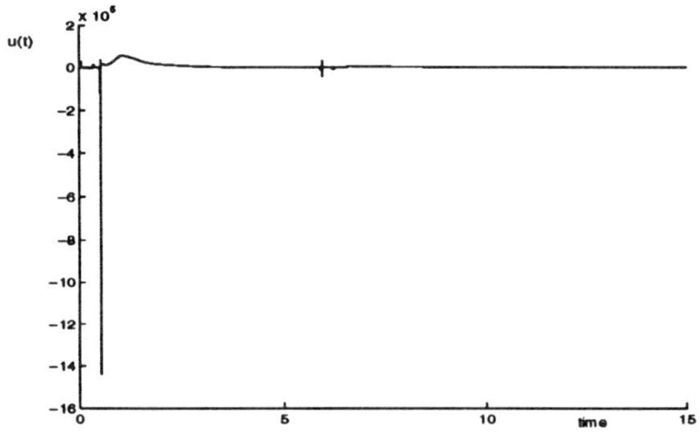

Fig 4.2: Control signal

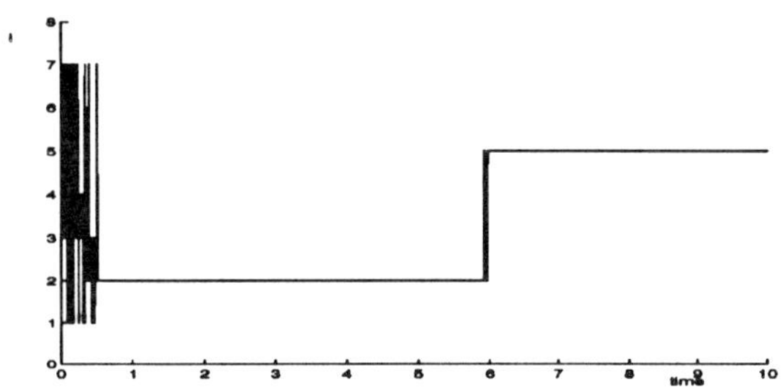

Fig 4.3: Index of the chosen beta-function

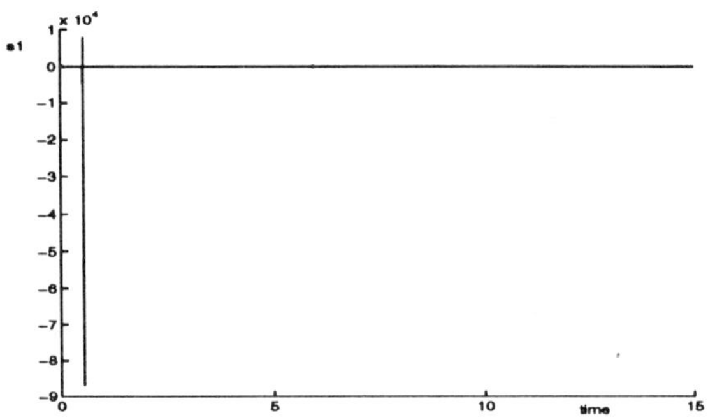

Fig 4.4: Control parameter s_1

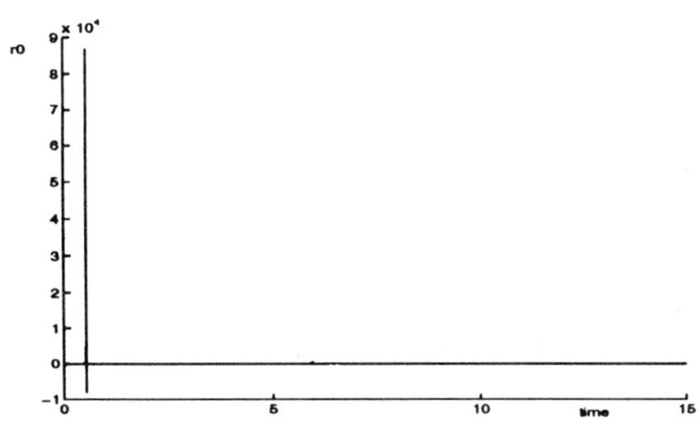

Fig 4.5: Control parameter r_0

Remark 4. From the comparation between the simulations 2 and 3, we see that the transient output is worse in Simulation 3 since in this case the jumps in the signals are more abrupt than in Simulation 2 when a switch in $\hat{\beta}$ occurs. It is due to that the controlability index $c(\hat{\beta})$ of the estimated model of the plant reachs values smaller when the hysteresis parameter is greater. Thus, in Simulation 3 the control parameters s_1 and r_0 reach values greater than in Simulation 2.

Remark 5. In Simulation 4 in the hysteresis the function $\hat{\beta}_j$ which has minimum value for $c(\hat{\beta}_j)$ is chosen. Thus, there are many switches in $\hat{\beta}$ and the transient output is worse than in Simulation 2 where the function $\hat{\beta}_j$ which has maximum value for $c(\hat{\beta}_j)$ is chosen and the hysteresis parameter is small. However, the output signal is better in Simulation 4 than in Simulation 3 where the function $\hat{\beta}_j$ which has maximum value for $c(\hat{\beta}_j)$ is chosen and the hysteresis parameter is large.

Remark 6. The control parameters and the signals y(t) and u(t) also present discontinuities at the time instants at which the dead-zone function s(t) is discontinue. Because of this the transient signals u(t) and y(t) are worse in Simulation 2 than in Simulation 1 since the unmodelled dynamics bound is larger in Simulation 2 than in Simulation 1. Then s(t) presents more discontinuity instants in Simulation 2 than in Simulation 1.

5. Conclusions

A robust adaptive stabilizer for a first-order hybrid plant subject to unmodelled dynamics and possibly not having stable inverse has been proposed without assuming any a priori knowledge on the plant parameters or high-frequency gain sign. The adaptive stabilizer consists of a parameter estimation algorithm and two additive terms in the adaptive control law. One of those terms is associated with prescribed pole-placement in the (non adaptive) stabilization of a known perfectly modelled plant. The second control term is used to compensate for the presence of discretized signals in the modelled part of the hybrid system in such a way that their dynamics is annihilated except for an associated residual term present in the adaptation error. The uniform boundedness of all the signals in the loop is ensured and the global asymptotic stability of the closed-loop system is proved in the absence of unmodelled dynamics.

ACKNOWLEDGEMENTS
The author is very grateful to DGICYT by its partial support of this work through Project PB93 - 0005 .

REFERENCES

[1] A. Feuer and A. S. Morse, "Adaptive control of single-input single-output linear systems", *IEEE Trans. Automat. Contr.*, Vol. AC-23, pp. 557-569, 1978.

[2] A. S. Morse, "Global stability of parameter-adaptive control systems", *IEEE Trans. Automat. Contr.*, Vol. AC-25, no. 3, pp. 433-439, 1980.

[3] K. S. Narendra, Y. H. Lin and L. S. Valavani, "Stable adaptive controller design, Part II: Proof of stability", *IEEE Trans. Automat. Contr.*, Vol. AC-25, no. 3, pp. 440-448, 1980.

[4] P. A. Ioannou and K. Tsakalis, "A robust direct adaptive controller", *IEEE Trans. Automat. Contr.*, Vol. AC-31, no. 11, pp. 1033-1043, 1986.

[5] A. S. Morse, "A three-dimensional universal controller for adaptive stabilization of any strictly proper, minimum phase system with relative degree not exceeding two", *IEEE Trans. Automat. Contr.*, Vol. AC-30, pp. 1188-1191, 1985.

[6] R. D. Nussbaum, "Some remarks on a conjecture in parameter adaptive control", *IEEE* Syst. *Contr. Lett.*, Vol. 3, no. 5, pp. 243-246, 1983.

[7] R. D. Mudgett and A. S. Morse, "Adaptive stabilization of linear systems with unknown high frequency gains", *IEEE Trans. Automat. Contr.*, Vol. AC-30, pp. 549-554, July 1985.

[8] G. Kreisselmeier and M. C. Smith, "Stable adaptive regulation of arbitrary n-th order plants", *IEEE Trans. Automat. Contr.*, Vol. AC-31, pp. 299-305, April 1986.

[9] P. De Larminat, "On the stabilization condition in indirect adaptive control", *Automatica*, Vol. 20, pp. 793-795, Nov. 1984.

[10] R. Lozano, A. Osorio and J. Torres, "Adaptive stabilization of nonminimum phase first-order continuous-time systems", *IEEE Trans. Automat. Contr.*, Vol. AC-39, no. 8, pp. 1748-1751, 1994.

[11] T. H. Lee and K. S. Narendra, "Stable discrete adaptive control with unknown high-frequency gain", *IEEE Trans. Automat. Contr.*, Vol. AC-31, no. 5, pp. 477-479, 1986.

[12] K.S. Narendra and A. M. Annaswamy, "A new adaptive law for robust stabilization without persistent excitation", *IEEE Trans. Automat. Contr.*, Vol. Ac-32, no. 2, pp. 134-135, Feb. 1987.

[13] K. S. Narendra, I. H. Khalifa and A. M. Annaswamy, "Error models for stable hybrid adaptive systems", *IEEE trans. Automat. Contr.*, 1992, Vol. AC-30, no. 4, pp. 339-347, April 1985.

[14] M. De la Sen, "Multirate digital adaptive control", *Computers and Mathematics with Applications*, Vol. 11, no. 12, pp. 1193-1210, 1985.

[15] M. De la Sen, "Multirate hybrid adaptive control", *IEEE Trans. Automat. Contr.*, Vol. AC-31, pp. 582-586, 1986.

[16] R. Middleton, G. C. Goodwin, D.J. Hill and D. Q. Mayne, "Design issues in adaptive control", *IEEE Trans. Automat. Contr.*, Vol. 33, no. 1, pp. 50-58, January, 1988.

[17] P.T. Kabamba and S. Hara, "Worst-case analysis and design of sampled-data control systems", *IEEE Trans. Automat. Contr.*, Vol. 38, no. 9, pp. 1337-1357, Sept. 1993.

[18] B. Martensson, "Remarks on adaptive stabilization of first-order nonlinear systems", *Syst. Contr. Lett.*, Vol. 17, pp. 1-7, 1987.

[19] A.C. Hicks and S. Townley, "On exact solutions of differential equations arising in universal adaptive stabilization", *Syst. Contr. Lett.*, Vol. 20, pp. 117-129, 1993.

Appendix A. *Proof of Lemma1*

(i) Define the positive function $V:=\tilde{\theta}^T P^{-1}\tilde{\theta}$, all $t \geq 0$. Since $\frac{d}{dt}(P^{-1}) = b\phi\phi^T$, one gets from (25.a)-(25.b) and (25.g) for all $t \geq 0$:

$$\begin{aligned}\dot{V} &:= 2\tilde{\theta}^T P^{-1}\dot{\tilde{\theta}} + \tilde{\theta}^T \frac{d}{dt}(P^{-1})\tilde{\theta} = \\ &= b\left\{2\tilde{\theta}^T\phi[-\tilde{\theta}^T\phi + \eta_f] + (\tilde{\theta}^T\phi)^2\right\} = \\ &= b[2(\eta_f - e)e + (\eta_f - e)^2] \leq b[\overline{\eta}_f^2 - e^2] \leq \\ &\leq \begin{cases} -\chi \leq 0 & \text{if} \quad |e| > \mu\overline{\eta}_f \\ 0 & \text{if} \quad |e| \leq \mu\overline{\eta}_f \end{cases}\end{aligned} \tag{A.1}$$

with

$$\chi := \frac{g\,(\mu^2-1)\,f^2}{\mu^2(1+\gamma\phi^T P\phi)} \leq \frac{g\,s\,(\mu^2-1)\,e^2}{\mu^2(1+\gamma\phi^T P\phi)} = \frac{\mu^2-1}{\mu^2} b\, e^2 \tag{A.2}$$

since $|e| \geq |f| \Rightarrow s\,e^2 = f\,e \geq f^2$ and $0 \leq (\mu^2-1)s^2/\mu^2 \leq 1$ from (25.g). Now $V(t_2) \leq V(t_1) \leq V(0) < \infty$, all t_1, t_2 with $t_2 \geq t_1 \geq 0$ and $\|\tilde{\theta}\| \in L_\infty$ and $\|\hat{\theta}\| \in L_\infty$. From (25.b), $P \in L_\infty$ and then from (25.d) to (25.e), $\hat{\beta} \in L_\infty$. Also, $b \in L_\infty$ from (25.c) since $gs \leq g_2$ so that $\hat{\theta}' \in L_\infty$ from (25.h). This completes the proof of (i).

(ii) Note from (A.1) that

$$\infty > V(0) \geq V(0) - V(t) = \int_0^t \frac{g\frac{\mu_2-1}{\mu_2}f^2(\tau)}{1+\gamma(\tau)\phi^T(\tau)P(\tau)\phi(\tau)}d\tau$$

all $t \geq 0$ (A.3)

This implies that $\bar{f} \in L_2$ since $0 < g\frac{\mu_2-1}{\mu_2} < g_2 < \infty$.

It is now proved by contradiction that $|\dot{V}|$ is uniformly bounded. The unboudedness of such a time-derivative implies the existence of a sequence $\{t_i\}$ of nonnegative real numbers such that $t_{i+1} = t_i + \varepsilon_i \geq t_i + \varepsilon$, $\varepsilon > 0$ such that $\lim_{t_i \to \infty}|\dot{V}(t_i)| = \infty$ and $|\dot{V}(t)| \geq |\dot{V}(t_i)|$; $t \in [t_i, t_i + \varepsilon_i]$ since it is zero on $\mathbf{I_1}$. Thus,

$$\infty > V(t_i) - V(t_i + \varepsilon_i) = \int_{t_i}^{t_i+\varepsilon_i}|\dot{V}(\tau)|\,d\tau \geq \varepsilon_i|\dot{V}(t_i)| \geq \varepsilon|\dot{V}(t_i)|$$

Then, a contradiction follows since $\infty \geq \varepsilon|\dot{V}(t_i)| > \infty$ since e is positive and the sequence $|\dot{V}(t_i)|$ diverges on R_0^+. Now, by using (A.1),

$$\left|\dot{V}\right| \geq b(e^2 - \bar{\eta}_f^2) = (b^{\frac{1}{2}}\left|e + \bar{\eta}_f\right|)(b^{\frac{1}{2}}\left|e - \bar{\eta}_f\right|) \Rightarrow$$

$$\Rightarrow (b^{\frac{1}{2}}\left|e \pm \bar{\eta}_f\right|) \in L_\infty \cap L_1$$

Also, since $\bar{f} \in L_\infty \Rightarrow (1+\gamma\phi^T P\phi)^{-1} \in L_\infty$ and then $f \in L_\infty$. Also,

$$\int_0^\infty \bar{f}^2(\tau)\, d\tau \geq \operatorname*{Inf}_{\tau\geq 0}(1+\gamma_1 K_1 \|\phi(\tau)\|_2^2)^{-1} \int_0^\infty f^2(\tau)\, d\tau$$

This implies that $f \in L_2$ since $(1+\gamma\phi^T P\phi)^{-1} \in L_\infty$ and $\bar{f} \in L_2$ provided that $\|\phi\|$ is bounded. A similar reasoning with $\operatorname*{Inf}_{\tau\geq 0}\{f^2(\tau)\} > 0$ on $\mathbf{I_2}$ can be used to prove that $(1+\gamma\,\phi^T P\phi)^{-1} \in L_2$ provided that $\mathbf{I_1}$ is of nonzero measure. From (25.a), (25.c), (25.g) and (25.h) one gets since $s = f / e$ on $\mathbf{I_2}$

$$\dot{\hat{\theta}} = \frac{gsP\phi e}{1+\gamma\phi^T P\phi} = \begin{cases} \dfrac{gP\phi\bar{f}}{\sqrt{1+\gamma\phi^T P\phi}} & \text{for } t \in I_2 \\ 0 & \text{for } t \in I_1 \end{cases} \tag{A.4}$$

but $(1+\gamma\phi^T P\phi)^{-\frac{1}{2}}\|gP\phi\bar{f}\| < \infty, \quad t \in I_2$ since $\bar{f}^2 \in L_\infty \Rightarrow \bar{f} \in L_\infty$. Thus, $\left\|\dot{\hat{\theta}}\right\| \in L_\infty$ since $\dot{\hat{\theta}}$ is uniformly bounded on $\mathbf{I_1}$. This completes the proof of (ii).

(iii)-(iv) Note from (25.b), (25.i) and (A.4) that

$$\left\|\dot{\hat{\theta}}\right\|_2^2 = -b\,\mathrm{tr}(\dot{P})\, e^2 \leq -b\,\mathrm{tr}(\dot{P})\, f^2 =$$

$$= b^2\phi^T P^2\phi\, f^2 = -\mathrm{tr}(\dot{P})\,\bar{f}^2 = b\,\phi^T P^2\phi\,\bar{f}^2$$

since $\bar{f}^2 = bf^2 \leq b\,e^2$ and $b^{\frac{1}{2}}|e|$ is uniformly bounded on R_0^+. Also, one has from (ii) that $\bar{f} \in L_\infty$ so that

$$\int_t^t \left\|\dot{\hat{\theta}}\right\|_2^2 d\tau \leq -K\int_0^t \mathrm{tr}(\dot{P})\, d\tau = K[\mathrm{tr}\,P(0) - \mathrm{tr}\,P(t)] \leq 8K_0 < \infty$$

for some positive real constant K_0. Thus, $\left\|\dot{\hat{\theta}}(t)\right\|_2^2$ is integrable on $[0,\infty)$ so that $\left\|\dot{\tilde{\theta}}\right\|_2 = \left\|\dot{\hat{\theta}}\right\|_2 \in L_2$, $\dot{\tilde{\theta}}$ converges to zero and $\hat{\theta}$ to a bounded limit as t tends to infinity This proves (iii) and the first part of (iv) since $\left\|\dot{\tilde{\theta}}\right\| \in L_\infty$ implies that $b\,e^2$ and $\left|\mathrm{tr}\,(\dot{P})\right|$ are uniformly bounded on R_0^+ so that they are bounded on R_0^+ and $\left|\mathrm{tr}(\dot{P})\right| = \left\|b^{\frac{1}{2}}P\phi\right\|_2^2$ is square-integrable on R_0^+. Also, $\left\|b^{\frac{1}{2}}P\phi\right\|_2^2$ is bounded and integrable on R_0^+. Thus, $\left\|b^{\frac{1}{2}}P\phi\right\| \in L_\infty \cap L_2$. Since $\mathrm{tr}(\dot{P})$ is integrable on R_0^+ and of constant definite sign, it converges asymptotically to zero and then P converges asymptotically to a constant matrix what fully proves (iv).

(v) follows since the convergence of P and $\hat{\theta}$ implies that of $\hat{\beta}$ from (25.d) and (25.e). Since $\delta_0 > 0$ in (25.d), the number of switches of $\hat{\beta}$ is finite.

(vi) is addressed as follows. Since the plant is controllable, there exists a positive real constant δ such that $c(\beta) := |b_0 a_1 - b_1| \geq \delta > 0$ and (22.b) is solvable in s_0 and r_1. Under parameter estimation with eventual parameter modification, eqns. (25.a)-(25.b): $c(\hat{\beta}) = \left|\hat{b}_0'\hat{a}_1' - \hat{b}_1'\right| = \left|\hat{f} + \hat{v}^T\hat{\beta} + b^2 p_1^T\hat{\beta} p_3^T\hat{\beta}\right|$ with $\hat{f} := \hat{b}_0\hat{a}_1 - \hat{b}_1$; $\hat{v} := b[\hat{b}_0 p_3 + \hat{a}_1 p_1 - p_2]$ (Note that $c(\hat{\beta}) = \left|\hat{f}\right|$ if b=0; i. e., $t \in I_1$). Since there is no parameter estimation on $\mathbf{I_1}$, the time instants within this set can be excluded from the controllability considerations which holds provided that the estimated model is controllable on all $t \in I_2$ Now, direct calculations lead to

$$0 < \delta \leq c(\beta) = |b_0 a_1 - b_1| = \left|\hat{f} + \hat{v}^T\beta + b^2 p_1^T\beta\, p_3^T\beta\right| \leq$$

$$\leq (\left|\hat{f}\right| + \|\hat{v}\| + b^2\|p_1\|\,\|p_3\|)\max\{1, \|\beta\|^2\}$$

for some positive real constant δ and $\hat{f}$, $\hat{v}$, p_1 and p_3 cannot be simultaneously zero[10] from Step 2 of the estimation algorithm. Thus, the switching rule of (25.d) and (25.e) always guarantees that if $\hat{\beta}(t^-) = \hat{\beta}_i(t^-)$, some $i \in \{1,2,\ldots,7\}$ all $t \in R_0^+$, then

$$\hat{\beta}(t^+) = \begin{cases} \hat{\beta}_i(t^+) & \text{if} \quad (1+\delta_0)\, c(\hat{\beta}_i(t^+)) > 0 \\ \hat{\beta}_j(t^+) & \text{if} \quad c(\hat{\beta}_i(t^+)) = 0 \text{ and } (1+\delta_0)\, c(\hat{\beta}_j(t^+)) > 0 \\ & \quad \text{some } j \in \{1,2,\ldots,7\}\ j \neq i \end{cases}$$

and such a $j \in \{1,2,\ldots,7\}$ allowing the switching of $\hat{\beta}$ always exists impliying $c(\hat{\beta}(t^+)) > 0$ from the controllability of the estimated plant model. Thus, (22.b) is uniquely solvable in the controller parameters when substituting the plant parameters by their modified estimates. As a result $\hat{r}_1$ and $\hat{s}_0$ are bounded, since the modified estimates, eqns. (25.h)-(25.i), are bounded from (i), and eventually converge to a limit under the conditions of (iv). The parameters $\hat{n}_{(.)}$ and $\hat{m}_{(.)}$ are also bounded (and eventually converge) by using the boundedness (and eventual convergence) of $\hat{\theta}$ and $\hat{\theta}'$ from (i) and (iii). The proof has been completed. ***

AIAA–97–3460

ON THE DIRECTIONAL DEPENDENCE AND NULL SPACE FREEDOM IN UNCERTAINTY BOUND IDENTIFICATION

K.B. Lim*, D.P. Giesy†
NASA Langley Research Center
Hampton, Virginia

Abstract

In previous work, the determination of uncertainty models via minimum norm model validation is based on a single set of input and output measurement data. Since uncertainty bounds at each frequency is directionally dependent for multivariable systems, this will lead to optimistic uncertainty levels. In addition, the design freedom in the uncertainty model has not been utilized to further reduce uncertainty levels. The above issues are addressed by formulating a min-max problem. An analytical solution to the min-max problem is given to within a generalized eigenvalue problem, thus avoiding a direct numerical approach. This result will lead to less conservative and more realistic uncertainty models for use in robust control.

1 Introduction

In applying multivariable robust control analysis and synthesis techniques to linear, time-invariant systems, as in for example [1], a set of plants as defined by a nominal and uncertainty model are required a priori. Nominal models are usually associated with a single "best" model, although what is considered "best" is debatable. Mathematical models derived from first principles are typically used as nominal models or sometimes identified from system identification experiments. In some cases where the physical conditions are not accurately or reliably known due to causes unknown or when simple models are desirable, it makes sense to require that the set of plants in question at least satisfy model validation conditions [2]-[6] with respect to available input and output measurement data.

A recently proposed approach [4, 5, 6] to obtaining uncertainty models from measurement data is based on searching for the smallest uncertainty set that will validate all available data. The approach assumes a single set of input and output measurement data. For multivariable systems, this results in a particular input direction at each discrete frequency which in turn leads to optimistic uncertainty bounds. In addition, the need to admit model validating solutions implys a sufficient degree of freedom in uncertainty structure selection. This remaining freedom in a certain null space has not been utilized to reduce the minimum norm uncertainty bounds. In this paper, we investigate both issues. The problem formulation leads to a min-max problem and the conditions for a solution is given.

We begin by first considering the standard $P-\Delta$ system shown in figure 1. Let the overall structured uncertainty be defined by the block diagonal matrices

$$\Delta = \text{diag}(\Delta^1, \ldots, \Delta^\tau); \quad \Delta^i \in C^{m_i \times n_i},\ i = 1, \ldots, \tau \quad (1)$$

and the set of all block diagonal and stable, rational transfer function matrices be given by

$$\mathcal{D} = \{\Delta(\cdot) \in RH_\infty :\ \Delta^i(s_o) \in C^{m_i \times n_i}, \forall s_o \in \bar{C}_+\} \quad (2)$$

where τ and $\bar{C}_+$ denote the number of uncertainty blocks and the closed right-half plane, respectively [7]. The output error is given by

$$e_y := y - \tilde{y} = y - \mathcal{F}_u(P, \Delta)u \quad (3)$$

where the output, input, and upper linear fractional transformation of (P, Δ), are denoted by y, u, and $\mathcal{F}_u(P, \Delta)$, and are variables in the z-domain so that they are complex vectors containing both gain and phase information.

For model validation [4, 5, 6] with respect to input and output data u and y, the output error is set to zero so that from (3)

$$y - P_{22}u = P_{21}\Delta(I - P_{11}\Delta)^{-1}P_{12}u \quad (4)$$

Observe from the left hand side of (4) that if the given nominal model P_{22} and input cancels the measured outputs (to within white noise), then $\Delta = 0$ validates the model. This is the case where a single

*Aerospace Technologist, Guidance & Control Branch, Flight Dynamics & Control Division, k.b.lim@larc.nasa.gov
†Aerospace Technologist, Guidance & Control Branch, Flight Dynamics & Control Division, d.p.giesy@larc.nasa.gov

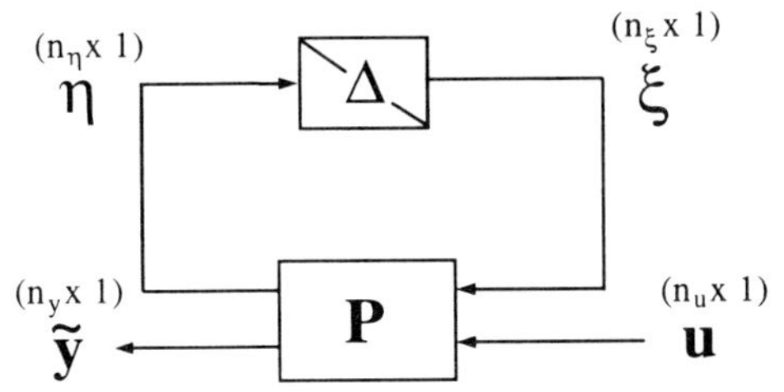

Figure 1: $P - \Delta$ structure in standard form.

"optimal" model can be found such that the output errors are perfectly uncorrelated to the inputs. In this ideal case, a single "true" model is available and robustness issues with respect to model error do not exist. Otherwise, note that the nominal model error is directionally dependent on the inputs. Hence, with a different (directional) choice of inputs, the mininum norm model validating uncertainty computed will also be different.

2 Problem Formulation

To address the directional dependence of the mininum norm model validating uncertainty with respect to the test input, we assume that n_u sets of inputs and corresponding outputs at each frequency are available and are arranged in the matrix forms

$$U := [u_1, \ldots, u_{n_u}]; \quad Y := [y_1, \ldots, y_{n_u}] \tag{5}$$

These sets of inputs and outputs can be formed from segments of a long identification data or independent identification experiments. Assuming a linear time-invariance system, a general input

$$u = U\alpha \tag{6}$$

will result in output

$$y = Y\alpha \tag{7}$$

so that the error equation for a general input signal is

$$e_y(\alpha, \Delta) = Y\alpha - \mathcal{F}_u(P, \Delta)U\alpha \tag{8}$$

Note that α is a complex coefficient vector which help to span the n_u dimensional vector space. The approach for handling the directional dependence is to search for the largest possible mininum norm model validating uncertainty over all possible test inputs.

Consider model validating a linear combination of input/output data set, i.e.,

$$e_y(\alpha, \Delta) = 0; \quad \forall\alpha \tag{9}$$

From (8), this condition is

$$P_{21}\Delta(I - P_{11}\Delta)^{-1}P_{12}U\alpha = (Y - P_{22}U)\alpha; \quad \forall\alpha \tag{10}$$

Using the earlier approach and making the assumptions as given in [4, 5], the Δ can be unscrambled to the form

$$\Delta x = y; \quad \forall\alpha \tag{11}$$

where

$$x(\alpha, \phi) := B\alpha + P_{11}\phi \tag{12}$$
$$y(\alpha, \phi) := A\alpha + \phi \tag{13}$$

and

$$A = P_{21}^{+}(Y - P_{22}U) \tag{14}$$
$$B = P_{12}U + P_{11}P_{21}^{+}(Y - P_{22}U) \tag{15}$$

$\phi \in \mathcal{N}(P_{21})$, where $\mathcal{N}(\cdot)$ denotes the null space. The variables depend on the arguments, α and ϕ, and in the remaining development, these arguments will not be explicitly used for simplicity.

To obtain lower bounds on the uncertainty components, partition in a conformal manner with respect to the uncertainty blocks the following:

$$x = \text{col}\left(x^1, \ldots, x^\tau\right); \quad y = \text{col}\left(y^1, \ldots, y^\tau\right) \tag{16}$$

Due to the block diagonality of Δ,

$$\Delta x = y,\ \forall\alpha \quad \Leftrightarrow \quad \Delta^i x^i = y^i, \forall(i, \alpha) \tag{17}$$

and in addition

$$\bar{\sigma}(\Delta) = \max_i \bar{\sigma}(\Delta^i), \quad \forall\alpha \tag{18}$$

Consider the following Lemmas as given in [4, 5].

Lemma 1:
For each component, $x^i \neq 0$,

$$\bar{\sigma}(\Delta^i) = \|\Delta^i\|_2 \geq \frac{\|y^i\|_2}{\|x^i\|_2}, \quad \forall(i, \alpha). \quad \Box \tag{19}$$

Lemma 2:
Let $y^i \in C^{m_i}$, $x^i \in C^{n_i}$, $x^i \neq 0, \exists\Delta^i \in C^{m_i \times n_i}$: $\Delta^i x^i = y^i$, $\bar{\sigma}(\Delta^i) = \frac{\|y^i\|_2}{\|x^i\|_2}$. $\Box$

The first lemma gives a lower bound on the maximum singular value for the ith block diagonal uncertainty and the input/output data defined by α. In order to model validate the input/output relationship as defined by (17) for a given input vector function, x^i, and an output vector function, y^i, the smallest normed uncertainty is given by the lower bound in Lemma 1. Of course if there are several input/output data sets (corresponding to different α's) to model validate, then the largest lower bound with respect to α is needed to validate all data given. The second lemma shows that the ith component uncertainty, Δ^i, can be chosen to achieve this lower bound. Thus Lemma 2 ensures that a minimum normed uncertainty exists which will validate the given data. Again, if there are several input/output data sets the largest bound with respect to α is necessary.

The problem is summarized in the following min-max problem:

$$\textit{Problem 1} : \max_{\alpha}\left(\min_{\phi\in\mathcal{N}(P_{12})} J(\alpha,\phi) := \frac{\|y(\alpha,\phi)\|_2}{\|x(\alpha,\phi)\|_2}\right) \quad (20)$$

The physical interpretations for α and ϕ are clear. The variable α represents all possible input disturbances in the system identification experiment whereas the variable ϕ denotes the remaining freedom in the assumed uncertainty. Obviously, a larger number of uncertainty components in Δ will generally give greater freedom in ϕ as realized by an increased dimension of the null space, $\mathcal{N}$.

3 Solutions

3.1 General Case

In this section we consider the general min-max problem defined in Problem 1. To avoid dealing with square roots, we consider the equivalent min-max of the ratio of norms squared. The problem considered is given as follows:

$$\max_{\alpha}\left(\min_{\psi} J(\alpha,\phi(\psi)) := \frac{\|A\alpha+N\psi\|_2^2}{\|B\alpha+D\psi\|_2^2}\right) \quad (21)$$

where

$$\begin{aligned} N^H A &= 0 && (22)\\ N^H N &= I && (23)\\ D &= P_{11}N && (24)\end{aligned}$$

The matrix N denotes an orthogonal basis for the null space $\mathcal{N}$ and (22) shows its orthogonality with respect to the range space of the pseudo-inverse P_{21}^{+}. The free parameter ψ has the same dimension as the null space and is related to ϕ by

$$\phi = N\psi \quad (25)$$

Let

$$\begin{aligned} A &:= A_r + jA_i && \in C^{r\times n} && (26)\\ N &:= N_r + jN_i && \in C^{r\times m} && (27)\\ B &:= B_r + jB_i && \in C^{s\times n} && (28)\\ D &:= D_r + jD_i && \in C^{s\times m} && (29)\\ \alpha &:= \alpha_r + j\alpha_i && \in C^{n\times 1} && (30)\\ \psi &:= \psi_r + j\psi_i && \in C^{m\times 1} && (31)\\ &&&&& (32)\end{aligned}$$

It can be shown that

$$\begin{aligned} \|A\alpha+N\psi\|_2^2 &= \xi^T\tilde{P}^T\tilde{P}\xi := a(\alpha,\psi) \ge 0 && (33)\\ \|B\alpha+D\psi\|_2^2 &= \xi^T\tilde{Q}^T\tilde{Q}\xi := b(\alpha,\psi) \ge 0 && (34)\end{aligned}$$

where the real column vector variable is

$$\xi := \begin{Bmatrix} \alpha_r \\ \alpha_i \\ \psi_r \\ \psi_i \end{Bmatrix} \quad (35)$$

and

$$\tilde{P} = [\ \tilde{P}_1 \quad \tilde{P}_2\ ] = \begin{bmatrix} A_r & -A_i & N_r & -N_i \\ A_i & A_r & N_i & N_r \end{bmatrix} \quad (36)$$

$$\tilde{Q} = [\ \tilde{Q}_1 \quad \tilde{Q}_2\ ] = \begin{bmatrix} B_r & -B_i & D_r & -D_i \\ B_i & B_r & D_i & D_r \end{bmatrix} \quad (37)$$

The positive definiteness of $\tilde{Q}^T\tilde{Q}$ is assumed from a physical argument. Observe that if $b = 0$ and $a \ne 0$, then J, the lower bound on the norm of model validating uncertainty is unbounded, which is to say a model validating uncertainty do not exist. Hence we assume that $b > 0\ \forall\xi \ne 0$ so that $\tilde{Q}^T\tilde{Q} > 0$ or equivalently, $\tilde{Q}$ has full column rank. The cost function is then written as

$$J(\alpha,\psi) = \frac{a(\alpha,\psi)}{b(\alpha,\psi)} \ge 0 \quad (38)$$

As figure 2 illustrates, a solution to the min-max problem lies at a saddle point. The particular type of saddle point we seek must possess the following properties: it should be stationary with respect to ξ, local maximum with respect to α, and local minimum with respect to ψ. Specifically,

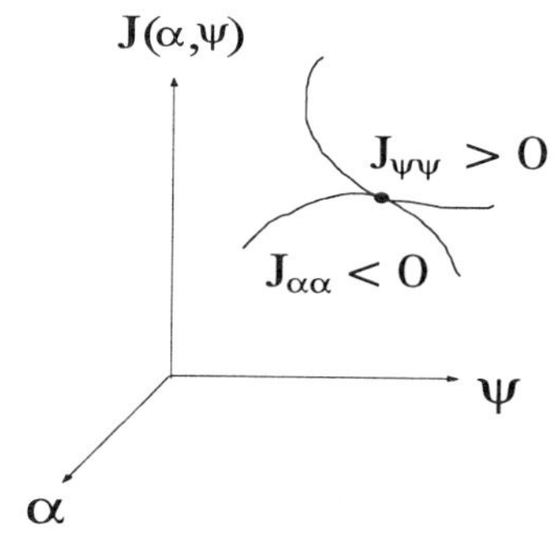

Figure 2: Geometry of a min-max solution: saddle point.

$$\frac{\partial J}{\partial \xi^T} = 0 \Leftrightarrow \frac{\partial J}{\partial \alpha^T} = 0,\ \frac{\partial J}{\partial \psi^T} = 0 \quad \text{stationarity (39)}$$

and

$$\frac{\partial}{\partial \alpha}\left(\frac{\partial J}{\partial \alpha^T}\right) \quad < \quad 0 \qquad \text{local maximum wrt } \alpha \qquad (40)$$

$$\frac{\partial}{\partial \psi}\left(\frac{\partial J}{\partial \psi^T}\right) \quad > \quad 0 \qquad \text{local minimum wrt } \psi \qquad (41)$$

By taking partial derivatives, it can be shown that the stationary condition reduce to

$$[\tilde{P}^T\tilde{P} - J\tilde{Q}^T\tilde{Q}]\xi = 0 \qquad (42)$$

It can also be shown that the local maximum and minimum conditions reduce to the sign definiteness conditions

$$\tilde{P}_1^T\tilde{P}_1 - J\tilde{Q}_1^T\tilde{Q}_1 \quad < \quad 0 \qquad (43)$$

$$\tilde{P}_2^T\tilde{P}_2 - J\tilde{Q}_2^T\tilde{Q}_2 \quad > \quad 0 \qquad (44)$$

Finally, it follows that the above negative and positive definiteness conditions in (43) and (44) respectively, is equivalent to the condition

$$\bar{\sigma}(\tilde{P}_1 L_1^{-T}) < J < \underline{\sigma}(\tilde{P}_2 L_2^{-T}) \qquad (45)$$

where L_i are square, nonsingular matrices derived from the singular value decompositions of $\tilde{Q}_i$ [8], i.e,

$$\tilde{Q}_i = U_i\Sigma_i V_i^T \qquad L_i = V_i\Sigma_i, \quad i = 1, 2. \qquad (46)$$

Note that $\tilde{Q}_1$, and $\tilde{Q}_2$ will have full column rank matrices since $\tilde{Q}$ was earlier assumed to have full column rank on physical grounds. To recapitulate, the min-max problem defined by Problem 1 reduces to finding α and ψ, such that J is stationary with respect to both α and ψ, i.e., the equality (42) holds, and the curvature of J with respect to α is negative and is positive with respect to ψ, i.e., the inequality (45) holds.

To satisfy the conditions defined in (42) and (45), first note that both J and ξ are unknowns in (42) so that this equation can be treated as a real, symmetric generalized eigenvalue problem. The functional dependence of both a and b and hence J on ξ is not constraining as evident from premultiplying (42) by ξ^T and solving for J. Hence a simple and direct solution algorithm follows. First, find all stationary points by solving for all the eigenvalues and corresponding eigenvectors for the eigenvalue problem in (42)

$$J^k := \left(\frac{a}{b}\right)_k, \quad \xi^k = \left\{\begin{array}{c} \alpha_r^k \\ \alpha_i^k \\ \psi_r^k \\ \psi_i^k \end{array}\right\}, \qquad (47)$$

where $k = 1, \ldots, 2(n_u + \dim(\mathcal{N}))$. Second, check to see if any of the λ_k's lie within the inequality bound (45). The results are summarized in the following:

Proposition 1:
Let J^k, ξ^k be the eigenpair of (42). Then the eigenpair is a solution to the min-max problem in (20) only if J^k satisfy the inequalities (45). □

If more than one eigenvalue lie within the inequality bound given in (45), this indicates that there is more than one saddle point and the smallest eigenvalue should be chosen since the square root of these eigenvalues directly correspond to uncertainty lower bounds. From a computational standpoint, the real, symmetric generalized eigenvalue problem that needs to be solved is of dimension $2(n_u+\dim(\mathcal{N}))$, which will not likely be a problem for most practical applications.

3.2 Special Case, $\phi = 0$

Consider the special case of $\phi = 0$ or when $\mathcal{N}(P_{21}) = \{0\}$. The mini-max problem reduces to the single optimization problem:

$$\textit{Problem 2}: \ \max_{\alpha} J(\alpha, 0) := \frac{||y(\alpha,0)||_2}{||x(\alpha,0)||_2} = \frac{||A\alpha||_2}{||B\alpha||_2} \qquad (48)$$

Since Δ must be finite in satisfying the model validation condition (9) for all nonzero α, $B^H B$ must be positive definite. With the nonsingular (and hence unique) coordinate transformation

$$L^T\alpha = \beta \qquad (49)$$

the optimization problem reduces to

$$\max_{\alpha} J(\alpha, 0) \quad = \quad \max_{\alpha} \frac{||A\alpha||_2}{||B\alpha||_2} \qquad (50)$$

$$= \quad \max_{\beta} \frac{||AL^{-T}\beta||_2}{||\beta||_2} \qquad (51)$$

$$= \quad \bar{\sigma}(AL^{-T}) \qquad (52)$$

The above development is summarized in the following.

Proposition 2:
The largest lower bound of the uncertainty with respect to all possible linear combinations of a set of n_u input/output data is given as

$$\bar{\sigma}(\Delta) \geq J(\bar{\alpha}, 0) = \bar{\sigma}(AL^{-T}). \ \square \qquad (53)$$

Consider the special case where a specific lower bound is used instead of any linear combination. For example, let δ_j denote the jth column of an n_u by n_u identify matrix. Then the jth input vector corresponds to $\alpha = \delta_j$ so that

$$u_j = U\delta_j; \quad y_j = Y\delta_j \tag{54}$$

From (49) the corresponding β is

$$\beta_j = L^T \delta_j \tag{55}$$

so that from Proposition 2

$$\frac{\|y(\delta_j, 0)\|_2}{\|x(\delta_j, 0)\|_2} = \frac{\|AL^{-T}\beta_j\|_2}{\|\beta_j\|_2} \leq \bar{\sigma}(AL^{-T}) \leq \bar{\sigma}(\Delta); \quad \forall \quad j = 1, \ldots, n_u \tag{56}$$

The expression on the left hand side of (56) denotes an uncertainty bound based on a single input/output data and is given in the earler works [4, 5].

3.3 Structured Uncertainty

To compute norm bounds for unstructured uncertainty, the maximum singular value in (53) or the eigenvalues in (47) need to be computed for general and special case respectively. However, for structured uncertainty case, eigenvector ξ^k and the maximizing principal vector $\bar{\beta}$ is needed to compute individual component uncertainty bounds for the general and special case respectively.

For the special case, the norm bound on the ith component uncertainty is given from (19) as

$$\bar{\sigma}(\Delta^i) \geq \frac{\|[AL^{-T}\bar{\beta}]^i\|_2}{\|\bar{\beta}^i\|_2} \tag{57}$$

The superscript i in the right hand side of (57) denotes the ith component of the column vectors. The principal vector, $\bar{\beta}$, comes from the singular value decomposition

$$AL^{-T} = P \cdot \text{diag}\left(\bar{\sigma}(AL^{-T}), \ldots\right) \cdot \left[\bar{\beta}, \ldots\right]^T \tag{58}$$

Similarly, for the general case, the norm bound on the ith component uncertainty is also given from (19) as

$$\bar{\sigma}(\Delta^i) \geq \frac{\|y^i(\alpha^k, \phi(\psi^k))\|_2}{\|x^i(\alpha^k, \phi(\psi^k))\|_2} \tag{59}$$

where the subscript k in (59) denotes the smallest eigenpair solution that satisfies the min-max inequality bounds.

4 Conclusion

The issue of directional dependence of multivariable uncertainty models is addressed along with the uncertainty design freedom beyond model validation. The problem is formulated as a search for a saddle point in a finite dimensional, complex space. Numerical and experimental validation and applications of the proposed solution is underway. It is expected that these results will lead to less conservative and more realistic uncertainty models for use in robust control.

References

[1] Balas, G.J., Doyle, J.C., Glover, K., Packard, A., and Smith, R., *μ-Analysis and Synthesis Toolbox*, MUSYN Inc., Minneapolis, MN, 1994.

[2] Smith, R.S., Doyle, J.C., "Model Validation: A connection between robust control and identification," *IEEE Transactions on Automatic Control*, vol 37, No 7, July 1992, pp.942-952.

[3] Kumar, A., and Balas, G.J., "A scaling approach to model validation in the μ framework," 1995 *American Control Conference*, pp.693-697.

[4] Lim, K.B., Balas, G.J., and Anthony, T.C., "Minimum-norm model validating identification for robust control," AIAA Paper No. 96-3717.

[5] Lim, K.B., "Closed form solution for minimum norm model validating uncertainty," 1997 *American Control Conference*, Paper No. 97050.

[6] Lim, K.B., Cox, D.E., Balas, G.J., and Juang, J-N., "Validation of an experimentally derived uncertainty model," AIAA Paper No. 97-0244.

[7] Stein, G., and Doyle, J.C., "Beyond singular values and loop shapes," *Journal of Guidance, Control, and Dynamics*, vol.14, No.1, Jan-Feb 91, pp.5-16.

[8] Golub, G.H., and Van Loan, C.F., *Matrix Computations*, The johns Hopkins University Press, 1983, p.89.

AIAA-97-3463

OPTIMAL PASSIVE DAMPER PLACEMENT METHODOLOGY FOR INTERFEROMETERS USING INTEGRATED STRUCTURES/OPTICS MODELING[1]

Sanjay S. Joshi,[2] Mark M. Milman,[3] and James W. Melody[4]
Jet Propulsion Laboratory
California Institute of Technology
Pasadena, California 91109

ABSTRACT

Interferometers collect the light from stars at two different points in space using telescopes or mirrors (apertures). They then direct this light to a common point to produce interference fringes. Sensitivity of the instrument is a function of separation of telescopes. A crucial requirement for this system is that the distance the light travels from one aperture to the interference point, called the optical pathlength, must match the distance the light travels from the other aperture to the interference point to within a few nanometers. The difference in optical pathlengths is called optical path difference (OPD). Clearly, with any type of structural disturbance, such as that caused by attitude control actuators, the nanometer requirement will not be met due to structural vibration. The fundamental approach to meeting this requirement involves three vibration attenuation layers: vibration isolation, structural quieting, and active optics. The strategy is to isolate vibrating machinery at the point of attachment, to add damping to the structure, and finally, to actively control certain optical components in order to reach the requirement. This paper focuses specifically on passive structural damping with the goal to lower the control demand on active optics. In order to develop a methodology for passive damper placement for interferometers, several analytical tools/methods must be integrated. These are (1) an integrated structural/optics interferometer model, (2) passive damper models, (3) damper placement strategy, and (4) optimization metric analysis. In this paper, we discuss each of these analytical tools/methods by applying them to a representative interferometer. The goal is to suppress errors in OPD caused by random noise disturbances on one point of the structure. An integrated structures/optics model is created of an interferometer. Since an unlimited number of dampers is infeasible and a large number of possible damper positions exist, we explore the use of combinatorial optimization of damper locations. In particular, simulated annealing optimization is chosen as a way to optimize over many possible combinations.

1 INTRODUCTION

Both NASA and the European Space Agency are currently studying the feasibility of placing large baseline interferometers in space to perform detailed astrometric measurements and also to image distant solar systems. Interferometers collect light from stars at two different points in space using telescopes or mirrors (apertures). They then direct this light to a common point to produce interference fringes. The visibility of these fringes provides information that allows astrometry and imaging. Sensitivity of the instrument is a function of separation of telescopes. A

---

[2]Staff Engineer.Member AIAA.Corresponding Author: JPL Mail Stop 198-326, Phone: (818) 354-0541 Email: sanjay.s.joshi@jpl.nasa.gov

[3]Senior Engineer.

[4]Staff Engineer.

crucial requirement for this system is that the distance the light travels from one aperture to the interference point, called the optical pathlength, must match the distance the light travels from the other aperture to the interference point to within a few nanometers (1×10^{-9} meters) . The difference in optical pathlengths is called optical path difference (OPD). Many interferometer designs studied to date place the two collecting apertures on either end of a long flexible structure. Clearly, with any type of structural disturbance, such as that caused by attitude control actuators, the nanometer requirement will not be met due to structural vibration.

The NASA Interferometry Technology Program evolved from a system design for the Focused Mission Interferometer (FMI): a space-based, 30m baseline, partial aperture telescope concept [Laskin & San Martin (1989)]. Using analytical models of the spacecraft and the disturbance environment, this effort predicted 1 micron (RMS) on-orbit fringe position motion in the unattenuated spacecraft environment. This fringe motion is a factor of 100 above the required level of 10 nm (RMS). This discrepancy inspired the layered vibration attenuation strategy. Figure 1 presents the fundamental approach in terms of the three vibration attenuation layers: vibration isolation, structural quieting, and active optics. The strategy is to isolate vibrating machinery at the point of attachment, to add damping to the structure, and finally, to actively control certain optical components in order to reach the 10 nm requirement.

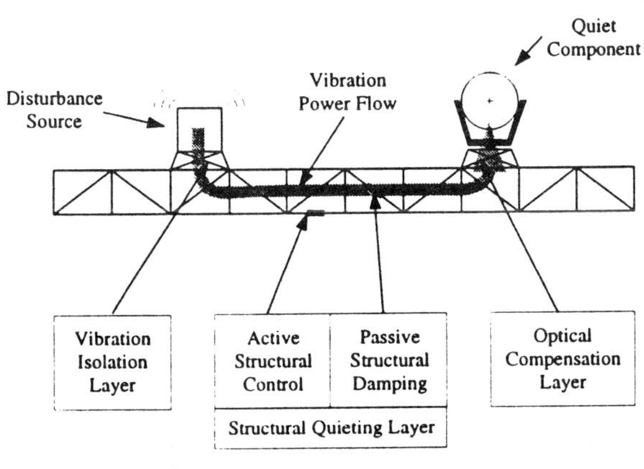

Figure 1. Interferometry Technology Program's layered control approach.

Several papers have addressed issues of structural isolation and active optical control for interferometers [Melody & Neat (1997)] [Spanos, et. al. (1995)]. This paper focuses specifically on passive structural damping with the goal to lower the control demand on active optics. In order to carry out a study on structural damping for interferometers, several analytical tools/methods must be integrated. These are (1) an integrated structural/optics interferometer model, (2) passive damper models, (3) damper placement strategy, and (4) optimization metric analysis. In this paper, we discuss each of these analytical tools/methods by applying them to a representative interferometer. The goal is to suppress errors in OPD caused by random noise disturbances on one point of the structure. For this purpose, an integrated structures/optics model is created of an interferometer. Since an unlimited number of dampers is infeasible and a large number of possible damper positions exist, we explore the use of combinatorial optimization of damper locations. In particular, simulated annealing optimization is chosen as a way to optimize over many possible combinations.

2 INTERFEROMETER MODEL

A finite element model of a representative two-truss interferometer structure was created using the Integrated Modeling of Optical Systems (IMOS) tool developed at JPL. IMOS allows finite element models of structures to be built with integrated optics attached. This model was based on a previous interferometer model described in [Melody (1996)]. It models a free-flying, 10 meter baseline, dual-star feed interferometer. The model includes two trusses on which the collecting apertures are mounted, a rigid central bus node modeled after the NASA Space Shuttle ASTRO-SPAS carrier, and attachment points for the optical elements. The ASTRO-SPAS node is connected to an optics node via a rigid body element. The two trusses are composed of six-degree-of-freedom beam elements. The ASTRO-SPAS node is connected to the trusses by four rigid body elements to each of the four corners of each truss. The geometry of the interferometer structure is shown in figure 2. The length of each truss is 4 meters. It is partitioned into 4 one meter long bays. The cross section of each bay is $0.28 \times 0.28\ m^2$. Each truss is composed

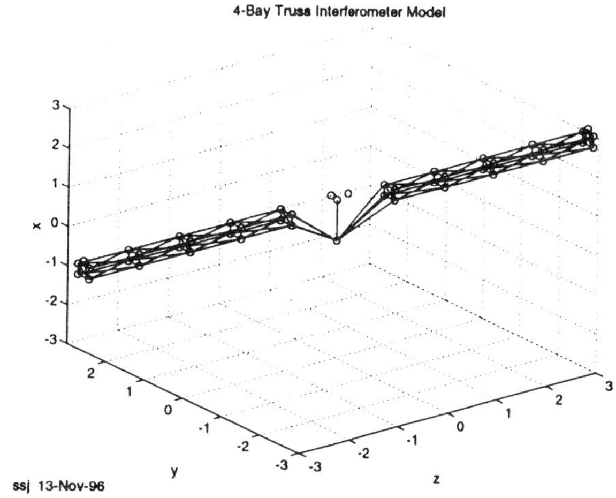

Figure 2. A two-truss interferometer based on ISIS [1].

of 57 beam elements [Milman (1995)]. The first four non rigid-body modes of the interferometer structure are shown in figure 3.

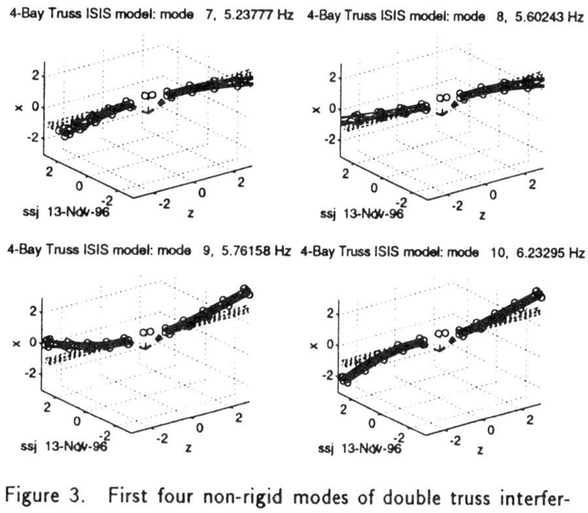

Figure 3. First four non-rigid modes of double truss interferometer.

From the double truss interferometer model, standard mass and stiffness matrices associated with a finite-element model were obtained,

$$M\frac{d^2x}{dt^2} + Kx = F \tag{1}$$

The model contained 408 degrees of freedom in physical space. This resulted in physical mass, M, and stiffness, K, matrices of size [408 × 408].

Of the 408 degrees of freedom, several were dependent DOF's. The dependent degrees of freedom resulted from rigid body element connections. After removing these dependent degrees of freedom, 162 degrees of freedom remained, resulting in reduced mass and stiffness matrices, M_r and K_r, of [162 × 162]. This size model was still too large to be analyzed repeatedly. As a result, a modal transformation was made to lower dimensional space for analysis of modal frequencies and damping. The transformation matrix, T [162 × 50], was composed of 50 eigenvectors, ν, of $M_r^{-1}K_r$.

$$T \triangleq [\nu_1, \nu_2, \ldots, \nu_{50}] \tag{2}$$

A reduced order system was then produced as

$$x \triangleq Ty; \qquad T^tMT\frac{d^2y}{dt^2} + T^tKTy = T^tF \tag{3}$$

where the superscript t refers to a matrix transpose. The new matrices, $\overline{M} \triangleq T^tMT$ and $\overline{K} \triangleq T^tKT$, are of size [50 × 50] and $\overline{F} \triangleq T^tF$ is of size [50 × 1].

3 PASSIVE DAMPERS

Passive dampers were incorporated in the FEM model as viscous dampers based on the Honeywell 'D-strut' [Neat, et. al. (1992)]. A viscous damping force is created along the axial direction that is proportional to the axial displacement rate of the beam element. In this article, it is assumed that the new passive damping element and the original beam element have the same stiffness, although significant advantage can be gained by optimizing these parameters as well. A new model with a passive damping element, denoted by the subscript i, is written as (in physical space)

$$M\frac{d^2x}{dt^2} + (K + \delta K_i)x = Fu + D_i d_i \qquad i = 1 \ldots N \tag{4}$$

where D_i is the influence vector associated with an axial force between two nodes of the structure, d_i is a damper actuator force as in [Chu & Milman (1992)], and δK_i is the change in stiffness associated with the change in elastic modulus of the damper beam element. Recall in this report, $\delta K \triangleq 0$. The force d_i is modeled as a linear velocity feedback so

$$d_i = -k_w v_i \qquad v_i = D^T \frac{dx}{dt} \tag{5}$$

where k_w is a scalar used to change the value of damping. By substituting (5) into (4), we obtain a damping matrix, C,

$$M\frac{d^2x}{dt^2} + (k_w D_i D_i^t)\frac{dx}{dt} + (K + \delta K_i)x = Fu \tag{6}$$

where $C \triangleq k_w D_i D_i^t$.

4 STATE SPACE REPRESENTATION

In this section, we create a state space model of a disturbance input to the ASTRO-SPAS node and a measurement output of optical path difference (OPD). The ASTRO-SPAS node is the most likely place of structural disturbance since it is the place where attitude control system actuators such as thrusters will be placed.

The reduced order dynamic system with viscous damping is represented as

$$\overline{M}\frac{d^2y}{dt^2} + \overline{C}\frac{dy}{dt} + \overline{K}y = \overline{F}u \tag{7}$$

where $\overline{M}$ is the reduced mass matrix, $\overline{C}$ is the reduced damping matrix, $\overline{K}$ is the reduced stiffness matrix, and $\overline{F}$ is the reduced disturbance influence vector for the ASTRO-SPAS node. By denoting a new state, p, as

$$p \triangleq \begin{bmatrix} y \\ \frac{dy}{dt} \end{bmatrix}, \tag{8}$$

(7) may be written in state variable form as

$$\frac{dp}{dt} = \begin{bmatrix} 0 & I \\ (-\overline{M}^{-1}\overline{K}) & (-\overline{M}^{-1}\overline{C}) \end{bmatrix} p + \begin{bmatrix} 0 \\ -\overline{M}^{-1}\overline{F} \end{bmatrix} u \tag{9}$$

The quantity of interest is the optical pathlength difference (OPD). The transformation matrix, H, that takes the reduced order displacement vector, y, to the OPD is created using the Controlled Optics Modeling Package (COMP) [Redding (1992)]. This is represented as

$$OPD = \begin{bmatrix} H & 0 \end{bmatrix} p \tag{10}$$

The H matrix is based on a sensitivity analysis of the optical pathlength with respect to the change in optical mounting positions and orientations. A nominal light ray trace from collecting apertures to the interference point is shown in figure 4. Before arriving at the interference point, light rays trace a path through several optical elements including siderostats, beam compressors, and steering mirrors. Rewrite (9) and (10) as

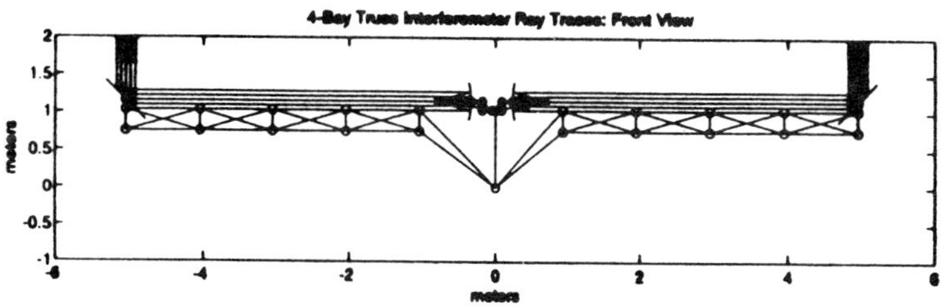

Figure 4. Nominal ray trace from apertures to interference point.

$$\frac{dp}{dt} = \tilde{A}p + \tilde{B}u \tag{11}$$

$$OPD = \tilde{C}p \tag{12}$$

where the definitions of $\tilde{A}$, $\tilde{B}$, and $\tilde{C}$ are clear.

5 OPTIMIZATION FOR DAMPER PLACEMENT

Since an unlimited number of dampers is infeasible and a large number of possible damper locations exist, we explore the use of combinatorial optimization of damper locations.

5.1 Optimization Metric

Let the system input, $u(t)$, be a Gaussian, white noise force input with zero mean and standard deviation, $\sigma_\nu = 1$ Newton. This is meant to simulate a very general disturbance input. Specific sources of disturbances, such as reaction wheels [Hasha (1986)], have particular power spectral densities that may be realized as white noise through a linear filter. This filter can be then incorporated into the state space description of the system (11). As a result of the Gaussian input, the output OPD is also a zero mean Gaussian process. (A Gaussian random variable through a linear filter is Gaussian.) The steady state variance of OPD, σ^2_{OPD}, is related to the steady state variance of the white, Gaussian noise input, σ^2_ν, as

$$\sigma^2_{OPD} = \tilde{C}Q\tilde{C}^* \tag{13}$$

where Q is the solution to the Lyapunov equation

$$Q\tilde{A} + \tilde{A}^*Q = -\tilde{B}\sigma^2_\nu\tilde{B}^* \tag{14}$$

This becomes a scalar measure of the disturbance transmission from input to output. The damper placement objective becomes

$$\min_{DamperLocation} [\sigma_{OPD}] \tag{15}$$

for a given $\sigma_\nu = 1$ Newton.

5.2 Simulated Annealing Optimization

Since the number of possible combinations for damper placement is prohibitively large, global optimization is infeasible. Therefore, we use a heuristic simulated annealing technique for optimization. Simulated annealing has previously been used in general case damper placement problems in [Chu & Milman (1992)] and [Chen, et. al. (1991)]. Most common combinatorial optimization routines follow an iterative improvement strategy in which a trial solution cost is compared to the previous low combination cost. If the new cost is lower, then the trial solution is deemed the new low-cost combination. This method will find local minima, but has the disadvantage of being stuck in a particular local minimum even though other local minima may provide a lower overall cost. The simulated annealing strategy differs in that is occasionally accepts a trial solution as the new low-cost solution even though the new cost has actually increased. This occasional acceptance of a non-improving trial combination allows the current solution to jump to other local minima. The probability of accepting a non-improving solution is governed by a so called Boltzmann probability function

$$P = e^{-\Delta E/\theta} \tag{16}$$

where ΔE is the change in cost from the previous low-cost combination and θ is a free-parameter known as the pseudo-temperature. θ is slowly decreased as the algorithm proceeds. It is to be noted that as the temperature decreases, the probability of accepting a non-improving solution also decreases. This forms a convergence pattern shown in figure 6 in which initially many non-improving solutions are accepted resulting in increased overall cost, but eventually the search settles to a particular local minimum region where only lower cost combinations are accepted. Note also that large increases in energy (large positive ΔE), prohibit the acceptance of a non-improving trial combination. In the IMOS simulated annealing routine, a new trial combination is created by perturbing one element in the current low-cost combination. The element which is perturbed is selected by a uniform random number generator. This element is changed to a random value in the space of all possible damper locations again by a uniform random number generator.

6 RESULTS

6.1 Vibration Attenuation Results

In order to compare results, a nominal model was created by assuming zero dampers and .1% damping in all modes of the model. This value was used as a reference in order to express performance in dB.

$$dB \triangleq 20log_{10}\left(\frac{\sigma^{WithDampers}_{OPD}}{\sigma^{NoDampers}_{OPD}}\right) \tag{17}$$

Note that σ_{OPD} with any number of dampers will be less than σ_{OPD} with zero dampers. Therefore, in terms of dB defined above, better performance will have a more negative dB value. Figure 5 shows the results. All dampers have a value, k_w,

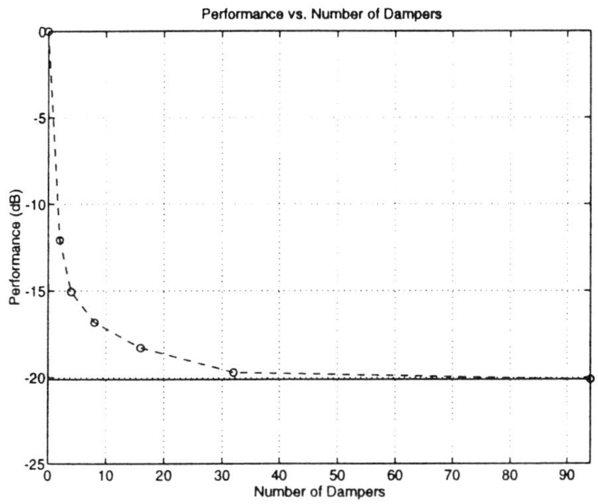

Figure 5. Performance vs. number of dampers. Circles represent calculated data points.

of 7.005×10^4 Ns/m . This value is representative of the values of the Honeywell dampers built for previous structures work done at JPL [Neat, et. al. (1992)]. The maximum number of damper locations is 94. Therefore, the performance value at 94 dampers of $-20.10dB$ forms a floor for the performance function. This indicates that an order of magnitude improvement in vibration attenuation could be achieved using these dampers. The minimum number of damper locations is 0. Therefore, the performance value of $0dB$ forms a ceiling for the performance function. We see that by placing dampers at optimal locations, we obtain significant performance value by using only a few dampers. Indeed, by using only 2 dampers, we obtain over half of the maximum performance available. This result becomes clear when we examine actual damper locations.

6.2 Four Damper Placement

Figures 6 shows a representative simulated annealing optimization run for the optimal placement of four dampers, each with $k_w = 7.005 \times 10^4$ Ns/m . In addition, .1% modal damping is assumed in the model. An initial damper combination is chosen at random. Figure 7 shows the

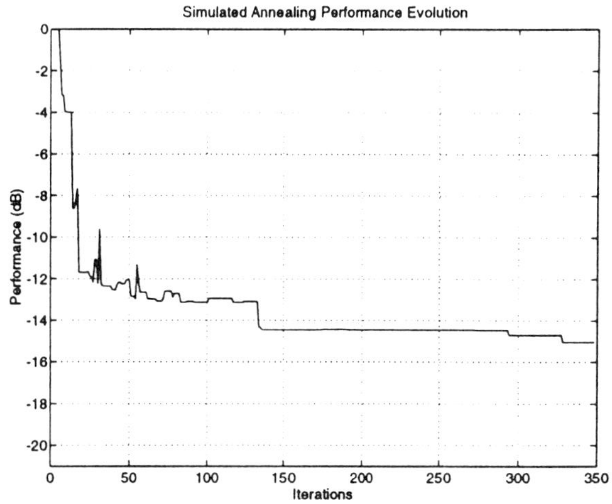

Figure 6. Simulated annealing convergence for four damper placement.

initial and final damper locations as a result of the optimization. The dampers tend to migrate

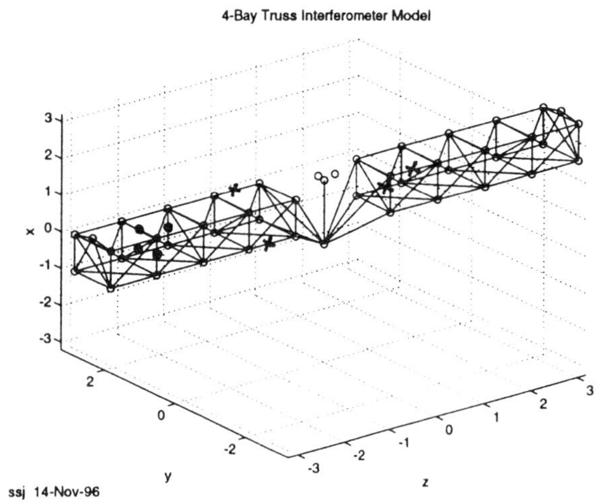

Figure 7. Four damper placement optimization. Initial damper locations are shown as closed circles. Final damper locations are shown as stars. Note that the interferometer is not drawn to actual dimension as in figure 2 in order to more clearly show damper positions.

to the base of the trusses. This indicates that dampers are placed at points of high axial stress of low modes.

7 EFFECT OF VIBRATION ISOLATION

It has been shown [Spanos, et. al. (1995)] that vibration isolation tends to act as a low-pass filter. Therefore, we may adjoin a low-pass filter to our model to study damper placement in the presence of vibration isolation (to first order). A low-pass chebyshev filter with cut-off frequency of 10 Hz was appended to the input of the model. Results of the analysis were extremely similar to those results in the previous section. This indicates that in this case, the damper placement works most on low modes whose effect is not curtailed by the low-pass "isolation" filter.

8 CONCLUSIONS AND FUTURE WORK

Damper optimization has been explored for interferometer structures. An integrated structures/optics model has been created. Using a particular scalar disturbance rejection performance metric, consisting of the standard deviation of OPD as a result of noise input at the interferometer base, it was shown damper placement can significantly reduce OPD error. Simulated annealing optimization was used to optimize the placement of a finite number of dampers. It was shown that a small number of dampers give significant performance enhancement. Optimal damper placement tended to place dampers near the base of the interferometer trusses. This result indicates that dampers are placed at locations of high axial stress of low modes. For future interferometer structures that contain several trusses and several disturbance sources, this methodology should provide valuable information for analysis of passive damper placement.

Future work involves producing this analysis for the Micro-Precision Interferometer testbed models [Melody & Neat (1997)] and then physically implementing passive damping on the testbed.

ACKNOWLEDGMENT

We are pleased to acknowledge Dr. Gregory W. Neat of JPL for many helpful discussions and suggestions.

This work was performed at the Jet Propulsion Laboratory, California Institute of Technology, under contract with the National Aeronautics and Space Administration.

REFERENCES

[Melody (1996)] Melody, J.W., "ISIS Integrated Model and RCS Thruster Disturbance Analysis," *JPL Interoffice Memorandum 341-96-174 ITP (Internal Document)* , May 20, 1996.

[Chu & Milman (1992)] Chu, C.-C. and Milman, M.H., "Eigenvalue Error Analysis of Viscously Damped Structures Using a Ritz Reduction Method," *AIAA Journal*, December, 1992.

[Milman & Chu (1994)] Milman, M.H. and C.-C. Chu, "Optimization Methods for Passive Damper Placement and Tuning," *Journal of Guidance, Control, and Dynamics*, July-August, 1994.

[Chen, et. al. (1991)] Chen, G.-S., Bruno, R.J., and Salama, M., "Optimal Placement of Active/Passive Members in Truss Structures Using Simulated Annealing," *AIAA Journal*, August, 1991.

[Neat, et. al. (1992)] Neat, G.W., O'Brien J.F., Lurie, B.J., and Garnica, A.,"Joint Langley Research Center/ JPL CSI Experiment," *15th Annual AAS Guidance and Control Conference*, February 8-12, 1992.

[Milman (1995)] Milman, M. (ed.), "Integrated Modeling of Optical Systems User's Manual: Release 2.0," JPL D-13040 (Internal Document), November 15, 1995.

[Redding (1992)] Redding, D., Needels, L., Wallace, K., and Levine, M., "Controlled Optics Modelling Package User Manual," JPL D-9816 (Internal Document), June 1, 1992.

[Laskin & San Martin (1989)] Laskin, R. A. and San Martin, M.,"Control Structure System Design of a Spaceborne Interferometer," *Proceedings of the AAS/AIAA Astrodynamics Specialist Conference*, Stowe, VT, 1989.

[Spanos, et. al. (1995)] Spanos, J.T., Rahman, Z., and Blackwood, G.,"A Soft 6-Axis Active Vibration Isolator," *American Control Conference*, pp. 412-416, Seattle, WA, June, 1995.

[Melody & Neat (1997)] J. W. Melody, G. W. Neat, "Integrated Modeling Methodology Validation Using the Micro-Precision Interferometer Testbed: Closed Loop Results", Proceedings of the 1997 American Control Conference, Albuquerque, NM, June, 1997.

[Hasha (1986)] Hasha, M. D.,"Reaction Wheel Mechanical Noise Variations," *LMSC EM SSS 218*, 1986.

AIAA-97-3464

Dynamics and Control of a Reaction Mass Actuator with Centering Springs

Tae W. Lim* and Kenneth A. Lieber†
Department of Aerospace Engineering
University of Kansas
Lawrence, KS 66045

Abstract

Instead of using an electronic centering force to provide a reaction mass actuator (RMA) with asymptotic stability, centering springs are installed in the RMA and their influence on the dynamics and control of the RMA is investigated. Typically, RMA frequency response characteristics (force output due to voltage input) with the electronic centering force resemble those of a high-pass filter. Because of this, the usable frequency range of an RMA is limited to the range above the break frequency of the RMA unless actuator dynamics is accounted for. The investigation in this paper shows that the absence of the voltage drop to produce the electronic centering force permits the RMA to have a flat frequency response both above and below the natural frequency of the actuator. This allows effective vibration control both below and above the actuator break frequency. Both analytically and experimentally, the dynamic characteristics of an RMA with centering springs are presented. It is also demonstrated both analytically and experimentally that the RMA can be employed to control effectively a cantilevered beam whose natural frequency is lower than that of the RMA using a simple proportional-derivative (PD) feedback controller.

Introduction

Space-realizable actuators for vibration suppression of flexible space structures have been investigated extensively in the last decade. In order for an actuator to be space-realizable, the actuator must not be attached to the ground and control force should be generated by reacting against an inertial mass. The devices that produce control force by accelerating a moving mass along a linear track are referred to as reaction mass actuators (RMA's) and they include proof-mass actuators (PMA's)[1-5] and linear motors.[6-8]

Since the introduction of the PMA prototypes by researchers at the University of Virginia for NASA Langley Research Center,[1] the dynamics and control aspects of the PMA's have been investigated by many researchers. Haviland et al.[2] investigated a linear controller based on the optimum response referred to as the limiting performance. The limiting performance provides an optimum response history of the actuator to reduce the initial vibration of the structure in the minimum settling time while meeting the proof-mass displacement and control force constraints. A linear quadratic regulator controller was then designed to produce the limiting performance response as closely as possible. Politansky and Pilkey[3] studied a suboptimal controller using a proof-mass relative position command. Instead of commanding control force directly, by controlling the relative position of the proof-mass, they showed that the displacement and force constraints could be met while achieving good performance.

Zimmerman et al.[4] investigated the dynamic characteristics and control of the PMA using a microprocessor based digital controller. By investigating the actuator dynamics both analytically and experimentally, they have found that the transfer function of the actuator force output to voltage command input resembles essentially a high pass filter. Also, they have concluded that the actuator dynamics can be ignored if the transfer function break frequency (and therefore the PMA natural frequency) is lower than the natural frequency of the structure to be controlled. Subsequently, Zimmerman and Inman[5] further investigated the nature of the interaction between the control law, the structure and the actuator. Again, they suggested that the break frequency of the actuator should be below the first natural frequency of the structure for best performance. In these investigations,[4,5] for asymptotic stability, an electronic centering force using a local displacement feedback control was incorporated. The input voltage drops to produce the electronic centering force. This voltage drop attributes to making the transfer function of the actuator force output to input voltage command resemble a high-pass filter.

This high-pass filter dynamic characteristics of the actuator limits practically the usage of the actuator below the transfer function break frequency. The goal of this research is to find an alternative to remove this limitation. As a viable alternative to provide centering force, an RMA installed with physical springs is examined. The dynamic characteristics and control of an RMA are investigated when centering springs are

---

* Assistant Professor, Member AIAA.

† Graduate Research Assistant. Currently with Cessna Aircraft Company. Member AIAA.

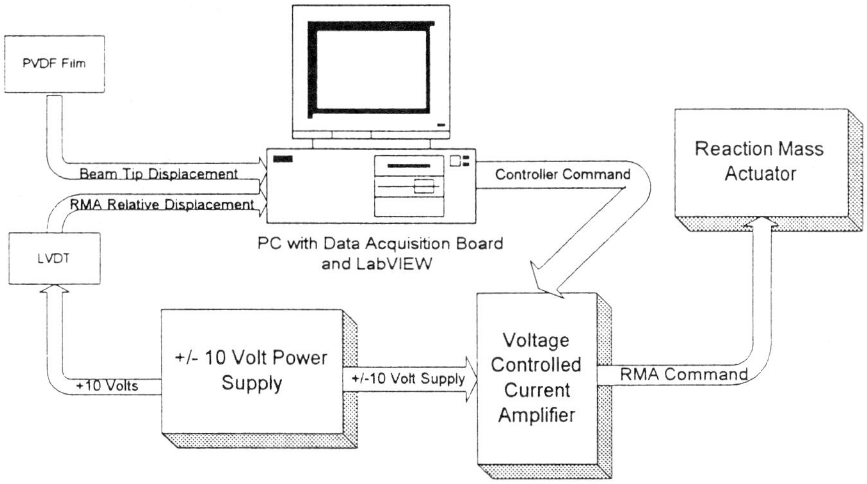

Fig. 1 Schematic of the reaction mass actuator system

installed. It was found that, by using centering springs, the actuator transfer function exhibited flat frequency response and adequate vibration control capability could be achieved below the break frequency of the actuator transfer function. Both analytical and experimental studies were conducted to support this conclusion. A vibration suppression experiment of a cantilevered beam with an RMA is presented.

Reaction Mass Actuator System Description

The reaction mass actuator system used in this research consists of three major components: an RMA with a displacement sensor, a cantilevered beam with a strain sensor, and a 486 personal computer (PC) running data acquisition and control software. Other peripheral devices such as power supply and voltage controlled current amplifier are also employed. A schematic diagram in Fig. 1 depicts the arrangement. Each component is described briefly in this section. Detailed information on the RMA system used for this research is available in Ref. 9.

Reaction Mass Actuator

The picture of the reaction mass actuator employed in this study is shown in Fig. 2. The RMA consists of a dumbbell shaped moving reaction mass which includes a permanent magnet in the middle section and a stationary housing which contains permanent magnet, fixed coil, and bearings supporting the moving mass. The moving mass weighs 2.2 lb. A schematic diagram of the RMA is shown in Fig. 3. A linear variable differential transformer (LVDT) is attached to the RMA and is employed to measure the relative displacement of the reaction mass with respect to the housing. An LVDT consists of a round bar shaped core and a cylinder shaped sleeve. The core slides inside the sleeve. To measure the relative displacement, the core is attached to the moving reaction mass and the sleeve to the RMA housing. The LVDT has a linear output range of $\pm$ 1 inch. The proof-mass actuator investigated previously[1-5] is free to move inside the housing. In order to provide asymptotic stability, an electronic centering force via a local position feedback has been used.[4-5] Unlike its predecessor, the RMA studied in this paper incorporates a couple of centering springs as indicated in Figs. 2 and 3.

Fig. 2 Reaction mass actuator

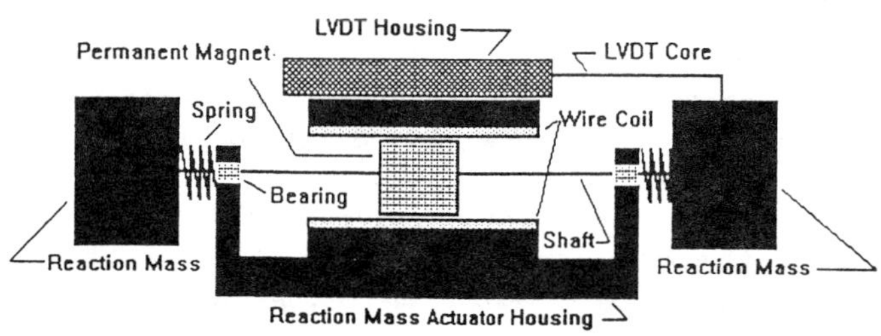

Fig. 3 Schematic and major parts of the RMA

Cantilevered Beam with Strain Sensor

The RMA is mounted at the free end of a cantilevered beam. The beam is employed to conduct vibration suppression control using the RMA. The beam is made of aluminum and has dimension of 3 in (height), 0.25 in (width) and 34 in (length). A polyvinylidene fluoride (PVDF) film sensor is attached at the root of the beam. The PVDF film sensor is used to measure the root strain of the cantilevered beam. By measuring the root strain of the beam, the corresponding tip displacement of the beam is approximated and used to suppress the beam vibration. The PVDF film sensor requires no external power supply and provides large amplitude signal which is sufficient to be used directly without amplification. This sensor consists of a 12mm by 30mm active area printed on a 15mm by 40mm substrate.

Controller Hardware and Software

A 486 PC is used to run the data acquisition software, to implement the control law, and to produce actuator command. LabVIEW for Windows software[10] is employed for this purpose. A 16 channel data acquisition board resides in the PC and the board has 16-bit analog-to-digital (A/D) and digital-to-analog (D/A) converters. This digital control setup allows easy implementation of various control algorithms and feedback signals. A detailed description of the software development environment is available in Ref. 9. Also, a voltage controlled current amplifier was employed to amplify the low current output from the data acquisition board.

Description of RMA Characteristics

In order to employ effectively the RMA described previously for the vibration suppression of the cantilevered beam, static and dynamic characteristics of the RMA are investigated. Salient features of the RMA are described herein.

Static Characteristics

Even when the RMA is turned off, i.e., no current in the coil, due to the magnetic force interaction between the magnets in the housing and in the moving mass, the relationship between the static force applied to the moving mass and the resulting displacement exhibits a nonlinear behavior. To quantify the relationship, the RMA was placed on a table top pointing the moving mass travel direction horizontally to eliminate the effect of the gravity. Then, force was applied in the direction of the moving mass travel by adding weights via a string and a roller. At one end of the string weights were added while the other end was attached to the moving mass. Measurements of the moving mass displacement were made through the LVDT sensor. Figure 4 shows the resulting force versus displacement relationship of the RMA. The curve resembles a cubic polynomial. The stiffness of the RMA increases as the absolute displacement increases. The range of the moving mass displacement is about ± 0.3 in and the corresponding force range is about ± 1.5 lb. A straight line was fitted to the measurement data using a least squares solution as shown in the figure. The resulting RMA stiffness was 4.08 lb/in.

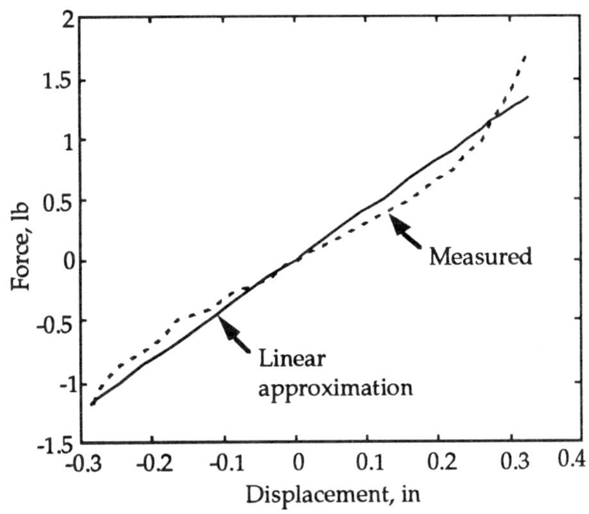

Fig. 4 Static force versus displacement relationship of the RMA

Dynamic Characteristics

The objectives of examining actuator dynamic characteristics are to incorporate them in the controller design process and to assess the limitations of the actuator. Previously, researchers[4,5] employed a local proportional feedback to provide an electronic centering force. This was used to help improve the stability of the RMA and prevent the moving mass from hitting the housing walls. The input voltage to the actuator coil drops to produce this electronic centering force. This voltage drop attributes to making the transfer function of the actuator force output to input voltage command resemble a high-pass filter. The actuator dynamics can be ignored if the transfer function break frequency is lower than the natural frequency of the structure to be controlled. Also, it has been suggested that the break frequency of the actuator should be below the first natural frequency of the structure for best performance. This high-pass filter dynamic characteristic of the actuator limits practically the usable frequency range of the actuator. The influence of using physical centering springs instead of electronic ones on the dynamic characteristics of the RMA is examined herein both analytically and experimentally.

Figure 5 shows a dynamic model of the RMA attached to a fixed base. In Fig. 5, m_p is the mass of

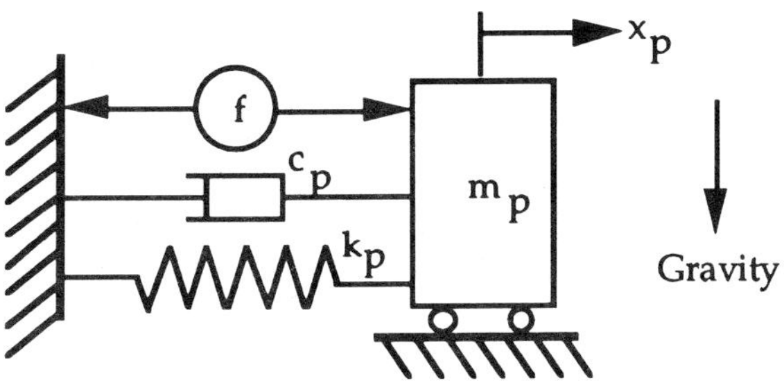

Fig. 5 Dynamic model of the RMA

the moving mass; x_p is the moving mass displacement; c_p is the magnetic damping force generated by the permanent magnets; k_p is the effective spring stiffness due to the physical springs installed in the RMA and the permanent magnets; and f is the net control force generated by the coil. Unlike the PMA investigated previously, the RMA in this study, due to the difference in its construction, has stiffness (k_p) and damping (c_p) without having voltage applied to the coil.

An experimentally obtained frequency response of the moving mass displacement to voltage command from the computer, i.e., voltage input to the power amplifier, is shown in Fig. 6. The displacement was measured by the LVDT sensor and the voltage command was a random signal with a Gaussian distribution. The transfer function exhibits the characteristics of a second order low pass filter. As shown in Fig. 6, a second order transfer function was fitted to the measured frequency response and it was identified as

$$\frac{X_p^d(s)}{V(s)} = \frac{G_v G_d / m_p}{s^2 + 2\zeta_p \omega_p s + \omega_p^2} \tag{1}$$

where V is the voltage command from the computer; X_p^d is the LVDT sensor signal measuring the moving mass displacement; ω_p is the undamped natural frequency of the RMA, i.e., $\omega_p = \sqrt{k_p/m_p} = 4.26\text{Hz}$; and ζ_p is the damping ratio which was identified as 0.25. The quantity G_v in the numerator of transfer function represents the combined gain from the current amplifier and the coil and the gain was identified to be 0.04 and G_d is the LVDT sensor gain (8.54 volts/in).

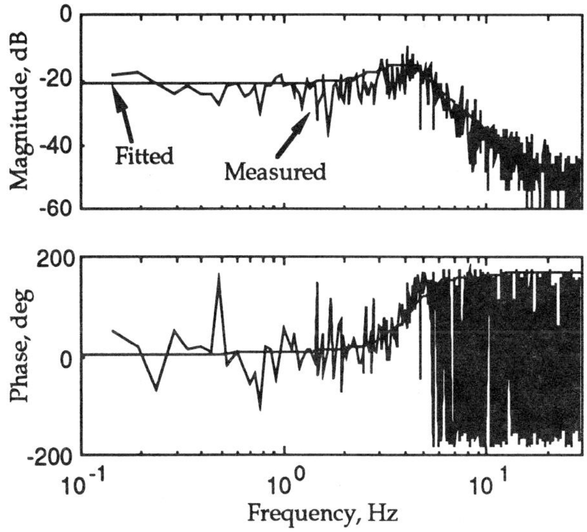

Fig. 6 Reaction mass actuator frequency response

The transfer function for the actuator dynamics can be obtained analytically by examining the equation of motion for the reaction mass as

$$F(s) = G_v V_n(s) \tag{2}$$

where F is the net electromagnetic control force generated by the coil; V_n is the net voltage applied to the coil. Due to the voltage drop from the back-emf damping effect, i.e., the electromagnetic force interaction between the coil and the permanent magnets, the net voltage across the coil becomes

$$\begin{aligned} V_n(s) &= V(s) - c_e V_e(s) \\ &= V(s) - c_e s X_p(s) \end{aligned} \tag{3}$$

where V_e is the voltage drop due to the back-emf damping. Since this net control force F is applied to the reaction mass along with the damping and spring forces as indicated in Fig. 5, we obtain

$$F(s) = m_p s^2 X_p(s) + c_p s X_p(s) + k_p X_p(s) \tag{4}$$

By substituting Eq. (3) into (2) and using Eq. (4), a transfer function can be obtained as

$$\frac{X_p(s)}{V(s)} = \frac{G_v}{m_p s^2 + (c_p + G_v c_e)s + k_p} \tag{5}$$

By combining Eqs. (4) and (5), we obtain the transfer function for the actuator control force output to the voltage command to the RMA current amplifier as

$$\frac{F(s)}{V(s)} = G_v \frac{m_p s^2 + c_p s + k_p}{m_p s^2 + (c_p + G_v c_e)s + k_p} \quad (6)$$

In an effort to determine the amount of damping attributable to the permanent magnets (c_p) and the back-emf effects (c_e), impulse hammer tests were conducted. The test results indicated that passive damping coefficient (c_p) was comparable to the combined damping coefficient ($c_p + G_v c_e$). This leads to the conclusion that the voltage drop due to back-emf damping is negligible, i.e., $c_p >> G_v c_e$. In other words, the damping due to back-emf effects is negligible. Therefore, the transfer function in Eq. (6) is practically a constant, G_v, and exhibits flat frequency response both below and above the break frequency of the actuator. From Eq. (5), the transfer function describing the LVDT sensor signal measuring the reaction mass displacement due to voltage command is obtained as

$$\frac{X_p^d(s)}{V(s)} = \frac{G_v G_d}{m_p s^2 + c_p s + k_p} = \frac{G_v G_d / m_p}{s^2 + (c_p / m_p)s + k_p / m_p} \quad (7)$$

where $X_p^d(s) = G_d X_p(s)$. This transfer function matches well the experimentally obtained transfer function shown in Eq. (1).

Modeling of a Cantilevered Beam with the RMA

The RMA described previously was employed to conduct active vibration suppression of a cantilevered beam. This section describes the modeling process of the beam itself and the beam with the RMA.

Modeling of the Cantilevered Beam

The cantilevered beam used in this study was a 3 in x 0.25 in (height x width) stock aluminum bar with 34 inches in length. The length of the beam was adjusted so that the resulting first bending frequency was lower than the RMA natural frequency of 4.26 Hz. The parasitic mass of the RMA was added at the free end of the beam as a lumped tip mass. The parasitic mass include the mass of the housing, LVDT sensor, and mounting brackets which were used to attach the RMA to the beam.

A finite element model of the beam with the tip mass was constructed and eigenvalue analysis was conducted to compute the natural frequencies and mode shapes of the beam. Resulting first mode natural frequency was 3.05 Hz. The second mode natural frequency was 32.8 Hz. Only the first mode was employed in the controller design process since the second mode frequency was sufficiently separated from the first.

Modeling of the Cantilevered Beam with the RMA

Using the first mode frequency and mode shape coefficient obtained for the beam, the model of the beam with an RMA was constructed. A schematic diagram of the beam with the RMA is shown in Fig. 7. In the figure, m_t is the tip mass, i.e., the parasitic mass of the RMA and y_n is the beam tip displacement.

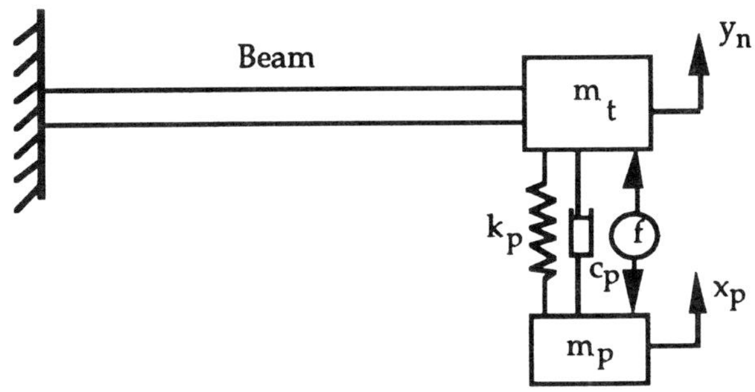

Fig. 7 A schematic diagram of the beam with the RMA

The equation of motion describing the beam is written as

$$\ddot{q}_1 + 2\zeta_1\omega_1\dot{q}_1 + \omega_1^2 q_1 = \phi_{1n}\left[-k_p(y_n - x_p) - c_p(\dot{y}_n - \dot{x}_p) + f\right] \quad (8)$$

where q_1 is the modal displacement of the first mode; ω_1 is the first mode natural frequency (ω_1=3.05 Hz); ζ_1 is modal damping ratio of the beam (ζ_1=0.005); f_{1n} is the unit generalized mass normalized mode shape coefficient at the tip of the beam for the first mode (ϕ_{1n}=11.06). Using the first mode, the beam tip displacement is written as

$$y_n = \phi_{1n} q_1 \quad (9)$$

By substituting Eq. (9) into Eq. (8) and rearranging, we obtain

$$\ddot{q}_1 + (2\zeta_1\omega_1 + c_p\phi_{1n}^2)\dot{q}_1 + (\omega_1^2 + k_p\phi_{1n}^2)q_1 = k_p\phi_{1n}x_p + c_p\phi_{1n}\dot{x}_p + \phi_{1n}f \quad (10)$$

From Fig. 7, the equation of motion describing the RMA can be obtained as

$$m_p\ddot{x}_p - c_p(\dot{y}_n - \dot{x}_p) - k_p(y_n - x_p) = -f \quad (11)$$

Substitute Eq. (9) into Eq. (11) and rearrange to obtain

$$m_p\ddot{x}_p + c_p\dot{x}_p + k_px_p = k_p\phi_{1n}q_1 + c_p\phi_{1n}\dot{q}_1 - f \quad (12)$$

By combining Eqs. (10) and (12), we obtain the equations of motion describing the beam with the RMA as

$$\mathbf{M}\ddot{\mathbf{x}} + \mathbf{C}\dot{\mathbf{x}} + \mathbf{K}\mathbf{x} = \hat{\mathbf{b}}f \quad (13)$$

where

$$\mathbf{M} = \begin{bmatrix} 1 & 0 \\ 0 & m_p \end{bmatrix}, \quad \mathbf{C} = \begin{bmatrix} 2\zeta_1\omega_1 + c_p\phi_{1n}^2 & -c_p\phi_{1n} \\ -c_p\phi_{1n} & c_p \end{bmatrix},$$

$$\mathbf{K} = \begin{bmatrix} \omega_1^2 + k_p\phi_{1n}^2 & -k_p\phi_{1n} \\ -k_p\phi_{1n} & k_p \end{bmatrix},$$

$$\mathbf{x} = \begin{Bmatrix} q_1 \\ x_p \end{Bmatrix}, \text{ and } \hat{\mathbf{b}} = \begin{Bmatrix} \phi_{1n} \\ -1 \end{Bmatrix}$$

The equations are coupled due to the inherent damping and stiffness characteristics of the RMA. Define a state variable

$$\mathbf{s} = \left\{\mathbf{x}^T \ \dot{\mathbf{x}}^T\right\}^T = \left\{q_1 \ \ x_p \ \ \dot{q}_1 \ \ \dot{x}_p\right\}^T \quad (14)$$

Then, a state space equation is obtained as

$$\dot{\mathbf{s}} = \mathbf{A}\mathbf{s} + \mathbf{b}V \quad (15)$$

where V is the voltage command to the RMA and

$$\mathbf{A} = \begin{bmatrix} \mathbf{0} & \mathbf{I} \\ -\mathbf{M}^{-1}\mathbf{K} & -\mathbf{M}^{-1}\mathbf{C} \end{bmatrix} \text{ and } \mathbf{b} = G_v\begin{Bmatrix} \mathbf{0} \\ \mathbf{M}^{-1}\hat{\mathbf{b}} \end{Bmatrix}$$

This equation will be used to design controller in the next section.

Controller Design

The block diagram describing the feedback control structure employed in this study is shown in Fig. 8. Since feedback signals from PVDF and LVDT sensors are available, controllers using either PVDF or LVDT sensor will be investigated.

PVDF Feedback Controller Design

Using the PVDF sensor, the tip deflection of the beam was measured. This tip deflection measurement was then employed to close the loop using proportional-derivative (PD) gains as indicated in Fig. 8. When this feedback structure is used, sensor hardware requirement

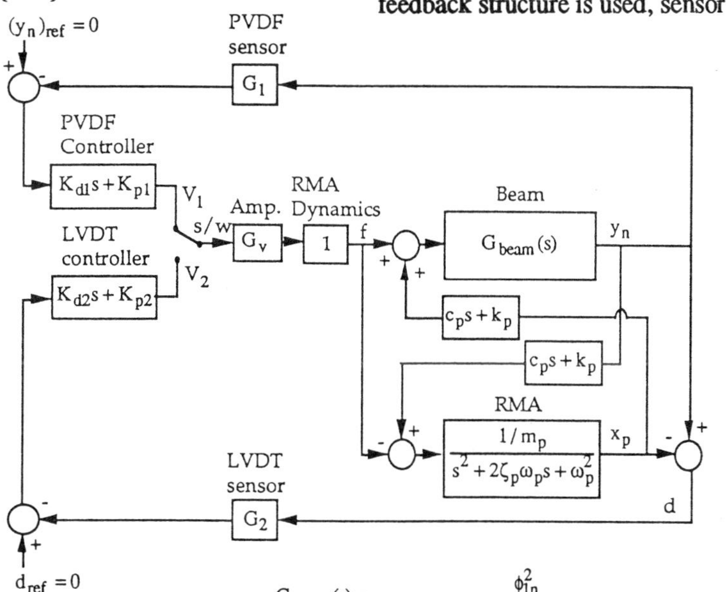

$$G_{beam}(s) = \frac{\phi_{1n}^2}{s^2 + (2\zeta_1\omega_1 + c_p\phi_{1n}^2)s + (\omega_1^2 + k_p\phi_{1n}^2)}$$

Fig. 8 Block diagram of the feedback control system

can be minimized since the PVDF sensor does not require separate signal conditioning. Using the same state variables and system equation in Eqs. (14) and(15), the voltage command to the RMA becomes

$$V_1 = -G_1\left[K_{p1}\phi_{1n} \;\; 0 \;\; K_{d1}\phi_{1n} \;\; 0\right]s = \mathbf{-K_1}s \tag{16}$$

where K_{p1} and K_{d1} are the proportional and derivative gains, respectively; and G_1 is the PVDF sensor gain of 4V/in. By substituting Eq. (16) into Eq. (15), we obtain the closed-loop system equation as

$$\dot{s} = (\mathbf{A} - \mathbf{bK}_1)s = \mathbf{A}_{CL_1}s \tag{17}$$

where $\mathbf{A}_{CL_1}$ is the closed-loop A matrix for the PVDF feedback.

The gains (K_{p1} and K_{d1}) must be designed to augment damping of the structure as much as possible while satisfying constraints on the reaction mass relative displacement. To help design the gains, root locus plots were employed. Initially, the influence of the proportional (displacement) and derivative (rate) gains was investigated individually as shown in Fig. 9. The displacement gain provides little improvement of the beam damping. The rate gain increases the beam damping but the RMA goes unstable if the gain goes beyond -4. After trial and error, the gains of $K_{p1} = 85$ and $K_{d1} = -2.5$ were selected. The beam damping ratio achieved with this gain setting was 8.7%.

LVDT Feedback Controller Design

Using the LVDT sensor, the relative motion of the RMA moving mass was measured. This measurement was then employed to close the loop using proportional-derivative (PD) gains as indicated in Fig. 8. Using the same state variables and system equation in Eqs. (14) and(15), the voltage command to the RMA becomes

$$V_2 = -G_2\left[K_{p2}\phi_{1n} \;\; -K_{p2} \;\; K_{d2}\phi_{1n} \;\; -K_{d2}\right]s = \mathbf{-K_2}s \tag{18}$$

where K_{p2} and K_{d2} are the proportional and derivative gains, respectively; and G_2 is the LVDT sensor gain of 8.54V/in. By substituting Eq. (18) into Eq. (15), we obtain the closed-loop system equation as

$$\dot{s} = (\mathbf{A} - \mathbf{bK}_2)s = \mathbf{A}_{CL_2}s \tag{19}$$

where $\mathbf{A}_{CL_2}$ is the closed-loop A matrix for the LVDT feedback.

To help design the gains, root locus plots were employed. Initially, the influence of the proportional and derivative gains was investigated individually as shown in Fig. 10. The displacement gain increases the beam damping until the beam locus goes unstable at K_{p2}=-103. The rate gain does not influence the beam damping but the RMA becomes unstable if the gain goes beyond -0.22. Since beam damping was not achieved from the rate gain, the rate gain was set to zero. As a compromise between performance (beam damping) and constraints (relative displacement), the displacement gain of -75 was selected. The beam damping ratio achieved with this gain setting was 19.2%.

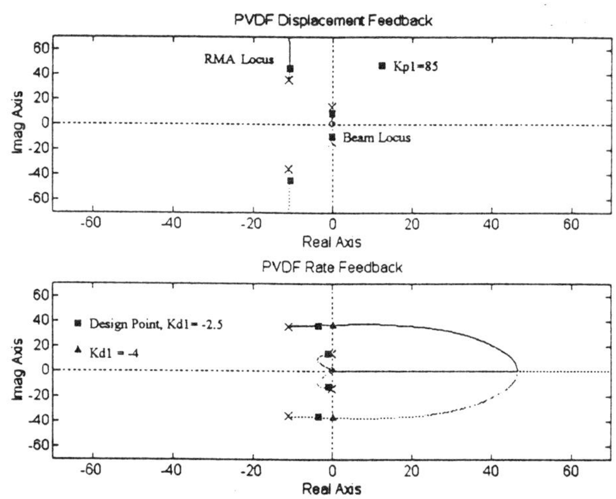

Fig. 9 Root locus of the PVDF feedback controller

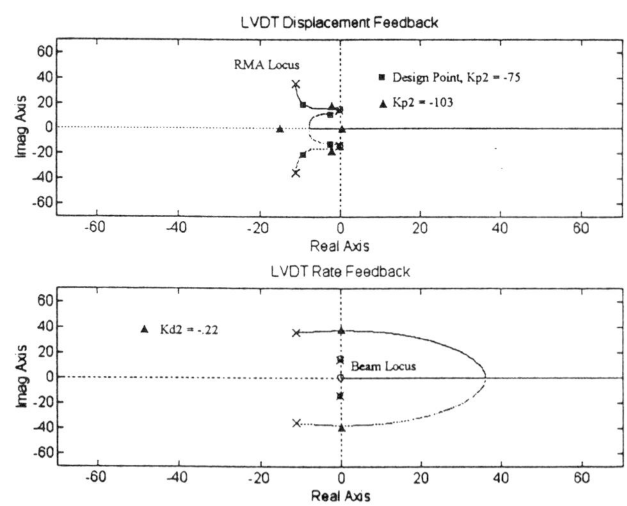

Fig. 10 Root locus of the LVDT feedback controller

This LVDT feedback structure is unique in that the direct measurement of structural vibration is not required. It is conceivable to make a self-contained RMA that has its own power supply, sensor, and controller. In other words, an RMA with the LVDT feedback controller will be able to be attached to any arbitrary location of the structure where vibration damping is desired.

Experimental Results

The controllers designed were implemented using the hardware described previously. As an initial test, an open-loop response was measured. This was then followed by the implementation and testing of both the LVDT and PVDF feedback control systems. The test sequence consists of a 10 second sinusoidal excitation at 2.65 Hz followed by a free-decay response for the open-loop case and the appropriate controlled responses for the active control cases.

Open Loop Response

To examine the accuracy of the modeling, an open-loop response experiment was conducted first and the results were compared with the analytical prediction. As indicated in Fig. 11, the beam was excited for 10 seconds by the RMA which was commanded using a sinusoidal signal of 2.65 Hz and the resulting motion was allowed to decay freely afterwards. Simulation and experiment are in good agreement indicating that the modeling of the RMA system was accurate.

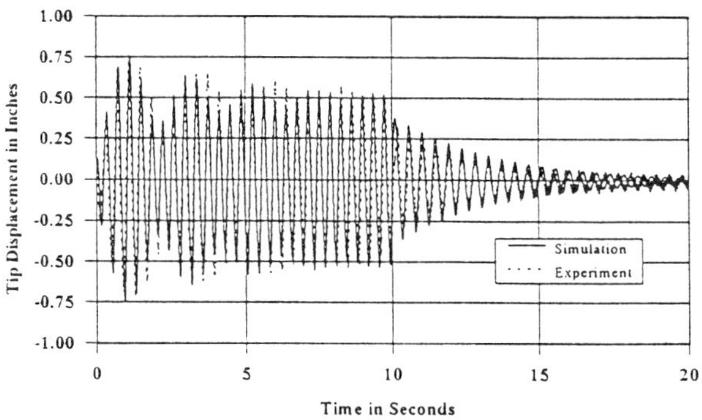

Fig. 11 Open-loop response (simulation vs. experiment)

Closed-Loop Response Using PVDF Controller

In this case, the PVDF film signal measuring root strain was fed back to the RMA using the PD (proportional and derivative) controller which was designed earlier. The controller gains were implemented in LabVIEW. The major advantage of this controller is that the experimental setup is simple as the PVDF sensor requires no external power supply or signal conditioning. Figure 12 shows the closed-loop response of the system with the previously selected proportional and derivative gains. The experimental time history matches well with the predicted response.

Closed-Loop Response Using LVDT Controller

In this case, the LVDT was used to feed back the relative displacement of the moving mass to the RMA via a proportional feedback controller. Figure 13 shows the resulting performance of the controller. Unlike the PVDF feedback controller case, as can be seen from the figure, the vibration of the beam does not damp out completely. This small residual motion can be attributed to a small amount of friction in the RMA and LVDT sensor. This friction allows the RMA to stop while the beam vibrates with small amplitude. Since the RMA relative displacement is the only measured signal for the controller, there is no information available about the beam deflection once the RMA stop moving. To eliminate the residual vibration, therefore, the beam deflection will also have to be measured.

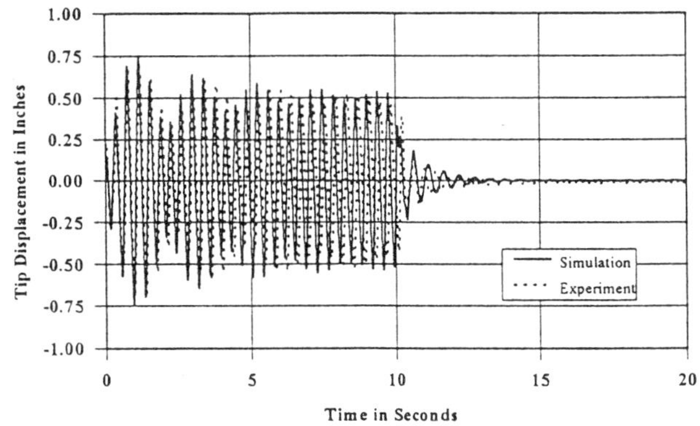

Fig. 12 Closed-loop response using PVDF feedback

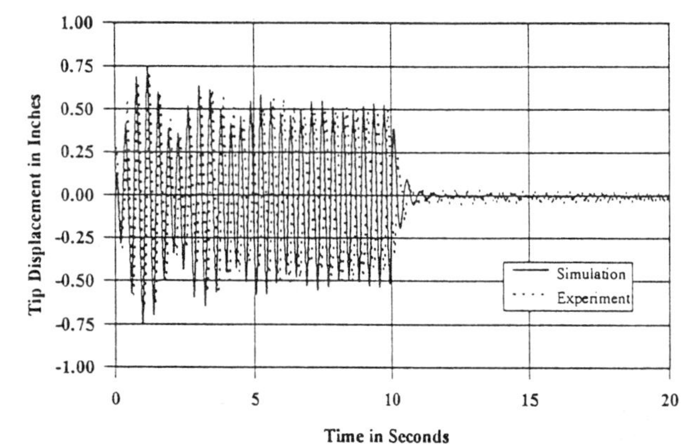

Fig. 13 Closed-loop response using LVDT feedback

Concluding Remarks

Both analytically and experimentally, it has been demonstrated that the RMA with centering springs can be used to suppress structural vibration of frequency less than the break frequency of the actuator. Performance of two controllers using PVDF and LVDT sensors has been investigated for the vibration suppression control. The PVDF sensor feedback is effective in that its hardware requirement is minimal and it provides good performance. Damping augmentation produced by the LVDT sensor feedback was higher but, due to the friction of the LVDT sensor and RMA, the LVDT controller produced a residual vibration of the beam. To maintain the higher damping ratio of the LVDT feedback and to eliminate the residual vibration, a combined controller using both LVDT and PVDF should be examined.

The fact that the RMA can be used to control structural frequency less than the RMA break frequency is significant since it expands the usable frequency range of the RMA. For future large flexible spacecraft, the natural frequencies of the important modes are expected to decrease as the on-orbit construction or deployment process continues for an extended period of time to achieve the desired size. Under the current design philosophy, the RMA's which were designed for earlier smaller flight configurations will become useless once the structural frequencies fall below the break frequency of the RMA. Additional RMA's may have to be supplied for each new configuration. With the extended frequency range of the RMA, the performance maintenance will be possible by adjusting the gains of the controller for new lower structural frequencies. This capability will help reduce the number of RMA's for future flexible spacecraft and weight while achieving comparable vibration suppression capability for the changing configurations of the spacecraft.

References

1. Pilkey, W. D. and Haviland, J. K., "Large Space Structure Damping Design," Final Report, Department of Mechanical and Aerospace Engineering, University of Virginia, Charlottesville, VA, 1983.

2. Haviland, J. K., Lim, T. W., Pilkey, W. D., and Politansky, H., "Control of Linear Dampers for Large Space Structures," *Journal of Guidance, Control and Dynamics*, Vol. 13, No. 2, 1990, pp. 234-240.

3. Politansky, H. and Pilkey, W. D., "Suboptimal Feedback Vibration Control of a Beam with a Proof-Mass Actuator," *Journal of Guidance, Control and Dynamics*, Vol. 12, No. 5, 1989, pp. 691-697.

4. Zimmerman, D. C., Horner, G. C., and Inman, D. J., "Microprocessor Controlled Force Actuator," *Journal of Guidance, Control and Dynamics*, Vol. 11, No. 3, 1988, pp. 230-236.

5. Zimmerman, D. C. and Inman, D. J., "On the Nature of the Interaction Between Structures and Proof-Mass Actuators," *Journal of Guidance, Control and Dynamics*, Vol. 13, No. 1, 1990, pp. 82-88.

6. Sulla, J. L., Juang, J.-N., and Horta, L. G., "Use of a Velocity Command Motor as a Reaction Mass Actuator," *Journal of Guidance, Control and Dynamics*, Vol. 15, No. 1, 1992, pp. 159-164.

7. Lindner, D. K., Celano, T. P., and Ide, E. N., "Vibration Suppression Using a Proofmass Actuator Operating in Stroke/Force Saturation," *Journal of Vibration and Acoustics*, Vol. 113, Oct. 1991, pp. 423-433.

8. Lindner, D. K., Zvonar, G. A., and Borojevic, Dusan, "Performance and Control of Proof-Mass Actuators Accounting for Stroke Saturation," Journal of Guidance, Control, and Dynamics, Vol. 17, No. 5, 1994, pp. 1103-1108.

9. Lieber, K., "Vibration Suppression with a Reaction Mass Actuator," M.S. Thesis, University of Kansas, Lawrence, KS, 1995.

10. *LabVIEW 3.0 User's Manual*, National Instruments, Inc., Austin, TX, 1994.

AIAA-97-3466

DEVELOPMENT AND UTILIZATION OF A MULTIVARIABLE ROOT LOCUS FLIGHT CONTROL DESIGN TOOL

Michael Graves II*
Wyle Laboratories
Arlington, Virginia 22202

Brett Newman**
Department of Aerospace Engineering
Old Dominion University
Norfolk, Virginia 23529

Abstract

New results are summarized which extend the Evans root locus design technique to fully coupled, two-channel control law architectures. The new results are based upon various root migration plots, and the underlying framework concerning how the plots influence one another, for the closed-loop poles and zeros in each channel, as key system gains are varied. A new computerized tool is presented that allows efficient construction of these root migration plots. This control design software program is coded in MATLAB® with a graphical user interface. Instead of sequentially closing each feedback loop, the software facilitates closure of all loops simultaneously, allowing efficient assessment of stability and performance levels in all channels. The new tool is used to explore the additional benefits in tailoring of handling qualities when using a two-channel, two-directional crossfeed control law architecture for lateral-directional flight dynamics. Utilization of this additional design freedom to tailor handling qualities, or increase performance levels and stability margins, is highly desirable.

Introduction

Design solutions for modern aircraft have reached a degree of maturity where reliance upon flight control systems, to satisfy an array of mission requirements, is common place and often necessary.[1] For example, in advanced fighter concepts, the vehicle is expected, under the commands from the pilot, to rapidly transition between pre-stall and post-stall flight conditions. Sluggish responses with severe coupling in the roll/pitch/yaw axes are a direct result of operating in this high angle of attack and high angular rate environment. Multivariable control laws are necessary to augment the vehicle characteristics to exhibit well coordinated responses extolling high levels of damping, quickness, and robustness, when contrasted against established handling quality criteria.[2]

Restoring exceptional handling qualities to the vehicle requires a method to synthesize multiple feedback and crossfeed paths. These signal paths must work in a coordinate fashion, providing operation at near maximum performance levels with stability margins approaching minimum requirement limits. Desirable attributes of any proposed design technique should include 1) clear and concise relationships between the design parameters and resulting closed-loop features, 2) system architecture and compensation which are amenable to implementation and modification, and 3) capacity to address multiple interacting system responses and inputs.

Design methods based upon classical single-loop concepts,[3-6] and extensions to multivariable settings with sequential loop closure techniques,[7] provide many of these attributes. For example, relationships between the design parameters and closed-loop properties can be portrayed graphically providing the flight control designer with a tool rich with information. Knowledge underlying the role of a particular feedback loop, or the effect from a specific dynamic compensation feature, upon the closed-loop properties is advantageous. These concepts also facilitate simple compensation and architecture. The main drawback of these techniques is their inability to address fully coupled interacting loops in an efficient manner.

Utilization of sequential loop closure concepts permits limited applicability of these techniques to multi-loop settings. In particular, for 2×2 flight control problems, a common architecture is a two-channel, one-directional crossfeed control law. In the lateral-directional axes, for example, this architecture consists of roll rate feedback to aileron, yaw rate feedback to rudder, and an aileron to rudder interconnect. When these loops are closed sequentially, the designer must anticipate effects from the current loop closure upon channels already closed and subsequent channels yet to be closed. Consideration of a two-channel, two-directional crossfeed control law architecture is a challenging task with this sequential design approach. However, utilization of this additional design freedom to tailor handling qualities, or increase performance levels and stability margins, is highly desirable, and advanced vehicle concepts may require this.

Ref. 8 provided new multivariable results for the Evans root locus design technique for such applications. This framework is based upon various root migration plots for the closed-loop poles and zeros in each channel, as key system gains are varied. The closed-loop pole plot is uniquely interesting in that poles "track moving zeros". Monitoring of these plots, and the underlying framework of how the plots influence one another, allows simultaneous consideration of closed-loop stability and performance characteristics in all channels, as candidate control laws are formulated.

* Cost Analyst. Masters Student. Member AIAA.
** Assistant Professor. Senior Member AIAA.

Further, the classical nature of the framework does not leave the designer in the dark concerning necessary modifications for improvement, but rather suggests, with the graphical information, what type of compensation is necessary for system improvement.

Development of a computerized design and analysis tool, which implements the new multivariable generalization of the Evans root locus concept, is one goal of the paper. With such a tool, common sequentially designed control system architectures, as well as fully coupled generalizations, can be designed within an overall framework allowing simultaneous features to be assessed, while still retaining an efficient graphical medium. A second goal is utilization of this tool to explore the potential of a fully coupled, two-channel lateral-directional flight control law. The extra design freedom allotted by such an architecture offers increased potential for tailoring of handling qualities.

Multi-Loop Root Locus Concepts

Consider a 2×2 system whose dynamics are to be augmented with feedback such as an aircraft with inherently poor flying qualities, or

$$y(s) = G(s)u(s) \tag{1}$$

$$y(s) = \begin{bmatrix} y_1(s) \\ y_2(s) \end{bmatrix}, \; u(s) = \begin{bmatrix} u_1(s) \\ u_2(s) \end{bmatrix}, \; G(s) = \begin{bmatrix} g_{11}(s) & g_{12}(s) \\ g_{21}(s) & g_{22}(s) \end{bmatrix}$$

In Eq. (1), G(s) represents the transfer function matrix between response vector y(s) and input vector u(s). Likewise, $g_{ij}(s)$ denotes the transfer function between individual response $y_i(s)$ and input $u_j(s)$. Further, suppose a two-channel feedback, two-directional crossfeed architecture is utilized, or

$$u(s) = K(s)\{y_c(s) - y(s)\} \tag{2}$$

$$y_c(s) = \begin{bmatrix} y_{1c}(s) \\ y_{2c}(s) \end{bmatrix}, \; K(s) = \mathcal{K} \begin{bmatrix} k_{11}(s) & k_{12}(s) \\ k_{21}(s) & k_{22}(s) \end{bmatrix}$$

In Eq. (2), $y_c(s)$ denotes the response command vector and K(s) represents the compensator transfer matrix. Further, $k_{ij}(s)$ represents individual compensation transfer functions and $\mathcal{K}$ is an overall compensator gain. Fig. 1 depicts the closed-loop system.

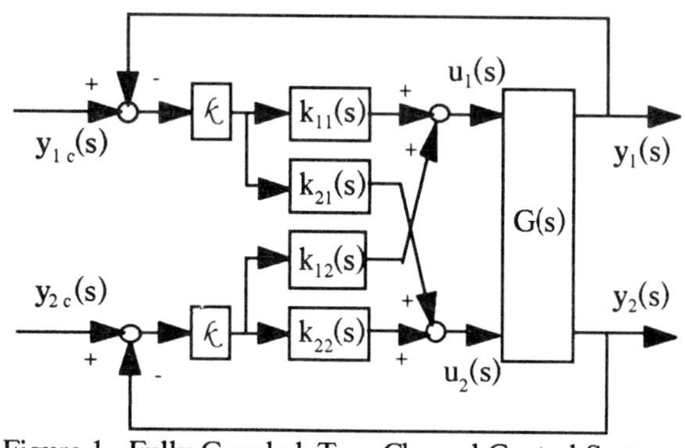

Figure 1. Fully Coupled, Two-Channel Control System

Closed-loop responses due to the commands are governed by

$$y = G[I + KG]^{-1}K\, y_c = G_{cl}\, y_c \tag{3}$$

where the explicit dependency upon complex frequency is dropped for conciseness. Individual closed-loop transfer functions can be expressed as follows

$$g_{11cl} = \frac{\mathcal{K}(k_{11}g_{11}+k_{21}g_{12}) + \mathcal{K}^2(\tilde{k}\tilde{g})}{\Delta} \quad \text{(main channel)}$$

$$g_{12cl} = \frac{\mathcal{K}(k_{12}g_{11}+k_{22}g_{12})}{\Delta} \quad \text{(cross channel)} \tag{4}$$

$$g_{21cl} = \frac{\mathcal{K}(k_{11}g_{21}+k_{21}g_{22})}{\Delta} \quad \text{(cross channel)}$$

$$g_{22cl} = \frac{\mathcal{K}(k_{12}g_{21}+k_{22}g_{22}) + \mathcal{K}^2(\tilde{k}\tilde{g})}{\Delta} \quad \text{(main channel)}$$

$$\Delta = 1+\mathcal{K}(k_{11}g_{11}+k_{12}g_{21}+k_{21}g_{12}+k_{22}g_{22})+\mathcal{K}^2(\tilde{k}\tilde{g})$$

$$\tilde{g} = g_{11}g_{22}-g_{12}g_{21}$$

$$\tilde{k} = k_{11}k_{22}-k_{12}k_{21}$$

In Eq. (4), $\tilde{g}(s)$ and $\tilde{k}(s)$ denote the plant and compensator coupling factors, respectively.

Consider Δ(s) in Eq. (4) which governs the closed-loop poles. The root locus relationship corresponding to these poles can be written as

$$1+\mathcal{K}\frac{\gamma}{\phi}+\mathcal{K}^2\frac{\psi}{\phi} = 0 \tag{5}$$

where

$$\phi = d_k d_g \tag{6}$$

$$\gamma = \mathcal{K}_{11} n_{k_{11}} n_{g_{11}} + \mathcal{K}_{12} n_{k_{12}} n_{g_{21}} + \mathcal{K}_{21} n_{k_{21}} n_{g_{12}} + \mathcal{K}_{22} n_{k_{22}} n_{g_{22}} = \frac{d_k d_g}{\mathcal{K}} \operatorname{tr}[KG]$$

$$\psi = \frac{\mathcal{K}_{11}\mathcal{K}_{22} n_{k_{11}} n_{k_{22}} - \mathcal{K}_{12}\mathcal{K}_{21} n_{k_{12}} n_{k_{21}}}{d_k} \times \frac{n_{g_{11}} n_{g_{22}} - n_{g_{12}} n_{g_{21}}}{d_g} = \frac{d_k d_g}{\mathcal{K}^2} \det[KG] = \psi_k \psi_g$$

Here, $\mathcal{K}_{ij} n_{k_{ij}}(s)$ and $n_{g_{ij}}(s)$ denote the numerator polynomials of $k_{ij}(s)$ and $g_{ij}(s)$, while $d_k(s)$ and $d_g(s)$ are the denominator polynomials. Finally, note that ψ(s) and φ(s) are the transmission zero and characteristic polynomials for the open-loop system K(s)G(s), and ψ(s) can be split into compensator and plant transmission zero polynomials, $\psi_k(s)$ and $\psi_g(s)$.

From Eq. (5), observe the complexities introduced by the system multivariableness. A two-channel feedback system results in a 2nd order polynomial relationship in $\mathcal{K}$ governing the closed-loop poles, rather than a linear relationship for scalar-loop feedback systems. Further, there are now two open-loop "numerator" polynomials which determine the loci behavior: ψ(s) and γ(s). Finally, these numerator polynomials are complicated functions of the individual open-loop numerator polynomials.

Eq. (5) has another more useful interpretation. Consider rewriting Eq. (5) as

$$1+\kappa\frac{\eta}{\phi}=0 \ , \ 1+\kappa\frac{\psi}{\gamma}=0 \qquad (7)$$

$$\eta=\gamma+\kappa\psi$$

Eq. (7) indicates the closed-loop pole behavior is governed by a "two-layered root locus." In other words, with increasing overall gain κ, the open-loop poles originating from $\phi(s)$ migrate towards "moving" open-loop zeros $\eta(s)$. These zeros originate at $\gamma(s)$ and migrate towards $\psi(s)$ as the system loop gain also increases.

Turning attention to the closed-loop numerators in Eq. (4), the main channel zeros are governed by the following relationships:

$$1+\kappa\frac{\psi}{\gamma_{11}}=0 \ , \ 1+\kappa\frac{\psi}{\gamma_{22}}=0 \qquad (8)$$

where the polynomials $\gamma_{11}(s)$ and $\gamma_{22}(s)$ are defined as

$$\gamma_{11}=\kappa_{11}n_{k_{11}}n_{g_{11}}+\kappa_{21}n_{k_{21}}n_{g_{12}} \qquad (9)$$

$$\gamma_{22}=\kappa_{12}n_{k_{12}}n_{g_{21}}+\kappa_{22}n_{k_{22}}n_{g_{22}}$$

With increasing overall gain κ the main channel zeros originating from $\gamma_{11}(s)$ and $\gamma_{22}(s)$ migrate towards $\psi(s)$. Note $\gamma(s)$ in Eq. (6) is simply the addition of $\gamma_{11}(s)$ and $\gamma_{22}(s)$, or

$$\gamma(s)=\gamma_{11}(s)+\gamma_{22}(s) \qquad (10)$$

The cross channel zeros, and the roots of $\gamma_{11}(s)$, $\gamma_{22}(s)$, $\gamma(s)$, and $\psi(s)$ are all invariant as the overall gain κ is adjusted. However, these important roots are dependent upon the relative gains between the cross channels and main channels. The cross channel zeros in Eq. (4) are governed by the following relationships:

$$1+\frac{\kappa_{12}}{\kappa_{22}}\frac{n_{k_{12}}}{n_{k_{22}}}\frac{n_{g_{11}}}{n_{g_{12}}}=0 \qquad (11)$$

$$1+\frac{\kappa_{21}}{\kappa_{11}}\frac{n_{k_{21}}}{n_{k_{11}}}\frac{n_{g_{22}}}{n_{g_{21}}}=0$$

Using Eq. (9), roots of $\gamma_{11}(s)$ and $\gamma_{22}(s)$ are given by

$$1+\frac{\kappa_{21}}{\kappa_{11}}\frac{n_{k_{21}}}{n_{k_{11}}}\frac{n_{g_{12}}}{n_{g_{11}}}=0 \qquad (12)$$

$$1+\frac{\kappa_{12}}{\kappa_{22}}\frac{n_{k_{12}}}{n_{k_{22}}}\frac{n_{g_{21}}}{n_{g_{22}}}=0$$

Finally, the governing relationships for the roots of $\gamma(s)$ and $\psi(s)$ are obtained form Eqs. (6) and (10),

$$1+\frac{\kappa_{22}}{\kappa_{11}}\frac{\gamma_{22}/\kappa_{22}}{\gamma_{11}/\kappa_{11}}=0 \qquad (13)$$

$$\left\{1-\frac{\kappa_{12}\kappa_{21}}{\kappa_{22}\kappa_{11}}\frac{n_{k_{12}}n_{k_{21}}/d_k}{n_{k_{11}}n_{k_{22}}/d_k}\right\}\psi_g=0$$

Fig. 2 summarizes the above relationships and illustrates how the corresponding root migration plots influence one another, ultimately yielding the closed-loop transfer functions, as well as closed-loop stability and performance levels. The compensator matrix K(s) in Eq. (2) has five "degrees of freedom": κ, κ_{11}, κ_{12}, κ_{21}, and κ_{22}, aside from the freedom available within each compensation term $n_{k_{ij}}(s)$. In Fig. 2 for root locus design applications, these parameters are reformulated in terms of the five "degrees of freedom" κ, κ_{12}/κ_{22}, κ_{21}/κ_{11}, κ_{22}/κ_{11}, and specification of any single κ_{ij} value.

The development here opens new possibilities for conventional-based multivariable flight control design. Eqs. (7)-(13) allow the designer to simultaneously monitor closed-loop stability and performance in all the feedback channels as the system gain and relative gains are varied. This design strategy is intended to relieve some of the burdens encountered with sequential loop closure design strategies, as well as extend the conventional-based Evans design technique to fully coupled control system architectures. Even with the results presented here, the multivariable design problem is formidable. Selection of actuator/sensor pairs and compensation still confront the designer. However, rather than conducting the design as a sequential process, the "design knobs" illustrated in Fig. 2 can be used to tailor system behavior under a unifying perspective. Critical to this effort will be development of a software design tool which portrays these concepts in an efficient and friendly manner.

Software Tool Description

The design tool is built around the theoretical framework summarized in Fig. 2 from the previous Section. The tool must be capable of displaying the various root migration plots with a capacity to easily modify the compensation and regenerate the graphical results. The tool has been coded in the encompassing analysis package known as MATLAB®, version 4.2. A Graphical User Interface (GUI) format has been utilized to decrease design iteration cycle time and to facilitate ease of use. The intent is to execute the tool on desktop computational hardware.

The GUI format consists of an array of edit boxes, slider bars, and push buttons which allow the user to

- designate files containing the plant dynamics to be loaded into the workspace
- specify compensator dynamics to be loaded into the workspace either by file or keyboard input
- adjust compensator relative gain and overall gain values by slider bar or keyboard input
- stipulate gain vectors for root loci generation either by file or keyboard input
- adjust window plot sizing by keyboard input
- refresh graphical results following design modifications by click input.

After launching the software tool, the file name containing the plant dynamics is inputted, and the code automatically loads the model data into the workspace. Next, unity gain compensator dynamics are inputted and automatically loaded into the workspace. The overall

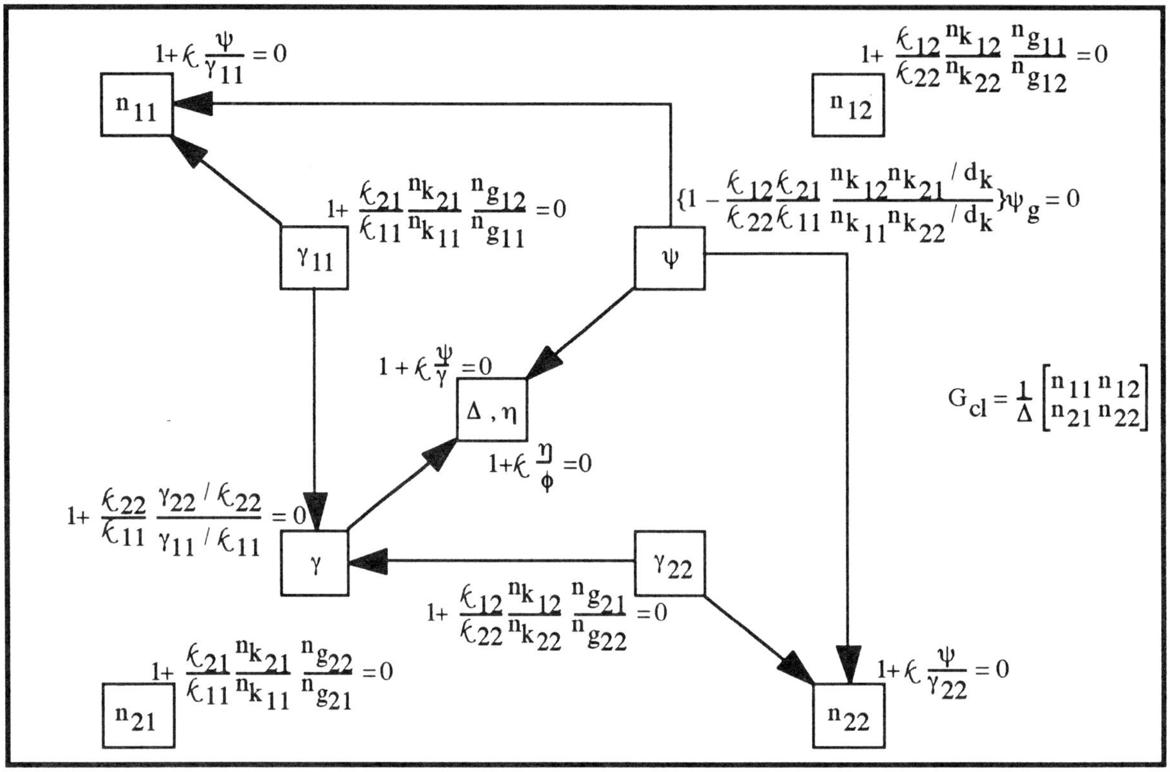

Figure 2. Relationships Governing Two-Channel, Two-Directional Crossfeed Control Systems

and relative gain values are then specified. After the refresh button is activated, the tool uses the inputted data to generate the applicable root migration plots. Closed-loop results are assessed by the user and, if necessary, modifications are made to the compensator gain values with the slider bars and the plots are regenerated with the refresh button. Fig. 3 displays the design tool in action.

It is important for the reader to understand that the control design software tool does not automatically develop the final, or for that matter, the initial control law with no inputs from the user. The tool assists the designer in efficiently generating and portraying the detail results within the framework of Fig. 2. To be an effective tool, the designer must have a grasp of conventional-based design aspects and the underlying relationships indicated in Fig. 2. Input from the designer who understands the control problem, the vehicle physics, and the associated trades is highly desirable, and leads to a problem-driven control law solution, not a design technique-driven control law solution.

Although many of the existing root locus features within MATLAB could be used directly, additional software for root locus computations was necessary. In particular, computations for the closed-loop pole migration (see Eq. (7)) were nontrivial and required new root locus code development. In generating the loci for these poles, calls to the root solver at each gain value required updating of the zero locations (i.e., "moving zeros"). Basic functions, such as gain vector generation and loci sorting, were also necessary in this step. To avoid unnecessary computations and the associated run time penalties, bypass logic was designed into the tool. Modifications to the compensator are monitored, and only those plots that are affected by these changes are updated when the refresh button is activated. Unaffected plots are bypassed.

Airframe Dynamics

Early results from exploratory programs addressing advancement of air-combat technologies have generated considerable interest for the development of a highly maneuverable airframe which can operate across a broad spectrum of aerodynamic attitude flight conditions.[9] The operational environment for this vehicle will lead to unsatisfactory dynamic characteristics and degraded handling qualities. For example, high angle of attack/side slip conditions will lead to unconventional airflow phenomena, with resulting affects upon airframe stability and control derivatives. These derivatives may not have their usual numerical values, and may even experience sign reversal. As a result, the vehicle will exhibit unfamiliar dynamics to the manual control loop. This type of vehicle is a prime candidate for, and may necessarily require, a multivariable flight control system. Therefore, a high angle of attack lateral-directional airframe application is chosen to demonstrate the new design technique and software tool presented in the previous Sections.

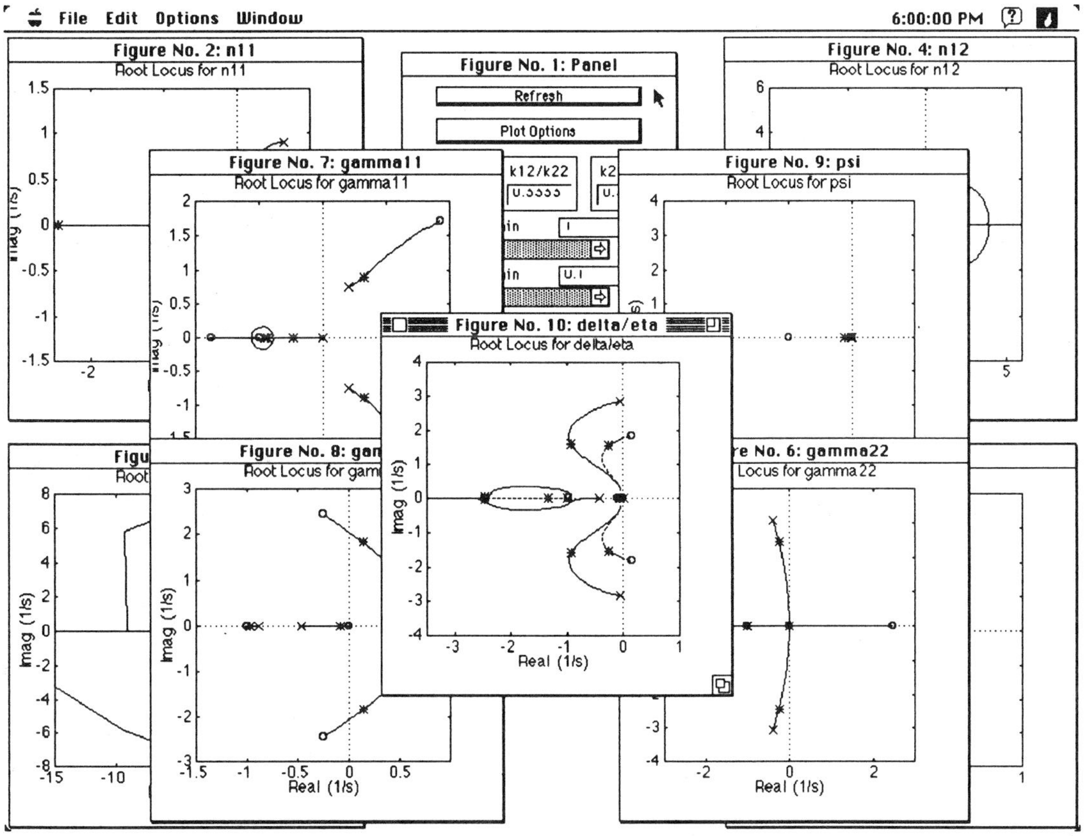

Figure 3. Two-Loop Control Design Tool In Action

Tab. 1 contains state space and transfer function data for the lateral-directional dynamics of a generic, highly maneuverable aircraft at Mach 0.6, 35,000 ft altitude, and moderately high angle of attack.[10] Figs. 4-7 show the roll rate and yaw rate responses to right-wing down (aileron) and nose right (rudder) pulse commands of 0.01 rad acting over 1 s. At this flight condition, the aircraft dynamics exhibit several objectionable characteristics including low dutch-roll damping ($\zeta_{dr} = 0.021$), slow roll response ($1/\tau_r = 0.43$ 1/s), undesirable dutch-roll contamination in all responses, and several unconventional turn entry characteristics. In particular, for roll right input, Fig. 5 indicates the airframe has a slight initial proverse-adverse yaw combination before the yaw rate builds up from the turn (i.e., nonminimum phase zeros in $n_{g12}(s)$). Further, for yaw right input, Fig. 7 shows delayed response reversal in yaw rate (i.e., nonminimum phase zeros in $n_{g11}(s)$). Finally, for yaw right input, Fig. 6 depicts adverse roll (i.e., nonminimum phase zero in $n_{g21}(s)$). Even though this last characteristic is common to nearly all conventional aircraft, and the "mil-specs" do not specifically address this issue, assume here that this behavior is undesirable from a manual control perspective. In summary, the inherent airframe dynamics appear unacceptable for manual control by the pilot and need improvement.

Preliminary Design

Objectives of the flight control system are to correct the characteristics noted above, as well as provide satisfactory levels of robustness to modeling errors with feedback architecture as simple as possible.

To augment dutch-roll damping and roll time constant, the "first cut" design will close the stability axis yaw rate to rudder and stability axis roll rate to aileron loops, or

$$K(s) = \kappa \begin{bmatrix} -\kappa_{11} & 0 \\ 0 & \kappa_{22} \end{bmatrix}, \quad \kappa_{ii} > 0 \qquad (14)$$

where static compensation is evaluated initially. Figs. 8-11 show root migration behavior for $n_{11}(s)$, $n_{22}(s)$, $\gamma(s)$, $\eta(s)$, and $\Delta(s)$. Results for $n_{12}(s)$, $n_{21}(s)$, $\gamma_{11}(s)$, $\gamma_{22}(s)$, and $\psi(s)$ are not applicable here since these roots are fixed for the trial control law. Also, Figs. 12-13 show selected responses for similar input conditions as in Figs. 4-5.

First consider the relationships between Figs. 8-11, recalling the framework displayed in Fig. 2. Fig. 10 shows how the roots of $\gamma(s)$ vary as κ_{22}/κ_{11} increases. The migration paths originate and terminate at the roots of $n_{g11}(s)$ and $n_{g22}(s)$, and for a value of $\kappa_{22}/\kappa_{11} = 0.3$ the roots are located at -0.092, -0.16±j2.0 1/s. Fig. 11 shows how the roots of $\Delta(s)$ and $\eta(s)$ vary as κ increases. For the roots of $\eta(s)$, the migration paths

Table 1. Lateral-Directional Aircraft Model

$$\dot{x}(t) = Ax(t) + Bu(t) \qquad y(s) = G(s)u(s)$$
$$y(t) = Cx(t)$$

$$y = \begin{bmatrix} r_s \text{ (stability axis yaw rate, rad/s)} \\ p_s \text{ (stability axis roll rate, rad/s)} \end{bmatrix}$$

$$x = \begin{bmatrix} \beta \text{ (side slip, rad)} \\ p \text{ (roll rate, rad/s)} \\ r \text{ (yaw rate, rad/s)} \\ \phi \text{ (roll angle, rad)} \end{bmatrix} \quad u = \begin{bmatrix} \delta_r \text{ (rudder, rad)} \\ \delta_a \text{ (aileron, rad)} \end{bmatrix}$$

$$\left[\begin{array}{c|c} A & B \\ \hline C & \end{array}\right] = \left[\begin{array}{cccc|cc} -0.0868 & 0.215 & -0.977 & 0.0539 & 0.0179 & 0 \\ -32.3 & -0.374 & 2.40 & 0 & 6.66 & 6.35 \\ 1.06 & -0.0406 & -0.0809 & 0 & -1.18 & 1.71 \\ 0 & 1 & 0.220 & 0 & 0 & 0 \\ \hline 0 & -0.215 & 0.977 & 0 & & \\ 0 & 0.977 & 0.215 & 0 & & \end{array}\right]$$

$$n_{g_{11}} = -2.58(s-0.401 \pm j0.752)(s+0.912)$$
$$n_{g_{12}} = 0.305(s+1.76)(s-1.85 \pm j1.71)$$
$$n_{g_{21}} = 6.25(s+3.21e{-6})(s+2.05)(s-2.44)$$
$$n_{g_{22}} = 6.57(s+3.21e{-6})(s+0.393 \pm j3.06)$$
$$d_g = (s-0.00477)(s+0.427)(s+0.0595 \pm j2.85)$$
$$\psi_g = -18.9s(s+0.0281)$$

originate and terminate at the roots of $\gamma(s)$ (from Fig. 10) and $\psi_g(s)$ (essentially two roots at the origin). With an excess of one root, the complex conjugate roots form a break-in point at -2 1/s, one root heads toward the origin while the other moves off to $-\infty$ along the real axis. For a value of $\mathcal{K} = 1$, the roots are located at -0.094, -0.78±j1.8 1/s. For the roots of $\Delta(s)$, the migration paths originate and terminate at the roots of $d_g(s)$ and $\psi_g(s)$. With an excess of two roots, the complex conjugate roots move off towards $-\infty$, parallel with the real axis. Note the roots of $\Delta(s)$ do not necessarily follow standard root locus rules. For example, the root originating at -0.43 1/s initially moves deeper into the left-half plane, but reverses it's direction of travel around -3 1/s and heads back towards the origin. In general, however, the roots of $\Delta(s)$ "track" the moving roots of $\eta(s)$. For a value of $\mathcal{K} = 1$, the roots are located at -0.081, -2.3, -1.3±j2.7 1/s. Fig. 8 shows how the roots of $n_{11}(s)$ vary as $\mathcal{K}$ increases. The migration paths originate and terminate at the roots of $n_{g11}(s)$ and $\psi_g(s)$, and for a value of $\mathcal{K} = 1$ the roots are located at -2.4, 0.046±j0.52 1/s. Finally, Fig. 9 shows how the roots of $n_{22}(s)$ vary as $\mathcal{K}$ increases. The migration paths originate and terminate at the roots of $n_{g22}(s)$ and $\psi_g(s)$, and for a value of $\mathcal{K} = 1$ the roots are located at -3.2e-6, -1.8±j2.5 1/s.

Now consider the design aspects of Figs. 8-11. From. Fig.11, dutch-roll damping and roll quickness have improved significantly, which is also seen in Figs. 12-13. However, the manner in which this augmentation takes place is undesirable. The migration paths in Fig. 11 indicate, for high gain, the dutch-roll mode becomes the dominate closed-loop mode while the augmented roll mode is suppressed. For the lateral-directional dynamics, a dominate mode characterized by a real root is more desirable. Further, the spiral mode is augmented in an unfavorable manner as follows. The

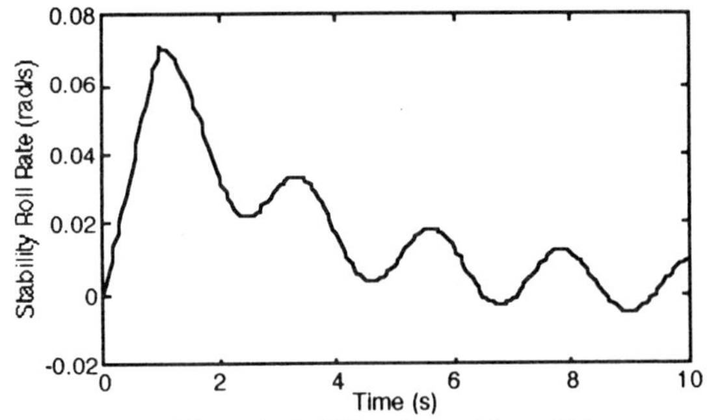

Figure 4. Roll Rate From Aileron Pulse

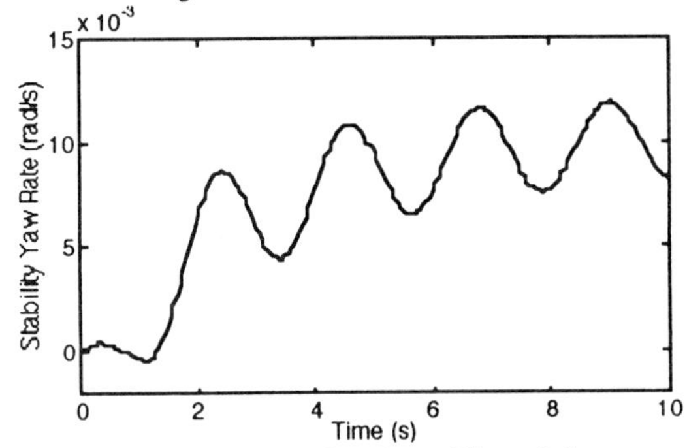

Figure 5. Yaw Rate From Aileron Pulse

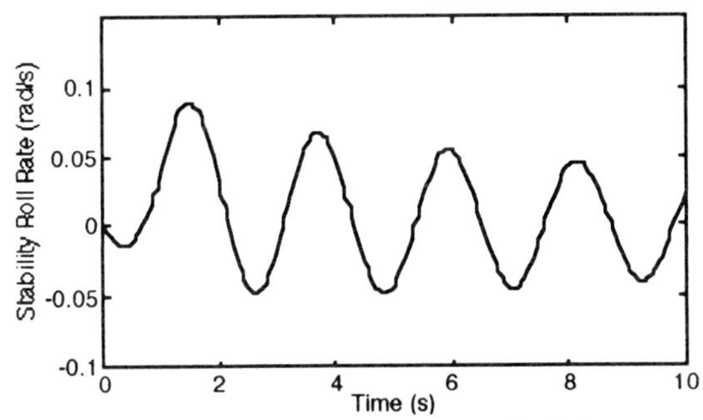

Figure 6. Roll Rate From Rudder Pulse

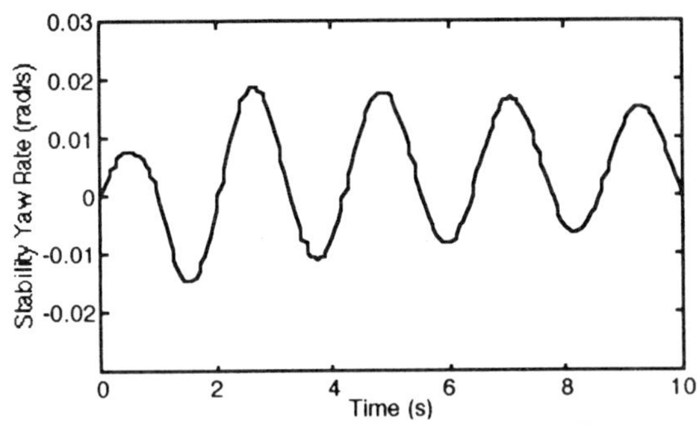

Figure 7. Yaw Rate From Rudder Pulse

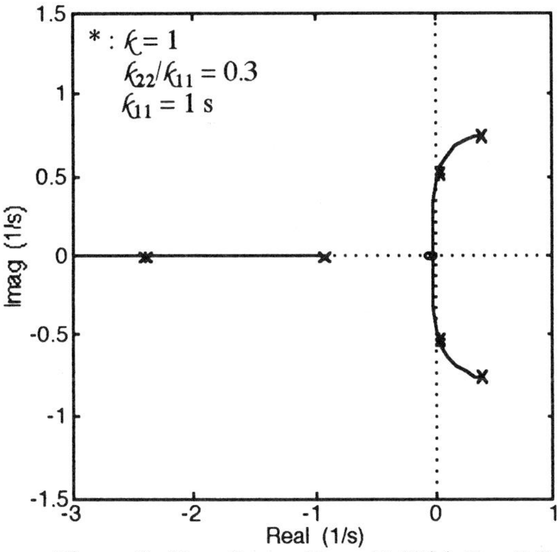

Figure 8. Root Locus For n11 With Eq. (14)

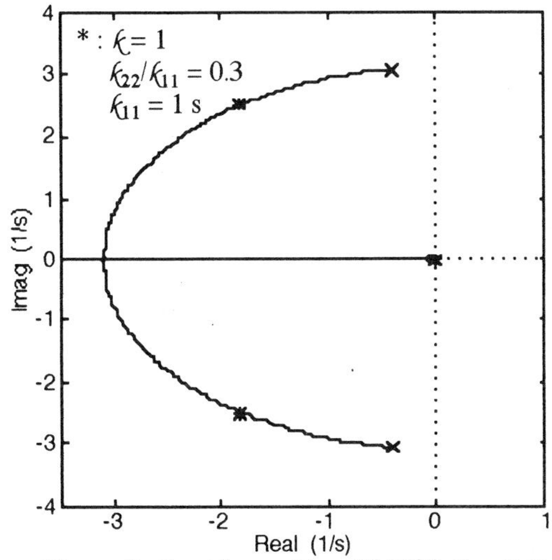

Figure 9. Root Locus For n22 With Eq. (14)

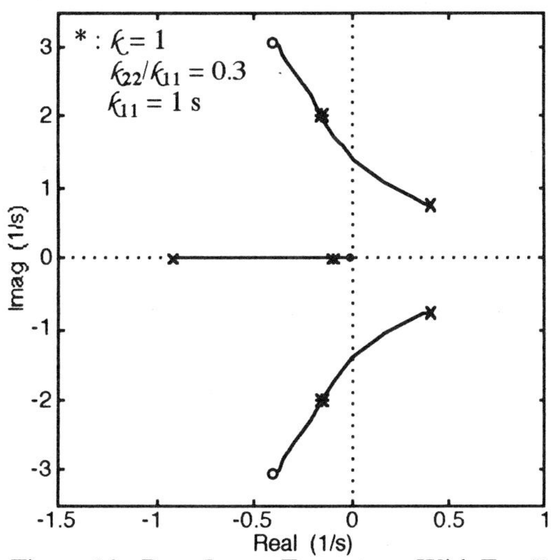

Figure 10. Root Locus For gamma With Eq. (14)

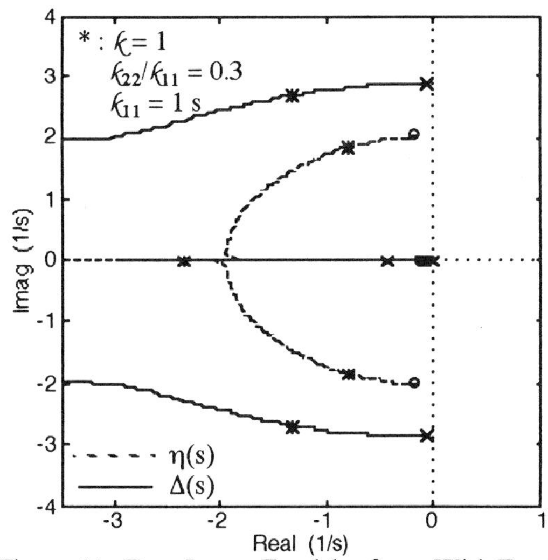

Figure 11. Root Locus For delta & eta With Eq. (14)

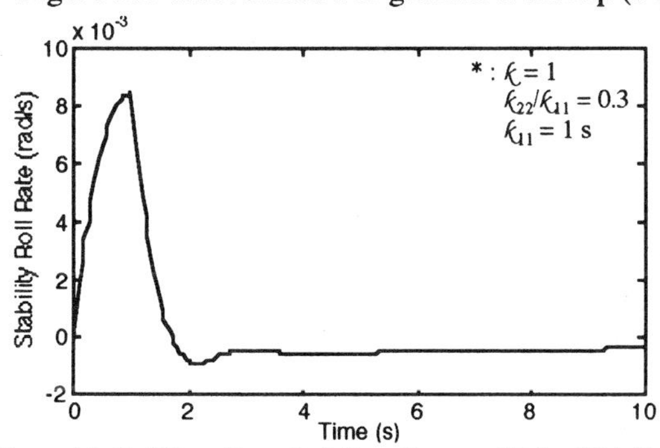

Figure 12. Roll Rate From Roll Rate Command Pulse With Eq. (14)

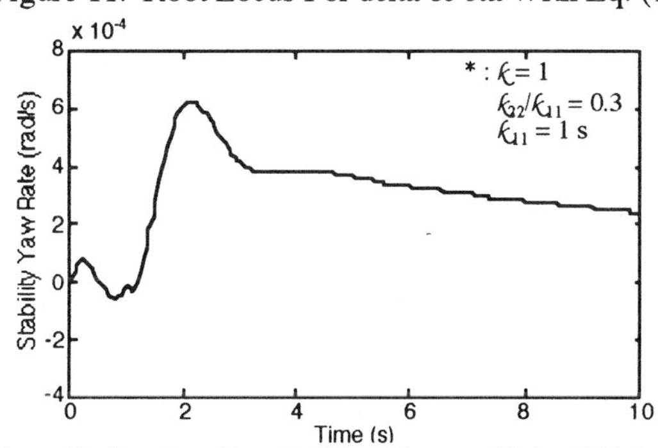

Figure 13. Yaw Rate From Roll Rate Command Pulse With Eq. (14)

"moving zero" in Fig. 11 migrating towards the origin combines with another "moving zero" already near the origin (which is pre-determined from $\gamma(s)$ in Fig. 10) to form complex conjugate roots which tend towards the roots of $\psi_g(s)$. This migration off the real axis allows the spiral mode to move through the origin to become stable. The consequence is seen in Fig. 13, the steady yaw rate from the turn drops off noticeably around 4 s. The yaw damper loop is taking the vehicle out of the steady turn. A control strategy which leaves the natural spiral behavior (essentially an integrator) untouched is more desirable.

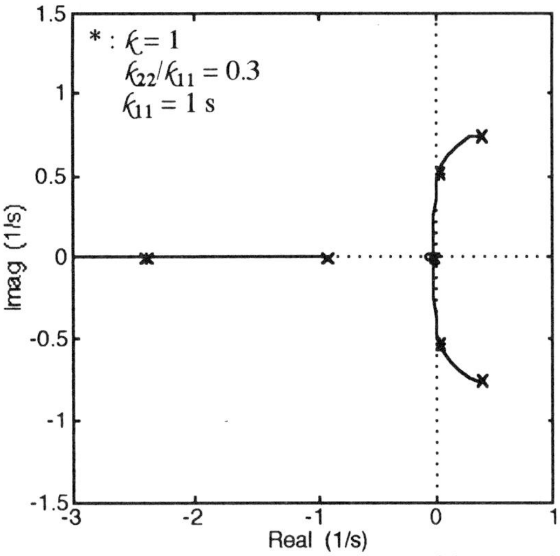

Figure 14. Root Locus For n11 With Eq. (15)

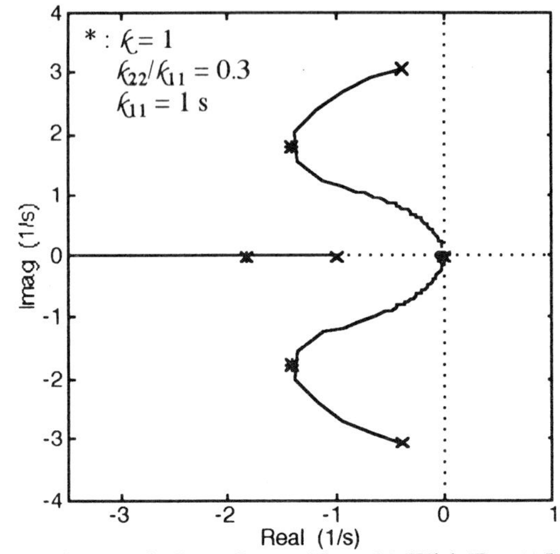

Figure 15. Root Locus For n22 With Eq. (15)

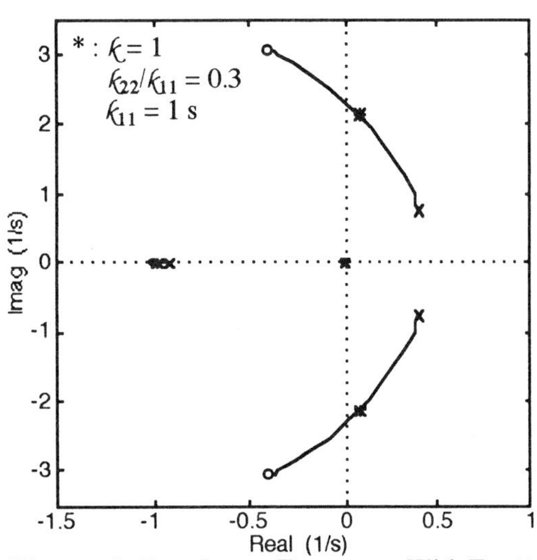

Figure 16. Root Locus For gamma With Eq. (15)

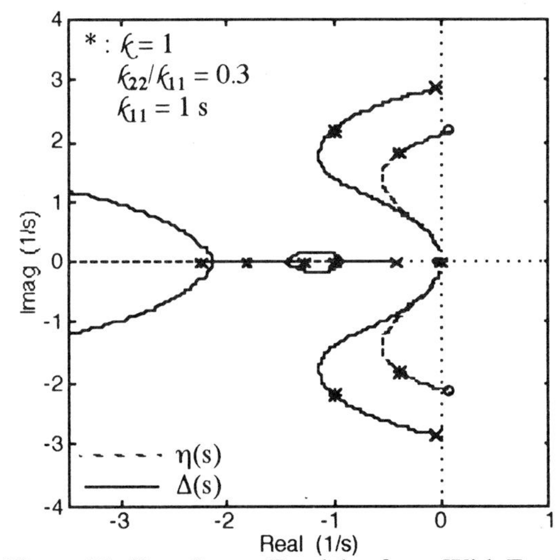

Figure 17. Root Locus For delta & eta With Eq. (15)

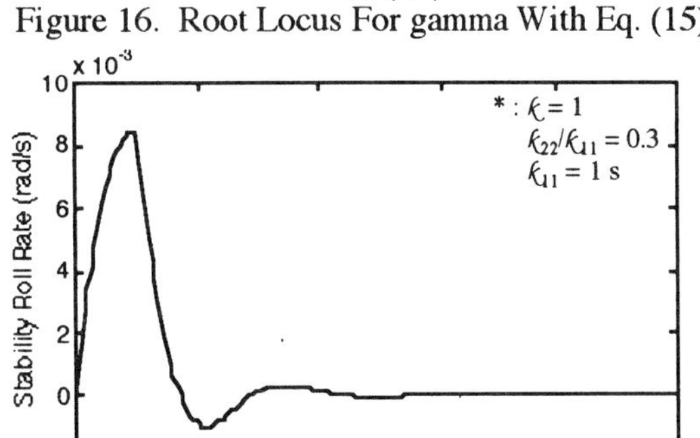

Figure 18. Roll Rate From Roll Rate Command Pulse With Eq. (15)

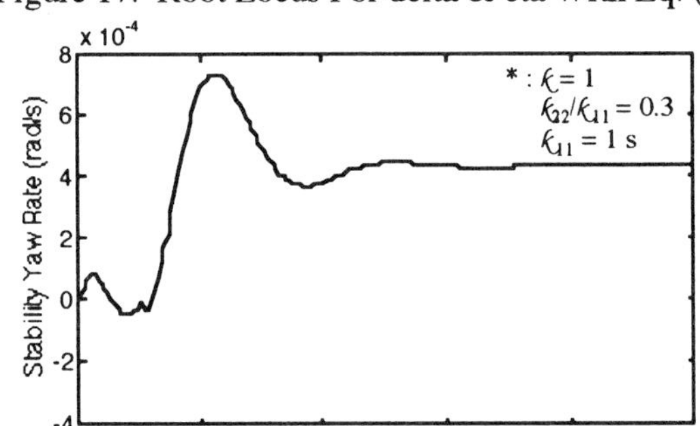

Figure 19. Yaw Rate From Roll Rate Command Pulse With Eq. (15)

With regards to the closed-loop numerators, most of the augmentation is acceptable. In Fig. 9, the complex conjugate roots follow the dutch-roll poles in Fig. 11 and the root near the origin is always close to the spiral root. Similarly, in Fig. 8 the real root follows the roll mode in Fig. 11. These trends reduce participation of the dutch-roll and spiral modes in the roll rate/roll rate command transfer function and roll participation in the yaw rate/yaw rate command transfer function. These characteristics are all desirable. However, the complex conjugate roots in Fig. 8, although slightly improved, remain in the right-half

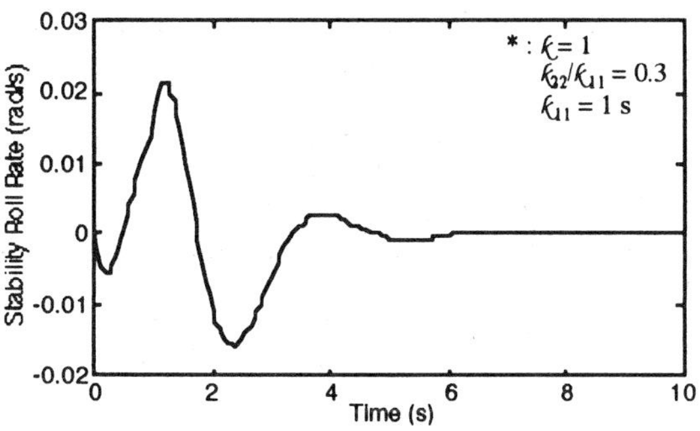

Figure 20. Roll Rate From Yaw Rate Command Pulse With Eq. (15)

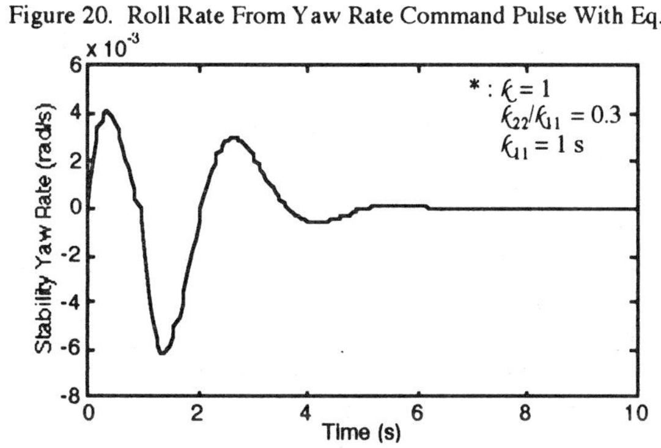

Figure 21. Yaw Rate From Yaw Rate Command Pulse With Eq. (15)

plane, leading to delayed yaw rate response reversal.

After careful evaluation of the undesirable features associated with the spiral mode augmentation near the origin in Fig. 11, and "back of the envelope" root locus sketches using the multi-loop framework in Fig. 2, a washout filter is inserted into the yaw damper loop, or

$$K(s) = \kappa \frac{1}{s+1} \begin{bmatrix} -\kappa_{11} s & 0 \\ 0 & \kappa_{22}(s+1) \end{bmatrix}, \quad \kappa_{ii} > 0 \qquad (15)$$

Figs. 14-17 show the root migration results, and Figs. 18-21 show the time responses. Without going into excessive description, inclusion of the washout filter introduces an additional root at the origin and constrains one root to remain near -1 1/s in Fig. 16. These new root characteristics for $\gamma(s)$ introduce an additional root at the origin, and one near -1 1/s, for $\eta(s)$ in Fig. 17. The new root at the origin blocks movement of the spiral mode and influences the dutch-roll roots to turn towards the origin. The new "moving zero" near -1 1/s moves off to $-\infty$ and draws the roll mode and compensator pole deeper into the left-half plane. These two roots are excess roots, and must migrate towards infinity. For high gain, the augmented roll mode will be the dominant response. Fig. 19 indicates the vehicle will now hold a steady turn. Also, Figs. 14-15 indicate the transfer function mode cancellation behavior is unaltered by the washout filter.

Improved Design

Attention is now focused on correcting the poor turn entry characteristics depicted in Figs. 18-21. In particular, the initial proverse-adverse yaw combination in Fig. 19, the adverse roll behavior in Fig. 20, and the delayed yaw rate response reversal in Fig. 21 are of concern. The source of these characteristics are the nonminimum phase closed-loop transfer function zeros within $n_{12}(s)$ and $n_{21}(s)$ (i.e., $n_{g12}(s)$ and $n_{g21}(s)$ in Tab. 1), and within $n_{11}(s)$ (i.e., augmented $n_{g11}(s)$ in Fig. 14).

During turn entry, studies have shown most pilots have a preference for mild adverse yaw where the vehicle nose follows the turn.[2] The control strategy, therefore, is to use the stability roll rate to rudder loop (i.e., aileron to rudder crossfeed) to modify the numerator characteristics of $n_{12}(s)$ from a single pair of complex conjugate nonminimum phase roots to a single real nonminimum phase root. To lessen, or remove altogether, the initial adverse roll during yaw commands, additional control strategy will be to utilize the yaw rate to aileron loop (i.e., rudder to aileron crossfeed) to remove the real nonminimum phase root within $n_{21}(s)$.

Rough, "back of the envelope" root locus sketches using the multi-loop framework in Fig. 2 indicate a feasible solution using gain feedback for these new channels, as well as the correct sign of the feedback gain. The new control law is

$$K(s) = \kappa \frac{1}{s+1} \begin{bmatrix} -\kappa_{11} s & \kappa_{12}(s+1) \\ \kappa_{21}(s+1) & \kappa_{22}(s+1) \end{bmatrix}, \quad \kappa_{ij} > 0 \qquad (16)$$

Figs. 22-30 show the root migration results, and Figs. 31-34 show the time responses.

Fig. 23 shows how the roots of $n_{12}(s)$ vary as κ_{12}/κ_{22} increases. The migration paths originate and terminate at the roots of $n_{g12}(s)$ and $n_{g11}(s)$, and for a value of $\kappa_{12}/\kappa_{22} = 0.33$ the roots are located at -1, 1.4, $-1.3 \pm j1.4$ 1/s. Observe how the nonminimum phase complex conjugate roots form a break-in point at 3.5 1/s, one root heads towards the origin and the other travels to the left-half plane via $\pm\infty$. The response in Fig. 32 has the desirable adverse yaw behavior. Fig. 24 shows how the roots of $n_{21}(s)$ vary as κ_{21}/κ_{11} increases. The migration paths originate and terminate at the roots of $n_{k11}(s)n_{g21}(s)$ and $n_{k21}(s)n_{g22}(s)$, and for a value of $\kappa_{21}/\kappa_{11} = 0.4$ the roots are located at -3.2e-6, -0.46, -2.8, 5.3 1/s. Note the nonminimum phase root is blocked by roots at the origin and must migrate to the left-half plane through $\pm\infty$. For the noted gain value the root does not quite make the full migration. However, Fig. 32 indicates a much less significant adverse roll characteristic.

With these new channels, the intermediate roots of $\gamma_{11}(s)$ and $\gamma_{22}(s)$ in Figs. 26-27 come into play. The roots of $\gamma_{11}(s)$ can influence the yaw rate to yaw rate command numerator characteristics and the closed-loop poles by modifying the origin and terminal points of the migration paths. Further, the roots of $\gamma_{22}(s)$ can influence the roll rate to roll rate command numerator characteristics and the closed-loop poles in the same manner. The designer's pencil sketches indicate the following significant features: high values for κ_{12}, κ_{21}, or both, will move the dutch-roll migration path towards the imaginary axis, and high values for k_{21} move the nonminimum phase complex roots for $n_{11}(s)$ deeper into the right-half plane. To achieve improved cross channel turn entry characteristics, dutch-roll damping must be traded off, and increased yaw rate response reversal must be tolerated. Comparison of Figs. 22, 25, and 30 with Figs. 14, 15, and 17 verify these trends. Here there is sufficient dutch-roll damping to tradeoff, but increased reversal in the yaw response is a significant concern.

Improvement of the yaw rate to yaw rate command dynamics shown here is a difficult control problem to solve. Two strategies the designer can explore include 1) additional filtering in $k_{11}(s)$ and $k_{21}(s)$, or in $k_{11}(s)$, $k_{22}(s)$, $k_{12}(s)$, and $k_{21}(s)$, to modify the roots of $\gamma_{11}(s)$ and $\psi(s)$ to eliminate the nonminimum phase zeros in $n_{11}(s)$ or 2) to consider alternate control effectors for generating yaw moments which are immune to aerodynamic phenomena at high angle of attack flight conditions. Risk associated with strategy 1 includes sophisticated, possibly right-half plane or near $j\omega$ axis compensation which is highly sensitive to model uncertainty, and which must be scheduled throughout the high angle of attack flight envelope. On the other hand, strategy 2 would suffer risk from new and untried technologies such as thrust vectoring or exotic forebody control schemes.

As a final point, Fig. 32 indicates the vehicle can no longer sustain a steady turn since the yaw rate begins to drop off beyond 4 s. The new control system cross channels have again stabilized the spiral mode in Fig. 30. To prevent this behavior, the low frequency content of the signal path along $k_{21}(s)$, $g_{12}(s)$, and around the yaw damper loop must be omitted. Inclusion of a washout filter for $k_{21}(s)$ should be considered.

None of the possible improvements to the control law in Eq. (16) discussed above are considered here. However, the main point not to be overlooked is the conventional-based design technique allows the designer to grasp the key control problem issues. This knowledge can then be used to formulate a sound solution strategy which facilitates implementation in a multiply redundant architectural environment often requiring scheduling over a wide spectrum of flight conditions and adjustment during flight test.

Conclusions

Extension of the Evans root locus design technique to fully coupled, two-channel control law architectures, while still retaining clear and concise relationships between the design parameters and resulting closed-loop features, and the ability to generate simple, low order control law solutions, has been described here. These results are significant because they offer a means to achieve the performance benefits of a highly integrated feedback compensator generated from conventional-based strategies. This approach to feedback synthesis fosters confidence in the control law augmentation role.

References

1. Bryson, A. E., "New Concepts in Control Theory, 1959-1984," *Journal of Guidance, Control, and Dynamics*, Vol. 8, No. 4, July-August, 1985, pp. 417-425.
2. U.S. Dept. of Defense, "Military Standard: Flying Qualities of Piloted Aircraft," MIL-STD-1797A, January, 1990.
3. Nyquist, H., "Regeneration Theory," *Bell System Technical Journal*, Vol. 11, January, 1932, pp. 126-147.
4. Bode, H. W., *Network Analysis And Feedback Amplifier Design*, Van Nostrand, New York, New York, 1945.
5. Evans, W. R., "Control System Synthesis by Root Locus Method," *AIEE Transactions Part II*, Vol. 69, 1950, pp. 66-69.
6. Ogata, K., *Modern Control Engineering*, Prentice-Hall, Englewood Cliffs, New Jersey, 1970.
7. McRuer, D., Ashkenas, I., and Graham, D., *Aircraft Dynamics And Automatic Control*, Princeton University Press, Princeton, New Jersey, 1973.
8. Newman, B., "Multivariable Flight Control Design Using Root Locus Techniques," Proceedings of the AIAA Guidance, Navigation, and Control Conference, Baltimore, Maryland, August, 1995, pp. 1157-1166.
9. Lynch, U. H. D., Ettinger, R. C., Pelt, J. V., and Skow, A. M., "Tactical Evaluation of the Air-to-Air Combat Effectiveness of Supermaneuverability," WRDC-TR-90-3035, Wright-Patterson AFB, Ohio, 1990.
10. McRuer, D. T., Myers, T. T., and Thompson, P. M., "Literal Singular-Value-Based Flight Control System Design Techniques," *Journal of Guidance, Control, and Dynamics*, Vol. 12, No. 6, November-December, 1989, pp. 913-919.

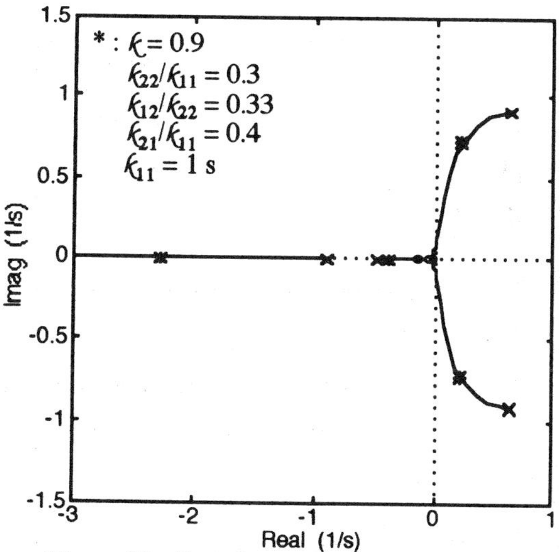

Figure 22. Root Locus For n11 With Eq. (16)

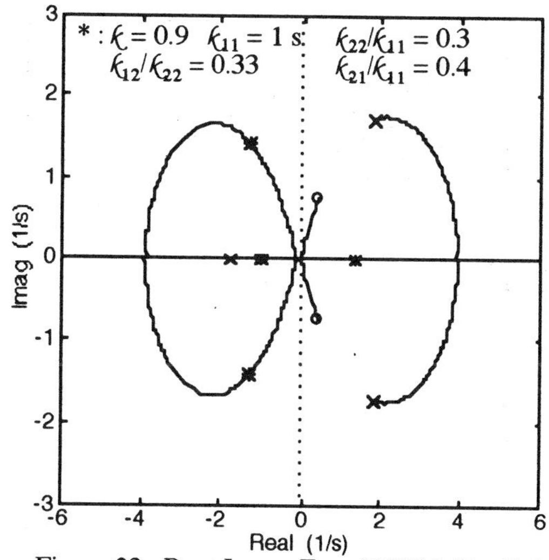

Figure 23. Root Locus For n12 With Eq. (16)

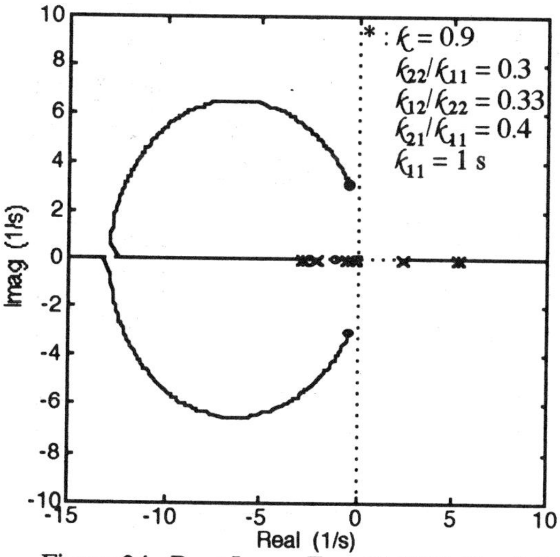

Figure 24. Root Locus For n21 With Eq. (16)

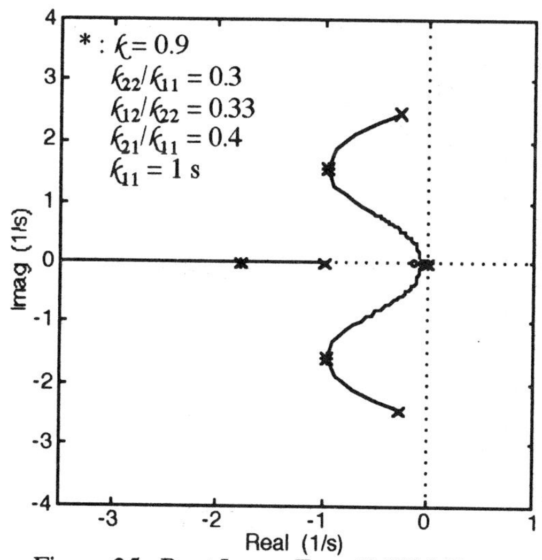

Figure 25. Root Locus For n22 With Eq. (16)

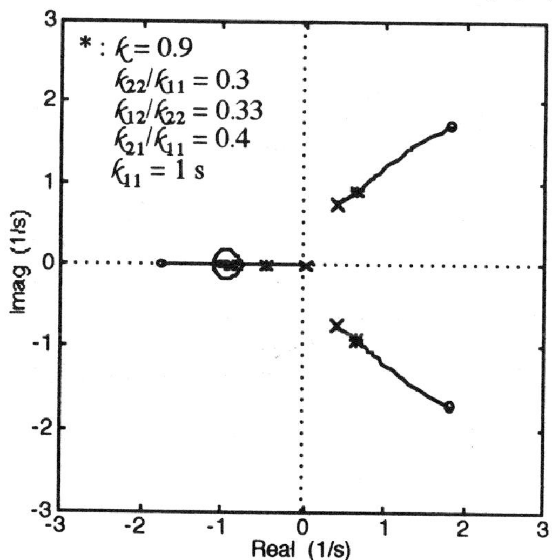

Figure 26. Root Locus For gamma11 With Eq. (16)

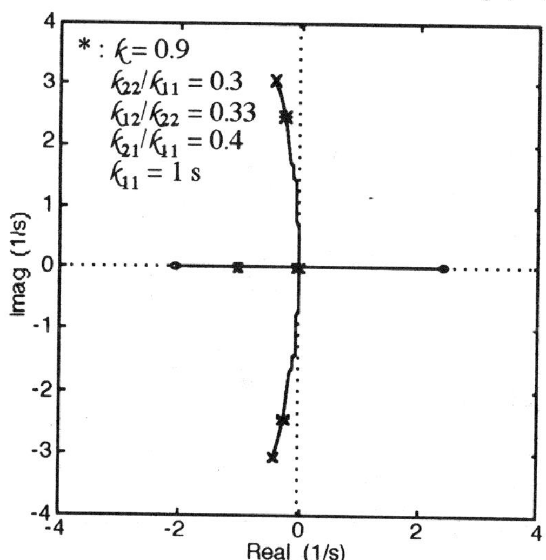

Figure 27. Root Locus For gamma22 With Eq. (16)

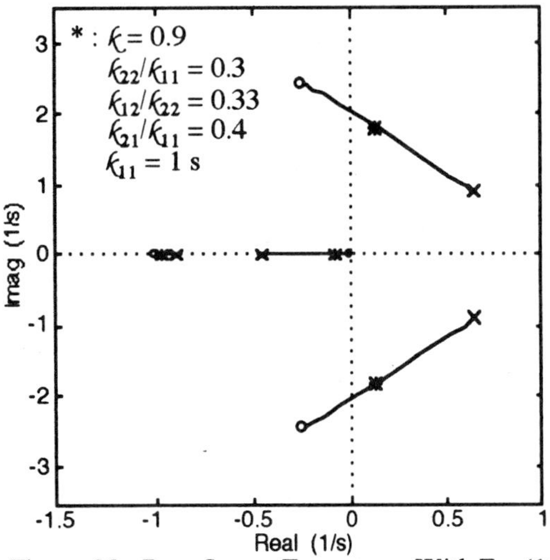

Figure 28. Root Locus For gamma With Eq. (16)

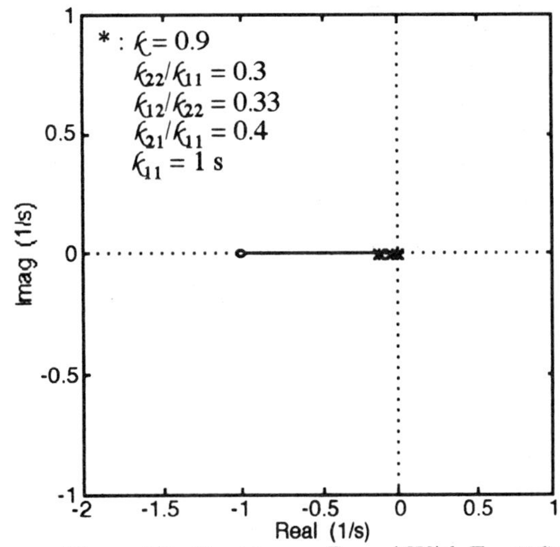

Figure 29. Root Locus For psi With Eq. (16)

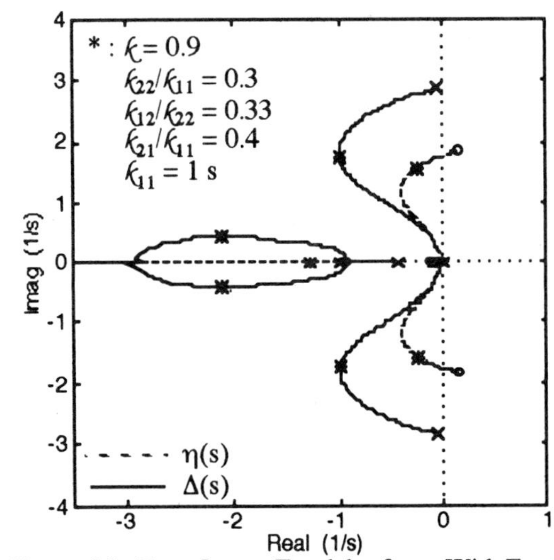

Figure 30. Root Locus For delta & eta With Eq. (16)

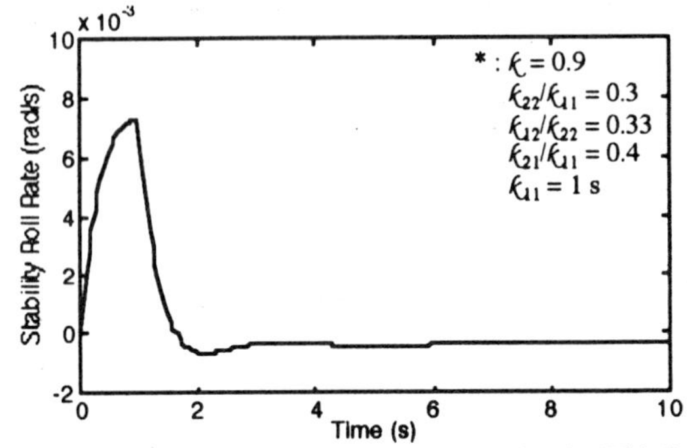

Figure 31. Roll Rate From Roll Rate Command Pulse With Eq. (16)

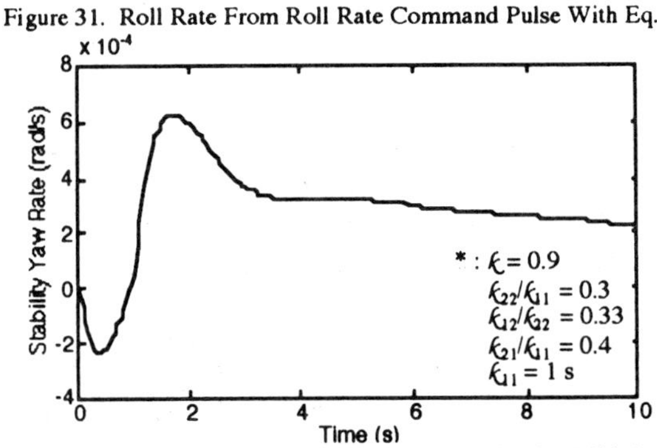

Figure 32. Yaw Rate From Roll Rate Command Pulse With Eq. (16)

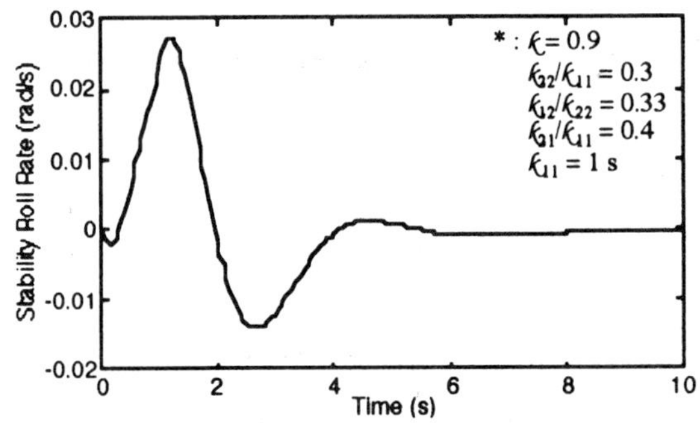

Figure 33. Roll Rate From Yaw Rate Command Pulse With Eq. (16)

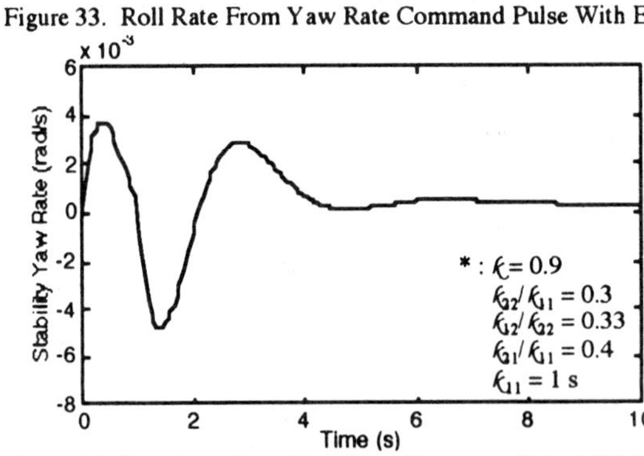

Figure 34. Yaw Rate From Yaw Rate Command Pulse With Eq. (16)

AIAA-97-3467

AVDS : A FLIGHT SYSTEMS DESIGN TOOL FOR VISUALIZATION AND ENGINEER-IN-THE-LOOP SIMULATION

Steven J. Rasmussen
Director of Technology
Artificial Horizons Inc.
San Francisco, CA

Stephen G. Breslin, PhD
Senior Engineer Aerospace Systems
Industrial Control Centre
University of Strathclyde
Glasgow, UK

ABSTRACT

This paper describes how a new engineering simulation tool, Aviator Visual Design Simulator (AVDS), enhances the flight control system (FCS) design process through the provision of data visualization and engineer-in-the-loop simulation. First the FCS design process is discussed with an emphasis on current practices and shortcomings. Next, there is a discussion of improvements to the FCS design process using AVDS. Included in this discussion are the benefits of data visualization, engineer-in-the-loop simulation, and local distributed interactive simulation (LDIS). There is a detailed description of AVDS including FCS and aircraft interactive simulation development, data visualization, synthetic environment, simulation points-of-view, instrumentation, data logging, and LDIS capability. Following this there are examples of the use of AVDS in FCS design and flight test, and flight systems research. Finally there is a discussion of planned additions to AVDS.

Introduction

The provision of manual flight control for the current generation of military aircraft represents a formidable challenge to the control systems designer, often requiring innovative and complex solutions. This has led to a marked increase in the cost of the flight control system design process. AVDS is an engineering simulation tool which has the potential for reducing this cost by providing an interactive environment for the rapid development and evaluation of aircraft and flight control systems. The tool was developed as a collaborative effort between Artificial Horizons Inc. (AHI) and the US Air Force's Wright Laboratory (WL). They combined their experience with interactive distributed simulations and FCS development to develop an environment which allows FCS designers to interact with their designs in high fidelity simulations before the design leaves the designer's desk. AVDS is the result of this collaboration. AVDS is capable of interactive simulation as well as data visualization, and distributed interactive simulation. Designers are able to define and integrate high fidelity aircraft and FCS models into AVDS and then interactively perform simulated flight tests over the full flight envelope of the models. Whether interactive simulation or data visualization, the simulated environment includes: user definable terrain, user definable instrumentation, and adjustable points of view, from inside and outside of the craft. An example of the AVDS graphical environment is shown in Figure 1. Aircraft and system models can be distributed over the local Ethernet, or the aircraft can be used as part of a local distributed interactive simulation (LDIS). AVDS gives the designer the ability to test the efficacy of new systems early in the design process, which can lead to reduction in costs and improved system capabilities. This paper describes the development of AVDS and includes examples which demonstrate its capabilities and versatility.

FCS Design Process

Current Practice

The design of the FCS for a modern aircraft is a highly labor intensive and lengthy process, as depicted in Figure 2. The process begins with the design of control laws for linear models of the aircraft at discrete operating conditions which are then simulated on a nonlinear model of the aircraft. The simulation stages used in the testing of control laws fall into a number of categories, including batch simulations on the linear and nonlinear model, real-time pilot-in-the-loop simulations, and real-time hardware and aircraft-in-the-loop simulations. The batch simulations are usually conducted using one of several available control systems design packages at the desk of the control engineer. The output from such packages is in the form of strip charts which record the transient response of the aircraft to different inputs. Upon satisfactory completion of this stage, the control laws are then 'flown' by a pilot on a

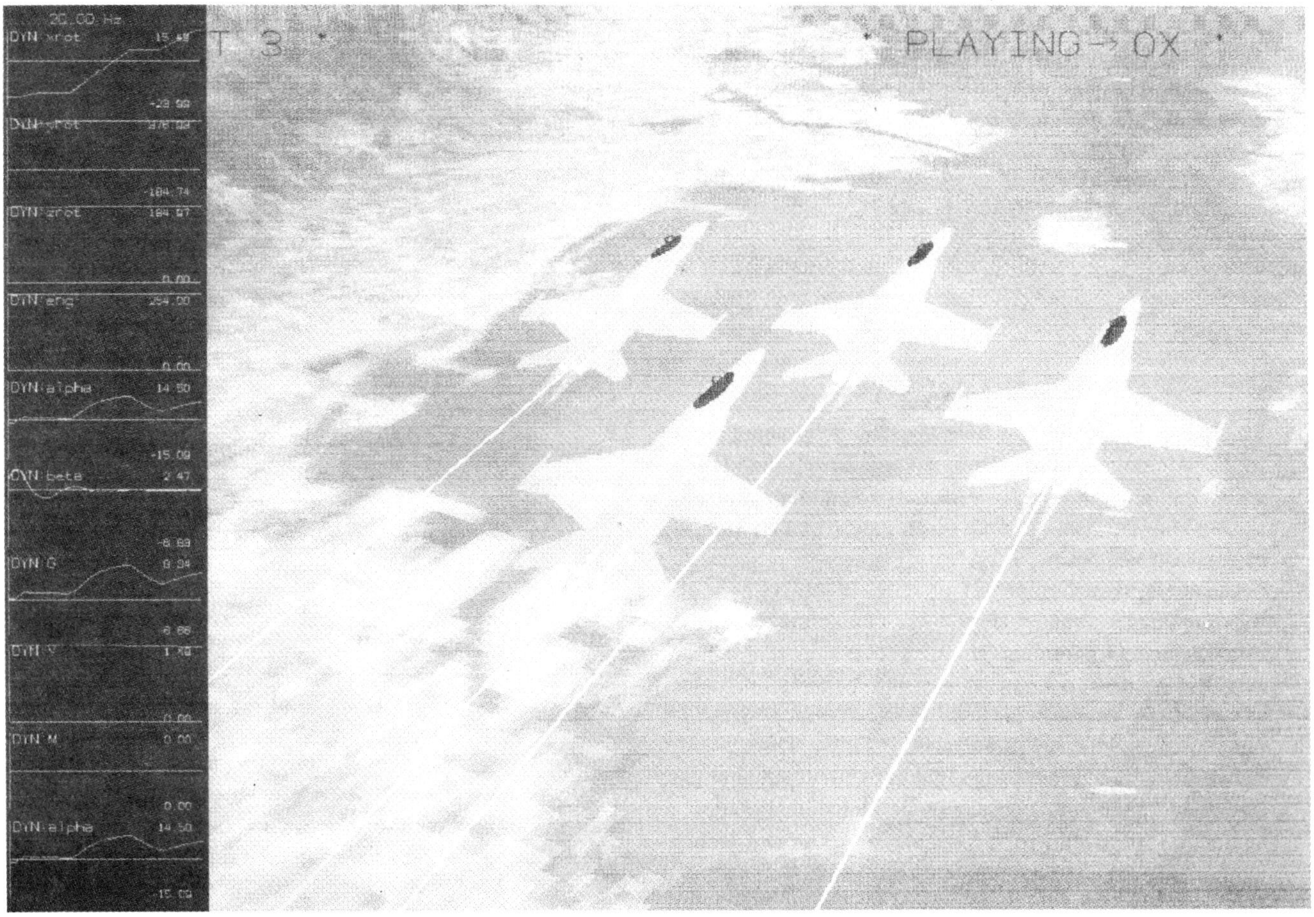

Figure 1 AVDS Graphical Environment

full-scale cockpit simulator to evaluate performance and handling qualities. Control engineers and test pilots iterate the design until they achieve the optimum combination of performance and handling qualities. The aircraft systems are then integrated and tested on hardware-in-the-loop simulations and finally on a prototype aircraft, interfaced with the simulation to drive the aircraft actuators, computers and hydraulic systems.

AVDS in the FCS Process

The FCS process will be improved by the availability of a simulation tool such as AVDS in three distinct areas :

i. visualization and replay of results of batch simulations on the linear and nonlinear models, interactive AVDS simulations, hardware or pilot-in-the-loop simulations, and flight tests.

ii. introduction of an additional simulation stage - 'engineer-in-the-loop' simulations.

iii. inclusion of an engineer-in-the-loop distributed simulation. This is several simulations that can interact in the same virtual environment and can be one or many aircraft.

Visualization

The results of simulations and flight tests have been traditionally recorded on 2-D strip charts which require a large degree of skill and experience from the control engineer to be interpreted correctly. The insight that it takes to understand a three dimensional aircraft maneuver from interpreting Euler angles plotted on strip charts takes experience. Add to this other pertinent aircraft information such as angular rates, angle of attack, sideslip angle and location of the aircraft in 3-D space, and the problem can be overwhelming. The effect that a visualization tool, such as AVDS, has on this problem is remarkable. While visualizing data, visible effects from all of the parameters are integrated and displayed simultaneously. This not only gives the designer clear insight into the state of the aircraft over time, but also allows the designer to observe subtle phenomena that might be missed interpreting strip

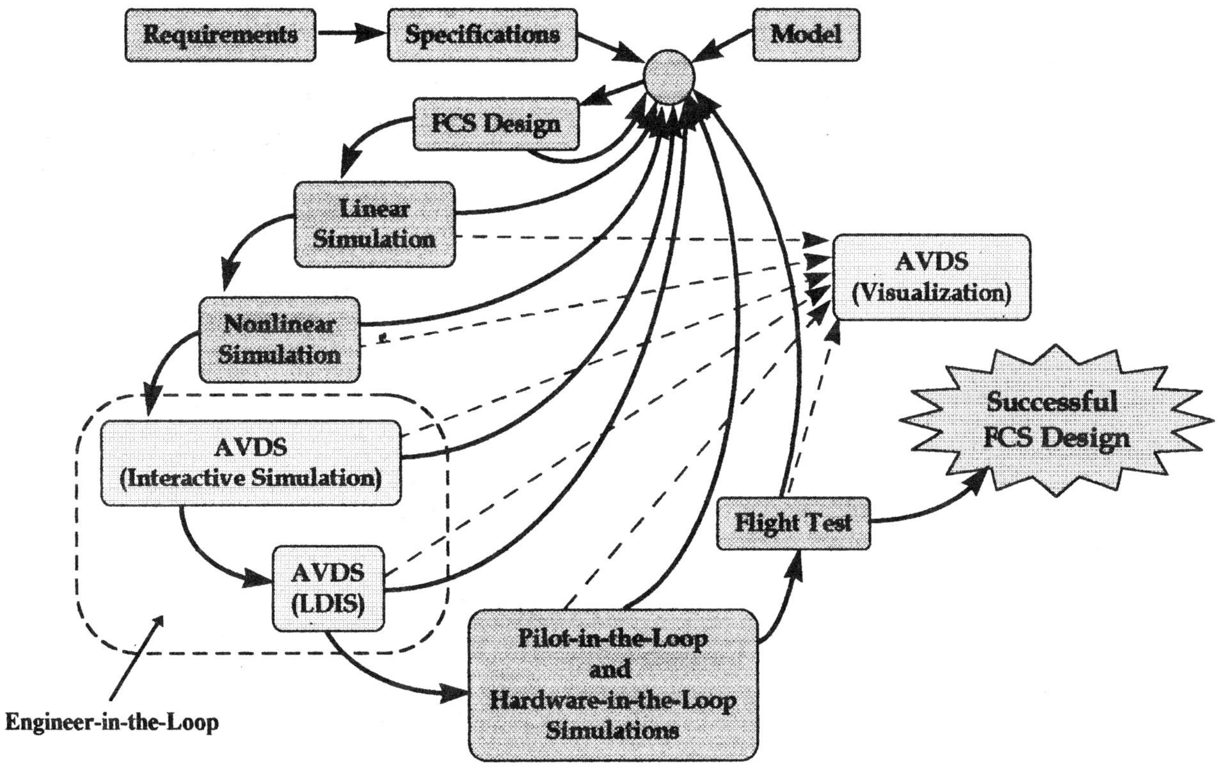

Figure 2 FCS Design Process

charts. AVDS provides a visualization facility so that the actual aircraft response can be examined in parallel with the strip charts providing much greater insight into the transient response of the aircraft and allowing data that may not produce visible effects to be monitored.

Interactive Simulation

AVDS can also be used as an interactive simulation, so that control engineers can 'fly' their control laws and thus begin to evaluate the performance and handling qualities and iterate on their designs at their desks before taking up expensive test pilot and pilot-in-the-loop simulator time. While it is true that AVDS does not have all of the visual and tactile cues that exist in a pilot-in-the-loop simulator, AVDS give the engineer a clearer insight into the operation and effects of the designed FCS. When FCS designers 'fly' their FCS in AVDS, they get an insight through the 'feel' of the simulation. For example, the designer can experience what happens when a pitch command is given or when a positive C* command is given. Using interactive simulation, the designer has a reference tool when, after a pilot-in-the-loop simulation, the pilot suggests that were problems with the aircraft handling qualities. With a description of the problems the designer can fly the design in AVDS and then redesign and then check the result against the original design before sending it back for pilot-in-the-loop simulation.

Currently the engineer can use joystick, mouse, and keyboard inputs to interact with their simulation. There is also the ability to insert canned inputs such as steps, doublets, etc. The configuration of AVDS software gives the user the ability to customize the method of interacting with the simulation.

LDIS

Using LDIS, engineers have the ability to visually simulate air-to-air combat. Simulations can be running side by side with a number of engineers flying different aircraft, maneuvering independently and trying to gain weapons firing position on opposing aircraft. In this scenario, guidance and control engineers and weapons engineers can evaluate the performance of the aircraft and the weapons as a whole system. LDIS facilities can also be used to evaluate formation flight algorithms. As algorithms are developed for automatic formation flight, simulations are required that have independent formation flight management software distributed across many different workstations.

AVDS DESCRIPTION

Development Environment

AVDS combines high fidelity graphical simulation capability with simulation development and analysis tools to produce a sophisticated environment for

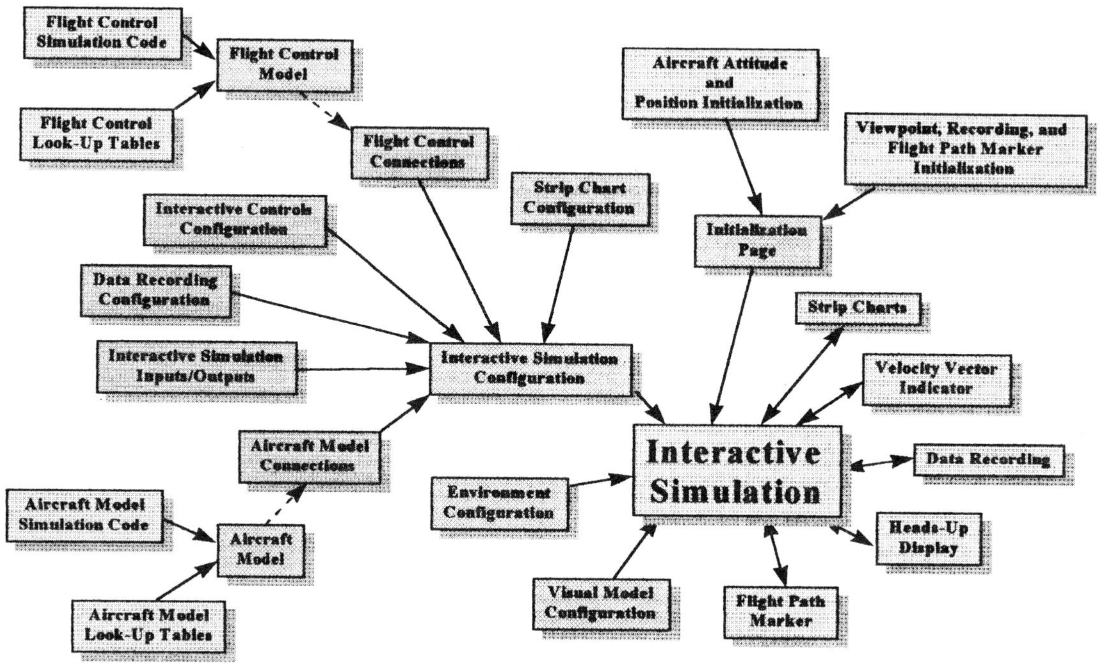

Figure 3 AVDS Interactive Simulation

development and analysis of dynamics models. Functional tools are incorporated to allow the designer to interact with the simulation of a given dynamic model. These simulations can be recorded and then animated. The ability to animate recorded data aids the designer in the visualization of dynamics output from any given simulation. AVDS includes a suite of tools that allow the designer to develop and use distributed interactive simulations. This suite of tools simplifies the use of distributed simulations so the designer can concentrate on the models to be simulated.

Entity Types

AVDS was designed to be an aircraft systems, especially FCS, design tool. But AVDS has potential as an multidisciplinary design tool which could allow rapid prototyping of conceptual aircraft, or other types of entities, and provide a uniform environment for systems development and testing. The designer defines an entity's dynamics by linking its dynamic simulation code to AVDS. The graphical representation of the entity is described through the definition of groups of polygons.

Simulation and Analysis

Included in AVDS is a facility to add separate aircraft and flight control system models and the ability to use three dimensional look-up tables in those models. Also included is the ability to interconnect the aircraft model, the flight control system model, and simulation environment variables. The addition of analysis tools such as strip charts, a flight path marker, and a velocity vector indicator give the designer much better insight into the dynamics of the simulation. AVDS allows selection of many viewpoints, or camera modes, that give the designer the ability to view the simulation from any vantage-point

Development History

Using the LDIS capabilities of AHI's Aviator software, engineers at WL built a number of tools that allowed them to put together interactive engineering simulations. This led to a Cooperative Research and Development Agreement (CRDA) between AHI and WL. Through this agreement, the tools required by the FCS designer to make optimal use of the proposed simulation development environment were defined and implemented. AVDS is currently being used by the U.S. Air Force for research into various FCS applications.

AIRCRAFT AND FCS INTERACTIVE SIMULATION

Capability

The designer is able to define and incorporate very sophisticated aircraft and flight control system simulations into AVDS. Figure 3 is an overview of the modules that make up AVDS interactive simulation development. There are also built-in aircraft simulations that consist of models that have full 6 degrees of freedom equations of motion with linearized aerodynamic coefficients. These aerodynamic coefficients are scheduled based on appropriate aerodynamic variables such as Mach number, dynamic pressure, and angle-of-attack.

Interactive simulation mode gives the user the capability to simulate highly complex aircraft and flight control system models in real time, while representing the vehicle dynamics graphically and allowing the user to interactively pass commands to the simulated aircraft. The aircraft and FCS models consist of two parts, model structure and model parameters and are discussed in the next section.

While in interactive simulation mode the user has access to all of the AVDS camera modes as well as the velocity vector indicator, flight path marker, and strip charts. During the interactive simulation, any of the output variables can be recorded and then saved to a file. This recorded information can be used for in-depth analysis with other software packages or animated in AVDS using the simulation visualization mode.

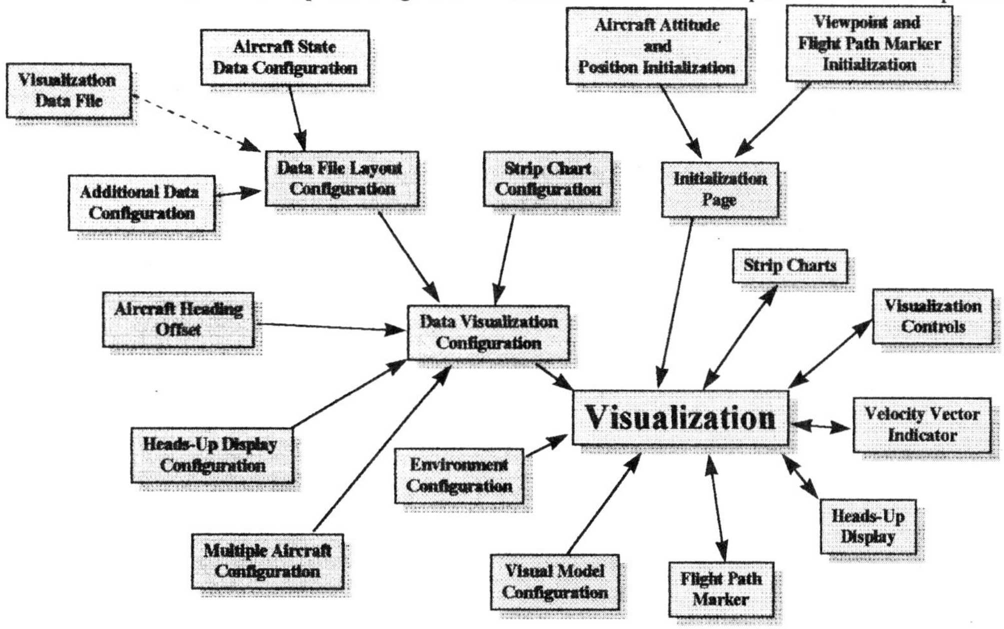

Figure 4 AVDS Visualization

Aircraft and FCS Models

Aircraft and FCS models are made up of two parts, the structure and the parameters. Model structures consist of executable subroutines that define how the model parameters are used to produce the dynamics required to simulate the aircraft or FCS. The model structure can either be a user defined structure or an existing AVDS structure. To incorporate a user defined structure, the user must compile and it with the supplied AVDS libraries. This is accomplished with the assistance of a makefile. AVDS models must be linked to C code, but can be written in any language that can be linked to C. AVDS comes with model structures built in. These structures allow the user to simulate a single-engine jet aircraft, with or without thrust vectoring, or a four engine jet transport aircraft.

Model parameters are used to cause the model structure to produce dynamics that mimic those of a given aircraft. The model parameters are specified in a parameter file and read into AVDS at the time the interactive simulation begins. This allows the user to change model parameters at any time between simulation runs without the need to recompile the program. The parameters are specified in the parameter file in the form of look-up tables. These parameters can be scheduled based on any of the aircraft's dynamic variables, e.g. dynamic pressure, Mach number, altitude. The look-up tables can have up to three dimensions and have associated bias and scale factor parameters.

VISUALIZATION MODE - *PLAYBACK*

Visualization or 'Playback' mode allows the user to animate the time history of an entity's linear and rotational motions. Figure 4 is an overview of the modules that make up the AVDS Visualization Mode. Time history data are contained in a user provided text file. The columns of the file represent variables to be animated, the rows represent time slices of the data. AVDS loads the file of data into memory and then animates that data using timing information specified in the data. Playback mode is controlled through the use of commands from the keyboard. During playback, the action can be displayed real time, faster than real time, frozen, or slower than real time, forwards or backwards.

In addition to animating the dynamics, additional data columns provided in the user text file can be displayed synchronously on strip charts. While in playback mode, the user is able to view the action from all of the AVDS camera modes, and also display the flight path and velocity vector of the vehicle.

SYNTHETIC ENVIRONMENT

Graphics Capability

AVDS's graphical environment is displayed through the use of polygons. Geographical coloring is accomplished by applying satellite derived terrain coloring to these polygons. AHI's ability to produce a very detailed graphics environment was made possible by building a graphical library specifically for the GX family of frame buffers available in Sun Microsystems UNIX workstations. With the advent of better graphics adapters AHI is in the process of porting AVDS to the OpenGL to make AVDS a cross-platform program.

Aircraft Icons

The visual representation of the aircraft, or icon, is defined by groups of polygons and contained in a separate initialization file. There are predefined types for these groups so that algorithms in AVDS can identify the object described by a particular set of polygons. For example, if the group representing a control surface should be damaged during the simulation then the modeled control inputs that the surface represents can be degraded or neutralized. The user can modify existing or define new aircraft icons.

Terrain

The terrain that AVDS simulates is based on digitized US Geological survey maps. The resolution of the simulated terrain is completely adjustable from simple to very detailed polygonal representation. User furnished terrain data can be converted to the AVDS terrain format and used in AVDS.

Visibility

Simulated visibility is adjustable from one mile to 35+ miles. Visibility is simulated by dropping polygons that are out of range and adding sheets of mist to the display.

Lighting

AVDS has simulated sunlight that causes all entities to be lighted and causes reflections off of all entities. When the point of view is directed towards the simulated sun, glare obscures the observer's visibility.

POINT-OF-VIEW

Camera Modes

There are three camera modes available in AVDS. They are the Flight Cam, Chase Cam, and Tower Cam modes.

A camera mode determines the dynamics of the graphical viewpoint of the user. AVDS allows the user to view the simulation from any angle and any distance, but how the user's viewpoint changes during the simulation is a function of the camera mode. In AVDS, the viewpoint can tied directly to the center of gravity of the aircraft, tied loosely to the center of gravity, or fixed with respect to the earth.

Pilot View

Included in AVDS is the pilot's point-of-view. The pilot's position relative to the aircraft is user definable.

Flight Cam

In Flight Cam mode, the viewpoint is tied directly to the center of gravity of the aircraft. This gives the user the ability to closely monitor the angular perturbations of the aircraft. In essence, the aircraft appears as if it were hanging by a string tied to its center of gravity, as the outside terrain passes by. Flight camera mode can be invoked so that the viewpoint remains fixed with respect to the aircraft.

Track Cam

In Track Cam mode, the viewpoint is tied loosely to the center of gravity of the aircraft. This gives the user the ability to closely monitor the linear and angular perturbations of the aircraft. In this mode the viewpoint appears to be flying in formation with the aircraft.

Tower Cam

In Tower Cam mode, the viewpoint is fixed with respect to the earth. This gives the user the ability to view the flight path of the simulation/playback. Track camera mode provides situational awareness by allowing the entire area of operations to be viewed at one time.

INSTRUMENTATION

Strip Charts

Ten strip charts are available for graphical display of simulation outputs. The information presented on the strip charts can be any of either the simulation outputs or the playback outputs. The strip charts are automatically scaled so that the maximum and minimum values on the strip charts are the maximum and minimum values of the output data. There is a zero reference line on each strip chart is readjusted each time the charts are rescaled.

Velocity Vector

The velocity vector indicator is available in the interactive simulation and playback modes. The velocity vector indicator is a black pointer that emanates from the aircraft's center of gravity. The length of the vector changes with the total velocity of the aircraft. The angles that the vector makes with the aircraft are based on the angle of attack and the sideslip angle.

Axes
Earth fixed, body fixed and stability axes are available for display in any of the AVDS modes. The use of these axes are useful in tracking aircraft orientation and as aids in understanding aircraft dynamic variables.

Heads-Up-Display
During any of the modes of operation of AVDS, a heads-up-display (HUD) can be displayed so that simulation parameters can be viewed. This HUD includes an attitude ladder for entity orientation and display of other parameters, when available, such as altitude, angle of attack, forward airspeed as well as others.

While in stealth observer mode or simulation playback mode, the HUD is attached to the locked-on entity. The parameters displayed are the dead reckoned values and any other appropriate parameters contained in the AVDS broadcast packets or simulation playback files.

Flight Path Markers
The flight path of the aircraft, or entity, can be highlighted using a white marker line, a 'smoke' trail. The marker consists of individual line segments that connect the points in space covered by the aircraft. During operations with multiple entities, flight paths of all entities are displayed.

Wing-Tip Markers
Wing-tip markers are colored lines that depict the time history of wing-tip positions. These markers are helpful in monitoring aircraft maneuvers. Wing tip positions are defined by the user so wing-tip markers can be positioned to emanate from any portion of the aircraft.

STEALTH OBSERVER MODE

AVDS includes a built-in stealth observer mode. During this mode, any of the entities being broadcast in a LDIS can be locked-on and observed. During initialization of this mode, all entities of a LDIS are displayed in a selection menu. Once selected the entity is lock-on, all of the AVDS viewpoints and instruments can be used to monitor the entity. To change entities the observer can either toggle through the entities using a selection key or move back to stealth observer initialization and make a new selection from the menu.

Data Logging

AVDS is capable of logging all data produced by a single entity during interactive simulation. During the interactive simulation, AVDS stores the required data into memory at multiples of the simulation time step. At the end of the simulation, the data is written to a file on disk. There is a configuration page that gives the designer the ability to select desired parameters for logging. The limitation of logging only one entity will be replaced, in the future, with the ability to log data from all of the other entities in a LDIS simultaneously.

DISTRIBUTED SIMULATION

AVDS Broadcast Packets
During a LDIS, AVDS generates, broadcasts, receives, and interprets TCP/IP packets on the local Ethernet segment. These packets are analogous to DIS Protocol Data Units (PDU) and contain information concerning the type of entity, its owner, it's dynamic state and configuration information, as shown in Table 1

AVDS Packet Broadcasting

Packets are broadcast by each computer running AVDS for all entities produced by that machine. Each computer receives packets from others and displays the entities within the field of view. To join a LDIS, an entity need only begin broadcasting Ethernet packets. This is automatic when an AVDS simulation, or playback, session begins.

Broadcasted AVDS packets are available to any computer on the local Ethernet segment. Packets are broadcasted whenever computation time allows, but not

AVDS version	craftcap checksum	home airport	xyz center of mass	angular velocities
craftid	lockon craftid	linear accelerations	pilot's view direction	collision hostid
collision craftid	craft type	linear velocities	render mask	engine level
ground type	ground slope	current fuel	current warhead	ground height
xyz rotations	xyz position			

Table 1 AVDS Broadcast Packet Parameters

faster than a set rate. For work accomplished to date, a broadcast rate of 10Hz seems to be fast enough for most simulations.

Simulation Compliance

AVDS packets contain AVDS version specific information that forces each participant to have a compliant version of AVDS. The visual representation of each entity and other identifying information is defined in a file called *craftcap*. This file is kept in a location that all participants in the session can access. Each time the craftcap file is changed, AVDS compiles it and derives a checksum of the compiled code. This craftcap checksum is part of AVDS broadcasted packets. All participants in a LDIS are forced to use the same craftcap file, as verified by the checksum. Because of this, simulation compliance can be enforced by having the distributed simulations comply with configuration information in the current craftcap. By sending information on object collisions, weapons lock-on, own object collision damage, current fuel amount, and warhead value, the sending participant informs the receiver what was damaged and how it was damaged. The receiver then determines what effect this type and level of damage has on the receiver's entity.

AVDS is supplied with a toolbox containing libraries and examples that give the designer the ability to develop simulation programs that are able to participate in a LDIS with AVDS. This toolbox also gives the designer the ability to adapt existing simulations, interactive or batch, for communications with AVDS.

AVDS EXAMPLES

Interactive Simulations

One way that AVDS can be used is to adapt large, high fidelity batch simulations to run interactively. This is accomplish by modifying the batch simulation to accept inputs interactively and adding the appropriate function calls to the toolbox libraries to communicate with AVDS on the local Ethernet segment. AVDS is used as the graphical output of the, now interactive, simulation by receiving and interpreting the broadcasted packets, on a separate computer.

In the same way that a batch simulation can be adapted to be interactive, other high fidelity simulations can be adapted, or developed, to operate in the AVDS virtual environment. This gives the designer the capability to develop simulations where entities are controlled from various locations and using various control mechanisms. For example, a distributed simulation could be developed where an aircraft model running in AVDS on one workstation lands on a ship model running on another workstation. This type of distributed simulation may be necessary because of the complexity of the models or due to the need for human control of both models.

If a FCS designer desires to simulate formation flight, AVDS can be run on separate workstations and the participants will automatically be in the same virtual environment. If the designer wants to test a formation flight algorithm based on radar contact, the algorithms can be developed and the necessary connections can be made using the AVDS modeling facility. If the designer desires to base the formation flight algorithms on some other type of sensor, the AVDS Distributed Simulation Toolbox can be used to model this connection

FCS Development Example

One of the first uses of the AVDS capabilities was interactive simulation and data replay for an unmanned research vehicle (URV) (Rasmussen and Houpis, 1994). A LDIS was used to run the aircraft dynamics on one workstation and display the simulation on another, using AVDS's observer mode. This interactive simulation setup allowed designers to check out the FCS and allowed the pilot to get a better understanding of the FCS before the actual flight test. Playback mode was utilized at each step of the process including analysis of the post-flight test data.

Managed Distributed Simulations

In order to better quantify the effects of mechanical failures onboard civilian airliners, WL engineers developed a 'pilot'-in-the-loop simulation using AVDS (Rasmussen and Sheldon 1993). This simulation used a rudimentary cockpit with simplified mechanical controls. Visual display was a 19 inch monitor attached to a workstation running AVDS in observer mode. The aircraft dynamics and fault management software were hosted on a different workstation. The simulation management software allowed researchers to interject modeled FCS failures into the simulation at any time. This software was developed using the AVDS distributed simulation toolbox so it could participate in an AVDS LDIS. The mechanical controls inputs were fed into the workstation running the dynamics and management software

The simulation consisted of set scenarios where the aircraft was initially flying straight and level and then a simulated mechanical failure was commanded by the test director. The pilot flew the aircraft from a viewpoint outside and behind so he could have a better indication of the dynamics of the simulation. During the failure, the pilot was expected to first gain control of the aircraft, and then cause the aircraft to turn 180 degrees and descend 5,000 feet. This was required to show that the pilot would be able to control the aircraft. During the duration of the simulation, all pertinent dynamics data and failure data was recorded for post simulation analysis and replay.

This type of simulation could easily be expanded to include entities such as other aircraft and missiles. The distribution of the simulation could have been changed so that the management software and the aircraft dynamics were hosted on separate workstations.

AVDS FUTURE IN FCS DESIGN

Since AVDS is a new software package, there are many planned improvements.

Computer Aided Design Software Interfaces
There are many CAD packages that facilitate the development of aircraft models and FCSs, such as Mathematica, MATLAB , and MATRIX$_X$. Interfaces to this type of CAD software will increase the ability of the FCS design engineer to develop low cost quick-look engineer-in-the-loop simulations.

Documentation Tools
While AVDS can save data in text format for use in separate data analysis programs it does not contains post-simulation data analysis tools such as those for data plotting and statistical analysis.

Simulation Analysis Tools
Tools are needed to help the FCS designer understand the health of the simulation. These tools will present the designer with information on subjects such as display frame rate, network latency, network traffic congestion, and dynamics computation rate.

Maneuver Profiles
Since there are many aircraft maneuvers that are commonly used in aircraft FCS design and testing, AVDS will have a library of common maneuvers that can be used in bench mark tests.

Interface to CAD Drawing Packages - Currently entity icons must be defined using individual polygons, future versions of AVDS will have an interface to common solid geometry CAD software.

Atmospheric Phenomena
Phenomena such as winds, turbulence, clouds, and changing time of the day are not now included in AVDS.

AVDS and DIS
AVDS was not designed to participate in large DIS exercises, but, AVDS performs most of the basic functions required of a DIS compliant simulation. AVDS can be made DIS compliant through changing the AVDS broadcast packet structure to DIS compliant PDUs, and using DIS compliant environmental databases. AVDS's built-in stealth observer mode and data logging capability can easily be made DIS compliant. The computer platforms that AVDS operates on can make it a relatively inexpensive DIS compliant environment.

CONCLUSION

An engineer-in-the-loop simulation tool can play an important role in the development of new aerospace systems. Whether used for an engineer's evaluation of new formation flight algorithms, air to air combat simulations, or any distributed simulation, AVDS has the potential to improve system performance and reduce development costs. Further, AVDS can be used to develop sophisticated engineer-in-the-loop simulations that utilize many existing complex batch simulations to form a single virtual simulation of the aerospace system.

AVDS is commercial interactive simulation and visualization software that provides the capability to visualize data and rapidly develop engineer-in-the-loop simulations. Using AVDS, designers can evaluate their designs before moving those designs to more expensive and time consuming simulations. Since AVDS is being developed as a cross-platform software environment, FCS designers will be able to integrate it into their FCS design process on their existing equipment.

BIBLIOGRAPHY

Rasmussen, S. J. and Houpis, C. H., 1994, "Develoment, Implementation and Flight Test of a MIMO Digital Flight Control, System for an Unmanned Research Vehicle Designed Using Quantitative Feedback Theory", *Dynamic Systems and Control 1994*, The American Society of Mechanical Engineers, New York, New York, pp. 145-150.

Rasmussen, Steven and Sheldon, Stuart, 1993, "Using Non-Real-Time Simulation to Investigate Transport Aircraft Survivability", *Proceedings of the 1993 Transport Aircraft Survivability Conference - Enhancing Survivability of Civil and Military Transport Aircraft Through the Application of Technology*, Air Warfare Section, Combat Survivability Division, American Defense Preparedness Association, pp. 102-115.

RESOURCES

Mathematica - Wolfram Research Inc., 100 Trade Center Drive, Champaign, IL 61820-7237, USA
http://www.wolfram.com

MATLAB - The Math Works Inc., 24 Prime Park Way, Natick, MA 01760-1500, USA
http://www.mathworks.com

MATRIX$_X$ - Integrated Systems Inc., 201 Moffett Park Drive Sunnyvale, CA 94089, USA
http://www.isi.com

AIAA-97-3468

MODERN FIXED STRUCTURE CONTROL DESIGN PART I: GAIN ADJUSTMENT TO IMPROVE HANDLING QUALITIES

Piero Miotto, James D. Paduano, Eric Feron
Massachusetts Institute of Technology, Cambridge, Massachusetts 02139

John J. Burken
NASA Dryden Flight Research Center, Edwards, California

Abstract

The problem of gain adjustment in an existing control law is cast into the framework of modern control theory, which results in a real-diagonal block structure of gains to be adjusted. The desire to recover performance in the form of input-output properties, as well as handling qualities, is introduced as a mixed cost functional. LMI-based solutions for H_∞ and H_2 based optimization are discussed, as well as a Quasi-Newton approach. The handling qualities of the F/A-18 HARV during refueling serves as a motivating example for the method. Five gains in the existing flight control system are adjusted to recover the performance and handling qualities of the standard F/A-18.

Introduction

Design of flight control laws for modern aircraft involves many steps which have not traditionally been directly addressed by control theory. In this two-part paper we discuss how some of these steps can be treated in the context of modern methods. The specific areas that we will address are: 1) modifying existing control structures for a new plant, 2) incorporating handling qualities criteria in the control design process, and 3) gain scheduling. We will briefly motivate these goals before introducing our approach.

In developing a complex engineering system, one of the main assets one draws upon is past experience. Practical flight control systems are typically modifications of previous control structures, modified in their particulars for the specific aircraft being controlled. Thus methods which allow a proven control structure to be optimized for a new plant are valuable. Further, if flight test indicates that a control system has inadequacies due to errors in the aerodynamic model, it is of interest to make simple changes to the existing control law, which has already undergone extensive development and validation. In both of these instances, fixed structure controller optimization (i.e. changing the gain tables in an existing or prototype controller) can be used to advantage.

Another issue that must be addressed by flight controls is handling qualities. Experience exists to help the designer choose feedbacks, weight stick inputs, and provide 'force-feel' feedback to the pilot — all with the goal of improving the pilots opinion of the aircraft's ability to accomplish its mission. Methods also exist to analyze the flying qualities of a control system, and much effort has been expended in improving these analyses. Less work has been done to explicitly include handling qualities in the design optimization step. If one proposes to automate the modification of gains in an existing flight control structure, maintaining or improving handling qualities is a central concern.

Finally, the flight control design (or modification) process must result in a control law which is valid over a wide range of operating conditions, which cause variations in the plant dynamics that are invariably represented using tabulated empirical coefficients. This has been quite successfully accomplished in the past using linear control law design at fixed operating points, followed by gain scheduling within the context of a fixed control structure. Much of the complexity of modern control systems lies in the details of scheduling and blending various feedback gains and filters, to cope with the disparate requirements of landing, up-and-away flight, acquisition, and tracking.

The approach described in this paper is to retain these interconnections, blends, gain schedules, and stick shaping filters, because they represent the accumulated knowledge of many years of practical design experience, and because it is costly to re-code these when problems are found late in development. We will focus on optimizing the gain tables and other numerical particulars of the control law, which are simple to change and do not require revalidation of the software. We will also show how to incorporate handling qualities requirements in the optimization[11], and how to perform the optimization over a range of operating conditions, so that a gain schedule results.

The fixed structure control problem has been addressed by numerous authors in the past[1,3,4,5,6,7]. Automating the gain scheduling procedure has also been addressed recently in the context of modern control[14,15,16,17,19] under the moniker 'Linear Parameter Varying' (LPV) control theory. We rely heavily on this theory to put the classical problem of fixed-structure controller optimization into the context of modern control theory, and to motivate our gain scheduling procedure. Modern control theory is well suited to defining desired closed-loop performance; we will combine these tools with a cost related to the handling qualities of the system to form the overall optimization problem. By doing this, we address the desire to improve the gains of a given control law, over a large operating envelope, while maintaining handling qualities.

This paper is organized as follows. First we describe a problem to motivate our procedure. Then we put this problem into a 'fixed-structure LPV' framework, to show how the practical and theoretical problem are related. Handling qualities will be incorporated into the LPV optimization problem, as required by our example problem. Finally, we apply the approach to our example problem and discuss the implications. In Part II of the paper, the method is applied to a large-envelope gain scheduling problem.

Problem Statement: Handling Qualities Improvement of F/A-18 HARV Refueling

Refueling is a critical flight control task because precision tracking is necessary. In order to refuel successfully and with a reasonable pilot work load, level 1 flying qualities (Cooper Harper rating from 1 to 3) are required. Refueling of the F/A-18 HARV was performed by two pilots during two different flights, twice per flight. In all the cases the pilot rating was 8 or 9. From the flight data shown in Fig.1 it is evident that a tendency to PIO exists and that the pilot work load is high. A flight engineering summary comments: *with the aircraft being sluggish in the longitudinal stick direction, it was discovered that the pilot gains to control the longitudinal axis were much less than anticipated.* Both pilots had to reduce their gains and be very gentle on the stick in order to perform the task.

The same maneuver is performed regularly on a standard F/A-18 and does not present the same problem. The control law in both cases is the same; the difference lies primarily in the mass and inertia characteristics. The HARV's mass and inertia characteristics are very different from those of an F/A-18. The moment of inertia around the y-body axis is almost doubled in the HARV with respect to the F/A-18, due to the weight of the thrust vectoring, spin chute, and associated ballast in the nose (see Tab.1). It is thus not surprising that the aircraft has poor performance in certain (suprisingly, very limited) regions of the flight envelope.

Our goal, then, is to re-optimize the gains of this control system to recover the input-output characteristics and handling qualities of the standard F/A-18. The desired input-output characteristics will be defined as a 'target plant', and the handling qualities will be defined based on the Neal-Smith criterion[19].

	F/A-18	F/A-18 HARV
Total Mass	1142.48 slugs	988.01 slugs
CG Percent Chord	22.88 %	21.50 %
CG Buttline	0.0 in	0.0 in
CG Waterline	102.96 in	103.47 in
Mom. Of Inertia Ix	23466.9 slug*ft^2	17178.3 slug*ft^2
Mom. Of Inertia Iy	181600.7 slug*ft^2	119208.2 slug*ft^2
Mom. Of Inertia Iz	196669.8 slug*ft^2	132354.5 slug*ft^2
Prod. of Inertia Ixz	-1741.85 slug*ft^2	-2148.64 slug*ft^2

Tab.1 HARV vs F/A-18 Mass and Inertia Characteristics

Definition of the target plant and of the controller parameters to re-design

Since in-flight refueling performance is good for the standard F/A-18, we will use the closed loop transfer function from stick input to pitch rate of the F/A-18 as our target. We will attempt to redesign the gains of the longitudinal CAS so that the transfer function from stick input to pitch rate of the HARV is similar to that of the standard F/A-18.

Tab.1 lists the mass and inertia characteristics of the F/A-18 HARV and the F/A-18; the latter is our target plant. The bode diagrams of the HARV and the F/A-18 are shown in Fig.2. The HARV's reduced crossover frequency is evident. Defining the bandwidth as the frequency at which the phase is -90 deg, the bandwidth of the HARV is about 7 rad/sec while that of the F/A-18 is about 10 rad/sec.

Another way to stress the difference between the standard F/A-18 and the HARV is to look at the Neal-Smith criterion for longitudinal handling quality. This method assumes that, given a certain value of pilot time delay, the pilot's lead, lag, and gain *adapt* to minimize the droop and the peak amplification of the

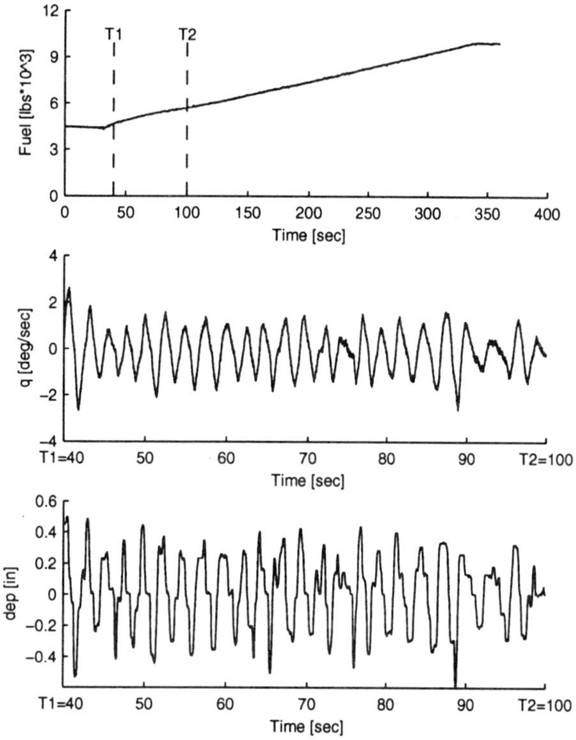

Fig.1 HARV Refueling Task Flight Data, 20,000 ft M=0.6. Fuel weight, pitch stick input (*dep*), and pitch rate (*q*). Pilot continuously corrects pitch angle throughout. The frequency of the pilot inputs was about 3.2 rad/sec.

closed-loop frequency response. Furthermore the phase of the pilot-vehicle transfer function $T_{\theta-\theta_c}(s)$ is forced to be *-90 deg* at the specified task bandwidth (*BW*). The Neal-Smith carpet plot is then computed. This is a valuable tool to measure the robustness of the closed loop transfer function with respect to changes in the required bandwidth and different droops. Fig.3 shows the Neal-Smith carpet plot for the F/A-18 and Fig.4 that of the F/A-18 HARV. The main difference is that while the F/A-18 is almost always in level 1 for bandwidths between 2.5 and 3.5 rad/sec, the HARV always has level 2 flying qualities. The HARV flying qualities also degrade rapidly as the bandwidth becomes higher.

Based on this discussion, our target plant will have two characteristics. First, it will have the same pitch rate response as the F/A-18 at 20,000 ft and Mach 0.60. Tab. 2 shows the poles and zeros of the target pitch response. Note that the phugoid poles have degenerated into two real poles, one of which is slightly unstable but within the limits of level 1 flying qualities. The phugoid poles are almost canceled by zeros and thus do not play an important role in the pitch-rate response.

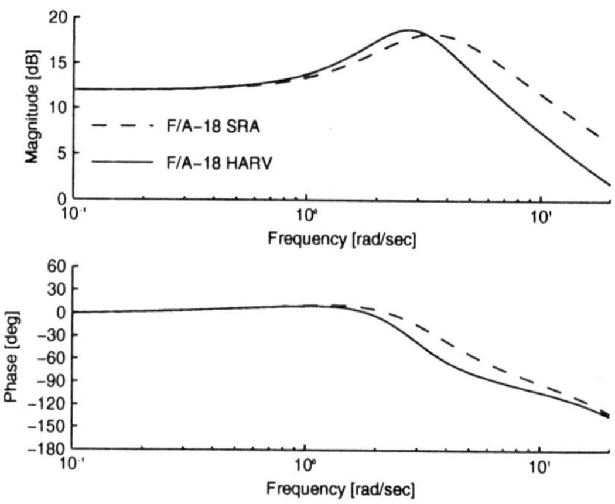

Fig.2 Transfer Function from longitudinal stick to pitch rate: Standard F/A-18 vs F/A-18 HARV

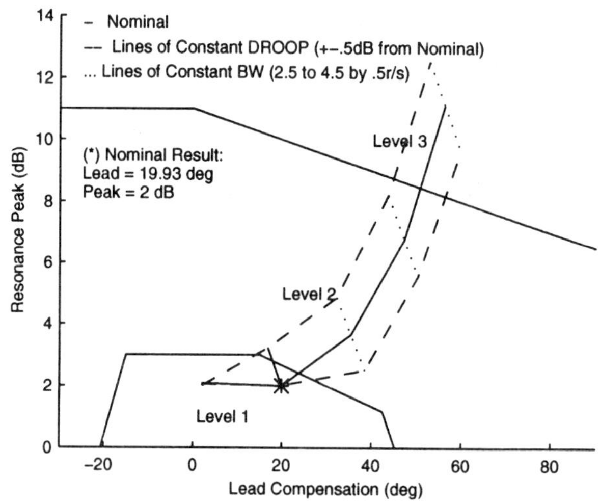

Fig.3 F/A-18 Neal/Smith Carpet Plot: Nominal *BW = 3* rad/sec, Pilot Delay *t = 0.3* sec

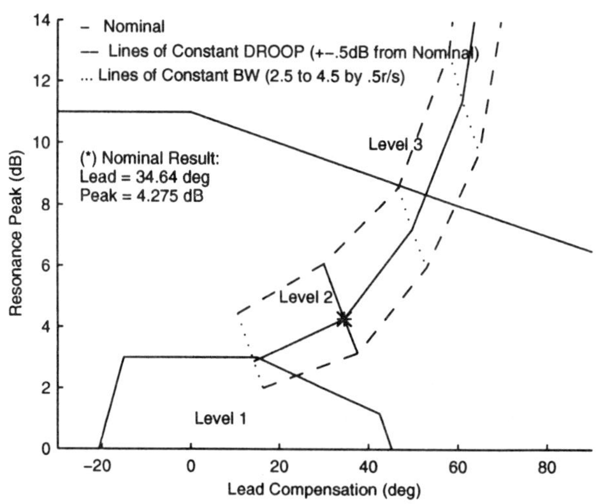

Fig.4 F/A-18 HARV Neal/Smith Carpet Plot: Nominal *BW = 3* rad/sec, Pilot Delay *t = 0.3* sec

Poles	Zeros
-61.6501 +/- 84.8776I	-5.6374 +/- 82.7068I
-14.2553 +/- 31.1079I	-24.2142
-21.8051	-6.2880
-5.5266	-2.6788
-4.2952	-1.6921
-2.2043 +/- 2.6588i *	-0.9364 +/- 0.5412I *
-0.9096 +/- 0.5227I	-0.0120 *
-0.0113 *	0 *
0.0027 *	
Gain = -6.2824e+004	

Tab. 2 Target Plant Transfer Function. Starred elements indicate plant poles and zeros.

The second characteristic of our target plant is that it has the Neal-Smith handling qualities of the F/A-18.

Once the target plant has been defined the next step is to create an interconnection structure that describes the model matching problem and select which gains and/or time constants need to be redesigned. The standard F/A-18 CAS is a PI compensator in the longitudinal channel. The controlled variable is pitch rate and the feedback variables are pitch rate, normal acceleration, and angle of attack. The refueling task is performed at low angle of attack and in this flight regime the α feedback is not used. We chose to adjust five gains to obtain the required performance; these are listed in Tab.3. We will redesign these gains to achieve the target performance. Although the gains listed are functions of flight condition, in this example we will optimize a point design (Part II discusses gain scheduling).

Gain	Description	Nominal
PK15	Stick Gain	0.3823
PK18	Pitch rate proportional gain	0.3801
PK16	Normal acceleration proportional gain	1.3380
PK21	Pitch rate integral gain	0.2486
PK19	Normal acceleration integral gain	1.3380

Tab.3 Longitudinal CAS gains. Dynamic pressure *qc = 261 [psf]*, static pressure *ps = 980 [psf]*.

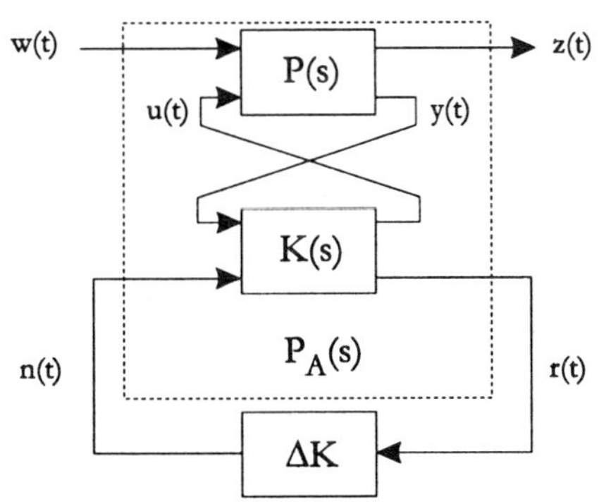

Fig.5 The Fixed Structure Control Problem

It is easily shown using root locus methods that the normal acceleration proportional gain (PK16) acts on the natural frequency of the short period mode, leaving the damping almost unaltered, while the pitch rate proportional gain (PK18) changes the damping maintaining the same natural frequency. The stick gain (PK15) regulates the gain of the response. As we will see shortly these gains give us enough degrees of freedom to reach a good matching with the target transfer function.

Approach: H_∞ Fixed Structure Control

Our problem, and many like it, is characterized in Fig.5. A linear controller $K_N(s)$ contains key gains and time constants that need to be adjusted. We isolate these key parameters and break the controller into a fixed part $K(s)$ and a variable part described by a diagonal matrix ΔK, $\Delta K \in R^{m\times m}$ where m is the number of parameters that characterize the controller. Depending on the structure of the controller ΔK can also contain *repeated real* elements.

The transfer function $P_A(s)$ of Fig.5 includes the plant augmented with the fixed part of the controller. It can also include weightings and target dynamics, as we will discuss shortly. The augmented plant is described by the following equations:

$$\begin{aligned}\dot{x}(t) &= Ax(t) + B_1 w(t) + B_2 n(t)\\ z(t) &= C_1 x(t) + D_{11} w(t) + D_{12} n(t)\\ y(t) &= C_2 x(t) + D_{21} w(t) + D_{22} n(t)\end{aligned} \qquad (1)$$

The fixed structure control problem is then stated in the following way:

Find a diagonal internally stabilizing controller ΔK, $\Delta K \in R^{m\times m}$, such that the n-norm $\|T_{zw}(s)\|_n$ is less than γ, where γ is some prescribed performance level. If such a controller exists it is called a γ suboptimal controller.

Depending on the norm that we minimize, this will be an H_∞ or an H_2 control problem.

To use this control problem to achieve our desired goals, we need to introduce a *model matching* structure to compute $T_{zw}(s)$. Fig.6 shows how this is done — if we minimize the norm of the transfer function from w to z, then the overall closed loop system will be as close as possible to the target plant $P_T(s)$ in terms of the chosen norm. The weighting function $W(s)$ is used to emphasize the frequency region where model matching is more important. The target plant can be defined in many different ways depending on the nature of the problem.

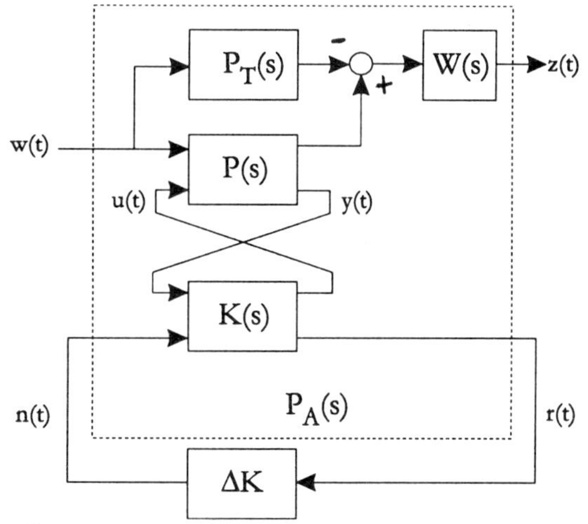

Fig.6 The H_∞ model matching problem

In some practical situations it is desirable to add to the norm based cost functional one or more additional terms that take into account other desired characteristics of the closed loop system. In our example, we add a term describing the distance of the Neal-Smith plot from a desired Neal-Smith plot. Our approach is to minimize the H_∞ norm of the transfer function $T_{zw}(s)$ of Fig.6, while making the Neal-Smith plot as close as possible to that of the target plant. The cost that we wish to minimize is thus:

$$J = \left\| T_{zw} \right\|_\infty + \frac{1}{5}\sum_{i=1}^{5}\sqrt{(P_i - P_{ti})^2 + (R_i - R_{ti})^2} \quad (2)$$

The handling quality metric in the above expression is the average distance between the target and the actual Neal-Smith plot calculated at the following five values of task bandwidth: 2.5, 3.0, 3.5, 4.0, 4.5 rad/sec. P_i is the pilot lead compensation in rad and R_i is the resonance peak calculated from the Neal-Smith criteria at the bandwidth (BW_i) (see Figs. 3 and 4). The subscript t indicates that the variable refers to the target plant.

This completes the description of our approach to the stated problem. Several comments are in order before we describe the technical details and results. First, we have adopted a modern framework, which allows additional robustness and performance considerations to be included. It also allows us to tap into the large body of knowledge that has been built up around this framework in recent years. In particular, this formulation can be modified to a Linear Parameter Varying (LPV) control problem, which will allow gain scheduling to be addressed. No consideration has yet been given to computational complexity; rather we have formulated a practical problem in the context of current methods. That this approach leads to a non-convex optimization problem (as we will see shortly) indicates that such problems are important to pursue.

Solution Procedure: Algorithms for solution of fixed-structure optimizations

A closed form solution of the fixed structure control problem does not exist. Fortunately fast computation and efficient numerical algorithms are widely available, and sufficient to solve our problem. We will present two algorithms: one based on an LMI formulation of the control problem and the other on Newton's method. Other algorithms, such as genetic algorithms or neural networks, could be used; in general it is good practice to solve non-convex problems using different methods to help insure a good solution.

Solution via LMIs

The LMI solution of the fixed structure control problem is based on two important lemmas: *The Bounded Real Lemma* and *The Maximum H_2 Norm Lemma.* These two lemmas can be used to formulate the fixed structure control problem as a bilinear matrix inequality. With some simple matrix algebra we can define the closed loop transfer function of the system described in (1) as follows:

$$\begin{aligned} \dot{\underline{x}}(t) &= A_{CL}\underline{x}(t) + B_{CL}\underline{w}(t) \\ \underline{z}(t) &= C_{CL}\underline{x}(t) + D_{CL}\underline{w}(t) \end{aligned} \quad (3)$$

With the assumption that the feedthrough term $D_{22} = 0$, we have:

$$\begin{aligned} A_{CL} &= A + \sum_{i=1}^{q} F_i k_i \qquad & B_{CL} &= B_1 + \sum_{i=1}^{q} G_i k_i \\ C_{CL} &= C_1 + \sum_{i=1}^{q} H_i k_i \qquad & D_{CL} &= D_{11} + \sum_{i=1}^{q} N_i k_i \end{aligned} \quad (4)$$

where: $F_i = \left[i^{th} \text{ column of } B_2\right] * \left[i^{th} \text{ row of } C_2\right]$

$$G_i = \left[i^{th} \text{ column of } B_2\right] * \left[i^{th} \text{ row of } D_{21}\right]$$

$$H_i = \left[i^{th} \text{ column of } D_{12}\right] * \left[i^{th} \text{ row of } C_2\right]$$

$$N_i = \left[i^{th} \text{ column of } D_{12}\right] * \left[i^{th} \text{ row of } D_{21}\right]$$

and $\Delta K = diag\left(k_1, k_2, k_3, \ldots, k_q\right)$.

In this way we have expressed the closed loop state space matrices as linear functions of the controller parameters k_i that we need to identify.

The solution of the H_∞ fixed structure control problem via LMIs is based on the Bounded Real Lemma, which states that, for a LTI system given by:

$$\begin{aligned} \dot{\underline{x}}(t) &= A\underline{x}(t) + B\underline{u}(t) \\ \underline{y}(t) &= C\underline{x}(t) + D\underline{u}(t) \end{aligned} \quad (5)$$

The H_∞ norm of the transfer function from u to y, $\|G\|_\infty < \gamma$ if and only if there exists $P \geq 0$ such that:

$$\begin{bmatrix} A^T P + PA & PB \\ B^T P & -\gamma^2 I \end{bmatrix} + \begin{bmatrix} C^T \\ D^T \end{bmatrix} \begin{bmatrix} C & D \end{bmatrix} \leq 0 \quad (6)$$

The importance of this lemma is that it transforms the H_∞ control problem into a finite dimensional linear matrix inequality. If we substitute the closed loop state space model (3) and (4) in (6), we get:

$$\begin{bmatrix} \left(A^T + \sum_{i=1}^{q} F_i^T k_i\right)P + P\left(A + \sum_{i=1}^{q} F_i k_i\right) & P\left(B_1 + \sum_{i=1}^{q} G_i k_i\right) & C_1^T + \sum_{i=1}^{q} H_i^T k_i \\ \left(B_1^T + \sum_{i=1}^{q} G_i^T k_i\right)P & -\gamma I & D_{11}^T + \sum_{i=1}^{q} N_i^T k_i \\ C_1 + \sum_{i=1}^{q} H_i k_i & D_{11} + \sum_{i=1}^{q} N_i k_i & -\gamma I \end{bmatrix} < 0 \quad (7)$$

The variables in the matrix inequality (7) are the positive definite matrix P and the controller coefficients k_i.. Unfortunately the matrix inequality (7) is nonlinear in the variables P and k_i, and thus we have a non-convex problem with no simple solution. For a *given* controller ΔK the matrix inequality is affine in γ and P and defines a convex set in P; efficient interior points methods from convex optimization are now available to compute the smallest feasible γ and the associated positive definite matrix P (this is the analysis problem). On the other hand, for a given matrix P the matrix inequality is affine in γ and the controller parameters k_i; once again we can use interior point methods to find the smallest feasible γ and the associated controller ΔK. The idea is then to find an optimal solution solving iteratively two linear matrix inequalities: the first in γ and P for a given ΔK, and the second in γ and ΔK for a given P. We have the following two step procedure:

1st LMI given a controller ΔK^* we can find the minimum γ for which there exist a positive definite matrix P that verifies (7);

2nd LMI given a positive definite matrix P^* (solution of the first LMI), find the minimum γ and the associated controller ΔK for which (7) is satisfied.

The algorithm is visualized in Fig.7. In order to start the algorithm we need to have an initial stabilizing controller, ΔK_0^*. Given an initial controller we solve the LMI (6) and find the matrix P_0^* that minimizes γ. At this point we go on iterating between the two LMIs. The algorithm guarantees that the value of γ will not increase between iterations, but there is no guarantee that a global minimum will be found. The first LMI also assures that the closed loop system will be stable.

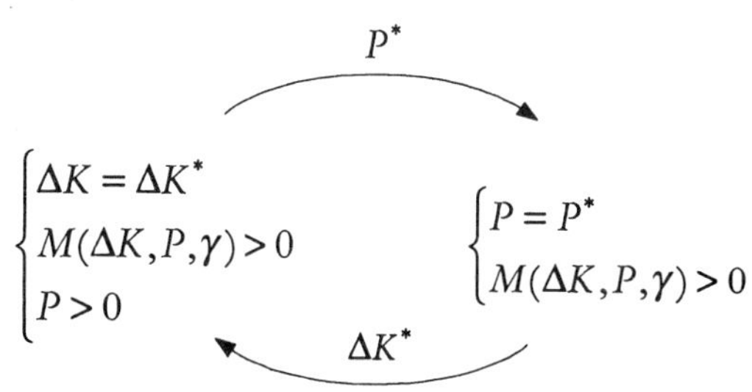

Fig.7 Iteration between Linear Matrix Inequalities

It is worth mentioning that it is not necessary to express the closed loop matrices as described in Eq. (4) to set up the LMI problem. The matrix ΔK can simply be kept in its diagonal form. Our formulation becomes convenient in Part II, when we solve the fixed structure *gain scheduling* problem.

The LMI can also be set up for the case when an H_2 norm is used instead of an H_∞ norm. In this case the property we rely on is that the H_2 norm of the transfer function from u to y, $\|G\|_2 < \gamma$ if and only if $D = 0$ and:

$$Trace\left(CPC^T\right) \leq \gamma^2 \quad (8)$$

$$\text{subject to}: \quad \begin{bmatrix} A^T P + PA & PB \\ B^T P & -I \end{bmatrix} \leq 0$$

$$P > 0$$

Again the norm criterion has been transformed into a finite dimensional linear matrix inequality. Using the closed loop state space model (3) and the maximum H_2 norm lemma, our H_2 Fixed Structure Control Problem can be stated in the following way:

$$minimize \ \text{Trace}\left[\left(C_1 + \sum_{i=1}^{q} H_i k_i\right) P \left(C_i^T + \sum_{i=1}^{q} H_i^T k_i\right)\right] \quad (10)$$

subject to

$$\begin{bmatrix} \left(A^T + \sum_{i=1}^{q} F_i^T k_i\right)P + P\left(A + \sum_{i=1}^{q} F_i^T k_i\right) & P\left(B_1 + \sum_{i=1}^{q} G_i k_i\right) \\ \left(B_1^T + \sum_{i=1}^{q} G_i^T k_i\right)P & -I \end{bmatrix} < 0 \qquad (11)$$

$$P > 0$$

Again we do not have a convex problem, and we define a two step procedure and iterate between two LMIs. Given an initial controller we can find the positive definite matrix P that minimizes the trace in (10) and satisfies the inequalities (11). We can then use this matrix P and solve for a new controller that minimizes the trace (10) and satisfies the inequalities (11). Using this iterative scheme until a minimum of the H_2 norm is reached or the maximum number of iteration is exceeded it is possible to find a controller that locally minimizes the H_2 norm of the transfer function from w to z, but there is no guarantee that a global minimum will be found.

Solution via Newton search

Besides the LMI approach both the H_2 and H_∞ fixed structure control problem can be efficiently solved using a search procedure. The method we selected is the 'BFGS Quasi-Newton' method. An algorithm that implements this technique is available in Matlab and it is straightforward to implement the fixed structure control optimization problem.

The main advantage with respect the LMI based solution is speed of convergence. The reduction in the cost obtained at each iteration between the two LMIs is always very small; in order to solve a problem involving 5 controller parameters and 14 states, we found that ~1000 iterations are required, requiring 10 hours on a workstation. The same problem can be solved in a matter of minutes using a Newton search.

Another advantage of the Newton search method is that it allows us to augment to cost with additional terms to account for other important performance specification, such as the handling qualities cost described in the previous section. In the LMI framework it is more difficult, if not impossible, to add additional elements to the cost. Additional terms in the cost must be added as LMI constraints.

Results

To apply our method to the HARV redesign problem, we must create the interconnection structure that describes the model matching problem. For this task a block diagram description of the system is of great help. Fig.8 shows the Simulink block diagram used to calculate the linear model of the actual plant augmented with the target plant. The longitudinal CAS is broken at the points where the stick gain, as well as the proportional and integral gains of the feedback, are introduced. This is shown in Fig. 9. Once the interconnection is defined, we can then calculate the linear state space model as defined in Eq. (1). The Matlab function *linmod* was used to obtain such a model.

In order to show the behavior of the LMI algorithm we will start by minimizing a pure H_∞ cost functional. We will compare the LMI optimal solution with that obtained via a Quasi-Newton search. In the Quasi-Newton search we need to calculate the H_∞ norm of the transfer function T_{zw}. Several methods are available to calculate the H_∞ norm of a linear system. We can use the bounded real lemma and solve an LMI problem, we can try to solve a Riccati equation, or form the Hamiltonian associated with the Riccati equation and find the value of γ for which one eigenvalue of M is purely imaginary. Note that if we find the minimum γ with any of the above methods, then the closed loop system of the plant and the modified controller is guaranteed to be stable.

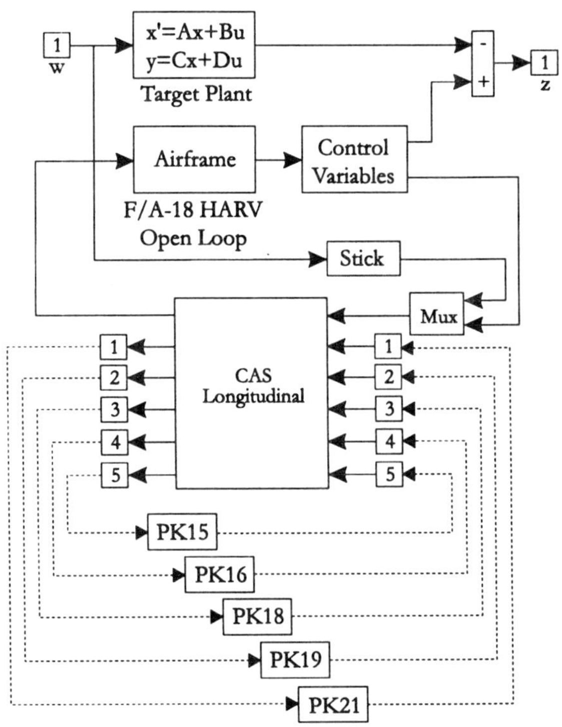

Fig.8 Fixed Structure H_∞ interconnection: Simulink block diagram

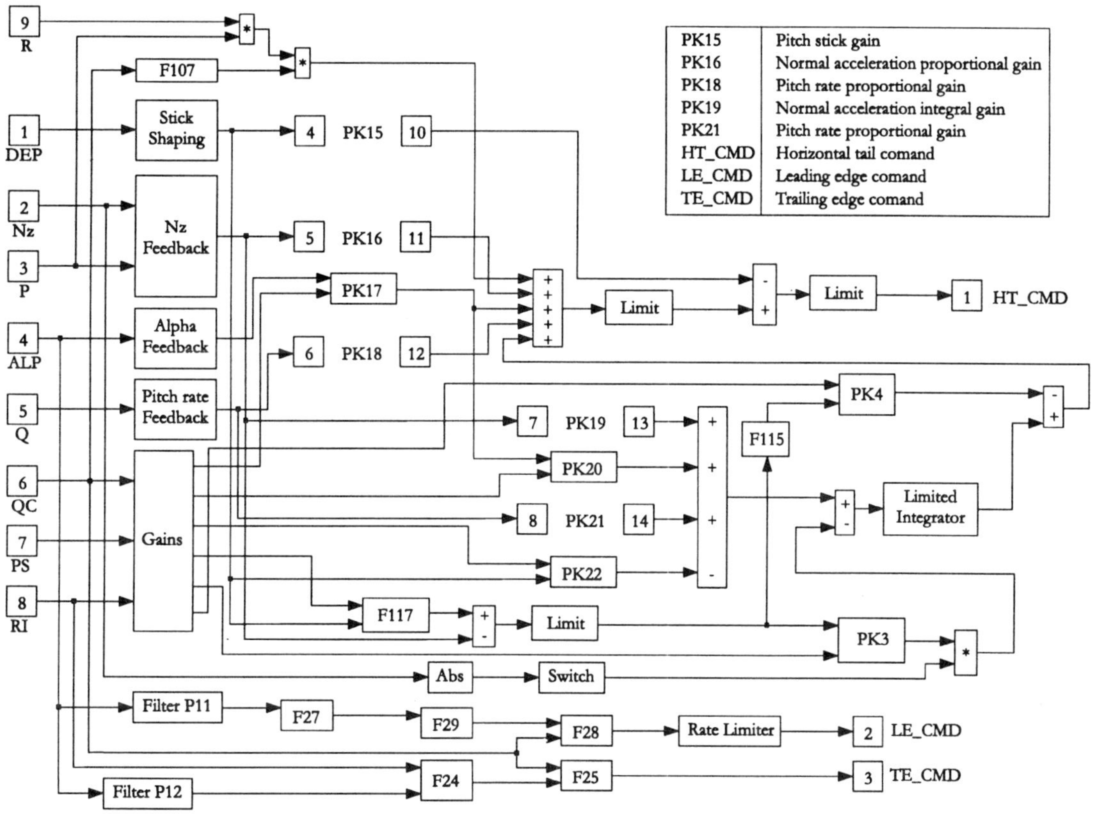

Fig. 9 Longitudinal CAS Opened at the PI gains

The most obvious way to check the performance of the fixed structure controller is to look at the cost achieved at the end of the search. We can decide to stop when a certain value is reached or go on until a minimum is found.

The initial value of the cost was: $\gamma = 4.1316$. The minimum value of gamma that was achieved with the Quasi-Newton search was: $\gamma = 0.1639$. While the minimum at the end of the LMI iterations was: $\gamma = 0.1653$. In both cases the search was started using the gains at their nominal values. The optimum solutions found with the two algorithms are listed in Tab.4.

Gain	LMI Gain	Newton Gain	Nominal Gain
PK15	0.6683	0.6646	0.3823
PK16	2.2488	2.1034	1.3380
PK18	0.6182	0.6144	0.3801
PK19	1.0195	0.9217	1.3380
PK21	0.3190	0.3496	0.2486
γ	0.1653	0.1639	4.1316

Tab.4 Comparison between nominal and new CAS gains

Fig.10 shows how the gains converge to the optimal solution while iterating between the two LMIs. After about 100 iterations the cost has almost reached its minimum value and the gains are stabilized at their final value.

Fig.11 shows a contour plot of the H_∞ norm as a function of the proportional gains PK16 and PK18. The steep increase in the norm divides the plane into stable and unstable regions. It is well known that the H_∞ norm goes to infinity when the plant becomes unstable. This kind of plot is very useful to check the solution: we can clearly see that in terms of PK16 and PK18 the solution represents a global minimum. Similar plots for different combinations of gains were plotted, and all showed that the solution was a global minimum with respect to the parameters plotted.

Finally we tested the Quasi-Newton search in the case of a multi-objective optimization problem. The cost that we wish to minimize is described in Eq. (2). It is the sum of the H_∞ norm (or the H_2 norm) of the transfer function T_{zw} and the handling quality metric. In this case we eliminated from the design the two integral gains. They don't play an important role in the model matching problem and they will not be used in the gain scheduling problem discussed in the second part of this paper.

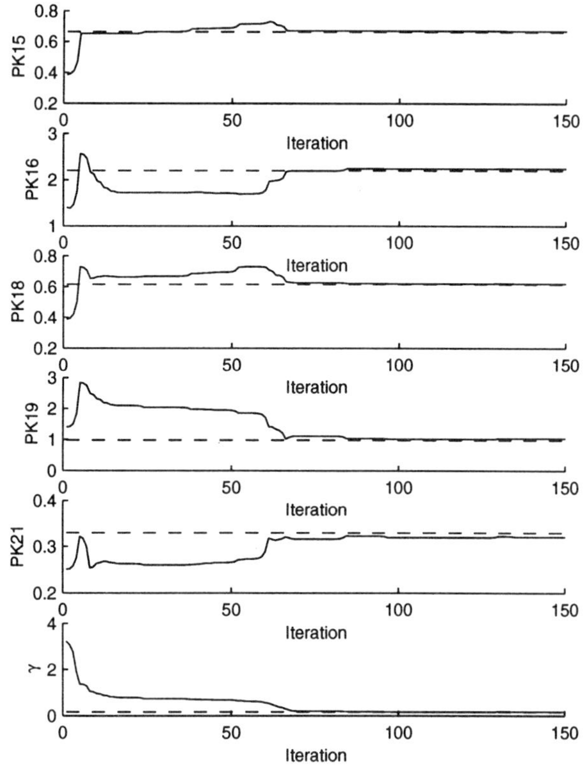

Fig.10 Convergence of the CAS gains during LMI optimization. Dashed lines: Quasi-Newton optimum gains

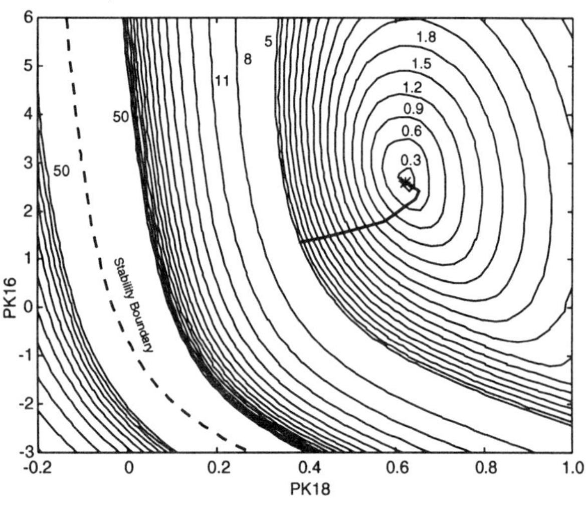

Fig.11 H_∞ norm vs proportional gains

Gain	H_2 Problem	H_∞ Problem	Nominal
PK15	0.5974	0.6100	0.3823
PK16	2.4150	2.3097	1.3380
PK18	0.5633	0.5831	0.3801
norm, γ	0.2919	0.2756	-

Tab.5 Longitudinal CAS: solution of the multi-objective optimization problem.

Fig.12 shows the Bode diagrams of the HARV, the F/A-18, and the HARV with the new gains from the H_∞ multi-objective optimization. The target plant has been reached to a good approximation — the differences at very low frequency do not affect the behavior of the plant. In this case a weighting function on the performance output z of Fig.8 was also inserted to emphasize the importance of a good match between 0.1 and 15 rad/sec.

Fig.13 shows the Neal-Smith carpet plot of the HARV with the modified flight control law. The plot is very close to that of the F/A-18 (see Fig.4), and solid level 1 flying qualities for a 3 rad/sec task bandwidth are expected.

Time domain nonlinear simulations

The modified gains were implemented in a full nonlinear simulator. The model consists of separate simulink C-functions that model the aerodynamics, the engine, the nonlinear equations of motion, the sensors, and the actuators. This model was developed from the existing F/A-18 flight simulator at NASA Dryden, and it replicates almost perfectly the highly sophisticated NASA simulator.

Pitch rate pulse responses were used to demonstrate the performance of the modified CAS. The response of the nominal control law (dashed line) and of the modified one (thick line) are compared in Fig.14. In the same plot is shown the response to the same impulse input of the F/A-18 (thin line).

The difference in the trim values of pitch angle and angle of attack reflects the differences in the mass and inertia characteristics of the F/A-18 HARV with respect to the standard F/A-18. Looking at the time domain traces we can see how the modified control

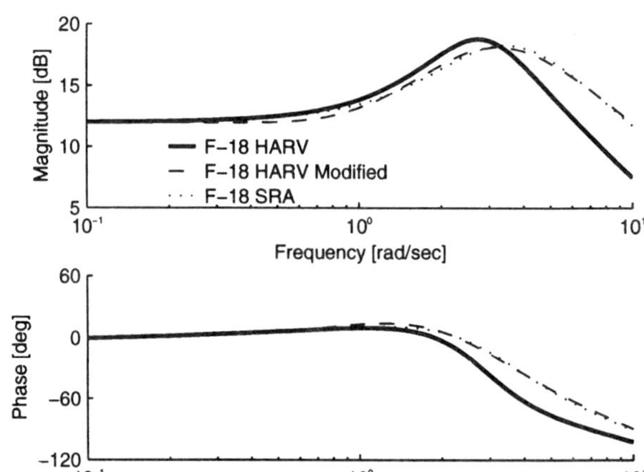

Fig.12 F/A-18 HARV Modified, Target, and Standard Transfer Functions

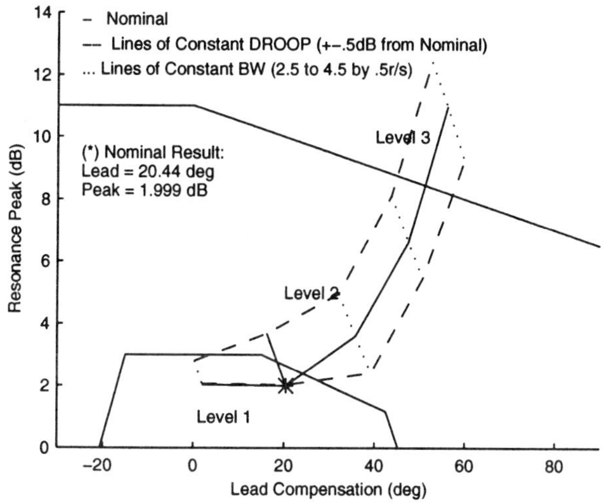

Fig.13 Modified HARV: Neal-Smith Carpet Plot

law has a response very similar to the F/A-18. It is particularly important to look at the trace of the pitch rate. Note that the response of the HARV with the unmodified control law is slower than the F/A-18. The modification to the control law corrects this aspect almost perfectly. Another important result is the increased damping of the response; it is evident from time domain data that a lower tendency to oscillate exists when the modified control law is used.

The stabilator deflection and rate time histories for both control laws are plotted in Fig.15. Here the main concern is rate saturation due to the increased gains in the modified 701E CAS control law. From the plot we can see that the modified CAS makes a stronger use of the horizontal tail but the rate saturation level is very similar to the standard CAS.

Tab.6 shows some of the characteristics of the pitch rate response of the three aircraft. Time to peak and the rise time of the F/A-18 HARV with the new set of gains are very close to those of the F/A-18.

Summary

Through a realistic example using the F/A-18 flight control system, we have shown how the important problem of fixed-structure gain adjustment can be accomplished using modern methods. The framework of block perturbations and norm-based objectives is an excellent way to develop generalized solutions to the problem. However, the resulting class of design problems is non-convex and therefore requires a DK-style iteration on LMIs. A Newton approach is found to be two orders of magnitude faster, and more versatile for mixed costs. A mixed cost is introduced which allows handling qualities, in terms of the Neal-Smith criterion, to be retained or improved during gain adjustments.

The generalized framework for gain adjustment is specifically tailored to allow gain scheduling to be addressed. In the second part of this paper, we draw upon the theory of linear parameter varying systems to develop a gain schedule for the F/A-18.

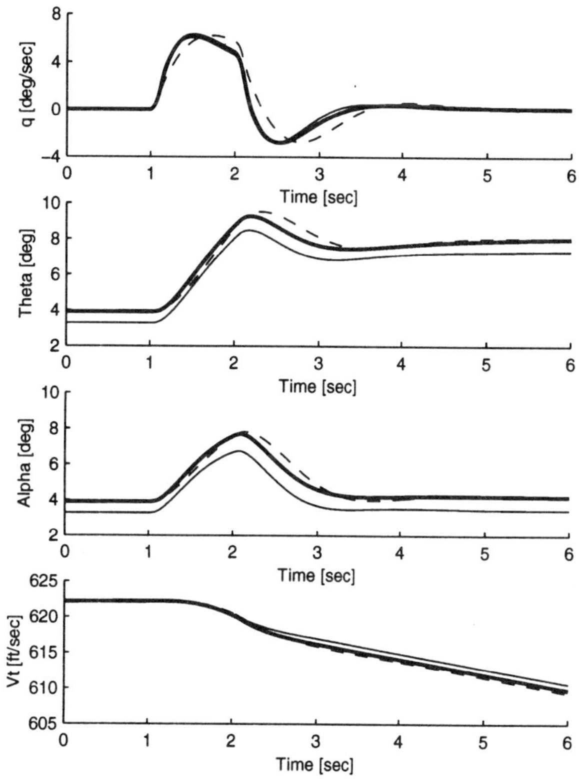

Fig.14 Time domain response to pitch impulse
Solid: F/A 18, Dashed: HARV, Thick: Modified HARV

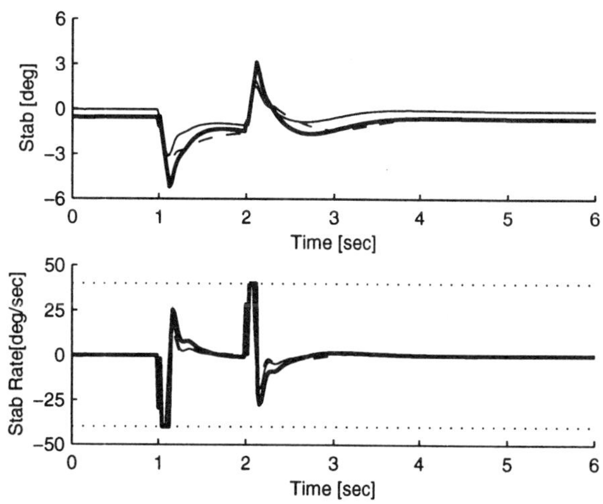

Fig.15 Stabilator Deflection and rate
Solid: F/A 18, Dashed: HARV, Thick: Modified HARV

	F/A-18	F/A-18 HARV New	F/A-18 HARV
Peak Overshoot	6.0356	6.2449	6.2004
Time to Peak	1.5014	1.5075	1.7206
Time at 10 %	1.0597	1.0675	1.0740
Time at 90 %	1.3172	1.3330	1.4678
Rise Time*	0.2575	0.2655	0.3938

* (Time at 90%) - (Time at 10%)

Tab.6 Time domain response characteristics

Acknowledgments

This work was supported by NASA Cooperative Agreement NCC-4001, John Burken, Technical Monitor. This support is gratefully acknowledged. The authors would also like to thank Patrick Stoliker, Keith Wichman and Joseph Pahle for technical assistance and Matlab handling qualities software.

References

[1] J. M. Maciejowski, Multivariable Feedback Design, Addison-Wesley, 1989, pp. 325-354.

[2] S. Boyd, L. El Ghaoui, E. Feron, and V. Balakrishnan, Linear Matrix Inequalities in System and Control Theory, SIAM Studies in Applied Mathematics, 1994.

[3] R. A. Hyde, H_∞ Aerospace Control Design, Advances in Industrial Control, Springer-Verlag, 1995, pp. 103-119.

[4] R. A. Hyde. Implementation complexity reduction of H_∞-optimal controllers using model matching and parameter optimization. Technical Report TR R71-92, DLR Oberpfaffenhofen, Institut fur Dynamik der Flugsysteme, December 1991.

[5] J.M. Edmunds, Control System Design and Analysis using Closed-loop Nyquist and Bode Arrays, *International Journal of Control,* 1979, Vol.30, No. 5, pp. 773-802.

[6] G. Kreisselmeier and R. Steinhauser, Systematic Control Design by Optimizing a Vector Performance Index, *Proc. IFAC Symp. On Computer Aided Design of Control Systems*, Zurich, 1979, pp. 113-117.

[7] N. Halyo, D.D. Moerder, J.R. Broussard, and D.B. taylor, A Variable-Gain Output Feedback Control Design Methodology, NASA Contractor Report 4226, 1989.

[8] R.J. Adams, J.M. Buffington, A.G. Sparks, and S.S. Banda, "Robust Multivariable Flight Control", *Advances in Industrial Control,* Springer-Verlag, 1994.

[9] "Military Specification, Flying Qualities of Piloted Airplanes", MIL-F/A-8785C, August 1980.

[10] "Military Standard, Flying Qualities of Piloted Vehicles", MIL-STD-1797A, January 1990.

[11] J.T. Bosworth and T.H. Cox, A design procedure for Handling Qualities Optimization of the X-29A Aircraft, NASA Technical Memorandum 4142, 1989.

[12] R.E. Bailey and R.E. Smith, Analysis of Augmented Aircraft Flying Qualities Through Application of the Neal-Smith Criterion, *AIAA Guidance and Control Conference*, Albumquerque, New Mexico, 1981.

[13] M.V. Cook and M.J. Rycroft, Aerospace Vehicle Dynamics and Control, Oxford University Press, 1994, pp.103-124, S.A. Carr and M.J. Grimble, "Comaprison of LQG, H_∞ and Classical Designs for Pitch Rate Control of an Unstable Military Aircraft.

[14] A. Packard, Gain Scheduling via Linear Fractional Transformations, *Systems and Control Letters*, October 1993.

[15] P. Apkarian and P. Gahinet, A Convex Characterization of Gain-Scheduled H_∞ Controllers, *IEEE Trans. Automat. Contr.* Vol. 40, May 1995, pp. 853-864.

[16] P. Apkarian, J. Biannic, and P. Gahinet, Self-Scheduled H_∞ Control of Missile via Linear Matrix Inequalities, *Journal of Guyidance, Control, and Dynamics*, Vol.18 pp.532-538, May 1995.

[17] J. Biannic, P. Apkarian, and W.L. Garrard, Parameter Varying Control of a High Performance Aircraft, *AIAA Guidance, Navigation, and Control Conference*, San Diego CA, July 1996.

[18] L. Lee and M. Spillman, Control of Slowly Varying LPV Systems: An Application to Flight Control, *AIAA Guidance, Navigation, and Control Conference*, San Diego CA, July 1996.

[19] T. P. Neal and R. E. Smith, "An In-Flight Investigation to Develop Control System Design Criteria for Fighter Airplanes", AFFDL-TR-70-74, June 1970.

AIAA-97-3470

LP SYNTHESIS OF A LATERAL FLIGHT CONTROL SYSTEM FOR A TRANSPORT AIRCRAFT

Gilles Ferreres Isabelle Dardenne

ONERA / CERT (DERA) B.P. 4025 F31055 TOULOUSE CEDEX, FRANCE

SUPAERO, Av. E. Belin, F31055 TOULOUSE CEDEX, FRANCE

e-mail: ferreres@cert.fr - dardenne@cert.fr

A simple method is first proposed for the synthesis of dynamic feedback or feedforward controllers, which satisfy classical time-domain specifications, such as rise time, overshoot, decoupling or rejection time of external disturbances. The only assumption on the plant is to be linear time invariant. The controller is obtained as the solution of a Linear (or Quadratic) Programming problem. When synthesizing a feedforward controller, a multi-model synthesis is moreover possible.

The method is then applied to the synthesis of a lateral flight control system for a transport aircraft. In this problem indeed, as in many industrial control problems, closed loop performance is primarily specified in terms of the time-domain response to a reference input, which here corresponds to a pilot requirement.

1 - Introduction

In many industrial control problems, closed loop performance is often primarily specified in terms of the time-domain response, either to a reference input, or to an external disturbance. Several solutions are available for taking into account these specifications.

As a first example, in modal control, the eigenstructure is chosen to adequately achieve time-domain specifications.[1] As an other example, in H_∞ control, it is possible on the one hand to tune the rise time or rejection time through the choice of the closed loop bandwidth.[2] On the other hand, the H_∞ controller naturally ensures the decoupling.[3]

A new method is proposed in this paper, which uses an affine representation of the closed loop response as a function of the control design parameters.[4] The only assumption on the plant is to be linear time invariant. The (feedback or feedforward) controller is then simply obtained as the solution of a Linear or Quadratic Programming (*LP or QP*) problem. When synthesizing a feedforward controller, a multi-model synthesis is moreover possible.

The method is then applied to the synthesis of a lateral flight control system for a transport aircraft. We focus on the time-domain responses to a step on the sideslip angle or on the roll rate output (these correspond indeed to typical pilot requirements). A rise time and a maximal overshoot are specified in each case. Saturations on the actuator rate outputs signals are also taken into account and our aim is then to minimize the coupling on the sideslip angle.

The paper is organized as follows. The basis of the method is presented in section 2. The design procedures are then detailed in section 3. The method is extended in section 4 to the multi-model synthesis of a feedforward controller. The flight control system is synthesized in section 5. Concluding remarks end the paper.

2 - The basis of the method

Introduction

Consider the augmented plant P of figure 1, where y_r represents the reference input, d an external (typically unmeasured) disturbance, u the control input and y the measured output. We would like to shape the time-domain response to an input signal $y_r(t)$ or $d(t)$ on the output w. The transfers H and K respectively denote the dynamic feedforward and feedback controllers.

For the sake of simplicity, the method is presented for the case of scalar inputs and outputs. It is however straightforward to extend the method to the case of vector inputs and outputs.

An affine representation of the closed loop response

Results 1 and 2 propose an affine representation of the closed loop response as a function of the control design parameters.

Result 1 *The feedback controller K of figure 1 is fixed. The feedforward controller H is parametrized as :*

$$H(z) = \sum_{i=0}^{n} \theta_i \, z^{-i} \tag{1}$$

A fixed reference input signal $y_r(t)$ is applied to the closed loop. Let $w^0(t)$ the corresponding closed loop response (on output w), which is obtained with a unit feedforward controller ($H = 1$). Using then the feedforward controller of equation (1), the closed loop response to the signal $y_r(t)$ can be written as :

$$w(t) = \sum_{i=0}^{n} \theta_i \, w^0(t-i) \tag{2}$$

Remarks 1 :
(i) for the sake of simplicity, the feedforward controller was parametrized as a Finite Impulse Response (FIR) filter in the above result. Other parametrizations are nevertheless possible. As a matter of fact, a great deal of work has been devoted to the problem of building orthonormal bases of filters, which generalize the classical Laguerre or Kautz bases :[5,6] and included references.
(ii) If the reference input signal is known in advance in the real time application, a non causal feedforward controller H can then be handled in equation (1) (Predictive control).

Before stating the second result, the following Lemma, which is extracted from,[4] is needed.

Lemma 1 *Consider the augmented plant of figure 1 (with $H = 1$) and note e the input vector $e = [d, \, y_r]^T$. Partition the transfer matrix P as :*

$$P = \begin{bmatrix} P_{11} & P_{12} \\ P_{21} & P_{22} \end{bmatrix} \tag{3}$$

where P_{11} (resp. P_{22}) is the transfer between e (resp. u) and w (resp. y). Consider the closed loop transfer $F_l(P,K)$ between input e and output w, obtained by connecting the controller K to the augmented plant P :

$$w = F_l(P,K)e = \left[P_{11} + P_{12}K(I - P_{22}K)^{-1}P_{21}\right] e \tag{4}$$

*The closed loop is assumed to be well posed *. Let $Q = K(I - P_{22}K)^{-1}$. Then :*

$$K = (I + QP_{22})^{-1}Q \tag{5}$$

As a matter of fact, many design specifications, which are not convex in the space of controllers K, are nevertheless convex in the space of closed loop transfer matrices Q.[4] As a consequence, the idea is to synthesize a transfer matrix Q satisfying the specifications. The corresponding controller K is then obtained using the above Lemma.

Result 2 *The feedforward controller H of figure 1 is now fixed. The closed loop transfer $F_l(P,K)$ between the input vector e and output w is parametrized as :*

$$F_l(P,K) = P_{11} + P_{12}QP_{21} \tag{6}$$

*ie P_{22} is strictly proper or Q is such that $det(I + Q(s)P_{22}(s)) \neq 0 \quad \forall Res > 0$.[7,8]

where :

$$Q = \sum_{i=1}^{n} \theta_i \, Q_i \tag{7}$$

and the transfers Q_i are a priori fixed. A given input signal $e(t)$ is applied to the closed loop. Let :

$$\begin{aligned} w^0(t) &= P_{11}\, e(t) \\ w^i(t) &= P_{12} Q_i P_{21}\, e(t) \end{aligned} \tag{8}$$

With reference to equations (6-7), it is then possible to write the closed loop response to the input signal $e(t)$ as :

$$w(t) = w^0(t) + \sum_{i=1}^{n} \theta_i \, w^i(t) \tag{9}$$

Remarks 2 :
(*i*) In a converse way, after the control design parameters θ_i have been computed, it suffices then to use Lemma 1 so as to obtain the corresponding controller K (see also section 5).
(*ii*) Here again, the issue is now to parametrize the transfer matrix Q as an affine combination of fixed filters Q_i (see equation (7)). A first solution is to choose a large base of filters, so as to have a sufficient amount of degrees of freedom (see e.g.[4] and remarks 1). The order of the resulting controller K will be however very large with this method. See section 5 for an alternative method.
(*iii*) It is also possible to simultaneously synthesize the feedback and feedforward controllers, so as to obtain a TDF (Two Degrees of Freedom) control law : see section 5.

3 - Design procedures

Obtaining LP and QP problems

In results 1 and 2, the time-domain response of the closed loop was obtained as an affine combination of the design parameters θ_i, namely $w(t) = w^0(t) + \phi^T(t)\theta$, where $w^0(t)$ and $\phi(t)$ were a priori computed. In a vector form, let $W = [w(1), \ldots, w(N)]^T$ the values of the output signal $w(t)$ at $t = 1, \ldots, N$. W can be written as $W = W^0 + \Phi^T\theta$, where the vector W^0 and the matrix Φ are a priori known.

Using the above affine representation of the closed loop response, we would like to translate the time-domain specifications into, either a Linear Programming problem:

$$\min_{A\theta \leq b} c^T\theta \tag{10}$$

or, more generally, a Quadratic Programming problem :

$$\min_{A\theta \leq b} (c^T\theta + \theta^T Q\theta) \tag{11}$$

The matrix Q must be positive definite, so as to obtain a convex optimization problem. Generally speaking, it is possible to consider l_2 and l_∞ time-domain objectives :

• l_∞ *constraints* : these can be generically written as $w(t) \leq \overline{w}(t)$ or $w(t) \geq \underline{w}(t)$ for $t \in [N_1,\ N_2]$. Consider as an example the case of a unit step on the reference input. A l_∞ constraint may typically correspond to a specification on the maximal allowable overshoot D (i.e. $w(t) \leq (1 + D)$ for all $t \geq 0$) or the rising time (e.g. $0.95 \leq w(t) \leq 1.05$ for all $t \geq N$).

• l_∞ *objectives* : we would like e.g. to minimize the value of α satisfying for all $t \in [N_1,\ N_2]$:

$$-\alpha \leq w(t) - w_r(t) \leq \alpha \tag{12}$$

Where $w_r(t)$ is typically a reference trajectory. This problem classically corresponds to an augmented LP problem. Let :

$$\begin{aligned} W &= [w(N_1), \ldots, w(N_2)]^T = W^0 + \Phi^T\theta \\ W_r &= [w_r(N_1), \ldots, w_r(N_2)]^T \end{aligned} \tag{13}$$

Define then :

$$\begin{aligned} x &= \begin{bmatrix} \theta \\ \alpha \end{bmatrix} \\ c &= [0, \ldots, 0, 1]^T \\ A &= \begin{bmatrix} \Phi^T & -1_m \\ -\Phi^T & -1_m \end{bmatrix} \\ b &= \begin{bmatrix} W_r - W_0 \\ W_0 - W_r \end{bmatrix} \end{aligned} \tag{14}$$

1_m denotes the unit column vector, of size $m = N_2 - N_1 + 1$. The minimization of α in equation (12) is equivalent to the minimization of $c^T x$, under the constraint $Ax \leq b$.

• l_2 *objectives* : these correspond to the minimization of a criterion of the form $c^T\theta + \theta^T Q\theta$. Suppose e.g. that the quadratic error between the response w and the reference trajectory w_r is to be minimized, namely :

$$J = \sum_{i=N_1}^{N_2} (w(i) - w_r(i))^2 \tag{15}$$

Note that :

$$(w(t) - w_r(t))^2 = (\phi^T(t)\theta + w^0(t) - w_r(t))^2 \tag{16}$$

so that J can be rewritten as :

$$J = \tilde{W}^T\tilde{W} + 2\tilde{W}^T\Phi^T\theta + \theta^T\Phi\Phi^T\theta \tag{17}$$

where $\tilde{W} = [w^0(N_1) - w_r(N_1), \ldots, w^0(N_2) - w_r(N_2)]^T$.

Design procedures

Synthesis of a dynamic feedforward controller:

- *Step 1* : choose an affine parametrization of the feedforward controller :

$$H = \sum_{i=1}^{n} \theta_i H_i \tag{18}$$

- *Step 2* : let $H = H_i$. Compute the closed loop response $w^i(t)$ to the reference input signal $y_r(t)$.

- *Step 3* : when using the feedforward controller of equation (18), the closed loop response $w(t)$ is obtained as :

$$w(t) = \sum_{i=1}^{n} \theta_i w^i(t) \tag{19}$$

  Following section 3.1, it is then possible to translate the time-domain specifications in terms of l_∞ constraints and l_2 and l_∞ objectives. A LP or QP problem is obtained.

- *Step 4* : the θ_i's are directly obtained as the solution of the LP or QP problem, which is to be solved.

Remarks 3 :
(*i*) Standard routines are available in Matlab for solving the LP or QP problems. Note however that more efficient algorithms may be used if the size of the problem (i.e. the dimension of matrix A in the constraint $A\theta \leq b$) is very large :[9] and included references. This may be especially the case when considering MIMO systems with a small sampling period.
(*ii*) Strictly speaking, it is not necessary to handle a parametric model of the closed loop system. It suffices to be able to measure the closed loop response to a given reference input signal.

The procedure for synthesizing a dynamic feedback controller essentially follows the one for synthesizing a dynamic feedforward controller.

Synthesis of a dynamic feedback controller :

- *Step 1* : choose an affine parametrization of $Q = K(I - P_{22}K)^{-1}$ (see Lemma 1) :

$$Q = \sum_{i=1}^{n} \theta_i Q_i \tag{20}$$

  The closed loop system is then obtained as $P_{11} + P_{12}QP_{21}$.

- *Step 2* : Let a given input signal $e(t)$. With reference to equation (8), compute the closed loop responses $w^0(t)$ (which corresponds to $Q = 0$) and $w^i(t)$ (which corresponds to $Q = Q_i$) for i=1,...,n.

- *Step 3* : when using the affine parametrization of Q (see equation (20)), the closed loop response $w(t)$ is obtained as :

$$w(t) = w^0(t) + \sum_{i=1}^{n} \theta_i w^i(t) \tag{21}$$

  Here again, use section 3.1 so as to translate the time-domain specifications into a LP or QP problem.

- *Step 4* : the θ_i's are first obtained as the solution of the LP or QP problem, which is to be solved. The corresponding transfer matrix Q is computed. See section 5 and[4] for the obtention of the state-space model of K from the state-space model of Q (see equation (5)).

Remark 4: results 1 and 2 and the above design procedures can be easily extended to the continuous-time case : as done in section 5, it suffices indeed to handle the values of the signals at each point of a time domain gridding.

4 - Multi-model synthesis of the feedforward controller

It is rather straightforward to extend the design procedure of a dynamic feedforward controller to the multi-model case. It suffices indeed to repeat steps 2 and 3 of the above procedure for each model. Let $i \in [1, n]$ the index of the model. The corresponding time-domain constraints $A_i\theta \leq b_i$ are stacked into an augmented linear constraint $A\theta \leq b$, with :

$$A = \begin{bmatrix} A_1 \\ \dots \\ A_n \end{bmatrix} \quad and \ b = \begin{bmatrix} b_1 \\ \dots \\ b_n \end{bmatrix} \tag{22}$$

Using the same method, note that it is possible to take simultaneously into account time-domain specifications, which correspond to different signals on the reference input y_r.

Result 3 *Assume first that the same time-domain constraints are required for each of the n closed loop systems $T_i(z)$ ($T_i(z)$ corresponds to the closed loop transfer between y_r and w in figure 1). If these time-domain constraints are satisfied by the same feedforward controller $H(z)$ for each of the $T_i(z)$, they are then satisfied on the polytope of closed loop systems, whose vertices are the T_i's :*

$$T(z) = \sum_{i=1}^{n} \lambda_i T_i(z) \ with \ \lambda_i \geq 0 \ and \ \sum_{i=1}^{n} \lambda_i = 1 \tag{23}$$

Sketch of proof : Let $A_1\theta \leq b_1$ and $A_2\theta \leq b_2$ the time-domain constraints for the first and second closed loop models $T_1(z)$ and $T_2(z)$. Let $T(z) = \lambda T_1(z) + (1-\lambda)T_2(z)$, with $\lambda \in [0, 1]$. Note simply that :

$$\begin{aligned} w(t) &= T(z)y_r(t) \\ &= \lambda(T_1(z)y_r(t)) + (1-\lambda)(T_2(z)y_r(t)) \end{aligned} \tag{24}$$

As a consequence, the affine constraints corresponding to the time-domain specifications on $T(z)$ can be written as :

$$(\lambda A_1 + (1-\lambda)A_2)\theta \leq \lambda b_1 + (1-\lambda)b_2 \tag{25}$$

As a consequence, if the solution vector θ satisfies both $A_1\theta \leq b_1$ and $A_2\theta \leq b_2$, it also satisfies the above equation.

5 - Synthesis of the flight control system

Preliminaries

Two problems are to be considered :

- The closed loop poles of $F_l(P, K) = P_{11} + P_{12}QP_{21}$ are a priori fixed as the poles of the augmented plant P and the poles of the closed loop transfer matrix Q.
- Figure 2 illustrates a way to compute the controller $K = (I + QP_{22})^{-1}Q$ from the closed loop transfer $Q = K(I - P_{22}K)^{-1}$. As a consequence, the order of the controller is necessarily equal to $n_P + n_Q$, where n_P (resp. n_Q) is the order of the augmented plant P (resp. the order of Q).

On the one hand, we would like to minimize the complexity of the control law. On the other hand, it is not possible to keep the same poles for the open loop and closed loop systems : in our aircraft model, there are indeed nearly unstable poles and badly damped modes. These two problems are solved as follows :

- An initial estimated state feedback (also called an observer based controller) K_{nom} is introduced in the augmented plant as seen figure 3. K_{nom} is to be chosen so as to obtain satisfactory poles and decoupling properties for the new augmented plant.

With such a nominal controller, the closed loop transfer $F_l(P, K)$ between the input vector e and the output w is parametrized as:

$$F_l(P, K) = T_1 + T_2 Q T_3$$

where T_1, T_2 and T_3 are stable maps depending on the plant. This parameterization is based on the so-called Q-parameterization of the closed loop input/output maps achievable with controllers wich stabilize the systems.[7,8] In particular, Doyle[10] has given a nice interpretation of this Q-parameterization when the nominal controller is an estimated state feedback. Figure 3 shows how the free transfer matrix Q is connected.

- A static transfer matrix Q is used, so that the order of the controller is equal to the order of the augmented plant P. Note that the dimensions of Q are the same as the dimensions of K.

Aircraft model and performance specifications

Aircraft model

A lateral model of a rigid transport aircraft is considered. Two lateral control surfaces, namely one aileron δ_p and one rudder δ_r, are used to pilot this aircraft. The measured outputs include the lateral acceleration n_y, the roll and yaw rates p and r and the roll angle ϕ.

The aircraft model is a linear time invariant transfer matrix of order 6 (a fourth aerodynamical model and a first order actuator on the rudder and the aileron).

When building the augmented plant of figure 1, additional outputs are considered in the design. These regulated outputs, which are not used by the control law, are the derivative $\dot{\Phi}$ of the roll angle Φ with respect to time and four outputs which correspond to the two actuator outputs and rate outputs.

Performance specifications

The inputs and outputs of the augmented plant of figure 1 are defined as follows :

- The exogeneous input vector e contains the reference inputs β_c and Φ_c on the sideslip and roll angles β and Φ.

- The control input vector u contains the actuator inputs δ_p, δ_r.

- The regulated output vector w contains :
  - the 2 actuator outputs.
  - the 2 actuator rate outputs.
  - the sideslip and roll angles β and Φ and the derivative $\dot{\phi}$ of ϕ with respect to time.

- The measured output vector y finally contains the lateral acceleration n_y, the roll and yaw rate p and r and the roll angle ϕ.

We now present the design specifications. As a first point, we consider the step response of a first order filter, which represents our first reference input signal. This filtered step is applied to the reference sideslip angle β_c. The specifications are :

- The sideslip angle β must track the reference input signal with a rise time of less than 6 seconds [2], an overshoot of less than 30 % and a zero static error.

- the roll angle must remain between two extremal values :

$$\phi_{min} \leq \phi(t) \leq \phi_{max} \quad \forall t \tag{26}$$

- The steady-state value ϕ_s of the roll angle must satisfy $\phi_s = -\beta_s$, where β_s is the steady-state value of the sideslip angle.

As a second point, a step is now applied to the roll rate output p. The specifications are essentially the same as above:

- The roll rate output must track the reference input signal with a rise time of less than 6 seconds and a zero static error.

[2]The rise time t_r is defined here as $0.95 \leq y(t) \leq 1.05$ for all $t \geq t_r$, if $y(t)$ is the unit step response of the closed loop.

- the sideslip angle must remain between two extremal values :

$$-\epsilon \leq \beta(t) \leq \epsilon \quad \forall t \tag{27}$$

The value ϵ is either fixed, or to be minimized.

The following specification is also requested for all two above responses (so as to take into account saturation constraints) :

- The actuator outputs must remain between - 30 degrees and + 30 degrees.
- The actuator rate outputs must remain between $\pm$ A degrees/s. The value A is here again either fixed, or to be minimized.

Synthesis of the flight control system

As a preliminary, an initial estimated state feedback K_{nom} and an initial static feedforward H_{nom} are synthesized. A modal method is used to ensure satisfactory closed loop poles and nearly perfect decoupling with the feedback K_{nom}. On the other hand, the static feedforward is used to respect the steady state design specifications (see above) :

$$\begin{bmatrix} \beta \\ \phi \end{bmatrix} = \begin{bmatrix} 1 & 0 \\ -1 & 1 \end{bmatrix} \begin{bmatrix} \beta_c \\ \phi_c \end{bmatrix} \tag{28}$$

The LP method is then applied to the augmented plant of figure 4. Note in this figure that reference inputs y_r are included in the measured outputs y, so that a TDF (Two-degrees-of-freedom) control law is synthesized : on the one hand, the feedback controller K and the feedforward controller H are simultaneously synthesized. On the other hand, K and H are two transfer matrices in figure 4, which share the same dynamics.

12 parameters θ_i are to be optimized, namely 8 for the feedback part [3] of the control law and 4 for its feedforward part [4]. As a static transfer matrix Q is chosen, the order of the control law remains equal to the order of the augmented plant, namely 6.

[3] $8 = 4 * 2$, since the aircraft model contains 2 inputs and 4 outputs.

[4] $4 = 2 * 2$, since the aircraft model contains 2 inputs and there are 2 reference inputs.

We would like to study the trade-off between the magnitude A of the actuator rate outputs signals $w_{actuator\ rate}(t)$ and the coupling ϵ on the side-slip angle. To this aim, a value of A is fixed and the minimization of the coupling ϵ is chosen as the optimization criterion. The other design specifications are expressed as l_∞ constraints. The minimization of ϵ leads to the introduction of a new parameter α :

$$-\alpha \leq \beta(t) \leq \alpha \tag{29}$$

which is to be minimized. This optimization problem can be recast into an augmented LP problem, as exposed in section 3.

Note that a continuous-time controller is synthesized, and that the time-domain responses are gridded with a sampling period of 0.05 seconds. The length of these responses is 10 seconds for the first simulation (step response on β) and 20 seconds for the second simulation (step response on p). In the augmented LP problem, the size of matrix A (in the constraint $A\theta \leq b$) is (11709 * 13). The problem is solved using Matlab in 5 and 10 minutes.

Various controllers are synthesized for several values of the magnitude A. The coupling ϵ is minimised in each case. figure 5 presents the trade-off between these quantities. As expected, the less coupling is required, the more the actuator are sollicitated.

As a final point, we present in figures 6 and 7 the closed loop responses for the controller, which corresponds to the star "*" in figure 5 ($\epsilon = 0.01$). These figures present the time-domain responses on the measured outputs β, p, r and ϕ and on the actuator outputs and rate outputs (which are noted respectively u_1, u_2, vs_1, vs_2).

7 - Conclusion

We proposed a new and simple method for synthesizing feedforward and feedback controllers :

- *Feedforward control* : it is possible to reshape the time domain response to a known reference input signal, depending on the specifications.

It is moreover possible to synthesize a unique feedforward controller, which verifies the time-domain specifications for a polytope of closed loop systems. As the major advantages of the method, the design procedure appears easy to implement, it takes into account (directly and in a non conservative way) the time-domain specifications, and a multi-model synthesis is possible.

- *Feedback control* : it is now possible to shape the time domain response, either to a reference input signal, or to an external disturbance signal. Our design procedure essentially uses the idea of.[4] These authors have showed that many time and frequency domain specifications are convex in the space of the closed loop transfer matrices. Our contribution was to emphasize that the general convex optimization problem, which is to be solved in,[4] boils down to a simple Linear (or Quadratic) Programming problem, which is thus much easier to solve.

As an application example, we have considered the case of a transport aircraft. We have especially studied the trade off between the coupling on the sideslip angle and the magnitude of the rate output actuators signals. Note however that other optimization criteria could have been taken into account, so as to study other trade-offs between the time-domain specifications.

Note also that it would be moreover possible to introduce in our method other specifications, especially in the frequency domain. The new optimization problem typically becomes then a LMI problem.[11] This renders especially possible to study in a systematic way the trade-off between a robust stability requirement (namely robust stability in the face of high frequency bending modes, in the context of a flexible transport aircraft) and a performance requirement, which typically corresponds to the time-domain response to a reference input (pilot requirement) or to an exogeneous perturbation (wind disturbance).

References

[1] K.M. Sobel, E.Y. Shapiro, and A.N. Andry. Eigenstructure assignment. *International Journal of Control*, 59(1):13–57, 1994.

[2] G. Ferreres, V. Fromion, G. Duc, and M. M'Saad. Application of real / mixed μ computational techniques to a H_∞ missile autopilot. *International Journal of Robust and Nonlinear Control*, 6(8):743–769, 1996.

[3] R.Y. Chiang, M.G. Safonov, K. Haiges, K. Madden, and J. Tekawy. A fixed H_∞ controller for a supermaneuverable fighter performing the Herbst maneuver. *Automatica*, 29(1):111–127, 1993.

[4] S. Boyd, C. Barrat, and S. Norman. Linear controller design : limits of performance via convex optimization. *Proceedings of the IEEE*, 78(3):529–574, March 1990.

[5] P.M.J. Van den Hof, P.S.C. Heuberger, and J. Bokor. System identification with generalized orthonormal basis functions. *Automatica*, 31(12):1821–1834, 1995.

[6] B. Ninness and F. Gustafsson. Orthonormal bases for system identification. *Proceedings of the ECC*, pages 13–18, 1995.

[7] B. A. Francis. A course in H_∞ control theory. *Lecture Notes in Control and Information Sciences*, 88, 87.

[8] M. Vidyasagar. *Control System Synthesis : A Factorization Approach.* Cambridge, MA: MIT Press, 85.

[9] J.C. Moreno Banos and M. Papageorgiou. A linear programming approach to large-scale optimal control problems. *IEEE Transactions on Automatic Control*, 40(5):971–977, 95.

[10] J.C. Doyle. *Matrix Interpolation Theory and Optimal Control.* PhD thesis, Univ. Calif. Bekerley, CA, 1984.

[11] G. Ferreres and I. Dardenne. LP synthesis of a lateral flight control system for a highly flexible aircraft. *in preparation.*

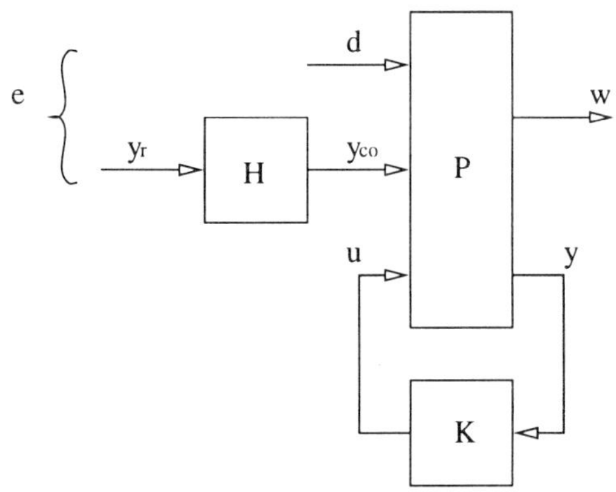

Figure 1: Augmented plant.

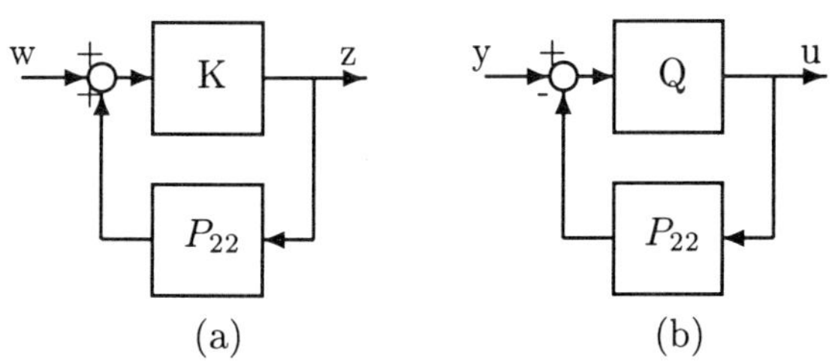

Figure 2: Realization of $z = Qw$ (a) and $u = Ky$ (b).

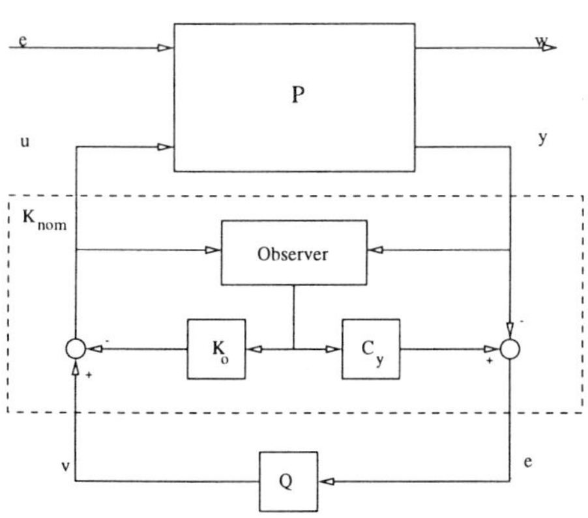

Figure 3: Q-parametrization for estimated state feedback nominal controller. K_o is a state feedback gain and $y = C_y x$ where x is the plant state.

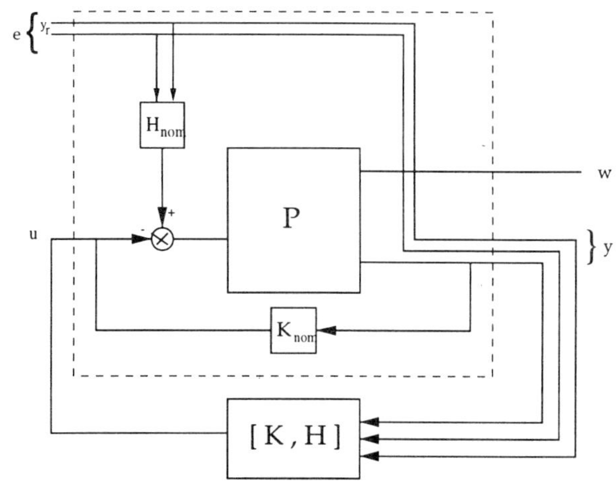

Figure 4: Augmented Plant with initial controller K_{nom} and feedforward H_{nom}

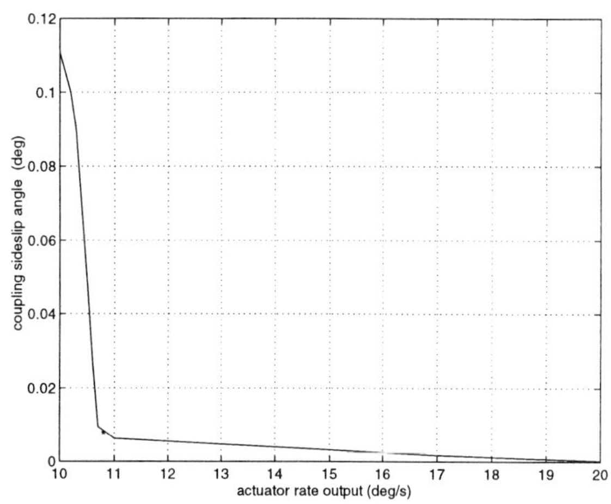

Figure 5: Coupling ϵ on the sideslip angle as a function of the magnitude of the actuator rate outputs.

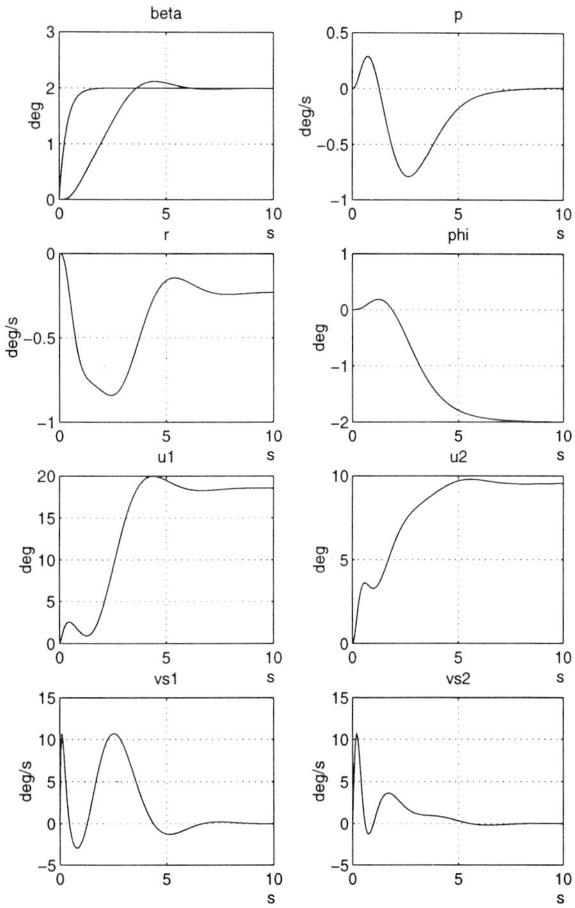

Figure 6: Step response on the sideslip angle.

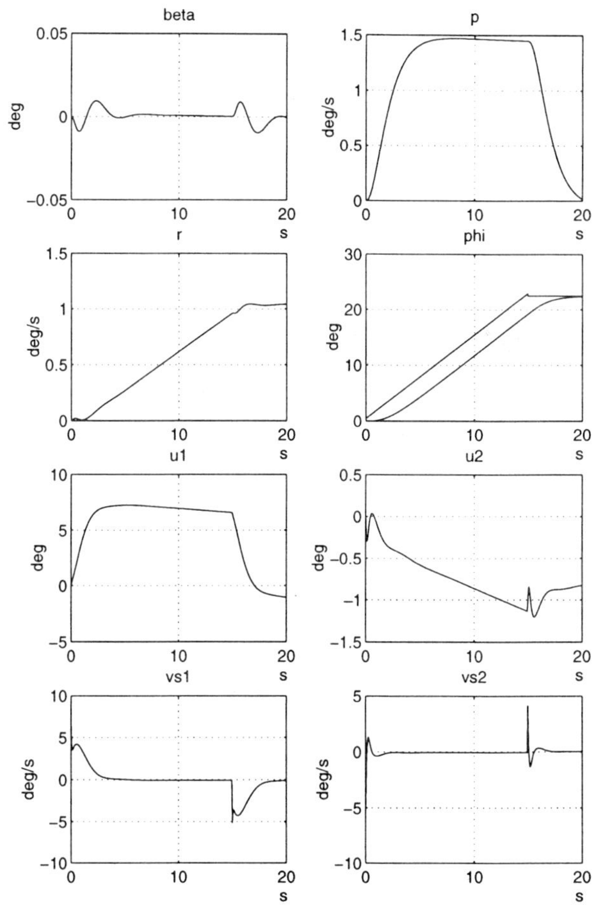

Figure 7: Step response on the roll rate output.

AIAA-97-3471

A NEW APPROACH TO INTERCEPTIBILITY ANALYSIS OF 3-DIMENSIONAL PPNG LAW

Jae-Hyuk Oh*
Department of Aeronautics and Astronautics,
Massachusetts Institute of Technology,
Cambridge, MA 02139, USA

In-Joong Ha†
Automatic Control Research Center,
School of Electrical Engineering,
Seoul National Univ., Seoul 151-742, Korea

Abstract – A lot of effort has been recently devoted to the interceptibility analysis of the pure proportional navigation guidance (PPNG). Nonetheless, there are still significant discrepancies between those analysis results and the realistic interceptibility of the PPNG law. In this paper, a vector analysis approach based on the so-called 'LOS plane' is presented to investigate the interceptibility of the 3-dimensional PPNG law against targets maneuvering with time-varying normal accelerations. The analysis result is not only much less restrictive than those presented in the prior works but also sufficient to explain well the high interceptibility of the PPNG law in practice. Our approach is expected to enable us to take fully into account more general pursuit situations.

NOMENCLATURE

(X_I, Y_I, Z_I)	Reference coordinate system
(X_L, Y_L, Z_L)	LOS coordinate system
(X_M, Y_M, Z_M)	Missile body coordinate system
(X_T, Y_T, Z_T)	Target body coordinate system
$\mathbf{i}_*, \mathbf{j}_*, \mathbf{k}_*$	Unit vectors corresponding to X_*, Y_*, Z_* axes, respectively ($* = \mathbf{L}, \mathbf{M}, \mathbf{T}$)
$\mid \mathbf{V} \mid$	Euclidean norm of vector $\mathbf{V}$
$\mathbf{V_M}$ ($\mathbf{V_T}$)	Missile (target) velocity vector
$\mathbf{A_M}$ ($\mathbf{A_T}$)	Missile (target) acceleration vector
A_{ym} (A_{zm})	Yaw(pitch) acceleration of missile
A_{yt} (A_{zt})	Yaw(pitch) acceleration of target
v_m (v_t)	$\mid \mathbf{V_M} \mid$ ($\mid \mathbf{V_T} \mid$)
θ_L, ψ_L	Euler angles from inertial frame to LOS frame
θ_m, ψ_m	Euler angles from LOS frame to missile body frame
θ_t, ψ_t	Euler angles from LOS frame to target body frame
N	Navigation constant
r	Misdistance between missile and target
ρ	v_t/v_m
$\mathbf{\Omega_L}$	Angular velocity vector of LOS
$\mathbf{\Omega_M}$	Angular velocity vector of missile relative to LOS
$\mathbf{\Omega_T}$	Angular velocity vector of target relative to LOS
$\dot{\lambda}_x, \dot{\lambda}_y, \dot{\lambda}_z$	X_L-, Y_L-, Z_L- components, respectively, of LOS angular velocity vector
$s\theta_*$, $c\theta_*$	$\sin\theta_i$, $\cos\theta_i$ ($* = m, t$)
$s\psi_*$, $c\psi_*$	$\sin\psi_i$, $\cos\psi_i$ ($* = m, t$)

*Postdoctoral Associate, email: hyuk@mit.edu
†Professor

1 INTRODUCTION

The class of PNG mainly consists of PPNG and True PNG (TPNG): the acceleration command of PPNG law is normal to the missile velocity vector, whereas that of TPNG law is normal to missile-to-target line-of-sight (LOS) vector. Although TPNG is analytically more tractable and hence has generated a lot of papers [1-4], the application of TPNG is confined within exo-atmospheric environment because the command of TPNG law includes axial acceleration or deceleration. On the other hand, PPNG has been widely recognized as one of the most efficient and practical guidance laws. A lot of effort has been recently devoted to the interceptibility analysis of the PPNG. Nonetheless, there are still significant discrepancies between those analysis results and the realistic interceptibility of the PPNG law.

Considering a nonmaneuvering target, Becker[5] derived a closed-form solution of the dynamic equations of the 2-dimensional PPNG pursuit situations and suggested a condition for an intercept to occur on the basis of the solution. Using the phase plane method, Guelman[6] first analyzed the interceptibility of the PPNG law against a target maneuvering with constant normal acceleration. Recently, it was shown in the reference[7] that Guelman's analysis result with slight extensions holds also in case of a target with time-varying normal acceleration. The approach used in these two works is so complicated that they could consider only the 2-dimensional pursuit situations in which the speed of the missile is at least squared root 2 times as fast as that of the target. And the conditions for the target interception, presented in these two works, still need many redundant restrictions on the initial states and the navigation constant. On the other hand, Ha, Hur, et al.[8] used a Lyapunov-like approach to analyze the interceptibility of PPNG law against the target maneuvering with time-varying normal acceleration. The

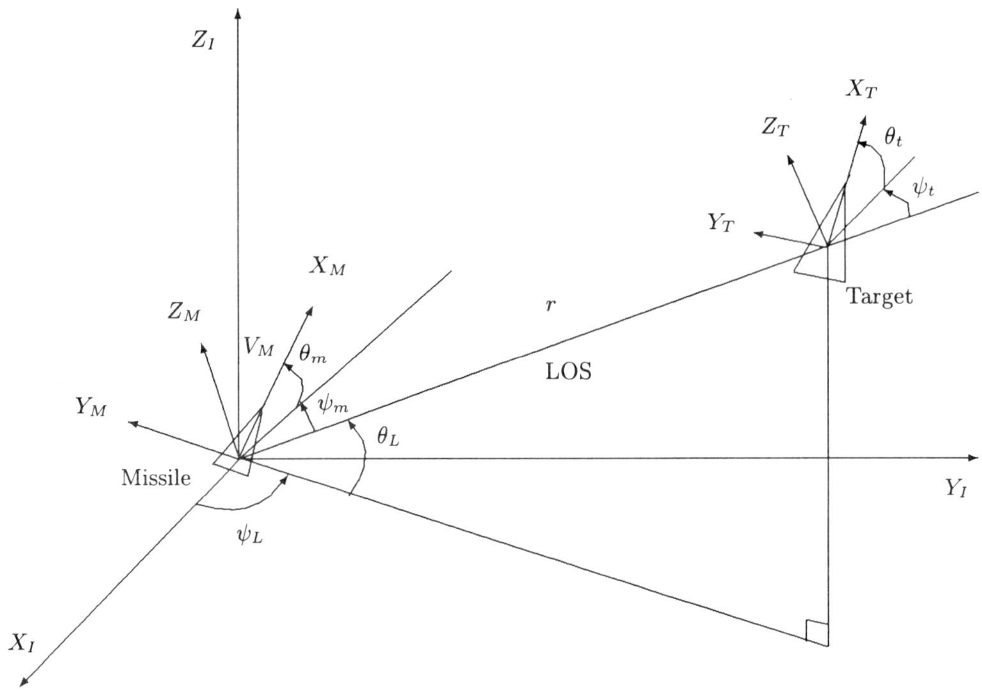

Figure 1: Pursuit geometry in 3-dimensional space

approach used in this work is mathematically more general than those presented in other works and, hence, it can be applied to the general 3-dimensional pursuit situations. Actually, Song and Ha[9] analyzed recently the performance of the 3-dimensional PPNG law using the same approach. But the conditions presented in these two works on the PPNG law were practically stringent.

This paper presents a new condition under which a missile guided by the 3-dimensional PPNG law can always intercept a target maneuvering with time-varying normal acceleration. Additionally, we provide some properties of the PPNG law, which are very useful in the analysis and design of PPNG-like missile guidance laws. In fact, the presented condition is nothing but a slight restriction that the navigation constant should be larger than 1 in case that the speed of the missile is at least squared root 2 times as fast as that of the target. No other restriction is necessary. Our result can explain very well the high interceptibility of the PPNG law in practice. Our present result is based on a vector analysis approach using a so-called 'LOS plane' which is a 2-dimensional subspace defined on the LOS frame. This approach is so powerful that we can not only handle the nonlinear 3-dimensional pursuit dynamics with ease but also expect to present various results on the performance analysis of the 3-dimensional PPNG Law in the near future.

2 Preliminaries

We first describe mathematically the 3-dimensional pursuit situation where a missile guided by the PPNG law pursues a target maneuvering randomly with time-varying accelerations. To simplify the equations of the pursuit situation, we assume that the missile and target are point masses and that the autopilot and seeker dynamics of the missile are fast enough to be neglected. We further assume that the angle-of-attack is small enough to be neglected and that the speeds and of the missile and target are constant. The 3-dimensional pursuit situation under these assumptions is depicted in Fig.1 and can be represented by the following equations using the well-known classical principles of dynamics[10].

$$\frac{d\mathbf{L}}{dt} = \mathbf{V_T} - \mathbf{V_M} = \dot{r}\,\mathbf{i_L} + \mathbf{\Omega_L} \times \mathbf{L} \quad \text{(1a)}$$

$$\begin{aligned} \mathbf{A_T} &= a_{yt}\,\mathbf{j_T} + a_{zt}\,\mathbf{k_T} \\ &= \mathbf{\Omega_L} \times \mathbf{V_T} + \mathbf{\Omega_T} \times \mathbf{V_T} \end{aligned} \quad \text{(1b)}$$

$$\begin{aligned} \mathbf{A_M} &= a_{ym}\,\mathbf{j_M} + a_{zm}\,\mathbf{k_M} \\ &= \mathbf{\Omega_L} \times \mathbf{V_M} + \mathbf{\Omega_M} \times \mathbf{V_M}, \end{aligned} \quad \text{(1c)}$$

where

$$\begin{aligned}
\mathbf{L} &= r\mathbf{i_L},\ \mathbf{V_T} = v_t\mathbf{i_T},\ \mathbf{V_M} = v_m\mathbf{i_M} \quad (2a)\\
\mathbf{\Omega_L} &= \dot{\psi}_L\ s\theta_L\ \mathbf{i_L} - \dot{\theta}_L\ \mathbf{j_L} + \dot{\psi}_L\ c\theta_L\ \mathbf{k_L}\\
&= \dot{\lambda}_x\ \mathbf{i_L} + \dot{\lambda}_y\ \mathbf{j_L} + \dot{\lambda}_z\ \mathbf{k_L} \quad (2b)\\
\mathbf{\Omega_M} &= \dot{\psi}_m\ s\theta_m\ \mathbf{i_M} - \dot{\theta}_m\ \mathbf{j_M}\\
&\quad + \dot{\psi}_m\ c\theta_m\ \mathbf{k_M} \quad (2c)\\
\mathbf{\Omega_T} &= \dot{\psi}_t\ s\theta_t\ \mathbf{i_T} - \dot{\theta}_t\ \mathbf{j_T}\\
&\quad + \dot{\psi}_t\ c\theta_t\ \mathbf{k_T}. \quad (2d)
\end{aligned}$$

In what follows, yaw and pitch accelerations of the missile are assumed to be equal to their commands since the dynamics of the autopilot are in general much faster than those of the guidance loop. The missile acceleration generated by the 3-dimensional PPNG law are given as follows[9].

$$\mathbf{A_M} = a_{ym}\mathbf{j_M} + a_{zm}\mathbf{k_M}, \quad (3)$$

where

$$\begin{aligned}
a_{ym} &= -N\ v_m\ \dot{\lambda}_y\ s\theta_m\ s\psi_m + N\ v_m\ \dot{\lambda}_z\ c\theta_m\\
a_{zm} &= -N\ v_m\ \dot{\lambda}_y\ c\psi_m. \quad (4)
\end{aligned}$$

Note that the above 3-dimensaional PPNG law can be directly reduced to the following planar PPNG law if we consider just only pitch plane, that is, we let $\psi_m = 0,\ \dot{\lambda}_z = 0$.

$$a_{ym} = 0, \quad a_{zm} = -N\ v_m\ \dot{\lambda}_y. \quad (5)$$

Now, by the definition of coordinate systems, it follows that

$$\begin{aligned}
\mathbf{i_M} &= c\theta_m c\psi_m \mathbf{i_L} + c\theta_m s\psi_m \mathbf{j_L} + s\theta_m \mathbf{k_L} \quad (6a)\\
\mathbf{j_M} &= -s\psi_m \mathbf{i_L} + c\psi_m \mathbf{j_L} \quad (6b)\\
\mathbf{k_M} &= -s\theta_m c\psi_m \mathbf{i_L} - s\theta_m s\psi_m \mathbf{j_L} + c\theta_m \mathbf{k_L}. \quad (6c)
\end{aligned}$$

Furthermore, the above relationships still hold with index letters T and t substituting for M and m respectively.

From now on, we call the 2-dimensional subspace (Y_L, Z_L) of the LOS coordinate system (X_L, Y_L, Z_L) as 'LOS plane'. Note that an arbitrary 3-dimensional vector $\mathbf{A}$ can be divided into two components; one belongs to the LOS plane and the other is perpendicular to the LOS plane. Then, we denote by $\mathbf{A_{LP}}$ its component which belongs to the LOS plane. Now, we define 2-dimensional vectors $\mathbf{M}, \mathbf{T}, \mathbf{D}$ by

$$\begin{aligned}
\mathbf{M} &\triangleq \frac{\mathbf{V_{M_{LP}}}}{v_m},\\
\mathbf{T} &\triangleq \frac{\mathbf{V_{T_{LP}}}}{v_m},\\
\mathbf{D} &\triangleq \frac{\mathbf{V_{M_{LP}}} - \mathbf{V_{T_{LP}}}}{v_m}, \quad (7)
\end{aligned}$$

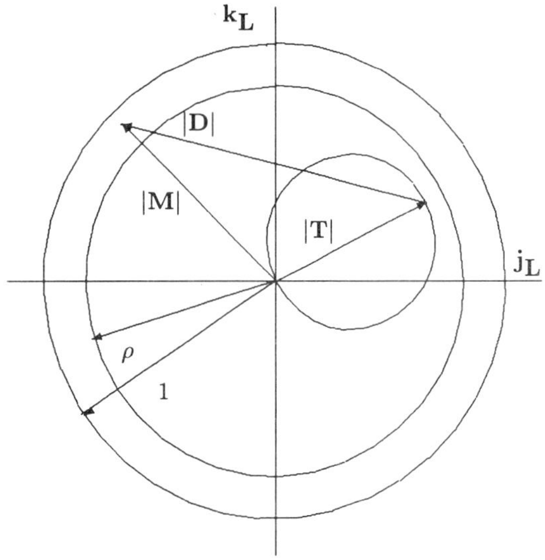

Figure 2: $\mathbf{M}, \mathbf{T}, \mathbf{D}$ on the LOS Plane

respectively. Obviously, the vectors $\mathbf{M}, \mathbf{T}$, and $\mathbf{D}$ belong to LOS plane as depicted in Fig.2. Note from (1a), (2a), (6), and (7) that the following properties hold.

$$\begin{aligned}
\mathbf{V_M} &= v_m\ (c\theta_m c\psi_m \mathbf{i_L} + \mathbf{M}),\\
\mathbf{V_T} &= v_m\ (\rho c\theta_t c\psi_t \mathbf{i_L} + \mathbf{T}),\\
\mathbf{V_T} - \mathbf{V_M} &= \dot{r}\mathbf{i_L} - v_m\mathbf{D} \quad (8)
\end{aligned}$$

On the other hand, by (1a), (2a), (6), (7), and (8), these vectors can be represented by the following equations.

$$\begin{aligned}
\mathbf{M} &= c\theta_m s\psi_m \mathbf{j_L} + s\theta_m \mathbf{k_L},\\
\mathbf{T} &= \rho c\theta_t s\psi_t \mathbf{j_L} + \rho s\theta_t \mathbf{k_L},\\
\mathbf{D} &= -\frac{r\dot{\lambda}_z}{v_m}\mathbf{j_L} + \frac{r\dot{\lambda}_y}{v_m}\mathbf{k_L}. \quad (9)
\end{aligned}$$

From (7) and (9), the following equations hold;

$$\begin{aligned}
|\mathbf{M}|^2 &= 1 - |c\theta_m c\psi_m|^2,\\
|\mathbf{T}|^2 &= \rho^2 - \rho^2 |c\theta_t c\psi_t|^2. \quad (10)
\end{aligned}$$

As will be shown in the next section, the introduction of the concept 'LOS Plane' and the vectors $\mathbf{M}, \mathbf{T}$, and $\mathbf{D}$ facilitates greatly our analyses. Moreover, we also expect that our approach will be very helpful to the interceptibility analysis with more general pursuit situations including, for an instance, the case of $v_m < \sqrt{2}v_t$.

3 MAIN RESULTS

In this section, we present the interceptibility analysis of the 3-dimensional PPNG law. Before going further, we assume that the following properties hold throughout in this paper;

$$c\theta_m(0)c\psi_m(0) > 0, \tag{11}$$
$$v_m > \sqrt{2}v_t \geq 0 \quad \text{i.e.} \quad 0 \leq \rho < \frac{1}{\sqrt{2}}. \tag{12}$$

The property in (11) implies that the initial heading of the missile is toward the target. Thus, it always holds in practice.

The following Theorem 1 states that a missile guided by the 3-dimensional PPNG law with any navigation constant $N > 1$ always intercepts a randomly maneuvering target.

Theorem 1 *Suppose that the navigation constant N is chosen so that*

$$N > 1 \tag{13}$$

Then, the 3-dimensional PPNG law in (4) guarantees that,

$$\begin{aligned} & c\theta_m(t)c\psi_m(t) \\ & \geq min\{c\theta_m(0)c\psi_m(0), \sqrt{1-\rho^2}\} \\ & > 0, \qquad \forall t \geq 0. \end{aligned} \tag{14}$$

and that there exists a finite capture time t_f such that

$$r(t_f) = 0. \tag{15}$$

□

This Theorem 1 shows that the condition (13) guarantees that the missile will always head toward the target during flight and finally intercept the target within a finite time. Here, it should be noted that the intercept will occur regardless of the normal acceleration profile of the target. This theorem can be regarded as the final solution to the PPNG interceptibility analysis problem in case of (12), which was dealt with in the references[6, 7]. This is because (*i*) the restrictions on initial heading errors and initial misdistance, which were indispensable in the prior works, are no longer necessary in our theorem, (*ii*) besides, our theorem does not depend on the target acceleration any more, and (*iii*) our condition on the navigation constant, (13), is not only very simple but also smaller than any others provided so far. Actually, (13) is the sufficient and necessary condition for the target interception to occur regardless of the target-maneuvering profile and the initial state values of the considered pursuit situation. Moreover, it should be noted that our analysis takes the whole nonlinear dynamics of 3-dimensional pursuit situation into account.

To prove Theorem 1 step by step, we provide the following two lemmas. Note that we don't need to assume (12) in stating all the following lemmas. This fact means that the following lemmas are general enough to be used in the analysis and design of other PPNG-like missile guidance laws.

Lemma 1 *Suppose that (13) holds. Then, the 3-dimensional PPNG law in (4) guarantees (14).*
Proof) *The following identities can be derived easily from (1a), (2a), (3), (4), (6a), (7)-(8), and (9);*

$$\mathbf{V_M} \cdot \mathbf{L} = r v_m c\theta_m c\psi_m \tag{16}$$
$$\begin{aligned} \mathbf{A_M} \cdot \mathbf{L} &= -r(s\psi_m a_{ym} + s\theta_m c\psi_m a_{zm}) \\ &= N v_m r(s\theta_m \dot{\lambda}_y - s\psi_m c\theta_m \dot{\lambda}_z) \\ &= N v_m^2 \mathbf{M} \cdot \mathbf{D} \end{aligned} \tag{17}$$

$$\begin{aligned} & \mathbf{V_M} \cdot (\mathbf{V_T} - \mathbf{V_M}) \\ & = \mathbf{V_M} \cdot (\dot{r}\mathbf{i_L} - v_m \mathbf{D}) \\ & = v_m(\dot{r} c\theta_m c\psi_m - v_m \mathbf{M} \cdot \mathbf{D}). \end{aligned} \tag{18}$$

By these equations with (7) and (13),

$$\begin{aligned} & \frac{d}{dt} c\theta_m c\psi_m \\ & = \frac{d}{dt} \frac{\mathbf{V_M} \cdot \mathbf{L}}{v_m r} \\ & = \frac{r(\mathbf{A_M} \cdot \mathbf{L} + \mathbf{V_M} \cdot (\mathbf{V_T} - \mathbf{V_M}))}{v_m r^2} \\ & \quad - \frac{\dot{r}\mathbf{V_M} \cdot \mathbf{L}}{v_m r^2} \\ & = \frac{v_m}{r}(N-1)\mathbf{M} \cdot \mathbf{D} \\ & \geq \frac{v_m}{r}(N-1)|\mathbf{M}|(|\mathbf{M}| - |\mathbf{T}|) \\ & > 0, \qquad \textit{whenever} \ |\mathbf{M}| > |\mathbf{T}|. \end{aligned} \tag{19}$$

Moreover, since $|\mathbf{T}| \leq \rho$, we have

$$\begin{aligned} \frac{d}{dt} c\theta_m c\psi_m &\geq \frac{v_m}{r}(N-1)|\mathbf{M}|(|\mathbf{M}| - \rho) \\ &> 0, \quad \textit{whenever} \ |\mathbf{M}| > \rho. \end{aligned} \tag{20}$$

By this along with (11), we see that (13) guarantees (14). □

This Lemma 1 which plays a very important role in this paper shows that the only thing needed to keep the missile heading toward the target is simply just taking the navigation constant larger than 1. This is a very important property because the missile guided by PNG should always head toward the target during flight.

Since Lemma 1 assures (14), the rest of our proof is to prove the existence of a finite capture time t_f satisfying (15). First, the following lemma is very useful in calculating maximal misdistances.

Lemma 2 *In addition to (13), suppose that the following inequality holds;*

$$|\mathbf{M}(t)| \geq |\mathbf{T}(t)|, \qquad \forall t \in [t_0, t_1]. \tag{21}$$

Then, the 3-dimensional PPNG law in (4) guarantees that

$$\max_{t\in[t_0,t_1]} r(t) \leq r(t_0)\exp[\frac{\rho}{(N-1)(1-\rho)} \times (\min(c\theta_m(t_1)c\psi_m(t_1), \rho) - c\theta_m(t_0)c\psi_m(t_0))] \tag{22}$$

Proof) *By (1a) and (6), we can write*

$$\dot{r} = v_m(\rho c\theta_t c\psi_t - c\theta_m c\psi_m) \tag{23}$$

This along with Lemma 1 implies that

$$\frac{\dot{r}}{v_m \rho} \leq 1. \tag{24}$$

Define the set S by

$$S \triangleq \{ t \in [t_0, t_1] \mid |\mathbf{M}(t)|^2 - |\mathbf{T}(t)|^2 \geq 1 - \rho^2, \; c\theta_t(t)c\psi_t(t) \geq 0\}. \tag{25}$$

Then, since $|\mathbf{M}(t)|$ and $|\mathbf{T}(t)|$ are continuous functions of time, there exist at most a countably infinite number of intervals I_i $_{,i=1,2,\cdots}$ such that

$$I_i = [b_i, e_i], \quad b_i \leq e_i < b_{i+1} \tag{26}$$

$$I_i \cap I_j = \emptyset \quad whenever \quad i \neq j \tag{27}$$

$$\cup_{i=1}^{\infty} I_i = S \tag{28}$$

By (10), (23), and (25), we see that

$$\begin{cases} \dot{r}(t) \geq 0, \; \forall t \in S \\ \dot{r}(t) < 0, \; \forall t \in [t_0, t_1] - S \end{cases} \tag{29}$$

and hence that

$$r(b_1) \leq r(t_0) \tag{30}$$

$$\begin{cases} r(b_i) \leq r(t) \leq r(e_i), & \forall t \in I_i \subset S \\ r(b_{i+1}) < r(t) < r(e_i), & \forall t \in (e_i, b_{i+1}) \\ & \subset [t_0, t_1] - S \end{cases} \tag{31}$$

Since $|\mathbf{M}(t)| \leq 1$ is always satisfied and, on the other hand, the definition of S in (25) implies that

$$|\mathbf{T}(t)| \leq \sqrt{|\mathbf{M}(t)|^2 - (1-\rho^2)}, \quad \forall t \in S \tag{32}$$

we can easily see that

$$\begin{aligned} &|\mathbf{M}(t)|(|\mathbf{M}(t)| - |\mathbf{T}(t)|) \\ &\geq \frac{(1-\rho^2)|\mathbf{M}(t)|}{|\mathbf{M}(t)| + |\mathbf{T}(t)|} \\ &= \frac{(1-\rho^2)}{1 + \sqrt{1 - \frac{1-\rho^2}{|\mathbf{M}(t)|^2}}} \\ &\geq 1 - \rho, \quad \forall t \in S, \end{aligned} \tag{33}$$

This along with (19), (21), and (24) implies that

$$\begin{aligned} &\frac{d}{dt} c\theta_m c\psi_m \\ &\geq \begin{cases} \frac{(N-1)v_m(1-\rho)}{r(t)}, & if\; t \in S \\ 0, & if\; t \in [t_0, t_1] - S \end{cases} \\ &\geq \begin{cases} \frac{(N-1)(1-\rho)}{\rho}\frac{\dot{r}(t)}{r(t)}, & if\; t \in S \\ 0, & if\; t \in [t_0, t_1] - S \end{cases} \end{aligned} \tag{34}$$

By this along with (30) and (31),

$$\begin{aligned} &c\theta_m(e_k)c\psi_m(e_k) - c\theta_m(t_0)c\psi_m(t_0) \\ &\geq \frac{(N-1)(1-\rho)}{\rho}(\sum_{i=1}^{k}\int_{I_i}\frac{\dot{r}(t)}{r(t)}dt) \\ &\geq \frac{(N-1)(1-\rho)}{\rho}(\sum_{i=1}^{k}\ln(\frac{r(e_i)}{r(b_i)})) \\ &\geq \frac{(N-1)(1-\rho)}{\rho}\ln(\frac{r(e_k)}{r(b_1)}) \\ &\geq \frac{(N-1)(1-\rho)}{\rho}\ln(\frac{r(e_k)}{r(t_0)}), \\ &\forall k = 1, 2, \cdots. \end{aligned} \tag{35}$$

This along with (21) implies that

$$r(e_k) \leq r(t_0) \times \tag{36}$$

$$\exp(\frac{\rho(c\theta_m(e_k)c\psi_m(e_k) - c\theta_m(t_0)c\psi_m(t_0))}{(N-1)(1-\rho)}),$$

$$\forall k = 1, 2, \cdots. \tag{37}$$

On the other hand, note from (23) that

$$\dot{r} \leq 0, \qquad whenever\; c\theta_m(t)c\psi_m(t) \geq \rho. \tag{38}$$

Now, (22) is the direct consequence of (31), (36), and (38). □

Now, we are ready to prove Theorem 1.

Proof of Theorem 1 : *By Lemma1 along with the given conditions, (14) holds. Now, choose a positive constant* η *such that*

$$\rho < \eta < \frac{1}{\sqrt{2}}. \tag{39}$$

Then (12) assures the existence of η *satisfying (39). From now, we will first show by contradiction that there exists a finite constant* $t_1 \geq 0$ *such that*

$$c\theta_m(t_1)s\psi_m(t_1) \geq \sqrt{1-\eta^2} \tag{40}$$

$$t_1 \leq t_2 \triangleq \frac{r(0)}{v_m}\exp(\frac{\rho\sqrt{1-\eta^2}}{(N-1)(1-\rho)}) \times \frac{\sqrt{1-\eta^2}}{(N-1)\eta(\eta-\rho)}. \tag{41}$$

To do so, suppose that

$$c\theta_m(t)c\psi_m(t) < \sqrt{1-\eta^2}, \quad {}^\forall t \in [0, t_2]. \tag{42}$$

Then the following inequality is the direct consequence of (10), (39), and (42);

$$|\mathbf{M}(t)|(|\mathbf{M}(t)| - |\mathbf{T}(t)|) > \eta(\eta-\rho) > 0, \quad {}^\forall t \in [0, t_2]. \tag{43}$$

By this along with Lemma 2,

$$\max_{t\in[0,t_2]} r(t) \leq r(0)\exp(\frac{\rho\sqrt{1-\eta^2}}{(N-1)(1-\rho)}). \tag{44}$$

This along with (19), (24), and (43) implies that

$$\frac{d}{dt}c\theta_m c\psi_m|_t > \frac{(N-1)v_m\eta(\eta-\rho)}{r(0)} \times \exp(-\frac{\rho\sqrt{1-\eta^2}}{(N-1)(1-\rho)}), \quad {}^\forall t \in [0, t_2]. \tag{45}$$

By this along with (11),

$$\begin{aligned} & c\theta_m(t_2)s\psi_m(t_2) \\ & > c\theta_m(0)c\psi_m(0) + \frac{(N-1)v_m\eta(\eta-\rho)}{r(0)} \times \exp(-\frac{\rho\sqrt{1-\eta^2}}{(N-1)(1-\rho)})\, t_2 \\ & = c\theta_m(0)c\psi_m(0) + \sqrt{1-\eta^2} \\ & > \sqrt{1-\eta^2}. \end{aligned} \tag{46}$$

This contradicts (42). Thus, there exists a finite constant $t_1 \geq 0$ *satisfying (40) and (41). Besides, by Lemma 1 and 2 along with (40) and (41), we have*

$$c\theta_m(t)c\psi_m(t) \geq \sqrt{1-\eta^2}, \quad {}^\forall t \geq t_1, \tag{47}$$

$$r(t_1) \leq r(0)\exp(\frac{\rho\sqrt{1-\eta^2}}{(N-1)(1-\rho)}). \tag{48}$$

Now, by (23) and (47), we see that

$$\dot{r}(t) \leq v_m(\rho - \sqrt{1-\eta^2}), \quad {}^\forall t \geq t_1. \tag{49}$$

This with (39), (47), and (12) implies

$$\dot{r}(t) < -v_m(\frac{1}{\sqrt{2}} - \rho) < 0, \quad {}^\forall t \geq t_1. \tag{50}$$

Thus, by this with (48), it is immediate that there exists a finite capture time t_f *satisfying (15). In detail, by (41) and (49), we have*

$$t_f \leq \frac{r(0)}{v_m}\left(\frac{\sqrt{1-\eta^2}}{(N-1)\eta(\eta-\rho)} + \frac{1}{\sqrt{1-\eta^2}-\rho}\right) \times \exp(\frac{\rho\sqrt{1-\eta^2}}{(N-1)(1-\rho)}). \tag{51}$$

□

Here, it should be noted that (51) can be much more alleviated by considering separately the initial conditions, that is, if we assume that

$$c\theta_m(0)c\psi_m(0) \geq \frac{1}{\sqrt{2}}, \tag{52}$$

then, by (23) along with Lemma 1, we have

$$\dot{r}(t) < 0, \quad {}^\forall t \geq 0, \tag{53}$$

$$t_f \leq \frac{r(0)}{v_m(1/\sqrt{2}-\rho)}. \tag{54}$$

Finally, from Lemma 1, we can easily derive the following corollary.

Corollary 1 *If* $N = 1$ *and* $|c\theta_m(0)c\psi_m(0)| > \rho$, *then the 3-dimensional PPNG law in (4) guarantees the existence of a finite capture time satisfying (15).*
Proof) *By the third inequality in (19) with* $N = 1$, *we have*

$$c\theta_m(t)c\psi_m(t) = c\theta_m(0)c\psi_m(0), \quad {}^\forall t \geq 0. \tag{55}$$

Then, by this with (23) and $|c\theta_m(0)c\psi_m(0)| > \rho$, *it is obvious that*

$$\dot{r} < -v_m(c\theta_m(0)c\psi_m(0) - \rho) < 0. \tag{56}$$

The remaining part of this proof is now immediate. □

4 Conclusion

In this paper, we have proven that a missile guided by the 3-dimensional PPNG law can always intercept a target maneuvering with time-varying normal acceleration if the navigation constant is chosen to be larger than 1. This analysis result can explain very well the high interceptibility of the PPNG law in practice. In our study, we have considered fully the nonlinear dynamics of the 3-dimensional pursuit situations in which the speed of the missile is at least squared root 2 times as fast as that of the target.

Our present result is based on a vector analysis approach using a so-called 'LOS plane'. This approach is so powerful that we can handle the nonlinear 3-dimensional pursuit dynamics with ease by this approach. Moreover, we also expect that we can present various results on the performance analysis of the 3-dimensional PPNG Law in the near future. Actually, some properties of the PPNG law, which are very useful in the analysis and design of PPNG-like missile guidance laws, are also presented in this paper.

References

[1] M. Guelman,"The Closed-Form Solution of True Proportional Navigation," *IEEE Transactions on Aerospace and Electronic Systems*, AES-12, pp.472–482, July 1976.

[2] P.J.Yuan and J.S.Chern, "Solutions of True Proportional Navigation for Maneuvering and Non-Maneuvering Targets," *Journal of Guidance, Control, and Dynamics*, Vol.15, pp268–271, Jan.-Feb. 1992.

[3] D..Ghose, "True Proportional Navigation With Maneuvering Target," *IEEE Transactions on Aerospace and Electronic Systems*, AES-30, pp229–237, Jan. 1994.

[4] C.-D.Yang and C.-C.Yang, "Analytical Solution of Three-Dimensional Realistic True Proportional Navigation," *Journal of Guidance, Control, and Dynamics*, Vol.19, pp569–577, May-June 1996.

[5] Klaus Becker,"Closed-Form Solution of Pure Proportional Navigation," *IEEE Transactions on Aerospace and Electronic Systems*, AES-26, pp.526–533, May 1990.

[6] M. Guelman,"Proportional Navigation with a Maneuvering Target," *IEEE Transactions on Aerospace and Electronic Systems*, AES-8, pp.364–371, May 1972.

[7] S. N. Ghawghawe and D. Ghose, "Pure Proportional Navigation Against Time-Varying Target Maneuvers," *IEEE Transactions on Aerospace and Electronic Systems*, AES-32, pp.1336–1347, Oct. 1996.

[8] I. J. Ha, J. S. Hur, M. S. Ko, and T. L. Song,"Performance Analysis of PNG Laws for Randomly Maneuvering Targets," *IEEE Transactions on Aerospace and Electronic Systems*, AES-26, pp.713–721, Sep. 1990.

[9] S. H. Song and I. J. Ha, "A Lyapunov - Like Approach to Performance Analysis of 3-Dimensional Pure PNG Laws," *IEEE Transactions on Aerospace and Electronic Systems*, AES-30, pp.349–358, Jan. 1994.

[10] D. T. Greenwood, *Principles of Dynamics*, Prentice-Hall.

[11] M. Guelman,"A Qualitive Study of Proportional Navigation," *IEEE Transactions on Aerospace and Electronic Systems*, AES-7, pp.637–643, July 1971.

AIAA-97-3472

ANALYTICAL SOLUTION OF MISSILE TERMINAL GUIDANCE

Tetsuya Takehira*, Nguyen X. Vinh† and Pierre T. Kabamba‡

The University of Michigan
Ann Arbor, MI 48109-2118, U.S.A.

Abstract

A new guidance law, which combines pursuit guidance and proportional navigation is proposed. This guidance law depends on two parameters that determine the relative importance of pursuit guidance and proportional navigation. Numerical simulations of the nonlinear equations of motion suggest that the parameters of this law can be chosen to reduce the peak value of the missile acceleration. When the engagement ends in a tail chase, and linearization is valid, the linearized equations of motion lead to a confluent hypergeometric equation. This equation is solved in closed form, in the general case where the target performs maneuvers such that its heading angle is a polynomial function of time. The analytic solution based on linearization and the numerical simulation of the nonlinear equations show good agreement.

Nomenclature

a, b	= parameters in normalized guidance law
C_i	= constants of integration
f	= forcing function
k_i	= coefficients of forcing function
m	= degree of polynominal
n	= velocity ratio
R	= relative distance
r	= normalized relative distance
t	= flight time
V_M	= missile velocity
V_T	= target velocity
z	= modified relative distance
α_i	= coefficients in polynominal of β_p
β	= line-of-sight(LOS) angle
β_p	= particular solution for β
Γ	= normal acceleration
θ	= missile flight-path angle
θ_T	= target flight-path angle
λ, μ	= parameters of guidance law
τ	= normalized flight time
$M(\cdot,\cdot,\cdot)$	= confluent hypergeometric function, or Kummer's function
$U(\cdot,\cdot,\cdot)$	= confluent hypergeometric function of the second kind
$(\cdot)_0$	= initial value

*Doctoral Candidate, Department of Aerospace Engineering, Student Member AIAA

†Professor, Department of Aerospace Engineering, Member AIAA

‡Professor, Department of Aerospace Engineering

1 Introduction

This paper considers the standard problem of guiding a craft or a missile on a collision course to a maneuvering target. Among the known guidance laws, the simplest one is pursuit guidance which constantly directs the velocity of the missile towards the moving target. This law only requires knowledge of the missile-target line-of-sight. Another practical guidance law is proportional navigation, in which the missile detects any rotation of the line-of-sight and applies a turning rate proportional to this rotation. The relative advantages and disadvantages of these guidance laws are well known and presented in many texts[1]-[8]. Various aspects of proportional navigation are topics of discussion in the current literature[9]-[16].

Guelman[9] obtained a closed-form solution of true proportional navigation for a nonmaneuvering target. Shukla et al.[10] derived a general form of linearized solution to proportional navigation. Various solutions for pure proportional navigation were given by Mahapatra et al.[11] and Becker[12]. Cochran et al.[13] obtained new closed-form solution in terms of elliptic functions and integrals. Yuan et al.[14]-[15]

also presented closed-from solutions of true proportional navigation for both maneuvering and nonmaneuvering targets and solution of generalized proportional navigation. Balakrishnan [16] introduced a class of proportional navigation laws through an approximation of time-to-go and a transformation of state variables.

In this paper we propose a guidance law which is a combination of pursuit and proportional navigation, with the aim of retaining the effectiveness of pure pursuit and at the same time reducing the lateral acceleration, which is a characteristic feature of proportional navigation. After formulation of the problem, the effectiveness of the control law is tested numerically for various maneuvers of the target. It is shown that by adjusting two parameters which control the turning rate of the missile, it is possible to reduce the peak acceleration to an acceptable level. Furthermore, in the final and most important phase of the interception, linearization of the dynamics leads to a confluent hypergeometric equation which is solvable analytically when the heading angle of the target is a polynominal function of a dimensionless independent variable.

2 Equations of Motions

For the case of planar pursuit-evasion, we have the equations

$$\frac{dR}{dt} = V_T \cos(\beta - \theta_T) - V_M \cos(\beta - \theta), \quad (1)$$

$$\frac{d\beta}{dt} = \frac{1}{R}\left[-V_T \sin(\beta - \theta_T) + V_M \sin(\beta - \theta)\right], \quad (2)$$

where R and β are the relative distance between target and missile positions and line-of-sight angle, and θ and θ_T are missile and target flight path angle, respectively. We use the flight time t as the independent variable. The geometry of the interception is shown in Fig.1.

To control the missile towards the intercept point, we use the following new guidance law

$$\frac{d\theta}{dt} = \lambda \frac{d\beta}{dt} + \mu \sin(\beta - \theta), \quad (3)$$

where λ and μ are constant parameters. The first term on the right hand side in this guidance equation represents proportional navigation, and the second term represents pursuit guidance. Therefore this guidance law is constructed by combining the two fundamental guidance laws through the use of the two parameters λ and μ.

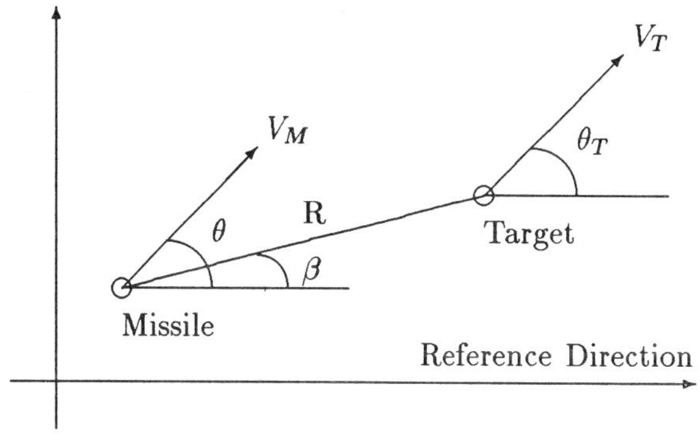

Figure 1: Geometry of Interception.

To normalize the equations, the following non-dimensional variables are defined for the case of constant speed of the target. We have

$$\begin{aligned} \tau &= \frac{V_T}{R_0} t, \\ r &= \frac{R}{R_0}, \\ n &= \frac{V_M}{V_T}, \\ a &= \lambda, \\ b &= \frac{R_0}{V_T} \mu, \end{aligned} \quad (4)$$

where R_0 is the initial relative distance, τ is the normalized time, r is the normalized relative distance, a and b are parameters of the normalized guidance law and n is the velocity ratio.

Using these non-dimensional variables, the normalized equations are

$$\frac{dr}{d\tau} = \cos(\beta - \theta_T) - n \cos(\beta - \theta), \quad (5)$$

$$\frac{d\beta}{d\tau} = \frac{1}{r}\left[-\sin(\beta - \theta_T) + n \sin(\beta - \theta)\right], \quad (6)$$

$$\frac{d\theta}{d\tau} = a \frac{d\beta}{d\tau} + b \sin(\beta - \theta). \quad (7)$$

3 Effectiveness of Guidance Law

Although we have restricted the analysis to the case of constant speed for the target, the nonlinear equations (5)~(7) allow its maneuverability through the arbitrary change in the heading $\theta_T(\tau)$ as a function of the dimensionless time, and the possibility of modulating the speed ratio $n(\tau)$ for the missile. Of course, the turning rate of the missile is dictated by the guidance law (7), now written in dimensionless form with two arbitrary constants a and b. For this paper, we use a constant value for n.

To show the effectiveness of the guidance law, and the influence of the parameters a and b on its performance, we consider the complete non-linear system and generate the intercept trajectory numerically.

Without loss of generality, the reference direction is taken along the initial target velocity. Hence, we have at the initial time

$$\theta_T(0) = 0, \quad \beta(0) = \beta_0, \quad r(0) = 1, \quad \theta(0) = \theta_0. \tag{8}$$

In the first two examples, we use a constant speed ratio $n = 3.0$, and $\beta_0 = 1/20$, while varying θ_0 from 0 to $\pi/2$. In the last example, we use $n = 1.701$

a / In the simplest case, we keep constant the heading of the target. Hence

$$\theta_T = 0. \tag{9}$$

Figure 2 shows the intercept trajectories for various initial headings of the missile, using $\theta_0 = 0, \pi/6, \pi/3, \pi/2$.

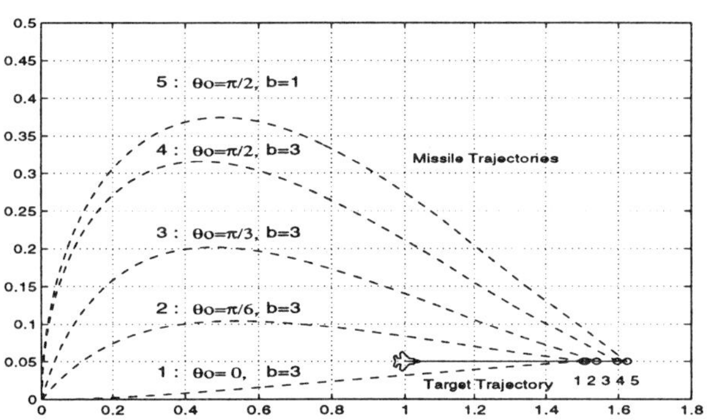

Figure 2: Intercept Trajectories for Non-Maneuvering Target.

As expected, for the same set of values $a = 3.0$, $b = 3.0$, the tail chase case $\theta_0 = 0$, is the fastest, and as θ_0 increases, it takes longer time for the interception. The interception points are denoted in order of increasing time and range in the figure. The normal acceleration Γ, in normalized form, is

$$\Gamma = n\frac{d\theta}{d\tau} = -\frac{an}{r}\sin(\beta - \theta_T) + n(\frac{an}{r} + b)\sin(\beta - \theta). \tag{10}$$

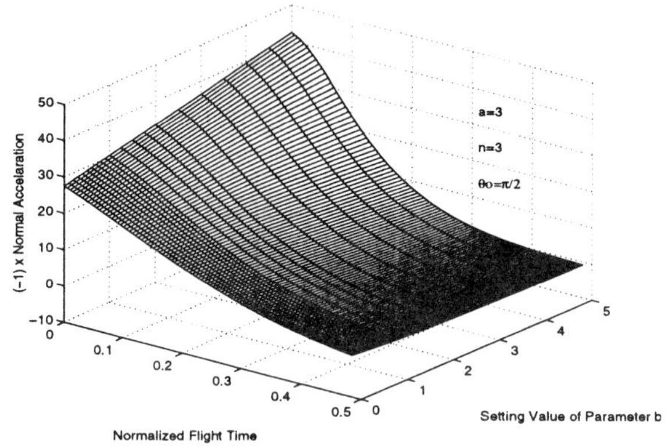

Figure 3: Time History of Normal Accelerations Γ vs. Setting Values of Parameter b.

It is highest at the initial time, and hence its peak value increases when θ_0 increases. To reduce it to an acceptable value for implementation in missiles with low turning rate, as indicated by Fig.3, we can decrease the value of parameter to $b = 1.0$, at the expense of a longer time for interception. This is achieved in Trajectory 5 in Fig.2 for the case of $\theta_0 = \pi/2$, with a lower initial turning rate than Trajectory 4.

b / In this case, we allow the target to have a constant turning rate by using

$$\theta_T = c\tau. \tag{11}$$

Figure 4 shows the missile trajectories with the initial values $\theta_0 = 0$ and $\theta_0 = \frac{\pi}{2}$

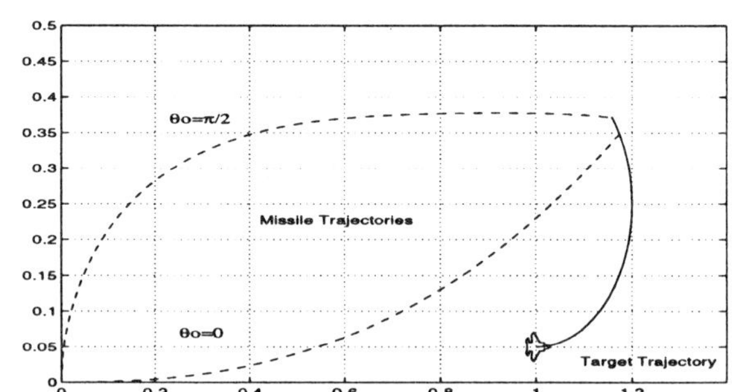

Figure 4: Circular Evasive Trajectory.

For the value of $c = 5.0$, both intercept trajectories are achieved with $a = 3.0$, $b = 3.0$, $n = 3.0$.

c / In the last example, we consider a sinusoidal change in the heading of the target in the form

$$\theta_T = c\sin\omega\tau. \tag{12}$$

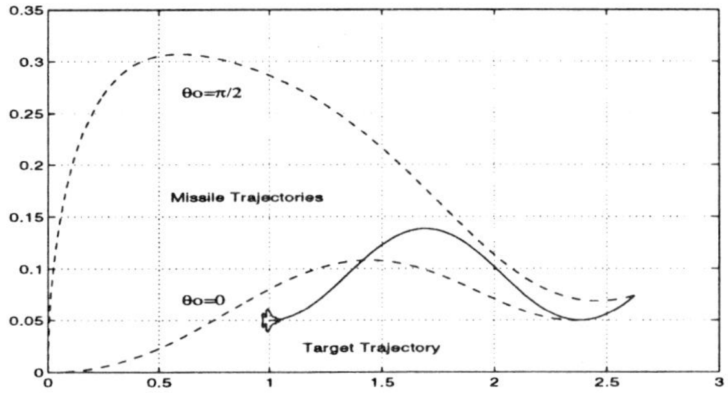

Figure 5: Sinusoidal Evasive Trajectory.

The values used for this evasive trajectory are $c = 0.2$ and $\omega = 4.5$. Both intercept trajectories are achieved with $a = 3.0$, $b = 3.0$, $n = 1.701$.

We have experimented with a variety of evasive maneuvers for the target, and it has been found that it is always possible to select the parameters a and b to have successful interception with an acceptable acceleration for the missile. Furthermore, the guidance law tends to tail chase, and hence in the terminal phase, the linearization of equations (5)~(7) is valid.

4 Terminal Guidance

The control law used in this guidance tends to align the velocity of the missile with the line-of-sight and with the velocity of the target. Hence, in the final phase of the homing, the two angles $(\beta - \theta)$ and $(\beta - \theta_T)$ are reasonably small. This leads to the linearized system for terminal guidance :

$$\frac{dr}{d\tau} = (1-n), \tag{13}$$

$$\frac{d\beta}{d\tau} = \frac{1}{r}\left[-(\beta-\theta_T)+n(\beta-\theta)\right], \tag{14}$$

$$\frac{d\theta}{d\tau} = a\frac{d\beta}{d\tau}+b(\beta-\theta). \tag{15}$$

The equation for r is decoupled and upon integration, we have the linear relation

$$r = 1-(n-1)\tau. \tag{16}$$

For the integration of the other two equations, we change the independent variable from τ to r to have the new system

$$\frac{d\beta}{dr} = -\frac{1}{r}\beta+\frac{n}{(n-1)r}\theta-\frac{1}{(n-1)r}\theta_T, \tag{17}$$

$$\frac{d\theta}{dr} = -\frac{a(n-1)+br}{(n-1)r}\beta+\frac{an+br}{(n-1)r}\theta-\frac{a}{(n-1)r}\theta_T. \tag{18}$$

For the integration of this system, it is convenient to use a new independent variable

$$z = \frac{br}{n-1}. \tag{19}$$

We notice that since r varies from 1 to *zero*, this new variable monotonically decreases from $b/(n-1)$ to *zero*. By changing to z, Equations (17) and (18) can be rewritten

$$\frac{d\beta}{dz} = -\frac{1}{z}\beta+\frac{n}{(n-1)z}\theta-\frac{1}{(n-1)z}\theta_T, \tag{20}$$

$$\frac{d\theta}{dz} = -\frac{a+z}{z}\beta+\frac{an+(n-1)z}{(n-1)z}\theta-\frac{a}{(n-1)z}\theta_T. \tag{21}$$

By eliminating θ between these two equations, we obtain a linear second order differential equation with a forcing function for the variable β

$$z\frac{d^2\beta}{dz^2}+[2-\frac{an}{n-1}-z]\frac{d\beta}{dz}+\frac{1}{(n-1)}\beta = \frac{1}{n-1}(\theta_T-\frac{d\theta_T}{dz}). \tag{22}$$

Upon obtaining the solution for β, the guidance angle θ for the missile is obtained from

$$\theta = \frac{(n-1)}{n}[\beta+z\frac{d\beta}{dz}+\frac{1}{n-1}\theta_T]. \tag{23}$$

The linear equation (22) is a confluent hypergeometric equation[17]. Including the forcing function on the right hand side, its general solution is

$$\begin{aligned}\beta(z) &= C_1 M(-\frac{1}{n-1},\frac{2(n-1)-an}{n-1},z)\\ &+ C_2 U(-\frac{1}{n-1},\frac{2(n-1)-an}{n-1},z)\\ &+ \beta_p,\end{aligned} \tag{24}$$

where $M(\cdot,\cdot,\cdot)$ and $U(\cdot,\cdot,\cdot)$ are linearly independent Kummer's solutions of the homogeneous equation, and $\beta_p(z)$ is a particular solution. Using this solution into Eq.(23), we have the solution for θ

$$\begin{aligned}\theta(z) &= C_1\frac{n-1}{n}[M(-\frac{1}{n-1},\frac{2(n-1)-an}{n-1},z)\\ &- \frac{z}{2(n-1)-an}M(1-\frac{1}{n-1},1+\frac{2(n-1)-an}{n-1},z)]\\ &+ C_2\frac{n-1}{n}[U(-\frac{1}{n-1},\frac{2(n-1)-an}{n-1},z)\\ &+ \frac{z}{n-1}U(1-\frac{1}{n-1},1+\frac{2(n-1)-an}{n-1},z)]\\ &+ \frac{n-1}{n}[\beta_p+z\frac{d}{dz}\beta_p+\frac{1}{n-1}\theta_T].\end{aligned} \tag{25}$$

The constants of integration C_1 and C_2 are determined by the initial conditions on β and θ.

For a maneuvering target, when the heading θ_T is an arbitrary function of τ, we model it as a polynominal in τ, and ultimately through the change of independent variable by Eqs.(16) and (19), we have θ_T as a polynominal in the independent variable z. In this case, the forcing function $f(z)$, which is the right hand side of Eq.(22) is also a polynominal in z and the particular function $\beta_p(z)$ in the solutions (24) and (25) for β and θ can be easily obtained by identification.

As an example, we consider the case of a third-degree polynominal

$$f(z) = k_0 + k_1 z + k_2 z^2 + k_3 z^3 = \frac{1}{n-1}(\theta_T - \frac{d\theta_T}{dz}). \tag{26}$$

The particular solution of the non-homogeneous equation (22) is assumed to be a polynominal of the same degree

$$\beta_p(z) = \alpha_0 + \alpha_1 z + \alpha_2 z^2 + \alpha_3 z^3, \tag{27}$$

where the coefficients α_i $(i = 0, 1, 2, 3)$ are to be found. Upon substituting into Eq.(22) and equating the coefficients of the same power in z, we have the solution

$$\begin{aligned} \alpha_3 &= -\frac{k_3}{A+3}, \\ \alpha_2 &= \frac{3(B+2)\alpha_3 - k_2}{A+2}, \\ \alpha_1 &= \frac{2(B+1)\alpha_2 - k_1}{A+1}, \\ \alpha_0 &= \frac{B\alpha_1 - k_0}{A}, \end{aligned} \tag{28}$$

where

$$\begin{aligned} A &= -\frac{1}{n-1}, \\ B &= \frac{2(n-1) - an}{n-1}. \end{aligned} \tag{29}$$

It is not difficult to show that if the forcing function is a polynominal of degree m, then in the solution for $\beta_p(z)$

$$\beta_p(z) = \alpha_0 + \alpha_1 z + \alpha_2 z^2 + \alpha_3 z^3 + \ldots + \alpha_m z^m, \tag{30}$$

we first calculate the coefficient α_m

$$\alpha_m = -\frac{k_m}{A+m}, \tag{31}$$

and for any other coefficient α_q $(q = 0, \ldots, m-1)$, we use the recursive formula

$$\alpha_q = \frac{(q+1)(B+q)\alpha_{q+1} - k_q}{A+q}. \tag{32}$$

As an example, we consider the case of a constant turning rate for the target as given in Eq.(11). Then, in terms of z, we have

$$\theta_T = c(\frac{1}{n-1} - \frac{z}{b}). \tag{33}$$

The solution $\beta_p(z)$ is also a linear function and the coefficients α_1 and α_0 are easily calculated from Eqs.(31) and (32). With $n = 1.701$, initial values $\theta_0 = 0$ and $\beta_0 = 1/20$ and using again $a = 3$, $b = 3$, we generate two evasive trajectories for $c = 0.06$ and $c = -0.06$. We calculate the intercept trajectories, first by integrating the non-linear equations, and then by using the analytical solution of the linearized system. The two solutions are in excellent agreement with each other, as seen in Fig.6.

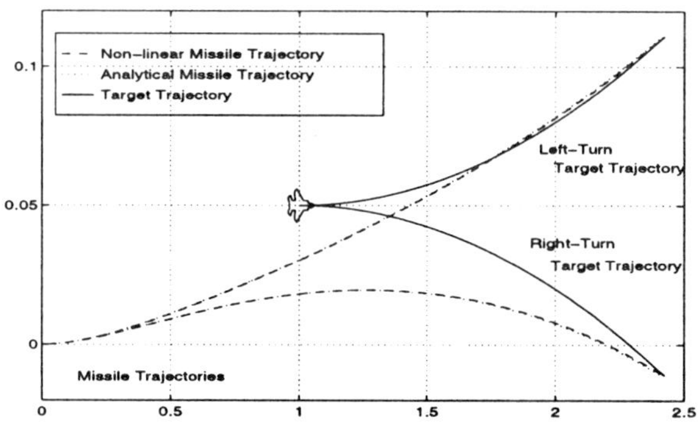

Figure 6: Circular Evasive Trajectory.

In the next example, we model the heading change of a sinusoidal evasive trajectory as a polynominal in τ.

$$\theta_T(\tau) = 0.38\tau^3 - 0.827\tau^2 + 0.4\tau, \tag{34}$$

After changing the independent variable, again, using the value $n = 1.701$, $a = 3$, $b = 3$ with the trinominal in z

$$\theta_T(z) = -0.01407z^3 - 0.08880z^2 - 0.12010z - 0.00928, \tag{35}$$

for the heading of the target, the analytical solution obtained is in perfect agreement with the numerical solution of the non-linear system. This is shown in Fig.7.

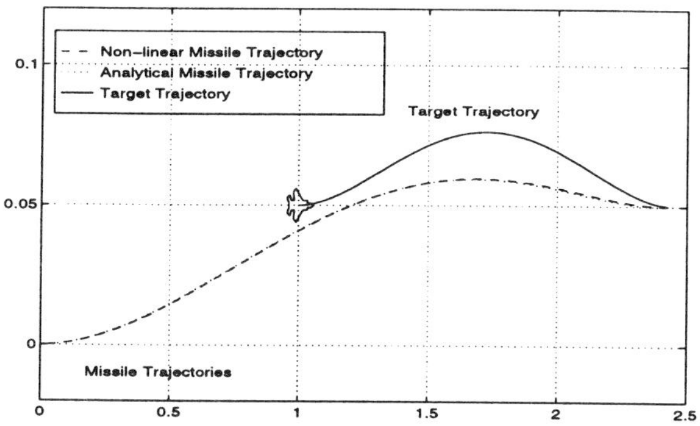

Figure 7: Sinusoidal Evasive Trajectory.

5 Conclusion

In this paper, a new guidance law, which combines pursuit guidance and proportional navigation has been proposed. This guidance law depends on two parameters that determine the relative importance of pursuit guidance and proportional navigation. Numerical simulations of the nonlinear equations of motion have been presented and suggest that the parameters of this law can be chosen to reduce the peak value of the missile acceleration. An analysis of the terminal portion of the trajectory has been presented, leading to a confluent hypergeometric equation. A closed form solution of this equation has been obtained in the general case where the target performs maneuvers such that its heading angle is a polynomial function of time. The analytic solution based on linearization and the numerical simulation of the nonlinear equations show good agreement.

References

[1] Locke, A. S., *Guidance*, D.Van Nostrand Co., Princeton, 1955, Chap.12.

[2] Howe, R. M., "Guidance," *System Engineering Handbook*, edited by R. E. Machol, W. P. Tanner Jr., and S. N. Alexander, McGraw-Hill, 1965, Chap.19.

[3] Pastrick, H. L., Seltzer, S. M., and Warren, M. E., Guidance Laws for Short-Range Tactical Missiles, *Journal of Guidance and Control*, Vol.4, No.2, March-April 1981, pp.98-108.

[4] Gonzalez, J., New Methods in the Terminal Guidance and Control of Tactical Missiles, *NATO AGARD Lecture Series*, Vol.101, 1979, pp.3-1~3-23.

[5] Kabamba, P. T., *Guidance and Navigation Lecture Notes*, The University of Michigan, 1995.

[6] Vinh, N. X., *Integral Performance of Supersonic Aircraft. Technical Report(4)*, The University of Michigan, 1994.

[7] Zarchan, P., *Tactical and Strategic Missile Guidance Second Edition*, AIAA, 1994.

[8] Lin, Ching-Fang, *Modern Navigation, Guidance, and Control Processing*, Prentice-Hall, 1991, Sec. 6.5.

[9] Guelman, M., The Closed-Form Solution of True Proportional Navigation, *IEEE Transactions on Aerospace and Electronic Systems*, Vol.12, No.4, July 1976, pp.472-482.

[10] Shukla, U. S., and Mahapatra, P. R., Generalized Linear Solution of Proportional Navigation, *IEEE Transactions on Aerospace and Electronic Systems*, Vol.24, No.3, May 1988, pp.231-238.

[11] Mahapatra, P. R. and Shukla, U. S., Accurate Solution of Proportional Navigation for Maneuvering Targets, *IEEE Transactions on Aerospace and Electronic Systems*, Vol.25, No.1, January 1989, pp.81-89.

[12] Becker, K., Closed-Form Solution of Pure Proportional Navigation, *IEEE Transactions on Aerospace and Electronic Systems*, Vol.26, No.3, May 1990, pp.526-533.

[13] Cochran Jr., J. E., No, T. S., and Thaxton, D. G., Analytical Solutions to a Guidance Problem, *Journal of Guidance, Control and Dynamics*, Vol.14, No.1, Jan.-Feb. 1991, pp.117-122.

[14] Yuan., P. J. and Chern, J. S., Solutions of True Proportional Navigation for Maneuvering and Nonmaneuvering Targets, *Journal of Guidance, Control and Dynamics*, Vol.15, No.1, Jan.- Feb. 1992, pp.268-271.

[15] Yuan., P. J. and Hsu, S. C., Exact Closed-Form Solution of Generalized Proportional Navigation, *Journal of Guidance, Control and Dynamics*, Vol.16, No.5, Sep.- Oct. 1993, pp.963-966.

[16] Balakrishnan, S. N., Analytical Missile Guidance Law with a Time-Varying Transformation, *Journal of Guidance, Control and Dynamics*, Vol.19, No.2, Mar.-Apr. 1996, pp.496-499.

[17] Abramowitz, M. and Stegun, I. A., *Handbook of Mathematical Functions*, National Bureau of Standard, 1964.

AIAA-97-3473

A VSS GUIDANCE LAW FOR AGILE MISSILES

by

Mario Innocenti‡, Fabio Pellegrini†, Francesco Nasuti*
Department of Electrical Systems and Automation, University of Pisa, 56126 Pisa, Italy

Abstract

A guidance law based on application of sliding mode control is proposed in the present paper. The structure of the guidance extends the current literature by adding a range term in the sliding manifold equation. Under the hypothesis of off-boresight capability, the sliding mode structure provides a guidance law with potential for maneuvering and acquiring the target, when the missile is initially moving away from it. Existence and reachability conditions are presented for the case of maneuvering target and missile acceleration with normal component and normal as well tangential. Numerical simulation is used to validate the proposed structure against conventional ones.

I. Introduction

The problems of missile guidance and, in general, the intercept problem have been addressed over the years by a variety of researchers using several approaches. Traditionally, most guidance schemes are based on the principle of Proportional Navigation *(PN)* [1], [2], [3], where missile steering is achieved by controlling its velocity variation in a manner proportional to the rate of change of the line of sight (LOS). In addition to providing satisfactory performance, *PN* becomes an optimal guidance law under some simplifying assumptions on missile velocity and response, target maneuvering characteristics, and decreasing range rate [4]. Depending mainly on the direction of commanded acceleration A_{mc}, different variants exist such as pure proportional navigation *PPN* [5], where A_{mc} is normal to the missile velocity, and true proportional navigation *TPN* [6], where A_{mc} is normal to *LOS*. In both, cases the magnitude of A_{mc} is proportional to the *LOS* rate. Other improvements include a modified *TPN*, with commanded acceleration proportional to the product between *LOS* rate and closing speed, the ideal *PN* and generalized true *PN*, where again the direction of commanded acceleration was taken in a different way [7], [8].

In order to compensate for maneuvering targets, proportional navigation was modified to yield an augmented *PN* guidance, where the commanded acceleration is a linear function of the target velocity changes as well [3]. Optimal control theory has also been used to improve *APN* both in two dimensions as well as in three dimensions, when system's dynamics become influential [9], [10].

Alternate methodologies were proposed recently. Reference [11] presents a good account of singular perturbations theory as applied to guidance and navigation problems. Game theoretic methods are used in [12], where the use of a state vector transformation enables the differential game strategy to be treated as a linear problem. Neural networks are introduced in [13], [14] where the NN architectures improve the optimal control problem solution, and feedback linearization has been proposed [15], which allows intercept over a wider field of view compared to standard proportional navigation.

In the present paper, the main interest is in a guidance strategy most appropriate for a system characterized by high angle of attack maneuvering capabilities, and, possibly with the generation of impulsive and discontinuous acceleration commands. To this end, a control methodology based on variable structure theory is suggested along the line of [16], and extended to encompass situations where the missile is flying away from the target. A new sliding manifold is presented; conditions for the existence and reachability of the sliding conditions are determined in a differential geometry framework, and validated using a 2D scenario.

Preliminary conditions for the existence of the solution in the case of variable missile velocity are also proposed, as a first step for the incorporation of direct attitude control within a guidance structure.

---

‡ Associate Professor, Associate Fellow AIAA
† Graduate Student
* Doctoral Candidate

II. Discontinuous Guidance Law

We refer in the following to a standard two-dimensional scenario shown in Figure 1. The baseline guidance law has a *PPN* form for the commanded acceleration given by Eq. (1), where V_c is the closing speed, $\dot{\sigma}$ is the LOS rate of change and N the proportional navigation constant

$$A_{mc} = N\ V_c \dot{\sigma} \tag{1}$$

the kinematic equations in polar form are given by

$$\begin{cases} \dot{R} = V_o \cos(\gamma_o - \sigma) - V_m \cos(\gamma_m - \sigma) \\ \dot{\sigma} = \dfrac{V_o \sin(\gamma_o - \sigma) - V_m \sin(\gamma_m - \sigma)}{R} \end{cases} \tag{2a}$$

$$\begin{cases} \ddot{R} = R\dot{\sigma}^2 + A_m \sin(\gamma_m - \sigma) - A_o \sin(\gamma_o - \sigma) \\ \ddot{\sigma} = \dfrac{-2\dot{\sigma}\dot{R} + A_o \cos(\gamma_o - \sigma) - A_m \cos(\gamma_m - \sigma)}{R} \end{cases} \tag{2b}$$

$$\begin{cases} \dot{\gamma}_m = \dfrac{A_m}{V_m} \\ \dot{\gamma}_o = \dfrac{A_o}{V_o} \end{cases} \tag{2c}$$

Defining a state vector:
$\underline{x} = [R \quad \sigma \quad \gamma_m \quad \gamma_o \quad A_o]^T \in \Re^5$, and input vector $\underline{u} = A_m \in \Re^1$, Eqs. (2a) and (2c) can be written in affine form (nonlinear system linear in the controls)

$$\dot{\underline{x}} = f(\underline{x}) + g(\underline{x})\underline{u} \tag{3}$$

As pointed out in the introduction, we are interested in the definition of a guidance law for a system capable of maneuvering and steering at high angles of attack. This specification leads to a kinematic model represented by a nonlinear uncertain system. In addition, the presence of propulsive commands for attitude and angle of attack control may require discontinuous control strategies. Such requirements will be addressed in a variable structure control framework.

Variable structure control has been described in the former Soviet literature since the early sixties, see, for example, [17] among others. Invariance of VSC to a class of disturbances and parameter variations was first developed by Drazenovic [21]. In the past two decades, a large amount of research has been performed in the area by the international community [18], [19].

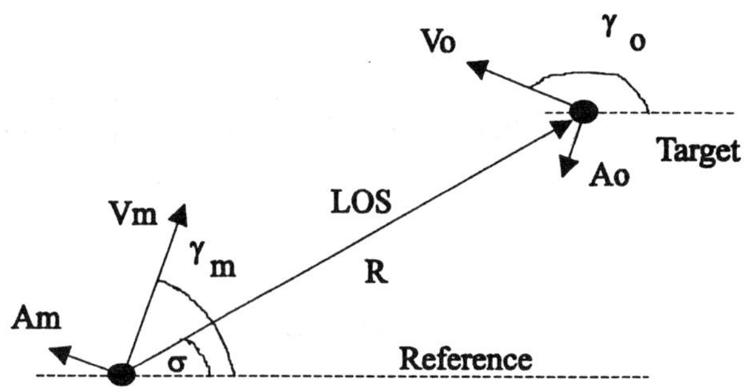

Figure 1. Intercept Scenario

.The essential feature of a variable structure controller is that it uses nonlinear feedback control with discontinuities on one or more manifolds (sliding hyperplanes) in the state space, or error space, in the case of model following control. This type of methodology is attractive in the design of controls for nonlinear, uncertain, dynamic systems with uncertainties and nonlinearities of unknown structure as long as they are bounded and occurring within a subspace of the state space [17]. The basic feature of VSC is the sliding motion. This occurs when the system state continuously crosses a switching manifold because all motion in its vicinity is directed towards the sliding surface. When the motion occurs on all the switching surfaces at once, the system is said to be in the "sliding mode" and then the original system is equivalent to an unforced completely controllable system of lower order. The design of a variable structure controller consists of several steps: the choice of switching surfaces, the determination of the control law and the switching logic associated with the discontinuity surfaces (usually fixed hyperplanes that pass through the origin of the state space). To ensure that the state reaches the origin along the sliding surfaces, the equivalent reduced order system, along the sliding surface must be asymptotically stable. This requirement defines the selection of the switching hyperplanes (sometimes called the "existence" problem), which is completely independent of the choice of control laws. The selection of the control law is the so-called "reachability" problem. It requires that the system be capable of reaching the sliding hypersurface from any initial state. The control law necessary during sliding has been defined as "equivalent control" in the literature.

One of the early attempts to formulate a guidance law using sliding modes can be found in Ref. 16, where Switched Bias Proportional Navigation (*SBPN*) is introduced. This interesting approach leads to a guidance strategy which contains an additional term, known as bias, compared to a standard PN, and it is used to improve robustness with respect to a class uncertainties in target maneuvers and speed variations. The main assumptions regarding the validity of *SBPN* are standard kinematic guidance conditions, with the addition of a bounded, but otherwise unknown target acceleration $A_t < \alpha$. The switching hyperplane chosen in [16] is simply the *LOS* rate, i.e. $s = \dot{\sigma}$. The choice, coupled with the assumption of a speed advantage by the missile compared to the target, guarantees intercept. The actual guidance law in [16] is derived by direct application of Lyapunov's stability theory, with positive definite quadratic Lyapunov function

$$V = \frac{1}{2} s^2$$

Asymptotic stability requires,

$$\dot{V} < 0, \quad for \quad s \neq 0$$

resulting in a steering command

$$A_{mc} = \frac{1}{\cos(\gamma_m - \sigma)} [-N \dot{R} \dot{\sigma} + W \; Sgn(\dot{\sigma})] \qquad (4)$$

Further analytical details, and results, can be found in [16].

The original guidance law described by Eq. (4) can be modified in order to improve intercept conditions, taking advantage of the freedom allowed by variable structure control methods. An obvious choice for sliding surface is one, which contains the main intercept variables, that is *LOS* rate and Range. Define

$$s = KR + \dot{\sigma} = \tilde{R} + \dot{\sigma} \qquad (5)$$

where K is a normalization parameter (note that K=0 implies Eq. (4)), for instance it could be chosen as $K=1/R(0)$ where $R(0)$ is the initial range value. Selecting a Lyapunov function as before

$$V = \frac{1}{2}\left[\tilde{R}^2 + 2\tilde{R}\dot{\sigma} + \dot{\sigma}^2\right] > 0 \qquad (6)$$

imposing asymptotic stability of the sliding condition

$$\dot{V} = s\dot{s} < 0, \qquad \forall s \neq 0$$

we obtain a modified guidance law of the form

$$A_{mc} = \frac{-(K'+2)\dot{\sigma}\dot{R} + (1-K')KR\dot{R} + WSgn[KR + \dot{\sigma}]}{\cos(\gamma_m - \sigma)} \qquad (7)$$

As an example of performance comparison between the two guidance laws given by Eqs. (4), and (7), let us consider two scenarios:

A. Target with variable Acceleration (engagement 2, Ref. [16])

In this engagement the starting conditions are a range of 4,500 m, 20 deg *LOS* angle, 10 deg. heading error, missile velocity 500 m/sec, and target velocity 300 m/sec. Commanded acceleration and maneuver evolution are shown in Figs. 2a, and 2b, with the continuous line corresponding to Eq. (7), and dashed line from [16]. The reduced amount of required acceleration in the modified guidance law is evident from the figures.

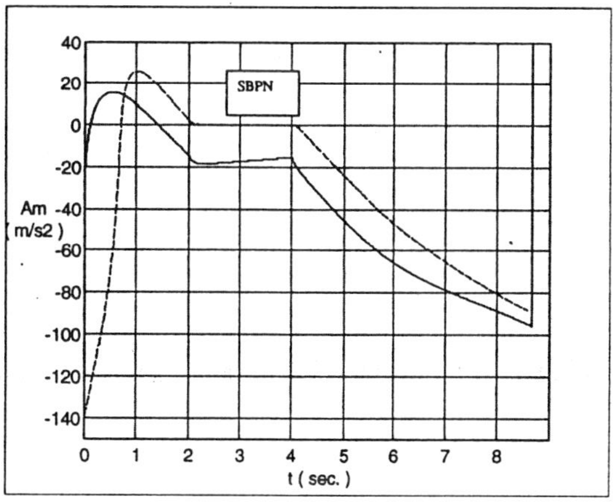

Figure 2a. Commanded Acceleration Comparison

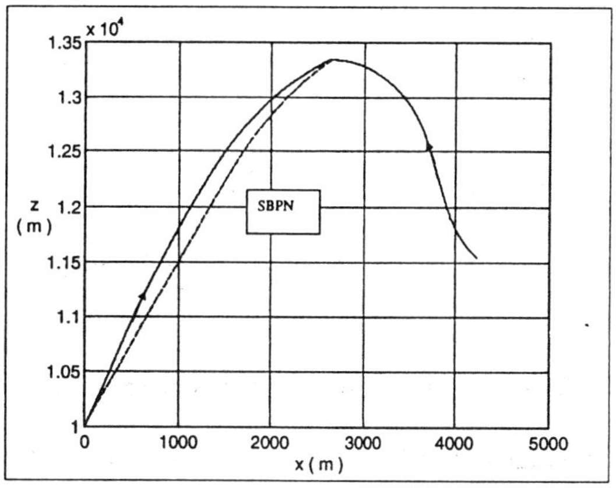

Figure 2b. Trajectory Comparison Scenario A

B. Target executing a 2D "Herbst Maneuver"

In this case, target speed and flight path angle are derived from a two dimensional approximation of the maneuver, the missile speed and flight path angle are set to 0.8 M, and zero degrees respectively. The starting altitude is 33,000 ft, and the two vehicles are closing in on each other from an initial distance of about 12,000 feet. The results of the engagement in Figs 3a, and 3b show the modified guidance law requiring about 30% less commanded acceleration for target intercept.

III. Off-Boresight Guidance *(OBGL)*

Recent development in aircraft maneuverability has had a major impact on missile technology. It is conceivable that future missile platforms will operate at high angle of attack regimes in several regions of the flight envelope, and in different missions (air-to-air, air-to-ground).

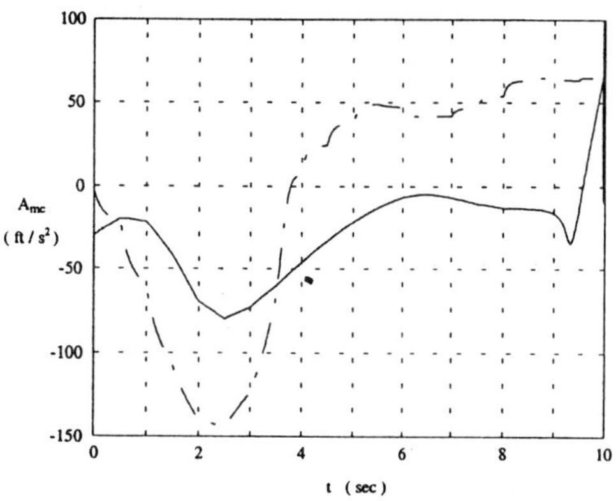

Figure 3a. Commanded Acceleration Comparison

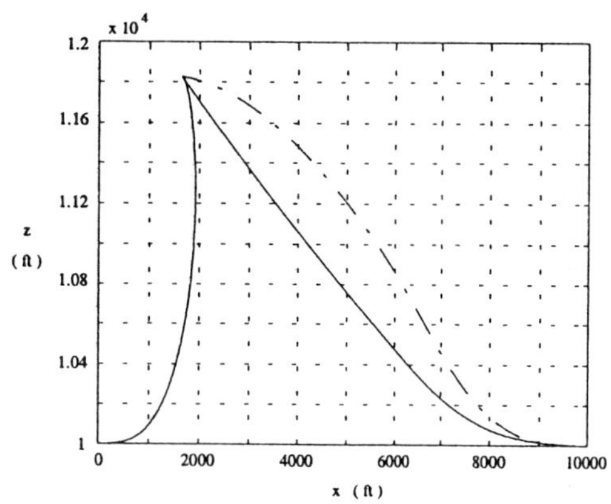

Figure 3b. Trajectory Comparison Scenario B

In this respect, it is important to investigate guidance laws capable of steering the vehicle, in a controlled fashion, through post-stall. The problem has been recently investigated in Refs. [12] and [15]. The former addresses the high angle of attack flight, by formalizing the guidance problem in a differential game framework. No information on the achieved angle of attack is present however, nor whether high alpha was considered a constraint in the differential game set-up. Reference 15 does not address high angle of attack directly, however it presents a guidance strategy capable of intercepting a target when the starting engagement conditions consist of a missile "moving away" from the target itself. The approach used in [15] is based on feedback linearization, and produces a guidance strategy that depends on the knowledge of the target acceleration.

A limit of about 70° on the look angle $\lambda_l = \gamma_m - \sigma$ was identified via simulation, with no other justification except for a generic comment on lack of sufficient energy or commanded acceleration.

Here we address a somewhat similar problem using a sliding mode approach similar to the one described in Section II. The basic concept of the proposed guidance structure is the missile capability of generating fast rotations of the look angle by effectively acting on the attitude using reaction jets as an additional control input. Once this is achieved, then a traditional guidance law, for instance proportional navigation, or a strategy given by Eq. (7) would lead to intercept.

From the standard intercept scenario shown in Fig. 1, it is necessary to achieve an ideal missile flight path angle capable of allowing intercept and given by

$$\gamma_{mid} = \sigma + \sin^{-1}\left(\frac{V_o \sin(\gamma_o - \sigma)}{V_m}\right) \tag{8}$$

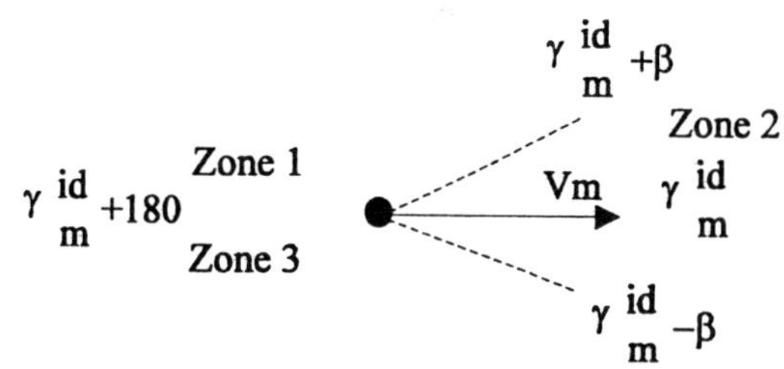

Figure 4. Geometry of Off-Boresight Guidance

If during the maneuver $\gamma_m \neq \gamma_{mid}$, then additional propulsive control is necessary in order for the missile to re-acquire an intercept condition, assuming of course a constant missile velocity $V_m > V_o$. Such situation is depicted in Fig. 4. If at the current instant the missile flies within zones 1 or 3, then a relay-type corrective action at maximum acceleration $\mp|A_{m\max}|$ is taken. This may lead to a high angle of attack situation provided the turn time is short enough, or turn rate is high enough. The choice of zone separation depends on the angle β shown in the Figure. This implies a specification of seeker characteristics and other design details that are beyond the scope of the present work. For our purposes, a selection such as in [15] is sufficient, where

$$\beta > \sin^{-1}\left(\frac{V_o}{V_m}\right) \tag{9}$$

Note that in our case Eq. (9) is merely taken as a limit on the region, whereas in [15] the condition is necessary for the feedback linearization guidance law to exist.

A sliding hyperplane for the proposed guidance is selected so as to guarantee intercept triangle conditions as in Eq. (8).
This choice is given by

$$s=\gamma_m-\sigma+\sin^{-1}\left(\frac{V_o\sin(\gamma_o-\sigma)}{V_m}\right) \tag{10}$$

Using the notation in [18], Eq. (3) can be written as:

$$f(\underline{x})=[V_o\cos(x_4-x_2)-V_m\cos(x_3-x_2)]\frac{\partial}{\partial x_1}+$$
$$\frac{V_o\sin(x_4-x_2)-V_m\sin(x_3-x_2)}{x_1}\frac{\partial}{\partial x_2}+\frac{x_5}{V_o}\frac{\partial}{\partial x_4}$$
$$g(\underline{x})=\frac{1}{V_m}\frac{\partial}{\partial x_3}$$

Existence and reachability of sliding motion can be proved for Eq. (10) using a differential geometric approach [18], which formalizes Utkin's equivalent control method [17]. The equivalent control u_{eq} is defined as that control law which satisfies ideal sliding conditions $s=\dot{s}=0$, and it is computed by zeroing the time derivative of $s(\underline{x})$ with respect to the vector field given by Eq. (3). When the equivalent control is applied during sliding, the system's dynamics would follow the switching manifold in an asymptotically stable fashion.

With the above definitions, using Lie algebra notation [20], denoting with $\langle a,b\rangle$ the inner product,

$$L_{f+gu_{eq}}s=\langle ds,f+gu_{eq}\rangle=0, s(\underline{x})=0 \tag{11a}$$

$$u_{eq}=\frac{L_f s}{L_g s}=-\left[\frac{\partial s}{\partial \underline{x}}g\right]^{-1}\frac{\partial s}{\partial \underline{x}}f \tag{11b}$$

In Eq. (11a), ds represent the gradient of s

$$ds=\frac{\partial}{\partial x_3}+\left\{\left[1-\left(\frac{V_o\sin(x_4-x_2)}{V_m}\right)^2\right]^{-.5}\frac{V_o}{V_m}\right.$$
$$\left.\cos(x_4-x_2)-1\right\}\frac{\partial}{\partial x_2}+$$
$$\left\{-\left[1-\left(\frac{V_o\sin(x_4-x_2)}{V_m}\right)^2\right]^{-.5}\frac{V_o}{V_m}\cos(x_4-x_2)\right\}\frac{\partial}{\partial x_4} \tag{12}$$

Define $\Sigma_s(\underline{x}):=\ker[ds(\underline{x})]$ as the sliding distribution associated with s($\underline{x}$), then Eq. (11a) can be rewritten as [18]

$$f+gu_{eq}\big|_{s=0}\in\ker[ds(\underline{x})]=\Sigma_s \tag{13}$$

Since we can write

$$\Sigma_s(\underline{x})=\alpha\frac{\partial}{\partial x_1}+\beta\frac{\partial}{\partial x_2}+\gamma\frac{\partial}{\partial x_3}+\delta\frac{\partial}{\partial x_4}+\sigma\alpha\frac{\partial}{\partial x_5}$$

from $\langle ds,\Sigma_s\rangle=0$, a basis for $\Sigma_s(\underline{x})$ is

(10)

$$\Sigma_s(\underline{x})=$$
$$span\left[\frac{\partial}{\partial x_1}\quad\frac{\partial}{\partial x_5}\quad\frac{\partial}{\partial x_2}-c_1(\underline{x})\frac{\partial}{\partial x_3}\quad\frac{\partial}{\partial x_4}-c_2(\underline{x})\frac{\partial}{\partial x_3}\right]$$

where

$$c_1(\underline{x})=\left\{\left[1-\left(\frac{V_o\sin(x_4-x_2)}{V_m}\right)^2\right]^{-0.5}\frac{V_o}{V_m}\cos(x_4-x_2)-1\right\}$$
$$c_2(\underline{x})=\left\{-\left[1-\left(\frac{V_o\sin(x_4-x_2)}{V_m}\right)^2\right]^{-0.5}\frac{V_o}{V_m}\cos(x_4-x_2)\right\}$$

Using Eq. (13), after some algebra the equivalent control is found to be

$$u_{eq}=A_o\cos(\gamma_o-\sigma)\frac{1}{\sqrt{1-\left(\frac{V_o}{V_m}\sin(\gamma_o-\sigma)\right)^2}} \tag{14}$$

Eq. (14) can be shown to correspond to a well-defined equivalent control, since *Lemma 1* in [18] is satisfied locally on the sliding manifold. In addition, local (global in our particular case) existence of sliding motion is guaranteed by choosing $u^-(\underline{x})$, $u^+(\underline{x})$ that satisfy

$$u^-(\underline{x})<u_{eq}(\underline{x})<u^+(\underline{x})$$

From Eq. (14), a sufficient condition based on an assumed ratio $V_o=0.99V_m$, yields $|A_{m\_\max}|\le 7|A_o|$ and the control law takes a relay form

$$u=A_{mc}=|A_{m\_\max}|\operatorname{sgn}(s) \tag{15}$$

[note that zone 1 and 3 in Fig. 4 can switch depending on the value of γ_m].

In order to verify the proposed guidance scheme; let us consider some examples:

A. Target behind Attacker at constant Speed

Commanded acceleration and trajectories profiles are shown in Fig. 5. The continuous line corresponds to the proposed guidance scheme, the dotted line to a standard *PN*, and the dash-dotted line to *SBPN*. The latter strategies become active only after about 6 seconds, when the intercept triangle conditions are achieved. Even in this case the *OBGL* requires the minimum amount of A_{mc}.

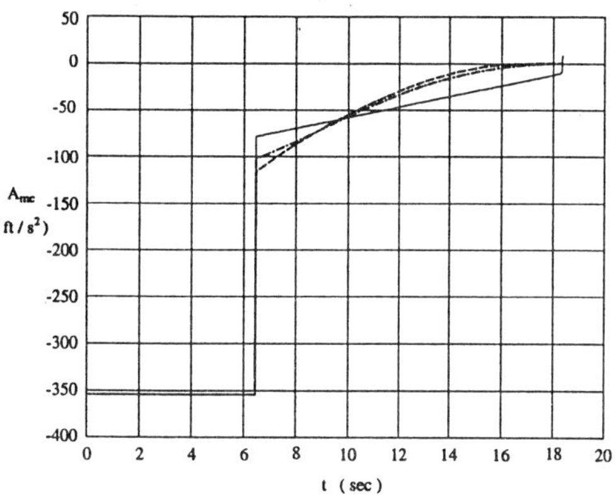

Figure 5a. Commanded Acceleration Comparison

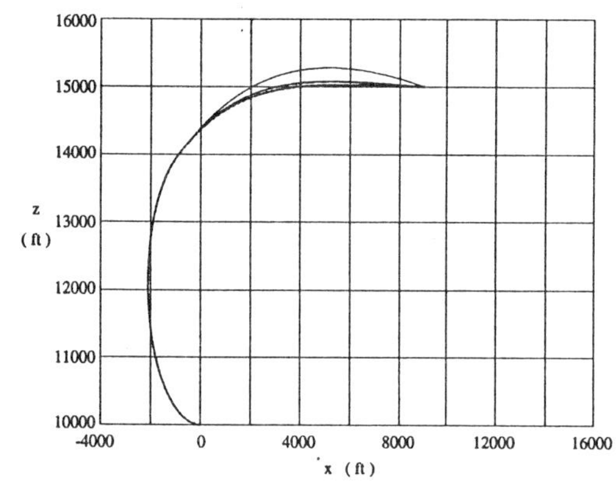

Figure 5b. Trajectory Comparison Scenario IIIA

B. Target in front performing an evasive "Herbst" Maneuver

In this example, the target is ahead and it slows down suddenly in order to acquire a firing advantage. As before, the guidance law operates at its maximum until triangle intercept, and then it may follow a conventional strategy. *PN* and *SBPN* schemes are unable to perform intercept until about 4.5 seconds. The peaks present in the commanded acceleration plots are due to numerical approximations in the code, and can be easily eliminated. Fig. 6 presents the engagement results, with the top graph showing the commanded Acceleration time histories while Fig. 7 shows a pictorial view of the scenario.

An interesting point of comparison is between the guidance law proposed by the present work and the so-called feedback linearization guidance *(FLGL)* introduced in [15], since both claims off-boresight capabilities. If we consider the triangle intercept geometry of Fig. 8, some considerations can be made.

As described in [15], *FLGL* operates with two different strategies depending on the missile leading or lagging the target.

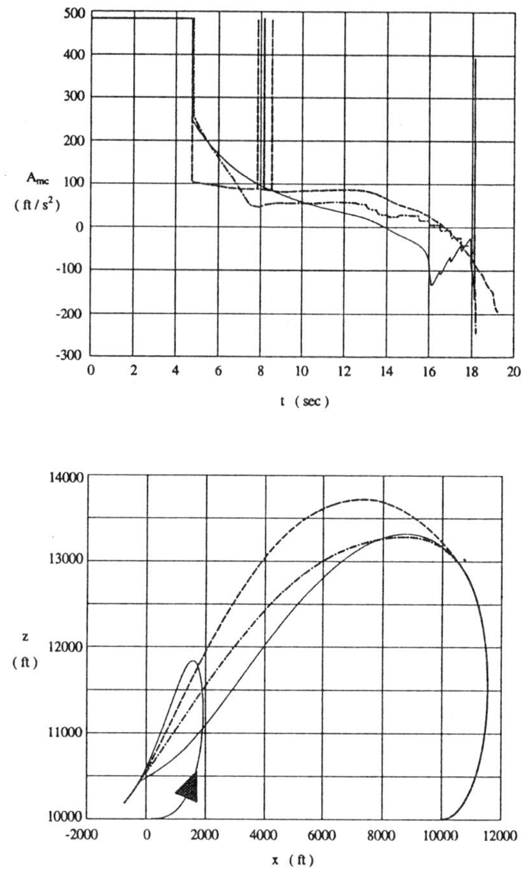

Figure 6. Guidance Performance: Engagement IIIB

Figure 7. Graphical Simulation of Engagement IIIB

This requires a guidance structure that changes depending on the missile velocity being above the line of sight, denoted by M-T, or below it. In the former case *FLGL* uses Eqs. (9a), (9c), and (9g), in the latter Eqs. (9a), (9d), and (9h), all of them written in the reference. *OBGL*, on the other hand, uses a switching strategy that depends on the definition of the sliding manifold described in Eq. (10) of the present paper.

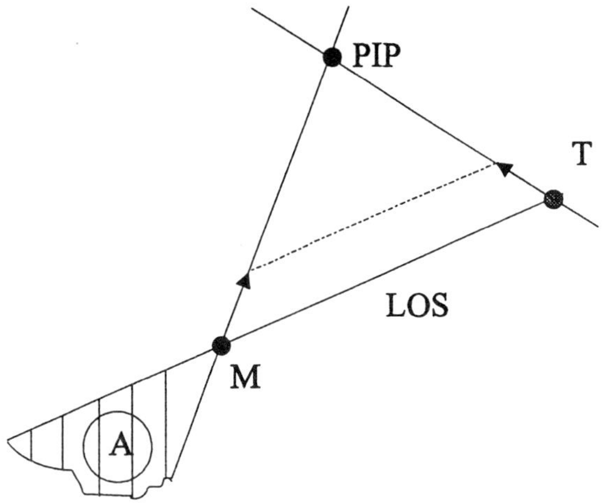

Figure 8. Intercept Triangle

In practice, *OBGL* operates as a relay giving plus or minus maximum commanded acceleration depending on the missile velocity being on the right or the left of the sliding surface denoted M-PIP in Fig. 8, where PIP stands for Predicted Intercept Point. This fact leads to the two guidance laws giving opposite commands whenever the missile velocity lies in the sector indicated by the dashed area A, which means that in such situations *OBGL* would provide a clockwise rotation of the velocity vector, whereas *FLGL* would command a counterclockwise rotation, with a potentially larger intercept time. It must be noted that the size of sector A increases as the target velocity increases (for a given missile speed).

The second consideration deals with the actual implementation of the guidance strategy in terms of commanded acceleration. If we look at the time history in Fig. 9 of Ref. [15], it is evident that *FLGL* resulting from Eqs. (9a), (9d), and (9h) of the same reference produces a saturated commanded value to the maximum assumed for this scenario, that is 100g, until the missile enters the intercept cone (after which *FLGL* reverts to proportional navigation). This physically obvious situation is instead a direct result of *OBGL* as from Eq. (15), since variable structure control gives a relay strategy with the commanded acceleration set to its maximum absolute value, without going through a complex feedback linearizing procedure, and providing all the potential robustness characteristics not present in a plant inversion approach.

The above facts are validated via simulation and shown in Figs. 9 and 10. Figure 9 represents the engagement evolution and commanded acceleration profile comparison for the scenario described in section V and Fig. 6 of Ref. [15]. Figures 10a and 10b differ from the previous one in that the initial range is set to 3000 meters (instead of 1500 meters), and the target heading angle is 140 degrees (as supposed to -20 degrees), the intercept cone angle β is 20 degrees in both cases.

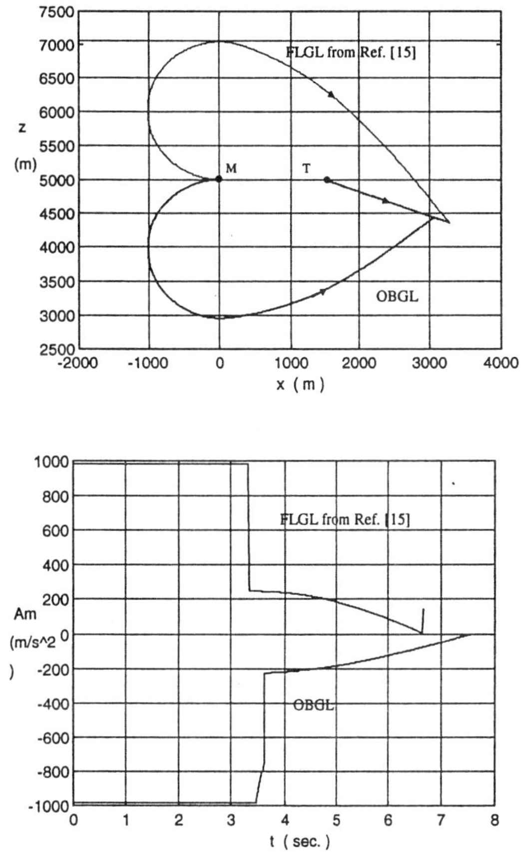

Figure 9. Trajectory and Performance Comparison between *OBGL* and *FLGL* [15], Example 1

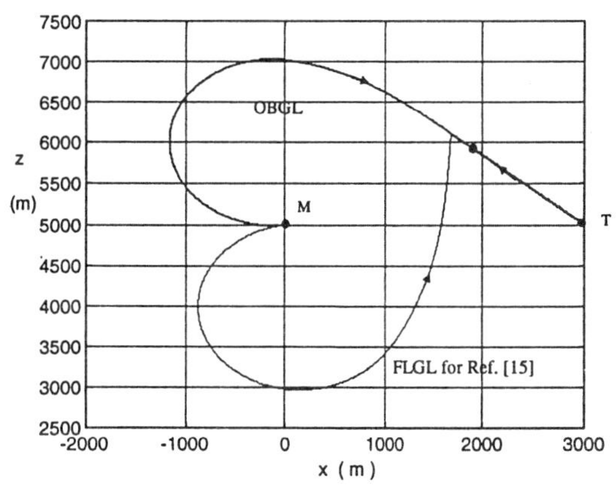

Figure 10a. Performance Comparison between *OBGL* and *FLGL* [15], Example 2

From the analysis of the above figures, although no formal validation has been performed at this time, we can see how the proposed guidance yields earlier intercept and a spatial evolution consistent with the predicted direction of motion of the target, conditions which become increasingly evident if the target velocity is higher.

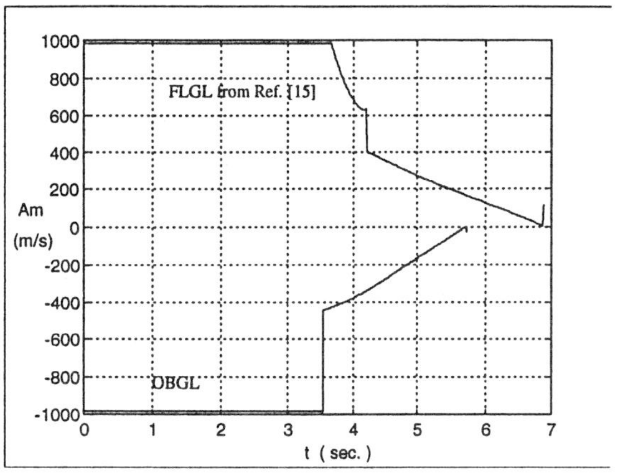

Figure 10b. Trajectory Comparison between *OBGL* and *FLGL* [15], Example 2

IV. Conditions in the Presence of Variable Speed

Traditionally, the majority of guidance laws assume constant modulo missile velocity. In the case of a missile that experiences high angle of attack conditions, however, there is a considerable speed variation (decrease) due to increased drag and stronger maneuverability requirements, leading to a tangential acceleration in addition to the (normal) commanded acceleration.

Conditions for the existence of an equivalent control in the presence of speed variations can be found, following the procedure described in Section III. Starting from the kinematic description of the intercept as in Eq. (3), the system now has a state vector which contains missile velocity as well as tangential acceleration and given by $\underline{x} = [R \quad \sigma \quad V_m \quad \gamma_m \quad \gamma_o \quad A_x \quad A_o]^T \in \Re^7$, and a control vector consisting of the normal acceleration $\underline{u} = A_z$. We now have

$$f(\underline{x}) = [V_o \cos(x_5 - x_2) - x_3 \cos(x_4 - x_2)]\frac{\partial}{\partial x_1} + \frac{V_o \sin(x_5 - x_2) - x_3 \sin(x_4 - x_2)}{x_1}\frac{\partial}{\partial x_2} + x_6 \frac{\partial}{\partial x_3} + \frac{x_5}{V_o}\frac{\partial}{\partial x_5}$$

$$g(\underline{x}) = \frac{1}{x_3}\frac{\partial}{\partial x_4}$$

From above, the equivalent control can be found to be

$$u_{eq} = A_z = A_o \cos(\gamma_o - \sigma)\frac{1}{\sqrt{1 - \left(\frac{V_o}{V_m}\sin(\gamma_o - \sigma)\right)^2}} - \frac{V_o}{V_m}\sin(\gamma_o - \sigma)\frac{1}{\sqrt{1 - \left(\frac{V_o}{V_m}\sin(\gamma_o - \sigma)\right)^2}} A_x$$

After which, bounds on u_{eq} are found as previously described.

Currently, variable speed guidance is being addressed, with the inclusion of the dynamic component of the system in the general model. From a qualitative standpoint, however, performance compare favorably with the results presented in [12]. As an example, let us consider an air-to-ground engagement given by a stationary target, and a missile moving with initial speed V_m, heading angle $\gamma_m = 60$ deg, *LOS* angle equal to 270 deg, and a maximum acceleration of about 22g. *OBGL* guidance operates in this engagement with a switching strategy up to a decision angle β =15 deg. after which proportional navigation (N=4) takes over. Figures 11a and 11b show the intercept trajectories corresponding to different speed profiles, and the time history of these profiles.

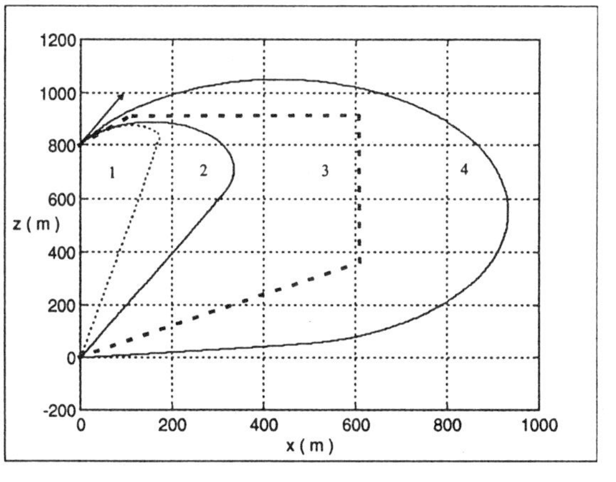

Figure 11a. Qualitative Trajectory of *OBGL* Guidance with variable Speed

The dashed line (line 3) corresponds to the trajectory envelope and velocity profile described in Ref. [12]. In that work, modeling included controlled missile dynamics, and intercept time was of the order of 5 seconds. The proposed guidance results in trajectories labeled as 1, 2, and 4.

Cases 2 and 4 are obtained with the constant speed condition corresponding to a low value (Mach 0.55), and a high value (Mach 1) respectively. As shown in Figure 11a, trajectories can fall inside or outside the envelope shown in [12], but in both cases with longer intercept times (see Figure 11b.).

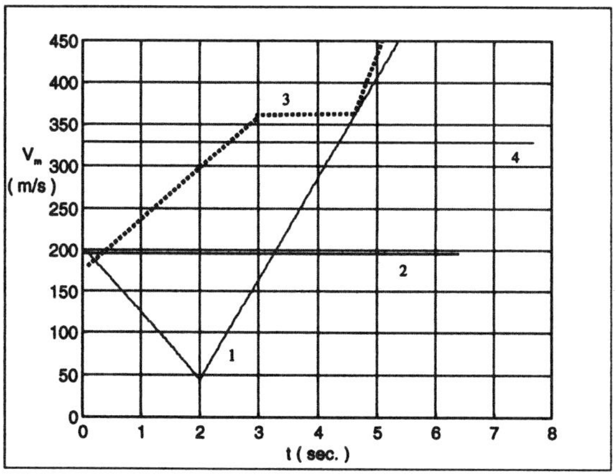

Figure 11b. Qualitative Performance of *OBGL* Guidance with variable Speed

This is due to the fact that speed is constant in magnitude, and *OBGL* has no control over the missile attitude dynamics, as opposed to the strategy described in [12].

If, however we hypothesize the capability of speed variation, indicative of a loss of energy due to the missile entering a controlled high angle of attack turn, then *OBGL* can improve drastically its performance both in terms of spatial envelope shown by curve 1 in Figure 11a, and intercept time (comparable now to that obtained in [12]). Clearly the above result must be taken as a qualitative example only, obtained by artificially introducing a speed reduction from Mach 0.55 to Mach 0.2 and then a subsequent acceleration to reach a final speed as in [12]. Nevertheless results indicate that the presence of an attitude control capable of flying the vehicle through high alpha can be cast into *OBGL* as an effective speed variation with beneficial effects on performance.

Conclusions

A new guidance strategy based on variable structure methodology was proposed, and extended to guarantee intercept in cases where the missile and target move away from each other. The guidance law stability was proved in the framework on nonlinear geometric theory, and it incorporates the capability of addressing a flight regime that includes high angles of attack. The proposed guidance was tested in several planar scenarios, with non-maneuvering as well as maneuvering targets. It was also compared with other guidance strategies showing favorable results, and a simpler implementation structure.

Acknowledgement

The present work was partially supported by the U.S. Air Force, WL/AMFC under Grant F08630-94-0001. Mr. Frederick A. Davis, WL/MNAV served as Technical Monitor. The authors wish to thank Dr. Alberto Landi for the useful comments and suggestions regarding the differential geometry approach to variable structure control.

References

[1] Martaugh, S.A., Criel, H.E., "Fundamental of Proportional Navigation ", *IEEE Spectrum*, Vol. 3, No. 6, 1966, pp. 75-85.

[2] Cloutier, J.R., Evers, J.H., Feeley, J.J., "Assessment of Air-to-Air Missile Guidance and Control Technology", *IEEE Control Systems Magazine*, October 2989, pp. 27-34.

[3] Zarchan, P., *Tactical and Strategic Missile Guidance*, Progress in Aeronautics and Astronautics, 1990.

[4] Kreindler, E., "Optimality of Proportional Navigation", *AIAA Journal*, Vol. 11, No. 6, 1973,pp. 878-880.

[5] Cochran, J.E., No, T.S., Thaxton, D.G., "Analytical Solutions to a Guidance Problem", *AIAA Journal of Guidance, Control, and Dynamics*, Vol. 14, No. 1, 1991.

[6] Rao, M.N., "New Analytical Solutions for Proportional Navigation", *AIAA Journal of Guidance, Control, and Dynamics*, Vol. 16, No. 3, 1993.

[7] Yuan, P.J., Chern, J.S., "Idela Proportional Navigation", *AIAA Journal of Guidance, Control, and Dynamics*, Vol. 15, No. 5, 1992.

[8] Yang, C.D., Hsiao, F.B., Yeh, F.B., "Generalized Guidance Law for Homing Missiles", *IEEE-TR-AES*, Vol. AES-25, No. 2, 1989.

[9] Bryson, A.E., Ho, Y.C., *Applied Optimal Control*, Hemisphere Publishing, New York, 1975.

[10] Lin, C.F., *Modern Navigation, Guidance and Control Processing*, Vo. 6, Prentice Hall, 1992.

[11] Calise, A., "Singular Perturbations and Time Scales in Guidance, Navigation and Control of Aerospace Systems: A Survey", Proc. AIAA Guidance, Navigation and Control Conference, Baltimore, MD, August 1995.

[12] Menon, P.K., Chatterji, G.B., "Differential Game Based Guidance Law for High Angle of Attack Missiles", Proc. AIAA Guidance, Navigation and Control Conference, San Diego, CA, July 1996.

[13] Balakrishnan, S.N., Biega, V., "A New Neural Architecture for Homing Missile Guidance", Proc. American Control Conference, Seattle, WA, June 1995.

[14] Balakrishnan, S.N., Shen, J., "Hamiltonian Bases Adaptive Critics for Missile Guidance", Proc. AIAA Guidance, Navigation and Control Conference, San Diego, CA, July 1996.

[15] Bezik, S., Rusnak, I., Gray, W.S., "Guidance of a Homing Missile via Nonlinear Geometric Control Methods", *AIAA Journal of Guidance,*

Control, and Dynamics, Vol. 18, No. 3, 1995, pp. 441-448.

[16] Babu, K.R., Sarma, I.G., Swamy, K.N., "Switched Bias Proportional Navigation against Highly Targets", *AIAA Journal of Guidance, Control, and Dynamics*, Vol. 17, No. 6, 1994, pp. 1357-1363.

[17] Utkin, V., *Sliding Modes and their Application to Variable Structure Systems*, MIR, Moscow, 1978.

[18] Sira-Ramirez, H., "Differential Geometric Methods in Variable-Structure Control", *International Journal of Control*, Vol. 48, No. 4, 1988, pp. 1359-1390.

[19] Innocenti, M., Thukral, A., " A Sliding Mode Missile Pitch Autopilot Synthesis for High Angle of Attack Maneuvering", *IEEE-TR-CST*, 1997, to appear.

[20] Isidori, A., Nonlinear Control Systems, 2nd Edition, 1989, Springer-Verlag.

[21] Drazenovic, B., "The Invariance Conditions in Variable Structure Systems," *Automatica*, Vol. 5, pp. 287-295, 1969.

AIAA-97-3474

HOMING GUIDANCE WITH TERMINAL ANGULAR CONSTRAINT AGAINST NONMANEUVERING AND MANEUVERING TARGETS

Byung Soo Kim* and Jang Gyu Lee†
Automatic Control Research Center & School of Electrical Engineering
Seoul National University, Seoul 151-742, Korea
Hyung Seok Han‡
Dept. of Electrical and Electronic Engineering
Soonchunhyang University, Asan 336-745, Korea
and
Chan Gook Park§
Dept. of Control and Instrumentation Engineering
Kwangwoon University, Seoul 139-701, Korea

Abstract

The guidance problem with terminal angular constraint is addressed. The guidance goal in the problem is to null miss distance and to achieve a desired impact attitude angle, simultaneously. In order to fulfill the guidance goal, a new homing guidance law, a modified proportional navigation, is proposed. Its performance is analyzed by employing a Lyapunov-like function with the sliding mode control methodology. The efficiency of the proposed guidance law is shown by proving the optimality in the sense that it minimizes the integral of the square of the pursuer acceleration. To account for randomly maneuvering targets, the modified PN with target maneuver compensation is also proposed. Its performance is analyzed by a similar approach. We show that the guidance command is bounded if the target executes a constant acceleration.

I. Introduction

For the past few decades, there have been extensive studies on the guidance laws for intercepting maneuvering and nonmaneuvering targets.[1-4] Among the existing guidance laws, the proportional navigation(PN) and its variants may be the most popular because of their ease of implementation and efficiency.[5] Recently, firm theoretical bases on high interceptibility of the PN have been revealed.[6-8] In Ref. 6, Ha, et al. performed the interceptibility analysis of the conventional PN by the Lyapunov method. The method has been shown to be powerful not only for analyzing the pursuit kinematics but also for designing new guidance laws.[7] The recent result based on vector analysis approach well explained the high interceptibility of the PN.[8]

There have also been efforts to improve the performance of the PN against highly maneuvering targets, so that many variants of the PN are found in the open literature.[6,7,9]

In general, the main objective of homing systems for guided weapons is to generate suitable commands so that terminal miss distance becomes zero. In some cases, however, zero miss distance does not always guarantee the completion of a mission. A typical example is the guidance of a missile or an underwater vehicle with directive warhead.[10,11] In these engagements, the final attitude angle of the pursuer should have a preferred value to achieve a proper penetration. Another example is the destruction of a ballistic missile by a interceptor missile, where head-on impact is required.[12] The guidance goal in the cases is to achieve a desired impact attitude angle as well as to null the miss distance. It is well known that the existing guidance laws such as the conventional PN cannot expect an acceptable performance in terms of the impact attitude angle.[2]

Despite its necessity in actual engagements, the guidance law with terminal angular constraint has not been actively reported in the open literature. Kim and Grider[13] firstly derived optimal and suboptimal guidance laws for the guidance goal. York and Pastrick[14] matched Kim and Grider's to a first-order autopilot. Ryu and Cho[15] and Kim, et al.[10] derived real-time guidance algorithms for anti-tank or anti-ship missiles. In addition, they matched the guidance law to high-order autopilot[15] and the height control was studied.[10]

In the above mentioned work, it is assumed that the target is slow compared with the pursuer. It is valid in the anti-tank or the anti-ship missiles. However, the target speed cannot be neglected in a certain engagement such as the anti-submarine warfare. For this reason, their guidance laws may have narrow launch envelope according to the initial conditions when the target speed is relatively high.

On the other hand, Idan, et al.[12] and Glizer[16] solved the problem in a more general framework.

---

* Doctoral Student.
† Professor. Senior Member.
‡ Assistant Professor.
§ Assistant Professor.

They provided closed-form solutions by using the Pontryagin minimum principle. However, their approaches do not seem to be adequate for real-time applications.

The previous work on the guidance problem with terminal angular constraint is either lack of the robustness against the variation in the initial condition or has difficulties in implementation. In this paper, a modified proportional navigation is proposed. The proposed guidance law has a simple form and makes it easy to implement the homing systems. A special Lyapunov function with the sliding mode control methodology by Babu, et al.[9] is employed for the performance analysis. By a rigorous approach, it is shown that a pursuer guided by our guidance law can impact a target with a desired attitude angle when the target exists in front of the pursuer over the whole engagement. We also show that the terminal acceleration of the pursuer is zero. Further, it is shown that the modified PN is optimal in the sense that it minimizes the integral of the square of the pursuer acceleration while producing zero terminal miss distance and zero impact angle error.

To account for highly maneuvering targets, the modified PN with target maneuver compensation is proposed. By a similar method, its performance is analyzed, so that an analytic capturability condition on randomly maneuvering target is derived. In addition, the boundedness of the guidance command is shown when the target acceleration is constant. The proposed guidance law is also easily implemented with the available measurement systems. It requires an estimate of the current target acceleration, while the previous one in Ref. 12 requires the future behavior of the target.

II. Preliminaries

Pursuit Kinematics

In order to simplify the equations of the pursuit situation, it is assumed that the pursuer and the target are point masses moving in a plane and that the seeker and the autopilot dynamics are perfect. We further assume that the pursuer and the target velocities are constants and the pursuer has a speed advantage over the target. It is also assumed that the angle-of-attack of the pursuer is small. These assumptions are generally made in the performance analysis of guidance laws. Under the above assumptions, pursuer-target engagement model of Fig. 1 can be represented by the following nonlinear differential equations.

$$\dot{\gamma}_t = A_t/V_t \tag{1}$$

$$\dot{\gamma}_m = A_m/V_m = A_c/V_m \tag{2}$$

$$\dot{r} = V_m(\rho\cos\theta_t - \cos\theta_m) \tag{3}$$

$$r\dot{\sigma} = V_m(\rho\sin\theta_t - \sin\theta_m) \tag{4}$$

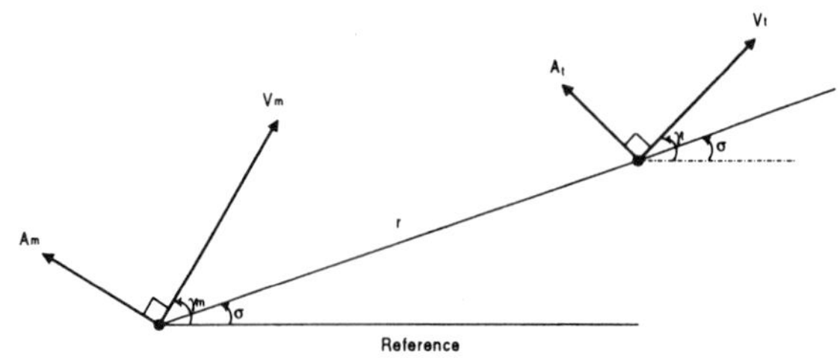

Fig. 1 Pursuit geometry.

where

$\theta_t = \gamma_t - \sigma$, $\theta_m = \gamma_m - \sigma$, $\rho = V_t/V_m$

A_m, A_c: lateral and commanded acceleration of the pursuer

A_t : lateral acceleration of the target

V_m, V_t : total velocity of the pursuer and the target

r, $\dot{r}$: relative range and range rate to the target

γ_m, γ_t : attitude angle of the pursuer and the target

σ, $\dot{\sigma}$: LOS angle and its rate.

Terminal Angular Constraint

In this paper, terminal angular constraint means a restriction on the pursuer's attitude angle at the time of capture. In order to define the constraint, introduce a function γ_d. Let $\gamma_d(t)$ be a desired impact attitude angle of the pursuer computed at time t assuming that the target makes no further evasive motion from the current time. It is a constant when the target does not maneuver over the whole engagement. However, it may be time-varying when the target maneuvers. By the definition of γ_d, the terminal angular constraint can be formally written as follows:

$$\gamma_m(t_f) = \gamma_d(t_f), \tag{5}$$

where t_f is final time such that $r(t_f) = 0$. It is obvious that the angular constraint (5) with the definition of γ_d is a generalization of the previous ones addressed in Ref. 10 and 13-15. Our definition can be applied to randomly maneuvering targets, while the previous formulations can be used only for the cases where the target does not maneuver or the future behavior of the target is known in advance.

Although $\gamma_d(t)$ can have an arbitrary value, it is usually represented by the target attitude angle in the practical applications. As an example, $\gamma_d(t) = \gamma_t(t) \pm \pi/2$ for perpendicular impact and $\gamma_d(t) = \gamma_t(t) - \pi$ for head-on impact. Then, the terminal angular constraint becomes

$$\gamma_m(t_f) = \gamma_t(t_f) \pm \pi/2,$$

in case of perpendicular impact. Similarly, in case of head-on impact, the terminal angular constraint is

$$\gamma_m(t_f) = \gamma_t(t_f) - \pi .$$

These special cases are important in actual applications and is dealt with later in detail.

Desired LOS Angle

Let $\sigma_d(t)$ be the desired impact LOS angle computed at time t assuming that the target makes no further evasive motion. It is implicitly defined by the following trigonometric equation and inequality:

$$\rho \sin(\gamma_t - \sigma_d) - \sin(\gamma_d - \sigma_d) = 0 , \qquad (6)$$

$$|\gamma_d - \sigma_d| < \pi/2 . \qquad (7)$$

It is easily shown that there exists unique σ_d satisfying (6) and (7) for given γ_t and γ_d. Conversely, if γ_t and σ_d are given, there exists unique γ_d satisfying (6) and (7).

By the above analysis, we know that $\sigma_d(t)$ can be computed by the desired impact attitude angle $\gamma_d(t)$. Now, it is claimed that the terminal angular constraint (5) is satisfied if the LOS angle at the final time is $\sigma_d(t_f)$. The reasoning is explained herein. Let us consider some conditions such that

$$\lim_{t \to t_f} r(t)\dot{\sigma}(t) = 0 , \qquad (8)$$

$$\lim_{t \to t_f} \sigma(t) = \sigma_d(t_f) , \qquad (9)$$

$$|\theta_m(t)| \le \Phi < \pi/2 , t \ge 0 . \qquad (10)$$

where Φ is a positive constant. If the assumptions (8) and (9) are true, it holds that

$$\begin{aligned}
&\lim_{t \to t_f} r\dot{\sigma} \\
&= \lim_{t \to t_f} V_m \left[\rho \sin(\gamma_t(t) - \sigma(t)) - \sin(\gamma_m(t) - \sigma(t)) \right] \\
&= V_m \left[\rho \sin(\gamma_t(t_f) - \sigma_d(t_f)) - \sin\left(\lim_{t \to t_f} \gamma_m(t) - \sigma_d(t_f) \right) \right] \\
&= 0
\end{aligned}$$

By comparing the above equation with (6), (10) and the definition of $\sigma_d(t_f)$ implies that

$$\lim_{t \to t_f} \gamma_m(t) = \gamma_d(t_f) .$$

Hence, it is clear that the terminal angular constraint is satisfied when the conditions (8)-(10) are fulfilled. Let us investigate the physical meaning of the conditions. The first equation (8) is well known condition for collision of two objects moving with constant speeds.[5] The second condition (9) means that the LOS angle becomes $\sigma_d(t_f)$ as the pursuer approaches to the target. The third condition (10) quite weak, which simply requires the angle between the pursuer velocity vector and the LOS to be kept acute over the whole engagement.[6,7] Unless it is satisfied, the pursuer may lose the target due to the physical limitation of the seeker field of view.

As mentioned above, the design of guidance systems for perpendicular or head-on impact is typical application of our theory. In these cases, we can express the desired impact LOS angle in terms of the target attitude angle. For perpendicular impact, $\gamma_d(t) = \gamma_t(t) \pm \pi/2$. Then, it is easily shown by (6) and (7) that

$$\sigma_d(t) = \gamma_t(t) \pm \left(\pi - \tan^{-1}(\rho^{-1}) \right) .$$

Similarly, we can express σ_d in term of the target attitude angle for head-on impact, i.e., $\sigma_d(t) = \gamma_t(t) - \pi$ when $\gamma_d(t) = \gamma_t(t) - \pi$. Hence, it is not necessary to solve the trigonometric equations (6) and (7) in these important applications.

Time Derivative of Impact LOS Angle

From (6) and (7), it is seen that $\gamma_d(t)$ and $\sigma_d(t)$ satisfy

$$\gamma_d(t) = \sigma_d(t) + \sin^{-1}\left\{ \rho \sin(\gamma_t(t) - \sigma_d(t)) \right\} . \qquad (11)$$

Differentiating (11) with respect to t gives

$$\begin{aligned}
&\dot{\gamma}_d \sqrt{1 - \rho^2 \sin^2(\gamma_t - \sigma_d)} - \dot{\gamma}_t \rho \cos(\gamma_t - \sigma_d) \\
&= \dot{\sigma}_d \left(\sqrt{1 - \rho^2 \sin^2(\gamma_t - \sigma_d)} - \rho \cos(\gamma_t - \sigma_d) \right)
\end{aligned} \qquad (12)$$

It is already addressed that desired impact attitude angle computed at time t is usually given by $\gamma_d(t) = \gamma_t(t) + \Omega$, where Ω may be any constant. This implies that $\dot{\gamma}_d(t) = \dot{\gamma}_t(t)$. Also note that $\sqrt{1 - \rho^2 \sin^2(\gamma_t - \sigma_d)} - \rho \cos(\gamma_t - \sigma_d) \neq 0$ since $\rho < 1$. Thus, it is seen from (12) that

$$\dot{\sigma}_d(t) = \dot{\gamma}_t(t) . \qquad (13)$$

Interceptibility of PNG Law

There have been extensive studies on the performance analysis of PN. However, the recent results by Oh and Ha[8] can be regarded as the final solution to the interceptibility analysis of the PN. They showed by LOS plane analysis that the pursuer guided by the pure PNG law can intercept a randomly maneuvering target if the target exist in front of the pursuer at the time of launch. Their result is introduced here with some notational modifications in order to analyze the proposed guidance laws.

Lemma:[8] If $|\theta_m(0)| < \pi/2$ and $0 \le \rho < 1/\sqrt{2}$. The navigation gain chosen such that $N(t) \ge N' > 1$. Then the pure PNG law $A_c(t) = N(t) V_m \dot{\sigma}(t)$ guarantees against a randomly maneuvering target that $\cos\theta_m(t) \ge \min\left\{ \cos\theta_m(0), \sqrt{1 - \rho^2} \right\}$, $t \ge 0$ and that there exists a finite time t_f' such that

$$r(t_f')=0.$$

III. Modified PN against Nonmaneuvering Targets

Guidance Law Derivation and Performance Analysis

In order to apply the sliding mode control methodology to the guidance problem, a switching surface is to be chosen. Before the selection of switching surface, it should be noted that there exist two guidance errors in the problems we are interested in. The heading error is the only guidance error in the conventional problem where zero miss distance is the guidance goal. However, in the guidance problem with terminal angular constraint, the guidance goal is transformed into achieving zero miss distance and a desired impact attitude angle, simultaneously. It is well known that the miss distance is nulled by nulling heading error. It is also shown in the previous section that the terminal angular constraint is satisfied by nulling LOS error, i.e., $|\sigma-\sigma_d|$. In this sense, the heading error and the LOS error together constitute the guidance errors. They are simultaneously nulled for fulfilling the guidance goal. Hence, the switching surface should include theses two factors. The switching surface of this section is chosen such that

$$s(t)=\dot{\sigma}(t)+\frac{\lambda V_m}{r(t)}(\sigma(t)-\sigma_d). \tag{14}$$

where λ is a positive constant. In (14), the first term is responsible for the heading error and the second term is responsible for the LOS error. Note that σ_d is a constant since the target does not maneuver. In order to satisfy the sliding condition, let us consider the time derivative of a Lyapunov function $\frac{1}{2}s^2(t)$.

$$\frac{1}{2}\frac{d}{dt}s^2=\ddot{\sigma}+\frac{\lambda V_m}{r}\dot{\sigma}-\frac{\lambda V_m}{r^2}\dot{r}(\sigma-\sigma_d). \tag{15}$$

Since the target does not maneuvers, $\ddot{\sigma}$ is given by

$$\ddot{\sigma}=(-2\dot{r}\dot{\sigma}-A_m\cos\theta_m)/r \tag{16}$$

The guidance law for guaranteeing the sliding condition is

$$A_c=[(\lambda V_m+K-2\dot{r})\dot{\sigma}+(K-\dot{r})\lambda V_m(\sigma-\sigma_d)/r + \omega\,\mathrm{sgn}(s)]/\cos\theta_m \tag{17}$$

where $K>0$ and $\omega>0$. It can be shown by inserting (17) with (16) to (15) that if (10) is true the following inequality holds.

$$\frac{1}{2}\frac{d}{dt}s^2<-\frac{\omega}{r}|s|<0,\ \text{for}\ s\neq 0\ \text{and}\ r>0. \tag{18}$$

Thus, the sliding condition is satisfied by the guidance command in (17) whenever $r>0$. However, r may become zero before the sliding surface is reached. To avoid the problem, let us choose the gain K and ω such that

$$K\geq\max\{K'|\dot{r}|,\ K'V_m(1-\rho)\}\ \text{and}\ \omega\geq\mu K, \tag{19}$$

where K' and μ are positive constants. Then, it holds that

$$\frac{1}{2}\frac{d}{dt}s^2\leq-\frac{K}{r}(s^2+\mu|s|)\leq\frac{K'\dot{r}}{r}(s^2+\mu|s|)\ \text{for}\ r>0.$$

It is trivial that $s(t)=0, t\geq 0$ when $s(0)=0$. We assume $s(0)\neq 0$. By solving the above differential inequality, we obtain that

$$|s(t)|\leq-\mu+(|s(0)|+\mu)(r(t)/r(0))^{K'}\ \text{for}\ r>0.$$

Hence, it holds that

$$s=0\ \text{when}\ r\in(0,\bar{r}] \tag{20}$$

where $\bar{r}$ is a constant such that

$$\bar{r}\geq r(0)\left(\frac{\mu}{|s(0)|+\mu}\right)^{1/K'}.$$

This shows that with the choice of (19) the sliding surface is reached before the interception occurs.

The proposed guidance law guarantees the completion of the guidance goal if the target exists in front of the pursuer over the whole engagement. It is summarized by the following theorem.

Theorem 1: Suppose that (10) is satisfied and $0\leq\rho<1/\sqrt{2}$. The guidance gains are selected such that

$$\lambda>1+\rho,\ K\geq\max\{K'|\dot{r}|,\ K'V_m(1-\rho)\},\ \text{and}\ \omega\geq\mu K \tag{21}$$

where λ, K' and μ are positive constants. Then, the guidance law (17) guarantees the existence of a finite time t_f such that

$$\lim_{t\to t_f}r(t)=0, \tag{22}$$

and

$$\lim_{t\to t_f}\gamma_m(t)=\gamma_d. \tag{23}$$

Proof) See Appendix A.

Convergence Property of Guidance Command

Since any vehicle has a finite maximum acceleration, an excessive guidance command results in saturation. The command saturation at the final phase may cause severe miss distance. Furthermore, large acceleration makes a high angle-of-attack which causes a discrepancy between the velocity vector and the actual heading of the pursuer. Observing the previous formulation, we know that the proposed guidance law makes the velocity vector achieve a desired angle. However, the actual heading of pursuer becomes much different from the desired value when the angle-of-attack is high. In this way, large acceleration command at the final phase may severely degrade the guidance performance in terms of both miss distance and impact attitude angle.

Now, it is shown that the slight modification of guidance gain results in zero terminal guidance command. Suppose that one of the guidance gain λ is chosen such that

$$\lambda \geq \kappa(1+\rho). \tag{24}$$

where κ is a constant and $\kappa > 1$. The above guidance gain is somewhat larger than that of (21). Suppose that all the assumptions of Theorem 1 is satisfied and that λ is chosen by (24). Then, by the same arguments used in proving (46), there exists a finite time $t_r \in [0, t_f)$ such that

$$\frac{\dot{\sigma}(t)}{\sigma(t)-\sigma_d} = -\frac{\lambda V_m}{r(t)} \leq \frac{\kappa \dot{r}(t)}{r(t)}, \quad t_r \leq t < t_f.$$

By solving the above differential inequality, it is seen that

$$|\sigma(t)-\sigma_d| \leq |\sigma(t_r)-\sigma_d| \left(\frac{r(t)}{r(t_r)} \right)^{\kappa}, t_r \leq t < t_f. \tag{25}$$

Then, the following properties can be easily shown by observing (25).

Property 1.1) $\lim_{t \to t_f} |\sigma(t)-\sigma_d| = 0$.

Property 1.2) $\lim_{t \to t_f} \frac{\sigma(t)-\sigma_d}{r(t)} = 0$.

Proof) It is obvious from (25) that

$$\frac{|\sigma(t)-\sigma_d|}{r(t)} \leq |\sigma(t_r)-\sigma_d| r(t)^{\kappa-1} / r^{\kappa}(t_r), \quad t_r \leq t < t_f.$$

Property 1.3) $\lim_{t \to t_f} \dot{\sigma} = 0$

Proof) It is immediate consequence of Property 1.2) and (14).

Property 1.4) $\lim_{t \to t_f} A_c(t) = 0$

Proof) It is clear from (17) that $A_c = (\lambda V_m - \dot{r})\dot{\sigma} / \cos\theta_m$, $t_r \leq t < t_f$. This with Property 1.3) implies the convergence of guidance command.

The following corollary can be easily proved with the above properties and Theorem 1.

Corollary 1: Suppose that (10) is satisfied and $0 \leq \rho < 1/\sqrt{2}$. The guidance gains are selected such that

$$\lambda \geq \kappa(1+\rho), K \geq \max\{K'|\dot{r}|, K'V_m(1-\rho)\}, \quad \omega \geq \mu K \tag{26}$$

where λ, κ, K' and μ are positive constants and $\kappa > 1$. Then, the guidance law (17) guarantees the existence of a finite time t_f such that

$$\lim_{t \to t_f} r(t) = 0,$$

$$\lim_{t \to t_f} \gamma_m(t) = \gamma_d$$

Furthermore, the guidance command converges to zero as the pursuer comes close to its target, i.e., $\lim_{t \to t_f} A_c(t) = 0$.

Relation between Previous Result and Proposed Guidance Law

It is well known that the conventional PN is similar to the optimal guidance law along the collision course. The optimal guidance gain which minimizes the integral of the square of the pursuer acceleration is three. A similar relation holds between the proposed guidance law and the optimal guidance law in the guidance problem with terminal angular constraint. The optimal guidance for the problem is derived by Kim and Grider.[13] It is represented by the following equation when hard constraints at final time are imposed.

$$A_c^* = \frac{6}{b(t_f-t)^2} y + \frac{6}{b(t_f-t)} \dot{y} + \frac{2}{t_f-t} V_m(\gamma_m - \gamma_d), \tag{27}$$

where y and $\dot{y}$ are lateral displacement and its rate, respectively. The lateral displacement is defined $y = r\sin(\sigma - \gamma_d)$ and $b = \cos\gamma_{m_0}$. The above guidance law minimizes $\int A_c^2 dt$ and satisfies the terminal constraints, i.e., $y(t_f) = 0$ and $\gamma_m(t_f) = \gamma_d$. In the following, it is shown that our guidance law (17) has a similar form along the collision course if the pursuer is sufficiently fast and the gain of switching term is small. Under these assumptions, it holds that

$$-\dot{r} \approx V_m, \quad \sigma_d \approx \gamma_d \text{ and } \cos\theta_m \approx 1.$$

The guidance gains are selected such that

$\lambda = \kappa(1+\rho) \approx \kappa$,

$K = \max\{K'|\dot{r}|, K'V_m(1-\rho)\} \approx K'V_m$, and

$\omega = \mu K \approx \mu K' V_m$.

It is obvious that the above selection satisfies (26). Then, (17) becomes

$$A_c \approx (\kappa + K' + 2)V_m\dot{\sigma} + \kappa(K'+1)V_m^2(\sigma - \sigma_d)/r + \mu K' V_m \operatorname{sgn}(s)$$

By the small angle assumption, the above equation is written by

$$A_c \approx \frac{\{\kappa + K' + \kappa(K'+1) + 2\}}{(t_f-t)^2} y + \frac{\kappa + K' + 2}{t_f - t}\dot{y} + \mu K' V_m \operatorname{sgn}(s). \tag{28}$$

On the other hand, since $\dot{y} \approx -V_m\gamma_m$ along the collision course when the pursuer is sufficiently fast,

(27) can be represented as follows.

$$A_c^* \approx \frac{6}{(t_f - t)^2} y + \frac{4}{t_f - t} \dot{y}. \tag{29}$$

Comparing (28) and (29), it is seen that the proposed guidance law is similar to the optimal guidance law along the collision course if the navigation gains are selected such that

$$K' = 1 \text{ and } \kappa = 1 + \varepsilon$$

with sufficiently small $\varepsilon > 0$ and $\mu > 0$. Hence, along the collision course, our guidance law (17) can be approximated to the optimal guidance law with $\omega \operatorname{sgn}(\cdot)$ for guaranteeing the sliding condition. However, it should be noted here that all the reasoning are valid along the collision course. Under the general engagements far from the collision course, these two schemes are much different.

IV. Modified PN against Maneuvering Targets

Random Maneuver

The switching surface of this section is chosen by

$$s(t) = \dot{\sigma}(t) + \frac{\lambda V_m}{r(t)} (\sigma(t) - \sigma_d(t)). \tag{30}$$

where λ is a positive constant. Note that the above switching surface is same to the nonmaneuvering target case except σ_d is a function of time. The definition and the computation of $\sigma_d(t)$ was addressed in the second section. Let us define a guidance law such that

$$A_c = [(\lambda V_m + K - 2\dot{r})\dot{\sigma} + (K - \dot{r})\lambda V_m(\sigma - \sigma_d)/r - \lambda V_m \dot{\gamma}_t + A_t \cos\theta_t + \omega \operatorname{sgn}(s)] / \cos\theta_m \tag{31}$$

where $K > 0$, $\omega \geq \mu K$, and μ is a positive constant. The above guidance law is same to the one proposed in Theorem 1 if target maneuver is negligible. By the similar way used in the previous section, the following properties can be proved under the condition (10).

Property 2.1) $\frac{1}{2}\frac{d}{dt}s^2 < -\frac{\mu K}{r}|s| < 0$, for $s \neq 0$ and $r > 0$.

Proof) The expression for $\ddot{\sigma}$ is derived from (4).

$$\ddot{\sigma} = (-2\dot{r}\dot{\sigma} - A_m \cos\theta_m + A_t \cos\theta_t)/r \tag{32}$$

Differentiating $\frac{1}{2}s^2$ with respect to t and using (13), (31) and (32),

$$\frac{1}{2}\frac{d}{dt}s^2 = \frac{s}{r}[-Ks - \omega \operatorname{sgn}(s)]$$

Suppose that $s > 0$. Then, the following inequality holds.

$$\frac{1}{2}\frac{d}{dt}s^2 \leq \frac{s}{r}[-Ks - \mu K].$$

The similar reasoning holds when $s < 0$. Thus, it can be concluded that

$$\frac{1}{2}\frac{d}{dt}s^2 < -\frac{\mu K}{r}|s| < 0, \text{ for } s \neq 0 \text{ and } r > 0.$$

■

Property 2.2) If $K \geq \max\{K'|\dot{r}|, K'V_m(1-\rho)\}$, $K' > 0$, then it holds that

$$s = 0 \text{ whenever } r \in (0, \bar{r}] \tag{33}$$

where $\bar{r}$ is a constant such that

$$\bar{r} \geq r(0)\left(\frac{\mu}{|s(0)| + \mu}\right)^{1/K'}.$$

Proof) The proof is same to the arguments used in proving (20). ■

The guidance law in (31) guarantees the completion of the guidance goal if the target exists in the seeker field of view over the whole engagement. It is summarized by the following theorem.

Theorem 2 : Suppose that;

A1) The geometry of pursuer and target satisfy

$$|\theta_m(t)| < \theta_s, \; t \geq 0, \tag{34}$$

for some constant $\theta_s \in (0, \cos^{-1}\rho)$.

A2) The guidance gains are selected such that

$$\lambda > 0, \; K \geq \max\{K'|\dot{r}|, K'V_m(1-\rho)\}, \; \omega \geq \mu K, \tag{35}$$

where λ, K' and μ are positive constants.

A3) The acceleration of the target is a piecewise continuous function.

Then, the guidance law (31) guarantees the existence of a finite time t_f such that

$$\lim_{t \to t_f} r(t) = 0, \tag{36}$$

$$\lim_{t \to t_f} \gamma_m(t) = \lim_{t \to t_f} \gamma_d(t). \tag{37}$$

Proof) See Appendix B.

Comparing Theorem 1 and 2, it is concluded that (17) is a special case of (31). Hence, the guidance law (31) guarantees the completion of the guidance goal whether the target maneuvers or not.

Let us see the physical meaning of (34). Generally, the impact attitude angle may be satisfied when the pursuer has a substantial speed advantage over the target. For example, $\rho = 1/500$ is assumed in the simulation condition of Ref. 13. Thus, in these engagements it is reasonable to suppose that $\rho \leq 1/2$, *i.e.*, the pursuer is at least two times faster than the target. On the other hand, because of a physical limitation, the seeker field of view generally does not exceed about 60°. In order words, the seeker cannot see a target when $|\theta_m| \geq 60^\circ$. With such consideration, it is concluded that (34) can be satisfied if a target stays within the seeker field of view.

Constant Maneuver

The guidance law (31) makes the pursuer impact the target with a desired attitude angle against randomly maneuvering targets. However, the boundedness of the command is not guaranteed with the guidance law except for the case that the target maneuver is negligible. In this section, it is shown that a simple modification of (31) guarantees the boundedness of the command when target maneuver is modeled by

$$A_t(t) = c \cdot V_t, \quad t \geq 0,$$

where c is a constant. The sliding surface in this case is

$$s(t) = \dot{\sigma}(t) - \dot{\sigma}_d(t) + \frac{\lambda V_m}{r}(\sigma(t) - \sigma_d(t)), \quad (38)$$

where λ is a positive constant. Note that the switching surface given by (38) is same to (14) and (30) except the adoption of $\dot{\sigma}_d$. Let us consider the following guidance command.

$$\begin{aligned} A_c &= [(\lambda V_m + K - 2\dot{r})\dot{\sigma} + \lambda V_m (K - \dot{r})(\sigma - \sigma_d)/r \\ &+ c(V_t \cos\theta_t - \lambda V_m - K) + \omega \operatorname{sgn}(s)] / \cos\theta_m \end{aligned} \quad (39)$$

where $K > 0$, $\omega \geq \mu K$, and μ is a positive constant. The guidance law (39) is similar to (31) except some gains and the definition of switching surface. In order to analyze the performance of (39), the following properties are proved. It is assumed here that there exists a finite time t_f such that $r(t_f) = 0$ and the condition (10) is valid.

Property 3.1) $\frac{1}{2}\frac{d}{dt}s^2 < -\frac{\mu K}{r}|s| < 0$, for $s \neq 0$ and $r > 0$.

Proof) Differentiating $s^2/2$ with respect to t and using (13), (32), and (39)

$$\frac{1}{2}\frac{d}{dt}s^2 = \frac{s}{r}[-Ks - \omega \operatorname{sgn}(s)]$$

Suppose that $s > 0$. Then, the following inequality holds.

$$\frac{1}{2}\frac{d}{dt}s^2 \leq \frac{s}{r}[-Ks - \mu K].$$

It becomes

$$\frac{1}{2}\frac{d}{dt}s^2 \leq -\frac{K}{r}(s^2 + \mu s).$$

The similar reasoning holds when $s < 0$. Thus, it can be concluded that

$$\frac{1}{2}\frac{d}{dt}s^2 < -\frac{K\mu}{r}|s| < 0, \text{ for } s \neq 0 \text{ and } r > 0. \quad ■$$

Property 3.2) If $K \geq \max\{K'|\dot{r}|, K'V_m(1-\rho)\}$, $K' > 0$, there exists a finite time t_r

$$s(t) = 0 \text{ for } t_r \leq t < t_f$$

where t_r satisfies $r(t_r) \geq r(0)\left(\frac{\mu}{|s(0)| + \mu}\right)^{1/K'} > 0$.

Proof) The proof is same to the arguments used in proving (20). ■

Furthermore, suppose that λ is chosen such that

$$\lambda \geq \kappa(1 + \rho) \quad (40)$$

where $\kappa > 1$. Then, the following holds.

$$\frac{\dot{\sigma}(t) - \dot{\sigma}_d(t)}{\sigma(t) - \sigma_d(t)} = -\frac{\lambda V_m}{r(t)} \leq \frac{\kappa \dot{r}(t)}{r(t)}, \quad t_r \leq t < t_f$$

By solving the above differential inequality, it is seen that

$$|\sigma(t) - \sigma_d(t)| \leq |\sigma(t_r) - \sigma_d(t_r)| \left(\frac{r(t)}{r(t_r)}\right)^{\kappa}, t_r \leq t < t_f \quad (41)$$

Then, the following properties can be easily shown by observing (41), (38) and (39).

Property 3.3) $\lim_{t \to t_f} |\sigma(t) - \sigma_d(t)| = 0$, $\lim_{t \to t_f} \frac{\sigma(t) - \sigma_d(t)}{r(t)} = 0$

Property 3.4) $\lim_{t \to t_f} \dot{\sigma}(t) = c$

Property 3.5) $\lim_{t \to t_f} A_c(t) = \lim_{t \to t_f} c(V_t \cos\theta_t - 2\dot{r})$

By Property 3.5), it is concluded that the guidance command is bounded with the selection of gains of (40). The following theorem summarizes the above reasoning.

Theorem 3 : Suppose that;

A1′) The geometry of pursuer and target satisfy

$$|\theta_m(t)| < \theta_s, \; t \geq 0,$$

for some constant $\theta_s \in (0, \cos^{-1}\rho)$.

A2′) The navigation gains are selected such that

$$\lambda \geq \kappa(1+\rho), \; K \geq \max\{K'|\dot{r}|, K'V_m(1-\rho)\}, \; \omega \geq \mu K,$$

where λ, κ, K' and μ are positive constants and $\kappa > 1$.

A3′) The target are assumed to turn with a constant acceleration such that

$$A_t = V_t \cdot c$$

where c is a constant real number.

Then, the guidance law (39) guarantees the existence of a finite time t_f such that

$$\lim_{t \to t_f} r(t) = 0$$

$$\lim_{t \to t_f} \gamma_m(t) = \lim_{t \to t_f} \gamma_d(t).$$

Furthermore, $\lim_{t \to t_f} A_c(t) = \lim_{t \to t_f} c(V_t \cos\theta_t - 2\dot{r})$.

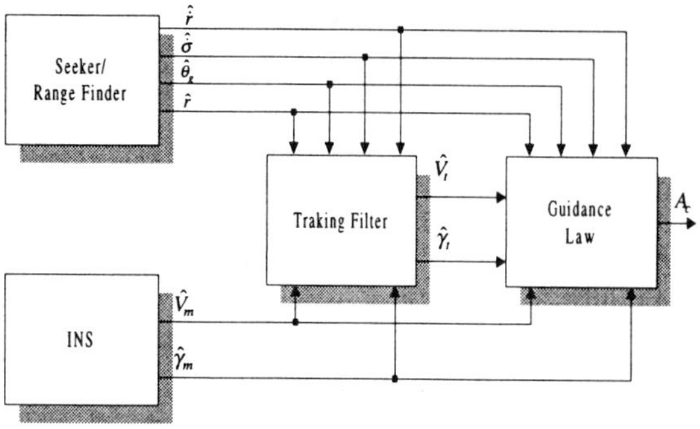

Fig. 2 Schematic diagram of guidance system.

Proof: It is immediate consequence of Property 3.1)-3.5). ■

Implementation of Guidance Laws

Note that required variables for the implementation of the modified PN in this paper are same to the previous ones in Ref. 10 and 13-15. In practice, active seeker with range finger, built-in INS, and tracking filter are generally available for missiles or torpedoes. It is clear that all of the required variables can be obtained with the above sensors. The schematic diagram of whole guidance system is presented in Fig. 2.

Since the implementation of the associated control switching is necessarily imperfect, the discontinuous control $\mathrm{sgn}(\cdot)$ in (31) and (39) should be smoothed.[17] It is approximated by Fig. 3 in this paper. The positive constant δ of Fig. 3 should be well chosen to achieve a suitable trade-off between guidance system bandwidth and overall performance.

V. Simulation Results

The simulation study in this section demonstrates the robustness of the modified BPN against variation in the initial condition, and the convergence of the terminal pursuer acceleration with the guidance gains of Corollary 1. For the implementation of $\mathrm{sgn}(\cdot)$, δ of Fig. 3 is set to 0.002.

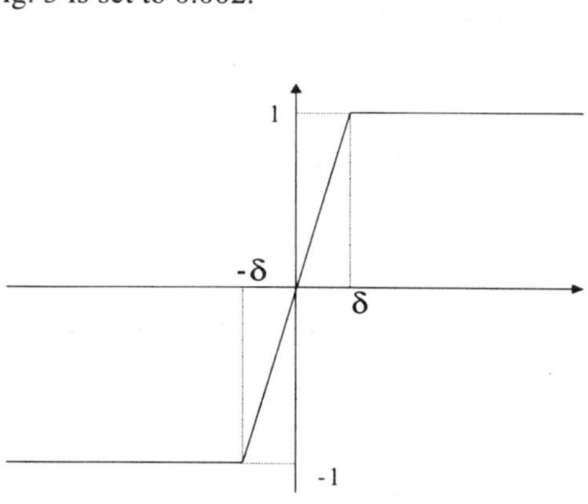

Fig. 3 Implementation of $\mathrm{sgn}(\cdot)$.

Let us first see the robustness of the modified PN against variation in the initial attitude angle of pursuer. The guidance gains are selected by (26) with $\kappa = 1.1$, $K' = 0.9$, and $\mu = 0.02$. It is obvious that the above selection satisfies the assumption of Corollary 1. The initial engagement geometry is given by

$$\sigma(0) = -30^\circ,\ r(0) = 500\mathrm{m},\ \text{and}\ \gamma_m(0) = \gamma_t(0) = 0^\circ.$$

The pursuer and the target velocities are assumed to be constants such that

$$V_m = 45\,\mathrm{m/sec}\ \text{and}\ V_t = 15\,\mathrm{m/sec}.$$

The desired impact attitude angle is assumed to be perpendicular to the target attitude angle. The optimal guidance law of (27) by Kim and Grider[13] is employed for performance comparison. By repeated trials, it is known that the pursuer can perpendicularly impact the target when b belongs to $[0.5446, 0.6018]$. Outside the interval, the impact angle error or the miss distance becomes excessive. In this section, b is selected as the center of the interval, i.e., $b = 0.5732$. Fig. 4(a) represents the trajectories of pursuer and target under the above conditions. The modified PN and the optimal guidance law together completes the guidance goal. However, the optimal guidance law fails to intercept the target when the initial attitude angle of pursuer is slightly changed to 10°. The modified PN still completes the guidance goal. Fig. 4(b) shows this situation. This simple example demonstrates clearly that the modified PN is better than the optimal guidance law in terms of launch envelope.

Theoretically, it is proved in Corollary 1 that the suitable selection of guidance gains guarantees zero terminal guidance command. However, there may a performance degradation since an approximation is used to implement $\mathrm{sgn}(\cdot)$. Let us see the effect of this approximation. The guidance gains in the following simulation is same to the above one of except that K' is changed to 1.2. Three cases are tried with different κ 's. The simulation results are presented in Fig. 5. Fig. 5(a) shows the trajectories of pursuer and target, and Fig. 5(b) represents the pursuer acceleration profiles. Corollary 1 states that κ 's greater than one produce zero acceleration of the pursuer at final time. By observing Fig. 5(b), it is concluded that the statement is still valid even if the $\mathrm{sgn}(\cdot)$ is replaced with a saturation function of Fig. 3. When κ is 1.1 or 1.5, the pursuer acceleration is very close to zero at final time. However, it becomes a nonzero value when $\kappa = 0.5$. It is also observed from Fig. 5(a) that the flight time becomes shorter as κ decreases. If κ is chosen appropriately, the various trajectories adequate for given missions can be obtained.

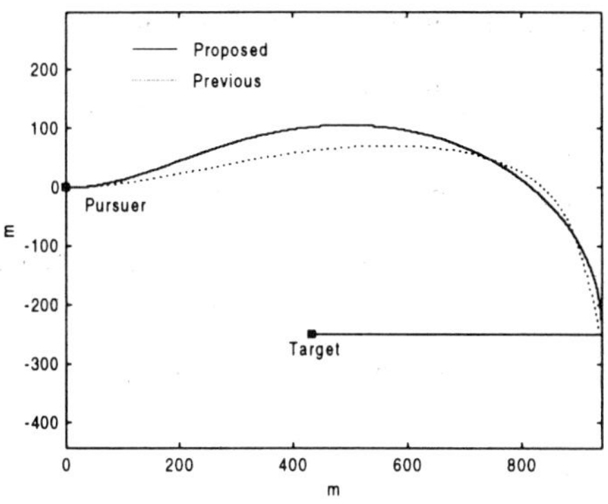

(a) $\gamma_m(0) = 0^\circ$

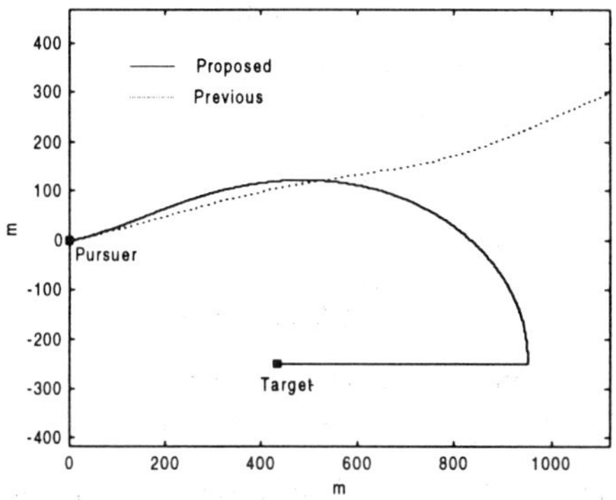

(b) $\gamma_m(0) = 10^\circ$

Fig. 4 Trajectories of pursuer and target.

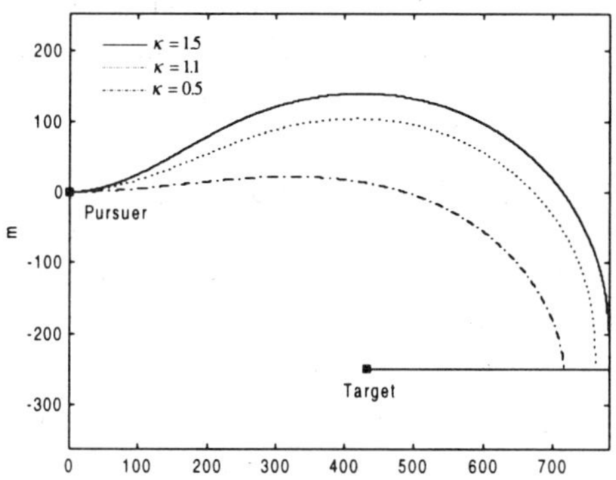

(a) Trajectories of pursuer and target.

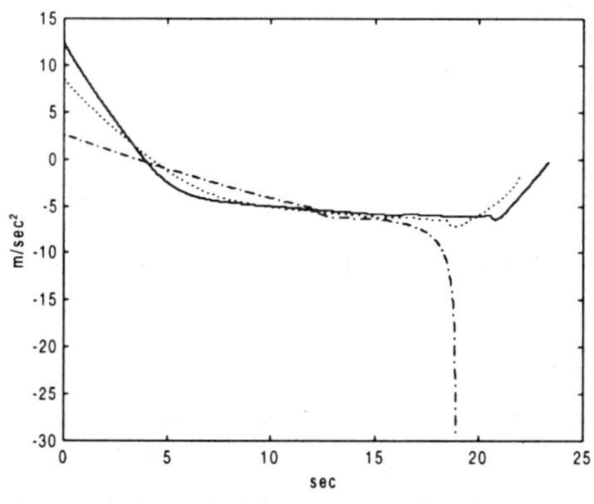

(b) Acceleration profiles.

Fig. 5 Modified PN with various κ.

Let us consider a maneuvering target with constant acceleration. In this simulation, the initial attitude angle of the target to be 30deg. The turn rate of target is $10^\circ/\text{sec}$. The initial LOS angle and the attitude angle of the pursuer are -30deg and -10deg, respectively. The simulation results are presented in Fig. 6. In the figure, the solid line represents the guidance law of (39) and the dotted line represents the guidance law of (17). The guidance law of (39) shows a good performance against maneuvering target. The acceleration at final phase in this case is not so large. Whereas, the acceleration of pursuer becomes excessive at final phase with the guidance law of (17), which may cause severe miss distance and impact angle error.

As another simulation scenario, let us consider a target evasive maneuver represented by Fig. 7. In this simulation, the target and the pursuer velocities are $45\,\text{m/sec}$ and $20\,\text{m/sec}$, respectively. The initial geometry satisfies

$$\sigma(0) = -30^\circ,\ \gamma_m(0) = -10^\circ, \text{ and } \gamma_t(0) = 45^\circ.$$

The absolute value of target acceleration is assumed to be $1.75\,\text{m/sec}^2$ which corresponds to the turn rate of $5^\circ/\text{sec}$. In order to access the performances of the guidance laws in realistic environments, the pursuer and the target dynamics are modeled as second-order time-invariant linear systems with the natural frequency of $10\,\text{rad/sec}$ and the damping ratio of 0.8. The seeker and tracking filter dynamics are also modeled as first-order linear systems with time constant of 0.1. The maximum acceleration of pursuer is assumed to be $23.6\,\text{m/sec}^2$ which corresponds to the turn rate of $30^\circ/\text{sec}$. Under the above conditions, the guidance laws of (17) and (31) are compared with the guidance gains such as $\kappa = 1.1$, $K' = 0.9$, and $\mu = 0.01$. The two guidance laws altogether make the pursuer impact the target. It is seen that the guidance law of (17) is better than that of (31) in terms of capture time. However, the impact angle error of (17) is somewhat larger as shown in Table 1. The miss distance of (31) is also slightly smaller than that of (17).

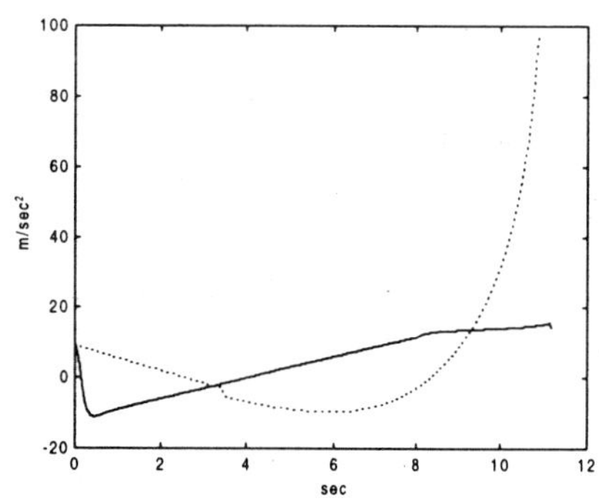

Fig. 6 Acceleration profiles.

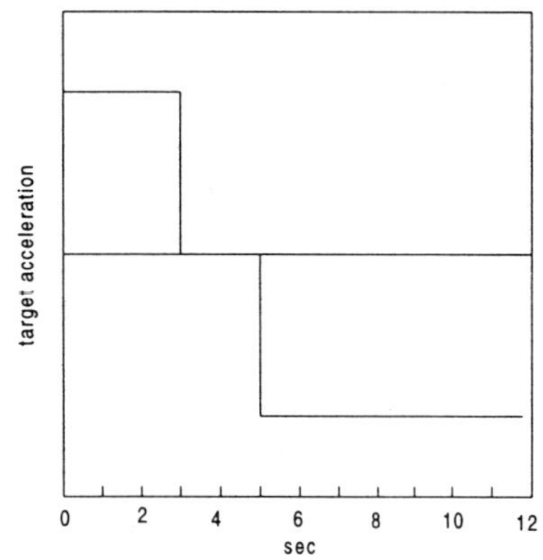

Fig. 7 Target acceleration time history.

Table 1 Simulation results for randomly maneuvering target.

	Final Time[sec]	Miss Distance[m]	Impact Angle Error[deg]
Guidance Law in (17)	15.4	0.69	8.29
Guidance Law in (31)	17.8	0.14	2.37

By these results, it is known that the guidance law of (31) with target maneuver compensation reduces the guidance errors if the target acceleration is accurately estimated.

VI. Conclusions

This paper presents modified PNG laws for the guidance problem with terminal angular constraint. Randomly maneuvering target as well as nonmaneuvering target is assumed in the performance analysis. The sliding mode control methodology is employed to analyze the proposed guidance laws. The relation between the modified PN and the optimal guidance law is also investigated. The optimal guidance gains of the modified PN which minimize the integral of the square of the pursuer acceleration are obtained. The proposed guidance laws are easily implemented with an active seeker and a built-in INS. Further, the estimate of the current target acceleration is sufficient even against maneuvering targets.

Acknowledgment

The authors gratefully acknowledge financial support from the Automatic Control Research Center and the Agency for Defense Development.

Appendix A: Proof of Theorem 1

Let us first prove (22) by contradiction. Suppose that

$$r(t) > 0, \quad t \ge 0. \tag{42}$$

Note that (42) with the assumptions of theorem guarantees (18). Without losing generality, assume that $s(0) > 0$. This with (18) implies that

$$\dot{s}(t) < -\frac{\omega}{r(t)} \le -\frac{\mu K' V_m (1-\rho)}{r(t)}, \quad s(t) \ne 0 \text{ and } t \ge 0$$

which can be written as

$$s(t) < -\mu K' V_m (1-\rho) \int_0^t 1/r \, d\tau + s(0), \; s(t) \ne 0 \text{ and } t \ge 0 \tag{43}$$

We can obtain the bound of r by (3) and (42) such that

$$0 < r(t) \le (V_m + V_t)t + r(0), \quad t \ge 0.$$

From this inequality, it is derived that

$$\int_0^t \frac{1}{r} d\tau \ge \int_0^t \frac{1}{(V_m + V_t)\tau + r(0)} d\tau, \quad t \ge 0, \tag{44}$$

Combining (43) and (44), it is easily seen that there exists $t_1 \in [0, \infty)$ such that

$$s(t) = 0, \quad t \ge t_1,$$

where $t_1 \le r(0)\left\{\exp\left(s(0)\frac{1+\rho}{\mu K'(1-\rho)}\right) - 1\right\} \Big/ (V_m + V_t)$.

Then, the guidance command becomes

$$A_m(t) = (\lambda V_m - \dot{r})\dot{\sigma}/\cos\theta_m, \quad t \ge t_1.$$

Noting that $\dot{r} < V_t$ by (10), it is seen that the above equation and the assumptions of theorem satisfies the assumptions of Lemma. Hence, there exists a finite time t_2 such that $r(t_2) = 0$. It contradicts to (42). This with the continuity of r together implies that there exists a finite time t_f such that $\lim_{t \to t_f} r(t) = 0$.

The next step is to prove that

$$\lim_{t \to t_f} \int_{t_3}^t \frac{1}{r(\tau)} d\tau = \infty, \text{ for any } t_3 \in [0, t_f). \tag{45}$$

In order to show this property, introduce the following inequality

$$\dot{r}(t) \ge -(V_m + V_t), \quad t \ge 0$$

which is an immediate consequence of (3). By integrating both sides, it is seen that

$$-r(t) = \lim_{\tau \to t_f} \int_t^\tau \dot{r}(\xi) d\xi \ge -(V_m + V_t)(t_f - t), \quad 0 \le t < t_f.$$

This implies that

$$0 < r(t) \le (V_m + V_t)(t_f - t), \quad 0 \le t < t_f.$$

From the above relation,

$$1/r(t) \ge 1/(V_m + V_t)(t_f - t), \quad 0 \le t < t_f$$

which implies that $\lim_{t \to t_f} \int_{t_3}^t \frac{1}{r(\tau)} d\tau = \infty$, for any $t_3 \in [0, t_f)$.

Let us prove (23) by the above properties. Observing (20), it is easily shown that (21) with (22) guarantees the existence of a finite time $t_4 \ge 0$ such that

$$s(t)=0 \text{ for } t_4 \le t < t_f . \quad (46)$$

Thus,

$$s(t)=\dot{\sigma}(t)+\frac{\lambda V_m}{r(t)}\left(\sigma(t)-\sigma_d\right)=0, \quad t_4 \le t < t_f .$$

It means that

$$\sigma(t)=\sigma(t_4)\exp\left[-\lambda V_m \Gamma(t,t_4)\right] + \exp\left[-\lambda V_m \Gamma(t,t_4)\right]\int_{t_4}^{t}\frac{\lambda V_m}{r}\sigma_d \exp\left[\lambda V_m \Gamma(\varsigma,t_r)\right]d\varsigma ,$$

$$t_4 \le t < t_f \quad (47)$$

where $\Gamma(\xi,t_4)=\int_{t_4}^{\xi}\frac{1}{r(\tau)}d\tau$. It holds by (45) and L'Hospital's rule that

$$\lim_{t\to t_f}\exp\left[-\lambda V_m \Gamma(t,t_4)\right]\int_{t_4}^{t}\frac{\lambda V_m}{r}\exp\left[\lambda V_m \Gamma(\varsigma,t_4)\right]d\varsigma = \lim_{t\to t_f}\frac{\frac{\lambda V_m}{r}\exp\left[\lambda V_m \Gamma(t,t_4)\right]}{\frac{\lambda V_m}{r}\exp\left[\lambda V_m \Gamma(t,t_4)\right]}=1$$

It is easily seen from (47) with the above equation and (45) that

$$\lim_{t\to t_f}\sigma(t)=\sigma_d . \quad (48)$$

It also holds by (48) and (14),

$$\lim_{t\to t_f} r(t)\dot{\sigma}(t)=0 .$$

This equation, (10) and (48) finally guarantees (23).

Appendix B: Proof of Theorem 2

Proof : It is easy to show that A1) implies that there exists a finite time t_f such that

$$\lim_{t\to t_f} r(t)=0 .$$

The remaining procedure is similar to the proof of Theorem 1. ■

References

[1]Siouris, G. M., "Comparison between Proportional and Augmented Proportional Navigation," *Nachrichtententechnische Zeitschrift,* 1974, pp. 278-280.

[2]Becker, K., "Closed-form Solution of Pure Proportional Navigation," *IEEE Transactions on Aerospace and Electronic Systems*, Vol. AES-26, No. 3, 1990, pp. 526-533.

[3]Ghose, D., "Capture Region for True Proportional Navigation Guidance with Nonzero Miss-Distance," *Journal of Guidance, Control, and Dynamics*, Vol. 17, No. 3, 1994, pp. 627-628.

[4]Yuan, P. J., and Hsu, S. C., "Exact Closed-Form Solution of Generalized Proportional Navigation," *Journal of Guidance, Control, and Dynamics*, Vol. 16, No. 5, 1993, pp. 963-966.

[5]Zarchan, P., *Tactical and Strategic Missile Guidance*, AIAA, Inc., 1994, Washington, pp. 25-43.

[6]Ha, I. J., Hur, J. S., Ko, M. S., and Song, T. L., "Performance analysis of PNG laws for randomly maneuvering targets," *IEEE Transactions on Aerospace and Electronic Systems*, Vol. AES-26, No. 5, 1990, pp. 713-721.

[7]Song, S. H., and Ha, I. J., "A Lyapunov-like approach to performance analysis of 3-dimensional pure PNG laws," *IEEE Transactions on Aerospace and Electronic Systems*, Vol. AES-30, No. 1, 1994, pp. 238-248.

[8]Oh, J. H, and Ha, I. J., "A New Approach to Interceptibility Analysis of 3-Dimensional PPNG Law," *Proceedings of 1997 AIAA Guidance, Navigation and Control Conference*, New Orleans, LA, 1997, to appear.

[9]Babu, K. R., Sarma, I. G., and Swamy, K. N., "Switched Bias Proportional Navigation for Homing Guidance against Highly Maneuvering Targets," *Journal of Guidance, Control, and Dynamics*, Vol. 17, No. 6, 1994, pp. 1357-1363.

[10]Kim, E., Cho, H., and Lee, Y., "Terminal Guidance Algorithms of Missiles Maneuvering in the Vertical Plane," *Proceedings of 1996 AIAA Guidance, Navigation and Control Conference*, San Diego, CA, 1996.

[11]Giovannetti, N., and Signorini, M., "Impact with Angular Constraint for a modern LWT: Validation by Means of Numeric Simulations," *Proceedings of the 1995 Undersea Defense Technology Conference*, Cannes, France, 1995, pp. 365-369.

[12]Idan, M., Golan, O. M., and Guelman, M., "Optimal Planar Interception with Terminal Constraints," *Journal of Guidance, Control, and Dynamics*, Vol. 18, No. 6, 1995, pp. 1273-1279.

[13]Kim, M., and Grider, K. V., "Terminal Guidance for Impact Attitude Angle Constrained Flight Trajectories," *IEEE Transactions on Aerospace and Electronic Systems*, Vol. AES-9, No. 6, 1973, pp. 852-859.

[14]York, R. J., and Pastrick, H. L., "Optimal Terminal Guidance with Constraints at Final Time," *Journal of Spacecraft and Rockets*, Vol. 14 , No. 6, 1977, pp. 381-382.

[15]Ryu, C. K., and Cho, H. J., "Optimal Guidance with Terminal Impact Angle and Control Constraint (in Korean)," *Proceedings of the 1992 Korean Automatic Control Conference,* Seoul, Korea, 1992, pp. 601-606.

[16]Glizer, V. Y., "Optimal Planar Interception with Fixed End Conditions: Closed-Form Solution," *Journal of Optimization Theory and Applications*, Vol. 88, No. 3, 1996, pp. 503-539.

[17]Slotine, J. E., and Li, W., *Applied Nonlinear Control*, Prentice-Hall, Inc., 1991, Englewood Cliffs, pp. 276-309.

AIAA-97-3475

Optimal Thrust Vector Control of Tactical Missiles

Der-Ren Taur* and Jeng-Shing Chern**
Chung Shan Institute of Science and Technology
P . O . BOX 90008-6-7
Lungtan, Taiwan 32526, ROC

The open loop optimal trajectories are analyzed for a point-mass model of tactical missiles with tail control enhanced by thrust vectoring capability. Care has been taken to model the missile aerodynamics in the high angle of attack flight regime executing post-stall maneuvers. The aim of this paper is to study the optimal turn laws for short range dogfight missiles to achieve maximum final speed in a given time with final flight path angle constraint during the post launch phase. Various problems such as the agile turn of a tactical surface-to-air missile (SAM) to engage very low altitude target, positioning of a SAM on a predicted line-of-sight (LOS) for beam-riding (BR) system, super agile turn of a air-to-air missile (AAM) for a large off-boresight angle targeting or to engage targets in the rear hemisphere, etc., are presented. In addition, the mechanism of the optimal thrust-angle programs for the agile missile will also be highlighted by the analytic solutions without considering missile aerodynamics.

Introduction

During the post launch phase of tactical missiles, dynamic pressures of the missile are often too low for aerodynamic controls to make an agile turn. Because the conventional aerodynamic control surfaces may no longer be effective at extreme angles of attack, alternate propulsive control effectors such as thrust vector control (TVC) will have to be employed for the next-generation agile missiles [1-3]. For instance in air-to-air missions [4], in aircraft close-in-combat scenarios, a large off-foresight angle targeting capability, or the ability to engage targets in the rear hemisphere, is a significant advantage. Another launch scenario may involve the post-stall maneuvering to engage very low altitude target for vertically launched SAM [5]. The third scenario is to position a SAM on a predicted LOS for BR system when the aiming angle is large .In all these cases, the missile will encounter the high angle of attack flight regime and therefore the thrust vectoring capability is necessary to maintain control over the missile during the agile turns.

In this paper, the open loop optimal trajectories are analyzed for a point-mass model of an agile missile with tail control enhanced by TVC. The aim of this study is to obtain the optimal turn laws to achieve maximum final speed in a given time with final flight path angle constraint during the post launch phase. Various problems such as the agile turn of a SAM to engage very low altitude target, positioning of a SAM on a predicted LOS for BR system at large aiming angle, super agile turn of an AAM for a large off-boresight angle targeting or to engage targets in the rear

---

* Associate Scientist ,Member AASRC
** Scientist, Associate Fellow AIAA, Permanent Member AASRC, Member AIAA Committee on Standards

hemisphere, etc., are presented.

The vehicle model is a generic high performance short range tactical missile with tail control and thrust vectoring capability. During the post launch phase, the main thrust-angle direction is used as the control variable in this study to provide controllable flight in the post-stall region .Care has been taken to model the missile aerodynamics in the high angle of attack flight regime [6] .The missile agile turn is confined in a vertical pitch-plane for surface-to-air missions, but in a horizontal yaw-plane for air-to-air missions instead. Besides, the main thrust is constant in magnitude but the mass of the missile is time varying.

The numerical solutions of the proposed optimal control problems have been solved by using a highly stable numerical scheme, the multiple shooting method[7]. The initial guesses of the adjoint variables and state variables required for solving the two point boundary value problem (TPBVP) are obtained from the analytic solutions solved by neglecting the missile aerodynamics. In addition, the mechanism of the optimal thrust-angle programs for the agile missile will also by highlighted by the solutions obtained without considering missile aerodynamics.

Missile Model

A point-mass model for missile flight over flat earth is used to represent the missile trajectory dynamics in the present study. The normalized equations of motion for the two-dimensional mathematical model of the missile flight are as follows:

$$\dot{\hat{x}} = \hat{u} \quad , \quad \dot{\hat{y}} = \hat{v}$$

$$\dot{\hat{u}} = \frac{T\cos\theta}{w} - \frac{N(\alpha)\sin\theta}{w} - \frac{A(\alpha)\cos\theta}{w} - \cos\sigma_0$$

$$\dot{\hat{v}} = \frac{T\sin\theta}{w} + \frac{N(\alpha)\cos\theta}{w} - \frac{A(\alpha)\sin\theta}{w} - \sin\sigma_0$$

$$w = \overline{M}(t)g \quad , \quad \hat{x} = \frac{x}{g} \quad , \quad \hat{y} = \frac{y}{g}$$

$$\hat{u} = \frac{u}{g} \quad , \quad \hat{v} = \frac{v}{g} \qquad (1)$$

where x is the down range of the missile in m, y the altitude in m for S-A missions or the cross range for A-A missions, u the missile velocity component along the x coordinate in m/s, v the missile velocity component along the y coordinate in m/s , σ_0 the aiming angle between x_I and y_E coordinates in degree (see Fig.1) , α the angle of attack of the missile in degree, $\overline{M}$ the mass of the missile in kg, g the acceleration of gravity in m/s^2. The parameter T is the main thrust magnitude along the missile body axis, A the aerodynamic axial force along the missile body axis, N the normal force perpendicular to the missile body axis. All are in Newton. Furthermore, θ is the attitude of the missile body axis, the trajectory control variable in the present problem.

Figure 1 illustrates the missile coordinate systems and the variables of interest. The missile equations of motion are expressed in terms of two coordinates. The coordinate system x_E-y_E is used to define the earth fixed frame where the coordinate x_E is in the local horizon plane, while x_I-y_I is used to define the position and attitude of the missile body with respect to some inertial frame. For BR system, the x_I coordinate is along the predicted LOS direction, but for A-A mission the initial velocity vector of the missile is along the x_E direction. However, for the S-A low altitude mission, the coordinates x_I-y_I and x_E-y_E are coincident.

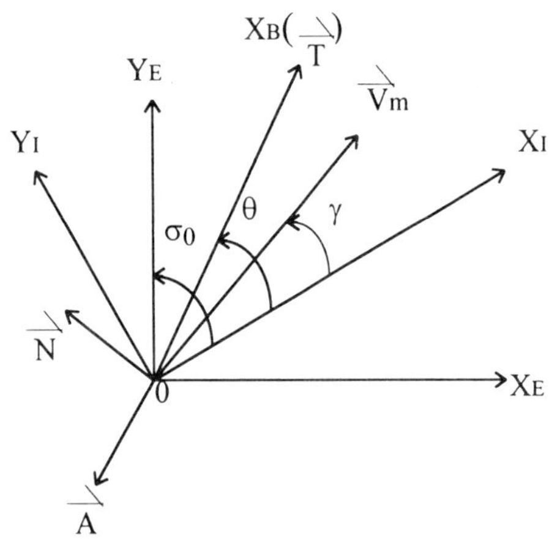

Fig.1 Missile coordinate systems

The form of the missile aerodynamic forces and the definition of other related variables are given by the following expressions:

$$N = QSCN(M,\alpha)$$
$$A = QSCA(M,\alpha) \quad (2)$$

where CN is the normal force coefficient, and CA is the axial force coefficient, all given as functions of mach number M and angle of attack α. The variable S is the reference area. Mach number M, dynamic pressure Q, total velocity Vm, and angle of attack α are defined as:

$$M = v_m / ss \quad , \quad Q = \frac{1}{2}\rho V_m^2$$

$$v_m = g\left(\hat{u}^2 + \hat{v}^2\right)^{\frac{1}{2}} \quad , \quad \alpha = \theta - \gamma$$

$$\gamma = Tan^{-1}\left(\frac{\hat{v}}{\hat{u}}\right) \quad (3)$$

where the variable ss is the speed of sound, ρ the air density, γ the flight path angle of the missile.

The high angle of attack aerodynamic coefficients of the missile below the transonic region are approximately modeled as follows[6]:

$$CN(\alpha) = CN_{\frac{\pi}{2}} \sin\alpha$$
$$CA(\alpha) = CA_0 + a_1\alpha + a_2\alpha^2 + a_3\alpha^3 + a_4\alpha^4 + a_5\alpha^5 \quad \text{for } \alpha \le 90^\circ$$
$$= CA_0 - (CA_0 - CA_\pi)\sin\frac{9}{7}(\alpha - 90^\circ) \quad \text{for } 160^\circ > \alpha > 90^\circ$$
$$= CA_\pi \quad \text{for } \alpha > 160^\circ \quad (4)$$

where $CN_{\pi/2}$ is the normal force coefficient at angle of attack $\alpha=90^\circ$, CA_0 is the axial force coefficient at angle of attack $\alpha=0^\circ$, CA_π is the axial force coefficient at angle of attack $\alpha=180^\circ$, and CA is taken positive when directed towards the base.

On the other hand, the aerodynamic coefficients in the supersonic region (M>1.1) are modeled as follows:

$$CA(\alpha) = CA_0^1 \cos^2\alpha \quad \text{for } \alpha \le 90^\circ$$
$$= CA_\pi^1 \cos^2(180^\circ - \alpha) \quad \text{for } \alpha > 90^\circ$$

$$CN(\alpha) = CN_{\frac{\pi}{2}}^1 \sin\alpha \quad (5)$$

Variational Formulation

The aim of this study is to maximize the final velocity of the missile with final flight path angle constraint in a given time, the final position of the missile is constrained for S-A missions but free for A-A mission. The cost functional can be expressed as:

$$J = MIN\left[-\hat{u}(t_f)\right] \tag{6}$$

Using the minimum principle, we introduce the adjoint vector $\vec{\lambda}$ to form the Hamiltonian function as follows:

$$H = \lambda_x \hat{u} + \lambda_y \hat{v}$$

$$+\lambda_u\left(\frac{T\cos\theta}{w} - \frac{N\sin\theta}{w} - \frac{A\cos\theta}{w} - \cos\sigma_o\right)$$

$$+\lambda_V\left(\frac{T\sin\theta}{w} - \frac{N\cos\theta}{w} - \frac{A\sin\theta}{w} - \sin\sigma_o\right) \tag{7}$$

The optimal value of the control variable θ is determined from the Pontryagin minimum principle ,i.e., minimize the Hamiltonian with respect to θ .Then

$$\frac{\partial H}{\partial \theta} = 0 \tag{8}$$

For this unconstrained case, the optimal control is given implicitly by the following formula

$$\lambda_u(-T\sin\theta + A\sin\theta - A_\theta\cos\theta$$
$$-N\cos\theta - N_\theta\sin\theta)$$
$$+\lambda_v(T\cos\theta - A\cos\theta - A_\theta\sin\theta$$
$$-N\sin\theta + N_\theta\cos\theta) = 0 \tag{9}$$

where the derivatives A_θ and N_θ are given by

$$A_\theta = QS\frac{\partial CA}{\partial\alpha} = QSCA_\alpha \quad , N_\theta = QS\frac{\partial CN}{\partial\alpha} = QSCN_\alpha \tag{10}$$

On the other hand, Euler-Lagrange equations provide the differential system of the adjoint variables

$$\dot{\lambda}_x = \frac{-\partial H}{\partial \hat{x}} = 0$$

$$\dot{\lambda}_y = -\frac{\partial H}{\partial \hat{y}} = 0$$

$$\dot{\lambda}_u = -\frac{\partial H}{\partial \hat{u}}$$
$$= -\left[\lambda_x + \frac{\lambda_u}{w}(-A_u\cos\theta - N_u\sin\theta)\right.$$
$$\left. + \frac{\lambda_v}{w}(-A_u\sin\theta + N_u\cos\theta)\right]$$

$$\dot{\lambda}_v = -\frac{\partial H}{\partial \hat{v}}$$
$$= -\left[\lambda_y + \frac{\lambda_u}{w}(-A_v\cos\theta - N_v\sin\theta)\right.$$
$$\left. + \frac{\lambda_v}{w}(-A_v\sin\theta + N_v\cos\theta)\right] \tag{11}$$

where the derivatives A_u, A_v ,N_u ,and N_v are given by

$$A_u = \frac{\partial A}{\partial \hat{u}} = \rho g^2 SCA\hat{u}$$
$$+ QSCA_\alpha g^2\frac{\hat{v}}{V_m^2}$$

$$A_v = \frac{\partial A}{\partial \hat{v}} = \rho g^2 SCA\hat{v}$$

$$-QSCA_\alpha g^2 \frac{\hat{u}}{V_m^2}$$

$$N_u = \frac{\partial N}{\partial \hat{u}} = \rho g^2 SCN \hat{u}$$

$$+QSCN_\alpha g^2 \frac{v}{V_m^2}$$

$$N_v = \frac{\partial N}{\partial \hat{v}} = \rho g^2 SCN \hat{v}$$

$$-QSCN_\alpha g^2 \frac{\hat{u}}{V_m^2} \qquad (12)$$

The optimal solution to the proposed problem is governed by Eqs.(1) for the state variables, and Eqs.(11) for the adjoint variables. The resulting boundary conditions are

Initial conditions:

$$\hat{x}(t_0) = 0 \qquad ,\hat{y}(t_0) = 0$$
$$\hat{u}(t_0) = \hat{V}_0 \cos\gamma_0 \ ,\hat{v}(t_0) = \hat{V}_0 \sin\gamma_0 \qquad (13)$$

Final conditions:

$$\hat{x}(t_f) = free \quad , \quad \hat{y}(t_f) = y_f \ \ or\ free$$

$$\hat{u}(t_f) = max \quad , \quad \hat{v}(t_f) = 0 \qquad (14)$$

where V_0 is the initial total velocity and γ_0 the initial flight path angle of the missile. The above boundary value problems will be solved for fixed final time t_f.

Analytic Solutions without Missile Aerodynamics

From Eqs.(9) and (10), it is clear that the optimal control θ^* do not possess an analytic solution due to the complexity of the missile aerodynamics. However during the post launch phase the dynamic pressure of the missile is quite low and the aerodynamic forces can be neglected in a first approximation. After neglecting the missile aerodynamics in Eq.(9), we obtain the optimal control θ^* as the following "bilinear tangent law [8]"

$$Tan\ \theta^* = \frac{\lambda_v}{\lambda_u} = \frac{at+b}{at+d} \qquad (15)$$

Due to the simplicity of the bilinear tangent law, the optimal control problems proposed in this study can be solved analytically. In fact, the analytic solutions not only can provide the initial guesses of the state and adjoint variables to solve the TPBVP numerically with missile aerodynamics, but also can give insight to highlight the mechanism of the optimal turn laws of the agile missile.

Maximum velocity transfer to a given line in pitch-plane

The basic problem considered here is to find the optimal thrust-angle programs to maximize the missile velocity along a given line(x_I coordinate) in pitch-plane with final flight path angle constraint in a given time. The associated TPBVP is defined as follows:

$$\dot{\hat{x}} = \hat{u}, \qquad \hat{x}(t_0) = 0, \qquad \hat{x}(t_f) = free$$

$$\dot{\hat{y}} = \hat{v}, \qquad \hat{y}(t_0) = 0, \qquad \hat{y}(t_f) = \hat{y}_f$$

$$\dot{\hat{u}} = \hat{a}\cos\theta - \cos\sigma_0, \quad \hat{u}(t_0) = \hat{V}_0\cos\gamma_0$$
$$\hat{u}(t_f) = \max, \qquad \hat{a} = T/W$$

$$\dot{\hat{v}} = \hat{a}\sin\theta - \sin\sigma_0, \quad \hat{v}(t_0) = \hat{V}_0\sin\gamma_0,$$
$$\hat{v}(t_f) = 0$$

$$\dot{\lambda}_x = 0, \qquad \lambda_x(t_f) = 0$$

$$\dot{\lambda}_y = 0, \qquad \dot{\lambda}_v = -\lambda_y$$

$$\dot{\lambda}_u = -\lambda_x, \qquad \lambda_u(t_f) = -1 \qquad (16)$$

With $\lambda_x=0$, it follows that $\lambda_u=-1$ throughout the flight, so the optimal control law becomes a " linear tangent law " :

$$\tan\theta = \tan\theta_0 - c_1 t, \quad \text{where} \quad c_1 = -\lambda_y \qquad (17)$$

For the case of constant mass, the above differential equations are readily integrated with the linear tangent law, using θ as the independent variable instead of t, one obtains

$$\hat{u} = \frac{\hat{a}}{c_1}\ell n\left(\frac{\sec\theta_0 + \tan\theta_0}{\sec\theta + \tan\theta}\right) + \frac{\cos\sigma_0}{c_1}\left(\tan\theta - \tan\theta_0\right) + \hat{V}_0\cos\gamma_0$$

$$\hat{v} = \frac{\hat{a}}{c_1}\left(\sec\theta_0 - \sec\theta\right) + \frac{\sin\sigma_0}{c_1}\left(\tan\theta - \tan\theta_0\right) + \hat{V}_0\sin\gamma_0$$

$$\hat{x} = \frac{\hat{a}}{c_1^2}\left[\left(\sec\theta_0 - \sec\theta\right) - \tan\theta\ \ell n\left(\frac{\sec\theta_0 + \tan\theta_0}{\sec\theta + \tan\theta}\right)\right] + \frac{\hat{V}_0\cos\gamma_0}{c_1}(\tan\theta - \tan\theta_0) + \frac{\cos\sigma_0}{2c_1^2}\left(\sec^2\theta - \sec^2\theta_0\right) - \frac{\cos\sigma_0\tan\theta_0}{c_1^2}\left(\tan\theta - \tan\theta_0\right)$$

$$\hat{y} = \frac{\hat{a}}{2c_1^2}\left[\left(\tan\theta_0 - \tan\theta\right)\sec\theta_0 - \left(\sec\theta_0 - \sec\theta\right)\tan\theta - \ell n\left(\frac{\tan\theta_0 + \sec\theta_0}{\tan\theta + \sec\theta}\right)\right] + \frac{\hat{V}_0\sin\gamma_0}{c_1}(\tan\theta_0 - \tan\theta) + \frac{\sin\sigma_0}{2c_1^2}(\sec^2\theta - \sec^2\theta_0) - \frac{\sin\sigma_0\tan\theta_0}{c_1^2}\left(\tan\theta - \tan\theta_0\right) \qquad (18)$$

The above formula can be directly applied to surface launched BRM with different predicted LOS directions, and the constants θ_0 and c_1 are determined by the final boundary conditions $\hat{v}(t_f) = 0$ and $\hat{y}(t_f) = 0$ The optimal thrust-angle programs for different LOS pointing directions with missile initial speed V_0=50m/s and launched perpendicular to the local horizon plane (see Fig.1) are given in Figure 2.

On the other hand, for the application to vertically launched low altitude SAM, the analytic solutions of the state variables are given by

$$\hat{u} = \frac{\hat{a}}{\hat{c}_1}\ \ell\text{n}\ \left(\frac{\sec\theta_0 + \tan\theta_0}{\sec\theta + \tan\theta}\right)$$

$$\hat{v} = \frac{\hat{a}}{\hat{c}_1}\left(\sec\theta_0 - \sec\theta\right) + \frac{1}{c_1}\left(\tan\theta - \tan\theta_0\right) + \hat{V}_0$$

$$\hat{x} = \frac{\hat{a}}{c_1^2}\Big[(\sec\theta_0 - \sec\theta) - \tan\theta \ell n\left(\frac{\sec\theta_0 + \tan\theta_0}{\sec\theta + \tan\theta}\right)\Big]$$

$$\hat{y} = \frac{\hat{a}}{2c_1^2}\Big[(\tan\theta_0 - \tan\theta)\sec\theta_0 - \left(\sec\theta_0 - \sec\theta\right)\tan\theta - \ell n\left(\frac{\tan\theta_0 + \sec\theta_0}{\tan\theta + \sec\theta}\right)\Big] + \frac{\hat{V}_0}{C_1}\left(\tan\theta_0 - \tan\theta\right) + \frac{1}{2c_1^2}\left(\sec^2\theta - \sec^2\theta_0\right) - \frac{\tan\theta_0}{c_1^2}\left(\tan\theta - \tan\theta_0\right) \tag{19}$$

where the constants θ_0 and c_1 are determined by the final boundary conditions $\hat{v}(t_f) = 0$ and $\hat{y}(t_f) = \hat{y}_f$. Besides, the optimal thrust-angle programs for different target altitudes $\hat{y}_f$ with V_0=50m/s are given in Figure 3.

Maximum velocity transfer of an agile turn in horizontal-plane

Another related important problem considered here is to find the optimal turn laws to maximize the final velocity of the missile in a horizontal agile turn with final flight path angle constraint in a given time. The associated TPBVP is defined as follows:

$$\dot{\hat{x}} = \hat{u}, \qquad \hat{x}(t_0) = 0, \qquad \hat{x}(t_f) = free$$

$$\dot{\hat{y}} = \hat{v}, \qquad \hat{y}(t_0) = 0, \qquad \hat{y}(t_f) = free$$

$$\dot{\hat{u}} = \hat{a}\cos\theta, \qquad \hat{u}(t_0) = \hat{V}_0\cos\gamma_0, \qquad \hat{u}(t_f) = \max$$

$$\dot{\hat{v}} = \hat{a}sin\theta, \qquad \hat{v}(t_0) = \hat{V}_0\sin\gamma_0, \qquad \hat{v}(t_f) = 0$$

$$\dot{\lambda}_x = 0, \qquad \lambda_x(t_f) = 0$$

$$\dot{\lambda}_y = 0, \qquad \lambda_y(t_f) = 0$$

$$\dot{\lambda}_u = -\lambda_x, \qquad \lambda_u(t_f) = -1$$

$$\dot{\lambda}_v = -\lambda_y \tag{20}$$

With λ_x=0,it follows that λ_u=-1 throughout the flight. Furthermore, with λ_y=0 , it follows that λ_v= constant, so the optimal control law becomes a "constant thrust direction profile":

$$tan\,\theta = tan\,\theta_0 \tag{21}$$

For constant mass, the differential equations are readily integrated with the constant direction law

$$\hat{u} = \hat{a}\cos\theta_0 t + \hat{V}_0\cos\gamma_0$$
$$\hat{v} = \hat{a}\sin\theta_0 t + \hat{V}_0\sin\gamma_0$$
$$\hat{x} = \frac{1}{2}\hat{a}\cos\theta_0 t^2 + \hat{V}_0\cos\gamma_0 t$$
$$\hat{y} = \frac{1}{2}\hat{a}\sin\theta_0 t^2 + \hat{V}_0\sin\gamma_0 t \tag{22}$$

where the constant thrust direction θ_0 is determined by the final boundary condition $\hat{v}(t_f) = 0$

$$\theta_0 = \sin^{-1}\left(\frac{-\hat{V}_0 \sin\gamma_0}{t_f \hat{a}}\right) \qquad (23)$$

By using the above formula, the optimal thrust-angle programs for different aspect angles of the missile in a horizontal-plane for agile turn is given in Figure 4. It is interesting to note that the optimal thrust-angle direction is programmed to cancel the velocity component perpendicular to the initial LOS at the end of $t=t_f$ .Therefore, for the heading reversal maneuver of the missile, the optimum thrust-angle is just opposite the velocity direction initially until the missile speed towards zero $V_m=0$. Then the thrust-angle is at the same direction as the missile velocity vector. In other words, the angle of attack is 180° at first then switches to zero after $V_m=0$.

Numerical Computation Result

Maximum velocity transfer for SAM in pitch-plane

The basic problem considered here is to find the optimal thrust-angle programs in pitch-plane to maximize the final velocity of a SAM to engage very low altitude target in a given time with final flight path angle constraint. The associated boundary conditions are given by

$$\hat{x}(t_0) = 0, \qquad \hat{x}(t_f) = free$$

$$\hat{y}(t_0) = 0, \qquad \hat{y}(t_f) = \hat{y}_f$$

$$\hat{u}(t_0) = \hat{V}_0 \cos\gamma_0, \qquad \hat{u}(t_f) = \max$$

$$\hat{v}(t_0) = \hat{V}_0 sin\gamma_0, \qquad \hat{v}(t_f) = 0$$

$$\lambda_x(t_f) = 0, \qquad \lambda_u(t_f) = -1 \qquad (24)$$

The TPBVP described above are solved numerically by using the multiple-shooting algorithm in which the independent variable interval $[0,t_f]$ is broken into sub-intervals. In our application examples it is difficult to obtain estimates for the variables which are within the domain of attraction for the algorithm. In fact, the initial guess values of the state and adjoint variables are obtained from the analytic solutions without missile aerodynamics in the preceding section.

For the numerical computation, we introduce different initial speed V_0, different initial flight path angle γ_0, and different target altitude y_f to obtain the optimal strategies for the missile to achieve the maximum final velocity parallel to the x coordinate. The final time t_f is chosen to be two seconds for tactical requirement. A set of results for this problem is presented in Figs.5 through 16 . Fig.5 shows the optimum thrust-angle programs for different target altitude $\hat{y}_f$ with initial conditions given by V_0=50m/s and $\gamma_0=90^{\circ}$ .Fig.6 presents the optimal speed profiles for the same initial conditions. On the other hand, the optimum thrust-angle programs and speed profiles of the missile without considering missile aerodynamics for the same boundary conditions are given in Figs.7 and 8. From these results, it is clear that the optimum speed profiles without aerodynamics (see Fig. 8) are quite close to those with aerodynamics (see Fig. 6). That means the aerodynamic effect is not significant for this scenario because the dynamic pressures are quite low at the beginning of the trajectory profiles in the subsonic region. However, at the end of the trajectories the optimum thrust-angles with aerodynamics (see Fig. 5) are much less than those without aerodynamics (see Fig. 7) because the aerodynamic forces are active in the transonic region. From Figs. 6 and 8, we can also observe that the final speeds of the missile are roughly the same for different target altitudes. In other words, the achieved final speed of the vertically launched missile seems independent of the target altitude for this low altitude mission. Besides, from the optimum thrust-angle programs (see Figs. 5 and 7) we notice that the thrust-angle direction is programmed to cancel the velocity component of the missile perpendicular to the x coordinate. Hence the lower the target altitude is , the more the attitude of the missile points to the negative side (with respect to the x_I coordinate) at the initial phase of the trajectories. On the contrary, the lower the target altitude is, the more the attitude of the missile points to the positive side at the final phase of the trajectories. This is just the optimal turn laws for the low altitude scenario because for the lower target altitude the optimal trajectory is overshooting the reference target altitude and for slightly higher target altitude the optimal trajectory is undershooting the reference target altitude . This phenomenon can be easily seen from the optimal trajectory profiles shown in Fig. 9. Furthermore, the optimum angle of attack profiles of the missile are given in Fig. 10. From Fig. 10, we see that the angles of attack of the missile are in the post-stall region at the initial phase of the trajectories and converge to the small angle of attack region within half of the given final time.

Another important observation is that for different initial speeds of the vertically launched missile the optimum speed profiles are roughly the same for the low altitude mission but for the case y_f=100m. Since the lower the missile initial speed is, the less the angle of attack of the missile has during the flight. Figs.11 through 14 represent the optimum angle of attack profiles and speed profiles for missile initial speeds V_0=5m/s and V_0=25m/s respectively. Finally, for different missile launch directions shoot-forward has superior optimum final speed to that of shoot-backward mission as shown is Figs. 15 and 16 for the optimum speed profiles with target altitude y_f=10m and y_f=100m respectively.

Maximum velocity transfer for BRM in pitch-plane

Another important class of the related problem in pitch-plane is to find the optimum thrust-angle programs to position a SAM on a predicted LOS for BR system to achieve maximum final speed in a given time with final flight path angle constraint. The aiming angle σ_0 of the given LOS is defined in Fig. 1 and the associated boundary conditions are given by

$$\begin{aligned}
\hat{x}(t_0) &= 0, & \hat{x}(t_f) &= free \\
\hat{y}(t_0) &= 0, & \hat{y}(t_f) &= 0 \\
\hat{u}(t_0) &= \hat{V}_0 \cos\gamma_0, & \hat{u}(t_f) &= \max \\
\hat{v}(t_0) &= \hat{V}_0 \sin\gamma_0, & \hat{v}(t_f) &= 0 \\
\lambda_x(t_f) &= 0, & \lambda_u(t_f) &= -1
\end{aligned} \tag{25}$$

For the numerical experiment, we are interested in the variation of different aiming angle σ_0 ,different initial speed

V_0, and different initial flight path angle γ_0 to study the optimum turn laws for the missile to reach the maximum final speed along the given LOS directions. Another set of results for this optimization problem is presented in Figs. 17 through 28. Fig. 17 shows the optimum thrust-angle programs for different aiming angle σ_0 with initial missile speed V_0=50m/s and the missile launched perpendicular to the local horizon plane (see Fig. 1). Fig. 18 presents the optimum speed profiles for the same launch conditions. On the other hand, the optimum thrust-angle programs and speed profiles of the missile without considering aerodynamics for the same launch conditions are given in Figs. 19 and 20. From all these results, it is obvious that the optimal solutions without missile aerodynamics are quite close to those with aerodynamics. In other words, the aerodynamic effect is not significant for this type of mission. From Figs.18 and 20, we also observe that the final speed of the missile decreased as the aiming angle σ_0 increased due to the fact that the angle of attack of the missile increased as the aiming angle increased. This can be proved by the optimum angle of attack profiles of the missile as shown in Fig.21 . Another important fact is that for different initial speed V_o of the missile, the final speeds are roughly the same because the lower the speed is the less the angle of attack of the missile has. The related optimum speed profiles and angle of attack profiles of the missile with different initial speeds V_0=5m/s and V_0=25m/s are given in Figs. 22 through 25 respectively . The third fact is that for a given LOS reference direction the speed profiles for the shoot-forward launch directions of the missile are superior to those for the shoot-reward directions as shown in Figs. 26 and 27 for the optimum speed profiles respectively. Finally, it is worthy noting that the optimal turn laws for the BRM are slightly different from those of the SAM for low altitude mission because now only the over-shooting optimal trajectory profiles (see Fig . 28) are observed. From Figs. 17 and 19, we learn that the optimum thrust-angle directions are programmed to cancel the missile velocity component perpendicular to the reference LOS. Therefore, the larger the aiming angle σ_0 is the more the attitude of the missile points to the negative side at the initial phase of the trajectories. However, at the end of the optimal trajectories the attitudes of the missile all point to the positive side due to the over-shooting character. Besides, we also see that the angles of attack of the missile are in the post-stall region at the initial phase of the trajectories as shown in Fig.21.

Maximum velocity transfer for AAM in horizontal-plane

The third important application of the thrust vectoring capability of the missile is the super agile turn for AAM in modern dogfight scenarios. The basic problem considered here is to find the optimal turn laws in the horizontal-plane for the missile to turn its heading agily which maximize the final velocity of the tactical missile in a given time with final flight path angle constraint. The associated boundary conditions are given by

$$\hat{x}(t_0) = 0, \qquad \hat{x}(t_f) = free$$

$$\hat{y}(t_0) = 0, \qquad \hat{y}(t_f) = free$$

$$\hat{u}(t_0) = \hat{V}_0 \cos\gamma_0, \qquad \hat{u}(t_f) = max$$

$$\hat{v}(t_0) = \hat{V}_0 \sin\gamma_0, \qquad \hat{v}(t_f) = 0$$

$$\lambda_x(t_f) = 0, \qquad \lambda_y(t_f) = 0,$$

$$\lambda_u(t_f) = -1 \tag{26}$$

For this new numerical experiment, we launch the missile from various different initial flight path angle γ_0 , different initial speed V_0 ,and different flight altitude h to study the optimum thrust-angle programs for the missile to make an agile turn with maximum final speed along the initial LOS direction. In Fig.1, the initial LOS direction is along the x_I coordinate. A different set of results for this optimization problem is presented in Figs . 29 through 41. Figs. 29 and 30 show the optimal thrust-angle programs and speed profiles for different initial flight path angle γ_0 with V_0=0.835 Mach and h=10^4 ft. It is obvious from these results that the higher the aspect angle is the less the final speed of the missile has. This is due to the fact that for higher aspect angle the angle of attack of the missile is larger as can be shown from the optimum angle of attack profiles for the missile in Fig. 31. Since for large angle of attack the main thrust of the missile can not be used to increase its speed. Another important observation is that for beam-attack and shoot-back missions of the missile the lower the initial speed V_0 is the more the final speed of the missile has. However, for shoot-forward mission of the missile the higher the initial speed is the more the final speed of the missile has. This can be proved by the optimum speed profiles for different initial speed V_0=0.835 Mach(see Fig. 30),V_0=0.6 Mach, and V_0=0.4 Mach as shown in Figs. 32 and 33 respectively. Besides the effect of the variation of the initial speed V_0 and initial flight path angle γ_0 of the missile, we are also interested in the influence of the flight altitude h on the performance of the thrust vectoring capability of the missile. Figs.34 through 36 represent the optimal speed profiles of the missile with V_0=274.4m/s for different flight altitude h=2×10^4 ft, h=3×10^4 ft, and h=5×10^4 ft respectively. Comparing the optimum speed profiles for different flight altitudes (see Figs. 30,34-36), we notice that the optimum speed profiles are roughly the same. This implies that the effect of the thrust vectoring capability seems independent of flight altitude. In other words, the optimum speed profile in dense air condition can be approximately represented by the optimum speed profile without considering missile aerodynamics since at high flight altitude the aerodynamic force can be neglected. Fig. 37 shows the optimum speed profiles without aerodynamics which are almost the same as those in Fig. 36 for the flight altitude h=5×10^4 ft. In addition, the optimum thrust-angle programs without aerodynamics (see Fig . 38) are also close to those at flight altitude h=5×10^4 ft (see Fig . 39) .Hence, we conclude that the optimal solutions at high altitude can be approached by the optimal solutions without considering missile aerodynamics. Furthermore, from the optimal thrust-angle programs without aerodynamics (see Fig. 38) we also observe that the optimal thrust-angle directions are programmed to cancel the missile velocity component perpendicular to the initial LOS direction. Therefor for heading reversal maneuver, the optimum thrust-angle of the missile is opposite to the missile velocity at the initial phase of the flight until the missile speed towards zero V_m=0 .Then the thrust-angle is at the same direction as the missile velocity vector which is aligned with the initial LOS. In other words, the angle of attack of the missile is 180° at first then switches to zero after V_m=0 .However, the optimal turn laws of the missile with aerodynamics are not so simple especially in the dense air condition (see Fig.29). Nevertheless, the optimum thrust-angle directions with aerodynamics are also programmed to cancel the missile velocity component perpendicular to the initial LOS exactly the same way as those without missile aerodynamics. Finally, the optimal trajectory profiles in dense air condition with different aspect angles at flight altitude $h=10^4$ ft and initial speed V_0=0.835 Mach are given in Fig .40. For comparison, the optimal trajectory profiles without missile aerodynamics are given in Fig . 41.

Conclusions

The open loop optimal trajectories are analyzed for a point-mass model of modern dogfight missiles with tail control enhanced by thrust vectoring capability. The optimal control theory and multiple shooting numerical method are invoked to solve the TPBVP of the optimal trajectories with the initial guesses of the solutions obtained from the analytic solutions without considering missile aerodynamics. The optimal turn laws for high performance missiles executing post-stall maneuvers to achieve maximum final speed in a given time with final flight path angle constraint during the post launch phase have been obtained .Various problems such as the agile turn of a SAM to engage very low altitude target, positioning of a SAM on a predicted LOS for BR system, super agile turn of a AAM for a large off-boresight angle targeting or to engage targets in the rear hemisphere, etc., are solved. These studies show that for the SAM the optimal turn laws are quite close to the analytic solutions without considering missile aerodynamics and the optimal speed profiles are independent of the initial speed of the missile. On the other hand, for AAM the optimal thrust-angle programs with aerodynamics are quite different from the analytic solutions without considering missile aerodynamics but the optimal speed profiles are roughly the same. In other words, the performance of the thrust vectoring capability of the AAM is approximately independent of the flight altitude of the missile. Besides, the optimal trajectory profiles for the very low altitude ASM show the over-shooting character but for slightly higher target altitude engagement the trajectory profiles become under-shooting. However, for BRM the optimal trajectory profiles always show the over-shooting behavior. It is also interesting to point out that during heading reversal agile turn of the AAM, the optimal trajectory profile shows the main thrust reversal maneuver. In addition, for the AAM the lower the initial speed V_0 the missile is the more the final speeds are for the beam-attack and shoot-back scenarios. But the higher the initial speed of the AAM is the more the final speed is for the shoot-forward missions instead.

References

1. Wassom, S..R., Faupell, L.C., and Perley, T.,"Integrated Aerofin/Thrust Vector Control for Tactical missiles," *Journal of Propulsion and Power,* NO.7. 1991, PP.374-381
2. Innocenti, M., and Tukral;. A.," Simultaneous Reaction Jet and Aerodynamic Control of Missile systems," *Proceedings of the AIAA Guidance, Navigation, and Control Conference*, 1993,(AIAA paper 93-3739).

3. Wise, K.A., and Broy, D.J., "Agile missile Dynamics and Control," *Proceedings of the AIAA Guidance, Navigation, and Control Conference*, 1996 , (AIAA paper 96-3912) .
4. Nasuti , F . , and Innocenti , M . , "Missile Trajectory Optimization with Agility Issues , " *Proceedings of the AIAA Guidance, Navigation, and Control Conference*, 1996 , (AIAA paper 96-3730) .
5. Jonsson, H-O., and Malamberg, G., "Optimal Thrust Vector Control for vertical launch of Tactical Missiles," *Journal of Guidance, Control, and Dynamics*, Vol.5, No.1, 1982, PP.17-21.
6. Aiello, G.F., and Bateman, M.C., "Aerodynamic Stability Technology for Maneuverable Missiles. Vol.1 Configuration Aerodynamic Characteristics," AFFDL-TR-76-55, 1976.
7. Stoor, J., and Bulersch, R., Introduction To Numerical Analysis, Springer-verlag, New York, 1980.
8. Bryson, A.E., and Ho, Y. C., Applied Optimal Control, Hemisphere, New York, 1975.

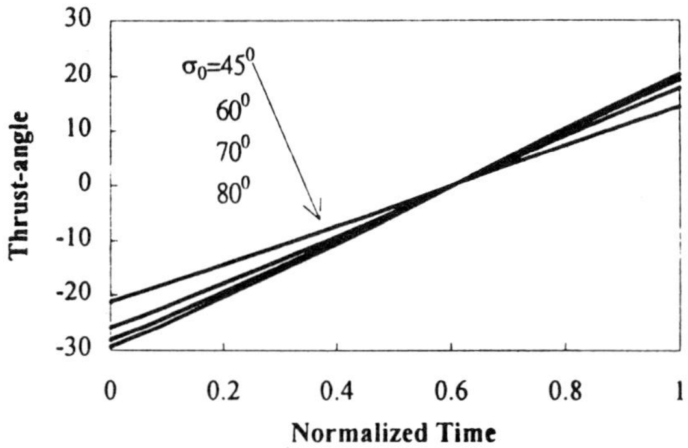

Fig.2 Analytic solutions of the optimal thrust-angle programs for BRM .

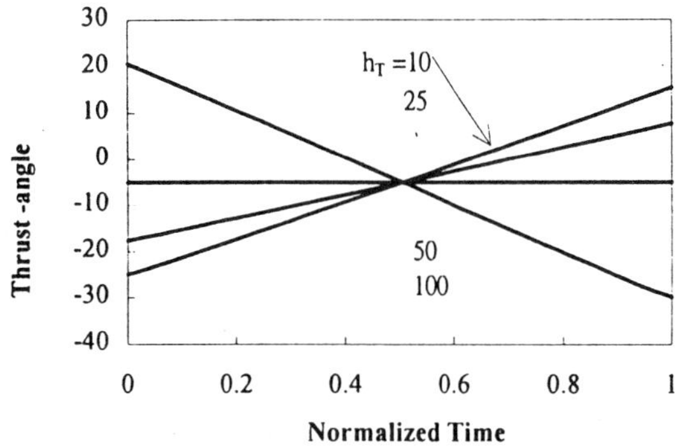

Fig.3 Analytic solutions of the optimal thrust-angle programs for SAM .

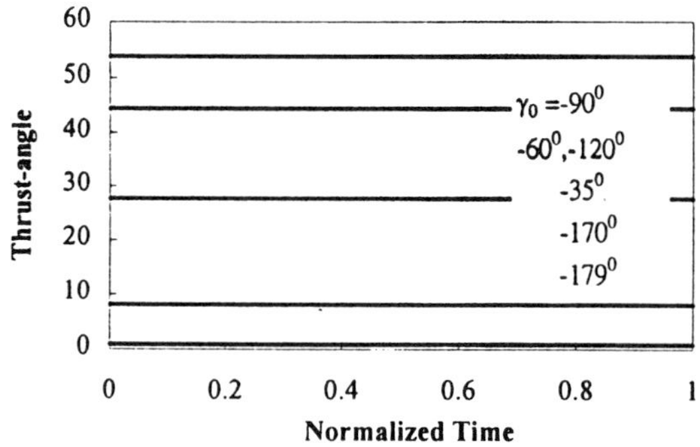

Fig.4 Analytic solutions of the optimal thrust-angle programs for AAM .

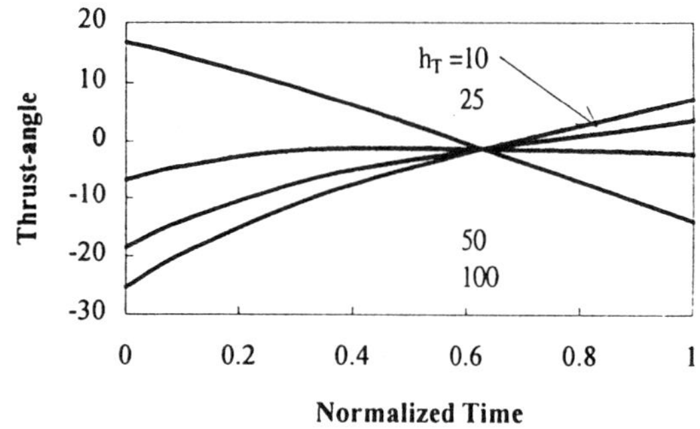

Fig.5 The optimal thrust-angle programs for SAM .

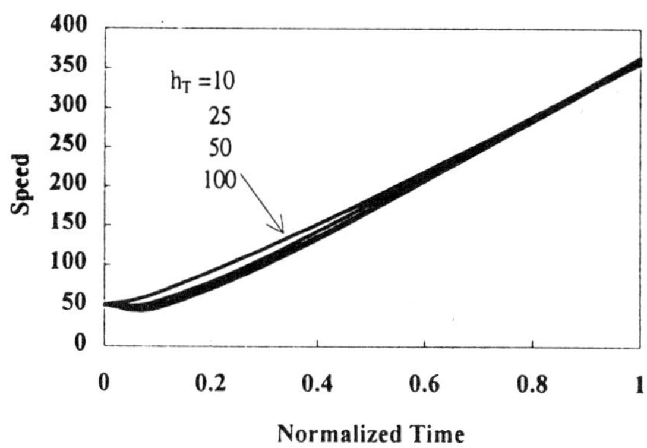

Fig.6 The optimal speed profiles for SAM .

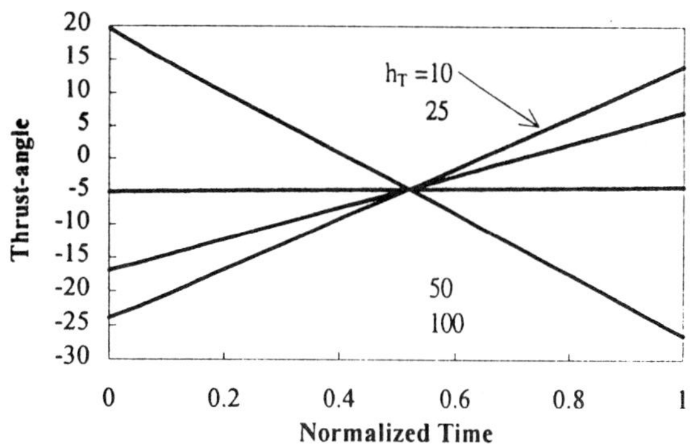

Fig.7 The optimal thrust-angle programs for SAM(without aero).

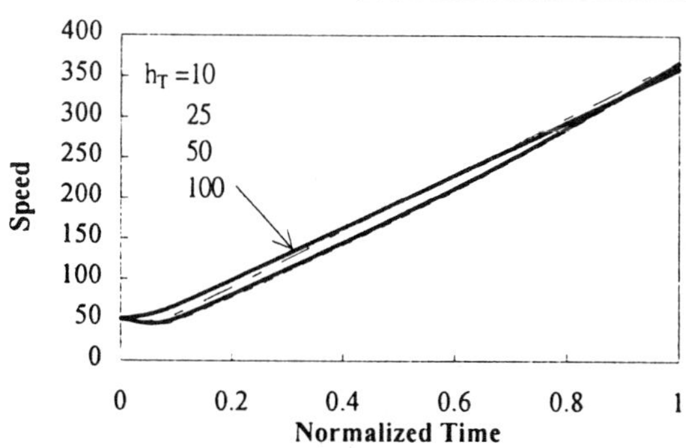

Fig.8 The optimal speed profiles for SAM(without aero).

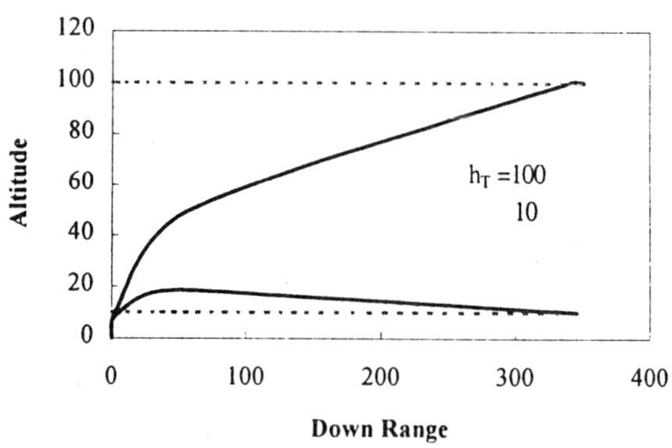

Fig.9 The optimal trajectory profiles for SAM .

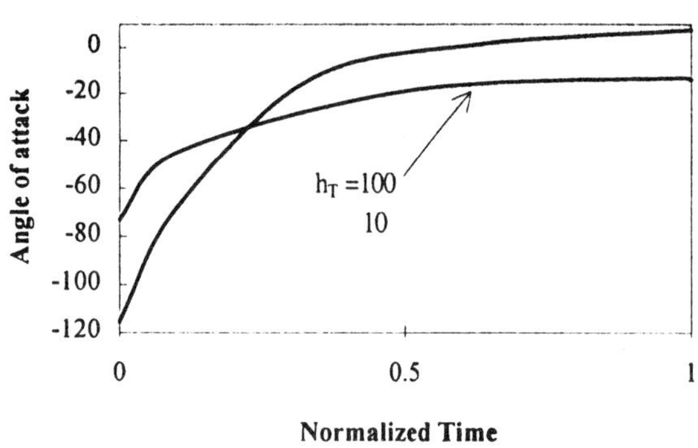

Fig.10 The optimal angle of attack profiles for the SAM .

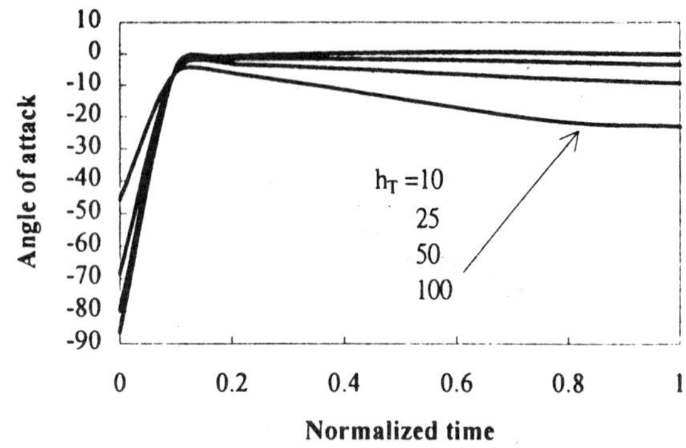

Fig.11 The optimal angle of attack profiles for the SAM(v0=5m/s).

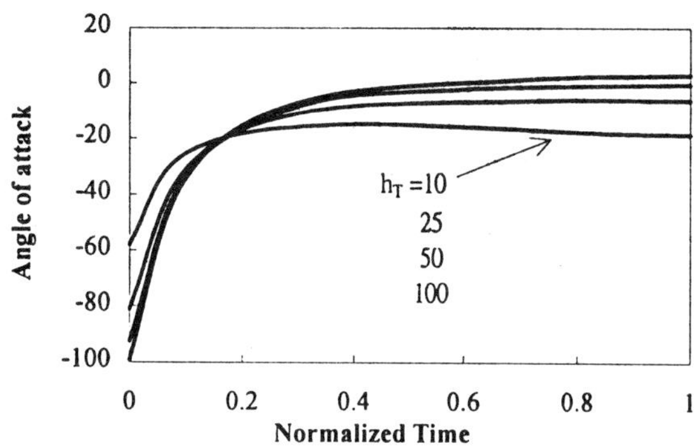

Fig.12 The optimal angle of attack profiles for the SAM(v0=25m/s).

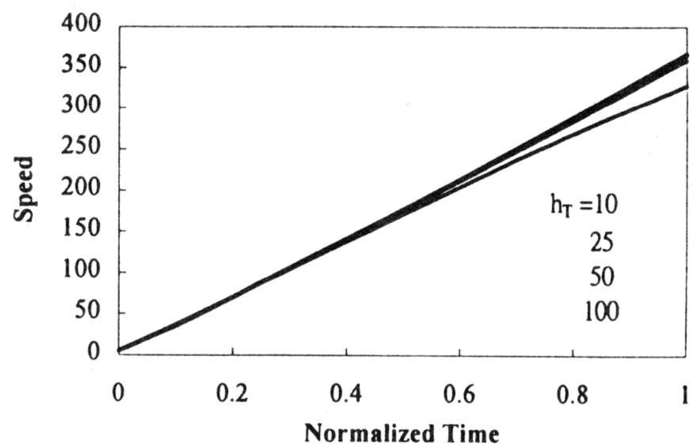

Fig.13 The optimal speed profiles for the SAM(v0=5m/s).

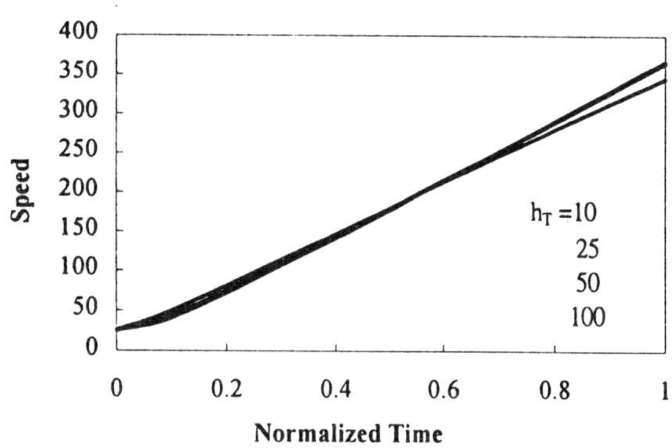

Fig.14 The optimal speed profiles for the SAM(v0=25m/s).

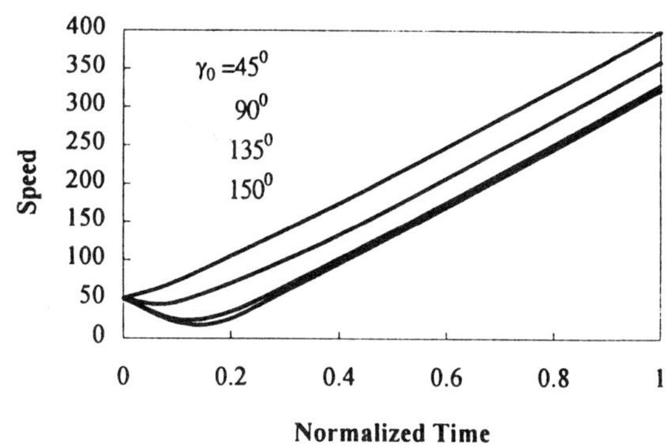

Fig.15 The optimal speed profiles of the SAM for different launch directions(ht=10m).

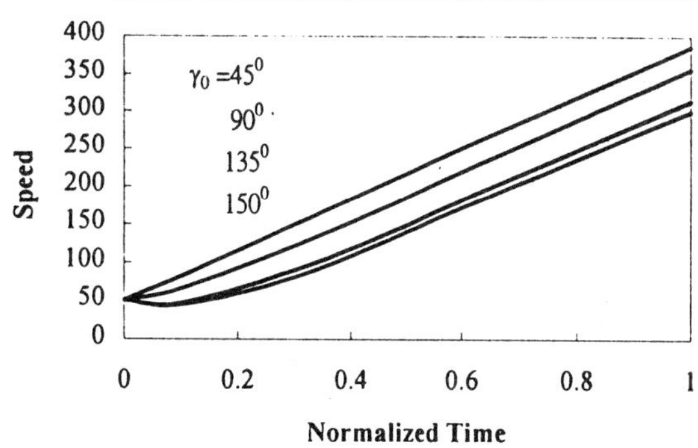

Fig.16The optimal speed profiles of the SAM for different launch directions(ht=100m).

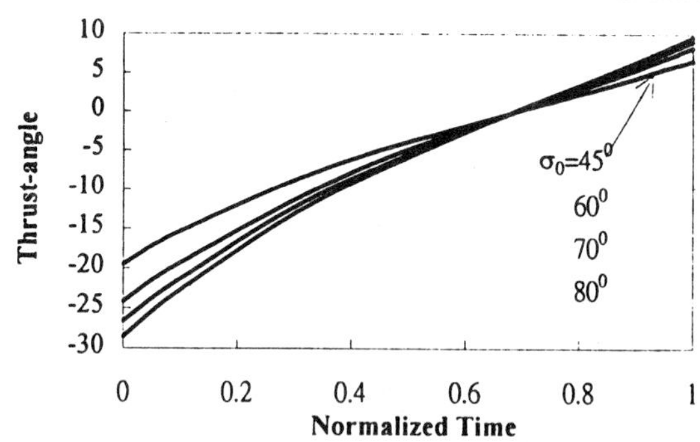

Fig.17 The optimal thrust-angle programs for the BRM.

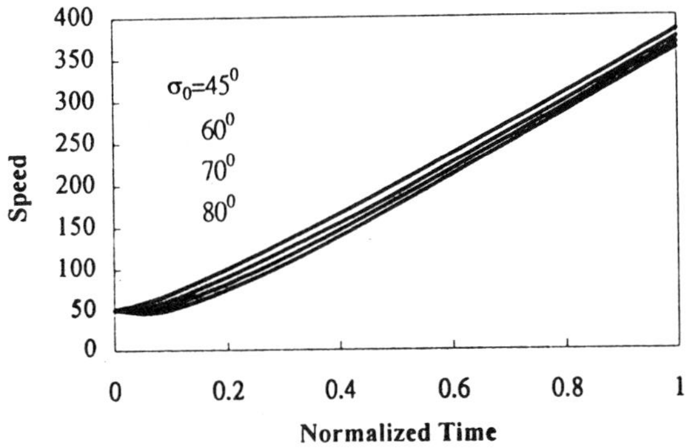

Fig.18 The optimal speed profiles for the BRM.

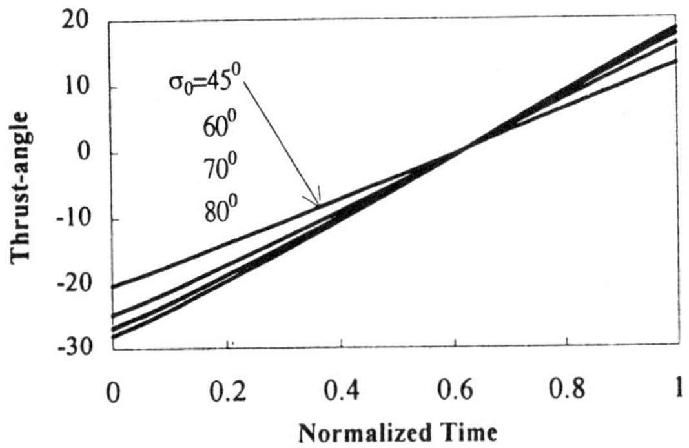

Fig.19 The optimal thrust-angle programs for the BRM(without aero).

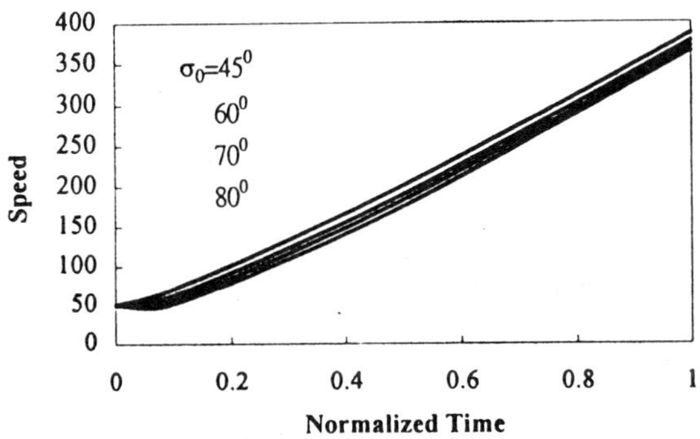

Fig.20 The optimal speed profiles for the BRM(without aero).

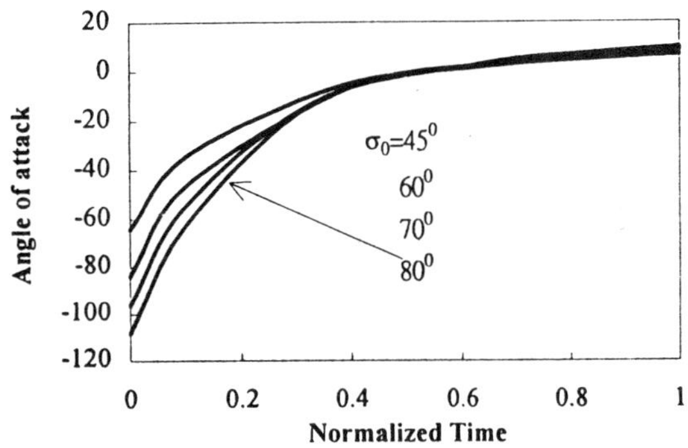

Fig.21 The optimal angle of attack profiles for the BRM.

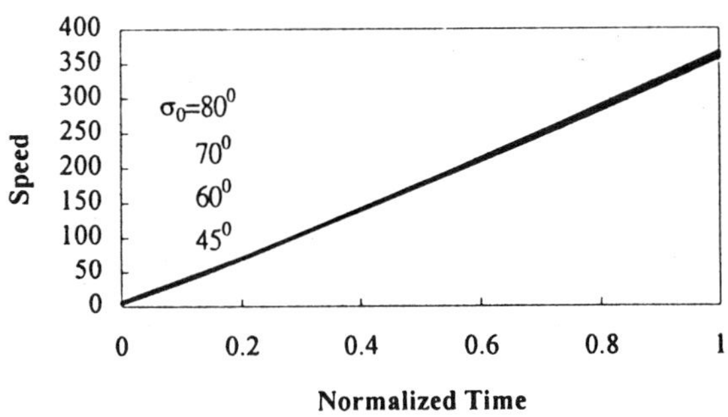

Fig 22 The optimal speed profiles for the BRM(v0=5m/s).

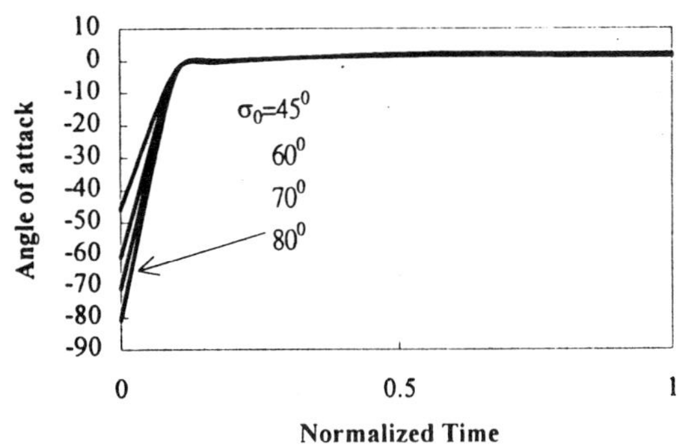

Fig.23 The optimal angle of attack profiles for the BRM(v0=5m/s).

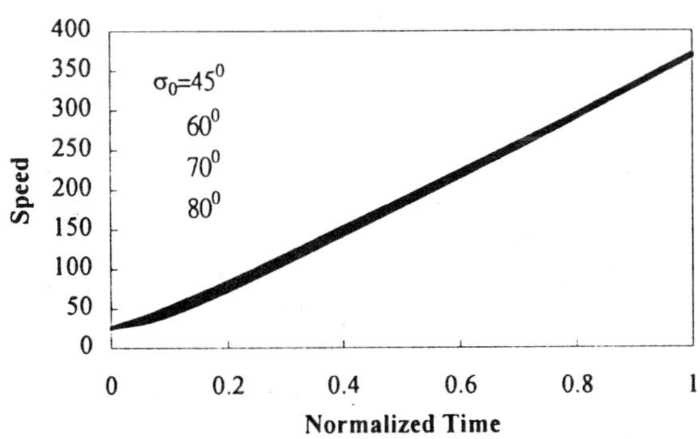

Fig.24 The optimal speed profiles for the BRM(v0=25m/s).

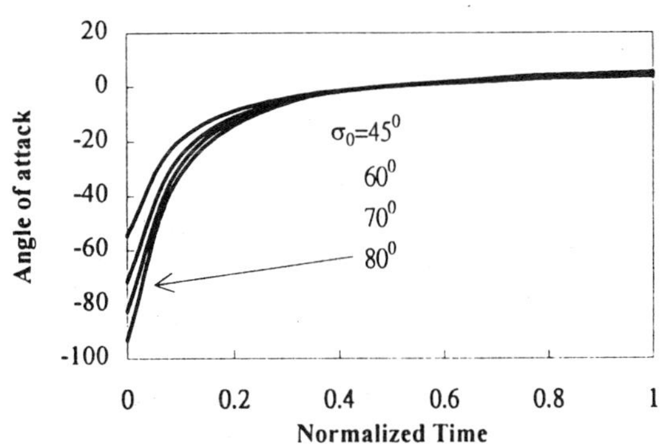

Fig.25 The optimal angle of attack profiles for the BRM(v0=25m/s).

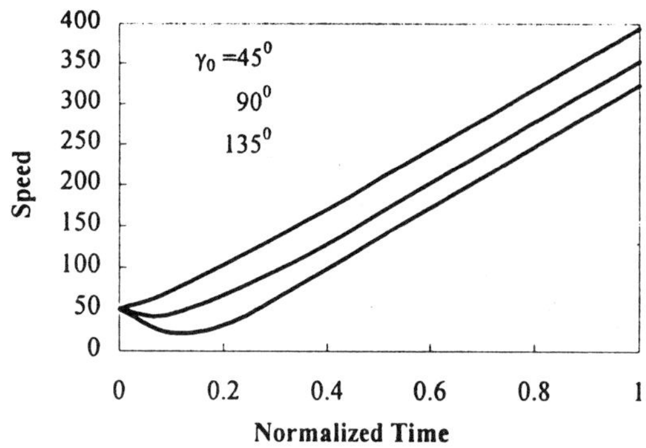

Fig.26 The optimal speed profiles of the BRM for different launch directions(aiming angle=80).

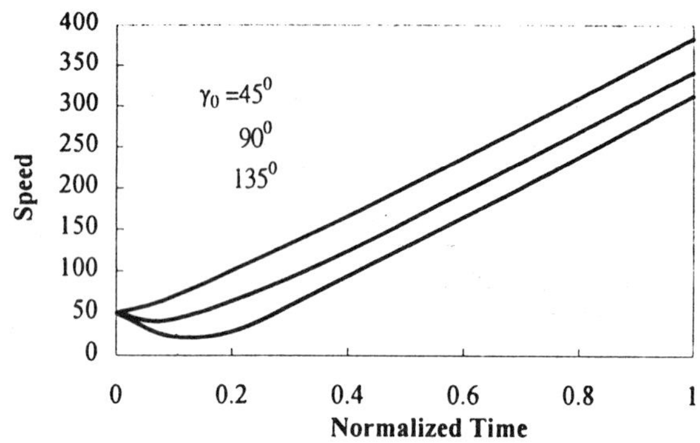

Fig.27 The optimal speed profiles of the BRM for different launch directions(aiming angle=45).

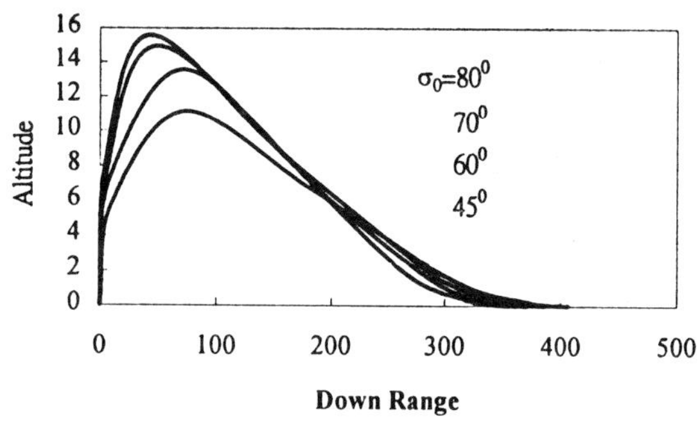

Fig.28 The optimal trajectory profiles for the BRM.

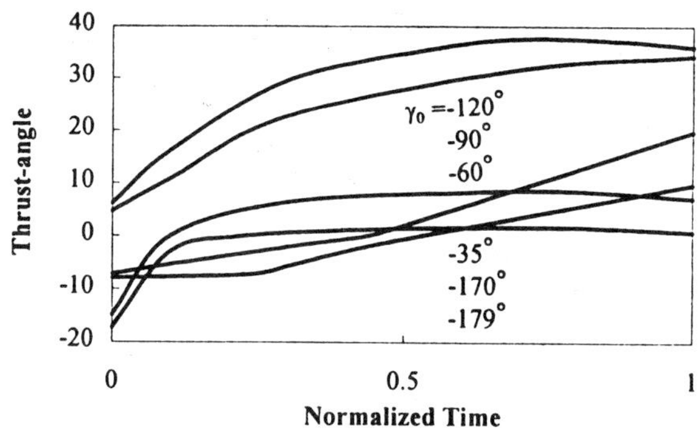

Fig.29 The optimal thrust-angle programs for the AAM.

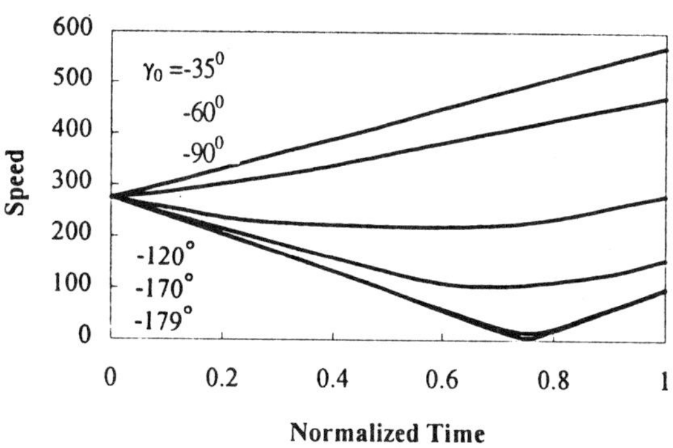

Fig.30 The optimal speed profiles for the AAM.

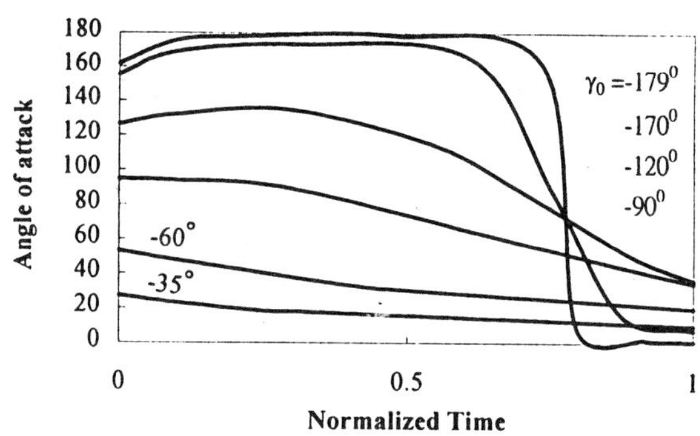

Fig.31 The optimal angle of attack profiles for the AAM.

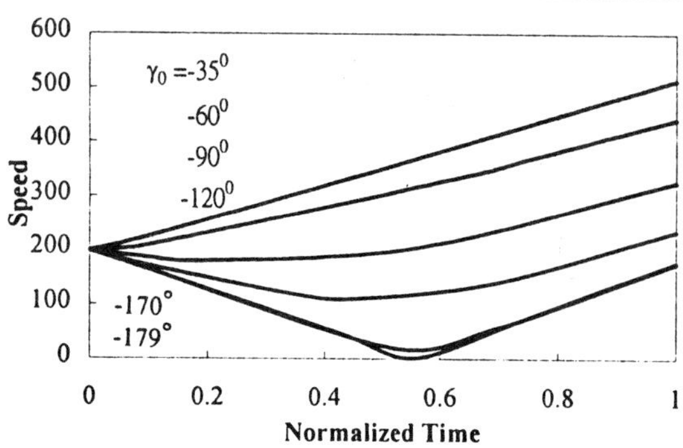

Fig.32 The optimal speed profiles for the AAM(v0=0.6mach).

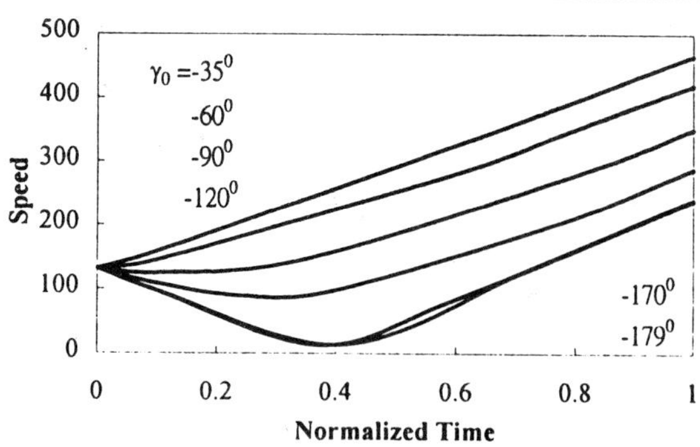

Fig.33 The optimal speed profiles for the AAM(v0=0.4mach).

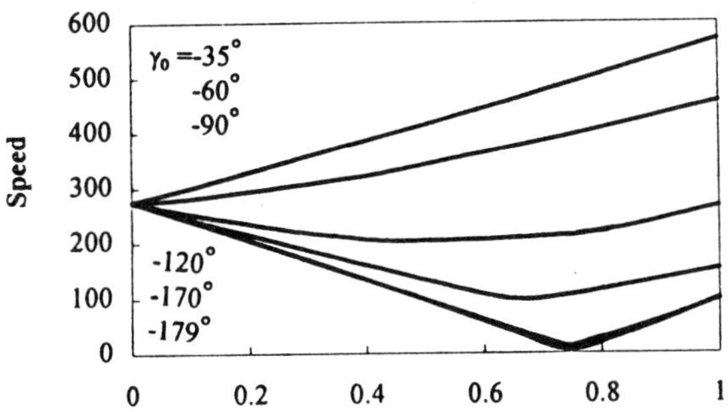

Fig 34 The optimal speed profiles for the AAM(h=20000ft).

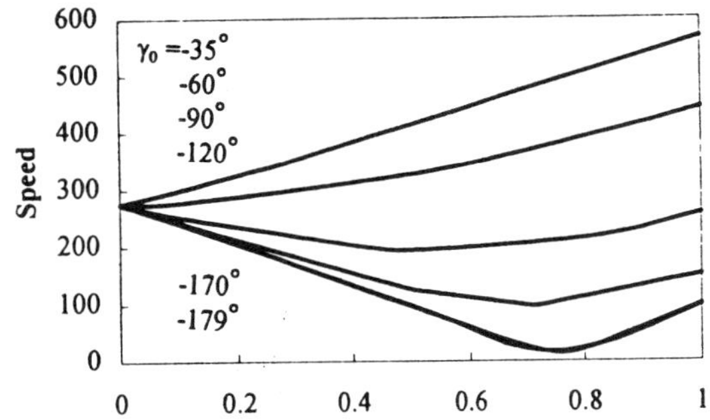

Fig.35 The optimal speed profiles for the AAM(h=30000ft).

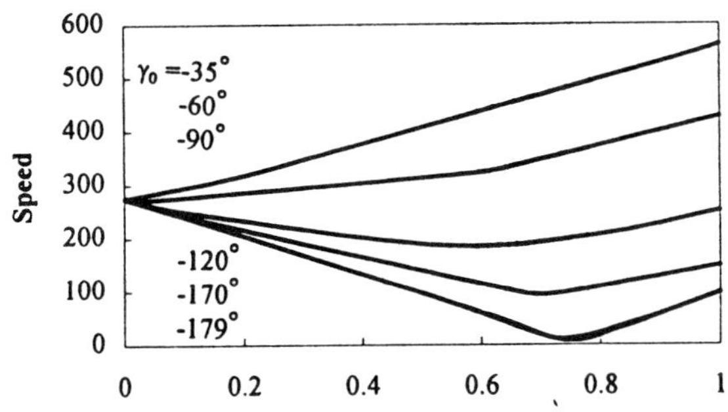

Fig.36 The optimal speed profiles for the AAM(h=50000ft).

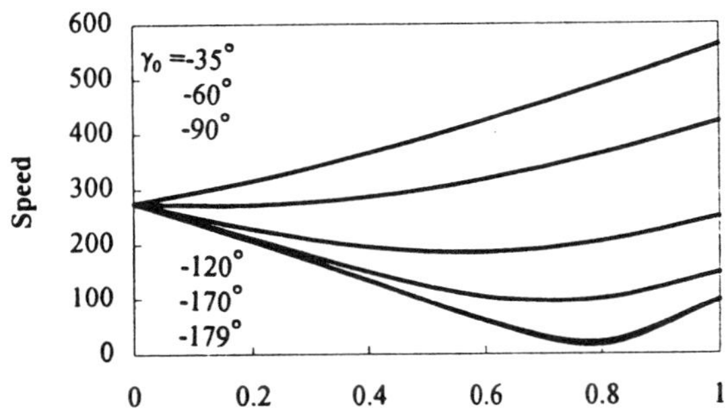

Fig.37 The optimal speed profiles for the AAM(without aero).

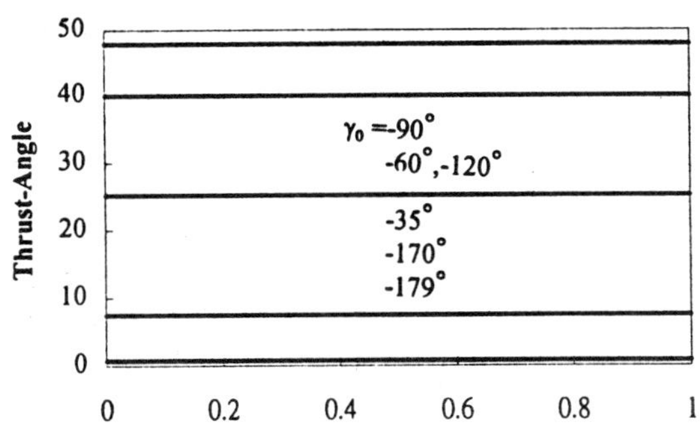

Fig.38 The optimal thrust-angle programs for the AAM(without aero).

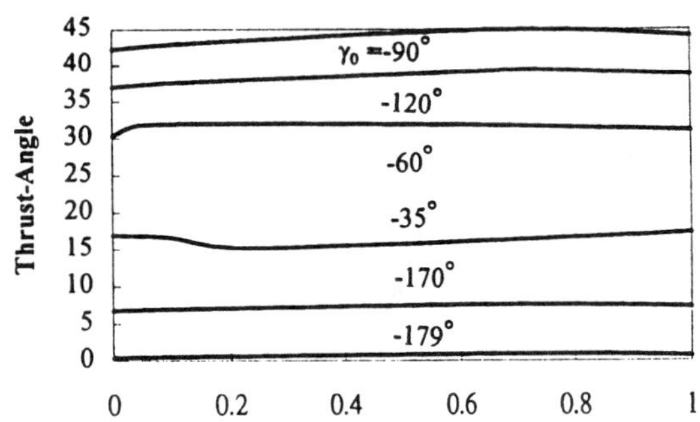

Fig.39 The optimal thrust-angle programs for the AAM(h=50000ft).

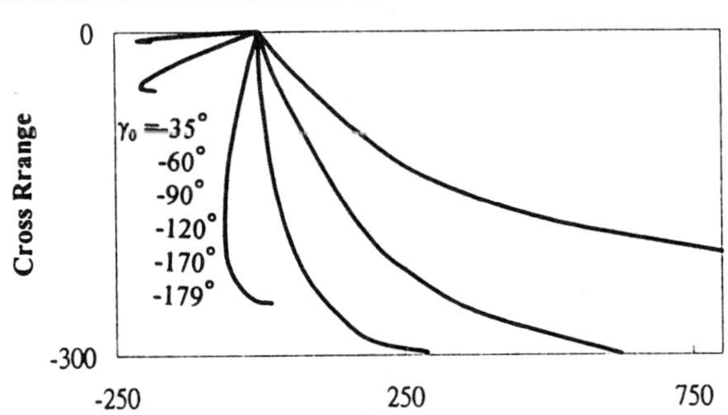

Fig.40 The optimal trajectory profiles for the AAM(h=10000ft).

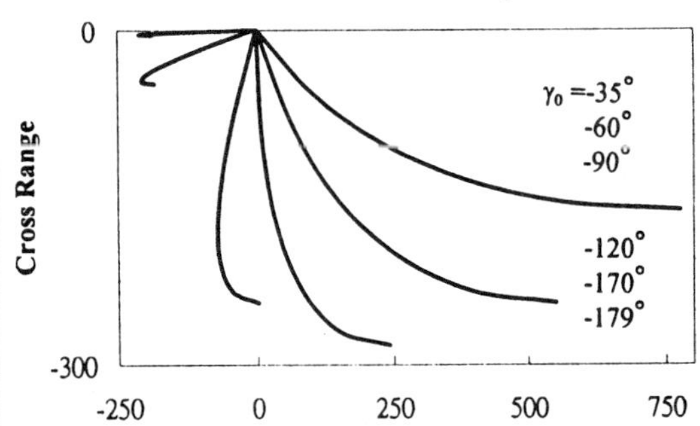

Fig 41 The optimal trajectory profiles for the AAM(without aero).

AIAA-97-3834

An Optofluidic Control System for Tactical Missiles and Guided Munitions

Dewey Benson[1]
AlliedSignal Aerospace

Bill Ryan[1]
AlliedSignal Aerospace

Dr. Ted Drzewiecki[2]
Defense Research

ABSTRACT

An optofluidic control system for a short range, hit-to-kill, interceptor was designed and demonstrated in a lab environment. The primary benefits of an optofluidic system are cost and simplicity. Most missiles and other guided munitions have expensive radar or optic tracking equipment, inertial sensors, and guidance computers[1]. The optofluidic approach requires no electronics on the missile, keeping the expensive subsystems on the ground for reuse. Control is achieved using attitude control thrusters and angle-of-attack sensors on the vehicle and an optical tracker on the ground. Guidance commands are relayed to the missile through a fiber-optic link. The key to the optofluidic control system is converting the light signals carried by the fiber-optic link directly into fluidic commands without the use of on-board electronics. The design effort included detailed and extensive 6-degree-of-freedom (6DOF) simulations, plus a hardware-in-the-loop test of the optofluidic control system coupled to a missile flight simulation. Both the 6DOF simulations and tests showed that the control system exceeded all design objectives.

INTRODUCTION

The drive to make products more cost effective has also had an impact on military procurement. Low cost missiles and smart munitions mean that the military can afford more rounds for training purpose and deploy more systems to the field on limited budgets. The control system described in this paper uses no on-board electronics, eliminating the need for an on-board battery. Instead, the gas source used to supply the thrusters also supplies power to the fluidic circuits which convert optic command signals into thruster responses. The result is a much simplified control system, which achieves higher levels of reliability and low 'throw away' cost. The first application of this technology was for the Small Low-Cost Interceptor Device (SLID)[2], a short range, hit-to-kill, tactical interceptor sponsored by the Defense Advanced Research Projects Agency (DARPA).

Because the SLID program is on-going, the sensitivity of the actual vehicle performance requires the use of normalized scales and time constants on all data presented in this paper. The principal objective of SLID is to defend lightly armored vehicles against both conventional anti-armor rounds and smart munitions with a per-round cost of less than $10,000. Because the SLID interceptor is gun launched, the vehicle is referred to as a projectile.

OPTOFLUIDIC CONTROL CONCEPT

The schematic for the SLID vehicle control system is shown in Fig. 1. This vehicle control system uses only fluidic (pneumatic) and passive optical devices to perform the necessary control functions, resulting in a simple, low-cost, and completely jam-proof command guidance capability. The projectile and threat positions are tracked by a ground based optical tracker. All guidance computations are performed in the fire control computer on the ground. Signals are sent via optic fiber to change the projectile angle-of-attack (AOA). The fluidic portion of the controller converts the optical signal into a pressure signal. Additional fluidic amplifiers boost this pressure signal to a level sufficient to drive high-power pneumatic thrusters.

A pair of thrusters is used for each of the two lateral control axes to change the vehicle's angle-of-attack. The projectile maneuvers by using lateral aerodynamic forces created by changing the angle-of-attack of the vehicle. Angle-of-attack is determined by the differential pressure across the nose of the vehicle, which is sensed by optical differential pressure transducers in the projectile nose. The optical pressure signals are returned down the fiber-optic command fink to the fire control computer. Fig. 2 shows how the entire system would fit into the 0.37 meter long (15 inch) SLID projectile.

The optofluidics on board the projectile are a subsystem of the entire guidance and control (G&C) system, Fig. 3. The optical tracker senses the positions of the threat and projectile in spherical coordinates of azimuth, elevation, and range. Positions are passed to the Kalman filter algorithm where a complete set of threat and projectile state estimates, such as position, velocity, and acceleration, are calculated. From the two sets of states, the guidance laws calculate the magnitude and direction that the projectile should accelerate to intercept the threat. The acceleration commands are compared with the projectile acceleration estimates derived from the Kalman filters. Within the accel control block of Fig. 3, the acceleration error is used to create an AOA command. The AOA signals, which are fed back down the fiber-optic link, are compared with the AOA commands, and then modified based on the roll position of the projectile, to create thruster commands for the two

[1] Members technical staff, Aerospace Equipment Systems Division, Tempe, Arizona, USA. E-mail: dewey.benson@alliedsignal.com., bill.ryan@alliedsignal.com.
[2] President, Rockville, Maryland, USA.E-mail: 73172.642@compuserve.com

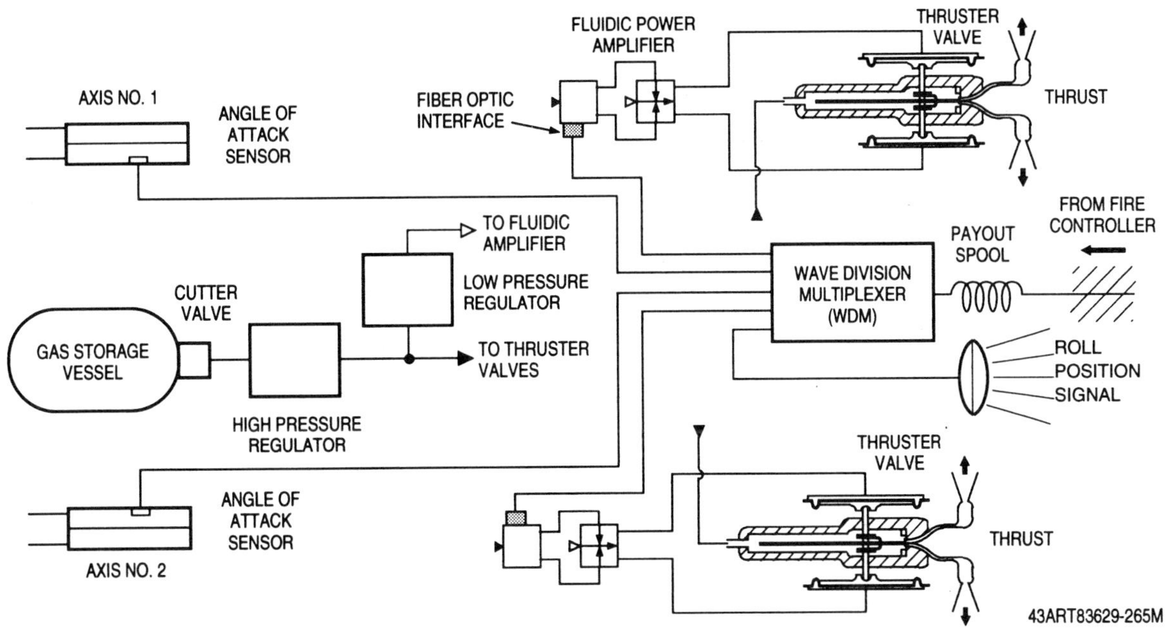

Fig. 1 SLID controller schematic.

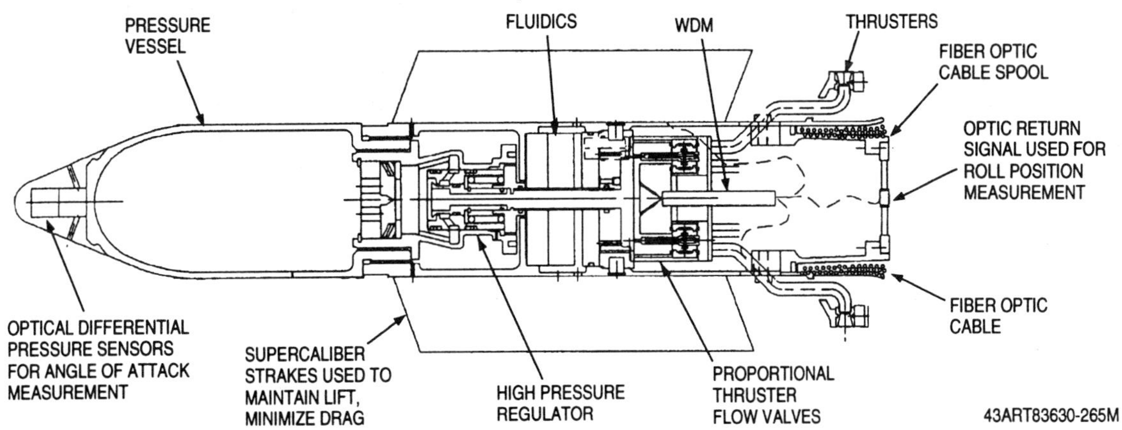

Fig. 2 Projectile.

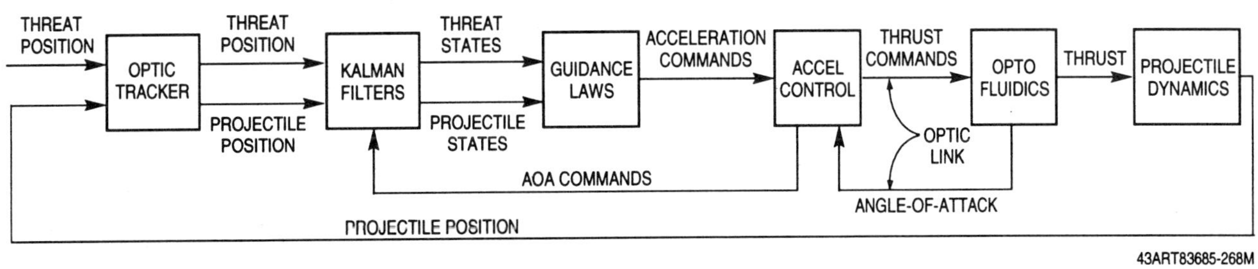

Fig. 3 Optofluidics integrated into complete guidance and control system.

lateral axis of the projectile. Roll position is determined by emitting a portion of the fiber-optic signals through a polarizing lens and out the tail of the projectile, shown in Fig. 1. The polarized light is directed rearward where the ground based optic tracker can receive the signal. A second rotating polarizing lens on the tracker determines the projectile roll position to within 5 degrees.

The thruster commands are sent up the fiber-optic link to the projectile where the optofluidics operate the thrusters. The total traffic on the fiber-optic link include four wavelengths sent from the ground to the projectile, two thruster command signals and two pressure sensor signals. The two carrier wavelengths sent to the projectile pressure sensors are modulated by the differential pressure on the projectile nose and returned back down the fiber-optic link.

THE CONTROL DESIGN PROBLEM

Conventional on-board guidance systems have both angular rate sensors and accelerometers, none of which are used in the optofluidic system. The computationally intense algorithms are handled by the fire control computer on the ground. In order to achieve this simple, low cost design, several issues had to be resolved: 1) A robust control algorithm had to be developed. 2) The AOA measurements must be suitable for control purposes. 3) Adequate acceleration estimates must be generated by the Kalman filters. The control design, therefore, focuses on the Kalman filter, accel control, optofluidics, and the projectile dynamics blocks in Fig. 3. Since the guidance algorithms are rather conventional, the guidance laws block of will not be discussed in Fig. 3 detail.

The AOA control concept performed almost as well as more complex approaches, such as using rate and acceleration sensors, but provided the advantage of allowing very simple devices on-board the vehicle. Fig. 4 shows a simplified block diagram of the AOA control laws and vehicle dynamics in the pitch plane, which forms the basis of the Kalman filter for the projectile. Lead-lag compensation is used in the AOA loop to provide a 'pseudo' rate signal to stabilize the projectile. In the case of the SLID projectile, the vehicle was aerodynamically 'trans-stable' as determined both analytically and by multiple wind tunnel tests. Fig. 5 shows a locus of all possible vehicle poles. Depending on the vehicle's AOA, velocity, and atmospheric conditions, the projectile will have either a pair of lightly damped but stable poles, or a pair of real poles, one of which is unstable. Fig. 6 demonstrates the effect of the lead compensation by placing two zeros on the real axis. The compensation stabilizes the unstable open loop pole, plus adds damping to the vehicle response. The locus of closed loop poles with the AOA compensation in place, Fig. 7, indicates only small variations in performance, despite the large open loop pole variations. The acceleration loop is closed using a simple lag command filter to slow the acceleration command response and decouple the acceleration loop from the AOA loop. The acceleration open loop margins are 80 degrees phase margin and 6 dB gain margin. Fig. 8 shows the well behaved time domain response of the vehicle to an acceleration command.

The next critical element of the design is the integrity of the AOA sensor data. During wind tunnel testing, data was taken at the vehicle AOA sensor ports with the vehicle in different roll angles (phi) and various angles of attack (alpha or beta). The subsonic results of Fig. 9 show consistent and nearly linear differential pressures across the vehicle nose displayed in terms of the airstream dynamic pressure, Cp. The AOA port pressure method of attitude control automatically adjusts the vehicle AOA for differences in air speed and atmospheric pressure to give similar accelerations for a given AOA command. That is, when atmospheric pressure is low, a greater AOA is required to give the same differential nose pressure as achieved at high atmospheric pressure. The result is the same lateral acceleration at any atmospheric pressure. Natural compensation is also true for crosswinds. A crosswind is sensed across the nose, causing the vehicle to "weather cock" into the wind, and resulting in a straight ground.

One concern when using lead-lag compensation is sensor noise. To address this issue, pressure data was taken from the nose ports at 1 kHz sampling rate during wind tunnel testing. The data showed noise magnitudes to be only 0.1% of full scale at AOA loop cross-over frequency. To further investigate the sensitivity of AOA noise on hit performance, the raw, unfiltered AOA measurements were used in the feedback within the accel control block. The result was approximately a 5% reduction in intercept percentage, compared to the performance when using filtered AOA sensor signals which reduced the noise level by a factor of 10.

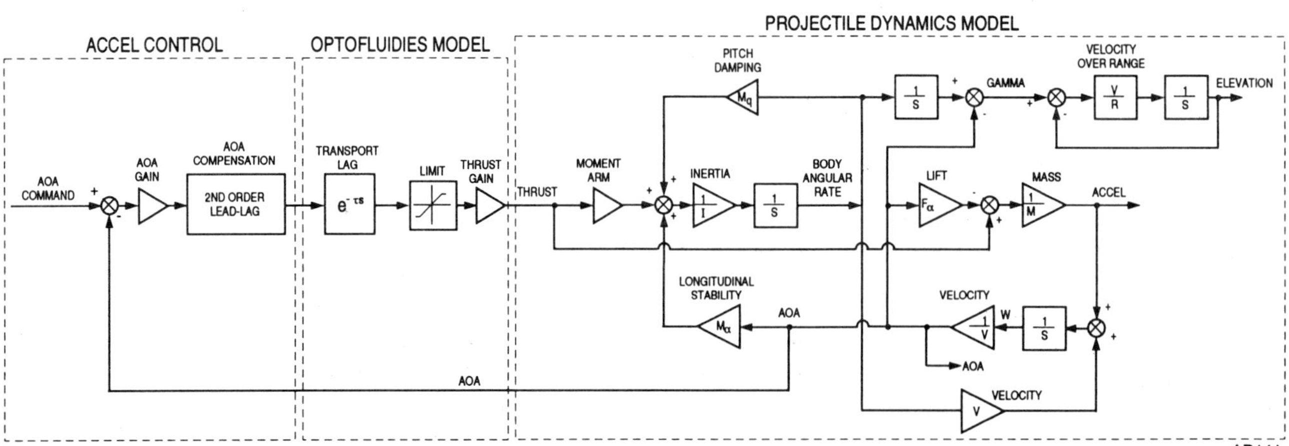

Fig. 4 Kalman filter model.

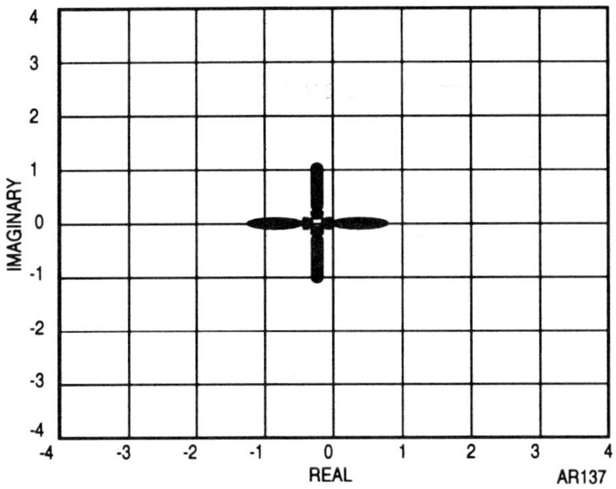

Fig. 5 Vehicle open-loop poles.

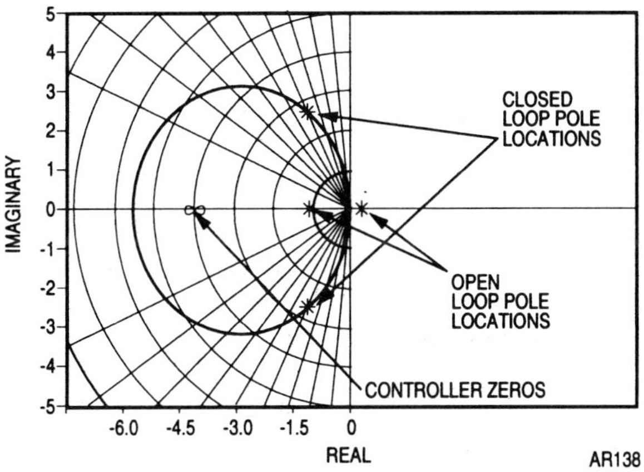

Fig. 6 AOA loop root locus.

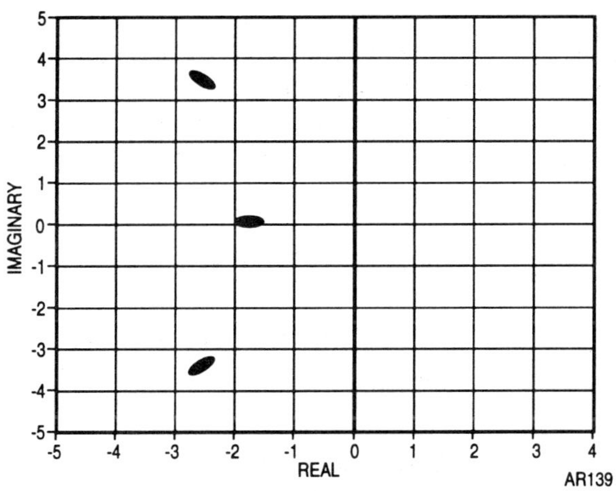

Fig. 7 Closed AOA loop pole variation.

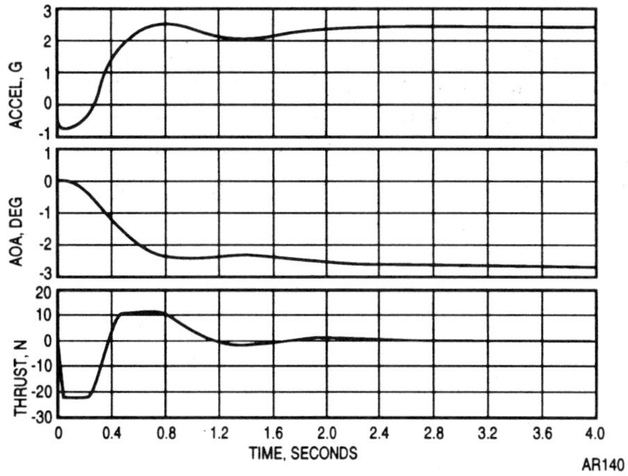

Fig. 8 Time domain response

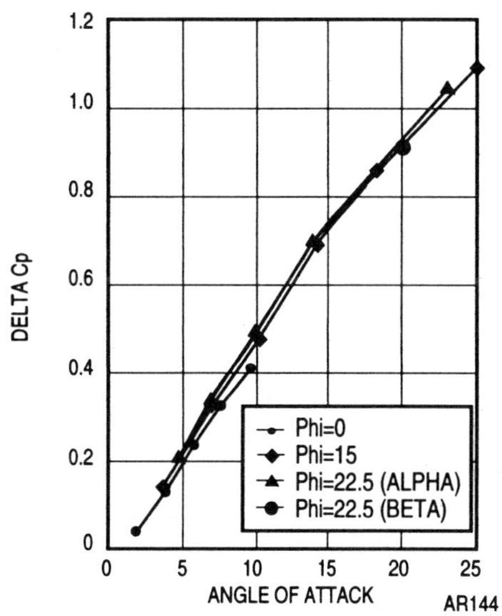

Fig. 9 Nose tip AOA port measurements at various roll angles (phi).

The final major design concern is the performance of the Kalman filters for estimating the projectile lateral accelerations. Fig. 4 presents the model used in the filter to determine acceleration in the vertical plane, which uses a common representation of the vehicle dynamics[3]. The optical tracker on the ground measures the elevation at which the projectile is observed and the filter produces vehicle lateral acceleration in the vertical plane. The model is discretized and used in a discrete Kalman filter implementation. Note that the model is non-linear, since the range value used in the calculation of elevation is always changing and a thrust command limit is used. The use of non-linear functions means that the optimality of the Kalman estimation cannot be guaranteed. On the other hand, trying to use a model with significant errors can lead to divergence of the filter[4]. The filter estimation accuracy is shown in Fig. 10, where the filter implemented with the embedded projectile model is compared to the vehicle acceleration from the 6DOF model and the estimate output from a simple three state Kalman filter, which is commonly used when no specific knowledge of the vehicle is available. The figure shows good tracking between the 6DOF model output and the embedded-model filter estimates. Without the additional knowledge of the embedded model, the simple three state filter has an additional 40 degrees of phase loss at the guidance loop crossover frequency.

RESULTS

SLID system performance was evaluated by extensive use of high fidelity system model and Monte Carlo analysis of intercepts. In total, there are 45 discrete random variables used in the system model, including tracking errors and platform disturbances, launcher pointing and tip-off errors, projectile mass properties and aerodynamic variations, atmospheric variations, and manufacturing errors such as strake misalignment and thruster positional errors. Twelve random noises, such as sensor noise, where also included in the system model. Over 10,000 simulations were run in developing the system and characterizing intercept performance. Simulations were run against seven different surrogate threats and all results showed excellent projectile performance and high probabilities of intercept. A single axis of the optofluidic control system was built and tested in a hardware in-the-loop facility. A flight simulation of the projectile making an intercept was run in parallel with the hardware test. Thruster commands from the guidance and control algorithms running in the simulation were sent to the control hardware along the fiber-optic link. The thruster outputs were measured during testing and returned to the projectile dynamics model running as a real-time simulation. Fig. 11 shows the simulated intercept of a threat making a constant g maneuver during the hardware-in-the-loop tests.

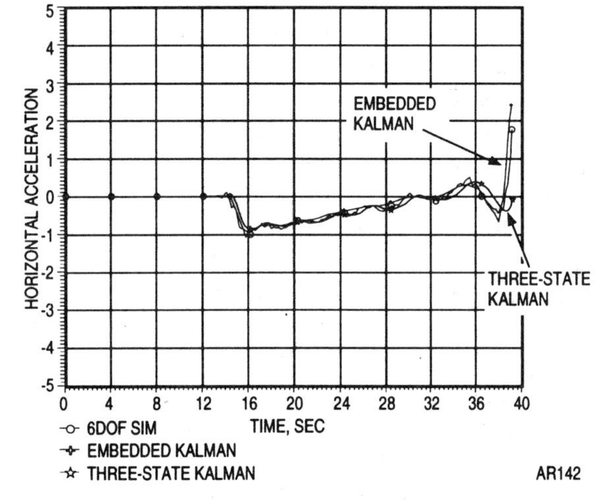

Fig. 10 Comparison of Kalman filter performance.

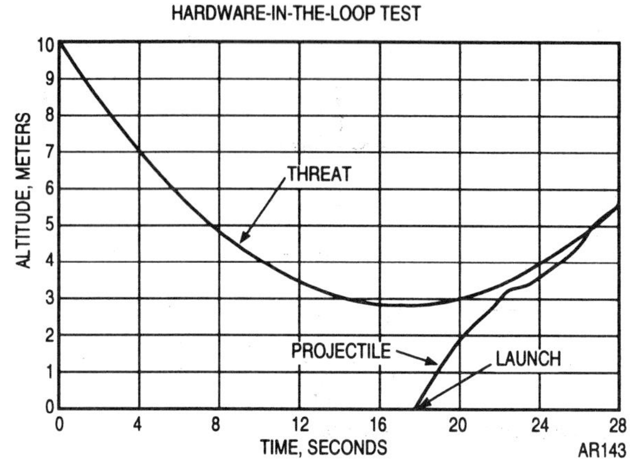

Fig. 11 Hardware-in-the-loop test showing real-time intercept.

CONCLUSION SUMMARY

The information presented in this paper provides, in a generic form, all the elements necessary to develop an optofludic AOA control law for a typical missile or guided munition. The SLID close range interceptor program was used as an example application.

An optofluidic control system was successfully demonstrated in a hardware-in-the-loop test environment. Extensive and detailed 6DOF simulations indicated the system to be robust for the SLID interceptor application based on aerodynamic and sensor noise data collected during wind tunnel testing. The optofluidic concept shows particular promise for achieving an accurate, low cost per round control system. Optofludic controls are equally applicable to a system with pneumatic or hydraulic fin actuators instead of thrusters, and would be suitable for a wide variety of tactical missiles and guided munitions.

REFERENCES:

[1] Lin, D.F., Modern Navigation, Guidance, and Control Processing, Vol. II, Prentice Hall, 1991.

[2] Drzewiecki, T.M., Waltz, E., Sutton, T. Kiernan, R.D., Joyce, K.E., "Small, Low-Cost Interceptor Device, Concept Refinement Program", ARDEC, Report ARFSD-CR-93008, June 1993

[3] Blakelock, J.H., "Automatic Control of Aircraft and missiles", Wiley & Sons, 1991

[4] Sorenson, H.W. "Kalman Filtering: Theory and Application", IEEE Press, 1985.

AIAA-97-3476

Efficient Flight Control System Assembly for Parameter Optimization Design Schemes

Mark F. Costello * Robert R. Schulz †
Department of Civil and Mechanical Engineering
United States Military Academy
West Point, New York 10996

August 11,1997

Abstract

Computer aided design packages used for aircraft flight control system analysis model the control laws as a set of filters and non linear modeling elements. Because the underlying assembly algorithm permits the elements to be connected arbitrarily, the control structure is open. This work shows that when starting from the common CAD GUI form of the control system, the FCS assembly process is not trivial from a computational standpoint. In fact, for practical aircraft system models, FCS assembly can be more computationally intensive than computing eigenvalues and time simulation of the system. The paper proposes an efficient assembly procedure to be used in conjunction with embedded parameter optimization FCS design schemes.

Introduction

Development of flight control laws for modern aircraft is generally an evolutionary process which consists of a cycle of analysis and testing beginning with some form of a baseline design. The cycle continues until required system performance is attained by satisfying a set of specifications. It is typical for aircraft flight control systems (FCS) to evolve over a period of several years. Flight control laws are generally communicated through block diagrams. Most computer aided design control system software packages contain algorithms to realize a control system comprised of a set of filters and non-linear modeling elements arranged in an arbitrary manner. As an example see Reference [1]. Computer aided design (CAD) FCS tools are being developed using application specific performance specifications [2]. For example, Yudilevitch, Tischler, Levine, and Lin have developed a FCS tool based on multi-criterion parameter optimization using military flight control specifications. A common feature of these techniques is the need to optimize design parameters of the control system to better meet performance objectives. Optimization of the design parameters requires computation of performance specifications of the closed loop system. In general, this dictates that the control system dynamic model be reassembled for each candidate set of control system parameters. For rigid fixed structure algorithms, this step is trivial but for user-friendly open structure algorithms where the design parameters are nestled deep in the control law block diagram this step becomes more difficult.

Consider the generic high hinge offset helicopter model described in the results sectionas well as in Figures 9 and 10. This system consists of a 39th order base airframe model and a 43rd order control system and is representative of the model texture required for detailed flight control system design of flexible aircraft. Figure 1 shows the relative computational burden of assembling the FCS for the high order helicopter model. From Figure 1, it can be seen that flight control system assembly requires 4 times more CPU time than computing the closed loop system eigenvalues and about half the computation time of a 4 second system time simulation. The work described here attempts to reduce the FCS assembly CPU time by segregating all design parameters into a diagonal constant gain output feedback matrix in order to accelerate subsequent FCS assembly as design parameters are varied.

Assembly Procedure

The uncontrolled plant is assumed to be given by

*Assistant Professor
†Instructor, U.S. Army Aviator

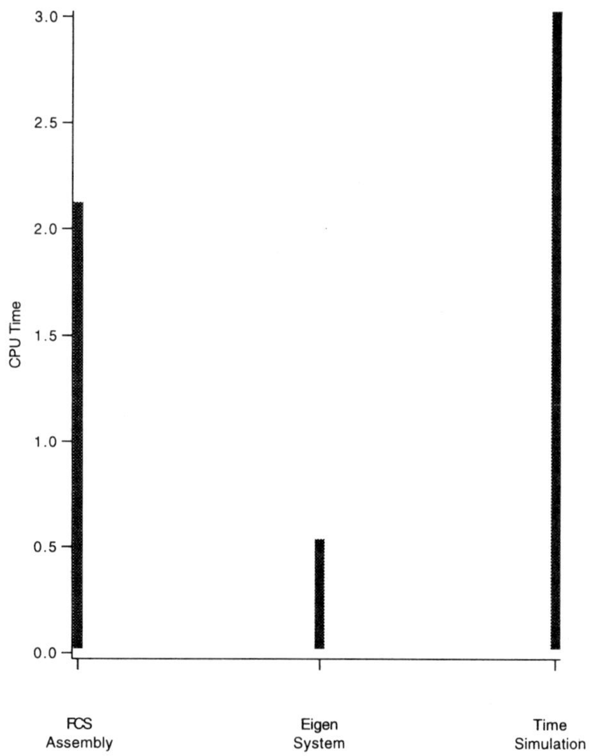

Figure 1: CPU Time Comparison

a linear system in state variable form,

$$\dot{x}_P = A_P x_P + B_P u_P \tag{1}$$

$$y_P = C_P x_P + D_P u_P \tag{2}$$

where, x_P is the state vector of the plant, u_P is the vector of actuator commands, and y_P is the vector of plant outputs. The order of the plant is n_P while the system has i_P inputs and o_P outputs.

The control system is comprised of n_F filters with an open structure which is defined with user input data. Each filter is a multiple input and single output filter given in polynomial form. Inputs to a filter can consist of system inputs, filter outputs, plant states, and plant state derivatives. Input data for each filter consists of the following: order, numerator and denominator coefficients, number of inputs, input identifiers, and gain values of each input. Other data necessary to define the control system is: number of variable control system parameters, filter number of each variable control system parameter, and the input identifier of each variable control system parameter.

To facilitate seperation of variable control system parameters of an individual filter, each filter is initially considered in phase variable block diagram form. The input to each numerator and denomenator filter coefficient is an output of the individual filter realization. Likewise, the output to each numerator and denomenator filter coefficient is an input of the individual filter realization. In this way all numerator and denomenator filter coefficients are separated from the filter dynamics matrix into a diagonal output feedback matrix. A filter of order n has $2n+3$ inputs and outputs and the corresponding system matrices are populated with 0's and 1's only. If the inputs and outputs are ordered with numerator coefficients first, denominator coefficients second, and external inputs and outputs last, then the filter state space realization consists of sparse matrices with the following fill.

$$a_{i,i+1} = 1 \qquad i = 1, n-1 \tag{3}$$

$$b_{n,2n+2} = 1 \tag{4}$$

$$\begin{aligned} c_{i,i} &= 1 \qquad i = 1, n \\ c_{n+1+i,i} &= 1 \qquad i = 1, n \end{aligned} \tag{5}$$

$$\begin{aligned} d_{n+1,2n+2} &= 1 \\ d_{2n+2,2n+3} &= 1 \\ d_{2n+3,n+1} &= 1 \\ d_{2n+2,n+2+i} &= 1 \quad i = 1, n \end{aligned} \tag{6}$$

Note that a matrix in lower case refers to an individual filter realization matrix and not a matrix of the complete control system. Using the variable parameter input data each numerator and denominator coefficient can be tagged as variable or fixed and appropriate bookkeeping can be performed to create separate input and output matrices for the variable and fixed parameters of the filter. Then the initial individual filter realization can be expressed as,

$$\dot{x}_s = a_s x_s + b_{s1} u_e + b_{s2} u_v + b_{s3} u_f \tag{7}$$

$$y_e = c_{s1} x_s + d_{s11} u_e + d_{s12} u_v + d_{s13} u_f \tag{8}$$

$$y_v = c_{s2} x_s + d_{s21} u_e + d_{s22} u_v + d_{s23} u_f \tag{9}$$

$$y_f = c_{s3} x_s + d_{s31} u_e + d_{s32} u_v + d_{s33} u_f \tag{10}$$

where x_s is the state vector of the kth filter, u_e and y_e are the external input and output of the filter, u_v and y_v are the internal variable parameter inputs and outputs of the filter, and u_f and y_f are the internal fixed parameter inputs and outputs. The internal parameter inputs and outputs are related by,

$$u_v = k_v y_v \tag{11}$$

$$u_f = k_f y_f \tag{12}$$

Subsequent to initially realizing the filter, the parameters of the filter which are fixed are eliminated with the substitution $u_f = k_f y_f$. Then each filter is written as,

$$\dot{x}_s = \tilde{a}_s x_s + \tilde{b}_e u_e + \tilde{b}_v u_v \tag{13}$$

$$y_e = \tilde{c}_e x_s + \tilde{d}_{11} u_e + \tilde{d}_{12} u_v \tag{14}$$

$$y_v = \tilde{c}_v x_s + \tilde{d}_{21} u_e + \tilde{d}_{22} u_v \tag{15}$$

where,

$$\tilde{a}_s = a_s + b_{s3} k_f (I - d_{s33} k_f)^{-1} c_{s3} \tag{16}$$

$$\tilde{b}_e = b_{s1} + b_{s3} k_f (I - d_{s33} k_f)^{-1} d_{s31} \tag{17}$$

$$\tilde{b}_v = b_{s2} + b_{s3} k_f (I - d_{s33} k_f)^{-1} d_{s32} \tag{18}$$

$$\tilde{c}_e = c_{s1} + d_{s13} k_f (I - d_{s33} k_f)^{-1} c_{s3} \tag{19}$$

$$\tilde{c}_v = c_{s2} + d_{s23} k_f (I - d_{s33} k_f)^{-1} c_{s3} \tag{20}$$

$$\tilde{d}_{11} = d_{s11} + d_{s13} k_f (I - d_{s33} k_f)^{-1} d_{s31} \tag{21}$$

$$\tilde{d}_{12} = d_{s12} + d_{s13} k_f (I - d_{s33} k_f)^{-1} d_{s32} \tag{22}$$

$$\tilde{d}_{21} = d_{s21} + d_{s23} k_f (I - d_{s33} k_f)^{-1} d_{s31} \tag{23}$$

$$\tilde{d}_{22} = d_{s22} + d_{s23} k_f (I - d_{s33} k_f)^{-1} d_{s32} \tag{24}$$

Equations 13 through 15 have effectively segregated all variable parameters of an individual filter into a diagonal output feedback matrix.

The individual filters are then assembled into an overall state space system where the individual filters are uncoupled from one another and the plant. The overall initial realization can be written as,

$$\dot{x}_{CS} = A_I x_{CS} + B_{I1} u_E + B_{I2} u_{VI} \tag{25}$$

$$y_E = C_{I1} x_{CS} + D_{I11} u_E + D_{I12} u_{VI} \tag{26}$$

$$y_{VI} = C_{I2} x_{CS} + D_{I21} u_E + D_{I22} u_{VI} \tag{27}$$

where x_{CS} is the state vector for the entire control system, u_E and y_E are the vector of all filter inputs and outputs, and u_{VI} and y_{VI} are the input and output feedback vectors related by $u_{VI} = K_I y_{VI}$. It should be recognized that the filter definitions concatenated into equations 25, 26, and 27 are uncoupled.

The filters are coupled together using the basic filter data, in particular, the input identifiers, $u_{k,j}$, and gain values, $K_{k,j}$. With this data the following coupling matrices can be directly formed.

$$u_E = K_1 y_E + K_2 \delta + K_3 x_P + K_4 \dot{x}_P + K_5 u_{VE} \tag{28}$$

where, u_{VE} is the input from a variable filter gain input, and δ is the external input vector to the overall dynamic system. Also,

$$u_{VE} = K_E \, y_{VE} \tag{29}$$

$$y_{VE} = G_1 y_E + G_2 \delta + G_3 x_P + G_4 \dot{x}_P \tag{30}$$

The final control system model is arrived at by substituting equation 28 into equations 25, 26, and 27. Also, the variable parameters, u_{VE}, are appended onto u_{VI} to form the complete vector of variable parameters, u_{OF}. The final equations are,

$$\dot{x}_{CS} = J_1 x_{CS} + J_2 \delta + J_3 x_P + J_4 \dot{x}_P + J_5 u_{OF} \tag{31}$$

$$y_{CS} = M_1 x_{CS} + M_2 \delta + M_3 x_P + M_4 \dot{x}_P + [M_5 \;\; M_6] u_{OF} \tag{32}$$

$$y_{OF} = L_1 x_{CS} + L_2 \delta + L_3 x_P + L_4 \dot{x}_P + L_5 u_{OF} \tag{33}$$

$$u_{OF} = K_{OF} y_{OF} \tag{34}$$

where,

$$J_1 = A_I + B_{I1} K_1 M_1 \tag{35}$$

$$J_2 = B_{I1}(K_2 + K_1 M_2) \tag{36}$$

$$J_3 = B_{I1}(K_3 + K_1 M_3) \tag{37}$$

$$J_4 = B_{I1}(K_4 + K_1 M_4) \tag{38}$$

$$J_5 = \begin{bmatrix} B_{I2} + B_{I1} K_1 M_5 & B_{I1}(K_5 + K_1 M_6) \end{bmatrix} \tag{39}$$

$$Q_1 = \Gamma M_1 \tag{40}$$

$$Q_2 = \Gamma M_2 \tag{41}$$

$$Q_3 = \Gamma M_3 \tag{42}$$

$$Q_4 = \Gamma M_4 \tag{43}$$

$$Q_5 = \Gamma [M_5 \;\; M_6] \tag{44}$$

$$L_1 = \begin{bmatrix} C_{I2} + D_{I21} K_1 M_1 \\ G_1 M_1 \end{bmatrix} \tag{45}$$

$$L_2 = \begin{bmatrix} D_{I21}(K_2 + K_1 M_2) \\ G_2 + G_1 M_2 \end{bmatrix} \tag{46}$$

$$L_3 = \begin{bmatrix} D_{I21}(K_3 + K_1 M_3) \\ G_3 + G_1 M_3 \end{bmatrix} \tag{47}$$

$$L_4 = \begin{bmatrix} D_{I21}(K_4 + K_1 M_4) \\ G_4 + G_1 M_4 \end{bmatrix} \tag{48}$$

$$L_5 = \begin{bmatrix} D_{I22} + D_{I21} K_1 M_5 & D_{I21}(K_5 + K_1 M_6) \\ G_1 M_5 & G_1 M_6 \end{bmatrix} \tag{49}$$

$$M_1 = [I - D_{I11} K_1]^{-1} C_{I1} \tag{50}$$

$$M_2 = [I - D_{I11} K_1]^{-1} D_{I11} K_2 \tag{51}$$

$$M_3 = [I - D_{I11} K_1]^{-1} D_{I11} K_3 \tag{52}$$

$$M_4 = [I - D_{I11} K_1]^{-1} D_{I11} K_4 \tag{53}$$

$$M_5 = [I - D_{I11} K_1]^{-1} D_{I12} \tag{54}$$

$$M_6 = [I - D_{I11} K_1]^{-1} D_{I11} K_5 \tag{55}$$

A byproduct of working with the scalar block diagram form of the control system data is that the matrix K_{OF} is diagonal.

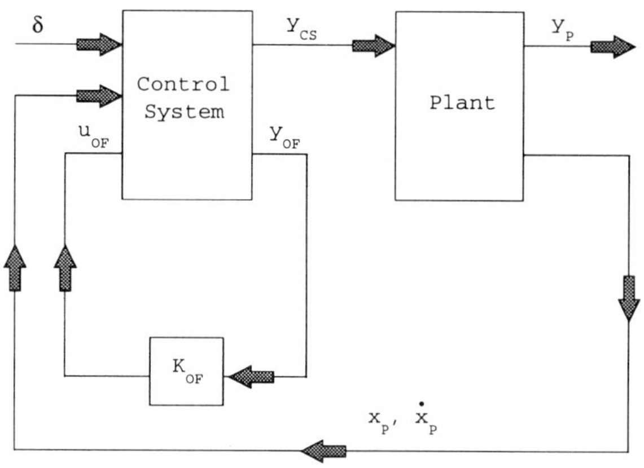

Figure 2: Physical Plant and Control System Connection

Physically the plant and control system are connected as shown in Figure 2. To efficiently apply a parameter optimization based design strategy it is necessary to separate the combined system plant and the variable parameters so that the closed loop system can be computed for a given set of variable parameters with a simple loop closure eliminating the need to completely recompile the entire plant and control system combination. This is accomplished by substituting $u_{AC} = \Gamma y_{CS}$ into equations 1 and 2. The matrix Γ restricts the control system output vector to consist of only those control system filter outputs which are plant inputs. Corresponding to Figure 3, the combined plant and control system equations are,

$$\dot{x}_P = H_1 x_P + H_2 x_{CS} + H_3 \delta + H_4 u_{OF} \tag{56}$$

$$\dot{x}_{CS} = F_1 x_P + F_2 x_{CS} + F_3 \delta + F_4 u_{OF} \tag{57}$$

$$y_P = T_1 x_P + T_2 x_{CS} + T_3 \delta + T_4 u_{OF} \tag{58}$$

$$y_{CS} = R_1 x_P + R_2 x_{CS} + R_3 \delta + R_4 u_{OF} \tag{59}$$

$$y_{OF} = S_1 x_P + S_2 x_{CS} + S_3 \delta + S_4 u_{OF} \tag{60}$$

where,

$$H_1 = [I - B_{AC} Q_4]^{-1} (A_{AC} + B_{AC} Q_3) \tag{61}$$

$$H_2 = [I - B_{AC} Q_4]^{-1} B_{AC} Q_1 \tag{62}$$

$$H_3 = [I - B_{AC} Q_4]^{-1} B_{AC} Q_2 \tag{63}$$

$$H_4 = [I - B_{AC} Q_4]^{-1} B_{AC} Q_5 \tag{64}$$

$$F_1 = J_3 + J_4 H_1 \tag{65}$$

$$F_2 = J_1 + J_4 H_2 \tag{66}$$

$$F_3 = J_2 + J_4 H_3 \tag{67}$$

$$F_4 = J_5 + J_4 H_4 \tag{68}$$

$$R_1 = Q_3 + Q_4 H_1 \tag{69}$$

$$R_2 = Q_1 + Q_4 H_2 \tag{70}$$

$$R_3 = Q_2 + Q_4 H_3 \tag{71}$$

$$R_4 = Q_4 H_4 + Q_5 \tag{72}$$

$$S_1 = L_3 + L_4 H_1 \tag{73}$$

$$S_2 = L_1 + L_4 H_2 \tag{74}$$

$$S_3 = L_2 + L_4 H_3 \tag{75}$$

$$S_4 = L_7 + L_4 H_4 \tag{76}$$

$$T_1 = C_{AC} + D_{AC} R_1 \tag{77}$$

$$T_2 = D_{AC} R_2 \tag{78}$$

$$T_3 = D_{AC} R_3 \tag{79}$$

$$T_4 = D_{AC} R_4 \tag{80}$$

It should be noted that with the exception of K_{OF}, all matrices of the combined plant and control system are independent of the variable control system parameters. The combined plant and control system equations need to be computed only once.

System Assembly

With the design parameters seperate from the fixed portion of the system, the final closed loop system can easily and rapidly be computed with a simple loop closure: $u_{OF} = K_{OF} y_{OF}$. The results are given by equations 81 and 82.

$$\dot{x}_{SY} = A_{CL} x_{SY} + B_{CL} \delta \tag{81}$$

$$y_{SY} = C_{CL} x_{SY} + D_{CL} \delta \tag{82}$$

where,

$$A_{CL} = \begin{bmatrix} H_1 + H_4 \nu_1 & H_2 + H_4 \nu_2 \\ F_1 + F_4 \nu_1 & F_2 + F_4 \nu_2 \end{bmatrix} \tag{83}$$

$$B_{CL} = \begin{bmatrix} H_3 + H_4 \nu_3 \\ F_3 + F_4 \nu_3 \end{bmatrix} \tag{84}$$

$$C_{CL} = \begin{bmatrix} T_1 + T_4 \nu_1 & T_2 + T_4 \nu_2 \\ R_1 + R_4 \nu_1 & R_2 + R_4 \nu_2 \end{bmatrix} \tag{85}$$

$$D_{CL} = \begin{bmatrix} T_3 + T_4 \nu_3 \\ R_3 + R_4 \nu_3 \end{bmatrix} \tag{86}$$

$$\nu_1 = [I - K_{OF} S_4]^{-1} K_{OF} S_1 \tag{87}$$

$$\nu_2 = [I - K_{OF} S_4]^{-1} K_{OF} S_2 \tag{88}$$

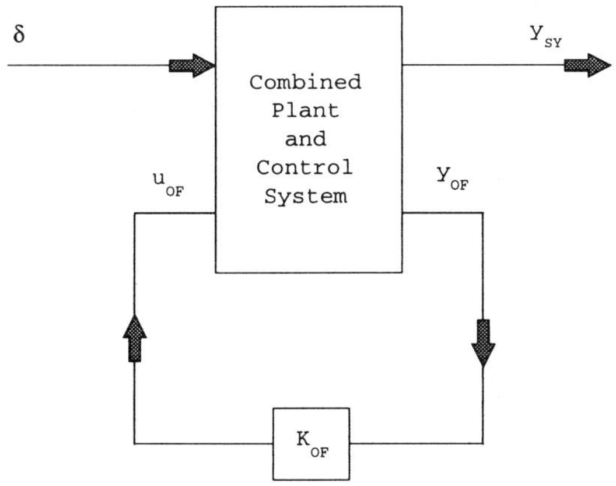

Figure 3: Computationally Efficient Design Parameter Connection

$$\nu_3 = [I - K_{OF}S_4]^{-1} K_{OF}S_3 \tag{89}$$

x_{SY} and y_{SY} are concatenations of x_P and x_{CS}, and y_P and y_{CS}, respectively. Note that since control system parameters in the feedback and feedforward path may be present in K_{OF}, both the closed loop poles and zeros can be affected by a modification in K_{OF}.

Results

In order to illustrate the computational efficiency of the control system realization method described above, a low and high order system are considered. The current method is contrasted against a standard open structure control system realization technique. The standard control system realization method used in this work is similiar to the method described above with one major exception. In the standard method, individual filters are directly realized in phase variable canonical form. The standard method does not separate variable and fixed control system parameters, and as such obviates the need for the bookkeeping associated with separated variable and fixed control system parameters. On the other hand the standard method must totally reassemble the plant and control system combination for each candidate set of variable control system parameters.

All Central Processing Unit (CPU) time and Floating Point Operations (FLOPS) data shown below was generated on a RISC 6000 machine running MATLAB. The CPU and and FLOPS data was obtained using the MATLAB cputime and flops commands, respectively [3]. The CPU time results generally showed a variance of approximately 10% and as such the results given are the average values for a sample size of 10.

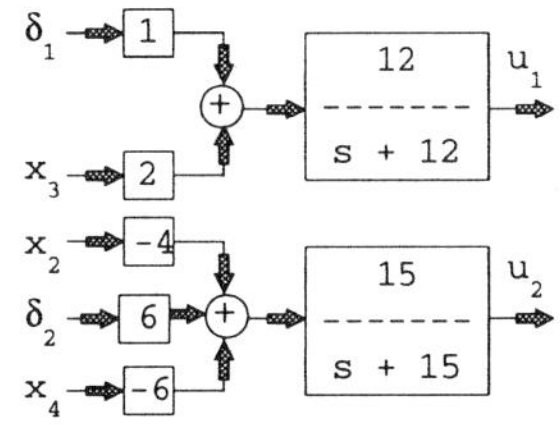

Figure 4: Low Order System Control System

Low Order System:

The low order state space system matrices are given by Equations through 93.

$$A_P = \begin{bmatrix} -0.117 & -0.00295 & -0.996 & 0.0386 \\ -5.2 & -1 & 0.249 & 0 \\ 1.54 & -0.0042 & -0.154 & 0 \\ 0 & 1 & 0 & 0 \end{bmatrix} \tag{90}$$

$$B_P = \begin{bmatrix} 0.02 & 0 \\ 0.337 & -1.12 \\ -0.744 & -0.32 \\ 0 & 0 \end{bmatrix} \tag{91}$$

$$C_P = \begin{bmatrix} 1 & 0 & 0 & 0 \\ 0 & 1 & 0 & 0 \\ 0 & 0 & 1 & 0 \\ 0 & 0 & 0 & 1 \end{bmatrix} \tag{92}$$

$$D_P = \begin{bmatrix} 0 & 0 \\ 0 & 0 \\ 0 & 0 \\ 0 & 0 \end{bmatrix} \tag{93}$$

As shown in Figure 4, the control system consists of two first order filters which take input from plant state feedback and external system inputs. Figures 5 and 6 show the CPU time and FLOPS to compute the closed loop system using the standard method described above and the new method with 2, 4, and 6 variable parameters. The variable parameters were selected randomly. Since the new method inherently requires computation of the combined plant and control system with variable parameters separated, CPU time is used before any loop closures occur. Notice that the standard method has a steeper slope than any of the variable parameter cases so that after a certain number of loop closures the new method becomes more computationally attractive. We notice for this example, the CPU break even point is

3 loop closures for all variable parameter configurations. Also note that the CPU time increases slightly as the number of variable parameters increases due to the loop closure matrices increase in size. It is also interesting to note that the overhead CPU time is approximately constant for all variable parameter cases. Corresponding to the same conditions as Figure 5, Figure 6 plots the FLOPS versus the number of loop closures. It is interesting to note that only the 2 variable parameter case requires less FLOPS than the standard method. After three loop closures the two variable parameter case requires less FLOPS than the standard method. Since the new method eliminates all bookkeeping after the initial system is constructed, the new method only requires FLOPS for a loop closure whereas the standard method requires both FLOPS and bookkeeping. These results were not sensitive to the parameters selected as variable.

High Order System:

The high order plant and control system combination consists of a 39th order plant and a 43rd order control system. The model is an accurate representation of a bearingless single main rotor helicopter with a conventional tail rotor in hover. The vehicle is 10,000 lbs. and has a main rotor radius of 20 feet which rotates at a speed of 35 rad/s. The baseline closed loop system has a spread in natural frequencies from 0-1075 rad/s. The plant models dynamics of the fuselage, main rotor, main rotor wake, drive system, and airframe elasticity. The plant matrices are not included here due to obvious space limitations. The control system structure is based on model following with body angular rate and attitude feedback. Figures 7 and 8 show the CPU time and FLOPS required to compute the closed loop system using the standard and new method with 5, 10, 15, and 20 variable parameters. Figures 9 and 10 show the FCS structure. As in the low order example, the variable parameters were selected randomly. The CPU time data is similiar to the low order example except that the loop closure break even point occurs slightly earlier. For 100 loop closures the new method is from 11.32-7.23 times faster than the standard method depending on how many variable parameters are considered. As in the low order example, CPU time increases slightly as the number of variable parameters increases due to the loop closure matrices increase in size and overhead CPU time is approximately constant for all variable parameter cases. In contrast to the low order example, FLOPS results shown in Figure 8 mirror the CPU results with the FLOPS break even point occuring after approximately one loop closure.

Conclusions

A method for separating variable control system parameters into a diagonal output feedback matrix for an open structure control system has been presented. The method is based on first considering all parameters of each individual control system filter as variable and subsequently closing the loop on all known fixed parameters. Through two examples, the method has been shown to offer a significant decrease in CPU time as the number of system evaluations increases. Results indicate that the method is most useful for high order plant and controller combinations where CPU time savings on the order of a factor of 10 are possible. This method may prove useful as a part of a parameter optimization based control system design tool where the cost function requires computation of the closed loop system for many trial sets of variable control system parameters.

References

[1] Grace A., Laub A., Little J., Thompson C., Control System Toolbox - For Use with MATLAB, The Math Works, Natick, MA, 1992.

[2] Yudilevitch G., Tischler M.B., Levine W.S., Lin C., "Rotorcraft Flight Control System Design Based on Multi-Criterion Parametric Optimization," Proceedings of the 1995 AIAA Guidance, Navigation, and Control Conference, Baltimore, MD, 1995.

[3] The Math Works, MATLAB - Reference Guide, The Math Works, Natick, MA, 1993.

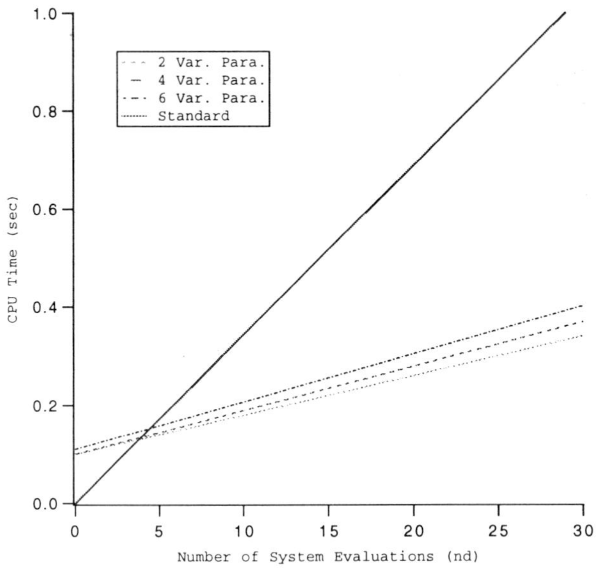

Figure 5: Low Order System - CPU Time versus Number of System Evaluations

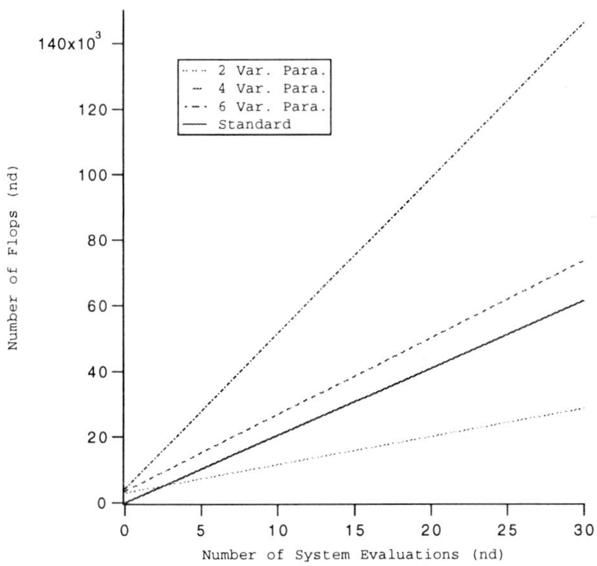

Figure 6: Low Order System - Flops versus Number of System Evaluations

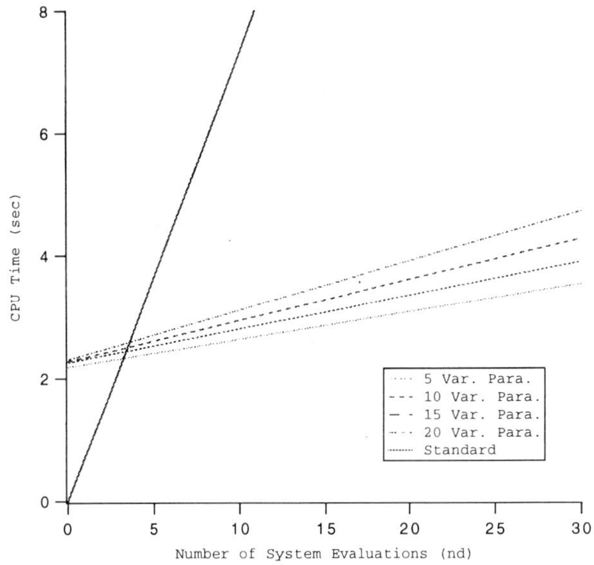

Figure 7: High Order System - CPU Time versus Number of System Evaluations

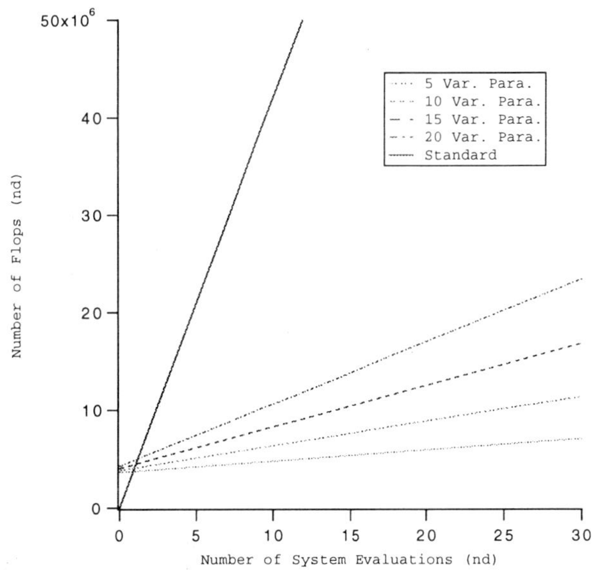

Figure 8: High Order System - Flops versus Number of System Evaluations

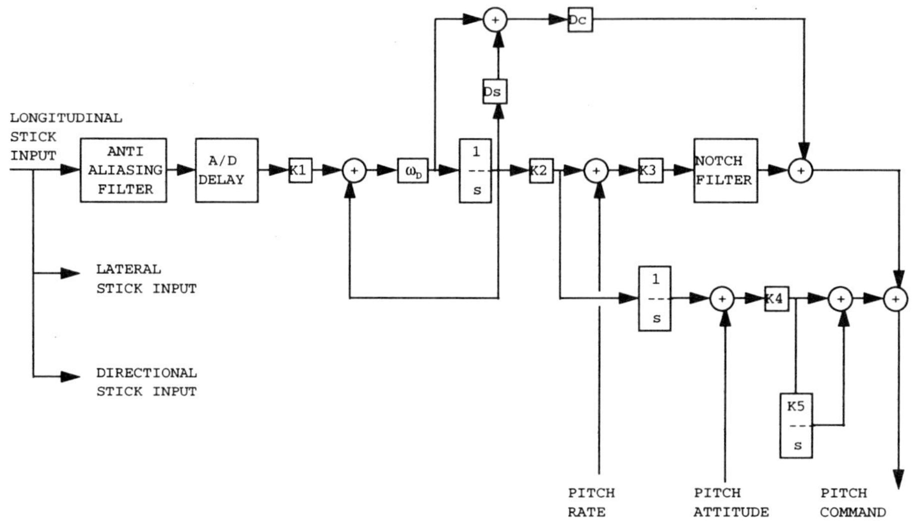

Figure 9: High Order Flight Control Laws

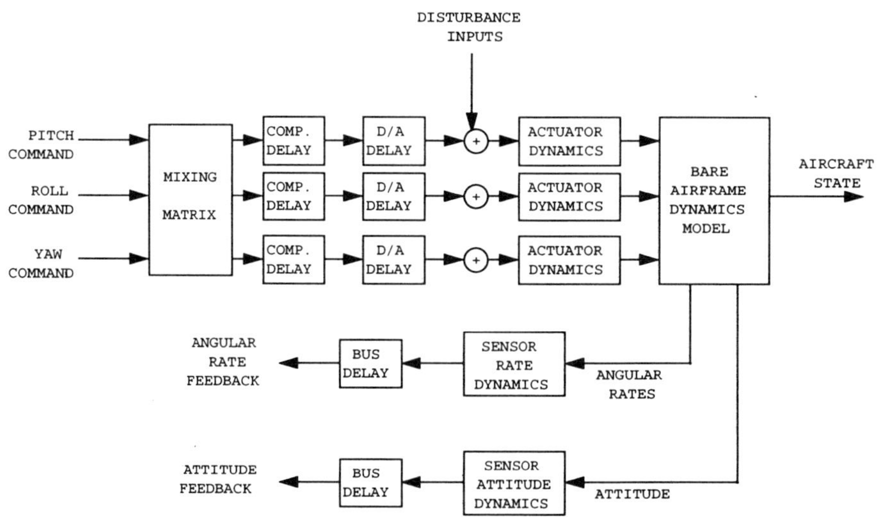

Figure 10: High Order Flight Control System - Actuator and Sensor Dynamics

AIAA-97-3477

STATIC AND DYNAMIC HYBRID OPTIMIZATION

Yan Hui *
Beijing Institute of Control Engineering, P.O.Box 2729, Beijing, 100080, P.R.China
Chen Shilu†
Northwestern Polytechnic University, Xi'an, 710072, P.R.China
Nan Ying‡
University of Minnesota, Minneapolis, MN55455

Abstract

The static and dynamic hybrid optimization (SDHO) is discussed in the paper. Treat the static variables as the state variables of optimal control. The derivatives of the static variables with respect to time is zero in the state equation. Introduce the static cost function to show the influence of the static variable on the performance function of SDHO. The necessary optimal conditions of SDHO have been constructed. The tradeoffs must be made in optimizing the static variables for SDHO. We can numerically solve SDHO by indirect method.

Introduction

Static and dynamic optimizations are usually studied, respectively[1-2]. The necessary optimal conditions are given by stationary points in static optimization. As to the constrained optimization, Kuhn-Tucker theory has been put forward. In dynamic optimization, the classic method is the calculus of variations.

* Doctor

† Professor, Senior Member AIAA

‡ Associate Professor , Visiting Researcher, Dept. of Aerospace Engineering and Mechanics

The development of the digit computer led to investigations of solving the problem. Pioneers in this field were Pontryagin and Bellman, who developed the Maximum Principle and Dynamic Programming, respectively. However, the performance functions are composed of static and dynamic variables in many problems. For example, the vehicle multidisciplinary optimization involves aerodynamics, propulsion, structure and trajectory, which include static and dynamic variables. The method to solve static and dynamic hybrid optimization (SDHO) is direct optimization. In Ref.[3][4], static variables were defined as design parameters. Combine the static with dynamic variables through discretizing the dynamic variable for optimization using nonlinear program. Another method is static and dynamic alternative optimization [5-6]. The static variables were firstly fixed, and the dynamic variables were optimized, then the static variables were optimized until the optimum performance function of SDHO was obtained.

The paper put forward indirect method to construct the optimal necessary conditions of SDHO. From the conditions, we can obtain the relations between static and dynamic optimization, and think it is necessary to combine static with dynamic optimization together.

SDHO

Since control variables vary with time in optimal control, we can not take the static variables as control variables to optimize in SDHO. So we introduce modifications to optimal control as follows:

Treat the static variables as the state variables of optimal control. The derivatives of the static variables with respect to time is zero in the state equation. Thus we can study SDHO on the basis of optimal control.

Similar to the terminal cost function in optimal control, we introduce the static cost function in the performance function of SDHO.

The Model of SDHO

If d is static variable, x is state variable, u is control variable, g(d) is static cost function, $\varphi(x_f, d_f, t_f)$ is terminal cost function and J is the performance function of SDHO, we desire to find the state and control vector such that

$$J = g(d) + \varphi(x_f, d_f, t_f) + \int_{t_0}^{t_f} L(x, d, u, t)dt \quad (1)$$

is minimized, subject to the state equation

$$\begin{cases} \dot{x} = f[x(t), d(t), u(t), t] \\ \dot{d} = 0 \end{cases} \quad (2)$$

the terminal equality constraints at the terminal time

$$N(x_f, d_f, t_f) = 0 \quad (3)$$

and the initial condition constraints

$$\begin{cases} x(t_0) = x_0 \\ h[d(t_0)] \geq 0 \end{cases} \quad (4)$$

where $h[d(t_0)]$ is the static variable should meet the constraint.

The admissible control inequality constraints

$$u_{min} \leq u \leq u_{max} \quad (5)$$

we now wish to derive the necessary condition for a minimum of the problem.

The Necessary Condition of SDHO

Consider the variational method to the solution of the optimization. We must change the constraint problem into unconstrained problem. Introduce the variables

$$\begin{cases} p = (p_1, p_2, \dots p_s)^T \\ z = (z_1, z_2, \dots z_r)^T \\ \sigma = (\sigma_1, \sigma_2, \dots \sigma_r)^T \end{cases} \quad (6)$$

Convert the inequality constraint to an equality constraints

$$\begin{cases} p^2 = h[d(t_0)] \\ \dot{z}^2 = (u_{max} - u)^T (u - u_{min}) \\ z(t_0) = z_0 \\ \dot{\sigma} = u \\ \sigma(t_0) = 0 \end{cases} \quad (7)$$

We adjoin, via the lagrange multiplier, constraints (2), (3), (4), (5) to equation (1) to obtain

$$J = g(d) + \nu^T \{h[d(t_0)] - p^2\} + \varphi(x_f, d_f, t_f) + \mu^T N(x_f, d_f, t_f) + \int_{t_0}^{t_f} \{L(x, d, \dot{\sigma}, t) - \lambda_d^T \dot{\sigma} + \lambda^T [f(x, d, \dot{\sigma}, t) - \dot{x}] + \xi^T [(u_{max} - \dot{\sigma})^T (\dot{\sigma} - u_{min}) - \dot{z}^2]\}dt \quad (8)$$

where $\lambda, \lambda_d, \nu, \mu, \xi$ are Lagrange multipliers. we take a first variation of the above performance function, and the necessary condition for a minimum requires

$$\delta J = 0 \quad (9)$$

We can obtain the necessary condition

a. The cannonic equations

$$\dot{\lambda} = -\frac{\partial H}{\partial x} \quad (10)$$

$$\dot{\lambda}_d = -\frac{\partial H}{\partial d} \quad (11)$$

where H is a Hamiltonian, and

$$H = L(x,d,\sigma,t) + \lambda^T f(x,d,\sigma,t) \quad (12)$$

b. Set the Hamiltonian with u=u* less than any other value of H with $u_{min} \leq u \leq u_{max}$

$$H(x,d,\lambda,u^*,t) = \min H(x,d,\lambda,u,t) \quad (13)$$

c. Two point boundary conditions

$$\lambda_d(t_0) = -\frac{\partial g'}{\partial d} \quad (14)$$

$$\lambda_d(t_f) = -\frac{\partial \varphi'}{\partial d} \quad (15)$$

$$\lambda(t_f) = -\frac{\partial \varphi'}{\partial x_f} \quad (16)$$

$$\frac{\partial g'}{\partial p} = 0 \quad (17)$$

where

$$\begin{cases} \varphi' = \varphi(x_f, d_f, t_f) + \mu^T N(x_f, d_f, t_f) \\ g' = g(d) + \nu^T [h(d(t_0)) - p^2] \end{cases} \quad (18)$$

d. The transversality condition

$$H(x_f, d_f, \lambda_f, u_f, t_f) + \frac{\partial \varphi'}{\partial t_f} = 0 \quad (19)$$

The equations (10)-(19) are the necessary optimal conditions of SDHO.

According to the stationary point conditions of static optimization, the right side of the equation (14) should be equal zero, but λ_d indicate the influence functions on J in the equation (11) in SDHO, which are usually not equal zero, that is, the tradeoffs must be made for SDHO in optimizing the static variables. On the other hand, SDHO can be numerically solved by two point boundary value problem (TPBVP). Because of the cannonic equation sensitivity to the initial value, SDHO can numerically be resolved by direct optimization[5,7].

Conclusion

In this paper, We have constructed the necessary optimal conditions of SDHO. The tradeoffs must be made in optimizing the static variables, and SDHO can numerically be resolved by TPBVP.

Reference

1. Bryson, A. E., and Ho, Y. C., Applied Optimal Control, Hemisphere, New York, 1975
2. Tabak, D., and Kuo, B. C., Optimal Control by Mathematical Programming, Prentice-Hall, Englewood-Cliffs, NJ, 1971
3. Hargraves, C. R., Paris, S. W., and Vlases, W. G., OTIS Past, Present and Future, AIAA 92-4530
4. Hull, D., G., Conversion of Optimal Control Problems into Parameter Optimization Problems, Journal of Guidance, Control, and Dynamics, Vol. 20, No. 1, 1997
5. Wang Zhigang, Yan Hui, Chen Shilu, and Nan Ying, Study of Integrated Optimization Method of Trajectory / Vehicle Overall Parameter, AAS 97-147
6. Nan Ying, Xiao Yelun, Yan Hui, and Wang Zhigang, A Common Numerical Calculation Method of Optimizing the Trajectories of Space Vehicles, AAS 97-148
7. Yan Hui, Chen Shilu, Wang Zhigang, Integrated Optimum Design of Ballistic Missile Using Parameterized Optimization, NPU/AIAA AFM 96-012, Xi'an, P.R.China, 1996

AIAA-97-3478

Fuel-Optimal Periodic Solutions to Hypersonic Cruise with Active Engine Cooling

Larry D. Dewell* and Jason L. Speyer†
University of California, Los Angeles
Michael Dinkelmann‡ and Gottfried Sachs§
Technische Universität München

Abstract

In this brief paper, it is shown that the active cooling of convectively-heated engine and airframe components in hypersonic vehicles may be approximately modeled and the associated questions of fuel-optimality are tractable using variational methods. Assuming a fuel-consumption which is linear with respect to throttle setting, it is shown that a first-order model of the cooling cycle results in a variational Hamiltonian which is piecewise linear in throttle, which is nonsmooth at a single interior equilibrium throttle setting. This special Hamiltonian control structure is exploited in the solution of the acceleration-constrained, fuel-optimal periodic cruise problem [2]. This work should be seen as a first step towards a more theoretical understanding of the coupling which exists between thermal-optimal and fuel-optimal hypersonic flight.

Nomenclature

a	speed of sound
C_D	drag coefficient
C_L	lift coefficient
$C_{T_{max}}$	maximum thrust coefficient
$C_{T_{min}}$	minimum thrust coefficient
D	drag
g	acceleration due to gravity
$g_{\max}$	load factor constraint
H	altitude
$\mathcal{H}$	Hamiltonian
h	enthalpy
$\mathcal{J}$	performance criterion
L	lift
M	Mach number
m_f	fuel mass consumed
P_r	Prandtl number
$\dot{q}_w$	heat flux
Re	Reynolds number
R_0	radius of the earth
r	horizontal range
S	throttle setting
S_b	reference area for aerodynamic forces
S_e	reference area for propulsive forces
St	Stanton number
T	thrust
t	time
v	velocity
W	vehicle weight
x	dynamic state vector
α	angle of attack
α_T	thrust offset angle
γ	flight path angle
λ	Lagrange multiplier vector
μ	Lagrange multiplier for constraint
ρ	atmospheric density
σ	trust-specific fuel consumption

*Currently with Lockheed Martin Missiles and Space; Member, AIAA.
†Prof., Mechanical and Aerospace Engineering Dept.; Fellow, AIAA
‡Dr.-Ing., Institute of Flight Mechanics and Flight Control; Member, AIAA
§Prof. Dr.-Ing., Director, Institute of Flight Mechanics and Flight Control; Associate Fellow, AIAA

1. Introduction

Problems related to the control of hypersonic velocity atmospheric vehicles present unique aspects inherent to the dynamic system and which are not shared in other aircraft systems. For example, the necessary usage of airframe surfaces for engine compression and expansion introduces strong aerodynamic and propulsive forces coupled into the dynamics. Many differences from traditional aircraft systems arise from the hypersonic flight envelope itself (typically $M > 6$) ; a representative such envelope is shown in Figure 1 [4]. In particular, the extremely high freestream dynamic pressure and compression of a large streamtube imply extreme convective heating of engine and airframe components, necessitating active cooling. For most hypervelocity vehicles, liquid hydrogen (H_2) is the only choice for fuel ; fortunately, this also implies on-board availability of a cryogenic coolant of high physical heat absorption. It is thus reasonable that the H_2 serves the dual purpose of coolant and fuel.

Fuel-optimal hypersonic cruise has been examined from several perspectives in recent years. For example, optimal cruise of a two-stage hypersonic vehicle was solved using a parameterized optimization technique [6]. Recent research has shown that hypersonic vehicles may be naturally amenable to periodic operation from a fuel-minimum point of view [1, 2]. In these studies, the fuel-optimal hypersonic cruise problem was posed as an optimal periodic control problem with a net vehicle acceleration constraint. Although aerothermodynamic heating was considered in these studies (a heating rate constraint was imposed on the optimal periodic solution of Chuang and Morimoto), no effort was made to model cooling cycle relative to fuel consumption. The fuel-optimal cruise problem with such cooling effects was studied by Sachs and Dinkelmann using a complex thermodynamic model of compression, combustion and expansion processes [7] . This problem was solved using direct numerical optimization techniques and did not focus on the periodic cruise problem.

The purpose of this study was to develop a simple model of fuel consumption in hypervelocity vehicles which accounts for active liquid H_2 cooling of engine/airframe components, and to study such a model from the perspective of fuel-optimal periodic cruise. The model used is meant to capture the essential feature of the cooling subsystem, while still allowing for a solution to the periodic problem by the calculus of variations, where the mathematical structure of the optimal control problem can be examined. In section 2, the model of the cooling subsystem is discussed. Section 3 reconsiders the hypersonic vehicle model previously studied in the context of fuel-optimal periodic cruise [2], and introduces the particular structure of the Hamiltonian for the cooling-cruise problem. Section 4 states the conditions for optimality of the optimal periodic control problem, and examines the question of convexity of the Hamiltonian with respect to throttle setting. Section 5 presents the numerical method and section 6 the results of the study. Finally, section 7 offers conclusions and directions for future work.

2. A Simplified Cooling Model

A simple schematic of the cooling cycle is illustrated in Fig. 2. Liquid H_2 at tank conditions (approximately 30 K) is passed through heat exchangers near high-temperature engine and airframe surfaces, which are assumed to be at their maximum allowable temperature. The gaseous H_2 cooling product cannot be returned to the tank, and thus must be either expelled into the atmosphere or injected into the SCRAM engine to produce thrust.

A simple model of the required fuel flow rate in the cooling cycle is developed based on the Eckert reference enthalpy method, which is used to evaluate approximately system cooling requirements [3]. In this method, the heat absorbed by the engine surfaces are assumed to result solely from the aerothermodynamic heating arising from the compression of a large freestream streamtube into the module area. Heat generated from the chemical combustion process is neglected. It should be noted that numerical comparisons of this model with a more complete thermodyanamic model [7] produced strong agreement.

The Eckert method predicts the heat transfer between the engine surface and the flow boundary layer as:

$$\dot{q}_w = St^* \cdot \rho^* \cdot v_e(h_{aw} - h_w), \tag{1}$$

where the asterisk indicates that the flow property is evaluated at an average reference condition which can be determined. Further, v_e is the flow velocity in the engine beyond the boundary layer, h_{aw} is the adiabatic wall enthalpy, and h_w is a selected design value of the static enthalpy at the surface. If the flow is assumed to be reasonably well modeled as turbulent, zero-pressure gradient flow over a constant–temperature flat plate, the Stanton number is given by:

$$St^* = \frac{0.0287}{(Pr^*)^{0.4} \cdot (Re_x^*)^{0.2}}, \tag{2}$$

then $\dot{q}_w$ can be written by:

$$\dot{q}_w = \frac{(0.0287)\rho^* v_e (h_{aw} - h_w)}{(Pr^*)^{0.4}(Re_x^*)^{0.2}}. \quad (3)$$

If it is further assumed that the freestream quantities may be used as the reference condition, the variation of transport properties with temperature is negligible, and if the enthalpy difference $(h_{aw} - h_w)$ can be replaced by $v_0^2/2$, the heat flux satisfies the proportionality:

$$\dot{q}_w \propto \frac{\rho_0 v_0^3}{(\rho_0 v_0)^{0.2}}. \quad (4)$$

The rate at which heat may be removed by flow of a coolant is expressed simply as:

$$\dot{q}_{fc} = \dot{m}_{fc} h_{fc}, \quad (5)$$

where h_{fc} is the physical heat absorption of the fuel, a property which measures the amount of heat absorbed by a unit mass of coolant (in this case H_2) in the transition from tank conditions to the maximum allowable gas temperature (usually about 1000 K). Equating equations (4) and (5) yields:

$$\dot{m}_{fc} = \frac{K\rho_0 v_0^2}{(\rho_0 v_0)^{0.2}} \quad (6)$$

This fuel flow rate $\dot{m}_{fc}$ may be injected into the burner to be used for combustion. In fact, proper design of a hypersonic vehicle dictates that at a reference flight condition, the fuel flow rate required to produce thrust should equal that required for active cooling. Thus, the constant K can be determined based on this reference flight condition. If, as in this study, the speed of sound is assumed a constant, this fuel flow rate may be expressed as a function of the Mach number as:

$$\dot{m}_{fc} = \frac{K_c \rho M^2}{(\rho M)^{0.2}} \quad (7)$$

For this study, the proportionality constant K_c of equation(7) was determined using a reference flight condition of $M = 12$ and $H = 30.48$ km. It is apparent from equation (7) that fuel is consumed regardless of any throttle setting commanded by pilot or controller. This fact will lead to an interesting structure for the fuel-optimal periodic control problem.

3. Hypersonic Vehicle Model and Constraint

The hypersonic vehicle of this study was derived from a generic hypersonic vehicle design study by NASA Dryden [8]. A schematic of the vehicle and the forces experienced during vertical-plane flight is illustrated in Fig. 3. Flight is assumed to be confined to the vertical plane over a nonrotating, spherical Earth. The equations of motion may be written as:

$$\frac{d\bar{H}}{d\bar{r}} = \tan\gamma \left(1 + \frac{\bar{H}}{\bar{R}_0}\right) \quad (8)$$

$$\frac{dM}{d\bar{r}} = \left\{\frac{gF_r(T\cos(\alpha + \alpha_T) - D - W\sin\gamma)}{Ma^2 W\cos\gamma}\right\}\left(1 + \frac{\bar{H}}{\bar{R}_0}\right) \quad (9)$$

$$\frac{d\gamma}{d\bar{r}} = \left\{\frac{gF_r(T\sin(\alpha + \alpha_T) + L - W\cos\gamma)}{M^2a^2 W\cos\gamma} + \frac{1}{\bar{H} + \bar{R}_0}\right\}\left(1 + \frac{\bar{H}}{\bar{R}_0}\right) \quad (10)$$

In this formulation, actual altitude H and range r have been normalized by the scalar F_r to produce the dynamic state $\bar{H}$ and the independent variable $\bar{r}$. As is indicated in Fig. 3, the thrust force T is assumed to have a nonzero thrust offset angle α_T due to exhaust expansion on the vehicle afterbody. The speed of sound a is considered to be a constant normalization factor for velocity.

The above model is implemented in the present study by specifying analytical relations between thrust, lift and drag and the dynamic states. In the following expressions, the various constant parameters associated with the polynomials were computed based on the baseline hypersonic vehicle model [8]. The aerodynamic forces are written in terms of dynamic pressure and the aerodynamic coefficients as:

$$T = ST_{\max} + (1 - S)T_{\min},$$
$$L = \frac{\rho}{2}M^2a^2S_bC_L,$$
$$D = \frac{\rho}{2}M^2a^2S_bC_D,$$
$$T_{\max} = \frac{\rho}{2}M^2a^2S_eC_{T\max},$$
$$T_{\min} = -\frac{\rho}{2}M^2a^2S_eC_{T_{\min}}.$$

The thrust model above assumes linear dependence on a throttle setting $S \in [0, 1]$ and includes an installation drag penalty term $T_{\min}$, where $C_{T_{\min}}$ is assumed a positive constant. The lift and drag coefficients are assumed to be related to a Mach number dependent lift-drag polar:

$$C_D = d_{00} + d_{01}M + (d_{10} + d_{11}M)\alpha + (d_{20} + d_{21}M)\alpha^2,$$

and lift coefficient is related to angle of attack as:

$$C_L = b_{00} + b_{01}\alpha + (b_{10} + b_{11}\alpha)M.$$

Maximum thrust coefficient is assumed to be related to M and α as:

$$C_{T\max} = C_{00} + C_{01}\alpha + (C_{10} + C_{11}\alpha)M + (C_{20} + C_{21}\alpha)M^2.$$

In addition to a model of vehicle dynamics, a propulsion system model is necessary for the solution of fuel-optimal trajectories. The fuel consumed for thrust production is modeled as

$$\frac{dm_f}{d\bar{r}} = \frac{S\sigma(T_{\max} - T_{\min})}{Ma\cos\gamma}(1 + \frac{\bar{H}}{\bar{R}_0}) \quad (11)$$

where σ is assumed to be linearly dependent on Mach number as :

$$\sigma = b_0 + b_1 M,$$

In order to obtain meaningful solutions to the fuel-optimal control problem, admissible controls C_L and S are constrained to satisfy the following inequality constraint at each instant:

$$C(\bar{H}, M, \gamma, C_L, S) = \sqrt{(T\cos(\alpha + \alpha_T) - D)^2 + (T\sin(\alpha + \alpha_T) + L)^2} - Wg_{\max} \leq 0, \quad (12)$$

where $g_{\max}$ is a (usually integer) constant which prescribes the maximum vehicle acceleration (load factor) as a multiple of the local acceleration of gravity.

4. Optimal Periodic Cooling-Cruise Problem

We can pose fuel-optimal *periodic* cruise problem as follows: find the initial states $\bar{H}(0)$, $M(0)$ and $\gamma(0)$, the piecewise continuous control histories $C_L(\cdot)$, $S(\cdot)$ and the period $\bar{r}_f > 0$ such that the local minimum of

$$\mathcal{J} = \frac{1}{\bar{r}_f}\int_0^{\bar{r}_f} \frac{dm_f}{d\bar{r}} d\bar{r} \quad (13)$$

is achieved, subject to inequality constraint (12) and state periodicity, as follows:

$$\bar{H}(\bar{r}_f) = \bar{H}(0),$$
$$M(\bar{r}_f) = M(0),$$
$$\gamma(\bar{r}_f) = \gamma(0).$$

The fuel-flow rate for the cooling cycle given by equation (7) is present regardless of the value of throttle setting S. Small values of throttle setting can be realized by injecting a portion of the cooling product into the combustion chamber, and expelling the difference into the atmosphere. In fact, there exists a *thermal equilibrium* throttle setting which we denote S_{eq}, for which all of the coolant product is used to produce thrust. For throttle setting greater than S_{eq}, fuel consumption is driven by equation (11). Thus, range-rate of fuel consumption may be written as:

$$\frac{dm_f}{d\bar{r}} = \mathcal{L} = \begin{cases} \frac{K_c\rho M^2}{(\rho M)^{0.2} a\cos\gamma}\left(1 + \frac{\bar{H}}{\bar{R}_0}\right), & S \leq S_{eq} \\ \frac{S(T_{max} - T_{min})\sigma}{Ma\cos\gamma}\left(1 + \frac{\bar{H}}{\bar{R}_0}\right), & S > S_{eq} \end{cases} \quad (14)$$

where, by equating coolant and thrust fuel flow rates, S_{eq} is given by:

$$S_{eq} = \frac{2K_c M}{S_e(\rho M)^{0.2} a^2 \sigma(C_{T_{max}} + C_{T_{min}})}, \quad (15)$$

In this model, increased combustion efficiency present in the combustion of hot hydrogen is neglected.

To aid in the examination of necessary optimality conditions, define the dynamic state vector $\mathsf{x} \in \mathbf{R}^3$ as:

$$\mathsf{x}(\bar{r}) = [\bar{H}(\bar{r}) \;\; M(\bar{r}) \;\; \gamma(\bar{r})]^T,$$

and let the vehicle dynamics (8-10) be written as[†]:

$$\dot{\mathsf{x}}(\bar{r}) = \mathsf{f}(\mathsf{x}(\bar{r}), C_L(\bar{r}), S(\bar{r})) \quad (16)$$

The first-order necessary conditions for local optimality are stated in terms of a scalar Hamiltonian function $\mathcal{H}$, defined as:

$$\mathcal{H}(\mathsf{x}, \lambda, C_L, S, \mu) = \mathcal{L}(\mathsf{x}, C_L, S) + \lambda^T(\bar{r})\mathsf{f}(\mathsf{x}, C_L, S) + \mu(\bar{r})C(\mathsf{x}, C_L, S) \quad (17)$$

where $\lambda(\bar{r}) \in \mathbf{R}^3$ and $\mu(\bar{r}) \in \mathbf{R}$. If the initial state x_0^*, the period $\bar{r}_f$ and the piecewise-continuous controls $C_L^*(\cdot)$ and $S^*(\cdot)$ minimize $\mathcal{J}$ subject to system dynamics, constraint (12) and dynamic state periodicity, then there exist Lagrange multiplier functions

[†] overdot notation denotes differentiation with respect to $\bar{r}$.

$\lambda(\cdot)$ and $\mu(\cdot)$ such that[†]:

$$(S^*(\bar{r}), C_L^*(\bar{r})) = \min_{C_L, 0 \le S \le 1} \{\mathcal{H}(\mathsf{x}^*(\bar{r}), \lambda(\bar{r}), C_L, S, \mu(\bar{r}))\} \tag{18}$$

$$\dot{\lambda}(\bar{r}) = -\mathcal{H}_{\mathsf{x}}^{*T}(\bar{r}) \;\; \forall \;\; \bar{r} \in [0, \bar{r}_f] \;\; ; \;\; \lambda(\bar{r}_f) = \lambda(0) \tag{19}$$

$$\mu(\bar{r}) \ge 0 \text{ and } \mu(\bar{r})C^*(\bar{r}) = 0 \;\; \forall \;\; \bar{r} \in [0, \bar{r}_f] \tag{20}$$

$$\left.\begin{aligned} &\text{If } (C_L^*, S^*)|_{\bar{r}=\bar{r}_i^+} \ne (C_L^*, S^*)|_{\bar{r}=\bar{r}_i^-}, \text{ then} \\ &\mathcal{H}^*(\bar{r}_i^+) = \mathcal{H}^*(\bar{r}_i^-) \text{ and } \lambda(\bar{r}_i^+) = \lambda(\bar{r}_i^-) \end{aligned}\right\} \tag{21}$$

$$\mathcal{H}^*(\bar{r}_f) - \mathcal{J}(\mathsf{x}_0^*, C_L^*(\cdot), S^*(\cdot), \tau) = 0 \tag{22}$$

In this statement of necessary conditions, the following notation was adopted:

$$\mathcal{H}^*(\bar{r}) \triangleq \mathcal{H}(\mathsf{x}^*(\bar{r}), C_L^*(\bar{r}), S^*(\bar{r}), \lambda^*(\bar{r}), \mu^*(\bar{r}))$$

The largest obstacle towards the use of these necessary conditions to solve for an extremal periodic cruise solution is the satisfaction of condition (18) for each $\bar{r} \in [0, \bar{r}_f]$. Suppose first that $C < 0$ - the constraint is not active. Inspection of equation (17) indicates that $\mathcal{H}$ is piecewise linear with respect to S for constant x, λ and C_L, with $\frac{\partial \mathcal{H}}{\partial S}$ discontinuous at $S = S_{eq}$. This implies that condition (18) may be simplified to:

$$(S^*, C_L^*) = \min_{S=0, S_{eq}, 1} \min_{C_L} \{\mathcal{H}(\mathsf{x}, \lambda, C_L, S)\} \tag{23}$$

The inner minimization of equation (23) is evaluated for fixed S by solving the stationarity condition of $\mathcal{H}$ with respect to C_L. $\mathcal{H}_{C_L} = 0$. Of course, for $S = S_{eq}$, this stationarity condition is augmented by the functional dependence of S_{eq} on C_L as:

$$\mathcal{H}_{C_L}(\mathsf{x}, \lambda, C_L, S) + \mathcal{H}_S(\mathsf{x}, \lambda, C_L)\frac{\partial S_{eq}}{\partial C_L} = 0.$$

Condition (23) implies that S exhibits a three-sided 'bang-bang' structure, such that for each $\bar{r}_i$ of discontinuous control, the Weierstrass-Erdmann conditions given in equation (21) must be satisfied.

On the other hand, consider satisfaction of (18) on an arc of active constraint, i.e., $C = 0$. In this case, $\frac{\partial \mathcal{H}}{\partial S}$ is explicitly a function of S, since $\mu > 0$ and C is of second order in S. Thus, condition (18) implies joint satisfaction of the equations:

$$\begin{aligned} \mathcal{H}_{C_L}(\mathsf{x}, \lambda, C_L, S, \mu) &= 0, \\ \mathcal{H}_S(\mathsf{x}, \lambda, C_L, \mu) &= 0, \\ C(\mathsf{x}, C_L, S) &= 0. \end{aligned}$$

[†]subscripts denote partial differentiation.

These equations may be reduced by one to a quadratic equation in S as:

$$\mathcal{A}(\mathsf{x}, \lambda, C_L)S^2 + \mathcal{B}(\mathsf{x}, \lambda, C_L)S + \mathcal{C}(\mathsf{x}, \lambda, C_L) = 0 \tag{24}$$

$$C(\mathsf{x}, S, C_L) = 0 \tag{25}$$

A control pair (S, C_L) may result from equations (24) and (25) for which S is interior to its control bound ; if not, then C_L is chosen to satisfy the constraint for fixed S. Unfortunately, the functional form of coefficients $\mathcal{A}$, $\mathcal{B}$ and $\mathcal{C}$ depend on whether or not $S \le S_{eq}$ or $S > S_{eq}$, which is of course not known *a priori*, since S is being solved for. This problem can be circumvented in the numerical solution of optimal control by assuming first the functional forms for $S > S_{eq}$ and finding all solutions (S_i, C_{Li}) to the combined quadratic / nonlinear equalities (24) and (25). Any solutions for which $S_i \le S_{eq}$ are eliminated, and the same procedure is performed assuming $S \le S_{eq}$.

5. The Numerical Method

An objective of this study was to obtain *exact*, extremal state and control histories to the optimal periodic cruise problem with engine cooling effects, through satisfaction of the first-order necessary conditions of optimality given in the previous section. This required numerical schemes for computation of constrained optimal control from conditions (12), (18) and (21), numerical integration of aircraft dynamics and Lagrange multiplier equation (19), and numerical solution of the periodic boundary value problem in state and costate. The numerical solution of optimal control has been outlined in the previous section. Numerical integration of the differential equations was carried out using a 4th-order Runge Kutta scheme with adaptive stepsize, with an accuracy of 10^{-9}. Instants of discontinuous control were located within the integration to the same order of accuracy. Solution of the two-point boundary value problem for fixed τ was performed by minimizing the cost function $\mathcal{J}_p$ given by:

$$\begin{aligned} \mathcal{J}_p(\mathsf{x}_0, \lambda_0) =& [\mathsf{x}(\bar{r}_f) - \mathsf{x}_0]^T Q_{\mathsf{x}} [\mathsf{x}(\bar{r}_f) - \mathsf{x}_0] \\ &+ [\lambda(\bar{r}_f) - \lambda_0]^T Q_\lambda [\lambda(\bar{r}_f) - \lambda_0], \end{aligned} \tag{26}$$

over the state/costate initial condition space $[\mathsf{x}_0^T \; \lambda_0^T]^T \in \mathbf{R}^{2n}$, where the matrices Q_{x} and Q_λ are generally diagonal with positive entries. This minimization was performed using the Davidon-Fletcher-Powell direction set method [5], with $\mathsf{x}(\bar{r}_f)$ and $\lambda(\bar{r}_f)$

Solution	cooling effects not considered [2]	cooling effects considered
optimal steady-state cruise	$J = 0.935$	$J = 0.935$
extremal periodic cruise ($g_{\max} = 5$)	$J = 0.844$	$J = 0.882$

Table 1: Range-average fuel consumption for different cruise solutions

computed via numerical integration using extremal $C_L^*(\cdot)$ and $S^*(\cdot)$. Since successful convergence of this method depends strongly on a fortunate choice of initial guess for state / costate initial condition, the cooling cruise problem was solved beginning with the extremal periodic solution of [2], and perturbing the fuel consumption model slowly to the cooling cruise problem by means of a scalar homotopy parameter.

6. Results and Discussion

The optimal steady-state cruise condition was not changed from the results of Dewell and Speyer [2] with the addition of engine / airframe cooling effects (see Table 1). That is, the best static cruise condition corresponded to a fully open throttle ($S = 1$) at $H = 48.2$ km and $M = 15.13$.

The extremal periodic cruise solution with consideration of engine cooling effects (Figs. 4-9), solved using the numerical method outlined in the previous section, was found to exhibit a range-averaged fuel consumption which was 6% less than the optimal static cruise solution and only 4.5% higher than the (less realistic) extremal periodic cruise solution without consideration of engine cooling effects. The altitude and Mach number behavior (Figs. 4 and 5) indicate the atmospheric 'skip' trajectory which is characteristic of extremal periodic cruise for hypersonic aircraft [2]. This skip trajectory is particularly apparent in Fig. 7, with the high-altitude quasi-ballistic phase and the low-altitude turn phase. The controls which realize this extremal periodic trajectory are shown in Fig. 8 and 9 ; in these figures, an enlargement of the range in the low-altitude flight regime is principally plotted, with an accompanying plot of the entire history. The cooling aspect of the problem is particularly evident in the plot of throttle setting in Fig. 9. In contrast to the results of [2] in which a zero throttle setting was optimal on a subarc of the trajectory, the lowest throttle setting appearing in the optimal solution is $S = S_{eq}$. That is, optimality dictates that *no portion* of the coolant product should be ejected into the atmosphere but should instead be used to impart kinetic energy to the vehicle.

A partial examination of second-order optimality conditions for the optimal periodic controls was performed on the extremal periodic cruise solution. Fig. 10 plots $\frac{\partial \mathcal{H}}{\partial S}$ for values of S slightly less and greater than S_{eq}, which are denoted by S_{eq-} and S_{eq+}, respectively. From this figure, arcs for which $S = S_{eq}$ exhibit strict convexity of $\mathcal{H}$ with respect to S. In fact, the equilibrium throttle setting becomes nonoptimal precisely when $\frac{\partial \mathcal{H}}{\partial S}(S_{eq-})$ becomes negative. Fig. 11 illustrates qualitatively the convexity of $\mathcal{H}$ with respect to the optimal control pair (S, C_L). In this figure, C_L is assumed unbounded, and S bounded. The figure is drawn for a given x, and the curved line indicates S_{eq} vs. C_L. The arrows in Fig. 11 indicate a direction of *increasing* Hamiltonian for control perturbations. The three solid dots indicate values of C_L which satisfy $\mathcal{H}_{C_L} = 0$ for the given value of S. The optimal throttle setting from this figure is S_{eq}, since both alternatives admit a control variation which can decrease the Hamiltonian. The equilibrium throttle setting becomes nonoptimal when it intersects the vertical dashed line which demarks the point of sign transition of $\frac{\partial \mathcal{H}}{\partial S}(S_{eq-})$. The positivity of $\mathcal{H}_{C_L C_L}$ for each stationary (S, C_L) pair of Fig. 11 (indicated by the large diverging arrows) was verified for the extremal periodic solution obtained.

7. Conclusions and Future Directions

In this paper we have examined together the fuel-optimal periodic cruise problem for hypersonic vehicles and the active cooling problem, with the goal of developing a simple model amenable to indirect methods, but which captures the essential features of the problem. With this model, the optimal throttle setting possessed an interesting three-sided bang-bang structure, in which the zero throttle setting was admissible. In the implementation of the Minimum Principle, however, such a coast phase never arose. Fuel improvement of 6% over best static flight was obtained for this problem, with a rather stringent aerodynamic load constraint of 5 times the acceleration of gravity. The calculus of variations approach allowed us to examine second-order optimality from the Legendre-Clebsch condition. The local minimum computed was found to satisfy this condition.

A cursory examination of the present cooling model reveals several deficiencies in the model which should be addressed in future studies. First, radiative cooling effects have not been addressed in this study. These effects are expected to be especially significant in the analysis of periodic cruise, due to the long, high-altitude phase of flight, in which radiative heat transfer to the surroundings would be significant. Second, the model studied here does not account for the airframe cooling and the therefrom resulting decrease in thrust-specific fuel consumption when burning the hot hydrogen, as opposed to the combustion of fuel at tank conditions. Inclusion of such effects is expected to significantly improve the fuel improvement of optimal steady-state flight.

Acknowledgement. This work was made possible by the generous support of the U.S. Fulbright scholarship fund and the German FORTWIHR program.

References

[1] C.-H. Chuang and H. Morimoto. Optimal periodic cruise for a hypersonic vehicle with constraints. In *AIAA Guidance Navigation and Control Conference*. American Institute of Aeronautics and Astronautics, 1996. Paper No. AIAA-96-3876.

[2] L.D. Dewell and J.L. Speyer. Fuel-optimal periodic control and regulation in constrained hypersonic flight. In *Proceedings of the AIAA Guidance, Nagivation and Control Conference*, 1996. Paper No. AIAA-96-3902.

[3] W. H. Heiser and D. T. Pratt. *Hypersonic Airbreathing Propulsion*. American Institute of Aeronautics and Astronautics, 1994.

[4] J. W. Hicks. Flight testing of airbreathing hypersonic vehicles. In *Space Course on Low Earth Orbit Transportation*, volume 2, pages 37:1–37:28. Technische Universität München, Germany, 1993.

[5] W. H. Press, B.P. Flannery, S.A. Teukolsky, and W.T. Vetterling. *Numerical Recipes*. Cambridge University Press, 1990.

[6] G. Sachs and M. Dinkelmann. Optimization of three-dimensional range and ascent trajectories of a two-stage hypersonic vehicle. *Space Technology*, 16(5/6):365–373, 1996.

[7] G. Sachs and M. Dinkelmann. Reduction of coolant fuel losses in hypersonic flight by optimal trajectory control. *AIAA Journal of Guidance, Control and Dynamics*, 19(6):1278–1284, 1996.

[8] D. White, A.H. Bowers, K.W. Iliff, G. Noffz, M. Gonda, and J. Menousek. Flight, propulsion, and thermal control of advanced aircraft and hypersonic vehicles. In *Handbook of Intelligent Control*. Van Nostrand Reinhold, 1992.

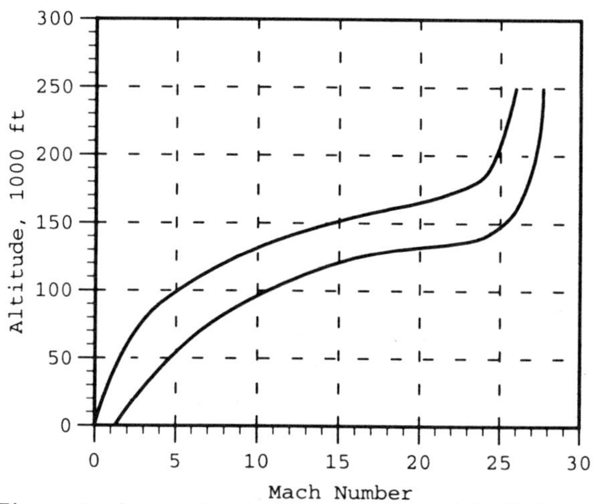

Figure 1: Approximate hypersonic vehicle flight envelope

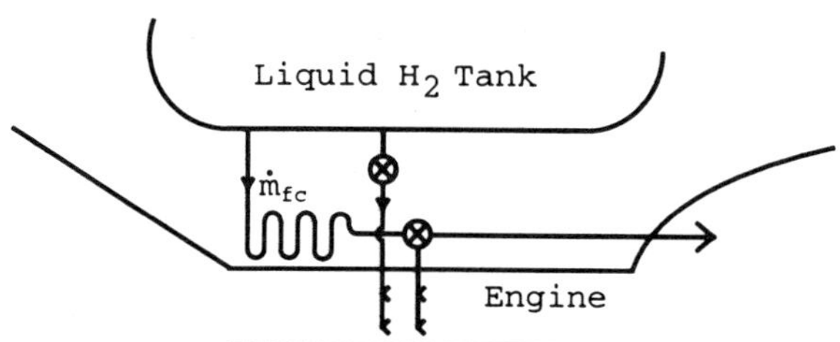

Figure 2: Schematic of cooling cycle

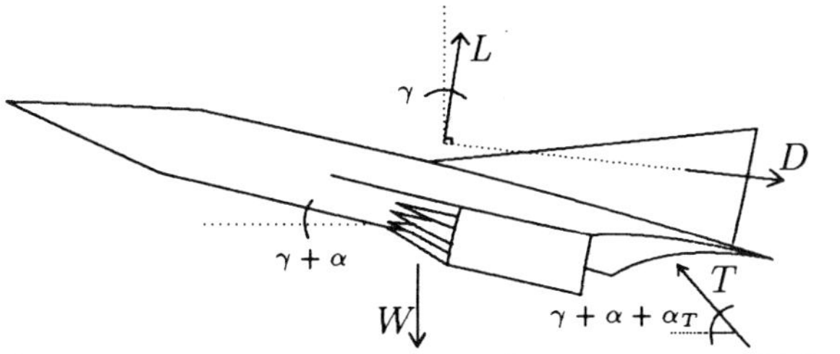

Figure 3: Free–Body Diagram of Hypersonic Vehicle

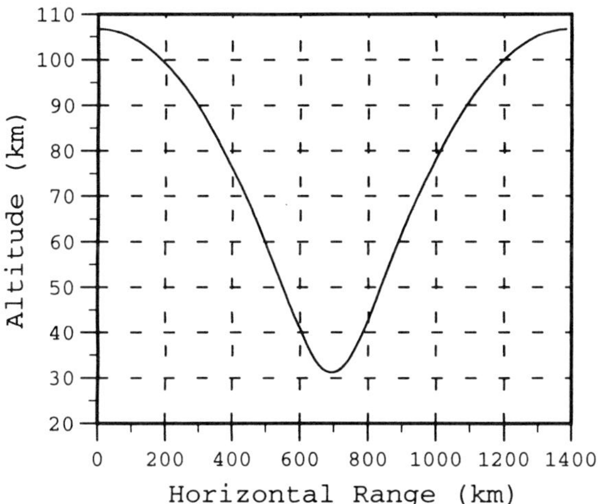

Figure 4: Extremal periodic solution : h vs. r

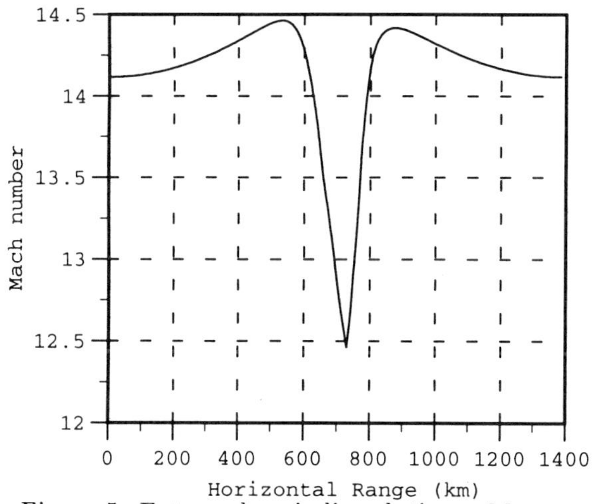

Figure 5: Extremal periodic solution : M vs. r

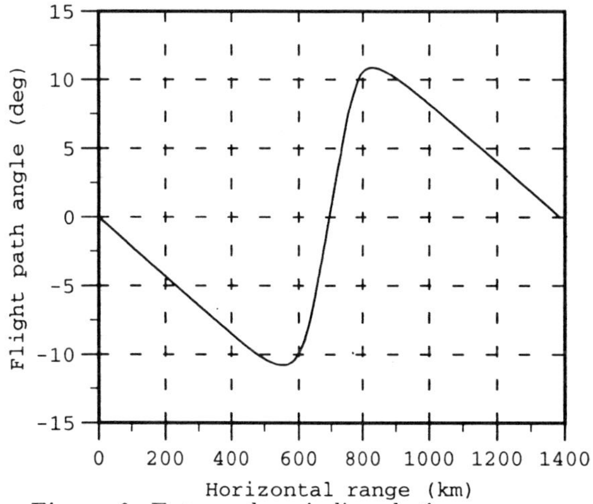

Figure 6: Extremal periodic solution : γ vs. r

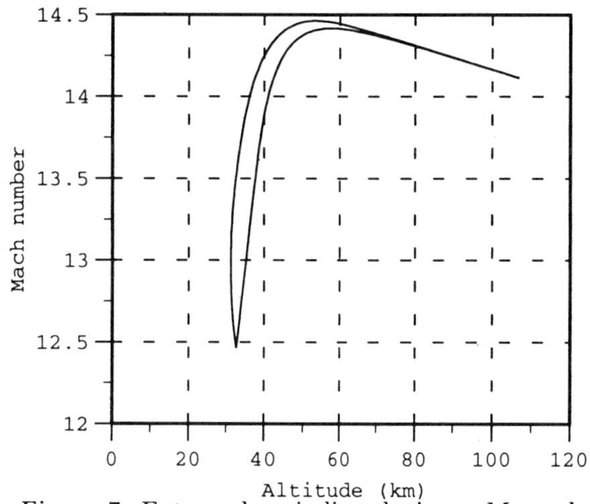

Figure 7: Extremal periodic solution : M vs. h

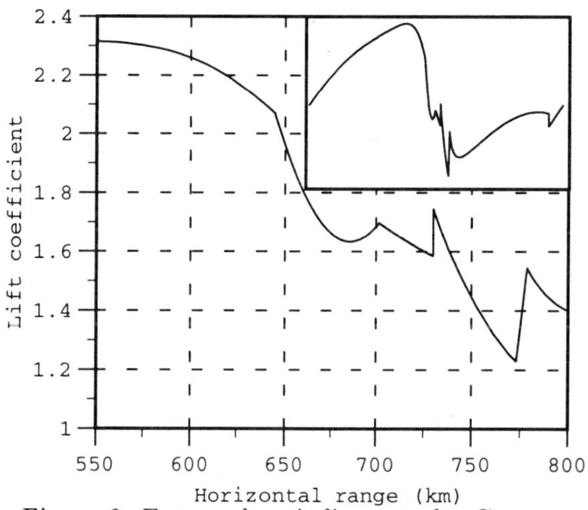

Figure 8: Extremal periodic control : C_L vs. r

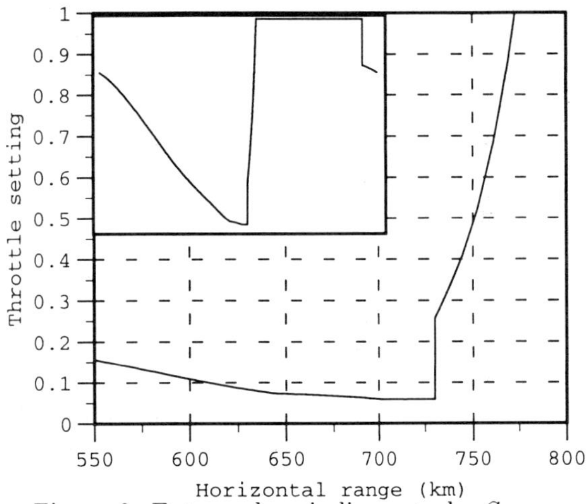

Figure 9: Extremal periodic control : S vs. r

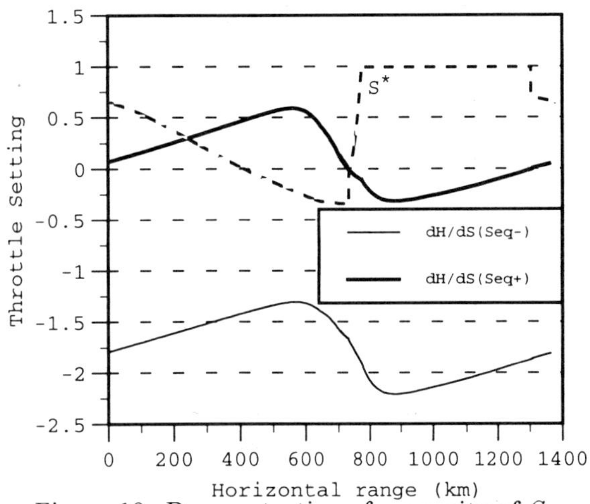

Figure 10: Demonstration of convexity of S_{eq}

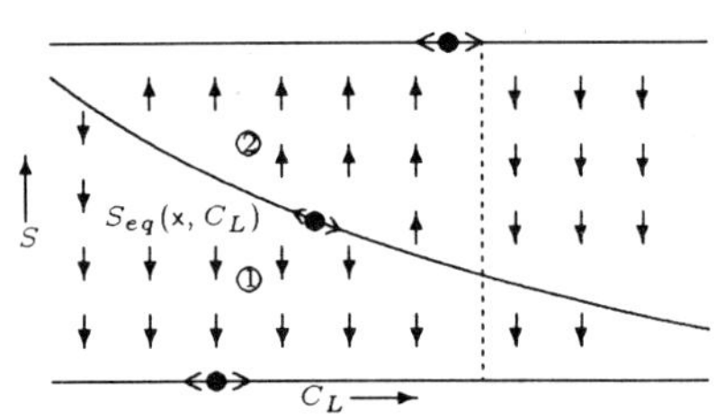

$$① \; \mathrm{sgn}\left\{\frac{\partial \mathcal{H}}{\partial S}\Big|_{S_{eq}^-}\right\} = \mathrm{sgn}\left\{\lambda_M \cos(\alpha + \alpha_T) + \frac{\lambda_\gamma}{M}\sin(\alpha + \alpha_T)\right\}$$

$$② \; \mathrm{sgn}\left\{\frac{\partial \mathcal{H}}{\partial S}\Big|_{S_{eq}^+}\right\} = \mathrm{sgn}\left\{\frac{aW\sigma}{F_r F_f g} + \lambda_M \cos(\alpha + \alpha_T) + \frac{\lambda_\gamma}{M}\sin(\alpha + \alpha_T)\right\}$$

Figure 11: Convexity of Cooling-Cruise Problem

AIAA-97-3479

LINEAR TIME VARYING APPROACH TO SATELLITE ATTITUDE CONTROL USING ONLY ELECTROMAGNETIC ACTUATION

Rafał Wiśniewski

Aalborg University, Department of Control Engineering,
Bajers Vej 7, DK-9220 Aalborg Ø, Denmark. raf@control.auc.dk

ABSTRACT

Recently small satellite missions have gained considerable interest due to low-cost launch opportunities and technological improvement of micro-electronics. Required pointing accuracy of small, inexpensive satellites is often relatively loose, within a couple of degrees. Application of cheap, lightweight, and power efficient actuators is therefore crucial and viable. This paper discusses linear attitude control strategies for a low earth orbit satellite actuated by a set of mutually perpendicular electromagnetic coils. The principle is to use the interaction between the Earth's magnetic field and the magnetic field generated by the coils. A key challenge is the fact that the mechanical torque can only be produced in a plane perpendicular to the local geomagnetic field vector, therefore the satellite is not controllable when considered at fixed time. Availability of design methods for time varying systems is limited, nevertheless, a solution of the Riccati equation gives an excellent frame for investigations provided in this paper. An observation that geomagnetic field changes approximately periodically when a satellite is on a near polar orbit is used throughout this paper. Three types of attitude controllers are proposed: an infinite horizon, a finite horizon, and a constant gain controller. Their performance is evaluated and compared in the simulation study of the realistic environment.

INTRODUCTION

The work is motivated by the Ørsted satellite mission. The Ørsted satellite is a 60 kg auxiliary payload scheduled to be launched by a MD-Delta II launch vehicle in the late 1997 into a 450 x 850 km orbit with a 96 degree inclination. The purpose of the Ørsted satellite is to conduct a research program in the discipline of the magnetic field of the Earth. When the satellite is firmly stabilized and the ground contact is established, an 8 m long boom is deployed. The boom carries the scientific instruments that must be displaced from the electro-magnetic disturbances present in the main body of the satellite.

Stabilization of the Ørsted satellite is accomplished by active use of a set of mutually perpendicular coils called magnetorquers. The interaction between external magnetic field of the Earth and the magnetic field generated in the magnetorquers produces a mechanical torque, which is used to correct the attitude. Magnetic control systems are relatively lightweight, require low power and are inexpensive, however they can only be successfully applied for the satellites on high inclined orbits. The Ørsted's coils are mounted in the x, y, and z facets of the main body. A maximum producible magnetic moment is 20 Am^2. The maximum mechanical torque produced by the magnetorquers is approximately $0.6 \cdot 10^{-3}$ Nm above the equator, and $1.2 \cdot 10^{-3}$ Nm above the Poles.

After boom deployment the nominal operation phase controller is activated. The satellite shall be three axis stabilized with its boom pointing outwards. Formally a coordinate system fixed in the satellite structure shall coincide with a reference coordinate system fixed in orbit. The pointing accuracy is required to be within 10 degrees in pitch, roll, and 20 degrees in yaw.

There is extensive literature covering satellite attitude control design. Most of the algorithms assume application of reaction wheels and/or thrusters for three axis stabilization, though. Attitude control with sole use of magnetorquers has the significant challenge that the system is only controllable in two axes at any time with the axes being perpendicular to the local geomagnetic field vector.

The number of internationally published papers on magnetic attitude control is still rather small. Most of them deal with momentum desaturation of reaction wheel systems. The available literature on magnetorquing for three axis stabilization of satellites includes Reference[1], where a configuration with two magnetic coils and a reaction wheel were analyzed. The problem of three axis control using only electromagnetic coils was addressed in Reference[2]. The local stabilization of the satellite was achieved via implementation of the infinite time horizon linear quadratic regulator. Another linear approach was given in Reference[3], where the linearized time varying satellite motion model was approximated by a linear time invariant counterpart. Three-axis stabilization with use of magnetic torquing of a satellite without appendages was treated in Reference[4], where sliding control law stabilizing a tumbling satellite was proposed. An approach for three-axis magnetic stabilization of a low earth near polar orbit satellite based on Lyapunov theory was presented in Reference[5].

The first part of this study addresses linearization technique for a LEO satellite motion. It is shown that a satellite on a near polar orbit actuated by a set of perpendicular magnetorquers may be considered as a periodic system. In the subsequent sections three types of controllers are design: an infinite horizon, a finite horizon, and a constant gain controller. Their performance is evaluated in the simulation study of the realistic environment.

SATELLITE LINEAR MODEL

The satellite considered in this study is modeled as a rigid body in the Earth gravitational field influenced by the aerodynamic drag torque and the control torque generated by the magnetorquers. The attitude is parameterized by the unit quaternion providing a singularity free representation of the kinematics. The details about mathematical modeling of a LEO satellite can be found in Reference[6]. In this paper only a satellite linear model will be investigated.

It has been already mentioned that the control torque, $\mathbf{N}_{ctrl}$, of the magnetic actuated satellite always lies perpendicular to the geomagnetic field vector, $\mathbf{b}$, furthermore a magnetic moment, $\mathbf{m}$, generated in the direction parallel to the local geomagnetic field has no influence on the satellite motion. This can be explained by the following equality

$$\mathbf{N}_{ctrl} = (\mathbf{m}_{\parallel} + \mathbf{m}_{\perp}) \times \mathbf{b} = \mathbf{m}_{\perp} \times \mathbf{b}, \qquad (1)$$

where $\mathbf{m}_{\parallel}$ is the component of the magnetic moment, $\mathbf{m}$, parallel to $\mathbf{b}$, whereas $\mathbf{m}_{\perp}$ is perpendicular to the local geomagnetic field.

Concluding, the necessary condition for power optimality of a control law is that the magnetic moment lies on a 2-dimensional manifold perpendicular to the geomagnetic field vector.

Consider the following mapping

$$\tilde{\mathbf{m}} \mapsto \mathbf{m} : \mathbf{m} = \frac{\tilde{\mathbf{m}} \times \mathbf{b}}{\| \mathbf{b} \|}, \qquad (2)$$

where $\tilde{\mathbf{m}}$ represents a new control signal for the satellite. Now, the magnetic moment, $\mathbf{m}$, is exactly perpendicular to the local geomagnetic field vector and control theory for a system with unconstrained input $\tilde{\mathbf{m}}$ can be applied. The direction of the signal vector $\tilde{\mathbf{m}}$ (contrary to $\mathbf{m}$) can be chosen arbitrary by the controller. From practical point of view, the mapping (2) selects the component of $\tilde{\mathbf{m}}$ which is perpendicular to the local geomagnetic field vector.

The linear model of the satellite motion is given in terms of the angular velocity and the first three components of the attitude quaternion. Linearization of the angular velocity is commonplace and based on the first order extension of the Taylor series. Linearization of the attitude quaternion is quite different due to the multiplicative transformation is needed to describe successive rotations. Consider two rotations, the first one from an orbit fixed coordinate system (OCS) to a reference coordinate system (RCS), the second from RCS to a satellite fixed coordinate system (SCS)

$${}^{S}_{O}\mathbf{q} = {}^{R}_{O}\mathbf{q}\, {}^{S}_{R}\mathbf{q} \qquad (3)$$

where ${}^{S}_{R}\mathbf{q}$, ${}^{R}_{O}\mathbf{q}$, ${}^{S}_{O}\mathbf{q}$ mean the quaternion mapping RCS to SCS, the quaternion OCS to RCS, and the quaternion transforming OCS to SCS, respectively. ${}^{S}_{R}\mathbf{q}$ can be considered as a small perturbation from the reference $[0\ 0\ 0\ 1]^T$. According to Reference[7]

$${}^{S}_{R}\mathbf{q} = \begin{bmatrix} e_1 sin\frac{\delta\phi}{2} \\ e_2 sin\frac{\delta\phi}{2} \\ e_3 sin\frac{\delta\phi}{2} \\ cos\frac{\delta\phi}{2} \end{bmatrix}, \qquad (4)$$

thus for small ϕ

$${}^{S}_{R}\mathbf{q} \approx \begin{bmatrix} \delta q_1 \\ \delta q_2 \\ \delta q_3 \\ 1 \end{bmatrix} \equiv \begin{bmatrix} \delta\mathbf{q} \\ 1 \end{bmatrix}. \qquad (5)$$

As the result, the linearized equation of motion is

$$\frac{d}{dt}\begin{bmatrix} \delta\Omega \\ \delta\mathbf{q} \end{bmatrix} = \mathbf{A}\begin{bmatrix} \delta\Omega \\ \delta\mathbf{q} \end{bmatrix} + \mathbf{B}(t)^c\tilde{\mathbf{m}}, \qquad (6)$$

where

$$\mathbf{A} = \begin{bmatrix} 0 & 0 & 0 & -6\omega_o\sigma_x & 0 & 0 \\ 0 & 0 & \omega_o\sigma_y & 0 & 6\omega_o\sigma_y & 0 \\ 0 & \omega_o\sigma_z & 0 & 0 & 0 & 0 \\ \frac{1}{2} & 0 & 0 & 0 & 0 & 0 \\ 0 & \frac{1}{2} & 0 & 0 & 0 & \omega_o \\ 0 & 0 & \frac{1}{2} & 0 & -\omega_o & 0 \end{bmatrix},$$

$$\sigma_x = \frac{I_y - I_z}{I_x},\ \sigma_y = \frac{I_z - I_x}{I_y},\ \sigma_z = \frac{I_x - I_y}{I_z},$$

$$\mathbf{B} = \begin{bmatrix} \frac{\mathbf{I}^{-1}}{\|\mathbf{b}\|}\begin{bmatrix} -b_y^2 - b_z^2 & b_x b_y & b_x b_z \\ b_x b_y & -b_x^2 - b_z^2 & b_y b_z \\ b_x b_z & b_y b_z & -b_x^2 - b_y^2 \end{bmatrix} \\ \begin{bmatrix} 0 & 0 & 0 \\ 0 & 0 & 0 \\ 0 & 0 & 0 \end{bmatrix} \end{bmatrix},$$

where ω_o is the orbital rate, and I_x, I_y, I_z are components on the diagonal of the inertia tensor $\mathbf{I}$. The matrix $\mathbf{B}(t)$ comes from the double cross product operation $-\mathbf{b}(t) \times (\mathbf{b}(t)\times)$.

Based on the mathematical model provided in this section linear attitude control concepts will be developed. Three controllers will be proposed: an infinite horizon, a finite horizon, and a constant gain controller.

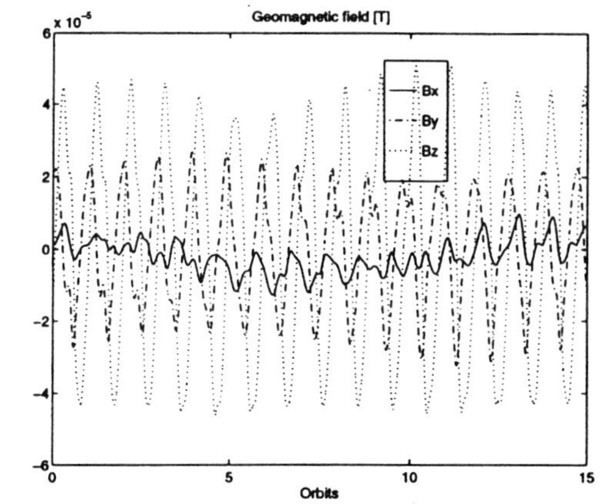

Figure 1: The geomagnetic field vector in RCS propagated by a 10th order spherical harmonic model during 24 h in May 1997.

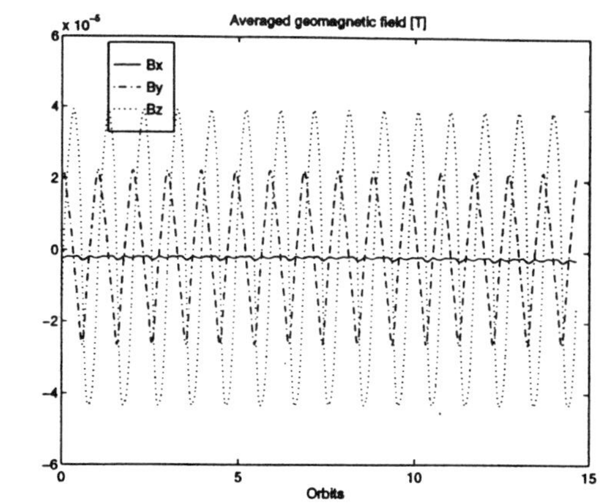

Figure 2: An averaged B-field vector in RCS. Compare with the realistic magnetic field of the Earth in Figure 1.

INFINITE HORIZON PERIODIC CONTROLLER

The geomagnetic field is essentially that of a magnetic dipole with the largest anomalies over Brazil and Siberia. The geomagnetic field in RCS, has large y and z components, while the x component is comparatively small. The rotation of the Earth is visible via fluctuations of the geomagnetic field vector's x component with frequency $1/24\ 1/hours$, see Figure 1. The following observation is used for the design of an attitude controller. The geomagnetic field on a near polar orbit is approximately periodic with a period $T = 2\pi/\omega_0$.

Due to periodic nature of the geomagnetic field, seen from RCS, the linearized model of the satellite can be considered as periodic. It is though necessary to find an ideally periodic counterpart of the real magnetic field of the Earth. This is done by averaging the geomagnetic field over a time interval reflecting a common period for the satellite revolution about the Earth and the Earth's own revolution. This interval for the Ørsted satellite corresponds to $N = 144$ orbits.

Furthermore, the geomagnetic field is parameterized by the mean anomaly M, since the geomagnetic field and the mean anomaly have the common period T

$$\mathbf{b}_{ave}(M) = \frac{1}{N}\sum_{i=1}^{N} \mathbf{b}(M + i * T). \tag{7}$$

An averaged B-field vector $\mathbf{b}_{ave}(M(t))$ is depicted in Figure 2.

The resultant linear periodic system is

$$\frac{d}{dt}\begin{bmatrix} \delta\Omega \\ \delta\mathbf{q} \end{bmatrix} = \mathbf{A}\begin{bmatrix} \delta\Omega \\ \delta\mathbf{q} \end{bmatrix} + \hat{\mathbf{B}}(M)^c\vec{\mathbf{m}}, \tag{8}$$

where $\hat{\mathbf{B}}(M)$ is given in Eq. (8) after substituting the symbol $\mathbf{B}(t)$ for $\hat{\mathbf{B}}(M)$, and the components of the vector $\mathbf{b}(t)$ for the components of $\mathbf{b}_{ave}(M)$.

The difference between the time varying matrix $\mathbf{B}(t)$ and the ideal periodic counterpart $\hat{\mathbf{B}}(M(t))$ used for the controller design is considered an additional external disturbance torque acting on the satellite.

The controller gain is calculated from the steady state solution of the Riccati equation. According to Theorem 6.3 in Reference[8], if the pair $(\mathbf{A}, \hat{\mathbf{B}}(M))$ is stabilizable then there exists a stabilizing symmetric periodic solution $\mathbf{P}_+(t)$ of the Riccati equation

$$\begin{aligned} -\dot{\mathbf{P}}_+(t) &= \mathbf{P}_+(t)\mathbf{A} + \mathbf{A}^T\mathbf{P}_+(t) \\ &- \mathbf{P}_+(t)\hat{\mathbf{B}}(M)\hat{\mathbf{B}}^T(M)\mathbf{P}_+(t) + \mathbf{Q}. \end{aligned} \tag{9}$$

The periodic solution of the Riccati equation, $\mathbf{P}_+(t)$ is computed from the periodic extension of the steady state solution $\mathbf{P}_\infty(t)$

$$\hat{\mathbf{P}}(t) = \begin{cases} \mathbf{P}_\infty(t) & \text{if } 0 \le t < T \\ \mathbf{0} & \text{otherwise} \end{cases} \tag{10}$$

$$\mathbf{P}_+(t) = \sum_{k=0}^{\infty} \hat{\mathbf{P}}(t - kT). \tag{11}$$

The solution $\mathbf{P}_\infty(t)$ is calculated using backward integration of the Riccati equation for an arbitrary positive definite final condition. This solution converges to the periodic solution. The matrix function $\mathbf{P}_\infty(t)$ corresponding to one orbital passage is stored in the computer memory, and then used for the subsequent orbits.

An example of the periodic matrix function $\mathbf{P}_+(t)$ is illustrated in Figure 3. $\mathbf{P}_+(t_0)$ at fixed time t_0 is a 6 x 6 positive definite matrix. The figure depicts the time history of $P_+(1,1)$, which is typical for the diagonal components. Off-diagonal components change their amplitudes between positive and negative values.

Again, the mean anomaly M can be used for parameterization of $\mathbf{P}_+(M)$, since both $\mathbf{P}_+(t)$ and $M(t)$ are

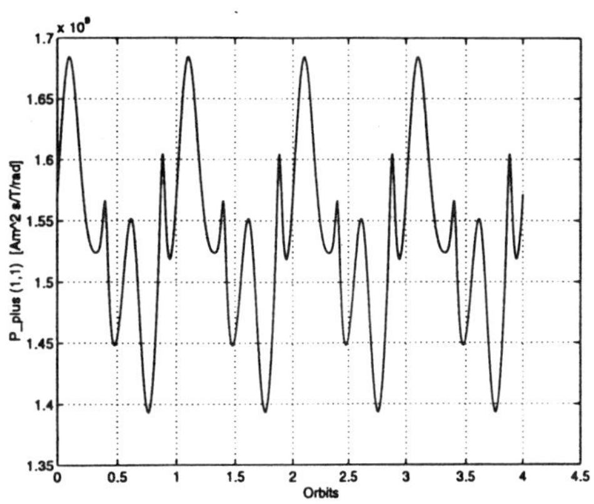

Figure 3: The time history of the (1,1) component of $\mathbf{P}_+$. Notice that $\mathbf{P}_+$ has a period equivalent to the orbit period.

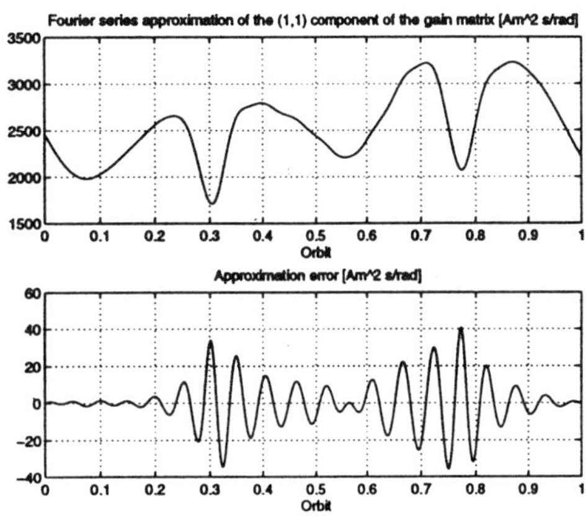

Figure 4: An approximation of the (1,1) component of the gain matrix $\mathbf{K}_+$ by 16th order Fourier series. The discrepancy between $\mathbf{K}_+$ and its Fourier approximation reaches 1.5 per cents at most.

T-periodic. Furthermore, the controller gain matrix is also T-periodic and parameterized with respect to M

$$\mathbf{K}_+(M) = -\mathbf{B}(M)\mathbf{P}_+(M). \tag{12}$$

Implementation

The mean anomaly dependent control gain matrix $\mathbf{K}_+(M)$ is computed off-line and stored in the computer memory. The control signal ${}^c\tilde{\mathbf{m}}(t)$ is then calculated according to

$$\tilde{\mathbf{m}}(t) = \mathbf{K}_+(M) \begin{bmatrix} \Omega_{SR}(t) \\ \mathbf{q}(t) \end{bmatrix}, \tag{13}$$

wehere Ω_{SR} is the angular velocity of the satellite relative to RCS, $\mathbf{q}(t)$ is the vector part of the ${}^S_R\mathbf{q}(t)$. Finally, the magnetic moment, ${}^c\mathbf{m}(t)$ is obtained by

$$\mathbf{m}(t) = \frac{\tilde{\mathbf{m}}(t) \times \mathbf{b}(t)}{\| \mathbf{b}(t) \|}. \tag{14}$$

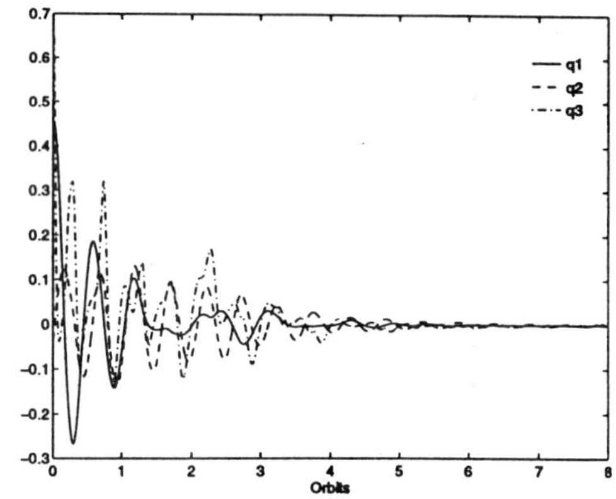

Figure 5: Performance of the infinite horizon controller for a satellite modeled as a linear object. The simulation is carried out for "ideally periodic" geomagnetic field. The initial attitude is 40 *deg* pitch, −40 *deg* roll and 80 *deg* yaw.

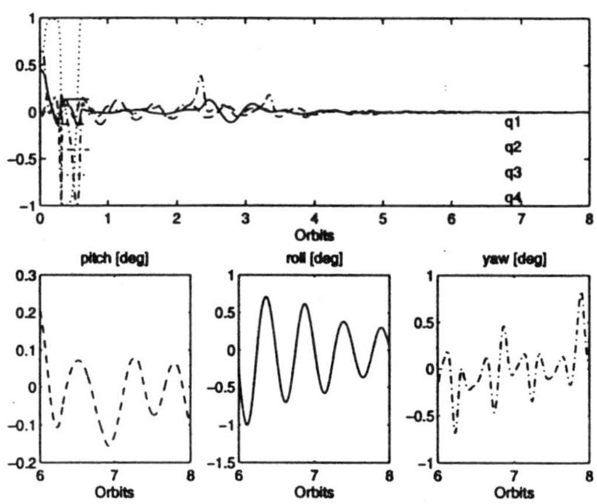

Figure 6: Performance of the infinite horizon controller for the Ørsted satellite on a circular orbit. The initial attitude is the same as in Figure 5. The steady state attitude error is below 1 *deg*.

Another option is to represent $\mathbf{K}_+(M)$ in terms of the Fourier coefficients, benefiting in a reduction of the data stored. A satisfactory approximation of the gain matrix $\mathbf{K}_+$ has been obtained with 16th order Fourier series, see Figure 4. The required capacity of computer memory given in floating point numbers is

$$Memory = \frac{T * No_{elemnts\ in\ \mathbf{K}_+} * Order_{Fourier\ series}}{Sampling\ time}.$$

For example with a sampling time of 10 sec and the orbital period 6000 sec , then 172800 floating point memory is required.

Simulation results of the infinite horizon attitude control are presented in Figure 5, 6 and 7. Figure 5 illustrates performance of the attitude controller for the linear model of the satellite motion with an ideally pe-

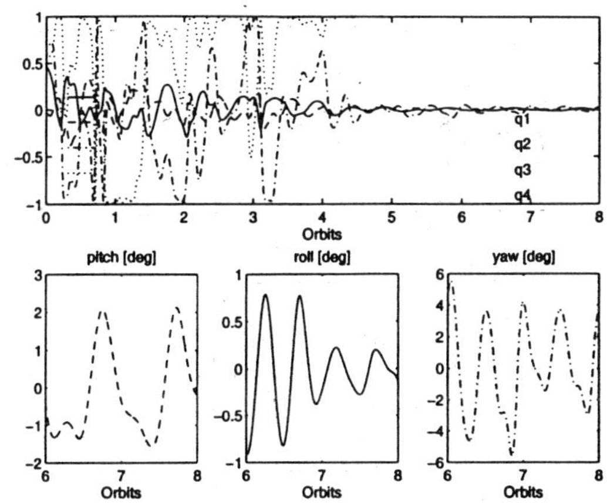

Figure 7: Performance of the infinite horizon controller for the Ørsted satellite on its elliptic orbit. The initial attitude is the same as in Figure 5. The satellite is influenced by the aerodynamic drag for normal solar activity. The attitude error is below 3 *deg* of pitch and roll. Yaw varies within 6 *deg*.

riodic geomagnetic field simulator. Figure 6 depicts performance of the infinite horizon attitude controller for the Ørsted satellite in a circular orbit. In Figure 7, the Ørsted satellite is simulated in its elliptic orbit. The satellite motion is affected by the aerodynamic drag for normal solar activity. Additionally, the realistic geomagnetic field is applied in both Figs. 6 and 7.

Disturbances due to eccentricity of the orbit and the aerodynamic drag torque act in the pitch direction. Therefore, both the first component of the quaternion and the first component of the angular velocity, which for small angles correspond to the pitch and pitch rate, are punished slightly more than the remaining components of the state. A diagonal weight matrix $\mathbf{Q}$ with the diagonal $[10\ 6\ 6\ 10\ 6\ 6]^T$ has been implemented for both linear and nonlinear models of the satellite. Initial values of the attitude are the same in both examples corresponding to 40 *deg* pitch, -40 *deg* roll, and 80 *deg* yaw.

The simulations show that the controller is stable for a wide range of operating points, also very much outside the reference. However, the performance of the infinite horizon controller is relatively poor outside vicinity of the reference, due to influence of the nonlinearities. This can be observed as large variations of the third and fourth component of the attitude quaternion, q_3 and q_4 in Figs. 6 and 7. The result of the disturbance torque due to a difference between the geomagnetic field and its periodic counterpart implemented in the attitude controller is the steady state attitude error in Figure 6. Performance of the infinite horizon attitude controller for the Ørsted satellite affected by the aerodynamic torque is illustrated in Figure 7. The satellite motion is influenced by a moderate aerodynamic drag torque corresponding to normal solar activity. The aerodynamic drag is equal $0.9\ 10^{-5}$ Nm at perigee. The attitude error is 3 *deg* of pitch and roll, whereas yaw angle varies within 6 *deg*.

A computational expense for the infinite horizon controller lies in the off-line numeric solution to the Riccati equation, but relatively large computer memory is required for keeping the gain data for one orbit. The controller gives a nonzero steady state error also for simulations without external disturbance torques. It is concluded that the infinite horizon magnetic controller is applicable for missions with low pointing requirements. The steady state performance could be improved by the finite horizon controller, which incorporates a realistic model of the geomagnetic field. This type of a control law is discussed in the next section.

FINITE HORIZON PERIODIC CONTROLLER

The linearized model of the satellite motion is only approximately periodic. There is a certain difference between the ideal periodic model of the geomagnetic field developed in the previous section, and the real magnetic field of the Earth. The controller performance could be improved by incorporating the time history of the real geomagnetic field into the controller structure. A new attitude controller based on a transient solution of the Riccati equation is therefore investigated.

The control algorithm is summarized as:

Procedure 1 *1. Calculate the time varying solution of the Riccati differential equation in the time interval* $t \in (\tau - T, \tau]$

$$-\dot{\mathbf{P}}(t) = \mathbf{A}^T\mathbf{P}(t)+\mathbf{P}(t)\mathbf{A}-\mathbf{P}(t)\mathbf{B}(t)\mathbf{B}^T(t)\mathbf{P}(t)+\mathbf{Q}(t) \tag{15}$$

with the final condition

$$\mathbf{P}(\tau) = \mathbf{P}_f. \tag{16}$$

2. *Implement controller*

$$\tilde{\mathbf{m}}(t) = -\mathbf{B}^T(t)\mathbf{P}(t)\begin{bmatrix}\Omega_{SR}\\ \mathbf{q}\end{bmatrix} \tag{17}$$

for $t \in (\tau - T, \tau]$.

3. *Calculate magnetic moment from the equation*

$$\mathbf{m}(t) = \frac{\tilde{\mathbf{m}}(t)\times\mathbf{b}(t)}{\|\mathbf{b}(t)\|}.$$

4. τ *becomes* $\tau + T$.

5. *go to 1.*

It was stated in Reference[6] that if the difference $\mathbf{F} = \mathbf{P}_f - \mathbf{P}(\tau - kT)$ is positive semidefinite for $k \geq 0$ then the procedure given above provides a stable control law (Theorem 3.4.2). The final condition, $\mathbf{P}_f$ is chosen sufficiently large such that $\mathbf{F}$ is positive semidefinite independently on the deviation of the geomagnetic field

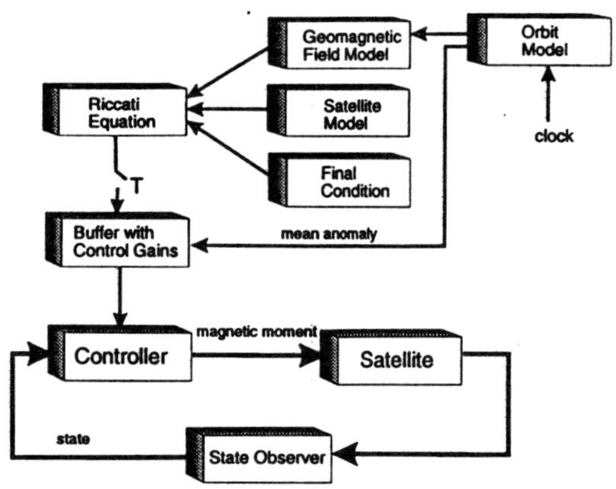

Figure 8: Attitude control system based on finite horizon control.

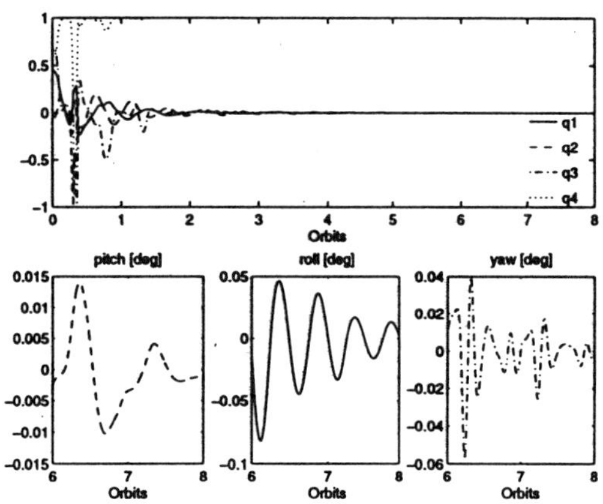

Figure 9: Performance of the quasi periodic receding horizon controller for the Ørsted satellite on circular orbit. The attitude converges asymptotically to the reference, i.e. ${}^{S}_{R}\mathbf{q} \rightarrow [0\ 0\ 0\ 1]^T$.

from its periodic model. It should be noted that the larger $\mathbf{P}_f$, the larger is the control torque. The controller shall comply with the power constraints imposed on the attitude control system, therefore the maximum value of the final condition shall be confined. The final condition is considered as a design parameter, that can be iterated by means of computer simulation.

The attitude control system based on the final horizon control is illustrated in Fig. 8.

The orbit model provides position of the satellite in orbit in terms of longitude, latitude and altitude. This information is used by the on board geomagnetic field model (here 10th order spherical harmonic model). The Riccati equation is computed for the subsequent orbit. The controller gain is computed and parameterized by the mean anomaly. The controller gain is stored in a buffer. This procedure is activated once per orbit. The control gain matrix is taken from the buffer on the basis of the mean anomaly associated with position of the satellite in orbit. The controller gain is updated ev-

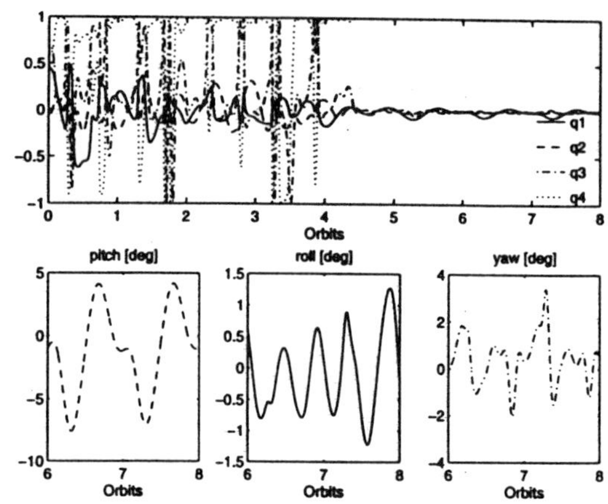

Figure 10: Performance of the quasi periodic receding horizon controller for the Ørsted satellite on the elliptic orbit. The satellite is influenced by the aerodynamic torque. Performance of the receding horizon is comparable with efficiency of the infinite horizon attitude controller in Fig. 7.

ery sampling cycle, and is implemented in the control loop.

Implementation

The quasi periodic receding horizon controller has been validated through computer simulation. The results are depicted in Figs. 9 and 10. Control parameters have been found empirically. Weight matrix $\mathbf{Q}$ has been set to 10 $\mathbf{E}_{6\times6}$* , and the final condition $\mathbf{P}_f$ has been calculated from the steady state solution: $\mathbf{P}_f = \mathbf{P}(\tau) = 2\mathbf{P}_\infty(\tau)$. Initial values of the attitude have again been assigned to 40 *deg* pitch, -40 *deg* roll and 80 *deg* yaw.

Fig. 9 depicts the Ørsted satellite motion on a circular orbit, i.e. there are no external disturbances. The satellite attitude is seen to converge asymptotically to the reference. The performance of the quasi periodic receding horizon controller for the satellite disturbed by the aerodynamic torque is comparable with the performance of the infinite horizon controller, see Figs. 10 and 7. This is due to the impact of the aerodynamic torque is seen to be much larger than the influence of the torque due to the discrepancy between the geomagnetic field and its periodic counterpart.

The steady state performance of the infinite horizon controller on a circular orbit is much better than infinite horizon, however they are seen to have the same performance for a satellite in an elliptic orbit effected by the aerodynamic drag. The computational burden for the finite horizon controller is heavy due to the Riccati equation shall be solved on board (alternatively uploaded to the on board computer during every ground

* $E_{6\times6}$ is the 6×6 unit matrix

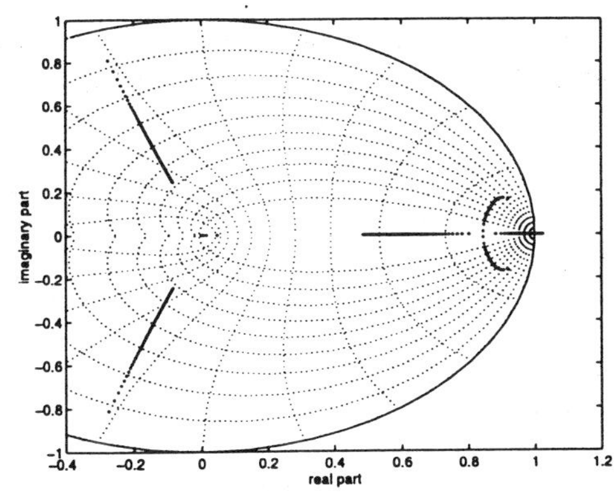

Figure 11: Locus of the characteristic multipliers $\lambda(\epsilon)$ for ϵ changing from 1 to 80 is evaluated for the closed loop system in Eq. (22).The satellite becomes unstable for $\epsilon = 52$, for $\epsilon = 18$ the largest characteristic multiplier is closest to the origin.

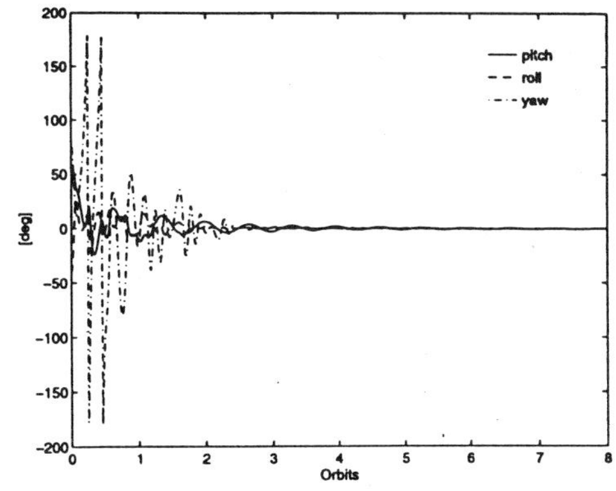

Figure 12: Performance of the constant gain controller for the Ørsted satellite on circular orbit, i.e. without external disturbances. The weigh matrix, $\mathbf{Q}$ has value $18 \cdot \mathbf{E}_{6\times6}$. Large amplitude of the yaw oscillations is encountered. The initial attitude is 40 *deg* pitch, −40 *deg* roll and 80 *deg* yaw.

station passage). Therefore, the infinite horizon controller is preferable for the missions like Ørsted with low pointing accuracy. The necessary computer power could be additionally limited if the constant controller was implemented, and had the same performance as the time varying controllers. This issue is addressed in the next section.

CONSTANT GAIN CONTROL

Computation of the infinite and finite horizon attitude controllers are tedious and difficult to implement on a real-time platform. A simple constant gain attitude controller could be an alternative. This type of controller is investigated in this section.

The design algorithm consists of replacing the time varying parameters of the satellite by its averaged values evaluated over a period of one orbit. This concept was first presented in Reference[3]. The theoretical basis of the method was given in Reference[6].

The time invariant counterpart of the time varying linearized satellite motion was

$$\frac{d}{dt}\begin{bmatrix}\delta\Omega\\ \delta\mathbf{q}\end{bmatrix} = \mathbf{A}\begin{bmatrix}\delta\Omega\\ \delta\mathbf{q}\end{bmatrix} + \mathbf{B}^c\tilde{\mathbf{m}}, \qquad (18)$$

where

$$\mathbf{B} = \frac{1}{T}\int_o^T \hat{\mathbf{B}}(M(t))dt, \qquad (19)$$

and T is the orbit period, and $\hat{\mathbf{B}}(M)$ is the control matrix in Eq. (8).

A linear quadratic regulator (LQR) is used for the constant gain controller design. The system is linear, time invariant and controllable thus a control law can be based on the solution of the steady state Riccati equation, see Reference[9]. The optimal control is given by

$$\tilde{\mathbf{m}} = -\mathbf{B}^T\mathbf{P}\begin{bmatrix}\delta\Omega\\ \delta\mathbf{q}\end{bmatrix}, \qquad (20)$$

where $\mathbf{P}$ satisfies the Riccati algebraic equation

$$\mathbf{PA} + \mathbf{A}^T\mathbf{P} - \mathbf{PBB}^T\mathbf{P} + \mathbf{Q} = \mathbf{0}. \qquad (21)$$

Once the control vector $\tilde{\mathbf{m}}$ in Eq. (20) is calculated, the magnetic moment, $\mathbf{m}$ is computed according to Eq. (2).

Stability of the control law in Eqs. (20) and (2) for the time varying linear model of the satellite in Eq. (6) is determined using Floquet theory, see Reference[10]. This check is necessary, since stability of the time varying system and its time invariant counterpart are not equivalent. The time invariant system is only the first order approximation of the time varying one, Reference[6]. Furthermore, the sensitivities of those systems are not equivalent neither, e.g. the disturbance torque acting on the satellite in the direction of yaw in a zone near the North nor South poles remains unaffected by the attitude controller (due to lack of controllability), whereas it can be arbitrarily damped by an LQ controller for the time invariant counterpart.

The following closed-loop system is considered for the Floquet analysis

$$\frac{d}{dt}\begin{bmatrix}\delta\Omega\\ \delta\mathbf{q}\end{bmatrix} = (\mathbf{A} - \hat{\mathbf{B}}^T(M(t))\mathbf{P})\begin{bmatrix}\delta\Omega\\ \delta\mathbf{q}\end{bmatrix}. \qquad (22)$$

As seen from Eqs. (21) and (22) stability of the closed-loop system is dependent on the weight matrix $\mathbf{Q}$. Figure 11 depicts locus of the characteristic multipliers for $\mathbf{Q}(\epsilon) \equiv \epsilon\mathbf{E}_{6\times6}$, where ϵ changes from 1 to 80 †. The

†The weight matrix $\mathbf{Q}(\epsilon)$ acts on the state space $[\delta\Omega\ \delta\mathbf{q}]^T$. $\delta\Omega$ is provided in rad/s and $\delta\mathbf{q}$ is given without units.

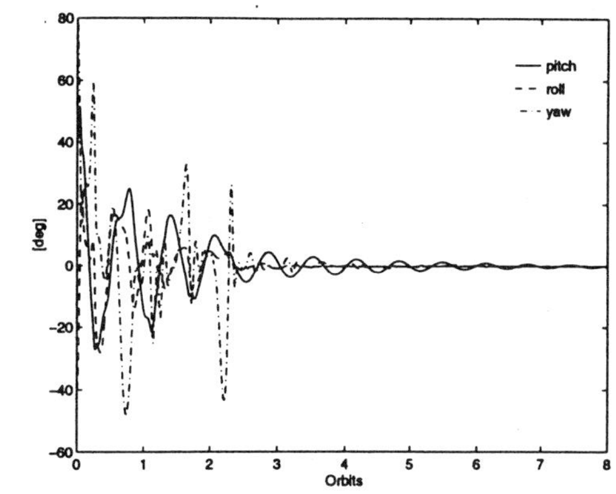

Figure 13: Performance of the constant gain controller for Ørsted satellite on circular orbit. The initial conditions are the same as in Figure 12. The diagonal weight matrix $\mathbf{Q}$ with diagonal $[18\ 18\ 90\ 18\ 18\ 90]^T$ is implemented. The amplitude of the yaw oscillation is reduced comparing with. Figure 12.

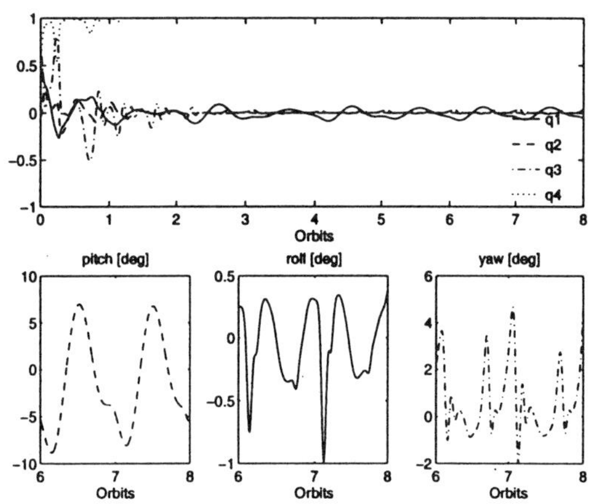

Figure 14: Performance of the constant gain controller for the Ørsted satellite on the elliptic orbit influenced by the aerodynamic drag. The initial conditions are as in Fig. 12. The resultant attitude is within 8 *deg*.

satellite becomes unstable for $\epsilon = 52$. For $\epsilon = 18$, the largest characteristic multiplier is closest to the origin.

Notice that the averaged geomagnetic field is implemented for the Floquet analysis. Therefore, an ultimate test is the simulation for the nonlinear model of the satellite in the realistic space environment.

Validation

The constant gain control demonstrated stability for the entire envelope of the expected satellite initial attitudes and angular velocities in the science observation mission phase. The control parameter: the weight matrix, $\mathbf{Q}$ has been found empirically. Its value has been set to $18 \cdot \mathbf{E}_{6\times 6}$. The simulation results for the Ørsted satellite on the circular orbit in Figure 12 show large amplitude of yaw oscillations. A new diagonal weigh matrix with diagonal $[18\ 18\ 90\ 18\ 18\ 90]^T$ is proposed. The amplitude of the yaw fluctuations is reduced, see Figure 13. The last Figure 14 illustrates the satellite motion on impact of the aerodynamic drag and the torque due to the eccentricity of the Ørsted orbit. The performance of the constant gain controller is very much the same as the infinite and finite horizon attitude controllers in Figs. 7 and 10. The attitude error is within 8 *deg*, which complies with required bond of ± 10 *deg* of pitch and roll, 20 *deg* of yaw.

CONCLUSION

This paper presented work on the magnetic attitude control based on the steady state and the transient solution of the Riccati equation. It was shown that a satellite on a near polar orbit actuated by a set of perpendicular magnetorquers could be described by a set of periodic differential equations. Three attitude controllers were designed: the finite horizon, the infinite horizon, and the constant gain controller. The control strategies presented were evaluated in the simulation study.

The performance of the designed controllers was comparable for a satellite in an elliptic orbit effected by the aerodynamic drag. The computer expense was smallest for the constant gain controller, therefore it was chosen for on board implementation in the Ørsted satellite. This controller was seen to be stable for a wide envelope of initial values of the attitude, though it had inherently low bandwidth, with time constants on the order of one orbit.

This work is believed to contribute to application of the theory of periodic linear systems to magnetic attitude control problem. It provides solution useful for small satellites with loose requirements on pointing accuracy.

ACKNOWLEDGEMENT

The support of this work by the Danish Ørsted Satellite Project and by the Faculty of Technology and Science at Aalborg University is greatly appreciated.

REFERENCES

1. Cavallo A & al 1993, A sliding manifold approach to satellite attitude control, in *12th World Congress IFAC, Sidney*.

2. Musser K & L Ward 1989, Autonomous spacecraft attitude control using magnetic torquing only, in *Flight Mechanics Estimation Theory Symposium, NASA*.

3. Martel F & al Sep. 1988, Active magnetic control system for gravity gradient stabilized spacecraft, in *Annual AIAA/Utah State University Conference on Small Satellites*.

4. Wisniewski R 1994, Nonlinear control for satellite detumbling based on magnetic torquing, in *Joint Services Data Exchange for Guidance, Navigation, and Control, Arizona.*

5. Wisniewski R & M Blanke Jun. 1996, Three-axis satellite attitude control based on magnetic torquing, in *13th IFAC World Congress, San Francisco, California.*

6. Wisniewski R Dec. 1996, *Satellite Attitude Control Using Only Electromagnetic Actuation*, Ph.D. thesis, Aalborg University.

7. Wertz J 1990, *Spacecraft Attitude Determination and Control*, Kluwer Academic Publishers.

8. Bittanti S, editor 1991, *The Riccati Equation*, Springer Verlag.

9. Kwakernaak H & R Sivan 1972, *Linear Optimal Control Systems*, Wiley.

10. Mohler R 1991, *Nonlinear Systems*, vol. Dynamics and Control, Prentice Hall.

AIAA-97-3524

PARAMETER IDENTIFICATION FOR THE INTERNATIONAL SPACE STATION USING NONLINEAR MOMENTUM MANAGEMENT CONTROL

Michael T. Carter* and Srinivas R. Vadali†
Texas A&M University, College Station, TX
and
Gregory E. Chamitoff‡
NASA Johnson Space Center, Houston, TX

A recursive least squares (RLS) identification algorithm to determine the inertia and aerodynamic parameters of the International Space Station has been applied to controlled maneuvers which seek the torque equilibrium attitude (TEA). Lyapunov stability theory is used to design the nonlinear momentum management control law. The RLS identification algorithm estimates the inertia and aerodynamic parameters from the attitude measurements and the control torques during the maneuver. This controller and its parameter estimation scheme are tested using a nonlinear rotational dynamics model of the International Space Station that includes external torques due to the gravity-gradient and aerodynamic drag. The aerodynamic drag moment is modeled such that the drag force and the center of pressure location are functions of the spacecraft attitude. Numerical results show that the TEA-seek maneuvers can generate sufficient excitation in the attitude during the maneuver such that the parameter estimates will converge to the true parameters.

Introduction

The International Space Station (ISS) is currently scheduled to be constructed over a five-year period beginning in June 1998 and will include at least 44 different configurations before its final assembly. For each of these configurations, the ISS moments of inertia will not be precisely known since the structure will never be pre-assembled for ground structural tests and since the ISS is not a single rigid body, but a system of interconnected bodies. The presence of multiple bodies does not necessarily mean that the time-varying moment of inertia cannot be calculated. The projected motion of discrete components, such as the solar arrays, the robotic arm, and attached payloads can be estimated. However, additional objects, such as the fuel/oxygen stored in partially-filled tanks, the cargo, and astronauts inside the space station, move in a somewhat arbitrary fashion. Thus, a precise estimate for the ISS inertia matrix can only be determined using on-line estimation techniques.

Attitude control systems for spacecraft in low-earth orbit (LEO) have been discussed in many papers. Predominantly, momentum-exchange actuators, such as the control moment gyroscopes (CMGs), have been employed in these control systems. Thus, momentum management techniques to prevent control torque and momentum saturation must be an integral part of the control design. In the absence of periodic disturbances, a spacecraft can remain at one of its torque-equilibrium attitudes (TEAs) without using CMG control torque. The CMG momentum will saturate if the spacecraft is held at any other arbitrary attitude for an extended period. The two most common attitude control laws for a satellite in LEO are TEA-seek control laws and attitude-hold control laws. Both of these control laws must include momentum management.

By linearizing the ISS rotational equation of motion about the TEA and uncoupling the roll/yaw motion from the pitch dynamics, Wie et al.[1] and Warren et al.[2] have designed a LQR control which seeks the TEA in the presence of constant and periodic disturbances. Wie also developed periodic disturbance rejection filters to suppress the effect of disturbances at multiples of the orbital frequency on the final attitude and momentum. [The atmospheric density varies at the orbital frequency due to the diurnal effect, while the external moment due to the solar arrays tracking the Sun varies at twice the orbital frequency.] Harduvel[3] evolved the LQR control design by including the three-axis coupled dynamics and an aerodynamic drag moment where the direction of the drag force is a function of the spacecraft attitude. Vadali and Oh[4] designed

* Graduate Student, Department of Aerospace Engineering. Student Member AIAA.

† Professor, Department of Aerospace Engineering. Associate Fellow AIAA.

‡ Flight Control Engineer, NASA/USA. Senior Member AIAA.

Lyapunov feedback control laws for the CMGs which provided three-axis nonlinear attitude control and momentum management of the space station under gravity gradient and constant disturbance torques.

Other researchers have studied how inertia uncertainty will affect the performance of the attitude control system. Parlos and Sunkel[5] have proposed using gain-scheduling between multiple LQR full state feedback designs to achieve the stability and performance required in the presence of mass properties variations. Each LQR control would be linearized about a different TEA, which would be determined by an unspecified on-orbit mass properties estimation algorithm. Byun et al.[6] developed a robust H_∞ control to satisfy attitude control and momentum management of the linearized, uncoupled roll/yaw and pitch dynamics of the ISS and to guarantee stability robustness for inertial variations up to 73%. Cristi et al.[7] has designed an adaptive Lyapunov attitude-hold control law with a recursive least squares (RLS) inertia estimate update law for a rigid body.

This paper will extend the previous work to a TEA-seek controller with momentum management that uses the RLS algorithm to estimate both the inertia and aerodynamic parameters of the ISS. The nonlinear rotational equations of motion for the ISS are modeled as a single rigid body in a circular LEO under the influence of the external gravity gradient and aerodynamic moments. An aerodynamic moment model is developed which will describe the drag force and the center of pressure location as functions of the attitude. Neither the diurnal effects on the atmospheric density nor the effects due to the articulating solar arrays are included in this paper.

A simple simulation of a spacecraft TEA-seek maneuver with momentum management is required to test the performance of the identification algorithm. Two nonlinear TEA-seek controllers are designed using Lyapunov stability theory to stabilize the motion about the TEA of a "gravity-gradient stable" spacecraft. Numerical studies show that a spacecraft under the influence of gravity gradient and aerodynamic drag moments as well as inaccurate inertia estimates will still converge to TEA using one of these control laws. The RLS algorithm will estimate the inertia matrix and aerodynamic parameters using the attitude measurements and the control history from these TEA-seek maneuvers. The TEA-seek maneuvers can generate sufficient excitation in the attitude measurements during the maneuver such that the parameter estimates will converge to the true parameters.

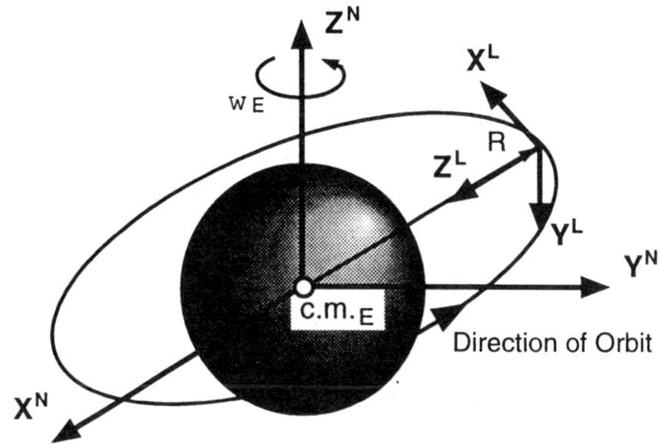

Figure 1. LVLH Reference Frame Relative to the Spacecraft Orbit

Rotational Equations of Motion

The following reference frames are used to describe the attitude motion of a spacecraft in circular orbit:

1. Inertial, **N** = An inertial reference frame whose origin is at the center of the Earth
2. LVLH (Local Vertical, Local Horizontal), **L** = A right-hand orthogonal reference frame whose origin is at the spacecraft center of mass (c.m.) and is always oriented such that the $Z^{\mathbf{L}}$ axis points at the center of the Earth and the $X^{\mathbf{L}}$ axis points in the same direction as the translational velocity vector for the spacecraft
3. Body Fixed, **B** = A right-hand orthogonal reference frame whose origin is at the spacecraft c.m. and whose orientation is fixed to the rigid spacecraft
4. Principal Axes, **P** = A right-hand orthogonal reference frame whose origin is at the spacecraft c.m. and whose orientation is fixed to the rigid spacecraft such that the inertia matrix in the principal axes reference frame is a diagonal matrix

Figure 1 illustrates how the LVLH reference frame is oriented relative to the inertial reference frame.

The nonlinear rotational equations of motion for a spacecraft in circular orbit using control moment gyros (CMGs) as the actuators are described by:

$$\dot{\underline{\beta}} = \tfrac{1}{2} H(\underline{\beta})(\underline{\omega} + n\underline{C}_2) \tag{1}$$

$$\begin{aligned} I\dot{\underline{\omega}} &= \underline{M}_{grav} + \underline{M}_{drag} - \underline{M}_{gyro} - \underline{u} \\ &= 3n^2 \underline{C}_3 \times I\underline{C}_3 + q\,C_D S\left(\underline{C}_1 \times \underline{\rho}_{cp}\right) \\ &\quad -\underline{\omega} \times I\underline{\omega} - \underline{u} \end{aligned} \tag{2}$$

$$\dot{\underline{h}} = -\underline{\omega} \times \underline{h} + \underline{u} \quad (3)$$

where $\underline{\beta}$ is the Euler parameter (quaternion) vector that describe the orientation of the body reference frame with respect to the LVLH reference frame, $\underline{\omega} = \underline{\omega}^{\mathbf{B}}_{\mathbf{B/N}}$ is the angular velocity of the body reference frame with respect to the inertial reference frame whose value is given in terms of the body reference frame, and $\underline{h}$ is the CMG angular momentum. The true inertia matrix I is assumed to be constant in the body reference frame. The control torque $\underline{u}$ is applied to the CMGs by electric motors.

The angular velocity of the body reference frame with respect to the LVLH reference frame is given by:

$$\underline{\omega}^{\mathbf{B}}_{\mathbf{B/L}} = \underline{\omega} + n\underline{C}_2 \quad (4)$$

$\underline{C}_1, \underline{C}_2$, and $\underline{C}_3$ are the first, second and third columns respectively of the coordinate transformation matrix from the LVLH reference frame to the body reference frame. n is the orbital mean motion, which is constant for circular orbits. The kinematics relationship between the Euler angle time derivative and the angular velocity will only be a function of the Euler angles if the body orientation with respect to the LVLH reference frame is described by Euler angle sets whose first rotation is about the pitch axis: 2-1-3, 2-3-1, 2-1-2, and 2-3-2. The $H(\underline{\beta})$ matrix in the Euler parameter time derivative is a 4 x 3 matrix which is a function of the Euler parameters:

$$H(\underline{\beta}) = \begin{bmatrix} -\beta_1 & -\beta_2 & -\beta_3 \\ \beta_0 & -\beta_3 & \beta_2 \\ \beta_3 & \beta_0 & -\beta_1 \\ -\beta_2 & \beta_1 & \beta_0 \end{bmatrix} \quad (5)$$

$\underline{M}_{grav}, \underline{M}_{drag}$, and $\underline{M}_{gyro}$ are the moments due to gravity gradient, aerodynamic drag and the gyroscopic effects. In the aerodynamic drag torque, the drag force:

$$\underline{F}_{drag} = q\, C_D S\, \hat{\underline{i}}_v \quad (6)$$

has been derived assuming the atmosphere is a continuum in LEO. C_D, S, and $\hat{\underline{i}}_v$ are the drag coefficient, reference area, and atmospheric velocity unit vector, while the dynamic pressure:

$$q = \tfrac{1}{2} \rho_a v_a^2 \quad (7)$$

is a function of the atmospheric density and the magnitude of the atmospheric velocity vector.

The torque equilibrium attitude, when zero control torque is employed, requires that the following two necessary conditions on the state variables are satisfied:

$$1.\ \underline{\omega} = -n\,\underline{C}_2 \quad (8)$$

$$2.\ \underline{h} = \alpha\,\underline{C}_2 \quad (9)$$

where α is any scalar. In addition, the spacecraft attitude must make the total external moment go to zero:

$$\begin{aligned} &\left(3\underline{C}_3 \times I\underline{C}_3 - \underline{C}_2 \times I\underline{C}_2\right) + \\ &\frac{1}{n^2} q\, C_D S\left(\underline{C}_1 \times \underline{\rho}_{cp}\right) = 0 \end{aligned} \quad (10)$$

The above attitude constraint can be divided into the following three necessary constraints:

$$\underline{C}_1^T I \underline{C}_3 = \frac{-qC_D S}{3n^2}\left(\underline{C}_3^T \underline{\rho}_{cp}\right) \quad (11)$$

$$\underline{C}_1^T I \underline{C}_2 = \frac{qC_D S}{n^2}\left(\underline{C}_2^T \underline{\rho}_{cp}\right) \quad (12)$$

$$\underline{C}_2^T I \underline{C}_3 = 0 \quad (13)$$

If the aerodynamic drag torque approaches zero, the spacecraft TEA is only a function of the attitude and the constant inertia matrix. Four out of the total twenty-four TEAs are stable when the external torque is only due to the gravity gradient. One stable TEA exists at $\underline{\beta} = [1 \;\; 0 \;\; 0 \;\; 0]^T$ if the body reference frame is aligned with the principal axes reference frame and the inertia matrix satisfies the following inequality constraint:

$$I_{22} > I_{11} > I_{33} \quad (14)$$

If the aerodynamic drag torque is non-zero, the spacecraft TEA is also dependent on the orbit, the orbital position, the spacecraft exterior geometry, and the average atmospheric conditions over the orbit.

Attitude-Dependent Aerodynamic Drag Torque

The aerodynamic drag torque is a function of attitude-dependent terms such as the drag-coefficient, the reference area and the center of pressure. However, these functions cannot be derived symbolically due to the complex geometry of most spacecraft. Instead, n discrete values of these parameters are assumed to be known for the corresponding n orientations. If interpolation functions can be defined over the entire orientation space, the parameter values can be calculated for any orientation (in terms of the Euler parameters) as the following linear combinations:

$$p(\underline{\beta}) = \sum_{j=1}^{n} p\left(\underline{\beta}_j\right) \psi_j(\underline{\beta}) \quad (15)$$

If the reference area is assumed to be the projected area on the Y-Z plane in the LVLH reference frame, (which is perpendicular to the aerodynamic velocity vector), the projected area for a given orientation can be defined using an ellipsoid interpolation. Assume that the aerodynamic reference frame, **A**, is defined as a fixed orthogonal reference

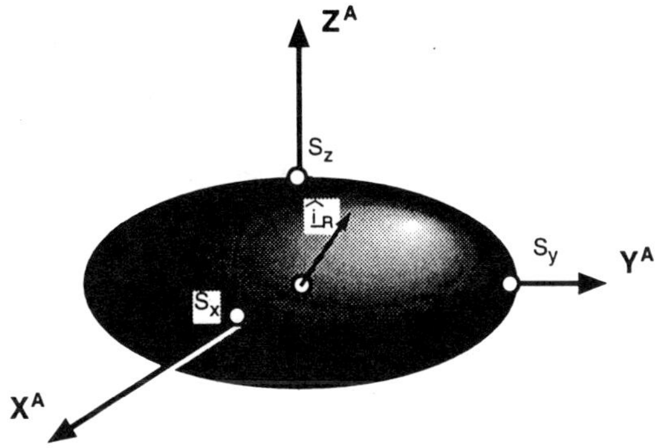

Figure 2. Projected Area Ellipsoid for the Aerodynamic Reference Frame

frame whose origin is at the spacecraft c.m. and whose orientation is fixed to the rigid spacecraft. Figure 2 shows how the ellipsoid will be aligned with the **A** reference frame. If three discrete values for the ellipsoid along the axes of this reference frame, $S_x, S_y,$ and S_z , are known when the **A** reference frame is aligned with the LVLH reference frame, the ellipsoid constraint equation will describe the ellipse that passes through these values:

$$\left(\hat{\underline{i}}_R^{\mathbf{A}}\right)^T S_1^{\mathbf{A}} \hat{\underline{i}}_R^{\mathbf{A}} = 1 \qquad (16)$$

where

$$S_1^{\mathbf{A}} = \text{diag}\left(\begin{bmatrix} \frac{1}{S_x^2} & \frac{1}{S_y^2} & \frac{1}{S_z^2} \end{bmatrix}\right) \qquad (17)$$

The ellipsoid constraint equation can be described for any spacecraft orientation as:

$$\underline{C}_1^T C_{B\to A}^T S_1^{\mathbf{A}} C_{B\to A} \underline{C}_1 = 1 \qquad (18)$$

Thus, the scalar projected area for a certain orientation can be calculated:

$$S(\underline{\beta}) = \frac{1}{\sqrt{\underline{C}_1^T S_1^B \underline{C}_1}} \qquad (19)$$

where

$$S_1^B = C_{B\to A}^T S_1^{\mathbf{A}} C_{B\to A} \qquad (20)$$

The coordinate transformation matrix $C_{B\to A}$ describes the orientation of the aerodynamic reference frame with respect to the body reference frame and is constant during the spacecraft maneuver.

The Euler parameters $\underline{\beta}_{\mathbf{A/L}}$ describe the orientation of the aerodynamic reference frame relative to LVLH. The center of pressure for the three discrete orientations described in Table 1 are assumed to be known. Using this information, the center of pressure for any orientation of the **A** reference frame can be calculated:

$$\underline{\rho}_{cp}^{\mathbf{A}} = -T_{cp}^{\mathbf{A}} C_{\mathbf{L\to A}} \hat{\underline{i}}_v^{\mathbf{L}} \qquad (21)$$

where $\hat{\underline{i}}_v^{\mathbf{L}}$ is the atmospheric velocity unit vector and where the aerodynamic orthogonal center of pressure matrix is defined as:

$$T_{cp}^{\mathbf{A}} = \left[\underline{\rho}_{cp}(\underline{\beta}_1) \;\; \underline{\rho}_{cp}(\underline{\beta}_2) \;\; \underline{\rho}_{cp}(\underline{\beta}_3)\right] \qquad (22)$$

The value of the center of pressure in the body reference frame is defined as:

$$\underline{\rho}_{cp}^{\mathbf{B}} = T_{cp}^{\mathbf{B}} \underline{C}_1^{\mathbf{B}} \qquad (23)$$

where

$$T_{cp}^{\mathbf{B}} = C_{B\to A}^T T_{cp}^{\mathbf{A}} C_{B\to A} \qquad (24)$$

Both the center of pressure location and the projected area for three orthogonal orientations are currently available for each ISS configuration[8]. The drag coefficient is assumed to be constant for all attitudes.

The previous aerodynamic drag moment in (2) is defined as:

$$\underline{M}_{drag} = q\, C_D S\left(\underline{C}_1 \times \underline{\rho}_{cp}\right) \qquad (25)$$

while the aerodynamic drag moment with the attitude-dependent center of pressure and projected

Table 1. Center of Pressure Location for Three Orthogonal Orientations

Center of Pressure	Description	$\underline{\beta}_{\mathbf{A/L}} = \underline{\beta}_i$	$C_{\mathbf{L\to A}}(\underline{\beta}_i)$
$\underline{\rho}_{cp}(\underline{\beta}_1)$	**LVLH** reference frame is aligned with **A** reference frame such that $X^{\mathbf{A}}$ and $X^{\mathbf{LVLH}}$ are aligned	$[1 \;\; 0 \;\; 0 \;\; 0]^T$	$\begin{bmatrix} 1 & 0 & 0 \\ 0 & 1 & 0 \\ 0 & 0 & 1 \end{bmatrix}$
$\underline{\rho}_{cp}(\underline{\beta}_2)$	-90° rotation about $Z^{\mathbf{A}}$ axis when **LVLH** reference frame is aligned with **A** reference frame such that $Y^{\mathbf{A}}$ and $X^{\mathbf{LVLH}}$ are aligned	$\begin{bmatrix} \frac{\sqrt{2}}{2} & 0 & 0 & -\frac{\sqrt{2}}{2} \end{bmatrix}^T$	$\begin{bmatrix} 0 & -1 & 0 \\ 1 & 0 & 0 \\ 0 & 0 & 1 \end{bmatrix}$
$\underline{\rho}_{cp}(\underline{\beta}_3)$	+90° rotation about $Y^{\mathbf{A}}$ axis when **LVLH** reference frame is aligned with **A** reference frame such that $Z^{\mathbf{A}}$ and $X^{\mathbf{LVLH}}$ are aligned	$\begin{bmatrix} \frac{\sqrt{2}}{2} & 0 & \frac{\sqrt{2}}{2} & 0 \end{bmatrix}^T$	$\begin{bmatrix} 0 & 0 & -1 \\ 0 & 1 & 0 \\ 1 & 0 & 0 \end{bmatrix}$

area is defined as:

$$\underline{M}_{drag}{}^{\mathbf{B}} = \frac{q C_D}{\sqrt{\underline{C}_1{}^T S_1^{\mathbf{B}} \underline{C}_1}} \left(\underline{C}_1 \times T_{cp}^{\mathbf{B}} \underline{C}_1 \right) \qquad (26)$$

Lyapunov TEA-Seek Control Design

Since the rotational equations of motion are nonlinear ordinary differential equations, Lyapunov's second method will be used to prove asymptotic closed-loop stability for the system. A Lyapunov function must be designed which is positive semidefinite such that it is only equal to zero at the equilibrium state. If the Lyapunov function time derivative is negative semidefinite such that it is only equal to zero at the equilibrium state, the asymptotic stability of the equilibrium state has been proven. If the Lyapunov function time derivative is zero at any other position, this state must be proven to only be an inflection point of finite duration after which the Lyapunov function will continue to decrease to the equilibrium state.

The desired equilibrium state for the TEA-seek control of a satellite in LEO requires that both the angular velocity of the spacecraft with respect to the LVLH reference frame and the CMG momentum error go to zero. If the external moment is only due to the gravity gradient torque, the following attitude potential function[9] can be defined:

$$\Phi = \tfrac{1}{2} n^2 \left\{ 3\left(\underline{C}_3{}^T I \underline{C}_3 - I_{\min} \right) + \left(I_{\max} - \underline{C}_2{}^T I \underline{C}_2 \right) \right\} \qquad (27)$$

This attitude potential function will be positive for any orientation and will be equal to zero at the four stable TEAs. The time derivative of this attitude potential function can be written as:

$$\begin{aligned} \dot{\Phi} &= -\underline{\omega}_{\mathbf{B/L}}{}^T \left\{ 3n^2 \underline{C}_3 \times I \underline{C}_3 - n^2 \underline{C}_2 \times I \underline{C}_2 \right\} \\ &= -\underline{\omega}_{\mathbf{B/L}}{}^T \left\{ 3n^2 \underline{C}_3 \times I \underline{C}_3 - \underline{\omega} \times I \underline{\omega} \right. \\ &\quad \left. - n I \underline{\omega} \times \underline{C}_2 \right\} \end{aligned} \qquad (28)$$

The following Lyapunov function for a stable TEA-seek control is proposed:

$$V = \Phi + \tfrac{1}{2} \underline{\omega}_{\mathbf{B/L}}{}^T I \underline{\omega}_{\mathbf{B/L}} + \tfrac{1}{2} k_1 \underline{h}^T \underline{h} + \tfrac{1}{2} k_2 \underline{h}^T \left(\underline{C}_1 \underline{C}_1{}^T + \underline{C}_3 \underline{C}_3{}^T \right) \underline{h} \qquad (29)$$

Since TEA can be maintained if $\underline{h} = \alpha \underline{C}_2$, the third term has been added to provide an additional gain to tune the momentum profile over the maneuver. The control law is defined as:

$$\underline{u} = Q\left(\underline{\omega}_{\mathbf{B/L}} - F \underline{h} \right) \qquad (30)$$

where Q is a positive definite gain matrix and the positive definite F matrix is given by:

$$F = k_1 I_{33} + k_2 \left(\underline{C}_1 \underline{C}_1{}^T + \underline{C}_3 \underline{C}_3{}^T \right) \qquad (31)$$

where I_{33} is the 3 x 3 identity matrix. Using this control law, the Lyapunov function time derivative will be a negative semidefinite function:

$$\dot{V} = -\left(\underline{\omega}_{\mathbf{B/L}} - F \underline{h} \right)^T Q \left(\underline{\omega}_{\mathbf{B/L}} - F \underline{h} \right) \qquad (32)$$

The Lyapunov function time derivative will only be zero when the angular velocity of the spacecraft with respect to the LVLH reference frame is $\underline{\omega}_{\mathbf{B/L}} = F \underline{h}$. Thus, if the Lyapunov function time derivative goes to zero when the spacecraft is not at the stable TEA, this only represents an inflection point for the Lyapunov function. When the Lyapunov function is zero, $\Phi = 0$, $\underline{\omega}_{\mathbf{B/L}} = \underline{0}$, and $\underline{h} = \underline{0}$. The above control law is locally asymptotically stable since this controller will find the nearest of the four stable TEAs. Since the inertia matrix does not explicitly appear in this control law, this TEA-seek controller will converge to the nearest TEA regardless of the inertia estimate error magnitude.

To bound the angular momentum, the following control law is proposed on the surface of this state magnitude constraint, $\|\underline{h}\|_2 = h_{\max}$:

$$\underline{u} = Q\left(\underline{\omega}_{\mathbf{B/L}} - k_1 \underline{h} \right) - \frac{\underline{h}^T Q \underline{\omega}_{\mathbf{B/L}}}{\underline{h}^T \underline{h}} \underline{h} \qquad (33)$$

This control law is still locally asymptotically stable at this state constraint if:

$$\left\| \underline{\omega}_{\mathbf{B/L}} \right\|_2 \le k_1 h_{\max} \qquad (34)$$

while the control remains on this surface. To smoothly transfer from the original TEA-seek control to this angular momentum constraint control, the following control law is proposed for the transition region, $\|\underline{h}\|_2 \in \left[h_{\max} - \Delta h,\ h_{\max} \right]$:

$$\underline{u} = Q\left(\underline{\omega}_{\mathbf{B/L}} - k_1 \underline{h} \right) - \lambda \frac{\underline{h}^T Q \underline{\omega}_{\mathbf{B/L}}}{\underline{h}^T \underline{h}} \underline{h} \qquad (35)$$

where the transition parameter is defined as:

$$\lambda = 1 - \frac{h_{\max} - \|\underline{h}\|_2}{\Delta h} \qquad (36)$$

Thus, the original control law in (30) has been transformed into three control laws over different regions of the CMG angular momentum magnitude space.

If both gravity-gradient and aerodynamic torques are applied to a spacecraft, the attitude potential function in (27) will not go to zero at TEA. Thus, any Lyapunov function which includes the above attitude potential function cannot be used to prove stability for this case. A Lyapunov control law with integral feedback is proposed for a satellite only under the influence of gravity-gradient moments. Integral feedback in the control law will bound the angular velocity and angular momentum, even if unmodeled external moments are applied. Afterwards, a

Lyapunov function without the attitude potential function will be used to study the effect of constant or attitude-dependent disturbance torques on this control law.

A second candidate Lyapunov function is proposed as:

$$V = \Phi + \tfrac{1}{2}\underline{\omega}_{\mathbf{B/L}}{}^T I \underline{\omega}_{\mathbf{B/L}} + \tfrac{1}{2}k_1 \underline{h}^T \underline{h} + \tfrac{1}{2}k_2\left[\int \underline{z}_1\, dt\right]^T IK\left[\int \underline{z}_1\, dt\right] \tag{37}$$

where

$$\underline{z}_1 = \underline{\omega}_{\mathbf{B/L}} - k_1 \underline{h} \tag{38}$$

k_1 and k_2 are positive scalars, while the matrices K and IK are positive definite. If the control law is defined as:

$$\underline{u} = I\left\{(q_1 + k_2)\left(\underline{\omega}_{\mathbf{B/L}} - k_1 \underline{h}\right) + k_2 K \int \left(\underline{\omega}_{\mathbf{B/L}} - k_1 \underline{h}\right) dt\right\} \tag{39}$$

the Lyapunov function time derivative can be written as:

$$\dot{V} = -(q_1 + k_2)\left(\underline{\omega}_{\mathbf{B/L}} - k_1 \underline{h}\right)^T I \left(\underline{\omega}_{\mathbf{B/L}} - k_1 \underline{h}\right) \tag{40}$$

where q_1 is a positive scalar. This Lyapunov function time derivative is negative semidefinite and will only go to zero when $\underline{\omega}_{\mathbf{B/L}} = k_1 \underline{h}$. At this state, the control law will reduce to:

$$\underline{u} = k_2 IK\left(\int \underline{\omega}_{\mathbf{B/L}}\, dt - k_1 \int \underline{h}\, dt\right) = \text{const.} \tag{41}$$

This is an inflection point for the Lyapunov function, but it is not obvious that the Lyapunov function will continue to decrease to zero.

If an appropriate set of gains in the control law can make $\int \left(\underline{\omega}_{\mathbf{B/L}} - k_1 \underline{h}\right) dt$ converge to zero at the same rate or faster than $\underline{\omega}_{\mathbf{B/L}} - k_1 \underline{h}$, the Lyapunov function time derivative will guarantee that the above control is locally asymptotically stable about the nearest TEA. By placing more emphasis on the gain k_2 rather than the gain q_1, the constraint:

$$\underline{\omega}_{\mathbf{B/L}} - k_1 \underline{h} = -K \int \left(\underline{\omega}_{\mathbf{B/L}} - k_1 \underline{h}\right) dt \tag{42}$$

will be more strongly enforced in the control law. Then, the Lyapunov function time derivative will only remain at zero if both the Lyapunov function and the control torque go to zero. When the Lyapunov function is zero, $\Phi = 0$, $\underline{\omega}_{\mathbf{B/L}} = \underline{0}$, $\underline{h} = \underline{0}$, and $\int \underline{\omega}_{\mathbf{B/L}} dt = k_1 \int \underline{h}\, dt$. Although the inertia matrix appears in the control law (39), this matrix is only required to be positive definite. However, the inertia estimate error magnitude can affect the rate of convergence for $\int \left(\underline{\omega}_{\mathbf{B/L}} - k_1 \underline{h}\right) dt$. The stability margin for this TEA-seek control law using a given set of gains must be tested for inertia uncertainty.

The term, $\underline{\omega} \times \underline{h}$, can be added to the control law:

$$\underline{u} = \underline{\omega} \times \underline{h} + I\left\{(q_1 + q_2)\left(\underline{\omega}_{\mathbf{B/L}} - k_1 \underline{h}\right) + k_2 K \int \left(\underline{\omega}_{\mathbf{B/L}} - k_1 \underline{h}\right) dt\right\} \tag{43}$$

to help the parameter identification by adding excitation in orthogonal directions from the current excitation. Unfortunately, it will create a second term in the Lyapunov function time derivative:

$$\dot{V} = -(q_1 + k_2)\left(\underline{\omega}_{\mathbf{B/L}} - k_1 \underline{h}\right)^T I \left(\underline{\omega}_{\mathbf{B/L}} - k_1 \underline{h}\right) - \underline{\omega}_{\mathbf{B/L}}{}^T \left(\underline{\omega} \times \underline{h}\right) \tag{44}$$

which is not negative semidefinite nor is its magnitude bounded by the first term. However, in numerical simulations, this term has been seen to have a negligible effect on the Lyapunov function going to zero.

If a constant or attitude-dependent torque $\underline{d}$ is added to the equations of motion such that the TEA for zero control torque still exists and is constant in the LVLH reference frame, the following Lyapunov function is used:

$$V = \tfrac{1}{2}\underline{\omega}_{\mathbf{B/L}}{}^T I \underline{\omega}_{\mathbf{B/L}} + \tfrac{1}{2}k_1 \underline{h}^T \underline{h} + \tfrac{1}{2}k_2\left[\int \underline{z}_1\, dt\right]^T IK\left[\int \underline{z}_1\, dt\right] \tag{45}$$

This is not a strict Lyapunov function since it can go to zero at orientations other than TEA. If the control law in (39) is used, the Lyapunov function time derivative is given by:

$$\dot{V} = -(q_1 + k_2)\left(\underline{\omega}_{\mathbf{B/L}} - k_1 \underline{h}\right)^T I \left(\underline{\omega}_{\mathbf{B/L}} - k_1 \underline{h}\right) + \underline{\omega}_{\mathbf{B/L}}{}^T \underline{f}_1 \tag{46}$$

where the total external moment in the LVLH reference frame is given by:

$$\underline{f}_1 = 3n^2 \underline{C}_3 \times I \underline{C}_3 - n^2 \underline{C}_2 \times I \underline{C}_2 + \underline{d} \tag{47}$$

When $\underline{f}_1 = \underline{0}$, the spacecraft is at the TEA for zero control torque. In a local region about the stable TEA, the magnitude of the first negative semi-definite term in the Lyapunov function time derivative will generally be greater than the second. Thus, the Lyapunov function will decrease in this local region. The Lyapunov function time derivative will only remain at zero if the spacecraft is at TEA when $\underline{\omega}_{\mathbf{B/L}} = \underline{0}$, $\underline{h} = \underline{0}$, and $\int \left(\underline{\omega}_{\mathbf{B/L}} - k_1 \underline{h}\right) dt = \underline{0}$. Thus, if attitude equilibrium is achieved using zero control torque, this control law has reached the TEA in the presence of constant or attitude-dependent torques. Then, this Lyapunov function shows that the control law in (39) can still be considered locally asymptotically stable with the addition of constant or attitude-dependent torques. If the initial attitude is outside this local region, the gains in the control law must be chosen carefully to constrain the attitude to repeatedly cross the local region around the nearest stable TEA until attitude equilibrium is finally achieved.

Linear Parameter Form of the Spacecraft Angular Acceleration

The spacecraft angular acceleration can be written as:

$$I\dot{\underline{\omega}} = 3n^2 \underline{C}_3 \times I\underline{C}_3 - \underline{\omega} \times I\underline{\omega} + \frac{q\,C_D}{\sqrt{\underline{C}_1{}^T S_1^{\mathbf{B}}\,\underline{C}_1}} \left(\underline{C}_1 \times T_{cp}^{\mathbf{B}}\,\underline{C}_1\right) - \underline{u} \tag{48}$$

The inertia matrix, the atmospheric density, the drag coefficient, and the discrete values for the center of pressure location and projected area are not precisely known. To simplify the identification algorithm, the angular acceleration must be rearranged such that the unknown parameters appear linearly. Appendix A gives examples of the linear element form of common vector/matrix operations that are used in this paper. If the six inertia elements are placed in the following inertia vector,

$$\underline{I}_{el} = \begin{bmatrix} I_{11} & I_{22} & I_{33} & I_{23} & I_{13} & I_{12} \end{bmatrix}^T \tag{49}$$

the inertia elements can be rearranged to appear linearly in each of the above terms:

$$I\dot{\underline{\omega}} = D_1(\dot{\underline{\omega}})\,\underline{I}_{el} \tag{50}$$

$$\tilde{\omega} I \underline{\omega} = D_3(\underline{\omega})\,\underline{I}_{el} \tag{51}$$

$$\tilde{C}_3 I \underline{C}_3 = D_3\left(\underline{C}_3(\underline{\beta})\right)\underline{I}_{el} \tag{52}$$

The aerodynamic parameters cannot be easily linearized due to the ellipsoid interpolation of the projected area. If the true projected area is the sum of the projected area estimate and the parameter estimate error (which can vary depending on the orientation),

$$S = \hat{S} + \Delta S(\underline{\beta}) = \hat{S}\left(1 + \frac{\Delta S(\underline{\beta})}{\hat{S}}\right) \tag{53}$$

the aerodynamic drag moment can be written as:

$$\begin{aligned} \underline{M}_{drag} &= \tfrac{1}{2} v_a{}^2 \hat{S}\,\underline{C}_1 \times \left(1 + \frac{\Delta S}{\hat{S}}\right) \rho_a C_D T_{cp}{}^{\mathbf{B}}\,\underline{C}_1 \\ &= \tfrac{1}{2} v_a{}^2 \hat{S}\,\underline{C}_1 \times N\,\underline{C}_1 \end{aligned} \tag{54}$$

The linear element form of the aerodynamic drag torque can be written as:

$$\underline{M}_{drag} = \tfrac{1}{2} v_a{}^2 \hat{S}\, D_4\left(\underline{C}_1(\underline{\beta})\right)\,\underline{N}_{el} \tag{55}$$

where the 9 x 1 aerodynamic parameter vector is given by:

$$\underline{N}_{el} = \begin{bmatrix} N_{11} & N_{12} & N_{13} \,\vdots\, N_{21} & N_{22} & N_{23} \,\vdots\, N_{31} & N_{32} & N_{33} \end{bmatrix}^T \tag{56}$$

Thus, the spacecraft angular acceleration with the unknown parameters appearing linearly is given by:

$$\begin{aligned} I\dot{\underline{\omega}} &= \left\{3n^2\, D_3\left(\underline{C}_3(\underline{\beta})\right) - D_3(\underline{\omega})\right\}\underline{I}_{el} \\ &\quad + \left\{\tfrac{1}{2} v_a{}^2 \hat{S}\, D_4\left(\underline{C}_1(\underline{\beta})\right)\right\}\,\underline{N}_{el} - \underline{u} \end{aligned} \tag{57}$$

Recursive Least Squares Algorithm

The least squares method will determine the parameter vector which minimizes the 2-norm of the error between the measured output and the model output for an input-output algebraic model. In addition, to avoid a nonlinear identification problem, the parameter identification model should have the unknown parameters appear linearly in the algebraic model:

$$\underline{y}_m = \Phi\underline{\theta} \tag{58}$$

where $\underline{y}_m$ is the model output vector, Φ is the known model matrix, and $\underline{\theta}$ is the unknown parameter vector. No additional terms should appear in this algebraic model.

The rotational equations of motion can be written as an input-output algebraic model:

$$\underline{u} = -\left[D_1(\dot{\underline{\omega}}_f) + D_3(\underline{\omega}) - D_3\left(\underline{C}_3(\underline{\beta})\right) \quad \tfrac{1}{2} v_a{}^2 \hat{S}\, D_4\left(\underline{C}_1(\underline{\beta})\right) \right] \begin{Bmatrix} \underline{I}_{el} \\ \underline{\eta}_{el} \end{Bmatrix} \tag{59}$$

In this case, the model output vector is the control torque and the parameter vector includes the unknown inertia and aerodynamic parameters. The model matrix is only a function of the Euler parameters, the angular velocity, the atmospheric velocity and the projected area estimate. Since the angular acceleration will not be measured, $\dot{\underline{\omega}}_f$ indicates that the angular acceleration is filtered from the angular velocity measurements. The least-squares parameter estimate is very sensitive to errors in the filtered angular acceleration values. Thus, assuming that the measurements are taken at evenly-spaced intervals, the following numerical derivative based on a polynomial approximation was used to estimate the angular acceleration in this paper:

$$\dot{\underline{\omega}}_f(z) = \frac{1}{72\Delta t} \cdot \frac{-3z^2 + 32z + 36 - 96z^{-1} + 31z^{-2}}{1 - 1/6\,z^{-2}}\,\underline{\omega}(z) \tag{60}$$

where z is the forward shift operator. This rational filter is obviously not causal. This is permissible since the filtering will be performed on a set of data off-line.

Assume that a previous parameter estimate is available. The recursive least squares (RLS) batch algorithm will determine the parameter vector that minimizes a weighted function of the model output error and the parameter change over each iteration:

$$V = \frac{1}{2}\left(\underline{\theta}_k - \underline{\theta}_{k-1}\right)^T W_1\left(\underline{\theta}_k - \underline{\theta}_{k-1}\right) + \frac{1}{2}\left(\underline{y}_k - \Phi\underline{\theta}_k\right)^T W_2\left(\underline{y}_k - \Phi\underline{\theta}_k\right) \tag{61}$$

where W_1 and W_2 are positive definite weighting matrices. The RLS estimate is defined as:

$$\underline{\theta}_k = \underline{\theta}_{k-1} + P^{-1}\Phi^T W_2\left(\underline{y}_k - \Phi\underline{\theta}_{k-1}\right), \quad P = W_1 + \Phi^T W_2 \Phi \tag{62}$$

To prevent numerical problems with the inverse, the Φ matrix may need to be column-normalized:

$$C = \Phi M \tag{63}$$

The RLS estimate using column scaling is given by:

$$\underline{\theta}_k = \underline{\theta}_{k-1} + MP^{-1}C^T W_2\left(\underline{y}_k - \Phi\underline{\theta}_{k-1}\right), \quad P = M^T W_1 M + C^T W_2 C \tag{64}$$

If the input excites all of the parameter modes, this iterative solution should converge to the true parameters with the appropriate choice of weighting matrices.

TEA-Seek Maneuvers for the Space Station

The final assembled configuration, Stage-44 Flight 19A, of the International Space Station is simulated in a circular orbit about a point-mass Earth model at an altitude of 278 km. Its orbital mean motion is 1.163e-3 rad/s and its orbital period is approximately 90 minutes. The orbital velocity is 7.74 km/s. The atmospheric density, ρ_a=4.980e-10 kg/m^3, is calculated using the 1976 Standard Atmospheric model from sea-level to 100 km and the 1962 Standard Atmosphere model from 100-500 km. The inertial and drag characteristics of the ISS are given by the ISS On-Orbit Assembly, Modeling and Mass Properties Databook.

The true inertia and aerodynamic parameter matrices for this ISS configuration are defined as:

$$I = \begin{bmatrix} 108.0 & 3.4 & 6.2 \\ 3.4 & 186.0 & 1.5 \\ 6.2 & 1.5 & 115.0 \end{bmatrix} 10^6 \text{ kg}-\text{m}^2 \tag{65}$$

$$N = \begin{bmatrix} 0.00 & 31.49 & -63.32 \\ -3.10 & 7.22 & -24.67 \\ -1.44 & 4.62 & 1.94 \end{bmatrix} 10^{-10} \text{ kg}-\text{m}^2 \tag{66}$$

For the following cases, the initial estimate for the inertia and aerodynamic parameter matrices are given by:

$$\hat{I}(t_0) = \begin{bmatrix} 118.5 & 0.0 & 0.0 \\ 0.0 & 193.9 & 0.0 \\ 0.0 & 0.0 & 101.0 \end{bmatrix} 10^6 \text{ kg}-\text{m}^2 \tag{67}$$

$$\hat{N}(t_0) = \begin{bmatrix} 0.00 & 4.75 & -9.56 \\ -0.47 & 1.09 & -3.72 \\ -0.22 & 0.70 & 0.29 \end{bmatrix} 10^{-10} \text{ kg}-\text{m}^2 \tag{68}$$

The parameter estimates are evaluated using the maximum and mean parameter estimate relative error, which are the maximum and mean values for the column magnitude maximum relative error. The inertia estimate and the aerodynamic estimate column magnitude maximum relative errors for column i are defined as:

$$\frac{\left\|I(:,i) - \hat{I}(:,i)\right\|_\infty}{\left\|I(:,i)\right\|_2}, \quad \frac{\left\|N(:,i) - \hat{N}(:,i)\right\|_\infty}{\left\|N(:,i)\right\|_2} \tag{69}$$

The initial maximum and mean inertia estimate relative errors are given by 12.19% and 8.72% respectively. The initial maximum and mean aerodynamic parameter estimate relative errors are given by 81.92% and 79.34% respectively.

The Euler angles are described using a 2-1-3 set (Pitch-Roll-Yaw Sequence). If the initial inertia estimate was correct, the TEA would be located at [0 0 0] deg. assuming that no aerodynamic moments are present. In this case, the TEA for the true inertia matrix is [60.05 -1.43 2.60] deg., if the gravity gradient is the only external moment. Since the space station must operate close to the true TEA, the initial Euler angle position for the following cases is assumed to be [50 0 0] deg. The initial spacecraft angular velocity in the LVLH reference and the initial CMG momentum are both zero.

Control Law A is the control law given in (30), (33) and (35), while Control Law B is the control law given in (43). The gains for Control Law A are given by k_1=2.01e-9, k_2=2.01e-9, and Q=1.0e-3 $\hat{I}$, where $\hat{I}$ is the current inertia matrix estimate. The gains for Control Law B are given by k_1=2.01e-8, k_2=1.0e-6, q_1=1.50e-9, and K=diag([1.0e-3 2.0e-3 1.0e-3]).

The maximum CMG angular momentum for the ISS is 18,980 N-m-s (or 14,000 ft-lb-sec). However, control systems for the ISS are required to work even if one CMG system fails. The maximum CMG angular momentum with one CMG system failure is 14,240 N-m-s (or 10,500 ft-lb-sec). In Control Law A, the angular momentum bounds are chosen as h_{max}= 14,000 N-m-s and Δh= 1000 N-m-s.

Two simulations will be studied. In the first case, Control Law A will be used when the external moments are only due to the gravity gradient. The six inertia parameters will be estimated at the end of each orbit using 150 measurements from that orbit. For the next orbit, the control system will make a

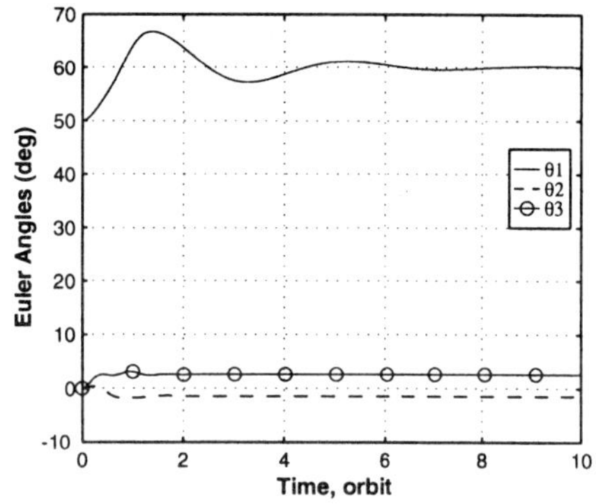

Figure 3. ISS Attitude (Control Law A)

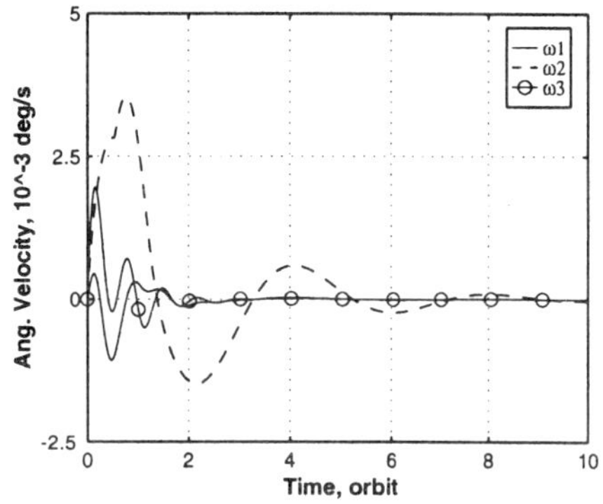

Figure 4. ISS Ang. Velocity (Control Law A)

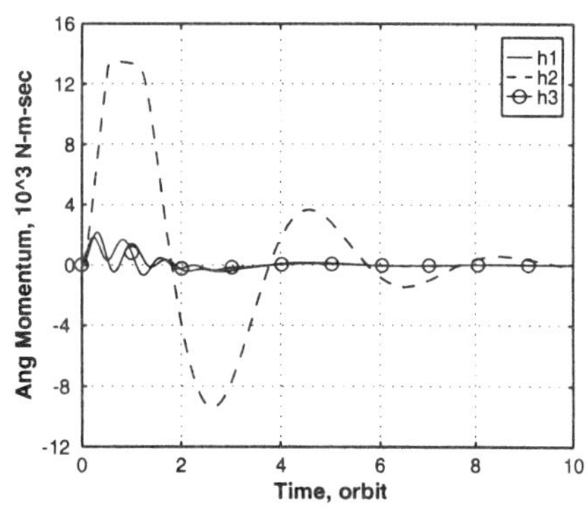

Figure 5. CMG Momentum (Control Law A)

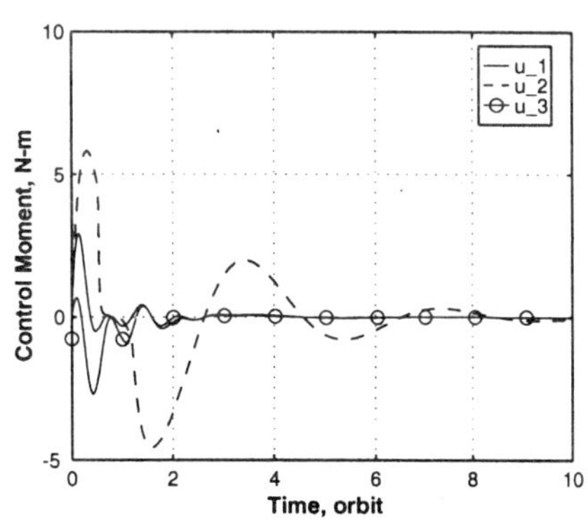

Figure 6. Control Torque (Control Law A)

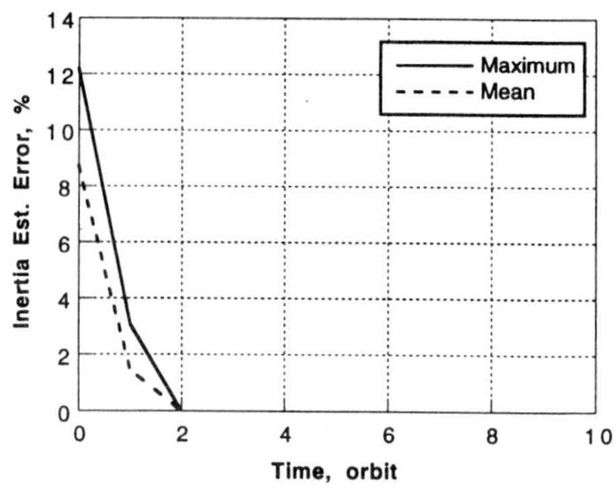

Figure 7. Case 1 Inertia Est. Relative Error

smooth transition from the previous inertia estimate to the new inertia estimate to prevent discontinuous jumps in the control torque. The second case will use Control Law B when the external moments are due to the gravity gradient and aerodynamic torques. In the same manner as described above, the six inertia meters and the nine aerodynamic parameters will be estimated at the end of each orbit using 150 measurements.

Figures 3, 4, 5, and 6 show the attitude, the angular velocity in the LVLH reference frame, the CMG angular momentum, and the control torque using Control Law A over 10 orbits. The attitude converges smoothly to the TEA, while the angular velocity, the angular momentum and the control go to zero. In the first two orbits, the CMG momentum nears its upper bound. The control law acts accordingly by reducing the u_2 component to prevent the angular momentum from exceeding this bound. The maximum magnitude for the angular momentum is 13,564 N-m-s. Figure 7 shows the inertia estimate relative error over this ten-orbit maneuver. After two orbits, the maximum inertia estimate relative error is given by 0.0074%, while the mean inertia estimate relative error is given by 0.0068%.

Figures 8, 9, 10, and 11 show the attitude, the angular velocity in the LVLH reference frame, the CMG angular momentum, and the control torque using Control Law B over 14 orbits. All of the Euler angles decay to the TEA in the mean, while the angular velocity, the angular momentum and the control approach zero. These oscillations are caused by the integral feedback in Control Law B. From the Euler angles at the final time in this simulation, the TEA with the addition of the aerodynamic moment has been identified as [27.54 6.31 -9.05] deg. The maximum magnitude for the angular momentum is 11,822 N-m-s. The angular velocity vector component magnitudes in Figure 9 are much higher than those in Figure 4, since the choice of gains is primarily concerned with momentum management. Consequently, the attitude response would converge

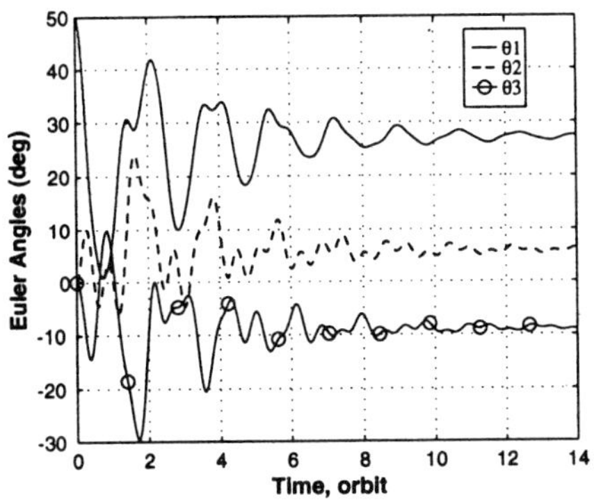

Figure 8. ISS Attitude (Control Law B)

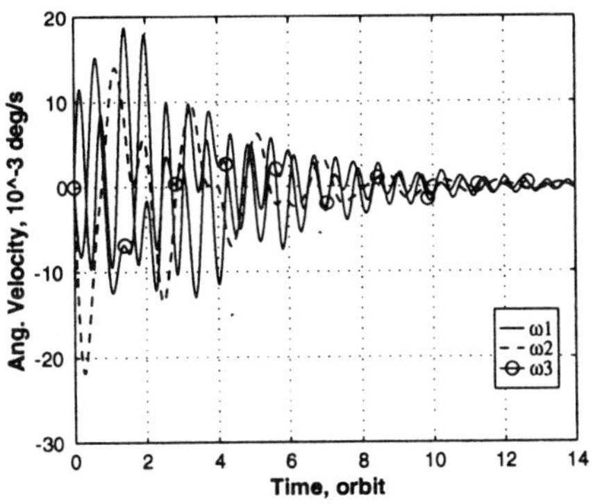

Figure 9. ISS Ang. Velocity (Control Law B)

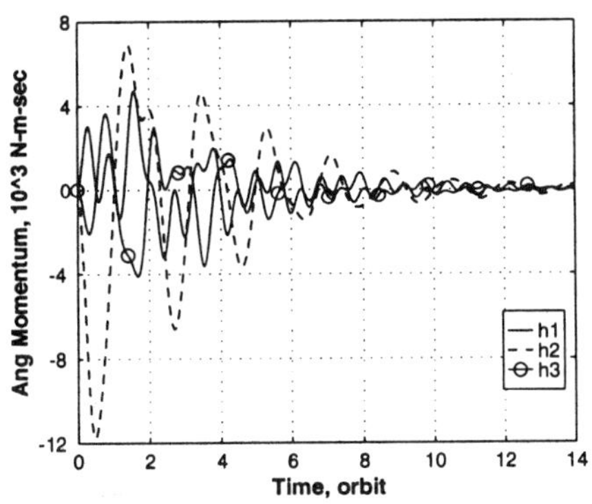

Figure 10. CMG Momentum (Control Law B)

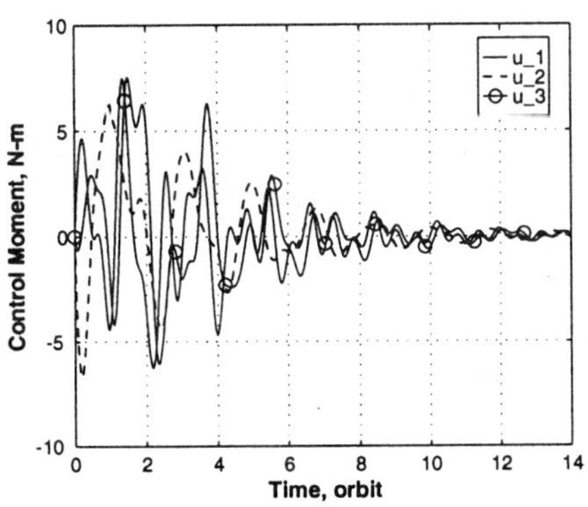

Figure 11. Control Torque (Control Law B)

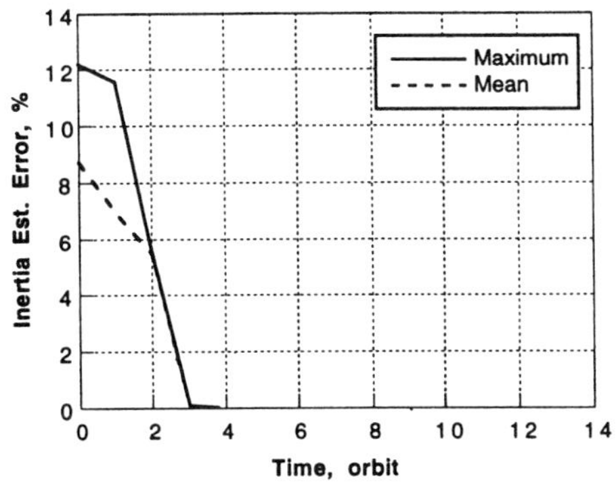

Figure 12. Case 2 Inertia Est. Relative Error

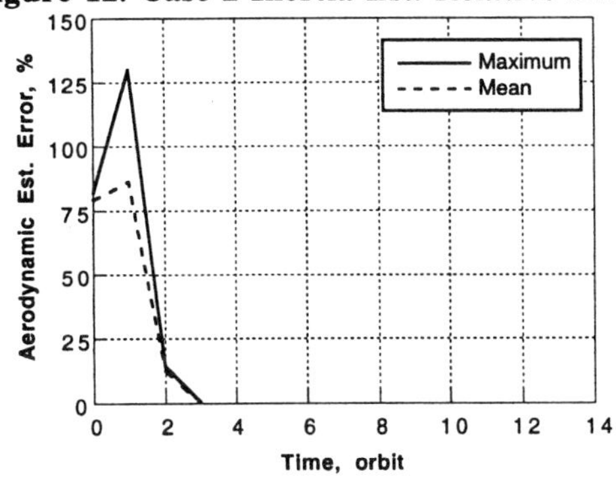

Figure 13. Case 2 Aerodynamic Estimate Relative Error

much quicker on the TEA if the momentum constraint was relaxed to allow a better choice of gains. Figures 12 and 13 show the inertia and aerodynamic parameter estimate relative errors respectively over this 14-orbit maneuver. After three orbits, the maximum and mean inertia estimate relative errors are 0.086% and 0.085%, while the maximum and mean aerodynamic parameter estimate relative errors are 0.161% and 0.142%. After fourteen orbits, the maximum and mean inertia estimate relative errors are 0.0009% and 0.0009%, while the maximum and mean aerodynamic parameter estimate relative errors are 0.0124% and 0.0071%.

The results from this parameter identification represent the best case scenario. Using Control Law B, the pitch Euler angle oscillates over 50 degrees before converging on the TEA. Typically, identification methods perform well if there is sufficient excitation in the maneuver. However, the actual ISS control system will constrain the motion of the space station to meet much tighter requirements on the attitude, angular velocity, and CMG momentum.

Conclusions

The RLS identification algorithm can accurately determine the inertia and aerodynamic parameters of the International Space Station using a nonlinear

TEA-seek control with momentum management. In addition, a new aerodynamic drag moment model has been proposed which includes the attitude effects on the projected area and the center of pressure location. Although the nonlinear Lyapunov control used in this paper worked well in the presence of aerodynamic disturbances and inertia uncertainty, the inertia and aerodynamic parameter identification algorithm must be studied in the future using the actual LQR control law that will be used on ISS.

Appendix A: Linear Element Forms of Matrix/Vector Operations

The following linear forms for matrix elements and vector elements inside a vector operator where:

$$\underline{A}_{el} = \begin{bmatrix} A_{11} & A_{22} & A_{33} & \vdots & A_{23} & A_{13} & A_{12} \end{bmatrix}^T \quad \text{(A.1)}$$

$$\underline{B}_{el} = \begin{bmatrix} B_{11} & B_{12} & B_{13} & \vdots & B_{21} & B_{22} & B_{23} & \vdots & B_{31} & B_{32} & B_{33} \end{bmatrix}^T \quad \text{(A.2)}$$

are given by:

1. Symmetric Matrix Elements in Matrix/Vector Multiplication

$$A\underline{x} = D_1(\underline{x})\,\underline{A}_{el}$$

$$= \left[\begin{array}{ccc|ccc} x_1 & 0 & 0 & 0 & x_3 & x_2 \\ 0 & x_2 & 0 & x_3 & 0 & x_1 \\ 0 & 0 & x_3 & x_2 & x_1 & 0 \end{array}\right] \underline{A}_{el} \quad \text{(A.3)}$$

2. Matrix Elements in Matrix/Vector Multiplication

$$B\underline{x} = D_2(\underline{x})\,\underline{B}_{el} = \left[\begin{array}{c|c|c} \underline{x} & \underline{0} & \underline{0} \\ \underline{0} & \underline{x} & \underline{0} \\ \underline{0} & \underline{0} & \underline{x} \end{array}\right]^T \underline{B}_{el} \quad \text{(A.4)}$$

3. Symmetric Matrix Elements in a Gyroscopic Matrix/Vector Operator

$$\tilde{x}A\underline{x} = D_3(\underline{x})\,\underline{A}_{el}$$

$$= \left[\begin{array}{cccccc} 0 & -x_2x_3 & x_2x_3 & x_2^2 - x_3^2 & x_1x_2 & -x_1x_3 \\ x_1x_3 & 0 & -x_1x_3 & -x_1x_2 & x_3^2 - x_1^2 & x_2x_3 \\ -x_1x_2 & x_1x_2 & 0 & x_1x_3 & -x_2x_3 & x_1^2 - x_2^2 \end{array}\right] \underline{A}_{el} \quad \text{(A.5)}$$

4. Matrix Elements in a Gyroscopic Matrix/Vector Operator

$$\tilde{x}B\underline{x} = D_4(\underline{x})\,\underline{B}_{el}$$

$$= \begin{bmatrix} \underline{0} & x_3\,\underline{x} & -x_2\,\underline{x} \\ -x_3\,\underline{x} & \underline{0} & x_1\,\underline{x} \\ x_2\,\underline{x} & -x_1\,\underline{x} & \underline{0} \end{bmatrix}^T \underline{B}_{el} \quad \text{(A.6)}$$

Acknowledgment

This research was performed in support of the Motion Control Systems group in the Mission Operations Directorate at the NASA Johnson Space Center. The work was sponsored by the United Space Alliance under the Dual-Use Academic Liason contract.

References

[1] Wie, B., Byun, K.W., Warren, V. W., Geller, D., Long, D., and Sunkel, J., "New Approach to Attitude/Momentum Control for the Space Station", *Journal of Guidance, Control, and Dynamics*, Vol. 12, No. 5, 1989, pp. 714-722.

[2] Warren, W.V., Wie, B., and Geller, D., "Periodic-Disturbance Accomodating Control of the Space Station for Asymptotic Momentum Management", *Journal of Guidance, Control, and Dynamics*, Vol. 13, No. 6, 1990, pp. 984-992.

[3] Harduvel, John T., "Continuous Momentum Management of Earth-Oriented Spacecraft", *Journal of Guidance, Control, and Dynamics*, Vol. 15, No. 6, 1992, pp. 1417-1426.

[4] Vadali, S.R., Oh, H.-S., "Space Station Attitude Control and Momentum Management: A Nonlinear Look", *Journal of Guidance, Control, and Dynamics*, Vol. 15, No. 3, 1992, pp. 577-586.

[5] Parlos, A.G., Sunkel, J.W., "Adaptive Attitude Control and Momentum Management for Large-Angle Spacecraft Maneuvers", *Journal of Guidance, Control, and Dynamics*, Vol. 15, No. 4, 1992, pp. 1018-1028.

[6] Byun, K.-W., Wie, B., Geller, D., Sunkel, J., "Robust H_∞ Control Design for the Space Station with Structured Parameter Uncertainty", *Journal of Guidance*, Vol. 14, No. 6, 1991, pp. 1115-1122.

[7] Cristi, R., Burl, J., Russo, N., "Adaptive Quaternion Feedback Regulation for Eigenaxis Rotations", *Journal of Guidance, Control, and Dynamics*, Vol. 17, No. 6, 1994, pp. 1287-1291.

[8] International Space Station On-Orbit Assembly, Modeling, and Mass Properties Databook: Design Analysis Cycle #2 Assembly Sequence, LESC-31166, Rev. D, Lockheed Engineering & Sciences, Co., May 1995.

[9] Hughes, P.C., *Spacecraft Attitude Dynamics*, John Wiley & Sons, New York, 1986, pp. 293-298.

NORMAL MODE ATTITUDE DETERMINATION ALGORITHM EXTENDED KALMAN FILTERING APPROACH

Andy Wu
Hughes Aircraft Company
Mail Station EO/E1/D125
P.O. Box 902
El Segundo, California 90245

Abstract

An attitude determination system has been designed for Hughes future geosynchronous orbit (GEO) satellites which employ a continuously running inertial rate sensor in conjunction with attitude sensors such as sun sensor, earth sensor, and/or star tracker. Real-time knowledge of spacecraft attitude with respect to an inertial reference frame is computed by numerically integrating gyro data through the relationship of kinematics equations. Attitude sensor data are processed to generate corrections to spacecraft attitude, gyro rate bias, and attitude sensor bias estimates. An extended Kalman filter based attitude determination algorithm is derived in this paper for a GEO satellite using static earth sensor assemblies (ESAs) and a two-axis fine sun sensor (FSS) as attitude sensors. A Matlab-based time domain simulation model is developed to evaluate the attitude determination performance. Simulation results show that precision attitude determination (less than ± 0.05 deg per axis) during normal mode operation is achieved using the selected attitude hardware and algorithms.

1.0 Introduction - Attitude Determination Algorithm Overview

The goal is to design an Attitude Determination and Control System (ADCS) which provides the spacecraft inertial attitude with accuracy better than ± 0.05 deg in roll, pitch, and yaw to meet the overall pointing requirement during the normal Earth pointing mode. The selected attitude hardware for precision attitude determination consists of (1) a three-axis gyro mounted on a radiator panel, (2) the two axis FSS mounted on the aft structure for sun viewing near sunset, (3) the static ESAs mounted at three places on the nadir structure to optimize Earth viewing angles, and (4) the Spacecraft Control Processor (SCP) which hosts the ADCS software and all sensor and actuator input/output (I/O) functions.

This paper describes the design and performance evaluation of precision attitude determination algorithms using the aforementioned attitude sensors.

A functional block diagram of the attitude determination algorithms is shown in Figure 1. The algorithms are partitioned into two major modules : attitude propagation and attitude estimator. Listed below are the nomenclature used throughout the paper:

$\omega^{B}_{I \leftarrow B}$: a 3x1 vector representing three angular rates of a Body reference frame with respect to an Inertial reference frame, where the vector is expressed in Body reference frame.

C^{O}_{B} : a 3x3 direction cosine matrix representing the attitude of an Orbital reference frame with respect to a Body reference frame.

Q^{O}_{B} : a quaternion presented by a 4x1 vector specifying the rotation of an Orbital reference frame with respect to a Body reference frame.

1.1 Attitude Propagation.

HIRU data is periodically sampled (every 32 msec.) into the SCP, compensated for known measurement errors such as input axis misalignments, converted into body reference frame, and corrected for gyro biases. The compensated gyro data is then used to numerically integrate a set of euler parameters (or quaternion) that specify the spacecraft attitude with respect to an orbital reference frame.

1.2 Attitude Estimator.

Because of the initial attitude error used in quaternion integration and the accumulated effect of both gyro measurement errors and integration truncation and round off errors, attitude determined by processing only gyro data will contain errors that grow with time. The gyro measurement errors include gyro random walk, bias instability, resolution, and scale factor error. The function of the attitude estimator is to use Earth/Sun sensor data to estimate the spacecraft attitude determination errors. In addition, the attitude estimator estimates the gyro biases to be compensated in gyro data processing submodule, thus reducing the rate of error growth between updates or in the absence of the Sun sensor and improving spacecraft rate stability.

At the end of each Earth/Sun sensor measurement cycle, the spacecraft attitude will be updated in accordance with equations summarized in Figure 2.2 using body attitude error estimates provided by the attitude estimator. The spacecraft roll, pitch, and yaw attitudes are then computed as shown in Figure 2.3.

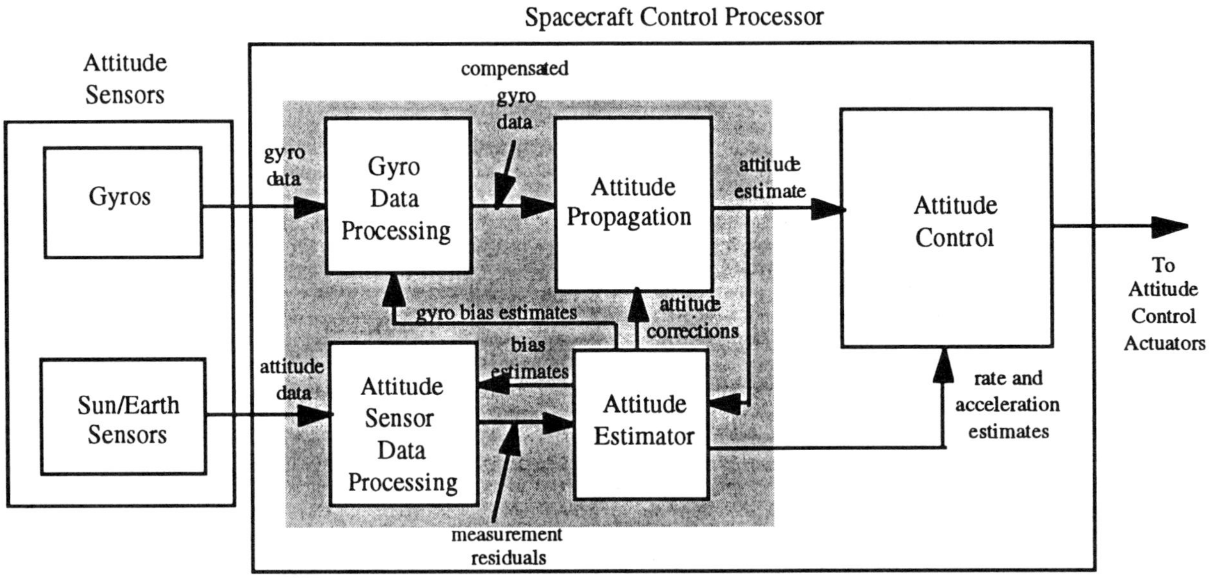

Figure 1. GEO Satellite Attitude Determination and Control Functional Block Diagram

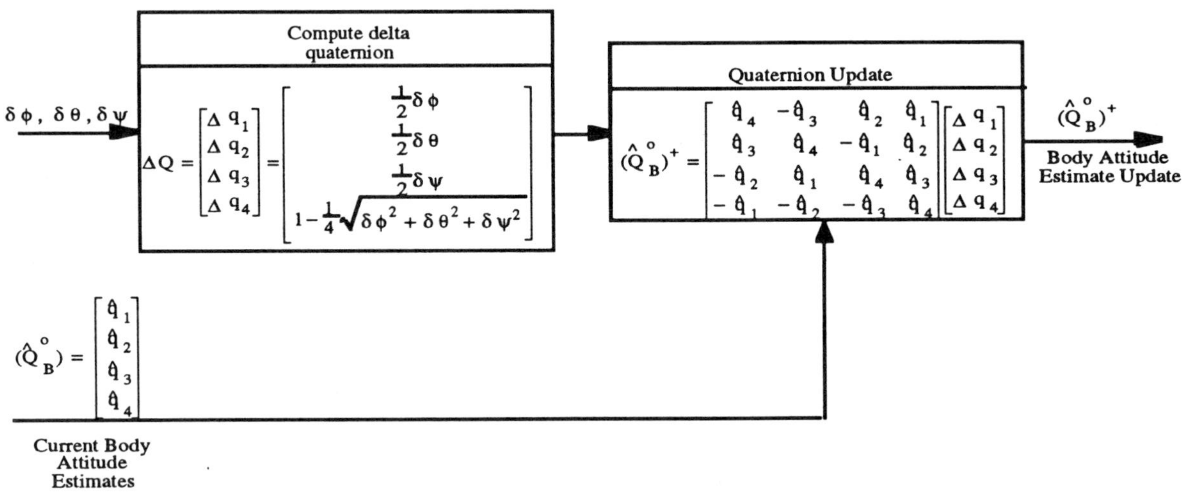

Figure 2.2 Quaternion Update

In the following, we derive the attitude estimator using Kalman filtering techniques, describe a time-domain simulation model using Matlab simulation language, and present the simulation results.

2.0 Derivation of Attitude Estimator

The application of Kalman filtering theory to attitude estimation requires : (1) to define the state variables to be estimated and their dynamics, (2) to express the measurements (or Earth/Sun observations) in terms of defined state variables, and (3) if necessary, to linearize the state equations and measurement equations.

2.1 State Variables and Dynamic Equations.

The dynamics of spacecraft attitude, represented by a quaternion Q^O_B, satisfies following well-known kinematics equations of motion :

$$\dot{Q}^O_B = \Omega\left(\omega^B_{O\leftarrow B}\right) * Q^O_B \tag{1}$$

$$\Omega\left(\omega^B_{O\leftarrow B}\right) = \begin{bmatrix} \frac{1}{2}\omega^B_{O\leftarrow B}(1) \\ \frac{1}{2}\omega^B_{O\leftarrow B}(2) \\ \frac{1}{2}\omega^B_{O\leftarrow B}(3) \\ 0 \end{bmatrix} \tag{2}$$

where "*" in Equation (1) denotes the quaternion multiplication [1,2]. The truth spacecraft attitude is obtained by integrating Equation (1) with given initial conditions and truth body rates, $\omega^B_{O\leftarrow B}$. Since the spacecraft attitude, $\hat{Q}^O_B$ is computed using a numerical integration of following differential equations :

$$\dot{\hat{Q}}^O_B = \Omega\left(\hat{\omega}^B_{O\leftarrow B}\right) * \hat{Q}^O_B \tag{3}$$

with

$$\hat{\omega}^B_{O\leftarrow B} = -\left[\hat{C}^O_B\right]\omega^O_{I\leftarrow O} + \hat{\omega}^B_{I\leftarrow B} \tag{4}$$

where

$$\omega^O_{I\leftarrow O} = \begin{bmatrix} 0 \\ -\omega_o \\ 0 \end{bmatrix};$$

ω_o = orbital rate ($\approx$ 15 deg/hr for a geosyncronized orbit)

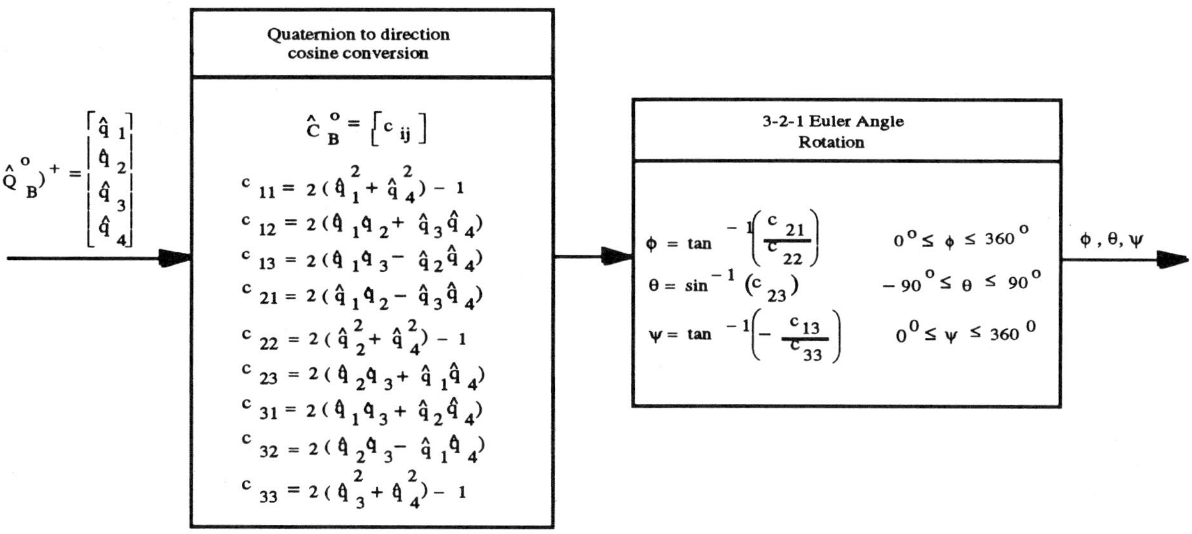

Figure 2.3 Body Attitude Computation

and $\hat{\omega}^B_{I\leftarrow B}$ are the measured body inertial rates provided by the HIRU, the computed spacecraft attitude, $\hat{Q}^O_B$ will be different from the truth spacecraft attitude, Q^O_B.

Let ΔQ^O_B denote the error quaternion satisfying :

$$Q^O_B = \left(\Delta Q^O_B\right) * \hat{Q}^O_B \tag{5}$$

with

$$\Delta Q^O_B = \left[\Delta q_1 \quad \Delta q_2 \quad \Delta q_3 \quad 1 - \sqrt{\Delta q_1^2 + \Delta q_2^2 + \Delta q_3^2}\right]^T;$$

and Q^O_B, $\hat{Q}^O_B$ satisfy Equations (1) and (3) respectively. By taking a time derivative of both sides of Equation (5), one obtains the following error quaternion dynamic equations :

$$\Delta\dot{Q}^O_B = \Omega\left(\hat{\omega}^B_{O\leftarrow B}\right) * \Delta Q^O_B - \Delta Q^O_B * \Omega\left(\hat{\omega}^B_{O\leftarrow B}\right) + \Omega(\Delta\omega) * \Delta Q^O_B \tag{6}$$

with

$$\Delta\omega = \Delta\omega^{B}_{I\leftarrow B} + [\delta C][\hat{C}^{O}_{B}]\,\omega^{O}_{I\leftarrow O} \tag{7}$$

$$\Delta\omega^{B}_{I\leftarrow B} = \omega^{B}_{I\leftarrow B} - \hat{\omega}^{B}_{I\leftarrow B} = \begin{bmatrix}\Delta\omega_x\\ \Delta\omega_y\\ \Delta\omega_z\end{bmatrix} \tag{8}$$

and δC satisfies :

$$C^{O}_{B} = (I + \delta C)\hat{C}^{O}_{B} \tag{9}$$

The gyro error vector, $\Delta\omega^{B}_{I\leftarrow B}$ defined in Equation (8) can be expressed as :

$$\Delta\omega^{B}_{I\leftarrow B} = \begin{bmatrix}\Delta\omega_x\\ \Delta\omega_y\\ \Delta\omega_z\end{bmatrix} = \begin{bmatrix}\Delta b_x\\ \Delta b_y\\ \Delta b_z\end{bmatrix} + \begin{bmatrix}n_{rwx}\\ n_{rwy}\\ n_{rwz}\end{bmatrix} \tag{10}$$

$$\text{with}\quad \begin{bmatrix}\Delta\dot{b}_x\\ \Delta\dot{b}_y\\ \Delta\dot{b}_z\end{bmatrix} = \begin{bmatrix}n_{rrwx}\\ n_{rrwy}\\ n_{rrwz}\end{bmatrix} \tag{11}$$

where n_{rw} are gyro angle random walk noises and n_{rrw} are gyro rate random walk noises. Based on Equations (6) and (11), the three attitude errors : $2\Delta q_1$, $2\Delta q_2$, $2\Delta q_3$ and three gyro bias errors : Δb_x, Δb_y, Δb_z are selected as state variables to be estimated.

2.2. Linearization of Dynamic Equations.

By substituting Equations (7), (8) and (10) into Equation (6), one notes that Equation (6) is a nonlinear function of defined states variables. To apply the Kalman filtering techniques, one needs to linearize the dynamic equations - Equation (6).

Linearization of Equation (6) can be obtained by using small angle approximations to a direction consine matrix $\hat{C}^{O}_{B}$,

$$\hat{C}^{O}_{B} \approx \begin{bmatrix}1 & -\hat{\psi} & \hat{\theta}\\ \hat{\psi} & 1 & -\hat{\phi}\\ -\hat{\theta} & \hat{\phi} & 1\end{bmatrix} \tag{12}$$

and an attitude error matrix δC,

$$\delta C \approx \begin{bmatrix}0 & -\delta\psi & \delta\theta\\ \delta\psi & 0 & -\delta\phi\\ -\delta\theta & \delta\phi & 0\end{bmatrix} \tag{13}$$

with

$$\begin{bmatrix}\phi\\ \theta\\ \psi\end{bmatrix} = \begin{bmatrix}\hat{\phi} + \delta\phi\\ \hat{\theta} + \delta\theta\\ \hat{\psi} + \delta\psi\end{bmatrix} \cong \begin{bmatrix}\hat{\phi} - 2\Delta q_1\\ \hat{\theta} - 2\Delta q_2\\ \hat{\psi} - 2\Delta q_3\end{bmatrix} \tag{14}$$

and $\hat{\omega}^{B}_{O\leftarrow B} \cong 0$. By substituting Equations (12) through (14) into Equation (6) and maintaining only the first-order terms, one obtains the following linear dynamic equations :

$$\begin{bmatrix}\delta\dot{\phi}\\ \delta\dot{\theta}\\ \delta\dot{\psi}\end{bmatrix} = \begin{bmatrix}0 & 0 & \omega_0\\ 0 & 0 & 0\\ -\omega_0 & 0 & 0\end{bmatrix}\begin{bmatrix}\delta\phi\\ \delta\theta\\ \delta\psi\end{bmatrix} - \begin{bmatrix}\Delta\omega_x\\ \Delta\omega_y\\ \Delta\omega_z\end{bmatrix} \tag{15}$$

2.3 Measurement Equations.

In the following we will derive the measurement residuals (Earth sensor measurement residuals and Sun sensor measurement residuals) and expressed them as a function of the above defined state variables.

2.3.1 Earth Sensor Measurement Residuals.

In Earth pointing mode, Barnes ESAs measure the spacecraft roll and pitch angles directly. The measured spacecraft roll and pitch angles, ϕ_m and θ_m are corrected for the biases first, and then subtracted from the estimated spacecraft roll and pitch angles, $\hat{\phi}$ and $\hat{\theta}$ to form the measurement residuals :

$$\begin{bmatrix}\Delta\phi\\ \Delta\theta\end{bmatrix} = \begin{bmatrix}-\hat{\phi} + \phi_m - \hat{\phi}_{mb}\\ -\hat{\theta} + \theta_m - \hat{\theta}_{mb}\end{bmatrix} \tag{16}$$

The Earth sensor biases, $\hat{\phi}_{mb}$, $\hat{\theta}_{mb}$ are provided by the attitude estimator, and the estimated spacecraft roll and pitch angles are computed using the current body attitude estimates, $\hat{C}^{O}_{B}$:

$$\hat{\phi} = \tan^{-1}\left(\frac{-\bar{r}^{B}_{e}(2)}{\bar{r}^{B}_{e}(3)}\right) \tag{17}$$

$$\hat{\theta} = -\sin\left(\bar{r}^{B}_{e}(1)\right) \tag{18}$$

$$\bar{r}^{B}_{e} = [\hat{C}^{O}_{B}]\begin{bmatrix}0\\ 0\\ 1\end{bmatrix}$$

where $\bar{r}^{B}_{e}$ is the estimated Earth unit vector expressed in spacecraft body reference frame.

2.3.2 Sun Sensor Measurement Residuals

Figure 2 shows the Sun sensor geometry, where the Sun sensor boresight is mounted in the roll/yaw plane at an angle, θ_m (= -70 deg), negative from the roll axis. When the Sun is within the sensor field-of-view (FOV), ± 32 deg as indicated in the figure, the sensor measures two angles , AZ_m and EL_m, azimuth and elevation angles in the FSS frame. Given the Sun Ephemeris data : sum azimuth and elevation angles, AZ_s and EL_s in the orbital frame, the Sun unit vector in the orbital frame can be computed as follows :

$$\tilde{r}_s^O = \begin{bmatrix} \cos(EL_s)\cos(AZ_s) \\ \sin(EL_s) \\ -\cos(EL_s)\sin(AZ_s) \end{bmatrix} \tag{19}$$

The Sun sensor measurement residuals are defined as the differences between the estimated Sun azimuth and elevation angles and the measured Sun azimuth and elevation angles in the FSS frame :

$$\begin{bmatrix} \Delta EL \\ \Delta AZ \end{bmatrix} = \begin{bmatrix} \tilde{EL} \\ \tilde{AZ} \end{bmatrix} - \begin{bmatrix} EL_m \\ AZ_m \end{bmatrix} \tag{20}$$

where the estimated Sun azimuth and elevation angles in the FSS frame are given by :

$$\tilde{AZ} = \tan^{-1}\left(\frac{-\tilde{r}_s^{ss}(3)}{\tilde{r}_s^{ss}(1)}\right) \tag{21}$$

$$\tilde{EL} = \sin^{-1}\left(\tilde{r}_s^{ss}(2)\right) \tag{22}$$

$$\text{with} \quad \tilde{r}_s^{ss} = \left[C_{ss}^B\right]\left[\hat{C}_B^O\right]\tilde{r}_s^O \tag{23}$$

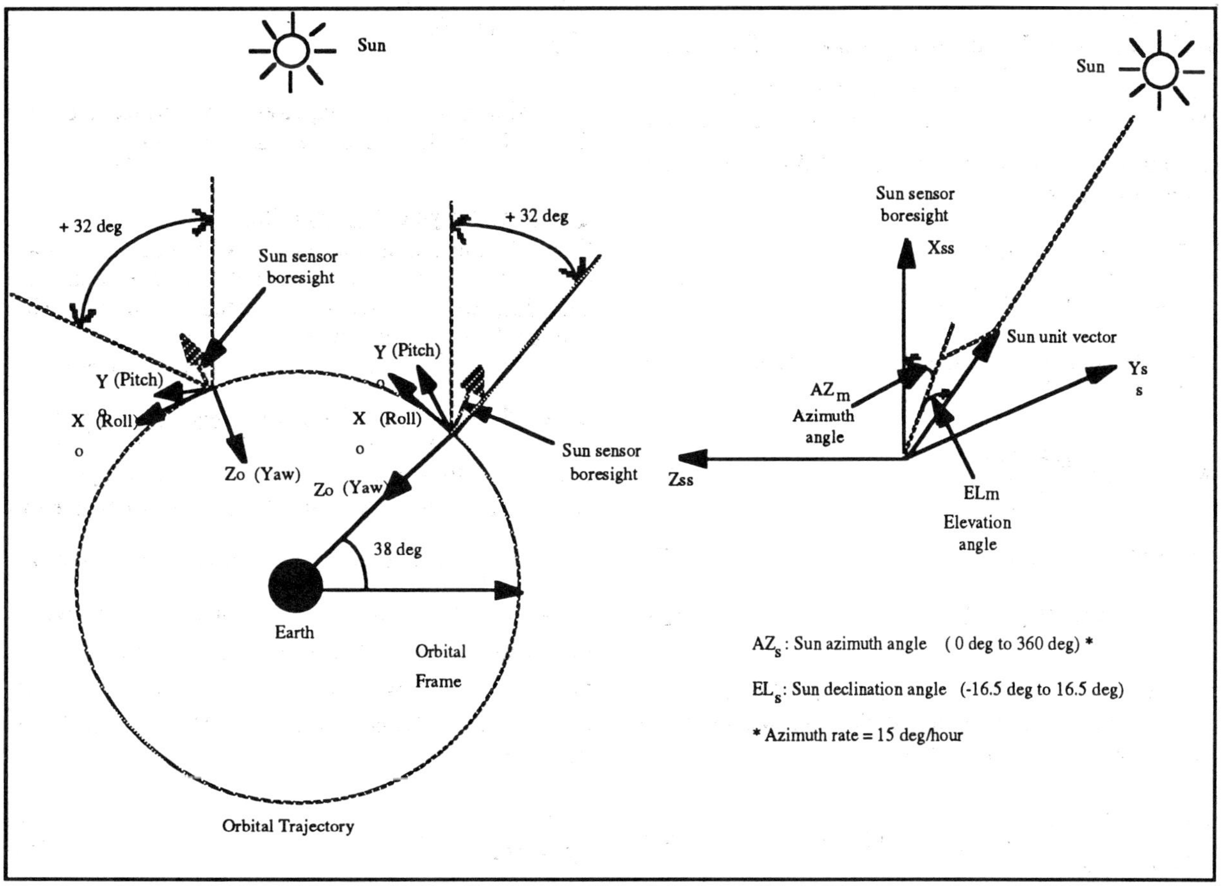

Figure 2. Sun Sensor Geometry

2.4 Linearization of Measurement Equations

2.4.1 Linearization of Earth Sensor Measurement Equations

Let $\Delta\phi_m$, $\Delta\theta_m$ denote the Earth sensor measurement errors, then the measured spacecraft roll and pitch angles, ϕ_m and θ_m are related to the truth spacecraft roll and pitch angles, ϕ and θ as follows :

$$\begin{bmatrix}\phi_m\\ \theta_m\end{bmatrix} = \begin{bmatrix}\phi - \Delta\phi_m\\ \theta - \Delta\theta_m\end{bmatrix} \tag{24}$$

Substituting Equations (14) and (24) into Equation (16) gives :

$$\begin{bmatrix}\Delta\phi\\ \Delta\theta\end{bmatrix} = \begin{bmatrix}\delta\phi - \Delta\phi_m\\ \delta\theta - \Delta\theta_m\end{bmatrix} \tag{25}$$

which are linear functions of state variables, $\delta\phi$ and $\delta\theta$.

2.4.2. Linearization of Sun Sensor Measurement Equations

Similarly, let ΔEL_m and ΔAZ_m denote the Sun sensor measurement errors, then the measured azimuth and elevation angles, EL_m and AZ_m are related to the truth azimuth and elevation angles, EL and AZ in the FSS frame as follows :

$$\begin{bmatrix}EL_m\\ AZ_m\end{bmatrix} = \begin{bmatrix}EL - \Delta EL_m\\ AZ - \Delta AZ_m\end{bmatrix} \tag{26}$$

where the truth azimuth and elevation angles in the FSS frame are given by :

$$AZ = \tan^{-1}\left(\frac{-r_s^{ss}(3)}{r_s^{ss}(1)}\right) \tag{27}$$

$$EL = \sin^{-1}\left(r_s^{ss}(2)\right) \tag{28}$$

$$\text{with} \quad r_s^{ss} = \left[C_{ss}^{B}\right]\left[C_{B}^{O}\right]\tilde{r}_s^{O} \tag{29}$$

By applying Taylor series expansion to Equations (27) and (28), the truth azimuth and elevation angles in the FSS frame can be expressed as:

$$\begin{bmatrix}EL\\ AZ\end{bmatrix} = \begin{bmatrix}\tilde{EL}\\ \tilde{AZ}\end{bmatrix} + [D]\,\delta r + O\left(\delta r^2\right) \tag{30}$$

$$D = \begin{bmatrix} 0 & \dfrac{1}{\cos(EL_m)} & 0 \\ -\dfrac{\sin(AZ_m)}{\cos(EL_m)} & 0 & -\dfrac{\cos(AZ_m)}{\cos(EL_m)} \end{bmatrix} \tag{31}$$

$$\delta r = r_s^{ss} - \tilde{r}_s^{ss} \tag{32}$$

where r_s^{ss}, $\tilde{r}_s^{ss}$ satisfy Equations (29) and (23) respectively. By substituting Equations (9), (13), (19), (23), (26), (29), (30), (31), and (32) into Equation (20) with

$$C_{ss}^{B} = \begin{bmatrix}\cos\theta_m & 0 & -\sin\theta_m\\ 0 & 1 & 0\\ \sin\theta_m & 0 & \cos\theta_m\end{bmatrix}$$

and truncating the higher-order terms, one obtains the linearized Sun sensor measurement equations :

$$\begin{bmatrix}\Delta EL\\ \Delta AZ\end{bmatrix} = \begin{bmatrix}-\sin(AZ_s) & 0 & -\cos(AZ_s)\\ \tan(EL_s)\cos(AZ_s) & -1 & -\tan(EL_s)\sin(AZ_s)\end{bmatrix}\begin{bmatrix}\delta\phi\\ \delta\theta\\ \delta\psi\end{bmatrix} + \begin{bmatrix}\Delta EL_m\\ \Delta AZ_m\end{bmatrix} \tag{33}$$

In obtaining the above equations, we utilize the facts that : $EL_s \cong EL_m$ and $AZ_s \cong AZ_m + \theta_m$.

2.5 Time Varying Kalman Filter

Given the linear dynamic equations - Equations (11) and (15), and linear measurement equations - Equations (25) and (33), we can design an attitude estimator using Kalman filtering techniques.

Let x be a 8x1 state vector : $x = \left[\delta\theta \;\; \Delta b_y \;\; \Delta\theta_e \;\vdots\; \delta\phi \;\; \Delta b_x \;\; \delta\psi \;\; \Delta b_z \;\; \Delta\phi_e\right]^T$ containing 3 attitude errors ($\delta\phi$, $\delta\theta$, $\delta\psi$), 3 gyro bias errors (Δb_x, Δb_y, Δb_z), and 2 Earth sensor bias errors ($\Delta\phi_e$, $\Delta\theta_e$), and y be a 4x1 measurement vector : $y = [\Delta\theta \;\; \Delta AZ \;\; \Delta\phi \;\; \Delta EL]^T$ when The Sun is within the sensor FOV and otherwise, y be a 2x1 measurement vector : $y = [\Delta\theta \;\; \Delta\phi\,]^T$. In state-space representation, Equations (11), (15), (25), and (33) becomes :

$$\dot{x} = A\,x + \omega \tag{34}$$

$$y = H\,x + n \tag{35}$$

where

$$A = \begin{bmatrix} 0 & -1 & 0 & & & & & \\ 0 & 0 & 0 & & & 0_{3x5} & & \\ 0 & 0 & 0 & & & & & \\ & & & 0 & -1 & \omega_0 & 0 & 0 \\ & & & 0 & 0 & 0 & 0 & 0 \\ & 0_{5x3} & & -\omega_0 & 0 & 0 & -1 & 0 \\ & & & 0 & 0 & 0 & 0 & 0 \\ & & & 0 & 0 & 0 & 0 & 0 \end{bmatrix}$$

where The Sun is within the sensor FOV, the measurement matrix H is given by :

$$H = \begin{bmatrix} 1 & 0 & 1 & 0 & 0 & 0 & 0 & 0 \\ -1 & 0 & 0 & c_{24} & 0 & c_{26} & 0 & 0 \\ 0 & 0 & 0 & 1 & 0 & 0 & 0 & 1 \\ 0 & 0 & 0 & c_{44} & 0 & c_{46} & 0 & 0 \end{bmatrix}$$

$$c_{24} = \tan(EL_s)\cos(AZ_s)$$
$$c_{26} = -\tan(EL_s)\sin(AZ_s)$$
$$c_{44} = -\sin(AZ_s)$$
$$c_{46} = -\cos(AZ_s)$$

otherwise,

$$H = \begin{bmatrix} 1 & 0 & 1 & 0 & 0 & 0 & 0 & 0 \\ 0 & 0 & 0 & 1 & 0 & 0 & 0 & 1 \end{bmatrix}$$

$$\omega = \begin{bmatrix} n_{rwy} & n_{rrwy} & n_{eb\theta} & n_{rwx} & n_{rrwx} & n_{rwz} & n_{rrwz} & n_{eb\phi} \end{bmatrix}^T;$$

$$n = \begin{bmatrix} n_\theta & \Delta AZ_m & n_\phi & \Delta EL_m \end{bmatrix}^T.$$

It is noted that two new states : the Earth sensor bias errors, $\Delta\phi_e$, $\Delta\theta_e$ are added to the original 6 states previously defined in Section 2.1 to improve attitude determination accuracy. Based on Equations (34) and (35), the discrete attitude estimator can be implemented as follows:

(a) Propagation : Between the Sun/Earth observations, the estimator propagates the covariance matrix P and the estimated state vector, $\hat{x}$:

$$P^-(t+\Delta t) = \phi(\Delta t)\, P^+(t)\, \phi^T(\Delta t) + Q_n(t) \quad (36)$$

$$\hat{x}^-(t+\Delta t) = \phi(\Delta t)\, \hat{x}^+(t) \quad (37)$$

where $\phi(\Delta t) = e^{A\Delta t}$ is the state transition matrix, $Q_n(t) = E\{\Gamma(\Delta t)\omega(t)\omega^T(t)\Gamma^T(\Delta t)\}$ is the process noise covariance matrix with $\Gamma(\Delta t) = \int_0^{\Delta t} e^{A\tau} d\tau$, and the covariance matrix P is defined as :

$$P^+(t) = E\left\{\left(x(t) - \hat{x}^+(t)\right)\left(x(t) - \hat{x}^+(t)\right)^T\right\}$$

(b) Correction : Upon the Sun/Earth observations at time t+Δt, the estimator computes the measurement matrix H as defined in Equation (35) and the time varying estimator gain matrix, $K_c(t+\Delta t)$:

$$K_c(t+\Delta t) = P^-(t+\Delta t) * H * \left(H * P^-(t+\Delta t) * H^T + R\right)^{-1} \quad (38)$$

where $R = E\{nn^T\}$ is the 4x4 measurement noise covariance matrix. It updates the covariance matrix, P and the estimated state vector, $\hat{x}$:

$$P^+(t+\Delta t) = (I_{8x8} - K_c(t+\Delta t) * H) * P^-(t+\Delta t) *$$
$$(I_{8x8} - K_c(t+\Delta t) * H)^T + K_c(t+\Delta t) * R * K_c(t+\Delta t)^T \quad (39)$$

$$\hat{x}^+(t+\Delta t) = \hat{x}^-(t+\Delta t) + K_c(t+\Delta t) * \Delta y(t+\Delta t) \quad (40)$$

$$\Delta y(t+\Delta t) = y(t+\Delta t) - H * \hat{x}^-(t+\Delta t) \quad (41)$$

and also updates the spacecraft attitude, gyro bias estimates, and the Earth sensor bias estimates :

$$\delta\theta = d\hat{x}(1);\ \delta\phi = d\hat{x}(4);\ \delta\psi = d\hat{x}(4);$$
$$\hat{b}_1 = \hat{b}_1 + d\hat{x}(5);\ \hat{b}_2 = \hat{b}_2 + d\hat{x}(2);$$
$$\hat{b}_3 = \hat{b}_3 + d\hat{x}(7);$$
$$\hat{\theta}_{mb} = d\hat{x}(3);\ \hat{\phi}_{mb} = d\hat{x}(8)$$

where

$$d\hat{x} = K_c(t+\Delta t) * \Delta y(t+\Delta t).$$

3.0 Simulation Model Description and Simulation Results

A time domain nonlinear simulation model coded with Matlab language was developed to evaluate the attitude determination performance. The model simulates : (1) the spacecraft orbital motion, (2) the truth spacecraft attitude, (3) the gyro noises, (4) the Sun/Earth sensor errors, and (5) the attitude determination algorithms - attitude propagation, attitude estimation, and attitude correction. Table 1 lists the assumptions made in the simulations including sensor noises/errors, initial conditions and filter processing rates.

Figure 3 shows the time responses of spacecraft roll, pitch, yaw attitude errors when the Earth sensors have bias errors. The gyro bias estimation errors and Earth sensor bias estimation errors are shown in Figures 4 and 5 respectively. The results indicate that the filter can accurately estimates the Earth sensor biases and meets the ± 0.05 deg. attitude performance accuracy requirements.

4.0 Conclusions

The normal mode attitude determination algorithms using data from continuously running gyros and Sun/Earth sensors for GEO mission has been described in this paper. A 8 state time varying filter bas been formulated and derived using Kalman filtering techniques. A computer simulation model simulating the basic functions of attitude determination algorithms, attitude sensors and spacecraft motion has been developed to evaluate the filter performance. Covariance analysis and time domain analysis have been performed for both time varying gain filter and fixed gain filter. The simulation results indicate that the attitude performance of 0.05 deg (3 sigma) requirement can be met with the selected attitude hardware and the fixed gain filter.

5.0 References

[1] : G. D. Niva, "The Use of Quaternion with an All-Attitude IMU," AAS 82-026, Annual Rocky Mountain Guidance and Control Conference, Jan. 30 - Feb. 3, 1982.

[2] : J. R. Wertz, "Spacecraft Attitude Determination and Control," D. Reidel Publishing Company, 1986.

Table 1. Attitude Determination Matlab Model Simulation Parameters

Earth Sensor Noise (1 sigma)	Sun Sensor Noise (1 sigma)	Gyro Noise (1 sigma)	Initial Errors	Filter Processing Rate
o Random noise : 0.01 deg @ 2Hz o Bias : 0.03 deg o Radiance errors : 0.016 deg with 7~14 days period o Thermal errors : 0.018 deg with 24 hrs period	o Random noise : 0.03 deg	o Angle random walk : 0.001 deg/sqrt(hr) o Rate random walk : 0.002 deg/hr*sqrt(hr) o Angle white noise : 0.054 arc sec/sqrt(Hz)	o Roll error : 0.1 deg o Pitch error : -0.2 deg o Yaw error : 0.57 deg o Gyro bias error : 0.06 deg/hr	o Gyro processing rate : 2 Hz o Filter processing rate : 1 Hz

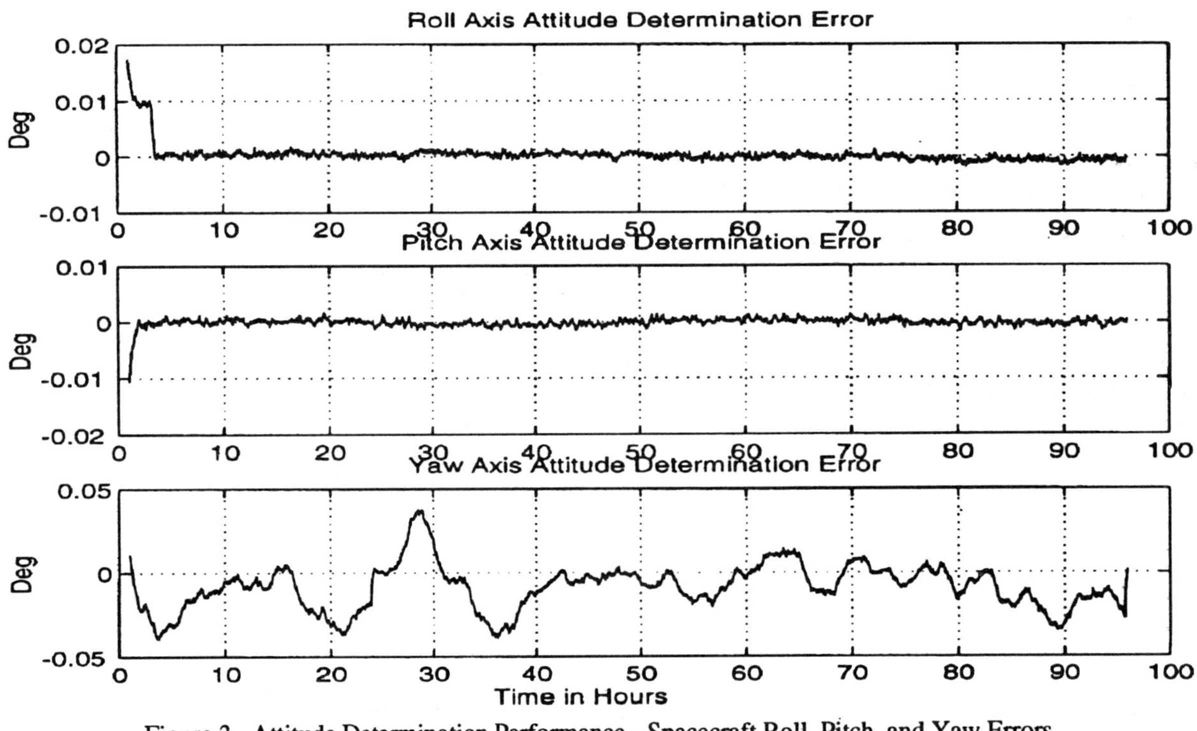

Figure 3. Attitude Determination Performance - Spacecraft Roll, Pitch, and Yaw Errors

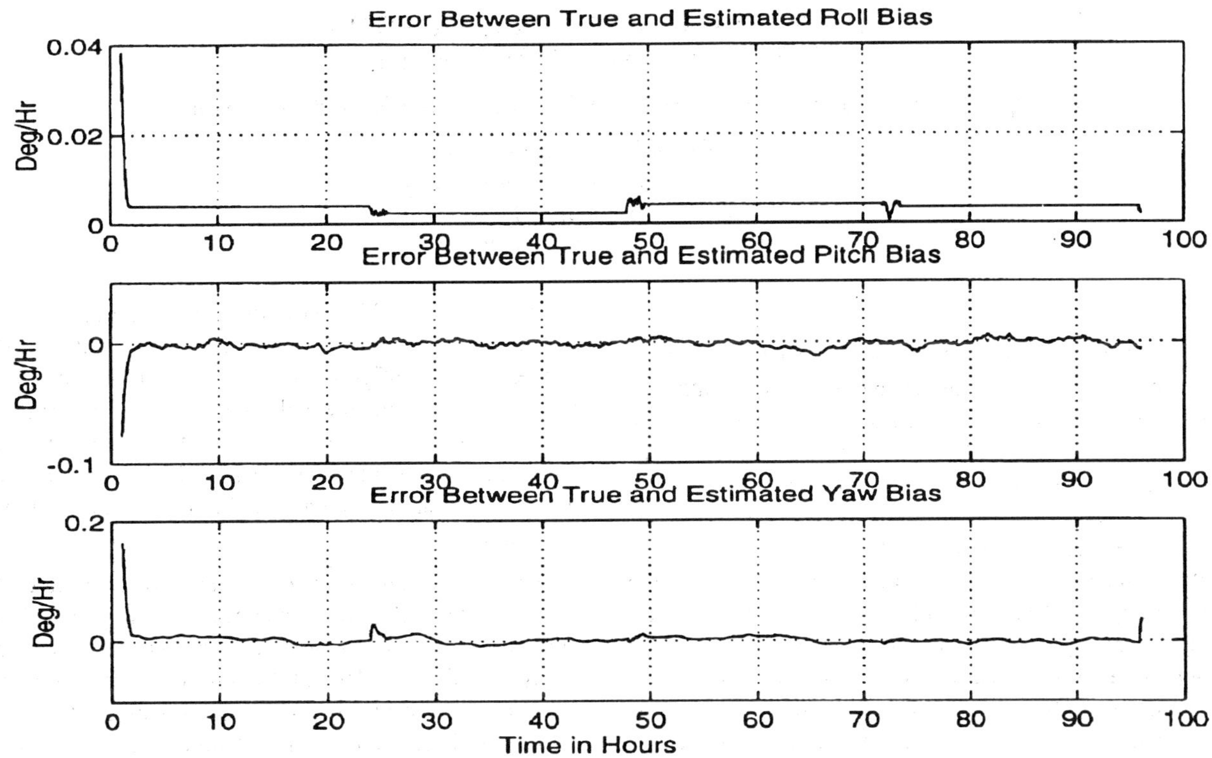

Figure 4. Attitude Determination Performance - Gyro Bias Errors

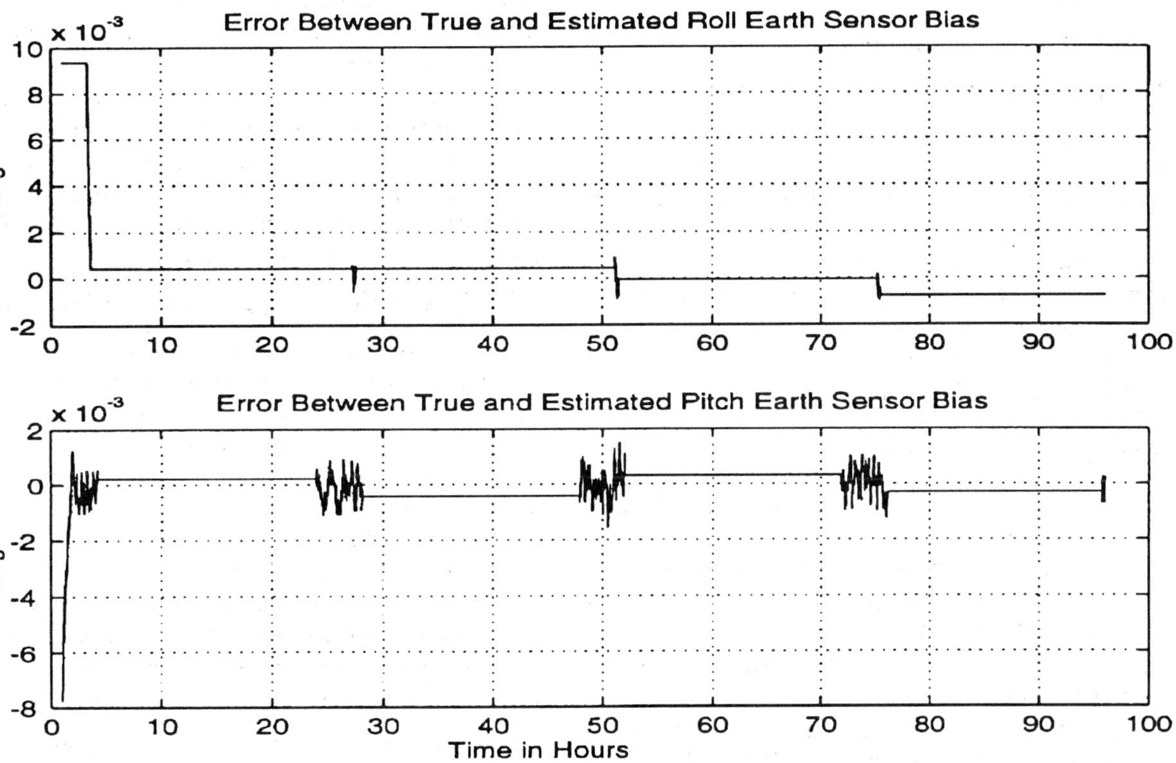

Figure 5. Attitude Determination Performance - Earth Sensor Bias Errors

AIAA-97-3526

A CONSTRAINT MONITOR ALGORITHM FOR THE CASSINI SPACECRAFT

Gurkirpal Singh*, Glenn Macala*, Edward C. Wong†, and Robert D. Rasmussen††

Jet Propulsion Laboratory
California Institute of Technology
4800 Oak Grove Drive
Pasadena, CA 91109-8099

Abstract

The Cassini spacecraft is commanded to turn from one point in space to another by commanding attitude, rate, and acceleration profiles, which the Attitude Controller is required to execute faithfully. The Constraint Monitor algorithm performs appropriate checks to ensure the legality and the realizability of the instantaneous attitude, rate, and acceleration commands. When the command is found to be unrealizable because of hardware limitations or illegal because it may enter a constraint space, Constraint Monitor modifies it appropriately such that the legality of the command is maintained. In doing so it ensures that the rate and acceleration are not outside the capabilities of the attitude control hardware and that certain sensitive spacecraft boresights are protected from exposure to bright objects.

1. Introduction

1.1. Mission and Science Objectives

The Cassini spacecraft, scheduled for launch in October 1997, will arrive at Saturn in 2004. On its way to Saturn, it will fly by Venus, Earth, and Jupiter to pick up the needed gravity assists. The spacecraft will be carrying a probe intended for delivery into the Titan atmosphere. The probe entry into the Titan atmosphere will occur about four months after Saturn arrival. The spacecraft will conduct a tour of the Saturnian system for approximately four years. Several close flybys of Titan and Saturn's icy satellites are planned. The nominal mission will conclude in the year 2008.

* Senior Technical Staff, Guidance and Control Analysis Group, Automation and Control Section

† Senior Technical Staff, Group Leader, Guidance and Control Analysis Group, Automation and Control Section, Member AIAA

†† Principal Technical Staff, Avionic Systems Engineering Section

Cassini science objectives include investigation of Saturn's atmospheric composition, winds and temperature, configuration and dynamics of the magnetosphere, structure and composition of the rings, characterization of the icy satellites, and Titan's atmospheric constituent abundance. The radar mapper will perform surface imaging and altimetry during each Titan flyby.

Cassini was originally one of the two spacecraft of Mariner Mark II series intended for multi-mission purpose: the CRAF (Comet Rendezvous and Asteroid Flyby) and Cassini. The CRAF mission was to follow a comet and conduct scientific investigations for 120 days and later flyby an asteroid. NASA budget constraints necessitated the cancellation of CRAF and descoping of Cassini spacecraft. Both the high and low precision scan platforms and their structural booms were deleted, as was the turn table which carried a fields and particles experiment. The spacecraft basebody assumed the role of the observation platform and the entire spacecraft now had to turn in order to do earth pointing, star tracking, and remote science pointing. This makes the detection and avoidance of spacecraft attitude constraints ever more difficult since movement of one boresight requires turning the entire spacecraft.

1.2. Spacecraft Configuration and Modeling

Cassini is a flexible spacecraft containing four structural appendages and three propellant tanks. The fully deployed Cassini spacecraft with the Huygens Titan probe is shown in Figure 1. The four long booms are a magnetometer boom and three radio and plasma wave science antennas. The core structure of the spacecraft houses the propulsion module which consists of two bipropellant tanks. At the bottom of the propulsion module is the lower equipment module, which supports three radio-isotope thermoelectric generators, a set of reaction wheels, and two articulable (two axis) main engines for large trajectory correction maneuvers.

At launch, the spacecraft total mass is approximately 5530 kg and the inertia approximately 8780,9050,3770 kg-m^2 about the S/C x, y, z axes. Towards the end of the Saturn tour (probe released), the mass and inertia will be reduced to 2400 kg and 6940, 5720, 3600 kg-

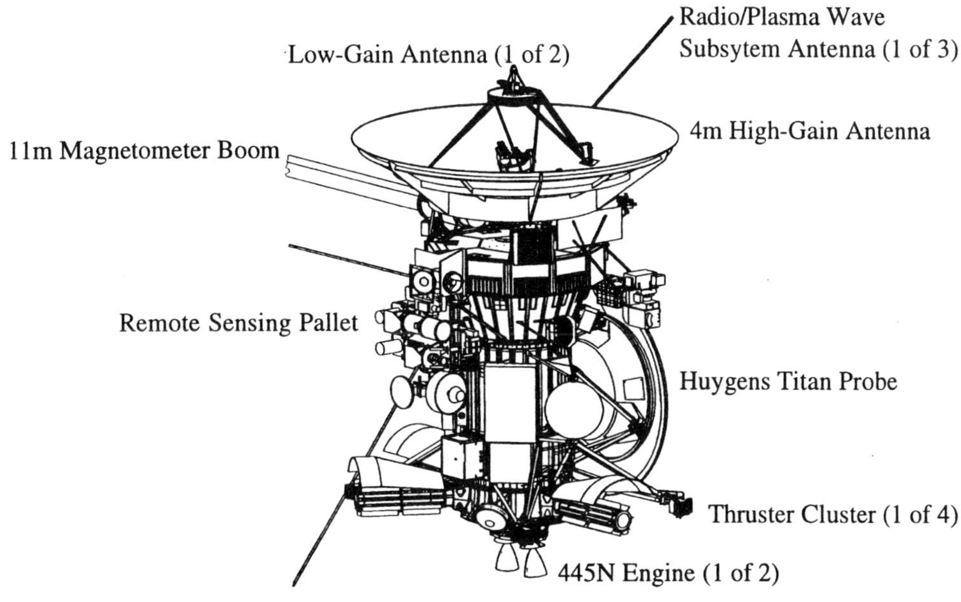

Figure 1. Cassini Spacecraft Configuration (Fully Deployed)

m^2. The magnetometer boom has a fundamental frequency at 0.65 Hz and damping between 0.2 and 1 %. The three radio and plasma wave antennas are much lower in mass and inertia, and have a frequency of 0.13 Hz or higher and a damping of 0.2 %.

At the beginning of the mission, the spacecraft propellant mass is heavier than the spacecraft dry mass. The bipropellant (Monomethyl Hydrazine (MMH) and Nitrogen Tetroxide (NTO)), consumed by the main engine, has a total launch mass of 3000 kg. The two bipropellant tanks are "stacked" along the spacecraft z-axis. Each of the two tanks contains an 8-panel propellant management device to alleviate propellant sloshing. The monopropellant (Hydrazine) launch mass is 132 kg. The monopropellant, consumed by the attitude control thrusters, is stored in a spherical tank off the S/C z-axis.

2. Attitude Control Functions

The primary functions of the Cassini Attitude and Articulation Control Subsystem (AACS) are attitude determination, attitude commanding, attitude control, spacecraft burn maneuvers, constraint violations detection and avoidance, fault protection, command processing, telemetry generation and data handling. The attitude control functional block diagram is shown in Figure 2.

Once the spacecraft attitude with respect to the J2000 celestial coordinate system has been initialized using the sun and the stars, the spacecraft attitude estimator maintains attitude knowledge using star identification and, when necessary, gyro outputs. The AACS software generates the required turn commands upon receiving the uplinked ground commands. The Attitude Controller follows the commanded attitude profile. The attitude profile can command a task as simple as pointing at Earth for downlink or more complicated tasks like surface mapping or probe tracking by the high gain antenna.

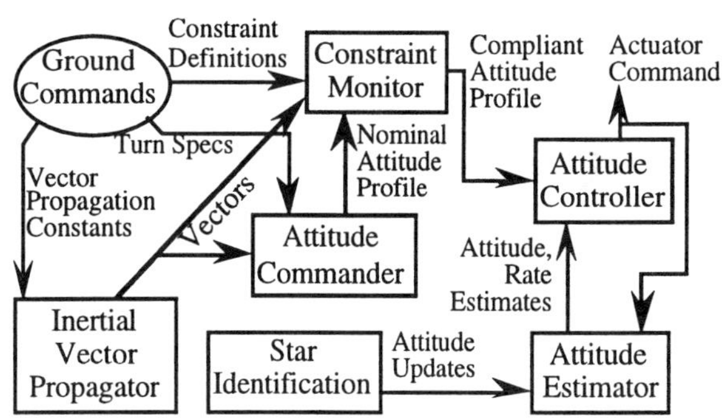

Figure 2. Attitude Control Functional Block Diagram

Large angle slews and coarse pointing are performed using the reaction control system (RCS) thrusters, and precision pointing and slewing for remote science observations by reaction wheels (RWA).

Constraint violations of turn commands are checked and an avoidance path is generated, if necessary, by onboard software prior to execution by the Attitude Controller. Profile commands generated by the Attitude Commander are continuously monitored by Constraint Monitor. By design, the attitude commanding and constraint monitoring tasks are not coupled. The actions of constraint monitor downstream of the nominal attitude profile generation do not effect the nominal attitude profile generation. Ideally, the commander-generated profile should never violate a constraint region if the ground command sequence is generated and checked properly. However, if the ground-commanded profile is in error, or must be abandoned due to a fault, and the attitude profile were to bring an instrument boresight or other sensitive aperture into a constraint region (e.g. an exclusion zone around Sun), the software would generate an alternative path to avoid the excluded region. It is the alternate path, traversed at an acceptable rate and acceleration, which is passed along to the attitude controller. The spacecraft is commanded to follow the alternate path until it can acceptably merge with the attitude commander-generated profile.

This paper considers the design of constraint detection and avoidance aspects of the software. The intent is to give the reader a clear understanding of the two steps involved in the process -- the detection of constraint violations and, once a violation has been detected, the steps involved in the avoidance process. The avoidance step is where an acceptable alternate path is generated.

In the following, underlined capital letters denote matrices, underlined lowercase letters denote vectors (in our usage, an n-vector is an n by 1 matrix), a ()′ implies transposition. The subscript b is used to denote a vector in S/C coordinates (as in $\underline{x}_b$). When no subscript is present, the vector is assumed to be in inertial coordinates. The symbol '×' implies vector cross product. A quaternion multiplication is implied whenever two quaternions appear together. In our usage the first three elements of a quaternion represent the vector component.

3. Constraint Monitor

The Constraint Monitor (CMT) function, which exists between the Attitude Commander (ACM) and the Attitude Controller (ACL) functions, implements part of the AACS fault protection functions. It determines what commanded motions are acceptable, when action should be taken to interrupt the ACM-commanded pointing profile so that an unacceptable motion is not realized, and determines how to re-join with the ACM-commanded motion, should the nominal path become acceptable again. Nominally, the Constraint Monitor should not alter the ACM-commanded motion. It is during a fault response that CMT is more likely to intervene and appropriately alter the ACM-commanded path. The Constraint Monitor function on-board Cassini

- ensures that the attitude commands received by the attitude controllers do not violate a celestial constraint.
- ensures that attitude commands passed to the controllers are smooth in attitude and rate, within tolerances (rate limited) and realizable (acceleration limited).

The two functions performed by the algorithm may be partitioned further as the Detection function and the Avoidance function. The Detection function provides the capability to detect constraint violations for ACM-commanded motions. The algorithm is capable of detecting imminent violations so that action may be taken well in advance. The Avoidance function generates the alternate path which satisfies all constraints when ACM-commanded path has been found to be in violation. The computational needs of the Avoidance algorithm are significantly greater than that of the Detection algorithm. In order to make CMT run in real-time, only the Detection algorithm is exercised every control cycle (125 millisec). The avoidance algorithm on the other hand executes in the background at a slower pace (once every 2 seconds). The end product of the Avoidance algorithm is an acceleration command which is to be applied until the next Avoidance computation 2 seconds later. The attitude and rate commands are needed every 125 millisec, however. The required commands are kinematic extrapolations, carried out every 125 millisec, which use the last commanded attitude and rate, and the latest avoidance acceleration computation.

The nature of all constraints enforced by CMT may be classified as celestial or dynamic. Flight rules restricting the pointing of spacecraft hardware at celestial objects (which can also include direction of incoming particles, e.g. at ring-plane crossing) are translated into celestial constraints enforced by the CMT. Enforcement of these constraints protect certain spacecraft fixed directions from exposure to celestial objects. There are two types of celestial constraints - non-timed and timed. A non-timed constraint may be expressed as follows.

- The angular separation between body vector B and celestial vector C shall never be less than Θ

A timed constraint, which protects certain spacecraft (S/C) axes from *extended* exposure to celestial objects, may be expressed as :

- The angular separation between body vector B and celestial vector C shall never be less than Θ for a time period greater than T. Time is accumulated at the rate of a 1 sec/sec inside the constraint to a maximum of T, and when outside the time is credited back (i.e. decremented) at a rate of R sec/sec to a minimum of zero.

Name	Celestial Object, C	Body Object, B	Half Cone Angle Θ (deg)	Max Allowable Time, T (sec)	Decay Rate R, (sec/sec)	Type	Drop/ Keep
const 1	SUN	B1	30	0	0	AVOID	KEEP
const 4	SUN	B2	30	0	0	AVOID	DROP
const 2	SUN	B4	85	300	0.01	DETECT	KEEP
const 3	SUN	B3	70	75	1.0	OFF	DROP
const 5	INERT 1	B5	30	45	3.0	DETECT	DROP

Table 1. Constraint Table Example

Celestial constraints define a conical region around the celestial vector. The body vector is required to stay out of the cone for non-timed constraints and for timed constraints it is required that the body vector not stay inside the cone for longer than the specified duration. CMT also enforces dynamic constraints to ensure that the acceleration and rates commanded by the ACM are not excessive. These constraints are specified as rate and acceleration ellipsoids in the spacecraft fixed coordinates. The sizes of these ellipsoids can be controlled via ground commands.

The celestial constraints enforced by CMT are defined by ground commands and stored in a table on board (Table 1). The table is of a fixed size (10 entries) and stores all relevant attributes of celestial constraints, which are, the constraint name (unique), the name of the body object, the name of the inertial object, the spatial extent (the half-cone angle of the constraint cone), the temporal extent (timed constraints only - the constraint is treated as a hard constraint when this attribute is zero), and the decay-rate (used in enforcing only the timed constraints). The constraint table need not be full. In addition, the ground commands also specify whether a constraint is to be avoided (avoidance action taken when the ACM-commanded path is about to violate the constraint) or only detected (no avoidance action taken but a violation is reported). A constraint can also be turned OFF (neither DETECT nor AVOID) The celestial and the body vector identifiers (C and B, respectively, in the examples above) and measures Θ, T, and R are commandable parameters. The last attribute of a constraint is the Drop/Keep flag which can have two possible values: DROP or KEEP. This allows CMT to ignore the constraint marked DROP in certain situations (Section 3.2.4). Any attribute of an existing constraint can be modified. Constraints can be removed from this table only by ground commands.

Each constraint is identified by a unique name. Notice that a constraint's inertial and body vectors are also referred to by *name*. The appropriate vector values are obtained from the Inertial Vector Propagator (IVP). Two separate tables in IVP maintain the appropriate values for the inertial and body vectors. The body vector table is not propagated with time but the inertial vector table is. In other words the positions and velocities of the objects stored therein are in general time-dependent. The vector name and epoch are supplied by CMT and the appropriate vector value is returned by IVP.

Once a constraint violation has been detected, the Avoidance function allows the ACM-commanded attitude, rate and acceleration to be modified such that the resulting attitude, rate and acceleration are not in violation of any celestial / dynamic constraints. The Avoidance machinery does not engage unless CMT is currently in Avoidance and / or an AVOIDable celestial constraint violation or a dynamic constraint violation has been detected. When "DETECT only" celestial constraints are violated, the ACM-computed attitude is not altered but such a violation is reported in telemetry. The CMT output is fed directly to the appropriate attitude controller as the attitude commands that the controller must follow.

3.1. Detection Function

This function provides the capability to anticipate AVOIDable celestial constraint and dynamic constraint violations. Two types of constraints are covered: AVOIDable celestial constraints and dynamic constraints. The celestial constraint checks are based on geometry and extrapolated attitude motion (this is an extrapolation based on the current ACM-commanded attitude, rate, and acceleration). The dynamic constraint checks are more "immediate" in nature.

3.1.1. Celestial Constraints

There is clearly the need for early detection of AVOIDable celestial constraint violations. We would like to start taking evasive action early enough so that the constraint cone is not entered for non-timed constraints and not entered for more than the allowed time for timed-constraints. This requires extrapolation of the current path to a point in time which is 2 background cycles (4 seconds) later. This allows sufficient time for the Avoidance algorithm to complete, should a violation be declared imminent. Constraint regions are always avoided by using maximum allowable acceleration in a direction which moves the offending body vector *radially* away from the celestial vector. Imminent violation of a non-timed constraint can therefore be detected early when the predicted (4 seconds later) celestial-body separation angle is less than the sum of the constraint angle and the maximum deceleration *stopping* distance. This stopping distance is the predicted angular motion in the direction of the constraint under the assumption that a radial acceleration is applied away from it until the rate towards the constraint is nulled. A closed-form solution for this distance is not possible. Instead a first order approximation is used. It works very well, especially for the relatively weak acceleration and rate capabilities of Cassini. The cross product of the celestial and the body vector (unit vectors $\underline{c}$ and $\underline{b}$, respectively) provides the direction (henceforth referred to as the "escape" direction) used in the evaluation of the stopping distance. Let the $\underline{\lambda}$ denote a unit vector in this direction.

$$\underline{\lambda} = Unit\,(\underline{c} \times \underline{b}),$$

where *Unit* is the normalization operator and predicted future values of $\underline{c}$ and $\underline{b}$ are used in this evaluation. The Avoidance function applies maximum available acceleration in the $\underline{\lambda}$ direction to move the body vector away from the constraint cone when a celestial constraint violation has occurred or is imminent (it is not necessarily time-optimal; it would be, if all axes had the same acceleration capability). Computation of the stopping distance requires the relative angular rate between the celestial and the body vectors (let the two rates be $\underline{v}$ and $\underline{\omega}$, respectively). The scalar rate which we are trying to annul by rotating about the escape direction, should the body vector get too-close to the constraint edge, is the projection of the differential rate onto the escape direction.

$$\omega = (\underline{v} - \underline{\omega})'\,\underline{\lambda}.$$

If the body vector is moving away from the celestial vector ($\omega < 0$) then there can not be an imminent violation. Consequently the stopping distance is set to zero, else it is computed as the angular distance moved such that the application of maximum deceleration about $\underline{\lambda}$ annuls the projected rate ω. The prediction requires the use of maximum available acceleration along the escape axis. The use of largest available acceleration might be optimistic since, while trying to escape, the escape direction may change as the offending body vector is pushed away from the celestial constraint it is trying to avoid. To avoid this optimism (and therefore future violations) and err on the conservative side, the smallest available acceleration (α) is used <u>only</u> in predicting violations. The exact rate and acceleration constraints are used in constraint avoidance. The stopping distance δ is therefore evaluated as

$$\delta = 0.5\,\omega^2/\alpha, \qquad \text{if } \omega > 0.$$

A violation is declared imminent if the predicted future c-b separation

$$\theta = Cos^{-1}(\underline{c}'\,\underline{b}) < \delta + \Theta.$$

To detect imminent violations of timed constraints, the additional time that a body vector can spend inside the timed constraint is also required. Note that it is legal to enter the constraint space here but not for more than the specified time. The time accumulation t is updated as follows. If the angle θ is smaller than the constraint angle Θ, implying that the body vector is predicted to be inside the constraint cone, the time must be accumulated. If the body vector is outside the constraint cone, the accumulated time must be reduced at the rate R specified in the constraint definition. In other words

$$\begin{aligned} t &= t + \Delta, && \text{if } \Theta > \theta, \\ &= t - R\,\Delta, && \text{if } \Theta < \theta, \end{aligned}$$

where Δ is the time elapsed since the last evaluation. This accumulation is bounded by 0 from below and Δ from above. At each cycle, the following question is asked. If an acceleration in the escape direction is applied, would the constraint be entered as the S/C decelerates to a zero rate in the escape direction. If not then there can not be a violation. If the constraint is predicted to be entered but can be exited in the time remaining (T - t) using the allowable rate and acceleration then there can not be a violation imminent, otherwise an imminent violation is declared and Avoidance machinery is invoked.

3.1.2. Dynamic Constraints

The dynamic constraints checks are relatively simple in nature. The rate and acceleration constraints are specified as S/C fixed ellipsoids. The ACM-commanded rate and acceleration commands must never lie outside their respective ellipsoids. A dynamic constraint violation is declared when the commands are outside, e.g. let $\{\alpha_x, \alpha_y, \alpha_z\}'$ be the acceleration command and (A_x, A_y, A_z) be the semi-major axes of the acceleration constraint ellipsoid, then the acceleration command is outside the acceleration ellipsoid (a violation) if

$$\{\alpha_x/A_x\}^2 + \{\alpha_y/A_y\}^2 + \{\alpha_z/A_z\}^2 > 1.$$

To detect imminent rate violations, the rate command at the next evaluation time is also estimated (assuming constant current acceleration) and verified against the rate ellipsoid. A dynamic violation is declared whenever the acceleration vector is outside the acceleration ellipsoid, or the current rate command or the predicted rate command is outside the rate constraint ellipsoid. There are two types of attitude controllers on-board, one which uses reaction wheels and the other which uses thrusters. The maximum rate and acceleration capabilities are different for each. Although the rate and acceleration constraints imposed by the algorithm are control-law specific, the fundamental behavior of the algorithm does not change when we transition from thrusters to reaction-wheels.

3.2. Avoidance Function

The Avoidance function is invoked when a violation is declared imminent by the Detection function. CMT stops calling this function only when ACM commands are in compliance of all constraints and CMT-propagated motion is close to the ACM-commanded motion. The purpose of this function is to modify the ACM-commanded attitude, rate, and acceleration commands such that the eventual attitude, rate, and acceleration commands (i.e. the commands relayed to the Attitude Controller) are in conformance with all constraints. The end product of the Avoidance evaluations is an acceptable acceleration command which will prevent the attitude from violating celestial constraints and the rate and acceleration commands from exceeding the hardware limits. This function executes at a slower pace (once every 16 Detection cycles). On initial entry into avoidance, the acceleration command is set to zero until the function returns with the correct acceleration command 2 seconds later. This delay is accounted for in the avoidance acceleration evaluation.

The corrective action taken is designed to be such that the constraint in question is not violated eventually. The corrective action consists of three steps, which may be repeated several times until the CMT-commanded motion merges with the ACM-commanded motion. The end goal is to always try to move towards a point in space which is not violating any constraints and is close to the ACM commanded path. When the ACM command is not in violation, the goal is of course the ACM command itself. The first action taken when a celestial constraint violation is declared imminent is the so-called Escape where an acceleration is commanded such that it moves the offending body vector radially away from the inertial vector. While in this mode, maximum available acceleration along the escape axis is used until the constraints are no longer in danger of imminent violations. When several constraints are to be "escaped" simultaneously, the appropriate escape directions are combined linearly to obtain the resultant escape direction. This action is maintained until the constraints in question are no longer in imminent violation. If at this point it is possible to move along a great circle towards the goal attitude and not violate any constraint in the immediate future (next 2 seconds), a Clear mode is declared where the motion is a great circle arc heading towards the goal. When this is not possible, the constraint or a collection of constraints have to be circum-navigated until it is possible to transition to the Clear mode. This intermediate mode is called the Circulate mode. While in this mode the CMT attitude is such that all body vectors remain outside their respective celestial constraints. The timed-constraints are allowed to be penetrated (under certain conditions) but non-timed constraints are generally followed at a fixed distance. This behavior is referred to as "...following the flow-field". While in Circulate mode, the attitude is continually checked for imminent violations (transition back to Escape) or possibility of movement towards the goal attitude (transition to Clear). The rate and acceleration constraints are never violated. The goal attitude that the CMT attitude is always striving for is such that it is in compliance of all AVOIDable celestial constraints. When there are no constraint violations, the attitude and rate goals coincide with the ACM commands. When one or more celestial constraint violations are present, the goal attitude is such that all celestial constraints (which are to be AVOIDed) are satisfied. The rate goal then is zero rate. This has been assumed since the attitude goal can exhibit rapid movement which may be discontinuous at times. The discontinuities are an artifact of trying to compute the attitude goal in a finite number of steps. It makes no sense therefore to compute a rate goal consistent with the variation in the attitude goal. Only the violated constraints are considered in determining the attitude goal. The attitude goal is not "close" to the ACM command necessarily but it does satisfy all violated constraints (only in rare cases have we found the algorithm to yield a non-compliant attitude: rapidly moving celestial constraints are one example). There may also be situations where no solution exists for a goal attitude. The algorithm is not interrupted in such an event since the rest of the CMT machinery would not allow the CMT attitude to acquire a non-compliant attitude.

Several cycles of Escape - Circulate - Clear (not necessarily in this order) may be required until the goal attitude is reached. Escape and Circulate are entered only when celestial constraint violations are imminent. When the CMT attitude is *far* from the goal it is trying to attain, largest possible rate is commanded until it is in the vicinity of it where a rate proportional to the distance from the goal is commanded.

3.2.1. Flow-Fields

It was noted earlier that while in the Circulate mode, the constraints have to be circum-navigated until a transition to either the Escape or Clear mode occurs. Computation of the direction in which to move around constraints is not straightforward and requires visualization of constraint cones in the Rodriguez parameter space (if $\underline{q}$ is the inertial-to-body attitude quaternion, then $\varepsilon(i) = q(i)/q(4)$, i = 1, 2, 3 are the Rodriguez parameters) - referred henceforth as the Q-space. The Rodriguez parameters form the basis of the Q-space. The constraint cones have the appearance of hyperboloids of one-sheet in the Q-space. The Q-space is obviously not finite but infinity here simply implies a π rotation. Attitude is a point in this space and the evolution of $\underline{\varepsilon}$, the "attitude", is required to stay outside a collection of hyperboloids.

In general the attitude $\underline{\varepsilon}$ is in compliance (i.e. outside the constraint) when the following quadratic form is negative, i.e.

$$\underline{\varepsilon}' \, \underline{U} \, \underline{\varepsilon} + 2 \, \underline{v}' \, \underline{\varepsilon} + w < 0 \quad \Leftrightarrow \quad \Theta < \theta,$$

where the matrix $\underline{U}$, vector $\underline{v}$, and scalar w are dependent on celestial and body vectors and the constraint angle. Let $\underline{c}$ be the constraint celestial vector in inertial coordinates, $\underline{b}_b$ be the constraint body vector *in S/C coordinates* and Θ be the constraint angle. Then it can be shown that

$$\underline{U} = \underline{c} \, \underline{b}_b' + \underline{b}_b \, \underline{c}' - \{\underline{c}' \, \underline{b}_b + Cos(\Theta)\} \, \underline{I},$$

$$\underline{v} = \underline{b}_b \times \underline{c},$$

$$w = \underline{c}' \, \underline{b}_b - Cos(\Theta),$$

where $\underline{I}$ is a 3 x 3 identity matrix. Attitude has to "flow" a certain way around a collection of such surfaces as $\underline{\varepsilon}$ makes its way towards the goal attitude. The direction ($\underline{c}+\underline{b}_b$) forms the symmetry axis of the constraint hyperboloid. It makes sense to pick a direction such that the flow around the constraints moves the attitude in positive sense about the ($\underline{c}+\underline{b}_b$) axis. This argument sets the flow-field direction in a relative sense. The correct orientation (i.e. one way or the other) is determined when the first imminent celestial constraint violation is declared.

3.2.2. The Goal Attitude

When CMT is in avoidance the attitude does not wander aimlessly, it is actually trying to move towards an attitude goal as it negotiates the constraints between the current attitude and the attitude goal. When the ACM-commanded attitude is in compliance with all AVOIDable celestial constraints, the goal attitude and rate are the ACM-commanded values. The goal attitude and rate are not close, necessarily, to the ACM attitude (it is, in single constraint situations). Rather, the algorithm provides an attitude which is in compliance with all AVOIDable celestial constraints. The solution process is iterative but the algorithm is exited after a fixed number of iterations, regardless of whether a goal attitude has been found or not. As a consequence of the fixed number of iterations, the alternate attitude can exhibit discontinuities.

The algorithm starts by accepting the ACM attitude and rate as the starting point for the goal-attitude iterations. At each instance of an AVOIDable celestial constraint violation (by the candidate goal attitude that is), the candidate goal attitude is updated such that the offending body vector (at the modified goal) lies slightly outside the edge of the constraint cone. The movement of the offending body vector is accomplished by rotating the body vector about the escape axis $\underline{\lambda}$. Two solutions are possible: either move the short distance about the positive escape direction or the long distance about the negative escape axis. The path taken (short / long) is such that the movement of one body vector will not push a previously adjusted body vector back inside it's constraint. The iterations to compute a goal attitude are repeated until there are no violations during an iteration or the iteration count exceeds the number of constraints under consideration (whichever comes first). The goal rate is zero whenever the goal attitude is not the ACM-commanded attitude.

3.2.3. Desired Direction of Motion

Once a goal attitude is available, the direction in which CMT should try to head towards is needed. The direction is a function of the current state, the goal state (attitude and rate), and the location of the constraints which are to be avoided. The desired direction of motion is prescribed in terms of a rate vector which CMT should try to attain by the end of next Avoidance cycle except in Escape mode, where the desired direction is actually a direction in which to accelerate in.

When in the Escape mode, the desired direction of travel is the weighted sum of all applicable escape directions and a maximum possible acceleration is commanded in this direction until either the constraints are no longer in imminent violation or the rate limit is reached.

When the constraints have been exited (i.e. an Escape is no longer needed), a transition to either the Circulate or the Clear mode takes place. These modes were described earlier. A Clear mode is trasitioned to if it is possible to move towards the goal attitude through a single-axis rotation and not be in imminent violation of constraints which may lie between the current attitude and the goal.

When far from the goal, maximum allowable rate is prescribed. In the vicinity of the goal, the prescribed rate ($\underline{\omega}^p{}_b$) becomes a linear function of the separation from the goal.

$$\underline{\omega}^p{}_b = \underline{\omega}^g{}_b + K\, \underline{q}^g\, \underline{q}^*, \qquad (1)$$

where $\underline{\omega}^g{}_b$ is the goal rate in S/C coordinates $\underline{q}^g$ is the goal attitude quaternion and $\underline{q}$ the current CMT-commanded attitude quaternion. The ()* is the conjugation operation. Note that only the vector part of the quaternion product $\underline{q}^g\, \underline{q}^*$ is used in (1). The gain K is such that stability of motion is guaranteed as CMT commanded attitude closes in on the goal attitude (more on this later) and the approach transients are acceptable.

When a transition from Escape to Clear mode is not possible it is only because a constraint violation may be *immediately* imminent had CMT tried to move towards the goal. A transition to Circulate mode is then declared until it is possible to move towards the goal (Clear mode). A desired rate vector has to be computed again. The desired rate magnitude is the largest possible when far from the goal, else the magnitude is the norm of $\underline{\omega}^p{}_b$ (1). The direction of the desired rate vector is a weighted sum of the so-called "circulation vectors". The weights bear an inverse square relationship to the distance from the edge of the constraints. Each constraint has a unique circulation vector which is defined such that it yields a minimum twist as the body vector follows the constraint boundary. Let $\underline{c}_b$ and $\underline{b}$ be the celestial and body vectors in the S/C coordinates now. Consider a perturbed attitude which is reached by following a sequence of two rotations: a rotation ψ about $\underline{c}_b$, followed by another rotation $-\psi$ about $\underline{b}$. It can be shown that the sequence yields a minimum twist offset between the starting and the perturbed attitude when ψ is small. The attitude perturbation (quaternion) can be shown to be:

$$\underline{q}^p = [\, (\underline{c}_b\text{-}\underline{b})\, s\, c + s^2(\underline{b}\times\underline{c}_b),\ c^2+s^2(\underline{c}_b{}'\,\underline{b})],$$

where $s = Sin(\psi/2)$, $c = Cos(\psi/2)$. Clearly, as ψ becomes small, the perturbation approaches:

$$\underline{q}^p = [\, (\underline{c}_b\text{-}\underline{b})\, \psi/2,\ 1],$$

which implies a rotation about the direction $\underline{c}_b$-$\underline{b}$ is S/C coordinates. This direction is the required circulation vector.

3.2.4. Avoidance Acceleration

Once the required rate vector has been computed in the previous step, all that is left to do is to compute the required acceleration command. Evaluations are different depending on the avoidance mode (Escape/Circulate/Clear) CMT happens to be in. An Escape mode implies that a constraint has been entered or is about to be entered and the commanded acceleration must move the violating body vector radially away from the celestial vector. The acceleration direction in this case is the sum of all applicable escape directions ($\underline{\lambda}s$). Largest possible acceleration is applied in this direction until the mode can be transitioned to Circulate or Clear or maximum rate is reached.

For non-Escape situations, the desired rate vector (1) is used to compute the acceleration command:

$$\underline{\alpha}^p{}_b = (\, \underline{\omega}^p{}_b - \underline{\omega}_b\,)\, /\, \Delta, \qquad (2)$$

where Δ is the avoidance cycle (2 seconds in this case), $\underline{\omega}^p{}_b$ is the prescribed rate (1) and $\underline{\omega}_b$ is the current CMT-commanded rate in S/C coordinates. It can be shown that, in the proximity of the goal attitude, the approach law (1), when used in (2) to evaluate the desired acceleration, results in a stable behavior when

$$0 < K\Delta < 4, \quad (\text{Cassini: } K = 0.05,\ \Delta = 2)$$

It is entirely possible that the acceleration suggested by (2) violates the CMT acceleration constraints, i.e. lies outside the acceleration ellipsoid. The direction of $\underline{\omega}^p{}_b$ must be preserved if possible. A geometric solution is possible sometimes[2] which allows the length of the $\underline{\omega}^p{}_b$ to be shortened such that $\underline{\alpha}^p{}_b$ lies exactly on the acceleration ellipsoid. When this is not possible the length of $\underline{\alpha}^p{}_b$ is shortened such that the acceleration command lies on the acceleration ellipsoid[2].

3.2.5. Avoidance End

Avoidance is declared over only when the ACM-commanded path is not in imminent violation and the CMT-commanded path (in attitude and rate) is *close* to the ACM-commanded path. The maximum rate and acceleration limits enforced by CMT are used to quantify *closeness* in this case.

Since sequences are checked rigorously on the ground before uplink, likelihood of a constraint violation during normal operations is very slight. Although chances are slight, we expect a dynamic constraint violation to be more likely to occur than a celestial constraint violation. It is not hard to construct impossible situations for CMT to handle. Pointing constraints can be constructed such that the CMT-commanded attitude never attains the CMT-computed attitude goal. Protection against such occurrences is provided by another attribute of the constraints (the DROP/KEEP column in Table 1). When a situation such as the one described here arises (i.e. away from the goal attitude for too-long) all non-Sun based constraints (i.e. constraints for which the inertial vector is not the S/C to Sun direction) and all Sun-based constraints whose DROP/KEEP attribute is DROP are marked

DETECT only (i.e. the type is changed to DETECT) by CMT. No ground intervention is needed in this step. The remaining constraints which have to be negotiated are therefore all Sun-based and it can be shown that CMT-commanded attitude should always be able to merge with the CMT-computed goal in this case.

4. Examples

The actions of the algorithm are demonstrated below via some examples. First consider a simple case where only one constraint is to be enforced. The constraint region and the path taken to avoid it are shown in Figure 3. The constraint on the left is a hard constraint and the one on the right a timed one which is allowed to be entered but for not more than 40 seconds. The constraint is defined as a cone of 30 degrees about the inertial Z-axis. The body vector to be kept out of it is the spacecraft x-axis. First consider a case where ACM commands a path which moves the spacecraft x-axis, initially outside, through the constraint to a final point outside the constraint. The two avoidance examples for this ACM command are shown in the top two sub-plots in Figure 3. The intersections of the spacecraft x-axis

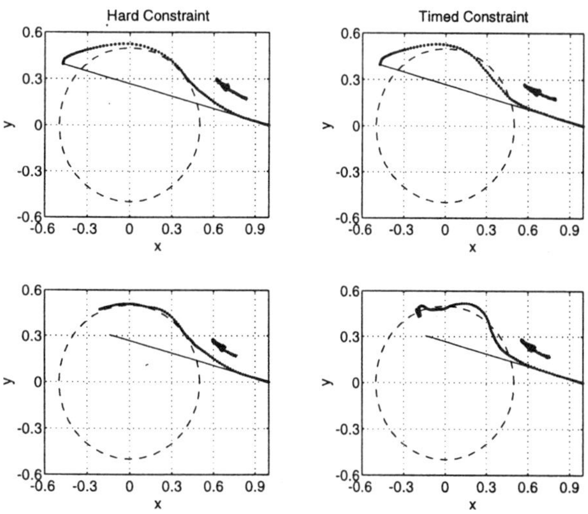

Figure 3. Single Constraint Avoidance

with the unit celestial sphere projected onto the inertial XY plane are shown here. The constraint edge is a dashed circle centered at (0,0) in this plane. Inside of this circle is the forbidden area. The ACM-commanded x-axis path goes from right to left and is shown as the thin solid line and the CMT action appears as a dot plotted every 2 seconds. The hard constraint is never entered and CMT rejoins with ACM after an Escape-Circulate-Clear sequence. The time constraint (top right) is indeed entered but for only 36 seconds (it is exited with 4 seconds to spare). It also rejoins with the ACM point later on.

Next consider the same example as above except that this time ACM commands a motion to leave the x axis at rest inside the constraint. In this case (the bottom plots), CMT seeks the goal attitude, just outside the constraint but close to the ACM point, and comes to rest there. Again, the hard constraint is never entered but the timed constraint is entered but exited in time (0.875 seconds to spare). Note that repeated attempts are made by CMT to approach the ACM commanded location inside the timed-constraint. This is of course perfectly legal as long as the spacecraft x-axis exits the timed-constraint in time.

Consider a more complicated example next, where three constraints have to be avoided. The constraint table is as shown in Table 2.

Name	C	B	Θ	T	R	Type	Dp/Kp
A	C1	B1	30°	0	0	AVOID	KEEP
B	C2	B1	45°	0	0	AVOID	KEEP
C	C3	B2	15°	50s	1.2	AVOID	KEEP

Table 2. Multiple Constraints Example - I

C1	0.0000	0.0000	1.0000
C2	0.4332	-0.2512	0.8661
C3	0.0000	-0.9662	0.2588
B1	-0.2418	0.6645	0.7071
B2	0.1228	0.6964	0.7071

Table 3. Vector Components for Constraints in Table 2

The constraint cone projections and projections of various axes on the unit J2000 celestial sphere are shown in Figure 4. Constraint regions are regions

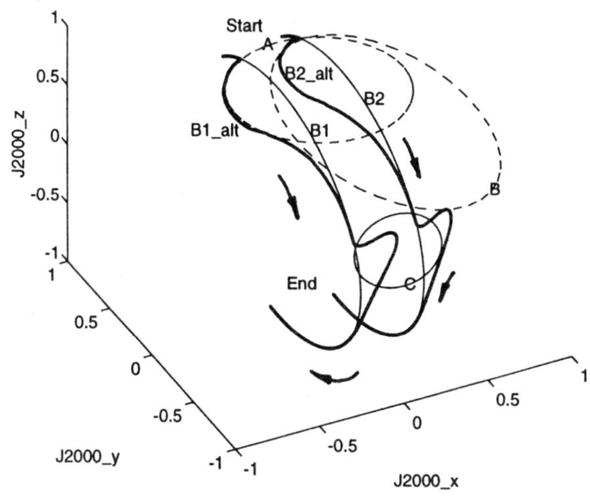

Figure 4. Multiple Constraint Avoidance (Table 2)

inside the dashed circles labeled A, B, C representing the three constraints A, B, C, respectively. The ACM-commanded paths, shown as thin arcs, begin at the top. The B1 path cuts across all three constraints but it is required to stay out of constraints A and B. B2 path also cuts across all three constraints but it is required to not remain in C for more than 50 seconds. The avoidance paths taken by vectors B1 and B2, shown as the thick line, are labeled B1_alt and B2_alt. The algorithm does its job by keeping B1 within 0.3 mrad and 18.3 mrad of violating constraints A and B respectively. Body vector B2 remained inside constraint C for 47 seconds, it exited with 3 seconds to spare. The time histories of the angular separation between the nominal ACM-commanded attitude (which is in violation some of the time) and the alternate path suggested by the CMT-commanded attitude (i.e. the separation between the thick and thin paths in Figure 4) and the CMT-mode are shown below in Figure 5. In this case CMT-commanded path merges with the ACM-commanded path at the 750 sec mark.

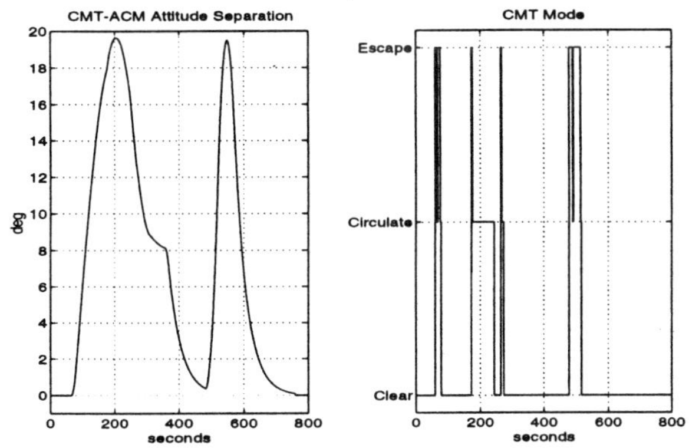

Figure 5. Attitude Separation and CMT Modes

As the final example consider a 2-constraint case where a complicated slew profile is commanded by ACM. The constraints in this case are as listed in Tables 4-6. Two variants are considered. In the first (Table 4), both constraints A and B are non-timed, and in the second (Table 6), Constraint B is allowed to be entered but not for more than 110 seconds. The two constraints share the same inertial vector.

Name	C	B	Θ	T	R	Type	Dp/Kp
A	C1	B1	90°	0	0	AVOID	KEEP
B	C1	B2	110°	0	0	AVOID	KEEP

Table 4. Multiple Constraints Example - II

C1	0.0000	0.0000	1.0000
B1	0.1710	0.0030	0.9848
B2	-0.3830	-0.3214	0.8660

Table 5. Vector Components for Tables 4, 6

The motion of the inertial vector C1 in S/C x-z plane is depicted (solid line) in Figure 6. The dashed lines in this figure denote the constraint boundaries (the edges closest to the inertial vector are marked A and B, respectively, for the two constraints A and B). The ACM-commanded path resembles a mosaic during which the inertial vector position is at the lower right corner (outside both constraints) at the start, it then zig-zags in and out of the constraint space, eventually returning back to the starting position.

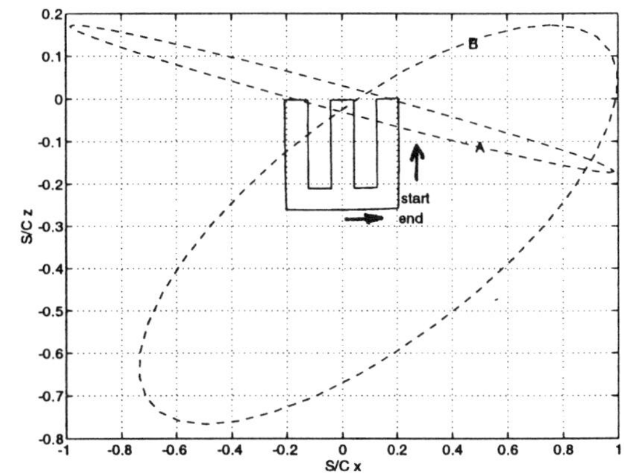

Figure 6. Constraints and the Commanded Path

The avoidance motion and the related time histories for this case are shown in the subplots on the left in Figure 7. The top figure depicts the constraint inertial vector path commanded by CMT in S/C x-z plane (dark line) superimposed on Figure 6. An 'x' and 'o' have been placed at 50 second intervals, respectively, along the ACM- and CMT-commanded paths. The minimum distance from the edges of the constraints in this case was 1 and 2.6 mrad, respectively. The time history of the angular separation between the ACM-commanded path and the CMT-commanded path is shown in the middle and the progression of CMT-mode appears at the bottom. ACM-commanded profile is rejoined soon after the 800 sec mark. As another variation of the example above, consider the following constraint table. It is exactly as in Table 4, except that Constraint B is now a

Name	C	B	Θ	T	R	Type	Dp/Kp
A	C1	B1	90°	0	0	AVOID	KEEP
B	C1	B2	110°	110s	1	AVOID	KEEP

Table 6. Multiple Constraints Example - III

timed-constraint allowed to be ventured into for not more than 110 seconds. The related motions and time histories appear as the subplots on the right in Figure 7. Note that Constraint B is entered this time but it is exited with 1.25 seconds to spare. ACM-commanded path remains inside Constraint B for about 120 seconds.

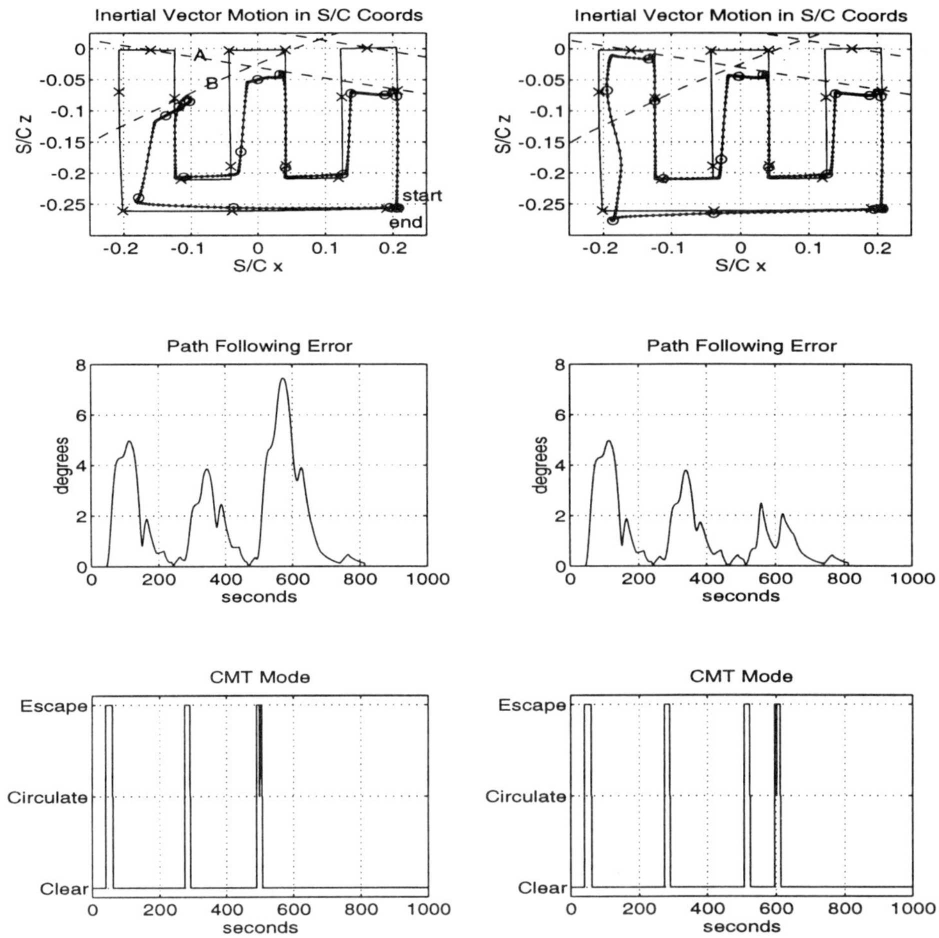

Figure 7. Multiple-Constraint Avoidance Example II

5. Summary

This paper presented the design of an autonomous avoidance algorithm. The algorithm is slated to fly on-board the Cassini spacecraft in October 1997. The algorithm does more than simply avoid pointing constraints. It prevents excessive rate and attitude commands from being sent to the attitude controllers. To our knowledge, autonomous constraint avoidance capability of the kind described here has never been designed or flown on an inter-planetary spacecraft. The algorithm is unique in the sense that it reacts to current S/C attitude relative to the constraints. No action is taken unless it is almost absolutely necessary to do so. The algorithm has been rigorously tested on many testbeds used to validate the Cassini Attitude Control Flight Software and it's performance has been very satisfying indeed.

6. Acronyms

S/C	Spacecraft
AACS	Attitude and Articulation Control Subsystem
ACM	Attitude Commander
ACL	Attitude Controller
CMT	Constraint Monitor
IVP	Inertial Vector Propagator

7. Acknowledgments

The work described in this paper was carried out by the Jet Propulsion Laboratory, California Institute of Technology, under a contract with the National Aeronautics and Space Administration.

8. References

[1] Rasmussen, R.D., Singh, G., Rathbun, D.R., and Macala, G.A., "Behavioral Model Pointing On Cassini Using Target Vectors", 18th Annual AAS Guidance and Control Conference, Keystone, Colorado, February 1-5, 1995

[2] Singh, G. and Macala, G.A., "Cassini Constraint Monitor Algorithm", IOM 3456-95-071, 21 December 1995

AIAA-97-3527

ENHANCED ATTITUDE CONTROL EXPERIMENT FOR SSTI LEWIS SPACECRAFT

Peiman G. Maghami

NASA Langley Research Center
Mail Stop 161
Hampton, Virginia 23681–0001

Abstract

The enhanced attitude control system experiment is a technology demonstration experiment on the NASA's small spacecraft technology initiative program's Lewis spacecraft to evaluate advanced attitude control strategies. The purpose of the enhanced attitude control system experiment is to evaluate the feasibility of designing and implementing robust multi-input/multi-output attitude control strategies for enhanced pointing performance of spacecraft to improve the quality of the measurements of the science instruments. Different control design strategies based on modern and robust control theories are being considered for the enhanced attitude control system experiment. This paper describes the experiment as well as the design and synthesis of a mixed H_2/H_∞ controller for attitude control. The control synthesis uses a nonlinear programming technique to tune the controller parameters and impose robustness and performance constraints. Simulations are carried out to demonstrate the feasibility of the proposed attitude control design strategy.

Introduction

The current practice in spacecraft attitude control design is based on a single-input/single-output (SISO) control strategy, wherein elementary and low bandwidth controllers are designed for each of the three-axes of the spacecraft separately. Typically, SISO attitude controllers are designed to be low bandwidth (gain stabilized) to avoid uncertain, and possibly harmful, interactions with the flexible modes of the spacecraft, such as those of solar arrays or antennas [1–2]. However, these controllers have limited performance because of their elementary structure and low bandwidth. Moreover, they are fairly hard to modify in their flight software implementation form because they are typically coded to follow a specific controller order and type. The enhanced attitude control system (EACS) experiment is a technology demonstration experiment on the NASA's small spacecraft technology initiative (SSTI) program's Lewis spacecraft to evaluate advanced attitude control strategies. The purpose of the EACS experiment is to evaluate the feasibility of designing and implementing robust multi-input/multi-output (MIMO) attitude control strategies for enhanced pointing performance of spacecraft to improve the quality of the measurements of the science instruments. This experiment is limited to controller designs for attitude control in the normal science mode which involves fine attitude pointing. The MIMO control designs would utilize robust and modern control theories to synthesize attitude control designs which can take better advantage of the control system hardware to provide higher authority controllers, i.e., phase or gain stabilized. Moreover, once an algorithm is incorporated on the spacecraft to implement a MIMO attitude control system (ACS) design, it would be very easy to replace controllers. The specific objectives of the experiment are as follows:

1. To develop a MIMO ACS algorithm and flight software, and implement this software within the SSTI/Lewis flight on-board computer software
2. To develop MIMO attitude control designs, based on robust and modern control theory
3. To conduct attitude control experiments by implementing the MIMO control designs (instead of the baseline normal mode controller) in the normal pointing mode
4. To evaluate the performance of MIMO attitude control designs by analyzing the telemetry data

An efficient algorithm, both in time and memory requirements, for the implementation of MIMO controllers has been developed. With this algorithm, the required memory and the number of floating point operations are linear functions of the number of states in the MIMO controller. The implementation software has been developed and incorporated within the on-board flight computer software. The enhanced attitude control system is implemented as an independent module within the ACS software. It should be noted that the various MIMO control designs are implemented by simply uploading the controller data sets into the MIMO routine, which also indicates the ease of replacing the controller with a new MIMO control design. Several synthesis techniques are being considered for the design of MIMO controllers for the EACS experiment. These include, H_2–based, H_∞–based, μ–based designs, as well as others. One of the techniques considered is a mixed H_2/H_∞ control synthesis approach, which is described in this paper. In this control synthesis approach, a nonlinear programming technique is used to tune the controller parameters of an H_2–based controller while imposing H_∞–based robustness constraints, as well as, additional performance constraints. The MIMO controllers are applied to a model of the Lewis spacecraft, and simulations are carried out to demonstrate the feasibility of the proposed attitude control design strategy.

Spacecraft Modeling

For the purpose of dynamics and control design and analysis, the task of modeling a spacecraft can be divided into the modeling of the spacecraft structure itself, and the modeling of the actuator components, sensor components, and the controller, all of which comprise the attitude control system.

Spacecraft structure modeling

Typically, the spacecraft structure can be modeled as a linear, time-invariant flexible system, which in turn can be represented by the following second-order dynamical equations:

$$M\ddot{x} + D\dot{x} + Kx = Bu + Hv \tag{1}$$

together with some set of measurement and performance output equations:

$$\begin{aligned} y_d &= C_{md}x, \quad y_v = C_{mv}\dot{x} \\ y_{pd} &= C_{pd}x, \quad y_{pv} = C_{pv}\dot{x} \end{aligned} \tag{2}$$

where M is the positive definite mass matrix; D is the positive semidefinite damping matrix; K is the positive semidefinite stiffness matrix; B is the input influence matrix; H is the disturbance input influence matrix; C_{md} and C_{mv} are the attitude and attitude rate measurement output influence matrices, respectively; C_{pd} and C_{pv} are attitude and attitude rate performance output influence matrices, respectively; x is a k x 1 vector of displacements; u is a m x 1 vector of inputs to the system; v is a e x 1 vector of disturbances to the system; y_d and y_v are the attitude and attitude rate measurement output vectors, respectively; and y_{pd} and y_{pv} are the attitude and attitude rate performance output vectors, respectively. Usually, a finite element analysis is used to obtain these matrices. In most cases, the number of displacements, k, is quite large and thus impractical to work with for general design and analysis purposes. To make the problem more tractable, the displacements vector x is transformed into modal coordinates using the transformation $x = \Phi r$, with r being a n x 1 vector of modal amplitudes and n << k. Here, only n significant modes are retained. The transformation matrix Φ contains n columns which are the eigenvectors associated with the n modes of interest of the flexible system. The equations for the system, in transformed coordinates, are:

$$\begin{aligned} M_r\ddot{r} + D_r\dot{r} + K_r r &= \Phi^T Bu + \Phi^T Hv \\ y_d = C_{md}\Phi r, \; & y_v = C_{mv}\Phi\dot{r} \\ y_{pd} = C_{pd}\Phi r, \; & y_{pv} = C_{pv}\Phi\dot{r} \end{aligned} \tag{3}$$

If normal modes are used, and their mode shapes have been normalized with respect to the mass matrix, and modal damping is assumed, then $M_r = I_{nxn}$, $K_r = diag(\omega_1^2, \omega_2^2, ..., \omega_n^2)$ and $D_r = diag(2\zeta_1\omega_1, 2\zeta_2\omega_2, ..., 2\zeta_n\omega_n)$, with w_i and ζ_i being the open-loop natural frequency and damping ratio values, respectively, for the ith mode.

By defining a new vector $x_s = [r^T \;\; \dot{r}^T]^T$, the second-order equations in eq. (3) can be rewritten into first-order form as:

$$\begin{aligned} \dot{x}_s &= A_s x_s + B_s u + B_d v \\ A_s &= \begin{bmatrix} 0 & I_{nxn} \\ -K_r & -D_r \end{bmatrix}; \\ B_s &= \begin{bmatrix} 0 \\ \Phi^T B \end{bmatrix}; \; B_d = \begin{bmatrix} 0 \\ \Phi^T H \end{bmatrix} \end{aligned} \tag{4}$$

The measurement and performance output equations can then be written as:

$$\begin{aligned} y &= \begin{bmatrix} y_d \\ y_v \end{bmatrix} = Cx_s \\ y_p &= \begin{bmatrix} y_{pd} \\ y_{pv} \end{bmatrix} = C_p x_s \\ C &= \begin{bmatrix} C_{md}\Phi & 0 \\ 0 & C_{mv}\Phi \end{bmatrix}; \\ C_p &= \begin{bmatrix} C_{pd}\Phi & 0 \\ 0 & C_{pv}\Phi \end{bmatrix}; \end{aligned} \tag{5}$$

Here, x_s is the plant state vector; A_s is the plant state matrix; B_s is the control input influence matrix; C and C_p denote the measurement output and performance output influence matrices, respectively.

Attitude Control System Modeling

In addition to modeling the spacecraft structure, or plant, the various components of the attitude control system (ACS) should be included to complete the spacecraft model. These components are the sensors, actuators and computer/data acquisition systems required to control the spacecraft. By including these components, significant real world effects such as actuator and sensor dynamics, noise, digital quantization and sampling time delays, can be accounted for in the overall analysis model. A typical ACS, depicted in block diagram form with the plant in the loop, is shown in Figure 1.

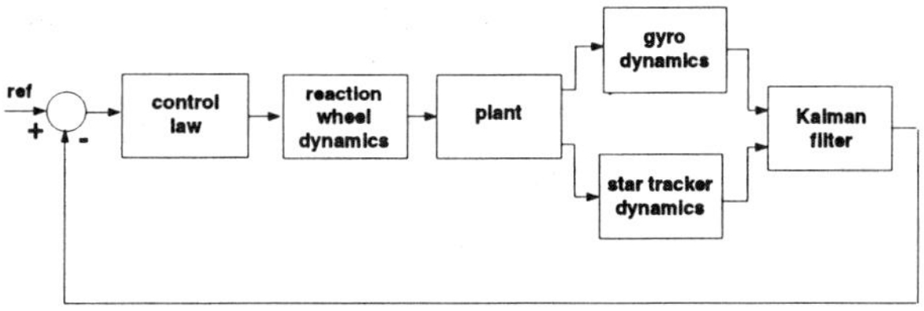

Figure 1. Typical ACS model components.

This block diagram represents a typical ACS simulation model for a spacecraft. The ACS consists of reaction wheels to provide the attitude maneuvering and control torques, with rate gyros and a star tracker, as sensors for measuring the spacecraft attitudes and attitude rates. The Kalman filter is used to estimate the vehicle's attitude from the sensor data, and the control law uses these estimated values, together with the attitude rate measurements from the gyros, to compute the appropriate torque commands. The reaction wheel dynamics may include nonlinear effects, like friction and stiction, limits on the input command voltages and digital voltage quantization, as well as the quantization effects on wheel RPM outputs due to the wheel's optical encoder. Each rate gyro senses the spacecraft attitude rates in one axis, and is modeled as a second-order system. Rate gyro units may be combined to comprise multi-axis gyro packages. To each gyro dynamic model output channel, random signals are added, which represent random drift walk and instrument noise. The modeling of the star tracker may include noise and alignment errors. Digital computer effects are modeled by performing the control law and Kalman filtering computations in discrete form, and using the appropriate sampling times for each.

The equations governing the dynamics of the ACS components are generally nonlinear, however, for the purpose of MIMO control designs a linear model which includes a linearized model of the wheel speed controller dynamics, a 2nd-order Pade approximation of a unit time delay of 1.024 sec (wheel speed controller time cycle), and a rational approximation of a zero-order hold at 1.024 sec, is used to represents the control system dynamics. These dynamics can be written in the form

$$\begin{aligned} \dot{x}_h &= A_h x_h + B_h \overline{u} \\ u &= C_h x_h + D_h \overline{u} \end{aligned} \tag{6}$$

where x_h is a vector of states for the ACS components; A_h, B_h, C_h, and D_h are the corresponding system matrices; and $\overline{u}$ is the vector of commanded control inputs. These equations can then be combined with the spacecraft dynamics equations in eqs. 4 and 5 to form the complete set of spacecraft system equations.

Control Synthesis

As mentioned earlier, several synthesis techniques are being considered in the EACS experiment for the design of MIMO controllers. These include, H_2–based, H_∞–based, μ–based, and other controllers. One technique considered is a mixed H_2/H_∞ approach, implemented through design optimization, which is described in this paper. In this approach, the controller is synthesized following H_2 methods while robustness constraints, to guarantee robustness against model uncertainties in the plant, input, and output, are imposed through H_∞–type constraints. The H_2 design follows the LQG/LTR (Loop Transfer Recovery) methodology to provide stabilizing controllers for the enhanced attitude control experiment. Loop shaping filters are implemented within the LQG/LTR framework to provide the ability to manipulate the loop gains. The design parameters include the characteristics of the process and measurement noise as well as the variables associated with the shaping filters. At the top level of this synthesis technique, nonlinear programming is used to optimize the design parameters to provide optimal pointing performance for the Lewis spacecraft while imposing H_∞–type robustness constraints, loop shaping constraints, as well as other design constraints. The details of the synthesis procedure is provided in the following sections.

LQG/LTR

In the LQG/LTR synthesis, the plant is assumed to have the form

$$\begin{aligned} \dot{x} &= Ax + Bu + \Gamma w \\ y &= Cx + v \end{aligned} \tag{7}$$

where x represents the augmented plant state vector defined as $x = \left[x_s^T, x_h^T\right]^T$, which is obtained by combining the spacecraft plant dynamics, given by eqs. (4) and (5), with the control system hardware dynamics, given by eq. (6), in series. The matrices A, B, and C are the augmented spacecraft state, input influence, and output influence matrices, respectively. The vectors w and v are the process and measurement noise vectors, respectively, and are modeled as zero-mean and uncorrelated white noises with covariances

$$E\{ww^T\} = W \geq 0\,;\; E\{vv^T\} = V > 0 \tag{8}$$

The optimal LQG controller is given by [3]

$$\begin{aligned} \dot{\hat{x}} &= (A - K_f - BK_c)\hat{x} + K_f y \\ u &= -K_c \hat{x} \end{aligned} \tag{9}$$

with

$$K_c = -R^{-1}B^T P_c\,;\; K_f = P_f C^T V^{-1} \tag{10}$$

where P_f and P_c are positive semidefinite matrices that satisfy the following algebraic Riccati equations.

$$\begin{aligned} A^T P_c + P_c A - P_c B R^{-1} B^T P_c + C^T Q C = 0 \\ P_f A^T + A P_f - P_f C^T V^{-1} C P_f + \Gamma^T W \Gamma = 0 \end{aligned} \tag{11}$$

If the triples $(A,\, B,\, Q^{1/2}C)$ and $(A,\, \Gamma W^{1/2},\, C)$ are stabilizable and detectable, positive semidefinite solutions to the Riccati equations in eq. (11) exist and the resulting controller is stabilizing, i.e., the closed-loop system is stable [3]. Loop transfer recovery is a way of designing the LQG controller such that the desirable robustness and performance characteristics of the full state feedback are recovered at the plant input or output. For a square plant, a two step procedure is followed to achieve LTR at the plant output [3].

1. Design a Kalman filter by choosing appropriate covariance matrices W and V until the return ratio of the filter $-C(sI-A)^{-1}K_f$ is satisfactory for the plant output.
2. Design an optimal regulator by setting $Q = I$ and $R = \rho I$, and choose a ρ small enough such that the return ratio of the compensated plant at the output resembles the return ratio of the filter $-C(sI-A)^{-1}K_f$ with reasonable accuracy over a desired range of frequencies.

It should be noted that in LQG/LTR the noise covariance matrices are taken as design parameters in order to shape the loop gain of the system, and they do not necessarily have any association with process or measurement noise characteristics. In practical applications, the degree of freedom provided by the element of the covariance matrices may not be sufficient to obtain a desired loop gain in particular ranges of frequency. In order to provide additional capability to adjust the loop gain, the spectral distribution of the process and measurement noises may be manipulated in desired frequency regions by augmenting the plant dynamics with additional noise dynamics. For the purpose of MIMO ACS designs for the Lewis spacecraft, the spectral distribution of both the process and measurement noises are manipulated. The spectral distribution of the process noise is manipulated to provide integral action in the loop gain which is critical for acceptable pointing performance. This is achieved by introducing the following dynamics

$$\begin{aligned} \dot{\eta} &= A_w \eta + B_w \bar{w} \\ w &= C_w \eta \end{aligned} \tag{12}$$

with $A_w = -0.0001\alpha$, $B_w = I_{3\times3}$, and $C_w = I_{3\times3}$. Here, $\bar{w}$ is a white noise with intensity W, and α is a 3x3 diagonal matrix, with positive elements that can be adjusted by the design process.

In order to ensure proper roll-off of the loop gain at the higher frequency ranges, to avoid destabilizing spill-over problems or high frequency noise pollution, the spectral distribution of the measurement noise is manipulated by introducing additional dynamics as follows.

$$\begin{aligned} \dot{\xi} &= A_v \xi + B_v \bar{v} \\ v &= C_v \xi + D_v \bar{v} \end{aligned} \tag{13}$$

with $A_v = -0.5\beta$, $B_v = I_{3\times3}$, $C_v = -12.5\beta\theta$, and $D_v = 25\theta$. Here, $\bar{v}$ is a white noise with intensity V, and β and θ are 3x3 diagonal matrices, with positive elements that can be adjusted by the design process to shape the roll-off.

The overall system dynamics may be written by combining eqs. (7), (12), and (13), to obtain

$$\begin{aligned} \dot{\bar{x}} &= \bar{A}\bar{x} + \bar{B}u + \bar{\Gamma}\psi \\ y &= \bar{C}\bar{x} + \Upsilon\bar{v} \end{aligned} \tag{14}$$

where

$$\begin{gathered} \bar{A} = \begin{bmatrix} A & \Gamma C_w & 0 \\ 0 & A_w & 0 \\ 0 & 0 & A_v \end{bmatrix}; \\ \bar{B} = \begin{bmatrix} B \\ 0 \\ 0 \end{bmatrix}; \; \bar{\Gamma} = \begin{bmatrix} 0 & 0 \\ B_w & 0 \\ 0 & B_v \end{bmatrix}; \\ \bar{C} = [C \quad 0 \quad CC_v]\,; \; \Upsilon = CD_v \end{gathered} \tag{15}$$

and

$$\psi \equiv \begin{Bmatrix} \bar{w} \\ \bar{v} \end{Bmatrix} \tag{16}$$

with

$$E(\psi\psi^T) = \Psi = \begin{bmatrix} W & 0 \\ 0 & V \end{bmatrix} \tag{17}$$

Now, the LQG/LTR based attitude control design is synthesized by using variables $\bar{A}$, $\bar{B}$, $\bar{C}$, $\bar{\Gamma}$, Ψ, and $\Upsilon V \Upsilon^T$, from eqs. (15) to (17), as A, B, C, Γ, W, and V, respectively, in eqs. (9) to (11), with $Q = I$, and $\rho = 10^{-8}$. Note that the design freedom is contained in the elements of the covariance matrices W and V, as well as the diagonal elements of matrices α, β, and θ, i.e., a new controller is synthesized for every combination of these parameters. As mentioned earlier, design optimization is used to choose these parameters so as to optimize the pointing performance of the spacecraft while satisfying robustness and performance constraints.

Robustness

The LQG/LTR controller can provide excellent performance if the model of the system is accurately known. However, this controller may have serious stability and performance issues if there is considerable uncertainty in the model. In typical spacecraft design, the issue of model uncertainty for the SISO controller is treated in the classical sense by imposing stability margins, such as gain and phase margins. In this paper, robust stability for the MIMO ACS controllers is addressed through the application of robust stability theory for various forms of uncertainty in the system model. Specifically, four types of uncertainties are considered, and they are (a) input multiplicative to accommodate uncertainties in the input of the plant, such as the reaction wheels; (b) output multiplicative to accommodate uncertainties in the output of the plant, such as the attitude measurement system; (c) additive to accommodate unmodelled high frequency spacecraft dynamics; and (d) parametric uncertainties to accommodate uncertainties in the modal frequencies of some of the flexible modes that are included in the design model. The conditions for robust stability of the system under these type of uncertainties is discussed in the following.

Uncertainties in the input

The uncertainties in the actuation system, such as those of the reaction wheel assembly or the controller, as well as general uncertainties in the spacecraft plant model, may be modeled as input multiplicative uncertainty, Δ_i, shown in Figure 2.

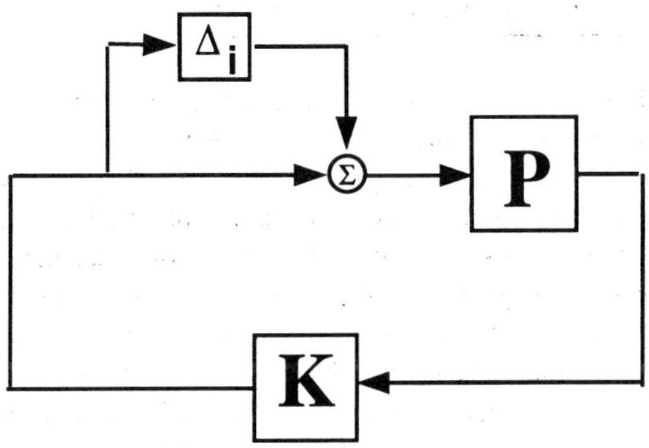

Figure 2. System with Input Multiplicative Uncertainty.

which is equivalent to

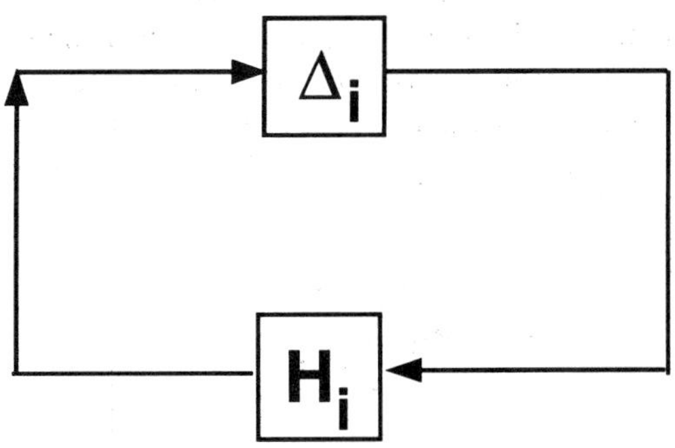

Figure 3. Equivalent System with Input Multiplicative Uncertainty.

where $H_i = (I + KP)^{-1}KP$, with P representing the plant (system in eq. (7) and K denoting the controller (system in eqs. (9) to (11)). The condition for robust stability of the feedback system shown in Figure 3 is established from the small gain theorem [3], which states that the closed-loop system is stable iff

$$||\Delta_i(jw)||_\infty < \frac{1}{||H_i(jw)||_\infty} \qquad (18)$$

From eq. (18), if it is required to allow an input multiplicative uncertainty $\Delta_i(jw)$ such that $||\Delta_i(jw)||_\infty \le \delta_i$, the condition for robust stability can then be written as

$$||H_i(jw)||_\infty < \frac{1}{\delta_i} \qquad (19)$$

Uncertainties in the Output

The uncertainties in the attitude measurement system, such as those in the rate gyros, star trackers, etc., as well as general uncertainties in the spacecraft plant model, may be modeled as output multiplicative uncertainty, Δ_o, shown in Figure 4.

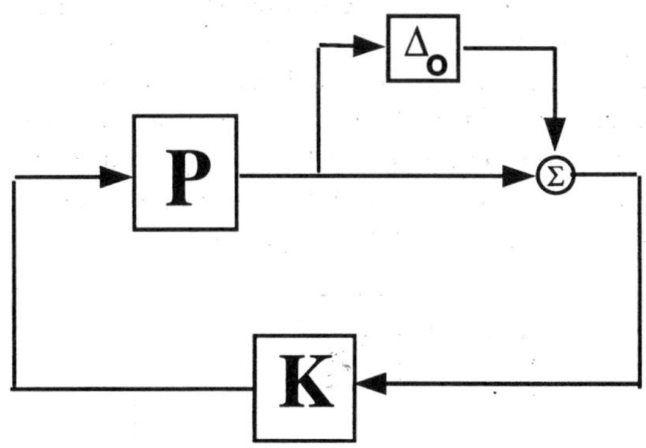

Figure 4. System with Output Multiplicative Uncertainty.

which is equivalent to

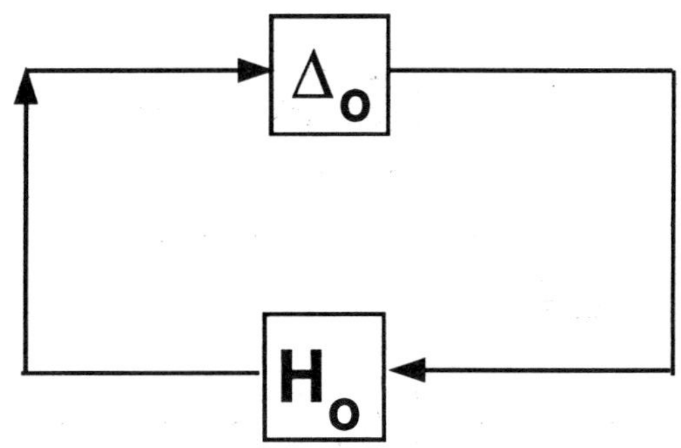

Figure 5. Equivalent System with Output Multiplicative Uncertainty.

where $H_o = PK(I + PK)^{-1}$. The condition for robust stability of the feedback system shown in Figure 5 is established from the small gain theorem [3], which states that the closed-loop system is stable iff

$$||\Delta_o(jw)||_\infty < \frac{1}{||H_o(jw)||_\infty} \qquad (20)$$

From eq. (20), if it is required to allow an output multiplicative uncertainty $\Delta_o(jw)$ such that $||\Delta_o(jw)||_\infty \le \delta_o$, the condition for robust stability can then be written as

$$||H_o(jw)||_\infty < \frac{1}{\delta_o} \qquad (21)$$

Unmodeled Dynamics

As mentioned earlier, the spacecraft structural model may include hundreds or thousands of flexible modes. However, realistically only a few significant low frequency modes are included in the control design model, i.e., the higher frequency modes are neglected. Moreover, the degree of uncertainty in the knowledge of the frequency and other modal attributes of the flexible modes increase as frequency increases. One option could be to neglect the higher frequency modes, but this could cause performance degradation and even instabilities in the form of spill over. The approach taken here is to model the neglected flexible dynamics along with associated uncertainties as additive plant uncertainty, Δ_a, as shown in Figure 6.

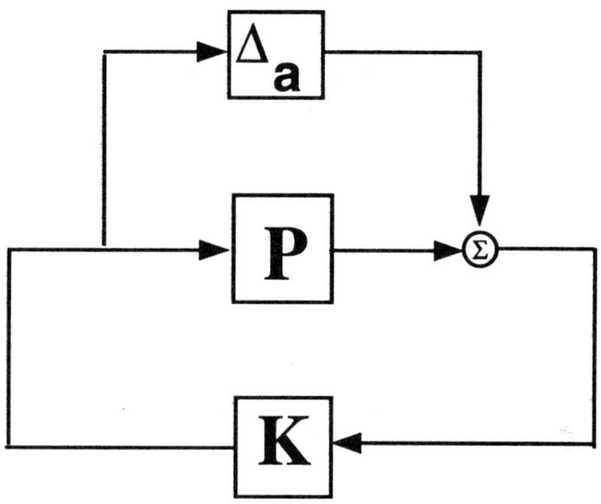

Figure 6. System with Additive Uncertainty.

which is equivalent to

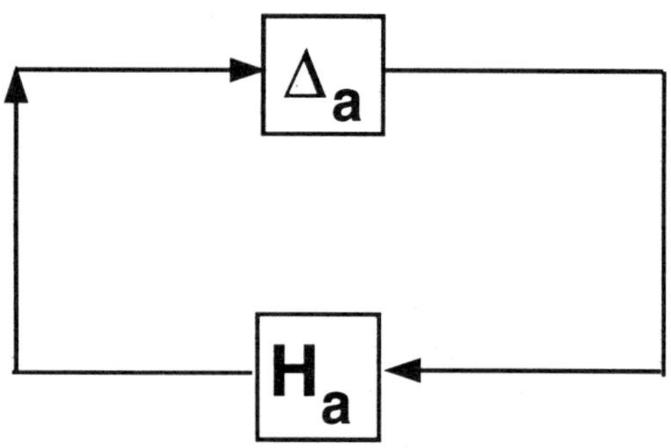

Figure 7. Equivalent System with Additive Uncertainty.

where $H_a = (I + KP)^{-1}K$. Let $\Delta_a(jw)$ be a stable, rational transfer function. Then, the condition for robust stability of the feedback system shown in Figure 7 is established from the small gain theorem [3], which states that the closed-loop system is stable iff

$$||\Delta_a(jw)||_\infty < \frac{1}{||H_a(jw)||_\infty} \tag{22}$$

Let the unmodeled dynamics along with associated uncertainties be such that the additive uncertainty $\Delta_a(jw)$ is bounded as follows

$$\bar{\sigma}(\Delta_a(jw)) \leq |\delta_a(jw)| \; ; \; \forall\omega \tag{23}$$

where $\delta_a(jw)$ is a scalar, rational, and stable transfer function. The condition for robust stability can then be rewritten as

$$||\delta_a(jw)H_a(jw)||_\infty < 1 \tag{24}$$

Uncertainty in modal frequencies

For the purpose of control design, only the first three flexible modes are included in the design model. This is chiefly due to the sampling rate of the control system, which limits the control authority over the flexible modes considerably. Nevertheless, because of uncertainties that may exist in the knowledge of the frequencies of these three modes (parametric uncertainties), it is desirable to design a MIMO ACS controller which provides robust stability in the presence of such uncertainties, i.e., the closed-loop system can tolerate a certain degree of uncertainty in these frequencies. In this paper, robust stability for uncertainties in the frequency of the first two flexible modes, which are most likely to have interactions with the ACS, is considered. Note that uncertainties in the mode shapes or modal damping ratios are not considered because they are not deemed as critical. The parametric uncertainties , Δ_m (see eq. (26), are modeled in a feedback configuration with an augmented plant as shown in Figure 8.

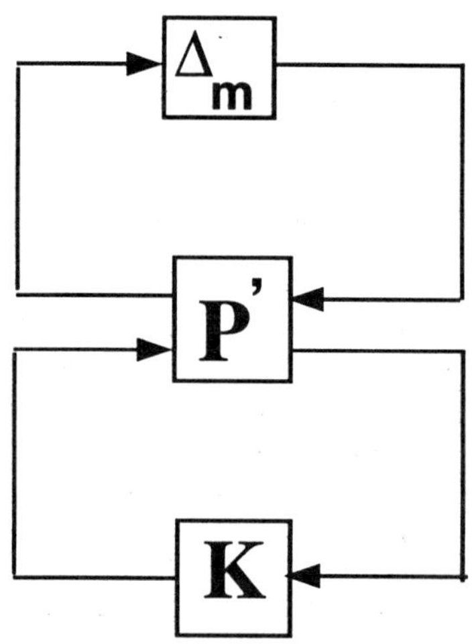

Figure 8. System with Parametric Uncertainty

Here P' represents an augmented plant with additional inputs and outputs to accommodate the two model frequency uncertainties (see eq. (27)). The configuration in figure 8

is equivalent to

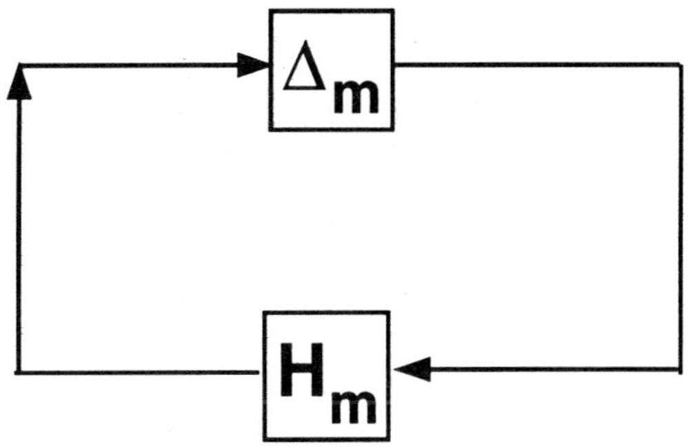

Figure 9. Equivalent System with Parametric Uncertainty

where

$$H_m = \hat{C}(sI - A_{cl})^{-1}\hat{B} \tag{25}$$

and

$$\Delta_m = \begin{bmatrix} \delta_{m1} & 0 \\ 0 & \delta_{m2} \end{bmatrix} \; ; \; \delta_{mi} \in \mathcal{C} \tag{26}$$

Here H_m represents the closed-loop transfer function around the parametric uncertainty. A_{cl} denotes the closed-loop state matrix; and

$$\hat{B} = \begin{bmatrix} 0 & 0 \\ \vdots & \vdots \\ 0 & 0 \\ 1 & 0 \\ 0 & 1 \\ \vdots & 0 \\ \vdots & \vdots \\ 0 & 0 \end{bmatrix} ; \; \hat{C} = \begin{bmatrix} 0 & \dots & 0 & -\omega_1^2 & 0 & \dots & \dots & 0 \\ 0 & \dots & 0 & 0 & -\omega_2^2 & 0 & \dots & 0 \end{bmatrix} \tag{27}$$

Note that the non-zero elements in $\hat{B}$ and $\hat{C}$ correspond to the state variables associated with the modal velocities and displacements of the first two flexible modes, respectively. The condition for robust stability of the feedback system shown in Figures 8 or 9 is given by Doyle [4], and may be stated as follows. The feedback system shown in Figure 9 is stable for Δ_m, $||\Delta_m||_\infty < \delta_m$, iff

$$\mu(H_m(jw)) < \frac{1}{\delta_m} \; ; \; \forall \omega \tag{28}$$

or

$$||H_m(jw)||_\mu < \frac{1}{\delta_m} \tag{29}$$

Since μ can not be directly computed, i.e., it is typically obtained by bounding it from above and below over a range of frequencies, direct implementation of the robustness condition given in eqs. (28) or (29) is costly and impractical in an optimization setting. Therefore, the condition in eqs. (28) or (29) is replaced by

$$||H_m(jw)||_\infty < \frac{1}{\delta_m} \tag{30}$$

Although, this condition is an upper bound for $\mu(H_m)$, and consequently may be more conservative than the previous one, it is computationally more tractable. Furthermore, since the modes are spatially independent (corresponding to two different principal axes of the spacecraft), and the plant and the controller tend to be diagonally dominant in this frequency range, it is expected that the condition in eq. (30) would not be too conservative. However, in actuality no matter what condition for robust stability is used (eqs. (28), (29), (30)), since the control authority over these two modes is fairly limited, the condition for robust stability tends to be conservative, i.e., $\delta_m \approx 2\zeta_i$. In other words, the maximum complex parameter uncertainty allowed would be approximately limited to 2 times the damping ratio of the flexible modes. Note that this limitation is due to the complex nature of uncertainty allowed and is not necessarily an indication of the real parameter uncertainty that can be tolerated by the system. This may be overcome by modeling uncertainties in real parametric form and using mixed μ analysis, but, as mentioned earlier, the computation of μ is costly, making its use in the nonlinear programming optimization prohibitive. Keeping in mind that the control authority is limited around the flexible modes, such that the controller would be rolling off before the first flexible mode, a more useful criteria for robust stability under parametric uncertainty may be established. Assume that it is desired to allow for 25% uncertainty in the first two flexible modes. Note that because of the limited control authority one need not be concerned with positive variations in these modes, but rather in variations that bring the frequency of these modes closer to the bandwidth of the controller. To do this, instead of the robust stability of $H_m(jw)$, robust stability of two perturbed models of the system is considered. Denote these perturbed models by $H_m^1(jw)$ ad $H_m^2(jw)$, with $H_m^1(jw)$ and $H_m^2(jw)$ corresponding to $H_m(jw)$ with the frequencies of the first two flexible modes decreased by 10% and 25%, each. Robust stability is then implied by requiring

$$\begin{aligned} \left\|H_m^1(jw)\right\|_\infty &< \frac{1}{\delta_m^1} \\ \left\|H_m^2(jw)\right\|_\infty &< \frac{1}{\delta_m^2} \end{aligned} \tag{31}$$

These conditions are intuitively expected to provide robust stability since the control authority is low around these modes, i.e., the controller is rolling off.

Controller Design via Optimization

As mentioned previously, the parameters of the controllers are designed via nonlinear programming techniques. In this approach, the parameters of the LQG/LTR synthesis approach are chosen to optimize the pointing performance of the system while satisfying a set of H_∞–type robust stability constraints along with other constraints to impose performance specifications or cost limitations. This optimization-based approach provides a systematic means of designing optimal MIMO controllers for the attitude control of the Lewis spacecraft. Although, a mixed H_2/H_∞ synthesis is pursued here, other synthesis techniques may be easily implemented as well. Furthermore, other performance and cost specifications may be readily implemented also. The design optimization problem is formulated as follows.

Using any linear, time invariant MIMO controller, the closed-loop system dynamics can be written as

$$\dot{\bar{x}} = \bar{A}\bar{x} + \bar{L}p; \quad y_p = \bar{C}_p\bar{x}; \quad u = \bar{C}_u\bar{x} \tag{32}$$

where $\bar{x}$ is the state vector for the closed-loop dynamics, $p(t)$ is a zero mean, white noise disturbance applied at the disturbance locations, y_p is the line-of-sight pointing error, $u(t)$ is the control vector, and $\bar{A}$, $\bar{L}$, $\bar{C}_p$ and $\bar{C}_u$ are the corresponding closed-loop system matrices. The steady state covariance matrix for the closed loop state, $\Sigma_{\bar{x}}$, is computed by solving the following Lyapunov equation [5]

$$\bar{A}\Sigma_{\bar{x}} + \Sigma_{\bar{x}}\bar{A}^T + \bar{L}\Sigma_p\bar{L}^T = 0 \tag{33}$$

where Σ_p is the covariance matrix for the disturbance noise, $p(t)$. The steady-state average control power is given as

$$P = \lim_{t\to\infty} Tr\{E\{u(t)u^T(t)\}\} = Tr[\bar{C}_u\Sigma_{\bar{x}}\bar{C}_u^T] \tag{34}$$

and the root-mean-square line-of-sight pointing error is

$$R = \lim_{t\to\infty}\left[Tr(E\{y_p(t)y_p^T(t)\})\right]^{\frac{1}{2}} = \left[Tr(\bar{C}_p\Sigma_{\bar{x}}\bar{C}_p^T)\right]^{\frac{1}{2}} \tag{35}$$

The control design optimization problem is posed as:

$$\min_{\{\bar{V},\bar{W},\alpha,\beta,\theta\}} P \tag{36}$$

s.t.

$$\begin{aligned}
&\|H_i(jw)\|_\infty < \frac{1}{\delta_i}\\
&\|H_o(jw)\|_\infty < \frac{1}{\delta_o}\\
&\|\delta_a(jw)H_a(jw)\|_\infty < 1\\
&\left\|H_m^1(jw)\right\|_\infty < \frac{1}{\delta_m^1}\\
&\left\|H_m^2(jw)\right\|_\infty < \frac{1}{\delta_m^2}\\
&R(1,1) \le e_{roll}\\
&R(2,2) \le e_{pitch}\\
&R(3,3) \le e_{yaw}\\
&\sigma_i\left(\bar{C}\left(j\omega_{orb}I - \bar{A}\right)^{-1}\bar{B}\right) \le \sigma_{orb\,;\,i=1,2,3}
\end{aligned} \tag{37}$$

Here, The first five constraints are H_∞–norm robust stability constraints. In the next three constraints, e_{roll}, e_{pitch}, and e_{yaw} represent the desired upper bound values for the spacecraft pointing error in roll, pitch, and yaw, respectively. The last three constraints are on the singular values of the closed-loop transfer function at orbital frequency to impose disturbance rejection requirements for environmental disturbances. The variables ω_{orb} and σ_{orb} denote the orbital frequency and upper bound value for the gain, respectively. This optimization-based synthesis technique is used in the next section to design MIMO ACS for the Lewis spacecraft.

Numerical Examples

The optimization-based mixed H_2/H_∞ synthesis technique described in the previous sections has been used to design MIMO ACS controllers for the Lewis spacecraft. A number of controllers were designed for various performance and cost specifications. In this section, the design and simulation of one of these controllers are discussed. As mentioned previously, the design variables available for optimization are contained in the elements of the covariance matrices W and V, as well as the diagonal elements of matrices α, β, and θ. The process noise covariance matrix W is expressed in terms of its Cholesky decomposition, i.e.,

$$W = L_w L_w^T \tag{38}$$

where the matrix L_w is a 3 x 3 lower triangular matrix. In this control design, the six nonzero elements of L_w are used as design variables. The measurement noise covariance matrix V is assumed to be diagonal, and its three diagonal elements are used as design variables. The nine design variables associated with the covariance matrices W and V, together with the nine design variables associated with the diagonal elements of matrices α, β, and θ, constitute the 18 design variables used in the design optimization. The design optimization was performed using the Automated Design Synthesis (ADS) program. Gradient computations were performed using finite difference approximations. An interior penalty function method of ADS was used to solve the nonlinear programming problems. In this method, the constrained optimization problem is transformed into an unconstrained problem through creation of a pseudo-objective function, which is the sum of the original objective function and an imposed penalty function (a function of the constraints [6]).

The parameters associated with the optimization constraints for stability robustness, pointing accuracy, loop shaping, etc., are described in the following:

1. The scalar parameter δ_i was chosen at 0.5 to provide input uncertainty levels comparable to the baseline design.
2. The scalar parameter δ_o was chosen at 0.5 to provide output uncertainty levels comparable to the baseline design.
3. The scalar transfer function $\delta_a(jw)$, which bounds the unmodeled dynamics of the plant as well as any of its potential perturbations or uncertainty, was chosen as

$$\delta_a(jw) = \frac{-\omega^2 + 2j\omega + 1}{-\omega^2 + 100j\omega + 2500} \tag{39}$$

   This proper transfer function would bound the unmodeled dynamics of the spacecraft plant.
4. The scalar parameters δ_m^1 and δ_m^2 were chosen at 0.005 to ensure the feasibility of the constraints (for parametric uncertainty) within the design optimization.
5. The pointing accuracy threshold for roll, e_{roll} was chosen at 0.025 to provide three fold improvement over the baseline design.
6. The pointing accuracy threshold for pitch, e_{pitch} was chosen at 0.025 to provide three fold improvement over the baseline design.
7. The pointing accuracy threshold for yaw, e_{yaw} chosen at 0.035 to provide comparable pointing performance to the baseline design.
8. The required attenuation at the orbital frequency, σ_{orb} was chosen at 15 dB to provide improved orbital disturbance rejection over the baseline design.

The design optimization resulted in a 44–order controller which satisfied all design, performance and robustness constraints. Two simulations were set up using the MATLAB v4.2/SIMULINK v1.3c simulation package on a SPARCsystem 600 Sun workstation, one for the designed MIMO ACS controller and the other for the baseline SISO ACS controller. The controller block in the simulation is implemented using the MIMO subroutine, developed by NASA Langley in C programming language, with the aid of the SIMULINK user-defined S-function. Both simulations used a 28-mode state space model of the Lewis spacecraft. However, no hardware dynamics (e.g., rate gyros, reaction wheels) were included in any of the simulations. Both simulations included full models (as they were available) of reaction wheels, rate gyros, and the star tracker. A Kalman filter was designed and used to estimate the vehicle's attitude from the sensor data. The reaction wheel dynamics included friction and stiction models, limits on the input command voltages and digital voltage quantization, as well as the quantization effects on wheel RPM outputs due to the wheel's optical encoder. To each gyro dynamic model output channel, random signals were added, which represent random drift walk and instrument noise. The modeling of the star tracker included noise and alignment errors. The spacecraft was subjected to environmental disturbances, which included gravity gradient torques, drag torques, magnetic unloading, etc. Each simulation was run for one orbit, with each orbit assumed to be 5996 seconds in duration.

Figures 10–12 show the spacecraft estimated attitude time histories in roll, pitch and yaw for the two simulations. These plots show the time histories from 500 seconds onwards to allow the start-up transient dynamics to die out. Figures 10 and 11 clearly demonstrate the improvement in attitude pointing in roll and pitch, where it is observed that peak to peak attitude response is reduced by a factor of 5 or more. There is no discernible change in yaw pointing as seen from figure 12. This is because the controller was mainly designed to improve roll and pitch pointing, as they effect science data considerably more. Figures 13–15 show the complete time histories of the wheel speeds for reaction wheels 1-3 for the two simulations. It is observed from these figures that the wheel speeds for the baseline SISO and MIMO controllers are fairly close to each other, particularly in steady-state, indicating that there is little penalty (in terms of power required) that is paid by the MIMO controller for providing a better pointing performance. It is also noted from these figures that the MIMO controller has higher frequency content, which indicates the increased bandwidth of the controller.

In conclusion, the optimal mixed H_2/H_∞ controller provided substantial improvements in the pointing performance of the Lewis spacecraft over those provided by the baseline controller. This improvement was achieved while satisfying all robust stability conditions as well as other design or performance specifications. Furthermore, the power required to achieve such an improvement in pointing performance is minimal and well within the capacity of the reaction wheels in normal pointing mode. It should be noted that the proposed mixed H_2/H_∞ approach follows the conventional design approach in that it satisfies the various robust stability conditions against input, output, parametric, and nonparametric uncertainties, individually. However, if it desired to analyze controller robustness in a collective sense, a mixed μ analysis [4] may be performed.

Concluding Remarks

The enhanced attitude control system experiment, which is a technology demonstration experiment on the NASA's small spacecraft technology initiative program's Lewis spacecraft to evaluate advanced attitude control strategies, has been described. The purpose of the experiment is to evaluate the feasibility of designing and implementing robust MIMO attitude control strategies for enhanced pointing performance of spacecraft to improve the quality of the measurements of the science instruments. Among the many control design strategies being considered for the experiment, a control synthesis technique based on mixed a H_2/H_∞ approach has been presented. The control synthesis uses a nonlinear programming technique to tune the controller parameters and impose robustness and performance constraints. Simulations were carried out to demonstrate the feasibility of the proposed attitude control design strategy, and have shown that the mixed H_2/H_∞ attitude control designs can provide substantial improvement in the pointing performance of the spacecraft over the conventional SISO designs, with minimal increase in required power, while maintaining reasonable degree of robustness against modeling and hardware uncertainties. An efficient algorithm, both in time and memory requirements, for the implementation of MIMO controllers has been developed. The algorithm has been incorporated within the on-board flight computer software.

Acknowledgment

The author would like to express his appreciation to Mr. Dean W. Sparks, for performing and providing the simulation results.

References

[1] GE Aerospace; "Pointing and Orbit Study Update (AFM T-9)" , NAS5–32500, May 22, 1992..

[2] J. R. Wertz, "Spacecraft Attitude Determination and Control," Chapter 18, Ed. by James R. Wertz, *Kluwer Academic Publishers*, Boston, MA, 1978.

[3] J. M. MacieJowski, ''Multivariable Feedback design,'' *Addison-Wesley Publishing Co.*, New York, NY, 1989.

[4] J. C. Doyle, ''Analysis of Feedback Systems with Structured Uncertainties,'' *IEEE Proc.*, Vol. 129, pt. D., No. 6, Nov. 1982, pp. 242–250.

[5] R. F. Stengel, ''Stochastic Optimal Control - Theory and Applications'' *John Wiley \& Sons, Inc.*, New York, NY, 1986.

[6] G. N. Vanderplaats, ''ADS–A Fortran Program for Automated Design Synthesis–version 1.10,'' Contractor Report 177985, NASA, 1985.

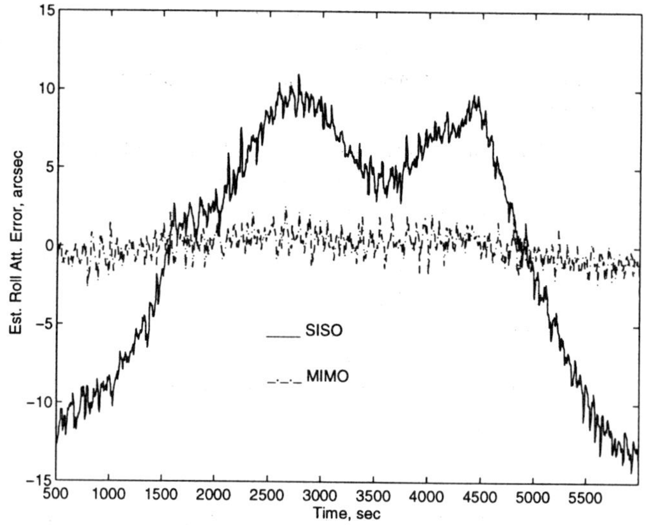

Figure 10. Time histories of estimated roll attitude for Lewis spacecraft

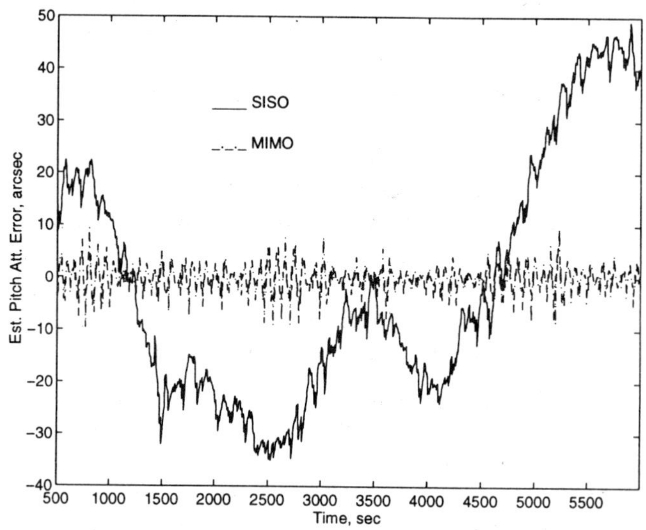

Figure 11. Time histories of estimated pitch attitude for Lewis spacecraft

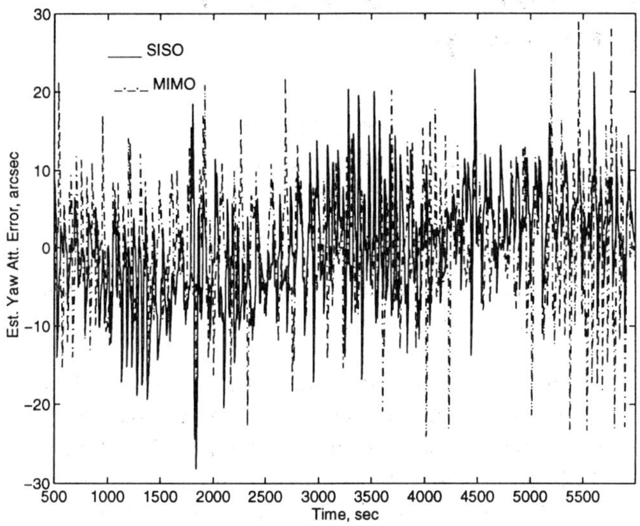

Figure 12. Time histories of estimated yaw attitude for Lewis spacecraft

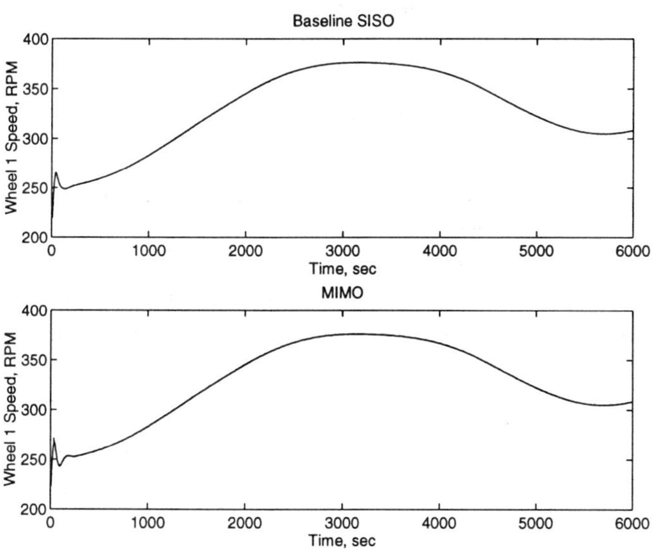

Figure 13. Time histories of the wheel speed for reaction wheel no. 1

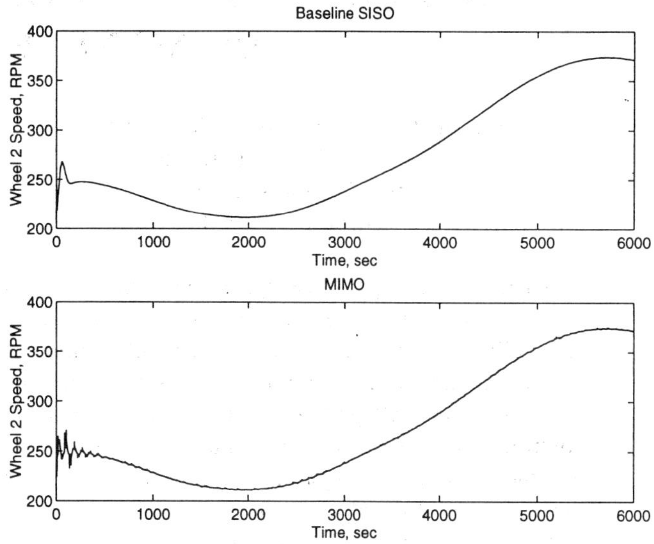

Figure 14. Time histories of the wheel speed for reaction wheel no. 2

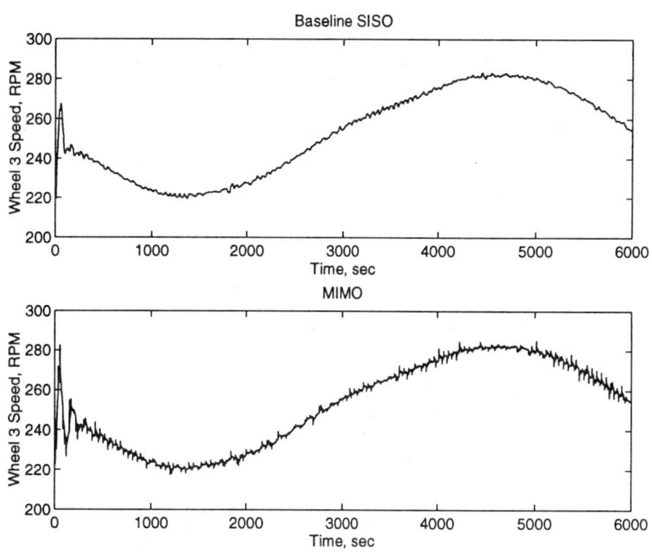

Figure 15. Time histories of the wheel speed for reaction wheel no. 3

AIAA-97-3528

INTELLIGENT CONTROL DESIGN AND SIMULATION USING NEURAL NETWORKS

Xu Feng, Ching-Fang Lin and Tie-Jun Yu
American GNC Corporation
9131 Mason Avenue
Chatsworth, CA 91311

Norm Coleman
Army Armament Research, Development and Engineering Center
Picatinny Arsenal, NJ 07806-5000

Abstract: This paper investigates the application of intelligent control to the XTE spacecraft. Specifically, an intelligent controller, the neural network controller, is studied and designed for the XTE spacecraft. The advantages of intelligent controllers over conventional controllers are analyzed. The simulation results show that the designed intelligent controllers provide a great improvement of the attitude control accuracy under the torque disturbance and achieve an accuracy of 4 arc-sec while the accuracy of the baseline controller is 20 arc-sec and the accuracy requirement is 30 arc-sec [1].

1. Introduction

Modern spacecraft systems are highly flexible, of high order, and highly coupled. At the same time, the spacecraft is required to perform rapid maneuvers, and achieve high-accuracy pointing in the presence of disturbances, uncertainties, and component failures. Accordingly, spacecraft control systems must be robust, adaptive and intelligent in order to meet this challenge. Currently, neural networks have been demonstrated to be the most promising and with potential techniques to design a robust, adaptive and intelligent control system.

The primary scientific objective of the X-ray Timing Explorer (XTE) satellite is to perform astronomical measurements in the X-ray portion of the electro-magnetic spectrum [1]. The XTE attitude control system consists of sensors (gyros, star trackers, sun sensors, magnetometers), actuators (reaction wheels, magnetic torquers), a digital computer, and supporting electronics [1].

To perform continuous X-ray sky surveys and to measure variations in energy and X-ray targets of interest, two science instruments, the All Sky Monitor (ASM) and the High Energy X-ray Timing Experiment (HEXTE), are gimbaled to allow single or two axis large angle rotation. The impact caused by such an angular motion of heavy instruments, along with thermal and aerodynamic disturbances, will create a serious problem in satisfying the pointing accuracy requirement of 30 arc-sec, 3σ. The pointing performance will be further worsened by the introduction of flexible-body dynamics.

The transfer function of the XTE spacecraft is given by

$$\frac{\theta_j(s)}{u_j(s)}=\frac{1}{I_j s^2}+\sum_i \frac{K_j / I_j}{s^2+2\varsigma\omega_i s+\omega_i^2}$$

where $j = X$, Y, or Z represents the roll, pitch, or yaw axis, respectively; θ is the attitude; u the control torque; I the moment of inertia; ω_i the i-th flexible mode frequency; K_i the modal gain for i-th flexible mode; and ζ the passive damping ratio assumed as 0.001.

Table 1 Model frequencies and gains of reduced-order XTE spacecraft (Yaw-axis)

$\beta = 0°$		$\beta = 45°$		$\beta = 90°$		$\beta = 150°$	
I = 2690 Kg-m^2		I = 2735 Kg-m^2		I = 2787 Kg-m^2		I = 2717 Kg-m^2	
ω_i (Hz)	K_i	ω_i (Hz)	K_i	ω_i (Hz)	K_i	ω_i (Hz)	K_i
0.80	0.4168	0.76	0.2023	0.72	0.0004	0.77	0.2781
2.33	0.0025	2.28	0.2040	2.29	0.4213	2.34	0.1513
3.36	0.0901	3.38	0.1216	3.38	0.1307	3.35	0.0944
4.18	0.1684	4.21	0.2178	4.23	0.2633	4.20	0.1935
14.85	0.0991	14.76	0.1089	14.68	0.1194	14.84	0.1010
29.80	−0.5459	29.65	−0.2991			29.77	−0.4361

The XTE system model contains 28~38 flexible modes for each axis. Model reduction is performed to reduce the order of the system model. Here, we only discuss the yaw-axis control design problem. Table 1 gives the modal frequencies and modal gains of reduced-order model for four different orientations of the XTE solar arrays.

2. Design of Neural Network Controller

The artificial neural network (ANN) technique is employed to design a neural controller of the XTE. In the control field, the ANN is more often utilized as a computation tool than as a control algorithm. Though the ANN possesses some important and promising attributes (to name a few, the ANN is a learning system, the ANN can be easily implemented in hardware, and the ANN has large scale, robust, and parallel computation power), the ANN has some inherent disadvantages to prevent it from becoming a popular control algorithm. One that is mostly criticized is that the ANN sometimes locks into a local equilibrium. Even worse, no effective learning algorithm exists which guarantees the convergence of the learning process.

Established results claim that any continuous function can be approximated arbitrarily well by a layered network with one hidden layer, where the hidden nodes represent either sigmoidal functions, or radial basis functions. For the XTE, we choose the popular three layer neural network structure, with one layer (one neuron) as the input layer, one layer (several neurons) as the hidden layer, and one layer (one neuron) as the output layer. Notice that since the XTE is a SISO plant, both the input layer and the output layer have only one neuron.

Generally speaking, neural networks are much better interpolators than extrapolators. It implies that if an off-line trained neural network controller is applied as control, it will work well only if it has been trained with adequate samples which include almost all of the circumstances that it will encounter later in the control. In most of the situations, this is an unrealistic assumption. On the contrary, a neural network controller with online learning capability can deal with unexpected situations quite well because of its learning ability. For the neural network to work well under any circumstances, on line learning is essential to keep tight control of the plant.

The neural network structure we used is shown in Figure 1. For the purpose of simplicity, four neurons in the hidden layer are illustrated. It is known that the structure of a single neuron is essential for the neural network to accomplish a specific task. Each neuron in Figure 1, except the input neuron, is built to share the similar structure. It should be noted that a threshold input is included in each neuron. The threshold of each neuron is found to be of critical importance for the neural network to achieve convergence through online learning. If no threshold is used, the neural network will not maintain stability no matter how the weight is adapted.

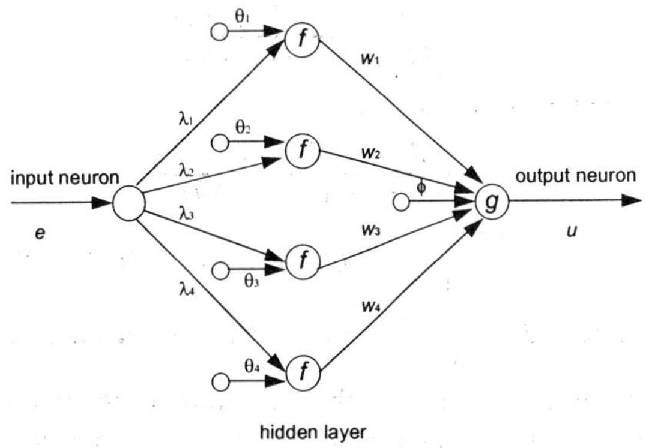

Figure 1 Structure of neural network

The output of the i-th neuron in the hidden layer is

$$h_i = f(\lambda_i e - \theta_i)$$

where e is the input of the neuron (here it is the tracking error), λ_i is the weight associated with the input, θ_i is the threshold related to the neuron, which is trained just as other weights during the learning phase, and $f(x)$ is the sigmoidal function with the nonlinear mapping,

$$f(x) = \frac{1-e^x}{1+e^x}$$

For the output neuron, the input-output relationship is described by

$$u = g(\sum_i h_i w_i - \phi)$$

where h_i is the i-th input of the neuron (which is the output of the hidden neuron), w_i is the associated weight, ϕ is the threshold, and $g(x)$ is the simple linear mapping,

$$g(x) = x$$

One important issue in neural network control is to choose an appropriate learning algorithm. The learning algorithm we used is explained as follows. Assume y to be the plant output and y_r the reference signal. The controller is expected to monitor the plant in such a way that the plant output will track the reference signal. Specifically, the following error is minimized,

$$J=\frac{1}{2}\sum_k[y_r(k)-y(k)]^2$$

The problem of most concern is how to online modify the weights of the neural network so that J is minimized. The error back propagation algorithm is used to update the neural weight. Suppose all weights in the neural network at time k are collected to construct a vector $W(k)$. Then by the chain rule,

$$\frac{\partial J}{\partial W(k)}=\frac{\partial J}{\partial u(k)}\frac{\partial u(k)}{\partial W(k)}$$

Again by the chain rule,

$$\frac{\partial J}{\partial u(k)}=\frac{\partial J}{\partial y(k+1)}\frac{\partial y(k+1)}{\partial u(k)}$$

where $\frac{\partial J}{\partial y(k+1)}=-[y_r(k+1)-y(k+1)]$ is the output error, and $\frac{\partial y(k+1)}{\partial u(k)}$ is the system gain at that time k which can be either obtained from experiments, or approximated online by the following relationship

$$\frac{\partial y(k+1)}{\partial u(k)}=\frac{y(k+1)-y(k)}{u(k)-u(k-1)}$$

$\frac{\partial u(k)}{\partial W(k)}$ can be established just as the standard error back propagation algorithm, in terms of the nonlinearity of the input-output feature of the neuron. Specifically, for the output neuron, it follows

$$\frac{\partial u(k)}{\partial w_i(k)}=h_i$$

$$\frac{\partial u(k)}{\partial \phi(k)}=-1$$

For the neurons in the hidden layer, we have

$$\frac{\partial u(k)}{\partial \lambda_i(k)}=\frac{\partial u(k)}{\partial h_i(k)}\frac{\partial h_i(k)}{\partial \lambda_i(k)}$$

$$\frac{\partial u(k)}{\partial \theta_i(k)}=\frac{\partial u(k)}{\partial h_i(k)}\frac{\partial h_i(k)}{\partial \theta_i(k)}$$

where it can be further calculated by

$$\frac{\partial h_i(k)}{\partial \lambda_i(k)}=\frac{2e\exp(-\lambda e-\theta)}{(1+\exp(-(\lambda e-\theta)))^2}$$

$$\frac{\partial h_i(k)}{\partial \theta_i(k)}=\frac{-2\exp(-\lambda e-\theta)}{(1+\exp(-(\lambda e-\theta)))^2}$$

In each step, the weight of the neural network is updated by the gradient descent algorithm

$$W(k+1)=W(k)-\eta\frac{\partial J}{\partial W(k)}$$

where η is a positive constant. To improve the stability of the learning, an alternative approach to update the weight is

$$\Delta W(k+1)=\alpha\Delta W(k)-\eta\frac{\partial J}{\partial W(k)}$$

where α is also a positive constant.

3. Analysis of the Neural Controller

Before proceeding to the simulation of the XTE model, we start with the simplest case. The neural network controller is applied to a first order model. This allows us to get some insight into the benefits and limitations of the neural network controller and facilitates further modification and improvement. The model considered is:

$$G(s)=1/(s+0.4)$$

For comparison, the initial weights of the neural network are chosen to be random numbers uniformly distributed in [0, 1]. Good performance is observed when the neural network described above is put into use. As shown in Figure 2, the system achieves the steady state within one period. It is also observed that as soon as the weights of the neural network converge, the output reaches its steady state. The transient process clearly manifests the learning characteristics of the neural network. Nevertheless, it seems that the number of the hidden neurons affects the transient dynamics greatly. The output reaches its steady state more quickly with more hidden neurons.

Then, we tested the neural network algorithm using the XTE model. In this case, it is found that either the neural network cannot stabilize the model, or it takes an extremely long period for the neural network to converge. What actually occurs in each separate simulation depends on the learning rate, i.e., α, η. If the learning rate is set to be too large, the closed loop becomes unstable. If the learning rate is set to be small enough, the learning does result in convergence, but it takes a long time to learn. An appropriate learning rate is critical for the neural network to converge and achieve reasonable transient performance, however, it is difficult to determine. We attribute this difficulty to the unstable pole in the model, i.e., the integrator. In order to achieve good performance for a system with an integrator, the controller must possess some kind of predictiveness. However, this is exactly what a neural network controller lacks: a neural network controller can only gain predictiveness through adequate learning.

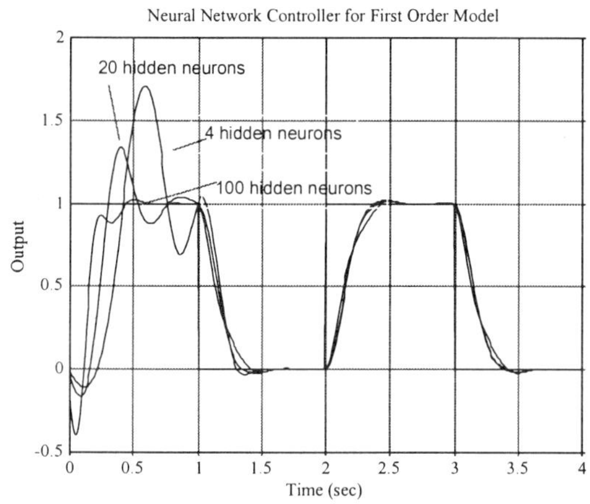

Figure 2 Neural network controller for first order model

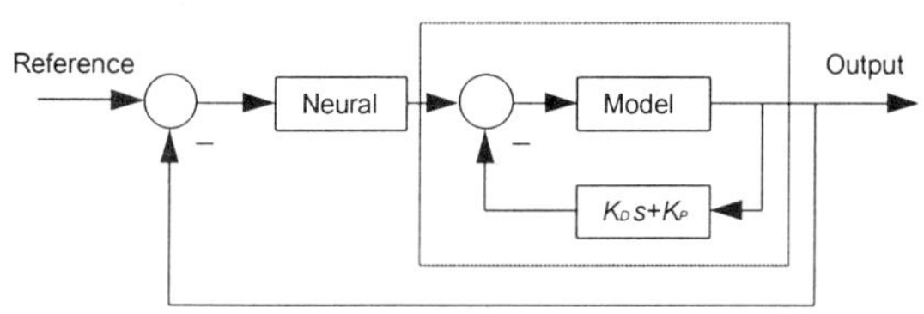

Figure 3 Modified structure

Starting with such considerations, an inner loop, which is actually a proportional derivative (PD) controller, is applied to stabilize the model. Then the neural network is used to track the reference signal. The structure of the controller is shown in Figure 3.

It is equivalent to applying the neural network controller to a modified model, as framed by the dashed line. Denote the former model as

$$G(s)=\frac{B(s)}{A(s)}$$

where $A(s)$ and $B(s)$ are polynomials in s. The controlled model now becomes

$$\tilde{G}(s)=\frac{\frac{B(s)}{A(s)}}{1+\frac{B(s)}{A(s)}(K_D s+K_P)}$$

$$=\frac{B(s)}{A(s)+K_D B(s)s+K_P B(s)}$$

With parameters K_D and K_p properly chosen, the equivalent model can be converted into a model which is much easier for neural network to control.

4. Simulation Software Development

In our simulation, LabVIEW by National Instruments Corporation is used as the simulation environment. LabVIEW is a graphical programming language in which programs are created in block diagram form. Graphic symbols, rather than textual language, are used to describe programming actions. It has extensive libraries of functions and subroutines for most programming tasks. Nevertheless, LabVIEW supports the utilization of various kinds of data acquisition boards, DSP boards, and A/D, D/A boards, which enable the implementation of real-time control systems within the LabVIEW environment when necessary.

To make simulation in LabVIEW more feasible, LabVIEW Code Interface Node (CIN) facilitates the link of external code written in a conventional programming language (usually in C) to LabVIEW. A CIN is a block diagram node associated with a section of source code written in a conventional programming language. The source code needs to be compiled first and then linked to LabVIEW to generate executable code. LabVIEW calls the executable code when the node executes, passing input data from the block diagram to the executable code, and returning data from the executable code to the block diagram. This feature makes it possible to incorporate complicated controllers, such as a neural network controller, a fuzzy controller, and a robust controller into the LabVIEW environment in a straightforward manner.

In our application, the controller CIN is programmed and compiled in WATCOM C and then linked to LabVIEW as a controller icon. In this way, offline simulation can be carried out to study the performance of the applied controller and subsequent refinement and modification can be made to improve the controller. When the controller turns out to be satisfactory, it can be conveniently converted into an online real time implementation.

In the simulation, the sampling period is set to be 0.05 second. Since the model is in a continuous form, the Runge-Kutta method is used to simulate the dynamics of the system.

5. Simulation Results

The neural controller is applied to the XTE model to examine the performance and effectiveness of intelligent controllers. First, dynamic responses with respect to step input are studied.

With respect to this consideration, the neural controller is utilized to close the control loop, as shown in Figure 4. The error between the output of the reference model and the output of the XTE is fed into the controller as the input, and the error rate is generated within the controller. Manipulations of these two error signals lead to the control signal, which is supposed to force the output of the XTE to follow the output of the reference model.

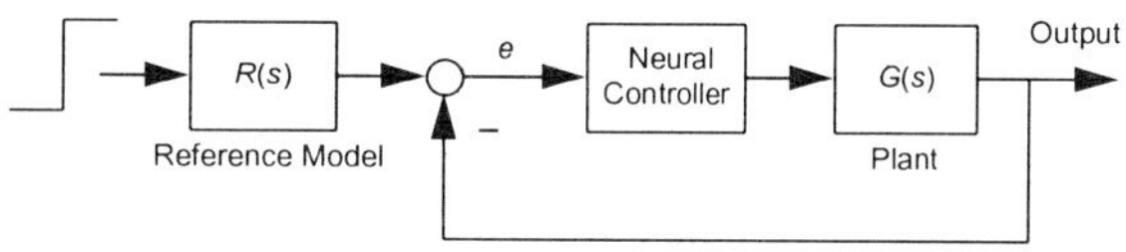

Figure 4 Neural controller for step response

Figure 5 through Figure 8 show the simulation results when the neural controller is applied to the XTE model. Four different orientations of the XTE aircraft, i.e., β = 0°, 45°, 90°, 150° are considered. Robustness of the neural controller can be concluded from the simulation results: although slight differences exist, the neural controller behaves equally well for all these four different orientations.

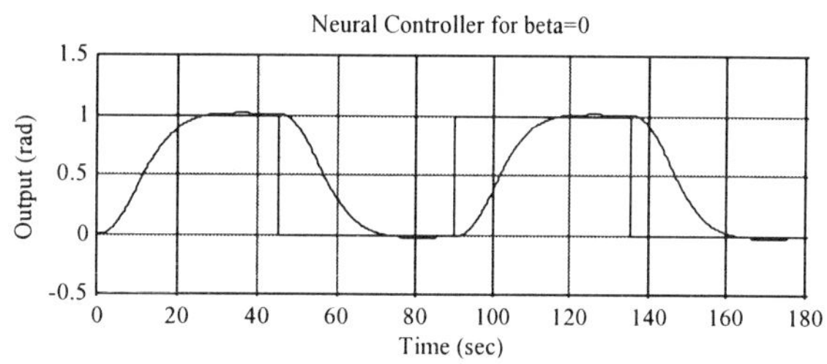

Figure 5 Step response for β = 0

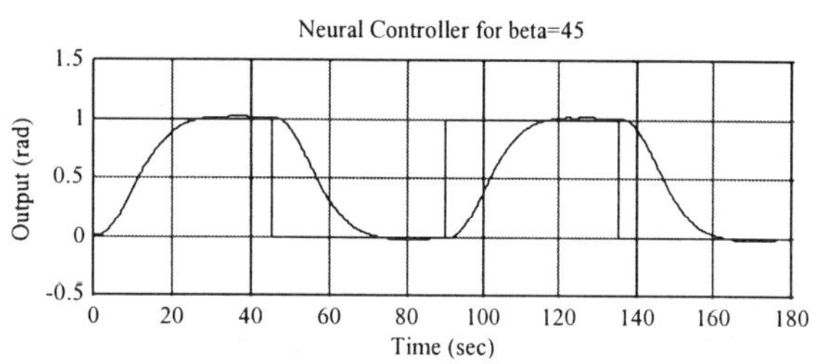

Figure 6 Step response for β = 45

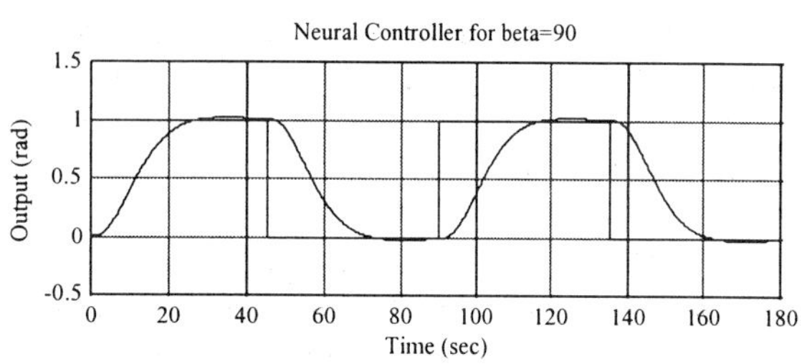

Figure 7 Step response for β = 90

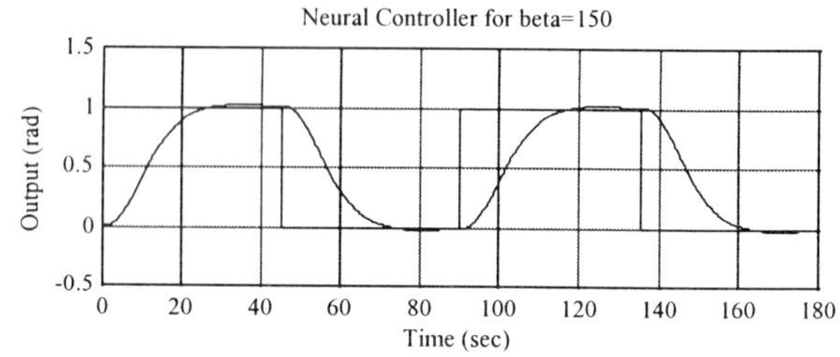

Figure 8 Step response for β = 150

A much more demanding requirement for the XTE controller is that the controller must reject the torque doublet disturbance as shown in Figure 9 in such a way that it will meet the 30 arc-sec 3σ pointing stability requirement. Due to this consideration, the same neural controller is combined with the configuration as shown in Figure 10, based on which the effect of the torque doublet disturbance on the control performance can be studied.

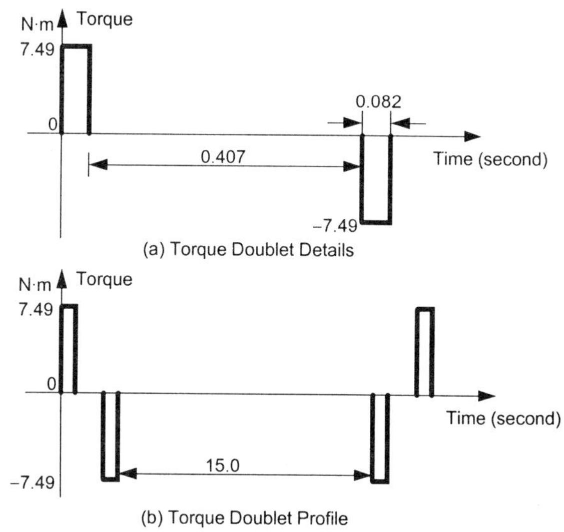

Figure 9 Torque doublet

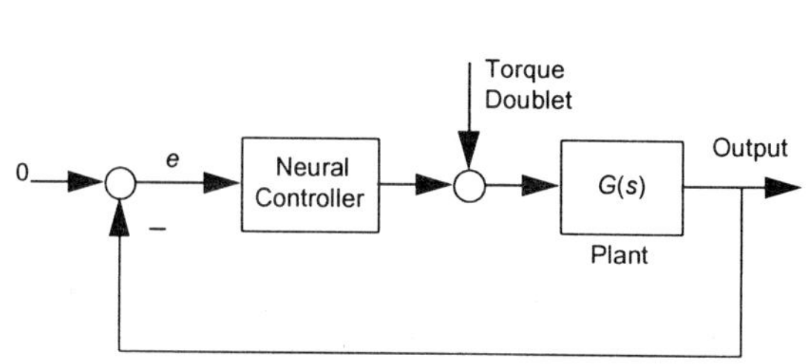

Figure 10 Neural controller for torque doublet rejection

Figure 11 through Figure 14 show the simulation results when the neural controller is applied to examine its ability to reject the torque doublet disturbance. Similarly, four different orientations of the XTE aircraft are considered. Note that different measurement unit from that in Figure 5 through Figure 8 is used here, specifically, *arc-sec* is used instead of *rad*. In addition to the robustness, excellent performance can also be observed from these curves: the output of XTE lies in [–4, 4] for all four orientations over the whole time period.

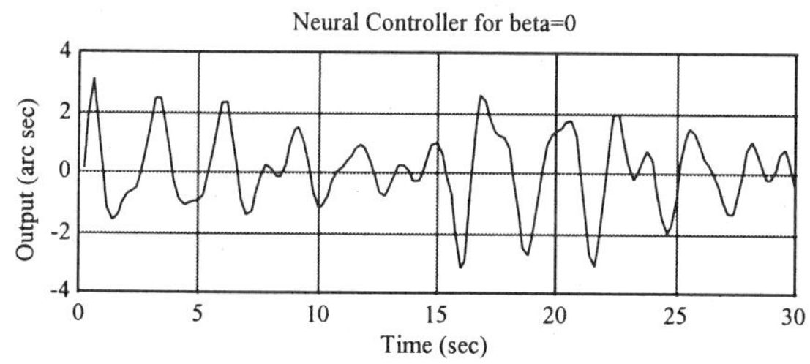

Figure 11 Response of torque doublet for β = 0

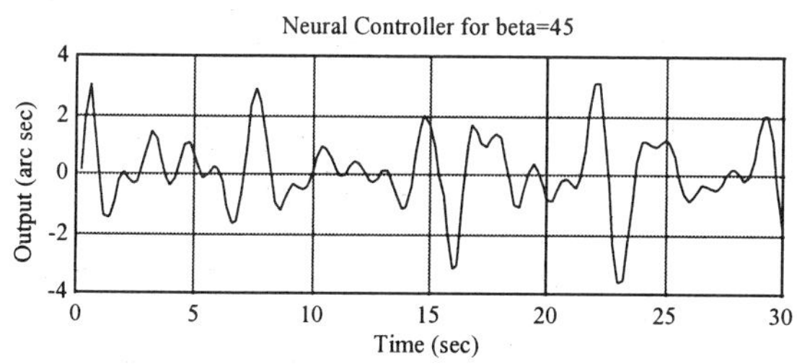

Figure 12 Response of torque doublet for β = 45

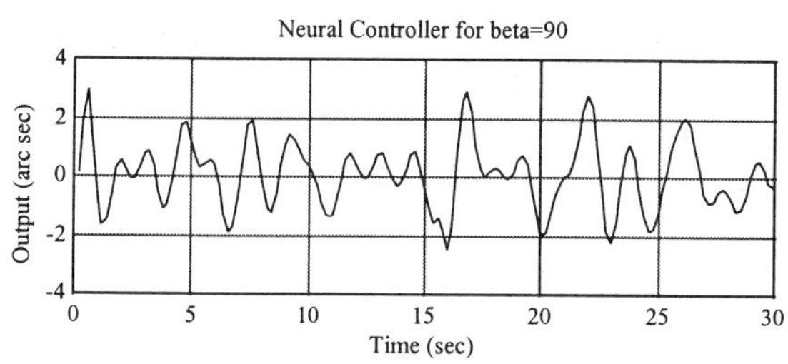

Figure 13 Response of torque doublet for β = 90

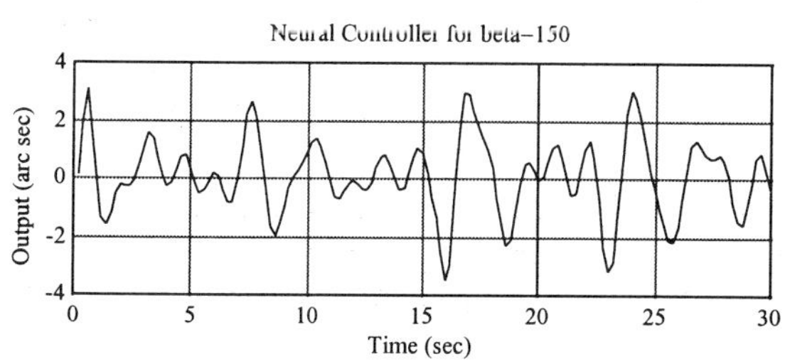

Figure 14 Response of torque doublet for β = 150

6. Conclusions

In this paper, the research focus is on exploring the application of the neural network techniques to spacecraft control, and developing, demonstrating and verifying advanced intelligent control techniques which can improve control accuracy. Specifically, the XTE spacecraft is chosen as a design plant and an intelligent controller, the neural controller, is designed. The software has also been developed to simulate and demonstrate the control performance. The simulation results showed that the designed intelligent controllers are robust and adaptive, work well at all flight conditions, and provide great improvement of the attitude control accuracy over the baseline controller.

References

[1] F. H. Bauer, M. D. Femiano, and G. E. Mosier. Attitude control system conceptual design for the X-ray timing explorer. *Proceedings of AIAA Guidance, Navigation, and Control Conference*, pp. 236-250, August 1992.

[2] E. Grant and B. Zhang. Neural network based reinforced learning. *Proceedings of IEEE Control and Decision Conference*, pp. 856-861, 1992.

[3] T.-H. Guo, J. Musgrave, and C.-S. Lin. Neural network based sensor validation for reusable rocket engines. *Proceedings of American Control Conference*, pp. 1367-1372, 1995.

[4] M. M. Gupta and D. H. Rao. *Neuro-Control Systems: Theory and Applications*. IEEE Press, 1994.

[5] C. F. Lin. *Advanced Control Systems Design*. Englewood Cliffs, NJ, 1994.

AIAA-97-3529

TRENDS IN GNC SYSTEMS FOR EARTH OBSERVATION MISSIONS

A.Tobias*, M.Aguirre, P.Silvestrin, J.Fuchs
Earth Observation Preparatory Programme - ESA/ESTEC
Noordwijk, The Netherlands

ABSTRACT

The trends in GNC systems for Earth Observation satellites are examined looking at the nine first candidates for ESA's Earth Explorer Programme of research oriented missions. These missions range from constellations of microsatellites to large (>2 tons) satellites. All satellites are 3 axis stabilised and two of them include passive stabilisation by gravity gradient or air drag augmented with active magnetic means. Pointing requirements are moderate and vary from 0.6 mrad to 50 mrad while attitude estimation requirements are more demanding and range from 0.01 mrad to 2 mrad and the most demanding attitude stability requirement is 0.001 mrad in 3s. Concerning orbit determination there are requirements for onboard position and velocity estimation with accuracy of 50 m and 0.1 m/s and restitution on ground to cm level. Some missions have no orbit maintenance requirements but one requires continuous compensation of the air drag to better than a fraction of a mN. From the point of view of implementation the synergy between payload and GNC, the wide utilisation of star imagers and of GPS/GLONASS receivers and the introduction of electric propulsion should be noted.

1. INTRODUCTION

The ESA strategy for Earth Observation includes two types of missions[1], the Earth Explorer of scientific research character, and the Earth Watch of operational character with emphasis on services. Examples of operational missions are the geostationary and polar meteorological systems being developed for Eumetsat. Several geostationary satellites of the Meteosat Second Generation (MSG) and polar satellites of the METOP series are under development. New ideas for Earth Watch missions are under study and include for instance a satellite system for the monitoring of coastal zones and related coastal oceans, sea ice and hinterland, and a system for the detection and monitoring of forest fires. The MSG satellites are spin stabilised as all their predecessors of the Meteosat series. The requirements on the GNC systems of the METOP satellites are within the capabilities of the systems used for the SPOT / ERS satellites.

The more scientifically oriented missions will be implemented within the Earth Explorer programme. Nine candidates[2] have been proposed for the first Earth Explorer missions which cover all scientific disciplines of Earth Observation and have been defined without a particular carrier spacecraft in mind. They form therefore a good sample to derive the trends in GNC systems for the immediate future Earth Observation missions.

This paper will introduce the nine Earth Explorer candidates, present the mission / system requirements that drive the GNC and outline the proposed implementation concepts.

2. THE GNC SYSTEMS OF THE EARTH EXPLORER CANDIDATES

The nine candidate missions can be classified in three groups concerning the satellite class:
- Medium-large missions: Radiation, Dynamics and Precipitation
- Intermediate missions with a single instrument onboard: Land, Chemistry and Topography
- Special missions: Gravity, Magnetometry and Atmospheric Profiling.

2.1 "Medium-large" missions

The Earth Radiation mission

The main objectives of this mission are the observation of clouds and aerosols and the understanding of the Earth radiation budget and fluxes. The satellite in a sun-synchronous orbit with 14:00 local time of the descending node,will carry the following instruments:
- a backscatter lidar called ATLID, 50 m vertical resolution,
- a cloud profiling radar, resolution 500 m vertical resolution, 1 km horizontal
- a cloud imager, 5 bands, 500 m spatial resolution
- a broadband radiometer

* Senior member AIAA

The altitude has been selected as low as possible to reduce the requirements on the lidar and the radar, aperture and antenna size and power, hence volume, mass, radiator area, etc. A resonant orbit at 411 km altitude is selected to revisit every two days during five years thousands of imaginary 50-by 50 km^2 cells around the globe.

The satellite, figure 1, has a mass of 1500 kg at launch and has to generate 2 kW of electric power. Orbit manoeuvres are considered to shift the ground track and thus change the set of cells observed. Concerning the GNC system:

- Attitude pointing requirements are not strict, 2.5 mrad, commensurate with the size of the grids of the climatological models.
- Attitude determination is driven by the vertical localisation requirements of the lidar observations, set at 50 m.
- Position determination (by on-ground post processing) must be as good as possible, meters, to leave most of the geolocalisation error allowance to attitude determination
- Attitude stability, 1 mrad/s, is driven by the observation quality of the cloud imager and the operation of the lidar.

Position determination is provided by a GPS/GLONASS receiver. The attitude determination system is based on gyros and star sensors. The error build up from satellite sensor to lidar line of sight is kept limited as required to obtain the 50 m vertical localisation accuracy. Reaction wheels (0.2 Nm and 20 Nms) and magnetorquers 150 Am^2 are used as actuators for attitude control and hydrazine thrusters for orbit maintenance and track drift adjustment. The number of thrusters is minimised as they are not used to produce torques, e.g. a B-dot law without thrusters is used for initial attitude acquisition and safe mode, figure 2.

The Atmospheric Dynamics mission

The objective of this Earth Explorer is to measure wind profiles in clear air from ground. The 2000 kg, 1.6 kW satellite, figure 3, will carry a Doppler wind lidar, based on CO_2 technology, called ALADIN, in a sun synchronous dawn dusk orbit of 480 km. The GNC requirements stem from:

- The need to separate in the total frequency shift of the backscattered signal the contributions of the spacecraft velocity and the Earth's rotation from the contribution of the wind proper. This translates into requirements for attitude determination to 0.035 mrad and velocity estimation to 0.1 m/s
- The vertical localisation of the lidar observations which implies good position determination, 50m
- The need to maintain stable pointing during the travel of the laser pulses.

The suggested implementation reuses the avionics of the ERS / SPOT platform, i.e. gyros and Earth and Sun sensors are used for atttitude determination. The high attitude determination accuracy imposed by the vertical localisation of the ALADIN observations is achieved by including a star sensor at instrument level. A GPS/GLONASS receiver provides position and velocity information. As actuators the standard SPOT / ERS reaction wheels and magnetotorques can be baselined. The stability requirements imply strict control of perturbations.

Other laser technologies are being discussed which could be less demanding for the satellite at the expense of reduced performance. An alternative implementation selected for further study consists of a full experiment in the International Space Station. The experiment will have to include navigation and attitude determination capabilities to cope with the limitations of the station for such demanding payloads.

The Precipitation mission

This mission is intended to measure precipitation by means of active, rain radar, and passive, microwave radiometer, instruments. The satellite is in a 56° inclination 500 km altitude orbit. The mass at launch is 2100 kg and the solar array has to generate almost 5 kW, equivalent to about 36 m^2 of solar array. The dimensions of the rain radar antenna are 6 x 5 m. The main reflector of the radiometer is an off-set parabola of 1.6 x 1.4 m^2 that rotates at more than 30 rpm. The above data give an idea of the internal and external torques to be considered in the GNC design. The satellite configuration is designed to minimise external torques, the radiometer has its own torque compensation mechanism.

Other GNC design drivers are:

- Attitude pointing: it is not demanding as the size of the model grids is large and errors of several mrad can be tolerated.
- Attitude estimation accuracy is driven by the localisation of the radar observations, resulting in 1.8 mrad in roll and 5 mrad in pitch and yaw.

Attitude estimation is performed by gyros updated by Earth and Sun sensors. A potential solution for attitude update in yaw is the use of the GPS/GLONASS receiver[4]. An accuracy of 3-5 mrad is expected. The same receiver is used for position determination. Reaction wheels and magnetotorquers are used as actuators.

2.2 Single instrument missions

These missions are implemented with medium size satellites carrying only one instrument. The performance of the GNC system can be fully used, e.g. satellite slew for aiming the instrument line of sight to the target, the full attitude estimation performance can be used because the alignment error propagation along the structure is minimised, compensation wheels at instrument levels can be avoided, etc.

The Land Surface Processes and Interactions mission

This Earth Explorer will provide observations in around 200 spectral bands 10 nm wide in the 450 to 2350 nm region and additional 1 micron wide bands in the thermal infrared region.of the spectrum with medium (50m) spatial resolution and high radiometric sensitivity. The mission will be implemented with a satellite in sun-synchronous orbit of 660 km of altitude and 11:00 of local time of descending node carrying an instrument called PRISM. The satellite must be able to observe selected sites within less than 3 days which at the selected orbit implies +/- 34° across track pointing. It is also required to be able to observe a site from several angles of view along the track (0°, +/- 30° or +/-45°, +/- 60°, +/-70° are typical values).

The sharing of the pointing functions between instrument and satellite is under study. If the instrument would have to perform both along and across track pointing, as it would be mandatory in a multi-instrument satellite- large image rotation problems would occur that would require large and complex derotation mechanisms. For observation quality reasons it is preferred that the along track pointing is performed by the satellite. This requires a compact, low inertia, and rigid satellite with fixed solar array. For such a satellite the across track pointing would also be feasible and in fact probably easier than the along track pointing. Figure 4 shows a possible satellite concept. The most demanding requirements on the GNC are derived from:

- The manoeuvrability requirements for possibly across and along track pointing. The satellite across track pointing requires roll manoeuvres while along track pointing requires combined roll and pitch manoeuvres in the short time imposed by the orbit velocity. This is not a straightforward task and has required to reduce the inertia of the satellite below 200 kgm^2 and to avoid flexibilities. In addition, yaw steering is required for image quality.
- The geolocalisation of the measurements to one pixel size or 0.050 mrad; though landmarks help to remove static and low frequency errors, the GNC has to retrieve the high frequency part of the error at frequencies $f > 0.05$ Hz.
- The overlapping of observations taken in different passes or from different angles in one pass which implies absolute pointing to better than 1 mrad.

The resulting implementation concept, figure 5, includes:

- A mission management function, based on time and navigation information, follows the site observation plan, activates the guidance generator and sets the spacecraft configuration, including the programming of the instrument (60 bands per site), the calibration operations, etc. The powerful onboard data handling with the capability of parallel-off line processing is also controlled by the GNC management function.
- The guidance generator that elaborates the slew profile and computes the guidance torque, i.e. the torque nominally required to follow the slew profile. This torque command is directly supplied to the reaction wheels in a feedforward path.
- The controller compensates for errors in the estimation and execution of the guidance torque, e.g. errors in the estimation of angles and rates, the knowledge of inertias, and other errors.
- Attitude estimation is performed by means of gyros and star sensors of wide field of view.

- A GPS/GLONASS receiver is used for time reference and for position determination,(the use of the GPS/GLONASS receiver as attitude sensor during the slew manoeuvers has also been considered with very promising results..
- Special measures are used to eliminate the effect of solar array and cryo-cooler induced perturbations.
- An alternative uses CMG to allow faster manoeuvering and reduce power consumption.

Figure 6 shows a pitch manoeuver from -60° to +60° The error is below 0.080 mrad which leaves margin for sensor and actuator errors.

Alternative satellite configurations are shown in figure 7. In these cases the across track pointing is performed by the instrument.

The Atmospheric Chemistry mission

This mission is focused on the study of ozone chemistry and chemistry - climate interactions with emphasis on the upper troposphere and lower stratosphere. A large number of species has to be observed which would require a very large payload. The implementation concept exploits the presence in orbit of other systems like METOP, with chemistry observation capabilities. The Atmospheric Chemistry Explorer, ACE, is therefore focused on the missing observations. A satellite carrying a single limb sounding instrument called MASTER is placed on the same orbit and ahead of METOP (sun-synchronous, 820 km altitude, 09:30 local time of descending node) so that MASTER scans at the limb from 0 to 50 km altitude the volume of atmosphere observed at nadir by the METOP instruments. To cover the sensitive polar regions it must be possible to depoint the system by +/- 10° in yaw. A clear advantage of single instruments satellites with respect to the large modular carriers advocated in the eighties is the possibility of using the platform services to simplify the payload. The +/-10° yaw scanning is provided by the satellite. The more demanding elevation scanning is subject of trade-off. The GNC drivers are:

- Accurate attitude pointing, 0.6 mrad, to use efficiently the scanning range of the limb sounder. Possible combinations of platform and instrument antenna scan are considered.
- Manoeuvrability, easy for yaw scanning, more difficult for the elevation scan.
- Orbit control to maintain formation flying with METOP.
- Accurate attitude determination, 0.060 mrad, for the vertical localisation of the observations
- Accurate position determination, 50 m, to leave most of the geolocalisation error margin to attitude determination.
- Sufficient attitude stability to preserve the observation quality of MASTER, 0.2 mrad/s below 1 Hz.

The above requirements lead to an implementation concept including gyros and star sensors for attitude determination and GPS/GLONASS based navigation. Orbit control is unfrequent and is based on a "boomerang" strategy: the ACE moves naturally around its nominal position on the METOP orbit from ahead and above, intersects the orbit behind its nominal position and moves below and ahead until it is boosted up which is needed at several month intervals. Reaction wheels are used for attitude control; their size and configuration depend on the system scanning strategy.

The Topography mission

Several system concepts have been considered. The simplest option for ice observations consists of a small satellite in dawn-dusk free decaying orbit of initially 560 km altitude. The payload is a laser altimeter. The driver for the GNC of the Topography Explorer is the measurement of ice topography, sea ice thickness and ice sheet elevation as it is based on beam limited altimetry of high horizontal resolution (100m) and vertical accuracy (2 cm) and repeatability (5 cm).

The key requirements on the GNC are:

- Position must be estimated with less than 4 cm radial and 0.5 cm horizontal error.
- The attitude estimation error must be below 0.01 mrad including transfer from altimeter to attitude reference.

The GNC includes gyros and star sensors for attitude determination. Integration of the instrument in the platform structure reduces attitude transfer errors. A dual-band GPS/GLONASS receiver is used for orbit determination. Reaction wheels and magnetotorquers are foreseen.

2.3 Special missions

These missions have very particular

requirements.

The Gravity Field and steady-state Ocean Circulation mission

This Earth Explorer will measure with high accuracy and spatial resolution the Earth's gravity field and its geoid. To achieve the required performance the satellite has to measure the weak gravity signal by flying in a very low orbit, around 260 km, with a very sensititive gravity gradiometer and a GPS/GLONASS receiver. The instruments are at the same time payload and part of the GNC. As payload the gradiometer provides the high resolution terms of the gravity field, as part of the GNC the gradiometer measures the linear and angular accelerations along and around three axes and allows to derive the satellite rates and in particular to measure directly the pitch rate, which is extremely important to derive certain components of the gravity tensor. The GPS/GLONASS receiver allows to derive the low resolution terms of the gravity field and provides time and position information for the GNC.

The low orbit altitude implies high air density and hence drag induced forces and torques. These disturbances are minimised by wise satellite design, figure 8, with small cross section and high symmetry, but still it is required to compensate the air drag to maintain the orbit, to reduce the dynamic range seen by the gravity gradiometer and indeed to cancel its effect within the gradiometer measurement bandwidth (0.005 to 0.1 Hz). The gravity signal could be corrupted not only by external, e.g. air drag, but also internal, e.g. sloshing, effect of moving elements, etc. perturbations. Errors in the reconstruction of the gravity field are also derived from the imperfect attitude pointing, hence allocation of signal to the reference axis, residual attitude rates and accelerations. The requirements on the GNC are established after error budget analysis which includes several iterations of top-down error allocation and bottom-up error budget closure. The resulting requirements are:

- Moderate for attitude pointing and attitude determination, better than 1.5 mrad, to minimise the measurement of gravity gradient terms on the wrong gradiometer axis.
- Very demanding for the determination of attitude rates, better than $5x10^{-8}$ rad/s
- Demanding on angular accelerations which must be lower than $5x10^{-7}$ rad/s^2 and have to be estimated with accuracy better than 10^{-9} rad/s^2
- Linear accelerations must not exceed $2x10^{-8}$ m/s^2 in the measurement bandwidth
- The orbit must be maintained.
- The GNC must not introduce perturbations, (this implies for instance that gyros are not used, reaction wheels are not desired and high Isp electric propulsion is preferred to eliminate sloshing).

The above requirements lead to the following implementation concept:

- The determination of attitude, angular rates and angular accelerations is based on the gravity gradiometer and star sensors.
- Position determination is based on a dual-band GPS/GLONASS receiver.
- Orbit maintenance and drag compensation along the velocity axis are achieved by electric propulsion. Two ion thrusters are used that consumme 30 kg of xenon during the 18 month mission. Sloshing is avoided. A special flow controller avoids the perturbations otherwise induced by all valve opening and closing actions.
- Drag compensation has to be implemented to reduce the effect along the satellite velocity, several mN. No drag compensation is required on the transverse axes due to the negligible effect of the low disturbances, figure 9, taking into account the capability of the instrument to further reduce those disturbances.
- Several options have been considered for attitude control which depend on the choice of gravity gradiometer, such as proportional helium or neon thrusters if an inductive superconductive gradiometer working at cryogenic temperature is used, impulsive cold gas N_2 thrusters, reaction wheels and electric propulsion if the boil-off gas of the cryostat is not available[3].

Figure 11 shows a potential GNC implementation concept.

A two level approach is proposed for orbit maintenance and drag compensation. To compensate the mean and the low harmonic terms of the air drag the guidance generator, based on the onboard navigation, issues the force commands in a feedforward path to the thrust controller. A closed loop based on the linear acceleration measurements of the gravity gradiometer is used to compensate the variations in the control bandwidth. Ion thrusters have been used in on/off mode for Eureca and are planned for Artemis. For the Gravity mission the utilisation will be different implying relatively large amplitude variations at low frequency with superimposed fast variations of low amplitude. There are strict requirements on thrust, including thrust orientation, accuracy and stability. Several ion thruster candidates have been successfully tested. Testing has been performed also for cold gas N_2 thrusters and reaction wheels with promising results.

The Magnetometry mission

The Magnetometry Explorer is devoted to the measurement of the Earth's magnetic field with high accuracy and spatial resolution. The mission includes two satelllites called Geomagnetic Space Observatories (GSO) in perfect polar orbit, 90° inclination of different altitude. The higher altitude satellite is the GSO-H which is at 600 km altitude for more than 5 years. The lower satellite is the GSO-L that flies below 250 km altitude for more than 6 months. The core payload consists of vector and scalar magnetometers, star imagers and GPS/GLONASS receivers and detectors for charged particles. The drivers for the GNC are:

- The accurate reference of vector magnetometer observations to the Earth centered frame.. A maximum 0.010 mrad error is established for attitude determination.
- The determination of position along the orbit to reduce the equivalent error in magnetic field (product position error times field gradient).
- Datation, as timing errors are equivalent to position errors.
- Magnetic cleanliness to avoid contamination of magnetometer observations.
- The high air drag perturbations for the GSO-L.

The above drivers lead to the following baseline concept:

- The GSO-H is mainly stabilised by gravity gradient. The boom required to separate the magnetometers from the satellite to achieve magnetic cleanliness is also used for attitude stabilisation. Air core coils are used to supplement passive stabilisation. The GSO-L is stabilised by air drag. Actuation is limited, magnetic cleanliness improved, field restitution accuracy is maximised.
- Attitude determination is achieved by means of wide field star imagers. Attitude transfer is achieved mounting the vector magnetometer and the star imager on an optical bench.
- Position determination and time reference are provided by a GPS/GLONASS receiver.

The Atmospheric Profiling mission

The objective of this mission is to provide profiles of temperature with 1 K accuracy, < 1 km vertical resolution from 8 to 30 km, and profiles of water vapour content with 10 % accuracy at lower altitudes. This is achieved by exploiting the delays experienced by the signals from GPS and GLONASS satellites that travel through the atmosphere before reaching an appropriate GPS/GLONASS receiver located on a low Earth orbit satellite. ESA is developing a GPS/GLONASS receiver[4] for atmospheric sounding, GRAS. GRAS could be embarked as passenger in suitable LEO satellites as already foreseen for METOP, or on dedicated microsatellites forming constellations.

In the case of microsatellites GRAS is also used to provide datation, position and attitude determination. The microsatellite, which could be called a GRAS-craft, is box shaped, smaller than the Ariane 5 ASAP class, and includes 2 directive sounding antennas on velocity and antivelocity sides and 4 navigation and attitude determination antennas on the zenith side. Position and velocity determination are typical tasks of the receiver. The associated performance requirements are twofold: onboard position and velocity estimation accuracy can easily support the acquisition of rising GPS and GLONASS satellites, while on ground position restitution shall support the derivation of the extra path of the occulting GPS and GLONASS satellites which is equivalent to an accuracy of 0.5 m.

GPS based attitude determination was first studied in ESA almost 10 years ago and algorithms have been improving continuously. The performance for a four antenna square pattern of 0.5 m side is estimated to be better than 0.1° since the "clean" satellite configuration allows to reduce the feared multipath effects to a bare minimum.

GRAS, figure 12, is a combined GPS/GLONASS receiver, with 16 dual-band channels that can be freely allocated to GPS or GLONASS satellites, to atmospheric sounding or navigation and attitude determination functions. The receiver is also capable of codeless tracking operation.

3. SUMMARY OF TRENDS

Misison classes

- All satellites are 3-axis actively stabilised. GSO-H and GSO-L use also passive stabilisation techniques.
- The missions with several instruments or conventional ERS/SPOT platforms, have no manoeuvrability requirements and only moderate performance requirements and may need augmentation at instrument level (Radiation, Dynamics and Precipitation),
- The missions as Land, Chemistry and Topography with a single instrument. This suggests an optimisation of the overall system to enhance performance at lower cost. Examples are the provision of across and along track pointing by the platform and the elimination of attitude sensors at instrument level. Standardisation of GNC systems can be intended for this class of missions targetting the highest performance requirements.
- There are thre special cases: Gravity, Magnetometry and Atmospheric Profiling.

Requirements

- Pointing requirements are not more demanding than in previous ESA missions. Attitude determination, including rates and accelerations, requirements are more demanding than so far.
- Requirements for manoeuvrability have emerged.
- The Gravity mission has very particular requirements.

Position determination

- Autonomous onboard position, velocity and time determination is baselined. A GPS/GLONASS recever is used. Real-time performances are in the order of 30 m., 0.1 m/s and 1 microsecond. After ground processing performance is dramatically increased as needed for the gravity, atmospheric profiling and magnetometry missions especially.

Attitude determination

- Attitude determination is in most cases performed with gyros and star imagers. Exceptions are the gravity mission which uses the gradiometer to replace the gyros, the magnetometry mission which does not need gyros for the GSO-H and the microsatellites of the atmospheric profiling constellation that use the sounding receiver GRAS also for attitude determination.
- The advent of wide field affordable star sensors is eliminating Earth and fine Sun sensors in nominal operations and relegated sun sensors to acquisition and back up modes.
- There is potential for GNSS (GPS/GLONASS) based attitude determination especially in the smaller satellites where the disturbing effect of multipath can be reduced.

Actuators:

- Chemical propulsion is used for orbit control in most missions. Exceptions are the Gravity mission that prefers electric propulsion, the GSO-H satellite and the simplest option of the Topography Explorer that do not require orbit maintenance. The advantage of electric propulsion for the Gravity Explorer is illustrated in figure 13.
- The use of magnetic means for attitude control in acquisition and safe modes may reduce to a minimum the number of thrusters.
- Ion thrusters can be used for continuous and accurate drag compensation.
- Electric propulsion is however not confirmed for more conventional missions. Advances in power generation and storage systems seem to be a pre-requisite for utilisation of electric propulsion in low Earth orbit.
- Reaction wheels are the most used actuator for attitude control in nominal modes. Exceptions are the GSO-H and most options of the Gravity mission.
- There is a trend towards manoeuvrability. The use of CMG is being considered for the Land mission.

Autonomy

- The Land Explorer will perform frequent fast and large slew manoeuvres outside ground control. The Gravity Explorer is continuously compensating air drag, i.e. maintaining its orbit.
- The Land mission uses the onboard GNSS time and position information to automatically sequence onboard a complex observation plan.

Verification

-It is worth mentioning the initimate relation

between GNC performance and overall mission return, and between payload (gravity gradiometer) and GNC for the Gravity Explorer. Early verification has been started in an appropriate test bed capable of providing increasing level of fidelity.
- Similar approach will probably be undertaken in the near future for other candidate Earth Explorers.

References

[1] ESA, 1994, Report of the Earth Observation User Consultation meeting, Noordwijk, ref. SP-1186.

[2] ESA, 1996, Reports for Assessment.The nine candidate Earth Explorer missions, ref. SP-1196

[3]M. Aguirre et. al. “Propulsion Systems for the Gravity Explorer Mission”2nd European Spacecraft Propulsion Conference, Noordwijk, 1997.

[4] P.Silvestrin et.al. “GNSS Receiver Design”3rd International Conference on Spacecraft GNC Systems. Noordwijk, 1996, ref. ESA SP-381.

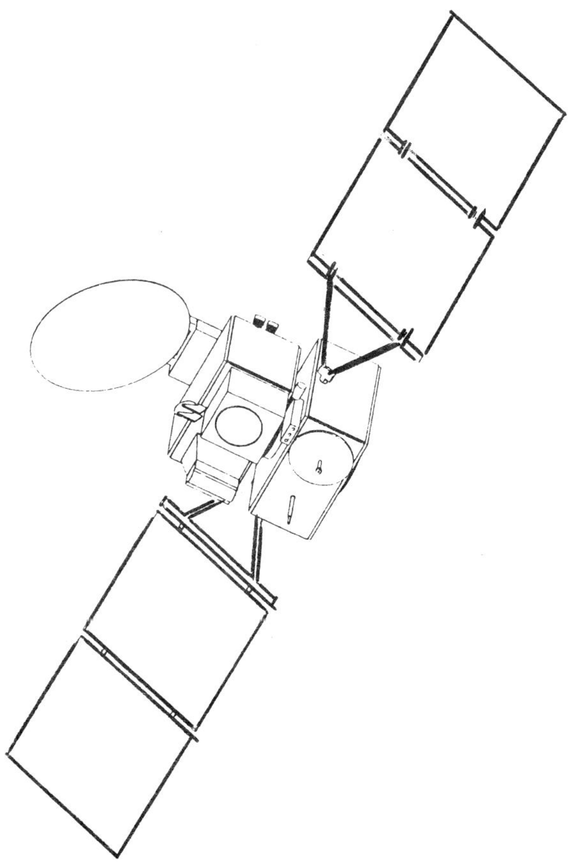

Figure 1. The Earth Radiation Explorer dominated by the 2m antenna of the Cloud Radar.

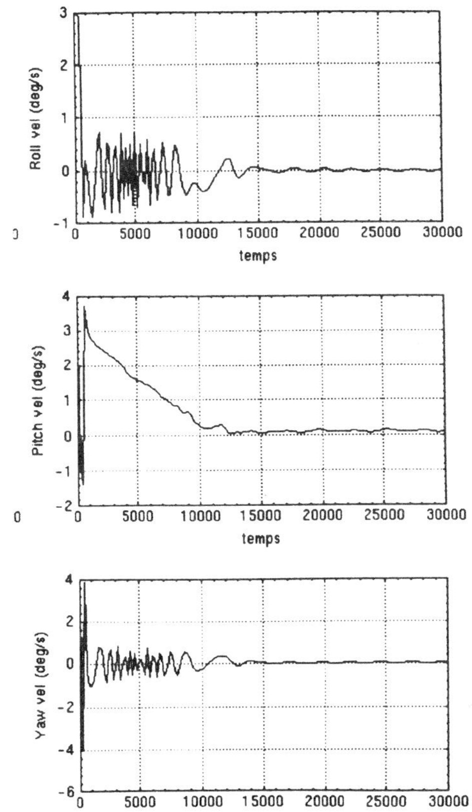

Figure 2. The B-dot law allows to stabilise the attitude in 3 orbits.

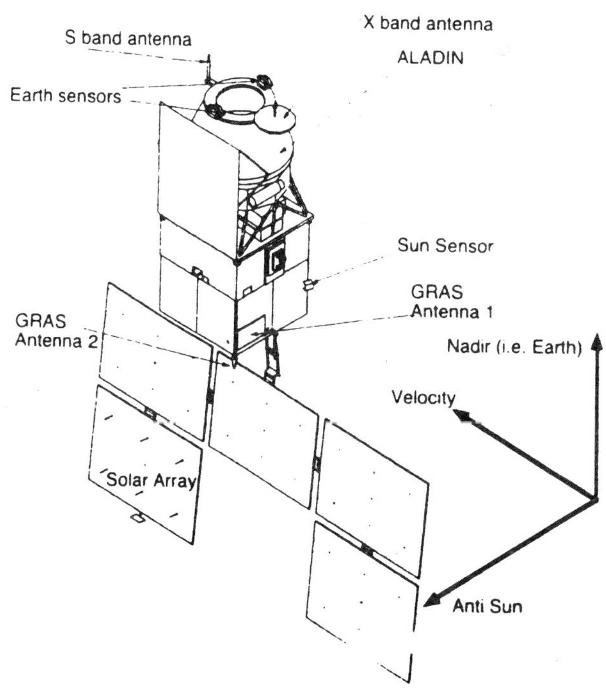

Figure 3. The Atmospheric Dynamics Explorer on a ERS/SPOT type bus.

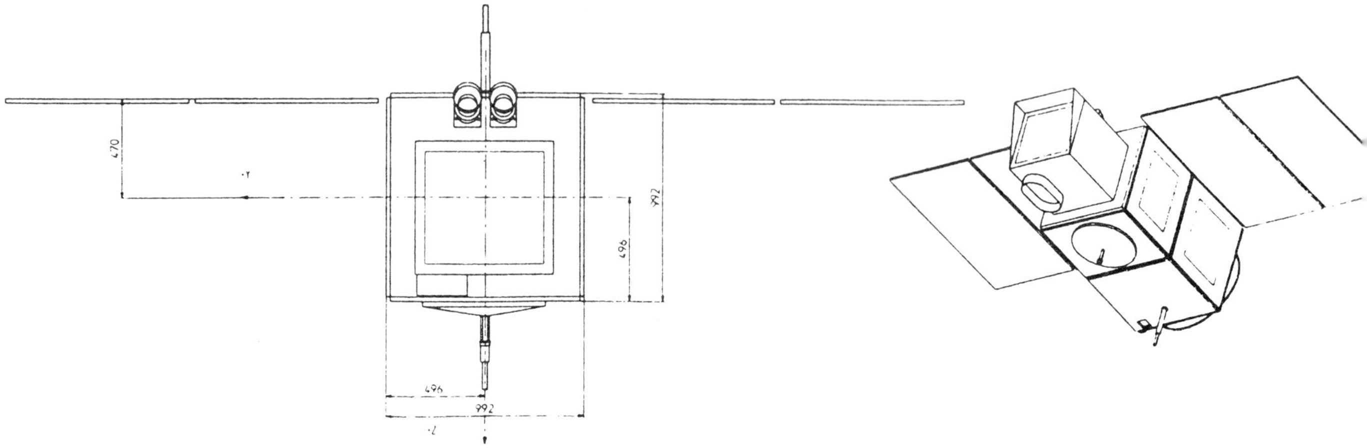

Figure 4. A possible Land Explorer concept for high manoeuvering capabilities

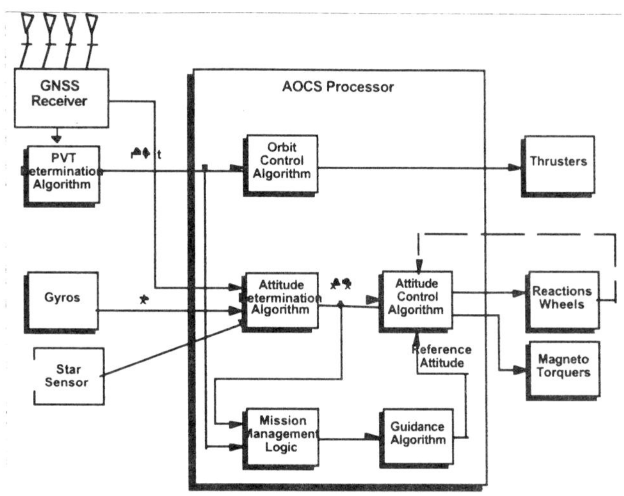

Figure 5. The GNC of the Land Explorer.

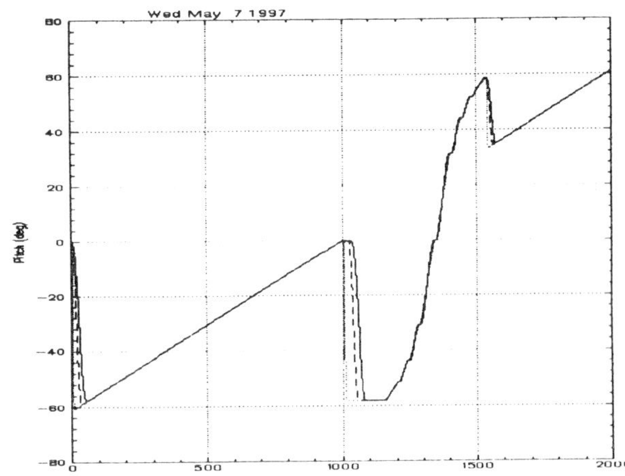

Figure 6. Response of the Land Explorer to large pitch slew manoeuvres

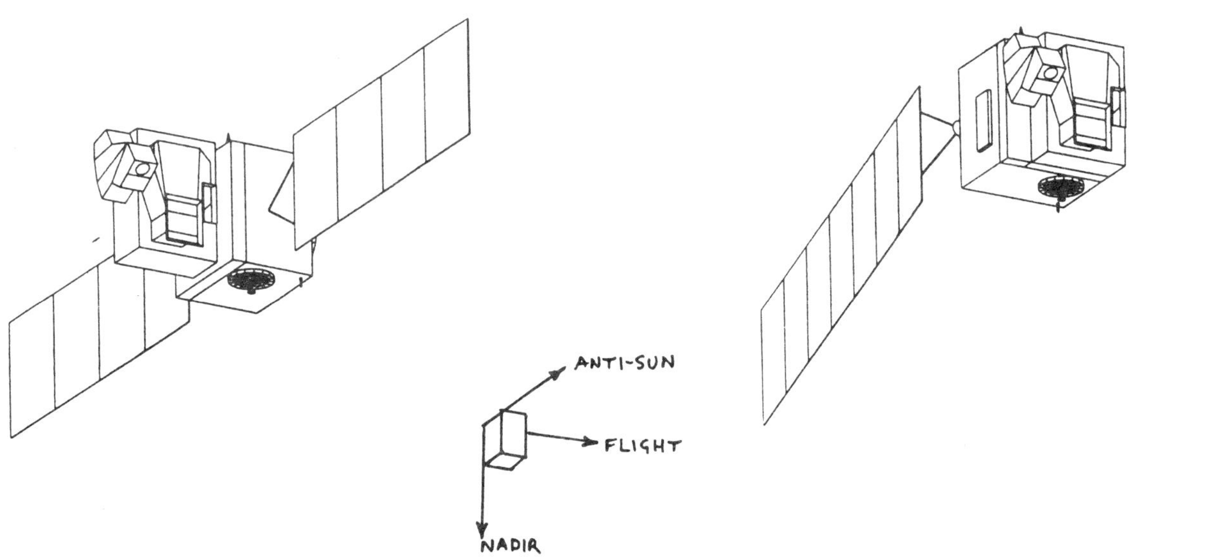

Figure 7. Land Explorer concepts with reduced manoeuvrability requirements.

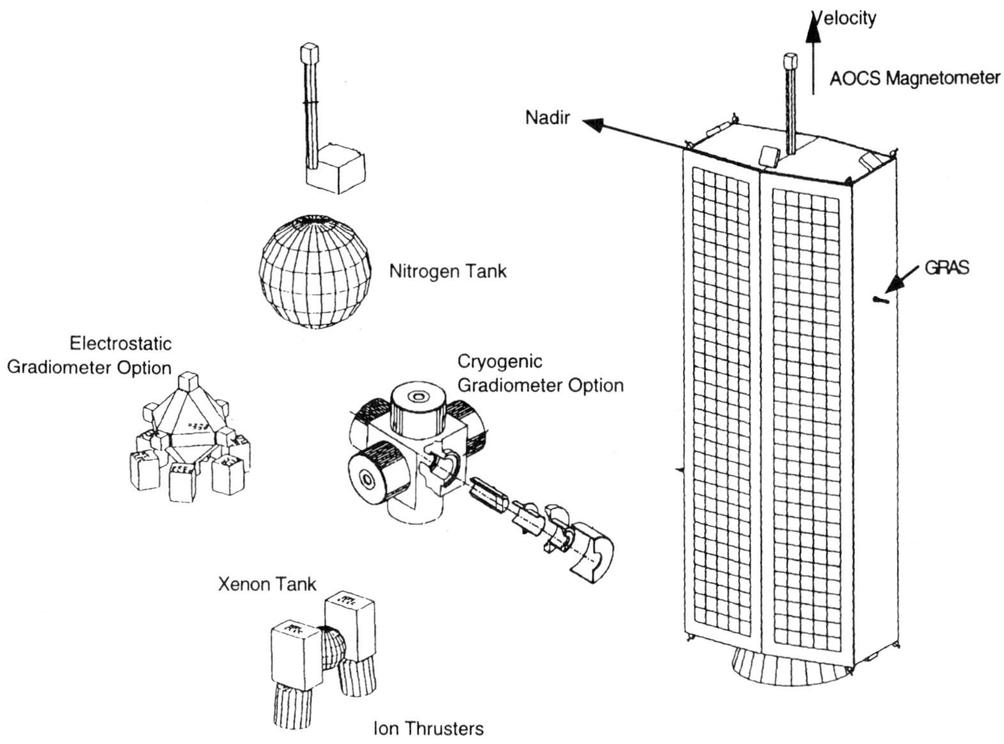

Figure 8. The Gravity Explorer. Small cross section, total symmetry, minimum disturbances.

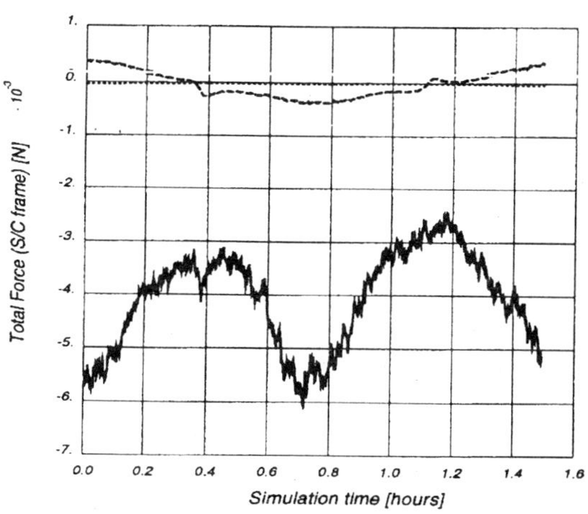

Figure 9. Gravity Explorer, air drag forces.

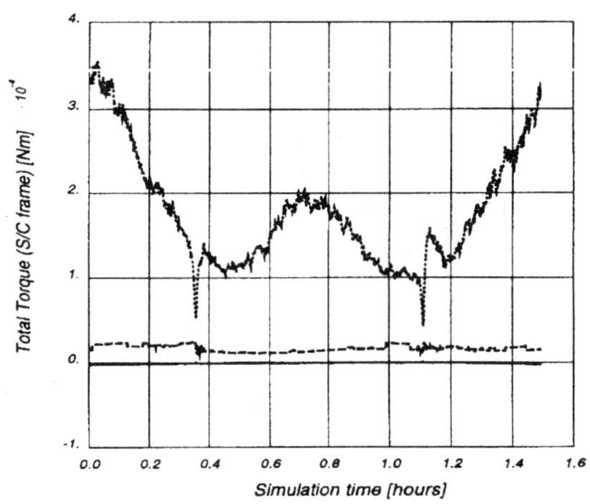

Figure 10. Gravity Explorer, air drag induced torque.

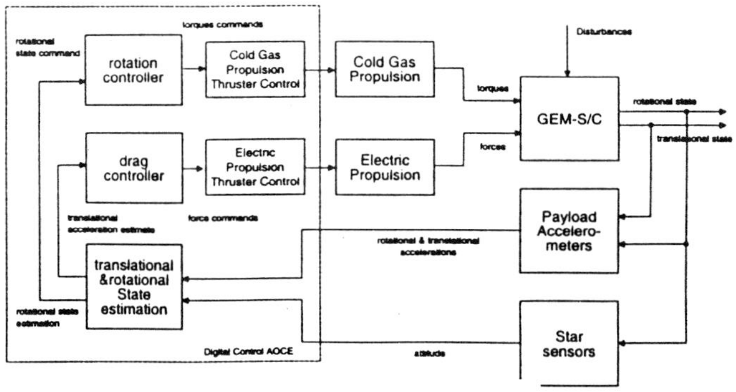

Figure 11. Gravity Explorer GNC concept.

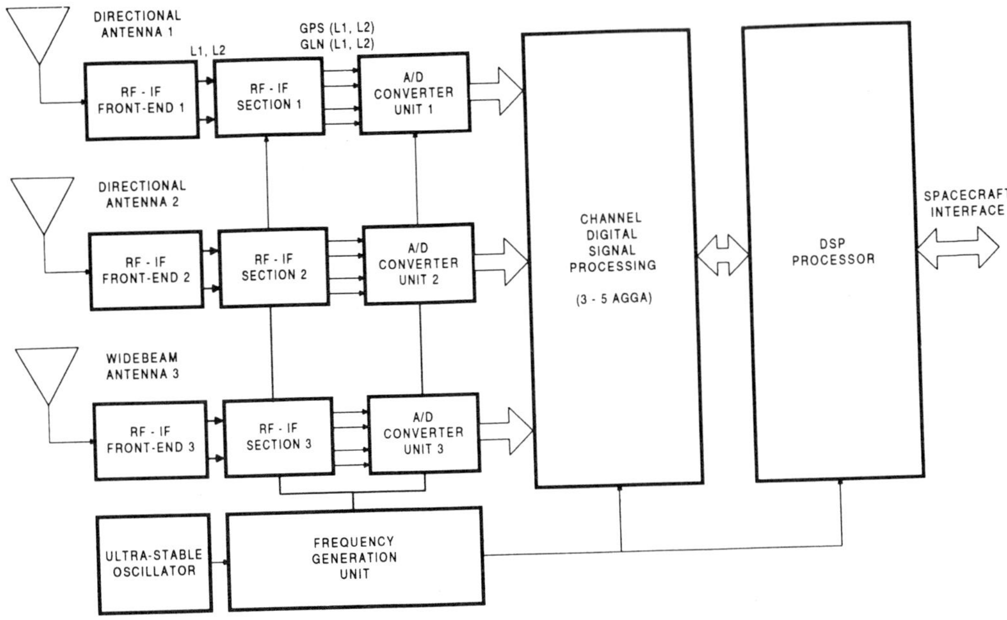

Figure 12. GRAS block diagram

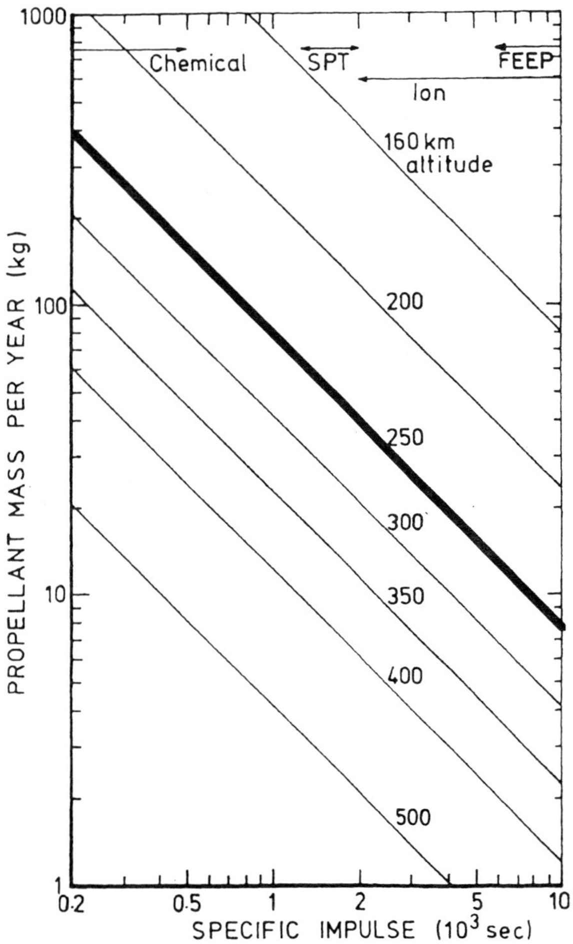

Figure 13. Ion propulsion choice for the Gravity Explorer.

AIAA-97-3530

ADAPTIVE ASYMPTOTIC TRACKING OF SPACECRAFT ATTITUDE MOTION WITH INERTIA MATRIX IDENTIFICATION

by

Jasim Ahmed, Vincent T. Coppola, Dennis S. Bernstein*
Department of Aerospace Engineering, University of Michigan
Ann Arbor, Michigan 48109-2118
(313) 764-3719, (313) 763-0578 (FAX)
{jasim, coppola, dsbaero}@engin.umich.edu

Abstract

The problem of a spacecraft tracking a desired trajectory is defined and addressed using adaptive feedback control. The control law, which has the form of a sixth-order dynamic compensator, does not require knowledge of the inertia of the spacecraft. A Lyapunov argument is used to show that tracking is achieved globally. A simple spin about the intermediate principal axis and a coning motion are commanded to illustrate the control algorithm. Finally, periodic commands are used to identify the inertia matrix of the spacecraft.

I. Introduction

The present generation of spacecraft require attitude control systems that provide rapid acquisition, tracking and pointing capabilities, while the equations that govern large angle maneuvers are coupled and nonlinear. As such, control system design must consider the nonlinear dynamics. Furthermore, since the mass properties of the spacecraft may be uncertain or may change due to fuel usage and articulation, it is necessary for the control system to be able to adapt changes in mass distribution. In this paper, we present a feedback control algorithm that achieves large angle tracking of velocity and attitude commands in spite of inertia uncertainty. Furthermore, we show how this tracking algorithm can be used to identify the spacecraft inertia matrix.

Many open-loop control strategies have been developed for large angle maneuvers[1,2]. However these open-loop schemes are generally sensitive to spacecraft parameter uncertainties, unexpected disturbances and initial attitude rates. In Refs. 3 and 4, global reorientation is achieved using body-fixed actuators without knowledge of the inertia of the spacecraft. The algorithms are based on decreasing the energy of the spacecraft until the desired orientation is achieved.

The attitude tracking problem has been considered in Ref. 5 using a locally convergent adaptive algorithm while adaptive feedback linearization is used in Ref. 6 to achieve tracking. However, the method of Ref. 6 requires measurements of the orientation and angular velocity of the spacecraft as well as angular acceleration. An adaptive tracking scheme has been developed that is globally valid except for a singularity[7]. A switching maneuver is used to avoid the singularity.

In the present paper, a spacecraft driven by reaction jets is commanded to track a desired trajectory using feedback control. The control algorithm requires no knowledge of the spacecraft inertia and re-

*Research supported in part by the Air Force Office of Scientific Research under grant F49620-95-1-0019 and the University of Michigan office of the Vice President for Research.

quires measurements of the orientation and angular velocity of the spacecraft. However, measurements of the angular acceleration are not required. The control law is globally valid, that is, free of singularities and has the form of a sixth-order dynamic compensator. The algorithm is adaptive in the sense of Ref. 8, p. 1 since it contains adjustable parameters and a mechanism for adjusting these parameters. Finally, the control algorithm is shown to be able to identify the spacecraft inertia matrix using a class of periodic commands.

We illustrate the tracking algorithm by commanding the spacecraft to perform a simple spin about its intermediate principal axis with the axis required to point in a given inertial direction. While the algorithm requires three torque inputs, we require no knowledge of the inertia except for the direction of the intermediate principal axis in the body frame. The problem of stabilizing a spacecraft about an intermediate principal axis without inertial pointing has been studied. For instance, in Refs. 9 and 10 the maneuver is performed using a single control torque. Next, a coning motion is commanded using the tracking algorithm. Finally, identification of the spacecraft inertia matrix is illustrated using a periodic command.

II. Equations of Motion

The spacecraft is modeled as a rigid body with actuators that provide body-fixed torques about three mutually perpendicular axes that define a body frame $\mathcal{B}$ with axes $X_{\mathcal{B}}$, $Y_{\mathcal{B}}$ and $Z_{\mathcal{B}}$. For each axis this assumption can be realized by employing a pair of actuators to produce equal and opposite forces perpendicular to the line joining the actuators. Note that the center of mass of the spacecraft does not need to be known. For $t \geq 0$, the equations of motion of the spacecraft are given by

$$J\dot{\Omega} = -\Omega^{\times} J\Omega + u, \tag{1}$$

$$\dot{\epsilon} = \frac{1}{2}\left(\epsilon^{\times}\Omega + \zeta\Omega\right), \tag{2}$$

$$\dot{\zeta} = -\frac{1}{2}\epsilon^{\mathrm{T}}\Omega, \tag{3}$$

where $\Omega = \Omega(t) \in \Re^3$ is the inertial angular velocity of the spacecraft with respect to an inertial frame $\mathcal{I}$,

$$J = \begin{bmatrix} J_{11} & J_{12} & J_{13} \\ J_{12} & J_{22} & J_{23} \\ J_{13} & J_{23} & J_{33} \end{bmatrix}$$

is the constant positive-definite inertia matrix of the spacecraft, both expressed in $\mathcal{B}$, $u = u(t) \in \Re^3$ is the vector of control torques, $(\epsilon, \zeta) = (\epsilon(t), \zeta(t)) \in \Re^3 \times \Re$ are the Euler parameters[11] representing the orientation of $\mathcal{B}$ with respect to an inertial frame $\mathcal{I}$ satisfying the constraint

$$\epsilon^{\mathrm{T}}\epsilon + \zeta^2 = 1,$$

and $(\dot{\ })$ denotes the derivative with respect to time t. The notation $a^{\times}$ for $a = [a_1\ a_2\ a_3]^{\mathrm{T}}$ denotes the skew-symmetric matrix

$$a^{\times} = \begin{bmatrix} 0 & -a_3 & a_2 \\ a_3 & 0 & -a_1 \\ -a_2 & a_1 & 0 \end{bmatrix}.$$

The rotation matrix $B = B(\epsilon, \zeta) \in \mathrm{SO}(3)$ relating $\mathcal{B}$ to $\mathcal{I}$ is given by

$$B = (\zeta^2 - \epsilon^{\mathrm{T}}\epsilon)I_{3\times 3} + 2\epsilon\epsilon^{\mathrm{T}} - 2\zeta\epsilon^{\times},$$

where $I_{3\times 3}$ is the 3×3 identity matrix. We assume that (ϵ, ζ) and hence B are known for all time $t \geq 0$. However, J is assumed to be unknown.

Let the desired rotational motion of the spacecraft be described by the attitude motion of a frame $\mathcal{D}$ whose orientation with respect to $\mathcal{I}$ is specified by the Euler parameters $(\xi, \mu) = (\xi(t), \mu(t)) \in \Re^3 \times \Re$ satisfying

$$\xi^{\mathrm{T}}\xi + \mu^2 = 1. \tag{4}$$

Let $D = D(\xi, \mu) \in \mathrm{SO}(3)$ be the corresponding rotation matrix given by

$$D = (\mu^2 - \xi^{\mathrm{T}}\xi)I_{3\times 3} + 2\xi\xi^{\mathrm{T}} - 2\mu\xi^{\times}, \tag{5}$$

and let $\nu = \nu(t) \in \Re^3$ denote the angular velocity of $\mathcal{D}$ with respect to $\mathcal{I}$ and expressed in $\mathcal{D}$ by

$$\nu = 2(\mu\dot{\xi} - \dot{\mu}\xi) - 2\xi^{\times}\dot{\xi}. \tag{6}$$

Let the time derivative of ν be denoted by $\dot{\nu} = \dot{\nu}(t) \in \Re^3$. Then $\dot{\nu}$ is given by

$$\dot{\nu} = 2(\mu\ddot{\xi} - \ddot{\mu}\xi) - 2\xi^{\times}\ddot{\xi}. \tag{7}$$

Let $(\varepsilon, \eta) = (\varepsilon(t), \eta(t)) \in \Re^3 \times \Re$ be the Euler parameters representing the orientation of frame $\mathcal{B}$ with respect to $\mathcal{D}$. Then (ε, η) satisfy

$$\varepsilon^{\mathrm{T}}\varepsilon + \eta^2 = 1. \tag{8}$$

The Euler parameters (ε, η) are related to (ξ, μ) and (ϵ, ζ) by the quaternion multiplication rule (Ref. 11, p. 17)

$$\begin{aligned} \varepsilon &= \mu\epsilon - \zeta\xi + \epsilon^{\times}\xi, && (9)\\ \eta &= \mu\zeta + \xi^{\mathrm{T}}\epsilon. && (10) \end{aligned}$$

The corresponding rotation matrix $C = C(\varepsilon, \eta) \in \mathrm{SO}(3)$ is given by

$$C = (\eta^2 - \varepsilon^{\mathrm{T}}\varepsilon)I_{3\times 3} + 2\varepsilon\varepsilon^{\mathrm{T}} - 2\eta\varepsilon^{\times} \tag{11}$$

and is related to B and D by

$$C = BD^{\mathrm{T}}. \tag{12}$$

The angular velocity $\omega = \omega(t) \in \Re^3$ of $\mathcal{B}$ with respect to $\mathcal{D}$ and expressed in $\mathcal{B}$ is then

$$\omega = \Omega - C\nu. \tag{13}$$

We assume that the desired maneuver is specified in terms of (ξ, μ), which are assumed to be C^2 functions, and that measurements of Ω and the attitude of $\mathcal{B}$ with respect to $\mathcal{I}$ are available. Using (5), (6) and (7) the quantities ν, $\dot{\nu}$ and D can be computed from (ξ, μ). Furthermore, since (ϵ, ζ) and B can be calculated from the measurements of the attitude of $\mathcal{B}$, it follows that C, ε, and η can be determined using (9), (10) and (12). The angular velocity ω is then determined using (13). The following problem expresses the requirement that the attitude and angular velocity of frames $\mathcal{B}$ and $\mathcal{D}$ should coincide asymptotically.

Tracking Problem: Let $\xi : [0, \infty) \rightarrow \Re^3$ and $\mu : [0, \infty) \rightarrow \Re$ be given C^2 Euler parameters satisfying (4) for all $t \geq 0$. Find a dynamic feedback control law of the form

$$\begin{aligned} \dot{\alpha} &= f(\hat{\alpha}, \omega, \varepsilon, \eta, \nu, \dot{\nu}), && (14)\\ u &= g(\hat{\alpha}, \omega, \varepsilon, \eta, \nu, \dot{\nu}), && (15) \end{aligned}$$

such that $C \rightarrow I_{3\times 3}$ and $\omega \rightarrow 0$ as $t \rightarrow \infty$.

Note that the attitude convergence condition $C \rightarrow I_{3\times 3}$ does not necessarily imply the angular velocity convergence condition $\omega \rightarrow 0$.

Using (8) and (11) it follows that $\varepsilon \rightarrow 0$ if and only if $C \rightarrow I_{3\times 3}$. Hence the tracking problem is solved if and only if $\omega \rightarrow 0$ and $\varepsilon \rightarrow 0$. Rewriting (1), (2) and (3) in terms of ω, ε and η we obtain

$$\begin{aligned} J\dot{\omega} &= -(\omega + C\nu)^{\times}J(\omega + C\nu)\\ &\quad + J(\omega^{\times}C\nu - C\dot{\nu}) + u, && (16)\\ \dot{\varepsilon} &= \frac{1}{2}\left(\varepsilon^{\times}\omega + \eta\omega\right), && (17)\\ \dot{\eta} &= -\frac{1}{2}\varepsilon^{\mathrm{T}}\omega. && (18) \end{aligned}$$

These equations describe the motion of the spacecraft with respect to $\mathcal{D}$. We observe that the tracking problem has been converted into an asymptotic stabilization problem for ω and ε in (16), (17) and (18).

III. Adaptive Control Law

In this section, we present a feedback control law that asymptotically tracks a desired maneuver and thus satisfies the requirements of the tracking problem. The control law is global in the sense that asymptotic tracking is achieved for arbitrary initial conditions.

We observe that the inertia parameters J_{ij}, where i, $j = 1, 2, 3$, appear linearly in equation (16). To isolate these parameters, we define a linear operator $L : \Re^3 \to \Re^{3\times 6}$ acting on $a = [a_1\ a_2\ a_3]^{\mathrm{T}}$ by

$$L(a) = \begin{bmatrix} a_1 & 0 & 0 & 0 & a_3 & a_2 \\ 0 & a_2 & 0 & a_3 & 0 & a_1 \\ 0 & 0 & a_3 & a_2 & a_1 & 0 \end{bmatrix}. \quad (19)$$

Letting

$$\alpha \triangleq [J_{11}\ J_{22}\ J_{33}\ J_{23}\ J_{13}\ J_{12}]^{\mathrm{T}},$$

it follows that

$$Ja = L(a)\alpha.$$

Equation (16) can now be rewritten in the form

$$J\dot{\omega} = F(\omega, C, \nu, \dot{\nu})\alpha + u, \quad (20)$$

where $F : \Re^3 \times \Re^{3\times 3} \times \Re^3 \times \Re^3 \to \Re^{3\times 6}$ is defined by

$$\begin{aligned} F(\hat{\omega}, \hat{C}, \hat{\nu}, \hat{\rho}) &\triangleq -(\hat{\omega} + \hat{C}\hat{\nu})^{\times} L(\hat{\omega} + \hat{C}\hat{\nu}) \\ &\quad + L(\hat{\omega}^{\times}\hat{C}\hat{\nu} - \hat{C}\hat{\rho}). \quad (21) \end{aligned}$$

Theorem 1: Assume that ν and $\dot{\nu}$ are bounded and let $K_1 \in \Re^{3\times 3}$, $K_2 \in \Re^{3\times 3}$ and $Q \in \Re^{6\times 6}$ be positive definite. Then the control law

$$\begin{aligned} \dot{\hat{\alpha}} &= Q^{-1}[F(\omega, C, \nu, \dot{\nu}) \\ &\quad + G(\omega, \varepsilon, \eta)]^{\mathrm{T}}[\omega + K_1\varepsilon], \quad (22) \\ u &= -[F(\omega, C, \nu, \dot{\nu}) + G(\omega, \varepsilon, \eta)]\hat{\alpha} \\ &\quad - (K_2K_1 + I_{3\times 3})\varepsilon - K_2\omega, \quad (23) \end{aligned}$$

where $G : \Re^3 \times \Re^3 \times \Re \to \Re^{3\times 6}$ is defined by

$$G(\hat{\omega}, \hat{\varepsilon}, \hat{\eta}) \triangleq -\frac{1}{2}L(K_1(\hat{\varepsilon}^{\times}\hat{\omega} + \hat{\eta}\hat{\omega})) \quad (24)$$

and F is given by (21), solves the tracking problem. Furthermore, $\hat{\alpha}$ is bounded for all $t \geq 0$ and $\dot{\hat{\alpha}} \to 0$ as $t \to \infty$.

Proof: Define σ, γ and β by

$$\begin{aligned} \sigma &= \omega + K_1\varepsilon, \quad (25) \\ \gamma &= \eta - 1, \quad (26) \\ \beta &= \alpha - \hat{\alpha}. \quad (27) \end{aligned}$$

Using (16), (17), (18), (22) and (23) we obtain the 13-dimensional system

$$\begin{aligned} J\dot{\sigma} &= [H(\sigma, \varepsilon, \gamma, \nu, \dot{\nu}) + M(\sigma, \varepsilon, \gamma)]\,\alpha + u, \quad (28) \\ \dot{\varepsilon} &= \frac{1}{2}\left[\varepsilon^{\times}\sigma - \varepsilon^{\times}K_1\varepsilon\right. \\ &\quad \left. + (\gamma + 1)(\sigma - K_1\varepsilon)\right], \quad (29) \\ \dot{\gamma} &= -\frac{1}{2}\varepsilon^{\mathrm{T}}(\sigma - K_1\varepsilon), \quad (30) \\ Q\dot{\beta} &= [H(\sigma, \varepsilon, \gamma, \nu, \dot{\nu}) + M(\sigma, \varepsilon, \gamma)]^{\mathrm{T}}\sigma, \quad (31) \end{aligned}$$

where $H : \Re^3 \times \Re^3 \times \Re \times \Re^3 \times \Re^3 \to \Re^{3\times 6}$ is defined by

$$H(\hat{\sigma}, \hat{\varepsilon}, \hat{\gamma}, \hat{\nu}, \hat{\rho}) \triangleq F(\hat{\sigma} - K_1\hat{\varepsilon}, C(\hat{\varepsilon}, \hat{\gamma} + 1), \hat{\nu}, \hat{\rho}), \quad (32)$$

$M : \Re^3 \times \Re^3 \times \Re \to \Re^{3\times 6}$ is defined by

$$M(\hat{\sigma}, \hat{\varepsilon}, \hat{\gamma}) \triangleq G(\hat{\sigma} - K_1\hat{\varepsilon}, \hat{\varepsilon}, \hat{\gamma} + 1), \quad (33)$$

and u is given by the feedback law

$$u = h(\sigma, \varepsilon, \gamma, \nu, \dot{\nu}, \beta), \quad (34)$$

where $h : \Re^3 \times \Re^3 \times \Re \times \Re^3 \times \Re^3 \times \Re^6 \to \Re^3$ is defined by

$$\begin{aligned} h(\hat{\sigma}, \hat{\varepsilon}, \hat{\gamma}, \hat{\nu}, \hat{\rho}, \hat{\beta}) &\triangleq -K_2\hat{\sigma} - \hat{\varepsilon} - [H(\hat{\sigma}, \hat{\varepsilon}, \hat{\gamma}, \hat{\nu}, \hat{\rho}) \\ &\quad + M(\hat{\sigma}, \hat{\varepsilon}, \hat{\gamma})]\,(\alpha - \hat{\beta}). \quad (35) \end{aligned}$$

It is observed that equations (28) - (31) are nonautonomous due to the presence of ν and $\dot{\nu}$. Under (34), we note that the origin $[\sigma^{\mathrm{T}}\ \varepsilon^{\mathrm{T}}\ \gamma^{\mathrm{T}}\ \beta^{\mathrm{T}}] = [0\ 0\ 0\ 0]$ is an equilibrium solution of the system (28) - (31).

Next, we show that under the control law (34), $\sigma \to 0$ and $\varepsilon \to 0$ as $t \to \infty$ for arbitrary initial conditions. To do this, consider the positive-definite candidate Lyapunov function $V : \Re^3 \times \Re^3 \times \Re^3 \times \Re^6 \to \Re$ defined by

$$V(\sigma,\varepsilon,\gamma,\beta) = \frac{1}{2}(\sigma^{\mathrm{T}}J\sigma + \beta^{\mathrm{T}}Q\beta) + \varepsilon^{\mathrm{T}}\varepsilon + \gamma^2. \quad (36)$$

Note that V is independent of time and is radially unbounded. The total time derivative of V along the trajectories of the system is given by

$$\begin{aligned}\dot{V}(\sigma,\varepsilon,\gamma,\beta) &= \Big[H(\sigma,\varepsilon,\gamma,\nu,\dot{\nu})^{\mathrm{T}}\sigma + M(\sigma,\varepsilon,\gamma)^{\mathrm{T}}\sigma \\ &+ Q\dot{\beta}\Big] - \sigma^{\mathrm{T}}K_2\sigma - \varepsilon^{\mathrm{T}}K_1\varepsilon.\end{aligned}$$

Using (31) we obtain the simplified expression

$$\dot{V}(\sigma,\varepsilon,\gamma,\beta) = -\sigma^{\mathrm{T}}K_2\sigma - \varepsilon^{\mathrm{T}}K_1\varepsilon, \quad (37)$$

which shows that $\dot{V}$ is negative semi-definite and not an explicit function of time. Since $V(\sigma(t),\varepsilon(t),\gamma(t),\beta(t)) \le V(\sigma(0),\varepsilon(0),\gamma(0),\beta(0))$ for all $t \ge 0$ and since V is radially unbounded, it follows that σ, ε, γ and β are bounded. Since by assumption ν and $\dot{\nu}$ are bounded and since α is constant it follows that $H(\sigma,\varepsilon,\gamma,\nu,\dot{\nu})$, $M(\sigma,\varepsilon,\gamma)$, $h(\sigma,\varepsilon,\gamma,\nu,\dot{\nu},\beta)$ and $\hat{\alpha}$ are bounded. The total time derivative of $\dot{V}$ along the trajectories of the system is given by

$$\begin{aligned}\ddot{V}(\sigma,\varepsilon,\gamma,\beta,t) &= -2\sigma^{\mathrm{T}}K_2J^{-1}\left[H(\sigma,\varepsilon,\gamma,\nu,\dot{\nu})^{\mathrm{T}}\alpha\right. \\ &+ h(\beta,\sigma,\varepsilon,\gamma,\nu,\dot{\nu}) \\ &+ \left.M(\sigma,\varepsilon,\gamma)^{\mathrm{T}}\alpha\right] - 2\varepsilon^{\mathrm{T}}K_1\left[\varepsilon^{\times}\sigma\right. \\ &+ \left.(\gamma+1)(\sigma - K_1\varepsilon)\right].\end{aligned}$$

Since σ, ε, γ, β, $H(\sigma,\varepsilon,\gamma,\nu,\dot{\nu})$, $M(\sigma,\varepsilon,\gamma)$ and $h(\sigma,\varepsilon,\gamma,\nu,\dot{\nu},\beta)$ are bounded and α is constant, it follows that $\ddot{V}(\sigma(t),\varepsilon(t),\gamma(t),\beta(t),t)$ is bounded for all $t \ge 0$. Using Theorem 5.4 of Ref. 8 we conclude that $\sigma \to 0$ and $\varepsilon \to 0$. Furthermore, since ν and $\dot{\nu}$ are bounded by assumption and $\sigma \to 0$ and $\varepsilon \to 0$ it follows from (31) that $\dot{\beta} \to 0$ and thus $\dot{\hat{\alpha}} \to 0$.

Because $\sigma \to 0$ and $\varepsilon \to 0$, it follows from (25) that $\omega \to 0$. Hence we conclude that (22) and (23) solves the tracking problem. □

The system (28) - (31) has two equilibrium points, namely $[\sigma^{\mathrm{T}}\ \varepsilon^{\mathrm{T}}\ \gamma^{\mathrm{T}}\ \beta^{\mathrm{T}}] = [0\ 0\ 0\ 0]$ and $[\sigma^{\mathrm{T}}\ \varepsilon^{\mathrm{T}}\ \gamma^{\mathrm{T}}\ \beta^{\mathrm{T}}] = [0\ 0\ -2\ 0]$, both of which correspond to the same orientation. The equilibrium point $[\sigma^{\mathrm{T}}\ \varepsilon^{\mathrm{T}}\ \gamma^{\mathrm{T}}\ \beta^{\mathrm{T}}] = [0\ 0\ 0\ 0]$ is stabilized using the control law (22) and (23).

We observe that the control law (22) and (23) is global in the sense that, for arbitrary initial conditions, $\omega \to 0$ and $\varepsilon \to 0$ as $t \to \infty$ which ensures that tracking is achieved asymptotically. The control law given by (23) is a sixth-order dynamic compensator. It is noted that the control law requires knowledge only of ω, ε and η and not of the inertia of the spacecraft.

The control law (22) and (23) is adaptive in the sense of Ref. 8, p. 1. The state $\hat{\alpha}$ represents adjustable parameters that, under certain conditions (see Section IV), converges to α. Equation (22) represents the mechanism for adjusting these parameters. The state $\hat{\alpha}$ is consequently termed the *adaptive parameter*. Although the time derivative of the adaptive parameter converges to zero as $t \to \infty$, $\hat{\alpha}$ does not necessarily converge.

We now consider the convergence of $\hat{\alpha}$ for periodic command signals.

Theorem 2: Assume that ν is periodic and define $W : [0,\infty) \to \Re^{3\times 6}$ by

$$W(t) \triangleq L(\dot{\nu}(t)) + \nu(t)^{\times}L(\nu(t)). \quad (38)$$

Under the control law given by (22) and (23), $\alpha - \hat{\alpha} \rightarrow \{\chi : W(t)\chi = 0 \text{ for all } t \geq 0\}$ as $t \rightarrow \infty$.

Proof: With σ, γ and β defined by (25), (26) and (27), respectively, we obtain the differential equations (28) - (31), where H and M are defined by (32) and (33), respectively, and $u = h(\sigma, \varepsilon, \gamma, \nu, \dot{\nu}, \beta)$, where h is defined by (35). Note that since ν is periodic and differentiable, $\dot{\nu}$ is periodic.

Consider the candidate Lyapunov function V defined by (36). Now V is C^1, positive definite, and radially unbounded, and $\dot{V}$ along the trajectories of the system is given by (37).

Let $x = \left[\sigma^{\mathrm{T}}\ \varepsilon^{\mathrm{T}}\ \gamma^{\mathrm{T}}\ \beta^{\mathrm{T}}\right]^{\mathrm{T}}$ and $E_0 = \{x : \dot{V}(x) = 0\}$. Let $x(t; x_0, t_0)$ denote the solution of the system (28) - (31) at time $t \geq t_0$ where $x(t_0; x_0, t_0) = x_0$. Let $L = \{(\hat{x}, \hat{t}) \in E_0 \times [0, \infty) : x(t; \hat{x}, \hat{t}) \in E_0 \text{ for all } t \geq \hat{t}\}$ and $N = \{x(t; \hat{x}, \hat{t}) : (\hat{x}, \hat{t}) \in L, t \geq \hat{t}\}$. Note that $N \subset E_0$. Using (8), (26) and (28) - (31) it follows that N is given by

$$N = \{(\sigma, \varepsilon, \gamma, \beta) : \sigma = 0, \varepsilon = 0, \gamma \in \{-2, 0\}, \beta \in G\},$$

where $G = \{\chi : W(t)\chi = 0 \text{ for all } t \geq 0\}$. It now folllows from Theorem 2.8 of Ref. 12 that $\alpha - \hat{\alpha} \rightarrow G$. □

The following corollary of Theorem 2 considers the special case in which ν is constant.

Corollary 1: Assume that ν is constant. Under the control law given by (22) and (23), $\alpha - \hat{\alpha} \rightarrow \{\chi : \nu^{\times} L(\nu)\chi = 0\}$ as $t \rightarrow \infty$.

Proof: Since ν is constant, Theorem 2 implies that $\alpha - \hat{\alpha} \rightarrow \{\chi : \nu^{\times} L(\nu)\chi = 0\}$. □

We now consider the case in which ν represents a constant spin about one of the principal axes which is equivalent to $\nu^{\times} L(\nu)\alpha = 0$. For such a command it is desirable that the control law satisfy $u \rightarrow 0$ as $t \rightarrow \infty$. The following result shows that the control law (22) and (23) has this property.

Corollary 2: Assume that ν is constant and satisfies $\nu^{\times} L(\nu)\alpha = 0$. Then, under the control law given by (22) and (23), $u \rightarrow 0$ as $t \rightarrow \infty$.

Proof: Under the control law given by (22) and (23), $\omega \rightarrow 0$ and $\varepsilon \rightarrow 0$ which implies that $u + \nu^{\times} L(\nu)\hat{\alpha} \rightarrow 0$ as $t \rightarrow \infty$. Since $\nu^{\times} L(\nu)\alpha = 0$, it follows from Corollary 1 that $\hat{\alpha} \rightarrow \{\kappa : \nu^{\times} L(\nu)\kappa = 0\}$. It thus follows that $u \rightarrow 0$ as $t \rightarrow \infty$. □

IV. Inertia Matrix Identification

In this section, we present a method for identifying the spacecraft inertia matrix. We first use Corollary 1 to identify the off-diagonal terms J_{12}, J_{23} and J_{31}.

Proposition 1: Let ν be constant. If $\nu = [0\ \nu_2\ 0]^{\mathrm{T}}$ where $\nu_2 \neq 0$ then, under the control law (22) and (23), $\hat{\alpha}_4 \rightarrow \alpha_4$ and $\hat{\alpha}_6 \rightarrow \alpha_6$ as $t \rightarrow \infty$. Furthermore, if $\nu = [0\ 0\ \nu_3]^{\mathrm{T}}$ where $\nu_3 \neq 0$ then, under the control law (22) and (23), $\hat{\alpha}_4 \rightarrow \alpha_4$ and $\hat{\alpha}_5 \rightarrow \alpha_5$ as $t \rightarrow \infty$.

Proof: From Corollary 1, under the control law (22) and (23), $\alpha - \hat{\alpha} \rightarrow \{\chi : \nu^{\times} L(\nu)\chi = 0\}$. Now $\nu^{\times} L(\nu)$ is computed to be

$$\nu^{\times} L(\nu) = \begin{bmatrix} 0 & 0 & 0 & \nu_2^2 & 0 & 0 \\ 0 & 0 & 0 & 0 & 0 & 0 \\ 0 & 0 & 0 & 0 & 0 & -\nu_2^2 \end{bmatrix}.$$

It thus follows that $\nu^{\times} L(\nu)\chi = 0$ if and only if $\chi_4 = 0$ and $\chi_6 = 0$ where $\chi = [\chi_1\ \chi_2\ \chi_3\ \chi_4\ \chi_5\ \chi_6]^{\mathrm{T}}$. Thus $\hat{\alpha}_4 \rightarrow \alpha_4$ and $\hat{\alpha}_6 \rightarrow \alpha_6$ as $t \rightarrow \infty$. If $\nu = [0\ 0\ \nu_3]^{\mathrm{T}}$ where $\nu_3 \neq 0$ then $\nu^{\times} L(\nu)$ is computed to be

$$\nu^{\times} L(\nu) = \begin{bmatrix} 0 & 0 & 0 & -\nu_3^2 & 0 & 0 \\ 0 & 0 & 0 & 0 & \nu_3^2 & 0 \\ 0 & 0 & 0 & 0 & 0 & 0 \end{bmatrix}.$$

It follows that $\hat{\alpha}_4 \rightarrow \alpha_4$ and $\hat{\alpha}_5 \rightarrow \alpha_5$ as $t \rightarrow \infty$. □

Hence the off-diagonal terms J_{12}, J_{13} and J_{23} can be identified by performing two constant tracking maneuvers. We now consider periodic maneuvers for identifying the entire inertia matrix.

Proposition 2: Let ν be periodic and let $W(t)$ be given by (38). Furthermore, let $0 \le t_1 \le t_2 \le \ldots \le t_n$ and suppose that

$$\operatorname{rank}\begin{bmatrix} W(t_1) \\ \vdots \\ W(t_n) \end{bmatrix} = 6. \tag{39}$$

Then, under the control law (22) and (23), $\hat{\alpha} \to \alpha$.

Proof: From Theorem 2 it follows that $\alpha - \hat{\alpha} \to \{\chi : W(t)\chi = 0 \text{ for all } t \ge 0\}$. However it follows from (39) that $\{\chi : W(t)\chi = 0 \text{ for all } t \ge 0\} = \{0\}$. Hence $\hat{\alpha} \to \alpha$. □

There are many signals that satisfy the conditions of Proposition 2. Consider, for example, the periodic signal

$$\nu(t) = [\sin t \;\; \sin 2t \;\; \sin 3t]^{\mathrm{T}}. \tag{40}$$

Then with $t_1 = 0$ and $t_2 = \pi/2$, we obtain

$$\begin{bmatrix} W(0) \\ W(\pi/2) \end{bmatrix} = \begin{bmatrix} 1 & 0 & 0 & 0 & 3 & 2 \\ 0 & 2 & 0 & 3 & 0 & 1 \\ 0 & 0 & 3 & 2 & 1 & 0 \\ 0 & 0 & 0 & -1 & 0 & -1 \\ -1 & -2 & 1 & 0 & 0 & 0 \\ 0 & 0 & 0 & -3 & 0 & 1 \end{bmatrix}.$$

Since the determinant of $\begin{bmatrix} W(0) \\ W(\pi/2) \end{bmatrix}$ is -64, it follows that

$$\operatorname{rank}\begin{bmatrix} W(0) \\ W(\pi/2) \end{bmatrix} = 6.$$

Hence under the control law (22) and (23), $\hat{\alpha} \to \alpha$. Thus the inertia matrix can be identified using a single periodic command signal.

V. Numerical Simulations

In this section we present simulations to illustrate tracking and identification of the spacecraft inertia matrix. The two tracking maneuvers considered are a simple spin about the intermediate principal axis and a coning motion. The maneuver given by (40) is used to identify the spacecraft inertia. The same initial conditions are chosen for the three maneuvers. The initial angular velocity is $\Omega = [0.4 \; 0.2 \; -0.1]^{\mathrm{T}}$ rad/sec, the initial orientation of the spacecraft is given by $\epsilon = [-0.1 \; 0.1 \; -0.1]^{\mathrm{T}}$, $\zeta = 0.9849$, and the initial value of the adaptive parameter $\hat{\alpha}$ in kg·m^2 is $\hat{\alpha} = [22 \; 18 \; 13 \; 1.6 \; 1.0 \; 1.3]^{\mathrm{T}}$. The gains are chosen to be $K_1 = 20 I_{3\times3}$, $K_2 = 5 I_{3\times3}$ and $Q = I_{6\times6}$.

First we command the intermediate principal axis of the spacecraft to point in a given inertial direction and the spacecraft to perform a simple spin about the intermediate principal axis. To do this, we assume that the direction of the intermediate principal axis is known with respect to $\mathcal{B}$, and, without loss of generality, we assume that the intermediate principal axis coincides with the $Y_{\mathcal{B}}$ axis of $\mathcal{B}$. We assume no knowledge of either the moments of inertia or the directions of the other two principal axes.

The maneuver is performed using the control law given by Theorem 1. In the body frame $\mathcal{B}$, let

$$J = \begin{bmatrix} 20 & 0 & 0.9 \\ 0 & 17 & 0 \\ 0.9 & 0 & 15 \end{bmatrix},$$

so that $Y_{\mathcal{B}}$ is the intermediate principal axis, let $\nu = [0 \; 1 \; 0]^{\mathrm{T}}$ rad/sec, let the initial orientation be $\xi = [0 \; 0 \; 0]^{\mathrm{T}}$ and $\mu = 1$, and let the X, Y and Z axes constitute the inertial frame. Then the desired motion is a spin about the intermediate principal axis, with the intermediate principal axis aligned with the Y axis of the inertial frame.

Applying the control law given by Theorem 1, we observe from Figures 1 and 2 that tracking is achieved. Figures 3 and 4 indicate that while J_{12} and J_{23} are identified in accordance with Proposition 1, the remaining entries of the inertia matrix are

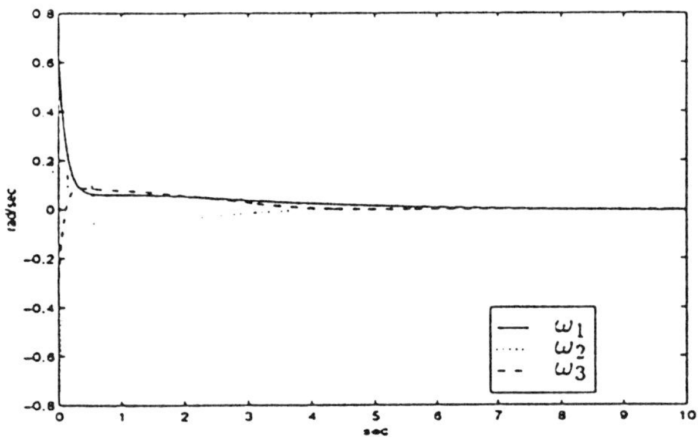

Figure 1: Relative angular velocity $\omega = [\omega_1\ \omega_2\ \omega_3]^{\mathrm{T}}$

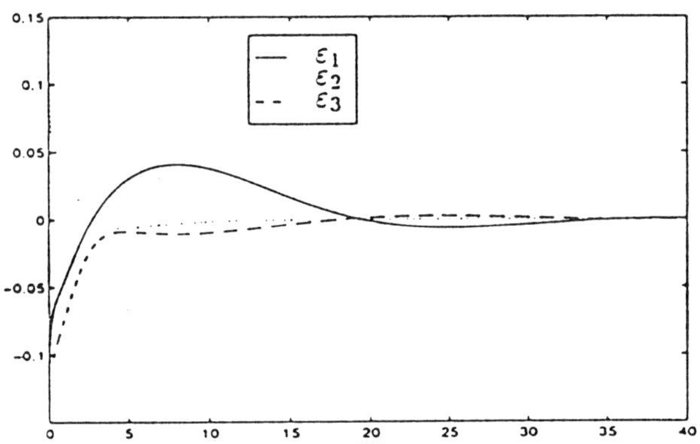

Figure 2: Relative Euler parameter $\varepsilon = [\varepsilon_1\ \varepsilon_2\ \varepsilon_3]^{\mathrm{T}}$

not. Figure 5 shows that Ω_2 converges to 1 rad/sec while Ω_1 and Ω_3 converge to 0 rad/sec. Thus the spacecraft approaches a simple spin about its intermediate axis.

Next we command the spacecraft to perform a specified coning motion. We use 3, 2, 1 Euler[13] angles ψ, θ and ϕ to represent the precession motion of the desired frame $\mathcal{D}$. Let $\psi(t) = 0.3t$ rad, $\theta(t) = -0.2778\pi$ rad and $\phi(t) = t$ rad. Then the desired motion is a coning maneuver with a precession rate of 0.3 rad/sec, a spin rate of 1 rad/sec and a coning angle of 40 degrees. The angular velocity ν

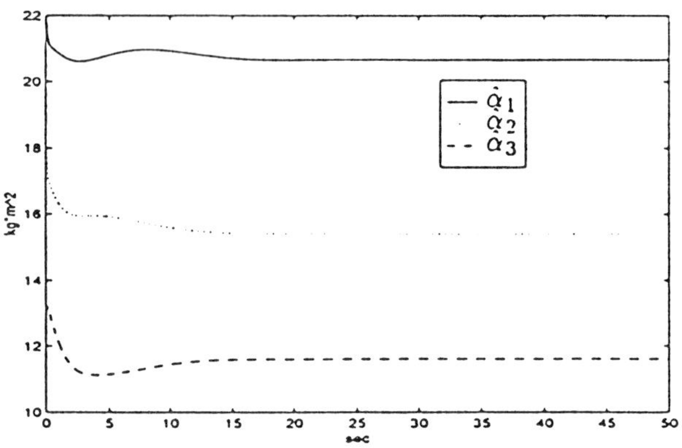

Figure 3: Adaptive parameters $\hat{\alpha}_1$, $\hat{\alpha}_2$ and $\hat{\alpha}_3$

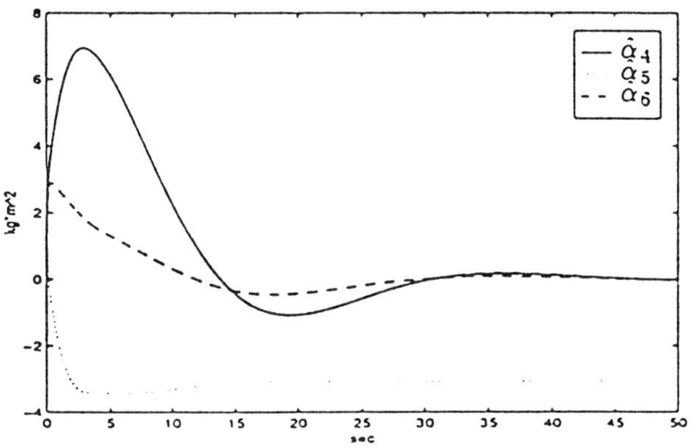

Figure 4: Adaptive parameters $\hat{\alpha}_4$, $\hat{\alpha}_5$ and $\hat{\alpha}_6$

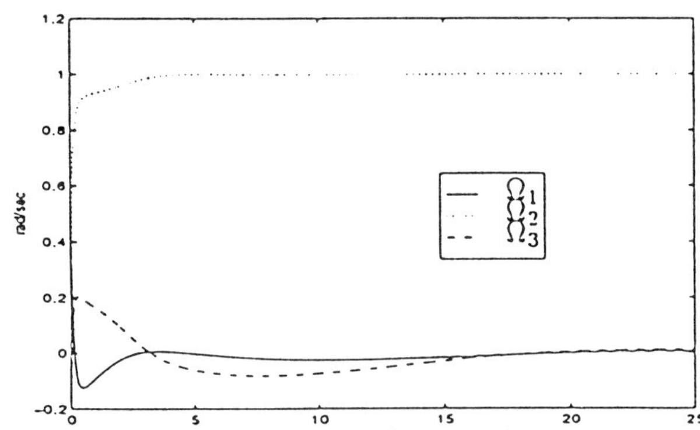

Figure 5: Angular Velocity of $\mathcal{B}$ with respect to Inertial $\Omega = [\Omega_1\ \Omega_2\ \Omega_3]^{\mathrm{T}}$

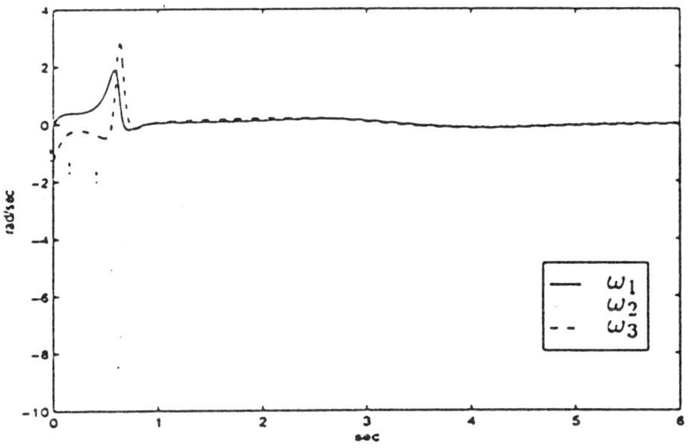

Figure 6: Relative angular velocity $\omega = [\omega_1\ \omega_2\ \omega_3]^T$

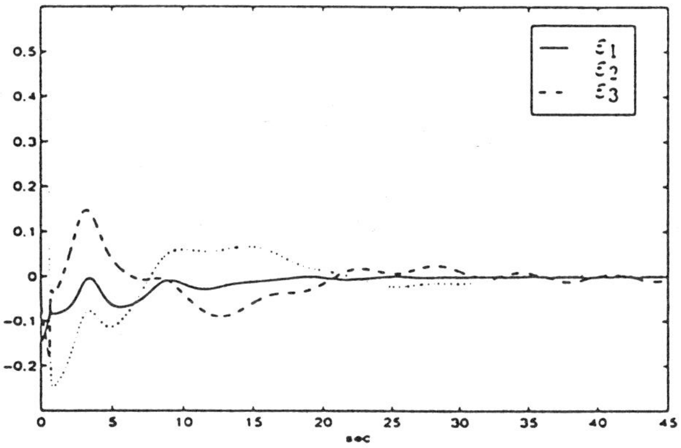

Figure 7: Relative Euler parameter $\varepsilon = [\varepsilon_1\ \varepsilon_2\ \varepsilon_3]^T$

and the initial values of the Euler parameters ξ and μ are computed from Ref. 11 using equation (24), p. 27, Table 2.1, p. 20 and equations (15) and (16), p. 18. The inertia matrix J of the spacecraft in kg·m^2 is given by

$$J = \begin{bmatrix} 20 & 1.2 & 0.9 \\ 1.2 & 17 & 1.4 \\ 0.9 & 1.4 & 15 \end{bmatrix}. \tag{41}$$

The controller given by Theorem 1 is used and it is observed from Figures 6 and 7 that $\omega \longrightarrow 0$ and $\varepsilon \longrightarrow 0$. However, note from Figures 8 and 9 that not all components of $\hat{\alpha}$ converge to α. Figure 5 shows the motion of the X_B axis of the spacecraft for a period of 100 seconds. The plots indicate that the desired coning motion is achieved.

To identify the inertia matrix, the maneuver $\nu(t) = [\sin t\ \sin 2t\ \sin 3t]^T$ rad/sec described in Section IV is chosen. The initial orientation of $\mathcal{D}$ is chosen to be $\xi = [0\ 0\ 0]^T$ and $\mu = 1$, and the inertia matrix is given by (41). Under the control law given by (22) and (23) we observe from Figures 11 and 12 that $\omega \longrightarrow 0$ and $\varepsilon \longrightarrow 0$. Furthermore, Figures 13 and 14 indicate that $\hat{\alpha} \longrightarrow \alpha$ in accordance with Proposition 2. It is noted from the numerical sim-

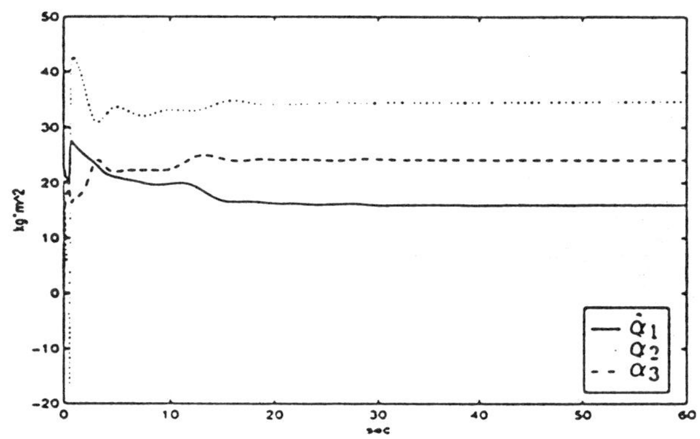

Figure 8: Adaptive parameters $\hat{\alpha}_1$, $\hat{\alpha}_2$ and $\hat{\alpha}_3$

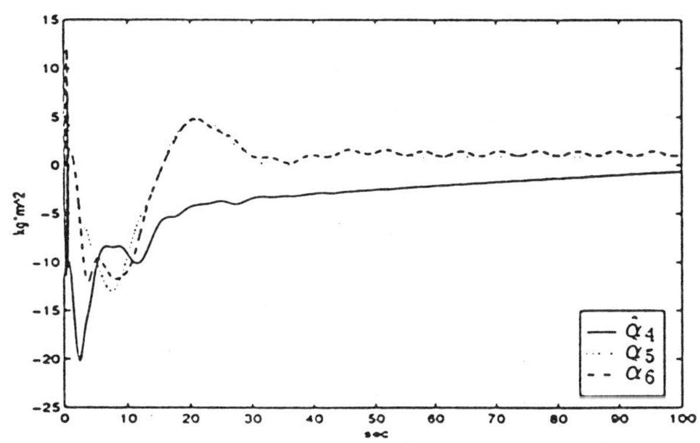

Figure 9: Adaptive parameters $\hat{\alpha}_4$, $\hat{\alpha}_5$ and $\hat{\alpha}_6$

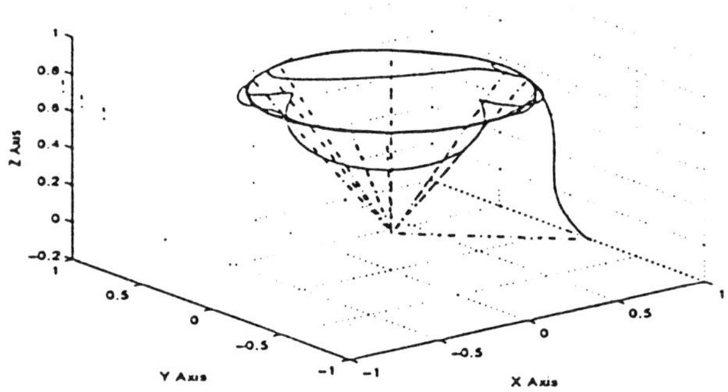

Figure 10: Motion of X_B axis

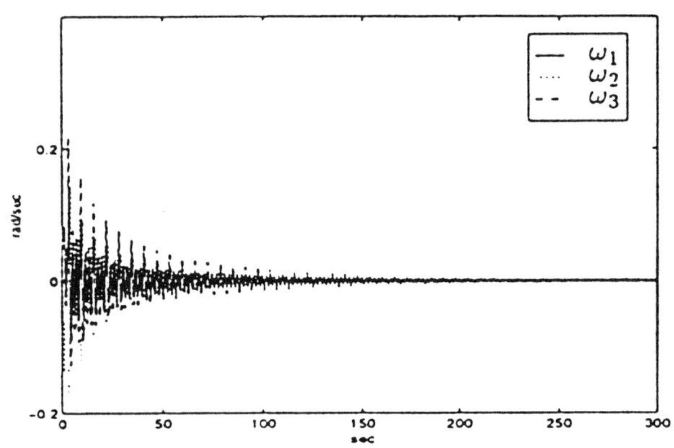

Figure 11: Relative angular velocity $\omega = [\omega_1\ \omega_2\ \omega_3]^T$

ulations that tracking is achieved rapidly whereas parameter identification takes much longer.

VI. Conclusions

An adaptive feedback control algorithm has been developed to provide global tracking of commanded spacecraft motion. The algorithm assumes no knowledge of the inertia of the spacecraft and is thus unconditionally robust with respect to this parametric uncertainty. It was shown using a

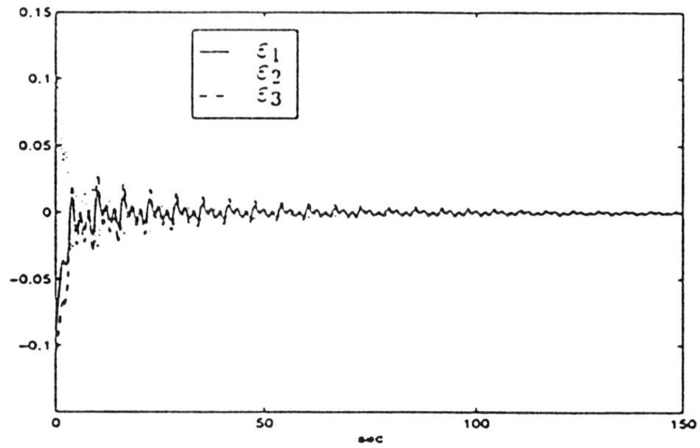

Figure 12: Relative Euler parameter $\varepsilon = [\varepsilon_1\ \varepsilon_2\ \varepsilon_3]^T$

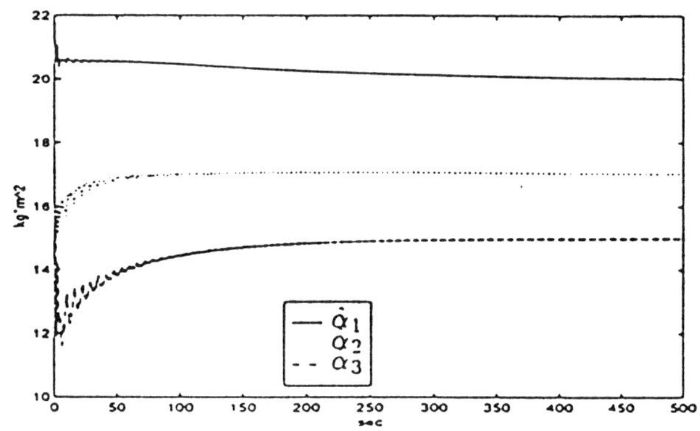

Figure 13: Adaptive parameters $\hat{\alpha}_1$, $\hat{\alpha}_2$ and $\hat{\alpha}_3$

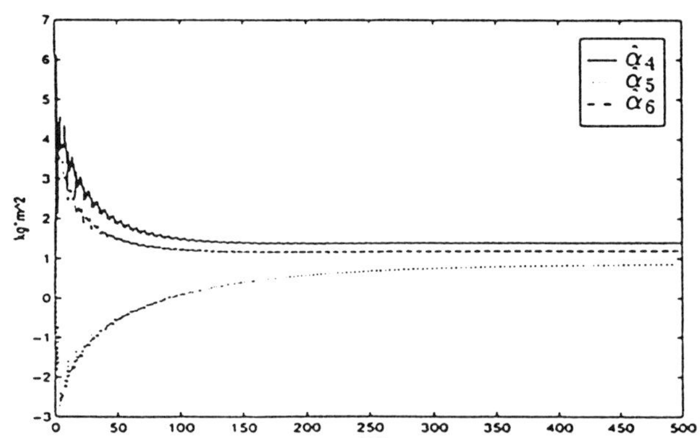

Figure 14: Adaptive parameters $\hat{\alpha}_4$, $\hat{\alpha}_5$ and $\hat{\alpha}_6$

Lyapunov argument that the attitude and angular velocity tracking error converge to zero. Furthermore, the control algorithm was used to identify the spacecraft inertia matrix. Numerical simulations illustrate tracking and identification of the inertia matrix.

References

[1]Junkins, J. L., and Turner, J. D., "Optimal Continuous Torque Attitude Maneuvers," *J. Guidance Contr.*, Vol. 3, no. 3, May-June 1980, pp 210-217.

[2]Skaar, S. B., and Kraige, L. G., "Large Angle Spacecraft Attitude Maneuvers Using an Optimal Reaction Wheel Power Criterion," *The Journal of Astronautical Sciences*, Vol. 32, no. 1, Jan.-Mar. 1984, pp. 47-61.

[3]Wie, B., Weiss, H., and Arapostathis, A., "Quaternion Feedback Regulator for Spacecraft Eigenaxis Rotations," *J. Guidance Contr.*, Vol. 8, no. 3, 1989, pp. 360-365.

[4]Joshi, S. M., Kelkar, A. G., and Wen, J. T.-Y., "Robust Attitude Stabilization of Spacecraft Using Nonlinear Quaternion Feedback," *IEEE Trans. Autom. Contr.*, Vol. 40, Oct. 1995, pp. 1800-1803.

[5]Singh, S. N., "Nonlinear Adaptive Attitude Control of Spacecraft," *IEEE Trans. Aerospace Elect. Sys.*, Vol. AES-23, No. 3, May 1987.

[6]Sheen, J. J., and Bishop, R. H., "Adaptive Nonlinear Control of Spacecraft," *Proc. Amer. Contr. Conf.*, Baltimore, MD, June 1994, pp. 2867-2871.

[7]Slotine, J. J. E., and Li, W., *Applied Nonlinear Control*, Prentice-Hall, NJ, 1989.

[8]Astrom, K. J. and Wittenmark, B., *Adaptive Control*, Addison-Wesley, Reading, MA, 2nd edition, 1995.

[9]Bloch, A. M., and Marsden, J. E., "Stabilization of Rigid Body Dynamics by the Energy-Casimir Method," *Systems & Control Letters*, 14, 1990, pp. 341-346.

[10]Aeyels, D., "On Stablization by Means of the Energy-Casimir Method," *Systems & Control Letters*, 18, 1992, pp. 325-328.

[11]Hughes, P. C., *Spacecraft Attitude Dynamics*, John Wiley & Sons, 1986.

[12]Narendra, K. S., and Annaswamy, A. M., *Stable Adaptive Systems*, Prentice Hall, 1989.

[13]Greenwood, D. T., *Principles of Dynamics*, Prentice-Hall, 1988.

AIAA-97-3532

EXTENSION OF A KOLMOGOROV ATMOSPHERIC TURBULENCE MODEL FOR TIME-BASED SIMULATION IMPLEMENTATION

John D. McMinn*
NASA Langley Research Center
Hampton, Va

Abstract

The development of any super/hypersonic aircraft requires the interaction of a wide variety of technical disciplines to maximize vehicle performance. For flight and engine control system design and development on this class of vehicle, realistic mathematical simulation models of atmospheric turbulence, including winds and the varying thermodynamic properties of the atmosphere, are needed. A model which has been tentatively selected by a government/industry group of flight and engine/inlet controls representatives working on the High Speed Civil Transport is one based on the Kolmogorov spectrum function. This report compares the Dryden and Kolmogorov turbulence forms, and describes enhancements that add functionality to the selected Kolmogorov model. These added features are: an altitude variation of the eddy dissipation rate based on Dryden data, the mapping of the eddy dissipation rate database onto a regular latitude and longitude grid, a method to account for flight at large vehicle attitude angles, and a procedure for transitioning smoothly across turbulence segments.

Symbols

A_n	Fourier series cosine coefficient
B_n	Fourier series sine coefficient
E	Kolmogorov energy spectrum
f_1	lower bound on frequency range of interest, Hz
f_2	upper bound on frequency range of interest, Hz
h	altitude, km
k_1	lower bound on wavenumber range of interest, rad/m
k_2	upper bound on wavenumber range of interest, rad/m
L	integral scale length, m
L_0	Kolmogorov outer scale length, m
u	longitudinal turbulence velocity, m/s
v	transverse turbulence velocity component, m/s
V	air relative vehicle velocity, excluding turbulence components, m/s
w	turbulence velocity component orthogonal to u and v, m/s
x	distance traveled along flight path, m
α	general Kolmogorov energy spectrum constant
α_p	Kolmogorov pressure energy spectrum constant, hPa^2s^2/m^2
γ	flight path angle, deg
$\Delta\rho$	density perturbation from mean
ΔP	pressure perturbation from mean
ΔT	temperature perturbation from mean
ε	eddy dissipation rate, m^2/s^3
ε_D	eddy dissipation rate derived from Dryden data, m^2/s^3
ε_0	eddy dissipation rate used in A_n and B_n calculation, m^2/s^3
Φ	Dryden energy spectrum
Φ_v	von Karman energy spectrum
ϕ	body axis roll angle, deg
θ	pitch attitude, deg
ψ_w	heading relative to wind, deg
Ω	wavenumber, rad/m
σ	RMS turbulence velocity, m/s

Abbreviations

3-D	three dimensional
2-D	two dimensional
lat.	latitude
long.	longitude
HSCT	High Speed Civil Transport
RMS	Root Mean Square

Subscripts

h	horizontal component
u	longitudinal component
v	transverse component
w	component orthogonal to u and v
0	indicates freestream value

Introduction

The development of any new aircraft requires the interaction of a wide variety of technical disciplines. Furthermore, when the vehicle being designed requires

* *Aerospace Engineer, Flight Dynamics and Control Division, Member AIAA*

the development of new technology in many of these disciplines, the level of interaction between the technology teams must be high to maximize efficiency in the resulting design. Such is the case with the technology development of the High Speed Civil Transport (HSCT) under the High Speed Research program.

In the case of the HSCT program it was recognized that different disciplines used different atmosphere models. In order to establish a common basis for comparison and minimize duplication of effort, an ad hoc group, including representatives from NASA and industry, has formed with the goal of adopting a model of atmospheric turbulence. The selected model will be used within the engine community as a basis for the evaluation of inlet/engine performance and in the flight controls area for the evaluation of the aircraft control system and handling qualities in piloted and batch computer simulations.

The model which has been proposed by the group is one suggested by Bill Tank of Boeing[1]. The model, like the von Karman model, is defined in the frequency domain, and for which the power has a -5/3 intensity roll off with increasing frequency. A key reason for the selection of the model is the capability to generate thermodynamic perturbations (density, pressure, and temperature) as well as wind gusts. A second reason for selecting the Tank model is the data used to parameterize it. Tank bases the model on data collected from a variety of sources where data was taken in flight as close as possible to the approximate cruise altitude of the HSCT.

This report reviews the Tank model, discusses implementation issues, and presents extensions made to the model to enhance its capability.

Background

The atmosphere has long been studied to better understand and predict its behavior, and for aircraft designers and operators, understanding the turbulent nature of the atmosphere is an area of concern. Current atmospheric theory[2] states that the turbulent energy of the atmosphere is distributed as a continuous function of wavenumber (spatial frequency) as shown in Figure 1. This Figure shows the zonal (east-west) wind, meridional (north-south) wind, and potential temperature power spectra (note: for clarity the meridional and temperature spectra have been shifted one and two decades to the right respectively). Notice that at the very long (planetary scale) wavelengths the power of turbulence decreases as the -3 power of decreasing wavelength (increasing frequency), and from about a wavelength of 500 km, the power fall off is roughly as the -5/3 power of frequency. Turbulence theory also states that the range where turbulence is fully three dimensional and isotropic extends from the very small scale (< 1cm) out to a scale of a few hundred meters[3].

In the range between a few hundred meters and 500 km the -5/3 law is still obeyed for turbulence in the near horizontal plane, though the turbulence cannot be classified as three dimensional (3-D) isotropic. Vertical turbulence has been found to obey a power fall off of -3 to -2 down to scales of 100 meters. It is expected to change to a -5/3 slope at even shorter scales[3].

In the 3-D region turbulence energy is not so much lost due to friction as it is transferred to shorter and shorter wavelengths. The rate at which energy is cascaded down in scale is called the eddy dissipation rate, ε. There is also evidence that supports the theory that energy is not only cascaded down in scale but also up. That the -5/3 law holds for scales out to hundreds of kilometers is a fortunate result of upscale transfer of two dimensional (2-D) turbulence energy. The impact this has on the modeling of turbulence for simulations is that models developed for 3-D, isotropic turbulence at short scales can be extended beyond the 3-D regime into the margins of the 2-D without changing the model.

In order to evaluate the performance of aircraft systems and flying qualities an appropriate, relatively simple, characterization of atmospheric turbulence is required. It is not sufficient to always assume a quiescent atmosphere or even that the thermodynamic properties of the atmosphere are solely deterministic functions of altitude.

Traditionally the Dryden turbulence model has been used whenever atmospheric (wind) turbulence was needed to add some sense of realism to a time-based simulation or to provide a disturbance input when evaluating systems in the frequency domain. The Dryden spectrum for the longitudinal, body axis velocity disturbance is given[4] as,

$$\Phi(\Omega)_u = \sigma^2 \frac{2L_u}{\pi} \frac{1}{1+(L_u\Omega)^2} \tag{1}$$

where Ω is the wavenumber, or frequency term, in units of radians per meter, σ represents the standard deviation of the velocity disturbance, and L_u is the integral scale of the longitudinal turbulence, which will be discussed in more detail later.

The spectra of the lateral and vertical components of Dryden body axis turbulence are given in the same reference by,

$$\Phi(\Omega)_{v,w} = \sigma^2 \frac{L_{v,w}}{\pi} \frac{1+3(L_{v,w}\Omega)^2}{\left[1+(L_{v,w}\Omega)^2\right]^2} \qquad (2)$$

with a similar definition of terms. Although the Dryden equations are more frequently used, they are not the best fit to the observed character of the atmosphere. Notice in equations 1 and 2, that the Dryden spectra fall off as the -2 power of frequency, not the -5/3 power as shown in Figure 1. The predominant use of the Dryden form stems from the ease of implementation within a time-based simulation. Such implementation usually involves forming a band limited, white noise filter equation by taking the spectral square roots of the spectrum equations. This is not so simple for an irrational -5/3 power fall off without approximating, but is relatively simple for the rational -2 power fall in equations 1 and 2. The resulting transfer functions[4] are given below,
longitudinal:

$$F_u(s) = \sigma_u \sqrt{\frac{2V}{\pi L_u}} \left(\frac{1}{s+\frac{V}{L_u}} \right) \qquad (3)$$

lateral and vertical:

$$F_{v,w}(s) = \sigma_{v,w} \sqrt{\frac{3V}{\pi L_{v,w}}} \left(\frac{s+\frac{V}{\sqrt{3}L_{v,w}}}{\left(s+\frac{V}{L_{v,w}}\right)^2} \right) \qquad (4)$$

Use of these equations with a band-limited, unit variance, white noise input and appropriate filter parameters yields Dryden aircraft-scale turbulence. For the purpose of classical rigid-body simulation and piloted assessment, the Dryden form is sufficient. When aircraft flexibility is an issue for vehicle flight assessment the frequency range of interest widens to include the higher frequency structural modes. At this point the difference between the -2 and the -5/3 power fall off can not be ignored.

The spectrum functions that better fit experimentally observed turbulence are the von Karman spectra as given in the same reference[4],

$$\Phi_v(\Omega)_u = \sigma^2 \frac{2L_u}{\pi} \frac{1}{\left[1+(1.339L_u\Omega)^2\right]^{\frac{5}{6}}} \qquad (5)$$

for the longitudinal, and

$$\Phi_v(\Omega)_{v,w} = \sigma^2 \frac{L_{v,w}}{\pi} \frac{1+\frac{8}{3}(1.339L_{v,w}\Omega)^2}{\left[1+(1.339L_{v,w}\Omega)^2\right]^{\frac{11}{6}}} \qquad (6)$$

for the lateral and vertical components.
Both the Dryden form and the von Karman are parameterized by two terms, the standard deviation of the turbulence, σ, and an integral scale length. The standard deviation is equal in magnitude to the square root of the area under a power spectrum plot. Note that increasing or decreasing σ in the equations 1, 2, 5, or 6 would have the effect of shifting up or down the amplitude of the spectrum, without distorting the shape of the plot. The Dryden spectra shown in Figure 2 were generated with the equation,

$$\Phi(\Omega) = \frac{10}{1+(0.5\Omega)^2} \qquad (7)$$

Referring to eqn. 1, this equation arises when $L_u = 0.5$m and $\sigma^2 = 10\pi$ m^2/s^2. The solid line in the upper plot on Figure 2 shows the typical Dryden spectral shape on a log-log plot. Note that the log-log plot is deceptive when attempting to characterize the area distribution by visual inspection. Although most of the area, and thus larger contributor to the integral of the spectrum, appears to lie under the flat (low frequency) portion of the power spectrum, that relation is not correct. As decade by decade integration of the spectrum would reveal, most of the turbulence power being described by the plot, is under the sloping portion of the curve.

The upper plot is a log-log plot and the lower a semi-log of the same data, with the solid lines defined by equation 7. The dashed line in Figure 2 represents the product of the power (or Dryden amplitude) and the axis frequency. Plotting this product as a function frequency gives a more realistic (less deceptive) graphical representation of the area distribution.
The knee, or the point where the spectrum has fallen to the half power point with increasing wavenumber (or frequency), is located at a wavenumber of 1/L. In equation 7, L has a value of 0.5. As shown in Figure 2, the knee is properly located at a wavenumber of $1/L_u$, or 2.0 Rad./m. For proper use, the wavenumbers of interest to the designer should be near $1/L_u$, so that the power associated with those wavenumbers is appropriate. Arbitrarily increasing the scale length, L, has the effect of shifting the knee of the spectrum to the left. If the knee moves left and σ is held constant, the spectrum will be shifted left and upward slightly to keep the integrated area constant. This shifting reduces the power associated with the higher frequencies. In order to both not affect the higher frequency power distribution

and move the knee to the left (for example, to include more low frequency power), both L and σ must increase. So it should seem intuitive now that the standard deviation and scale length should be related. Tables defining L and σ pairs, exist and give the pairs as a function of altitude, such as is shown in Table 1[5].

Despite being frequently used there is much controversy over specification of the scale length. According to Etkin, the determination of the scale length at altitude remains an unresolved difficulty. This stems from the uncertainty of how to determine L, whether from the integral of the velocity autocorrelation function or from a fit to a spectrum[6]. The matter is further confused by the tendency to think of the integral scale as a physical characteristic of the turbulence field, and is often confused with the 'outer scale' of turbulence, or the longest wavelength that could be expected. For example, in Figure 2 the integral scale is the reciprocal of the knee frequency, or 0.5, but inspection of the bell shaped area distribution indicates that significant turbulence intensity exists at longer scales (lower frequencies).

Additionally, the use of the Dryden or von Karman models provide turbulent winds, but not thermodynamic variations. The models do not usually account for the global variation of turbulence. In time-based simulations, where the Dryden model is used, the power fall off is not the observed -5/3 value, though some may argue that the Dryden model gives a better fit to the vertical gusts due their observed -3 to -2 character at scales above 100m as mentioned above.

The Tank Turbulence Model

In the light of these issues with using the Dryden and von Karman models, it is interesting to note that atmospheric scientists use a different model. Where the aerodynamicist uses the Dryden or von Karman models for characterizations of the turbulent atmosphere, atmospheric scientists tend to use the Kolmogorov energy spectrum[7]. The Kolmogorov spectrum function for longitudinal turbulence has the form,

$$E(k) = \alpha \varepsilon^{\frac{2}{3}} k^{\frac{-5}{3}} \tag{8}$$

where k is the spatial wavenumber, ε is the eddy dissipation rate, and α is a specified constant. In log space eqn. 8 is the equation for a straight line with a -5/3 slope. Notice that it matches the -5/3 character observed in Figure 1, and is similar to the von Karman form in the limit as wavenumber approaches infinity. In the 3-D range the transverse spectra are 4/3 times the longitudinal for an incompressible fluid as Kolmogorov showed[8].

The turbulence model suggested by Tank[1] for application to the HSCT, hereafter referred to as the Tank model, uses this Kolmogorov form for defining wind and thermodynamic turbulence. According to the Tank model the temperature spectrum has the same form as the longitudinal, the only difference being the constant α, whose units vary depending on the spectrum being defined. A description of the similarity of the temperature and velocity spectra is available[9]. The spectrum of atmospheric pressure perturbations associated with turbulence is given by Tank to be

$$E(k) = \alpha_p (P_0/T_0)^2 \ \varepsilon^{2/3} \ k^{-5/3} \tag{9}$$

where P_0 is the standard atmospheric pressure in hPa and T_0 is the standard Kelvin temperature.

The Tank model defines the temperature and pressure equations, leaving the the density perturbation to be determined from the equation of state. If,

$$(P_0 + \Delta P) = (\rho_0 + \Delta\rho) \ R \ (T_0 + \Delta T) \tag{10}$$

where ρ_0 is the standard atmospheric density, R is the gas constant, and the Δ symbol indicates a perturbation of an atmospheric quantity (temperature, T, or pressure, P), then the density perturbation can be solved for

$$\Delta\rho = \frac{(P_0 + \Delta P)}{(T_0 + \Delta T)R} - \frac{P_0}{RT_0} \tag{11}$$

Given these equations and the appropriate α value for the quantity being simulated, the Tank model reduces to one parameter, ε. This contrasts with conventional Dryden and von Karman which have two parameters (σ and L). The eddy dissipation rate, ε, characterizes the turbulence for whatever range of frequencies is of interest. To account for the global non-homogeneous nature of turbulence, Tank has defined ε, at the approximate HSCT cruise altitude, as shown in Figure 3. Inspection of the figure reveals that Tank has defined ε between 80 degrees north and 40 degrees south latitude over the oceans around the world.

The eddy dissipation rate can be linked to more familiar terms by integrating both sides of eqn. 8 from k_1 to k_2 and recognizing that the integral of the left side is equal to σ^2. This yields the following equation,

$$\varepsilon = \left(\frac{2\sigma^2}{3\alpha\left(k_2^{\frac{-2}{3}} + k_1^{\frac{-2}{3}} \right)} \right)^{3/2} \tag{12}$$

Here σ^2 represents the area under the spectrum defined in eqn. 8 and is the variance of the turbulence within the range k_1 to k_2. Integrating from k_1 to infinite wavenumber equation 12 reduces to

$$\varepsilon = \frac{\sigma^3}{(1.5\alpha)^{\frac{3}{2}} L_0} \qquad (13)$$

where L_0 is $1/k_1$ and represents the longest wavelength in the range of interest and is referred to as the outer scale length. This outer scale should not be confused with the integral scale length.

Tank Model Extensions

To implement and add functionality to a time-based simulation employing the Tank model, several features have been developed by the author in a series of MATLAB scripts. A variation of the eddy dissipation rate with altitude based on Dryden data has been calculated. The ε database has been modified. Methods to more properly account for flight at large attitude angles have been implemented. Also, a procedure for transitioning smoothly from one turbulence segment to the next has been developed.

Altitude Variation

The model proposed by Tank defined the eddy dissipation rate only at the estimated cruise altitude (~60 kft) as shown in Figure 3. To use the model in a broader range of the flight envelope a method for changing ε based on altitude was developed using eqn. 13 and data from reference 3.

Table 1[5] defines a set of standard deviations and integral scale lengths for light, moderate, and severe turbulence as a function of altitude for use in a Dryden turbulence model. Using this σ and L information for horizontal turbulence, with α (in eqn. 13) set equal to 0.15 as specified by Tank for longitudinal turbulence, eddy dissipation rates were estimated from the Dryden parameters. These estimated eddy dissipation rates are shown in Table 2.

Noting that Tank defines the lower stratosphere global mean epsilon to 0.00002 m^2/s^3 and observing that at an altitude of 18 km the light, Dryden turbulence data produced a nearly identical epsilon value (shaded in Table 2), it was accepted that the 'light' turbulence data best fit Tank's data. To support that claim, two time histories were evolved, one by the Tank method and the other with the Dryden filter. Sample time histories can be compared in Figure 4 and the corresponding power spectra in Figure 5. Notice that both models generate nearly the same low frequency character and that this observation is substantiated by the overlap of the power spectra at the low frequencies. Also note that the Tank model produces the higher amplitude high frequency turbulence, this is due to the difference in power fall off between the two models. As should be expected by looking at the power spectra the standard deviation of the Tank time history is higher than the Dryden (.2699 vs. .2158 m/s).

Accepting the similarity, the light turbulence ε data was normalized to unity at 18 km and is shown in the last column of Table 2. The resulting normalized variation with altitude is the altitude variation that has been added through eqn. 14,

$$\varepsilon(\text{lat., long., h}) = \varepsilon(\text{lat., long.}) * \varepsilon_D(h) \qquad (14)$$

where h is altitude in km.

Database Gridding

The ε data given by Tank[1] was only presented in the form of the isoline plot shown as Figure 3. For implementation within a computer simulation, the data was mapped into a regular latitude and longitude grid as shown in Table 3. The missing data over the continents was filled in with a constant value of .000021 corresponding to the global average ε value of .00002 plus an offset (the insignificant offset was used for bookkeeping purposes to identify the value as one not originally defined by Tank).

Large Angle Flight

To explain the remainder of the additions to the Tank model first requires the explanation of the time-based implementation method.

With the spectrum defined in eqn. 8, any of a number of methods could be used to evolve a time simulation. Tank suggests a method described by Fairall, White, and Thompson[3]. If the spectrum is discretized in M frequency bands Δk wide between k_1 and k_2, the strip area can be computed and equated as

$$\text{area} = 0.5 * (A_n^2 + B_n^2) \qquad (15)$$

where Δk is defined as numerically equal to k_1 (the lowest frequency of interest), and n ranges from 1 to M.

After solving for A_n and B_n, with a uniform random split between A_n and B_n, a Fourier series can be used to represent a turbulence field as follows

$$u(x) = \sum_{n=1}^{M} A_n \cos\left(\frac{2\pi x n}{L_0}\right) + B_n \sin\left(\frac{2\pi x n}{L_0}\right) \qquad (16)$$

Here L_0 is the segment length $1/k_1$. When the sequence above is evolved a one dimensional turbulence field in space is defined. Using Taylor's frozen field hypothesis the space field can be converted to a time-based sequence with the simple transformation

$$x = V * t \tag{17}$$

Incorporating this and exchanging Δk for $1/L_0$, eqn. 16 becomes

$$u(t) = \sum_{n=1}^{M} A_n \cos(2\pi\Delta ktn) + B_n \sin(2\pi\Delta ktn) \tag{18}$$

If evolved for a time longer than it takes to travel a distance L_0, as defined by eqn. 17, the field will begin to repeat. To avoid this a new set of A_n and B_n coefficients must be computed and used each time a distance L_0 is traversed.

Given this method of generating a turbulence field, several comments can be made. The frequency range of interest is defined in terms of spatial waves (if the vehicle velocity changes, the effective frequency changes). The turbulence field is evolved in precomputed segments. The vehicle state does not continuously update the model. The vehicle orientation has no effect on the sensed winds.

Dissipation Rate Scaling

To account for the fact that ε may change significantly during a segment that maybe as long as 100 km, ε scaling has been implemented. After computing the turbulence field with a fixed eddy dissipation rate, ε_0, the instantaneous value of ε can be used to form a scaling term,

$$\text{scale factor} = (\varepsilon/\varepsilon_0)^{1/3} \tag{19}$$

Equation 19 was derived through algebraic manipulation of equation 15 (with the strip area equal to the integral of equation 8) and observing the effect of ε on A_n and B_n. This scaling factor is applied directly to all the turbulence quantities. Figure 6 contains two traces of temperature as a function of time. The upper trace shows dissipation rate scaling as ε is arbitrarily increased by a factor of 4 over the 250 second run. The lower trace, which has been shifted down 2 degrees for clarity, is the same sequence without scaling. Notice how the magnitude of the perturbations increases in the upper trace and not in the lower.

Coordinate System Rotation

As in the case of Dryden turbulence, the Tank model as defined produces body axis turbulence for horizontally flying vehicles. This restriction is the result of the turbulence filters being derived for a turbulence field aligned along and normal to the vehicle velocity vector. Also, for horizontal flight, the vehicle altitude is nearly constant so the filter parameters, which are usually functions of altitude, do not change. To allow simulation of flight at large vehicle attitude angles, where there may be significant misalignment of the velocity vector and x-body axis, or large roll angle, a transformation has been defined.

To develop the axis transformation equations, the axis systems had to be defined. For ease of implementation it was desired that the output of the turbulence model should be body axis turbulence. It was also known that the longitudinal turbulence equation was specifically derived to generate turbulence along the velocity vector. The two remaining axes are only required to be normal to the longitudinal (and each other). For simplicity, in this implementation, one of these axes always lies in the local horizontal. The third axis points as required to remain normal to the local horizontal and the velocity vector. This choice has the benefit of aligning with the gravity and body axes in the case of standard horizontal flight.

To compute the transformation the heading 'off the wind' (Ψ_w), pitch attitude (θ), flight path angle (γ), and roll angle (ϕ) are required. Rotating from these turbulence axes to body axes involves first a rotation through Ψ_w about the local vertical axis. Second, is an upward rotation about the one axis lying in the local horizontal plane that is normal to the velocity vector, through the angle $\theta - \gamma$. The final rotation is about the x-body axis through the roll angle ϕ.

The turbulence to body axis transformation matrix can be written as the reverse order multiplication of the individual vector rotations as follows

$$T_{BT} = \begin{bmatrix} 1 & 0 & 0 \\ 0 & \cos\phi & \sin\phi \\ 0 & -\sin\phi & \cos\phi \end{bmatrix} \begin{bmatrix} \cos(\theta-\gamma) & 0 & -\sin(\theta-\gamma) \\ 0 & 1 & 0 \\ \sin(\theta-\gamma) & 0 & \cos(\theta-\gamma) \end{bmatrix} \begin{bmatrix} \cos\Psi_W & \sin\Psi_W & 0 \\ -\sin\Psi_W & \cos\Psi_W & 0 \\ 0 & 0 & 1 \end{bmatrix} \tag{20}$$

At each time step then the turbulence velocities are rotated into body axes with the following equation,

$$\begin{bmatrix} u \\ v \\ w \end{bmatrix}_{body} = \left[T_{BT}\right] \begin{bmatrix} u \\ v \\ w \end{bmatrix}_{turbulence} \tag{21}$$

Figure 7 shows the effect of the transformation in the case of a complete roll about the longitudinal body axis. The roll maneuver begins at the 10 second mark and is completed in 24 seconds. The top two traces are the turbulence time histories of the lateral components as generated in the turbulence axis system. The bottom two traces are the corresponding plots in body axes. Notice that the corresponding v and w plots are identical for the first and last 10 seconds in the traces. Between 10 and 34 seconds the roll occurs. Notice at the half way point the two v plots and two w plots are equal in magnitude but opposite in sign, as they should be for 180 degrees of roll.

Turbulence Segments

The model as described by Tank was developed for simulations with the simulation time duration defined a priori. The low frequency wave number was then computed as the quotient of the time duration and the flight velocity. This manner of implementation allows one set of Fourier coefficients to generate the entire turbulence field.

When the simulation time duration is not known, the low frequency wavenumber must be specified. If the duration of the simulation is long enough for the vehicle to travel a distance greater then the reciprocal of the selected low frequency wavenumber, hereafter called the segment distance, the turbulence field would begin to repeat. To avoid the repeating pattern, a new set of Fourier coefficients must be calculated each time a segment ends.

This changing of Fourier coefficients allows the likely possibility that a significant discontinuity could exist between segments. To minimize the discontinuity, an exponential decay from the last value of the previous segment is faired into the following segment according to,

$$u(t) = u'(t) + (\Delta u)\exp\left(\frac{f_2}{2}\Delta t\right) \qquad (22)$$

where u'(t) is the raw turbulence perturbation as calculated, Δu is the magnitude of the discontinuity between the segments, f_2 is the upper frequency limit, and Δt in this instance is the time since the discontinuity. The result of equation 23 is that the discontinuous jump is faired in over a few seconds in most cases. Figure 8 shows a turbulence time history sequence in its raw and faired form. In this example the minimum frequency of interest was 0.05 Hz, therefore a discontinuity occurs every 20 seconds. In Figure 8 they are visible in the raw sequence (dotted line) at the 260 and 280 second points. Notice that the faired sequence (solid line) does not follow the discontinuity, but quickly merges with the raw sequence shortly thereafter.

Summary

In this report Dryden and Kolmogorov based turbulence models have been reviewed with attention paid to the parameterization of the models. The specific Kolmogorov model proposed by Tank has been implemented as a series of MATLAB scripts. Features that add functionality to the Tank model have been developed, implemented and described. A variation of the eddy dissipation rate with altitude based on Dryden data has been calculated. The eddy dissipation rate database has been mapped onto a regular latitude and longitude grid and over-land values filled in. Methods to account for flight at large vehicle attitude angles have been implemented. Finally, a procedure for transitioning smoothly across turbulence segments has been developed.

References

1. Tank, William G.; Atmospheric Disturbance Environment Definition, NASA CR-195315, February 1994.

2. Nastrom, G.D.; Gage, K.S.: A Climatology of Atmospheric Wavenumber Spectra of Wind and Temperature Observed by Commercial Aircraft. J. Atmospheric Sciences, vol. 42, No. 9, May 1985, pp. 950-960.

3. Fairall, C.W.; White, A.B.; Thompson, Dennis W.; A Stochastic Model of Gravity-Wave-Induced Clear-Air Turbulence, J. Atmos. Sci. Vol. 48, No. 15, August, 1991.

4. Hoblit, Frederic M.; Gust Loads on Aircraft: Concepts and Applications, AIAA, Washington DC,1988

5. Johnson, D. L.; Terrestrial Environment (Climatic) Criteria for Use in Aerospace Vehicle Development, 1993 Revision; NASA TM 4511, August 1993.

6. Etkin, B.; Turbulent Wind and Its Effect on Flight, J. Aircraft Vol. 18, No. 5, May 1981.

7. Cornman, L.; Morse, C.; Cunning, G.; Real-time Estimation of Atmospheric Turbulence Severity From In-Situ Aircraft Measurements, AIAA 94-0268, 32nd Aerospace Sciences Meeting & Exhibit, Reno NV, January, 1994.

8.Gage, K.S.: Evidence for a k-5/3 Law Inertial Range in Mesoscale Two-Dimensional Turbulence. J. Atmospheric Sciences, vol. 36, October 1979, pp. 1350-1354.

9. Tennekes, H.; Lumley, J.L.; A First Course In Turbulence, The MIT Press, Cambridge 1972

Table 1. Dryden horizontal and vertical turbulence parameters as functions of altitude and severity.

	Light Turbulence		Moderate Turb.		Severe Turbulence		Scale Lengths	
Altitude km	Horizontal σ_h m/s	Vertical $\sigma_{v,w}$ m/s	Horizontal σ_h m/s	Vertical $\sigma_{v,w}$ m/s	Horizontal σ_h m/s	Vertical $\sigma_{v,w}$ m/s	Horizontal L_u km	Vertical $L_{v,w}$ km
1	0.17	0.14	1.65	1.36	5.70	4.67	0.832	0.624
2	0.17	0.14	1.65	1.43	5.80	4.75	0.902	0.831
4	0.20	0.17	2.04	1.68	6.24	5.13	1.04	0.972
6	0.21	0.17	2.13	1.69	7.16	5.69	1.04	1.01
8	0.22	0.17	2.15	1.69	7.59	5.98	1.04	0.98
10	0.22	0.17	2.23	1.73	7.72	6.00	1.23	1.10
12	0.25	0.18	2.47	1.79	7.89	5.71	1.80	1.54
14	0.26	0.19	2.62	1.91	6.93	5.05	2.82	2.12
16	0.24	0.21	2.44	2.10	5.00	4.31	3.40	2.6
18	0.22	0.21	2.21	2.07	4.07	3.81	5.00	3.34
20	0.23	0.20	2.26	1.99	3.85	3.38	8.64	4.41
25	0.27	0.21	2.71	2.09	4.34	3.34	12	6.56
30	0.37	0.24	3.73	2.39	5.6	3.59	28.6	8.88

Table 2. Epsilon values derived from Table 1 data.

Altitude, km	Light Turbulence ε, m^2/s^3	Moderate Turbulence ε, m^2/s^3	Severe Turbulence ε, m^2/s^3	Normalized variation, ε_D
1	5.53E-05	0.0506	2.09	2.77
2	5.10E-05	0.0467	2.03	2.56
4	7.21E-05	0.0765	2.19	3.61
6	8.34E-05	0.0871	3.31	4.18
8	9.59E-05	0.0895	3.94	4.81
10	8.11E-05	0.0845	3.50	4.07
12	8.13E-05	0.0784	2.56	4.08
14	5.84E-05	0.0598	1.11	2.93
16	3.81E-05	0.0400	0.34	1.91
18	2.00E-05	0.0202	0.13	1.00
20	1.32E-05	0.0125	0.06	0.66
25	1.54E-05	0.0155	0.06	0.77
30	1.66E-05	0.0170	0.06	0.83

Table 3. ε at regular latitude and longitude intervals (for proper scaling multiply values by 10^{-5} m^2s^{-3}).

Longitude (east), deg	Latitude, deg.												
	80	70	60	50	40	30	20	10	0	-10	-20	-30	-40
100	0.8	2.1	2.1	2.1	2.1	2.1	2.1	2	1.5	1	0.3	0.3	0.3
120	1	2.1	2.1	2.1	2.1	2.1	4	4	2	1.5	1	0.3	0.3
140	1	2.1	6	6	8	6	3	4	6	3	2.1	0.3	0.3
160	0.3	2.1	2.1	3	1.9	1.5	1.4	2	2	2	1.6	1	0.3
180	0.3	1	2.2	2.5	1	1.3	1.4	1.5	2	1.8	1	0.3	0.3
-160	0.3	2.1	2.1	1.5	1.3	1.8	2	1	2	1	1	0.3	0.3
-140	0.3	2.1	2.1	1	1.2	1.5	1.5	1.5	2	1	1	0.3	0.3
-120	1	2.1	2.1	2.1	2.1	1	1	1.8	2	1	1	0.3	0.3
-100	1	2.1	2.1	2.1	2.1	2.1	1	2	2	1	1	0.3	0.3
-80	1	2.1	2.1	2.1	2.1	2	2	2	2	1.5	1	0.3	0.3
-60	1	0.3	0.3	2	1.3	1	1	2	2.1	2.1	2.1	0.3	0.3
-40	1	2.1	2	1	2	1.5	1.2	2	2	1.5	1	0.3	0.3
-20	2	2	1.8	1.2	1.5	1.2	1.2	2	1.8	1	1	0.3	0.3
0	1.5	1.5	2	2	1	2.1	2.1	2.1	1.5	1	0.3	0.3	0.3
20	1	2	2.1	2.1	2.1	2.1	2.1	2.1	2.1	2.1	2.1	2.1	0.3
40	1.5	1.5	2.1	2.1	2.1	2.1	2.1	2.1	2.1	0.3	0.3	0.3	0.3
60	1.5	1.5	2.1	2.1	2.1	2.1	2	1.5	1	0.3	0.3	0.3	0.3
80	1.5	2.1	2.1	2.1	2.1	2.1	2.1	1.5	1	0.3	0.3	0.3	0.3

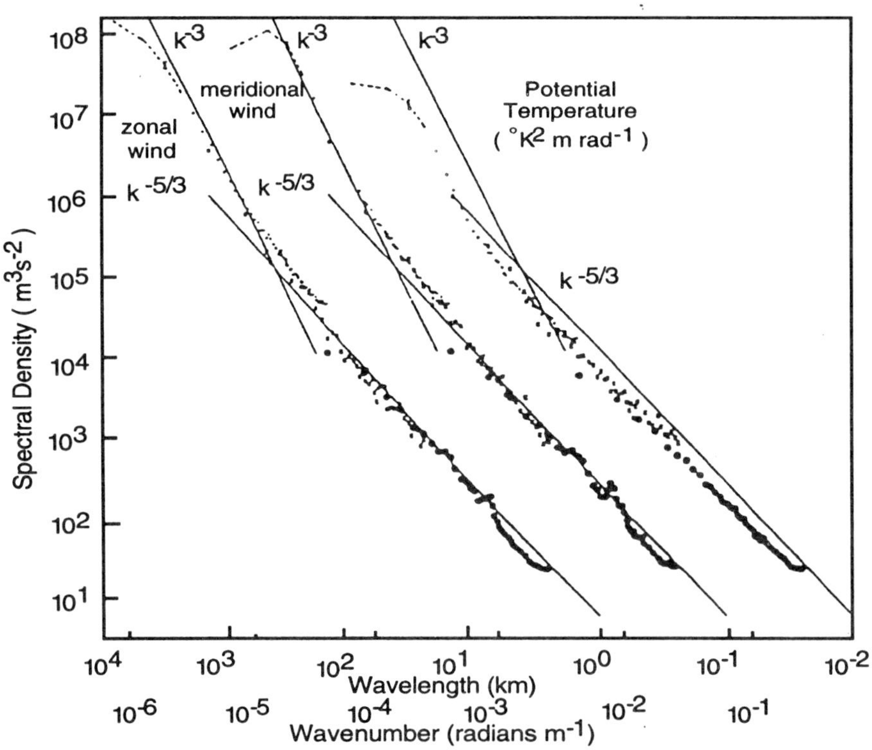

Figure 1. Wind and Potential Temperature Spectra as reported by Nastrom and Gage (1981). Note: for clarity, the meridional wind and temperature spectra have been shifted one and two decades to the right, respectively.

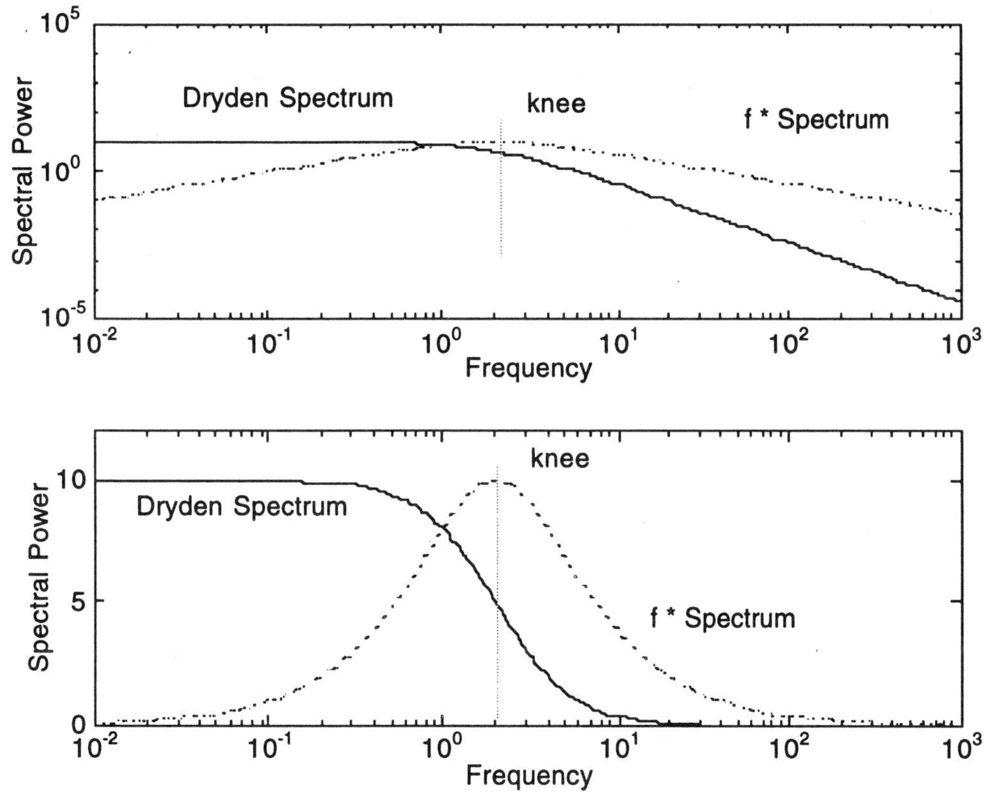

Figure 2. -Log and semi-log plots of Dryden Spectrum showing 'knee' frequency (rad./m).

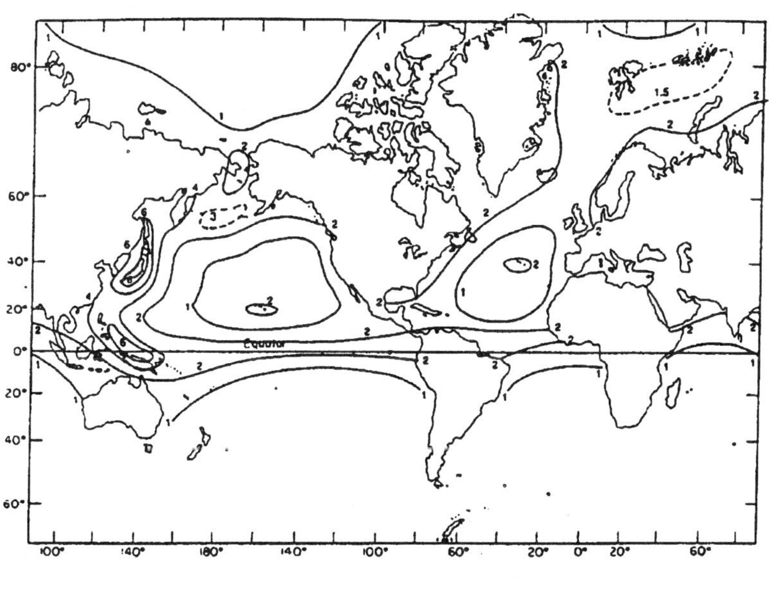

Figure 3.- Lower stratosphere, annual mean, eddy dissipation rate, x10^5 m^2/s^3, from reference 7.

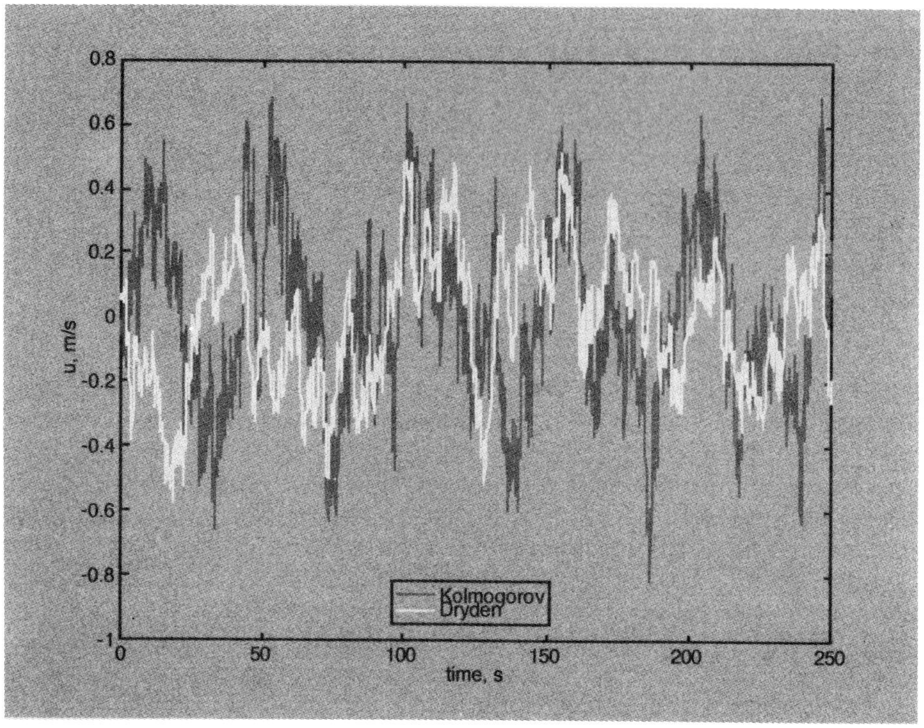

Figure 4.- Comparison of longitudinal turbulence time histories evolved by the Dryden filter method ($\sigma = 0.22$ m/s and $L_u = 5$km) and the Tank model (f_1=.0125 Hz, f_2 = 5 Hz, $\varepsilon = 2 \times 10^{-5}$ m^2/s^3).

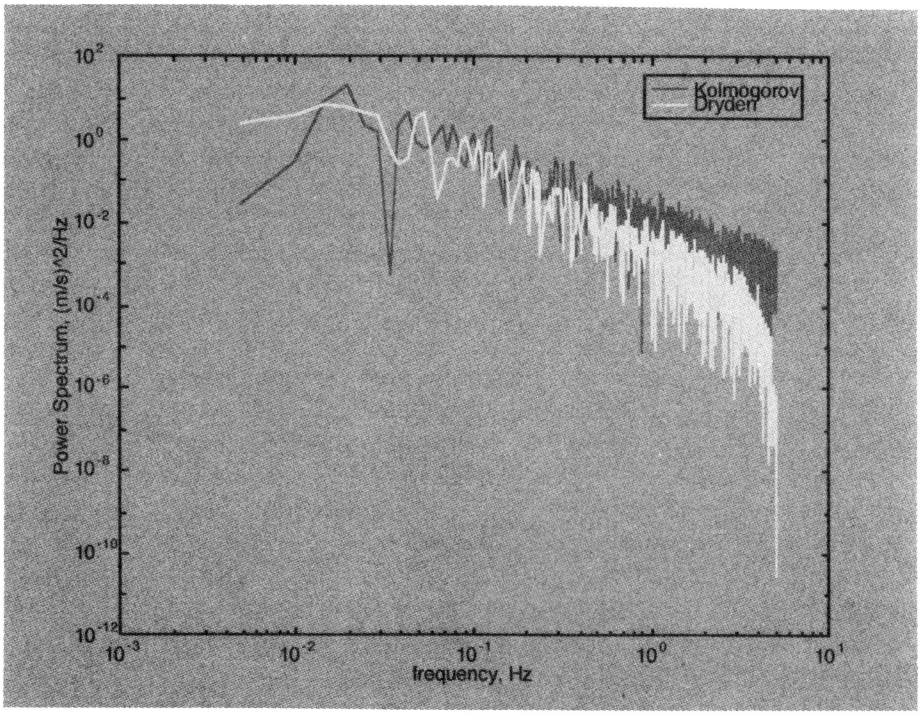

Figure 5.- Power spectra of time histories shown in Figure 4.

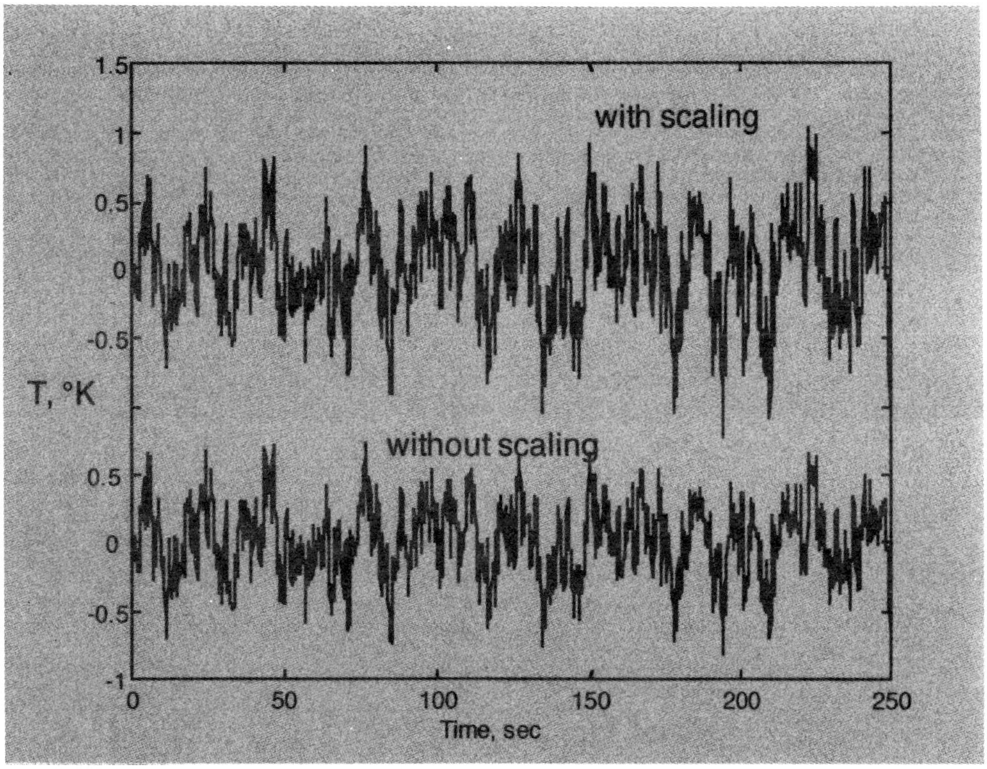

Figure 6.- Temperature histories with and without ε scaling.

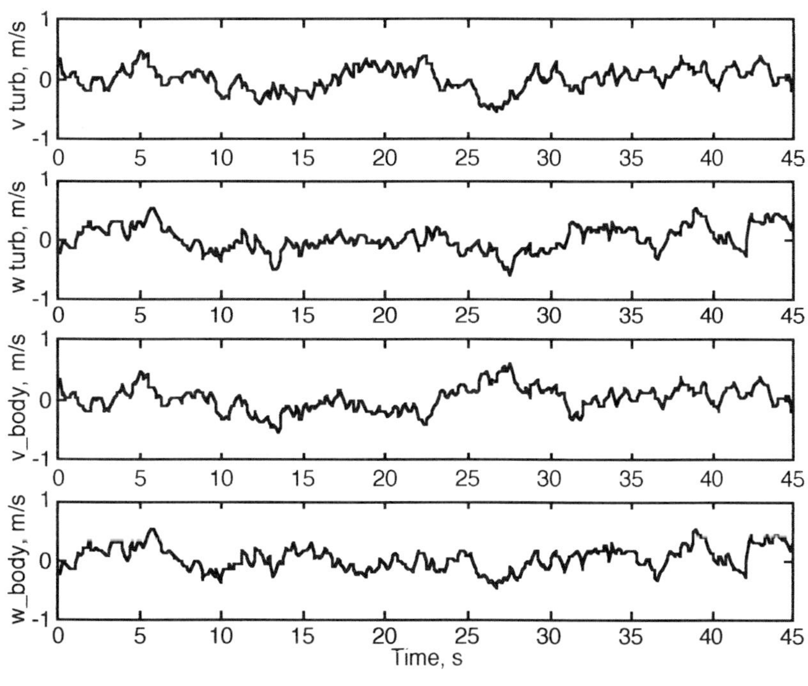

Figure 7.- Comparison of transverse turbulence in turbulence and body axes during 360 degree roll.

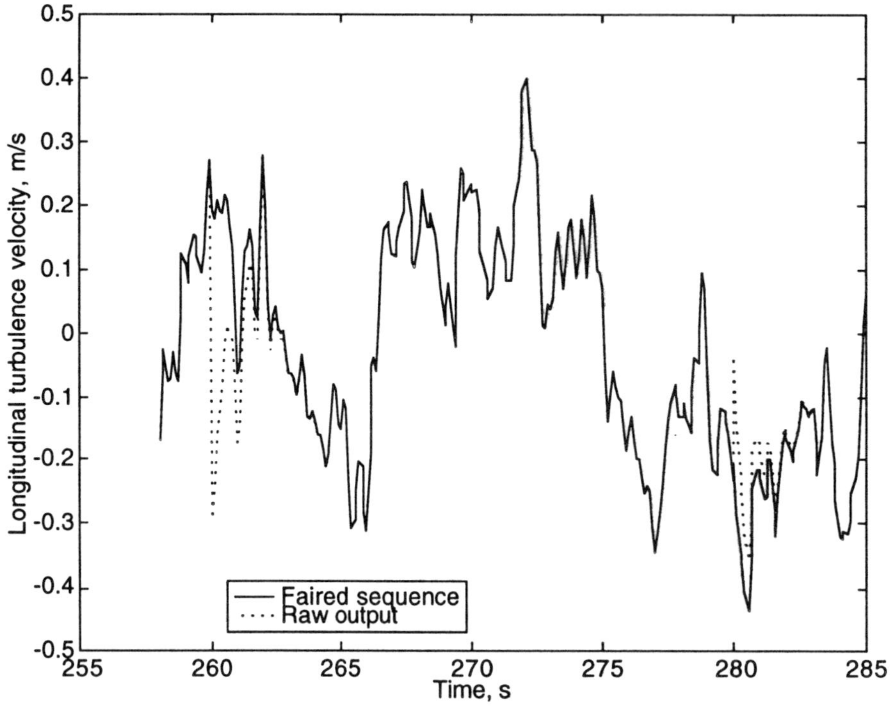

Figure 8.- Longitudinal turbulence time history showing the effect of 'fairing' between segments.

AIAA-97-3533

REUSABLE LAUNCH VEHICLE CONTROL IN SLIDING MODES

Yuri Shtessel and Christian Tournes

Electrical and Computer Engineering Department
University of Alabama in Huntsville
Huntsville, AL 35899
(205)890-6164, shtessel@ebs330.eb.uah.edu

Don Krupp

Guidance and Control Systems Division
NASA Marshall Space Flight Center
MSFC, AL 35812
(205)544-1812, Don.Krupp@msfc.nasa.gov

Abstract

Control of a reusable launch vehicle (RLV) is considered in sliding modes. The sliding mode controllers are designed for the RLV in two operational modes: re-entry (descending) mode and launch (ascending) mode. The decentralized sliding mode controllers are designed to guarantee robust de-coupled asymptotic tracking of the Euler angles' reference profiles upon the RLV model uncertainties and external disturbances. The angular rates command vector is formed in the outer loop in order to provide an asymptotic tracking of the Euler angles' reference profiles. This command vector is tracked in the internal loop via the smoothed sliding mode controllers. Effectiveness of the designed sliding mode controllers is confirmed via computer simulations of a vertical-takeoff, horizontal-landing wing-body single-stage RLV studied at NASA Marshall Space Flight Center.

Introduction

The current research in developing a next generation launch vehicle is focused on reusable launch vehicles (RLV). Reusable single-stage-to orbit launch vehicles developing in X-33 project will replace the aging Space Shuttles at the beginning of the next century[1]. One proposed configuration for a RLV vehicle is a vertical-takeoff, horizontal-landing wing-body single-stage vehicle (WB001)[2]. The reentry control problem consists in automatic controlling the atmospheric entry trajectory which can be precalculated or generated in current time by guidance algorithms. The launch control problem consists of automatic tracking the launch trajectory which as assumed to be optimally precalculated. For present launch systems the prescribed trajectory has to be remodeled for each mission, depending on the weight of the payload as well as the prescribed orbit and safety margins. From the control point of view it requires development of reliable, robust control algorithm that can automatically adjust to some changes in mission specifications (mass of payload, target orbit) and the operating environment (atmospheric perturbations, interconnection perturbations from the other subsystems of the launch vehicle, thrust deficiencies, failure scenarios). It is obvious that the errors in optimal launch trajectory tracking caused be various perturbations and uncertainties lead to the payload weight reduction launched to the orbit. From this point of view a robust feedback controller design improving the accuracy of tracking of the optimal trajectory is a critical issue for RLV. Various linear and nonlinear guidance and trajectory control algorithms[3,4] were proposed.

One of the effective control strategies successfully applied in nonlinear systems is the *Sliding Mode Control*[5-7]. The main advantage of the Sliding Mode Control is that the system's state response in the sliding surface remains insensitive to certain parameter variations, nonlinearities and disturbances. In this paper the WB001 RLV trajectory tracking with unmatched external and interaction disturbances and nonlinearities is addressed via the de-coupling robust sliding mode control[8-11].

Problem Formulation

The following mathematical model of a rigid RLV is taken.

1. Orientation equations of a rigid RLV are given in terms of Euler angles ψ, θ, φ

$$\begin{bmatrix} \dot{\varphi} \\ \dot{\theta} \\ \dot{\psi} \end{bmatrix} = R(\varphi,\theta,\psi) \begin{bmatrix} p \\ q \\ r \end{bmatrix} \qquad (1)$$

where

$$R(\varphi,\theta,\psi) = \begin{bmatrix} 1 & \tan\theta\sin\varphi & \tan\theta\cos\varphi \\ 0 & \cos\varphi & -\sin\varphi \\ 0 & \dfrac{\sin\varphi}{\cos\theta} & \dfrac{\cos\varphi}{\cos\theta} \end{bmatrix}$$

2. The equations of an RLV rotational motion are

$$\begin{cases} \dot{p} = f_1(p,q,r) + \dfrac{M_1}{J_{xx}} \\ \dot{q} = f_2(p,q,r) + \dfrac{M_2}{J_{yy}} \\ \dot{r} = f_3(p,q,r) + \dfrac{M_3}{J_{zz}} \end{cases} \quad (2)$$

where $\mathrm{J} = \begin{bmatrix} J_{xx} & J_{xy} & J_{xz} \\ J_{yx} & J_{yy} & J_{yz} \\ J_{zx} & J_{zy} & J_{zz} \end{bmatrix}$ is the inertia matrix;

$$\begin{cases} f_1(p,q,r) = \dfrac{1}{J_{xx}}\left[J_{xy}\dot{r} - qr\left(J_{zz} - J_{yy}\right) + J_{xz}pq\right] \\ f_2(p,q,r) = \dfrac{1}{J_{yy}}\left[-pr\left(J_{xx} - J_{zz}\right) - J_{xz}\left(p^2 - r^2\right)\right] \\ f_3(p,q,r) = \dfrac{1}{J_{zz}}\left[J_{xz}\dot{p} - pq\left(J_{yy} - J_{xx}\right) - J_{xz}qr\right] \end{cases} \quad (3)$$

p is a roll rate, q is a pitch rate, r is a yaw rate; $M = \{M_1, M_2, M_3\}^T$ is a vector of moments acting on an RLV.

3. The equations of an RLV translational motion are

$$\begin{cases} \dot{v}_x = -qv_z + rv_y - g\sin\theta + \dfrac{1}{m_0}F_x \\ \dot{v}_y = -rv_x + pv_z + g\cos\theta\sin\varphi + \dfrac{1}{m_0}F_y \\ \dot{v}_z = -pv_y + qv_x + g\cos\theta\cos\varphi + \dfrac{1}{m_0}F_z \end{cases} \quad (4)$$

where $v = \{v_x, v_y, v_z\}^T$ is a velocity vector of a center of gravity of the RLV in the body frame, m_0 is mass of the RLV, $F = \{F_x, F_y, F_z\}^T$ is a vector of forces acting to RLV.

Assuming a vector of a wind velocity is equal to zero a side slip angle β and angle of attack α are defined as follows:

$$\alpha = \tan^{-1}\left(\frac{v_z}{v_x}\right), \quad \beta = \tan^{-1}\left(\frac{v_y}{\sqrt{v_x^2 + v_z^2}}\right) \quad (5)$$

4. The altitude of the RLV is described as

$$\dot{R} = v_x\sin\theta - v_y\sin\varphi\cos\theta - v_z\cos\varphi\cos\theta \quad (6)$$

where R is a distance between the surface of the flat Earth to the RLV's center of gravity.

Two control modes of the RLV are considered: the reentry mode and the launch mode. The control algorithms in these two modes are completely different, because of the following reasons:

- in the reentry mode the control torque are created by deflections of the aerodynamic surfaces, and mass m_0 of the RLV is constant (no fuel consumption),
- in the launch mode the control torque are created by thrusters (rocket engines), and mass of the RLV varies because of fuel consumption.

Remark 1. In some papers, for example[11], the equations (1) are given in quaternion notation in order to avoid singularities occurring in an orbital flight. This is not a case considering in this paper.

Reentry Mode

In the reentry mode the desired Euler angles' trajectories are assumed to be generated in a real time by the pilot or the guidance system. The vector of moments M (in a body frame) is defined as follows:

$$M = \begin{bmatrix} M_{aero_x} \\ M_{aero_y} \\ M_{aero_z} \end{bmatrix} = \begin{bmatrix} \bar{q}S_{ref}bC_{l_{TOTAL}} \\ \bar{q}S_{ref}cC_{m_{TOTAL}} \\ \bar{q}S_{ref}bC_{n_{TOTAL}} \end{bmatrix} + \tilde{M}_{aero}; \quad (7)$$

aerodynamic moments acting on the rigid body of the RLV are described as follows:

$$\tilde{M}_{aero} = \mathrm{r}_{aero} \times \mathrm{F}_{aero},$$

$$\mathrm{F}_{aero} = \begin{bmatrix} F_{aero_x} \\ F_{aero_y} \\ F_{aero_z} \end{bmatrix} = \begin{bmatrix} \bar{q}S_{ref}C_{A_{TOTAL}} \\ \bar{q}S_{ref}C_{Y_{TOTAL}} \\ \bar{q}S_{ref}C_{N_{TOTAL}} \end{bmatrix},$$

$$\mathrm{r}_{aero} = \begin{bmatrix} r_{aero_x} \\ r_{aero_y} \\ r_{aero_z} \end{bmatrix} = \begin{bmatrix} -\left(x_{MRP} - x_{cg}\right) \\ y_{MRP} - y_{cg} \\ -\left(z_{MRP} - z_{cg}\right) \end{bmatrix}, \quad (8)$$

where $\bar{q}$ is dynamic pressure, S_{ref} is a reference area, b is a reference span, c is a reference length of the RLV's body, $C_{l_{total}}, C_{m_{total}}, C_{n_{total}}, C_{A_{total}}, C_{Y_{total}}, C_{N_{total}}$ are aerodynamic coefficients, $x_{MRP}, y_{MRP}, z_{MRP}$ are the coordinates of the moment reference point, x_{cg}, y_{cg}, z_{cg} are the coordinates of the center of gravity.

Aerodynamic coefficients in the formulas (7) and (8) are calculated through the data base given basic, elevator, aileron, body flap, tip fin, side slip (beta) and power-off aerodynamic coefficients

$$\begin{cases} C_{A_{TOTAL}} = C_{A_{BASIC}} + C_{A_{ELEVATOR}} + C_{A_{AILERON}} + C_{A_{BODY\ FLAP}} \\ \quad + C_{A_{TIP\ FIN}} + C_{A_{POW\ OFF}} \\ C_{N_{TOTAL}} = C_{N_{BASIC}} + C_{N_{ELEVATOR}} + C_{N_{AILERON}} + C_{N_{BODY\ FLAP}} \\ \quad + C_{N_{TIP\ FIN}} \\ C_{m_{TOTAL}} = C_{m_{BASIC}} + C_{m_{ELEVATOR}} + C_{m_{AILERON}} + C_{m_{BODY\ FLAP}} \\ \quad + C_{m_{TIP\ FIN}} \end{cases} \tag{9}$$

$$\begin{cases} C_{Y_{TOTAL}} = C_{y_{BETA}} + C_{Y_{ELEVATOR}} + C_{Y_{AILERON}} + C_{Y_{BODY\ FLAP}} \\ \quad + C_{Y_{TIP\ FIN}} \\ C_{l_{TOTAL}} = C_{l_{BETA}} + C_{l_{ELEVATOR}} + C_{l_{AILERON}} + C_{l_{BODY\ FLAP}} \\ \quad + C_{l_{TIP\ FIN}} \\ C_{n_{TOTAL}} = C_{n_{BETA}} + C_{n_{ELEVATOR}} + C_{n_{AILERON}} + C_{n_{BODY\ FLAP}} \\ \quad + C_{n_{TIP\ FIN}} \end{cases} \tag{10}$$

Assuming side slip angle $\beta = 0$ the aerodynamic coefficients $C_{l_{total}}, C_{m_{total}}, C_{n_{total}}$, which contribute to control torque, are approximated by linear functions

$$\begin{aligned} C_{m_{TOTAL}} &= C_{m,1} + C_{m,2}\delta_{Ele} + C_{m,3}\delta_{Ail} + C_{m,4}\delta_{BF} + C_{m,5}\delta_{TF} \\ C_{l_{TOTAL}} &= C_{l,1}\delta_{Ail} + C_{l,2}\delta_{TF} \\ C_{n_{TOTAL}} &= C_{n,1}\delta_{Ail} + C_{n,2}\delta_{TF} \end{aligned} \tag{11}$$

where $\delta_{Ele}, \delta_{Ail}, \delta_{BF}, \delta_{TF}$ are deflections of virtual aerodynamic surfaces: elevators, ailerons, body flap and tip fins correspondingly, $C_{i,j} = C_{i,j}(Mach, \alpha)$., and $Mach$ is a Mach number.

Aerodynamic coefficients $C_{A_{total}}$, $C_{Y_{total}}$, $C_{N_{total}}$ are assumed to be bounded disturbances. The component $\tilde{M}_{aero}$ of the vector of moments (7), (8) is assumed to be a bounded vector-disturbance .

The flight control surfaces (elevators, ailerons, body flap, and tip fins) are actuated by corresponding actuators. The following constrains to actuator deflections are imposed:

$$|\delta_{Ele}| \le 18^0, \quad |\delta_{Ail}| \le 18^0, \quad |\delta_{BF}| \le 18^0, \quad |\delta_{TF}| \le 60^0. \tag{12}$$

The body flap control action is fixed at the level $\delta_{BF} = -18^0$.The desired flying qualities in terms of Euler angles desired responses are defined by the following transfer functions

$$\frac{\theta_c}{\theta^*} = \frac{\omega_\theta^2}{s^2 + 2\xi_\theta \omega_\theta s + \omega_\theta^2}, \quad \frac{\psi_c}{\psi^*} = \frac{\omega_\psi^2}{s^2 + 2\xi_\psi \omega_\psi s + \omega_\psi^2},$$
$$\frac{\varphi_c}{\varphi^*} = \frac{\omega_\varphi^2}{s^2 + 2\xi_\varphi \omega_\varphi s + \omega_\varphi^2} \tag{13}$$

where $\theta^*, \psi^*, \varphi^*$ are guidance commands, $\theta_c, \psi_c, \varphi_c$ are command signals to actuators, $\xi_\theta = \xi_\psi = \xi_\varphi = 0.7$ and $\omega_\theta = 7\ rad/s$, $\omega_\psi = 1.2\ rad/s$, $\omega_\varphi = 0.3\ rad/s$.

Taking into account the equations (11), the angular rate and actuator equations are rewritten in a following format

$$\begin{cases} \begin{bmatrix} \dot{p} \\ \dot{q} \\ \dot{r} \end{bmatrix} = \begin{bmatrix} f_1(p,q,r) \\ f_2(p,q,r) \\ f_3(p,q,r) \end{bmatrix} + \begin{bmatrix} M_{1d} \\ M_{2d} \\ M_{3d} \end{bmatrix} + \begin{bmatrix} b_{11} & 0 & b_{13} \\ b_{21} & b_{22} & b_{23} \\ b_{31} & 0 & b_{33} \end{bmatrix} \begin{bmatrix} \delta_{Ail} \\ \delta_{Ele} \\ \delta_{TF} \end{bmatrix} \\ \begin{bmatrix} \dot{\delta}_{Ail} \\ \dot{\delta}_{Ele} \\ \dot{\delta}_{TF} \end{bmatrix} = \begin{bmatrix} 20 & 0 & 0 \\ 0 & 20 & 0 \\ 0 & 0 & 20 \end{bmatrix} \left\{ \begin{bmatrix} \delta_{Ail} \\ \delta_{Ele} \\ \delta_{TF} \end{bmatrix} + \begin{bmatrix} u_1 \\ u_2 \\ u_3 \end{bmatrix} \right\}, \end{cases} \tag{14}$$

where u_i are the control inputs of the actuators, $M_d = \{M_{1d}, M_{2d}, M_{3d}\}^T$ is a vector of disturbance moments reducing to rad/s.

The problem is to design the control laws $u_i \ \forall i = \overline{1,3}$ to provide the desired robust de-coupled tracking of prescribed Euler angles' trajectories for RLV in a reentry (descending) mode upon effect of external disturbances and plant uncertainties.

Launch Mode

Without throttling the rocket engines feedback velocity regulation cannot be achieved effectively. Therefore we will only concentrate on Euler angles reference profiles tracking. In the launch mode the desired Euler angles' trajectories are assumed to be precalculated off line and must be accurately tracked by a control system. In the equations of a translational motion a mass m_o is not constant any more and its evolution is described by the following equations

$$\dot{m}_f = \begin{cases} mdot, & if \ m_f > 0 \\ 0, & if \ m_f = 0 \end{cases}$$
$$m_0 = \begin{cases} m_{dry} + m_f, & if \ m_f > 0 \\ m_{dry}, & if \ m_f = 0 \end{cases} \tag{15}$$

where m_f is mass of fuel, m_{dry} is mass of the RLV without fuel, $mdot$ is a rate of fuel consumption.

The vector of moments $M=\{M_1, M_2, M_3\}^T$ in the equations (2) is defined as

$$M = M_{th} + \widetilde{M}_{aero} \quad (16)$$

where M_{th} is a vector of control (thrusters) moments, $\widetilde{M}_{aero}$ is a vector of aerodynamic moments, defining by the equations (8) and treating as a disturbance. The moments (torque) M^i_{th} $\forall i=1,3$ are produced by thrusters which deflection angles are actuated by actuators. No time lag between thrusters' deflection angles and corresponding thrusters' torque is assumed. The transfer functions of the actuators are taken as follows:

$$\frac{M^i_{th}(s)}{U_i(s)} = \frac{\omega_n^2}{s^2 + 2\xi\omega_n s + \omega_n^2} \quad \forall i=1,3 \quad (17)$$

where $u_i(t)$ are control commands to actuators, and $\xi = 0.7$, $\omega_n = 26.389\, rad/s$.

The problem is to design the control laws u_i $\forall i=\overline{1,3}$ to provide the desired robust de-coupled tracking of prescribed Euler angles' trajectories for RLV in a launch (ascending) mode upon effect of external disturbances and plant uncertainties.

Sliding Mode Control Strategy for RLV Trajectory Tracking

The back-stepping control[12] is combined with a sliding mode control technique. The following two-step algorithm of the sliding mode controller design for the RLV trajectory tracking in re-entry and launch modes is developed:

- **At the first step** the angular rates vector $\{p,q,r\}^T$ is considered as a virtual control for the Euler angles vector $\{\varphi,\theta,\psi\}^T$ describing by the equations (1). The angular rates command profile $\{p_c,q_c,r_c\}^T$ is identified to provide the desired evolution of Euler angles tracking errors. This evolution is described by the following system of de-coupled linear time-invariant homogeneous differential equations

$$\begin{bmatrix} \dot{\widetilde{\varphi}} \\ \dot{\widetilde{\theta}} \\ \dot{\widetilde{\psi}} \end{bmatrix} = K \begin{bmatrix} \widetilde{\varphi} \\ \widetilde{\theta} \\ \widetilde{\psi} \end{bmatrix}, \qquad K = \begin{bmatrix} k_1 & 0 & 0 \\ 0 & k_2 & 0 \\ 0 & 0 & k_3 \end{bmatrix} \quad (18)$$

where $\widetilde{\varphi} = \varphi_c - \varphi$, $\widetilde{\theta} = \theta_c - \theta$, $\widetilde{\psi} = \psi_c - \psi$ are Euler angles' tracking errors. The entries k_i $\forall\overline{1,3}$ of a gain-matrix K must be selected to provide the desired (given) eigenvalues placement to the system (18). Substituting the equations (1) in to the equations (18) the following angular rates command vector is identified

$$\begin{bmatrix} p_c \\ q_c \\ r_c \end{bmatrix} = R^{-1}(\varphi,\theta,\psi)\left\{ \begin{bmatrix} \dot{\varphi}_c \\ \dot{\theta}_c \\ \dot{\psi}_c \end{bmatrix} + K \begin{bmatrix} \widetilde{\varphi} \\ \widetilde{\theta} \\ \widetilde{\psi} \end{bmatrix} \right\},$$

$$R^{-1}(\varphi,\theta,\psi) = \begin{bmatrix} 1 & 0 & -\sin\varphi \\ 0 & \cos\varphi & \sin\varphi\cos\theta \\ 0 & -\sin\varphi & \cos\varphi\cos\theta \end{bmatrix} \quad (19)$$

- **At the second step** the corresponding sliding mode controllers are designed to robustly, accurately track the angular rates command profiles (19) in accordance with the equations (3) and (14) in the reentry mode, and the equations (2) and (3) in the launch mode.

Sliding Mode Controller Design for the Reentry Mode

The equations (3), (14) are employed for the description of the RLV orientation motion during the de-coupling sliding mode controller design for the reentry mode.

Implementing the sliding mode controller design algorithm the sliding surfaces are designed to provide the desired (given) evolution of the angular rates tracking errors in sliding modes. The angular rates tracking errors must be described by de-coupled linear time-invariant homogeneous differential equations with a desired (given) eigenvalues placement.

In order to design the sliding surfaces the vector relative degree is identified as $\bar{r} = \{2,\ 2,\ 2\}^T$ without zero dynamics. The corresponding vector-sliding surfaces is designed as follows

$$\begin{bmatrix} \sigma_1 \\ \sigma_2 \\ \sigma_3 \end{bmatrix} = \begin{bmatrix} \dot{e}_1 \\ \dot{e}_2 \\ \dot{e}_3 \end{bmatrix} + C^1 \begin{bmatrix} e_1 \\ e_2 \\ e_3 \end{bmatrix} + C^2 \begin{bmatrix} \int_0^t e_1 d\tau \\ \int_0^t e_2 d\tau \\ \int_0^t e_3 d\tau \end{bmatrix} = \begin{bmatrix} 0 \\ 0 \\ 0 \end{bmatrix}, \quad (20)$$

$$C^1 = \begin{bmatrix} c_1^1 & 0 & 0 \\ 0 & c_2^1 & 0 \\ 0 & 0 & c_3^1 \end{bmatrix}, \quad C^2 = \begin{bmatrix} c_1^2 & 0 & 0 \\ 0 & c_2^2 & 0 \\ 0 & 0 & c_3^2 \end{bmatrix}.$$

where $e_1 = p_c - p$, $e_2 = q_c - q$, $e_3 = r_c - r$ are the angular rates tracking errors. The coefficients $c_i^j\ \forall i = \overline{1,3}\ \forall j = \overline{1,2}$ are identified to provide the desired (given) eigenvalues placement of the angular rates evolution in the de-coupled sliding surfaces (20).

Remark 2. It is clear that the sliding surfaces (19) and (20) are nonlinear functions of Euler angles.

Remark 3. The system's (3), (14) motion in the sliding surface is de-coupled and robust to plant uncertainties and external disturbances.

Completing the sliding mode controller design the control function must be identified to provide the existence of the sliding modes in the sliding surfaces (20). These control functions must provide the asymptotic stability of the origin of the system (3), (14) in the σ -space. This motion is described as follows:

$$\dot{\sigma} = F(.) - BGu \tag{21}$$

where

$$\sigma = \{\sigma_1, \sigma_2, \sigma_3\}^T,\ u = \{u_1, u_2, u_3\}^T,$$

$$B = \begin{bmatrix} b_{11} & 0 & b_{13} \\ b_{21} & b_{22} & b_{23} \\ b_{31} & 0 & b_{33} \end{bmatrix},\ G = \begin{bmatrix} 20 & 0 & 0 \\ 0 & 20 & 0 \\ 0 & 0 & 20 \end{bmatrix}$$

$$F(.) = \begin{bmatrix} \ddot{p}_c + c_1^1 \dot{p}_c + c_1^2 p_c \\ \ddot{q}_c + c_2^1 \dot{q}_c + c_2^2 q_c \\ \ddot{r}_c + c_3^1 \dot{r}_c + c_3^2 r_c \end{bmatrix} - C^2 \begin{bmatrix} p \\ q \\ r \end{bmatrix} - C^1 \begin{bmatrix} f_1(p,q,r) + M_{1d} \\ f_2(p,q,r) + M_{2d} \\ f_3(p,q,r) + M_{3d} \end{bmatrix} - \begin{bmatrix} \dot{f}_1(p,q,r) + \dot{M}_{1d} \\ \dot{f}_2(p,q,r) + \dot{M}_{2d} \\ \dot{f}_3(p,q,r) + \dot{M}_{3d} \end{bmatrix} - \left(C^1 B + BG\right) \begin{bmatrix} \delta_{Ail} \\ \delta_{Ele} \\ \delta_{TF} \end{bmatrix} \tag{22}$$

Introducing a candidate to a Lyapunov function

$$V = \frac{1}{2}\sigma^T \sigma > 0 \tag{23}$$

its derivative is identified as follows:

$$\dot{V} = \sigma^T \dot{\sigma} = \sigma^T \left[F(.) - BGu\right] \tag{24}$$

Assuming the vector-function $F(.)$ is estimated by the function $\hat{F}(.)$, and introducing $\Delta(.) = F(.) - \hat{F}(.)$ such that $|\Delta(.)_i| \le L_i\ \forall i = \overline{1,3}$, the sliding mode controller is designed in the following format:

$$u = \hat{u}_{eq}(.) + \rho SIGN\,\sigma, \tag{25}$$

where

$\hat{u}_{eq}(.) = (BG)^{-1}\hat{F}(.)$ is an equivalent control estimate,

$$\rho = \begin{bmatrix} \rho_1 & 0 & 0 \\ 0 & \rho_2 & 0 \\ 0 & 0 & \rho_3 \end{bmatrix},\ SIGN\,\sigma = \begin{bmatrix} sign\,\sigma_1 \\ sign\,\sigma_2 \\ sign\,\sigma_2 \end{bmatrix}.$$

Substituting expression (25) to the formula (24) the following formula is obtained

$$\dot{V} = \sigma^T\left[\Delta(.) - \rho SIGN\,\sigma\right] = \sum_{i=1}^{3}\sigma_i\left[\Delta_i(.) - \rho_i sign\,\sigma_i\right] =$$

$$= \sum_{i=1}^{3}\left(\sigma_i \Delta_i(.) - \rho_i|\sigma_i|\right) \le \sum_{i=1}^{3}\left(|\sigma_i| L_i - \rho_i|\sigma_i|\right) =$$

$$= \sum_{i=1}^{3}|\sigma_i|\left(L_i - \rho_i\right) < 0 \tag{26}$$

The inequality (26) is valid if

$$\rho_i > L_i\ \forall i = \overline{1,3}. \tag{27}$$

It is easy to show that the origin $\sigma = 0$ in the σ - space is reached by the system's (3), (14) trajectory in a finite time, which can be estimated as

$$t_r \le \max_{i=1,2,3}\frac{|\sigma_i(0)|}{\rho_i - L_i} \tag{28}$$

After the sliding surface (20) has been reached the system's (3), (14) motion will be in this sliding surface.

Implementation of Sliding Mode Controller for the Reentry Mode

In order to simplify implementation of the sliding mode controller (25) it is assumed that the matrix B is diagonal. Avoiding the control chattering due unmodeled dynamics the sliding mode controller (25) is continuously implemented[13] as follows:

$$\begin{bmatrix} u_1 \\ u_2 \\ u_3 \end{bmatrix} = \begin{bmatrix} U_{1\max} & 0 & 0 \\ 0 & U_{2\max} & 0 \\ 0 & 0 & U_{3\max} \end{bmatrix} \begin{bmatrix} sat\dfrac{\sigma_1}{\varepsilon_1} \\ sat\dfrac{\sigma_2}{\varepsilon_2} \\ sat\dfrac{\sigma_3}{\varepsilon_3} \end{bmatrix},\ \varepsilon_i = 0.005\ \forall i = \overline{1,3} \tag{29}$$

where $sat\dfrac{(.)}{\varepsilon}$ is a saturation function with an ε - width of a linearity zone, $U_{1\max} = U_{2\max} = 18^0$, $U_{3\max} = 60^0$, $i = 1$ corresponds to the pitch channel, $i = 2$ corresponds to the roll channel, $i = 3$ corresponds to the yaw channel.

Remark 4. It is worth to notice that the complicated term $\hat{u}_{eq}(.)$ in the sliding mode controller is not included in the controller implementation (29). It can be done if $U_{i\max}$ $\forall i = \overline{1,3}$ are large enough to make the inequalities (27) valid.

The matrices C^1 and C^2 in the equation of the sliding surface (20) are selected as follows:

$$C^1 = \begin{bmatrix} 10 & 0 & 0 \\ 0 & 10 & 0 \\ 0 & 0 & 10 \end{bmatrix}, \quad C^2 = \begin{bmatrix} 25 & 0 & 0 \\ 0 & 25 & 0 \\ 0 & 0 & 25 \end{bmatrix} \tag{30}$$

Computer simulations of the reentry mode

The reentry mode was simulated for the WB001 RLV coming back to Earth with a payload in. The trim conditions are taken as follows:
altitude: 19530.9 m, Mach#: 2.0, angle of attack $\alpha = 9.0^0$, side slip angle $\beta = 0$.
The following parameters of the RLV were accepted:

$$\begin{cases} m_0 = 103756\,kg,\ x_{cg} = 38.85\,m,\ y_{cg} = 0.0\,m, \\ z_{cg} = 1.53\,m, \bar{q} = 17108.7\,N/m,\ S_{ref} = 391.2\,m^2 \\ J_{xx} = 1997922.0\,kg\,m^2,\ J_{yy} = 2766299668\,kg\,m^2 \\ J_{zz} = 28383800.1\,kg\,m^2, y_{MRP} = 0.0\,m, \\ J_{xy} = J_{yx} = J_{xz} = J_{zx} = J_{yz} = J_{zy} = 0.0, \\ b = 29.2\,m,\ c = 56.6\,m,\ x_{MRP} = 38.8\,m,\ z_{MRP} = 1.68\,m \end{cases} \tag{31}$$

The reference profiles for all three Euler angles are formed on line and tracked by the smoothed de-coupling sliding mode controller (20), (29) and (30) The results of the simulation are shown in fig. 1 - 7. It is obvious that the designed sliding mode controller provides robust, de-coupled extremely accurate tracking of reference Euler angles' profiles insensitive to external and interaction disturbances in the re-entry mode. No control chattering can be observed. The output tracking in the Pitch Channel was simulated separately with the designed smoothed sliding mode controller and with the PID controller, developed at NASA MSFC. The results of the simulation are shown in fig. 8 and 9. Comparison of these two controllers shows that the sliding mode controller (SMC) provides the perfect pitch angle tracking, and is insensitive to disturbances. On the other hand the PID controller tracks a prescribed pitch angle profile with a significant tracking error. Effect of disturbances to the pitch angle tracking error is significant as well.

Sliding Mode Controller Design for the Launch Mode

The equations (2), (3) and (16) were employed for the description of the RLV orientation motion during the de-coupling sliding mode controller design for the launch mode. In order to simplify the sliding mode controller design the dynamics of actuators (17) were neglected. Actuator dynamics are added during the simulation.

Implementing the sliding mode controller design algorithm the sliding surfaces are designed to provide the desired (given) evolution of the angular rates tracking errors in sliding modes. The angular rates tracking errors must be described by de-coupled linear time-invariant homogeneous differential equations with a desired (given) eigenvalues placement.

In order to design the sliding surfaces the vector relative degree is identified as $\bar{r} = \{1,\ 1,\ 1\}^T$ with no zero dynamics. It implies the sliding surfaces of zero order.

Applying the smoothed sliding mode controller design algorithm developing for the RLV reentry mode to the controller design for the RLV in the launch mode the following de-coupling control law is obtained

$$\begin{bmatrix} u_1 \\ u_2 \\ u_3 \end{bmatrix} = \begin{bmatrix} U_{1\max} & 0 & 0 \\ 0 & U_{2\max} & 0 \\ 0 & 0 & U_{3\max} \end{bmatrix} \begin{bmatrix} sat\frac{\sigma_1}{\varepsilon_1} \\ sat\frac{\sigma_2}{\varepsilon_2} \\ sat\frac{\sigma_3}{\varepsilon_3} \end{bmatrix}, \varepsilon_i = 0.005\ \forall i = \overline{1,3} \tag{32}$$

$$\begin{bmatrix} \sigma_1 \\ \sigma_2 \\ \sigma_3 \end{bmatrix} = \begin{bmatrix} e_1 \\ e_2 \\ e_3 \end{bmatrix} + C \begin{bmatrix} \int_0^t e_1 d\tau \\ \int_0^t e_2 d\tau \\ \int_0^t e_3 d\tau \end{bmatrix} = \begin{bmatrix} 0 \\ 0 \\ 0 \end{bmatrix}, \quad C = \begin{bmatrix} 5 & 0 & 0 \\ 0 & 5 & 0 \\ 0 & 0 & 5 \end{bmatrix}, \tag{33}$$

where $U_{1\max} = 30000\,Nm,\ U_{2\max} = 10000\,Nm,$ $U_{3\max} = 5000\,Nm$; $i = 1$ corresponds to the pitch channel, $i = 2$ corresponds to the roll channel, $i = 3$ corresponds to the yaw channel.

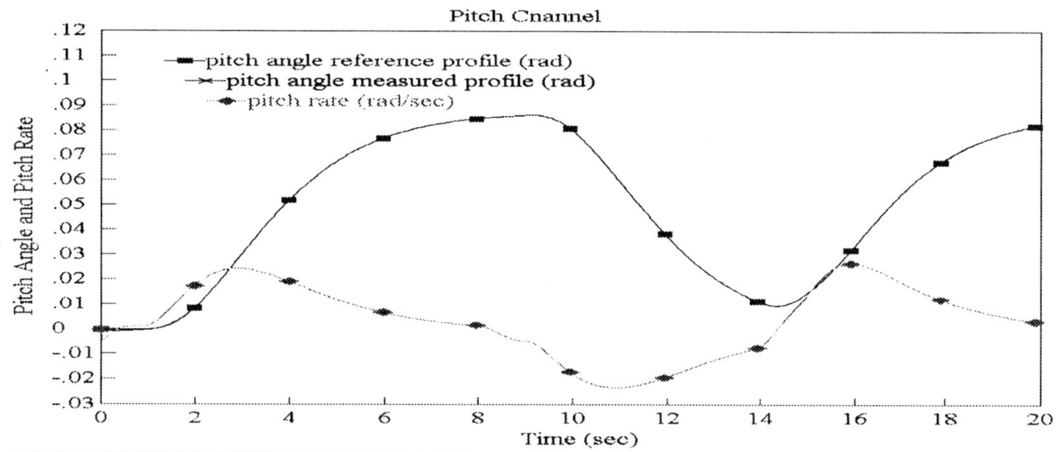

Fig. 1 Tracking of a pitch angle reference profile

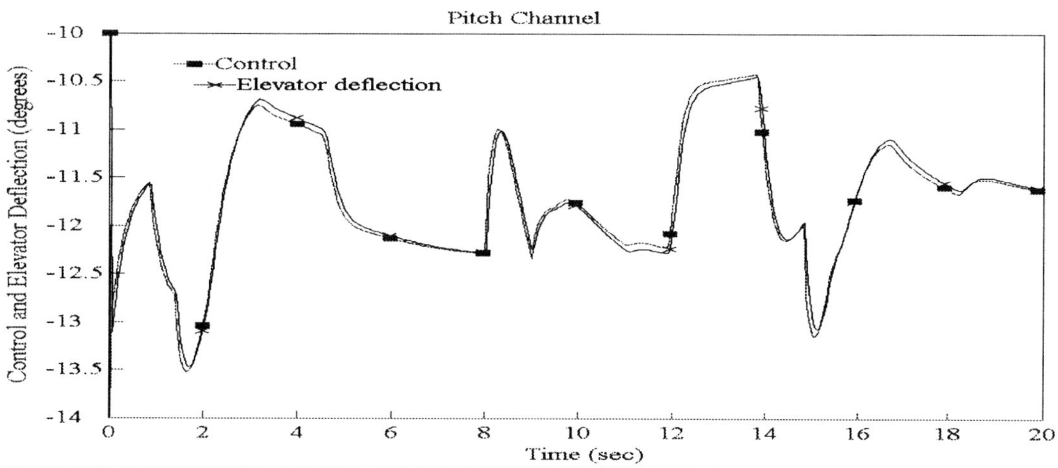

Fig. 2 Control signal and Elevator deflection

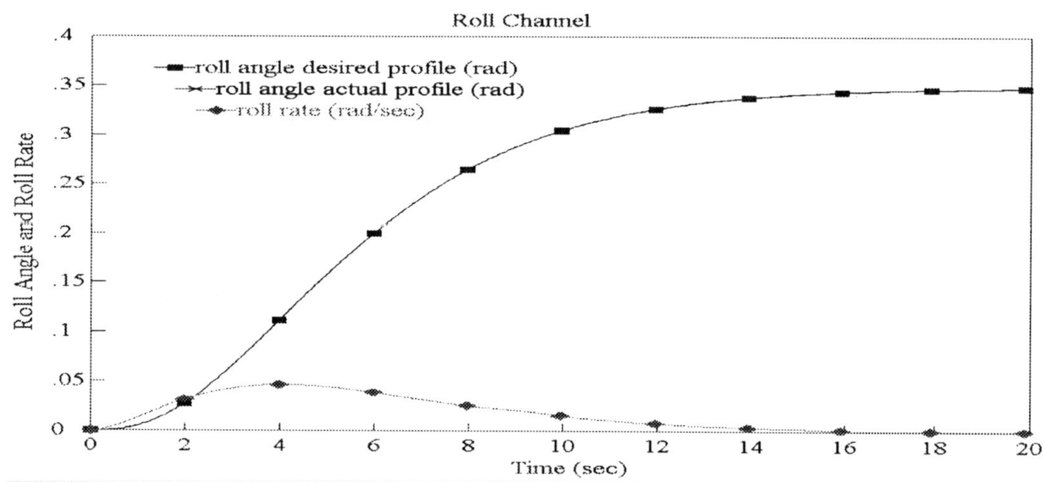

Fig. 3 Tracking of a roll angle reference profile

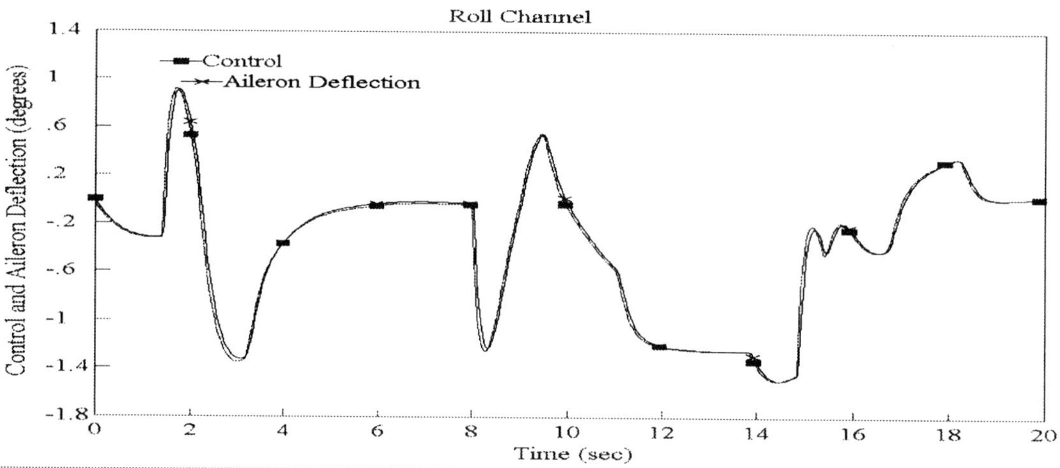

Fig. 4 Control signal and Aileron deflection

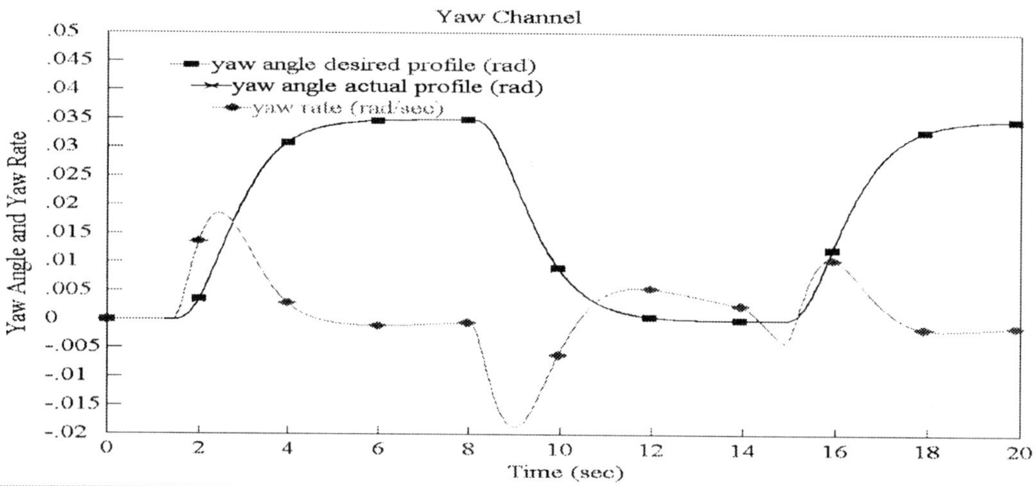

Fig. 5 Tracking of a yaw angle reference profile

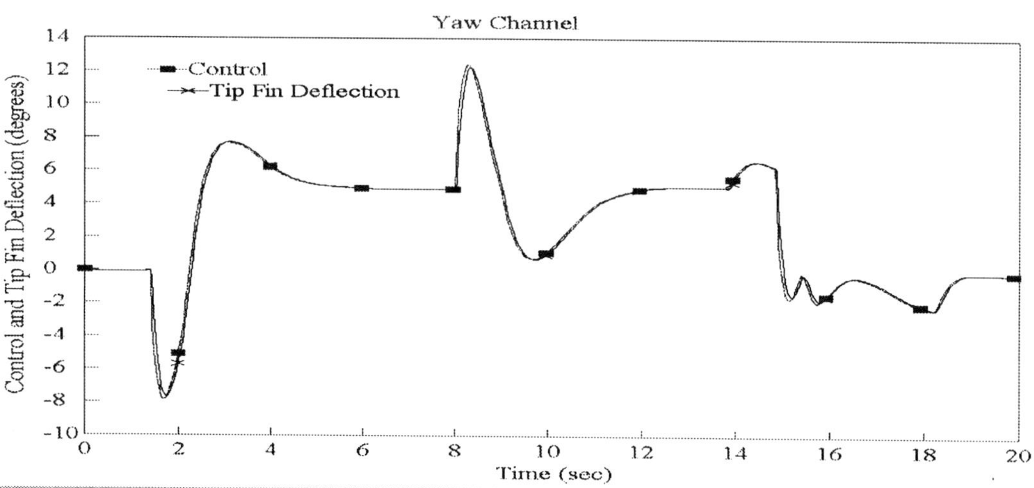

Fig. 6 Control signal and Tip fin deflection

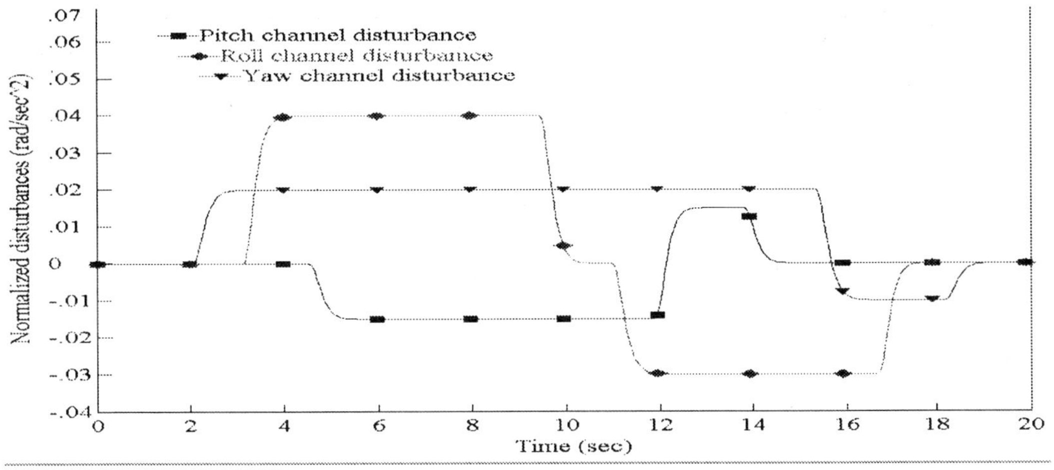

Fig. 7 External disturbances

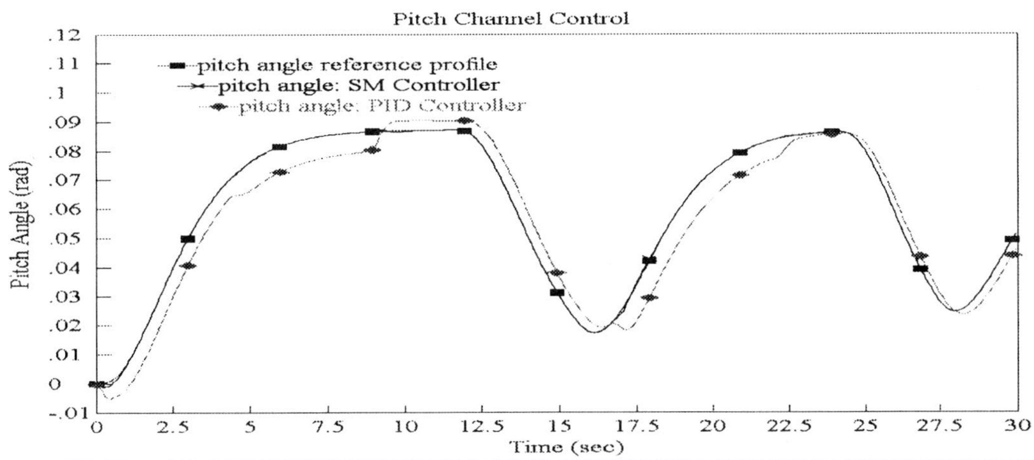

Fig. 8 Pitch angle reference profile tracking via SM and PID controllers

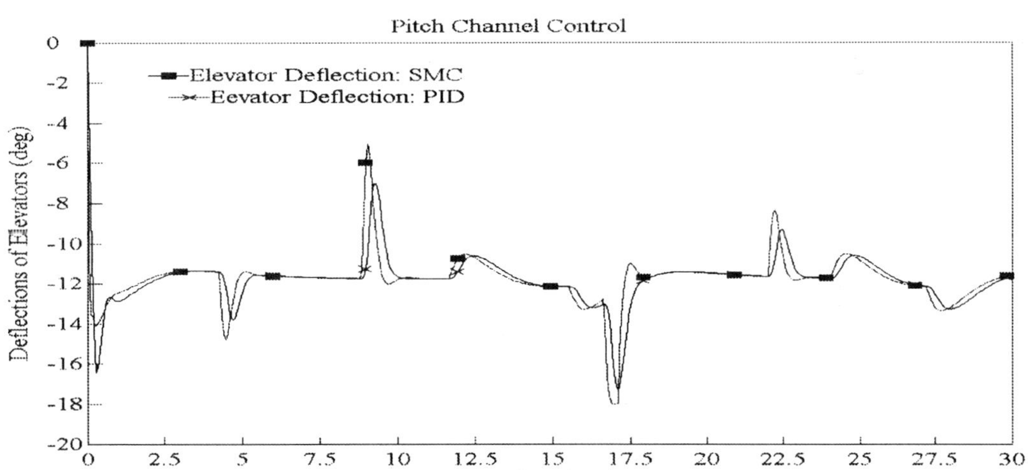

Fig. 9 Elevator deflections during pitch angle control via SM and PID controllers

Computer simulations of the launch mode

The reference profiles for all three Euler angles are formed on line and tracked by the smoothed de-coupling sliding mode controller (27), (33) and (34) Disturbance (aerodynamic) moments with magnitudes up to 10% of control moments are applied arbitrary to the angular rates equations during the simulation. The dynamics of actuators were simulated concerning expressions (17) as well. The results of the simulation are shown in fig. 10 - 13.

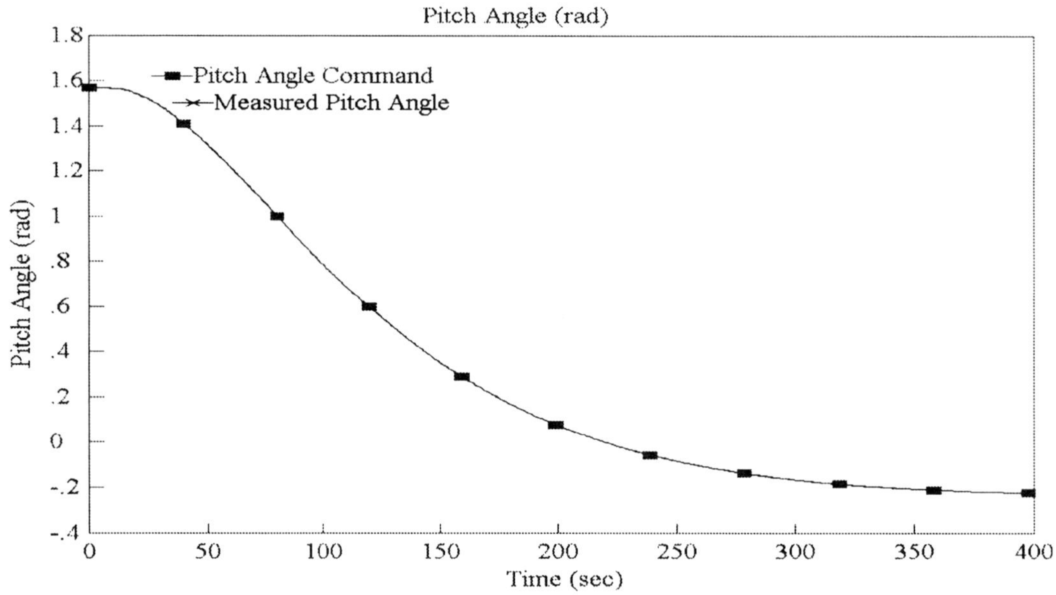

Fig. 10 Pitch Angle command tracking in the launch mode

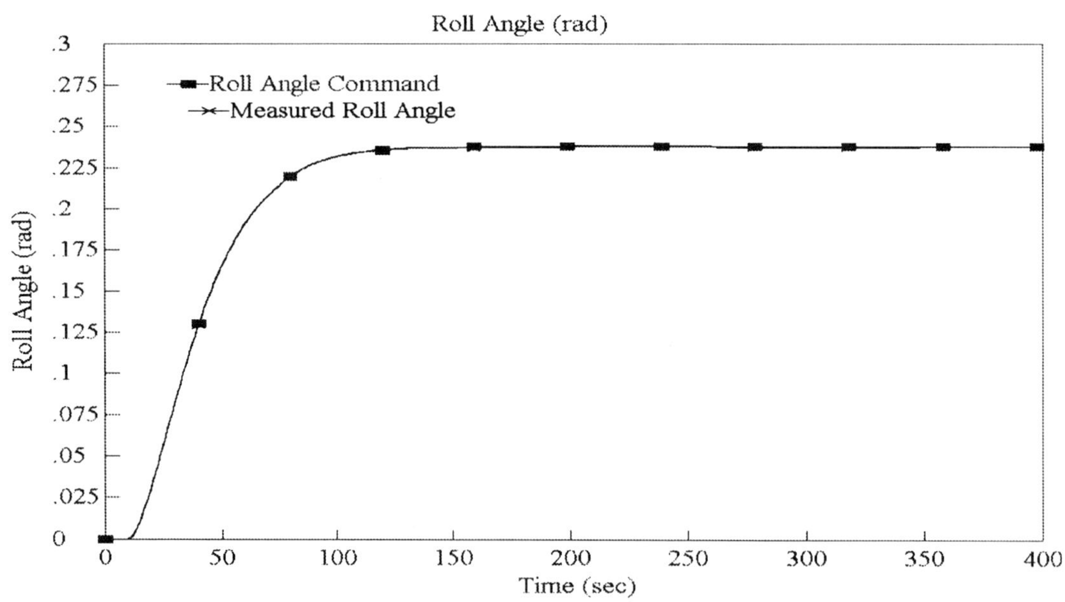

Fig. 11 Roll Angle command tracking in the launch mode

It is obvious that the designed sliding mode controllers provide robust, de-coupled extremely accurate tracking of reference Euler angles' profiles insensitive to external and interaction disturbances in the launch mode. No control chattering can be observed.

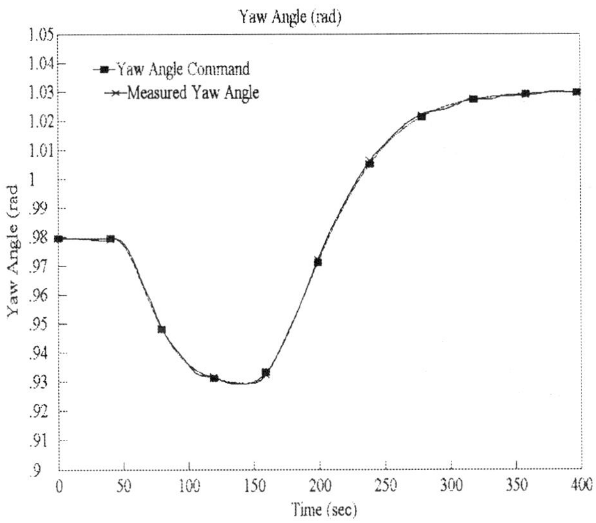

Fig. 12 Yaw Angle command tracking in the launch mode

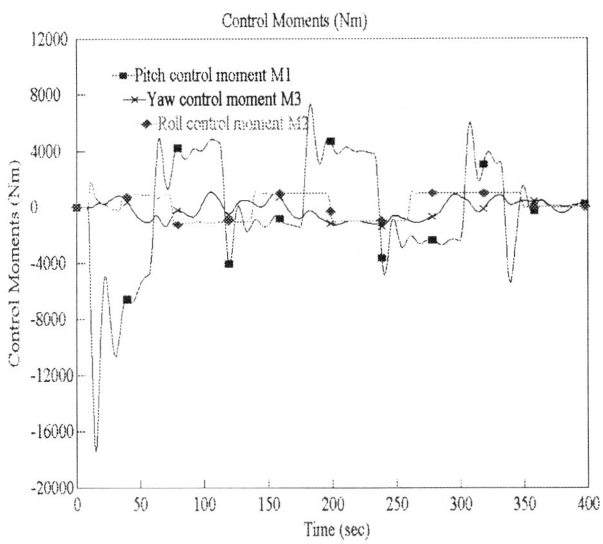

Fig. 13 Control moments in the launch mode

Conclusion

The smoothed sliding mode controllers providing the robust de-coupled high accuracy tracking of orientation reference trajectories for the re-entry and launch modes for the reusable launch vehicle are developed. Improving the reference trajectories following will permit to maximize the payload weight launching to the orbit and will increase accuracy and robustness of the landing phase of the RLV. The proposed sliding mode controllers are relatively simple and convenient for implementation.

References

[1] D. O. Stanley, W. C. Engelund, R. A. Lepsch, M. McMillin, K. E. Wruster, R. W. Powell, T. Guinta, and R. Unal, "Rocket-Powered Single-Stage Vehicle Configuration Selection and Design," *Journal of Spacecraft and Rockets*, No. 5, 1994, pp. 792-798.

[2] "Reusable Launch Vehicle Concept Study - Winged Body Concept," Final Technical Report, NASA, January 1995

[3] P. Lu, "Entry Guidance and Trajectory Control for Reusable Launch Vehicles," AIAA Paper 96-3700, *AIAA Guidance Navigation, and Control Conference*, San Diego, CA, 1996.

[4] P. Lu, "Nonlinear trajectory Tracking Guidance with Application to a Launch Vehicle," *Journal of Guidance, Control, and Dynamics*, No. 1 1996, pp. 99-106.

[5] Utkin, V. I. *Sliding Modes in Control Optimization*, Springer-Verlag, Berlin, 1992.

[6] DeCarlo, R. A., Zak, S. H. and Matthews, G. P. "Variable structure control of nonlinear multivariable systems: a tutorial," *IEEE Proceedings*, Vol. 76, 1988 pp. 212-232.

[7] Fernandez, B. R., Hedrick, K. J., *Control of multivariable non-linear systems by the sliding mode method*, International Journal of Control, 1987, No. 3. pp. 1019-1040.

[8] Shtessel, Y. B., Tournes, C., Non Linear Flight Control Problem on Decentralized Sliding Modes, *Proceedings of The Fourth IEEE Conference on Control and Its Applications*, Sept. 27-29, 1995, Albany NY. pp. 488-493.

[9] Y. Shtessel, "Decentralized Sliding Mode Control in Three-Axis Inertial Platforms", *Journal of Guidance, Control and Dynamics*, No. 4, 1995, pp. 773-781.

[10] C. Tournes, and Y. Shtessel, "Aircraft Control Using Sliding Mode Control," AIAA Paper 96-3692, *AIAA Guidance, Navigation, and Control Conference*, San Diego, CA, 1996.

[11] A. Cavallo, G. De Maria, F. Ferrara and E. Filippone, "A Trajectory and Attitude Control Strategy for the CRV/CRT Atmospheric Reentry," AIAA Paper 96-3703, *AIAA Guidance, Navigation, and Control Conference*, San Diego, CA, 1996.

[12] P. Kokotovic., "The joy of feedback: nonlinear and adaptive," *IEEE Control Systems Magazine*, No. 12, 1992, pp. 7-17.

[13] J. Slotine, Weiping Li, *Applied non Linear Control*, Prentice Hall, Englewood Cliffs, N J, 1991.

AIAA-97-3534

The attitude stabilization and trajectory tracking of reentry vehicle via variable-structure based control method *

Gao Jianping**, Chen Zongji

Department of Automatic Control
Beijing University of Aeronautics & Astronautics
Beijing 100083, P.R.China
Email:{zjchen@cs.sebuaa.ac.cn}

Abstract

In the paper some variable-structure-based control methods are presented to the attitude stabilization and trajectory tracking of reentry vehicle. The attitude stabilization and trajectory tracking are two main control issues of reentry vehicle, while reentry vehicle is a strong nonlinear plant which needs three attitude angles stablization to desired ones at first phase of reentry with the pertubations of aerodynamics parameters, and flight trajectory tracking while keeping the velocity and orbit angle stable at second phase of reentry. In this paper, we present a variable structure predictive method to the attitude stabilization, while the method can keep attitude angle decoupling tracking and also trade off the control power and performance. Simulation shows the closed loop system performance with the perturbations of aerodynamics parameters. To the problem of trajectory tracking, we present a PD(Proportional+Differential) algorithm combined with variable structure method, while in the inner loop variable structure method can be used for linearization and stabilization of velocity and orbit angle, and in the outer loop PD algorithm can guarantee the desired trajectory tracking.

** From Sept.1996 to July.1997, on leave from Beijing Unversity of Aeronautics and Astronautics and work as fellow in UNU/IIST. P.O.Box 3058, Macau, Email: gjp@iist.unu.edu

Keyword: VSC, attitude stablization, trajectory tracking.

1 Introduction

The attitude stabilization and trajectory tracking are two main control issues in the control of reentry vehicle. The difficulties lie in, the reentry vehicle is a strong nonlinear plant, which is difficult to be linearized by conventional linear approximation. Morever, it needs attitude angles stabilization to desired ones in the first phase of reentry with the perturbations of aerodynamics coefficients, and trajectory tracking in x,y,z coordinate in the second phase while keeping the velocity and orbit angle stable.

Feedback linearization[1] can be a promising approach to nonlinear control problem. However, this approach requires the exact modeling of nonlinear plant and internal zero dynamics exponential stable.

In these years, robust methods combined with feedback linearization, such as adaptive

[2][3], H^∞ and μ synthesis approaches [5][7] appear to tackle robust nonlinear decoupling problems, especially to the nonlinear aircraft manuevers with high angle of attack[4].

The variable-structure control is also a promising method to nonlinear problem, which has various applications to the attitude stabilization and trajectory tracking of robots and spacecraft [8]. The variable structure technique requires a switching surface which defines the desired dynamics relationships between variables to be constructed. The control can be designed to drive the system to the switching surface where in the switching surface, the states are invariant to any perturbations.[11]

In this paper, we present a variable-structure- predictive method which constructs the predictive switching surface instead of conventional one. Finally the method leads to trading off of decoupling performance and control power limits. We apply it to the attitude stabilization of reentry vehicle with bounded control powers. To the trajectory tracking, a PD(proportional+Differential) algorithm combined with variable structure method is used, where in the inner loop the variable structure method can be used to linearize and stabilize the velocity and orbit angle robustly, and PD algorithm can guarantee the desired trajectory tracking in the outer loop.

The paper is organized as following. Section 2 investigates the nonlinear problem and presents a variable-strucure-predictive method. We also address its robustness analysis in this section. In section 3, we apply the method to the attitude stabilization of reentry vehicle and show the simulation. The PD algorithm combined with variable-structure control method to the trajectory tracking of reentry vehicle is presented in section 4, while in section 5, we give the conclusion and future work outline.

2 The Variable structure predictive approach

2.1 Problem formulation

$$\dot{x} = f(x) + g(x)u, \quad y = h(x), \tag{1}$$

Where $x \in R^n, y \in R^m, u \in R^m$ and $f(\cdot), g(\cdot), h(\cdot)$ sufficiently smooth in x and t.

Assumption 2.1 y_i have strong relative degree $\gamma_i, i = 1, 2, \cdots, m$.

which means that $\forall x \in R^n, L_g h_i(x) = L_g L_f h_i(x) = \ldots = L_g L_f^{\gamma_i - 2} h_i(x) = 0$ and $L_g L_f^{\gamma_i - 1} h_i(x) \neq 0$.

Through standard state-dependant coordinate and input transformations, we may input/output linearize the plant (1) so that it takes the form:

$$\begin{array}{ll} \dot{\xi}_{i1} = \xi_{i2} & \vdots \\ \dot{\xi}_{i\gamma_i - 1} = \xi_{i\gamma_i} & \dot{\xi}_{i\gamma_i} = \alpha_i(\xi_i, \eta_i) + \beta_i(\xi_i, \eta_i)u \\ \dot{\eta}_i = \eta_i(\xi_i, \eta_i) & y_i = \xi_{i1}, i = 1, 2, \cdots, m \end{array} \tag{2}$$

Where

$$\begin{aligned} &\xi_i = (\xi_{i1}, \xi_{i2}, \cdots, \xi_{i\gamma_i})^T \\ &\eta_i = (T_{i\gamma_i+1}, T_{i\gamma_i+2}, \cdots, T_{in})^T \\ &\alpha_i(\xi_i, \eta_i) = L_f^{\gamma_i} h_i(x) \\ &\beta_i(\xi_i, \eta_i) = L_g L_f^{\gamma_i - 1} h_i(x), i = 1, 2, \cdots, m. \end{aligned}$$

Definition 2.1 The reference is $y_r(t) = (y_{r1}, y_{r2}, \cdots, y_{rm})^T$.

Definition 2.2 The decoupling matrix is:

$$\begin{aligned} &D = (L_g L_f^{\gamma_1 - 1} h_1(x), L_g L_f^{\gamma_2 - 1} h_2(x), \cdots \\ &L_g L_f^{\gamma_m - 1} h_m(x))^T \\ &L(x) = (L_1(x), L_2(x), \cdots, L_m(x))^T, \\ &L_i(x) = L_f^i c_i \\ &Y = (y_1^{\gamma_1}, y_2^{\gamma_2}, \cdots, y_m^{\gamma_m})^T \end{aligned}$$

Then:

$$Y = L(x) + D(x)u \tag{3}$$

Assumption 2.2 D is nonsingular in a neighbourhood area of (ξ, η).

Assumption 2.3 Zero dymanics $\dot{\eta}_i = \eta_i(0, \eta_i)$ is exponential stable.

2.2 Variable-structure-predictive approach

Considering a vectored switching surface as:

$$\begin{aligned} &S = (S_1, S_2, \cdots, S_m)^T = 0 \\ &S_i = g_{i0} z_i + g_{i1} e_i + g_{i2} \dot{e}_i + \cdots + \\ &g_{i\gamma_i - 2} e^{\gamma_i - 2} + e_i^{(\gamma_i - 1)} \end{aligned} \tag{4}$$

Where γ_i is the strong relative degree of y_i, $e = (y - y_r) = (e_1, e_2, \cdots, e_m)^T$ is the tracking error vector and $\dot{z}_i = e_i, i = 1, 2, \cdots, m$, and design of $g_{i0}, g_{i1}, \cdots, g_{i\gamma-2}$, ensures that the polynomial of e satisfying that $S_i = 0$ is *Horowitz.*

Differentiating S with respect of t, we get:

$$\dot{S}_i = g_{i0} e_i + g_{i1} \dot{e}_i + \cdots + g_{i\gamma_i - 1} e^{\gamma_i - 1} + e_i^{(\gamma_i)} \tag{5}$$

$$\dot{S} = Y - Y_r + zz \tag{6}$$

where,

$$\begin{aligned} &zz_i = (g_{i0} e_i + g_{i1} \dot{e}_i + \cdots + g_{i\gamma_i - 1} e^{\gamma_i - 1}), \\ &zz = (zz_1, zz_2, \cdots, zz_m)^T, \\ &Y_r = (y_{r1}^{(\gamma_1)}, y_{r2}^{(\gamma_2)}, \cdots, y_{rm}^{(\gamma_m)}) \end{aligned}$$

Let a performance index be:

$$J[u] = \frac{1}{2}[S^T(t+h) Q S(t+h) + u^T(t) R u(t)] \tag{7}$$

where $Q > 0, R > 0, h > 0$. Note that $S(t+h)$ is the predictive value of S at time $t + h$. We can expand the $S(t+h)$ in Tayor series:

$$\begin{aligned} &S(t+h) \approx S(t) + h\dot{S}(t) \\ &= S(t) + h(Y - Y_r + zz)) \\ &= S(t) + h(L(x) + D(x)u - Y_r + zz) \end{aligned} \tag{8}$$

To minimize the J, partial differentiate (7) with respect to u, and let $\frac{\partial J}{\partial u} = 0$ to get:

$$\frac{\partial J}{\partial u} = \frac{\partial S(t+h)^T}{\partial u} Q S(t+h) + Ru = 0 \tag{9}$$

$$Ru + hD^T Q[S + h(L(x) + D(x)u - Y_r + zz)] = 0 \tag{10}$$

$$\begin{aligned} &u = -h(R + h^2 D^T Q D)^{-1} D^T Q (S + \\ &h(L(x) - Y_r + zz)) \end{aligned} \tag{11}$$

Specially, if control weight matrix $R = 0$,

$$u = -h^{-1} D^{-1} [S + h(L(x) + zz - Y_r)] \tag{12}$$

Then:

$$Y = -h^{-1} S - zz + Y_r \tag{13}$$

From (6), we have:

$$\dot{S} + h^{-1} S = 0 \tag{14}$$

$S(t) = e^{-h^{-1} t} S(0)$, implies that with arbitary initial value $S(0)$, $t \to \infty, S \to 0$, Then $S_i = 0, i = 1, 2, \cdots, m$ holds, which means that variable-structure-predictive control introduced above could guarantee the desirable asymptotic decoupling performance while trading off the control power and performance.

2.3 Robustness analysis

In this section, we analyze the robustness of the control with perturbations. In case of perturbations, the output equations become:

$$Y = L(x) + \delta L(x) + [D(x) + \delta D(x)]u \quad (15)$$

Assumption 2.4 $D(x) + \delta D(x)$ keeps nonsigularity in the neighbourhood of (ξ, η)

Then we have the theorem.

Theorem 2.1 Suppose that following conditions holds:

1. $\exists N1 > 0, N2 > 0$ such that: $\|\delta D(x)\| \leq N1, \|\delta L(x)\| \leq N2$.
2. $\exists M > 0, \|D(x)^{-1}(L(x) + zz - Y_r)\| < M$

Then:
$\forall \epsilon > 0, \quad \exists h > 0$, such that with the control (12), the output errors of perturbed system satisfy:

$$\|e(t)\| = \|y - y_r\| < \epsilon$$

Proof: Introducing (12) into (15), to get:

$$Y = L(x) + \delta L(x) - [D(x)+ \delta D(x)]h^{-1}D^{-1}[S + h(L(x) + zz - Y_r)] \quad (16)$$

Noticing (6) we have:

$$\dot{S} + h^{-1}(I + \delta D(x)D(x)^{-1})S = \delta D(x)D(x)^{-1}[(L(x) + zz - Y_r)] - \delta L(x) \quad (17)$$

combining with assumptions,

$$S(t) = e^{-h^{-1}(I+\delta DD^{-1})t}S(0)+ \int_0^t e^{-h^{-1}(I+\delta DD^{-1})(t-\tau)} [\delta D(x)D(x)^{-1}(L(x) + zz - Y_r) - \delta L(x)]d\tau \quad (18)$$

$$\|S(t)\| \leq \|e^{-h^{-1}(I+\delta DD^{-1})t}S(0)\|+ (N1 * M + N2)\|\int_0^t e^{-h^{-1}(I+\delta DD^{-1})(t-\tau)}d\tau\| \quad (19)$$

$$\|S(t)\| \leq \|e^{-h^{-1}(I+\delta DD^{-1})t}S(0)\| +(N1 * M + N2)h\|[I - e^{-h^{-1}(I+\delta DD^{-1})t}\| \quad (20)$$

Given $\epsilon > 0$, we can select proper t, such that

$$\|e^{-h^{-1}(I+\delta DD^{-1})t}S(0)\| < \frac{1}{2}\epsilon$$

and proper h,

$$(N1 * M + N2)h\|[I - e^{-h^{-1}(I+\delta DD^{-1})t}]\| < \frac{1}{2}\epsilon$$

Then:

$$\|S(t)\| < \epsilon \quad (21)$$

From the definition of $S(t)$, the output errors of perturbed system $\|e(t)\|$ can enter ϵ neighhood also, which proves the claimed conclusion. □

3 The attitude stabilization of reentry vehicle with variable structure predictive decoupling method

The model of reentry vehicle in the first phase of reentry could be described as:

$$\dot{\Theta} = R^{-1}(\Theta)\omega = f_1(\Theta, \omega)$$
$$\dot{\omega} = f_2(\Theta, \omega) + Bu$$

Where $\Theta = (\gamma, \psi, \phi)^T$ are the three euler angles, (roll angle, bank angle and pitch angle) respectively. $\omega = (\omega_x, \omega_y, \omega_z)^T$ are the angular velocities in the related directions expressed in the body frame.

$$R^{-1} = \begin{bmatrix} 0 & \frac{sin\phi}{cos\psi} & \frac{cos\phi}{cos\psi} \\ 0 & cos\phi & -sin\phi \\ 1 & sin\phi tg\psi & cos\phi tg\psi \end{bmatrix}$$

$$B = \begin{bmatrix} \frac{1}{J_x} & 0 & 0 \\ 0 & \frac{1}{J} & 0 \\ 0 & 0 & \frac{1}{J} \end{bmatrix}$$

$$f_2(\theta, \omega) = \begin{bmatrix} \frac{1}{J_x}(\frac{qsd^2 C_{lp}\omega_1}{v}) \\ \frac{1}{J}[(J - J_x)\omega_x\omega_y + \frac{qsl^2 C_{NR}\omega_y}{2v}] \\ \frac{1}{J}[-(J - J_x)\omega_x\omega_z + \frac{qsl^2 C_{MQ}\omega_z}{2v}] \end{bmatrix}$$

C_{lp}, C_{NR}, C_{MQ} are the aerodynamics coefficients, which can be interpolated in tabular of height, mach number, and angle of attack. The design objective of robust attitude stabilization is to make three attitude angles track the desired ones $\Theta_r = (\gamma_r, \psi_r, \phi_r)^T$.

This is a typical robust decoupling problem of nonlinear plant. Here $C = [I, 0]$ is a constant matrix, which simplifies our problem.

Differentiating Θ with respect to t, we get:

$$\begin{aligned} \ddot{\Theta} &= \frac{\partial f_1}{\partial \Theta}\dot{\Theta} + \frac{\partial f_1}{\partial \omega}\dot{\omega} = \\ &\frac{\partial f_1}{\partial \Theta} f_1(\Theta, \omega) + \frac{\partial f_1}{\partial \omega}(f_2(\Theta, \omega) + Bu) \\ &\frac{\partial f_1}{\partial \Theta} R^{-1}(\Theta)\omega + R^{-1}(\Theta) f_2(\Theta, \omega) + R^{-1}Bu \end{aligned} \tag{22}$$

It is clear to see:

$$L(x) = \frac{\partial f_1}{\partial \Theta} R^{-1}(\Theta)\omega + R^{-1}(\Theta) f_2(\Theta, \omega)$$

and $D(x) = R^{-1}(\Theta)B$ is nonsingular and no zero dynamics exists, so the method presented in section 2 can be applied here. The required switching surface can be constructed like:

$$\begin{aligned} &S = G_0 z + G_1(\Theta - \Theta_r) + \dot{\Theta} - \dot{\Theta}_r = 0 \\ &\dot{z} = \Theta - \Theta_r \\ &G_0 = diag(g_{01}, g_{02}, g_{03}) \\ &G_1 = diag(g_{11}, g_{12}, g_{13}) \end{aligned}$$

The control can be:

$$\begin{aligned} u = &-h(R + h^2 D^T Q D)^{-1} D^T Q (S + \\ &h(L(x) - Y_r + zz) \end{aligned} \tag{23}$$

Where,

$$\begin{aligned} &Y_r = (\ddot{\gamma}_r, \ddot{\psi}_r, \ddot{\phi}_r) \\ &L(x) + zz = G_0 e + G_1 \dot{e} + \ddot{\Theta}, e = \Theta - \Theta_r \end{aligned}$$

We select

$$\begin{aligned} R &= diag(0.1, 0.1, 0.1), h = 0.05, \\ Q &= diag(1, 1, 1) \end{aligned}$$

and do the simulation. The initial values of attitude angle are

$$(\gamma_0, \psi_0, \phi_0)^T = (0.087, 0.087, -0.6167)^T$$

while references are

$$(\gamma_r, \psi_r, \phi_r)^T = (0, 0, -0.7207)^T$$

During this phase, the control power could be the limited jet thrust. With proper selection of R, the system performs good under bounded control power with the 25% disturbances of aerodymamics coefficients. **Fig 1** show the control torques and the decoupling attitudes with and without disturbances.

4 The tracjectory tracking of reentry vehicle

In the second phase of reentry, the velocity, orbit angle and the trajectory in x, y, z coordinate are main issues of consideration. So here we express state equation of the plant(reentry vehicle) in velocity frame,

$$\begin{aligned} m\dot{V} &= -mg\sin\theta\cos\sigma - C_x q S_m \\ mV\dot{\theta}\cos\sigma &= -mg\cos\theta + C_y q S_m \\ -mV\dot{\sigma} &= -mg\sin\theta\sin\sigma + C_z q S_m \end{aligned} \tag{24}$$

and two additional complementary equations:

$$\begin{aligned} \psi &= \sigma + \beta \\ \phi &= \theta + \alpha \end{aligned} \tag{25}$$

Where V, θ, σ are velocity, pitch orbit angle, and steering orbit angle, $\gamma, \psi, \phi, \alpha, \beta$ denote roll angle, bank angle, pitch angle, angle of attack and side-slip angle respectively. C_x, C_y, C_z represent the aerodynamics coefficents which can be calculated from the coefficients introduced in the section 3.

The requirements here are to track the desired trajectory in x, y, z coordinate while keeping the velocity and orbit angles stable at the same time.

From the requirements above, a control law can be designed like:

The variable structure control in the inner loop guarantees the stability of velocity and tracjectory angle, whereras the outer loop with PD control ensures the accurancy of trajectory tracking.

The switching surface in the inner loop can be constructed as:

$$S = \begin{bmatrix} g_1(V - V_c) + \dot{V} - \dot{V}_c + \dot{\gamma} \\ g_2(\sigma - \sigma_c) + \dot{\sigma} - \dot{\sigma}_c \\ g_3(\phi - \alpha - \theta_c) + \dot{\theta} - \dot{\theta}_c \end{bmatrix} = 0$$

$$\dot{S} = \begin{bmatrix} g_1(\dot{V} - \dot{V}_c) + \ddot{V} - \ddot{V}_c + \ddot{\gamma} \\ g_2(\dot{\sigma} - \dot{\sigma}_c) + \ddot{\sigma} - \ddot{\sigma}_c \\ g_3(\dot{\theta} - \dot{\theta}_c) + \ddot{\theta} - \ddot{\theta}_c \end{bmatrix} \tag{26}$$

$$= \begin{bmatrix} g_1(\dot{V} - \dot{V}_c) + \ddot{V} - \ddot{V}_c \\ g_2(\dot{\sigma} - \dot{\sigma}_c) - \ddot{\sigma}_c - \ddot{\beta} \\ g_3(\dot{\theta} - \dot{\theta}_c) - \ddot{\theta}_c - \ddot{\alpha} \end{bmatrix} + \begin{bmatrix} \ddot{\gamma} \\ \ddot{\psi} \\ \ddot{\phi} \end{bmatrix}$$

From the (22), we get:

$$\dot{S} = \begin{bmatrix} \dot{\gamma} - \dot{\gamma}_c - \ddot{\gamma}_c \\ \dot{\sigma} - \dot{\sigma}_c - \ddot{\sigma} - \ddot{\beta} \\ \dot{\theta} - \dot{\theta}_c - \ddot{\theta}_c - \ddot{\alpha} \end{bmatrix} + \frac{\partial f_1}{\partial \theta} R^{-1}\omega$$
$$+ R^{-1}(\theta)(f_2 + Bu)$$
$$= L^*(X) + \frac{\partial f_1}{\partial \theta} R^{-1}\omega + R^{-1}(\theta)(f_2 + Bu) \tag{27}$$

Where the $L^*(X)$ is the function of states and their differentials. We assume that these variables are measureable. So from the nonsingularity of $R^{-1}(\Theta)B$, the control can be:

$$u = -B^{-1}R(\theta)(L^*(X) + \frac{\partial f_1}{\partial \Theta} R^{-1}\omega + R^{-1}(\theta) f_2 + Ksgn(S)) \tag{28}$$

From the similar derivations in the section 2, the switching surface is reachable. which guarantees the $S = 0, \dot{S} = 0$, and implies the velocity V, the trajectory angle θ, and the steering angle σ can track the desired ones. The objective of the outer loop is to use the PD(proporitional+differential)control of tracking errors and make the errors approximate to zero.

The errors of trajectory tracking can be:

$$e = (e_1, e_2, e_3)^T = (x_c - x, y_c - y, z_c - z)^T$$

Let control be:

$$u = (\dot{V}_c, \dot{\sigma}_c, \dot{\theta}_c)^T = K_1 * e + K_2 * \dot{e} \tag{29}$$

Where K_1, K_2 are the proportional and differential constant matrix respectively.

Then we have the theorem:

Theorem 4.1 With the proper choice of K_1 and K_2 the control law (29)leads to $t \to \infty, e \to 0$.

Proof: From the $S = 0$, variable structure control laws in the inner loop leads to:

$$V \to V_c, \sigma \to \sigma_c, \theta \to \theta_c$$

The output of inner loop can be $\dot{x}, \dot{y}, \dot{z}$.

Let $X = (x, y, z)^T, X_c = (x_c, y_c, z_c)^T$.

$$\dot{X} = \begin{bmatrix} \dot{x} \\ \dot{y} \\ \dot{z} \end{bmatrix} = \begin{bmatrix} V\cos\sigma\cos\theta \\ V\cos\sigma\sin\theta \\ -V\sin\sigma \end{bmatrix} = F(V, \sigma, \theta) \tag{30}$$

$$\ddot{X} = \frac{\partial F(V, \sigma, \theta)}{\partial (V, \sigma, \theta)} (\dot{V}, \dot{\sigma}, \dot{\theta}) \tag{31}$$

So the close-loop equation can be:

$$\begin{bmatrix} \ddot{x} \\ \ddot{y} \\ \ddot{z} \end{bmatrix} = \ddot{X} = \frac{\partial F(V,\sigma,\theta)}{\partial (V,\sigma,\theta)} \begin{bmatrix} \dot{V} \\ \dot{\sigma} \\ \dot{\theta} \end{bmatrix} \approx \frac{\partial F(V,\sigma,\theta)}{\partial (V,\sigma,\theta)} \begin{bmatrix} \dot{V}_c \\ \dot{\sigma}_c \\ \dot{\theta}_c \end{bmatrix}$$

$$= \frac{\partial F(V,\sigma,\theta)}{\partial (V,\sigma,\theta)} (K1 \begin{bmatrix} x_c - x \\ y_c - y \\ z_c - z \end{bmatrix} + K2 \begin{bmatrix} \dot{x}_c - \dot{x} \\ \dot{y}_c - \dot{y} \\ \dot{z}_c - \dot{z} \end{bmatrix})$$

$$\frac{X}{X_c} \approx \frac{\frac{\partial F(V,\sigma,\theta)}{\partial (V,\sigma,\theta)} K_2 S + \frac{\partial F(V,\sigma,\theta)}{\partial (V,\sigma,\theta)} K_1}{S^2 + \frac{\partial F(V,\sigma,\theta)}{\partial (V,\sigma,\theta)} K_2 S + \frac{\partial F(V,\sigma,\theta)}{\partial (V,\sigma,\theta)} K_2}$$

The design of proper K_1 and K_2 can ensure that the poles of the transfer function matrix in the left half of complex plane, then $t \to \infty, X \to X_c$ is guaranteed. □

With different (V, σ, θ), $\frac{\partial F(V,\sigma,\theta)}{\partial(V,\sigma,\theta)}$ can have different eigenvalues, so different K_1, K_2 can be designed to ensure the stability of the transfer function matrix.

The control structure can be shown in **Fig 2**:

5 Conclusion and future work outline

In the paper we present a variable-structure based method to the attitude stabilization and trajectory tracking of reentry vehicle, and show a good performance under perturbations. Since the method needs the state information of the system which is difficult to get in real world, the state estimator design combined with the method presented above should be considered in the future work. Morever, the method presented doesn't consider the non-minimum phase problem which occurs sometimes in nonlinear plant. How to tackle the non-minimum phase and how to combine it with the method presented is still open.

References

1. A.Isidori, *Nonlinear Control System*. Springer-Verlag, second edition.1989.
2. Sastry S.S and lsidori A, *Adaptive contol of linearizable systems*, IEEE Trans. on Automat. Contr. 34(1989): 1123-1131.
3. Kanellakopoulos,I. eta-il.*Systematic design of adaptive controller for feedback linearizable systems*, IEEE Trans on Automat Contr., Vol AC-36(11): 1241-1251, 1991.
4. Hauser.J.E, Sastry.S and Meyer.G, *Nonlinear Control Design for Slightly Nonminimum Phase System:Application to V/STOL Aircraft*, Automatica, Vol 28,No.4 1992, pp.665-680.
5. Snell S.A et al, *Nonlinear inversion flight control for a supermaneuverable aircraft*. AIAA-90-3406-CP.
6. Asseo,S.J, *Decoupling of a class of nonlinear systems and its application to a aircraft control problem*. Journal of Aircraft 10,1973 739-747.
7. Lane.S.H and Stengel R.E, *Flight control design using nonlinear inverse dynamics*, Automatica, 1988.
8. S.N.Singh, *Asymptotically decoupled discontinuous control of system and nonlinear aircraft maneuver*. IEEE Transaction On Aerospace and Electronic Systems Vol 25, No 3 1989.
9. Hakan Elmali and Nejat Olgac, *Robust Output Tracking Control of Nonlinear MIMO System via Sliding Mode Technique*. Automatica Vol 1, pp 145-151,1992.
10. S.D.Brierlay and R.Loy.Champ, *Application of Sliding Mode Control to Air-Air Interception*, IEEE Transaction On Aerospace and Electronic Systems Vol 26, No 2 1989.
11. V.I.Utkins, *Variable Structure Controller Design for Nonlinear System*, Internatinal Journal of Control, Vol 40, 2, pp421-434,1984
12. J.J.E.Slotine and J.A.Coetsee, *Adaptive Sliding Controller Synthesis for Nonlinear System*. Int.J.Control, Vol 43, pp1631-1651 June,1986
13. J.J.E.Slotine and J.A.Coetsee, *An extended directly scheme for robust nonlinear control*, Automatica, Vol 27, pp247-255, 1991.

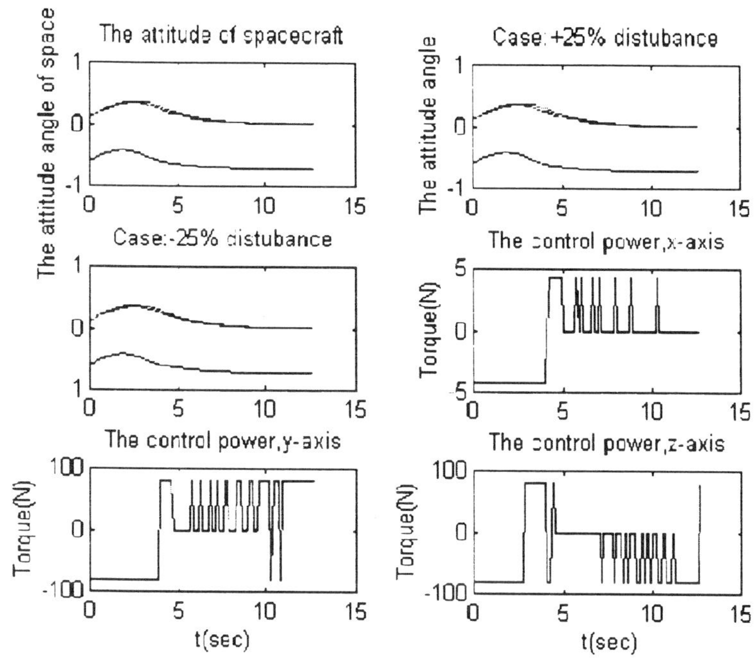

Fig 1, Three attitude angle tracking and related control power

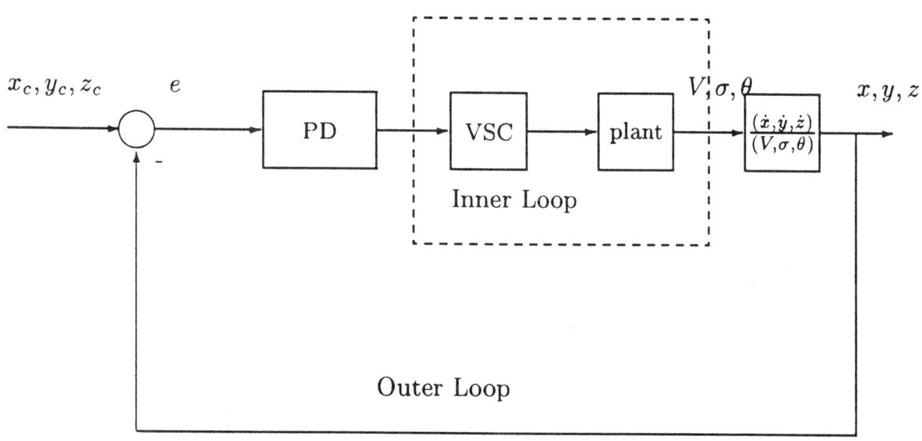

Fig 2, Structure of the control process

This article was processed using the LaTeX macro package with LLNCS style

AIAA-97-3535

MINIMUM HEATING RE-ENTRY TRAJECTORIES FOR ADVANCED HYPERSONIC LAUNCH VEHICLES

Robert Windhorst
Research Assistant
Santa Clara University
Santa Clara, California 95053

Mark Ardema*
Professor and Chairman of Mechanical Engineering
Santa Clara University
Santa Clara, California 95053

Jeffrey Bowles†
Aerospace Engineer
NASA Ames Research Center
Moffett Field, California

Abstract

Optimal re-entry trajectories are generated for reusable launch vehicles which minimize: (i) the heat absorbed at the vehicle surface, (ii) the lower surface temperature, and (iii) the heat absorbed by the internal structure. The approach uses the energy state approximation technique and a finite control volume heat transfer code coupled to a flight path integration code. These trajectories are compared to the optimal re-entry trajectory minimizing the integrated convective heat rate to determine which trajectory produces the minimum internal structural temperatures for a given thermal protection system. Three different thermal protection systems are considered: tile, blanket, and metallic.

Nomenclature

c = specific heat, btu/(lb*R)
dt = time step required by vehicle to travel between energy levels, sec
ε = emissivity
E = specific excess energy, ft
k = thermal conductivity, btu/(hr*ft*R)
L = total thickness of TPS
P = specific excess power, ft/sec
P_{surf} = pressure at vehicle surface, lb/ft^2
$\dot{\Phi}$ = optimization function
Φ = integrated optimization function
q = dynamic pressure, lb/ft^2
$\dot{Q}_{abs}$ = conductive heat rate per square foot, btu/(hr*ft^2)
Q_{abs} = total heat conducted per square foot, btu/ft^2
$\dot{Q}_{conv}$ = convective heat rate per square foot, btu/(hr*ft^2)
Q_{conv} = total heat convected per square foot, btu/ft^2
$\dot{Q}_{rad}$ = radiative heat rate per square foot, btu/(hr*ft^2)
Q_{rad} = total heat radiated per square foot, btu/ft^2
ρ = density, lb/ft^3
σ = Stefan-Boltzmann cons., btu/(hr*ft^2*R^4)
t = flight time, min.
T_{surf} = estimated temperature at vehicle surface, °F
T_i = temperature of TPS node i, °F
x = TPS depth measured from surface, inches

* Associate Fellow AIAA
† Member AIAA

Introduction

Current research and development of advanced reusable launch vehicles (RLVs) focuses on reducing the cost of access to space and the turn around time between missions. The RLV program proposes to reduce operating cost and turn around in part by redesigning the vehicle thermal protection system (TPS) for lighter weight, reusability, and increased durability relative to the space shuttle. There are two key interrelated aspects of launch vehicle TPS design: optimizing the vehicle re-entry trajectory for minimum internal structural temperatures for a given TPS, and evaluation of advanced, lightweight, reusable TPS materials. Although this report concentrates on the trajectory optimization problem, results are given for three different TPS concepts: tile, blanket, and metallic. The tile and blanket TPS concepts are similar to what is currently used on the space shuttle. The underside of the shuttle uses a tile TPS and the top side uses a blanket TPS. The metallic TPS, however, represents a new TPS concept designed for the goals mentioned above: lighter weight, reusability, and increased durability.

As a RLV descends into the atmosphere from orbit it undergoes extreme convective heat rates($\dot{Q}_{conv}$) due to its high kinetic energy. The magnitude and duration of $\dot{Q}_{conv}$ determine the thickness of the TPS required to prevent the vehicle internal structure temperature from exceeding its limit. In Reference 1 this problem is approached by using the energy state trajectory approximation [2-5] to minimize the required thickness of the TPS by minimizing the integrated convective heat rate (Q_{conv}) on the vehicle.

Sachs and Dinkelmann [6] minimize fuel consumption of a hypersonic vehicle subject to a heat absorption constraint (Q_{limit}). The vehicle is an air-breathing first stage of a two-stage launch system. Because the maximum speed is Mach 6, the kinetic energies and $\dot{Q}_{conv}$ of this vehicle are much lower than those for the re-entry of a single stage vehicle. The constraint on the heat absorbed by the vehicle used in Reference 6 is

$$\int_0^{t_f} \dot{Q}_{abs} dt \leq Q_{limit}$$

where $\dot{Q}_{abs}$ is the rate of heat absorbed by the innermost layer of the insulation and is determined by a finite difference heat transfer code much like the one used in the present paper. Since Q_{limit} is evaluated at the end of the mission, Sachs and Dinkelmann must solve a two point boundary value trajectory optimization problem. They found that the heat absorbed could be greatly reduced with only a small penalty in fuel consumption.

The approach used in this paper avoids the complexity of the two point boundary value problem by using the energy state approximation (ESA) technique [2-5]. By the ESA technique optimal trajectories minimizing a variety of different functions can be generated with a single integration of the equations of motion over the vehicle flight path. Furthermore, ESA techniques could be used to write an on-board algorithm that uses position feedback to generate optimal trajectories in real time [5]. This paper presents results of optimal trajectories minimizing: (i) the heat absorbed (Q_{abs}) by the RLV at its surface, (ii) the RLV estimated lower surface temperature, and (iii) the heat absorbed by the RLV internal structure, and compares them to results obtained by minimizing Q_{conv}, as is done in Reference 1, to determine which trajectory minimizes the internal structural temperatures of the RLV for a given thermal protection system. Also discussed are the assumed vehicle constraints on the flight trajectory, the method of generating approximate optimal trajectories, and the equations for calculating temperatures and conductive heat rates($\dot{Q}_{abs}$).

Methods

Vehicle Synthesis Code

The NASA AMES Hypersonic Aircraft Vehicle Optimization Code (HAVOC) generated all the optimal trajectories presented in this paper. HAVOC aides RLV design by synthesizing key vehicle elements such as geometry, aerodynamics, dynamics, propulsion, and structures into a single code. The integration of all these elements into a single code creates a versatile platform ideal for studying RLV optimal trajectories. Furthermore, the code contains built in subroutines which use energy state approximation (ESA) techniques to generate optimal trajectories minimizing time, fuel, Q_{conv}, Q_{abs}, or a weighted combination of these [1]. Although the optimal paths are determined from ESA methods, the trajectory integration in HAVOC uses a point mass model assuming a spherical rotating earth with no winds and a flight path angle rate of zero. The equations of motion are listed as equation (1) in Reference 2.

Constraints

There are seven vehicle constraints imposed on the flight trajectories. Each constraint along with its upper and/or lower limit is listed in Table 1. The first constraint is imposed on the dynamic pressure (q) of the vehicle. Three temperature constraints are imposed at three different locations on the vehicle body.

TSTAGN is the stagnation temperature constraint at the vehicle nose. Temperature lower surface (TLS) is the temperature constraint of a point located along the windward surface, centerline, and 1/3 the vehicle length from the nose, and temperature upper surface (TUS) is the temperature constraint of a point located along the leeward surface, centerline, and 1/3 the vehicle length from the nose. The load factor (SLDFAC) constraint limits the acceleration of the vehicle in the direction normal to the velocity vector. AN is the constraint on the vehicle's angle of attack, and FA is the constraint on the vehicle's flight path angle.

Constraint	Flight Parameter Constrained	TPS Type	Lower Limit	Upper Limit
q	Dynamic Pressure	all	20 psi	900 psi
TSTAGN	Stagnation temp. at vehicle nose	all	n/a	3000° F
TLS	Temperature lower surface	Tile Blanket Metallic	n/a n/a n/a	2400° F 2000° F 1800° F
TUS	Temperature upper surface	all	n/a	1200° F
SLDFAC	Load factor	all	0	2
AN	Vehicle angle of attack	all	-60°	+60°
FA	Vehicle flight path angle	all	-45°	+45°

Table 1. Vehicle Constraints Imposed on Flight Trajectory

Energy State Approximation

The ESA method of trajectory optimization generates an optimal trajectory minimizing the integral

$$\Phi = \int \dot{\Phi} \frac{dE}{P} \qquad (1)$$

where $\dot{\Phi}$, called the optimization function, is the vehicle parameter of interest, P is the specific excess power of the vehicle, and dE is the energy step [2-5]. This approach assumes that the energy state variable is on a slower time scale than the other trajectory and temperature variables. Substituting

$$\frac{dE}{P} = dt \qquad (2)$$

into equation (1) results in

$$\Phi = \int \dot{\Phi} dt \qquad (3)$$

Therefore by equation (3) the optimal trajectory generated using the ESA technique directly minimizes the integral of $\dot{\Phi}$, with respect to time. We note that E is preferable to t as an independent variable because its total change is the same for every trajectory.

Temperature Estimation

In addition to its optimal trajectory and vehicle dynamics routines, HAVOC contains a subroutine TEMP which estimates vehicle surface temperature (T_{surf}). HAVOC uses engineering approximations in its aerothermal subroutines to construct for each surface panel on the RLV $\dot{Q}_{conv}$ as a numerical function of angle of attack, altitude, and Mach number. HAVOC then calculates T_{surf} for a given surface panel using the following energy balance equation:

$$\dot{Q}_{conv} = \dot{Q}_{rad} \qquad (4)$$

$$\dot{Q}_{conv} = \varepsilon\sigma\left[T^4_{surf} - T^4_{amb}\right] \qquad (5)$$

$$T_{surf} = \left[\frac{\dot{Q}_{conv}}{\varepsilon\sigma} + T^4_{amb}\right]^{\frac{1}{4}} \qquad (6)$$

In this case T_{surf} is much greater than T_{amb}, so T_{amb} may be neglected. The absence of conduction terms in the surface energy balance equation slightly inflates T_{surf} and does not calculate the conductive heat rates ($\dot{Q}_{abs}$) of the TPS or the vehicle internal structure temperature.

Heat Transfer Analysis

The HAVOC subroutine HEATX receives inputs of time step, surface pressure, $\dot{Q}_{conv}$, and past nodal temperatures. HEATX outputs $\dot{Q}_{abs}$ and current TPS nodal temperatures. HAVOC integrates $\dot{Q}_{abs}$ with respect to time to find the total heat absorbed (Q_{abs}). HEATX calculations are made at a single point on the vehicle body. For this study this point is at the same location as the TLS temperature constraint, along the windward surface, centerline, and 1/3 the vehicle length from the nose.

HEATX solves the one dimensional transient heat equation,

$$\rho c \frac{\partial T}{\partial t} - \frac{\partial}{\partial x}\left[k \frac{\partial T}{\partial x} \right] = 0 \tag{7}$$

where

$$T=f(x,t) \qquad 0 \le x \le L \tag{8}$$

$$c=f(T) \qquad 255\,\text{min} \le t < \infty \tag{9}$$

$$k=f(T,P_{surf}) \tag{10}$$

where L is the thickness of the TPS and the RLV re-enters the atmosphere 255 minutes into its mission. Equation (7) is a linear partial differential equation with the initial condition

$$T(x,0)=70°F \tag{11}$$

and boundary conditions, at x=0

$$-k\frac{\partial T}{\partial x} = \dot{Q}_{conv} - \varepsilon\sigma\left[T^4 - T_{amb}^4\right] \tag{12}$$

and at x=L

$$-k\frac{\partial T}{\partial x} = 0 \tag{13}$$

The 70°F initial temperature arises from what is currently done with the space shuttle. Just before the shuttle leaves orbit to make its descent, it performs a series of 360° rolls in the sun. These rolls are intended to create an even temperature distribution around the shuttle TPS at approximately 70°F. The TPS surface boundary condition at x=0 imposes the law of conservation of energy, forcing the sum of $\dot{Q}_{conv}$, $\dot{Q}_{rad}$, and $\dot{Q}_{abs}$ at the surface to zero. The tank TPS boundary condition at x=L is adiabatic. The adiabatic boundary condition arises from the assumption of a dry tank wall. This assumption is justified by the fact the vehicle has used up all its fuel during ascent. Since the fuel tank is empty, any heat reaching the tank has no where to go and therefore can only be used in heating the tank structure.

HEATX generates the solution to equation (7) using numerical techniques. A 1-D section of the TPS is divided up into n number of nodes. Figure 1 illustrates a general nodal breakdown of a TPS section and the five different types of nodes: inner layer nodes, TPS surface boundary nodes, TPS tank boundary nodes, layer boundary nodes, and layer boundary nodes with an air or vacuum gap. Table 2 lists for each TPS the number of layers, layer thickness, layer material, layer weight, and number of nodes in each layer. Equation (7) is discretized using the control volume formulation, resulting in a set of n equations, one for each node, and n unknowns, the temperature of each node [7].

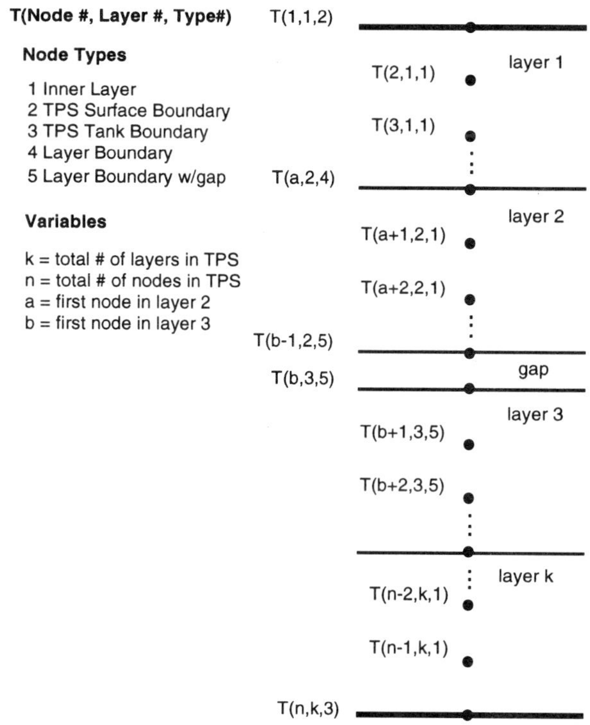

Figure 1. TPS Nodal Breakdown

TPS	Layer #	Material	# of Nodes	Depth Inches	Weight lb/ft^2
Tile	1	TUFI / AETB12	5	0.1	0.500
	2	AETB12	5	2	2.008
	3	RTV - 560	1	0.012	0.088
	4	Aluminum	1	0.08	1.167
	Total		**12**	**2.192**	**3.763**
Blanket	1	TABI	5	2	1.550
	2	RTV - 560	1	0.012	0.088
	3	Aluminum	1	0.08	1.167
	Total		**7**	**2.092**	**2.805**
Metallic	1	Inconel 617 Honeycomb Sandwich	5	0.29	0.628
	2	Q - Fiber Felt	5	3.5	1.020
	3	Titanium Honeycomb Sandwich	1	0.18	0.363
	4	Air (gap)	1	0.5	n/a
	5	Aluminum	1	0.08	1.167
	Total		**13**	**4.55**	**3.178**

Table 2. TPS Material and Nodal Breakdown

The discretized equation for each type of node is as follows:

1) Inner Layer Node

$$\frac{k_e}{\Delta x}[T_{i+1}-T_i]-\frac{k_e}{\Delta x}[T_i-T_{i-1}]=\frac{\rho c}{\Delta t}[T_i-T_{i,past}] \quad (14)$$

2) TPS Surface Boundary Node at x=0

$$-\frac{k_e}{\Delta x}[T_2-T_1]=\dot{Q}_{conv}-\varepsilon\sigma[T_1^4-T_{amb}^4] \quad (15)$$

3) TPS Tank Boundary Node at x=L

$$T_n-T_{n-1}=0 \quad (16)$$

4) Layer Boundary Node

$$\frac{k_e}{\Delta x}[T_{i+1}-T_i]-\frac{k_e}{\Delta x}[T_i-T_{i-1}]=0 \quad (17)$$

5) Layer Boundary Node with a gap

equation for gap upper surface node T_i

$$-\frac{k_e}{\Delta x}[T_i-T_{i-1}]=\varepsilon_e\sigma[T_i^4-T_{i+1}^4]+h_\infty[T_i-T_{i+1}] \quad (18)$$

equation for gap lower surface node T_{i-1}

$$-\frac{k_e}{\Delta x}[T_{i+2}-T_{i+1}]=\varepsilon_e\sigma[T_i^4-T_{i+1}^4]+h_\infty[T_i-T_{i+1}] \quad (19)$$

where k_e is the interface conductivity [7]. These equations are linearized, grouped into a matrix, and solved for the nodal temperatures.

Once the nodal temperatures are known, $\dot{Q}_{abs}$ may be calculated by (20). Equation (21) calculates $\dot{Q}_{abs}$ at the vehicle surface, while equation (22) calculates $\dot{Q}_{abs}$ at the vehicle structure. The subscript m in equation (22) denotes the layer boundary node between the aluminum structure layer and the layer just before it.

$$\dot{Q}_{abs}=-k\frac{\partial T}{\partial x} \quad (20)$$

$$\dot{Q}_{abs,surf}=-\frac{k}{\Delta x}[T_2-T_1] \quad (21)$$

$$\dot{Q}_{abs,tank}=-\frac{k}{\Delta x}[T_{m+1}-T_m] \quad (22)$$

Results

Minimum Time and Temperature Trajectories

Before considering minimum heating trajectories, minimum time and minimum estimated lower surface temperature trajectories were computed. These are of interest because, loosely speaking, minimum heating trajectories are a weighted sum of time and temperature trajectories. The optimization criteria for these two cases are as follows:

Minimum Time: Minimize $\int dt$

Minimum Temperature: $\min_{h}\,[T_{surf}]_{E=constant}$

As expected, the minimum time ESA trajectory follows the highest possible dynamic pressure (q) path, being limited by maximum stagnation temperature at Mach numbers higher than 15 and by maximum dynamic pressure below that (see Figure 2 and Table 3, next page). Table 3 shows that the descent time for this trajectory was 26 minutes and the maximum internal structure temperatures ranged from 267°F to 337°F for the three TPS concepts. Reference 1 shows that the descent time can be substantially reduced by bank angle chattering.

The minimum estimated lower surface temperature results are also shown on Figure 2 and Table 3. The surface temperature is estimated at a point 1/3 back from the nose on the windward side of the vehicle. This trajectory follows the minimum dynamic pressure constraint, has a descent time of 40 minutes, and gives a range of 271°F to 355°F for the maximum internal structural temperature. In Reference 1 contours of surface temperature for a RLV are plotted in the flight envelope. It was found, generally speaking, that there are regions of relatively low temperature near both the low and high q boundaries with higher values at intermediate values of q. The temperatures in the low q region, however, are lower than those in the high q region, which agrees with our results. This pattern of minimum time and temperature trajectories means that minimum heating trajectories will be on or near either the low q or high q boundaries.

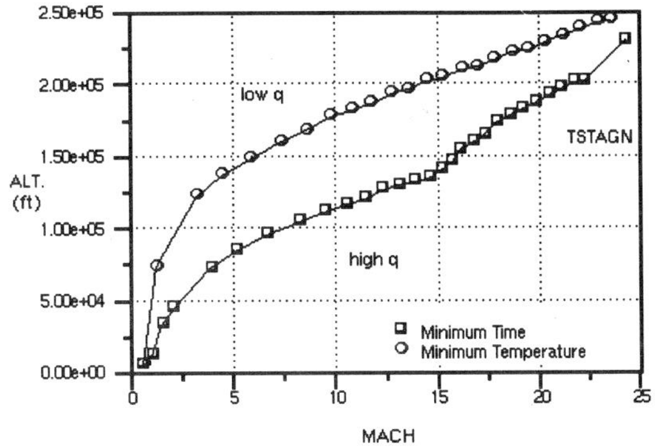

Figure 2. Altitude vs. Mach Number

Case	Flight Time @ Landing	Descent Time (Min.)	Traj. Type	Max Structure Temp. °F		
				Tile	Blanket	Metallic
Min. Time	281	26	high q	267	287	337
Min. Temp	295	40	low q	272	271	355
1	297	42	low q	271	267	353
2	288	33	low q	255	253	329
3	281	26	high q	267	287	337
3 *	271	16	high q	230		

* This case violates constraints on stagnation temperature and load factor.

Table 3. Temperature and Time Results of Optimal Trajectories

Minimum Heating Trajectories

The main purpose of the TPS is to limit the interior structural temperature to some maximum acceptable value. Since this value is not known until the end of the trajectory (indeed, the maximum structural temperature typically occurs after the vehicle has landed), the coupled problem of sizing the insulation and determining the optimal trajectory is by nature an iterative one. It is the goal of this paper to formulate and evaluate ESA criteria which give non-iterative approximations to the trajectory optimization part of this problem.

Three optimization criteria will be investigated in the following sections:

Case (1): Minimize $\int \dot{Q}_{abs} dt$ at the surface of the TPS, equation (21).

Case (2): Minimize $\int T_{surf}^{4} dt$, see equation (5).

Case (3): Minimize $\int \dot{Q}_{abs} dt$ at the internal structure, equation (22).

The integral in case (2) was the criterion used in Reference 1 and is directly proportional to Q_{conv}. Case (3) was the criterion used in Reference 6.

Figure 3 shows the trajectories in the flight envelope for the three cases, and Figures 4 and 5 show their time and estimated surface temperature histories, respectively. Note that Figures 4 and 5 are plotted using specific excess energy as the independent variable and illustrate that the total change in specific excess energy of each case is the same whereas the change in flight time is not. According to Figure 3 two of the trajectories, cases (1) and (2), are of the low q type and the other, case (3), is of the high q type. Figures 4 and 5 show that relative to the high q type, the low q trajectories have long descent times and low surface temperature. They also have high angle of attack and low drag. Table 3 shows, however, that all three cases give maximum interior structural temperatures that are very similar, with case (2) giving the best (lowest) value. Also shown on Figure 3 for reference is the space shuttle trajectory [8]; it is seen that this is of the low q type.

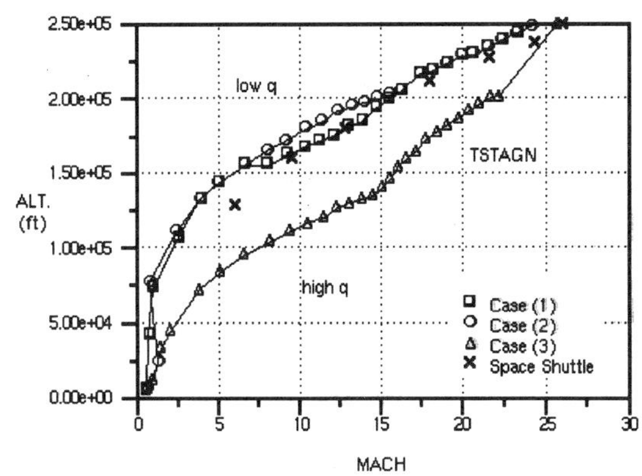

Figure 3. Altitude vs. Mach Number

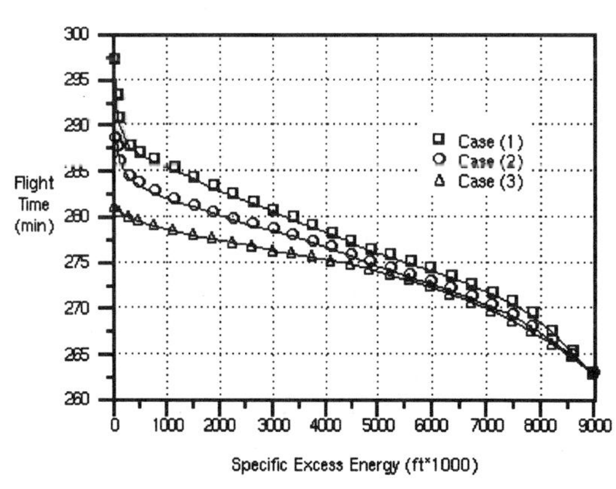

Figure 4. Flight Time vs. Specific Energy

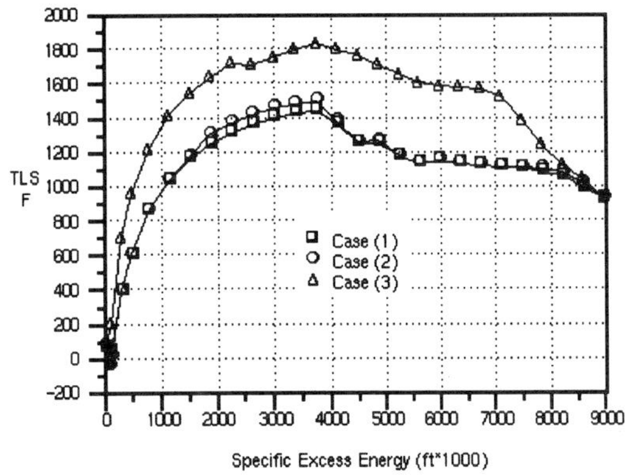

Figure 5. Estimated Surface Temperature vs. Specific Excess Energy

The time histories of the convected heat rates are shown on Figure 6. This figure again illustrates the different natures of the high q and low q trajectories. Notice the magnitude of $\dot{Q}_{conv}$ for the high q trajectory is much greater and lasts for a shorter time period relative to the magnitude of $\dot{Q}_{conv}$ for the low q trajectories.

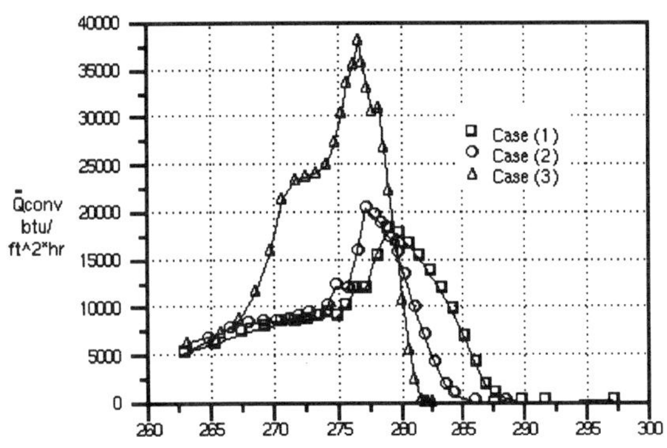

Figure 6. Convective Heat Rate vs. Flight Time

Figures 7-9 show the time histories of the heat absorbed at the vehicle surface for the three cases and the three TPSs. Although the shape of the curves on the figures are similar, the magnitude of the heat absorbed at the vehicle surface differs depending on the TPS concept used. Also according to the figures, at a certain point in the trajectory Q_{abs} at the vehicle surface peaks. Past this point the vehicle surface temperature is so high that $\dot{Q}_{rad}$ has exceeded $\dot{Q}_{conv}$ (see equ. 15). From then on the vehicle loses heat and surface temperatures cool. This process continues well after the vehicle has landed.

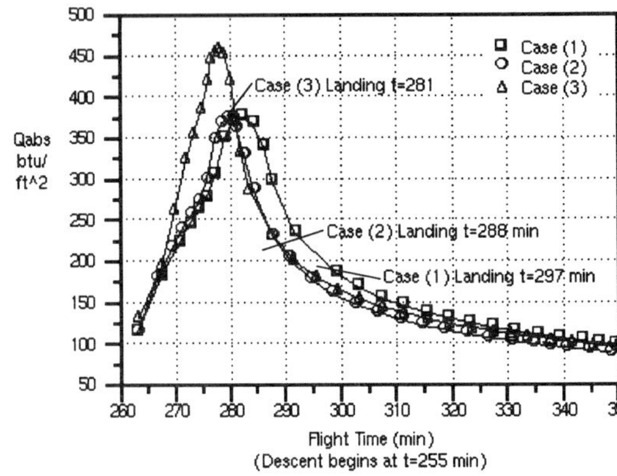

Figure 7. Heat Absorbed at the Vehicle Surface vs. Flight Time (Tile TPS)

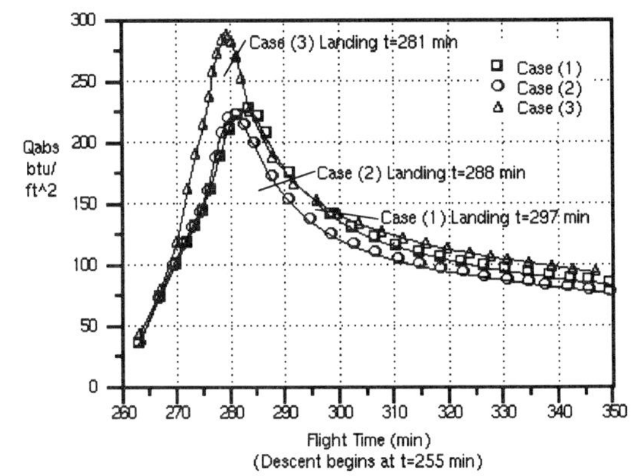

Figure 8. Heat Absorbed at the Vehicle Surface vs. Flight Time (Blanket TPS)

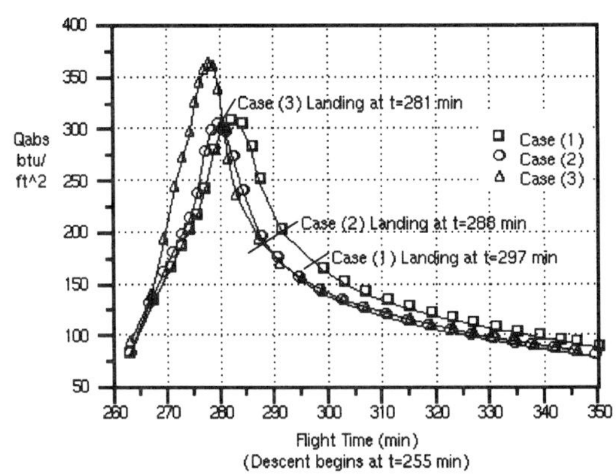

Figure 9. Heat Absorbed at the Vehicle Surface vs. Flight Time (Metallic TPS)

Figures 10-12 show the time histories of interior structural temperatures. It is seen that case (2) gives the lowest maximum temperature. Also of interest is the fact that the internal structure temperature peaks well after landing in all cases. This directly contrasts with the surface temperature which peaked with Q_{abs} earlier in the trajectory and is still cooling after landing.

This transient heating effect is illustrated in Figure 13. Notice that after the maximum surface temperature has been reached a heat pulse forms and begins traveling from the TPS surface to the vehicle structure. This heat pulse continues to increase the vehicle structure temperature even after the vehicle has landed.

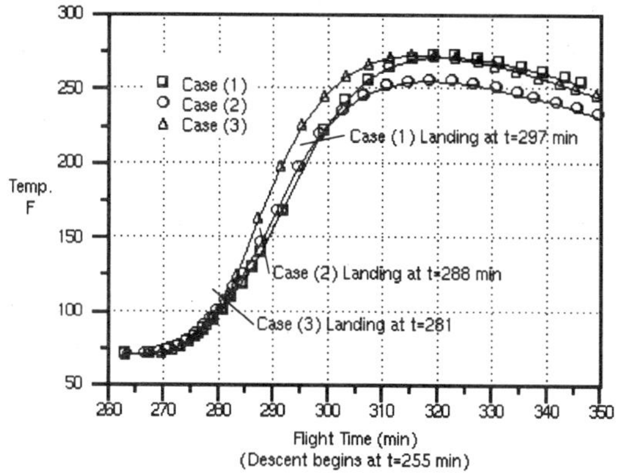

Figure 10. Vehicle Structure Temperature vs. Flight Time (Tile TPS)

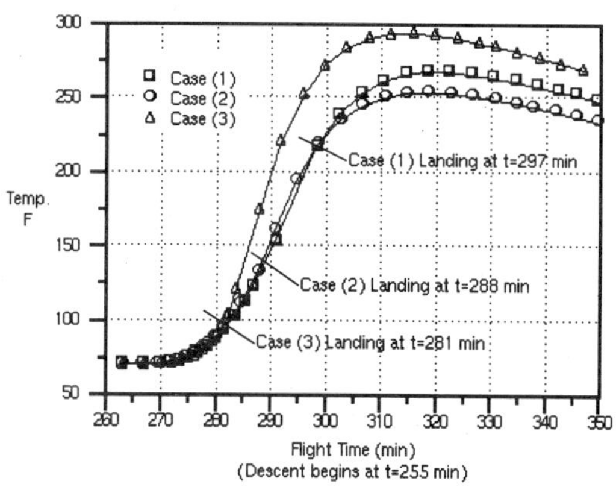

Figure 11. Vehicle Structure Temperature vs. Flight Time (Blanket TPS)

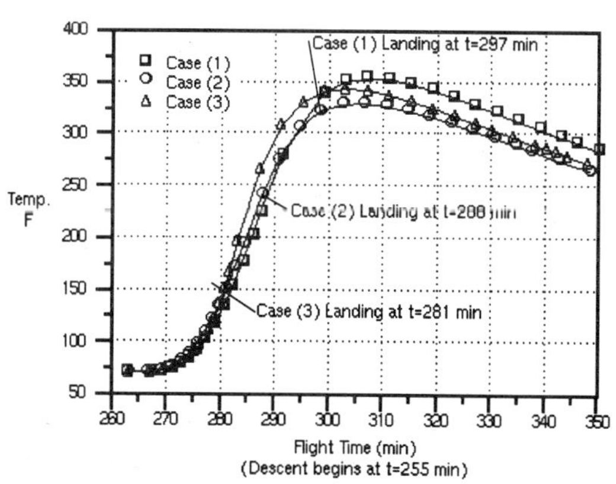

Figure 12. Vehicle Structure Temperature vs. Flight Time (Metallic TPS)

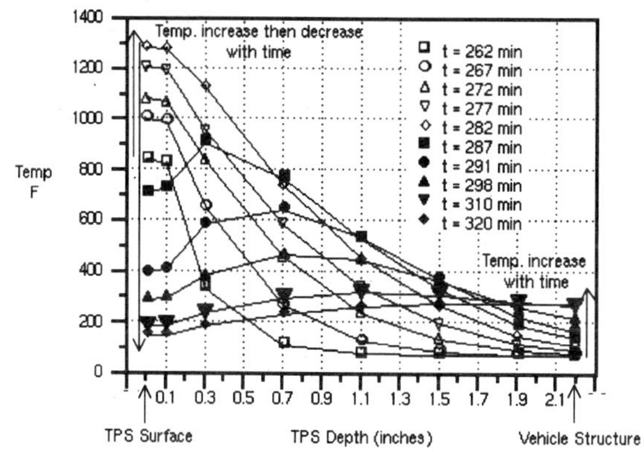

Figure 13. TPS Temperature Profiles at Different Flight Times (Tile TPS, Case (1))

Since it is desired to keep interior structural temperatures below 250°F, all the TPS would have to be re-sized to meet this objective, particularly the metallic system.

Other Options

The case (3) trajectory lies on the maximum nose cap stagnation temperature boundary of the flight envelope for the first portion of the descent. If this condition is relaxed from 3000°F to 4000°F, a significant improvement results. Case (3) and this case, labeled 3*, are compared on Figure 14. Case (3*) gives a trajectory which rapidly dives to the new nose cap stagnation temperature boundary (the load factor constraint was ignored for this dive) and remains in the high q region for the rest of the flight. Table 3 and Figure 15 show that this trajectory not only has minimum flight duration but the lowest internal structural temperature as well.

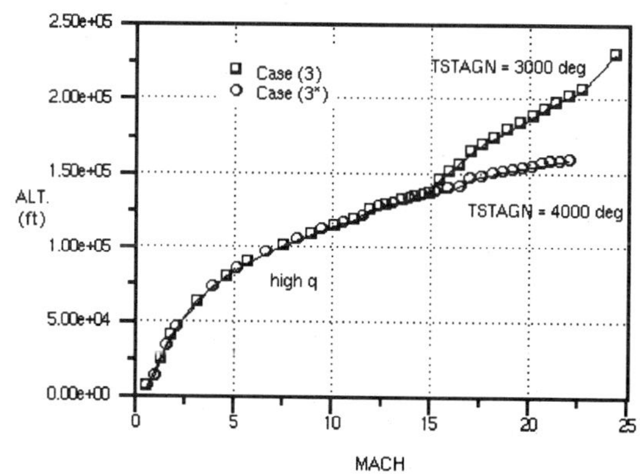

Figure 14. Altitude vs. Mach Number (Tile TPS)

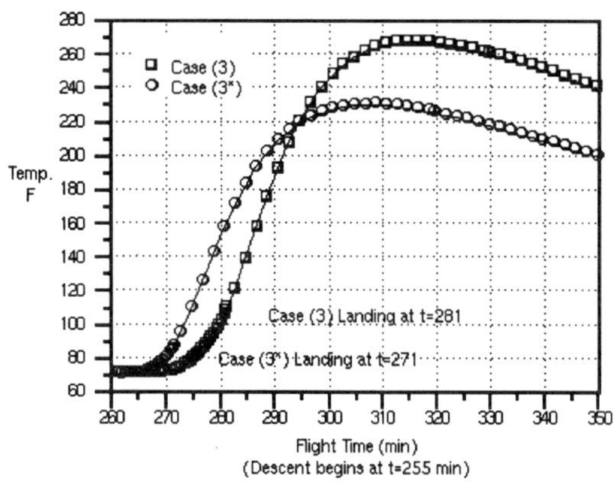

Figure 15. Vehicle Structure Temperature vs. Flight Time (Tile TPS)

The superiority of case (3*) is due to two factors: first, the amount of time the vehicle undergoes convective heating determines how far the heat will diffuse into the TPS. A short descent time and a TPS with low thermal diffusivity does not allow the heat to diffuse into the TPS very far. Constraining the heat near the surface of the TPS makes a shorter path for the heat to radiate back out of the vehicle instead of continuing to conduct further into the vehicle structure. Second, the high T_{surf} caused by a case (3*) optimal trajectory increases the heat rejection rate, $\dot{Q}_{rad}$, by T_{surf} to the fourth power, see equation (5). Thus, although the case (3*) trajectory induces higher heat absorption rates on the vehicle, it allows even higher heat rejection rates. The combined effect of the short descent time and increased efficiency of heat rejection, keeps the heat at the TPS surface, rejects the heat quickly, and reduces conduction into the vehicle structure, thereby minimizing the maximum structural temperature of the vehicle. These combined effects are illustrated in Figure 16, which shows TPS temperature profiles for each case at the time when peak surface temperatures occurred. Notice the case 3* profile has the highest surface temperature and the least amount of heat diffusion.

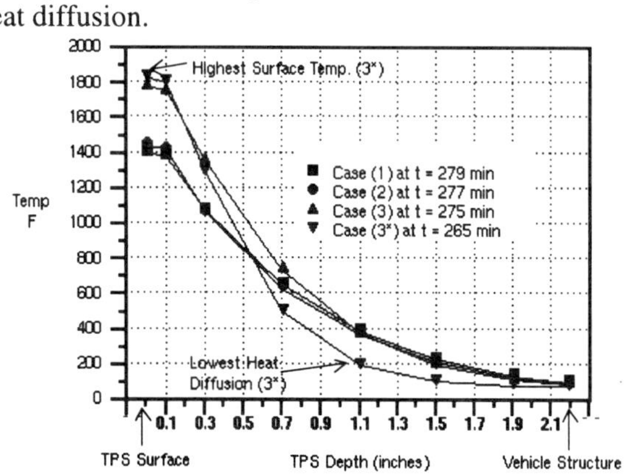

Figure 16. TPS Temperature Profile at Flight Time of Peak Surface Temperature (Tile TPS)

Chattering trajectories were also considered. In these trajectories, the vehicle rapidly banks left and right so that its flight path remains straight. This increases the vehicle lift and hence its drag. Although chattering drastically reduces descent time by increasing drag, it also increases $\dot{Q}_{conv}$, thereby increasing T_{surf}. Since the vehicle is constrained by TSTAGN already during the first half of it's descent, chattering does not significantly change the case (3) optimal trajectory unless the TSTAGN constraint is relaxed.

Trajectories were also generated minimizing the integral of T_{surf} with respect to time. This optimization integral is proportional to the integral of the steady state conductive heat rate of the vehicle, assuming the vehicle tank wall temperature remains constant.

$$\dot{Q}_{steady} = \frac{k}{L}\left[T_{surf} - T_{tank,wall}\right] \tag{23}$$

The integral T_{surf} dt optimal trajectory continually jumps between high q trajectories and low q trajectories. This is because T_{surf} and dt are now weighted equally, and low q trajectories minimize T_{surf} and high q trajectories minimize dt. Although jumping from one path to another is physically impossible, it is allowable by the ESA method. Constraining the trajectory with load factor limits eliminates the jumps but destroys the optimality of the trajectory. For example, a constrained integral T_{surf} dt trajectory that starts out as a low q trajectory is forced to remain a low q trajectory because of the load factor constraint. On the other hand, the integral T_{surf} dt trajectory could be forced to follow the high q trajectory at first and it would remain on that trajectory due to the load factor constraint.

Concluding Remarks

Control volume heat transfer equations have been coupled to the re-entry flight dynamics equations for the purpose of determining trajectories that minimize heat input and interior structural temperatures. The energy state approximation was used to obtain the near-optimal paths.

It was found that the trajectories were of two types: very low dynamic pressure (q) and very high q. The low q trajectories have lower surface temperatures and higher descent times than do the high q ones. The low q trajectories gave lower maximum structural temperatures, but the difference was very small. If TPS materials capable of withstanding higher temperatures are developed, then the high q trajectories become superior.

Finally, we note that the technique developed in this paper is non-iterative and has modest computational requirements. Thus it could be used as an on board guidance scheme to generate minimum heating trajectories in real time.

Acknowledgment

This research was supported by NASA Ames Research Center Grant NCC2-5165.

References

[1] H.C. Chow, M.D. Ardema, and J.V. Bowles, "Near Optimal Re-entry Trajectories for Reusable Launch Vehicles," AIAA Paper 97-3582, to be presented at the 1997 AIAA Atmospheric Flight Mechanics Conference.

[2] M. D. Ardema, J V. Bowles, and T. Whittaker, "Optimal Trajectories for Hypersonic Launch Vehicles," Dynamics and Control, Vol. 4, pages 337-347 (1994).

[3] M. D. Ardema, J V. Bowles, E. J. Terjesen, and T. Whittaker, "Approximate Altitude Transitions for High-Speed Aircraft," Journal of Guidance, Control, and Dynamics, Vol. 18, No. 3, pages 561-566.

[4] M.D. Ardema, J.V. Bowles, T. Whittaker, "Near-Optimal Propulsion-System Operation for an Air-Breathing Launch Vehicle," Journal of Spacecraft and Rockets, Vol. 32, No. 6, pages 951-956.

[5] M.D. Ardema, H. C. Chow, and J.V. Bowles, "Near-Optimal Operation of Dual Fuel Vehicles," Journal of Guidance, Control, and Dynamics, Vol. 19, No. 5, pages 1180-1182.

[6] G. Sachs and M. Dinkelmann, "Heat Input Reduction in Hypersonic Flight by Optimal Trajectory Control," AIAA Guidance Navigation and Control Conference, July 29-31, 1996 / San Diego, CA.

[7] Suhas V. Patankar, "Numerical Heat Transfer and Fluid Flow," Hemisphere Publishing Co. 1980.

[8] David R. Olynick and Tim Tam, "Trajectory Based Validation of the Shuttle Heating Environment," AIAA Thermophysics Conference, June 17-20, 1996 / New Orleans, LA.

AIAA-97-3536

NONLINEAR CONTROL USING NEURAL APPROXIMATORS WITH LINEAR CONTROL THEORY

J. Neidhoefer and K. Krishnakumar
Department of Aerospace Engineering and Mechanics
The University of Alabama, Tuscaloosa, AL 35487-0280
e-mail: jneidhoe@buster.eng.ua.edu

Abstract

This paper presents an approach for designing nonlinear, adaptive controllers using Neural Approximators (NA) in conjunction with linear control theory. The adaptive capabilities of these controllers are such that, real time control should be possible. Inspiration for the Neural Approximator approach was found largely in the methodology for "gain scheduling", however, the resultant Approximators are not discrete schedulers, but continuous approximations to the nonlinear mapping between local linear systems. The method is demonstrated here by applying it to two aircraft control problems. The first is a stability augmentation problem and the second is a guidance problem. The aircraft model used is a high-performance fully nonlinear model representative of modern day fighter aircraft.

1. Introduction

As detailed by Shamma and Athans in [1], the general idea behind gain scheduling is to break the control design process into two steps. First, one designs a set of "locally valid" (usually linear) controllers, and second a global nonlinear controller is designed by utilizing some "logical operator" to choose which local controller is the most valid. A "third step", now often used is to employ some type of tuning technique to the "chosen" local controller. This idea is not a new one. In the 1970's Lainiotis[2] proposed using multiple Kalman filters to improve the accuracy of state estimations. Athans used multiple models with MMAC switching (a probability based logical operator) for aircraft control[3]. Much more recently Narendra, Balakrishnan, and Ciliz have proposed using multiple models with switching [4]. In [4], how neural networks might be employed is also discussed. Krishnakumar's idea of Immunized Neural Systems [5][6] is based upon the idea of creating and combining several neural networks which are locally valid in some global hypersphere of uncertainty. For fast adaptation, Krishnakumar has discussed the idea of "innate immunity" and has proposed the use of "linear controllers with fuzzy schedulers" and also that "systems known *a priori* can be stored in look up tables"[7].

The two main limitations of gain scheduling as discussed by Shamma[1], are that 1) The scheduling variable must vary slowly, and 2) The scheduling variable should capture the nonlinearities of the plant. A third disadvantage of most gain scheduling methods is that we have no way of knowing how far from the equilibrium point (about which the system has been linearized) that the linear controller will remain valid for the nonlinear system. Two of the greatest advantages of modern neural networks are 1) They can generalize a function arbitrarily well utilizing a only few representative data points. and 2) They can model highly nonlinear functions.

This paper sets forth a technique for implementing Neural Approximators which utilize the strengths of artificial neural systems to overcome some weaknesses (limitations) in the general gain scheduling technique. NA can be thought of as high-dimensional curve fits which provide continuous information throughout a nonlinear plants operational envelope by providing a continuously changing linear approximation of the plant. The result is a very fast adapting controller architecture suitable for real-time control.

The rest of this paper is organized as follows: First the general NA technique is outlined. Neural Approximators are then applied to the stability augmentation of a high-performance aircraft model. Next, Neural Approximators are implemented on a much more complex guidance control system. Finally conclusions and future recommendations are given.

2. NA Design Methodology

2.1 Define The Hypershpere of Uncertainties: The first step in the design of a Neural Approximator is to specify the hypershpere of uncertainties. In other words, one must specify which uncertainties the system may encounter, and which of these should be negated by the controller.

Defining the hypersphere of uncertainties is most important in successfully implementing a NA based controller. It is important that the control system designer differentiates between the types of uncertainties (i.e. parametric, initial condition, environmental, etc.) and trains the approximator accordingly. This is important because this method will be impractical for specific problems in which uncertainty measurements, estimations, or approximations are unobtainable.

2.2 Define The Scheduling Variables: The next step is to define the "scheduling variables". Neural Approximators can be thought of as "continuous schedulers", and it will become apparent from the design methodology that the NA inputs will have the same restrictions as the "scheduling variables" of more traditional "scheduling or switching type" controllers. Thus the inputs to NA are termed "scheduling variables". As formalized by Shamma [1], the two main restrictions on the scheduling variables are that 1) The scheduling variables must vary slowly, and 2) The scheduling variables should capture the nonlinearities of the plant. A rigorous mathematical justification for these restrictions is presented in [12]-[14].

The scheduling variables are often very closely associated with the uncertainties. Sometimes the scheduling variables may be the uncertainties themselves. Usually the scheduling variables are some function of one or more of the measured uncertainties, and/or system states.

2.3 Define a Suitable Set of Equilibrium Points: The equilibrium points will represent the points at which the nonlinear plant system is linearized, and will also be closely related to the NA's neural net training set. The number and location (within the hypersphere) of equilibrium points is an important design consideration. Equilibrium points must be chosen to sufficiently capture the nonlinearities of the hypersurface.

Neural Approximators can be thought of as "pseudo-global" as the region of Approximator validity is actually designed by the choice of equilibrium points.

2.4 Linearize the System: Once the equilibrium points are defined, the nonlinear system is linearized at each point. Local linear stability augmenting, or tracking controllers are then designed (for neural gain approximation) or the linear representations of the nonlinear plant themselves are used (for neural state matrix approximation). The only restriction on the local linear systems is that their individual elements must be statistically correlated between equilibrium points. The necessity for correlation between elements arises from the fact that standard neural networks cannot generalize uncorrelated data. And it is the generalizing capabilities of neural networks that we are most interested in here.

An example of a way to ensure correlated data is to use pole placement in individual local controller designs and place the poles the same for each controller. Thus, as we move from one equilibrium point to another, the gains will change, but they will be correlated, and the response of the closed loop system will remain constant. This is the method used for the first example presented in this paper. LQR or LQG controllers could be used as well, but care must be taken in specifying the Q and R matrices such that the resultant controllers are correlated (i.e. keep the closed loop system a constant).

2.5 Train A NN Approximator: The key to the NA technique is as follows: *If multiple linear systems are correlated between equilibrium points, then one can train a neural network using input/output pairs consisting of the scheduling variables (inputs) and elements of the local linear systems (i.e. elements of a state matrix, or controller gains) (outputs).* The resultant network will provide an approximation of the nonlinear mapping between the local linear systems.

It should now be apparent why it is necessary for our training data to be correlated. If NN training data is uncorrelated, the best one can hope for is for the NN to *memorize* the data. If the NN memorizes the data, then the NN *generalization* will probably be meaningless (resulting in little or no robustness).

2.6 Analyze the Approximator: Analyzing the validity of a NA can be very application specific. Rigorous stability proofs of NA-based controllers would be highly desirable, however, to this point, validity assessment has been made through rigorous testing throughout the design region. The main objective of this analysis is to convince oneself that the NA provides an accurate, robust, and valid approximation in the designed region. Since we have the ability to linearize the system, we may be able to check the eigenvalues of (A-BK) over a fine mesh covering the design region. In the first example presented herein, this analysis method was used. If the closed loop system becomes unstable (in gain approximation), two things can be immediately tried, 1) train the NN to a lower training error, or 2) increase the stability margin of the linear controllers. In the second case, it is possible to determine the max and mean error and the standard deviation in the approximator output. These values can then be used to calculate the stability margin necessary for the linear controllers. The designer must also be careful when designing a controller for a real system, that there will be sufficient control power in the system to provide the gains specified by the control law.

3. Neural Gain Approximators for SAS

Neural gain approximators were implemented in a stabilizing controller for a 6 DOF aircraft model with nonlinear aerodynamics, and nonlinear kinematics. The aircraft model used, is a high-performance, supersonic vehicle representative of modern fighters. The aircraft's primary flight-control surfaces consist of horizontal stabilators capable of symmetric or differential deflection, conventional ailerons, and rudder. This model is described fully in [15]. For the purpose of this study, the model was somewhat modified. In the original model, the control surfaces are all deflected with rate limited actuators with first order dynamics. The engines in the original system were also rate limited and employed first order dynamics. For this study, the first order dynamics as well as rate limiting were removed from the engine and control surfaces, also, the differential stabilator deflection was set to zero, as it provided redundant roll control. Thus, our study used control surface deflections and total engine thrust directly as plant system control inputs.

The design objective of this control system was that it provide a valid stability augmenting control law.

3.1 Define The Hypersphere of Uncertainties:

Initially, the uncertainties were specified as being parameters specifying the flight regime. The different flight regimes were specified by the mach number and the altitude. These uncertainties were sufficient to provide plant system perturbations over a wide range of frequencies and amplitudes.

Now that we have specified the disturbances which define our hypersphere of uncertainty, we define our "scheduling variables" to be velocity and altitude.

3.2 Define Suitable Set of Equilibrium Points: The next step in the control system design was to come up with the trim conditions for the aircraft model in each of these flight regimes. The aircraft's flight envelope in terms of altitude and airspeed is well defined in [15], a portion of this envelope was well covered by incrementing airspeed from 500 fps to 1000 fps in increments of 100 fps, and incrementing altitude from 0 to 20,000 ft in increments of 5,000 ft (See Figure 1). After specifying the altitude and airspeed, a simplex routine was used for finding aircraft trim points.

3.3 Design a Linear Controller At Each of the Equilibrium Points: Once the "trim points" were found, the model was linearized about these points resulting in state space A, and B matrices. Now, it was necessary to create correlated linear controllers at each of the trim points. The first step in this process was to define a pole placement that would result in favorable aircraft handling characteristics. This was done by using handling quality tables for our particular flight case. Figure 3 shows the open loop poles at the trim points. A genetic algorithm was used to place the eigenvalues at the desired positions by allowing the GA to optimize the gains for pole placement. To insure correlation, a suitable set of gains were found at one trim point, and the GA was used to find perturbations about these gains for the rest of the trim points. This method was used as use of the pole placement routines in MATLAB did not ensure correlated gains.

3.4 Train a NN Approximator: A backpropagation algorithm for optimizing both weights and connections was used for optimizing the net of the approximator. Once again, it is here that the correlation between the gains in the linear controllers at different trim points became important to achieve a successful neural net training.

3.4 Analyze the Approximator: The NA was, in this case, a neural gain approximator and it was implemented in a SAS control system as shown in figure 2. In order to analyze the robustness of this controller, a fine mesh was created in the designed region (a finer mesh than in figure 1). And these points were then checked by running them through the controller, generating the gains, and finally plotting the eigenvalues of the closed loop system. Figure 3 contains the open loop eigenvalues as well as closed loop eigenvalues for several on training (points used in the NA training, i.e. figure 1) and off training points (points not used in NA training) The accuracy of the controller can be judged from the difference between the mean of the eigenvalue clusters, and the desired pole placement, and the robustness of the controller can be judged from the standard deviation of the clusters.

4. Neural Approximators for Guidance

In the second example, neural state matrix approximators were implemented in a guidance controller for the same aircraft model used in the previous example.

The design objective of this control system was that it provide a stable guidance control law.

4.1 Define The Hypersphere of Uncertainties: Once again, the uncertainties were specified as being parameters specifying the flight regime. The different flight regimes were specified by the mach number and the altitude. Therefore, once again we define our "scheduling variables" to be velocity and altitude.

4.2 Define Suitable Set of Equilibrium Points: The same set of equilibrium points as was used in the previous example was used for this study (See figure 1). See section 3.2 for a more detailed explanation of their selection.

4.3 Linearize the System At Each of the Equilibrium Points: Once the "trim points" were found, the model was linearized about these points resulting in state space A, and B matrices. The individual elements of the matrices were examined with a sensitivity study, and only those elements that were sensitive to variations in the scheduling variables were approximated by the NA. Thus two NA training data files were created, one for the A matrix, and one for the B matrix. After training, the NA were used to reconstruct linear approximations of the aircraft (dependent on the operating condition) by calculating the sensitive elements with the NA and using the mean values of the nonsensitive elements.

4.4 Train a NN Approximator: A backpropagation algorithm for optimizing both weights and connections was used for optimizing the net of the approximator. Once again, it is here that the correlation between the elements of the A and B matrices in the local linearized systems at different equilibrium points became important to achieve a successful neural net training.

4.5 Analyze the Approximator: The NA were, in this case, neural state matrix approximators and they were implemented in a guidance control system as shown in figure 4. In order to analyze the validity of the NA, a fine mesh of off training points was created in the designed region (a finer mesh than in figure 1). Over this fine mesh, the linearized system was approximated, and the open loop eigenvalues were plotted (similar to figure 3). As in figure 3, different clusters, represent the aircraft modes with cluster spread being representative of different operating conditions. A sufficiently small standard deviation in the clusters showed that the linear approximations were valid.

The basic idea of the controller shown in figure 4, is that if a linear approximation of the nonlinear plant is constantly updated and valid, then at each time step, a valid inverse control can be calculated. The controller was first used to control the linearized system. Figures 5 and 6 show the desired and actual aircraft paths through a climbing turn maneuver. At the time of this writing, work is underway to implement the guidance controller with the complex, 21 state fully nonlinear plant. Initial results are promising as can be seen in figures 7, 8, and 9.

5. Conclusions and Future Recommendations

During the course of this study, it was concluded that if a system is linearizable, then there is little advantage to using nonlinear local controllers when trying to implement a global nonlinear controller with scheduling or switching type control laws. If we can linearize the system then we can use linear controllers along with neural gain approximation to derive a psuedo-global control law.

It was also concluded that if multiple linear controllers designed for a given system are correlated between equilibrium points, then we can train a neural network using input/output pairs consisting of the scheduling variables (inputs) and the controller gains (outputs).

Most adaptive control techniques can only function when uncertainties vary slowly. This is partially due to the fact that most adaptation laws use gradient information to optimize and thus there is some transient time in which the controller is not optimal. This technique may over come this shortcoming of traditional adaptive control (the "slowly varying uncertainty" rule is also due to system response transients, and the NA approach does not have any advantage over traditional methods for this shortcoming).

This technique though similar to scheduling techniques is not a discrete scheduler. It should be applicable to any situation where a scheduler might work, however, it should work in a much more streamlined manner than traditional schedulers. The method of neural gain approximation has several advantages over traditional gain scheduling techniques. First, it negates the necessity of storing a large number of gain sets corresponding to equilibrium points of the system (NA is in a way, data-compression). Also, we should experience no computational time disadvantage, as running a single forward pass through a NN should not be significantly computationally slower than looking up values in a table, and then interpolating between the values. Another advantage, is that neural gain approximation can successfully extrapolate outside designed region (though this will vary from application to application).

A drawback of this technique is that we must be able to get correlated local systems. Linearization is one easy way of doing this, however, it may also be possible to employ a similar technique using parameterized nonlinear local controllers.

The most pressing area of extended future research should be the development of rigorous stability proofs for this type of controller. Though it is easy to conceptualize (using Lyapunov linear stability techniques along with the understanding of what a neural net is doing when it maps a higher dimensional surface) why this method is valid, rigorously proving this may be more challenging.

Acknowledgment

This material is based on work supported by the National Science Foundation under Grants No. ECS-9415351 and ECS-9113283.

REFERENCES

[1] J.S. Shamma, M. Athans, "Gain Scheduling: Potential Hazards and Possible Remedies", Control Systems Magazine, Vol. 12 No. 3, June 1992.

[2] D.G. Lainiotis, "Partitioning: A Unifying Framework for Adaptive Systems, I: Estimation, II: Control," *Proceedings of the IEEE*, 64: 1126-1143, 1182-1197, August 1976

[3] M. Athans et al., "The Stochastic Control of the F-8C Aircraft Using a Multiple Model Adaptive Control (MMAC) Method-Part I: Equilibrium Flight," *IEEE Transactions on Automatic Control*, AC-22: 768-780,1977

[4] K.S. Narendra, J. Balakrishnan, M. Ciliz, "Adaptation and Learning Using Multiple Models, Switching, and Tuning", IEEE Control Systems, June 1995.

[5] K. Krishnakumar, "Immunized Artificial Neural Systems", World Congress on Neural Networks, Portland, July 11-15, 1993

[6] K. Krishnakumar, "Immunized Neural Control: Concepts and Initial Implementation Results", BER Report #513-177, The University of Alabama, Tuscaloosa AL, 1992.

[7] K. Krishnakumar, A. Satyadas, J. Neidhoefer, "An Immune System Framework For Integrating Computational Intelligence Paradigms With Applications To Adaptive Control", Computational Intelligence (A Dynamic System Perspective), IEEE Press, 1995.

[8] J.J. Slotine, W. Li, "Applied Nonlinear Control", Prentice Hall, Englewood Cliffs, New Jersey 1991

[9] W.J. Rugh, "Analytical Framework for Gain Scheduling", Control Systems Magazine, Vol. 11, No. 1, Jan. 1991

[10] P. Cook, "Nonlinear Dynamical Systems", Prentice / Hall International , 1986

[11] K. Krishnakumar, "Immunized Neurocontrol: Concepts and Initial Results", COGANN-92, A Workshop on Combinations of GA and NN. June 6, 1992.

[12] J.S. Shamma, "Analysis and design of gain scheduled control systems," Ph.D. thesis, Dept. of Mechanical Engineering, M.I.T., 1988.

[13] J.S. Shamma and M. Athans, "Analysis of nonlinear gain scheduled control systems," *IEEE Trans. Auto Control*, vol. AC-35, no. 8, pp. 898-907,1990.

[14] J. S. Shamma and M. Athans, "Guaranteed properties of gain scheduled control of linear parameter-varying plants," *Automatica*, vol. 27, no. 3, pp. 559-564, May 1991.

[15] R. Brumbaugh, "Aircraft Model for the AIAA Controls Design Challenge", Journal of Guidance, Control, and Dynamics, Vol. 17 No. 4. July-August 1994.

[16] J. E. Bailey, K. Krishnakumar, A. Homaifar, "GPS-Based Automatic Landing System For The AH-64 Helicopter", BER Report No. 361-177, September, 1985.

FIGURES

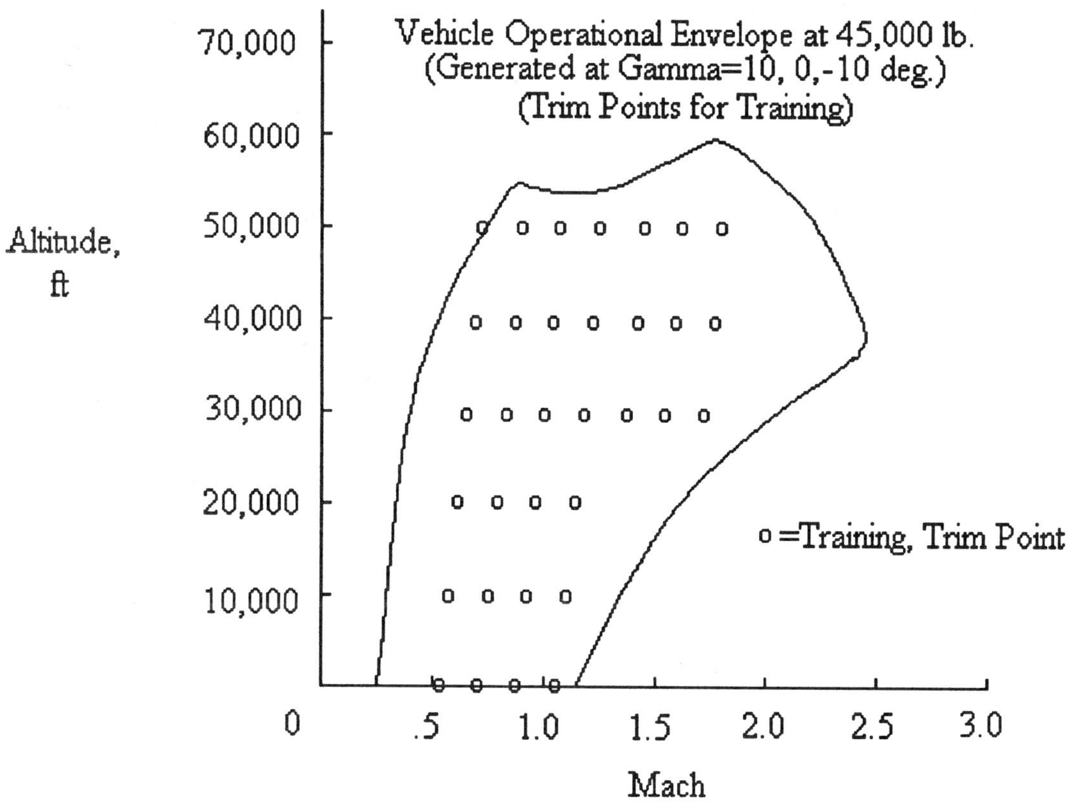

Figure 1. Vehicle Operational Envelope, and Equilibrium Points for NA

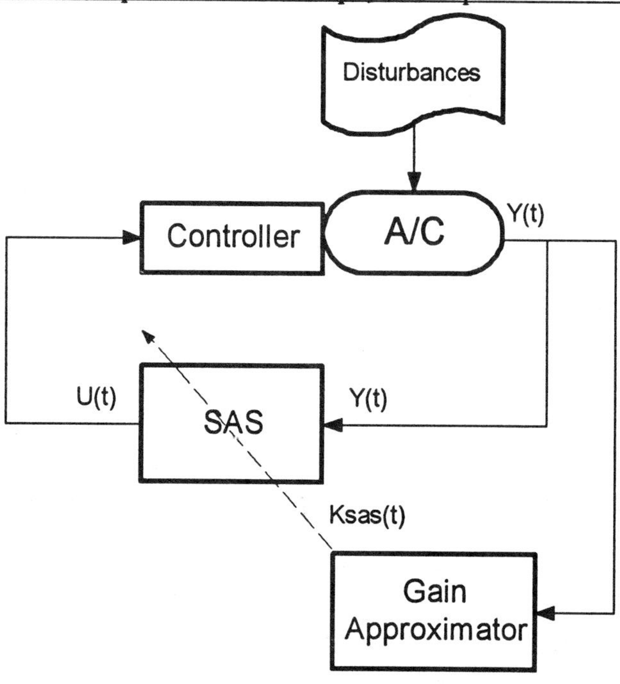

Figure 2. NA-Based Controller For SAS

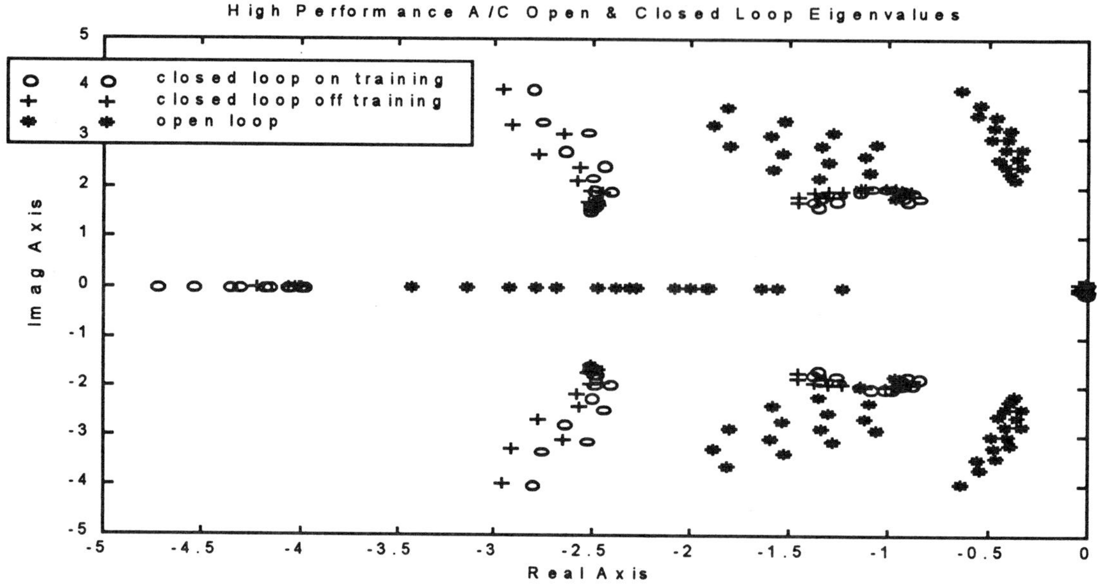

Figure 3. Open and Closed Loop Poles with SAS NA-Based Controller

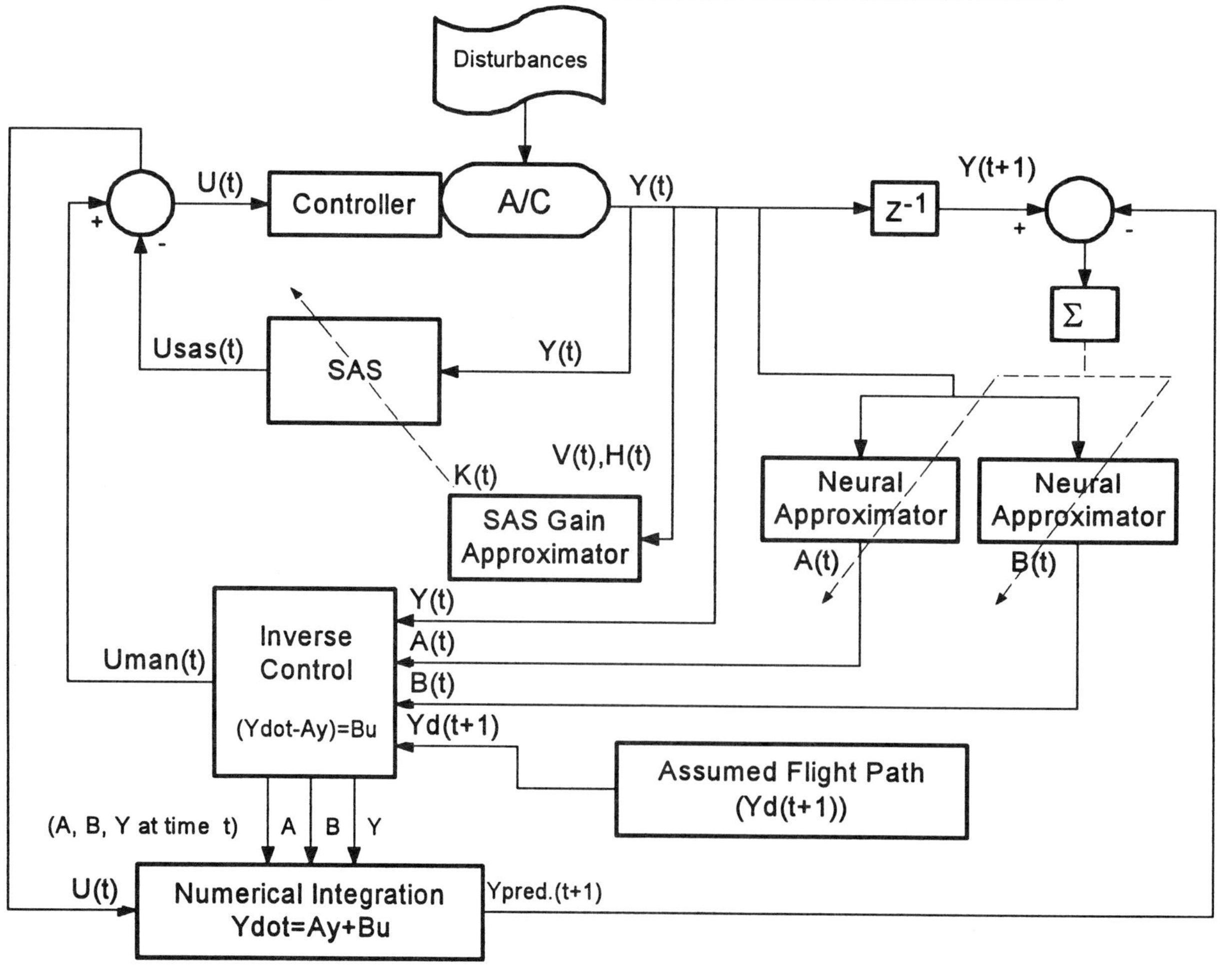

Figure 4. Level 1 Intelligent Controller

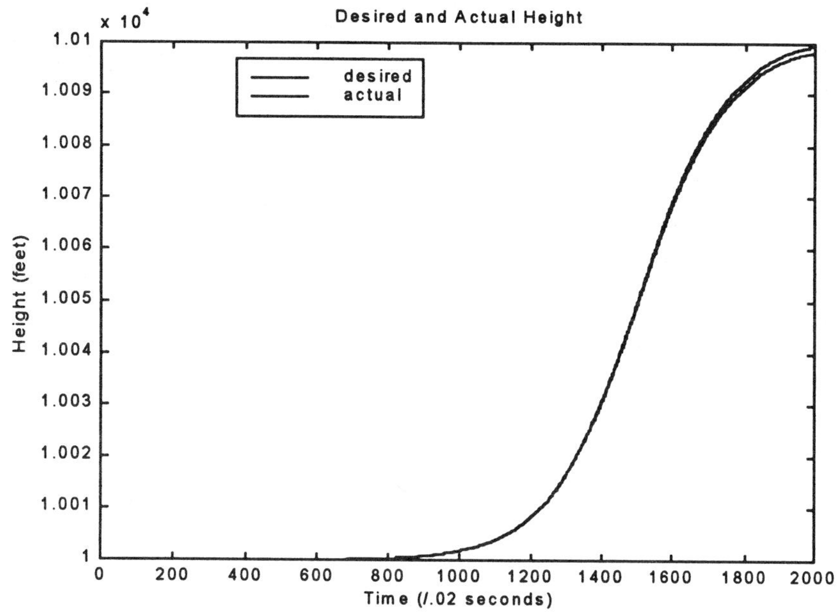

Figure 5. Desired and Actual Height NA-Based Guidance, Linear Approximated Plant

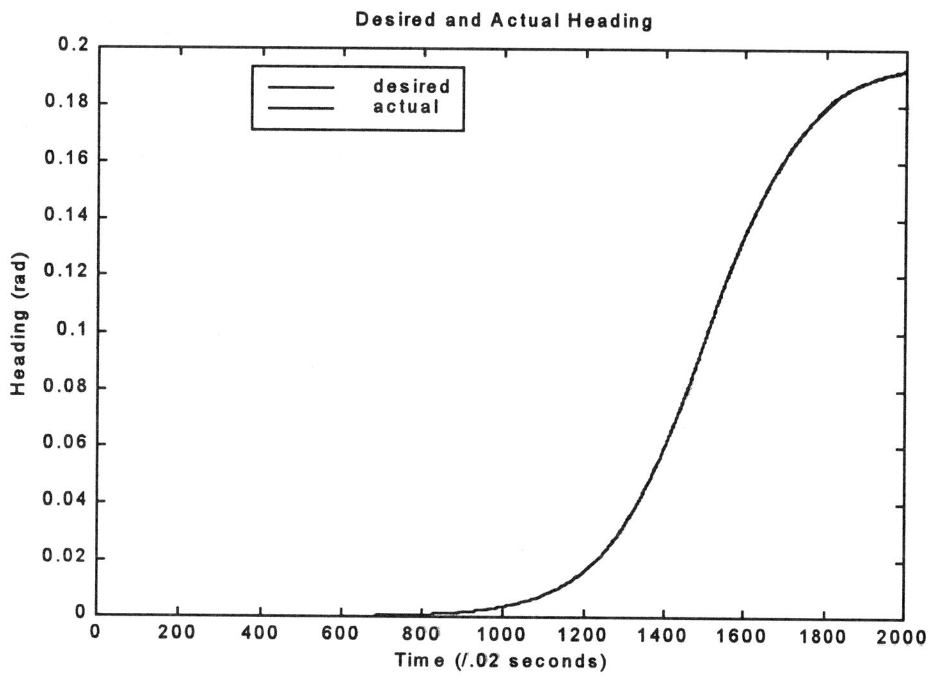

Figure 6. Desired and Actual Heading NA-Based Guidance, Linear Approximated Plant

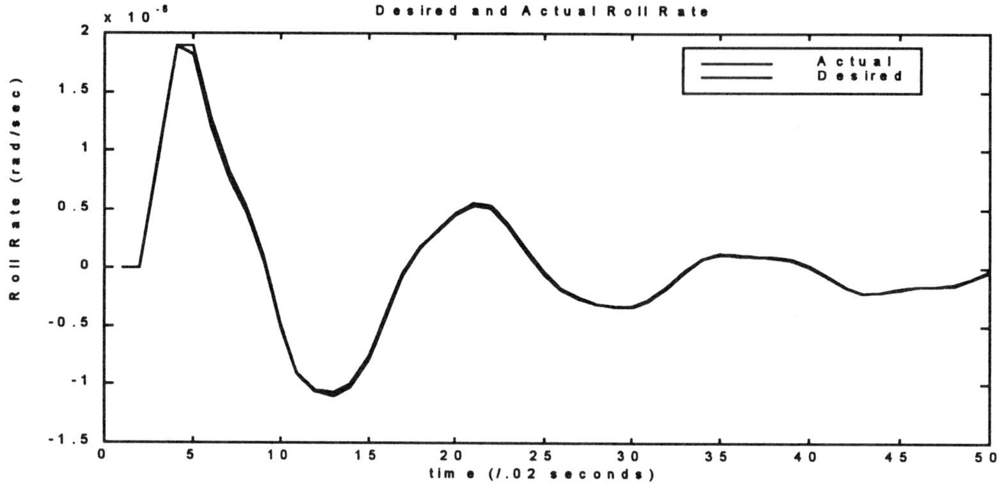

Figure 7. Desired and Actual Roll Rate NA-Based Guidance, Nonlinear Plant

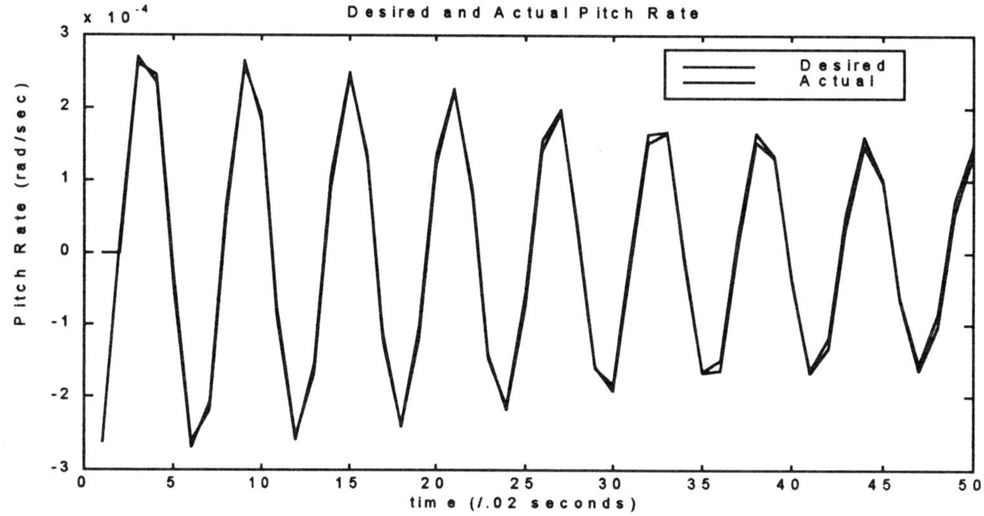

Figure 8. Desired and Actual Pitch Rate NA-Based Guidance, Nonlinear Plant

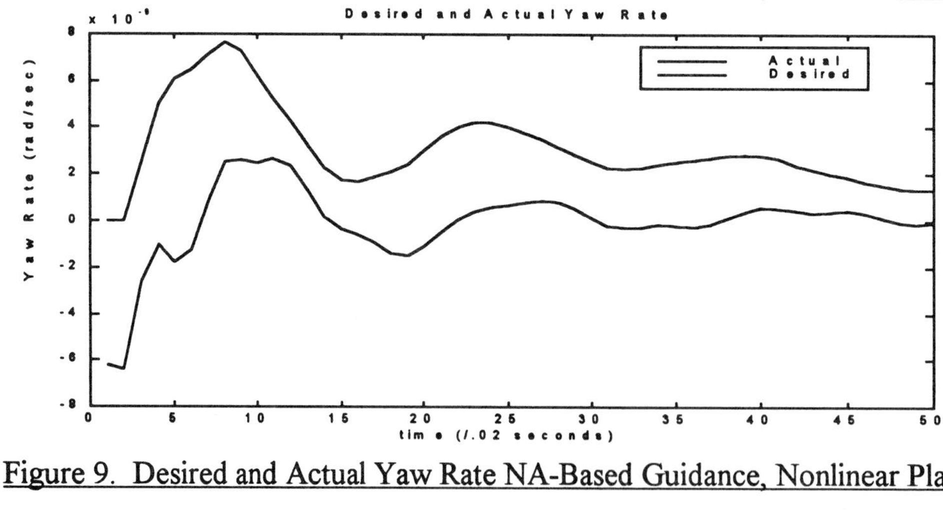

Figure 9. Desired and Actual Yaw Rate NA-Based Guidance, Nonlinear Plant

AIAA-97-3537

ADAPTIVE FLIGHT CONTROL AND ON-LINE LEARNING

Joseph J. Totah
Aerospace Engineer
NASA Ames Research Center
Moffett Field, California

Abstract

The objective of this paper is to present results from the evaluation of a modified direct adaptive tracking flight controller with on-line learning. The control architecture employs both pre-trained and on-line neural networks combined with an aerodynamic coefficient-based adaptive feedback path. The pre-trained neural network is used to represent the nominal, non-linear aircraft dynamics in the model inversion portion of the controller. The on-line neural network is used to learn, in real-time, the aircraft dynamics in- flight even if those dynamics change in-flight, as would be the case for damaged aircraft. Finally, the coefficient-based adaptive feedback path provides a certain amount of controller adaptivity to any changes in the aircraft dynamics. The aircraft model used for this evaluation is representative of the F-15 Advanced Control Technology for Integrated VEhicles (ACTIVE) aircraft. Three cases were evaluated: (1) nominal conditions, (2) loss of control power, and (3) loss of control power in the presence of atmospheric turbulence. The results were compared with the existing F-15 ACTIVE conventional mode controller in all cases. The modified direct adaptive tracking flight controller exhibited extremely desirable airframe stabilization characteristics for case (1) that do not degrade significantly for case (2) or (3) as does the conventional mode controller. It was concluded that this controller exhibits both stable and robust adaptive characteristics when subjected to mild and extreme loss of control power conditions, and in the presence of heavy atmospheric turbulence. The on-line neural network also exhibited extremely desireable characteristics that allowed operation within the framework of a real-time simulation environment. These results are considered a preliminary but necessary first step to validating the technology through proof-of-concept flight tests in the F-15 ACTIVE aircraft.

Senior Member

Introduction

In 1994 NASA awarded a joint NASA/McDonnell Douglas Aerospace (MDA) contract as an Advanced Concepts Program Research Project, entitled Intelligent Aircraft Control System. The objective of this project is to demonstrate a flight control concept that can learn aircraft stability and control characteristics using neural networks, and then utilize this information to optimize aircraft performance in nominal, off-nominal, and simulated failure conditions. The F-15 ACTIVE aircraft shown in figure 1 will be used for actual flight demonstration at NASA Dryden Flight Research Facility throughout this project.

The project is divided into three phases. Phase I focuses on the development and flight evaluation of pre-trained neural networks for mapping the very large aerodynamic stability and control derivative database that currently exists for the F-15 ACTIVE aircraft's six-degree-of-freedom simulation. Phase II employs advanced self-learning, or on-line neural network learning methods to train on estimates of the total aerodynamic coefficients in nominal, off-nominal, and simulated failure conditions. This will culminate in a Phase III flight demonstration of an adaptive flight controller that will employ on both the pre-trained and on-line neural networks to provide the best available estimate of the aircraft dynamics. The approach described above is similar to other efforts[1,2,3,4], with the following two exceptions: (1) this effort will include actual flight testing throughout a large portion of the envelope, and (2) the technology is also designed for rapid prototyping of new aircraft control systems.

The direct adaptive tracking control architecture developed by Kim and Calise[1] was implemented by NASA to study the performance of both pre-trained and on-line neural network learning methods in a real-time simulation environment. However, several modifications were made to this control architecture, namely: (1) the pre-trained radial-basis networks were replaced by a Levenberg-Marquardt (LM) network developed at NASA Ames[5]; (2) the model inversion portion of the controller was changed from a coefficient form to a derivative form; (3) controls were blended to enable all effectors; (4) the adaptive portion of the architecture was modified to regulate adaptivity using total moment coefficient error; and (5) the on-line radial basis networks were replaced by a Dynamic Cell Structure (DCS) network developed at NASA Ames[6].

The results presented in this paper evaluate the performance of the modified direct adaptive tracking controller against the performance of the existing conventional mode controller currently on the F-15 ACTIVE aircraft. DCS's on-line learning capability was also evaluated. Three cases were considered: (1) nominal conditions, (2) loss of control power, and (3) loss of control power in the presence of atmospheric turbulence. The results are considered a preliminary step to evaluating pre-trained and on-line network performance in a high fidelity piloted simulation environment prior to qualifying them for flight.

F-15 ACTIVE Simulation

An F-15 ACTIVE simulation was developed at NASA Ames, and it is modeled after configuration G of the U.S. Air Force's Short takeoff and Landing Maneuver Technology Demonstrator (S/MTD) program.[7] The model assumes no external loads present, gear/flaps retracted, and rotating vanes disabled (no thrust vectoring capability). The conventional mode controller was modeled in addition to the modified direct adaptive tracking controller to drive the ailerons, canards (symmetric and differential), stabilator (symmetric and differential), and rudder. Sensor and actuator dynamics were modeled for both controllers. A first order throttle/propulsion system was developed for both manual and auto throttle control, however they are not exact representations of the actual F-15 ACTIVE throttle logic and powerplant dynamics.

The equations of motion are based on perturbation theory, however most of the non-linear terms were retained in the gravitational and inertial portions of the translational axes. All of the non-linear terms were retained in the inertial portions of the rotational axes. A block diagram of the simulation is shown in figure 2.

Environmental conditions were also modeled to account for pressure, density, temperature, and speed-of-sound variations with altitude. This allowed for scheduling the aerodynamic derivatives, conventional control system gains, and nominal operating conditions throughout the flight envelope. The Dryden turbulence model was also implemented to provide an option for simulating light, medium, or heavy turbulence levels at altitudes below 10,000 feet.

The simulation was integrated with a high-fidelity graphics database, a simulator pod with simple visual and auditory feedback, a three-axis hand controller, throttle lever, and push-button switches to provide an excellent, low-cost fixed-base simulator for pilot-in-the-loop simulation, shown in figure 3.

Model validation was obtained by matching perturbed state time histories at various flight conditions to those generated by the full, non-linear, six-degree-of-freedom simulation conducted by the Air Force during the S/MTD program.

Direct Adaptive Tracking Controller

Controller

The design philosophy used to develop the direct adaptive tracking control architecture is described in detail by Kim.[8] Essentially, the controller is comprised of a command augmentation system (CAS) that has both an attitude orientation system and command augmentation logic. The attitude orientation system is an airframe stabilization system that accepts rate commands. This serves the function of stabilizing the airframe while following the commanded rates. The command augmentation logic is an outer loop function to track the pilot commands. The pilot commands normal acceleration with longitudinal stick, lateral acceleration with pedals, and roll rate with lateral stick. This design allows for normal acceleration and roll rate command tracking with automatic turn coordination.

The output of the attitude orientation system (U_{1pd}, U_{2pd}, and U_{3pd}) is summed with the adaptive portion of the controller (U_{1ad}, U_{2ad}, and U_{3ad}) to generate U_1, U_2, and U_3, which are then transformed into commanded aircraft body axis rotational accelerations. These accelerations are the inputs to the inverted model for calculating control surface deflections.

The direct adaptive tracking control architecture developed by Kim and Calise was implemented by NASA with the following modifications: (1) the pre-trained radial-basis networks were replaced by a LM network; (2) the model inversion portion of the controller was changed from a coefficient form to a derivative form; (3) controls were blended to enable all effectors; (4) the adaptive portion of the architecture was modified to regulate adaptivity using total moment coefficient error; and (5) the on-line radial basis networks were replaced by a DCS network. These modifications were all designed to enhance controller performance and improve neural network accuracies for further examination within the framework of a real-time simulation environment.

A flowchart for the controller is shown in figure 4. T1, T2, T3, and T4 defined in reference 5 are valid for all five modifications, with the following two exceptions. First, a typographical error was made in T3 of reference 5 (equation A.19 for $\dot{P}_c$). The correct equation is:

$$\dot{P}_c = U_1 - U_3 \sin\Theta - \dot{\Psi}\dot{\Theta}\cos\Theta \quad (1)$$

Second, modifications (4) and (5) described below require a revision of the control laws that comprise T2, which contain the adaptive components U_{1ad}, U_{2ad}, and U_{3ad}.

Modifications (1) and (2)

The method developed by Kim and Calise to perform the inversion in T4 was modified such that a single LM network was used to map stability derivatives rather than using radial basis networks to map total aerodynamic coefficients. The stability derivative mapping was so accurate that the LM network was essentially an exact representation of the aircraft aerodynamics, and thus the T4 inversion was considered to be an exact inversion.

Modification (3)

Additional modifications were made to T4 to account for symmetric canard position, which is scheduled with angle-of-attack (the same schedule as that of the conventional mode controller) and ailerons and differential canard position, which are blended as:

$$\delta_a = 2\delta_t \quad (2)$$

$$\delta_{dc} = -\delta_r \quad (3)$$

Modifications (4) and (5)

The adaptive portion of the control architecture was modified to regulate adaptivity using total moment coefficient error. The error is multiplied by euler accelerations derived from euler angle sensor data using simple numerical differentiation. The on-line radial basis networks used in the original design were replaced by a single DCS network developed at NASA-Ames.[6] The approach was also changed to train on derived total moment coefficients. New equations comprising T2 are given below detailing the coefficient-based adaptive feedback path. The aerodynamic stability derivatives denoted with an overstrike, such as $\overline{C}_{m_\alpha}$, signify that they are LM network estimates. Both modifications (4) and (5) are illustrated in figure 5.

$$U_1 = U_{1pd} + U_{1ad} \quad (4)$$

$$U_2 = U_{2pd} + U_{2ad} \quad (5)$$

$$U_3 = U_{3pd} + U_{3ad} \quad (6)$$

$$U_{1ad} = \left(C_l\big|_{derived} - C_l\big|_{LM} \right)\ddot{\Phi} \quad (7)$$

$$U_{2ad} = \left(C_m\big|_{derived} - C_m\big|_{LM} \right)\ddot{\Theta} \quad (8)$$

$$U_{3ad} = \left(C_n\big|_{derived} - C_n\big|_{LM} \right)\ddot{\Psi} \quad (9)$$

$$C_l\big|_{LM} = \overline{C}_{l_\beta}\beta + \frac{b}{2(U_0+u)}\left(\overline{C}_{l_p}p + \overline{C}_{l_r}r\right) + \overline{C}_{l_{\delta_a}}\delta_a + \overline{C}_{l_{\delta_t}}\delta_t + \overline{C}_{l_{\delta_{dc}}}\delta_{dc} + \overline{C}_{l_{\delta_r}}\delta_r \quad (10)$$

$$C_m\big|_{LM} = \overline{C}_{m_u}\frac{u}{(U_0+u)} + \overline{C}_{m_\alpha}\alpha + \frac{\overline{c}}{2(U_0+u)}\overline{C}_{m_q}q + \overline{C}_{m_{\delta_s}}\delta_s + \overline{C}_{m_{\delta_c}}\delta_c \quad (11)$$

$$C_n\big|_{LM} = \overline{C}_{n_\beta}\beta + \frac{b}{2(U_0+u)}\left(\overline{C}_{n_p}p + \overline{C}_{n_r}r\right) + \overline{C}_{n_{\delta_a}}\delta_a + \overline{C}_{n_{\delta_t}}\delta_t + \overline{C}_{n_{\delta_{dc}}}\delta_{dc} + \overline{C}_{n_{\delta_r}}\delta_r \quad (12)$$

$$C_l\big|_{derived} = \frac{1}{\overline{q}Sb}\begin{bmatrix} \dot{p}I_x - \dot{r}I_{xz} + \\ qr(I_z - I_y) - pqI_{xz} \end{bmatrix} \quad (13)$$

$$C_m\big|_{derived} = \frac{1}{\overline{q}S\overline{c}}\begin{bmatrix} \dot{q}I_y + rp(I_x - I_z) + \\ I_{xz}(p^2 - r^2) \end{bmatrix} \quad (14)$$

$$C_n\big|_{derived} = \frac{1}{\overline{q}Sb}\begin{bmatrix} \dot{r}I_z - \dot{p}I_{xz} + \\ pq(I_y - I_x) + qrI_{xz} \end{bmatrix} \quad (15)$$

$$U_{1pd} = K_{p\Phi}\left(\frac{\dot{\Phi}_c}{S} - \Phi\right) + K_{d\Phi}\left(\dot{\Phi}_c - \dot{\Phi}\right) \quad (16)$$

$$U_{2pd} = K_{p\Theta}\left(\frac{\dot{\Theta}_c}{S} - \Theta\right) + K_{d\Theta}\left(\dot{\Theta}_c - \dot{\Theta}\right) \quad (17)$$

$$U_{3pd} = K_{p\Psi}\left(\frac{\dot{\Psi}_c}{S} - \Psi\right) + K_{d\Psi}\left(\dot{\Psi}_c - \dot{\Psi}\right) \quad (18)$$

The DCS network architecture has 13 inputs that include aircraft states and control surface positions, and 3 outputs corresponding to derived roll, pitch, and yaw total moment coefficients. The initial network was trained on coefficient data throughout the entire flight envelope. Normalized coefficients are shown in figures 6-8, and maximum errors were held to less than 8%, as shown in figure 9. The pre-trained DCS network has 86 nodes which are held constant during active learning. Both Mach and altitude are scaled by subtracting the mean value and dividing by the standard deviation of the data range over the entire flight envelope. The DCS network parameters (defined in ref. 6) are as follows:

$$\Gamma\ recall\ metric \quad (19)$$

$$\alpha_{DCS} = 0.1 \quad (20)$$

$$\Theta_{DCS} = 0.8 \quad (21)$$

$$\varepsilon_{bmu} = 0.99 \quad (22)$$

$$\varepsilon_{N(u)} = 0.06\ (0.006\ for\ pre-training) \quad (23)$$

Results

The results presented in this paper evaluate the performance of the modified direct adaptive tracking controller against the performance of the existing conventional mode controller currently on the F-15 ACTIVE aircraft. DCS's on-line learning capability was also evaluated. Three cases were considered: (1) nominal conditions, (2) loss of control power (similar to the conditions considered by Kim and Calise in ref. 1), and (3) loss of control power in the presence of atmospheric turbulence. These results are a preliminary step to evaluating controller and neural network performance in a high fidelity piloted simulation environment prior to qualifying them for flight.

The maneuver chosen to illustrate controller and neural network performance was the same maneuver considered by Kim and Calise for an F-18 aircraft. Starting at Mo=0.6 and ho=10,000 ft, a roll rate command of 10 deg/sec is given until an 80 deg bank angle is obtained. Normal acceleration is maintained at 1.0g during the roll maneuver. This is followed by a 5.0g normal acceleration command that is maintained until the end of the maneuver. The throttle is held at 100%, resulting in a speed of M=1.1 after 30 seconds.

A baseline set of responses are presented for case (1) under nominal conditions. Next, case (2a) is examined for a 30% loss of control power. Case (2b) represents a 75% loss of control power. Finally, case (3) is the same as case (2b) with heavy atmospheric turbulence present.

Case (1)

Figure 10 depicts normal acceleration and roll rate response for the 30 second maneuver for both controllers under nominal conditions in order to establish baseline performance without the introduction of any modified aerodynamics. As shown in figure 10, there are differences between the two controllers. The most noticeable is the overshoot and time constant of normal acceleration for the conventional mode controller. This is due to the presence of a $-\frac{g}{v_t}(1-\cos\theta\cos\phi)$ term added to pitch rate in the longitudinal control law, which tends to improve responsiveness but decrease damping at large bank or pitch angles. The differences in roll rate response between the two controllers are not significant.

As mentioned in the previous section, the DCS network was initially trained using total moment coefficient data over the entire flight envelope. With DCS engaged in the active learning mode, the network learns actual derived total moment coefficient values calculated at each time step throughout the maneuver. Figure 11 is an error histogram that represents the absolute value of the on-line learning errors as a percentage of data range, plotted in ascending order for that maneuver. The performance of DCS in the active learning mode is nearly perfect for all three total moment coefficients.

Case (2a)

Figure 12 depicts controller responses to the same inputs for the same maneuver performed in Case (1). However, modified aerodynamics are now introduced that reduce the effectiveness of all control surfaces by 30%. Essentially, all of the aerodynamic control derivatives used in the equations of motion are multiplied by a factor of 0.7. The conventional controller normal acceleration response has noticeably more overshoot than shown in figure 10, and a roll rate response of approximately 8 deg/sec. On the other hand, the modified direct adaptive tracking controller performs quite well, with no noticeable difference in roll rate and normal acceleration response.

Figure 13 shows very similar error levels to those presented in figure 11. The aircraft adapted extremely well to the 30% loss of control power, which can be attributed to the remarkable performance of the coefficient-based adaptive feedback path described in the previous section. This produces Mach and altitude trajectories nearly identical to Case (1).

Case (2b)

Figure 14 depicts an extreme condition in which modified aerodynamics are now introduced that reduce the effectiveness of all control surfaces by 75%. For this case, all of the aerodynamic control derivatives used in the equations of motion are now multiplied by a factor of 0.25. The conventional controller normal acceleration response shown in figure 14 cannot maintain the commanded value at all, and the roll rate response can only achieve a value of 5 deg/sec. The modified direct adaptive tracking controller, however, again performs quite well. It shows only a slight wrinkle in normal acceleration and a slight overshoot in roll rate just prior to attaining the commanded values.

Figure 15 also shows very similar error levels to those presented in figure 11. Again, the remarkable performance of the coefficient-based adaptive feedback path described in the previous section produces Mach and altitude trajectories nearly identical to Case (1).

Case (3)

Figure 16 depicts the same condition described in Case (2b), except now the maneuver was performed in the presence of heavy atmospheric turbulence. The modified direct adaptive tracking controller performed extremely well in terms of achieving commanded normal acceleration and roll rate values. The conventional mode controller performed poorly, as it did in Case (2b).

Figure 17 shows slightly larger on-line learning errors that reach maximum values of about 3.5% during a very short portion of the maneuver.

Conclusion

The objective of this paper was to present results from the evaluation of a modified direct adaptive tracking flight controller with on-line learning. The control architecture employs both pre-trained and on-line neural networks, in addition to an aerodynamic coefficient-based adaptive feedback path.

The excellent results presented for the modified direct adaptive tracking controller can be attributed to the abundance of control power available with the F-15 ACTIVE aircraft along with the augmentation produced by the adaptive coefficient-based feedback portion of the controller. Maneuvers were performed at loss of control power conditions greater than 75%, and the modified direct adaptive tracking controller performed as poorly as the conventional mode controller because the control surfaces saturated.

The excellent results presented for DCS can be attributed to the design of perfect topology preserving networks described in reference 6. An illustration of this is shown in figure 18. Derived roll moment coefficient is plotted against DCS in the "active learning" mode for Case (3). It can be seen that DCS is tracking rolling moment coefficient quite closely prior to, and after, 9 seconds, which is when the lateral stick step input is removed.

Future plans are to perform similar studies in the presence of the changes to aerodynamic stability (rather than control) derivatives. Piloted simulation evaluations of the modified direct adaptive tracking controller with DCS are also planned prior to installing the networks on the F-15 ACTIVE aircraft flight processor. Ten flight tests are planned in the 3rd quarter of FY98 with the possibility of DCS running in "active learning" mode. Integration and flight testing of all three phases of the F-15 Intelligent Aircraft Control System program are planned to be completed in 1998.

References

[1] Kim, B. S. and Calise, A. J., "Nonlinear Flight Control Using Neural Networks", AIAA Paper 94-3646-CP, 1994.

[2] Chiang, C. and Youssef, H. M., "Neural Network Approach to Aerodynamic Coefficients Estimation and Aircraft Failure Isolation Design, AIAA Paper 94-3599-CP, 1994.

[3] Napolitano, M. R. , Naylor, S., Neppach, C., and Casdorph, V., "On-Line Learning Non-Linear Direct Neuro Controllers for Restructurable Control Systems", AIAA Paper 94-3643-CP, 1994.

[4] Barron, R. L., Cellucci, R. L., Jordan, P. R. III, and Beam, N. E., "Applications of Polynomial Neural Networks to FDIE and Reconfigurable Flight Control", IEEE Paper CH2881-1/90/0000-0507, 1990.

[5] Norgaard, M., Jorgensen, C., and Ross, J., "Neural Network Prediction of New Aircraft Design Coefficients", NASA TM 43188, 1997.

[6] Jorgensen, C. C., "Direct Adaptive Aircraft Control Using Neural Networks", NASA TM 47136, January, 1997.

[7] Zeh, J. M., Young, D. L., and Couture, N. J., "STOL and Maneuver Technology Demonstrator (S/MTD) Stability Derivatives", WL-TR-92-3055, 1992.

[8] Kim, B. S., "Nonlinear Flight Control Using Neural Networks", PhD. Thesis, Georgia Institute of Technology, School of Aerospace Engineering, 1993.

[9] Totah, J., "An Examination of Aircraft Aerodynamic Estimation Using Neural Networks", SAE Technical Paper Series 952036, 1995.

Figure 1. F-15 ACTIVE Aircraft

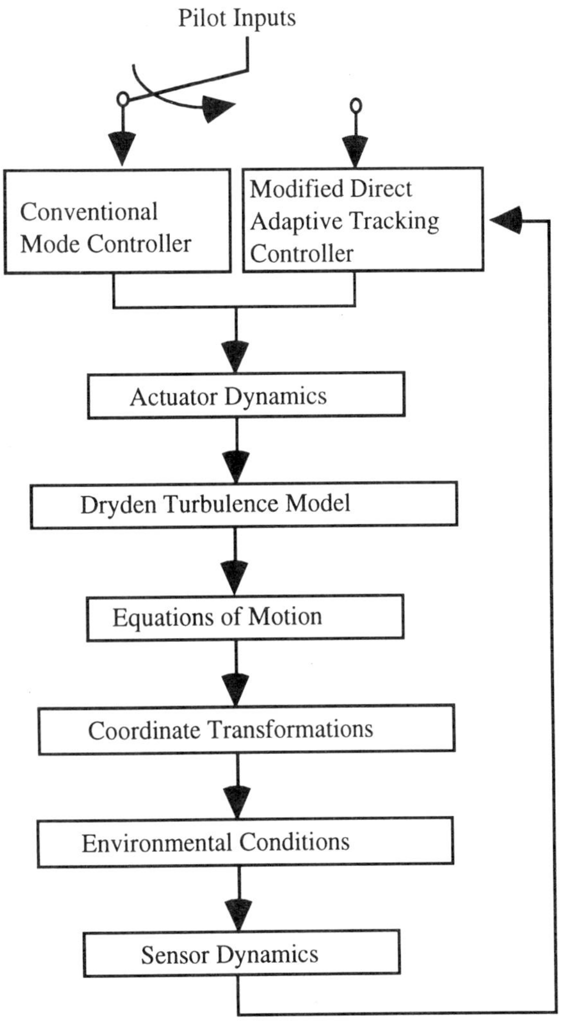

Figure 2. F-15 ACTIVE Simulation

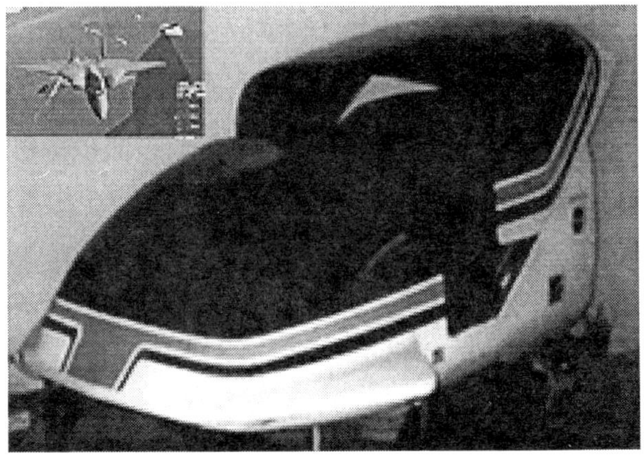

Figure 3. Simulator POD and F-15 ACTIVE Graphics.

$$\begin{bmatrix}\delta_{la}\\ \delta_{ln}\\ \delta_{pd}\end{bmatrix} \xrightarrow{T1} \begin{bmatrix}\dot{\Phi}_c\\ \dot{\Theta}_c\\ \dot{\Psi}_c\end{bmatrix} \xrightarrow{T2} \begin{bmatrix}U_1\\ U_2\\ U_3\end{bmatrix} \xrightarrow{T3} \begin{bmatrix}\dot{P}_c\\ \dot{Q}_c\\ \dot{R}_c\end{bmatrix} \xrightarrow{T4} \begin{bmatrix}\delta_t\\ \delta_s\\ \delta_r\end{bmatrix}$$

Figure 4. Modified Direct Adaptive Tracking Controller.

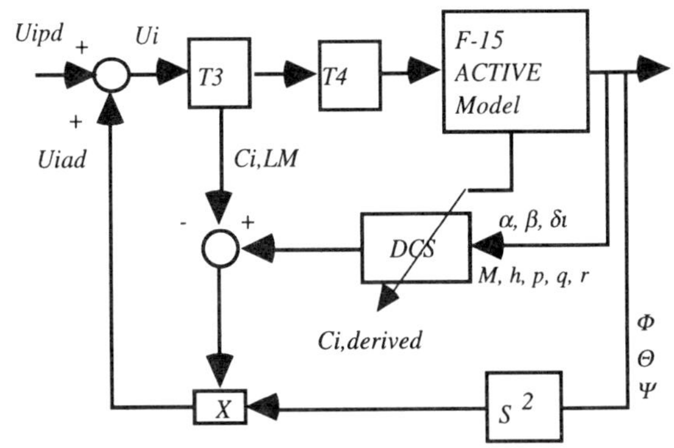

Figure 5. Modified Direct Adaptive Tracking Control Architecture with On-Line Learning.

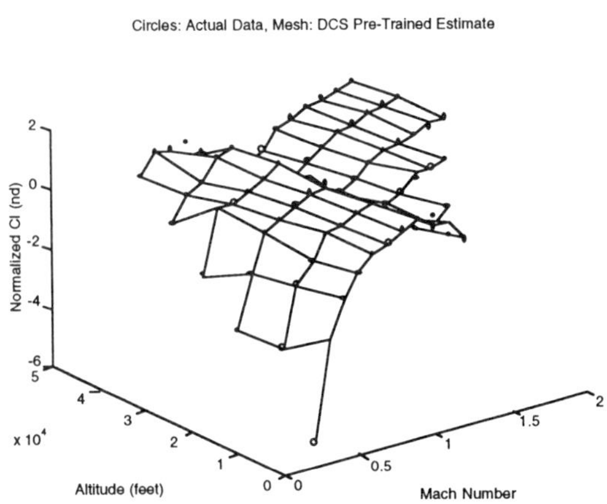

Figure 6. Cl Variation Versus Mach and Altitude.

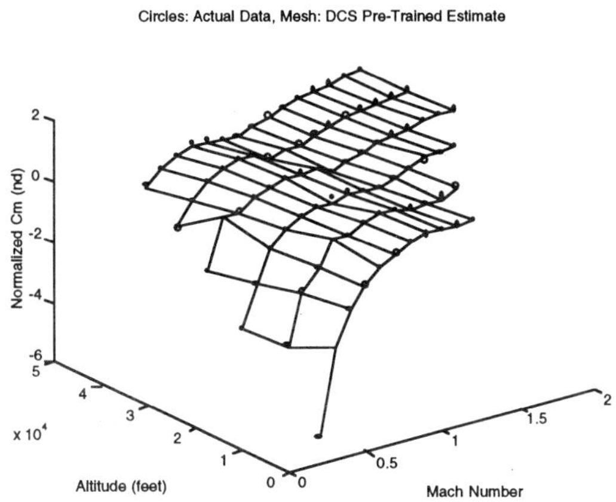

Figure 7. Cm Variation Versus Mach and Altitude.

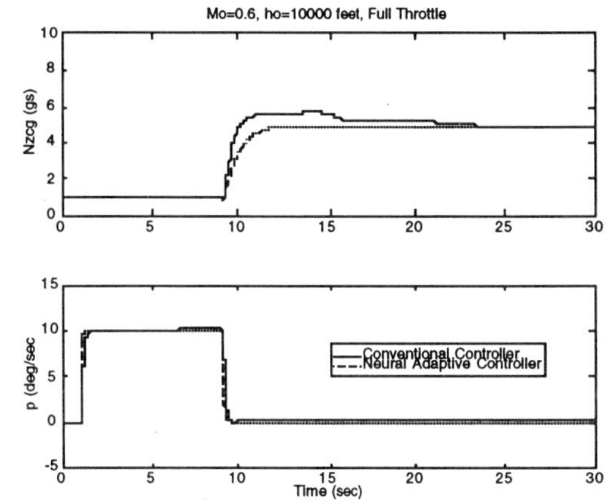

Figure 10. Normal Acceleration and Roll Rate Responses to Stick Inputs; Case (1): Nominal Conditions.

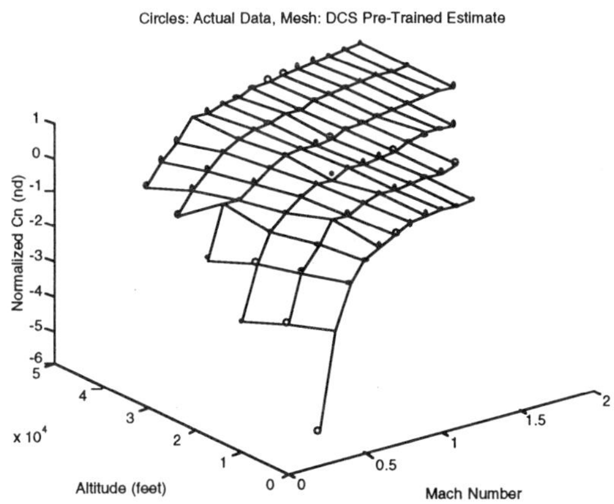

Figure 8. Cn Variation Versus Mach and Altitude.

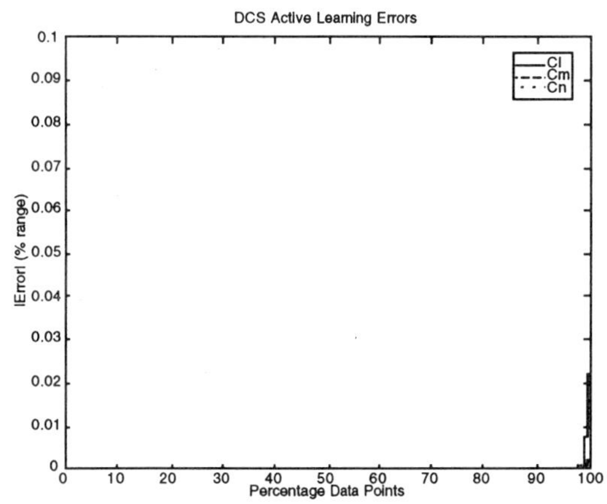

Figure 11. DCS Roll, Pitch, and Yaw Coefficient Error Histogram for Case (1).

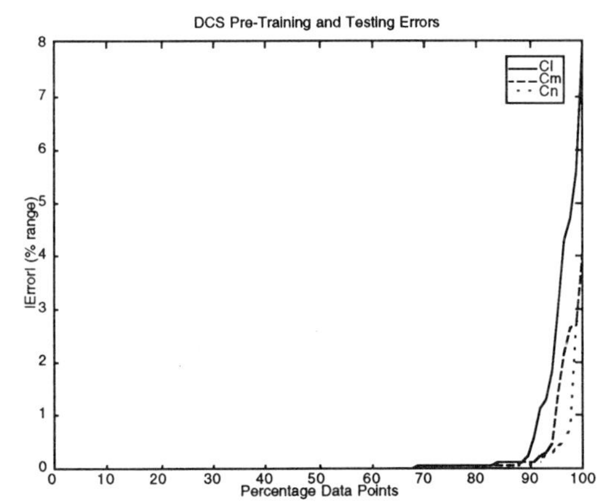

Figure 9. Initial DCS Network Error Histogram.

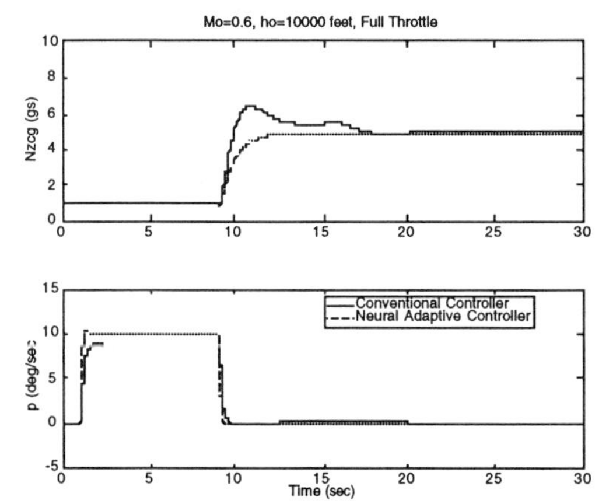

Figure 12. Normal Acceleration and Roll Rate Responses to Stick Inputs; Case (2a): 30% Loss of Effectiveness to all Control Surfaces.

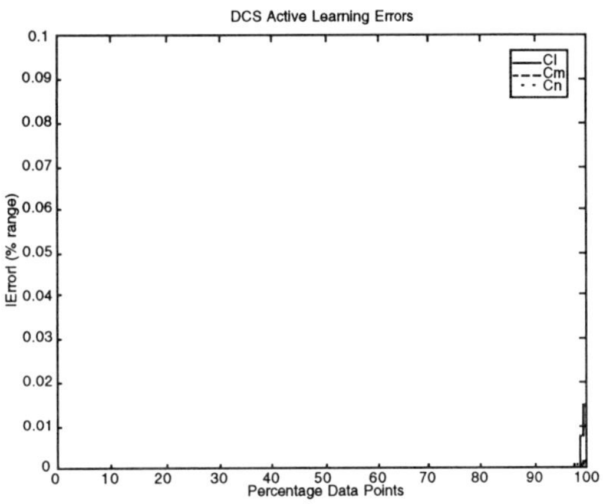

Figure 13. DCS Roll, Pitch, and Yaw Coefficient Error Histogram for Case (2a).

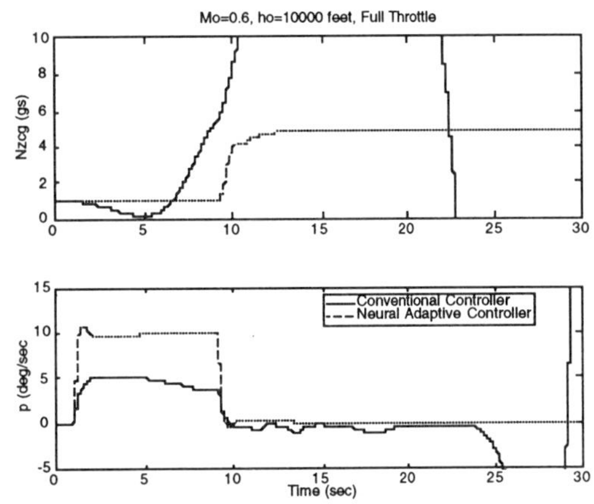

Figure 14. Normal Acceleration and Roll Rate Responses to Stick Inputs; Case (2b): 75% Loss of Effectiveness to all Control Surfaces.

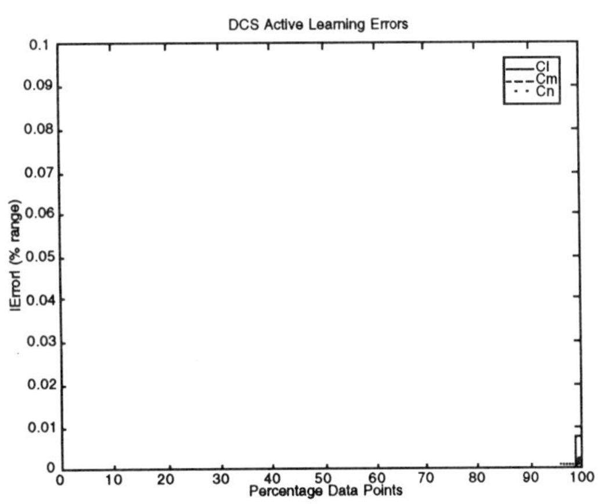

Figure 15. DCS Roll, Pitch, and Yaw Coefficient Error Histogram for Case (2b).

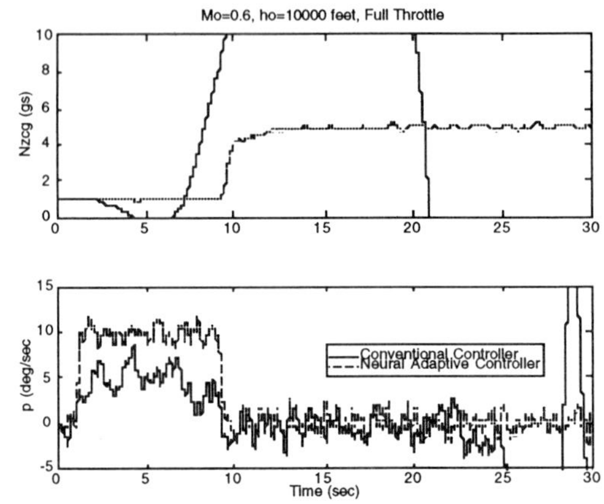

Figure 16. Normal Acceleration and Roll Rate Responses to Stick Inputs; Case (3): 75% Loss of Effectiveness to all Control Surfaces, Heavy Atmospheric Turbulence.

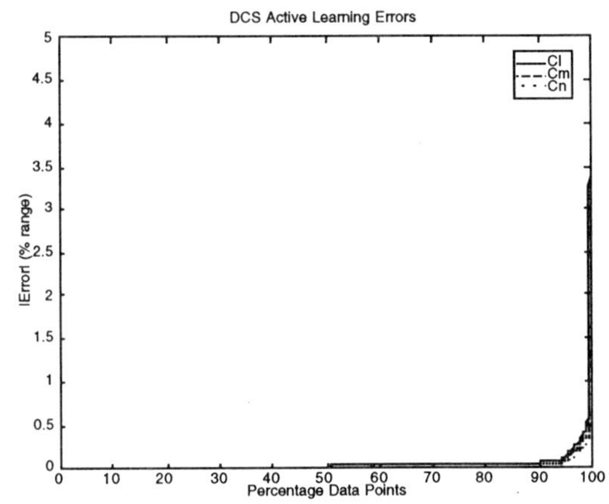

Figure 17. DCS Roll, Pitch, and Yaw Coefficient Error Histogram for Case (3).

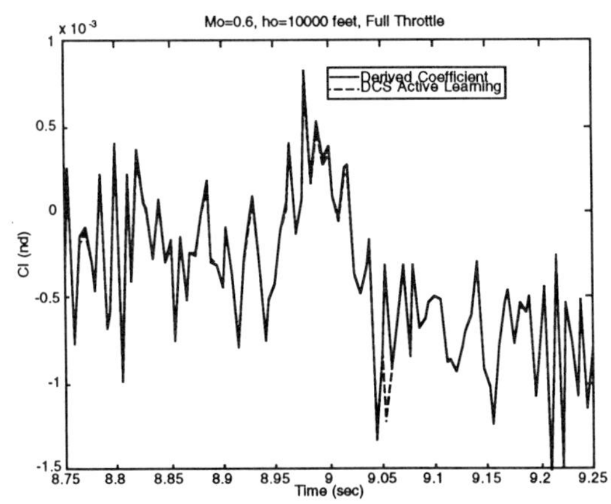

Figure 18. Time History of DCS During Active Learning of Cl for Case (3).

AIAA-97-3538

The Virtual Flight Data Recorder : a Neural Extension of Existing Data Recording Capabilities

Marcello R. Napolitano[*†], Dale A. Windon[*‡], José L. Casanova[*‡], Brad A. Seanor[*‡]
[*]Aviation Operation and Safety Center, Department of Mechanical and Aerospace Engineering
David R. Martinelli[§†]
[§]Staggers Transportation Center, Department of Civil and Environmental Engineering
West Virginia University, Morgantown, WV 26506/6106

Abstract

This paper presents the concept of a Virtual Flight Data Recorder (VFDR) for commercial airliners. Current FAA regulations mandate the recording of 11 dynamic parameters not including the aircraft control surface deflections. The absence of these data can be critical in the event of a crash investigation. This paper proposes the introduction of a scheme based on a Neural Network Simulator (NNS) and a Neural Network Reconstructor (NNR). The NNS is trained, using available flight data for the particular aircraft, to simulate any desired dynamic output recorded in current FDRs and then interfaced with an NNR The outputs of the NNR are the control surface deflections (or any other parameter of interest not recorded by the FDR) which minimize a performance index based on the differences between the available data from the FDR and the output from the NNS. The scheme has been introduced, tested, and validated with flight data from a B737-300 with an FDR with extended recording capabilities showing accurate reconstruction of the control surface deflections time histories. The VFDR can be considered as an alternative or interim solution for a very expensive retrofitting of all the FDRs installed on commercial airliners prior to 1989 as well as a valuable tool for crash investigations where loss of control surface effectiveness is believed to be a factor.

Symbols

a = acceleration, g's
CAS = Calibrated Air Speed, ft/sec
h = altitude, ft
J= cost function
k= discrete time index
L= lower bound of the modified sigmoid function
N_1 = Percentage of maximum fan RPM
p= pattern for the neural network input data
T= slope of the modified sigmoid function
t= time, sec
U= upper bound of the modified sigmoid function
δ = deflection of control surfaces, deg
μ = mean
θ=Euler pitch angle, deg
ϕ=Euler bank angle, deg
σ^2 = variance
ψ=Euler heading angle, deg
A= aileron
c= control surface command
E = elevator
LEF = Leading Edge Flaps
LES = Leading Edge Slats
R= rudder
TEF = Trailing Edge Flaps

[†] Associate Professor, AIAA Member
[‡] Graduate Research Assistant, AIAA Member

Introduction

Since the mid 1980's there has been an increase in Neural Network (NN) research induced by the introduction of the Back-Propagation algorithm for feed-forward NNs with supervised learning [1,2,3]. In more recent years there have been numerous studies focusing on applications of NNs to a variety of flight control problems. Particularly, NNs have been proposed for:

- fault-tolerant control systems (for actuator and sensor failure);
- autopilot and stability augmentation systems;
- aircraft parameter estimation from flight data.

In each of these applications, the proposed use of an NN is that of a controller or a state or parameter estimator. In addition to these applications, the interesting and, at times, unique capabilities of NNs allow for the design of alternative schemes for very specific problems.

In recent years, general and commercial aviation have been plagued by the occurrence of several crashes. The crash investigations conducted by the National Transportation Safety Board (NTSB) typically start with the inspection of the aircraft wreckage along with the analysis of the flight data recorder (FDR) and the cockpit voice recorder (CVR) tracks. In most instances a combination of these elements provides all the necessary clues for a successful investigation. Unfortunately, this is not always the case.

This paper is structured as follows. The next section describes the concept of a Virtual Flight Data Recorder (VFDR). Following sections review key properties of neural networks, describe the Neural Network Simulator (NNS) and Neural Network Reconstructor (NNR) concepts, and present the results of a proof-of-concept study with actual flight data from a B737-300. A final section follows with the conclusions.

The Virtual Flight Data Recorder

Federal Aviation Administration (FAA) rules mandate the aircraft FDR to record a certain number of dynamic time histories. These usually include airspeed, altitude, Euler angles, and linear accelerations. Often, additional time histories, such as those of engine-related parameters, are recorded by the airline for engine maintenance purposes. Presently, however, to the best of the authors knowledge no FAA rule dictates the mandatory recording of the deflections of the primary control surfaces although rule modifications on this matter are being considered.

Ref.[4] by the first and second authors proposes a NN-based reconstruction of the control surface deflections. The approach assumes the availability of a non-linear simulation code for the considered aircraft. This paper proposes a modification to that approach by removing a major drawback consisting, namely, in the assumed availability of a non-linear simulation code. The parallel non-linear simulation is replaced by a neural network simulator (NNS). For each aircraft type the NNS is trained off-line with a complete set of actual flight data from a single FDR with extended recording capabilities. Different NNs should be trained with flight data at each representative flight condition within the flight envelope (for example : take-off, climb, low-cruise, high-cruise, descent, approach, landing).

Once trained to a desirable level of fidelity, the NNS is interfaced with the neural network reconstructor (NNR)[4]. The NNR through an iterative minimization scheme calculates the control surface deflections (or any other desired dynamic parameter not recorded by the FDR for which a set of training data exists) which minimizes a performance index at any given discrete time step. The performance index is based on the weighted differences between the FDR and the NNS data. The combination of the NNS and the NNR leads to the concept of a Virtual Flight Data Recorder (VFDR). Figure 1 shows a block diagram of the VFDR scheme.

The VFDR concept may have interesting economic implications. The FAA is currently considering mandating that US airlines replace old FDRs with limited recording capabilities (which were installed on all aircraft built prior to 1989) with new FDRs with extended recording capabilities. The total cost of this retrofitting can be substantial for each major US airline. These costs include the direct cost for the new FDRs, their installation as well as the losses in revenue associated with the aircraft ground time. As of August 1996, 5,979 commercial aircraft are supposed to be retrofitted for a total cost estimated to be in the range of approximately $1.5 B.

The VFDR approach assumes the availability of a single aircraft equipped with an FDR with extended recording capabilities for each aircraft type operated by the airlines. The complete set of flight data for this FDR can then be used to train and regularly update the NNS to be interfaced with the NNR. The most immediate application for the VFDR is, of course, for use by the NTSB in the event of a crash. Other additional uses may include dynamic reconstruction of aircraft time histories in the event of "non-catastrophic, uncommanded" maneuvers as well as updating and improving the fidelity of existing flight simulators for pilot training.

Basic Principles of Neural Network Theory

A NN is an approximated mathematical model of the human brain's problem solving activities. A complete review of the NN theoretical principles beyond the purpose of this paper and is available in Refs.[1,2]. A feed-forward NN with supervised learning is utilized. For the purpose of this study, two of the key NN properties of interest are the following:

- applicability to non-linear systems due to their mapping capabilities. Ref.[5] shows that a feed-forward NN with at least one hidden layer is capable of approximating any non-linear function once sufficient training is provided.
- applicability to multi-variable systems. NNs are, by definition, multi-input, multi-output entities and this, naturally, leads to their application to multi-variable systems.

The off-line learning implemented to achieve the mapping between input and output data is performed through the application of the Extend Back-Propagation algorithm (EBPA)[6].

A Neural Network Simulator

The VFDR concept assumes the availability of a single complete set of flight data for the considered aircraft model obtained from a FDR with extended recording capabilities. The aircraft considered for this study is a Boeing 737-300. A complete set of dynamic data was provided by the NTSB for the purpose of developing an accurate NN simulator. The 14,000 sec. dynamic time histories received from the NTSB from an extended FDR equipped B737-300 consisted of the following parameters:

$$\phi,\ \theta,\ \Psi,\ a_n,\ a_x,\ a_y,\ CAS,$$
$$\text{Altitude},\ N_1,\ \delta_E,\ \delta_A,\ \delta_R,\ \delta_{TEF},\ \delta_{LEF},\ \delta_{LES}$$

Figure 2 shows an altitude and airspeed profile of the received flight data. These plots show the entire 14,000 sec. flight and the selection of the flight regions that were used for training and testing of the neural schemes.

Recall that the FDRs with which most of older commercial airliners are equipped only record a subset of these parameters. For the purpose of the reconstruction study, it was assumed that the following parameters are recorded by a typical FDR:

$$\phi,\ \theta,\ \phi,\ a_n,\ a_x,\ a_y,\ CAS,\ \text{Altitude}$$

The Boeing 737-300 features on each wing 2 trailing edge flaps, 2 leading edge flaps, and 2 leading edge slats. The deflections of the leading edge surfaces are scheduled with the deflection of the trailing edge flaps and the trailing edge flaps are scheduled as a function of airspeed. The aircraft also features flight spoilers and ground spoilers which were not activated during these flight tests. It should be said that these data were ideal for the purpose of training the NNS since they showed a comprehensive set of maneuvers.

The sampling frequencies varied among the different parameters from 1 samples/sec for N_1 to 8 samples/sec for the linear accelerations and the Euler angles. Therefore, prior to the NN training, the data underwent a cubic interpolation in order to have the same sampling frequency of 8 samples/sec for each time history. A 5th order digital Butterworth filter was then applied to the interpolated data to reduce the noise level. The cut-off frequencies were carefully selected to avoid removing significant portions of the aircraft rigid body dynamics; cut-off frequencies of 5 rad/sec and 3 rad/sec were used for the longitudinal and lateral-directional time histories respectively. Samples of raw FDR data and the equivalent smoothed and filtered data are shown in Figure 3.

Once the training data sets were smoothed and filtered, the next issue was the selection of the architecture for the NNS. As it is well known in the technical literature, there are several degrees of freedom in the selection of an NN architecture: number of hidden layers, number of processing elements (neurons) for each hidden layer, learning rates and momentum coefficients, selection of input data at the input layer, and the selection of the performance index at the output layer. The conclusions in Refs.[7,8] provided general guidelines for selecting the NNS architecture. Therefore, architectures with a single hidden layer, with a number of hidden neurons slightly higher than the number of input data, with low learning rates, were considered.

Different training sessions were conducted for different NNSs with different input and output combinations. The configurations that have been tested are listed in Table 1. The results presented in the paper are relative to Configuration 1. The set of data covered 14,000 sec. of flight. Excluding the take-off, climbing, descent, and landing phases, there were approximately 10,500 sec. of maneuvered flight. The training for the NNS was conducted using 7,500 sec. of data with the remaining 3,000 sec. being used for the testing of the NNS. The training consisted of submitting the 7,500 sec. training data to the NNS for a total of 250 iterations. The training was monitored by "freezing" the numerical architecture of the NNS every 10 iterations and by letting the "frozen" NNS go through the first 1,000 sec. of testing data. The following classical statistical parameters for the estimation error associated with the simulation were introduced:

$$\mu_{simul} = \frac{\sum_{i=1}^{N} (X_{simul} - X_{FDR})}{N} \tag{1}$$

$$\sigma^2_{simul} = \frac{\sum_{i=1}^{N}\left[(X_{simul} - X_{FDR}) - \mu_{simul}\right]^2}{N} \qquad \textbf{(2)}$$

where X indicates any of the NNS outputs, and N is the total number of data points for the testing phase (N=1,000 x 8 samples/sec = 8,000). Figures 4 and 5 show the statistical trend of the NNS error for the 250 iterations. Note that the initial learning rate for the NNS was set to 0.01; the rate was decreased to 0.005 after 100 training iterations and then to 0.001 after 150 iterations. An analysis of Figure 4 shows a small non zero asymptotic mean error for the longitudinal and vertical accelerations as well as a non negligible asymptotic bias for the roll Euler angle. A close look at Figure 5 reveals instead a desirable decreasing trend for the estimation error variances; however the presence of a persistent asymptotic variance for the roll angle estimation error was of some concern since it can lead to substantial error in the reconstruction of the rolling dynamics.

A careful analysis of the FDR ailerons data revealed some difference between the magnitudes of the left and right aileron recordings. Furthermore this difference was not consistent throughout the duration of the flight. In fact it was ranging between (1-4) degrees; therefore it could not be treated as a removable bias. Figure 6 shows a comparison of the raw left and right aileron deflections along with the magnitude of their difference for generic 500 sec. of the training data. Most likely this difference was due to a minor calibration problem for the right and left aileron channels of the data acquisition system. The raw data for the deflections of left and right elevators and spoilers were also checked for a similar problem showing no significant inconsistency.

While the occurrence of this type of problems may not be uncommon it is significant that the neural scheme in the NNS was able to detect this anomaly in the rolling dynamics as shown by the statistics of the estimation error for the bank angle in Figures 4 and 5.

It should be emphasized that the availability of an accurate simulation scheme (NNS) in the overall VFDR is only a necessary but not sufficient condition for an accurate reconstruction. Figures 7,8 provide a comparison between the NNS results and the actual data from the extended FDR for the first 500 sec. of the 3,000 sec testing phase. The figures show that the NN simulation is quite accurate despite the presence of small biases for each of the linear accelerations and Euler angles; some of these biases can not be noticed due to the scaling of the plots. However, based on the analysis of the statistics of the estimation error in Figures 4 and 5, it was believed that the neural simulation of the B737 dynamics was reliable enough to attempt the interface with the reconstruction scheme.

A Neural Network Reconstructor

For the purpose of reconstructing the deflections of the control surfaces, a multi-input, multi-output (MIMO) neural network reconstructor (NNR) was introduced. The inputs to the NNR are given by the NNS output data or by a subset of these data. The selected architecture of the NNR is shown in Table 2. The NNR outputs are the estimates of the deflections of the control surfaces at time instant 'k' which minimize the differences between the NNS time histories and those from the FDR at time instant 'k+1'. The process is iterative and the reconstruction scheme proceeds to the next time step only when the deflections of the control surfaces at time 'k', minimizing the FDR and NNS time histories at time 'k+1', are found by the NNR. For this initial study the attention was focused on the reconstruction of the lateral-directional control surfaces. Several configurations for the cost functions to be minimized for each of the output nodes at each iteration of the neural computation for each time step of the simulation were considered. The results presented below are relative to the following cost functions:

$$J_{lat}(k) = a_1\,(\phi_{FDR}(k+1) - \phi_{NNS}(k+1) \qquad \textbf{(3)}$$

$$J_{dir}(k) = b_1\,(\Delta\psi_{FDR}(k+1) - \Delta\psi_{NNS}(k+1)) \qquad \textbf{(4)}$$

where a_1=-10, b_1=10, and $\Delta\psi(k+1)=\psi(k+1)-\psi(k)$. The following statistical parameters for the reconstruction error were introduced:

$$\mu_{recon} = \frac{\sum_{i=1}^{N}(\delta_{x_{NNR}} - \delta_{x_{FDR}})}{N} \qquad \textbf{(5)}$$

$$\sigma^2_{recon} = \frac{\sum_{i=1}^{N}\left[(\delta_{x_{NNR}} - \delta_{x_{FDR}}) - \mu_{recon}\right]^2}{N} \qquad \textbf{(6)}$$

where x = A (ailerons), or R (rudder), and N is, again, the total number of discrete data points. Of course, in an application of the method to operating commercial aircraft, the FDR control surface deflections are not available and are the objectives of the reconstruction. Figures 9 and 10 show the reconstruction for the aileron and rudder deflection time histories during 800 sec. of the testing data. It can be seen that the reconstructed deflections are able to follow the actual deflections of the control surfaces within some approximation. However the error in the reconstruction of the aileron deflections should not be surprising due to the already discussed presence of a non negligible difference between the left and right aileron readings.

Overall, it can be said that the neural scheme recognizes substantial lateral-directional control surface deflections which may lead to an uncommanded maneuver or to a crash. Such analysis can, perhaps, provide a first immediate and inexpensive step in a crash investigation.

Conclusions

This paper has presented an approach for the reconstruction of control surface deflections which could be used for aircraft crash investigations since these parameters are not recorded by current FDRs for aircraft built prior to 1989. The methodology takes advantage of the capabilities offered by neural networks for non-linear mapping purposes. Specifically, the approach is based on the development of a neural network reconstructor which provides estimates of the control surface deflections minimizing weighted quadratic differences, at each computational time step, between the actual time histories from an FDR and the simulated time histories from a parallel non-linear simulation, also neural-network based. The method does not assume the availability of a simulation code for the considered aircraft. The process is iterative in that only when the weighted differences are minimized at time step 'k' the process proceeds to time step 'k+1'. This approach has been tested with actual B737-300 flight data provided by the NTSB. Its application could provide future investigators with a reconstruction of the control surface deflections which could furnish appropriate clues to focus only on a limited set of probable causes of a crash rather than conducting broad range, lengthy, and expensive investigations.

Acknowledgments

The authors wish to thank the NTSB for the B737-300 data provided for this study.

References

1 - Rumelhart, D. and McClelland, J. "*Parallel Distributed Processing*", MIT Press, Cambridge, Ma, 1986

2 - Narendra, K.S., Partasarathy, K. "Identification and Control of Dynamical Systems Using Neural Networks", IEEE Transactions on Neural Networks, Vol.1, No.1, 1990

3 - Hunt, K.J., Sbarbaro, D., Zbikowski, R., Gawthrop, P.J. "Neural Networks for Control Systems - A Survey", Automatica, Vol. 28, No. 6, pp.1083-1112, 1992

4 - Napolitano, M.R., Windon, D.A., "Neural Networks-Based Reconstruction of Flight Data for Aircraft Crash Investigations", Proceedings of the International Conference on Neural Networks, Washington, D.C., June 1996

5 - Hornik, K., Stinchcomb, M., White, H. "Multi-layer Feed-forward Networks Are Universal Approximators", Neural Networks, Vol. 2, pp. 359-366, 1989

6 - Napolitano, M.R., Chen C.I., Naylor, S. "Aircraft Failure Detection and Identification Using Neural Networks", AIAA Journal of Guidance, Control and Dynamics, Vol. 16, Number 6, Nov-Dec. 1993, pp.999-1009

7 - Neppach, C., "Application of Neural Networks in Sensor Failure Detection, Identification and Accommodation in a System Without Sensor Redundancy", MS Thesis, School of Mechanical and Aerospace Engineering, West Virginia University, WV, August 1994

8 - De Villiers, J., Barnard, E. "Back-Propagation Neural Nets with One and Two Hidden Layers", IEEE Transactions on Neural Networks, February 1993, Volume 4, pp. 136

Table I - Architectures of the NNS

Configuration #1	
Input Parameters :	FDR data for $\theta,\Phi,\Delta\Psi,a_x,a_n,a_y,\delta_E,\delta_A,\delta_R,\delta_{TEF},N_1$ at k-1,.., k-p
Data pattern (p) :	3
Total number of inputs :	33
Hidden layer neurons :	45
Outputs :	Simulation of $\theta,\Phi,\Delta\Psi,a_x,a_n,a_y$ at current instant k
Learning rates (training):	0.01 - 0.005
Parameters to be updated (PTBU):	1850 at each computational step
Configuration #2	
Input Parameters :	FDR data for $\theta,\Phi,\Delta\Psi,a_x,a_n,a_y,\delta_E,\delta_A,\delta_R,\delta_{TEF},N_1$ at k-1,.., k-p
Data pattern (p) :	5
Total number of inputs :	55
Hidden layer neurons :	60
Outputs :	Simulation of $\theta,\Phi,\Delta\Psi,a_n,a_y$ at current instant k
Learning rates (training):	0.005
Parameters to be updated (PTBU):	3860 at each computational step
Configuration #3	
Input Parameters :	FDR data for $\theta,\Phi,\Delta\Psi,a_x,a_n,\delta_E,\delta_A,\delta_R,\delta_{TEF},N_1$ at k-1,.., k-p
Data pattern (p) :	3
Total number of inputs :	30
Hidden layer neurons :	40
Outputs :	Simulation of $\theta,\Phi,\Delta\Psi,a_x,a_n$ at current instant k
Learning rates (training):	0.005
Parameters to be updated (PTBU):	3020 at each computational step
Configuration #4	
Input Parameters :	FDR data for $\theta,\Phi,\Delta\Psi,a_n,\delta_E,\delta_A,\delta_R,\delta_{TEF},N_1$ at k-1,.., k-p
Data pattern (p) :	5
Total number of inputs :	45
Hidden layer neurons :	50
Outputs :	Simulation of $\theta,\Phi,\Delta\Psi,a_x,a_n,a_y$ at current instant k
Learning rates (training):	0.005
Parameters to be updated (PTBU):	2774 at each computational step

Table II - Architecture of the NNR

Input Parameters :	FDR data θ,Φ,Ψ,a_n,a_y at time instant k-1,.., k-p
Data Pattern (*p*) :	2
Total number of inputs :	10
Hidden layer neurons :	10
Outputs :	reconstruction of δ_A, δ_R at current instant k
Learning rates :	0.5
Parameters to be updated (PTBU):	216 at each computational step

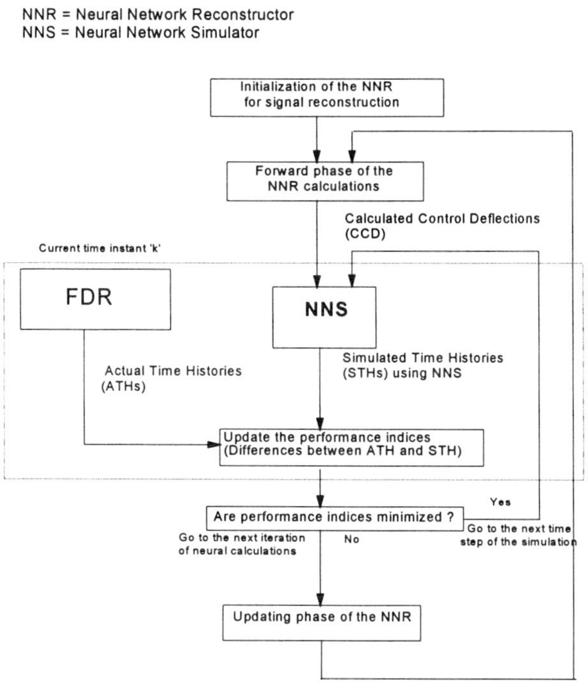

Figure 1: Block diagram of the Virtual Flight Data Recorder (VFDR)

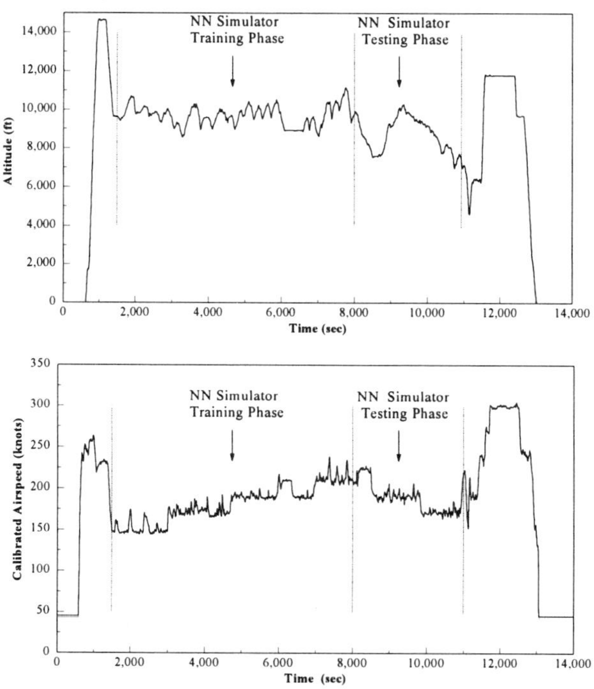

Figure 2: Altitude and airspeed profile of the B737 data

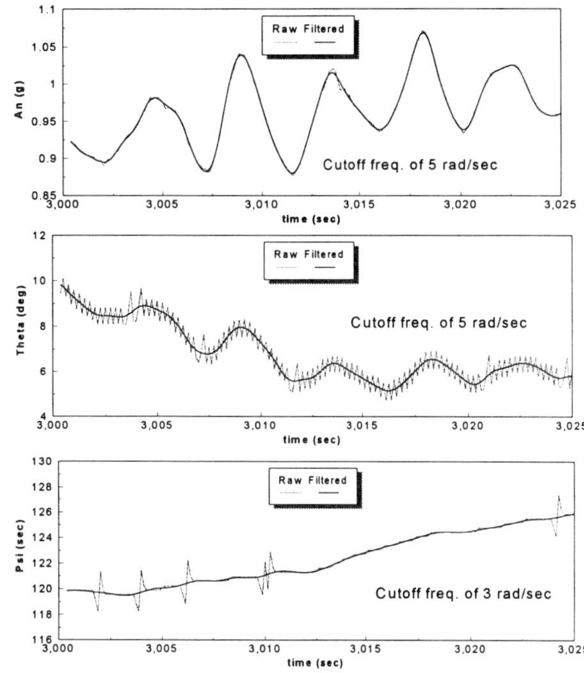

Figure 3: Samples of raw and conditioned B737 data

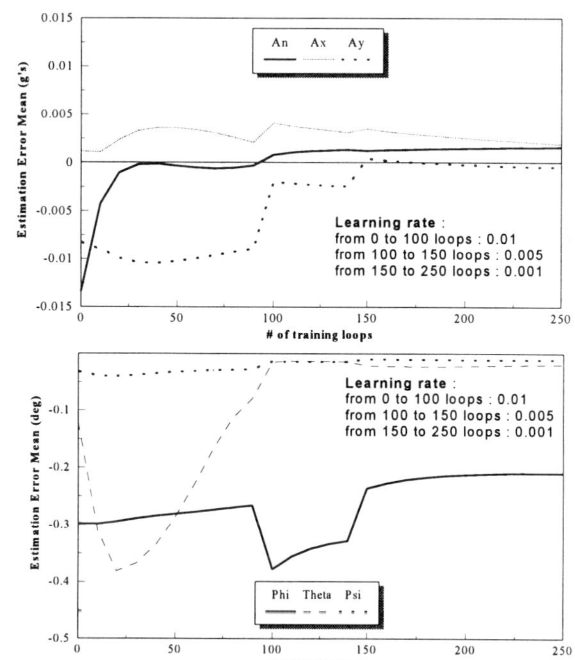

Figure 4: Mean of the NNS simulation error vs. number of training loops for the linear accelerations and the Euler angles

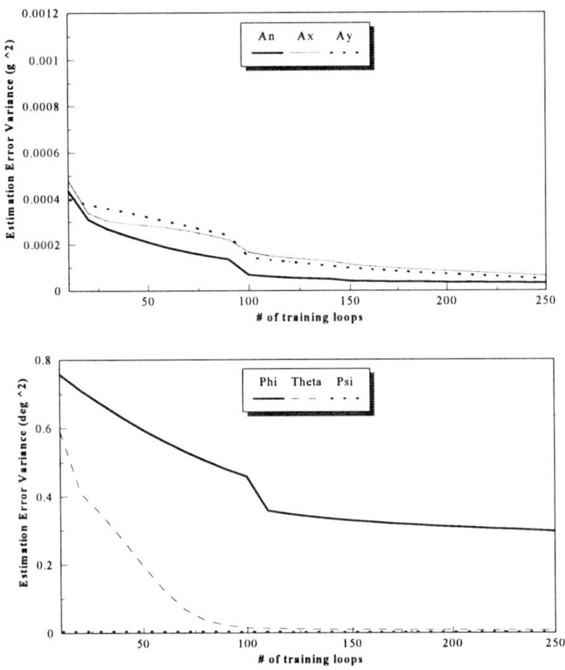

Figure 5: Variance of the NNS simulation error vs. number of training loops for the linear accelerations and the Euler angles

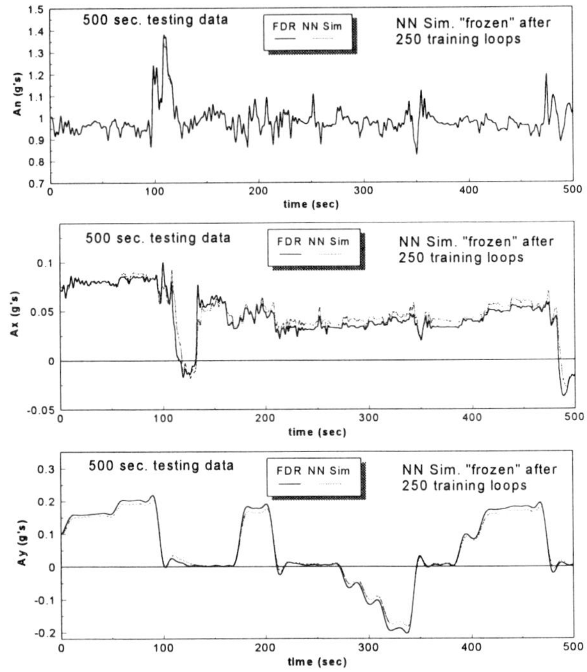

Figure 7: Comparison between NNS and FDR time histories for linear accelerations

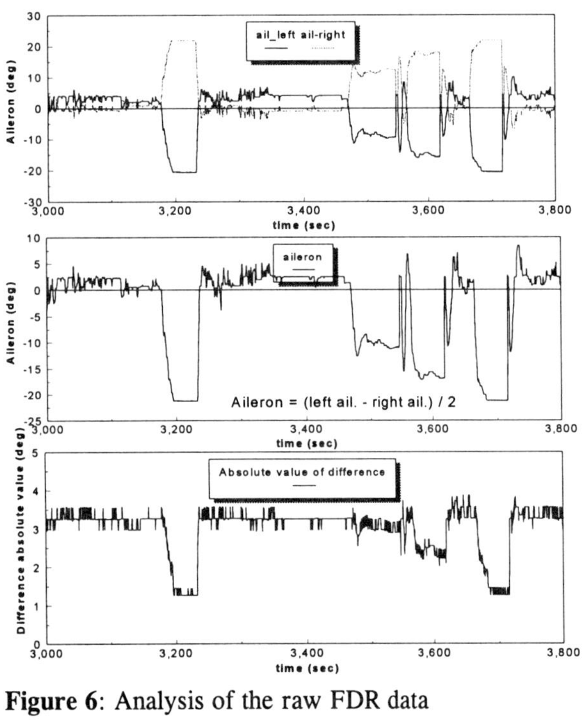

Figure 6: Analysis of the raw FDR data

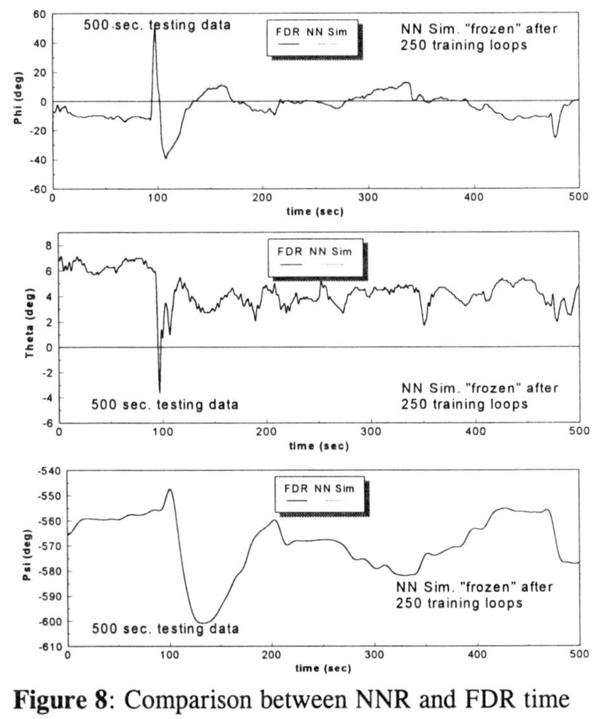

Figure 8: Comparison between NNR and FDR time histories for ailerons and rudder deflections

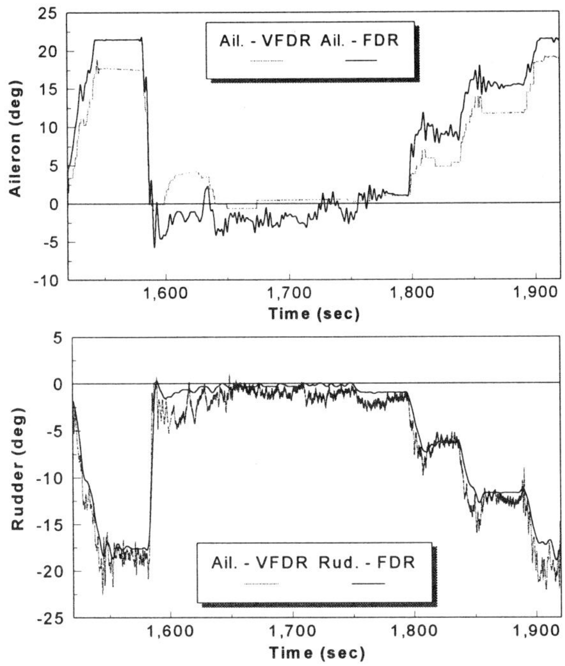

Figure 9: Comparison between NNR and FDR time histories for ailerons and rudder deflections

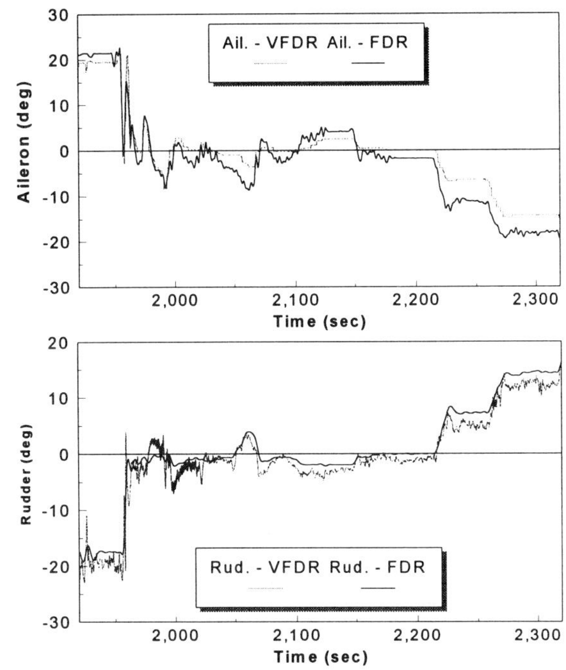

Figure 10: Comparison between NNR and FDR time histories for ailerons and rudder deflections

AIAA-97-3539

On-Line Learning Neural and Fuzzy Logic Controllers for Actuator Failure Accommodation in Flight Control Systems

Marcello R. Napolitano[*], Jose L. Casanova[+], Mario Innocenti[#]
[*+]Aviation Operation and Safety Center, Department of Mechanical and Aerospace Engineering
West Virginia University, Morgantown, WV 26506/6106
[#]Department of Electrical Systems and Automation
University of Pisa, Pisa, 56126 Italy

Abstract

This paper presents the results of a comparative study evaluating the performance of on-line learning neural and fuzzy logic controllers for fault-tolerant capabilities in the event of actuator failures. The objective is to design and evaluate both controller schemes through numerical simulations for the purpose of allowing the aircraft to regain and maintain a new equilibrium condition after substantial damage to a vital control surface. The robustness of the schemes is tested for elevator and aileron actuator failures. For each failure types a comparison is performed during a pre-programed maneuver. The two approaches are illustrated and tested through a non-linear, six degrees of freedom dynamic simulation using the mathematical model of an unmanned aerial vehicle. The study shows that both controllers provide desirable performance and robustness over the entire flight envelope.

Introduction

Several factors have induced in recent years an increased interest toward fault-tolerant flight control systems with capabilities for accommodating sensor and actuator failures. For military purposes it is clear that such a feature can increase the chances of survivability following a control surface failure in a battle or a dual combat scenario. In fact, studies on the casualties by the US Air Force during the Vietnam conflict have concluded that 70% of the aircraft losses could have been avoided if a flight control reconfiguration scheme was properly implemented in the flight control system. If the cost of new military aircraft along with their low procurement rate are also considered, a retrofit of the flight control system to introduce fault-tolerant capabilities may become appealing. Experimental flight testing using commercial jetliners has recently shown the benefits of fault-tolerant capabilities using differential thrust[1].

On a different side of the aviation industry, medium and high-technology, low weight, unmanned aerial vehicles and spacecrafts have been recently introduced for remote sensing purposes for scientific and military applications. It is obvious that a fault-tolerant control system is very desirable for these unmanned aircraft.

A full fault-tolerant flight control system is required to accomplish :

-Sensor failure, detection, identification, and accommodation (SFDIA);

-Actuator failure, detection, identification, and accommodation (AFDIA).

The study described in this paper is relative to the AFDIA task. Particularly the AFDIA task can then be divided into:

- actuator failure detection and identification (AFDI), which detects significant abnormalities and searches for the cause or a set of probable causes;
- actuator failure accommodation (AFA), which determines what actions should be taken to recover the impaired aircraft. Note that an actuator failure may imply a locked surface, a missing part of the control surface, or a combination of both. In increasing order of severity an actuator failure induces altered trim conditions, dynamic coupling, loss or deterioration of dynamic stability potentially leading to unrecoverable flight conditions. The final objective of the AFDIA task is to achieve, in increasing order of importance, a lower failure-induced handling qualities degradation, a lower mission abort rate, and a lower aircraft loss rate.

In trying to define the AFA task one must distinguish between reconfigurable and restructurable flight control systems. In reconfigurable flight control systems the types of control surface failures are classified and anticipated, and the relative change in control gains are calculated off-line and stored in the memory of the flight control computer, ready for on-line use whenever needed. On the contrary, in restructurable flight control systems, the reconstruction of the control law is performed on-line. In this case, the computational efficiency of the control algorithms and the availability of

---

[*] Associate Professor, AIAA Member
[+] Graduate Research Assistant, AIAA Member
[-] Associate Professor, Associate Fellow AIAA

sufficient computational power, as opposed to memory, can become a critical issue. Ref.[2] provides a list of different control algorithms, most of them linear, for both reconfigurable and restructurable flight control systems.

More precisely, the AFA task can be divided into the following phases:

Phase 1 - To regain control of the aircraft with a control law capable of handling non-linear dynamic and aerodynamic conditions before the aircraft enters unrecoverable flight conditions.

Phase 2 - To restructure the control laws for the continuation of the mission through recalculation and re-scheduling of the feedback gains, using a linear control approach, to reflect the new dynamic condition following the actuator failure.

In this study it was assumed that a successful AFDI had already been achieved. The principal author has previously introduced in Ref.[3] a neural network-based approach for the AFDI task. The failure detection is achieved by spotting substantial changes in the aircraft angular velocities following and type of actuator failure. Therefore, an estimation of the aircraft angular velocity from a neural estimation scheme would present an estimation error higher than a selected threshold. The selection of the threshold should be a trade-off between high detectability and low level of false alarm rate. For the identification part, the approach in Ref.[3] is based on the analysis of the cross-correlation functions between different angular velocities taking advantage of the different levels of dynamic coupling induced by any type of actuator failures.

This paper presents the results of two different approaches for Phase 1 of the AFA task: on-line learning Neural Network Controllers (NNCs) and fuzzy logic controllers (FLCs). The capabilities of both schemes allows the handling of the potentially non-linear conditions following actuator failures.

The paper is organized as follows. The next sections briefly introduce the NN and the FL-based control schemes. A section describes the aircraft model used in this study and it introduces the modeling of the actuator failures. Then, following sections describe the results of comparative simulations. A final section summarizes the paper with conclusions.

Basic Review of On-Line Learning Neural Network Controllers

The previous decade has witnessed an increase in NN research, mostly induced by the introduction of the Back-Propagation Algorithm (BPA) for feed-forward NNs with supervised learning[4,5]. In recent years NNs have been proposed for identification and control of linear and non-linear dynamic systems. Ref.[6] and, to a great extent, Ref.[7] provided a substantial background work for several NN applications and, clearly stated some of the necessary theoretical assumptions in order to have well formulated problems.

An aircraft flying at non-linear dynamic and aerodynamic conditions at any point in its flight envelope can be considered a time-varying, non-linear system with a certain degree of uncertainty on the exact modeling of the system as well as the system and measurement noises. Therefore, the control of such a system can be attempted with an adaptive scheme. For NN applicability to adaptive control systems, the following properties are important:

Applicability to non-linear systems - Ref.[8] shows that a feed-forward NN with at least one hidden layer is capable of approximating any non-linear function once enough training is provided.

Parallel distributed processing and hardware implementation - NNs have an inherent parallel architecture which, naturally, leads to parallel hardware implementations.

Learning - NNs can be trained using past recorded data or simulated data (off-line training) or current data (on-line learning).

Applicability to multi-variable systems - NNs, by definition, are multi-input, multi-output entities and this, naturally, leads to their application to multi-variable systems.

In this paper the authors refer to feed-forward neural networks with supervised learning with an error correction scheme. A critical design choice in the use of NNCs is on-line learning vs. off-line learning. Off-line learning implies that the NNCs have a "frozen" numerical architecture and do not take advantage of the learning which can be provided in on-line. On the other side for on-line learning NNCs points of particular concern are:

- the amount of time needed to acquire a substantial learning level;
- the complexity of the NNs.

Both these issues are related to the learning algorithm. To avoid the "classic" and well documented problems associated with the BPA (namely slow learning and local minima), the on-line learning for the neural controllers is performed using the Extended Back-Propagation Algorithm (EBPA) this algorithm is described with details in Refs.[3,9].

The design of on-line learning NNCs can also be complicated by the large number of degrees of freedom (number of hidden layers, number of neurons per hidden layer, size of input data pattern, magnitude of the learning rate and momentum coefficients) in the selection of the neural architectures. This issue has been addressed in Refs.[10,11]. Previous work coordinated by the first author[12] provided the following guidelines for the selection of suitable architectures for neural controllers using the EBPA:

- the performance of on-line learning neural controllers are generally robust to different number of neurons per hidden layer and a different input data pattern;
- medium to high learning rates, either constant or slowly decreasing with time, performed better for on-line learning purposes;
- simple single hidden layer architectures generally perform as well as more complicated and more computationally intensive multiple hidden layers architectures;
- a proper selection of the data to be presented as input and an appropriate choice of the cost functions are, by far, the most effective degrees of freedom in the selection of neural architectures; this makes the design of a NNC very problem-dependent.

In general, the conclusions in Ref.[12] relative to the EBPA were consistent with those in Refs.[10,11] relative to the BPA.

Basic Review of Fuzzy Logic Controllers

Fuzzy logic controllers (FLCs) provide an alternative approach to non-linear control problems. Unlike on-line learning NNCs, they have already been implemented in several commercial applications, mostly for ground transportation purposes[13,14].

Fuzzy Logic (FL) is a methodology based on conventional logic that permits the representation and manipulation of elements that may be true, false or anywhere in between. They are ruled based systems which encounters a problem as that of the translation of all possible solutions to fuzzy sets; therefore, their lack of flexibility in decision-making limits their performance[15].

In a fuzzy expert system there are four sequential processes, described in Figure 1: fuzzyfication, inference, composition, and defuzzification. The *fuzzification process* is extracted from a human's expert knowledge of the dynamic system, where the membership functions defined on input variables are applied to their actual values, such as the changes of longitudinal and lateral aerodynamic coefficients, to determine the degree of truth for each rule premise. The selection of these variables will be important in choosing an effective control action so that they cover the data to a reasonable degree. In the *inference process* the truth value for the premise value for the premise of each rule is computed and applied to the conclusion part of each rule. This results in one fuzzy membership to be assigned to each output variable for each rule. Finer partitioning is needed for better control performance, although computational efficiency and robustness decreases when using a high number of rules. The inference method applied in these studies is the Mandami's MIN inferencing, in which the output membership function is clipped off at a height corresponding to the rule's premise computed degree of truth. In the *composition process*, all of the fuzzy memberships assigned to each output variable are combined together to form a single fuzzy membership for each output variable. One composition method applied here is the MAX composition, where the combined output fuzzy membership is constructed by taking the point wise maximum over all of the fuzzy memberships assigned to the output variable by the inference rule. Finally, in the *defuzzification process*, the result of the composition process needs to be converted to a single number. The defuzzification method used in these studies is the Weighted Average method[15], which is only valid for symmetrical output membership functions. Figure 2 shows the fuzzy logic membership functions for the lateral and longitudinal controllers.

Aircraft Simulation and Modeling of Actuator Failures

The mathematical model for these studies is the model of a high altitude unmanned aircraft used for scientific missions within the NASA mission to Planet Earth program. Using provided aerodynamic and thrust data[16] a non-linear simulation code was developed [17,18,19]. The primary control surfaces consists of two differential horizontal stabilators, two differential rudders on the same vertical tail and a set of three independent ailerons per wing. The aircraft flight envelope is shown in Figure 3. Two different failure types were simulated when the aircraft is performing a maneuver at approximately 19,000 m and a velocity of 106 m/sec. Failure #1 consists of a stuck left elevator at +15 degrees with 50% of missing surface remaining while failure #2 consists of a stuck right aileron at +20 degrees with 50% of missing surface remaining. Failures occur at 30 seconds of the simulated flight with a time step for the simulation of 0.02 sec. According to the AFD scheme introduced in Ref.[3], the detection of a particular failure implies that an angular velocity estimation error parameter defined by:

$$ANGVELERR(k) = \frac{1}{2}[(p(k) - \hat{p}(k))^2 + (q(k) - \hat{q}(k))^2 + (r(k) - \hat{r}(k))^2] \tag{1}$$

will exceed a selected threshold providing, therefore, an indication that a failure has indeed occurred. The AFI task also introduced in Ref.[3] makes use of the cross-correlation function defined by:

$$R_{YX}(n) = E[Y(k)\ X(k+n)] \qquad \textbf{(2)}$$

for two given generic random processes Y(k) and X(k)[20]. The threshold values for the AFD and the AFI tasks are selected to reflect tradeoffs between quick detectability and identifiability vs. low false alarm rates. The linearized state variable models of the aircraft at nominal, failure #1 and failure #2 conditions are shown in Ref.[19]. The mathematical modeling of the actuator failures, including the event of a missing part of the control surface, is achieved through a set of closed form expressions for the non-dimensional stability and control derivatives as functions of the normal force coefficient of the failed surface[21]. A similar modeling can be introduced for failures to other control surfaces. The system noise is modeled as zero mean, white, Gaussian gust disturbance on the angle of attack and the sideslip angle. Therefore:

$$w = [\ \alpha_{gust}\ ,\ \beta_{gust}\ ]^T \qquad \textbf{(3)}$$

The sensor noise is modeled based on the studies from Ref.[22]. The standard deviations for the sensor noise are given in Tables I and II for the system and measurement noise respectively.

Comparative Studies

The neural and fuzzy control schemes required the use of all aircraft control surfaces to regain equilibrium for either failure #1 or failure #2 due to the coupling between the longitudinal and the lateral directional dynamics. Roll and pitch controllers were considered for this investigation; a yaw controller was not considered necessary due to the availability of two rudders on the aircraft. Studies on the settling times of the controllers were performed on the various parameters of the neural architecture to determine the "near-optimal" configuration for use and comparison in this study. These studies were based on input data window, number of hidden layers, neurons per layer, learning and momentum rates, and the composition of the performance indices. The final selected neural networks are shown in Table III. A similar procedure was developed for the fuzzy logic controllers. Studies for this investigation evaluated the effects of fuzzy subset distributions, membership subdivisions for inputs and outputs, ranges of the membership functions for the error and error rate. The final fuzzy architectures used for this investigation are given in Table IV.

For the following figures flight condition #29 of the flight envelope was considered. Figure 4a shows the AFD results for failure #1, where at t=30 sec. the spike for ANGVELERR allows the detection of a possible failure. With the use of the cross correlations shown in Figure 4, the identification task is able to validate the detection and determine an elevator failure. Once these two tasks are accomplished, the controllers start the accommodation of the impaired aircraft. A comparison between the NNC and FLC for the pitch rate is presented in Figure 5, showing that both controllers are able to regain equilibrium for the aircraft after the failure has been properly detected and identified. Since a longitudinal failure induces a coupling with the lateral dynamics, as shown in Figure 6. Figure 6 also shows the compensating right elevator deflection; it can be seen that the output of the fuzzy controller tends to oscillate more then the output of the neural network controller at this particular flight condition. Similarly, Figures 7 and 8 show the AFDIA results for failure #2. Also, to analyze the robustness of both controllers for the entire flight envelope, settling times were recorded for all 31 discretized flight conditions; they are reported in Table V and shown in Figures 9 and 10. For the longitudinal failure, the performance of both controllers are very similar while for the lateral failure the NNC outperforms the FLC in the mid-low section of the flight envelope. The opposite seems to happen around the higher portions of the flight envelope.

The computational effort required from the two schemes was quantified in terms of CPU time. Table VI gives a quantification of the computational effort for a 60 second simulation for both the longitudinal and the lateral controllers using a 90 MHZ Pentium PC with 32 Mb of RAM.

Conclusions

This paper has presented a comparison between Neural Network-based and a Fuzzy Logic-based schemes for the problem of actuator failure accommodation in a flight control system. The comparison is performed after the failure has been successfully detected using the actuator failure detection, identification, and accommodation schemes introduced in Ref.[3].

Both techniques have shown robustness over the entire flight envelope following a complete set of parametric studies that determined their "near-optimal" configurations. The neural network-based scheme is based on the robustness of on-line learning capabilities of the controllers while the fuzzy logic-based scheme is dependent on a fixed architecture developed from the designer experience. The difference in performance was not substantial; particularly, for the longitudinal failure it should be noted how for the low and mid points in the flight envelope the FL controller outperforms the NN controller; the opposite happens at the extreme points in the flight envelope. For the lateral failure instead the performance of the NN controller is better than the FL controller over most of the flight envelope,as shown in Figures 9 and 10. While the robustness of the FL controller is bounded, the NN controller performance are expected to improve with the on-line learning although this specific issue has not been addressed in this investigation. On the other side the FL approach is computationally more efficient.

In general it can be concluded that both approaches have similar potential and appealing capabilities with respect to conventional control methods. The performance and properties of both schemes, in addition to the feasibility of implementing them in dedicated parallel microprocessors, make them attractive alternatives to gain scheduling-based restructurable and/or reconfigurable control methods.

Acknowledgments

Support for the first two authors has been provided through the NASA Cooperative Agreement Award NCCW-0051.

References

1 - AVIATION WEEK, Summer and Fall 1995

2 - Napolitano, M. R., Naylor S., Neppach C., Casdorph V., "On-Line Learning Nonlinear Direct Neurocontrollers for Restructurable Control Systems," Journal of Guidance, Control and Dynamics Vol. 18 No. 1, Jan-Feb 1995 pp. 170-176

3 - Napolitano, M. R., Chen, C. I., Naylor, S., "Aircraft Failure Detection and Identification Using Neural Networks," Journal of Guidance, Control and Dynamics Vol. 16, No. 6, 1993, pp. 999-1008

4 - Rumelhart,D., and McClelland, J., Parallel Distributed Processing, MIT Press, Cambridge, MA, 1986

5 - Simpson, P. K., Artificial Neural System, Pergamon, Fairview Park, NY, 1990

6 - Narendra, K. S., Partasarathy, K. "Identification and Control of Dynamical Systems Using Neural Networks", IEEE Transactions on Neural Networks, Vol.1, No.1, March 1990

7 - Huang, C., Tylock, J., et al. "Failure-Accommodating Neural Network Flight Control". AIAA Guidance, Navigation, and Control Conference. August 1992

8 - Hornik, K., Stinchcombe, M., and White, H. "Multilayer Feed-forward Networks are Universal Approximators". Neural Networks. Vol. 2, 1989

9 - Chen, C. L., and Nutter R. S., "An Extended Back-Propagation Learning Algorithm by Using Heterogeneous Processing Units," Proceedings of the International Joint Conference on Neural Network (IJCNN '92), Baltimore, MD, June 1992

10 - Barnard, E., "Optimization for Training Neural Nets", IEEE Transactions on Neural Networks, March 1992, Volume 3, Number 3, pg. 232

11 - De Villiers, J., Barnard, E., "Back-Propagation Neural Nets with One and Two Hidden Layers", IEEE Transactions on Neural Networks, January 1993, Volume 4, Number 1, pg. 136

12 - Naylor, S. "On-Line Learning Non-Linear Neural Controllers for Restructurable Flight Control Systems", Master Thesis in Aerospace Engineering, Department of Mechanical and Aerospace Engineering, West Virginia University, August 1994

13 -Kosko, B. "Neural Networks and Fuzzy Systems: a Dynamical Approach to Machine Intelligence," Englewood Cliffs, NJ: Prentice-Hall 1992.

14 - Kosko, B. "Fuzzy Thinking: the New Science of Fuzzy Logic," 1993

15 - Ross, T. J., "Fuzzy Logic with Engineering Applications," McGraw-Hill Inc. 1995

16 - Aurora Flight Sciences, "Theseus Aerodynamic Data", Technical Report, March 1995

17 - Antoniewicz, R. F., Duke, E. L., Patterson, B. P., "Derivation and Definition of a Linear Aircraft Model", NASA Reference Publication 1207, August 1988

18 - Antoniewicz, R. F., Duke, E. L., Patterson, B. P.,"User's Manual for Interactive Linear, a Fortran Program to Derive Linear Aircraft Models", NASA Technical Paper 2835, Sept 1988

19 - Casanova, Jose L., "Design and Comparison of Neural Network and Fuzzy Logic Actuator Failure Schemes for Flight Control Systems", Thesis WVU MAE Dept., Morgantown, WV, December 1996

20 - Newland, D. E., Random Vibrations and Spectral Analysis, Longman Group, Essex, England, UK, 1984

21- Napolitano, M.R., Swaim, R. L., "New Technique for Aircraft Flight Control Reconfiguration", AIAA Journal of Guidance, Navigation and Control, Vol. 14, Number 1, February 1991, pp. 184-190.

22 - Moytka, P., W. Bonnics, et al., "The Evaluation of Failure Detection and Isolation Algorithms for Reconstructurable Control", NASA Contractor Report 177983, 1985

Table I - System Noise Standard Deviations

Parameter	Standard Deviation	Units
α_{gust}	1.74e-4	rad
β_{gust}	1.74e-4	rad

Table II - Measurement Noise Standard Deviations

Parameter	Standard Deviation	Units	Parameter	Standard Deviation	Units
p	2.45e-4	rad/s	a_x	7.65e-4	g's
q	2.45e-4	rad/s	a_z	7.65e-4	g's
r	2.45e-4	rad/s	a_y	7.65e-4	g's
ϕ	8.80e-4	rad	dp/dt	8.73e-4	rad/s^2
θ	1.75e-4	rad	dr/dt	8.73e-4	rad/s^2
α	8.80e-4	rad	$d\alpha/dt$	2.45e-4	rad/s
β	8.80e-4	rad	$d\beta/dt$	2.45e-4	rad/s
V_T	4.19e-4	m/s			

Table III - Data for the NN and FL Pitch Controller, NN and FL Roll Controller

	NN based AFDI scheme	
	Estimator	
Input Parameters	$p,q,r,\theta,\phi,\delta_A,\delta_E,\delta_R$,dp/dt,dq/dt,dr/dt	
Data Pattern	2	
Total number of inputs	22	
Number of Hidden Layer(s) (HL)	1	
Number of Neurons in HL(s)	25	
Outputs	3	
Learning Rates	0.2	
Momentum Coefficients	0.5	
Parameters to be Updated (PTBU)	737 at each computational step	
	NN based AFA scheme	
	Pitch	Roll
Input Parameters	q,θ,δ_E	p,q,dp/dt,dq/dt,θ,ϕ,δ_A
Data Pattern	2	1
Total number of inputs	6	7
Number of Hidden Layer(s) (HL)	1	2
Number of Neurons in HL(s)	20	5/7
Outputs	1	1
Learning Rates	0.6	0.1
Momentum Coefficients	0	0
Parameters to be Updated (PTBU)	224	129 at each computational step
	FL based AFA scheme	
	Pitch	Roll
Input Parameters	$\Delta\theta,\Delta q$	$\Delta\phi,\Delta p$
Range of the Inputs	$\Delta\theta = \pm$ 24(deg.)	$\Delta\phi=\pm$12(deg.)
	$\Delta q = \pm$ 12(deg./sec.)	$\Delta p = \pm$ 12(deg./sec.)
Membership Subdivisions	7	7
Outputs	1	1

Table IV - Fuzzy Associative Memory Tables for the Pitch and Roll Controllers

Pitch Controller

		Δθ (deg)						
		NB	**NM**	**NS**	**Z**	**PS**	**PM**	**PB**
ΔQ (deg/sec)	**NB**	NB	NB	NB	NM	NS	PS	PM
	NM	NB	NB	NM	NM	NS	PS	PM
	NS	NB	NB	NM	NS	Z	PM	PB
	Z	NB	NM	NS	Z	PS	PM	PB
	PS	NB	NM	Z	PS	PM	PB	PB
	PM	NM	NS	PS	PM	PM	PB	PB
	PB	NM	NS	PS	PM	PB	PB	PB

Roll Controller

		Δϕ (deg)						
		NB	**NM**	**NS**	**Z**	**PS**	**PM**	**PB**
ΔP (deg/sec)	**NB**	PB	PB	PB	PM	PS	NS	NM
	NM	PB	PB	PM	PM	PS	NS	NM
	NS	PB	PB	PM	PS	Z	NM	NB
	Z	PB	PM	PS	Z	NS	NM	NB
	PS	PB	PM	Z	NS	NM	NB	NB
	PM	PM	PS	NS	NM	NM	NB	NB
	PB	PM	PS	NS	NM	NB	NB	NB

Table V - Settling Times for the Longitudinal and Lateral Failure Accommodation

	Longitudinal Controller Failure Type #1		Lateral Controller Failure Type #2	
Flight Condition	NN Ts	FL Ts	NN Ts	FL Ts
F.C. #1	15.16	9.84	29.24	10.56
F.C. #2	16.74	11.34	30.32	14.3
F.C. #3	16.24	4.8	31.76	18.96
F.C. #4	15.18	10.62	17.92	6.44
F.C. #5	16.74	11.62	17.98	6.86
F.C. #6	15.16	5.1	23.9	9.8
F.C. #7	10	9.46	25.98	13.62
F.C. #8	5.94	4.26	27.1	21.68
F.C. #9	17.1	12.16	15.68	6.1
F.C. #10	11.74	5.48	15.86	6.44
F.C. #11	6.88	6.92	16.32	9.62
F.C. #12	3.16	4.26	17.42	16.48
F.C. #13	12.5	7.28	13.64	5.54
F.C. #14	7.02	5.52	14.32	7.84
F.C. #15	6.12	6.94	14.58	11.48
F.C. #16	6.88	5.28	15.68	13.22
F.C. #17	7.2	6.86	8.52	5.26
F.C. #18	6.86	7.92	12.52	7.66
F.C. #19	4.64	4.74	12.72	8.78
F.C. #20	7.3	6.52	13.72	15.54
F.C. #21	2.72	5.62	7.94	5.12
F.C. #22	5.52	6.84	8.14	7.6
F.C. #23	6.82	5.04	11.62	10.68
F.C. #24	8.74	6.36	12.9	21.52
F.C. #25	7.44	4.86	7.38	5.28
F.C. #26	7.54	6.12	7.7	7.66
F.C. #27	8.64	6.34	7.5	11.5
F.C. #28	7.74	8.02	7.0	6.68
F.C. #29	7.54	8.6	6.9	7.92
F.C. #30	9.3	8.4	6.56	4.84
F.C. #31	9.48	9.0	6.72	5.54

Table VI - Results of Computational Effort between NNC and FLC

	CPU time
Neural Network-based AFA scheme	
Pitch Controller	220.58 sec.
Roll Controller	220.25 sec.
Fuzzy Logic-based AFA scheme	
Pitch Controller	54.10 sec.
Roll Controller	53.72 sec.

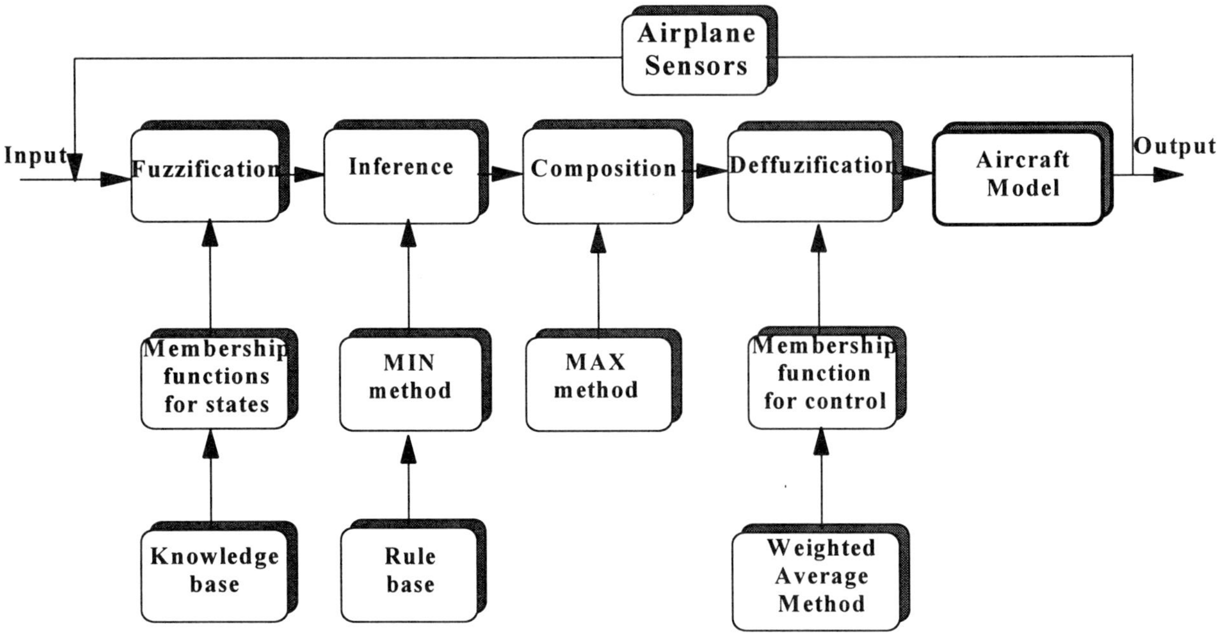

Figure 1: Fuzzy Logic Control System Block Diagram

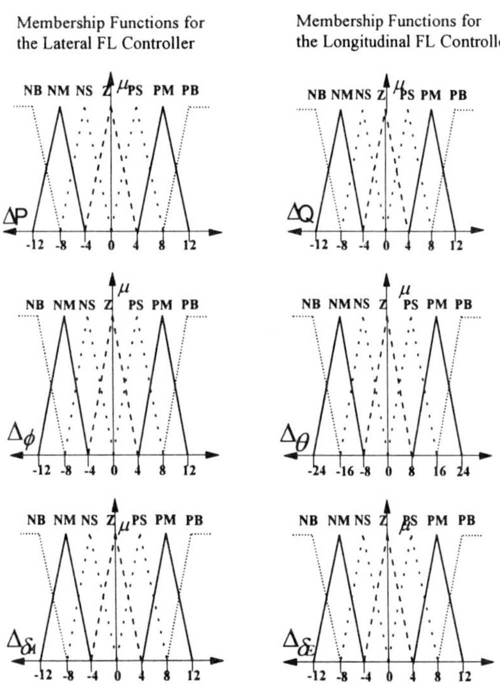

Figure 2: Fuzzy Logic Membership Functions for the Lateral and Longitudinal Controllers

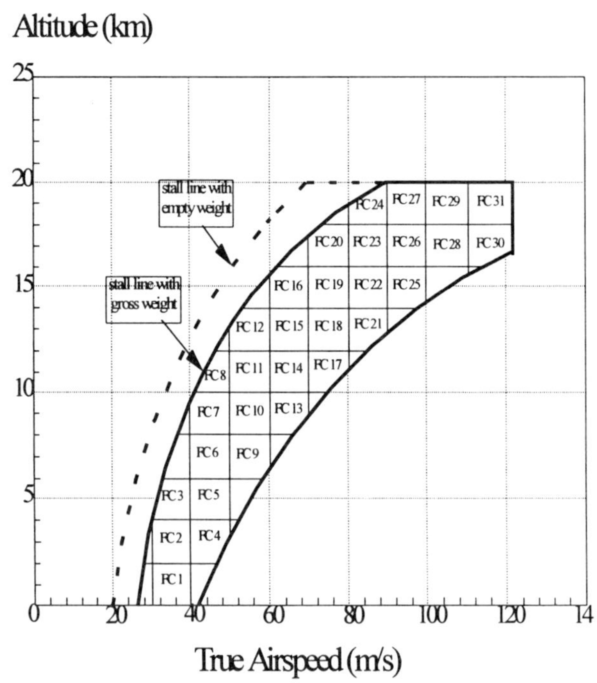

Figure 3: Aircraft Flight Envelope

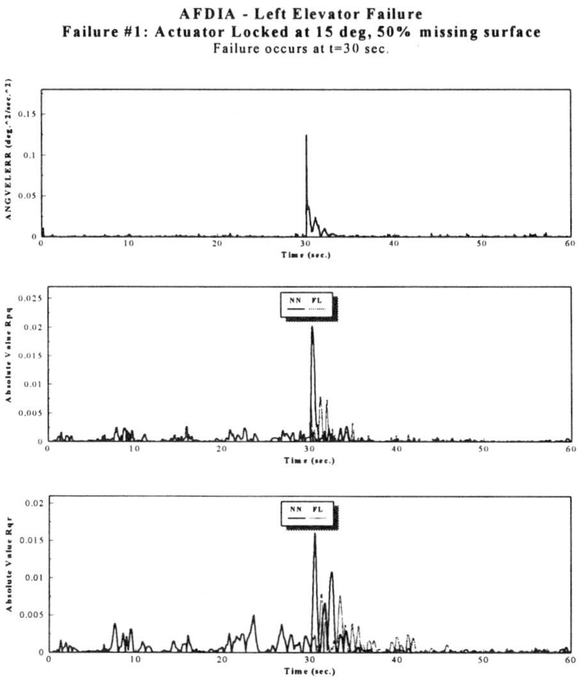

Figure 4: a) ANGVELERR for the AFD task, b) Rpq Cross-Correlation for the AFI task, c) Rqr Cross-Correlation for the AFI task

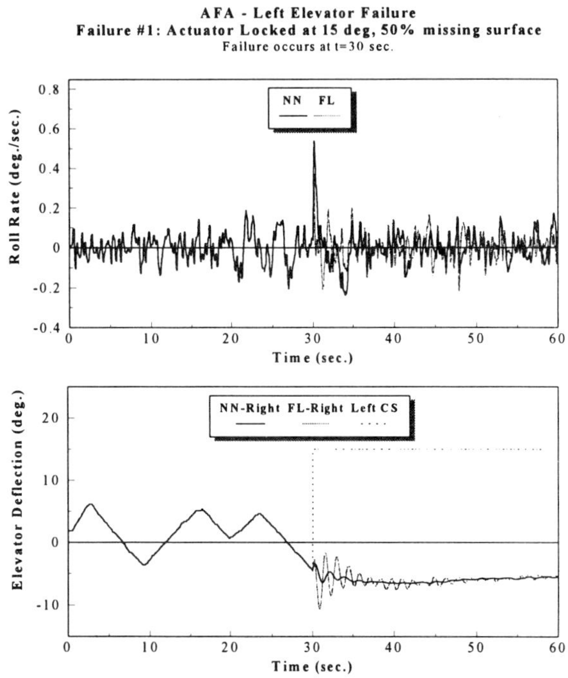

Figure 6: a) Roll Rate, b) Elevator Deflection

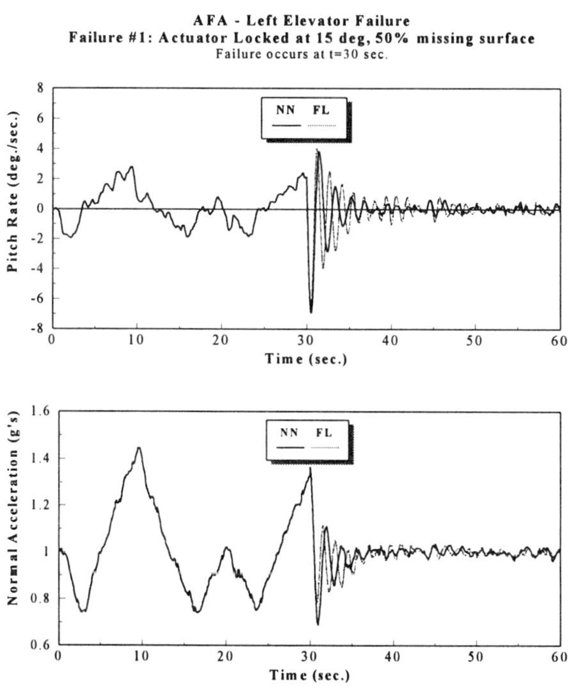

Figure 5: a) Pitch Rate, b) Normal Acceleration

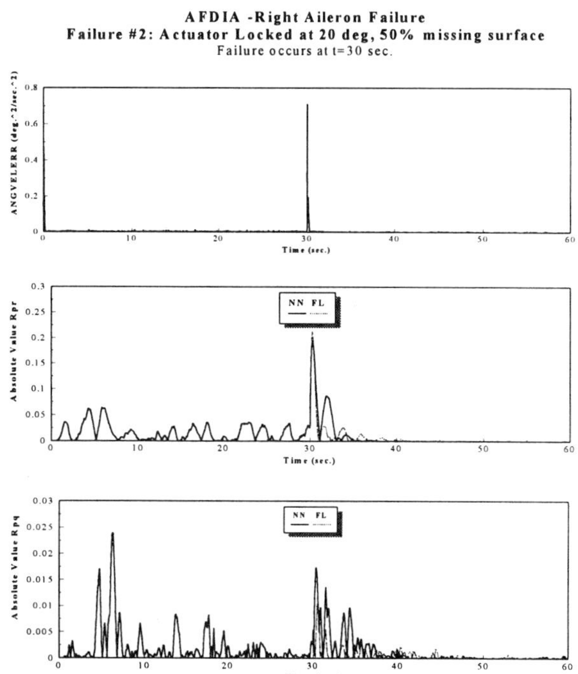

Figure 7: a) ANGVELERR for the AFD task, b) Rpq Cross-Correlation for the AFI task, c) Rqr Cross-Correlation for the AFI task

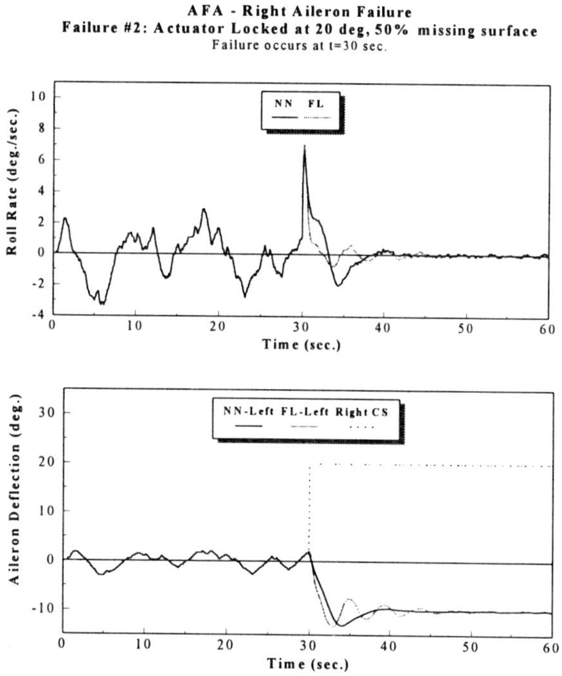

Figure 8: a) Roll Rate, b) Aileron Deflection

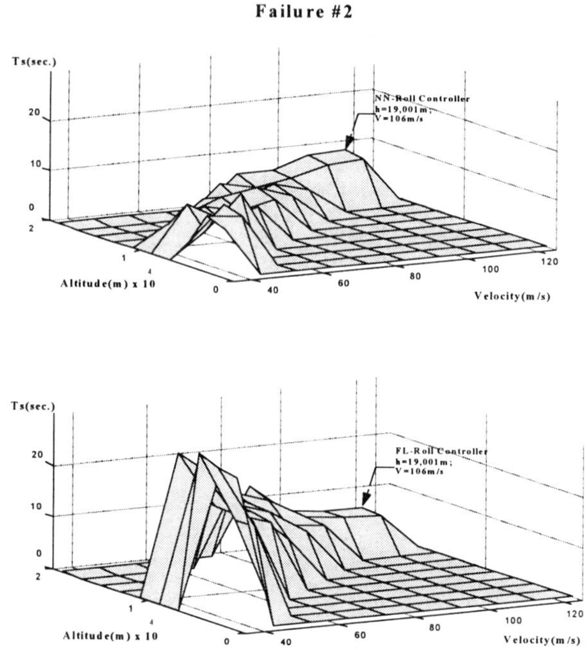

Figure 10: a) Flight Envelope Settling Times for the NN-Pitch Controller, b) Flight Envelope Settling Times for the FL-Pitch Controller

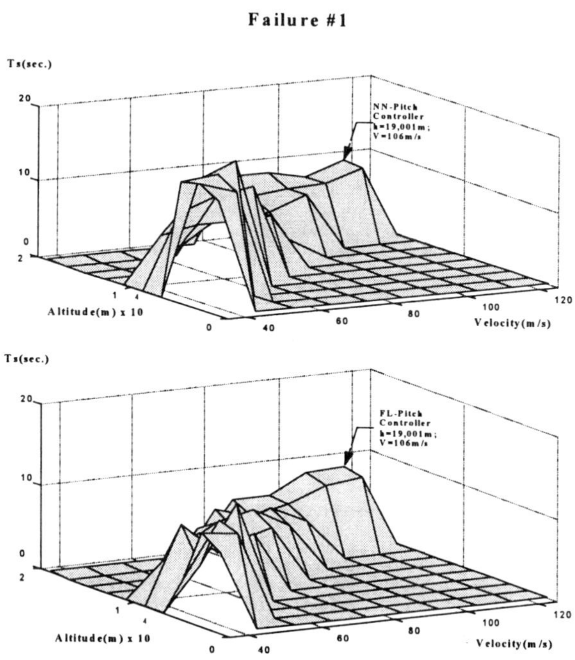

Figure 9: a) Flight Envelope Settling Times for the NN-Pitch Controller, b) Flight Envelope Settling Times for the FL-Pitch Controller

AIAA-97-3540

MULTILAYER NEURAL NETWORKS AND ADAPTIVE NONLINEAR CONTROL OF AGILE ANTI-AIR MISSILES

Michael B. McFarland* and Anthony J. Calise†
School of Aerospace Engineering
Georgia Institute of Technology
Atlanta, Georgia, USA

Abstract

Research has shown that neural networks can be used to improve upon approximate dynamic inversion for control of uncertain nonlinear systems. In one architecture, the neural network adaptively cancels inversion errors through on-line learning. Such learning is accomplished by a simple weight update rule derived from Lyapunov theory, thus assuring the stability of the closed-loop system. In this paper, the authors review this methodology and extend it to incorporate an important class of neural networks with a single hidden layer. An agile anti-air missile autopilot is subsequently designed using this control scheme. First, a control law based on approximate inversion of the nonlinear dynamics is presented. This control system is augmented by the addition of a multilayer neural network with on-line learning. Finally, numerical results from a nonlinear agile anti-air missile simulation demonstrate the effectiveness of the resulting autopilot.

Introduction

Advances in fighter aircraft technology continue to create challenges for designers of anti-air weapon systems. The introduction of low-observable aircraft has increased the need for small, lightweight missiles. This, in turn, has led to various control problems associated with air-breathing propulsion, asymmetric airframes, and greatly reduced aerodynamic control surface area. Similarly, the advent of supermaneuverable aircraft has motivated efforts aimed at expanding the flight envelope of a missile to include high-angle-of-attack conditions. Accordingly, some proposed next-generation missiles employ propulsive control effectors similar to those associated with supermaneuverable aircraft. These so-called "agile" missiles possess enhanced range as a result of their ability to execute propulsive heading changes during the boost phase of an intercept. Such weapons can be deployed during high-angle-of-attack maneuvers, and may engage targets in the rear hemisphere relative to the launch aircraft.

The dynamics of an agile missile flying in bank-to-turn mode at a high angle of attack are inherently nonlinear and may vary rapidly with time. Further, these dynamics are uncertain since aerodynamic data for vehicles operating under such conditions is difficult to obtain and may in fact be a poor approximation to actual flight conditions. These and other concerns have prompted researchers to look beyond classical methods which have historically dominated missile autopilot design to robust, nonlinear, and "intelligent" control.

Most nonlinear control techniques are based on linearizing the equations of motion by the application of nonlinear feedback. Known as feedback linearization, or dynamic inversion, these methods rely heavily on precise *a priori* knowledge of the plant dynamics. An early application of this theory to the missile autopilot design problem is found in Ref. 1, while Ref. 2 presents a more sophisticated approach involving variable structure control. More recently, neural networks have emerged as a means of explicitly accounting for uncertainties in plant dynamics. Their on-line learning and functional approximation capabilities make neural networks an excellent candidate for this application. Ref. 3 is one example in which neural networks are investigated for nonlinear control of missiles.

This paper concerns a neural network based approach to direct adaptive control of nonlinear systems which originated in Ref. 4 and was further developed in Ref. 5. In the proposed architecture, a simple dynamic inversion controller approximately linearizes the vehicle dynamics. This controller is augmented by a neural network which acts to improve the linearization by adaptively canceling inversion errors in real-time. The neural network implementation features a stable on-line learning algorithm derived from Lyapunov theory. While Ref's. 4 and 5 dealt specifically with fighter aircraft and helicopter applications, respectively, the work described in this paper extends these efforts while adapting them specifically to the agile missile control problem. The result is an autopilot which combines the

---

* Graduate Student. Student Member, AIAA.
† Professor. Fellow, AIAA.
Research sponsored in part by the U.S. Air Force Wright Laboratory under contract # F08630-95-1-0006.

best features of dynamic inversion, neural networks, and adaptive control to increase the effectiveness and versatility of tomorrow's missile systems.

The preliminary study presented in Ref. 6 revealed that neural networks are capable of attaining sufficiently high learning rates to make adaptation feasible even during the most demanding aerial engagements. In fact, previous work documented in Ref. 7 shows that the methodology considered here compares favorably to traditional gain-scheduled linear methods for this application. This paper extends the results of earlier research to accommodate the important class of multilayer neural networks with one sigmoidal hidden layer. These architectures are of particular interest because of their proven functional approximation capabilities. This theoretical extension is based on the work described in Ref. 8, which was specialized to the control of robotic manipulators. The development here expands this approach to address a more general class of nonlinear systems, including systems which depend nonlinearly on the input.

The paper begins with a detailed description of the proposed controller architecture, including some of the principal aspects of its Lyapunov-based proof of stability. The methodology in question is then applied to an agile anti-air missile autopilot design problem. First, the baseline control scheme consisting of an approximate inversion of the missile's six-degree-of-freedom nonlinear dynamics is presented. A multilayer neural network is then designed to offset residual nonlinearities, enhancing the performance this nonlinear controller. Numerical simulation results for an agile anti-air missile demonstrate the feasibility of this autopilot design technique. Finally, conclusions and future research directions are discussed.

Control Design Methodology

Consider a block-triangular nonlinear system with the following structure:

$$\begin{aligned} \dot{x}_1 &= f_1(x_1) + g_1(x_1)x_2 \\ \dot{x}_2 &= f_2(x_1, x_2, u) + d(t) \end{aligned} \tag{1}$$

where $x_1, x_2, u, d \in \Re^n$ and $g_1(x_1)$ remains nonsingular for all x_1. We assume that the external disturbance (d) is bounded by $|d| \le d^*$. For reference, a block diagram of this system is included in Fig. 1.

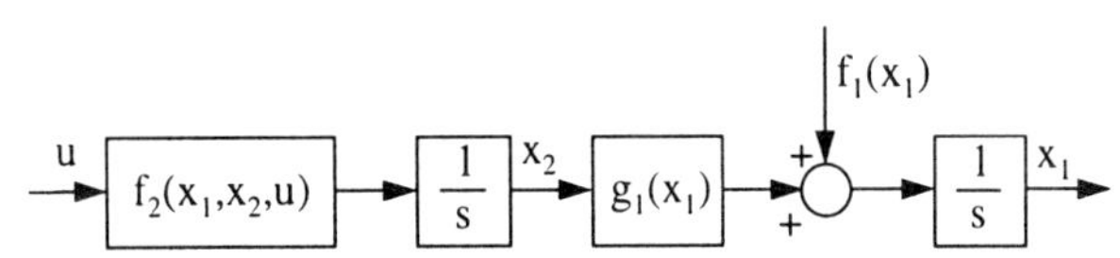

Figure 1: Open-Loop Nonlinear System

Given a reference trajectory defined by $x_{1c}(t)$ and its derivatives $\dot{x}_{1c}(t)$ and $\ddot{x}_{1c}(t)$, the objective is to achieve stable command following via full-state feedback.

Approximate Nonlinear Control

Suppose that the dynamics of the x_1 subsystem are well-known, but the nonlinearity f_2 is only known approximately. With the state error defined by $\tilde{x}_1 = x_{1c}(t) - x_1$, suppose furthermore that an asymptotically stabilizing controller $\alpha_1(\tilde{x}_1, t)$ for the $\tilde{x}_1$ error dynamics with x_2 as input is also known with an associated Lyapunov function V_1. One simple choice for designing $\alpha_1(\tilde{x}_1, t)$ is to use dynamic inversion. Note, however, that $\alpha_1(\tilde{x}_1, t)$ need not be an inverting controller. If, for example, the plant dynamics include so-called "beneficial" nonlinearities, these terms need not be canceled by the nonlinear feedback. This is described as "lean" nonlinear control in Ref. 9 with regard to integrator backstepping. Next, form a new control law as follows:

$$\bar{\alpha}_1(\tilde{x}_1, t) = \alpha_1(\tilde{x}_1, t) - g_1^T (\partial V_1 / \partial \tilde{x}_1)^T \tag{2}$$

This control ensures that the $\tilde{x}_1$ error dynamics are Input-to-State Stable (ISS) with respect to z, an error variable defined by

$$z = \bar{\alpha}_1(\tilde{x}_1, t) - x_2 \tag{3}$$

The ISS property allows us to address the problem of stabilizing the z dynamics independently of $\tilde{x}_1$. Note that when $d(t) \equiv 0$, these dynamics take the form

$$\dot{z} = \dot{\bar{\alpha}}_1 - f_2(x_1, x_2, u) \tag{4}$$

By redefining the input, the x_2 dynamics may be rewritten (in the absence of external disturbances) as follows:

$$\begin{aligned} \dot{x}_2 &= v \\ v &= f_2(x_1, x_2, u) \end{aligned} \tag{5}$$

where $v(t) \in R^n$ is a pseudo-control input. Provided that the mapping $f_2(x_1, x_2, u)$ is invertible and full-state feedback is available, the definition of v above provides a linearizing transformation of the control. The corresponding inverse transformation is written as

$$u = f_2^{-1}(x_1, x_2, v) \tag{6}$$

and must be computed in real-time when the control is implemented. If the mapping f_2 is known and its inverse is computed accurately, then the system is exactly linearized.

In the case of perfect nonlinear inversion, linear control theory may be used to construct a pseudo-control input which stabilizes the z dynamics. For this purpose, we propose a pseudo-control of the form:

$$v = K_2\tilde{x}_2 + \dot{\bar{\alpha}}_1(\tilde{x}_1, t) \tag{7}$$

The last term in Eq. (7) has the effect of the "command derivative" term in model-following control, and is necessary to achieve tracking of arbitrary smooth trajectories. For slowly varying commands, this term may be neglected.

Because f_2 is uncertain, and external disturbances are present, exact linearization is impossible. Eq. (6) is approximated as $\hat{u} = \hat{f}_2^{-1}(x_1, x_2, v)$, resulting in:

$$\dot{x}_2 = v + \Delta'(x_1, x_2, \hat{u}) + d \tag{8}$$

where

$$\begin{aligned}\Delta'(x_1, x_2, \hat{u}) &= f_2(x_1, x_2, \hat{u}) - \hat{f}_2(x_1, x_2, \hat{u}) \\ &= \Delta(\tilde{x}_1, z, x_{1c}, \dot{x}_{1c}, v)\end{aligned} \tag{9}$$

In Eq. (9), $\Delta' : R^n \times R^n \times R^n \rightarrow R^n$ is a nonlinear mapping representing the inversion error. The second set of dependencies in Eq. (9) denotes an equivalent representation of this inversion error (Δ) which will prove useful in the subsequent adaptive control development. Fig. 2 illustrates this approximate dynamic inversion.

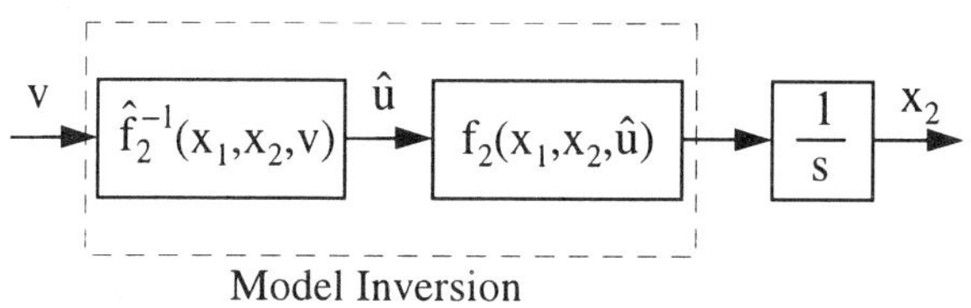

Figure 2: Approximate Dynamic Inversion

Neural Network Based Adaptation

In the case of non-zero inversion error (Δ) the control law in Eq. (7) must be augmented by an adaptive term, to be denoted by v_{ad}. This adaptive control is designed to counteract dynamic inversion errors by taking advantage of the documented functional approximation capabilities of multilayer neural networks such as the one depicted in Fig. 3. The output of a typical multilayer neural network with input $\bar{x} \in \Re^{N_1}$ takes the form

$$y_i = \sum_{j=1}^{N_2}\left[w_{ij}\sigma\left(\sum_{k=1}^{N_1} v_{jk}\bar{x}_k + \theta_{v_j}\right) + \theta_{w_i}\right] \tag{10}$$

$$i = 1, 2, \ldots, N_3$$

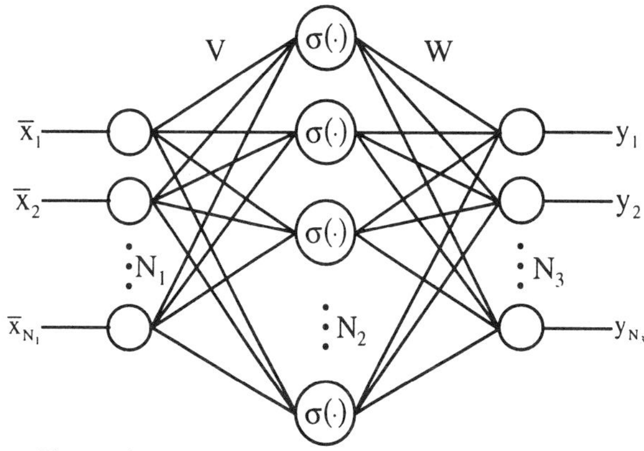

Figure 6: Neural Network With One Hidden Layer

where σ is the hidden-layer activation function, v_{jk} are the first-to-second layer interconnection weights, and w_{ij} are the second-to-third layer interconnection weights. The bias terms θ_{wi} and θ_{vj} represent thresholds. This architecture has N_1 inputs, N_2 hidden-layer neurons, and N_3 outputs. The functional form of the hidden-layer activation is a design parameter, but we will consider the case of sigmoidal activation functions:

$$\sigma(z) = \frac{1}{1 + e^{-az}} \tag{11}$$

We may express Eq. (10) in matrix form as

$$y = W^T\sigma(V^T\bar{x}) \tag{12}$$

where the thresholds are incorporated into the weight matrices as follows:

$$W^T = \left[\theta_{w_i} \middle| w_{ij}\right] \quad V^T = \left[\theta_{v_i} \middle| v_{ij}\right]$$

Here, we have defined

$$\begin{aligned} y^T &= \begin{bmatrix} y_1 & y_2 & \cdots & y_{N_3}\end{bmatrix} \\ \bar{x}^T &= \begin{bmatrix} 1 & \bar{x}_1 & \bar{x}_2 & \cdots & \bar{x}_{N_1}\end{bmatrix} \\ \sigma(z) &= \begin{bmatrix} 1 & z_1 & z_2 & \cdots\end{bmatrix}\end{aligned}$$

A neural network of the type depicted in Fig 3 is capable of approximating any smooth function to any desired accuracy, provided the number of hidden-layer neurons N_2 is sufficiently large. This implies that for continuous Δ and any $\varepsilon_N > 0$ there exists a finite N_2 and weight matrices W and V such that

$$\Delta = W^T\sigma(V^T\bar{x}) + \varepsilon(\bar{x}) \tag{13}$$

with $\|\varepsilon\| \le \varepsilon_N$. Therefore, Eq. (7) may be modified to account for inversion errors by introducing as an additional term a neural-network approximation of Δ. In this paper, an adaptive scheme will be employed to tune the weight matrices on-line so that the network requires no off-line "learning" phase. The pseudo-control v in Eq. (7) may be represented in terms of its various components as follows:

$$
\begin{aligned}
v &= v_0 - v_{ad} + \bar{v} \\
v_0 &= K_2 z + \dot{\bar{\alpha}}_1(\tilde{x}_1, t) \\
v_{ad} &= \hat{W}^T \sigma(\hat{V}^T \bar{x})
\end{aligned} \tag{14}
$$

where $\bar{v}$ is an additional pseudo-control term, to be defined. In the expression for v_{ad}, $\hat{W}$ and $\hat{V}$ represent adaptive estimates of constant weight matrices W and V which have in some sense "ideal" values.

Lyapunov stability theory will be applied to determine stable learning rules for the parameter estimates $\hat{W}$ and $\hat{V}$ as well as $\bar{v}$. First, we assume that the ideal weight matrices are bounded in the sense that $\|W\|_F \le \bar{W}$ and $\|V\|_F \le \bar{V}$, where $\|\cdot\|_F$ denotes the Frobenius norm, $\|A\|_F^2 = \mathrm{tr}\{A^T A\}$. When not specified otherwise, $\|\cdot\|$ indicates the Euclidean norm. Introducing more compact notation, we may also write

$$
Z = \begin{bmatrix} W & 0 \\ 0 & V \end{bmatrix} \text{ with } \|Z\|_F \le \bar{Z} \tag{15}
$$

The reference command x_{1c} and its first two derivatives are also assumed to be norm-bounded.

Observing the functional dependencies in Eq.'s (9) above and (26) to be presented later, we choose the network input

$$
\bar{x}^T = \begin{bmatrix} 1 & \tilde{x}_1^T & z^T & x_{1c}^T & \dot{x}_{1c}^T & \ddot{x}_{1c}^T & v_{ad}^T & \|\hat{Z}\|_F \end{bmatrix} \tag{16}
$$

which is bounded as follows:

$$
\|\bar{x}\| \le c_1 + c_2 \|\tilde{x}\| + c_3 \|\tilde{Z}\|_F \quad ; \quad c_i > 0 \tag{17}
$$

where we have introduced the partitioned vector $\tilde{x}^T = \begin{bmatrix} \tilde{x}_1^T & z^T \end{bmatrix}$ of state errors.

A key result which facilitates the extension of theory previously developed for simple linear-in-parameters neural networks to more complicated neural networks with one hidden layer has been described in Ref. 8. It involves the use of a Taylor series expansion of the hidden-layer output. First, define the error variables $\tilde{W} = \hat{W} - W$, $\tilde{V} = \hat{V} - V$, and $\tilde{Z} = \hat{Z} - Z$. Also, define the hidden-layer output error as follows:

$$
\tilde{\sigma} = \hat{\sigma} - \sigma = \sigma(\hat{V}^T \bar{x}) - \sigma(V^T \bar{x}) \tag{18}
$$

The Taylor series expansion of σ about $\hat{\sigma}$ may then be written as

$$
\sigma(V^T \bar{x}) = \sigma(\hat{V}^T \bar{x}) - \sigma'(\hat{V}^T \bar{x}) \tilde{V}^T \bar{x} + O(\tilde{V}^T \bar{x})^2 \tag{19}
$$

where $\sigma'(\hat{z}) = d\sigma(z)/dz\big|_{z=\hat{z}}$ and $O(\cdot)^2$ denotes those terms with order greater than one. These higher-order terms satisfy Eq. (20):

$$
\begin{aligned}
& O(\tilde{V}^T \bar{x})^2 = \left[\sigma(V^T \bar{x}) - \sigma(\hat{V}^T \bar{x})\right] + \sigma'(\hat{V}^T \bar{x}) \tilde{V}^T \bar{x} \\
& \Rightarrow \left\| O(\tilde{V}^T \bar{x})^2 \right\| \le c_4 + c_5 \|\tilde{V}\|_F + c_6 \|\tilde{V}\|_F \|\tilde{x}\| + c_7 \|\tilde{V}\|_F \|\hat{W}\|_F \\
& \qquad c_i > 0
\end{aligned} \tag{20}
$$

where the inequality follows from the definition of the sigmoid activation function and certain properties of vector norms.

Here, it is necessary to rewrite the z-dynamics as:

$$
\begin{aligned}
\dot{z} &= -K_2 z + \hat{W}^T \hat{\sigma} - W^T \sigma - (d + \varepsilon) - \bar{v} \\
&= -K_2 z + \tilde{W}^T \hat{\sigma} - \tilde{W}^T \tilde{\sigma} + \hat{W}^T \tilde{\sigma} - (d + \varepsilon) - \bar{v}
\end{aligned} \tag{21}
$$

Recalling Eq. (18), we may also write

$$
\dot{z} = -K_2 z + \tilde{W}^T (\hat{\sigma} - \hat{\sigma}' \hat{V} \bar{x}) + \hat{W}^T \hat{\sigma}' \tilde{V}^T \bar{x} + (w - \bar{v}) \tag{22}
$$

where $\hat{\sigma}' = \sigma'(\hat{V}^T \bar{x})$ and any extraneous terms have been collected into the disturbance signal w, defined in Eq. (23).

$$
\begin{aligned}
& w = \tilde{W}^T \hat{\sigma}' V^T \bar{x} + W^T O(\tilde{V}^T \bar{x})^2 - (d + \varepsilon) \\
& \Rightarrow \|w\| \le C_0 + C_1 \|\tilde{Z}\|_F + C_2 \|\tilde{Z}\|_F \|\tilde{x}\| + C_3 \|\tilde{Z}\|_F^2 \\
& \qquad C_i > 0
\end{aligned} \tag{23}
$$

The norm-bound on w follows from Eq.'s (17) and (20) along with the boundedness of d, e, and $\hat{\sigma}'$ by certain norm inequalities.

For the stability analysis, consider a typical choice of Lyapunov function candidate.

$$
V_2 = \frac{1}{2} z^T z + \frac{1}{2} \mathrm{tr}(\tilde{W}^T \Gamma_W^{-1} \tilde{W}) + \frac{1}{2} \mathrm{tr}(\tilde{V}^T \Gamma_W^{-1} \tilde{V}) \tag{24}
$$

The Lyapunov analysis is similar to that of Ref. 8, with modifications to account for the more general structure of the open-loop nonlinear plant as described in Eq 1. Based on this development, learning rules for the weight matrices are chosen as follows:

$$
\begin{aligned}
\dot{\hat{V}} &= \dot{\tilde{V}} = -\Gamma_V \left[\bar{x}(z^T \hat{W}^T \hat{\sigma}') + \lambda \|z\| \hat{V}\right] \\
\dot{\hat{W}} &= \dot{\tilde{W}} = -\Gamma_W \left[(\hat{\sigma} - \hat{\sigma}' \hat{V}^T \bar{x}) z^T + \lambda \|z\| \hat{W}\right]
\end{aligned} \tag{25}
$$

Note that the λ terms in Eq. 25 correspond to the e-modification found in adaptive control literature. These terms provide additional damping which helps to contain the growth of the parameter estimates. In addition, the pseudo-control term $\bar{v}$ in Eq. (14) is defined as

$$
\bar{v} = K_Z \left(\|\hat{Z}\|_F + \bar{Z}\right) \left(\|\tilde{x}_1\| + \|z\|\right) e_z \tag{26}
$$

where $e_z = z/\|z\|$ is a unit vector in the direction of z. The gain K_Z in Eq. (26) must be chosen such that

$$
K_Z > C_2 \tag{27}
$$

in order to effectively dominate nonlinear phenomena associated with parameter errors.

We then conclude that the trajectories of the closed-loop error system are uniformly ultimately bounded provided that in addition to $\lambda > C_3$ either of the following two inequalities hold:

$$\|\tilde{x}\| > \frac{(\lambda - C_3)C_4^2/4 + C_0}{\underline{\sigma}(K_2)} = b_r$$
$$\|\tilde{Z}\|_F > C_4/2 + \sqrt{C_4^2/4 + C_0/(\lambda - C_3)} = b_Z \tag{28}$$

where $C_4 = (\lambda\bar{Z} + C_1)/(\lambda - C_3)$ and $\underline{\sigma}(K_2)$ is the minimum singular value of the gain matrix K_2.

This result is a consequence of an extension of Lyapunov stability theory which requires that the Lyapunov derivative ($\dot{V}_2$) remain negative outside of some compact set. In fact, the quantities on the right-hand sides of the inequalities in Eq. (28) may be considered as practical bounds on the errors. This completes the development of the neural network based adaptive nonlinear control system. Fig. 4 depicts the z subsystem under this adaptive control architecture.

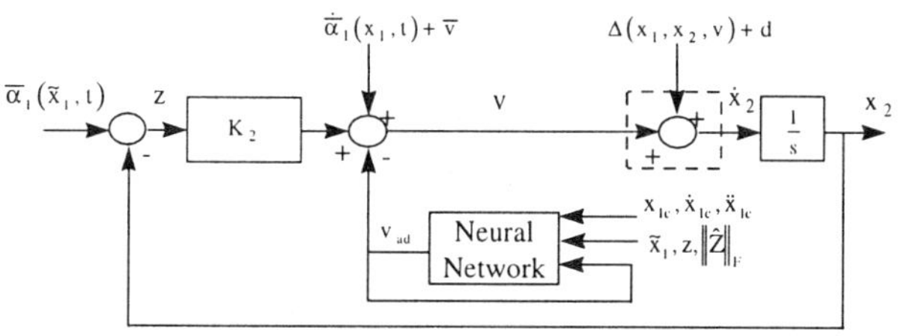

Figure 4: Adaptive Control Architecture

Note from Fig. 4 and Eq. (16) that the network output v_{ad} is also a network input. Thus, a critical assumption in the stability proof involves the existence of a fixed-point solution for the quantity v_{ad}. This condition is guaranteed to be satisfied, since sigmoidal activation functions are bounded in magnitude. Furthermore, when a stable fixed-point exists, a simple iterative scheme may be employed to compute the network output.

Missile Autopilot Application

In this section, the neural network-based adaptive nonlinear control methodology described above is used to design an autopilot for a next-generation anti-air missile as depicted in Fig. 5. This vehicle is described in greater detail in Ref's. 10 and 11. In the current scenario, the objective is to design a bank-to-turn autopilot which tracks external guidance commands in angle of attack and bank angle while holding sideslip near zero.

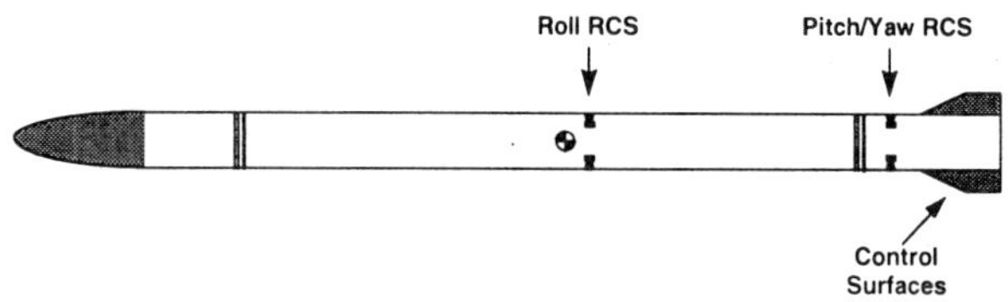

Figure 5: Agile Missile Configuration

Approximate Dynamic Inversion

The motion of a symmetric missile about its velocity vector may be described by the following equations.

$$\dot{V} = (a_x \cos\alpha + a_z \sin\alpha)\cos\beta + a_y \sin\beta$$
$$\dot{\alpha} = q - (r\sin\alpha + p\cos\alpha)\tan\beta + \frac{-a_x \sin\alpha + a_z \cos\alpha}{V\cos\beta} \tag{29}$$
$$\dot{\beta} = p\sin\alpha - r\cos\alpha - \frac{(a_x \cos\alpha + a_z \sin\alpha)\sin\beta - a_y \cos\beta}{V}$$

where V, α, and β denote airspeed, angle of attack, and sideslip angle. Here, a_x, a_y, and a_z are the body-axis components of acceleration (including gravitational effects) while p, q, and r are the body-axis angular rates. The moment equations have the form

$$\dot{p} = \frac{L}{I_{xx}}$$
$$\dot{q} = \frac{M}{I_{yy}} + \left(1 - \frac{I_{xx}}{I_{yy}}\right)pr \tag{30}$$
$$\dot{r} = \frac{N}{I_{yy}} - \left(1 - \frac{I_{xx}}{I_{yy}}\right)pq$$

where L, M, and N represent aerodynamic moments about the body axes. Finally, I_{xx} and I_{yy} are rolling and pitching moments of inertia, respectively, and are assumed to have nearly constant values.

In order to simplify roll control in bank-to-turn flight, we now introduce an aerodynamic bank angle (μ) about the velocity vector. The bank angle dynamics are described by Eq. (31).

$$\dot{\mu} = \left(\frac{p\cos\alpha + r\sin\alpha}{\cos\beta}\right) + \frac{(a_x \sin\alpha - a_z \cos\alpha)\tan\beta}{V} \tag{31}$$

The following development assumes the presence of a guidance law which commands angle of attack, sideslip angle, and bank angle. There are several possible alternative choices for the roll command, including body roll angle (ϕ), body roll rate (p), and stability-axis roll rate ($p_s = p\cos\alpha + r\sin\alpha$).

Introducing more compact notation,

$$x = \begin{Bmatrix} \alpha \\ \beta \\ \mu \end{Bmatrix} \quad \omega = \begin{Bmatrix} p \\ q \\ r \end{Bmatrix} \tag{32}$$

the last two of Eq's. (29) may be rewritten along with Eq. (31) as

$$\dot{x} = a_f + T(x)\omega \tag{33}$$

where T(x) and a_f are given by Eq. (34).

$$T(x) = \begin{bmatrix} -\cos\alpha\tan\beta & 1 & -\sin\alpha\tan\beta \\ \sin\alpha & 0 & -\cos\alpha \\ \cos\alpha\sec\beta & 0 & \sin\alpha\sec\beta \end{bmatrix}$$

$$a_f = \begin{Bmatrix} \dfrac{-a_x\sin\alpha + a_z\cos\alpha}{V\cos\beta} \\ \dfrac{-(a_x\cos\alpha + a_z\sin\alpha)\sin\beta + a_y\cos\beta}{V} \\ \dfrac{(a_x\sin\alpha - a_z\cos\alpha)\tan\beta}{V} \end{Bmatrix} \tag{34}$$

The autopilot design methodology presented above will now be applied to the system described by Eq. (33) and Eq. (30). First, body-axis angular rate commands which stabilize the aerodynamic angle error dynamics are computed as follows:

$$\overline{\omega}_c = T^{-1}(x)(K_x\tilde{x} + \dot{x}_c - a_f) \tag{35}$$

where we have used dynamic inversion. The positive definite gain matrix K_x is typically chosen to be diagonal. An integral term may also be included in Eq. (35), but has been omitted here for simplicity. No adaptive control terms are necessary at this stage of the design, since the nonlinear dynamics are sufficiently well-known. Eq. (35) makes use of available acceleration and velocity information, as well as estimates of angle of attack and sideslip angle which are constructed from inertial data. The accelerometer measurements are filtered and biased appropriately to account for gravitational effects. Note that the matrix T(x) may be inverted, since it is nonsingular for all α and β except $\beta = \pi/2$, which should not be encountered in bank-to-turn flight.

Next, a first-order filter is applied to the guidance commands ($\overline{\alpha}_c$, $\overline{\beta}_c$, and $\overline{\mu}_c$). Its outputs are the filtered commands and first two time derivatives.

$$x_c = \begin{Bmatrix} \alpha_c \\ \beta_c \\ \mu_c \end{Bmatrix} \quad \dot{x}_c = \begin{Bmatrix} \dot{\alpha}_c \\ \dot{\beta}_c \\ \dot{\mu}_c \end{Bmatrix} \tag{36}$$

At this point in the control design procedure, we turn our attention to the body-axis angular rates. First, a pseudo-control input, v, is defined by

$$\dot{\omega} = v$$
$$v = f(x, \omega, \hat{\delta}) \tag{37}$$

where the function $f(\cdot)$ refers to the right-hand-side of Eq. (30). The missile's body-axis angular rate dynamics must be approximately inverted to determine the required control input. For an approximate inversion, the body-axis moments are represented linearly:

$$L \approx \hat{L} = L_\beta\beta + L_p p + L_r r + L_{\delta p}\hat{\delta}_p + L_{\delta r}\hat{\delta}_r$$
$$M \approx \hat{M} = M_\alpha\alpha + M_q q + M_{\delta q}\hat{\delta}_q \tag{38}$$
$$N \approx \hat{N} = N_\beta\beta + N_p p + N_r r + N_{\delta p}\hat{\delta}_p + N_{\delta r}\hat{\delta}_r$$

The body angular rate dynamics of Eq. (30) are then rewritten as

$$\dot{\omega} = f(x, \omega, \hat{\delta}) = \hat{F}(x, \omega) + \hat{B}\begin{Bmatrix} \hat{\delta}_p \\ \hat{\delta}_q \\ \hat{\delta}_r \end{Bmatrix} + \Delta(x, \omega, \hat{\delta}) \tag{39}$$

where f is approximated linearly by introducing

$$\hat{F} = \begin{Bmatrix} \hat{F}_p \\ \hat{F}_q \\ \hat{F}_r \end{Bmatrix} \quad \text{and} \quad \hat{B} = \begin{bmatrix} \dfrac{L_{\delta p}}{I_{xx}} & 0 & \dfrac{L_{\delta r}}{I_{xx}} \\ 0 & \dfrac{M_{\delta q}}{I_{yy}} & 0 \\ \dfrac{N_{\delta p}}{I_{xx}} & 0 & \dfrac{N_{\delta r}}{I_{yy}} \end{bmatrix} \tag{40}$$

with

$$\hat{F}_p = \frac{L_\beta\beta + L_p p + L_r r}{I_{xx}} \tag{41}$$

$$\hat{F}_q = \frac{M_\alpha\alpha + M_q q}{I_{yy}} + \left(1 - \frac{I_{xx}}{I_{yy}}\right)pr \tag{42}$$

$$\hat{F}_r = \frac{N_\beta\beta + N_p p + N_r r}{I_{yy}} + \left(1 - \frac{I_{xx}}{I_{yy}}\right)pq \tag{43}$$

Performing the approximate dynamic inversion, the control input ($\hat{\delta}$) is then given by Eq. (44).

$$\hat{\delta} = \hat{B}^{-1}(v - \hat{F}) \tag{44}$$

The next step in the control design is the computation of the pseudo-control input (v). Here, the commands generated by Eq. (35) are filtered to obtain

$$\omega_c = \begin{Bmatrix} p_c \\ q_c \\ r_c \end{Bmatrix} \quad \dot{\omega}_c = \begin{Bmatrix} \dot{p}_c \\ \dot{q}_c \\ \dot{r}_c \end{Bmatrix} \tag{45}$$

which are used in Eq. (14) as follows:

$$\begin{aligned} v = {} & K_\omega\tilde{\omega} + \dot{\omega}_c - v_{ad} \\ & + K_Z(\|Z\|_F + \overline{Z})(\|\tilde{x}\| + \|\tilde{\omega}\|)\tilde{\omega}/\|\tilde{\omega}\| \end{aligned} \tag{46}$$

where v_{ad} is the adaptive control component included to cancel nonlinear inversion errors. As in the aerodynamic angle loops, K_ω is a positive definite gain matrix and the error variable $\tilde{\omega} = \omega_c - \omega$ has been introduced. The matrix K_Z must be chosen as described previously in Eq. (27).

In practice, the stability and control derivatives in Eq. (39) , along with trim values of the control inputs, could be scheduled as a function of flight condition. This, however, presumes that an accurate full-envelope model of the vehicle is available. In the current implementation, the moment coefficients are treated as constants, shifting the burden of gain scheduling to the neural network. The dynamic inversion control law may be simplified further by introducing additional approximations as simulation results warrant. The neural network would then be required to additionally compensate for any effects which are neglected by the dynamic inversion.

In the actual implementation, the control inputs computed by Eq. (44) are distributed among the missile's aerodynamic control surfaces and RCS thrusters using a suitable control allocation algorithm.

Neural Network Architecture

Substituting Eq. (35) into Eq. (33) and Eq. (44) into Eq. (39), the closed-loop error dynamics can be written as follows:

$$\begin{aligned} \dot{\tilde{x}} &= -K_x \tilde{x} \\ \dot{\tilde{\omega}} &= -K_\omega \tilde{\omega} + v_{ad} - \Delta(x, \omega, v) - \bar{v} \end{aligned} \tag{47}$$

A sufficiently large neural network of the type depicted in Fig. 3, is capable of approximately reconstructing the nonlinear inversion error (Δ) and may therefore be used to compute the adaptive contribution to the pseudo-control. The adaptive pseudo-control input in each channel is thus computed by Eq. 14 presented previously. Moreover, a stable learning rule is again given by Eq. (25).

Various neural network topologies and choices of inputs were considered in this study. The results presented correspond to a neural network with only five sigmoidal hidden-layer neurons. Each neuron has an internal activation potential of a = 0.01. Network inputs were chosen to include $\tilde{x}$, $\tilde{\omega}$, x_c, $\dot{x}_c$, v_{ad}, $\|Z\|_F$, and Mach number. Although it does not appear in Eq. 16, Mach number is included here because its variation affects the nonlinear inversion error. This does not alter the proof of stability, since for our purposes Mach number may be considered a bounded, slowly-varying parameter. The network inputs $\ddot{x}_c$ have been omitted in order to reduce the size of the network. Also note that since $\beta_c(t) \equiv \dot{\beta}_c(t) \equiv \ddot{\beta}_c(t) \equiv 0$, these particular inputs may be omitted without loss of generality. The learning rates Γ_W and Γ_V were chosen as diagonal matrices with nonzero elements equal to 400. Other parameter values used in the neural network implementation were $\lambda = 0.01$ in Eq. 25 and $K_Z = 0.5$, $\bar{Z} = 50$ in Eq. 26.

Simulation Results

Numerical results have been obtained using a neural-adaptive nonlinear autopilot in a nonlinear, six-degree-of-freedom (6-DOF) simulation of the agile anti-air missile featured in Ref. 11. A typical engagement geometry for this missile is the "head-on merge" scenario illustrated in Fig. 6. In this example, the missile and target intercept trajectories lie primarily in the local horizontal plane.

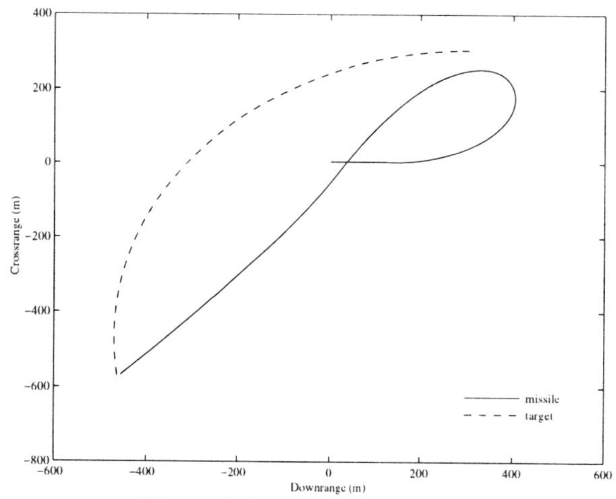

Figure 6: Intercept Trajectories

Fig. 7 compares responses of several autopilot designs to a 90 degree step command in angle of attack. These include a conventional linear gain schedule, the multilayer neural network autopilot described above, and an autopilot design that employs a linear-in-parameters neural network as investigated in Ref. 7.

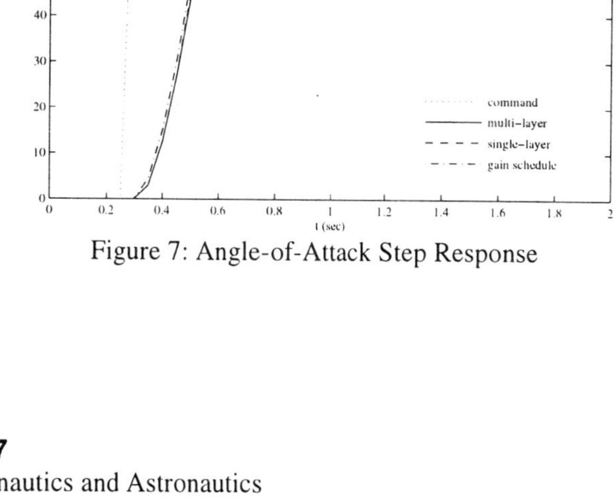

Figure 7: Angle-of-Attack Step Response

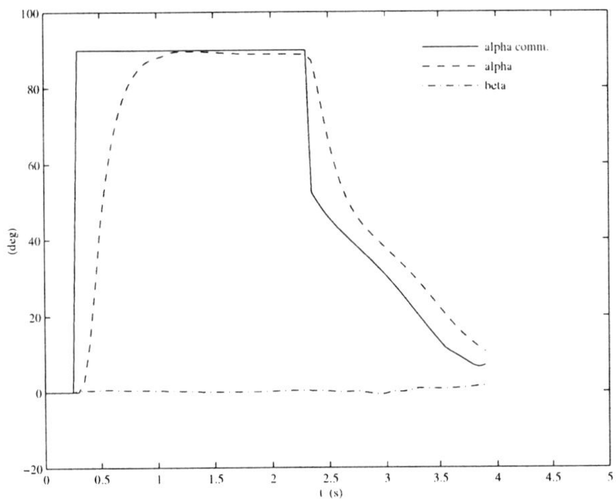

Figure 8: Angle of Attack and Sideslip Responses

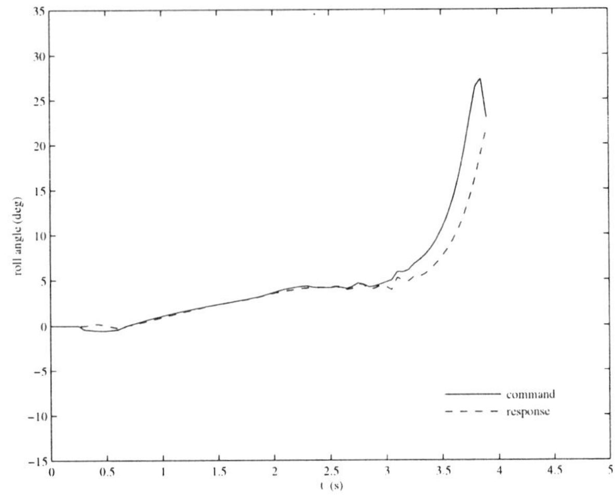

Figure 9: Roll Angle Response

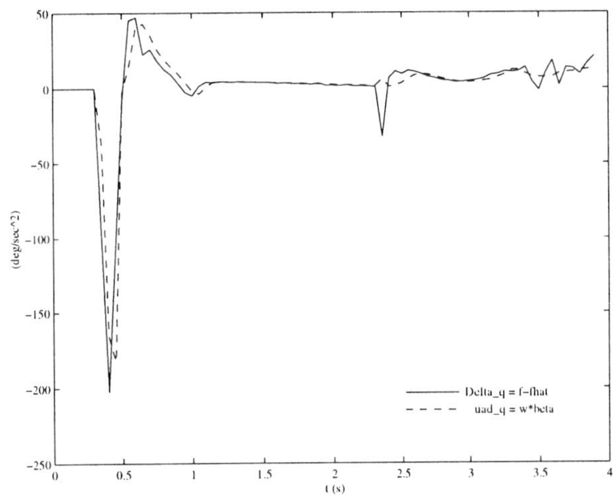

Figure 10: Pitch Rate Inversion Error Reconstruction

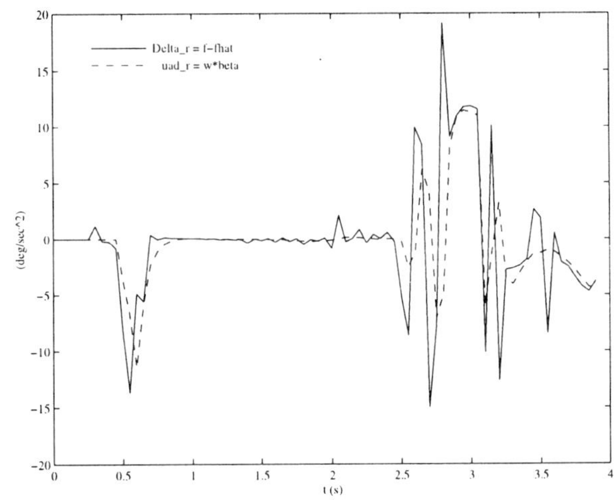

Figure 11: Yaw Rate Inversion Error Reconstruction

The simpler network required a learning rate of 10,000 to achieve desired performance, causing on-line learning interfere with the transient response. In this case, a network with one hidden layer more effectively parameterizes the inversion error, facilitating improved approximation without excessive learning. The multilayer neural network controller surpasses even the gain schedule in performance, with reduced knowledge of the system dynamics.

Moreover, the neural networks in these two controllers have approximately the same number of weights. Here, there are 15 inputs, 5 hidden layers, and 3 outputs, which (with thresholds) add up to a total of 98 adjustable parameters. The linear-in-parameters network had 10 inputs and performed a second-order polynomial approximation with 111 unknown coefficients. One disadvantage associated with the introduction of hidden-layer neurons is the increased complexity and conservatism of the stability proof, including the need for the additional high-gain term $\overline{v}$.

Fig. 8 shows the angle-of-attack and sideslip responses for the engagement depicted in Fig. 6. Excellent tracking is achieved, in spite of the fact that the aerodynamic coefficients were approximated as constants in computing the nonlinear inverse control. Turn coordination is preserved by maintaining zero sideslip throughout the intercept. The roll angle histories presented in Fig. 9 indicate that similar tracking performance is achieved in the roll channel.

Figures 10 and 11 illustrate the neural network's ability to reconstruct nonlinear inversion errors in the pitch rate and yaw rate dynamics. The neural network response illustrated in Fig. 11 indicates that the most severe inversion errors are associated with the yaw rate dynamics. These errors are caused by variations in the aerodynamic moment coefficients with Mach number and angle of attack, which were not modeled in the nonlinear control design. Based on the sideslip response in Fig. 8, however, it is unnecessary to modify the network for improved yaw rate error reconstruction.

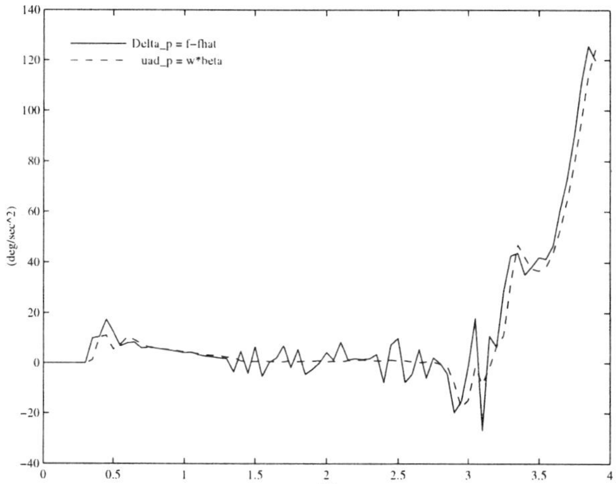

Figure 12: Roll Rate Inversion Error Reconstruction

Fig. 12 depicts the neural network reconstruction of roll rate inversion errors. Modifications which tend to improve the response of the network include adding hidden-layer neurons and increasing the learning rates. No significant improvements were observed, however, for greater than 10 hidden-layer neurons. Five neurons were ultimately used because this number provided a balance between network size and accuracy. Similarly, increasing the learning rates improves the network's approximation of Δ up to the point at which the weight update becomes too fast to be adequately represented in the discrete-time simulation.

Finally, Fig. 13 is included simply to illustrate the variation of the weight matrices over time, as captured by $\|\hat{V}\|_F$ and $\|\hat{W}\|_F$. Because the neural network provides only a local approximation of the inversion error, the weights do not approach steady-state values. Instead, learning occurs when conditions vary most rapidly.

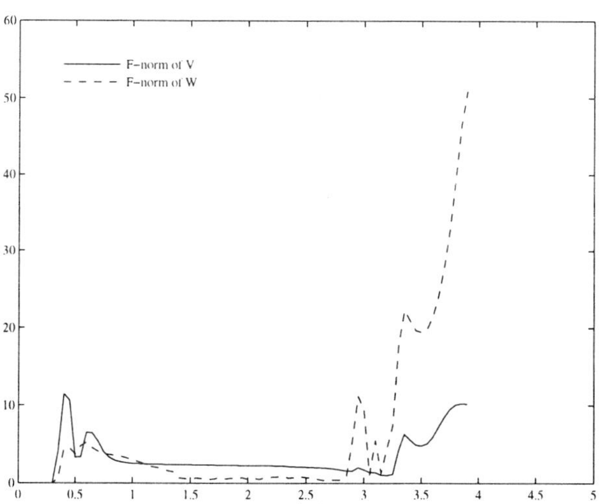

Figure 13: Neural Network Weight Update

Conclusions

A nonlinear bank-to-turn missile autopilot based on approximate dynamic inversion has been proposed. With the aid of a multilayer neural network, this autopilot tracks guidance commands in angle of attack and bank angle while holding sideslip angle near zero. A more traditional skid-to-turn autopilot design could be developed simply by replacing the bank angle state (μ) with roll angle (ϕ) and maintaining $\phi = \dot{\phi} = 0$.

The effectiveness of this control system has been demonstrated in a nonlinear 6-DOF agile anti-air missile simulation. Numerical results indicate distinct improvements over single-layer networks used in Ref. 7. Moreover, such improvements are achieved using fewer adjustable weights. The price of improved inversion error reconstruction is the need for additional design parameters λ, K_Z, and $\overline{Z}$. Lyapunov stability theory dictates only that these gains should be sufficiently large, but excessive conservatism in assigning their values may degrade performance. Nonetheless, we have demonstrated that neural networks enable an approximate nonlinear controller to efficiently adapt on-line to uncertain nonlinear aerodynamic phenomena. This is an especially important feature when, as in the missile autopilot example, these phenomena are difficult or impossible to accurately model for purposes of design and simulation.

Neural network based adaptive control of nonlinear systems is a maturing technology area which promises to be applicable to a wide range of systems. To date, challenging nonlinear control problems characterized by simultaneous non-affine dependence on plant inputs and nonlinearly parameterized uncertainties have been addressed successfully using these techniques. Meanwhile, progress has also been made in the area of robustness to unmodeled dynamics, as described in Ref. 12. Other current research is directed toward implementing an adaptive nonlinear control system which features multilayer neural networks along with robustness to unmodeled dynamics. Discrete-time implementation issues, including the use of recurrent neural networks, are also being actively investigated. Each of these advancements represents another step toward the development of a design technique powerful enough for tomorrow's demanding control problems.

References

1. Tahk, M., Briggs, M., and Menon, P. K. A., "Application of Plant Inversion via State Feedback to Missile Autopilot Design," *Proceedings of the 27th Conference on Decision and Control*, Austin, TX, 1986.

2. Innocenti, M., and Thukral, A., "Simultaneous Reaction Jet and Aerodynamic Control of Missile Systems," *Proceedings of the AIAA Guidance, Navigation, and Control Conference*, Monterey, CA, 1993, pp. 347-354.

3. McDowell, D. M., Irwin, G. W., and McConnell, G., "Online Neural Control Applied to a Bank-to-Turn Missile Autopilot," *Proceedings of the AIAA Guidance, Navigation, and Control Conference*, Baltimore, MD, 1995, pp. 1286-1294.

4. Kim, B. S., and Calise, A. J., "Nonlinear Flight Control Using Neural Networks," *AIAA Journal of Guidance, Control and Dynamics*, Vol. 20, No. 1, pp. 26-33.

5. Leitner, J., Calise, A., and Prasad, J. V. R., "Analysis of Adaptive Neural Networks for Helicopter Flight Controls," *Proceedings of the AIAA Guidance, Navigation, and Control Conference*, Baltimore, MD, 1995, pp. 871-879.

6. McFarland, M. B., and Calise, A. J., "Neural Networks for Stable Adaptive Control of Air-to-Air Missiles," *Proceedings of the AIAA Guidance, Navigation, and Control Conference*, Baltimore, MD, 1995.

7. McFarland, M. B., and Calise, A. J., "Neural-Adaptive Nonlinear Autopilot Design for an Agile Anti-Air Missile," AIAA-96-3914, *AIAA Guidance, Navigation and Control Conference*, San Diego, CA, 1996.

8. Lewis, F. L., Yesildirek, A., and Liu, K., "Multilayer Neural-Net Robot Controller with Guaranteed Tracking Performance," *IEEE Transactions on Neural Networks*, Vol. 7, No. 2, March, 1996.

9. Krstic, M., Kanellakopoulos, I., and Kokotovic, P., *Nonlinear and Adaptive Control Design*, John Wiley and Sons, New York, 1995.

10. Menon, P. K. A., *et al.*, *Integrated Control and Guidance Systems for High Angle of Attack Missiles*, Final Report, Contract No. F08630-94-C-0015, USAF Wright Laboratory (AFMC), February, 1995.

11. Wise, K. A., and Broy, D. J., "Agile Missile Dynamics and Control," AIAA-96-3912, *AIAA Guidance, Navigation and Control Conference*, San Diego, CA, 1996.

12. McFarland, M. B., and Calise, A. J., "Robust Adaptive Control of Uncertain Nonlinear Systems Using Neural Networks," *Proceedings of the American Control Conference*, 1997, *to appear*.

AIAA-97-3541

TERMINAL AREA GUIDANCE INCORPORATING HEAVY WEATHER

Jimmy Krozel, Ph.D.,
Tara Weidner, and George Hunter
Seagull Technology, Inc.
16400 Lark Ave, Los Gatos, CA 95032

Abstract

Inclement weather is a major cause of air transportation delays, and the majority of this cost is incurred in terminal areas. Even though current sensor technology, meteorological methods, and displays are capable of producing accurate and timely mesoscale nowcasts in the terminal area, controllers still find it difficult to vector aircraft around storm cells, maintain safety, and achieve efficient throughput. In short, the weather and airspace management information is available, but the lack of time to mentally solve this complex problem causes inefficient solutions and high workloads for air traffic controllers. In this paper, we model the weather avoidance problem with two algorithmic route planning methods, a grid-based Dijkstra's algorithm and a weighted regions problem algorithm, in order to synthesize arrival aircraft trajectories that avoid penetrating heavy weather. The goal is to create flexible models for the terminal area that provide a means to investigate how aircraft can efficiently be guided through and around weather.

Introduction

Inclement weather is a major cause of air transportation delays, and the majority of this cost is incurred in terminal areas — within 100 nmi from an airport. Furthermore, domestic airline schedules are currently arranged for visual meteorological conditions, so delays occur when storms force deviations from this nominal operating condition. This is especially true when weather impacts a major hub, because delays ripple throughout the entire system [2].

With scores of aircraft and storm cells simultaneously moving about the terminal airspace, the mental workload for an air traffic controller is high, and throughput is adversely affected. Even though current sensor technology, meteorological methods, and displays are capable of producing accurate and timely mesoscale nowcasts in the terminal area, controllers still find it difficult to vector aircraft around storm cells, maintain safety, and achieve efficient throughput. Controllers are left to somehow bridge this gap mentally, in real-time. In short, the weather and airspace management information is available, but the lack of time to mentally solve this complex problem causes inefficient solutions and high workloads for air traffic controllers. The result is that aircraft are delayed during heavy weather conditions as only a fraction of the airspace is utilized. To better understand this problem domain and to propose decision support tools, further engineering investigations are needed.

Related Work

While significant work has been done in using winds in trajectory synthesis and traffic scheduling, little work has been done in using storm data in the Air Traffic Control (ATC) process. For example, the NASA Ames CTAS (Center-TRACON Automation System) program imports the latest wind field forecast and uses it in scheduling arrival aircraft [1, 6, 7], but no attempt has yet been made at using storm data. In our previous research [8], weather intensity is integrated over candidate trajectories to determine the best route among a set of fixed candidate routes; this development, sponsored by the FAA TASS (Terminal Area Surveillance Systems) program office, is the foundation for the work in this paper.

Recent work performed by Dixon [4] deals with combining weather and traffic data into ATC automation tools; this work is also directly related to our investigation. Trajectories were constrained to constant altitude and fixed starting and stopping points. ATC restrictions and operations (*e.g.*, traffic spacing) were not considered. The purpose of the study was to evaluate the effects of weather on the theoretically derived aircraft deviations, and to begin to determine sensor and forecast requirements.

Presented at the AIAA Guidance, Navigation, and Control Conf., New Orleans, LA, Aug., 1997.

Variable route techniques, which is the basic focus of this paper, have primarily been developed for mobile robot applications in both 2D and 3D domains. In 2D domains, grid-based approaches are the most popular, due to their simple form and ease of implementation. Quadtree and weighted regions solutions are becoming more popular as further research identifies the efficiency benefits of these techniques. In 3D domains, grid and multi-grid based techniques have been investigated, but not to the degree of research that 2D methods have received. A review of many of these route planning techniques is given by Mitchell [10].

Pilot Survey

Pilot surveys, interviews, and analysis of heavy weather flight data are necessary to ascertain *what* determines the level of weather that pilots are willing to fly through and *how close* pilots are willing to fly to storm cells. Pilots (and airlines) often prefer a safety buffer between heavy weather and their aircraft. But this constraint reduces the effective airspace available to synthesize routes. Additionally, even if a route successfully avoids adverse weather cells, if routes do not include a sufficient factor of safety, pilots may invalidate our modeling techniques. Thus, pilot surveys are required to establish acceptable values for weighting cells in our solution approach – establishing the cost function for the weather avoidance problem.

To examine actual pilot preferences, we assess initial results of a NASA-funded Adverse Weather Pilot Survey effort, part of their Terminal Area Productivity (TAP) program. The survey responses analyzed were received from 100 very experienced commercial pilots logging an average of over 12,000 flight hours and 16 years of experience. Two questions from this general questionnaire are particularly relevant to this paper. The first, shown in Figure 1, assesses the importance of three attributes in the pilot's acceptance of a bad weather route; (a) overall weather penetration, (b) duration spent in significant weather, and (c) proximity to adjacent, more severe weather. The responses reveal that all three attributes are very important (Level 5) over 60% of the time (74%, 70% and 63% respectively). Indeed, when viewed in histogram form in Figure 2, there appears to be little difference between the three. However, "overall weather penetration" seems to be the most important, showing the largest importance ranking as well as the smallest response variation. Conversely, "proximity to more adjacent, more severe weather" ranked lowest and highest response variability.

Survey Question: How important are the following aspects in your acceptance of an adverse weather route?

(a) Overall weather penetration
(b) Duration spent in significant weather
(c) Proximity to adjacent, more severe weather

Figure 1. Question appearing on pilot survey.

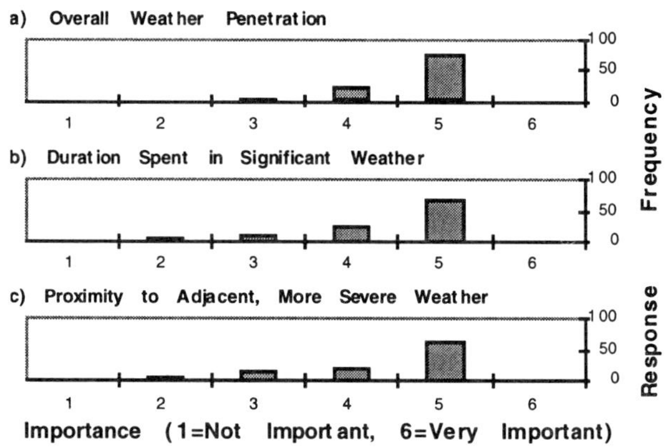

Figure 2. Results to survey question on aspects of weather penetration.

In the second question, shown in Figures 3 and 4, the pilots were asked to evaluate six alternate adverse weather routes. These routes represented possible paths around bad weather during arrival into Dallas-Ft. Worth Airport. Route A is the nominal arrival route but traverses the most severe weather, as shown by the radar return of Figure 3. Route F is the clearest weather route but requires flying an additional 10 miles. Pilots were asked to identify whether each route, if given as a clearance, would be acceptable as well as rank each route in preference order. Table 1 shows the proportion of pilots willing to accept each route and Figure 5 shows the rankings. The clearer weather routes E and F were significantly more acceptable to pilots. Route D is also marginally acceptable with 51% acceptance and most ranking it within their top 3 route choices. The remaining 3 routes would not be acceptable to 68% to 100% of the respondents, rejections growing in relation to the route's most severe level of weather penetration. The low variability on the ranking of routes A, B, C and D of Figure 5 indicates that there is high consensus on the ranking, unlike Routes E and F. Additionally, it is interesting to note that the results overwhelmingly show that pilots are willing to travel the two extra miles of Route D rather than penetrate the low level weather island of Route C. Unfortunately, the pilot comments raise some caution. Comments such as "Would rather hold", "Depends on how it looks

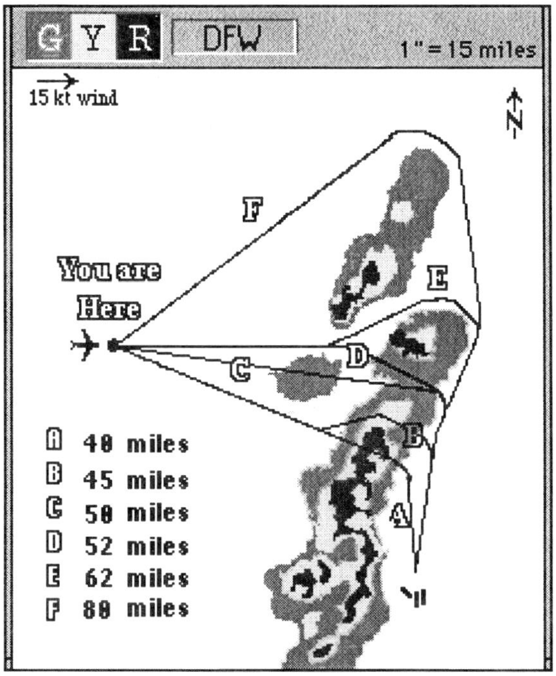

Figure 3. Pilot survey illustration.

	Would you reject this clearance?	Preference Rank (1-6) (1=Best Route)	Comments
Route A	Y N	______	______
Route B	Y N	______	______
Route C	Y N	______	______
Route D	Y N	______	______
Route E	Y N	______	______
Route F	Y N	______	______

Figure 4. Question appearing with illustration.

visually," "Only in an emergency" show that a pilot's response under real-life circumstances could differ. Additionally, comments such as "None of these would be acceptable", "Too close to red areas by company procedures", "Safety first, distance flown least important" may be overly simplistic responses that would differ under real conditions.

These survey results set the stage for this paper. The first question indicates that evaluations of pilot perceptions of route weather severity should include all three attributes, shown in Figure 2, in a cost function. The second question reveals the tradeoff of weather severity and distances helping to identify weights of specific attributes. Finally, pilot comments indicate the need to study actual pilot routing under adverse weather in order to derive a more complete picture of pilot preferences.

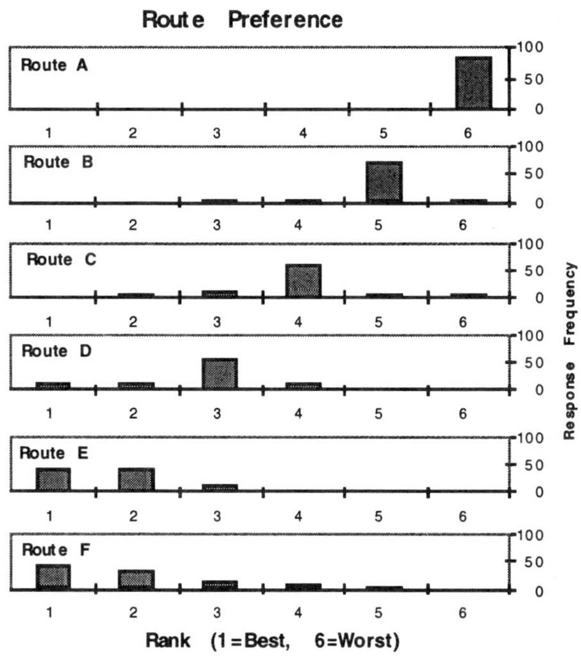

Figure 5. Results to survey question on aspects of weather penetration.

Table 1. "Would you reject this clearance?"

ROUTE	YES, REJECT	NO, ACCEPT
Route A	100 %	0 %
Route B	93 %	7%
Route C	68 %	32 %
Route D	49 %	51 %
Route E	7 %	93 %
Route F	5 %	95 %

Modeling

In this section, the models for fixed and variable routes, weather, and runway constraints are discussed.

Modeling Routes

In order to best exploit weather information in the terminal area, aircraft must have the flexibility to maneuver around weather. The best route can be selected from a set of fixed routes or from a variable route synthesis technique.

In a prior research effort [8], a database of fixed routes is assessed. Figure 6 shows the routes used for the Bridgeport metering fix arrival traffic in the Dallas-Ft.Worth (DFW) terminal area. Each route starts at a metering fix and ends at a runway. Fixed routes are

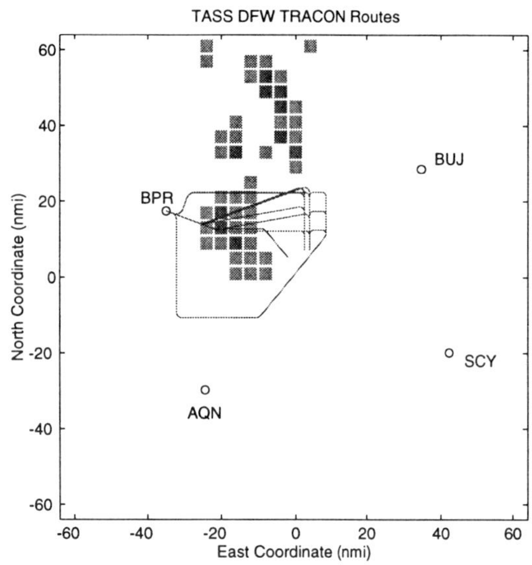

Figure 6. Fixed routes used for arrival routes from the Bridgeport BPR metering fix to the DFW runways (o denotes a metering fix).

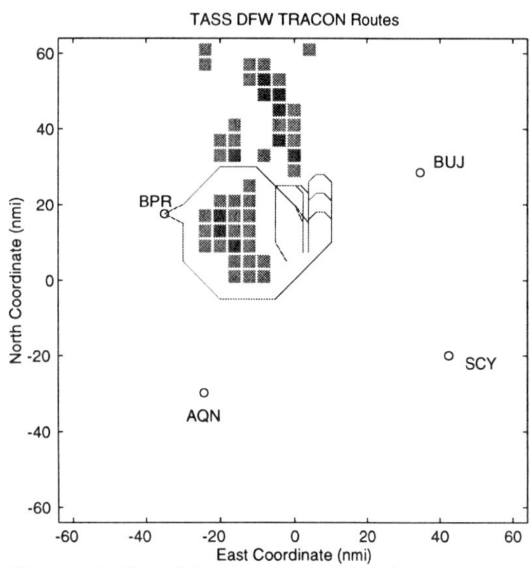

Figure 7. Variable routes that are designed to avoid weather cells (o denotes a metering fix).

defined by a set of fixed waypoints which are easy to define, implement, and select. In the case of selection, for example, the algorithm from our previous work in [8] tests each possible route, working from the fastest to the slowest route. The search is terminated when a route is found that is both conflict-free and below a weather intensity threshold. However, given the variety of weather patterns, it is difficult to know how many fixed routes are needed to produce good results. Fixed routes may or may not provide for good weather avoidance. For instance, as shown in Figures 6 and 7, many of the Bridgeport (BPR) metering fix routes could be covered by a weather pattern (Figure 6), which variable routes could negotiate (Figure 7).

Variable routes are defined by either a grid of waypoints, or a series of flight segments. Variable routes define a *range of paths* because there are multiple paths represented in the grid, or with flight segments with variable length. Variable routes provide a richer set of routes than fixed routes, so that the clear airspace can be exploited better. The drawback is that they require more computational resources to synthesize routes.

In addition to defining the horizontal path, each route also contains speed control information. The placement of these deceleration segments affects flight time; thus, the adjustment of them is referred to as *speed control.* Speed control fine-tunes the arrival time and traffic spacing. It is not, however, an effective tool for weather avoidance. This is because aircraft travel more than an order of magnitude faster than storms. Minor changes in the aircraft flight time have little effect on the storm-aircraft relative locations. In our research, speed control is primarily used to optimize aircraft-to-aircraft spacing. The primary tool for weather avoidance is the horizontal path selection, referred to as *path control.* Thus, the remainder of this paper will be primarily focused on path control techniques.

Modeling Weather

In general, weather can be modeled in 2D or 3D for the terminal airspace. Data in the form of Airport Surveillance Radar (ASR–9) Vfill is an example of a 2D form and the Next Generation Weather Radar (Nexrad) 3D grid is an example of a 3D form; see [5] for an operational comparison. The spatial resolutions of these two grids are similar. These data are typically registered as the ASR Vfill data (6 weather levels) or levels of radar reflectivity (0 - 60+ dBZ). A 3D grid allows ATC to vector aircraft both above (assuming that the storm does not extend too far above the terminal airspace) and around storms (note, however, vectoring aircraft beneath storms is problematic because storms can quickly drop in altitude). With a 2D grid, the routing algorithm has no knowledge of a storm's altitude extent, so it cannot exploit clear air that might be available above the storm. The tradeoffs are: (*i*) the Nexrad grid is 3D whereas the ASR-9 Vfill is 2D, (*ii*) Nexrad has a slower update rate due to a more timely scanning strategy, and (*iii*) Nexrad has greater storm data spatial resolution. In this paper, we only consider 2D grid-based data and one method of data weighting, which is described next.

For 2D weather data, the weather may be modeled in a grid, multi-grid, or a polygonal data structures. Data structures can use the original raw data, or the raw data can be passed through a filter which groups the data into 5 bins with set weights, w_i, $i=1,2,...,5$ as described by Table 2. With little or no weather data, the "no color" weight is defined to be $w_0 \equiv 1$. The values for all other weights w_i are determined through pilot surveys and engineering analysis so that the resultant variable routes are acceptable to pilots. The weights w_i define locally (within grid cells or within polygon regions) the penalty for traversing a heavy weather cell. Thus, a route π has a total cost C_π defined by a line integral:

$$C_\pi = \int_\pi w(s)ds \tag{1}$$

where ds is the differential element along the arc length of the route π.

Table 2. Weights for Weather Cell Data

Color	Data Range	Weight
No Color	0-19 dBZ	$w_0 \equiv 1$
Light Green	20-29 dBZ	w_1
Dark Green	30-39 dBZ	w_2
Yellow	40-49 dBZ	w_3
Red	50-59 dBZ	w_4
Magenta/Purple	60+ dBZ	w_5

In a uniform 2D grid form, each grid cell represents a square region of airspace in a 2D plane. Figure 8 illustrates weather data in grid form. Weather data are typically recorded in a grid form, so the benefit of using a grid form is that there are no additional data structures to be created. Also, as weather moves and weather data are updated, the data structure need not change in size or shape. The space complexity of the data is represented by the size of the grid, which remains a constant. For instance, the number n=XY grid cells for a X×Y grid. Figure 8 illustrates a 32×32 grid with n=1024 size. The drawback to using the grid form is that the variable route techniques derived in grid form require larger computation times; this topic will be discussed further later.

To reduce the size of the weather data structure, a multi-grid quadtree technique may be implemented [15, 16]. The motivation of using quadtrees is that this data representation requires less space complexity by taking advantage of aggregation of homogeneous cells of data. In the worst case, the quadtree will require the same amount of space as a uniform grid. (A quadtree technique was not implemented in our study.)

Weather data can also be processed to form polygonal regions of homogeneous data. Figure 9 shows an example of this approach. The motivation for the polygonalization technique is that it models

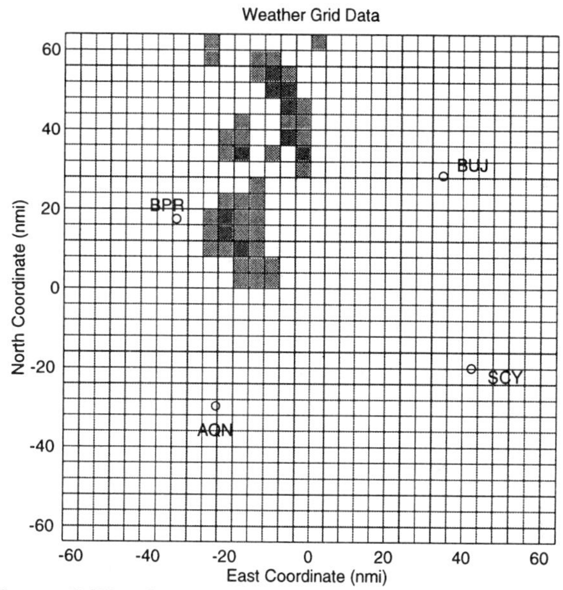

Figure 8. Weather data in a grid-based representation.

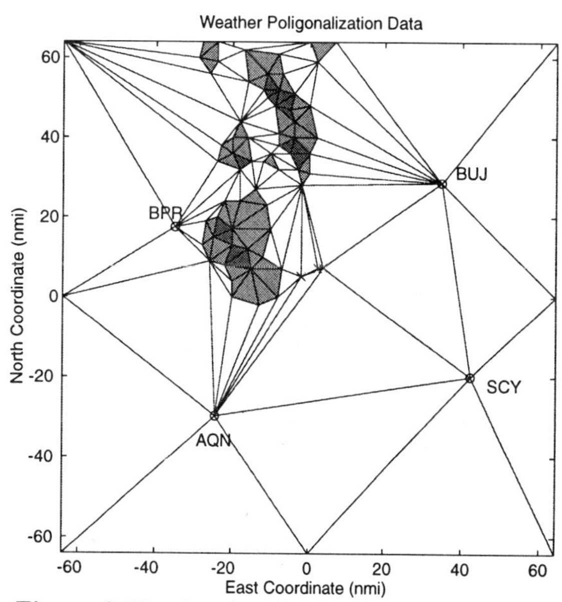

Figure 9. Weather data in a polygon representation.

weather cell shapes directly without digitization bias [10] or aliasing effects due to a fixed axis system. In this form, the space complexity is described by both the number of polygons p and the number of vertices v for all polygons. For example, in Figure 9, $p=137$ polygons and $v=73$ vertices. The polygonalization technique avoids modeling open spaces with no weather to model, and also provides for an efficient method of generating variable routes. The drawback, like the quadtree technique, is that a polygonalization must be computed each time weather data are updated.

Modeling Airport Runways

Fixed and variable routes terminate at airport runways. For variable routes, the runway is modeled with a grid-based or polygon-based model, consistent with the form of the weather model. In addition to the runway itself, heading constraints for the final approach to the runway are also imposed in the model. Heading constraints are enforced by requiring trajectories to terminate in-line with the runway. For grid-based techniques, this means that the grid axis system is oriented in-line and perpendicular to the runway and trajectories are not allowed to pass through grid cells on either side of the runway. For polygon techniques, polygon regions on both sides of the runway are disallowed for trajectory synthesis. Figures 10 and 11 illustrate the grid-based and polygon-based models used for a runway and the runway heading constraints. Figure 12 illustrates routes that are generated using the weighted regions problem [11] (discussed later); notice how the lower weights in the area near the final approach fix allow routes to turn to final approach in-line with the direction of the terminal runway. This technique also allows for the use of extended final approaches. "No Go" regions model the minimum final approach required.

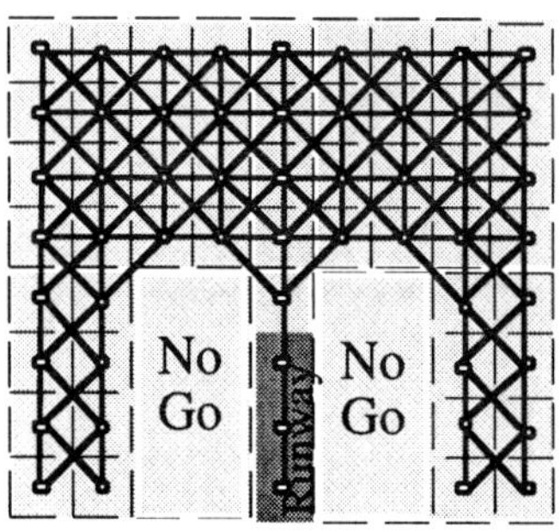

Figure 10. Runway model in a grid-based representation with No Go regions to enforce terminal heading constraints.

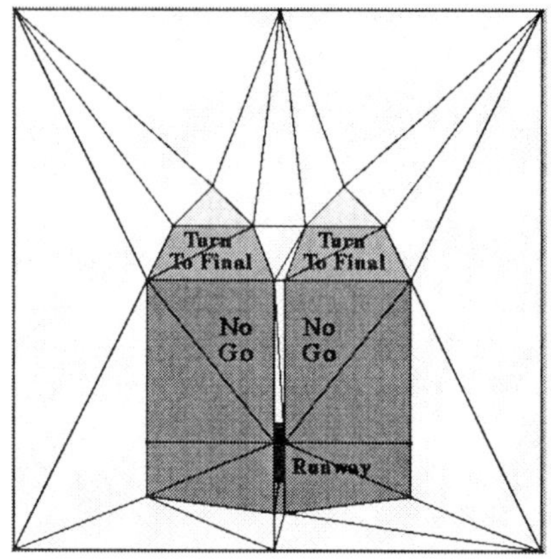

Figure 11. Runway model in a polygon representation with weighted No Go regions to enforce terminal heading constraints.

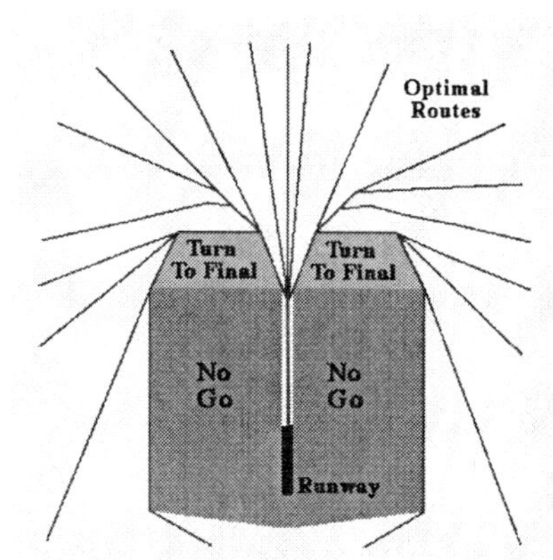

Figure 12. Optimal routes leading to the runway based on the polygonal representation and weighted regions problem solutions.

Algorithmic Techniques for Variable Routes

In this section, we present the algorithms that we use to construct variable routes. The techniques are based on the weather data being in grid, multi-grid, or polygonal form, as previously described.

In general, to construct variable routes, the weather data are abstracted into a graph where vertices of the graph represent grid, multi-grid, or polygonal cells of data. Edges of the graph represent the costs to travel between vertex locations. By searching the graph for a path that minimizes distance weighted by the exposure to weather, a solution path from the metering fix location to runway location can be found. For this discussion, a constant weather pattern with no motion will be used to explain the search methods, and later moving weather patterns will be considered.

Grid-Based Algorithms

One solution for variable routes uses a grid-based search method. Generally, the airspace is modeled with a graph G as an N-dimensional lattice (N=2,3). For a 2D airspace, the graph G might be a 4-neighbor, 8-neighbor, or 16-neighbor connected uniform grid, or a lattice that connects adjacent quadtree multi-grid cells. For a 3D airspace, the graph G might be a 6-neighbor or 26-neighbor connected lattice, or a lattice that connects adjacent octree cells. A flight path π from a start vertex v_0 to a runway terminal vertex v_k in a graph G is a non-null sequence of alternating vertices and edges, $\pi = (v_0, e_1, v_1, e_2, \ldots, e_k, v_k)$, such that the first vertex is v_0, the runway terminal vertex is v_k, and edge e_i connects vertex v_{i-1} to vertex v_i for $i = 1, 2, \ldots, k$, and all vertices and edges are distinct. Flight paths are also written as a sequence of vertices, without specifying edges, *e.g.*, $\pi = v_0 v_1 \ldots v_k$. Basically, grids or lattice networks are constructed by modeling a vertex of the graph G for each grid or multi-grid weather cell. The travel cost $c(e)$ defines a positive cost for each edge e in the graph G modeling the airspace. The path cost C_π for a flight path $\pi = v_0 v_1 \ldots v_k$ is defined to be:

$$C_\pi(v_0, v_k) = \sum_{e \in \pi} c(e). \qquad (2)$$

which is a discrete approximation of equation (1).

Next, we consider a search problem that identifies the solution for a variable route: find a least cost path in a simple graph G from a start vertex v_0 to a goal vertex v_k. An optimal path from v_0 to v_k in a graph G can be found by applying a graph searching technique such as Dijkstra's algorithm [3] or the A^* algorithm [12]. Alternatively, one can find the least cost path from many start vertices to one runway terminal v_k; the collection of these optimal paths constitute what is called an Optimal Path Map OPM [9]. The OPM, as shown in Figure 13, is a map that guides aircraft from arbitrary start locations (including metering fixes) to the runway located at v_k. For an OPM, Dijkstra's algorithm or the A^* algorithm is used to propagate a search from the runway terminal vertex v_k outward until optimal paths from one or all of the metering fixes (or beyond) are identified. In searching outward from v_k the intermediate results indicate the minimum cost solution paths from all possible start vertices to the terminal runway vertex v_k; additionally, the cost $C^*(v_0, v_k)$ of a minimum cost path from v_0 to v_k is recorded as an intermediate result. Figure 14 illustrates the optimal routes for the example case of Figure 13. For a given terminal runway vertex v_k, a *search pointer* is assigned to each edge e during a search which indicates an optimal direction to travel. The search pointers are a gradient field representation of the search results; a search

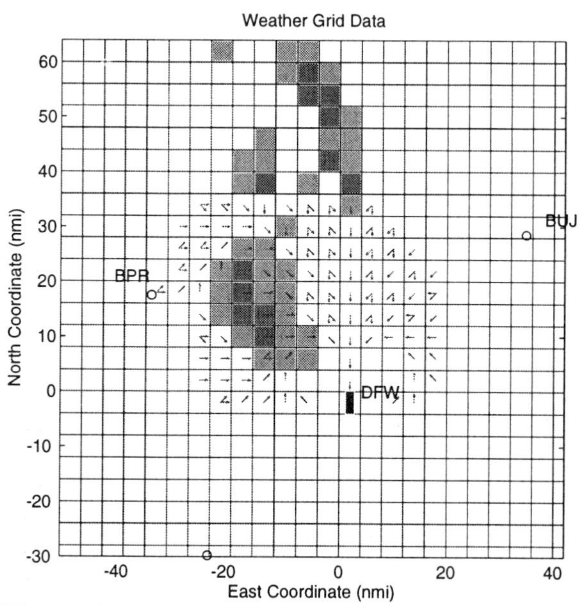

Figure 13. Optimal path maps can be used to create variable routes guiding an aircraft from one or more metering fixes to a runway.

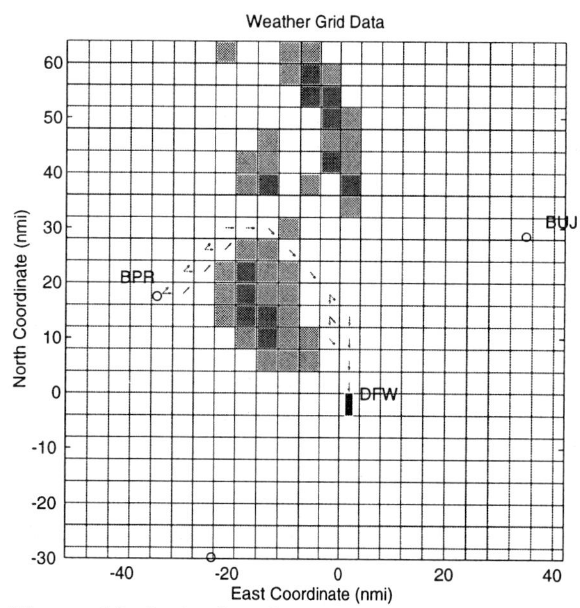

Figure 14. Optimal path maps can be used to create variable routes guiding an aircraft from one or more metering fixes to a runway.

pointer represents the unit vector in the direction of the gradient $\nabla C^*(v_0, v_k)$. Note that search pointers are not necessarily assigned to every edge e investigated during a search. Also, for a vertex v in the graph G, there may be more than one optimal path leading to the vertex v_k, thus, there may be more than one search pointer directed away from the vertex v, as is the case in Figure 14.

By assigning appropriate graph edge costs, flight path segments with heavy weather penalties can be modeled. If there is little or no weather, then the travel cost $c(e)=\delta$, where δ is the distance between grid cells, can be used to identify the minimum distance path. If there is heavy weather, then weather costs, the weights w_i as shown in Table 2, can be assigned to graph edges $c(e)=w_i\delta$ to identify optimal flight paths that minimize the exposure to heavy weather. For example, we have found from experimentation that the weather costs w_1=1.1 for light green, w_2=1.3 for dark green, w_3=1.7 for yellow, w_2=3 for red, and w_2=10 for magenta/red result in paths that "cut the corner" around green colored weather cells, but due to the extreme cost of traversing heavy weather, does not allow the path to penetrate red or magenta weather cells. Flight paths that are acceptable by pilots and controllers are achieved by assigning the appropriate weights w_i to each weather level; these weight assignments are part of the engineering solution to this problem.

Because aircraft have dynamics, scheduling, and restricted airspace constraints, the search algorithms must also be adjusted to reflect these constraints. With limiting turn radius constraints, the search algorithms can be adjusted to disallow solution paths that turn over a specified angle. Turning constraints are currently being implemented in our algorithm by allowing the search to proceed along the graph G with turns of 45° or less — this eliminates 4-neighbor grids, but 8-neighbor and 16-neighbor grids are still applicable. For scheduling and aircraft spacing constraints, the search algorithms can be adjusted to disallow or adjust speed control parameters for flight paths that result in scheduling and/or separation violations. Currently, we have not yet incorporated these constraints into our algorithm. Furthermore, grid-based search techniques easily handle the constraint of restricted airspace zones by weighting the airspace in a restricted airspace zone appropriately.

Polygon-Based Algorithms

A second method for generating variable routes is based on the polygonal form of weather data. The variable route is generated by solving a weighted regions problem [9, 11, 13, 14]. The weather intensity data define weighted polygon regions, as previously shown in Figures 9 and 11. Then, the *continuous* Dijkstra's algorithm is applied to identify the paths with minimal weighted length starting at arbitrary locations and ending at the terminal runway location. The weighted regions solution approach takes advantage of the fact that shortest paths are piecewise-linear and turn only at polygon edges (never within a weighted region). There is no digitization bias [10] imposed by a grid using a weighted regions solution approach. When an optimal path passes from one weighted region with weight w_1 to another adjacent weighted region with weight w_2, the path obeys Snell's Law of Refraction:

$$\frac{\sin\theta_1}{\sin\theta_2} = \frac{w_1}{w_2} \qquad (3)$$

where θ_1 defines the angle of incidence and θ_2 defines the angle of refraction, both defined relative to the normal of the polygon edge boundary. Figure 15 shows an example.

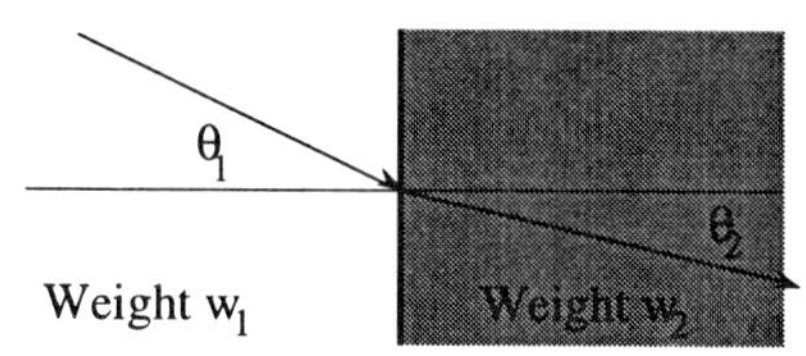

Figure 15. An optimal path through weighted regions obeys Snell's Law of Refraction.

The weighted regions problem can be applied to find optimal path maps and individual optimal routes leading to the runway. Figures 16 through 18 illustrate the application. Figure 16 illustrates the polygonalization of the weather data; note that some smaller clusters of weather data are filtered out prior to the polygonalization process. Figure 17 illustrates the OPM for multiple routes leading to the DFW runway. Figure 18 illustrates the optimal route from the Bridgeport metering fix to the DFW runway. In this example, we found that the weather costs w_1=1.1 for light green, w_2=1.3 for dark green, w_3=1.7 for yellow, w_2=3 for red result in paths that "cut the corner" around light green colored weather cells, but avoid heavy yellow and red colored weather cells.

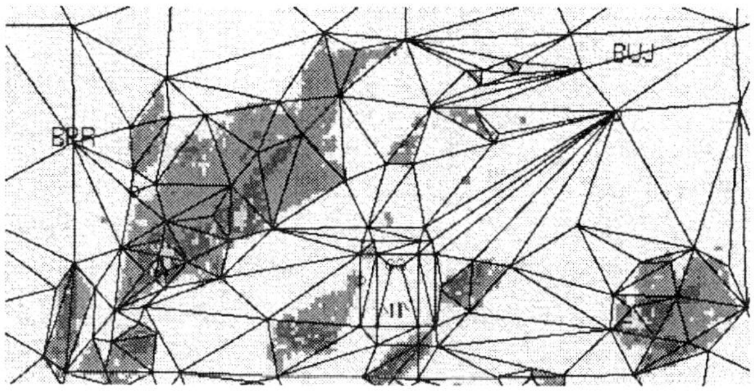

Figure 16. The weighted regions problem uses weather data in polygonal form to create variable routes guiding an aircraft to the runway.

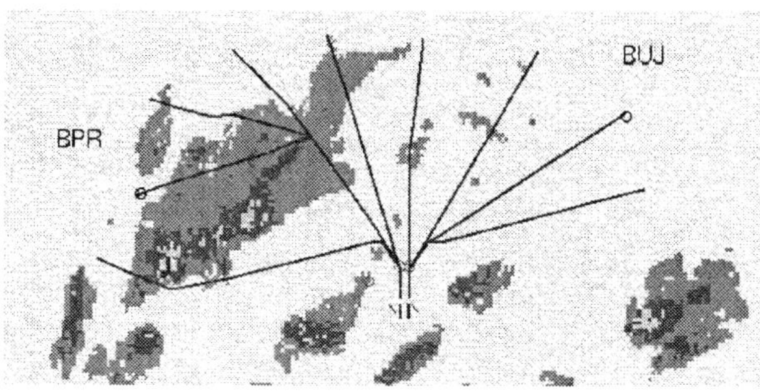

Figure 17. The Optimal Path Map OPM indicating multiple routes leading to the DFW runway.

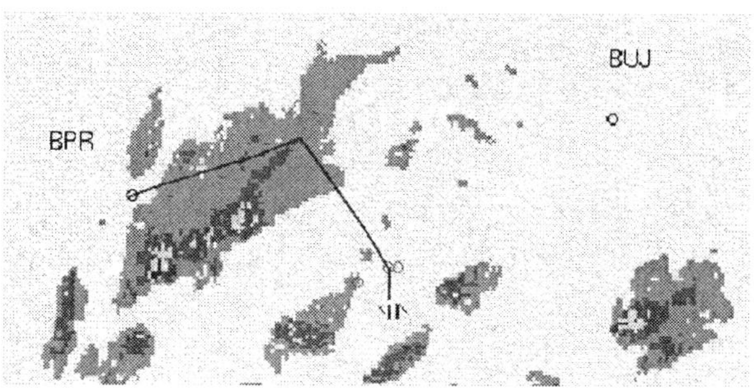

Figure 18. The optimal route solution from the BPR metering fix to the DFW runway.

Just as grid based techniques must be modified to enforce dynamics, scheduling, and restricted airspace constraints, the weighted regions search algorithm must also be adjusted to reflect these constraints. The weighted regions problem evaluates flight path turns only at region boundaries. To include turning radius constraints, the portion of the code that computes Snell's Law must be modified to eliminate turns that violate any turning radius constraints. These alterations have not yet been implemented in our code. Separation constraints due to nearby aircraft may also eliminate certain candidate flight paths or cause speed control adjustments for flight paths that result in scheduling and/or separation violations. Separation constraints have also not been implemented in our code. Finally, the weighted regions technique easily handles the constraints of restricted airspace zones by applying very high weights to restricted airspace zones.

Time Varying Weather

Clearly, weather moves, and further considerations must be given to identify how these grid-based and polygon-based methods must be adapted to account for moving weather patterns. Today, weather prediction is becoming more and more viable, especially within time horizons of 30 minutes or less, which is the case for terminal area guidance. For the sake of this discussion, assume that the weather pattern movement can be predicted by some appropriate mechanism.

The grid-based technique more easily incorporates the dynamics of moving weather compared to the polygon technique. The solution of the grid-based technique propagates a search from the runway location to the metering fixes. At each grid location, the graph G models the grid location with a vertex v_i and an adjacent grid location with a vertex v_{i+1}, and the cost of traversing between adjacent grid locations with the edge weight e_i. In the dynamic weather case, the edge weights are time varying, and the appropriate edge weight is determined by using a weather predictor. Next, because the search can arrive at a vertex at more than one time, the search must also carry along with it a time-to-go parameter T. The search space is thus increased from $X \times Y$ in the non-moving weather case to $X \times Y \times T$ in the moving weather case. To limit the size of the search space, heuristics can be applied to reduce the number of vertices in the search space that have multiple path pointers with multiple times-to-go; for instance, maintaining only the search pointers that have times-to-go within a given window, the lower-bounding time-to-go for the window being the no weather condition time-to-go. At this time, we have not yet incorporated these moving weather constraints into our grid-based algorithm.

In the polygon-based approach the incorporation of moving weather is more difficult. First of all, for the grid-based technique, when the weather moves, the underlying grid structure remains constant while the weights change within grid cells. But for the polygon-based approach, the polygonalization structure itself changes shape as the weights change and the weather moves. Currently, there is no approach published in the literature that deals with such polygon movements in the weighted regions problem. However, an approximation can be made. The approximation is to polygonalize the weather data and solve the weighted regions problem at discrete times t_k, t_{k-1}, ... , t_1, t_0 (where t_k is the current time and t_0 is the touchdown time) using the appropriate weather forecast for each time. Thus, an optimal path map OPM(t_i) is created for each time t_i, $i=0,1,...,k$. The aircraft is first advised to follow the route indicated by the optimal path map OPM(t_k), then at time t_{k-1} the aircraft should switch to following OPM(t_{k-1}), and follow it until the next discrete time period. A route to follow from the metering fix to the runway is created by appending each of these route segments from the appropriate OPM.

Computational Complexity of Algorithms

By studying the computational complexity of fixed and variable route methods, software design considerations can be explored. The algorithmic complexity of the fixed route method is $O(fw)$ where f is the number of fixed routes and w represents the number of waypoint locations used by the fixed routes. This is next compared to variable route methods. The algorithmic complexity of the grid-based variable route solution approach is $O(n \log n)$ where n represents the number of grid locations. The value of n is typically quite large in comparison to the number of fixed routes f or number of polygons p. However, note that grid techniques are not dependent on the number of weather cells, so they will not necessarily speed up when there is little weather nor slow down when there is heavy weather. The algorithmic complexity of the weighted regions solution approach is $O(p \log p)$ where p is the number of polygon regions. These two variable route methods are roughly comparable in complexity, but differ based on the weather cell characteristics. Applying a very fine grid models weather information quite well, but the number of grid locations will be very large and thus will require more computation time to establish routes. On the other hand, with weighted polygonal regions, the number of polygons increases with the number of weather cells, so the amount of computation time is dependent on the number of weather cells.

Summary

We have investigated the problem of guiding aircraft through heavy weather in the terminal area from the metering fixes to runways. From pilot surveys, we have found that solutions that pass through light green regions of weather are acceptable, and on the other extreme, passing through red or magenta regions are unacceptable. The synthesis of appropriate routes can be achieved using algorithmic route planning methods, including grid-based and polygon-based methods. The weights assigned to cells can be established by engineering analysis such that the types of routes resulting from these grid-based and polygon-based methods are acceptable to pilots and controllers. The two algorithms investigated were the grid-based Dijkstra's algorithm and the weighted regions problem algorithm. The grid-based approach has the benefit of a fixed grid structure which does not change for new weather patterns nor as the weather moves. The disadvantage for grid-based techniques is that they require more time to search for aircraft trajectories and can create a digitization bias based on the limiting connectivity between grid cells. From computational geometry, we also considered using the weighted regions problem to model the trajectory synthesis problem. The weighted regions problem allows for weather to be modeled with polygon regions. The disadvantage is that creating a polygonalization from grid-based weather data takes additional processing time; this would have to be performed every time the weather pattern moves. The advantage to this approach is that the total number of polygons used to model weather can be much smaller than the number of cells used in a fine grid, and consequently, finding good routes that avoid heavy weather can be done quickly. There is also no digitization bias for the polygon-based approach.

Acknowledgments

This research was funded by NASA Ames Research Center under contract NAS2-97055 for the Small Business Innovation Research (SBIR) program and under contract NAS2-13767 for the Terminal Area Productivity (TAP) program. The authors appreciate the program support of Dr. Karl D. Bilimoria and Dr. Rhonda Slattery at NASA. The authors acknowledge various technical contributions from Dr. Rich Bortins, Ms. Susan Dorsky, and Paul Davis of Seagull Technology, Inc. The authors thank Prof. J.S.B. Mitchell and Dr. Cristian Mata from State University of New York, Stony Brook, for providing the weighted regions problem software and advice on applying the theory to this weather avoidance application.

References

[1] Davis, T.J., Erzberger, H., and Green, S.M., "Design and Evaluation of an Air Traffic Control Final Approach Spacing Tool", *Journal of Guidance, Control, and Dynamics*, Vol. 14, No. 4, pp. 848-854, 1991.

[2] DeArmon, J.S., "Do Air Traffic Flow Problems Interact?", *IEEE Trans. on Control Systems Technology*, Vol. 1., No. 3, pp. 195-203, Sept., 1993.

[3] Dijkstra, E.W., "A Note on Two Problems in Connection with Graph Theory," *Numerische Mathematik*, Vol. 1, pp. 269-271, 1959.

[4] Dixon, M. and Weiner, G, "Automated Aircraft Routing Through Weather-Impacted Airspace", *Fifth Intern. Conf. on Aviation Weather Systems*, Vienna, VA, pp. 295-298, Aug., 1993.

[5] Dunbar, B.E., Mittelman, J.P., and Dixon, M.J., "ARSR/NEXRAD Operational Comparison", *Air Traffic Quarterly* , Vol. 2, No. 1, pp. 5-32, 1994.

[6] Erzberger, H., "Design Principles and Algorithms for Automated Air Traffic Management," *Knowledge-Based Functions in Aerospace Systems*, AGARD Lecture Series 200, NATO, Nov. 1995.

[7] Erzberger, H., Davis, T.J., and Green, S., "Design of Center-TRACON Automation System", *AGARD Guidance and Control Symposium on Machine Intelligence in Air Traffic Management*, Berlin, Germany, May, 1993.

[8] Hunter, G., Bortins, R., "Effects of Weather Forecasting Errors on Air Traffic Control Automation System Performance," *AIAA Guidance, Navigation, and Control Conf.*, Baltimore, MD, Aug., 1995.

[9] Mitchell, J.S.B., *Planning Shortest Paths*, Ph.D. Dissertation, Stanford University, Stanford, CA, Aug., 1986.

[10] Mitchell, J.S.B., "An Algorithmic Approach to Some Problems in Terrain Navigation," *Artificial Intelligence*, Vol. 37, pp. 171-201, Dec., 1988.

[11] Mitchell, J. S. B., "The Weighted Region Problem", *Journal of the Assoc. for Computing Machinery*, Vol. 38, No. 1, pp. 18-73, Jan., 1991.

[12] Nilsson, N.J., *Principles of Artificial Intelligence*, Tioga Pub. Co., Palo Alto, CA, 1980.

[13] Richbourg, R.F., Rowe, N.C., Zyda, M.J., and McGhee, R.B., "Solving Global, Two-Dimensional Routing Problems using Snell's Law and A* Search," *IEEE Intern. Conf. on Robotics and Automation*, Raleigh, NC, March/April, 1987.

[14] Rowe, N.C. and Richbourg, R.F., "An Efficient Snell's Law Method for Optimal-Path Planning across Multiple Two-Dimensional, Irregular, Homogeneous-Cost Regions," *Intern. Journal of Robotics Research*, Vol. 9, No. 6, pp. 48-66, Dec., 1990.

[15] Samet, H., "The Quadtree and Related Hierarchical Data Structures," *Assoc. for Computing Machinery, Computing Surveys*, Vol. 16, No. 2, pp. 187-260, June, 1984.

[16] Samet, H., *Applications of Spatial Data Structures*, Addison-Wesley, New York, NY, 1990.

FUZZY REASONING-BASED SEQUENCING OF ARRIVAL AIRCRAFT IN THE TERMINAL AREA

John E. Robinson III*, Thomas J. Davis* and Douglas R. Isaacson*
NASA Ames Research Center, Moffett Field, California 94035

Abstract

A fuzzy reasoning-based method for scheduling air traffic in the terminal area has been designed at the NASA Ames Research Center and operationally tested at the Dallas/Fort Worth Terminal Radar Approach Control (TRACON). The scheduling system sequences and assigns landing times to arrival aircraft by utilizing continuous updates of aircraft radar data and controller inputs. The scheduling algorithm contains a knowledge base which was refined during several thousand hours of controller-in-the-loop real-time simulations. This knowledge base applies fuzzy reasoning to evaluate propositions that consider both performance criteria and workload criteria, such as delay reduction and conflict avoidance. Operational test results show that the algorithm determines an efficient arrival plan to land aircraft in a manner acceptable to the air traffic controllers. This paper details the scheduling algorithm and presents data regarding its effectiveness in predicting the landing order of arrival aircraft.

Introduction

The nation's air traffic system has experienced significant growth over the past fifteen years that has resulted in substantial increases in delays at nearly every major airport. This increase in traffic has not, however, been met with many deployed and operationally sound technologies for aiding the air traffic management and controller workforce. Many efforts to alleviate these problems have focused on developing decision support systems that schedule air traffic on a real-time basis in order that the traffic may be better managed and controlled. The main objective has been to provide an accurate prediction of the future traffic situation such that accurate traffic management and control decisions can made early, thus resulting in more efficient traffic flow.

NASA Ames Research Center has been conducting research on algorithms and decision support tools for the extended terminal area for nearly twenty years. This research has resulted in the development of a suite of tools known as the Center/TRACON Automation System (CTAS).[1] CTAS elements include the Traffic Management Advisor (TMA) to aid enroute traffic management, the Descent Advisor (DA) to supply arrival descent clearance advisories and management, and the Final Approach Spacing Tool (FAST) to provide sequence, runway, heading, and airspeed advisories to TRACON air traffic controllers.[2-5]

Previous Work

In order to build air traffic decision support systems such as CTAS, researchers at various labs worldwide have experimented with a wide range of algorithms for scheduling arrival aircraft. All of the methods explored to date build a schedule or plan based on estimated times of arrival at a fixed point, typically a runway threshold or meter fix. In Germany, Volckers successfully developed and deployed the COMPAS system which utilized a branch-and-bound algorithm to schedule arrival aircraft and provide controllers with broad advisories on whether to expedite, maintain, or slow aircraft in order to meet an overall strategic traffic plan.[6] The inputs to this scheduling algorithm included estimated times of arrival at a metering fix and runway threshold based on models of aircraft performance and the arrival route structure. In France, Garcia developed and deployed the MAESTRO system based on similar algorithms and modeling as the COMPAS system. MAESTRO, however, included expert systems logic for determining the most desirable runways in order to maintain proper airport balancing.[7] Brinton developed a more sophisticated branch-and-bound algorithm which again scheduled aircraft at a single point for an early version of a terminal controller tactical advisory tool.[8] While this scheduler produced near-optimal delay minimization, its schedules were found to be too difficult for controllers to implement in the busy terminal area air traffic environment.

Meanwhile, Erzberger designed the scheduling algorithm that forms the basis for the TMA functionality in CTAS.[9] The TMA scheduling algorithm utilizes accurate aircraft, winds aloft, and air

---

* Research Scientist, Terminal Area Air Traffic Management Research Branch

traffic control models as inputs to a 4D trajectory synthesis generator. The trajectory synthesis algorithm generates highly accurate 4D trajectories and estimated times of arrival (ETAs). These ETAs are the inputs that the scheduling algorithm uses to produce scheduled times of arrival (STAs) at the runway threshold and at the meter fix or entry point into a TRACON. Controllers use the STAs to "meter" arrival traffic into the TRACON and thus produce the desired flow rate of traffic. Operational tests of the TMA scheduling system produced highly successful results at Fort Worth Center in mid-1996.[10]

All of the methods for scheduling arrival aircraft described above were attempted during the development phase of FAST; however, all were ultimately unsuccessful. While these methodologies produce favorable results for decision support tools that provide strategic planning, none encompass the necessary factors to be acceptable in a more tactical decision support environment. In order to provide the detailed tactical advisories directly to the air traffic controllers, it was learned through real-time, controller-in-the-loop simulations, that the scheduling algorithm must account for a wide range of factors including efficiency, controller workload, and adaptability.

Based on the above, an initial spatial constraint satisfaction algorithm was designed to schedule aircraft at multiple points along their trajectories, not just at the runway threshold or meter fix.[11] Like the TMA scheduling algorithm, this early FAST scheduling algorithm utilized accurate 4D trajectories as inputs. All merges within the extended terminal area were considered by breaking the large global sequencing problem of merging at the runway threshold into a network of smaller local sequencing problems of merges at intersections of trajectory segments. A refined and enhanced spatial constraint satisfaction concept remains the foundation of the present FAST sequencing algorithm and will be presented as the topic of this paper.

Motivation

As a tactical decision support tool, FAST supplies air traffic controllers with assigned runway, sequence number, heading and airspeed advisories for aircraft in the terminal area. The underlying premise of its design has been that all increases in airport capacity must only be achieved in a manner that is acceptable to the air traffic controllers. A subset of the FAST functionality, encompassing only the sequence number and assigned runway advisories, was recognized as the minimum required functionality of a tactical decision support tool intended for the terminal area. This version of FAST, called "Passive" FAST, underwent a successful operational test at the Dallas-Fort Worth TRACON in mid-1996.[12]

Sequence advisories alone cannot significantly increase delay savings or airport throughput. Brinton has shown that the delay benefits of an optimal sequence versus a first-come-first-serve (FCFS) sequence in the terminal area are minimal.[8] Moreover, sequence advisories do not provide enough information to enable the reduction of mean excess spacing at the runway threshold. However, sequence advisories decrease a controller's workload by increasing his situational awareness. For example, a controller recognizes that extra separation is needed between two aircraft in order to merge another aircraft not yet under his control, when the sequence numbers of the two aircraft are not consecutive. In addition, accurate schedule information about traffic inside the terminal area, but under the control of another controller, reduces the controller-to-controller communication during hand-offs.

The remainder of this paper describes the design methodology and effectiveness of the sequencing algorithm used by Passive FAST. The other primary functionality of Passive FAST, runway assignment, is discussed in Reference 13.

Controller Techniques in the Terminal Area

In order to fully understand the design requirements for a sequencing algorithm in the terminal area, common techniques employed by air traffic controllers must be understood. These practices describe situations where the controller departs from a delay-optimal or FCFS landing order. These departures should not be viewed as controller ineffectiveness. Rather, they are based on a deeper understanding of operational procedures and their impact on controller workload.

Using Standard Trajectory Segments

Air traffic controllers are responsible for defined partitions of airspace, known as sectors. Sector boundaries prohibit a controller from arbitrarily vectoring an aircraft along paths that may interfere with other aircraft not under his control. Instead, controllers conduct aircraft along a series of standard trajectory segments towards the airport. Figure 1 illustrates the typical arrival routes† to runway 17C at Dallas-Fort Worth (DFW) International Airport. Routes for low side arrivals begin with "short" segments, followed by "base" segments and a "final" segment. Similarly, routes for high side arrivals begin with "long"

† These routes reflect airspace procedures that were changed in late 1997 due to the addition of the new runway 17L at DFW.

segments, followed by "downwind" segments, followed by "base" segments and a "final" segment.

Trajectory segments limit the areas where aircraft from different traffic streams intersect and merge, and, in fact, describe the standard procedures for controlling the air traffic. For a controller, standard segments reduce the workload required to sequence the arrival traffic, because they eliminate the need to compare every aircraft with every other aircraft.

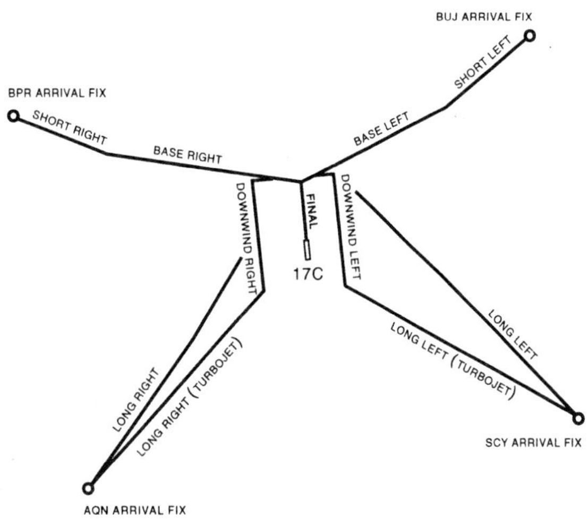

Fig. 1. Standard trajectory segments to Dallas/Fort Worth runway 17C

Allowing Overtakes

An overtake occurs when one aircraft passes alongside, over, or under another aircraft sharing the same trajectory segment. Overtakes in the terminal area are typically workload intensive for the controller and therefore undesirable in most cases. There are exceptions, however, and acceptable overtakes will generally need to meet three basic requirements.

First, an overtake must occur only in the airspace farthest from the airport. Nearly all arrival aircraft at a busy airport, such as Dallas/Fort Worth, are eligible for and capable of meeting the same set of standard clearances issued in the terminal area. Therefore, once inside the terminal area, all traffic is vectored to the same target speeds and altitudes, with little regard for aircraft type. Only near entry into the terminal area do traffic streams usually have speed or altitude differences large enough to warrant overtakes.

Second, an overtake must not demand excessive attention or controller-to-controller coordination. Rather, the overtake must be completed in a timely manner. If the overtake requires a long time to develop, the controller will need to focus extra attention on the situation in order to verify that the aircraft do not represent potential conflict.

Lastly, an overtake cannot require excessive aircraft vectoring. Regardless of delay savings, an overtake is avoided when it causes workload-intensive vectoring of either aircraft. For example, such vectoring would occur if no arrival slot exists ahead of the slower aircraft for the faster aircraft to occupy. In this situation, the controller would need to vector both aircraft - the faster aircraft to complete the overtake and the slower aircraft to achieve legal separation.

Merging In-trail Aircraft

Two aircraft are considered in-trail when they occupy the same trajectory segment with similar speeds and near-minimum legal separation. Similar to overtakes, splitting in-trail aircraft usually yields only minimal delay savings and substantial workload penalties.

For the controller, little effort is needed to keep such aircraft consecutive in the sequence all the way to the runway threshold. Therefore, the additional vectoring caused by merging another aircraft between them can significantly increase workload. For example, in order to merge an aircraft from a separate traffic stream between the in-trail aircraft, the controller needs to both vector the in-trail aircraft to gain an additional landing slot between them and vector the merging aircraft into that slot. Conversely, in order to place the aircraft behind the in-trail aircraft, the controller simply vectors it to that position.

Separating Stacked Aircraft

In the terminal airspace, near the final approach course, altitude often indicates an aircraft's place in the landing order. This is especially relevant for aircraft separated by altitude, or "stacked". While their current positions and ETAs to the runway threshold are nearly identical, the lower aircraft will be given preference. The controller's workload drives this decision. In order to sequence the higher aircraft ahead, the controller first needs to achieve minimum in-trail separation by vectoring the lower aircraft. Then, the controller needs to descend the higher aircraft and verify that no conflicts arise with the traffic below. Conversely, when the lower aircraft is sequenced ahead, the controller only needs to maintain minimum in-trail separation.

Equalizing Delay

The techniques previously described consist mainly of tactical methods used to sequence individual pairs of aircraft. It is also necessary to understand how controllers more generally sequence aircraft which are not in close proximity but who will be merging at some time in the future. Though the controller's cognitive process in making these longer term plans is not fully understood, a working model suggests that controllers attempt to ensure that aircraft occupying nearby slots in the landing sequence have similar levels of delay. A scheduler which is acceptable to controllers will try to equalize delay penalties to consecutive aircraft. In other words, a controller would be reluctant to implement an arrival plan which had base traffic intercepting the final approach course twenty miles from the runway threshold but downwind traffic intercepting only five miles from the runway threshold. Instead, controllers often identify a common "pivot" point along the final approach course where the different traffic streams will merge. This technique maintains comparable delay levels for all traffic streams, maximizes the controller's available options for vectoring aircraft, and thus increases the safety of the operation by decreasing his workload.

Criteria for Sequence Advisory Acceptability

It is also necessary to define qualitative measures of the acceptability of a system intended to provide sequence advisories to the air traffic controllers. Four criteria have been identified: controllability, speed of response, stability and flexibility.

Controllability of the Sequence Advisories

The arrival plan presented to the controllers must be achievable without the use of unnecessary or extreme vectors and the endurance of unacceptable workload levels. This is defined as the system's controllability. Implementation of a suggested arrival plan by the controller must not compromise the safety of the air traffic control system. Therefore, the arrival plan cannot, for the purpose of increasing delay savings, force aircraft into situations where their maneuverability or delay capacity are minimized. Moreover, the arrival plan cannot depart too drastically from the standard techniques used by the controller.[11]

Speed of Response of Sequence Advisories

The system must be quick to respond to controller decisions to change the relative order of aircraft in the arrival plan. No sequencing algorithm can be in agreement with all controllers 100% of the time, because there are many circumstances which the system cannot account for in its predictions. For example, a pilot may execute a clearance late or incorrectly, a pilot may vector around unknown weather, an aircraft may have unknown performance restrictions and individual controllers will have different styles. In light of these uncertainties, the system must be quick to recognize and respond when the controller is following a plan other than the recommended sequence.

Stability of Sequence Advisories

While a quick response time is desired, the system must also present a stable arrival plan to the controller. The relative orders of aircraft cannot be characterized by large or frequent jumps. Certainly, it is not possible to eliminate all sequence switches, especially for aircraft which are far from the airport or which are following non-standard routes. However, it is important that the number of these intrusive sequence number jumps be minimized. Controllers would like to be given a workable arrival plan, and then be allowed to execute the actions required to meet that plan.

Flexibility of Sequence Algorithm

Lastly, the system must be flexible. While the sequencing algorithm may be designed to be applicable to a generic terminal airspace, additional site-specific preferences may become necessary for the system to generate acceptable sequence advisories. The incorporation of these sequencing considerations into the knowledge base must be easy and straightforward. Similarly, the planning process of the automation system should be an extension of the controller's own planning process. An algorithm implemented in terms of criteria and decisions similar to those made by the human controller benefits from its ability to use the expert knowledge of those controllers in its design. Then, rather than optimizing a complex, non-intuitive cost function, the sequencing algorithm relates its decisions to the human controller's cognitive processes.

Passive FAST Sequencing Algorithm

This section describes the knowledge-based sequencing (KBS) algorithm which combines the trajectory based sequencing of the spatial constraint satisfaction method with fuzzy reasoning to produce acceptable sequence advisories for tactical control in the terminal area.[11]

Inputs to the Sequencing Algorithm

The KBS algorithm utilizes a 4D trajectory from each aircraft's current position to the runway threshold and an estimated time of arrival range for each aircraft to determine its relative order in the arrival plan.

Unlike time-based sequencing which schedules to a single point, the KBS algorithm considers each aircraft's complete trajectory to determine a satisfactory arrival plan. This requires an accurate 4D trajectory from each aircraft's most recently reported position to the runway threshold. These trajectories are produced by a trajectory generation engine which integrates point mass equations of motion along a 3D route in order to meet specific altitude, routing and airspeed constraints.[14] The horizontal route of the fastest trajectory represents the shortest path the aircraft is expected to take from its current position to the runway threshold, and the airspeed profile of the fastest trajectory represents the latest locations where required speed reductions will likely occur. The altitude profile approximates a standard terminal area profile by descending along fixed flight path angles to meet the altitude restrictions imposed at the final approach fix and runway threshold. The current aircraft state, winds aloft, atmospheric temperature and pressure profiles, aircraft flight characteristics and standard terminal area procedures are included in the integration of the equations of motion.

The time range over which the aircraft could arrive at the runway threshold is also used by the sequencing algorithm. This time range is computed by perturbing the fastest trajectory by its most extreme vector deviations and its earliest speed reductions. The estimated time of arrival range does not consider the influences of other traffic on the aircraft's route or airspeed and altitude profiles. Rather, the time range simply defines the upper and lower bounds on arrival time for the aircraft; this is the total delay capacity of the aircraft during normal operations.

Outputs from the Sequencing Algorithm

The essential output of the sequencing algorithm is the relative sequence of all aircraft in the arrival plan to their assigned runways. Sequence number advisories, indicating the relative order of aircraft destined to each runway, are displayed to the controller. As a result, it is possible to have aircraft assigned to different runways with the same sequence number advisory.

The sequence number advisories of aircraft throughout the terminal area are not decremented when an aircraft lands. Rather, after aircraft #1 lands, aircraft #2 becomes the next aircraft scheduled to land and so on until an adaptable maximum number is reached (typically 30). At this point, the sequence number advisories reset to a value of 1. By not decrementing the sequence numbers as aircraft land, a change in an aircraft's sequence number always indicates a change in that aircraft's relative position in the landing order.

The sequencing algorithm also computes a nominal scheduled time of arrival (STA) for each aircraft based upon the required separation behind the aircraft that it trails at the runway threshold. This nominal STA is defined as

$$STA_B^{AB} = \max(STA_A^{AB} + \bar{t}_{AB},\ TA_B^{earliest})$$

where, STA_A^{AB} and STA_B^{AB} are the STAs of Aircraft A and Aircraft B, respectively, given a relative sequence AB. $TA_B^{earliest}$ is the earliest time of arrival of Aircraft B and $\bar{t}_{AB}$ is the required separation time at the runway threshold.

The nominal STA does not account for additional delay used to resolve conflicts away from the runway threshold. Rather, it is used by the sequencing algorithm to compare estimates of the delay associated with different relative orders of aircraft.

Trajectory-Based Constraints

It was learned through extensive real-time simulations that to produce an acceptable arrival sequence it is necessary to consider all merges within the terminal area, not just the merge onto the final approach course. To accomplish this, the sequencing problem is broken into a network of common trajectory segments. These trajectory segments are first ordered and then merged to produce a consistent and acceptable sequence at the runway threshold.

Trajectory Segment Tree

A trajectory segment is the portion of a trajectory that falls within a defined segment of flight. Figure 2 shows several aircraft and their trajectories broken into trajectory segments, referred to as "long", "downwind", "base" and "final".

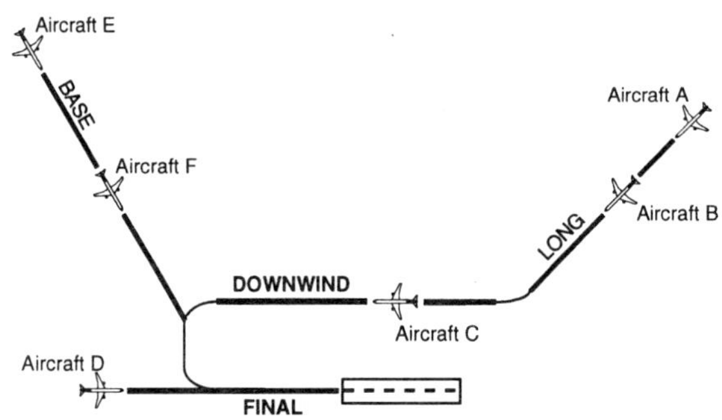

Fig. 2. Illustration of Merging Trajectory Segments

Determining the sequence of aircraft to a given runway consists of both ordering the aircraft on each trajectory segment and combining these relative orders into a consistent final sequence to that runway. Figure 2 depicts a situation where six aircraft merge together to land on the same runway. The trajectory segment tree shown in Figure 3 illustrates how this sequencing problem is broken into a series of smaller sequencing problems. The root of the tree represents the advised landing order. The branches of the tree symbolize merges that occur in the terminal airspace. These branches connect nodes which represent the relative sequence of aircraft on a particular trajectory segment.

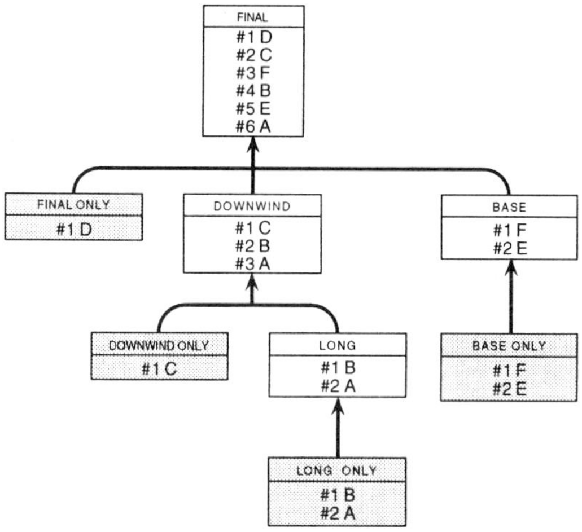

Fig. 3. Trajectory Segment Tree

The sequencing process begins at the leaves of the trajectory segment tree. A relative order is determined for the aircraft on each trajectory segment. This process of sequencing aircraft which share a common trajectory segment is referred to as "ordering." The ordering process considers many factors but its details are not needed to describe the trajectory-based sequencing methodology. The result is an ordered aircraft list for a particular leaf of the trajectory segment tree. This list represents the aircraft's relative order on its first trajectory segment.

Once the leaves are ordered, these leaves, and later branches, are repeatedly merged to create a relative order in each node. Thus, a node represents the relative order of aircraft on a particular trajectory segment which was formed by merging the earlier predictions of other leaves and branches. Like the ordering process, the merging process considers many criteria; the description of their details is deferred to a later section.

The merging process preserves the relative order of aircraft from preceding nodes and leaves. This constrains the number of possible sequences at the runway threshold by enforcing all sequence constraints encountered at earlier merge points of the trajectory.

Advantages of Trajectory-Based Constraints

This method of addressing the sequencing problem has several benefits. First, separate criteria can be specified for each trajectory segment which take into account specific methods controllers use to sequence aircraft on that segment, e.g. allowing or disallowing overtakes. Also, different criteria can be specified for the ordering and merging steps of the sequencing process. Therefore, aircraft already present on a trajectory segment can be sequenced differently from aircraft predicted to be on that trajectory segment at a future time.

Second, the relative order for each pair of aircraft is determined using information about when those aircraft are in close spatial proximity. The first set of relative orders is comprised of aircraft on a common trajectory segment - each aircraft's first trajectory segment. This set of decisions uses the current aircraft positions and short-term trajectory predictions. The next set of decisions combines the first set to produce the relative order on the next trajectory segment that the aircraft are approaching. These orders are determined from predictions made farther into the future. As the sequencing process continues, it is forced to use predictions that are farther into the future. Yet, at every merge, the sequencing process uses the earliest predicted information of that segment, hence the most accurate, that can be used to compare the aircraft.

Fuzzy Reasoning-Based Sequencing

This section details the knowledge base of the sequencing algorithm implemented by Passive FAST and successfully tested in the operational environment of Dallas-Fort Worth TRACON in mid-1996. The KBS algorithm combines trajectory-based sequencing with fuzzy reasoning to determine an acceptable arrival plan. The subsequent discussion assumes familiarity with fuzzy reasoning methods, specifically decision making based upon the fuzzification and defuzzification of linguistic propositions. These concepts and their inclusion in the KBS algorithm are discussed in detail in Reference 15. More general explanations of fuzzy set theory can also be found in References 16 and 17.

Spatial Sequence Constraints

Rather than sequence aircraft along each trajectory segment, the KBS algorithm sequences aircraft in larger groups called spatial constraint sets[‡]. A spatial constraint contains aircraft whose trajectories now or later will pass through the same section of airspace. There are three types of spatial constraints, GENERAL, FINAL and DEPENDENCY. For conciseness, only the specific sequencing criteria for GENERAL-type spatial constraints are presented in this paper. Refer to Reference 15 for an in-depth discussion of the sequencing criteria for FINAL-type and DEPENDENCY-type spatial constraints.

A GENERAL-type spatial constraint contains a set of aircraft from a particular trajectory segment, other than the FINAL trajectory segment. While a common trajectory segment is the primary attribute of these

[‡] Spatial Constraint Set is hereafter referred to as Spatial Constraint.

spatial constraints, other attributes are used to isolate a subset of aircraft from a single trajectory segment. For example, the spatial constraints, SHORT_LEFT_PROPS and SHORT_LEFT_JETS, subdivide the aircraft on the SHORT_LEFT trajectory segment.

Sequencing differences require FINAL trajectory segments to represent a separate type of spatial constraint. A FINAL-type spatial constraint is simply a specialized GENERAL-type spatial constraint that contains all aircraft from a single FINAL trajectory segment. Every arrival aircraft belongs to a single FINAL-type spatial constraint, and the aircraft's sequence number is its relative sequence on that spatial constraint.

Although two aircraft landing on different runways never physically merge, each can affect how the other is sequenced. Therefore, a DEPENDENCY-type spatial constraint links multiple FINAL-type spatial constraints in order to account for runway dependencies like staggered approaches to parallel runways or converging approaches to intersecting runways.

Fuzzy Set Antecedents

The KBS algorithm uses five fuzzy antecedents to build "if-then" propositions for its sequence decision-making logic of GENERAL-type spatial constraints. These antecedents evaluate numerous aircraft position and trajectory path distance quantities to determine an aircraft's levels of membership in the corresponding fuzzy sets. In order to be independent of airspace, distance values are normalized by the required separation distance. The normalized values of these quantities represent the number of landing slots. For example, a normalized distance value of 1.0 represents a savings of one full landing slot.

"A is ahead of B at their FCTS:" The primary criteria for ordering two aircraft on a GENERAL-type spatial constraint are the aircrafts' predicted positions at the first instance both belong to that spatial constraint. These positions are called First Common Time Steps (FCTS). The aircrafts' FCTS measure how far ahead or behind one aircraft is to another. Preference can be given to a FCFS ordering of the aircraft. The normalized separation at the FCTS, (NSD_{fcts}), is

$$NSD_{fcts} = \begin{cases} \dfrac{d^B_{fcts,ref} - d^A_{fcts,ref}}{\bar{d}_{AB}} & if\ (d^B_{fcts,ref} > d^A_{fcts,ref}) \\ 0 & if\ (d^B_{fcts,ref} \le d^A_{fcts,ref}) \end{cases}$$

where, $d_{fcts,ref}$ is the distance from an aircraft's FCTS to the spatial constraint's endpoint and $\bar{d}_{AB}$ is the required separation distance given the sequence AB.

A positive NSD_{fcts} indicates that Aircraft A is predicted to be closer to the runway threshold than Aircraft B at their FCTS. The membership function used to map the normalized separation at the FCTS to a proposition's firing strength is shown in Figure 4.

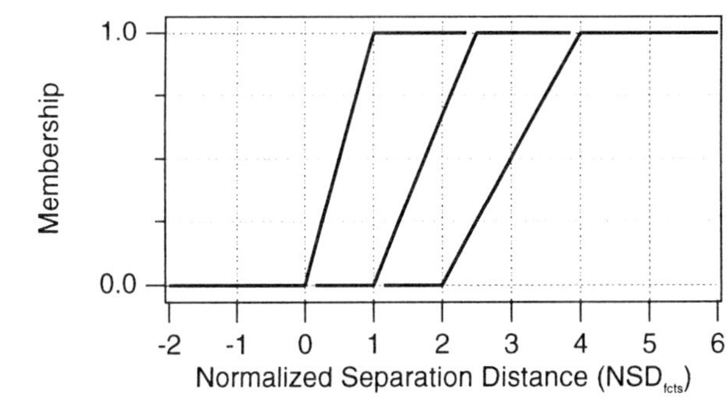

Fig. 4. Membership Functions for "Ahead at FCTS"

"A is ahead of B at their current positions:" The geometric order describes which aircraft is presently closest to the runway threshold along its predicted trajectory. This antecedent defines propositions which limit overtakes and give preference in the sequence to the aircrafts' current geometric order. The current normalized separation (NSD_{trk}) is

$$NSD_{trk} = \begin{cases} \dfrac{d^B_{trk,ref} - d^A_{trk,ref}}{\bar{d}_{AB}} & if\ (d^B_{trk,ref} > d^A_{trk,ref}) \\ 0 & if\ (d^B_{trk,ref} \le d^A_{trk,ref}) \end{cases}$$

where $d_{trk,ref}$ is the path distance from an aircraft's current position to its FCTS and $\bar{d}_{AB}$ is the required separation distance given the sequence of AB.

A positive NSD_{trk} indicates that Aircraft A is currently ahead of Aircraft B. The membership function used to map the current normalized separation to a proposition's firing strength is shown in Figure 5. For current normalized separation values less than 0.25, the aircraft are essentially considered tied, so the resulting membership is 0.0.

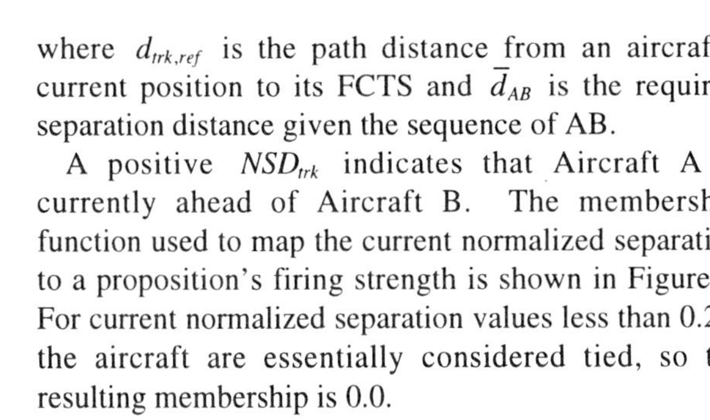

Fig. 5. Membership Function for "Currently Ahead"

"*A is faster than B at their FCTS:*" A ground speed difference at the aircrafts' FCTS is used to identify acceptable overtake situations. The membership function used to map a ground speed difference to a proposition's firing strength is shown in Figure 6. Since the error in computed ground speed is frequently more than 10 knots, ground speed differences less than 25 knots have a membership value of 0.

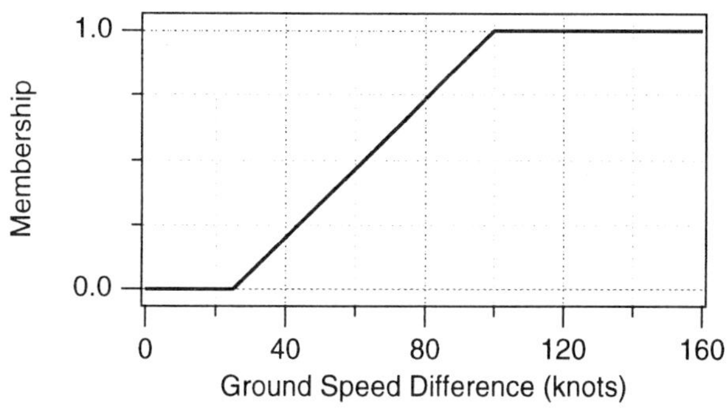

Fig. 6. Membership Function for "Faster at FCTS"

"*A is lower than B at their current positions:*" An altitude difference at the aircrafts' current radar track positions is used to define heuristics for ordering stacked aircraft. The membership function used to map the current altitude difference to a proposition's firing strength are shown in Figure 7.

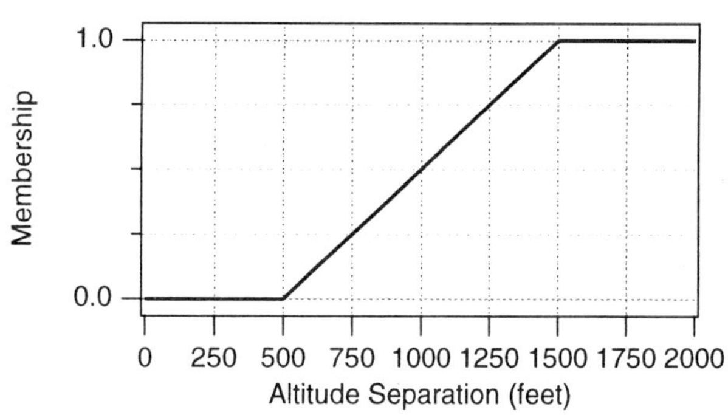

Fig. 7. Membership Function for "Currently Lower"

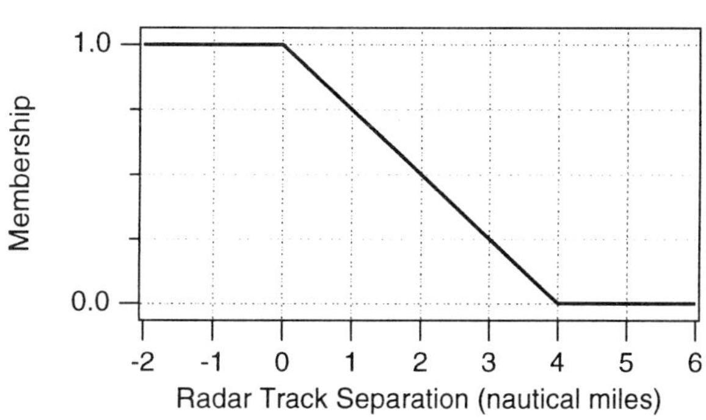

Fig. 8. Membership Function of "Close"

"*A and B are Close:*" The horizontal spatial separation of two aircraft is also used to determine if they are in close proximity. This antecedent is used in conjunction with the antecedent "Aircraft A is Faster than Aircraft B at their FCTS" to recognize a controller's decision to allow an overtake when none was initially being advised. The membership function used to map the current separation to a firing strength is shown in Figure 8.

Fuzzy Set Consequents

Fuzzy set consequents form the "then" part of an "if-then" relationship. The KBS algorithm defines a set of six standard consequents used in the expression of its fuzzy propositions. This set includes mathematical descriptions of marginally favored, slightly favored, somewhat favored, favored, quite favored and significantly favored. These linguistic labels and their associated fuzzy regions are illustrated in Figure 9.

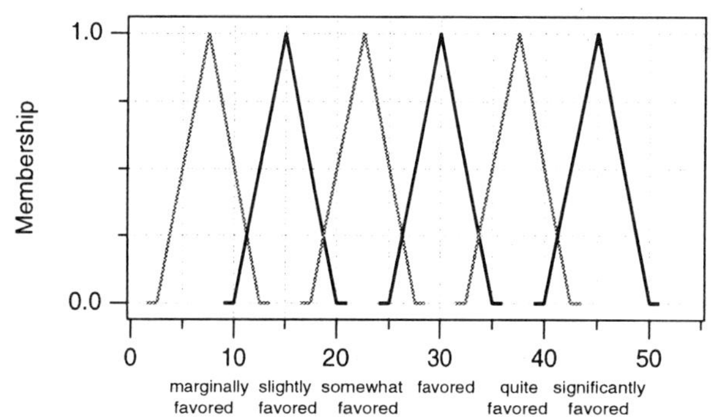

Fig. 9. Consequents Describing "FAVORED"

Early in the development of the fuzzy reasoning algorithm, each fuzzy proposition had its own consequent. Such complexity arose when the consequents were separately modified in each fuzzy proposition to fix specific deficiencies in the sequence advisories. Allowing a different consequent for each fuzzy proposition was later abandoned to improve maintainability. Instead, the consequent of each fuzzy proposition is chosen from the standard set. This bounds the number of permutations possible when the fuzzy propositions need modification.

By design, the consequents are symmetric. This attribute would not be appropriate for a fuzzy system whose output is a real-valued quantity. However, the output of this algorithm is simply an algebraic sign, positive for one relative sequence, negative for the other. Both the weight and the centroid of an asymmetric fuzzy region vary with membership value. Therefore, each rule's maximum strength relative to the other rules changes with membership value. This would needlessly increase the complexity of the KBS algorithm and decrease its maintainability.

Fuzzy Propositions for Ordering

Every aircraft belonging to the GENERAL-type spatial constraint is compared to every other aircraft belonging to that spatial constraint. Seven fuzzy propositions are evaluated in order to sequence each pair of aircraft. Since these seven propositions are independently evaluated for both aircraft of the pair, there are effectively fourteen different rules representing the decision making process. For two aircraft, "A" and "B," these fuzzy propositions are expressed as follows:

1. if (A is significantly ahead of B at their FCTS) then (A is significantly favored)
2. if (A is ahead of B at their FCTS) then (A is favored)
3. if (A is slightly ahead of B at their FCTS) then (A is slightly favored)
4. if (A is ahead of B at their current positions) then (A is slightly favored)
5. if (A is faster than B at their FCTS) then (A is marginally favored)
6. if (A is lower than B at their current positions) then (A is marginally favored)
7. if (A is close to B AND A is faster than B at their FCTS) then (A is slightly favored)

Aircraft occupy and thus are sequenced on GENERAL-type spatial constraints when they are farthest from the runway threshold. In this region, controllers, known as "feeder" controllers, are responsible for managing several traffic streams before transferring control of these aircraft to other controllers, known as "final" controllers. The final controllers are then responsible for merging this traffic onto the final approach courses. The feeder controllers must manage their traffic to supply multiple streams of aircraft which can be smoothly merged into a single stream. Therefore, feeder controllers execute a near FCFS sequence. Rules 1, 2 and 3 represent this preference by favoring the aircraft predicted to have the least path distance to fly to the runway threshold from the FCTS.

The few exceptions to a FCFS sequence occur when aircraft are in close proximity to each other. Therefore, several fuzzy propositions serve to constrain the problem by favoring a sequence based upon the aircrafts' current geometric relationships. This is accomplished by considering the aircraft's state at the current radar track. Rule 4 gives preference to the aircraft which has the least path distance to fly to the runway threshold from its current radar track location. Similarly, rule 5 gives preference to the lower aircraft, and rule 6 gives preference to the faster aircraft.

Finally, since ill-advised overtakes represent considerable workload for the controller, the system has been designed to be conservative in its prediction of overtakes. However, for the system's performance to remain acceptable, this approach was augmented to provide a quick response in situations where the initial prediction of no overtake was incorrect. Rule 7 represents logic intended to recognize and respond quickly to unadvised overtakes. If an overtake was not originally suggested by the system, but two aircraft are in close proximity at dissimilar ground speeds, rule 7 recognizes that an unadvised overtake is being executed by the controller and the aircraft are resequenced accordingly.

Fuzzy Propositions for Merging

Once the aircraft are ordered on their initial spatial constraints, the aircraft are repeatedly merged onto GENERAL-type spatial constraints closer to the runway threshold. This is analogous to traversing the network of spatial constraints by beginning at the leaves and ending at the root of the tree (Fig. 3).

The first aircraft of each spatial constraint immediately preceding the spatial constraint to be merged are compared to determine who will be next. Again, seven fuzzy propositions are evaluated for each aircraft in order to determine the relative order of each pair of aircraft. Like ordering, merging of GENERAL-type spatial constraints is well-described by a near-FCFS sequence. For this reason, the same decision making criteria were used in both situations. The aircraft chosen to be next is removed from the ordered list of its preceding spatial constraint, and appended to the ordered list of the merged spatial constraint. Moreover, the aircraft which immediately followed this chosen aircraft is now considered first from that preceding spatial constraint. The merging process continues by comparing this new set of aircraft first from the preceding spatial constraints. Again, that aircraft determined to be next is removed from its spatial constraint's ordered list, and appended to the merged spatial constraint's ordered list. The merging process is finished when the relative order of all aircraft on the merged spatial constraint has been determined.

Most importantly, the relative order of aircraft from each preceding spatial constraint is preserved in the relative order of aircraft on the merged spatial constraint. This approach is crucial for determining an acceptable final sequence of aircraft. It forces earlier decisions, made with near-term information, to constrain later decisions.

Operational Test Results

Following several thousand hours of real-time controller-in-the-loop simulations of the Passive FAST system, it was determined that an operational test was necessary to completely validate the results obtained in the laboratory.

A series of operational tests were conducted at the Dallas-Fort Worth TRACON during the period from January 1996 through July 1996. These tests included all major arrival rush periods, instrument flight rules (IFR) and visual flight rules (VFR), North and South Flow TRACON operations, two and three runway airport configurations and a variety of meteorological conditions.

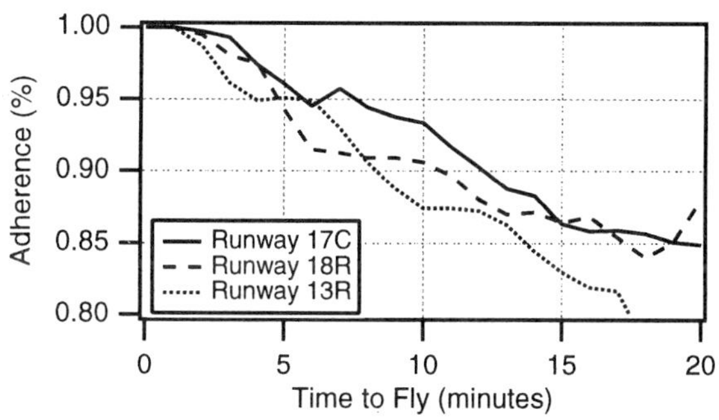

Fig. 10. Adherence to South Flow Sequence Advisories

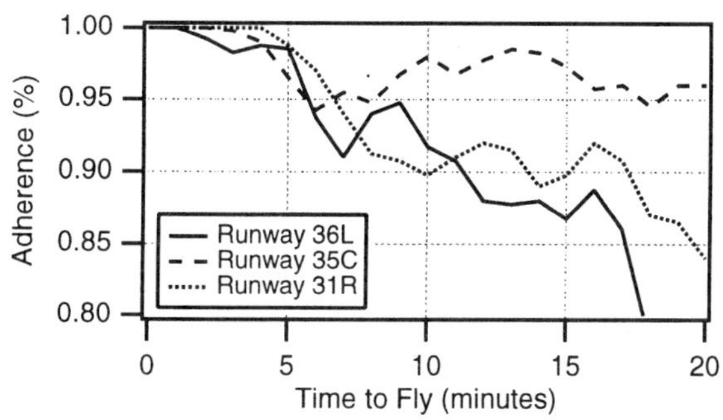

Fig. 11. Adherence to North Flow Sequence Advisories

A critical measure of the success of a sequencing algorithm is the degree to which the air traffic controllers adhere to the suggested arrival plan. Following the conclusion of the operational tests, several modifications were made to the sequencing algorithm. The algorithm documented in this paper represents this refinement of the algorithm used during the operational tests. These changes did not greatly increase the resultant adherence levels. Instead, they lessened the severity of unacceptable sequence advisories. Data collected during the operational tests were re-analyzed to quantify what fraction of the suggested arrival plans matched the actual aircraft landing orders. Figures 10 and 11 show the percentage of aircraft landing in the advised order as a function of time to fly for South Flow and North Flow airport configurations.

While these results seem to suggest a high level of adherence, these findings alone do not validate the algorithm. Controller comments during the operational test indicated that it is more important to guard against a poor sequence than to provide a perfect sequence. This is evident when a single ill-advised overtake during an hour-long, 100-aircraft arrival rush can make the difference between the system being judged acceptable or unacceptable. In the end, the controllers who evaluated Passive FAST during the operational test felt that any frustration caused by a rare ill-advised sequence advisory was greatly outweighed by the system's benefits of decreased workload and increased throughput.

Conclusions

A fuzzy reasoning-based algorithm for sequencing arrival air traffic in the terminal area has been developed at the NASA Ames Research Center and operationally tested at the Dallas-Fort Worth TRACON. Although developed using several thousand hours of real-time controller-in-the-loop testing with DFW traffic scenarios and nearly thirty hours of operational testing at the DFW TRACON, the algorithm is designed to be generic, and thus, applicable to other terminal airspaces.

The sequencing algorithm performs trajectory-based planning by breaking the whole sequencing problem into sets of smaller short, medium and long term decisions. This design explicitly allows for special treatment of non-FCFS controller preferences which cannot be considered by time-based scheduling methods. It also permits separate sequencing preferences to be defined for trajectory segments throughout all phases of the flight from the extended terminal area to the runway threshold. Through the use of fuzzy-reasoning, the algorithm directly incorporates the experience of expert controllers in its design. Its knowledge base is appropriately defined in terms of fuzzy propositions since the problem of sequencing arrival aircraft in the terminal area is inherently difficult to define. Further, fuzzy propositions allow a more direct connection between the controller's decision making process and the system's.

Data and controller comments gathered during the operational test verify that the system generates an arrival plan that is acceptable to the air traffic controllers. These results show that the system's advised landing order at the time the aircraft enter the terminal area was correct more than 90% of the time.

This advisory adherence steadily increases and converges to 100% as the aircraft gets closer to landing.

As a result of the successful operational test, Passive FAST will begin long-term operational testing at the DFW TRACON in July 1997.

References

[1] Erzberger H., Davis T.J., Green S.M., "Design of Center-TRACON Automation System,"ß Proceedings of the AGARD Guidance and Control Panel 56th Symposium on Machine Intelligence in Air Traffic Management, Berlin, Germany, 1993, pp. 11-2 – 11-12.

[2] Nedell W., Erzberger H., Neuman F., "The Traffic Management Advisor," Proceedings of the American Control Conference, San Diego, California, May 1990.

[3] Synnestvedt R.G., Swenson H., Erzberger H., "Scheduling Logic for Miles-In-Trail Traffic Management," NASA Technical Memorandum 4700, September 1995.

[4] Green S., Vivona R., "Field Evaluation of Descent Advisor Trajectory Prediction Accuracy," Proceedings of the AIAA Guidance, Navigation and Control Conference, San Diego, California, No. AIAA-96-3764-CP, July 1996.

[5] Davis T.J., Krzeczowski K.J., Bergh C., "The Final Approach Spacing Tool," Proceedings of the 13th IFAC Symposium on Automatic Control in Aerospace, Palo Alto, California, September 1994.

[6] Volckers U., "Arrival Planning and Sequencing with COMPAS-OP at the Frankfurt ATC-Center," Proceedings of the 1990 American Control Conference, San Diego, California, pp. 496-501.

[7] Garcia J., "MAESTRO - A Metering and Spacing Tool," Proceedings of the 1990 American Control Conference, San Diego, California, pp. 501-507.

[8] Brinton C.R., "An Implicit Enumeration Algorithm for Arrival Aircraft Scheduling," Presented at the 11th Digital Avionics Systems Conference, Seattle, Washington, October 1992.

[9] Erzberger H., "Design Principles and Algorithms for Automated Air Traffic Management," AGARD Lecture Series No. 200: Knowledge-Based Functions in Aerospace Systems, Madrid, Spain; Paris, France; San Francisco, USA, November 1995.

[10] Swenson H.N., Hoang T., Engelland S., Vincent D., Sanders T., Sanford B., Heere K., "Design and Operational Evaluations of the Traffic Management Advisor at the Fort Worth Air Route Traffic Control Center," To be published as a NASA Technical Memorandum.

[11] Bergh C.P., Krzeczowski K.J., Davis T.J., "TRACON Aircraft Arrival Planning and Optimization through Spatial Constraint Satisfaction," Air Traffic Control Quarterly, Volume 3, Number 2, pp. 117-138.

[12] Davis T.J., Isaacson D.R., Robinson J.E., den Braven W., Lee K.K., Sanford B., "Operational Test Results of the Passive Final Approach Spacing Tool," Presented at the 8th IFAC Symposium on Transportation Systems, Chania, Greece, June 1997.

[13] Isaacson D.R., "Knowledge Based Runway Assignment for Arrival Aircraft in the Terminal Area," To be presented at the AIAA Guidance, Navigation and Control Conference, New Orleans, Louisiana, August 1997.

[14] Erzberger H., Tobias L., "A Time-Based Concept for Terminal Area Traffic Management," Proceedings of the 1986 AGARD Conference, No. 410 on Efficient Conduct of Individual Flights and Air Traffic, Brussels, Belgium, 1986, pp. 52-1 – 52-14.

[15] Robinson J.E., Davis T.J., Isaacson D.R., "Fuzzy Reasoning-Based Sequencing of Arrival Aircraft in the Terminal Area," To be published as a NASA Technical Memorandum, October 1997.

[16] Yager R.E., Zadeh L.A. Eds., Introduction to Fuzzy Logic Applications in Intelligent Systems, Kluwer Academic Publishers, Norwell, MA, 1992.

[17] Cox E. Ed., The Fuzzy Systems Handbook, Academic Press Professional, New York, 1994.

KNOWLEDGE-BASED RUNWAY ASSIGNMENT FOR ARRIVAL AIRCRAFT IN THE TERMINAL AREA

Douglas R. Isaacson*, Thomas J. Davis*, and John E. Robinson III*
NASA Ames Research Center, Moffett Field, California 94035

Abstract

A knowledge-based system for scheduling arrival traffic in the terminal area, referred to as the Final Approach Spacing Tool (FAST), has been implemented and operationally tested at the Dallas/Fort Worth Terminal Radar Approach Control (TRACON) facility. Two types of controller advisories are generated by FAST: sequence number and runway assignment. The knowledge base for runway assignment employs a set of hierarchical rules and decision logic that evaluates both performance and workload criteria. This formulation is based on over 2,000 hours of controller-in-the-loop, real-time simulations. In the field tests, controllers had the option to accept or reject the FAST-generated runway assignments. Results indicate strong adherence to the advisories and increased capacity, with no significant impact on controller workload.

Introduction

Runway assignment of arrival aircraft is a tactical decision made by controllers. Strategic re-assignments, or allocations, assist in balancing controller workload and reducing delay, but are difficult for controllers because of the high workload already associated with the traffic load. As terminal area controllers become consumed with the task of separation, strategic runway allocation becomes neglected. During high workload periods, controllers simply assign runways to fill available landing slots when aircraft are well within TRACON airspace. This process of tactical adjustments to the arrival schedule requires coordination between controllers and ultimately leads to higher workload and increased delay. Any system which attempts to alleviate this problem must consider controller workload. This report describes such a system; the knowledge-based runway allocation algorithm for an air traffic automation tool called the Final Approach Spacing Tool (FAST).[1,2,3] FAST is the terminal area component of the Center/TRACON Automation System (CTAS).[4]

* Research Scientist, Terminal Area ATM Research Branch

The runway allocation algorithm in FAST attempts to achieve runway load balancing and increased capacity without increasing controller workload. Previous research focused on optimization of delay reduction. Optimization of a cost function requires that all relevant inputs be quantified. Therefore, if controller workload is considered important, it must be quantified. Brinton attempted to represent workload as a series of terms in a cost function, however, his implicit enumeration algorithm did not adequately address this issue and was found unacceptable by controllers.[5] The runway allocation function provided by FAST employs a knowledge base obtained through thousands of hours of simulation with expert controllers. This knowledge base incorporates controller preferences and workload into the runway allocation problem. By doing this, FAST emulates the decision patterns of expert controllers, while using accurate calculations of aircraft performance to reduce delay in a manner acceptable to controllers.

This paper begins by defining the motives and potential benefits of runway allocation. The runway assignment algorithm employed by FAST is described, as well as the rules used to create the knowledge base. Results of operational testing of FAST at the Dallas/Fort Worth TRACON facility are briefly discussed, followed by some concluding remarks.

Runway Allocation Motives and Benefits

As aircraft enter TRACON airspace, they are typically assigned by the controller to the closest runway to the arrival feeder gate. This default assignment defines an initial arrival plan. Adjustments are made to the arrival plan, as the aircraft approach the runway, by assigning aircraft to alternate runways. This process is known as runway allocation, and is the primary means of balancing arrivals to each runway and controller workload. Runway allocation decisions made by controllers in today's system are tactical adjustments to the arrival schedule prompted by near-term concerns. As traffic volume and workload increase, controllers have less time to evaluate potential runway allocations. Strategic runway allocation is lost at high controller workload levels. This void in strategic decision making could be filled by a decision support system. To develop such a system, it is necessary to comprehend the factors that necessitate runway allocation.

Runway Allocation Motives

Several factors can lead a controller to assign a new runway to an aircraft. Factors such as airline preference, shortest flight time, and controller preference dominate during periods of low demand; while workload reduction drives decision patterns during high demand periods.

Airline preferences are usually determined by parking gate location, and perceived taxi time to the gate. During rush periods, controllers are less likely to accommodate airline preferences, as this increases their workload due to increased coordination between controllers and more issuance of clearances if no slot is available on the desired runway.

Controller preferences stem from a variety of issues. For example, even at low workload, comfort is gained through knowing an aircraft has more vectoring options available if a problem arises: an aircraft may have more available airspace for maneuvering if assigned to one runway over another, with no significant impact on delay. During low workload periods, controllers have time to evaluate the arrival plan and make runway allocations which result in reduced delay. This is accomplished through coordination with other controllers and the ability to recognize available landing slots on alternate runways. However, because the default runway is usually the closest runway, this type of allocation is rare during low arrival rates.

As the number of aircraft handled by a controller increases, the number of required commands increases, as does the time required to issue these commands. It is essential for a controller to limit workload to a level at which the system is controllable and safe. The complexity of merging streams and insuring separation increases and at some level becomes the only task the controller can effectively perform. This means less time will be available to evaluate strategic runway allocations. Unfortunately, the controller with the highest workload has the least time to accurately evaluate the evolving problem. Decisions which affect workload are often made too late to be of significant benefit. The fundamental problems with strategic decision making by terminal area controllers are twofold: 1) the controllers having the necessary information to evaluate potentially beneficial decisions, have no time to do so and 2) evaluation of the problem itself becomes more complex and time consuming as the potential benefits increase.

Potential Benefits Of Automated Runway Selection

Decision support tools for runway assignment could fill the void left in strategic decision making during high workload periods. A number of benefits could be realized from automated runway assignment: workload balancing, increased throughput, and delay reduction.

By providing decision support for runway assignment, the terminal area controller has more time to perform the task of separating aircraft. With proper runway selection, less coordination between controllers would be necessary to accommodate late runway changes. However, the primary workload benefit is workload balancing. Extremely high workload for a single controller in the TRACON complicates the entire system, thus requiring increased coordination between controllers. Balancing the number of aircraft landing on each runway insures workload is evenly distributed among the final controllers. Balancing workload between controllers reduces both the workload of the busiest controller, and the coordination between controllers. Furthermore, runway balancing may reduce surface congestion, taxi time and departure delay.

Runway balancing is a methodology that attempts to provide each runway with adequate demand. The actual runway threshold throughput can only reach the airport capacity if demand meets or exceeds capacity on each arrival runway. It is not possible to consistently meet minimum separation on each runway if sufficient demand does not exist for each runway. Effective runway balancing insures sufficient demand exists for each runway at high arrival rates. Unbalanced runways can result in the demand for a given runway greatly exceeding the capacity, while another runway's capacity exceeds demand. Excess demand for a runway leads to flight time delay to land all the aircraft assigned to that runway. Runway balancing attempts to match a runway's demand with its capacity, thus eliminating delay due to capacity shortfall. Some additional flight time may be required to land on an alternate runway, but this additional time is often less than the required delay to land on the default runway. Furthermore, TRACON delay reduction through maximum utilization of all runways, could lead to increased TRACON acceptance rate from Center airspace. Increased acceptance rate translates into reduced delays in Center airspace during metering.

Requirements For An Operational System

For a runway assignment decision support system to be operationally acceptable in today's ATC environment, four requirements are realized: acceptable workload, schedule stability, trustability, and cost reduction.

The primary concern of controllers is safety. Safety can be linked to workload and thus any system that makes a controller feel on the edge of safe operation, due to high workload, is unacceptable. Controller workload is difficult to quantify; instead, controller acceptability is generally evaluated through surveys such as NASA-Ames' Controller Acceptance Rating Scale (CARS).[6]

The schedule advised by the decision support system must not appear variable to the controller. While minor instabilities may be acceptable in advising the sequence of aircraft in a "close-call" situation, instabilities in runway advisories are unacceptable. The runway advisories displayed to controllers handling aircraft as they first enter the TRACON can determine when and to whom aircraft are handed off (transfer of control) to next. Instability in runway advisories would lead to an increase in handoffs, controller workload and stress.

Some level of trust in the system must be gained in order to realize the potential benefits of the advised schedule. If a controller takes a "wait and see" attitude on a fully acceptable advisory, the advised solution may eventually become unacceptable. Trust is gained through familiarity with the system, stability, and observed benefits.

Finally, an ATC decision support tool must realize benefits. Benefits are largely achieved through increased throughput at the airport.

FAST Knowledge Based Runway Allocation

The foundation of the FAST runway allocation algorithm lies in the wealth of information provided by accurate 4D trajectories. FAST employs an extensive database of aircraft performance models, continuous radar updates, flightplan information and 3D weather predictions to produce accurate 4D trajectories, estimated times of arrival (ETAs), route deviation possibilities, and advisories to controllers. The inputs used by the FAST Knowledge Based Runway Allocation (KBRA) algorithm will be discussed in this section, followed by an overview of the KBRA algorithm. The rules and criteria used to implement the knowledge base developed for DFW TRACON are included in the Appendices.

KBRA Inputs

There are four inputs to the KBRA algorithm: airport configuration, 4D aircraft trajectories, available degrees of freedom for each aircraft, and the relative sequence of arrival aircraft. Each input is now briefly discussed.

Airport Configuration: An airport configuration is chosen by the traffic manager for each airport in the TRACON. The configuration is determined by wind direction and magnitude, visibility, traffic load, and various other factors such as ongoing runway maintenance. For FAST, the airport configuration defines: runways available for arrivals, default runway assignments, potential runways for each arrival traffic stream, and the runway allocation window for each runway.

Each traffic stream (stream class), consisting of aircraft of common engine type arriving through a given feeder gate, is mapped to a default runway. The default runway is usually the closest available runway to the traffic stream's feeder gate. The nature of an arrival rush may lead to a runway other than the closest being mapped as the default in order to strategically balance default runways for arrivals across all available runways.

In addition to mapping the default runway for each traffic stream, possible alternate runways are defined by the airport configuration. Due to standard operating procedures and workload considerations, not all runways available for arrivals will be defined as potential runways for all traffic streams. This reduces the scope of the runway allocation problem and prohibits allocations which would lead to high controller workload.

The runway allocation window for each runway defines a window in time for which aircraft are eligible for runway allocation by FAST. The times are referenced to the fastest time to the final approach fix of each runway. The allocation window is selected to terminate at a time corresponding to aircraft locations just outside the feeder gates, thereby providing stable runway assignment advisories that can be easily implemented early in the control of each aircraft.

Accurate 4D Trajectories: A trajectory generation engine produces accurate 4D trajectories by integrating point mass equations of motion along a horizontal route with specified altitude/speed constraints. The resultant 4D trajectory is broken into trajectory segments for use by the FAST sequencing and runway allocation algorithms.[7] Reducing each trajectory to a set of trajectory segments is analogous to how a controller approaches the merging aircraft problem. Trajectory segments simplify the problem for a controller because they dictate where a controller can expect an aircraft to be in the near future, and where aircraft need to merge with other aircraft onto the next trajectory segment. Figure 1 illustrates an aircraft and its trajectory broken into four trajectory segments: LONG, DOWNWIND, BASE and FINAL.

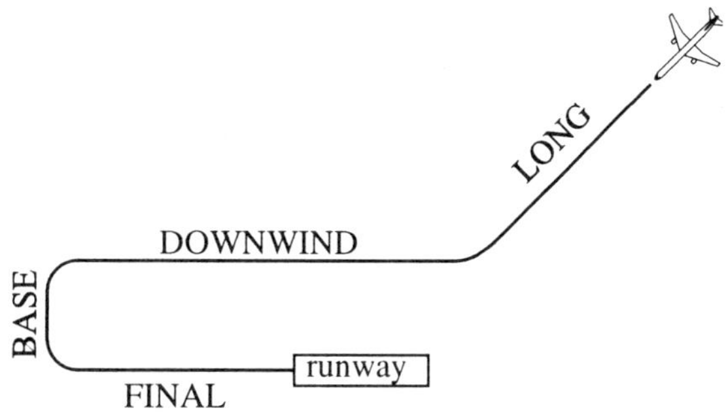

Figure 1. Division of Aircraft Trajectory into Segments

Available Degrees of Freedom: TRACON controllers employ degrees of freedom (DOF) for aircraft to avoid conflicts and merge aircraft streams within their airspace. Typical DOFs for arrivals to the TRACON include speed reductions, base extension and varying intercept angles. Figure 2 shows the base extension DOF.

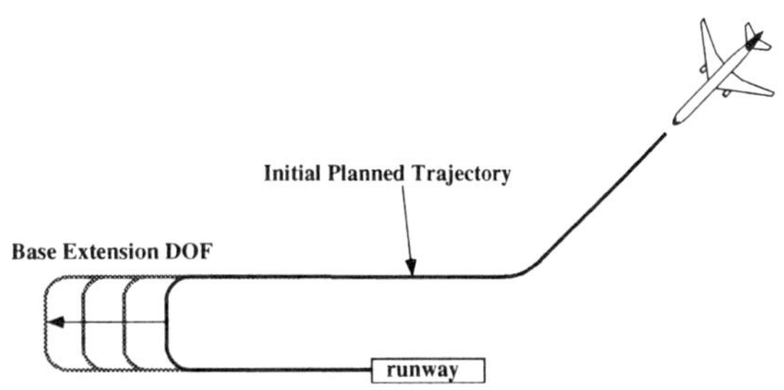

Figure 2. Use of Base Extension Degree of Freedom

The set of all available DOFs are defined for each aircraft based on location, aircraft type and airport configuration. The potential delay effects of each DOF are calculated by producing a set of ETAs corresponding to both full and no employment of each DOF available. Given the relative order between two aircraft determined by the sequencing logic, conflicts are resolved by adding delay to the trajectory of the trailing aircraft through employment of the appropriate DOFs available (as shown in Fig. 3).

Trajectory Segment Ordering for Deconfliction: The deconfliction algorithm requires ordered lists of all aircraft sharing common trajectory segments. That is, if an aircraft is to be checked for conflicts with the aircraft ahead on each trajectory segment, it is necessary to create a sequenced list of aircraft for each trajectory segment in the airspace. These lists of aircraft are referred to as "constraints". The list of aircraft belonging to each constraint is built as each aircraft's trajectory is dissected into trajectory segments. The sequence of the aircraft within each list is determined

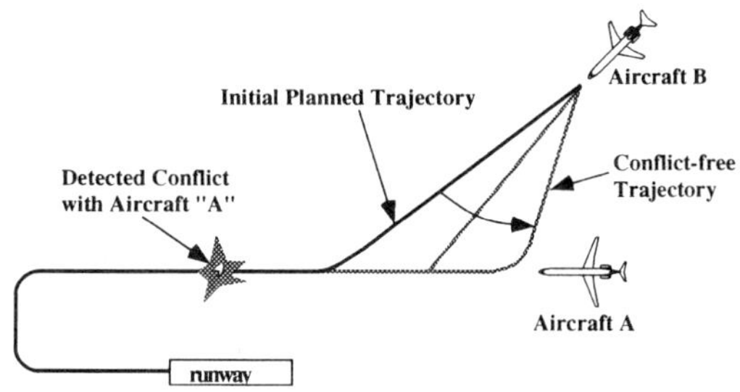

Figure 3. FAST Conflict Detection and Resolution

by the FAST knowledge-based sequencing (KBS) algorithm.[7] The result of the KBS is a sequence of aircraft for each FINAL constraint in the system. However, it is not sufficient to simply employ the sequences obtained from the sequencing for each FINAL trajectory segment in the deconfliction algorithm: some aircraft share segments prior to FINAL, but land on different runways or even at different airports, as shown in Figure 4 (Aircraft B and C).

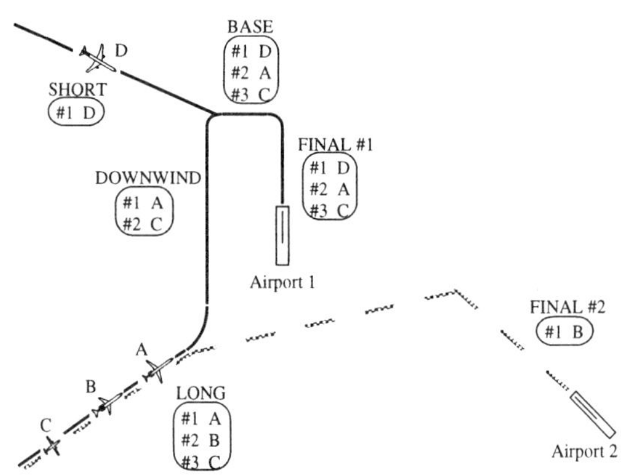

Figure 4. Trajectory Segment Ordering

Also shown in Figure 4, the relative sequence between two aircraft on FINAL trajectory segments is maintained across all segments. All aircraft sharing a given trajectory segment are included in the ordered list for that segment, regardless of which FINAL segment terminates each trajectory. This indicates that Aircraft C will absorb any necessary delay to insure separation with Aircraft B on the LONG segment and Aircraft A on the DOWNWIND, BASE and FINAL#1 segments. An important result of this process is that an aircraft will likely depend on, and therefore be de-conflicted from, different aircraft depending on the order for each trajectory segment.

KBRA Outputs

The output of the knowledge based runway allocation is an advisory to the controllers suggesting a runway for each aircraft. If the controller chooses to accept this advisory, the runway becomes the assigned runway for that aircraft.

KBRA Algorithm

This section describes the KBRA algorithm and its integration into the FAST architecture. First, an overview of how the runway allocation algorithm is integrated into the FAST update cycle will be presented, followed by descriptions of the individual components of the KBRA algorithm.

FAST updates the arrival plan every 6 seconds, operating asynchronously with the TRACON radar, which updates every 4.7 seconds. Within the FAST

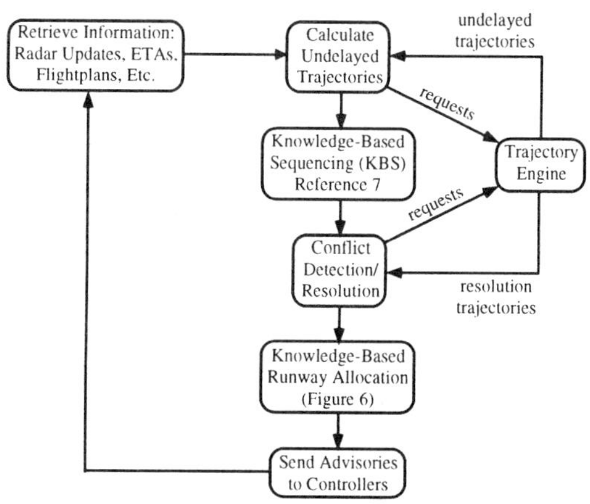

Figure 5. FAST Scheduler Cycle

update cycle, the following is achieved: sequences and STAs are calculated for each aircraft, potential runway allocations are evaluated and resulting sequence and runway advisories are sent to the controllers. As shown in Figure 5, evaluation of potential runway allocations occurs after sequencing and conflict detection/resolution of the arrival plan has occurred. The results of the knowledge-based sequencing and conflict detection/resolution are used in evaluation of runway allocations.

As shown in Figure 6, the runway allocation algorithm is divided into two cycles: the preliminary evaluation of all potential allocations, and the final determination of a single, most promising allocation.

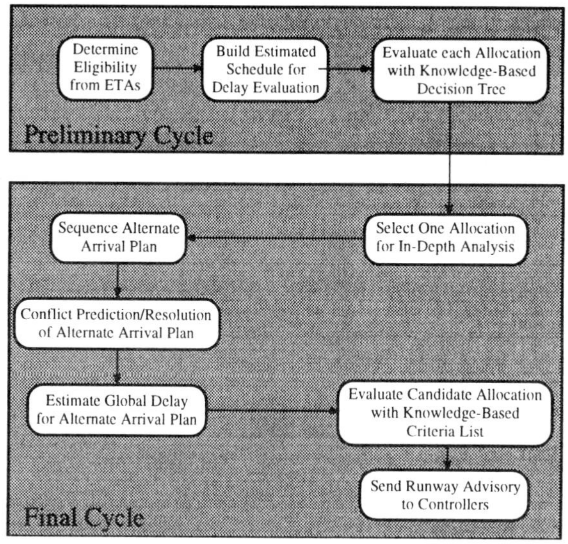

Figure 6. Knowledge Based Runway Allocation Flowchart

Preliminary Cycle

The purpose of the preliminary cycle is to reduce the set of possible tactical runway allocations to a manageable set and to quickly evaluate these allocations. Reduction of the set of all aircraft and potential allocations is first achieved through a test for runway allocation eligibility.

An aircraft's eligibility for runway allocation is largely determined by the airport configuration and its undelayed arrival time or ETA. Specifically the requirements for eligibility are:

1) The aircraft has an available alternate runway. This requires that more than one possible runway be defined in the airport configuration for the aircraft's stream class.

2) The undelayed time-to-fly to the available alternate runway is within the runway allocation window defined by the airport configuration. The allocation window is defined independently for each arrival runway in the configuration.

3) The runway assignment has not been "frozen." A frozen runway assignment indicates that only the controller is allowed to assign a new runway to the aircraft. A frozen runway results when a controller manually assigns a runway to the aircraft, or when the aircraft's time-to-fly to its assigned runway is less than the minimum defined by the runway allocation window.

An aircraft can be eligible for allocation to more than one runway. For this reason, the algorithm employs a runway-pair structure which defines an allocation as an aircraft from its currently assigned runway to an available alternate runway. One of the factors in evaluation of a potential allocation is its effect on total system delay. In preparation for evaluation by the knowledge-based rules of the preliminary cycle, the schedule for each aircraft to each available runway is estimated.

The delay savings for each allocation is estimated as the difference in the sum of expected time to fly for all aircraft in the system for both the allocation and non-allocation cases. At this point, it is important to define what is used as the expected time to fly.

A time known as the nominal scheduled time of arrival (STA) is used as the reference for expected time to fly. Nominal STA is defined as the later of an aircraft's undelayed ETA (fastest possible trajectory) or the arrival time corresponding to minimum separation with the aircraft sequenced one ahead on final approach at the runway threshold.

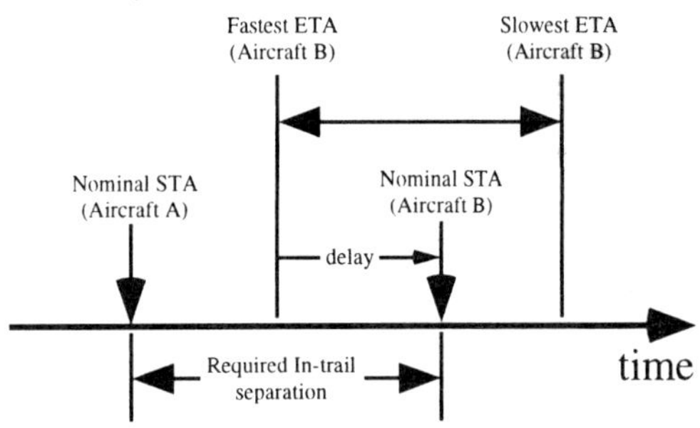

Figure 7. Nominal STA Derivation

As shown in Figure 7, Aircraft B would be in violation with Aircraft A if it were to meet its fastest ETA; therefore, its nominal STA is dependent upon the nominal STA of Aircraft A and the required minimum separation distance, governed by FAA regulations and based on aircraft type and winds aloft on final approach.

The nominal STA is used as the reference for two reasons: simplicity and accuracy. Simplicity is essential if the preliminary cycle is to achieve its goal of rapid evaluation of all potential allocations. Calculation of the nominal STAs require an established sequence on each final approach, each aircraft's ETA, and the weight class of each aircraft to determine required separation, as well as an assumption for ground speed on final approach. Each of these inputs has already been determined, and is easily accessible; leading to rapid estimation of total system flight time and delay savings for each allocation pair.

Producing an acceptable set of conflict-free 4D trajectories for all aircraft in the system is an extremely difficult task.[7] To produce such a solution set requires precise modeling of controller decision patterns, coordination and prioritization of tasks. Furthermore, it would require exactness in route deviation (and DOF) possibilities, and accurate modeling of how a controller employs these DOFs. This is largely accounted for in the FAST sequencing algorithm, however, it is too computationally intensive for the task described here. To simplify estimation of total flight time in the preliminary cycle, the allocation aircraft is sequenced First-Come-First-Served on the alternate runway. This allows for numerous estimations of delay savings to be made without revisiting the sequencing logic. For this reason, it is assumed that this model of the sequencing algorithm is sufficiently close to produce usable results in the preliminary cycle.

Knowledge-Based Decision Tree

Once the potential delay savings have been estimated in the preliminary cycle, each potential runway allocation is evaluated with a decision tree which incorporates the knowledge base of facility procedures, controller workload issues and delay reduction criteria. This decision tree determines if each potential allocation would achieve necessary delay benefits and would be acceptable to controllers. Figure 8 illustrates a simplified runway decision tree, showing a single thread of a series of branches. Each branch in the tree is based on the result of one of the rules presented in Appendix A. Individually these criteria represent simple, understandable ideas related to controller workload and delay reduction. In combination, however, they can adequately and efficiently represent the complex decision patterns of experienced terminal area controllers. For each potential allocation, a decision tree is traversed until a rule is evaluated which results in a decision to either allow further evaluation of the allocation (pass the preliminary cycle), or to remove it from the list of potential allocations (fail the preliminary cycle).

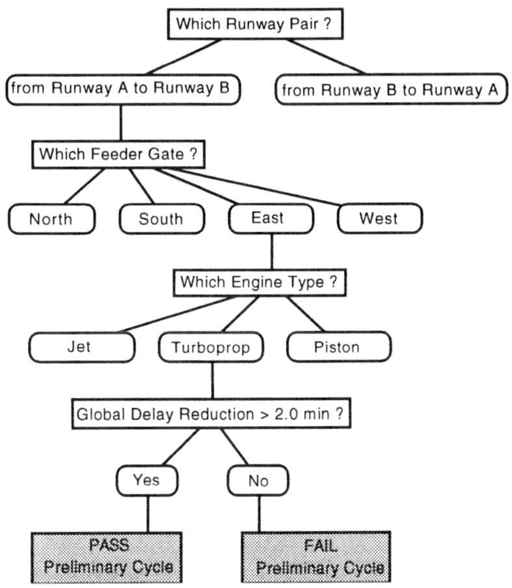

Figure 8. Sample Preliminary Evaluation Decision Tree

A list of potentially beneficial and acceptable allocations results from the preliminary evaluation cycle. This reduced set of allocations is further evaluated to insure delay benefits and controller acceptability in the final cycle.

Final Cycle

The final cycle of the FAST KBRA algorithm is essentially a preventative measure employed to fully evaluate the most promising allocation in the list of allocations which passed the short cycle rules. Inappropriate allocations are avoided by performing an in depth evaluation of one allocation pair per update cycle, as shown in Figure 9.

This greater depth of evaluation is accomplished by creating an alternate schedule which includes the candidate allocation aircraft to its alternate runway. The final cycle consists of five steps: selecting an allocation pair for evaluation, creating the alternate arrival plan, conflict detection/resolution for the alternate arrival plan, estimation of delay savings for the alternate arrival plan and evaluation of the candidate allocation pair with a knowledge-based criteria list.

The list of potentially beneficial and acceptable allocations from the preliminary cycle is sorted primarily by delay savings potential. Allocation pairs with similar delay savings potential are sorted based on

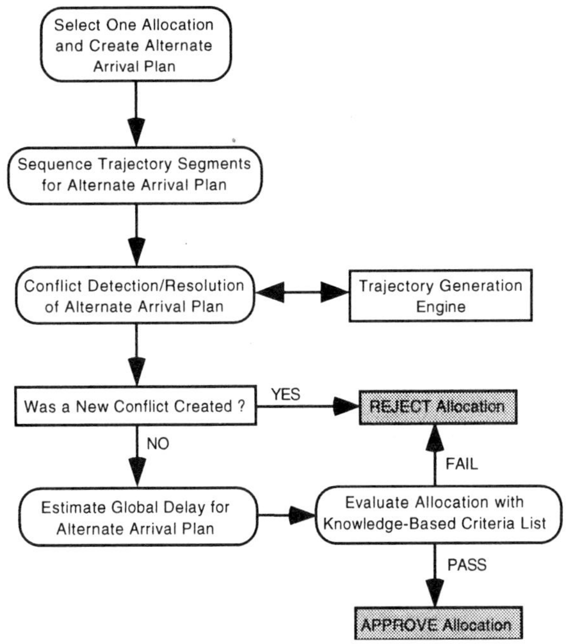

Figure 9. Final Cycle for Runway Allocation

elapsed time since the allocation pair was last selected for evaluation by the final cycle. Following sorting, the first allocation pair in the list is selected for in-depth evaluation.

An alternate arrival plan is created by revisiting the trajectory segment evaluation logic for the candidate allocation aircraft. Once the trajectory to the alternate runway has been dissected into segments, the knowledge based sequencing (KBS) algorithm sequences the alternate arrival plan in the same manner as was used for the current arrival plan.[7] Sequencing the alternate arrival plan produces a more realistic schedule than that which was used in the preliminary cycle. The preliminary cycle simply placed the candidate allocation aircraft first-come-first-serve on its alternate runway. Sequencing in the final cycle takes into account merging streams of aircraft and controller workload in its knowledge base. It is possible that the sequence on any trajectory segment in the system could change, not only the sequence on the segments which the allocation aircraft traverses.

As a minimum requirement for an allocation to be acceptable, it must not adversely affect the conflict resolution status of the aircraft. In other words, if the candidate allocation aircraft is conflict-free on the default runway, but predicted to be in conflict on the alternate runway, the allocation is assumed unacceptable from a controller workload standpoint. A conflict occurs when an aircraft does not have enough delay absorption capability, as defined in the route adaptation, to resolve a predicted separation violation. Regardless of any predicted delay savings (based on nominal STAs), FAST will not advise an allocation which, because of a new conflict, may unacceptably increase controller workload.

If an allocation passes the conflict resolution criterion, its delay savings must be recalculated. Because nominal STAs depend on the sequence of aircraft, they may differ from the STAs predicted in the preliminary cycle. For this reason, the delay savings estimate may not be the same for the final cycle as for the preliminary cycle. The new delay savings estimate is stored for further evaluation by the final cycle.

Evaluation of the candidate allocation pair by the final cycle is accomplished through a criteria list. This list is a set of criteria specific to the allocation being evaluated. A list exists for each runway pair/category combination. As shown by Figure 10, the criteria in the list are evaluated either until a criterion is not met, or until all criteria are satisfied. If all criteria are satisfied, the candidate runway allocation is advised to the controller. If any one of the criteria are not satisfied, the final determination cycle rejects the candidate allocation, and in most cases the current arrival plan is maintained.

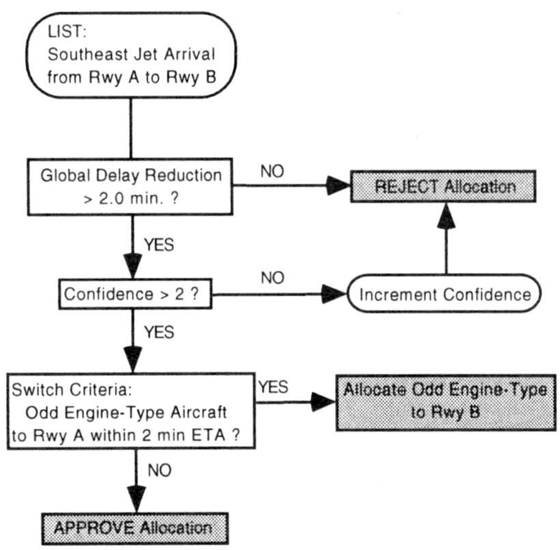

Figure 10. Example Final Cycle Criteria List

The criteria used in the lists are presented in Appendix B. However, the roles of two criteria should briefly be discussed here. The "confidence" criterion simply increments a counter for a given allocation pair, and will not let an allocation occur until the counter reaches an adaptation specified value. The confidence level criterion prevents improper allocations resulting from various errors (variable ETA or radar data, etc.). As a result of the confidence criterion, aircraft which enter the preliminary cycle's allocation time window first, are more likely to satisfy the confidence criterion before those with slightly later ETAs. This may not always lead to the most acceptable allocation.

When the candidate allocation satisfies the confidence criterion, "switch" criteria can be used to search for more appropriate allocations which would serve the same general purpose as the candidate allocation. If an aircraft is found which would yield a more acceptable allocation, the final cycle rejects the candidate allocation, and advises controllers of the more acceptable allocation.

Operational Test Results

FAST was evaluated operationally at the Dallas/Fort Worth TRACON during 1996.[8] FAST was operational for over twenty arrival rushes spanning the spectrum of nearly all traffic patterns encountered at DFW TRACON. During these arrival rushes, controllers evaluated FAST generated sequence and runway advisories for over 1200 arrivals. Both controller feedback and statistical trends indicate the expected benefits were realized.[8]

The TRACON traffic management coordinator (TMC) routinely increased the acceptance rate during FAST tests to levels above normal operations. This indicates that controllers were able to handle aircraft more efficiently; leading to more operations at similar workload. As shown in Figure 11, airport throughput was increased 9.3% during IFR operations and 13.3% during VFR operations with FAST.[8]

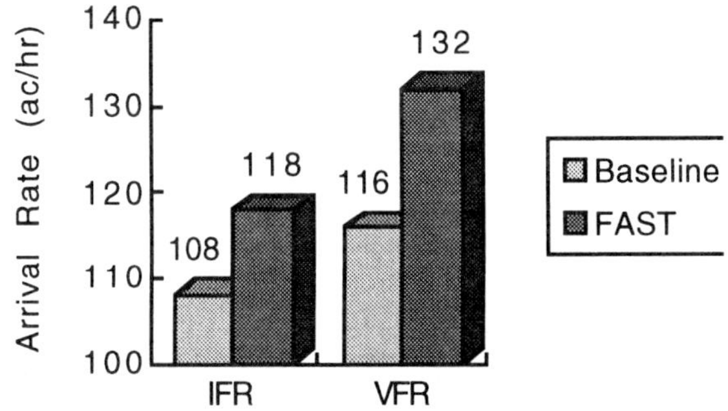

Figure 11. Comparison of mean airport throughput during peak portion of 11:15 am rushes.[8]

Comments from tower controllers were overwhelmingly positive. The departure queue backlog was reduced by

9% due to well-balanced runways and consistently-spaced arrivals. Tower operations logs showed an average increase of 28 operations per hour or approx. 13% (15 arrivals & 13 departures) during rushes in which FAST was operating. Taxi-in and taxi-out times were not impacted despite capacity increases.[8]

Controller feedback through post-rush evaluation forms, modeled after the NASA Task Load Index (TLX) scale, indicated no significant impact on overall controller workload due to the use of Passive FAST.[8] A modified Controller Acceptance Rating Scale (CARS) was used to gauge overall system acceptability. CARS responses indicated the prototype FAST system needed minor modifications to become fully acceptable. Comments from the CARS forms stated the major concerns with runway allocation performance of FAST were late runway allocations and response time to controller inputs. Late allocations could be avoided by sliding the runway allocation eligibility window further from landing time, thus improving perceived schedule stability. Response to controller input has been improved by employing faster computers for FAST.

Advisory adherence can be used to gauge acceptability of FAST-generated advisories. Controllers had the option to override any advisory judged to be unsafe, sub-optimal, high-workload or generally unacceptable for any reason.

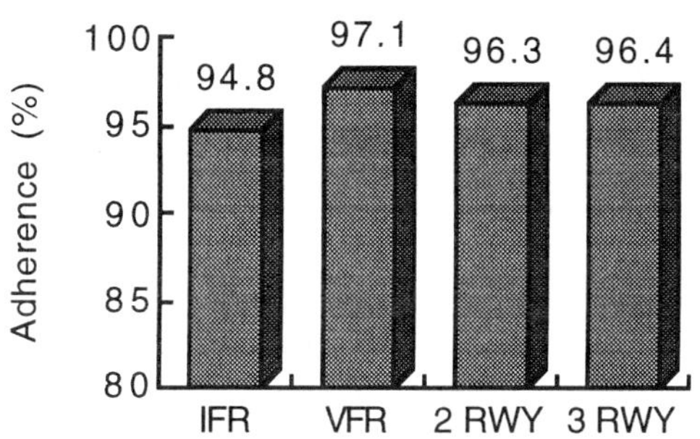

Figure 12: Adherence to FAST Runway Advisories

Non-adherence to a large number of advisories could lead to increased controller workload, reduced benefits, and distrust of the system. While controllers disagreed with some of the runway advisories, overall adherence to runway advisories was high, indicating trust in the system (Fig. 12).

Conclusions

A knowledge-based system for sequencing and assigning runways to arrival traffic to the terminal area has been developed and tested. The algorithms and knowledge base were developed through thousands of hours of controller-in-the-loop simulation. Field testing of FAST has demonstrated the ability of a limited number of rules to adequately model the runway assignment decision process. While this set of rules may need to be expanded for other airspaces (e.g. Chicago, New York), the ability to model the runway assignment decision process with a limited rule-base has been demonstrated.

The knowledge-based runway allocation (KBRA) algorithm uses the results of the knowledge-based sequencing algorithm, in a two-cycled approach. The preliminary cycle quickly evaluates all eligible re-assignments for potential benefits and controller acceptability. The preliminary cycle reduces the number of potential reassignments to a manageable number for further evaluation. The final cycle evaluates this reduced set and selects one re-assignment for in-depth evaluation. This candidate re-assignment is verified to achieve the required benefits and acceptability requirements established by the knowledge base.

Operational testing at Dallas/Fort Worth TRACON has demonstrated significant increases in airport capacity without adversely affecting controller workload. The tests also indicated reduced departure delays are possible due to runway balancing and reduced ground congestion. As a result of the benefits demonstrated through testing and evaluation of the prototype system, an operational version of FAST is now scheduled for deployment in 5 to 10 major U.S. airports.

References

1. Davis, T. J.; Krzeczowski, K. J.; Bergh, C.: "The Final Approach Spacing Tool", 13th IFAC Symposium on Automatic Control in Aerospace, Palo Alto, CA, Sept. 1994.
2. Davis, T. J.; Erzberger,H.; Green, S. M.; Nedell, W.: "Design and Evaluation of an Air Traffic Control Final Approach Spacing Tool", Journal of Guidance, Control, and Dynamics, Vol. 14, No. 4, July-August 1991, pp.848-854.
3. Krzeczowski, K. J.; Davis, T. J.; Erzberger, H; Lev-Ram, I; Bergh, C: "Knowledge-Based Scheduling of Arrival Aircraft in the Terminal Area", AIAA Guidance, Navigation, and Control Conference, Baltimore, MD, August, 1995.
4. Erzberger, H.; Davis, T. J.; Green, S. M.: "Design of Center-TRACON Automation System", Proceedings of the AGARD Guidance and Control Panel 56th Symposium on Machine Intelligence in Air Traffic Management, Berlin, Germany, 1993, pp. 11-2 - 11-12.

5. Brinton, C.: "An Implicit Enumeration Algorithm for Arrival Aircraft Scheduling", 11th Digital Avionics Conference, Seattle, WA, October 1992.
6. Lee, K.; Davis, T. J.: "The Development of the Final Approach Spacing Tool (FAST): A Cooperative Controller-Engineer Design Approach", NASA TM 110359, August 1995.
7. Robinson, J. E. III; Davis, T. J.; Isaacson, D. R.: "Fuzzy-Reasoning Based Sequencing of Arrival Traffic in the Terminal Area", Proceedings of the AIAA Guidance, Navigation, and Control Conference, New Orleans, LA, August 1997.
8. Davis, T. J.; Isaacson, D.R.; Robinson, J.E. III; denBraven, W.; Lee, K.K.;Sanford, B.: "Operational Field Test Results of the Passive Final Approach Spacing Tool", Proceedings of the IFAC 8th Symposium on Transportation Systems, Chania, Greece, June 1997.

Appendix A - Rules used for Preliminary Runway Allocation Evaluation

The rules in this appendix are used in the decision trees of the preliminary evaluation cycle. While the number of rules presented here is brief, they were found completely able to adequately model runway allocation decision patterns when used in combination. Air traffic systems and facility procedures are varied, and may require development of more rules to solve problems unique to each facility. Eight rules were used in developing the decision trees for the DFW TRACON. The logic behind each rule will be discussed, as well as its input, basis in controller decision patterns, and usage.

AC IN CATEGORY EXISTS: This rule determines if any aircraft in a specified runway category exist in the system. The runway category is determined from the feeder gate, airport configuration, destination, and engine type. The existence of an aircraft in the specified runway category could affect the acceptability of an allocation.

WHICH ENGINE: This rule determines the aircraft engine type: jet, turboprop, or piston. Certain allocations are preferred for a given engine type over others. This is due to TRACON routing, which separates aircraft by stream class as they arrive over the feeder gate.

FEEDER GATE RUSH: This rule is used to recognize when a large number of aircraft are to arrive through a single feeder gate and are scheduled to a single runway. Without allocation, a feeder gate rush leads to unacceptable workload and high delay. This can be avoided by recognizing this situation before this group of aircraft reach the feeder gate, and allocating from this group to alternate runways.

WHICH RUNWAY CATEGORY: This rule determines if the aircraft being evaluated by a decision tree is of the runway category specified by the decision tree. The factors determining runway category are often used in favoring one group of aircraft for allocation over another. This rule is commonly used in a similar fashion to the engine type rule, but can be expanded to employ runway categories that are dependent on airport configuration and common operating procedures.

ODD AC TYPE: It is common practice for controllers to stratify a traffic flow based on engine type. Controllers prefer to maintain stream consistency (engine type) whenever possible. Aircraft of similar engine type generally have similar performance characteristics. This leads to repeatability of commands and lowers the workload associated with maintaining separation in a traffic stream. This predicates allocations based on engine type. However, the engine type such an allocation is based on varies depending on the mix of aircraft in the stream from which an allocation will come. For this reason, it is necessary to recognize when an odd-type aircraft is in a traffic stream. An odd type aircraft is defined as an aircraft whose engine type consists of less than 43% of the stream being considered.

RUNWAY BUSY: This rule counts the number of aircraft assigned to a given runway within a specified time window of the evaluation aircraft's ETA on that runway, to determine if a runway will be busy at the time of the evaluation aircraft's ETA. This rule is linked to controller workload and is employed to incorporate controller workload in the knowledge base. Used alone, it could set a limit on the number of aircraft allowed to be assigned (by FAST) to a given runway. However, it is usually grouped to form numerous branches with varying delay threshold values, based on how busy a runway is. Furthermore, nesting of runway busy rules increases flexibility by allowing a "tradeoff" of delay to be used in determining if an allocation is acceptable. Unlike "feeder gate rush", "runway busy" counts all aircraft to a given runway in the specified time window, not just those arriving over a single feeder gate.

DELAY REDUCTION: Delay reduction is employed as a rule that usually is the final rule used to determine eligibility for the long cycle. It is not required that delay be reduced for an allocation to pass the preliminary cycle rules. If workload issues warrant an allocation, one could be made which increases delay: as long as the amount by which the delay is increased is less than that specified in the decision tree. Situations which could lead to such an allocation are captured in the decision tree by the previously described rules.

RUNWAY AVAILABLE WITH NO DELAY: While delay reduction is a measure of the savings in total flight time of all aircraft in the system, it is not necessarily the basis for controller preferred allocations. This rule determines if the allocation under evaluation would lead to delay for any aircraft assigned to the alternate runway. Regardless of prescribed configuration and default runway, situations exist in which controllers prefer an alternate runway over the default if such an allocation does not affect any other aircraft on the new runway (no added delay). This effectively removes any flight time difference from the delay estimation. For example, it may be preferred by the controller or the airline to land on the parallel runway as opposed to a diagonal runway due to perceived taxi time, if such an allocation does not add delay to any other aircraft.

Appendix B - Criteria used for Final Runway Allocation Determination

These criteria are used by the final cycle to evaluate the acceptability of the candidate allocation. Some of the criteria are very similar to the rules used in the preliminary evaluation decision trees. However, these criteria are evaluated following KBS sequencing of the alternate arrival plan. This could lead to different delay savings estimates and controller acceptability. Each of the five long cycle criteria, their input and usage will now be discussed.

USE RUNWAY IF DELAY REDUCED: The delay reduction threshold required for the long cycle can be explicitly specified in the adaptation, but is generally set to the same value as specified in the preliminary cycle. Used in this manner, this criterion simply verifies that the delay savings estimated in the preliminary cycle, are still realized following KBS sequencing of the alternate arrival plan.

USE RWY IF AVAIL WITH NO DELAY: Also similar to the rule used in the preliminary cycle, this criterion is generally used as a verification of delay estimates of the short cycle.

CHECK CONFIDENCE LEVEL: This criterion is used in a manner consistent with a low pass filter. Only those allocations which have consistently passed the previous criteria in the list achieve the required confidence, thus stabilizing the runway allocation process. Each time an allocation reaches the confidence criterion, its confidence level is incremented. Once an allocation achieves the required confidence level, this criterion is satisfied. The confidence level of allocations can be reset by the criteria if an allocation is advised.

"Switch" criteria: Switch criteria are used after an aircraft has been determined to meet the confidence level required, along with all other criteria employed. The purpose of a switch criteria is to search the specified traffic stream for a more suitable aircraft for allocation. Switch criteria are means of enforcing controller preferences in the long cycle for allocations which may be acceptable but not the most acceptable. Once a switch criteria has been reached in the criteria list, the candidate allocation has been deemed acceptable. For that reason, once a switch criterion has been reached, an allocation will occur, but the candidate allocation may be replaced with a similar allocation considered more acceptable to controllers. Two switch criteria are employed: one concerning runway category preference, and the other employing the odd engine type logic similar to that of preliminary cycle.

SWITCH RUNWAY WITH OTHER CATEGORY: This criterion searches the system for aircraft within the specified time window in the specified runway category which would be a more acceptable allocation than the one determined to meet all previous criteria in the list. This criterion can be used to model general controller decision patterns for runway allocation once it has been determined an allocation is necessary for a given time or runway slot. Choosing the most acceptable aircraft to fill that slot, or balance runways, can be achieved by defining preferred runway categories for each allocation pair considered.

SWITCH RWY WITH ODD AC TYPE: Again, this switch criteria is a means of insuring the most acceptable allocation of a given type is advised. Similar to the short cycle odd engine type logic, this criterion searches the stream of the candidate allocation aircraft for an aircraft of odd engine type, if the candidate aircraft is not of an odd type in the stream. This criteria is necessary due to the cyclical nature of the runway allocation process. The runway allocation cycle only occurs once every update cycle (6.0 seconds). This determines that allocations of similar benefit are evaluated in turn as they enter the allocation window. This fact, coupled with the confidence criterion, means the earlier an aircraft enters the allocation time window, the better chance it has of being allocated. Such behavior does not always lead to the most acceptable allocation. The switch criteria corrects for this cyclical nature by determining if there are any other aircraft in the system which would closely match the candidate allocation, yet be more acceptable to controllers.

AIAA-97-3545

Departure Trajectory Synthesis and the Intercept Problem

Michael A. Bolender *
G.L. Slater †
Department of Aerospace Engineering and Engineering Mechanics
University of Cincinnati
Cincinnati, OH 45221-0070

Abstract

Two areas of the departure problem in air traffic control are discussed. The first topic is the generation of climb-out trajectories to a fix. The trajectories would be utilized by a scheduling algorithm to allocate runways, sequence the proposed departures, and assign a departure time. The second area is concerned with finding horizontal trajectories to merge aircraft from the TRACON to an open slot in the en-route environment. Solutions are presented for the intercept problem for two cases: (1) the aircraft is traveling at the speed of the aircraft in the jetway, (2) the merging aircraft has to accelerate to reach the speed of the aircraft in the en-route stream. An algorithm is given regarding the computation of a solution for the latter case. For the former, a set of equations is given that allows us to numerically solve for the coordinate where the merge will occur.

Introduction

Departure scheduling is an important area of air traffic control. Currently, there are no automation tools available to assist the tower, departure, and sector controllers with the management of departure traffic. The current procedure of runway assignment and sequencing, administering departure clearances, and the tactical control of aircraft within the TRACON is largely a manual process. By taking advantage of current technology, the development of advanced automation systems that will assist air traffic controllers with the control of departures is being realized. One such system is Expedite Departure Path (EDP). EDP is the component of the Center–TRACON Automation System (CTAS) which will handle the scheduling and flow of departure traffic. EDP will provide departure and en–route controllers with advisories to assist in the movement of aircraft from the runway to the en–route stream.

The departure problem in air traffic control automation can be separated into three sub–problems. These are runway allocation and sequencing for departure, trajectory synthesis, and the en–route merge problem. The runway allocation and sequencing function will ideally take inputs from a trajectory synthesis program in order to produce an optimal and robust schedule. Miles–in–trail restrictions over a departure gate, runway capacities, and airspace constraints must all be considered in this implementation. The en–route merge problem will ultimately require a trajectory synthesis function to predict aircraft positions along a route, and the time to fly a time–optimal trajectory to a fixed point.

The purpose of this paper is to discuss two parts of the departure problem– the trajectory synthesis problem and the en–route merge or "intercept" problem. Denery and Erzberger[1] state that the management of traffic in constrained airspace requires a capability to accurately predict aircraft trajectories. Due to the abundance of literature which discusses trajectory synthesis, a brief discussion on this topic as it pertains to departures is included for completeness. The intercept problem is addressed since there presently lacks a means to assist controllers in merging aircraft departing the TRACON and entering en–route streams. It will be shown that horizontal trajectories can be computed which deliver an aircraft to a known moving position between aircraft flying in–trail.

Trajectory Synthesis

The trajectory synthesis portion will make up a large part of the structure of the departure scheduling software. Its importance lies in the fact that accurate departure fix crossing times are needed to produce schedules which are beneficial to the air traffic system. The trajectory synthesis is based entirely upon the trajectory synthesis program (TS) currently within

*Graduate Student. Student Member AIAA.
†Professor and Department Head. Associate Fellow AIAA.

CTAS. Stated simply, the TS uses point mass performance models of the jet and turboprop aircraft employed by the airlines. These models are then used, along with pilot procedures defined by airline operating handbooks, to calculate arrival times to the outer marker for final approach.[2]

Several simplifying assumptions are made which speed-up computation without sacrificing accuracy. Commercial aircraft typically do not partake in large amplitude maneuvers or encounter sudden changes in the flight path (e.g. zoom climbs, maximum performance turns). Therefore, assume that the rate of that the aircraft's flight path angle, γ_a, is "small", and the acceleration normal to the velocity vector is negligible. It is assumed that the total derivative of the wind velocity field with respect to time is approximately zero. The explicit time dependency of wind is neglected, and it is assumed that the spatial variation changes slowly. It is further assumed that there is no wind component in the vertical direction.

For completeness the simplified equations of motion which are used for determining a trajectory are presented below. The reader is referred to Slattery and Zhao[3] for a complete discussion.

$$\dot{V}_t = \frac{T-D}{W}g - g\gamma_a \quad (1)$$

$$\dot{\psi}_i = \frac{Lg\sin\phi}{WV_g} \quad (2)$$

$$0 = L\cos\phi - W \quad (3)$$

$$\dot{h} = V_t\gamma_a = V_g\gamma_i \quad (4)$$

$$\dot{x} = V_g \sin\psi_i \quad (5)$$

$$\dot{y} = V_g \cos\psi_i \quad (6)$$

Where V_t is the true airspeed, T is the thrust and is parallel to the velocity vector, D is the drag, and g is the acceleration due to gravity. The flight path angle, γ_a, is measured with respect to the moving air mass, and γ_i is measured with respect to the $x-y$ plane. The groundspeed, $\vec{V}_g$, is the vector sum of the true airspeed and the wind velocity and is given by

$$\vec{V}_g = \vec{V}_t + \vec{V}_w \quad (7)$$

The magnitude of the groundspeed, V_g, denotes the speed of the aircraft "over the ground." The groundtrack heading, ψ_i, is measured positive clockwise from North. The bank angle, ϕ, is positive when the aircraft is turning clockwise. The position of the aircraft is measured on a coordinate system where the x axis is positive East, and the y axis is positive North. The altitude is given by h, and the rate of climb is given by $\dot{h}$. The lift generated by the aircraft is L and the aircraft's weight is W.

The departure trajectory is determined by considering the trajectory as two independent components. The first component is the horizontal path. The horizontal path extends from the runway to the final waypoint on the departure route. The second is the vertical profile that is flown to a final or "control" altitude. The aircraft generally follows a pre-determined route through the TRACON, along which the speed profile is known from the vertical profile. Since the aircraft is assumed to have a "small" flight path angle, the equations of motion nearly decouple. This allows the horizontal path to be found independent of the vertical path.

Horizontal Path

The horizontal path flown will generally be one of the standard departure routes. At those airports located within Class B airspace and some within Class C, Standard Instrument Departure routes (SIDs) are used as the primary routes out of the airspace. The purpose of the SIDs is to separate the departure traffic from the arrivals by using different "corridors." A SID will typically contain several waypoints for the aircraft to fly until it reaches the VOR for transition to the jetway. The purpose of the horizontal synthesis is then to assign and compute such a route from the runway to the outer departure fix satisfying any necessary constraints along the route.

A typical horizontal trajectory consists of a series of straight line trajectory segments connected at "waypoints" by circular arc transitions. At each waypoint the aircraft undergoes a heading change toward the next point until the aircraft has merged into an en-route trajectory. At the waypoint there are several ways to model the turning segment of the trajectory. The first and most common approach (i.e. that which is used in most on-board flight management systems) is to construct the circular segment such that it is tangent to the two straight line paths, and does not intersect the waypoint (Fig. 1). This is called an "inside turn." In fact, this is the only type of turn used in our current analysis as other types of turns may force the trajectory to pass through the waypoint.

Vertical Profile

The vertical profile is the second component of the trajectory. The vertical profile consists of a number of trajectory segments. These segments are (1) an acceleration to a calibrated airspeed (CAS), (2) a climb at a constant CAS, (3) a climb at a constant Mach number, and (4) level flight at a constant CAS or Mach. For simplicity, it is assumed that the bank angle is zero. The vertical trajectory is governed by only two equations of motion, Eqns. 1 and 4. We will consider V_t and h to be states and the controls to be T and γ_i. In order to define a segment, any combination of two controls or states needs to be specified. Alternatively, any two relations that define the controls may be specified. The most commonly used are (1) constant rate-of-climb or flight path angle, (2) constant CAS or Mach, and (3) constant acceleration.

A typical profile that is currently flown in our simulation is shown in Fig. 2. The first segment is for the aircraft to accelerate to a calibrated airspeed of 250 kts (KCAS) while climbing at a fixed rate of climb beginning at approximately 1,000 ft above ground level (AGL). The aircraft then maintains 250 KCAS until 10,000 ft according to Federal Aviation Regulations (FARs). The aircraft then accelerates, again at a fixed climb rate, to a new climb CAS that is dependent upon the aircraft. This CAS is maintained until the aircraft reaches a climb Mach number (a typical altitude for this transition is usually above 27,000 ft). The aircraft continues climbing at a constant Mach until it reaches a cruise altitude.

Taken together, the vertical and horizontal trajectory approximates the aircraft's flight through the TRACON. The trajectory then may be used as an input to a scheduler as well as to provide tactical control once the aircraft is aloft. An example of a climb-out trajectory is shown in Fig. 3. The aircraft departed Denver International on January 8, 1997. The computed climb-out trajectory is also shown along with lines of constant Mach number and constant CAS for comparison. The aircraft was "picked-up" on radar at 12,300 ft at a climb CAS of 275 kts. The "tail" at the bottom of the trajectory was ignored as it is an artifact of the groundspeed filtering algorithm used by the radar. In the simulation, the aircraft climbs at 275 KCAS until it reaches a Mach number of 0.68. It continues at this Mach until it reaches its cruise altitude at flight level 330. The actual climb profile has the aircraft initially at 275 KCAS until it reaches Mach 0.6. The aircraft climbs at Mach 0.6 until it reaches approximately 23,000. At this altitude the aircraft accelerates to Mach 0.68. The new Mach is maintained until the aircraft reached its cruise altitude. The actual time to climb was approximately 14.1 min. The calculated time to climb was 13.8 min, a difference of 21 sec. The simulation assumed a climb rate of 1500 fpm. The actual climb rate started at 2000 fpm and decreased to 1000 fpm as the aircraft reached the end of its climb. The differences in the times to climb can be attributed to the differences in climb rate and the different speed–altitude profiles.

The Intercept Problem

One problem of interest in departure scheduling is to have an aircraft merge into or intercept a "slot". A slot may be defined to be a gap between two aircraft such that the merging aircraft may fly in–trail to the lead aircraft without violating separation constraints. A slot may also be a position behind a single aircraft such that the mandatory separation requirement is not violated. In either case, the current headings and speeds of all aircraft are known. The former situation may occur in the en–route environment where the sector controller is trying to merge traffic coming across different fixes. The latter is likely to occur in the terminal area where a departure controller is trying to provide miles–in–trail separation across a departure fix when aircraft are using radar departure procedures. In either situation, we want to be able to find the following: (a) the physical location of the slot at the instant the intercept occurs, (b) the time to intercept the slot, and (c) the horizontal trajectory that satisfies (a) and (b).

The intercept problem is broken down into two different cases. The first case is when the aircraft is traveling at the speed of the traffic. The second case is where the aircraft is traveling at lower speed than the en–route aircraft. The analysis that follows assumes that the intercept problem is independent of the trajectory problem discussed in the previous section. Also, only the horizontal intercept problem is discussed. The "3-D" intercept problem (i.e. the aircraft is merging while climbing to a cruise altitude) is left as the subject of a future paper.

For brevity, the solution for a particular geometry is presented. Using symmetry, the results derived here can be extended to other geometries. The coordinate system is chosen such that the merging aircraft is initially at the origin, with a known heading, $-180° < \psi_0 \leq 180°$, and airspeed, V_0. The route is assumed to be parallel to the x–axis. Traffic along this route is moving from left to right and has a speed V_T. The distance from the aircraft to the stream is known and denoted as y_f, which is assumed positive. Finally, the slot is assumed to have a known initial position on the route, x_{t0}. The intercept point, x_f, is where the aircraft merges into the stream. Define the distance traveled by the slot as S_s, and the distance traveled by the aircraft as S_a. Likewise, the time for the slot to travel from x_{t0} to x_f is T_s and the time for the aircraft to travel from the origin to the intercept point is T_a. The problem may be stated as follows: Given the initial conditions (ψ_0, x_{t0}, y_f, V_0 and V_T), find x_f, T_a, and the horizontal trajectory to deliver the merging aircraft.

The distance between the origin and the intercept

point, (x_f, y_f), is assumed to be large enough such that the optimal horizontal trajectory is a "turn-straight-turn" trajectory. As was shown by Erzberger and Lee,[4] a constant speed, turn-straight-turn trajectory is a "time-optimal" trajectory for aircraft guidance to a point with a specified final heading. Although there are four solutions to such a turn-straight-turn trajectory problem, only two are considered here. The merging aircraft will be required to make a "direct turn" to the route. A "direct turn" is defined to be a turn where the merging aircraft does not "cross over" the route before turning back to complete the merge. This is done in order to prevent conflicts with the en-route traffic. For the geometry presented in this analysis, the final turn will be a clockwise turn, as a counterclockwise final turn will require the aircraft to cross the route and then turn back to intercept the route.

Case 1: $V_0 = V_T$

The first case presented is the case where the aircraft is traveling at the same speed as the en-route stream. It is the more interesting of the two to analyze since we are able to determine functions which can be solved for the intercept point. The fact that the aircraft are traveling at the same speed means that the aircraft will travel the same distance in the same time. The analysis directly follows from McLean,[5] where the turn-straight-turn trajectories for minimum time flight to a point with a known heading are solved geometrically.

Counterclockwise-Straight-Clockwise

We will begin our analysis by studying the counterclockwise initial turn, with a goal of finding a function that can be solved, either numerically or analytically for the intercept point. The approach will be to find the total distance traveled by the aircraft, then to equate this to the distance traveled by the slot. The geometry for this problem is shown in Fig. 4.

To begin, some basic relationships are given that describe any general turn-straight-turn trajectory. We start with the relationships for the center of curvature of a turn arc, which is assumed to follow a circular path. McLean [5] defines a parameter S_i which describes the direction of a turn, where $S_i = +1$ if the turn is clockwise and $S_i = -1$ if the turn is counterclockwise. Also, R_1 and R_2 are the turn radii for the first and second turns respectively. The centers of the turns can be given in terms of a position, heading, and turn radius (or air/ground speed) as

$$a_i = x_i + R_i S_i \cos\psi_i \tag{8}$$

$$b_i = y_i - R_i S_i \sin\psi_i \tag{9}$$

where x_i, y_i, and ψ_i are arbitrary points on the turn circle. Two more parameters to be used are the distance between the centers of the circles, Q, and the length of the straight line path, D (see Fig. 4). The distance between the centers of the circles is given as

$$Q = \sqrt{(a_2 - a_1)^2 + (b_2 - b_1)^2} \tag{10}$$

McLean shows the tangent line between the circles to have a length of

$$D = \sqrt{Q^2 - (R_2 S_2 - R_1 S_1)^2} \tag{11}$$

For this particular case, where the first turn is counterclockwise and the second turn is clockwise, $S_1 = -1$, $S_2 = +1$, and $R_1 = R_2 = R$. Therefore, using Eqns. 8 and 9, the initial turn is centered at

$$a_1 = -R\cos\psi_0 \tag{12}$$

$$b_1 = R\sin\psi_0 \tag{13}$$

The center of the final turn circle is a function of the final point, x_f. The final heading is known (90 deg) as well as the final y coordinate, y_f. Applying Eqns. 8 and 9

$$a_2 = x_f \tag{14}$$

$$b_2 = y_f - R \tag{15}$$

The length of the straight line segment is found using Eqn. 11

$$D = \sqrt{Q^2 - 4R^2} \tag{16}$$

The tangent points on the initial turn circle are denoted by (x_2, y_2), and on the final turn circle by (x_3, y_3). The heading of the tangent path is ψ_t. The tangent points, (x_2, y_2) and (x_3, y_3), can be found from Eqns. 8 and 9 to be

$$x_2 = a_1 + R\cos\psi_t \tag{17}$$

$$y_2 = b_1 - R\sin\psi_t \tag{18}$$

$$x_3 = a_2 - R\cos\psi_t \tag{19}$$

$$y_3 = b_2 + R\sin\psi_t \tag{20}$$

The components of D are found by subtracting Eqn. 17 from Eqn. 19 and Eqn. 18 from Eqn. 20.

$$x_3 - x_2 = (x_f - a_1) - 2R\cos\psi_t \tag{21}$$

$$y_3 - y_2 = (b_2 - b_1) + 2R\sin\psi_t \tag{22}$$

From Fig. 4 it is seen that

$$x_3 - x_2 = D\sin\psi_t \tag{23}$$

$$y_3 - y_2 = D\cos\psi_t \tag{24}$$

Equating Eqn. 21 with Eqn. 23 and Eqn. 22 with Eqn. 24 gives

$$\begin{aligned} D\sin\psi_t &= x_f - a_1 - 2R\cos\psi_t && (25)\\ D\cos\psi_t &= \Delta b + 2Rsin\psi_t && (26) \end{aligned}$$

where $\Delta b = b_2 - b_1$.

These two equations in turn are solved for $\sin\psi_t$ and $\cos\psi_t$.

$$\begin{aligned} \sin\psi_t &= \frac{D(x_f - a_1) - 2R\Delta b}{D^2 + 4R^2} && (27)\\ \cos\psi_t &= \frac{2R(x_f - a_1) + D\Delta b}{D^2 + 4R^2} && (28) \end{aligned}$$

To solve for ψ_t, divide Eqn. 27 by Eqn. 28

$$\tan\psi_t = \frac{D(x_f - a_1) - 2R\Delta b}{2R(x_f - a_1) + D\Delta b} \quad (29)$$

Note the dependence of Eqn. 29 on x_f. It is this dependence that will make analytical solution of x_f impossible, and the solution must be done numerically. Also, a four quadrant inverse tangent should be used to find ψ_t. This is required to insure that the heading is in the correct quadrant.

Define the angles, $\Delta\psi_1$ and $\Delta\psi_2$ as the initial and final turn respectively:

$$\begin{aligned} \Delta\psi_1 &= \psi_t - \psi_0 + 2\pi S_1 k_1 && (30)\\ \Delta\psi_2 &= \psi_f - \psi_t + 2\pi S_2 k_2 && (31) \end{aligned}$$

where ψ_f is the heading of the route, and k_1 and k_2 are constants. The constants, k_i, are dependent upon the direction of the turn, and are evaluated as follows:[5]

$$k_1 = \begin{cases} 0 & S_1(\psi_t - \psi_0) \geq 0, \\ 1 & S_1(\psi_t - \psi_0) < 0 \end{cases} \quad (32)$$

and

$$k_2 = \begin{cases} 0 & S_2(\psi_f - \psi_t) \geq 0, \\ 1 & S_2(\psi_f - \psi_t) < 0 \end{cases} \quad (33)$$

Eqns. 32 and 33 are needed to obtain the correct "2π multiples" of the turn angles $\Delta\psi_1$ and $\Delta\psi_2$. By definition a counterclockwise turn will result in a final heading that is less than the initial heading. The turn angle in this case is negative. Likewise, a clockwise turn results in a positive turn angle.

The total angle through which the aircraft turns, $\Delta\psi$, is simply the sum of the absolute values of the two turn angles

$$\Delta\psi = |\Delta\psi_1| + |\Delta\psi_2| \quad (34)$$

The total distance flown by the aircraft is then $S_a = R\Delta\psi + D$. The distance that the slot has moved is $S_s = x_f - x_{t0}$. Equating S_a and S_s yields the following:

$$x_f - x_{t0} = R\Delta\psi + \sqrt{(x_f - a_1)^2 + (\Delta b)^2 - 4R^2} \quad (35)$$

Solving Eqn. 35 and Eqn. 29 simultaneously will yield a solution to the merge problem for the given initial conditions, given that the solution exists. Typically, a solution for a counterclockwise first turn will not exist for all initial headings. There may be instances where the aircraft will unable to "catch-up" with the slot, and an intercept will not be possible.

Clockwise-Straight-Clockwise

The second possible intercept path is when the initial turn is a clockwise turn. Again, our goal is to find solutions where the aircraft and the slot travel an equal distance for given initial conditions. The geometry is shown in Fig. 4 for this problem. However, the sense of the velocity vector will be opposite that in the figure, and the line segments Q and D will be parallel instead of crossing. Eqns. 8–11 will be used; however, since the initial turn is clockwise, $S_1 = 1$.

Using Eqns. 8 and 9, and using $S_1 = 1$, the center of the first turn is located at the point

$$\begin{aligned} a_1 &= R\cos\psi_0 && (36)\\ b_1 &= -R\sin\psi_0 && (37) \end{aligned}$$

The center of the final turn is the same as given in Eqns. 14 and 15. The distance between the centers of the turns is given by the expression for Q in Eqn. 10. The length of the straight line segment is found from Eqn. 11, with $S_1 = S_2 = 1$, which gives

$$D = \sqrt{(x_f - a_1)^2 + (\Delta b)^2} \quad (38)$$

Note that D has the same length as Q, which implies that the tangent path is parallel to the line segment connecting the centers of the turns.

The endpoints of the line segment that is tangent to the circles are:

$$\begin{aligned} x_2 &= a_1 - R\cos\psi_t && (39)\\ y_2 &= b_1 + R\sin\psi_t && (40)\\ x_3 &= a_2 - R\cos\psi_t && (41)\\ y_3 &= b_2 + R\sin\psi_t && (42) \end{aligned}$$

The components of D are

$$\begin{aligned} x_3 - x_2 &= (x_f - a_1) && (43)\\ y_3 - y_2 &= (b_2 - b_1) && (44) \end{aligned}$$

From Fig. 4, the components of D are

$$\begin{aligned} x_3 - x_2 &= D\sin\psi_t && (45)\\ y_3 - y_2 &= D\cos\psi_t && (46) \end{aligned}$$

So, equating Eqn. 45 with Eqn. 43 and Eqn. 46 with Eqn. 44 will give the sine and cosine of the tangent heading.

$$\sin\psi_t = \frac{x_f - a_1}{D} \tag{47}$$

$$\cos\psi_t = \frac{\Delta b}{D} \tag{48}$$

In turn, dividing these two expressions yields

$$\tan\psi_t = \frac{x_f - a_1}{\Delta b} \tag{49}$$

The heading, ψ_t is found by taking a four quadrant inverse tangent of Eqn. 49. The turn angle, $\Delta\psi$, is defined as before.

The total distance flown by the intercepting aircraft is

$$S_a = R\,\Delta\psi + D$$

$$S_a = R\,\Delta\psi + \sqrt{(x_f - a_1)^2 + (\Delta b)^2} \tag{50}$$

Equate Eqn. 50 to the distance traveled by the slot.

$$x_f - x_{t0} = R\,\Delta\psi + \sqrt{(x_f - a_1)^2 + (\Delta b)^2} \tag{51}$$

Collecting terms and solving for x_f yields the equation

$$x_f = \frac{a_1 + x_{t0} + R\,\Delta\psi}{2} - \frac{(\Delta b)^2}{2(x_{t0} + R\,\Delta\psi - a_1)} \tag{52}$$

Either Eqn. 51 or 52 will yield the intercept point for a clockwise initial turn. As with the counterclockwise initial turn solution, the solution for an initial clockwise turn typically will not exist for all values of ψ_0. Consider Fig. 5, which shows solutions for counterclockwise and clockwise initial turns. The aircraft and slot have a calibrated airspeed of 280 kts and the slot's initial position is -30 n.mi. Furthermore, the distance from the aircraft to the jetway, y_f, is 50 n.mi. From an initial heading of 0 to 27 degrees, a solution does not exist for a counterclockwise initial turn. On the other hand the clockwise initial turn solution exists across this range of initial headings. The heading at 27 degrees where the initial turn "switches" from a clockwise turn to a counterclockwise turn is significant. At this particular heading, the initial turn is degenerate (i.e. to complete the intercept no initial turn is required.) The aircraft in this case simply continues along its initial heading until it begins its turn to intercept the route. A discussion on the no turn heading follows in the next section.

The other boundaries that define whether the initial turn is clockwise or counterclockwise are not as easily determined. Fig. 5 defines one scenario. Note that solutions for each initial turn exist over a common range. The point where the initial turn switches direction in this case will be the heading where the distance traveled along both trajectories is identical. The other case is one where the range of solutions do not overlap. In each case, however, it is not possible to determine *a priori* these particular headings.

No First Turn Solution

The heading where the aircraft does not require an initial turn for the intercept is important because it defines a point on the switching curve boundary. For a given initial slot position, the heading for the no first turn solution is easily obtained. The geometry given in Fig. 6 is used in the following analysis.

Assume that the aircraft is initially at the origin with a heading ψ_{nt}. Define L to be the distance from the origin to the point where the aircraft begins its turn to intercept the route. The angle $\Delta\psi$ is simply $90° - \psi_{nt}$. The distance that the aircraft travels will be

$$S_a = L + R\,\Delta\psi \tag{53}$$

In order to find the length L, the point, (x_3, y_3), where L is tangent to the final turn circle is needed. Using Eqn. 9 with $S = 1$ and Eqn. 15 for the center of the turn circle gives

$$y_3 = y_f + R\,(\sin\psi_{nt} - 1) \tag{54}$$

Equating the expression for y_3 to the vertical distance $L\cos\psi_{nt}$ and solving for the distance L yields

$$L = \frac{y_f + R\,(\sin\psi_{nt} - 1)}{\cos\psi_{nt}} \tag{55}$$

The slot has to travel an identical distance as the aircraft for an intercept to occur (i.e. $S_s = x_f - x_{t0}$). Therefore, the intercept point needs to be determined. This is done solving Eqn. 8 for x_f where $a_i = x_f$, $S_i = 1$ and $x_i = x_3$

$$x_f = x_3 + R\sin\psi_{nt} \tag{56}$$

However, from Fig. 6, it is observed that $L\cos\psi_t = x_3$. Substituting this into Eqn. 56 gives us the intercept point as a function of the turn radius, the initial heading, and the length L

$$x_f = L\sin\psi_{nt} + R\sin\psi_{nt} \tag{57}$$

Substituting the expression for x_f into the distance equation

$$S_s = L\sin\psi_{nt} - x_{t0} + R\cos\psi_{nt} \tag{58}$$

where L is given by Eqn. 55.

Equate Eqns. 53 and 58 and solve for x_{t0}

$$x_{t0} = \frac{y_f + R(\sin\psi_{nt} - 1)}{\cos\psi_{nt}}(\sin\psi_{nt} - 1) + R(\cos\psi_{nt} - \Delta\psi) \quad (59)$$

Obviously, this equation is only valid for $-90° < \psi_{nt} < 90°$. Fig. 7 shows the solution of Eqn. 59. Note that this analysis is done for the aircraft and the slot traveling at the identical speeds. To find ψ_{nt} given x_{t0}, Eqn. 59 is solved numerically for the no turn heading given the initial position of the slot, and the speed of the aircraft.

Case 2: $V_T > V_0$

In this section, an iterative algorithm is described that generates horizontal trajectories for the case where the merging aircraft must increase its airspeed to that of the aircraft in the en-route stream. It is assumed that the merging aircraft makes its initial turn at its initial speed. The aircraft then accelerates only along the straight line segment of the trajectory. This assumption is only valid as long as the distance required to accelerate the aircraft is less than the length of D. The purpose of flying the trajectory this way is two-fold. First, it is computationally simpler. Second, keeping the aircraft's speed at the lower value during the initial turn means that the aircraft has a higher turn rate, in turn minimizing the time to turn to the heading of the tangent path. The optimality of this method has not been shown, but should be the subject of future research.

The trajectory is broken up into four parts: (1) the initial turn, which is made at the aircraft's initial speed; (2) an acceleration along the beginning of the straight line segment to the desired speed; (3) travel along the straight line segment at the new speed; and (4) a turn to the final heading. The algorithm uses the "newpsi" subroutine found in McLean[5] to get the total path length for a given intercept point, initial and final heading, and initial and final speed. The time to traverse the trajectory is then found by calculating the time to traverse each segment:

$$\begin{aligned} T_1 &= \frac{V_0 \Delta\psi_1}{R_1} \\ T_2 &= \frac{V_T - V_0}{a} \\ T_3 &= \frac{D}{V_T} - \frac{V_T^2 - V_0^2}{2aV_T} \\ T_4 &= \frac{V_T \Delta\psi_2}{R_2} \\ T_a &= T_1 + T_2 + T_3 + T_4 \end{aligned} \quad (60)$$

Where T_1 is the time to turn through the angle $\Delta\psi_1$, T_2 is the time to accelerate to the new airspeed, T_3 is the time to traverse the remainder of the straight line segment, and T_4 is the time to turn through an angle $\Delta\psi_2$. Note that if the aircraft cannot reach the new airspeed along the straight line segment, this algorithm is not valid.

Using the final position, x_f, the time for the slot to reach the intercept point is then

$$T_s = \frac{x_f - x_{t0}}{V_T} \quad (61)$$

Eqn. 60 and Eqn. 61 must be solved iteratively for x_f such that $T_a = T_s$. At the i_{th} iteration, let

$$T_{error} = T_a - T_s \quad (62)$$

If the absolute value of Eqn. 62 is greater than some tolerance, a new intercept point is calculated

$$x_f(i+1) = x_f(i) + V_T\, T_{error} \quad (63)$$

and the solutions repeated. Convergence is fairly rapid, when a solution exists. If the iteration reaches an upper limit, or the final point moves "too far" downrange, it is assumed that there is no solution for this problem.

The difficulty with this algorithm lies in the determination of the direction of the initial turn *a priori*. Depending upon the initial conditions, there are either two or three headings that determine the "switching curve." The most important heading, the no turn heading, is found using the method shown in the section above. The only difference will be that the aircraft and slot must fly paths that have equal times. The time is computed as described for the turn–straight-turn case, with $\Delta\psi_1$ set to zero. The initial heading is then found by searching over a range of slot initial positions. The heading for the "no turn" solution lies between -90 and 90 degrees.

The importance of the no turn heading is dependent upon the observation that any initial aircraft heading will turn towards the no turn heading, as long as the initial heading is between -90 and 90 degrees. However, the switching curve has either one or two additional boundaries, depending upon the initial conditions. For example, if at least one solution (either clockwise or counterclockwise initial turn) exists for all headings, for fixed initial conditions, there is an additional heading, which lies between 90 and 270 degrees, that defines the second half of the switching curve. Physically, this heading represents the point where the time to intercept for a clockwise initial turn and a counterclockwise initial turn are equal. Unfortunately, there is no way to determine this heading without examining all solutions for the given initial conditions.

The second situation is where there would be two headings in addition to the no turn heading that define the switching curve. The additional headings represent a range of headings for which no intercept is possible. Again, this is not something that can be predicted *a priori*, and the best practical solution is to compute both candidate trajectories to see if a solution exists.

Figure 8 shows representative trajectories for the case where there are two headings that define the switching curve. The initial conditions for this particular problem are $x_{t0} = -50$ n.mi., $y_f = 50$ n.mi., $V_0 = 280$ KCAS at 30,000 feet, and $V_T = 330$ KCAS at 30,000 feet. The aircraft is assumed to be able to accelerate at $0.1g$. The heading where there is a no first turn solution is -6 degrees. Likewise, the heading where the times are equal is 243 degrees. One can see that as the initial heading approaches the bifurcation heading, the trajectory length increases. As the initial heading passes through the bifurcation point and continues on towards 360 degrees, the trajectories become shorter.

Conclusions

The intercept problem and the departure trajectory problem have been addressed. Solutions have been found for the intercept problem for the case where the aircraft is traveling at the speed of the aircraft in the jetway, as well as the case where the aircraft has to accelerate to reach the velocity of the stream. An algorithm is given regarding the computation of a solution for the latter case. For the former, a set of equations is given that allows us to numerically solve for the coordinate where the intercept will occur. Currently, validation of the code that calculates the departure trajectory is in progress. Results are being compared for a data set taken from Denver International Airport on Jan. 8, 1997.

Acknowledgement

This research was supported by the Air Traffic Management Branch at NASA Ames Research Center under Grant NGT-2-52205 and Cooperative Agreement NCC2-950.

References

[1] Denery, D. and Erzberger, H., "The Center-TRACON Automation System: Simulation and Field Testing," TM-110366, NASA Ames Research Center, August 1995.

[2] Slattery, R., "Terminal Area Trajectory Synthesis for Air Traffic Control Automation," *Proceedings of the 1995 ACC Control Conference*, 1995, pp. 1206–1210.

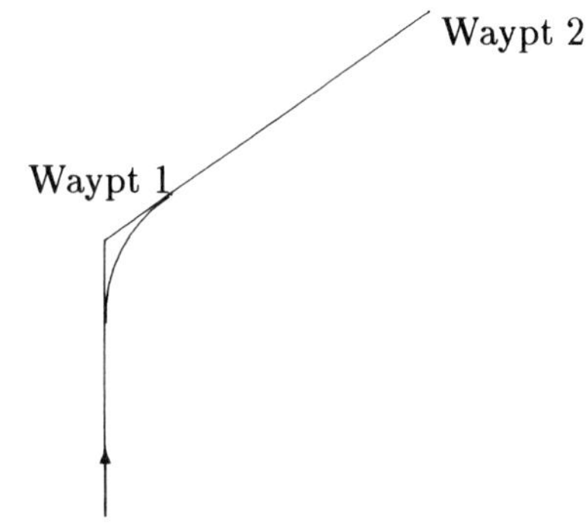

Figure 1: Route Intercept Geometry

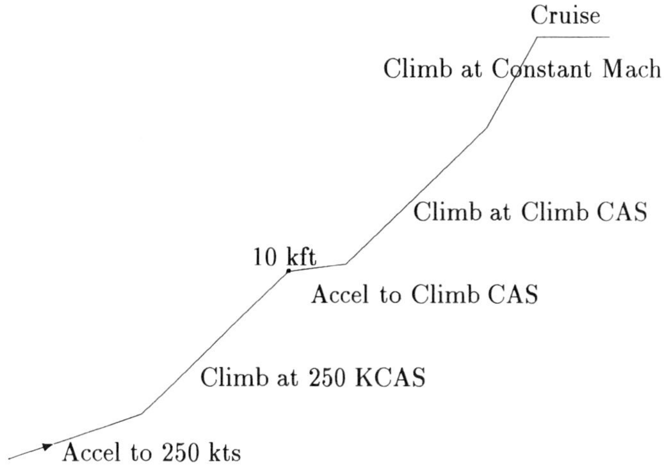

Figure 2: Climb Trajectory

[3] Slattery, R. and Zhao, Y., "Trajectory Synthesis for Air Traffic Automation," *Journal of Guidance, Control, and Dynamics*, Vol. 20, Mar./Apr. 1997, pp. 232–238.

[4] Erzberger, H. and Lee, H., "Optimal Horizontal Guidance Techniques for Aircraft," *Journal of Aircraft*, Vol. 8, July 1972, pp. 95–101.

[5] McLean, J., "A New Algorithm for Horizontal Capture Trajectories," TM–81186, NASA Ames Research Center, March 1980.

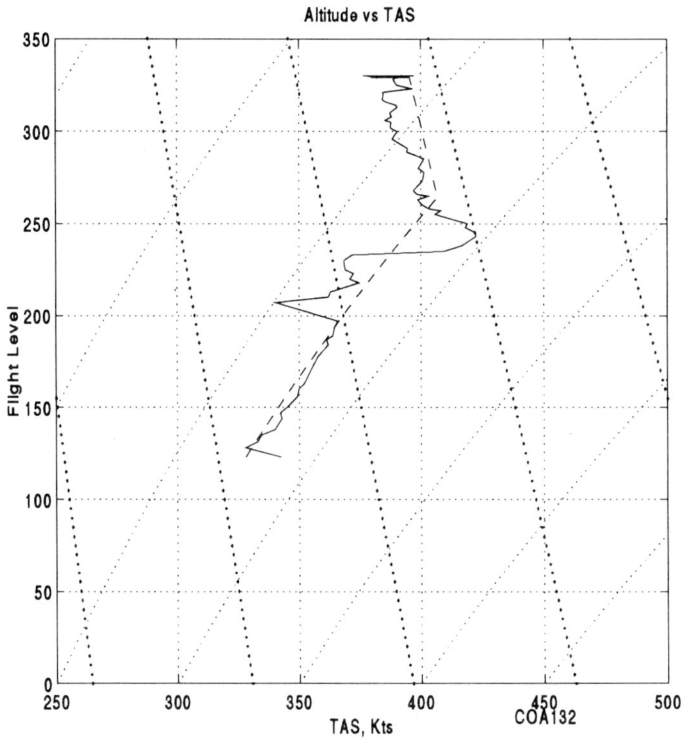

Figure 3: Example Climb-Out Trajectory

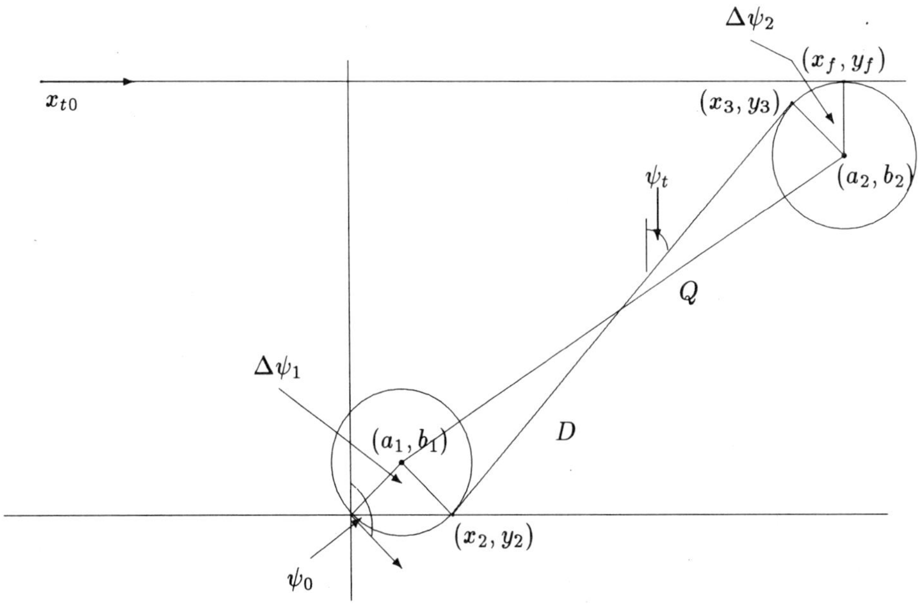

Figure 4: Geometry of a Horizontal Intercept Trajectory

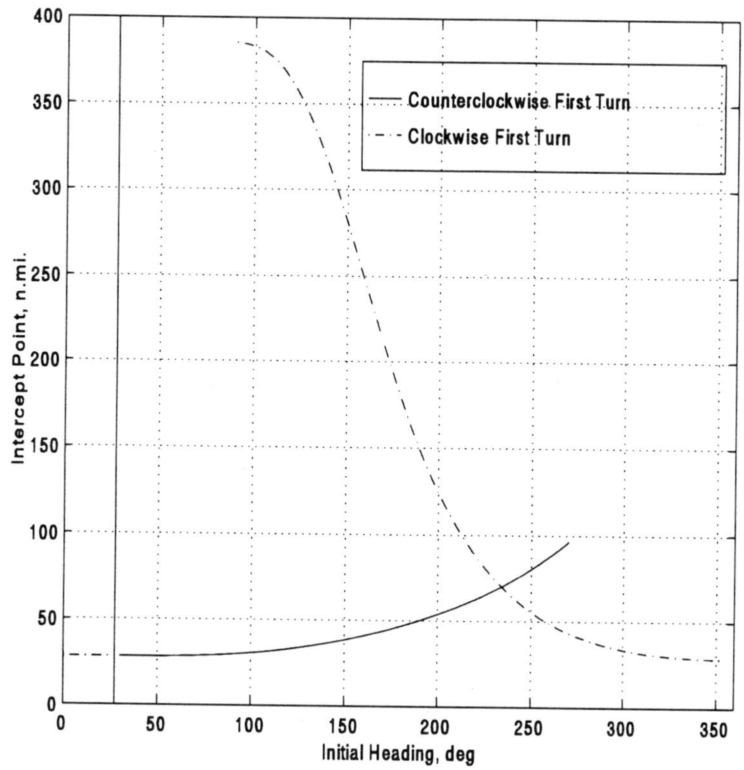

Figure 5: Intercept Points vs Initial Heading

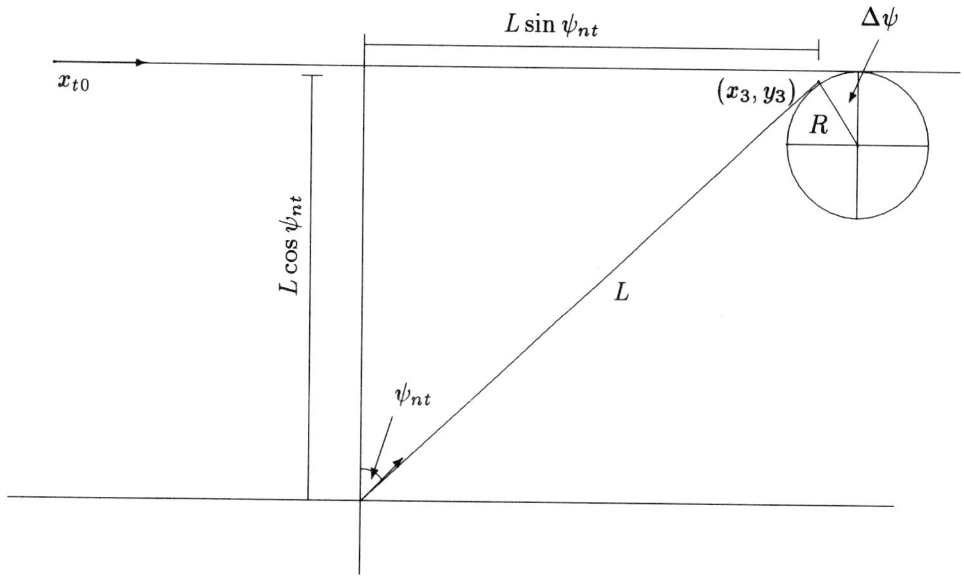

Figure 6: No First Turn Horizontal Trajectory

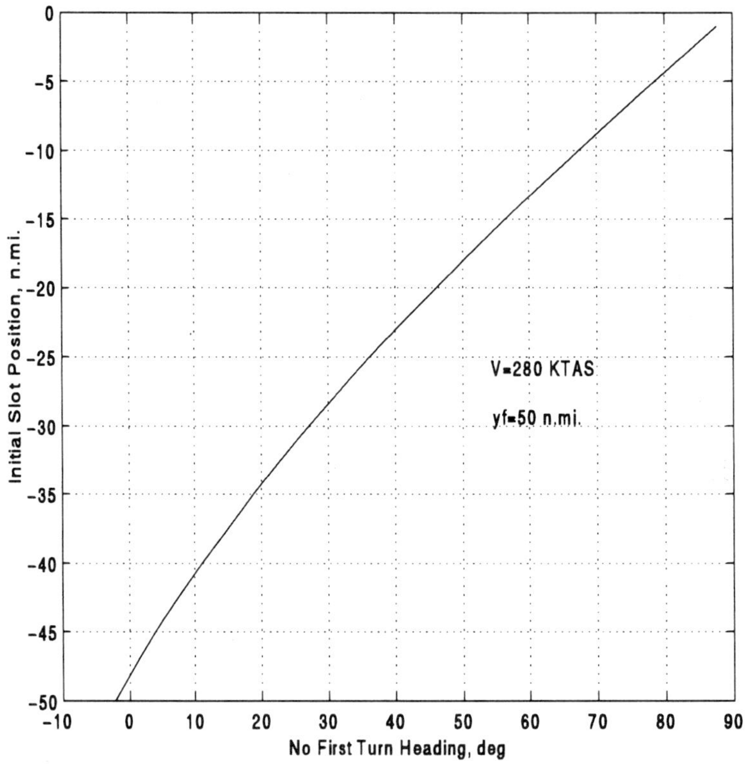

Figure 7: Initial Slot Position vs No First Turn Heading for Same Speed Case

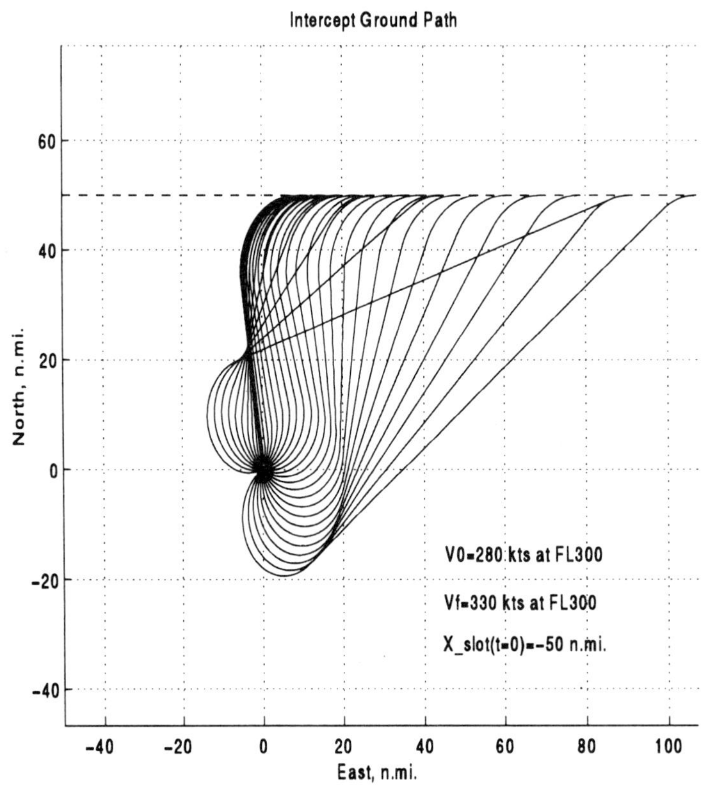

Figure 8: Variable Speed Horizontal Intercept Trajectories

AIAA-97-3546

Optimal Strategies for Free Flight Air Traffic Conflict Resolution*

By

P. K. Menon[1] and G. D. Sweriduk[2]
Optimal Synthesis
450 San Antonio Road, Suite 46
Palo Alto, CA 94306-4638

B. Sridhar[3]
NASA Ames Research Center
Moffett Field, CA 94035-1000

Abstract

Recent advances in navigation and data communication technologies make it feasible for individual aircraft to plan and fly their trajectories in the presence of other aircraft in the airspace. This way, individual aircraft can take advantage of the atmospheric and traffic conditions to optimally plan their paths. This capability is termed as the *free flight* concept. While the free flight concept provides new degrees of freedom to the aircraft operators, it also brings-in complexities not present in the current air traffic control system. In the free flight concept, each aircraft has the responsibility for navigating around other aircraft in the airspace. While this is not a difficult task under low speed, low traffic density conditions, the complexities of dealing with potential conflict with multiple aircraft can significantly increase the pilot's work load.

This paper presents the development of a conflict resolution algorithm based on the quasi-linearization method to enable the practical implementation of the free flight concept. The algorithm development uses nonlinear point-mass aircraft models, and incorporates realistic operational constraints on individual aircraft. The analytical framework can also incorporate information about ambient atmospheric conditions. Realistic conflict resolution scenarios are illustrated. Due to their speed of execution, these conflict resolution algorithms are suitable for implementation on-board aircraft.

1. Introduction

Recent advances in Global Positioning System (GPS) - based navigation techniques and satellite data communications technology have motivated the FAA and commercial air traffic carriers to consider the "free flight" concept as a major component of the future air traffic control system [1, 2]. Free flight is defined as safe and efficient flight operations under instrument flight rules in which the operators have the freedom to select their path and speed in real time. Thus, individual aircraft will be able to plan and execute their trajectories without direction from any external agents during most of their flight duration. The ability of individual aircraft to plan their trajectories can result in substantial improvements in operational efficiency, since the aircraft can take advantage of atmospheric and traffic conditions to minimize the delays and fuel consumption. This approach can also lead to optimal coordination between aircraft involved in hub/spoke modes of operation currently employed by most major air carriers. However, since different aircraft in a given airspace may have conflicting objectives, the desired trajectories synthesized by individual aircraft can generate conflicts with other aircraft trajectories. Successful implementation of the free flight concept will require the development of systematic methods for multi-aircraft conflict resolution. While conflict resolution strategies are obvious under low speed, low traffic density conditions, the problem can become complex as the traffic density and the aircraft speeds increase. This fact has motivated several research studies on various aspects of the free-flight conflict resolution problem at NASA and other research centers [3 - 10].

The focus of the research effort described in this paper is on the development of systematic methods for multiple aircraft conflict resolution. Conflict resolution algorithms employed in the present air traffic management automation tools are largely rule-based, and are designed to operate under the existing air traffic control procedures [11 - 15]. While these methods are adequate for resolving simultaneous conflicts between aircraft following standard jet routes, it is not clear how these methods can be generalized to handle multiple aircraft conflicts that can arise in the free flight environment. The emphasis of the research reported in this paper is in formulating and solving the conflict resolution problem as a trajectory optimization problem. As with other aerospace guidance problems, application of optimal control theory [16] and differential game theory [17] will result in more systematic methodologies for the solution of the conflict resolution problem. This fact has been demonstrated in several recent aerospace trajectory synthesis problems [18 - 22]. Unlike rule-based conflict resolution approaches, one of the advantages of the formalism presented here is that the same algorithms are independent of the conflict resolution geometry. Present research builds upon a previous research effort [5, 23] on the conflict resolution problem.

Conflict resolution methodologies developed in the present research assume that each aircraft trajectory involved in a potential conflict is given by a sequence of 4 dimensional (three position coordinates and time) waypoints. Nominal waypoint sequences are adjusted by the conflict resolution algorithm in order to synthesize conflict-free trajectories that optimize desired performance indices. The conflict resolution problem is

[1]Research Scientist, Associate Fellow AIAA
[2]Research Scientist, Member AIAA
[3]Research Scientist, Associate Fellow AIAA.

formulated both as a single objective optimization problem and as a multiple objective optimization problem [24, 25]. Various types of curves can be used to define the actual aircraft trajectory between the waypoints. The use of piecewise linear and cubic spline trajectories are both considered in the present research. Inter-aircraft conflict is defined through a conflict envelope in the form of an oblate spheroid.

The trajectory optimization formulation uses nonlinear point-mass models of aircraft with realistic aerodynamic and engine models [26, 27]. Two different cost functions are used in this study: a linear combination of total flight time and fuel consumption, and integral square of the perturbation from nominal trajectories. The latter function assumes that the nominal trajectories have already been optimized by each individual aircraft, and that the conflict resolution methodology should only perturb these trajectories by the smallest amount required to accomplish conflict resolution. The advantage of the latter formulation is that individual aircraft need not reveal their core trajectory planning strategies to other aircraft involved in the conflict resolution process. The constraints employed in the present formulation are: inter-aircraft distance, maximum and minimum airspeeds, maximum climb and descent rates, maximum and minimum altitude limits, maximum bank angle, maximum and minimum load factors, and maximum and minimum thrusts limits. Vehicle models, constraints, and trajectory parameterization schemes are discussed in Section 2.

Iterative numerical methods and semi-analytical guidance laws are developed based on trajectory optimization theory [16]. Iterative methods that formulate the conflict resolution problem as a single objective optimization problem, and as multiple objective optimization problem are given in Section 3. Semi-analytical guidance law development using quasi-linearization is presented in Section 4. Several conflict resolution scenarios are given in Sections 3 and 4 to illustrate the performance of these algorithms. In every case, it is shown that the methods produce useful conflict resolution strategies. Further, the results show that the conflict resolution algorithms developed in the present research are general enough to handle every type of conflict that can occur in the free flight air transportation environment. Conclusions from the present research effort and future research issues are given in Section 5.

2. Aircraft, Trajectory, and Conflict Envelope Models

Three major components of the conflict resolution problems analyzed in the present research are the aircraft models, trajectory parameterization schemes, and conflict envelope definitions. Efficiency of the conflict resolution algorithms can be greatly enhanced by using appropriate aircraft models and trajectory parameterization schemes. The computational efficiency of implementing aircraft operational constraints can be improved by employing appropriate transformations on the aircraft models. Careful definition of the inter-aircraft conflict envelopes can further improve the conditioning of the conflict resolution problem. Each of these issues are discussed in the following subsections.

2.1. Aircraft Models

Point-mass aircraft models are used in the present research. The point-mass aircraft model captures most of the dynamical effects encountered in civil aviation aircraft. These models are popular in aircraft performance work, and are used in practically every aircraft trajectory optimization problem discussed in the literature. The point-mass equations of motion are formulated with respect to a coordinate system shown in Figure 1.

The point-mass model assumes that the aircraft thrust is directed along the velocity vector, and that the aircraft always performs coordinated maneuvers. It further assumes a flat, non-rotating earth. These assumptions are reasonable for civil aviation aircraft operating within a range of 200 nautical miles. Since the conflict resolution generally occurs within this range, the fidelity provided by the point-mass model is adequate for formulating these problems. Note that it is feasible to develop point mass models valid for spherical earth approximations also. The point-mass aircraft equations describing aircraft flight are:

$$\dot{V}_i = \frac{(T_i - D_i)}{m_i} - g \sin \gamma_i, \quad \dot{m}_i = -Q_i \tag{1}$$

$$\dot{\gamma}_i = \frac{g}{V_i}\left(\frac{L_i \cos \phi_i}{g\, m_i} - \cos \gamma_i\right), \quad \dot{\chi}_i = \frac{L_i \sin \phi_i}{m_i V_i \cos \gamma_i} \tag{2}$$

$$\dot{x}_i = V_i \cos \gamma_i \cos \chi_i$$
$$\dot{y}_i = V_i \cos \gamma_i \sin \chi_i, \quad \dot{h}_i = V_i \sin \gamma_i \tag{3}$$

with i = 1, 2,...n being the aircraft under consideration. In these equations, and V_i is the ground speed, assumed to be equal to airspeed during the present research. T_i is the aircraft engine thrust, D_i is the drag, m_i is the aircraft mass, g is the acceleration due to gravity, γ_i is the flight path angle, L_i is the vehicle lift, ϕ_i is the bank angle, χ_i is the aircraft heading angle, x_i is the down range, y_i is the cross range, h_i is the altitude, and Q_i is the fuel flow rate. The fuel flow rate depends on the altitude, Mach number and the engine thrust. The aircraft drag is given in terms of the zero-lift drag coefficient, and induced drag coefficient. Both of these are specified as functions of Mach number and the aircraft configuration setting such as the deployment of flaps, speed brakes, and the landing gear. For the present research, it is assumed that the aircraft is in the cruise flight configuration. The control variables in the aircraft model are the load factor n_i controlled using the elevator, the bank angle ϕ_i controlled using a combination of rudder and ailerons, and the engine thrust T_i controlled using the throttle. Throughout the conflict resolution process, the control variables will be constrained to remain within specified limits.

For analytical convenience, the kinematic equations describing the aircraft position can be differentiated once with respect to time, and the remaining dynamic equations can be used to cast the point-mass aircraft model in an alternate form as:

$$\ddot{x} = U_1 \quad \ddot{y} = U_2, \quad \ddot{h} = U_3 \qquad (4)$$

U_1, U_2, U_3 are the three new control variables in the point mass model. The relationship between these control variables and the actual control variables is given by the expressions [28]:

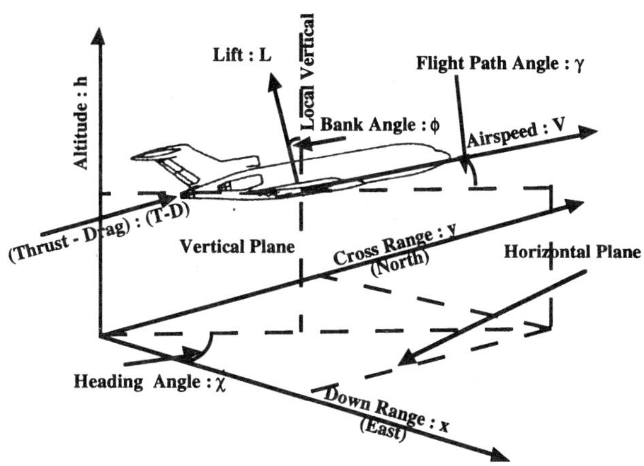

Fig. 1 . The Aircraft Coordinate System

$$\phi = \tan^{-1}\left[\frac{U_2 \cos\chi - U_1 \sin\chi}{\cos\gamma(U_3 + g) - \sin\gamma(U_1 \cos\chi + U_2 \sin\chi)}\right] \qquad (5)$$

$$n = \frac{\cos\gamma(U_3 + g) - \sin\gamma(U_1 \cos\chi + U_2 \sin\chi)}{g \cos\phi} \qquad (6)$$

$$T = \left[\sin\gamma(U_3 + g) + \cos\gamma(U_1 \cos\chi + U_2 \sin\chi)\right]m + D \qquad (7)$$

the heading angle χ and the flight path angle γ are computed as:

$$\tan\chi = \frac{\dot{y}}{\dot{x}} \qquad \sin\gamma = \frac{\dot{h}}{V} \qquad (8)$$

This latter form of aircraft point-mass model is useful for mapping geometric trajectory parameters in terms of the aircraft control variables, and vice-versa. Thus, constraints on the aircraft control variables can be readily transformed into constraints on the trajectory curvature. Such mappings are useful for efficient implementation of control constraints, as will be demonstrated in subsequent sections.

2.2. Trajectory Parameterization

Trajectory parameterization methods allow the description of individual aircraft trajectories using a small number of parameters. In addition to providing a compact description of the trajectories, trajectory parameterization methods permit the solution of the trajectory optimization problems as parameter optimization problems. Parameterized aircraft trajectories are described using a set of four-dimensional waypoints consisting of three position components and time, together with the specification of the type of curves joining them. Two types of trajectory parameterization schemes are used in the present research. These are piecewise linear trajectories and cubic spline trajectories. Piecewise linear trajectory parameterization is highly efficient from a computational standpoint. However, the piecewise linear trajectory parameterization implies abrupt changes in aircraft control variables at the waypoints. Since such abrupt control variations are not desirable, smooth trajectory segments will need to be introduced in the vicinity of the waypoints to ensure practical implementability. Note that these trajectory segments can introduce uncertainties in the aircraft position near the way points. It is assumed in the present research that these smooth trajectory segments can be introduced without adversely affecting the conflict resolution solutions.

On the other hand, the cubic spline parameterization produces smooth trajectories as well as smooth control settings. The disadvantage of this parameterization scheme is that it requires more computation time, and that it tends to introduce anomalies in the path by virtue of its fitting a fixed-order smooth curve through a given set of points. Due to these factors, although a few cubic spline parameterization cases were studied during the present research, the emphasis is on the use of piecewise linear trajectory parameterization.

2.3. Conflict Envelope Definition

According to the currently accepted definition [12], a conflict is said to occur if the altitude difference between an aircraft pair occupying the same down-range, cross-range coordinates is less than 2000 feet; or if the two aircraft are closer than 5 nautical miles while flying at the same altitude. It is convenient to conceptualize these requirements by imagining that each aircraft is centered inside a hypothetical conflict box with dimensions of 10 nautical miles length and breadth with 4000 feet thickness. Any other aircraft entering this hypothetical box can be considered as causing a conflict.

Although this definition of a conflict is simple to conceptualize, it can cause severe numerical difficulties due the corners present in such a conflict envelope. For instance, in certain kinematic conflict configurations, small motion of the aircraft can cause a large change in the conflict status. Such large changes can cause numerical instabilities in the conflict resolution algorithm. Alternate definitions can be developed to enforce the aircraft separation constraint without introducing such numerical discontinuities. Possibilities include quadratic surfaces [30] and super quadric surfaces [31, 32]. An oblate spheroidal conflict envelope is an example of a quadratic surface. The oblate spheroidal conflict envelope has an elliptical cross section in the vertical plane, and a circular cross section in the horizontal plane. In order to approximate the FAA definition of the conflict envelope, the semi-major axis of the elliptical cross section can be chosen as 5 nautical miles, and the semi-minor axis can be set at 2000 feet. A schematic view of the oblate spheroidal conflict envelope in the vertical plane is shown in Figure 2 for a pair of aircraft.

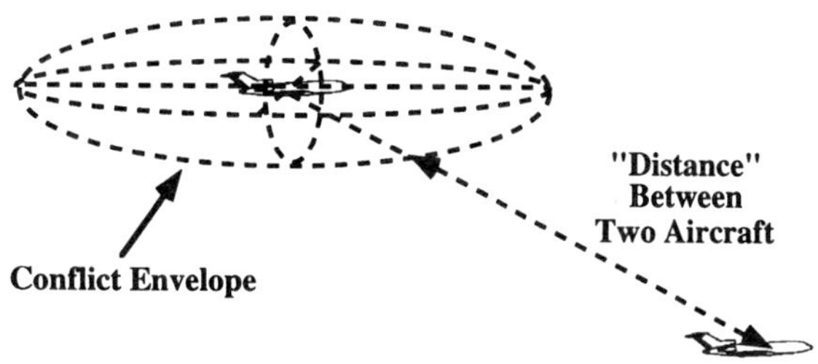

Fig. 2 . Oblate Spheroidal Conflict Envelope and the Inter-Aircraft Distance in the Vertical Plane

Figure 3 shows the definition of the oblate spheroid conflict envelopes for three aircraft in the horizontal plane. Note that the "inter-aircraft distance" is the distance from an aircraft's CG to another aircraft's conflict envelope.

Using this geometry, the distance $r_{i,j}$ between any two aircraft i, j can be defined using the expression:

$$r_{i,j} = \sqrt{\Delta x_{i,j}^2 + \Delta y_{i,j}^2 + \Delta h_{i,j}^2} - \sqrt{\frac{a^2 b^2 \left(\Delta x_{i,j}^2 + \Delta y_{i,j}^2 + \Delta h_{i,j}^2\right)}{a^2 \Delta h_{i,j}^2 + b^2 \left(\Delta x_{i,j}^2 + \Delta y_{i,j}^2\right)}}, \quad i \neq j \tag{9}$$

In this expression, a is the conflict envelope semi-major axis, b is the semi-minor axis, and $\Delta x_{i,j}$, $\Delta y_{i,j}$, $\Delta h_{i,j}$ are the components of the relative position vector between any two aircraft i and j. It can be verified that if a = b, the conflict envelope degenerates into a sphere. Further, it can be shown that $r_{i,j} = r_{j,i}$. A conflict can be defined as a situation in which any of the inter-aircraft distances $r_{i,j}$ fall below zero. Note that if n aircraft are involved in an air traffic control situation, a maximum of n(n-1)/2 conflicts can simultaneously arise. The resulting inter-aircraft distances can be represented as a symmetric matrix with zeros along the diagonal. Any conflict resolution methodology must ensure that all the off-diagonal elements of the inter-aircraft distance matrix remain above zero at all times.

A further refinement of the inter-aircraft conflict definition is to introduce a matrix norm in the formulation. The Frobinius norm or the Matrix 2 norm [33] is a candidate measure that can be used to express the inter-aircraft conflict in a compact form.

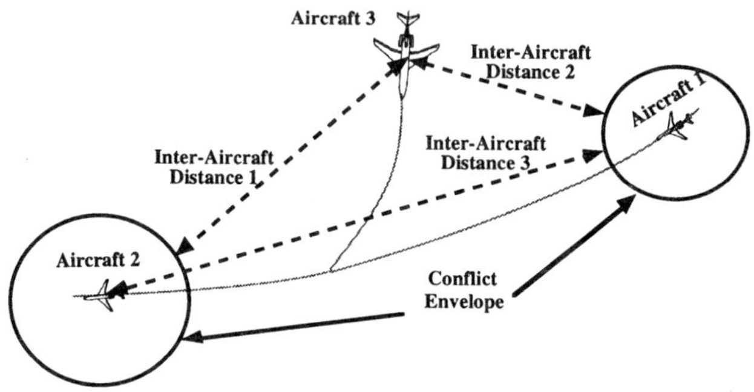

Fig. 3 . Definition of the "Inter-Aircraft Distance" in the Horizontal Plane

3. Direct Iterative Methods for Conflict Resolution

The conflict resolution process consists of modifying the aircraft nominal airspeed, altitude, down-range and cross range trajectory components to ensure that every aircraft stays out of the every other aircraft's conflict envelope while executing their flight plans. These trajectory perturbations must not significantly change the arrival times at the specified terminal point, and should not result in excessive maneuvering. Moreover, conflict resolution trajectory changes should be such that they do not lead to future conflicts. In some cases, it may be desirable to include fuel conservative measures in the conflict resolution performance objectives. Since the atmospheric perturbations and piloting techniques introduce uncertainties in the aircraft trajectories, the trajectory synthesis should be such that the worst-case perturbations in each aircraft trajectory still preserve the integrity of the conflict resolution process.

From the foregoing requirements, it can be observed that the conflict resolution problem can be formulated as a multi-participant trajectory optimization problem. This problem can be cast in several different forms by assigning roles to individual aircraft. For instance, it can be treated as a single objective trajectory optimization problem in which every aircraft performs conflict resolution maneuvers that optimize a defined performance index. Alternatively, the conflict resolution problem can be cast as a series of one-sided optimal control problems by considering a pair of aircraft at a time, with one aircraft operating along the nominal trajectory, and the other aircraft making the necessary trajectory adjustments required for conflict resolution. Finally, it can be treated as a multi-objective optimization problem in which each participant makes the least amount of trajectory deviations necessary for maintaining the inter-aircraft distance above a minimum specified value while simultaneously optimizing its performance criterion. Such approaches to the conflict resolution problem can be considered as a "cooperative conflict resolution approach". Other variations of the conflict resolution problem formulation are presented in the Section 4.

All these approaches require the definition of performance indices that reflect the operational requirements of the conflict resolution problem. Performance indices considered in the present research are linear combinations of flight time and fuel, and integral square of the deviations from the nominal flight paths. The constraints in the problem include the minimum permissible inter-aircraft separation and aircraft performance limits. Additionally, aircraft trajectories may be required to pass through a series of metering waypoints, as is required in the current air traffic control environment. The resulting problem is a state constrained optimal control problem [16, 17] that can be approached using well developed mathematical techniques.

The necessary conditions for the optimality of state constrained optimal control problems result in multi-point boundary-value problems. There are two major families of techniques useful for solving these problems. Firstly, the conflict resolution problem can be converted into a parameter optimization problem by parameterizing the trajectories. Trajectory parameterization methods described in the Section 2 are useful for this purpose. The resulting optimization methods are commonly known as *direct* methods. The second family of solution approaches consists of satisfying the necessary conditions for optimality by iteratively solving simplified versions of the multi-point boundary value problem. This latter family of approaches is termed as *indirect* methods in the literature. Indirect methods are especially suitable for the derivation of explicit guidance laws. Solutions to the conflict resolution problem via trajectory parameterization will be discussed in this section. Conflict resolution guidance law derivation using indirect methods will be considered in Section 4.

Two versions of the direct conflict resolution technique are discussed in the following subsections. In Section 3.1, the conflict resolution problem is formulated as a single objective optimization problem. The Sequential Quadratic Programming (SQP) method [34] is used to solve the resulting optimization problem. The conflict resolution problem is formulated and solved as a multi-objective optimization problem in Section 3.2. The goal attainment method [24, 25] is used to obtain the solutions to the multi-objective optimization problem.

The computational techniques are discussed in Sections 3.1 and 3.2, followed by a discussion of the numerical results in Section 3.3. Solutions are obtained for conflict situations involving two, three, four and six aircraft. All the conflict resolution solutions are obtained using piecewise linear trajectory parameterization. Comparisons between the single objective conflict resolution solutions, and multi-objective solutions are also given in Section 3.3.

3.1. Conflict Resolution as A Single Objective Optimization Problem

Single objective formulation of the conflict resolution problem considers the sum of individual aircraft flight times and fuel consumption as the performance index. Starting from a set of nominal waypoint sequences, the objective is to synthesize trajectories that resolve conflicts while optimizing the performance index. As a second part of the single objective optimization study, the performance index is re-cast as the sum of the integral deviations of each aircraft from their nominal trajectories. Inequality constraints are included for inter-aircraft distance, maximum and minimum velocities, maximum climb and descent rates, maximum and minimum altitudes, maximum roll angle, maximum and minimum load factors, and maximum and minimum thrust magnitudes. The nominal trajectories of each aircraft are specified in terms of a set of waypoints (x,y,h,t). The present formulation of the trajectory optimization problem allows the waypoints to be constrained in a any desired manner. Physical characteristics of each aircraft such as mass, drag coefficient, and fuel consumption rate also included in the study.

The algebraic transformations that relate thrust, load factor, and roll angle to the acceleration components discussed in Section 2 are included in the formulation so as to permit the imposition of aircraft performance constraints without including the aircraft equations of motion. Trajectories are parameterized using the piecewise linear parameterization scheme discussed in Section 2. The optimization algorithm adjusts the trajectory parameters to resolve any existing conflicts while minimizing cost and meeting the aircraft performance constraints. The optimization problem is currently set up to accommodate a maximum of ten aircraft, although this can easily be increased if necessary.

Optimal trajectories are computed using the SQP method. The sequential quadratic programming method has been found highly effective in solving a wide variety of constrained nonlinear programming problems [34]. The SQP method solves optimization problems of the form:

$$\begin{gathered}\text{minimize } f(x)\\ \text{subject to } g(x) \le 0\end{gathered} \tag{10}$$

by approximating the performance index by a quadratic function, and the constraints by linear functions. The SQP method attempts to imitate the Newton's method, which is known to converge to the optimal solution in the least number of steps for quadratic problems. In expressions (10), f(x) is a scalar objective function, g(x) is a vector of constraint functions, and x is an n-vector of optimization parameters. The scalar Lagrangian function [34] is formed by adjoining the constraints to the objective function with a vector λ of Lagrange multipliers:

$$L(x,\lambda) = f(x) + \lambda^T g(x) \tag{11}$$

The next step is to form a quadratic approximation of this function and to linearize the constraints about the current solution vector. The approximate optimization problem then becomes:

$$\text{minimize} \quad \tfrac{1}{2} p^T H p + \nabla f^T p \tag{12}$$

$$\text{subject to} \quad \nabla g^T p + g \le 0 \tag{13}$$

The vector p is the local approximation of the optimization parameters, H is the Hessian matrix and ∇ denotes the gradient operation with respect to x, evaluated at the current solution vector. This approximate problem (12), (13) is next solved using the Quadratic Programming (QP) algorithm [34]. The active set strategy is used for constraint satisfaction. The first step in solving the minimization problem is to find a feasible starting point. Such a point can often be found using a linear programming algorithm with the linearized constraints. Failing that, a search can be performed in the direction of decreasing constraint violations. The feasible point determines the set of active constraints. A search direction d is then calculated using the active constraints, and a one-dimensional line search is performed to minimize (12). A step is taken along the search direction d, with a step length α.

$$x_{i+1} = x_i + \alpha d \tag{14}$$

For quadratic problems it can be shown that the minimum occurs when $\alpha = 1$ as long as the constraints are not violated. Otherwise, smaller steps have to be taken and a new active set should be defined. However, due to the approximate nature of the QP problem, a more effective method is to use the solution of the QP problem as a direction and minimize a merit function which takes the inactive constraints also into consideration. The function employed in the present research is from Reference 35:

$$\psi(x) = f(x) + \sum_i r_i \max\{0, g_i(x)\} \tag{15}$$

where r_i are weighting parameters. Having found the appropriate step size, the new parameter vector x is found and a new approximate quadratic problem is formed and solved. The optimization toolbox of the MATLAB® software [36] is used in the present investigation for implementing the sequential quadratic programming algorithm.

An interesting aspect of the single objective conflict resolution method using the SQP algorithm is that the sensitivity of the cost function and constraints with respect to individual trajectory parameters can be readily computed using the gradient of the cost ∇f, Hessian matrix H, and the Jacobian matrix of the constraints ∇g. These sensitivities can be then used by individual aircraft to further negotiate changes to their path. For instance, large trajectory changes can be made in regions of low cost/constraint sensitivity, while trajectory perturbations must be limited in the high sensitivity regions. Such information can be valuable for the practical implementation of the conflict resolution method.

Numerical results using the single objective formulation of the conflict resolution problem will be given in Section 3.3. An alternate formulation of the conflict resolution problem is discussed in the following section.

3.2. Conflict Resolution As A Multiple Objective Optimization Problem

Single objective formulation of the conflict resolution problem is difficult to justify in the free-flight environment, because individual aircraft involved in the maneuvers may not be interested in optimizing a single performance index. For instance, some of the aircraft may place larger emphasis on flight time than fuel, while other aircraft may be interested in minimizing fuel consumption. Multiple objective optimization methods allow the use of more than one cost function, and are perhaps more realistic in representing the true nature of the multi-aircraft conflict resolution problem. Note that the multiple objective formulation automatically allows the simultaneous inclusion of cooperating and competing objectives, if any are present in the problem.

The cost function in the multi-objective conflict resolution problem is a vector. Several algorithms have been reported in the literature for the solution of multiple objective optimization problems [24]. Multiple objective optimization problems seek a set of "non-inferior" points as their solution. Non-inferiority implies that any further improvement in one objective would result in a degradation in another objective. Thus, all points in the solution set minimize the performance objectives to the maximum possible extent. The technique selected for use in the present research is the Goal Attainment method discussed in Reference 24. Here, the objectives and the constraints are all represented as goals to be satisfied, and the degree to which these goals are met are adjusted using weighting functions. The goal attainment formulation allows the imposition of inequality and equality constraints.

Mathematically, the goal attainment method can be stated as:

$$\text{minimize } \gamma, \text{ where } \quad f(x) - w\gamma \le f^* \tag{16}$$

Note that f(x) is a vector which includes the cost functions and the constraints. The goals are contained in the vector f, and the weights are in the vector w. The scalar γ is a measure of how far away a solution point is from the goals. Thus, the term $w\gamma$ can be considered as the slackness in the problem. Individual weights capture the fact that in many design problems, a goal need not be met exactly and one is more interested in arriving at an optimal trade-off between conflicting objectives. The weighting vector defines the direction from the goal point to the feasible region. As γ is varied, the feasible region changes size, until the constraint boundaries converge to a solution point. One of the advantages of the goal attainment formulation is that it can be solved as a sequential quadratic programming problem. A difference arises, though, in the merit function used for the line search. There are again several possibilities, but the one employed in the present research is from Reference 25:

$$\psi(x) = \max_{i} \left\{ \frac{f_i(x) - f_i^*}{w_i} \right\} \qquad (17)$$

Conflict resolution trajectories involving two, three, four and six aircraft have been generated using the goal attainment method. Some of these will be illustrated in Section 3.3.

3.3. Conflict Resolution Results

Conflict resolution methods developed in Sections 3.1 and 3.2 are evaluated in a sample conflict scenario in this section. The results for single objective and multiple objective formulations corresponding a six aircraft conflict will be given here.

This conflict scenario is generated by including six merging aircraft, five of them flying at 30,000 feet , and the sixth aircraft merging from 32,000 feet altitude. Such a conflict can arise during merging operations near an air traffic control center.

The initial aircraft speeds are:

V_1 = 780 ft/s, V_2 = 733 ft/s, V_3 = 548 ft/s, V_4 = 708 ft/s, V_5 = 586 ft/s, V_6 = 490 ft/s.

The aircraft parameters used in this conflict scenario are:

Mass = 185,000 lb_m, Drag coefficient (C_D) = 0.05, Wing area (S) = 1700 ft^2

Fuel consumption rate is assumed be a quadratic function of airspeed, with the constant of proportionality being 0.3×10^{-6} lb_m- sec /ft^2. The performance constraints imposed on the conflict resolution problem are:

340 ft/sec < Airspeed < 900 ft/sec
0.9 g's < Load Factor < 1.1 g's
1,000 lb_f < Thrust < 48,000 lb_f
20,000 ft < Altitude < 40,000 ft
-3,000 ft/min < Rate of Climb < 1,000 ft/min

The arrival order at the end of the maneuvers are assumed to be specified. Nominal trajectories for the aircraft are given Figure 4. It may be observed from the inter-aircraft distance plot that the conflicts between aircraft occur right from the beginning, and by 300 seconds, every aircraft is in conflict with the rest. Note that the conflict resolution strategies are far from obvious in this situation.

Conflict resolution using single objective formulation is given in Figure 5. Square of the deviations from nominal trajectories is used as the performance index in these optimization runs. Note that the methodology has modified each aircraft trajectory sufficiently to completely resolve the conflicts. Trajectory corrections have been introduced in both vertical plane and horizontal plane. Conflict resolution using multiple objective formulation is shown in Figure 6. Although it is rather difficult to gauge the differences between the two approaches, an analysis of the sensitivities of the trajectories with respect to the constraints show that the multiobjective formulation results in a more robust solution.

4. Closed-Loop Guidance Law for Conflict Resolution

The previous section discussed iterative techniques for conflict resolution. The conflict resolution problem will be formulated as a closed-loop guidance problem in this section. Guidance laws are developed based on the assumption that the nominal paths are the desirable paths, and that any perturbation from these paths should only be made to achieve conflict resolution. The guidance law derivation for two aircraft will be illustrated in this paper. Generalization to multiple aircraft conflicts is a future research item. From a guidance law derivation point-of-view, the aircraft involved in a potential conflict can act in the three following fashions.

a) All aircraft involved in the potential conflict cooperate with each other for resolving the conflict. This would be the case when every aircraft is aware of every other aircraft in the vicinity, no aircraft has any priority over other aircraft, and all aircraft have comparable maneuverabilities. The resulting conflict resolution process will be termed here as *cooperative conflict resolution.*

b) Some aircraft involved in the potential conflict continue to fly nominal paths, while others take on the responsibility for resolving the conflicts. This situation can arise because of three distinct reasons. Firstly, some of the aircraft may be significantly be more agile than other aircraft. In this case, the agile aircraft may be able to resolve conflicts much faster, and more efficiently than non-agile aircraft. The second mechanism for such a scenario is when the aircraft are assigned a priority ordering, such that certain aircraft are allowed to stay on their nominal trajectories, while the others are made responsible for conflict resolution. The third situation where this might occur is when some of the aircraft are not aware of other aircraft in the vicinity, and hence continue on their nominal paths, while the remaining aircraft have to take responsibility for conflict resolution. These scenarios can be termed as *noncooperative* conflict resolution.

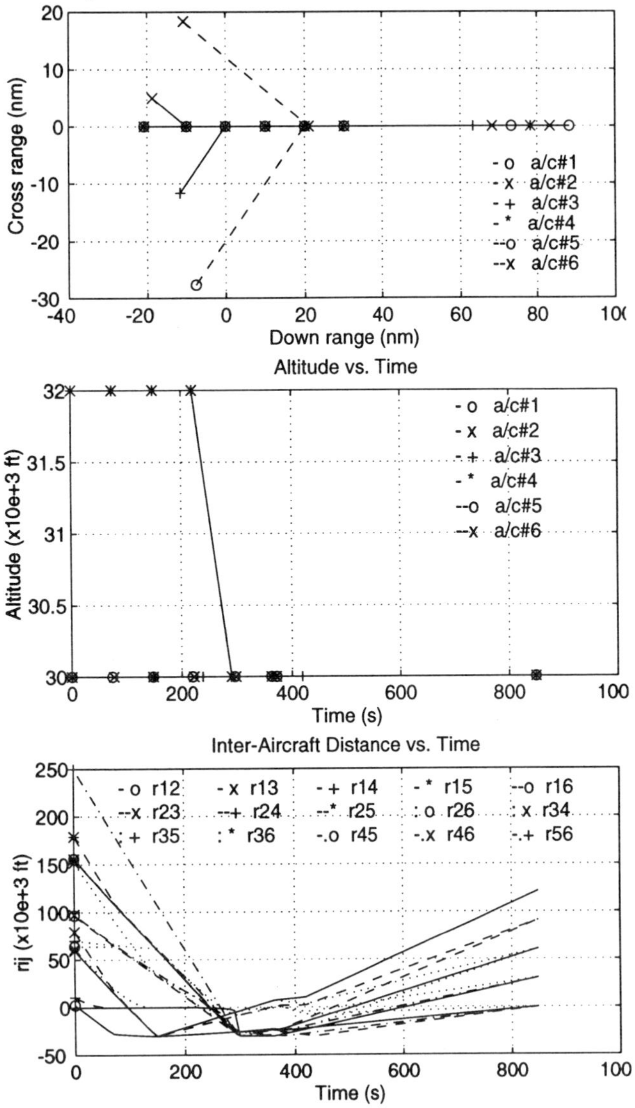

Fig. 4. Nominal Trajectories for Six Aircraft Conflict Resolution

c) The third operational mode corresponds to the case of some of the aircraft in the airspace maneuvering in such a way as to increase the potential for conflict. This situation may arise whenever some of the aircraft are not aware of other aircraft in the vicinity, and may maneuver in such a way as to increase the conflicts. Alternatively, in order to assure guaranteed conflict resolution, each properly equipped aircraft in the vicinity may decide to maneuver under the assumption that the unequipped aircraft in the vicinity may alter their trajectories in such a way as to cause conflicts. This case will be termed here as *adversarial* or *competitive* conflict resolution. Note this conflict scenario has a solution only if the conflict-resolving aircraft have superior data gathering and maneuvering capabilities when compared with the adversarial aircraft.

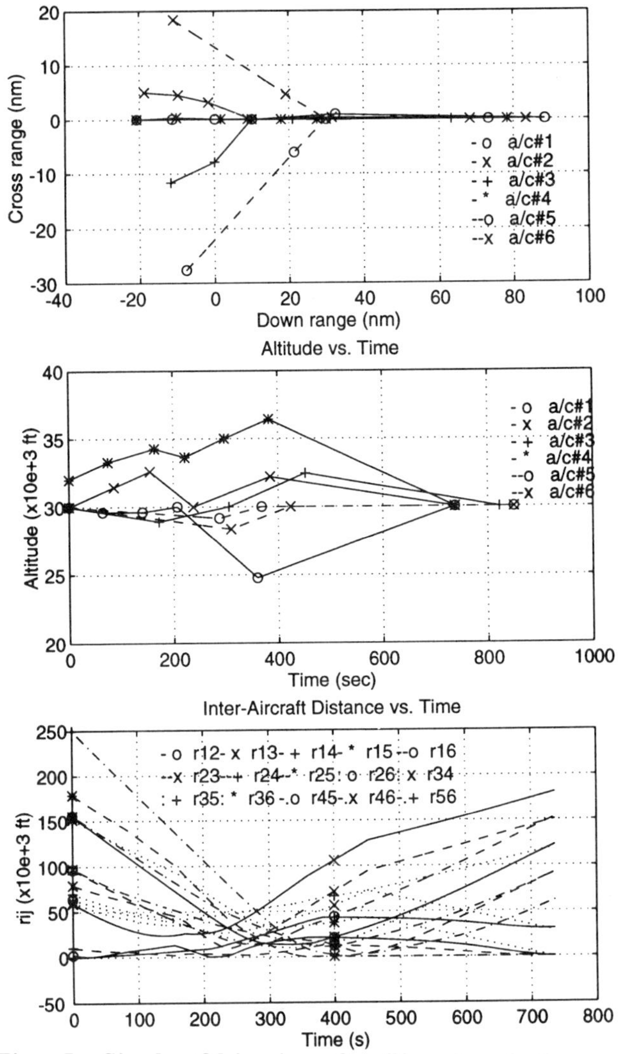

Fig. 5. Single Objective Conflict Resolution For the Six Aircraft Geometry

Characterizing the conflict resolution problem in terms of these categories serves to capture the degree of optimism of the aircraft involved in a given conflict scenario. Thus, the first mode corresponds to fully optimistic conflict resolution process, while the second is more neutral. The third formalism is pessimistic, and will lead to highly conservative conflict resolution maneuvers.

Guidance law development will assume that each aircraft in the airspace has selected a nominal waypoint sequence based on some other criteria. The criteria for nominal trajectory selection may include the minimization of flight time and fuel consumption to reach a selected set of final conditions while satisfying aircraft and environmental constraints. They may also be based on prevailing atmospheric conditions, and the need for coordinating operations with other aircraft in the air transportation system. Conflict resolution guidance laws introduce perturbations on these nominal waypoints to resolve the conflicts. Conflict resolution perturbations can be introduced in several different ways. For instance, the trajectory perturbations can be based on a set of rules capturing the maneuvering strategies in every given conflict situation.

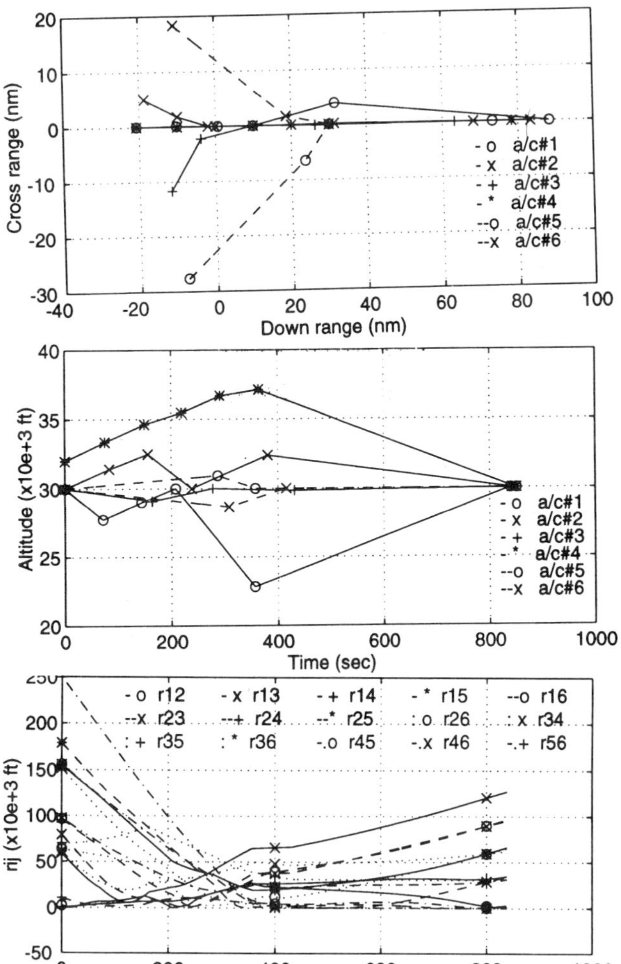

Fig. 6. Multiple Objective Conflict Resolution For the Six Aircraft Geometry

Alternatively, one may cast the trajectory perturbation problem as an optimal control problem [16] that minimizes the deviations from the nominal trajectories while achieving conflict resolution. In this framework, penalties can be placed on the magnitudes of acceleration, velocity, and position components. This approach is followed in the present research.

In order to formulate the conflict resolution problem in this manner, perturbations in each aircraft trajectory are first expressed as:

$$\delta x_1 = x_1 - \bar{x}_1,\ \delta y_1 = y_1 - \bar{y}_1,\ \delta z_1 = z_1 - \bar{z}_1 \qquad (18)$$

$$\delta x_2 = x_2 - \bar{x}_2,\ \delta y_2 = y_2 - \bar{y}_2,\ \delta z_1 = z_2 - \bar{z}_2 \qquad (19)$$

An over-bar denotes nominal values. Differentiating the expressions (18) and (19) twice with respect to time leads to the perturbed equations of motion:

$$\begin{aligned} \delta\ddot{x}_1 &= \ddot{x}_1 - \ddot{\bar{x}}_1 = u_1, \quad \delta\ddot{x}_2 = \ddot{x}_2 - \ddot{\bar{x}}_2 = u_2 \\ \delta\ddot{y}_1 &= \ddot{y}_1 - \ddot{\bar{y}}_1 = v_1, \quad \delta\ddot{y}_2 = \ddot{y}_2 - \ddot{\bar{y}}_2 = v_2 \\ \delta\ddot{z}_1 &= \ddot{z}_1 - \ddot{\bar{z}}_1 = w_1, \quad \delta\ddot{z}_2 = \ddot{z}_2 - \ddot{\bar{z}}_2 = w_2 \end{aligned} \qquad (20)$$

where u_1, u_2, v_1, v_2, w_1, w_2 are the perturbations in the aircraft acceleration components along down range, cross-range and altitude directions. These quantities can be related to perturbations in actual control variables of the aircraft using the expressions derived in Section 2. The reference values of the velocity and position components along the nominal trajectory can be determined using the given waypoint sequence.

Deviations from nominal trajectory can be characterized in terms of the Euclidean norm of the perturbations on velocity and position perturbations. The position perturbations are not included in the performance index used in the present research. Thus, a performance index of the form:

$$\min_{u_1,u_2,v_1,v_2,w_1,w_2} \frac{1}{2}\int_0^{t_f} \Big[\alpha_1^2\left(\delta\dot{x}_1^2 + \delta\dot{y}_1^2 + \delta\dot{z}_1^2\right) + \alpha_2^2\left(\delta\dot{x}_2^2 + \delta\dot{y}_2^2 + \delta\dot{z}_2^2\right) + \frac{1}{\beta_1^2}\left(u_1^2 + v_1^2 + w_1^2\right) + \frac{1}{\beta_2^2}\left(u_2^2 + v_2^2 + w_2^2\right) \Big] dt \qquad (21)$$

is used for the conflict resolution problem. Weighting factors α_1, β_1, α_2, β_2 can be used to adjust the relative importance of velocity and acceleration perturbations. They can also be used to cast the problem as cooperative, noncooperative or adversarial conflict resolution problem. For instance, using equal weights for both aircraft will result in cooperative conflict resolution, while using very large weights for one of the aircraft will produce the noncooperative conflict resolution solution. Changing the signs of the terms for one of the aircraft in the performance index will produce the adversarial conflict resolution solution. For the sake of illuminating the guidance law development, the present research will employ equal weighting factors for both aircraft. In this case, the weighting factors on the perturbed velocity components can be used to scale the weighting factor on the acceleration components. Other conflict resolution guidance laws will be more fully explored during future research.

Since the conflict resolution perturbations must begin and end on the nominal trajectories, they must satisfy the boundary conditions:

$$\begin{aligned} \delta x_1(o) = \delta\dot{x}_1(o) = \delta x_2(o) = \delta\dot{x}_2(o) = 0 \\ \delta x_1(t_f) = \delta\dot{x}_1(t_f) = \delta x_2(t_f) = \delta\dot{x}_2(t_f) = 0 \end{aligned} \qquad (22)$$

The minimum inter-aircraft distance requirement between the two aircraft at the time of conflict can be expressed as an interior-point equality constraint of the form:

$$\begin{aligned}&(\bar{x}_1+\delta x_1-\bar{x}_2-\delta x_2)^2\\&+(\bar{y}_1+\delta y_1-\bar{y}_2-\delta y_2)^2+(\bar{z}_1+\delta z_1-\bar{z}_2-\delta z_2)^2\\&-(\bar{x}_1-\bar{x}_2)^2-(\bar{y}_1-\bar{y}_2)^2-(\bar{z}_1-\bar{z}_2)^2\,\big|_{t=t_c} = R_{min}^2\end{aligned} \quad (23)$$

Or:

$$\begin{aligned}&(\delta x_1-\delta x_2)^2+(\delta y_1-\delta y_2)^2+(\delta z_1-\delta z_2)^2\\&+2(\bar{x}_1-\bar{x}_2)(\delta x_1-\delta x_2)\\&+2(\bar{y}_1-\bar{y}_2)(\delta y_1-\delta y_2)+2(\bar{z}_1-\bar{z}_2)(\delta z_1-\delta z_2)\,\big|_{t=t_c}\\&= R_{min}^2\end{aligned} \quad (24)$$

A conflict is assumed to have been detected at time t_c, and R_{min} is the desired additional separation distance between the two aircraft at the point-of-closest approach. It is assumed that : $0 \le t_c \le t_f$.. In the case of n conflicting aircraft, the optimal trajectories will have to satisfy n(n-1)/2 constraints of this form.

Additionally, in order to ensure that the separation between the aircraft does not decrease after t_c, the relative velocity along the direction of minimum separation will be required to satisfy the additional interior-point constraint:

$$\begin{aligned}&(\delta x_1-\delta x_2)(\delta\dot{x}_1-\delta\dot{x}_2)+(\delta y_1-\delta y_2)(\delta\dot{y}_1-\delta\dot{y}_2)\\&+(\delta z_1-\delta z_2)(\delta\dot{z}_1-\delta\dot{z}_2)+(\dot{\bar{x}}_1-\dot{\bar{x}}_2)(\delta x_1-\delta x_2)\\&+(\bar{x}_1-\bar{x}_2)(\delta\dot{x}_1-\delta\dot{x}_2)+(\dot{\bar{y}}_1-\dot{\bar{y}}_2)(\delta y_1-\delta y_2)\\&+(\bar{y}_1-\bar{y}_2)(\delta\dot{y}_1-\delta\dot{y}_2)+(\dot{\bar{z}}_1-\dot{\bar{z}}_2)(\delta z_1-\delta z_2)\\&+(\bar{z}_1-\bar{z}_2)(\delta\dot{z}_1-\delta\dot{z}_2)\,\big|_{t=t_c} = 0\end{aligned} \quad (25)$$

In the interests of maintaining simplicity, for this initial study, the conflict envelope has been assumed to spherical. The guidance law derivation along one of the coordinate axes will be illustrated in the following.

4.1. Conflict Resolution Guidance Law in the Down-Range Direction

In order to limit the complexity of the derivations in this initial research, it will be assumed that the conflict resolution problem can be solved independently along each coordinate direction by generating the identical amounts of separation along each axis. For instance, a performance index of the form:

$$\min_{u_1,u_2} \frac{1}{2}\int_0^{t_f}\left[\left(\delta\dot{x}_1^2+\delta\dot{x}_2^2\right)+\frac{1}{\beta^2}\left(u_1^2+u_2^2\right)\right]dt \quad (26)$$

is used in the down range direction. Similar performance indices can be can used along the cross range and altitude directions. The interior point equality constraints are then expressed as:

$$\left[(\delta x_1-\delta x_2)^2+2(\bar{x}_1-\bar{x}_2)(\delta x_1-\delta x_2)\right]\Big|_{t=t_c} = r_{min}^2 \quad (27)$$

$$\left[\begin{aligned}&(\delta x_1-\delta x_2)(\delta\dot{x}_1-\delta\dot{x}_2)+(\dot{\bar{x}}_1-\dot{\bar{x}}_2)(\delta x_1-\delta x_2)\\&+(\bar{x}_1-\bar{x}_2)(\delta\dot{x}_1-\delta\dot{x}_2)\end{aligned}\right]\Bigg|_{t=t_c} = 0 \quad (28)$$

where r_{min} is the additional separation to be generated along the down range direction. Note that the second interior point constraint further simplifies to $(\delta\dot{x}_1-\delta\dot{x}_2)=0$, if $(\dot{\bar{x}}_1-\dot{\bar{x}}_2)=0$ at $t=t_c$.

Optimal control problems with interior-point constraints can be treated as two distinct optimal control problems, the first one defined in the interval $[0,t_c]$ with the interior point constraints as the terminal conditions, while the second optimal control problem is defined in the interval $[t_c, t_f]$ with interior point constraints as the initial conditions.

The solution to the second problem will be discussed first because it is easier to solve. Define the variational Hamiltonian [16] as

$$\begin{aligned}H=\frac{1}{2}&\left[\left(\delta\dot{x}_1^2+\delta\dot{x}_2^2\right)+\frac{1}{\beta^2}\left(u_1^2+u_2^2\right)\right]\\&+\lambda_1u_1+\lambda_2u_2+\lambda_3\delta\dot{x}_1+\lambda_4\delta\dot{x}_2\end{aligned} \quad (29)$$

The Euler-Lagrange Equations are of the form:

$$\dot{\lambda}_1=-\delta\dot{x}_1-\lambda_3,\quad \dot{\lambda}_3=0,\quad u_1=-\beta^2\lambda_1 \quad (30)$$

$$\dot{\lambda}_2=-\delta\dot{x}_2-\lambda_4,\quad \dot{\lambda}_4=0,\quad u_2=-\beta^2\lambda_2 \quad (31)$$

The first two sets of costate equations and optimality conditions can be solved in closed-form as:

$$\lambda_1=a_1e^{\beta t}+b_1e^{-\beta t},\quad u_1=-\beta^2\left(a_1e^{\beta t}+b_1e^{-\beta t}\right) \quad (32)$$

$$\lambda_2=a_2e^{\beta t}+b_2e^{-\beta t},\quad u_2=-\beta^2\left(a_2e^{\beta t}+b_2e^{-\beta t}\right) \quad (33)$$

where a_1, b_1, a_2, b_2 are the arbitrary constants in the solution. The expressions for optimal control are next substituted in the equations of motion and integrated to yield:

$$\begin{aligned}\delta\dot{x}_1&=c_1-\beta\left(a_1e^{\beta t}-b_1e^{-\beta t}\right)\\\delta\dot{x}_2&=c_2-\beta\left(a_2e^{\beta t}-b_2e^{-\beta t}\right)\end{aligned} \quad (34)$$

$$\begin{aligned}\delta x_1&=d_1+c_1t-\left(a_1e^{\beta t}+b_1e^{-\beta t}\right)\\\delta x_2&=d_2+c_2t-\left(a_2e^{\beta t}+b_2e^{-\beta t}\right)\end{aligned} \quad (35)$$

where c_1, c_2, d_1, d_2 are arbitrary constants in the solution. Using these expressions, the costates λ_3, λ_4 can be found to be:

$$\lambda_3 = -c_1,\quad \lambda_4 = -c_4 \quad (36)$$

The arbitrary constants a_1, b_1, c_1, d_1, a_2, b_2, c_2, d_2 can be evaluated using the given boundary conditions. Applying the boundary conditions at t_c and t_f yields the expressions:

$$a_1\left(1-e^{\beta(t_f-t_c)}\right)-b_1\left(1-e^{-\beta(t_f-t_c)}\right) = \frac{\delta\dot{x}_1(t_f)-\delta\dot{x}_1(t_c)}{\beta} \quad (37)$$

$$a_2\left(1-e^{\beta(t_f-t_c)}\right)-b_2\left(1-e^{-\beta(t_f-t_c)}\right) = \frac{\delta\dot{x}_2(t_f)-\delta\dot{x}_2(t_c)}{\beta} \quad (38)$$

$$a_1\left[1+\beta(t_f-t_c)-e^{\beta(t_f-t_c)}\right] + b_1\left[1-\beta(t_f-t_c)-e^{-\beta(t_f-t_c)}\right] = \delta x_1(t_f)-\delta x_1(t_c)-[t_f-t_c]\,\delta\dot{x}_1(t_c) \quad (39)$$

$$a_2\left[1+\beta(t_f-t_c)-e^{\beta(t_f-t_c)}\right] + b_2\left[1-\beta(t_f-t_c)-e^{-\beta(t_f-t_c)}\right] = \delta x_2(t_f)-\delta x_2(t_c)-[t_f-t_c]\,\delta\dot{x}_2(t_c) \quad (40)$$

$$c_1=\delta\dot{x}_1(t_c)+\beta(a_1-b_1),\; d_1=\delta x_1(t_c)+(a_1+b_1) \quad (41)$$

$$c_2=\delta\dot{x}_2(t_c)+\beta(a_2-b_2),\; d_2=\delta x_2(t_c)+(a_2+b_2) \quad (42)$$

These expressions can be solved to obtain the trajectory perturbations required to restore the aircraft to the nominal trajectories after conflict resolution.

In order to handle the state constrained optimal control problem between the initial time and the point of closest approach, define

$$\Phi(t_c)=\mu_1\left[(\delta x_1-\delta x_2)^2+2(\bar{x}_1-\bar{x}_2)(\delta x_1-\delta x_2)\,|_{t=t_c} - r_{min}^2\right] + \mu_2\left[(\delta x_1-\delta x_2)(\delta\dot{x}_1-\delta\dot{x}_2) + (\dot{\bar{x}}_1-\dot{\bar{x}}_2)(\delta x_1-\delta x_2)+(\bar{x}_1-\bar{x}_2)(\delta\dot{x}_1-\delta\dot{x}_2)\,|_{t=t_c}\right] \quad (43)$$

μ_1 and μ_2 are two undetermined constants. The condition (4.28) imposes constraints on the value of the costates at $t = t_c$ as:

$$\lambda_1(t_c)=\mu_2\left[(\delta x_1-\delta x_2)+(\bar{x}_1-\bar{x}_2)\,|_{t=t_c}\right] \quad (44)$$

$$\lambda_2(t_c)=-\mu_2\left[(\delta x_1-\delta x_2)+(\bar{x}_1-\bar{x}_2)\,|_{t=t_c}\right] \quad (45)$$

implying: $\lambda_1(t_c)=-\lambda_2(t_c)$. This leads to the result:

$$(a_1+a_2)e^{\beta t_c}=-(b_1+b_2)e^{-\beta t_c} \quad (46)$$

Similarly,

$$\lambda_3(t_c)=\mu_1\left[2(\delta x_1-\delta x_2)+2(\bar{x}_1-\bar{x}_2)\,|_{t=t_c}\right] + \mu_2\left[(\delta\dot{x}_1-\delta\dot{x}_2)+(\dot{\bar{x}}_1-\dot{\bar{x}}_2)\,|_{t=t_c}\right] \quad (47)$$

$$\lambda_4(t_c)=-\mu_1\left[2(\delta x_1-\delta x_2)+2(\bar{x}_1-\bar{x}_2)\,|_{t=t_c}\right] - \mu_2\left[(\delta\dot{x}_1-\delta\dot{x}_2)+(\dot{\bar{x}}_1-\dot{\bar{x}}_2)\,|_{t=t_c}\right] \quad (48)$$

which implies $\lambda_3(t_c)=-\lambda_4(t_c)$. Substituting for these quantities:

$$(a_1+a_2)-(b_1+b_2)=\frac{-\delta\dot{x}_2(0)-\delta\dot{x}_1(0)}{\beta} \quad (49)$$

These equations can be used to obtain:

$$(b_1+b_2)=\frac{\delta\dot{x}_2(0)+\delta\dot{x}_1(0)}{\beta\left(1+e^{-2\beta t_c}\right)}$$

$$(a_1+a_2)=-\left[\frac{\delta\dot{x}_2(0)+\delta\dot{x}_1(0)}{\beta\left(1+e^{-2\beta t_c}\right)}\right]e^{-2\beta t_c} \quad (50)$$

The terminal constraints yield the following relationships:

$$(\delta x_1-\delta x_2)\big|_{t=t_c}=-(\bar{x}_1-\bar{x}_2)\pm\sqrt{(\bar{x}_1-\bar{x}_2)^2+r_{min}^2}\,\Big|_{t=t_c} \quad (51)$$

$$(\delta\dot{x}_1-\delta\dot{x}_2)\big|_{t=t_c} = \frac{-(\dot{\bar{x}}_1-\dot{\bar{x}}_2)\left[-(\bar{x}_1-\bar{x}_2)\pm\sqrt{(\bar{x}_1-\bar{x}_2)^2+r_{min}^2}\right]}{\pm\sqrt{(\bar{x}_1-\bar{x}_2)+r_{min}^2}}\Bigg|_{t=t_c} \quad (52)$$

From closed-form solutions,

$$(\delta\dot{x}_1-\delta\dot{x}_2)\big|_{t-t_c} = c_1-c_2-\beta(a_1-a_2)e^{\beta t_c}+\beta(b_1-b_2)e^{-\beta t_c} \quad (53)$$

$$(\delta x_1-\delta x_2)\big|_{t=t_c} = d_1-d_2+(c_1-c_2)t_c-(a_1-a_2)e^{\beta t_c}-(b_1-b_2)e^{-\beta t_c} \quad (54)$$

Next substituting for the arbitrary constants in terms of the boundary conditions,

$$(\delta\dot{x}_1-\delta\dot{x}_2)\big|_{t-t_c}=[\delta\dot{x}_1(0)-\delta\dot{x}_2(0)] + (a_1-a_2)\beta\left[1-e^{\beta t_c}\right]-(b_1-b_2)\beta\left[1-e^{-\beta t_c}\right] \quad (55)$$

$$(\delta x_1-\delta x_2)\big|_{t=t_c}=[\delta x_1(0)-\delta x_2(0)] + [\delta\dot{x}_1(0)-\delta\dot{x}_2(0)]t_c + (a_1-a_2)\left[1+\beta t_c-e^{\beta t_c}\right] + (b_1-b_2)\left[1-\beta t_c-e^{-\beta t_c}\right] \quad (56)$$

These two equations can be solved for (a_1-a_2) and (b_1-b_2). They can then be combined with the expressions for (a_1+a_2) and (b_1+b_2) to yield the arbitrary constants a_1, b_1,

a_2, b_2. These can then be used to determine the remaining constants in the problem c_1, d_1, c_2, d_2.

This completes the computation of the perturbations required to resolve conflicts. These perturbations can be added to the nominal paths to yield the conflict resolved trajectories. Note that the perturbation equations are solved from the current time to the final time in every guidance interval.

These perturbations correct the nominal waypoints to form the new waypoint sequence. Note that the addition of the perturbations to the nominal may not always resolve the conflict. In highly nonlinear geometries, the conflict resolution computations will have to be repeated several times to generate conflict-free waypoint sequences. In these cases, it is beneficial to multiply the perturbations by an attenuating factor before adding to the nominal trajectories. The resulting algorithm will turnout to be the quasi-linearization technique [16, 38] or the technique of invariant embedding. Quasilinearization is a second-order technique and is known to converge rapidly to the optimal solution.

The conflict resolution algorithm is next evaluated for a set of conflict scenarios involving two aircraft. These simulation results are given in the next subsection.

4.2. Simulation Results

A two-aircraft conflict scenario is chosen to demonstrate the guidance law performance. In all cases, the weighting factor β is chosen to be 10^{-03}. The conflict resolution case considered is that of two aircraft with crossing trajectories. Figure 7 illustrates the down range-cross range trajectories, altitude profiles and inter-aircraft distances before and after the conflict resolution maneuvers. The trajectories before conflict resolution are denoted by dashed lines, and the trajectories after conflict resolution are denoted using dashed lines. Since cooperative conflict resolution is assumed, both aircraft perform conflict resolution maneuvers as can be observed in the down range - cross range plot. Due to the assumption that equal amounts of conflict resolution maneuvers will be made in the down range, cross range and altitude directions, the aircraft also perform maneuvers in the vertical plane. It can be observed from the inter-aircraft distance shown in Figure 7 that each aircraft has adjusted its trajectory to resolve the conflict.

In the present example, the guidance problem was solved as a quasilinearization problem. In the quasilinearization formulation, the trajectory perturbations generated by the guidance law are attenuated before being added to the nominal trajectories. The updated nominal trajectory is then used to compute the conflict resolution trajectory perturbations. Iterative computations are continued until the conflict is resolved. In the present case, a perturbation factor of 0.2 is used, and the solution converged in three iterations.

The simulation results presented in this subsection illustrate the performance capabilities of the conflict resolution guidance law. In each case, the guidance law resolves the conflicts using modest control magnitudes. As indicated elsewhere in this chapter, the guidance law can be extended to include multiple aircraft executing widely different nominal trajectories. Moreover, inequality constraints can be imposed on the control magnitudes to improve the implementability of the present guidance laws. The ride-quality of these conflict resolution guidance laws can be further improved by introducing control rate limits. This will produce smooth control settings, resulting in nearly jerk-free trajectories. Finally, the present formulation can be extended to the spherical earth case. The kinematic trajectory models of the form given in Reference 4 can be readily adapted for this purpose.

The derivations presented in this chapter, together with the simulation results amply demonstrate the feasibility of developing on-board implementable guidance laws for conflict resolution in free flight.

5. Conclusions

This paper documented the results of a research study that examined development of advanced algorithms for air traffic conflict resolution. The research was motivated by the free flight concept currently being considered for implementation as a part of the national air traffic control system. In the free flight concept, individual aircraft will be allowed to plan and execute their trajectories without direction from any external agents during most of their flight duration. However, since each aircraft in the airspace may have conflicting objectives, the desired trajectories synthesized by individual aircraft can produce conflicts with other aircraft trajectories. While conflict resolution strategies are obvious under low speed, low traffic density conditions, the problem can become complex as the traffic density and the aircraft speeds increase.

The conflict resolution problem was formulated and solved as a multi-participant optimal control problem in the present research. The problem was formulated under the assumption that the objective of each aircraft involved in a potential conflict is to optimize its performance while maintaining adequate separation between other aircraft in the vicinity. Point-mass aircraft models and oblate spheroidal conflict envelopes were used in this study. Point mass models included thrust, load factor, and bank angle as the control variables. Inequality constraints were imposed on the aircraft altitude, climb/descent rate, airspeed, and control magnitudes. The oblate spheroidal conflict envelope provided a smooth approximation to the FAA mandated separation between aircraft. The conflict envelope had a 5 nautical mile semi-major axis, and a 2000 feet semi-minor axis.

Two distinct approaches for solving the conflict resolution problem were advanced. In the first approach, the aircraft trajectories were approximated using piecewise linear curves to convert the trajectory optimization problems into parameter optimization problems. In the second approach, the aircraft equations of motion were linearized about the nominal trajectories, and state-constrained multi-point boundary value problems were formulated. Parameterized conflict resolution problems were solved using single objective sequential quadratic programming method, and a multiple objective goal attainment method. Conflict resolution involving six aircraft were demonstrated. State constrained multi-point boundary value problems were solved in closed-form to generate conflict resolution guidance laws. Performance of the guidance laws for several conflict scenarios were demonstrated.

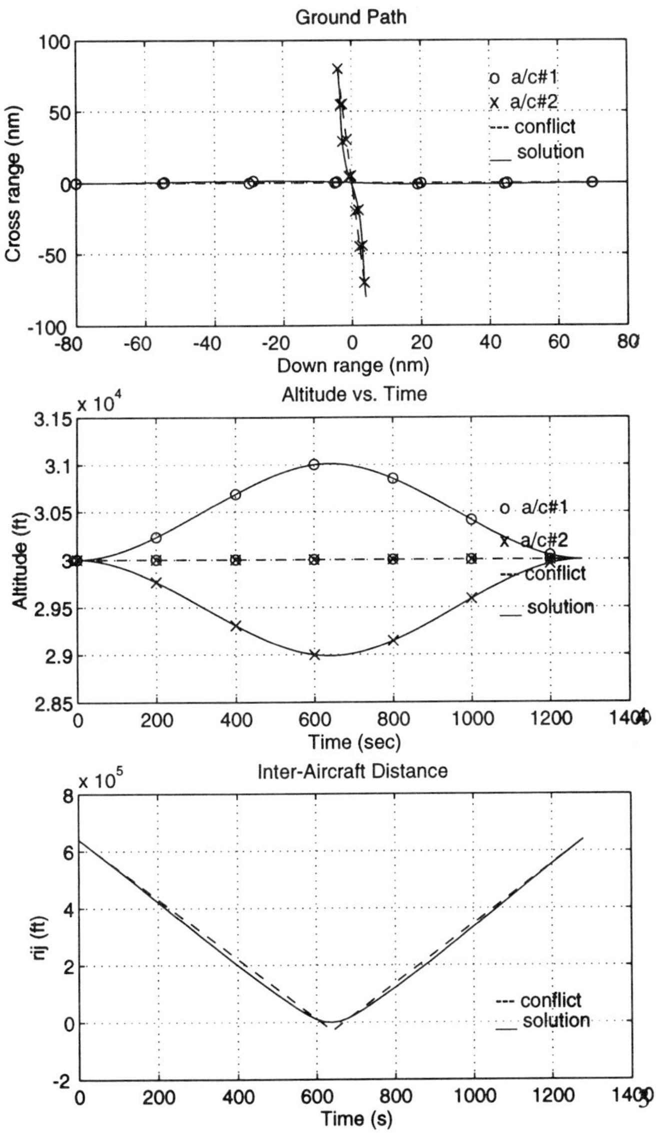

Fig. 7. Conflict Resolution Trajectories Using The Quasilinearization Guidance Law: Crossing Trajectories

Acknowledgement

This research was supported under NASA Ames Research Center Contract No. NAS1-14333.

6. References

1. Final Report of the RTCA Task Force 3: *Free Flight Implementation*, RTCA Inc., October 26, 1995.
2. Special Report on Free Flight, *Aviation Week and Space Technology*, July 31, 1995.
3. Bilimoria, K., Sridhar, B., and Chatterji, G., "Effects of Conflict Resolution Maneuvers and Traffic Density on Free Flight", *AIAA Guidance, Navigation, and Control Conference*, July 29-31, 1996, San Diego, CA.
4. Chatterji, G., Sridhar, B., and Bilimoria, K., "En-route Flight Trajectory Prediction for Conflict Avoidance and Traffic Management", *AIAA Guidance, Navigation, and Control Conference*, July 29-31, 1996, San Diego, CA.
5. Menon, P. K., "Control Theoretic Approach to Air Traffic Conflict Resolution", *Guidance, Navigation, and Control Conference*, August 9 -11, 1993, Monterey, CA.
6. Pritchett, A. R., and Hansman, R. J., "Experimental Study of Collision Detection Schema Used by Pilots During Closely Spaced Parallel Approaches", *AIAA Guidance, Navigation, and Control Conference*, July 29-31, 1996, San Diego, CA.
7. Slattery, R. A., "Terminal Area Trajectory Synthesis for Air Traffic Control Automation", *Proceedings of the 1995 American Control Conference*, Seattle, WA.
8. Krozel, J., Mueller, T., and Hunter, G., "Free Flight Conflict Detection and Resolution Analysis", *AIAA Guidance, Navigation, and Control Conference*, July 29-31, 1996, San Diego, CA.
9. Pujet, N., and Feron, E., "Flight Plan Optimization in Flexible Air Traffic Environments", *AIAA Guidance, Navigation, and Control Conference*, July 29-31, 1996, San Diego, CA.
10. Adams, M., Kolitz, S., and Odoni, A., "Evolutionary Concepts for Decentralized Air Traffic Flow Management", *AIAA Guidance, Navigation, and Control Conference*, July 29-31, 1996, San Diego, CA.
11. Erzberger, H., "CTAS: Integrated Automation Tools for the Center and TRACON", *The Journal of Air Traffic Control*, Oct. -Dec. 1990, Vol. 32, No. 4, pp. 90 - 91.
12. Erzberger, H., and Nedell, W, "Design of Automated System for Management of Arrival Traffic", NASA TM 102201, June 1989.
13. Erzberger, H., "CTAS: Computer Intelligence for Air Traffic Control in the Terminal Area", NASA Technical Memorandum 103959, July 1992.
14. Davis, T. J., Erzberger, H., Green, S. M., and Nedell, W., "Design and Evaluation of an Air Traffic Control Final Approach Spacing Tool", *Journal of Guidance, Control, and Dynamics*, Vol. 14, No. 4, July-Aug. 1991, pp. 848-854.

15. Davis, T. J., Erzberger, H., and Green, S. M., "Simulator Evaluation of the Final Approach Spacing Tool", NASA TM 102807, May 1990.
16. Bryson, A. E., and Ho, Y. C., *Applied Optimal Control*, Hemisphere, New York, 1975
17. Isaacs, R., *Differential Games*, R. E. Krieger, New York, 1965.
18. Menon, P. K. and Cheng, V. H. L., "Minimum-Exposure Near Terrain Flight Trajectories for Rotorcraft", Chapter 9 in *Control and Dynamic Systems* Edited by C. T. Leondes, Vol. 52, pp. 391 - 433.
19. Menon, P. K., Kim, E., and Cheng, V. H. L., "Helicopter Trajectory Planning Using Optimal Control Theory", *Proceedings of the 1988 American Control Conference*, Atlanta, GA, June 15 - 17, pp. 1440-1447.
20. Cheng, V. H. L., "Obstacle-Avoidance Automatic Guidance: A Concept Development Study", *AIAA Guidance, Navigation, and Control Conference*, Aug. 15 - 17, 1988, Minneapolis, MN.
21. Pekelsma, N. J., "Optimal Guidance with Obstacle Avoidance for Nap-of-the-Earth Flight", NASA CR 177515, December 1988.
22. Menon, P. K., Chatterji, G. B., and Sridhar, B., "Vision-Based Optimal Obstacle Avoidance Guidance for Rotorcraft", *Proceedings of the AIAA Guidance, Navigation, and Control Conference*, New Orleans, LA, August 1991.
23. Menon, P. K., "Air Traffic Conflict Resolution Using Modern Control Theory", *Optimal Synthesis* Report Prepared for NASA Ames Research Center under Subcontract No. 02804230, October 1992.
24. Gembicki, F. W., "Vector Optimization for Control with Performance and Parameter sensitivity Indices", *Ph.D Dissertation*, Case Western Reserve University, Cleveland, OH, 1974.
25. Brayton, R. K., Director, S. W., Hachtel, G. D., and Vidigal, L., "A New Algorithm for Statistical Circuit Design Based on Quasi-Newton Methods and Function Splitting", IEEE Transactions on Circuits and Systems, Vol. CAS-26, September 1979, pp. 784-794.
26. Nelson, R. C., *Flight Stability and Automatic Control*, McGraw-Hill Book Company, San Francisco, 1989.
27. Etkin, B., *Dynamics of Atmospheric Flight*, John Wiley, New York, NY, 1982.
28. Menon, P. K., "Short-Range Nonlinear Feedback Strategies for Aircraft Pursuit-Evasion", *Journal of Guidance, Control and Dynamics*, Vol. 12, January - February 1989, pp. 27 - 32.
29. Gerald, C. F., Applied Numerical Analysis, Addison Wesley, Reading, MA, 1978.
30. Kreyszig, E., *Advanced Engineering Mathematics*, Wiley, New York, New York, 1993.
31. Anon, *MATLAB® User's Manual*, The MathWorks, Inc., Natick, MA, 1994.
32. Barr, A. H., "Super Quadric Surfaces", *IEEE Transactions on Computer Graphics and Applications*, 1981.
33. Golub, G. H., and Van loan, C. F., *Matrix Computations*, The John Hopkins University Press, Baltimore, Maryland, 1989.
34. Gill, P. E., Murray, W. and Wright, M. H., *Practical Optimization*, Academic Press, New York, NY, 1981.
35. Dantzig, G., *Linear Programming and Extensions*, Princeton University Press, Princeton, NJ, 1963.
36. Grace, A., *Optimization Toolbox for MATLAB*, The MathWorks, Inc., Natick, MA, 1992.
37. Green, W., Swanborough, G., and Mowinski, J., *Modern Commercial Aircraft*, Crown Publishers, New York, NY, 1987.
38. Balakrishnan, A. V., and Neustadt, L. W. (Editors), *Computing Methods in Optimization Problems*, Academic Press, New York, NY, 1964.

AIAA-97-3547

Deterministic Resolution of Two Aircraft Conflict in Free Flight

Yiyuan Zhao*
University of Minnesota, Minneapolis MN 55455
Robert L. Schultz†
Honeywell Technology Center, 3660 Technology Drive, Minneapolis, MN 55418

Abstract

This paper studies strategies for resolving two aircraft conflicts in a free flight environment for air traffic management. In a free flight environment, pilots of individual aircraft can change flight paths in real time. As a result, resolving conflicts among aircraft becomes a key priority for controllers. In this paper, a point-mass level flight aircraft model is used. Resolution of a two-aircraft conflict is formulated as deterministic nonlinear optimal control problems. Performance indices are selected to minimize deviations of both aircraft from their originally-intended flight paths or to minimize banking angles, subject to a minimum separation requirement. In addition, terminal constraints are included that force each aircraft to return to its original path after conflict avoidance maneuvers. In an unrestricted region, it is assumed that aircraft are originally flying with a constant heading. In general, the two aircraft would turn the same way (left or right) by roughly the same amount to avoid a conflict. When the intersecting angle is small, however, one aircraft needs to take a path stretching to avoid the other aircraft. In the terminal region where aircraft converge to the same gate, one aircraft needs to stretch its flight paths to avoid the other aircraft. If the two aircraft have the same speed, either aircraft can perform the path stretching. If the two aircraft have different speeds, the slower aircraft should perform the path stretching.

Introduction

The current air traffic control[1] (ATC) system provides sufficient separation among all aircraft through pre-defined routes and flight procedures. In this ATC infrastructure, lateral separations are guaranteed by choices of different flight routes, vertical separations are achieved by different flight levels, and longitudinal separations are maintained through speed control.

Recently, a concept of free flight is proposed for future air traffic control[2]. In a free flight environment, aircraft operators can change flight paths in **real time**. Air traffic controllers are only supposed to intervene when situation warrants, to resolve potential conflicts among aircraft. It is argued that a free flight environment enables individual aircraft to minimize their operating costs and thus improves the economic benefits of the current system. Clearly, a key component in the free flight study is the determination of proper strategies for conflict resolution.

A conflict occurs between two aircraft if the relative separation distance between the two falls below a specified minimum. Minimum separation requirements depend on aircraft type, accuracy of aircraft position determination, and air traffic management structure. In the current air traffic control system, any two aircraft during en route flight must be separated by 5 mni horizontally, and 1000 or 2000 ft in the vertical direction depending on aircraft flight levels.

Successful conflict resolution strategies depend on many elements such as methods of aircraft position determinations and air-to-air, air-to-ground communications. From a system automation point of view, conflict resolution process consists of aircraft trajectory predictions, conflict detections and estimation of conflict likelihood, and synthesis of conflict resolution strategies.

Trajectory predictions determine the most likely aircraft paths from the current time to a certain future time based on measurements of current aircraft positions, filed aircraft flight intentions, and aircraft dynamics. Conflict detections need to compare flight paths of many aircraft to determine if any two aircraft may fall below specified minimum separation standards over the predicted trajectories. Because all trajectory predictions contain errors that normally grow with time, the likelihood of a potential conflict needs to be estimated to reduce the possibility of false alarms. Finally, involved aircraft must perform necessary resolution maneuvers to avoid a potential conflict.

Efforts have been made at NASA Ames Research Center to design the first operational air traffic control automation system called Center-TRACON Automation System, or CTAS[3-6]. As a result, mature methods of trajectory predictions exist for on-board flight management[7] and air traffic control automation[8-9] for the current ATC infrastructure. Conflict detections can be made by comparing pairwise trajectories of two selected aircraft[10]. Ref. 11 presents an efficient method for estimating the likelihood of a potential conflict. This method can be used in both the current ATC system and in a free flight environment. Over the last few decades, air traffic controllers have accumulated a mature set of methods for resolving potential aircraft conflicts in the existing ATC system[12]. These

* Associate Professor, Dept. of Aerospace Eng. and Mechanics, Member AIAA
† Senior Research Fellow, Member AIAA

methods are incorporated into CTAS[6,13]. In addition, studies were made to use optimal control theories for conflict resolution in the descent phase[14].

Similar problems exist for conflict resolutions in free flight. Ref. 15 studied great circle paths and sensitivities of ground speed for free-flight path predictions. In a free flight environment, aircraft operators have more degrees of freedom to resolve potential conflicts. In general, pilots can use altitude change, speed change, and/or horizontal maneuvers to avoid a potential conflict. Ref. 16 examined the effectiveness of each control freedom in resolving a potential conflict, and laid a foundation for the current work. Ref. 17 also discussed the relative effectiveness of avoidance maneuvers.

Various approaches are used to study conflict resolution strategies. Refs. 18-23 used optimization and game theory. These work determine patterns of optimal maneuvers for conflict resolutions. Factors that influence the minimum separation standards are discussed in Ref. 24. Ref. 25 used a genetic algorithm approach for conflict resolutions.

Extensive research has been conducted for conflict-free robot motion planning[26,27]. It is not clear to what extend and how that results in robot motion planning can be used directly to aircraft conflict resolutions because of the unique aircraft dynamics. Refs. 28 and 29 used the potential field idea of robot motion planning for aircraft conflict resolutions. Potential field methods model obstacles as repulsions and desired flight paths as attractions. It seems that these methods can be used in real time and for multiple aircraft conflict resolution. On the other hand, these methods may not maintain a strict minimum separation requirement at times. In Ref. 30, a different method of robot motion planning is used for aircraft conflict resolution.

Extensive efforts have been made on the Traffic Alert and Collision Avoidance System (TCAS)[31-33]. TCAS is an airborne system that actively detects the intrusion of nearby aircraft, and recommends emergency advisories for pilots to avoid an impending collision. The response time is on the order of 30 seconds. Therefore, TCAS is a safety net built to ward off unforeseeable errors. In comparison, Automated En Route ATC project (AERA) investigates automated algorithmic conflict detection and resolution for today's ATC system during the en route phase. AERA consists of a system of logics that breaks a multiple aircraft conflict into a series of pairwise conflicts and eliminate any possible domino's effect in the process[34].

This paper determines optimal patterns of two aircraft conflict resolutions in free flight. Specifically, resolutions of potential conflicts are formulated as nonlinear optimal control problems. Both aircraft involved in a potential conflict are assumed to be on level flight, and aircraft banking angles are used as controls. Performance indices are selected to minimize deviations of the two involved aircraft from their originally intended flight paths, or to minimize banking angles, subject to the minimum separation requirement. In addition, terminal constraints are included that force both aircraft to return to their original flight paths after the potential conflict is avoided.

Conflict Resolution in Free Flight

For the purpose of conflict detection and resolution, aircraft trajectories need to be predicted over a certain period of time $[t_0, t_f]$. Let $\vec{r}_1(t)$ and $\vec{r}_2(t)$ represent the actual positions of two aircraft in the three-dimensional space as functions of time $t \in [t_0, t_f]$. A conflict occurs if

$$\|\vec{r}_1(t_c) - \vec{r}_2(t_c)\| \leq r_{\text{req}} \tag{1}$$

at some point $t_c \geq t_0$, where t_0 is the initial time of trajectory predictions, r_{req} is the minimum required separation between two aircraft, and $\|\cdot\|$ is a certain vector norm.

The current ATC standards specify the minimum separation requirement between any two aircraft. If two aircraft are in the same flight level, the geometrical distance between them must be larger than or equal to 5 nmi. If the two aircraft are roughly in the same horizontal location, their vertical separation must be larger than or equal to 2000 ft. These separation standards are caused by aircraft turbulent wake and error margins of radar measurements. They constitute the protected zones of each aircraft with today's technology. The horizontal separation distance requirement becomes 3 nmi when aircraft are within 40 nmi from radar antennas, and the vertical separation requirement becomes 1000 ft for aircraft below 29,000 ft.

If $\vec{r}_1^{\,o}(t)$ and $\vec{r}_2^{\,o}(t)$ denote predicted aircraft trajectories based on known or assumed pilot intentions, we have

$$\vec{r}_1(t) = \vec{r}_1^{\,o}(t) + \vec{r}_1^{\,i}(t) + \Delta\vec{r}_1(t) \tag{2}$$

$$\vec{r}_2(t) = \vec{r}_2^{\,o}(t) + \vec{r}_2^{\,i}(t) + \Delta\vec{r}_2(t) \tag{3}$$

where $\Delta\vec{r}_1$ and $\Delta\vec{r}_2$ are errors of trajectory predictions, $\vec{r}_1^{\,i}(t)$ and $\vec{r}_2^{\,i}(t)$ are possible trajectory variations caused by unintended pilot maneuvers or errors in the assumption of pilot intentions. In the current study, we assume that both pilots are cooperating and the pilot intentions are well known. In other words,

$$\vec{r}_1^{\,i}(t) = 0 \quad \vec{r}_2^{\,i}(t) = 0 \tag{4}$$

Define

$$e(t) \triangleq \|\Delta\vec{r}_1(t) - \Delta\vec{r}_2(t)\| \tag{5}$$

it is clear from Eqs. (1) through (3) that

$$\|\vec{r}_1(t) - \vec{r}_2(t)\| \leq \|\vec{r}_1^{\,o}(t) - \vec{r}_2^{\,o}(t)\| + e(t) \tag{6}$$

$$\|\vec{r}_1(t) - \vec{r}_2(t)\| \geq \|\vec{r}_1^{\,o}(t) - \vec{r}_2^{\,o}(t)\| - e(t) \tag{7}$$

Therefore, a potential conflict exists if

$$\|\vec{r}_1^o(t) - \vec{r}_2^o(t)\| + e(t) \leq r_{\text{req}} \tag{8}$$

or

$$\|\vec{r}_1^o(t) - \vec{r}_2^o(t)\| \leq r_{\text{req}} - e(t) \tag{9}$$

and there is no potential conflict (a conflict is avoided) if

$$\|\vec{r}_1^o(t) - \vec{r}_2^o(t)\| - e(t) \geq r_{\text{req}} \tag{10}$$

or

$$\|\vec{r}_1^o(t) - \vec{r}_2^o(t)\| \geq r_{\text{req}} + e(t) \tag{11}$$

Trajectory predictions always contain errors. In general, trajectory prediction errors grow with time, especially in the longitudinal direction. After a certain amount of "lookahead" time (e.g. 20 to 30 minutes), trajectory prediction errors become so large that predicted trajectories can no longer be used meaningfully.

Trajectory prediction errors can either be described through deterministic worst case bounds or as a stochastic process. In a deterministic framework, for example[24],

$$e = k_0 + k_1 t + k_2 t^2 \tag{12}$$

Due to limited space, it is assumed that $e = 0$ in the current paper. This condition simplifies the following analysis and essentially assumes that perfect trajectory predictions are made. These results are valid for a short time span and can serve as useful guides in general.

Multiple aircraft conflict consists of several aircraft pairs that are predicted to fall below the minimum separation requirement. It is expected that many multiple aircraft conflicts can be resolved through pairwise resolutions. Research work on multiple aircraft conflict resolution is in progress.

Two Aircraft Conflict Detection and Resolution

In the current air traffic control system, lateral separation en route is achieved through assigned routes, while vertical separation by different flight levels. Speed change is the main tool for maintaining sufficient longitudinal separation en route. During the final descent to approach over the extended terminal area, speed change is also the main tool for maintaining separation, followed by altitude changes and path stretching.

In the proposed free flight system, all three controls can be used for conflict resolution: speed change, altitude change, and heading change (path stretching). However, some controls are more effective than others. During en route flight, heading change is most effective in producing long term miss distances, while altitude change (climb/descent) can provide sufficient vertical separation in a very short time. Speed change is the least effective due to high flight speeds at cruise (Ref. 16).

In this paper, we examine conflict resolutions using heading control only. Point-mass equations of motion for an aircraft in level flight with a constant speed are

$$\dot{x} = V \sin \Psi \tag{13}$$

$$\dot{y} = V \cos \Psi \tag{14}$$

$$\dot{\Psi} = \frac{g}{V} \tan \phi \tag{15}$$

Roll motion for a commercial transport has a time constant on the order of a few seconds. Therefore, the roll (or bank) angle ϕ can be treated as the control.

There are two basic types of potential conflicts. If two aircraft try to converge to the same waypoint or runway at very close arrival times, the conflict is said to be a **goal conflict**. If the separation between two aircraft drops below a certain minimum standard, the conflict is said to be a **loss of minimum separation**.

Goal conflicts require one or both aircraft to reschedule their arrival at the waypoint or runway. If a goal conflict is not resolved, there will necessarily be a loss of minimum separation sometime in the future. On the other hand, aircraft involved in a loss of minimum separation conflict en route may be able to resolve their conflict without affecting scheduled times of arrival. In the en route phase of a free flight, loss of minimum separation is the main concern. In the transition region or extended terminal area, however, goal conflicts must be resolved before separation conflicts can be resolved permanently.

En Route Conflict

We consider the detection and resolution of conflict involving two aircraft at flight levels of no more than 2000 ft apart during en route flight. At the time of conflict detection $t_0 = 0$, both aircraft are flying at constant speeds and following constant heading angles. From Eqs. (13) and (14), we have

$$x_1(t) = x_1(t_0) + V_1 \sin \Psi_1^0 \, t \tag{16}$$

$$y_1(t) = y_1(t_0) + V_1 \cos \Psi_1^0 \, t \tag{17}$$

$$x_2(t) = x_2(t_0) + V_2 \sin \Psi_2^0 \, t \tag{18}$$

$$y_2(t) = y_2(t_0) + V_2 \cos \Psi_2^0 \, t \tag{19}$$

The separation distance between the two aircraft as a function of time is

$$d(t) = \sqrt{(x_1 - x_2)^2 + (y_1 - y_2)^2} \tag{20}$$

The smallest separation can be found by differentiating this distance with respect to time and determining the minimum value. While details are omitted, the predicted closest approach distance and the corresponding time at closest approach are given by

$$d_{\min} = \frac{1}{N} \left| (x_1^0 - x_2^0)(\cos \Psi_1^0 - \sigma \cos \Psi_2^0) \right.$$

$$- (y_1^0 - y_2^0)(\sin \Psi_1^0 - \sigma \sin \Psi_2^0)| \quad (21)$$

$$t_{\min} = -\frac{1}{N^2 V_1} \left[(x_1^0 - x_2^0)(\sin \Psi_1^0 - \sigma \sin \Psi_2^0) + (y_1^0 - y_2^0)(\cos \Psi_1^0 - \sigma \cos \Psi_2^0) \right] \quad (22)$$

where

$$N = \sqrt{1 + \sigma^2 - 2\sigma \cos(\Psi_1^0 - \Psi_2^0)}$$

In the above, $(\ )^0$ represent conditions at t_0, and

$$\sigma \triangleq \frac{V_2}{V_1} \quad (23)$$

Therefore, a conflict is detected if both of the following two conditions are satisfied.

$$d_{\min} \leq r_{\text{req}} \quad (24)$$

$$t_{\min} > 0 \quad (25)$$

where r_{req} is the minimum longitudinal separation requirement (5 nmi). Eq. (25) insures that the predicted closest approach distance is in the future.

These results provide a useful tool for conflict detection en route, if both aircraft are assumed to fly fixed courses at constant speeds. In reality, trajectory predictions will contain errors. However, these simple relations can be used for a short time interval repeatedly. The denominator

$$1 + \sigma^2 - 2\sigma \cos(\Psi_1^0 - \Psi_2^0) > (1 - \sigma)^2 \quad (26)$$

is always positive except when the two aircraft are flying on top of each other at all times ($\sigma = 1$ and $\Psi_1^0 = \Psi_2^0$).

We now study strategies for conflict resolutions. For good numerical properties, we normalize all distances by $\frac{V_1^2}{g}$ and all times by $\frac{V_1}{g}$:

$$\bar{x}_1 = \frac{x_1}{V_1^2/g} \quad \bar{y}_1 = \frac{y_1}{V_1^2/g} \quad (27-1)$$

$$\bar{x}_2 = \frac{x_2}{V_1^2/g} \quad \bar{y}_2 = \frac{y_2}{V_1^2/g} \quad (27-2)$$

$$\tau = \frac{t}{V_1/g} \quad \bar{r}_{\text{req}} = \frac{r_{\text{req}}}{V_1^2/g} \quad (27-3)$$

Representing the differentiation with respect to τ by prime,

$$\frac{d(\ )}{d\tau} \triangleq (\ )' \quad (28)$$

we have

$$\bar{x}_1' = \sin \Psi_1 \quad (29)$$

$$\bar{y}_1' = \cos \Psi_1 \quad (30)$$

$$\Psi_1' = \tan \phi_1 \quad (31)$$

$$\bar{x}_2' = \sigma \sin \Psi_2 \quad (32)$$

$$\bar{y}_2' = \sigma \cos \Psi_2 \quad (33)$$

$$\Psi_2' = \frac{1}{\sigma} \tan \phi_2 \quad (34)$$

The avoidance of a predicted conflict can be formulated as an optimal control problem. We seek to determine optimal aircraft maneuvers for both aircraft that deviate the least from their original flight paths while avoiding the conflict, and return to their original paths at a **fixed** final time τ_f. Specifically,

$$\min_{\phi_1, \phi_2} \ I = \int_{\tau_0}^{\tau_f} \left[K_1(\Psi_1 - \Psi_1^0)^2 + K_2(\Psi_2 - \Psi_2^0)^2 \right] d\tau \quad (35)$$

subject to equations of motion in Eqs. (29) through (34), the following path constraints for conflict avoidance and bank angle limit

$$(\bar{x}_1 - \bar{x}_2)^2 + (\bar{y}_1 - \bar{y}_2)^2 \geq \bar{r}_{\text{req}}^2 \quad (36)$$

$$|\phi_1| \leq \phi_{\max} \quad (37)$$

$$|\phi_2| \leq \phi_{\max} \quad (38)$$

and terminal constraints

$$\Psi_1(\tau_f) = \Psi_1^0 \quad (39)$$

$$\Psi_2(\tau_f) = \Psi_2^0 \quad (40)$$

$$[\bar{x}_1(\tau_f) - \bar{x}_1^0] \cos \Psi_1^0 = [\bar{y}_1(\tau_f) - \bar{y}_1^0] \sin \Psi_1^0 \quad (41)$$

$$[\bar{x}_2(\tau_f) - \bar{x}_2^0] \cos \Psi_2^0 = [\bar{y}_2(\tau_f) - \bar{y}_2^0] \sin \Psi_2^0 \quad (42)$$

These terminal constraints specify that both aircraft return to their corresponding original courses. K_1 and K_2 are positive weightings in the performance index.

Strictly speaking, another terminal constraint is needed to insure that the separation distance between the two aircraft increase at the end of the solution interval. Mathematically, at $\tau = \tau_f$,

$$\frac{d}{d\tau} \sqrt{(\bar{x}_1 - \bar{x}_2)^2 + (\bar{y}_1 - \bar{y}_2)^2} > 0 \quad (43)$$

This constraint can often be verified by checking the solutions and by properly selecting the final time. As a result, this condition is not directly enforced in the optimization process.

Numerical solutions of the above problem are obtained using a *Sequential Gradient Restoration Algorithm* developed in Ref. 37. Before any solutions are computed, a conflict must have been detected using Eqs. (21) and (22).

Numerous cases are computed. In the examples presented below, both aircraft have a cruise speed of $V_1 = 450$ knots, except in the example where effect of speed difference is examined. Both aircraft are about 15 nmi from the predicted conflict point. The maximum allowed bank angle is assumed to be $\phi_{\max} = 40^o$. Avoidance maneuvers are allowed to start as soon as a conflict is detected.

1. **Effect of intersecting angle** is studied by calculating the optimal trajectories for $\theta = \Psi_2^0 - \Psi_1^0 = 90^o, 30^o, 120^o$.

1a. We first assume $\theta = 90^o$ with $\Psi_1^0 = 90^o$ and $\Psi_2^0 = 180^o$. Using Eqs. (21) and (22), a collision is predicted at 120 seconds later. Figures 1 and 2 pertain to specified final time of 180 seconds and 240 seconds, respectively. In both cases, aircraft turn **in the same way** (left or right) to follow straight lines toward some target points until the potential conflict is passed. These roughly constant heading histories are caused by the performance index choice in Eq. (35). With $t_f = 240$ seconds, both aircraft also follow straight lines after the potential conflict has passed. As the required time to complete the whole maneuver becomes shorter, as with $t_f = 180$ seconds, aircraft are forced to return to their original headings sooner.

1b. Next, we assume $\theta = 30^o$ with $\Psi_1^0 = 90^o$ and $\Psi_2^0 = 120^o$. A potential conflict is predicted at 120 seconds later. Because of the smaller intersecting angle, the two aircraft come into each other's protected zone sooner than in the above case. As a result, one aircraft must delay its arrival at the predicted conflict point. Figure 3 shows the optimal solution with $t_f = 180$ seconds, where one aircraft performs path stretching, while the other aircraft performs just a little heading change. We solved the same problem with a different set of initial guesses. A similar result occurs (Figure 4), except now aircraft two performs the path stretching maneuver. Therefore, with a small intersecting angle, one aircraft must make the sacrifice through path stretching to avoid the potential conflict with another aircraft. With equal speeds, either aircraft can perform the path stretching.

1c. Finally, we assume $\theta = 120^o$ with $\Psi_1^0 = 90^o$ and $\Psi_2^0 = 240^o$. Figures 5 and 6 show the optimal avoidance maneuvers with $t_f = 200$ seconds and $t_f = 240$ seconds. In these solutions, the two aircraft perform similar rolling motions to avoid the conflict. Aircraft trajectories can be approximated by two straight lines connected at target points. Each aircraft has its own target point. Again, the two aircraft turn the same way (left or right) by about the same amount.

2. **Effect of speed difference** is studied by repeating all of the above cases with $V_2 = 0.8V_1$. Figure 7 presents the optimal trajectories for $\theta = 90^o$, while Figures 8 and 9 show the cases of $\theta = 30^o$ and $\theta = 120^o$, respectively. Conclusions are the same as in the case of equal speeds with one important exception. When the intersecting angle is small, it is always the slower aircraft that makes the path stretching (regardless of different initial guesses).

Goal Conflict

We now consider a scenario where two aircraft are trying to turn onto the same route. Mathematically, a route with a specified heading can be described by a position and a heading

$$x_f, y_f, \Psi_f \tag{44}$$

A goal conflict occurs if the two aircraft try to pass through the same point (x_f, y_f) in close times. It is another form of loss of minimum separation conflict.

We formulate the following problem for optimal goal conflict avoidance. For a specified final time τ_f, determine optimal controls ϕ_1 and ϕ_2 from

$$\min_{\phi_1,\phi_2} \; I = \int_{\tau_0}^{\tau_f} [K_{u1}\phi_1^2 + K_{u2}\phi_2^2]\, d\tau \tag{45}$$

where K_{u1} and K_{u2} are positive weightings, subject to equations of motion in Eqs. (29) through (34), conflict-free path constraints in Eqs. (36) through (38), and the following terminal constraints,

$$\Psi_1(\tau_f) = \Psi_f \tag{46}$$

$$[\bar{x}_1(\tau_f) - \bar{x}_f]\cos\Psi_f = [\bar{y}_1(\tau_f) - \bar{y}_f]\sin\Psi_f \tag{47}$$

$$\Psi_2(\tau_f) = \Psi_f \tag{48}$$

$$\bar{x}_2(\tau_f) = \bar{x}_f \tag{49}$$

$$\bar{y}_2(\tau_f) = \bar{y}_f \tag{50}$$

In the terminal constraints, Eqs. (46) and (47) direct aircraft one to turn to the route with specified heading, while Eqs. (48) through (50) direct aircraft two to turn to the same route, **as well as** provide a meaningful ending to the optimization program. Please note that the performance index in Eq. (45) minimizes bank angles during the maneuver.

In the following example, initial conditions of aircraft one are: $x_1^0 = 0$, $y_1^0 = 0$, and $\Psi_1^0 = 90^o$. Initial conditions of aircraft two are $x_2^0 = 15$ nmi, $y_2^0 = 15$ nmi, and $\Psi_1^0 = 180^o$. The two aircraft have the same speed: $\sigma = 1$. The final specified route begins at $x_f = 20$ nmi and $y_f = -5$ nmi, with a heading of $\Psi_f = 135^o$. If no avoidance maneuvers are taken, a conflict is predicted when the two aircraft enter the route.

Figure 10 shows the optimal avoidance trajectories. The aircraft must resolve the potential conflict before coming to the same route. Actually, goal conflict dictates that one aircraft must take an effort in delaying its arrival at the route through path stretching. Because these aircraft normally roll in certain directions to get onto the route, the path stretching is accomplished by the chosen aircraft to roll in a direction opposite to its roll into the route.

Conclusions

Based on the above results, the following conclusions can be made. These conclusions are directly related to the choices of performance indices in Eqs. (35) and (45).

1. If the intersecting angle is large, optimal avoidance maneuvers for two aircraft involved in a potential conflict are similar. In particular, the two aircraft turn the same way (left or right) to avoid the potential conflict. Each aircraft's trajectory roughly consists of two straight line paths connected by a target point. For an approximate strategy of conflict resolution, therefore, one only needs to determine a target point for each aircraft. This is the same as inserting an intermediate waypoint.
2. If the intersecting angle is small, optimal avoidance maneuvers call for one aircraft to delay its arrival at the predicted conflict point while the other aircraft follows a two-straight line path. When the two aircraft have the same speeds, either aircraft can perform the path stretching to avoid the other aircraft. When the speeds are not equal, however, the slower aircraft should stretch its path to delay its arrival at the predicted conflict point.
3. When two aircraft try to converge onto the same route with a specified heading, one aircraft has to slow down to stay away from the other aircraft. Similar conclusions hold as in the case of small intersecting angles.

Acknowledgments

This work was supported by NASA Ames under Contract NAS2-14290.

References

[1] Nolan, Michael S., *Fundamentals of Air Traffic Control*, Wadsworth Pub., Belmont, CA, 1990.

[2] RTCA, "Free Flight Implementation," Final Report of RTCA Task Force 3, Washington D. C., October 26, 1995.

[3] Erzberger, H., Davis, T. J., and Green, S., "Design of Center-TRACON Automation System," presented at the AGARD Guidance and Control Symposium on *Machine Intelligence in Air Traffic Management*, 11-14 May 1993, Berlin, Germany.

[4] Erzberger, H., "Design Principles and Algorithms for Automated Air Traffic Management," AGARD Lecture Series No. 200 on *Knowledge-Based Functions in Aerospace Systems*, Madrid, Paris, and San Francisco, Nov. 1995.

[5] Davis, T. J., Erzberger, H., Green, S. M., and Nedell, W., "Design and Evaluation of Air Traffic Control Final Approach Spacing Tool," *Journal of Guidance, Control, and Dynamics*, Vol. 14, No. 4, July-August, 1991, pp. 848-854.

[6] Bergh, C. P., Krzeczowski, K. J., and Davis, T. J., "TRACON Aircraft Arrival Planning and Optimization through Spatial Constraint Satisfaction," *Air Traffic Control Quarterly*, Vol. 3, No. 2, 1995, pp. 117-138.

[7] Erzberger, H., "Automation of On-Board Flight-path Management," NASA TM 84212, December 1981.

[8] Erzberger, H., and Tobias, L., "A Time-Based Concept for Terminal-Area Traffic Management," NASA TM 88243, April 1986.

[9] Slattery, R. A., and Zhao, Y., "Trajectory Synthesis for Air Traffic Automation," *Journal of Guidance, Control, and Dynamics*, Vol. 20, No. 2, March-April, 1997.

[10] Erzberger, Heinz, and Engle, Laurie, "Conflict Detection Tool," Addendum to NASA TM-102201, October 1989.

[11] Paielli, Russell A., and Erzberger, Heinz, "Conflict Probability Estimation for Free Flight," *Journal of Guidance, Control, and Dynamics*, Vol. 20, No. 3, May-June, 1997, pp. 588-596.

[12] Lee, H. Q., "An Advisory System for Predicting and Resolving Airspace Violations Based on Four-Dimensional Guidance Techniques," *Proceedings of AIAA Guidance, Navigation and Control Conference*, August 1987, Monterey, CA, AIAA Paper No. 87-2525, pp. 1197-1209.

[13] Slattery, Rhonda, and Green, Steve, "Conflict-Free Trajectory Planning for Air Traffic Control Automation," NASA TM-108790, January 1994.

[14] Menon, P. K. A., "Air Traffic Conflict Resolution Using Modern Control Theory," report of *Optimal Synthesis*, 908 E. Meadow Drive, Palo Alto, CA 94303, October 15, 1992.

[15] Chatterji, G., Sridhar, B., and Bilimoria, K., "En-Route Flight Trajectory Prediction for Conflict Avoidance and Traffic Management," AIAA GNC, August 1996, San Diego, CA, *AIAA 96-3766*.

[16] Bilimoria, K., Sridhar, B., and Chatterji, G., "Effects of Conflict Resolution Maneuvers and Traffic Density on Free Flight," presented at the *the AIAA Guidance, Navigation and Control Conference*, July 29-31, 1996, San Diego, CA.

[17] Morrel, J. S., "The Mathematics of Collision Avoidance in the Air," *Journal of Navigation*, Vol. 11, No. 1, pp. 18-28, Jan. 1958.

[18] Merz, Antony W., "Optimal Aircraft Collision Avoidance," *Proceedings of the Joint Automatic Control Conference*, 1973, pp. 449-454,

19 Merz, A. W, "Maximum-miss aircraft collision avoidance," Lockheed Missiles and Space Co, Palo Alto, CA, USA, *Lecture Notes in Control & Information Sciences*. Vol. 151, 1991, pp. 271-280. *Proceedings of the 3rd Workshop on Control Mechanics*,

20 Friedman, Moshe F, "Decision Analysis and Optimality in Air Traffic Control Conflict Resolution. I. Optimal Timing of Speed Control in a Linear Planar Configuration," Univ of Miami, Coral Gables, FL, USA. *Transportation Research* Part B-Methodological, Vol. 22B, No. 3, Jun 1988, pp. 207-216.

21 Friedman, M. F., "Decision Analysis and Optimality in Air Traffic Control Conflict Resolution: II. Optimal Heading (Vectoring) Control in a Linear Planar Configuration," *Transportation Research*, Vol. 25B, No. 1, pp.39-53, 1991.

22 Krozel, Jimmy, Mueller, Tysen, and Hunter, G., "Free Flight Conflict Detection and Resolution Analysis," presented at the *the AIAA Guidance, Navigation and Control Conference*, July 29-31, 1996, San Diego, CA.

23 Tomlin, C., Pappas, G. J., and Sastry, S., "Conflict Resolution for Air Traffic Management: A Case Study in Multi-Agent Hybrid System," Memorandum No. UCB/ERL M96/38, July 1996, Electronics Research Lab., University of California, Berkeley.

24 Holt, J. M., and Marner, G. R., "Separation Theory in Air Traffic Control System Design," *Proceedings of the IEEE*, Vol. 58, No. 3, pp. 369-376. March, 1970.

25 Alliot, Jean-Marc. Gruber, Herve. Joly, Georges. Schoenauer, Marc, "Genetic algorithms for solving air traffic control conflicts," *Proceedings of the 9th Conference on Artificial Intelligence for Applications*, Orlando, FL, USA, 1993, pp. 338-344.

26 Latombe, Jean-Claude, *Robot Motion Planning*, Kluwer Academic Publishers, 1991.

27 Canny, J. F., *The Complexity of Robot Motion Planning*, The MIT Press, 1988.

28 Eby, M. S., "A Self-Organizational Approach for Resolving Air Traffic Conflicts," *the Lincoln Laboratory Journal*, Vol. 7, No. 2, 1994, pp. 239-254.

29 Miura, Akira. Morikawa, Hiroyuki. Mizumachi, Moriyuki, "Aircraft collision avoidance with potential gradient - ground-based avoidance for horizontal maneuvers, " *Electronics & Communications in Japan*, Part Iii: Fundamental Electronic Science. Vol. 78, No. 10, Oct 1995, pp. 104-114.

30 Chen, Y.-B., and Inselberg, "Conflict Resolution for Air Traffic Control," Tech. Rep., USC-CS-93-543, Computer Science Department, University of Southern California, 1993.

31 Williamson, T. and Spencer, N. A., "Development and Operation of the Traffic Alert and Collision Avoidance System (TCAS)," *Proceedings of the IEEE*, Vol. 77, No. 11, pp. 1735-1744, Nov., 1989.

32 Ball, D. W., Altman, S. I., and Wood, M. L., "TCAS: Maneuvering Aircraft in the Horizontal Plane," *the Lincoln Laboratory Journal*, Vol. 7, No. 2, pp. 295-312, 1994.

33 Harman, W. H., "TCAS, A System for Preventing Midair Collisions," *the Lincoln Laboratory Journal*, Vol. 2, No. 3, pp. 437-457, 1989.

35 Niedringhaus, W. P., "Maneuver Option Manager: Automated Simplification of Complex Air Traffic Control Problems," *IEEE Transactions on Systems, Man, and Cybernetics*, Vol. 22, No. 5, Sept./October 1991, pp. 1047-1057.

36 Wangermann, J. P. and Stengel, R.F., "Principled Negotiation Between Intelligent Agents: A Model for Air Traffic Management," AIAA GNC, San Diego, August, 1996.

37 Miele, A., Damoulakis, J. N., Cloutier, J. R., and Tietze, J. L., "Sequential Gradient Restoration Algorithm for Optimal Control Problems with Nondifferential Constraints," *Journal of Optimization Theory and Applications*, Vol. 13, No. 2, Feb. 1974, pp. 218-255.

Fig. 1 90 deg Intersection, t_f = 180 s

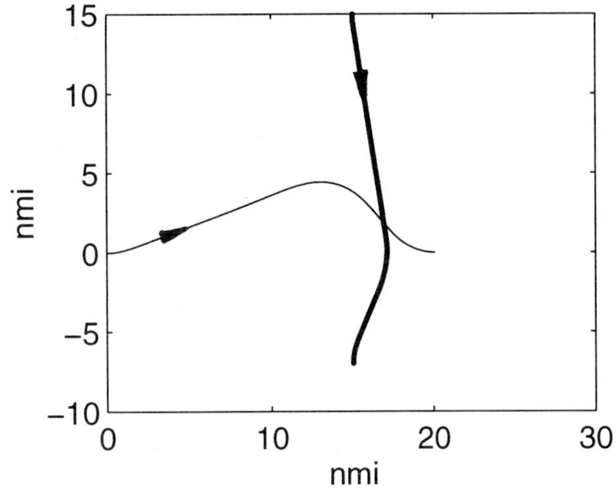

Fig. 2 90 deg Intersection, t_f = 240 s

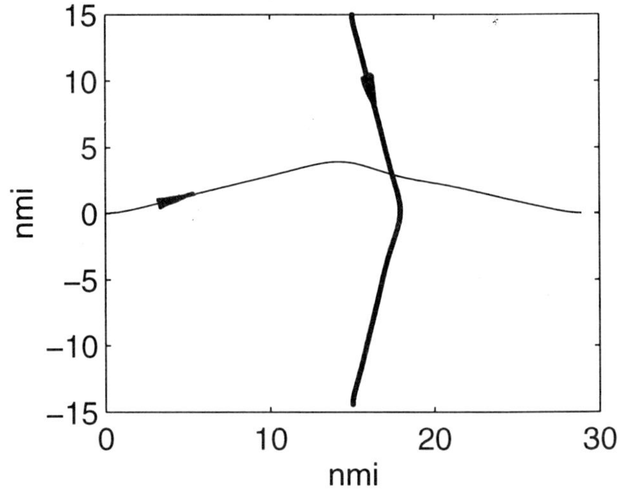

Fig. 3 30 deg Intersection, t_f = 180 s

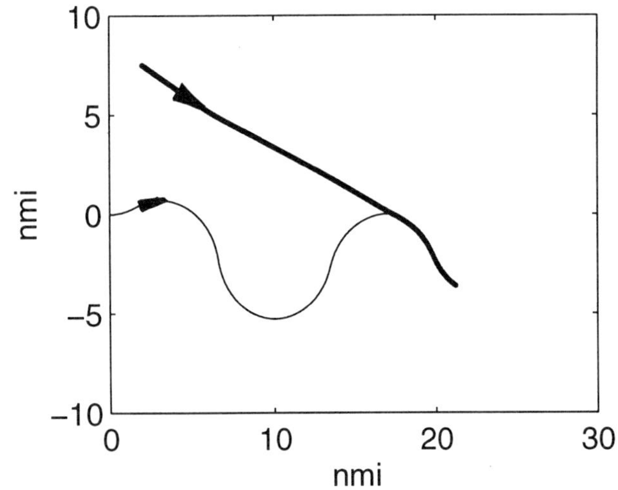

Fig. 4 30 deg Intersection, t_f = 180 s, II

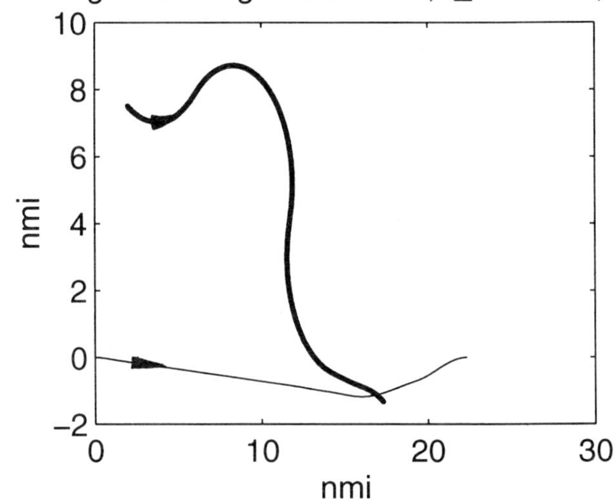

Fig. 5 120 deg Intersection, t_f = 200 s

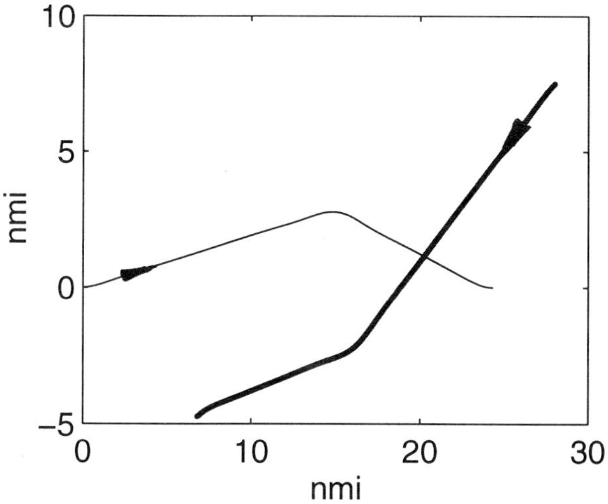

Fig. 6 120 deg Intersection, t_f = 240 s

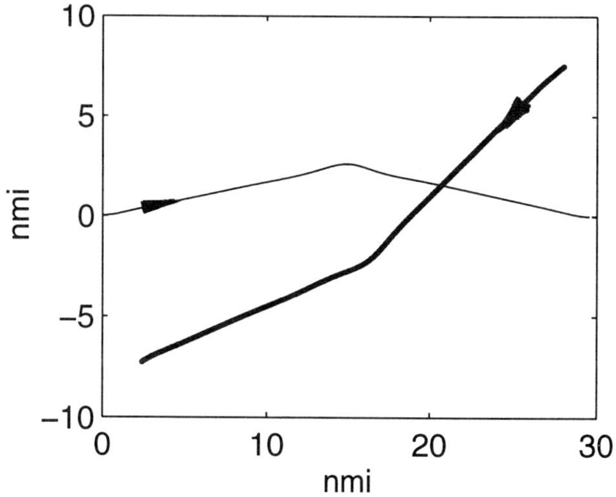

Fig. 7 90 deg Intersection, V2/V1 = 0.8

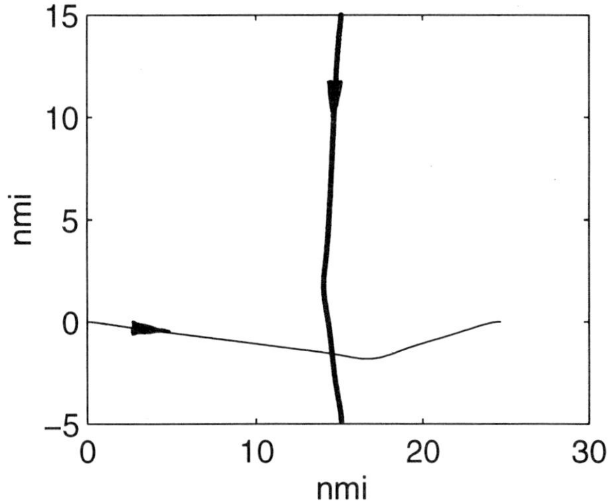

Fig. 8 30 deg Intersection, V2/V1 = 0.8

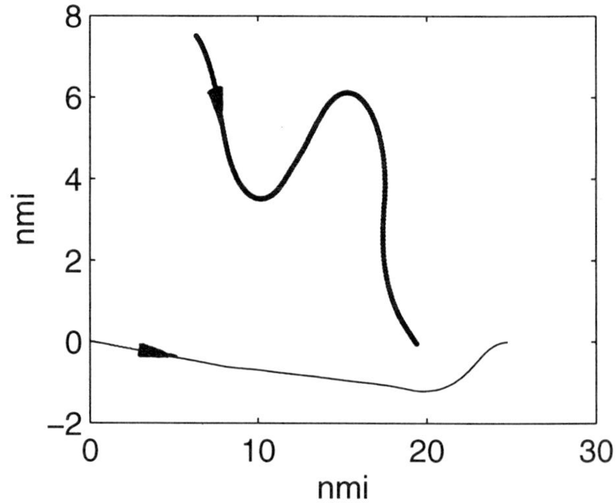

Fig. 9 120 deg Intersection, V2/V1 = 0.8

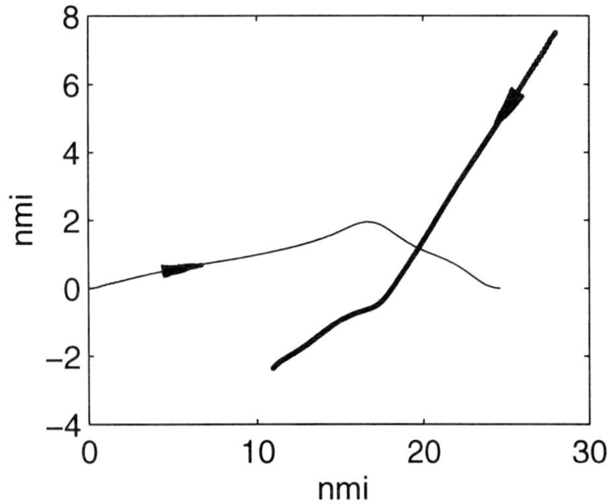

Fig. 10 Converging Conflict, t_f = 220 s

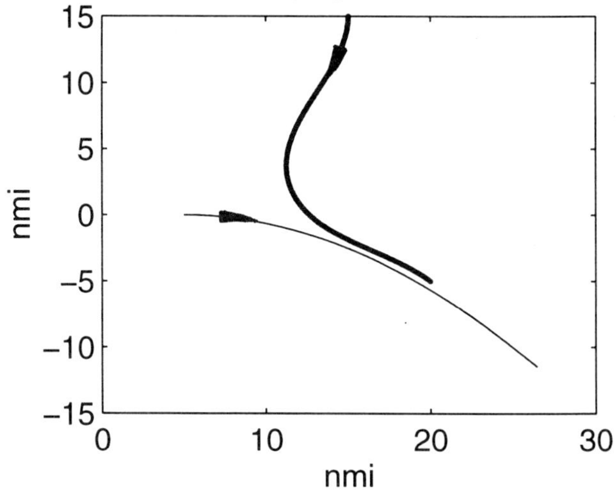

AIAA-97-3548

Fiber Optic Gyro Development for Precision Guidance and Navigation Applications

Joung C. Ha
Wright Laboratory, Avionics Dir.
Wright-Patterson AFB, Ohio

Ed Kanegsberg and Ralph A. Patterson
Litton Guidance and Control Systems Division
Woodland Hills, California

Abstract

In support of objectives to exploit recent technical advances in fiber optic gyroscope technology, the US Air Force, with support from the US Navy, is conducting an exploratory development program to develop and demonstrate an affordable, high accuracy, strapdown, solid-state gyroscope for precision navigation and guidance applications. The precision fiber optic gyro (PFOG) must demonstrate high accuracy performance under high dynamic environments, potential for high reliability (>100,000 hours) and low cost in a small, low power, lightweight package requiring no periodic maintenance. Performance goals for PFOG are 0.001 degrees per hour bias, 0.0005 degrees per square-root hour random walk and 5 parts per million scale factor.

Litton Guidance and Control Systems has been awarded a contract for the design, development, test and demonstration of a PFOG instrument. This paper describes the overall PFOG program and provides a review of Litton's technical approach including test results. A breadboard model PFOG has been fabricated and is undergoing performance and developmental testing. Progress toward meeting the PFOG performance goals is presented.

Introduction

The defense community has a need for smaller, more reliable inertial sensors for platforms with a precision strike capability. Fiber optic gyroscope technology can be utilized to meet this need. The scope of the present program, sponsored by the U.S. Air Force with support from the U. S. Navy, is an exploratory development effort to design, develop, test and demonstrate an affordable, high accuracy, strapdown, solid-state, precision fiber optic gyro (PFOG) for use in precision navigation and guidance applications. The technical and design approach for the PFOG have been previously presented[1]. Litton's qualifications to conduct this development effort involve a synergy among the establishment of a tactical grade Interferometric Fiber Optic Gyro (IFOG) product now in production, development of navigation grade IFOG designs over a fifteen year period, development and implementation of automated production processes, and a continuing IRAD program of IFOG and component design development.

The approach which offers the greatest potential for meeting the future needs of high-accuracy guidance and navigation applications is the IFOG. Recent technical advances at Litton have supported this approach [2,3].

Objectives

The objectives of the present phase of the PFOG program are to demonstrate the ability of IFOG technology to meet the requirements of precision strike navigation applications. These include low cost, small size, light weight, high reliability and high performance. The demonstration is being done by developing a PFOG breadboard design, building and testing a PFOG breadboard instrument to provide design validation data and establish required improvements. A brassboard specification, a production plan and a radiation survivability assessment will provide a detailed data base for the initiation of the next phase of the program.

Applications

The precision fiber optic gyro program is to develop a new solid state optical rate sensor which achieves performance in the 0.001 deg/hr range over full military environment. This is an order of magnitude better performance than the best existing fiber gyros and compares favorably with the very best ring laser gyroscopes, which unlike fiber gyros, require high voltage power supplies and heavy, low-coefficient-of-thermal-expansion, precision machined glass blocks. This new fiber optic sensor will have potential application to many aerospace platforms - both military and civilian - requiring highly reliable, light weight, lower cost inertial navigation equipment, including manned aircraft, unmanned aerial vehicles, and cruise missiles.

Technical Approach

The technical approach to this phase of the PFOG program consists of detailed component trade studies, the design, build, test and iteration of a PFOG breadboard as required to meet the PFOG performance objectives. A cooperative program between the PFOG program efforts and Litton's continuing IRAD component development tasks[2] as well as synergy with other Litton fiber optic gyro programs is a key to this effort. The component trade studies have established the initial PFOG breadboard optical design and flowdown to component requirements. Components which can achieve these requirements will continue to be developed on Litton's IRAD program and provided to the PFOG breadboard for test and evaluation within the breadboard architecture. Rigorous functional testing as well as developmental testing on the breadboard will be utilized to determine any component performance deficiencies as well as refinements to the optical design and or flowdown requirements. Iterations of this process will continue through the program resulting in the development of a brassboard specification and a production plan for the next phase of the program. A radiation survivability assessment of the final breadboard iteration will also be completed.

Design approach

The optical architecture of Litton's navigation grade IFOG is being upgraded to provide the performance needed for PFOG. The PFOG architecture [1] utilizes the same factory processes as Litton's existing production IFOGs and therefore lends itself to low cost production. Figure I is a block diagram of Litton's navigation grade IFOG optical architecture. The long wavelength, broadband optical source provides optical energy for the gyro. The source optical power output is being improved significantly to achieve the random walk goal of 0.0005 degrees per square-root hour. Its wide optical bandwidth will be maintained to minimize bias uncertainty of the instrument due to polarization nonreciprocity [4], and ensure low random walk. The stability of the output wavelength versus temperature will be improved to meet the 5 parts per million (ppm) scale factor error requirement. Less than 10 ppm scale factor residual in a dynamic thermal environment was previously demonstrated with an IFOG utilizing this source [4,6]. Polarization and pigtail cross coupling performances of the Multifunction Integrated Optical Chip (MIOC) are being upgraded to PFOG requirements though parallel Litton IRAD efforts. Other IRAD efforts include coil and fiber design to reduce sensitivities to thermal environment.

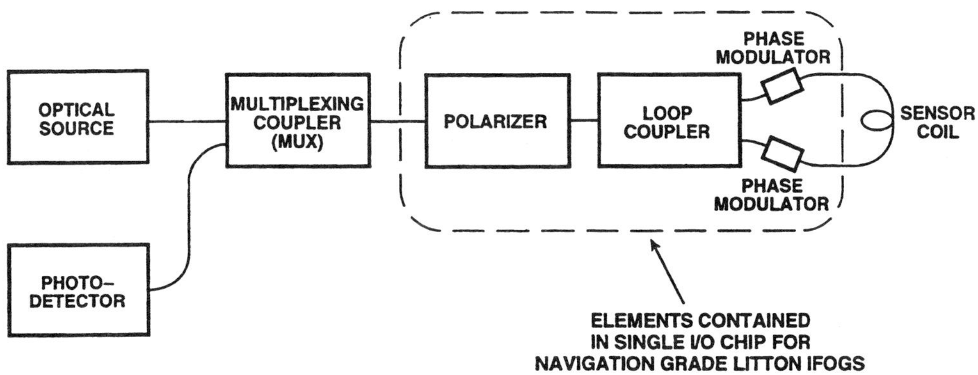

Figure 1. IFOG Block Optical Architecture Used in Litton Inertial Navigation Systems

Breadboard

The most important tool in the PFOG development effort is the breadboard gyro. It has been designed to allow the evaluation of individual components and their effects on PFOG performance. The first iteration breadboard package is shown in Figure 2 and features a small fiber coil of less than three inches in diameter. Testing of the PFOG breadboard is being accomplished to evaluate components and error sources through full characterization tests and developmental testing. The results of the breadboard tests and analysis will be used to iterate the breadboard design and the component requirements as required. The PFOG breadboard will then be upgraded and testing repeated.

Program status

The detailed design trade-off studies have been completed and the results incorporated into the design of the first iteration of the breadboard PFOG. A preliminary design review was successfully completed and optical breadboard components were fabricated/procured and put through component level screening tests to ensure compliance with the design requirements. The components were assembled into the optical breadboard, were integrated with the electronics, test site and software and testing begun.

Figure 2. Photograph of Precision Fiber Optic Gyro Breadboard Package

Preliminary test results for angle random walk and bias stability were very encouraging, though all requirements are not yet demonstrated. Figure 3 is a plot showing angle random walk of 0.0007 degrees/root-hour. Angle random walk was measured by collecting bias data at constant temperature and analyzing the data using an Allan variance technique which is implemented in the computer program called 'AutoFit'[7]. These results demonstrate the potential to achieve the PFOG program angle random walk goals. This result agrees well with an analytical model. As lower loss components and a higher optical power source become available within the next few months, the angle random walk will meet PFOG goals.

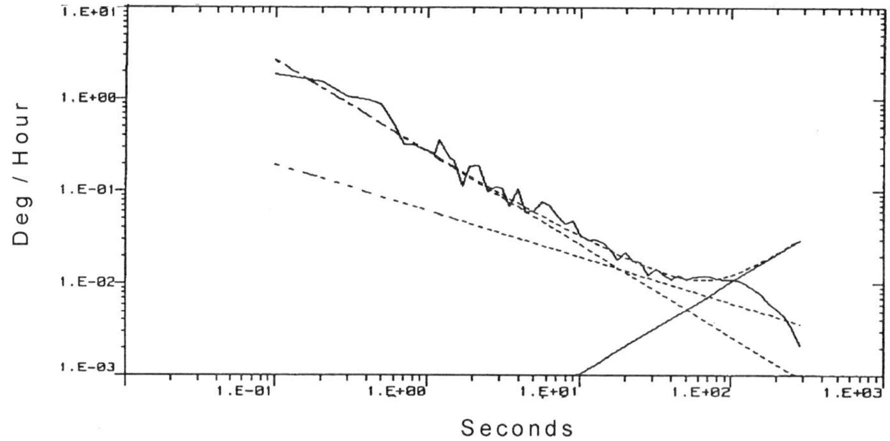

Figure 3. Autofit Alan Variance Analysis Angle Random Walk (ARW) = 0.00072 $°/\sqrt{Hr}$

The bias stability over time was measured with the PFOG breadboard mounted in a computer controlled oven. Figure 4 is a plot of a constant temperature run. Smoothing times have been made long to average the random walk variations and expose any systematic bias changes.

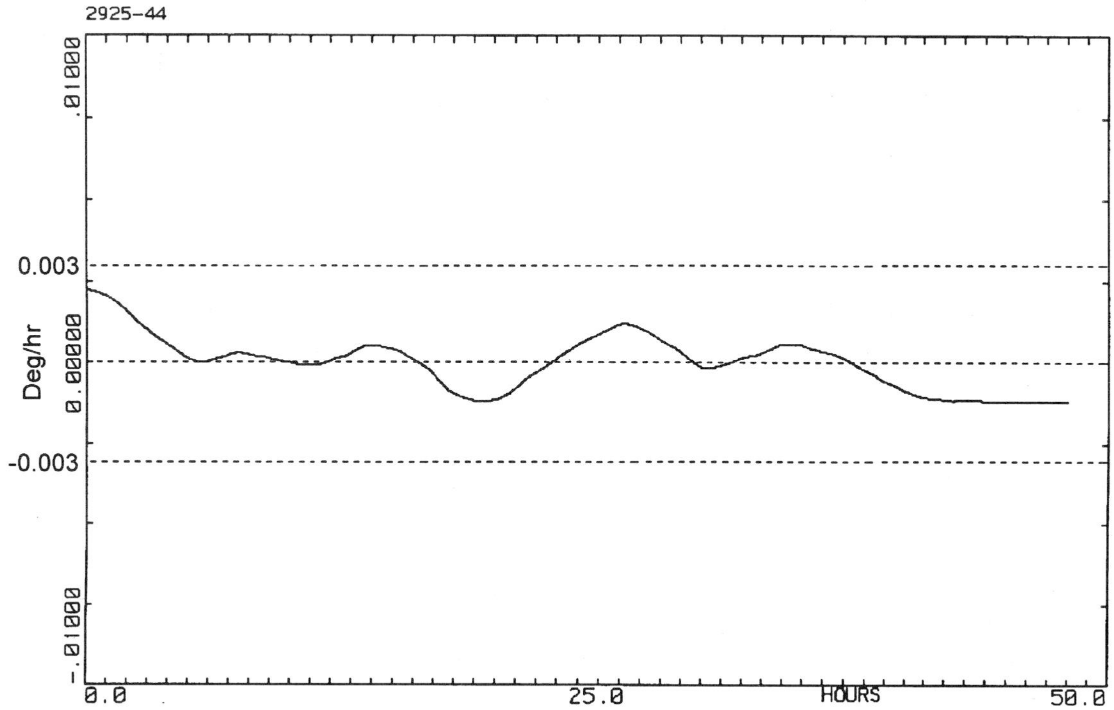

Figure 4. 48-hour constant temperature bias stability (4-hour filtering) sigma=0.00081 deg/hr (PFOG 3-sigma limits of 0.003 deg/hr shown)

Due to some bias instabilities associated with the higher optical power in this instrument, a redesign of the original source architecture was suggested. Analysis shows that this modified approach will not only meet the PFOG performance goals but make it more probable that an even smaller diameter coil can be incorporated into the next iteration of the PFOG. This modified approach was presented at a successful critical design review. The modified source is currently being assembled and testing is expected by late summer, 1997.

Summary

The U.S. Air Force, with support from the U.S. Navy, and its contractor Litton Guidance and Control Systems Division are conducting an exploratory development program to satisfy the needs of the precision guidance and navigation applications. The PFOG will provide a solid state, high reliability solution for many future applications. A program of detailed design, build, test and design iteration activities has been developed and is underway. A critical design review was held for the design approach. Detailed performance characterization tests are underway.

Acknowledgments

This work is supported by the United States Air Force, Wright Laboratories, Avionics Directorate, with support from the United States Naval Air Systems Command, PMA-209. The authors wish to express their appreciation to John Cappi, Amado Cordova, Eric Goldner, Leo Lam, Dave Libman, Tom McLean, John Rahn, Dave Rozelle, James Steele, Charles Thorsen, Fernando Torres and Tak Yoneyama for their contributions to the design, fabrication and test of the breadboard PFOG.

References

1. J. C. Ha and R. A. Patterson, “Precision Fiber Optic Gyro Development”, presented at the 23rd Joint Services Data Exchange for Guidance, Navigation and Control , Orlando, FL (1996)

2. A. Cordova , R.A. Patterson, L. Lam, J. Rahn, and D. Rozelle, “ Recent Advances in Navigation Grade Interferometric Fiber Optic Gyroscopes”, presented at the 23rd Joint Services Data Exchange for Guidance, Navigation and Control , Orlando, FL (1996)

3. A. Cordova , R. Patterson, E. Goldner, D. Rozelle, “Interferometric fiber optic gyroscope with inertial navigation performance over extended dynamic environments.”, SPIE 2070 Fiber Optic and Laser Sensors XI (1993) pp. 164-180.

4. R.A. Patterson, D.M. Rozelle, A. Cordova and E.L. Goldner, "Inertial Navigation IFOG Performance ver Static and Dynamic Environments," presented at the 22nd Joint Services Data Exchange for Guidance, Navigation and Control, Scottsdale, AZ (1994).

5. G. Pavlath, "Fiber Optic Gyro Evolution: Concept to Product," presented at the AIAA Guidance and Control Conference, Hilton Head, SC (1992).

6. J. DeFato and R.A. Patterson, "Fiber Optic Gyro Development for Navy Shipboard Inertial Reference System Application," Proceedings of the Seventeenth Biennial Guidance Test Symposium, 78 (1995).

7. D. Sargent and B. 0. Wyman, "Extraction of Stability Statistics from Integrated Rate Data," AIAA Guidance and Control Conference, Danvers, MA (1980).

AIAA-97-3550

VERSATILE MICROMECHANICAL VIBRATION SENSOR AND APPLICATIONS

Jonathan Bernstein

and

Christopher Trainor*

The Charles Stark Draper Laboratory

Cambridge, MA 02139

Abstract

Micromechanical vibration sensors have been designed and are being fabricated which cover a sensing range from 3 Hz to 10 kHz. The primary application is for an air-dropped unattended ground sensor. The sensor is housed in a hermetically sealed robust package to ensure function in harsh environments. Special attention is paid to the lowest possible noise floor resulting in a sensor which approaches geophone performance at a much smaller size and weight. Applications include: canard turbulence indicator, tank locator, ground-based battlefield acoustics, miniature geophones for nuclear test ban monitoring, underground facility locator (both during construction and operation), smart skins, health monitoring of space structures and micro g environment control, directional hydrophones for sonobuoys and intelligent sensor modules. This paper discusses the current development of the Draper Laboratory vibration sensor and its applications.

Introduction

Micro-Electro-Mechanical Systems (MEMS) have been at the forefront of sensor technology for several years and will continue to provide breakthroughs in sensor technology in the years to come. MEMS technology achieves miniaturization of sensors and actuators using fabrication tools and processes originally developed for integrated circuits. These well-known techniques include photolithography, thin film deposition and etching, and diffusion and oxidation of silicon wafers. In addition, many new fabrication techniques have been added in order to provide the unique materials and mechanical structures needed for electromechanical transducers.

MEMS processes have been used to create almost every sensor imaginable, although not all have reached the stage of commercial viability. Silicon sensors are commercially available for measuring strain, temperature, light, pressure, sound, magnetic fields, acceleration, electrical power, and rotation rate. Some of these sensors include on-chip electronics, resulting in an entire system on a chip. ASICs and mixed signal technology have been developed at Draper for vibration sensors.

We distinguish vibration sensors from accelerometers because DC output accuracy (bias, drift) is not of primary concern; the noise floor of the instrument is the main focus. Although commercial silicon accelerometers are widely available, no available MEMS sensor is capable of measuring vibration levels at the nano-g/$\sqrt{\text{Hz}}$ level. Measurement of sub-μg vibration is necessary in numerous applications, most notably seismic localization of vehicles and underground facilities. The performance of the micromachined vibration sensor is predicted to approach geophone sensitivity levels, with much smaller size and weight.

Figure 1 shows a concept drawing of the sensor which is a silicon proof mass suspended by silicon springs over a glass substrate chip. Future versions of this sensor will have a symmetric top glass electrode as well as the bottom sense electrode which will allow force rebalance to be used.

A summary of the predicted specifications of the vibration sensor are given in Table 1. Calculated specifications for sensors designed with three resonant frequencies (500 Hz, 1 kHz, and 10 kHz) are given. Specifications for a small commercial geophone, GS-14-L9 are included for reference [1].

Comparison of Capacitive Vibration Sensor with Geophones

Conventional geophones are widely used for seismic measurements and have the advantages of being self powered, low noise, and a well established technology. They operate on the principle of a moving coil cutting through a magnetic field which generates a voltage. The total system power required to read out a geophone vs. a capacitive microsensor should be compared. Low power CMOS electronics are well suited to read out capacitive microsensors, hence it is not necessarily true that a geophone system will use less power than a capacitive sensor system. Commercial geophones are relatively large and heavy compared to micromachined sensors and they cannot measure frequencies down to DC.

---

* Associate Director, Instruments Division, Member AIAA. Presented as Paper 97-3550 at The AIAA Guidance, Navigation, and Control Conference, New Orleans, LA, August 11-13, 1997.

Geophones measure velocity and operate properly only *above* their resonant frequency, hence they do not measure low frequencies unless they are made very large and heavy. Capacitive vibration sensors measure *below* their resonant frequency, hence they can measure down to DC if the electronics allow it. The CMOS ASICs developed for this project allow a user selectable lower frequency limit by changing an RC time constant.

Brownian motion noise from the sensor damping is an important noise source for micromachined sensors due to the small mass and narrow capacitive sense gaps. This noise term can be minimized by packaging the sensor in a vacuum package which routinely yields sensors with a Q of 10^4 to 10^5. The Brownian motion of a mass-spring damper system with Q's of 1 and 10^4 are shown in Figure 2. This sensor has a resonant frequency of 500 Hz and a proof mass of 22 mg (5 mm X 5 mm X 0.38 mm). The proof mass is suspended by springs above a capacitor electrode on a glass substrate. The Brownian motion noise is reduced by a factor of $\sqrt{Q}$ in a vacuum. Motion at the resonance would be objectionable in an open loop system, but can be removed by force rebalancing. Note that the Brownian motion is only on the order of a few times 10^{-15} m for the high Q case. This is small enough that electronic noise determines the noise floor of the sensor.

Prototype Fabrication and Test Results

Prototypes of the vibration sensor were fabricated using silicon wafers bonded to glass substrate wafers. The proof mass and springs are etched from the single crystal silicon wafer using an ICP (Inductively Coupled Plasma) deep reactive ion etch. Test devices are shown in Figures 3 and 4 after the etch step which proceeds through the entire silicon wafer. In live devices metal electrodes are located under the proof mass for a capacitive pickoff.

Figure 5 shows test data on a pre-prototype device. The magnitude and phase are given. A sensitivity of 120 mV/g was achieved with a resonant frequency of 1 kHz. An on-chip JFET source-follower was used as buffer amplifier.

Applications

There are many applications for vibration sensors in the fields of aeronautics, astronautics, undersea, and land based measurements. Some of these are summarized below.

Undersea Applications

Pressure Gradient Hydrophones

Many acoustic sensing tasks are currently performed using arrays of hydrophones, such as towed arrays and sonobuoys. A single hydrophone can measure an acoustic pressure but cannot determine the direction of incidence of an acoustic wave without array processing. In the case of a straight line towed array there is always a cone of ambiguity which must be resolved by bending the array or taking more data after turning the ship. A set of 3 orthogonal neutrally buoyant vibration sensors of sufficient sensitivity can measure the pressure gradient [2]. The vibration sensors presented here should have a sufficiently low noise floor to be useful in undersea acoustic gradient sensing.

Smart Skins

For many undersea applications it is desirable to mount sensors in the hull of a ship. In this application the size and weight of the sensor are very important giving the micromachined sensors an advantage.

Composite skins have advantages over traditional metal structures, however, monitoring the health of the composites is more difficult. Embedded sensors and actuators may be able to use acoustic (vibration) emission and transmission to characterize a healthy structure and warn of changes prior to failure. The small size and high sensitivity of these sensors makes them ideal for this application.

Space Applications

Micro-gravity environments are achieved in space which may be useful for a variety of applications, such as growing perfect crystals or separation of large biomolecules. Optical satellites and radar antennae require precision pointing and are vulnerable to errors due to vibration and rotation. Measurement and control (or compensation) for residual vibration requires sensors which can measure below 1 μg.

Battlefield Acoustics

Ground-based vibration and air acoustic sensing can be used to locate soldiers, snipers, artillery, tracked and wheeled vehicles, and artillery. Small, sensitive MEMS sensors can be useful in detecting the magnitude, frequency and direction of incidence of seismic waves.

Underground Facility Locating

In response to overwhelming U.S. air superiority, some nations are burying their vulnerable military targets under concrete, mountains, and civilian structures such as hospitals. Seismic signature analysis can be used to locate these structures during construction or operation. Networks of widely distributed miniature seismic sensors, linked by R.F., can be used to find such facilities or track mobile missiles even under cloud cover.

Seismic arrays can also be used to monitor compliance with nuclear test ban treaties.

Health Monitoring of Dynamic Structures

Ball bearings are used in numerous dynamic structures critical to the D.O.D. such as many inertial instruments. Vibration sensors could be used as a ball bearing health monitoring device to detect changes or degradation in instrument performance. Turbo pump vibration detection and health monitoring in space applications have also been recommended as applications.

Conclusions

A new micromachined vibration sensor has been presented as well as preliminary test data and specifications. Advanced micromachining techniques are used to fabricate the proof mass assemblies including through wafer plasma etching. Calculations indicate a self noise level for a vacuum packaged sensor of 15 $ng/\sqrt{Hz}$, dominated by electronics noise. Improvements to the electronics can drop the noise floor still further. At this level of sensitivity the micromachined devices can compete with geophones which are orders of magnitude larger and heavier. The micromachined vibration sensors can also read out arbitrarily low frequencies which the smaller geophones cannot do.

A wide range of applications are expected for this device such as seismic localization, battlefield acoustics, micro-gravity measurement and control, noise and vibration reduction, smart skins, composite health monitoring, and pressure gradient sonar measurements.

Acknowledgments

The authors would like to thank Draper Laboratory for supporting this research with an IR&D grant, Jerry Sohn for developing the through wafer ICP etch, and Technicians Doug Preble, Michele Lind, Connie Cardoso, and Brenda Hugh for excellent work developing a new fabrication process.

References

1. Geospace Inc., GS-14 geophone data sheet, 1993, 7334 N. Gessner Road, Houston, TX 77040.

2. Abraham, B., in "Acoustic Particle Velocity Sensors: Design, Performance, and Applications," AIP Conference Proceedings 368, pp. 189-201.

Table 1. Preliminary Specifications for Micromachined Vibration Sensor

	Units	500 Hz Sensor	1 kHz Sensor	10 kHz Sensor	GS-14-L9 Geophone
Upper Frequency Limit	Hz	400 Hz	800 Hz	8 kHz	2 kHz
Lower Frequency Limit	Hz	3	3	3	27
Self Noise Floor (electronics noise)	ng/Rt-Hz	16.5	33	330	*1
Sensitivity w/ unity gain buffer	V/g	0.49	0.245	0.025	23.6 V/m/s
Maximum g-level (open loop)	g's	0.3	1.2	120	*2
Sensor Chip Size	mm X mm	7 X 7	7 X 7	7 X 7	n.a.
Package Volume	cm^3	0.41	0.41	0.41	5.43

*1 Output and noise are specified in velocity instead of acceleration
*2 Measurement limited by displacement of proof mass (coil).

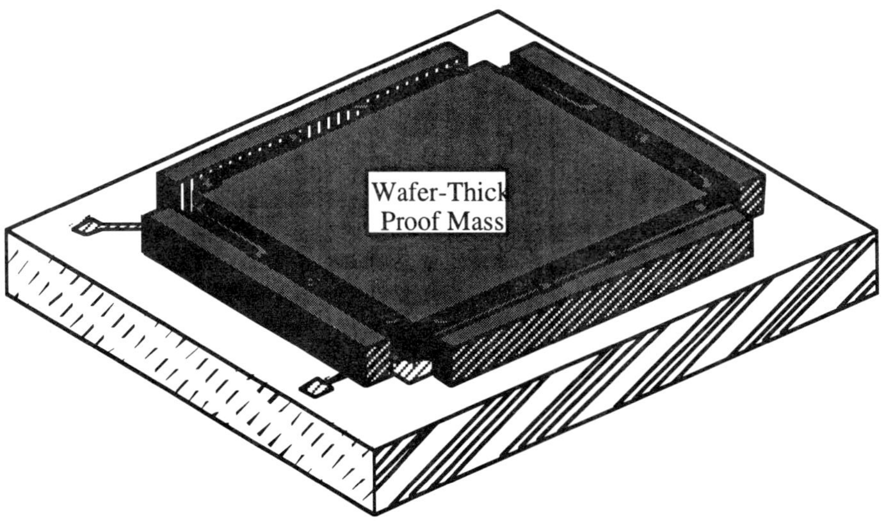

Figure 1. Capacitive micromachined vibration sensor using silicon springs and wafer thick proof mass.

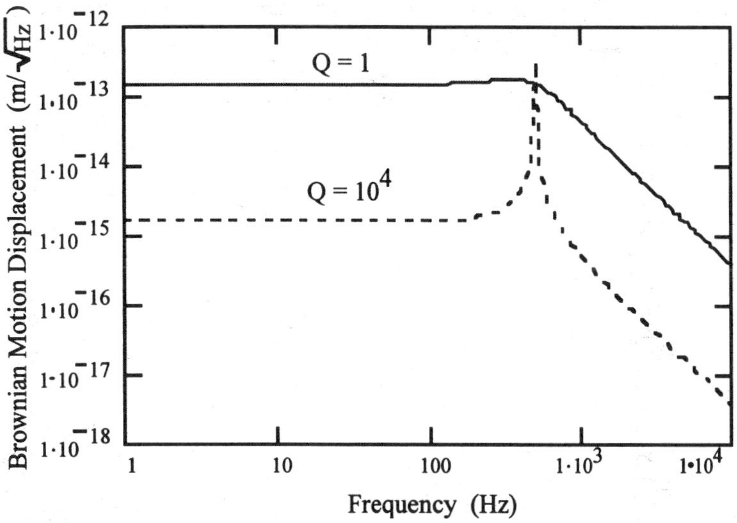

Figure 2. Brownian motion induced displacement of the proof mass of a micromachined vibration sensor for $Q = 1$ and 10^4.

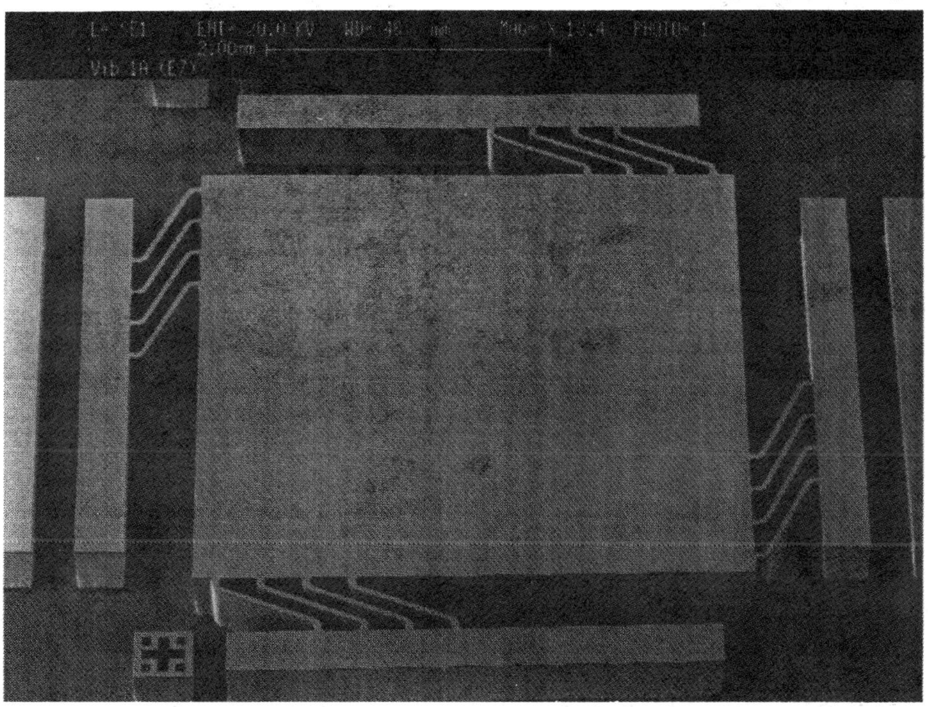

Figure 3. Vibration Sensor chip with 3 mm proof mass, at intermediate fabrication step. Structure is reactive ion etched through the thickness of a silicon wafer.

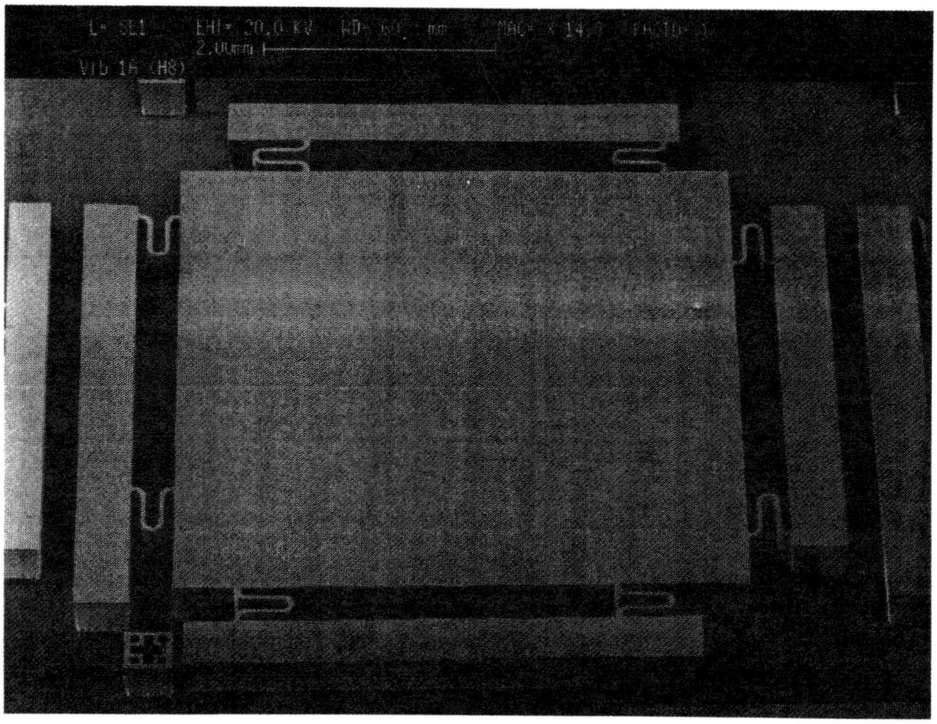

Figure 4. Vibration Sensor test chip with 5 mm proof mass, at intermediate fabrication step.

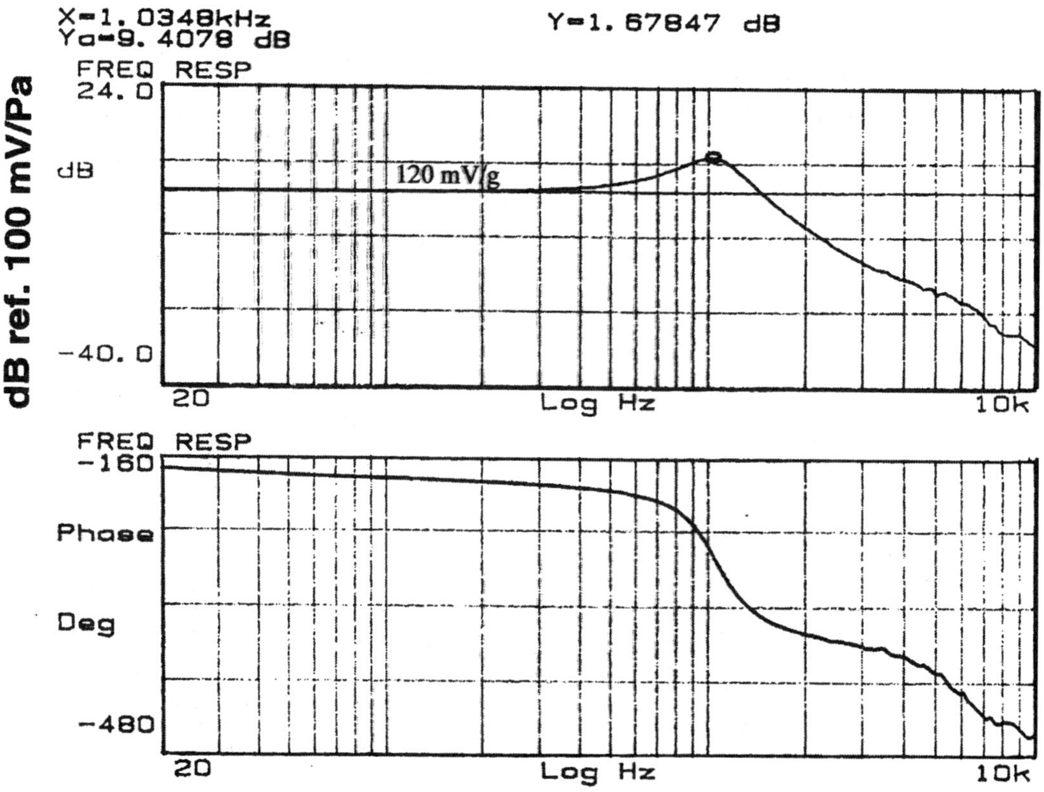

Figure 5. Frequency response of pre-prototype vibration sensor. Resonance is at 1 kHz, and sensitivity is 120 mV/Pa using unity-gain buffer amplifier.

AIAA-97-3551

Testability of Micromachined Silicon Rate Sensors

R. Jaffe, J.Heiertz, S. Ventresco
Boeing
AIAA Guidance, Navigation, and Control Conference
August 11-13 1997
New Orleans, LA

Abstract

In 1993, Boeing of Anaheim, CA, and the Charles Stark Draper Laboratories (Draper) of Cambridge, MA, entered into an alliance to bring to market micromachined silicon inertial angular rate sensors. The intended initial application for the sensors is in automotive skid control/antilock braking systems. The use of the sensors in an automotive safety system, and the military markets, demand a low-cost sensor design that can be built in high volume with high quality. The success of the Boeing/Draper venture is dependent on establishing high-yield, low-cost fabrication and test processes for the micromachine product design.

This paper describes the necessary micromachine test processes established at Boeing to address the automotive market requirements, and will review the approach developed to test micromachined devices.

Introduction

Micromachined silicon sensors will have tremendous cost advantage over many competing technologies. Using batch processing techniques that are typical in today's semiconductor industry, thousands of identical sensors, less than a few square millimeters in area, can be fabricated simultaneously. These single crystal silicon resonating micromechanical rate sensors are extremely rugged, inherently balanced, and physically small.

Boeing is presently in pilotline, producing production intent sensors. Several hundred sensor modules from the pilotline effort will be used in automotive qualification tests during the second half of 1997 to validate the sensor and electronic design for full production. These modules are currently under test at a number of military and commercial test facilities. Production ramp up to full production after design validation is scheduled to occur in the summer of 1998, with production parts becoming available in early 1999.

Automotive Application Description

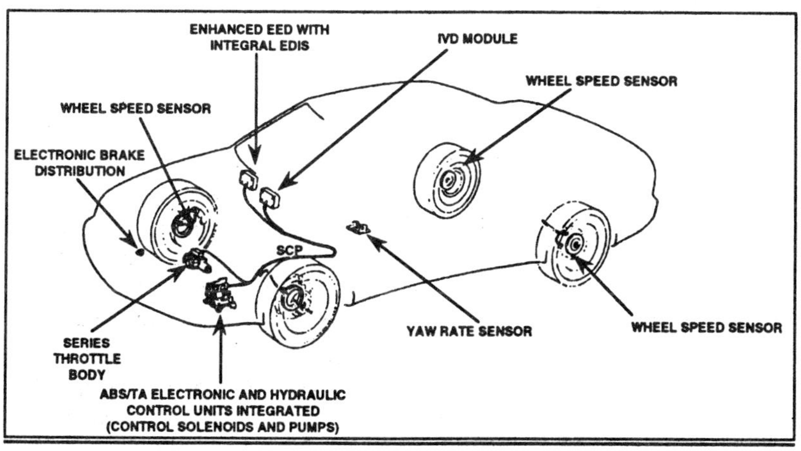

Figure 1. ABS Systems

Consumer demand for Antilock Braking Systems (ABS), as well as other automobile safety systems, is increasing. ABS system sophistication has also been increasing to take advantage of additional performance improvements and new cost effective technological developments. Many automobiles sold today have ABS as standard equipment. Future ABS systems utilizing yaw rate sensors and accelerometer inputs will outperform today's standard ABS systems.

Standard ABS uses wheel speed sensors to detect when one or more wheels of either the front or rear wheel pair are near lock-up, and then pulses the front or rear brake hydraulic pressure to avoid lock-up and keep brakes operating at peak efficiency in slippery conditions or during emergency braking maneuvers. (reference Figure 1)

Split coefficient surfaces, such as road surfaces that have a combination of dry, wet, oily, or icy conditions, can degrade the efficiency of ABS. On split coefficient surfaces, if one wheel of a front or back wheel pair sees a slip condition during braking, the ABS will act by modulating the brake pressure. The second wheel of the wheel pair will then have a reduced braking capability and the total braking efficiency of the system will be effected.

An ABS implemented with a Yaw Rate Sensor (YRS), lateral accelerometer, and steering wheel input sensor can actively adjust the braking system to deliver the optimal setting for every possible road condition (without undesired effects to the vehicle dynamics), and also tell if the vehicle is following the driver's intended path. The system can help compensate for vehicle yaw torques induced by braking on split coefficient surfaces, and can detect oversteer or understeer due to tire slip and apply one or more of the brakes to bring the car back onto the proper line. ABS's incorporating these additional sensors will provide increased vehicle control and in most cases be transparent to the vehicle operator.

Military Application Description

Inertial sensors have long been integral to flight control and navigation functions for military platforms. Advances in electronics and sensor technology have allowed the use of inertial sensors to smaller and less expensive platforms. Micromanchined technology is the latest in this trend towards, more rugged, lighter, less expensive, smaller, lower power inertial measurement units (IMUs). Micromachined technology is applicable to a wide range of military applications. Guided projectiles is an example of an application that is now feasible with the use of micromachined technology, previous technologies could not fit within the projectiles, nor would they be cost effective. Micromachined sensors with their small size and rugged nature are perfect for these applications. Another ideal application is used as an IMU in guided munitions. In these applications, cost and reliability are the primary advantages micromachined based systems provide. Micromachined based sensors are ideally suited for any application where, size, weight, cost or power are considerations.

Product Description

Sensor Principal of Operation

The principle of operation for the rate sensors is based on the effect of Coriolis forces. The resonating structure is composed of two elements driven electrostatically with opposite oscillatory phases. When the device is rotated about its sensing axis, the elements experience small changes in their motion due to Coriolis forces. The elements, (or proof masses) oscillate slightly out of their original plane of motion by an amount proportional to the rotation rate. Capacitive measurement of the mass deflection allows the determination of the rotational rate. Macroscopic tuning-

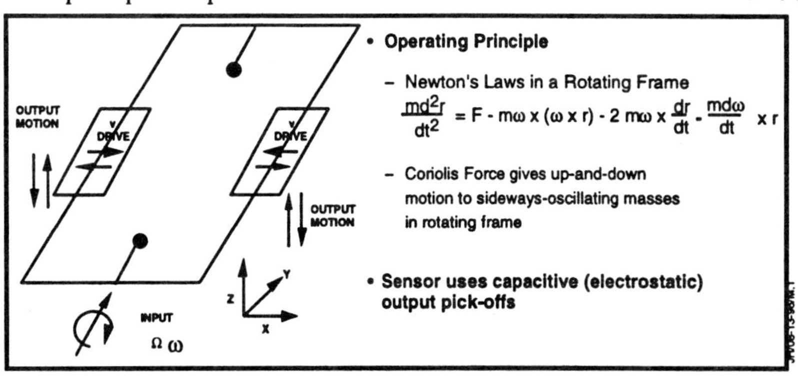

Figure 2. How They Work: Device Concept Evaluation - Rotation Sensor

fork gyroscopes that take advantage of this effect have been used for many years and are well understood. (reference Figure 2)

Device Fabrication, Block Diagram, and Module Physical Layout

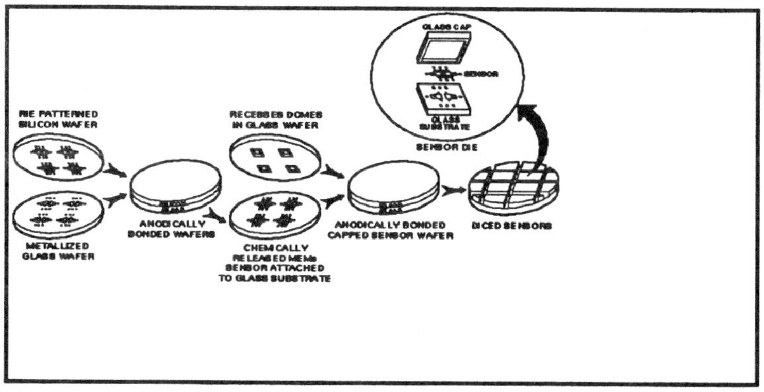

Figure 3. Wafer Level Fabrication

The micromachined structures are fabricated using a dissolved wafer process with boron diffusion and dry etch process steps utilized to define the final dimensions. This structure is then anodically bonded to a glass substrate, that has been prepared with a metal deposition step to provide electrical contacts for the rate sensor's operation. (reference Figure 3)

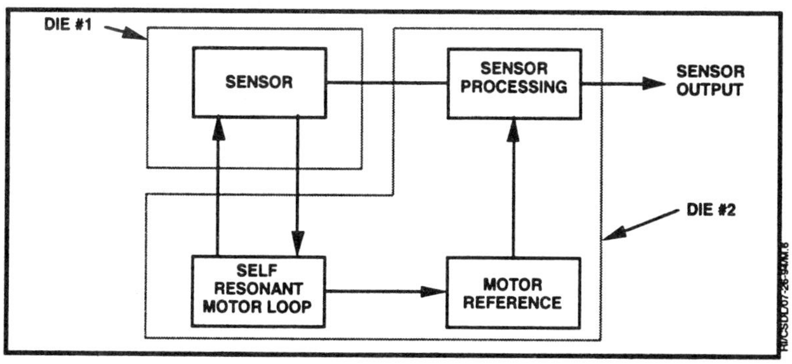

Figure 4. Micromachined Gyro Block Diagram

Finished sensor modules will include the packaged sensor, a CMOS Application Specific Integrated Circuit (ASIC) containing the motor drive loop and sense processing electronics, and individual discretes for power conditioning and EMI protection. (reference Figure 4)

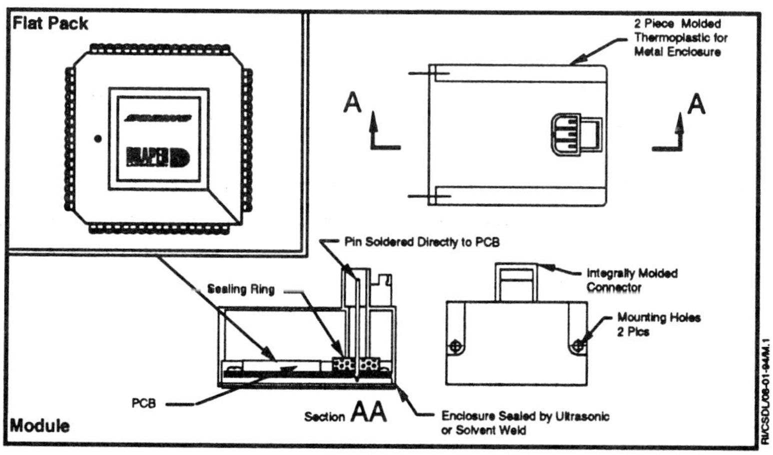

Figure 5. Automotive Packaging

The ASIC will be packaged in a standard flat pack configuration. The packaged sensor and ASIC will be integrated into a common plastic package, similar to other automotive electronic modules. (reference Figure 5)

Testability

Requirements

The design development and the potential YRS unit cost has been driven by customer performance requirements, the available process capability at Boeing, and the YRS testability. Boeing has developed a Micromachined Gyro performance specification to address the requirements of several potential automotive and military customers. A summary of the operational characteristics and significant performance requirements can be found in Figure 6. Scale Factor and Bias numbers include the variation expected over life and operating temperatures. Bias refers to the gyro's output, scaled into units of angle per unit time, in the absence of any inertial angular input. Scale factor is the conversion factor from the gyro's output (in this case volts) to units of angle per time. Scale Factor is adjustable to meet individual customers' full scale dynamic range needs.

Requirement	Automotive Limit	Military Goal (single axis)
Size	40x40x60 mm	.5 cubic inches
Weight	<150 g	<100 g
Supply Voltage	5±0.25 V	5±0.25 V
Power Consumption	<300 milliwatts	<200 milliwatts
Temperature Range	-40 °C to +85 °C	-55 °C to +85 °C
Full Scale Input	50 to 100 deg/sec (adjustable)	1500 deg/sec
Accuracy (over life, temperature)		
Scale Factor	±3 %	.1 %
Bias	±3 °/sec	+.1 °/hr
Bandwidth	100 Hz 3 dB	100 Hz 3 dB
Noise	0.5 °/sec rms	.1 deg/rt-hr

Figure 6. Performance Requirements

The ability to test and characterize the final performance of the YRS module in a cost effective manner is a critical factor in creating a viable product. Due to the performance requirements of the customer, a final acceptance rate test is still required to verify that device performance conforms to the customer specification.

Relationships of Device Dimension to Sensor Performance

The mechanical dimensions of a micromachined device directly affects the inertial performance of a final yaw rate sensor. The DRMG micromachined sensor operation can be modeled by standard equations of motion for a second order system, similar to a mass/spring/damper system studied in most elementary physics courses. Spring constants, damping constants, and proof masses establish the natural response of the system. In a micromachined sensor, the in-plane and out-of-plane spring constants are set by the dimensions of the flexing elements, and the damping constants are related to the pressure the device is run under and the limits of the device material. (reference Figure 7)

$$(ms^2+b_xs+c_x)x=F_d$$
$$(ms^2+b_ys+c_y)y + (2m\Omega)sx=F_s$$

m = *Mass*	Ω = *Angular rate input*
b = *Damping constant*	F_d = *Force in drive axis*
c = *Spring constant*	F_s = *Force in sense axis*

Figure 7. Simplified Second Order Equations of Motion

Uncontrolled process variation leads to device dimension errors that effect the yield of acceptable sensors. For example, the inertial rate scale factor is effected by both the in-plane motor mode and the out-of-plane sense mode, and the magnitude of the pickoff capacitances. Lithographic control, and process etch control determine the variability of these parameters. Developing the device design, setting limits on the acceptable frequency values and capacitances, and establishing the proper process controls to achieve the proper device values with minimum variation, has been the focus of the technology transfer and development at Boeing over the past three years. Capacitance information can be measured by straightforward standard electrical measurements. Modal frequency measurements can be collected by open loop transfer function measurements using standard network analyzers. Statistical processing of the information is used to establish yield numbers for producibility. (reference Figure 8)

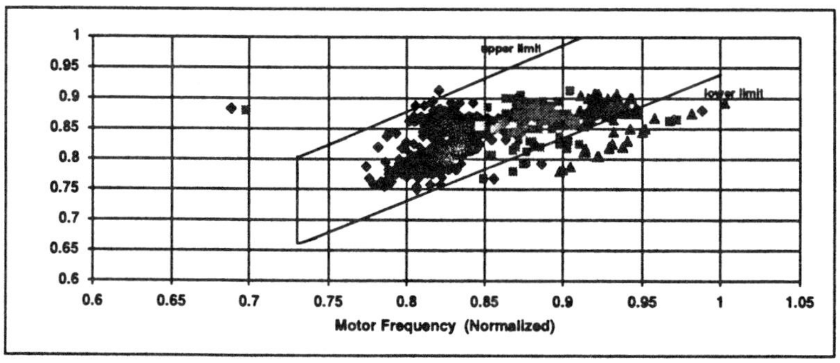

Figure 8. Frequency Separation Results and Limits

Boeing Test Sequence

A series of test steps and information feedback paths have been implemented over the course of prototype development to develop the ability to measure basic device parameters (electrical, mechanical, and inertial) at the wafer probe level. Performance characteristics are then verified by integrating the sensor into a complete instrument package and performing controlled rate and thermal evaluation testing. Data from actual instrument performance testing are then fed back and correlated to both the process control sequence and to the wafer probe test level. This iterative loop is used to correlate the appropriate device parameters with inertial level performance. (reference Figure 9)

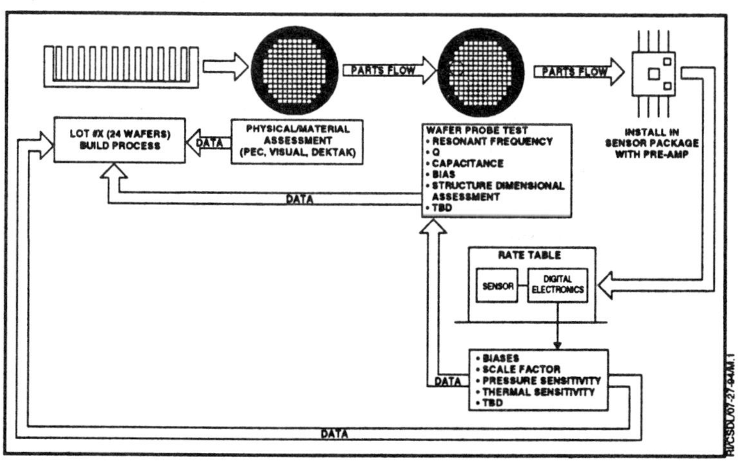

Figure 9. Development Test Sequence

Initial device development at Draper involved fabrication of a few devices at a time. During the initial process transfer to Boeing, development focused on scaling the Draper process up to a wafer level, and then to a wafer batch process. At present, each Boeing fabrication lot contains 25 wafers that are processed simultaneously through most of the fabrication sequence.

Initial tests at Boeing during development were designed to characterize the process and the device sensitivity. Testing was used to identify process associated defects, process repeatability, device failure modes, and functionality of devices fabricated using the developing batch techniques. Testing was performed in a manual probe station and required approximately three days to functionally test a sampling of devices on each 3" wafer containing several hundred devices. Typical measurements included measuring device resonant frequencies, resonant mode quality factors, device drive and pickoff capacitances, and device operational rate bias.

Present probe testing employs automation techniques on an optimized test sequence, with improved electronics, on wafers with approximately double the original device density. Test time now requires approximately seven hours to do 100 percent testing on each wafer. Devices are functionally operated as they are in the intended product, and characterized for their operating modes and capacitances. The automation has been accomplished by using National Instruments Lab View software, and by interfacing the standard test equipment used for device characterization to a data collection computer.

In production, a fully automated probe test station and a computer controlled test switching matrix will be used to perform the minimum number of critical tests as defined during development testing. An automated station has already been procured and is undergoing acceptance test in the production test laboratory, and programming of the test routines planned for production is in progress.

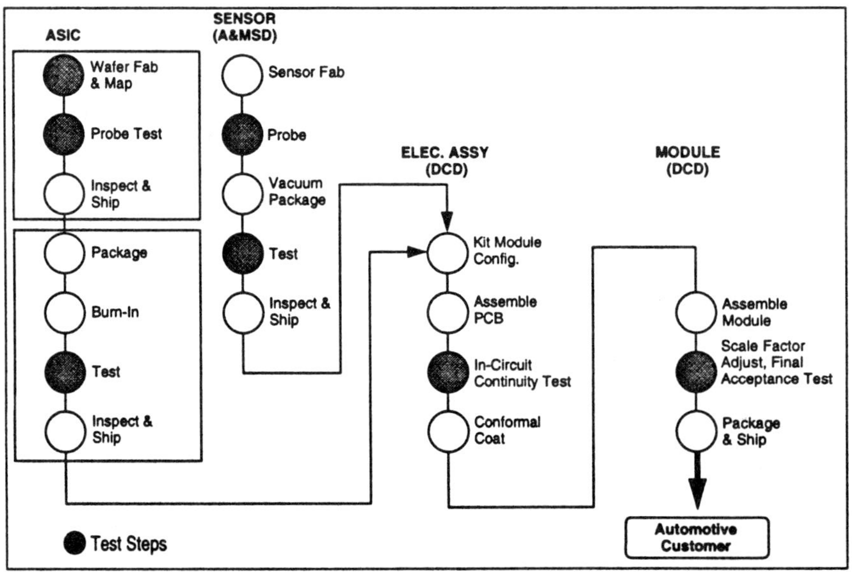

Figure 10. Manufacturing Process and Test Sequence

Critical device testing will take place in process and at probe test (reference Figure 10). Post-package testing will identify die that failed the packaging process. After assembly, in-circuit test will identify printed circuit board faults and non-functional discrete components. Module Final Acceptance testing will check the final inertial characteristics, the Built In Test functions, and calibrate the inertial scale factor for each particular customer.

Database/SPC Activities

The sensor development database has been important for correlating device performance back to the dimensional characteristics. The database, in conjunction with the system requirements, is instrumental in making process yield estimates and establishing acceptance requirements for the individual characterization tests.

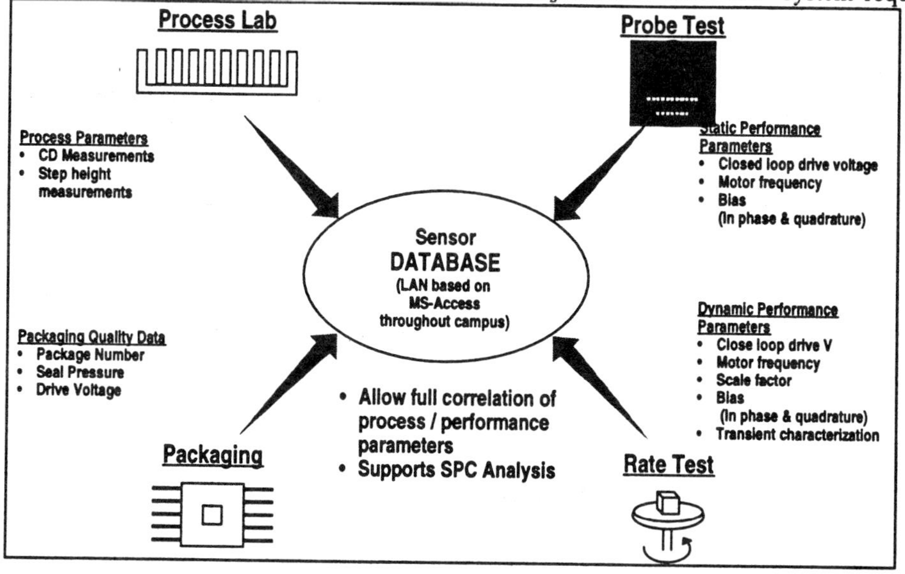

Figure 11. Sensor Fabrication Data Base

The present Database has been created in Microsoft Access 2.0. Tables in the database have been created for each process/test area, and queries have been generated to link the fabrication to final function and to estimate inertial quality of sensors at the earliest test level possible.

Data entry is still either manual, such as at the rate site, or via Labview to spreadsheet to database format, such as at probe test. Automatic data entry directly from the test sites' data collection systems is being developed and will be in place by the end of the present pilotline activities. (reference Figure 11)

Statistical Process Control (SPC) activities to date have been centered around the fabrication area. Device performance, backed by Scanning Electron Microscope and Wyco profilometer measurements, have been used to feedback device yield to the process area.

Examples of Performance

The following plots exhibit the typical DRMG sensor performance over temperature. (reference Figure 12 through 15).

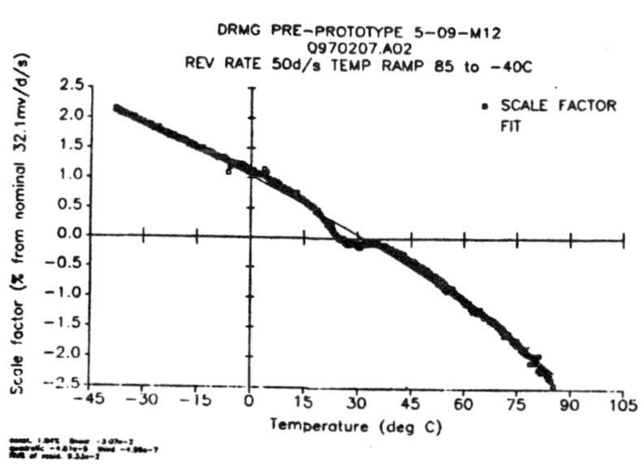

Figure 12. Uncompensated Scale Factor Performance Across Temperature (85° to -40°C)

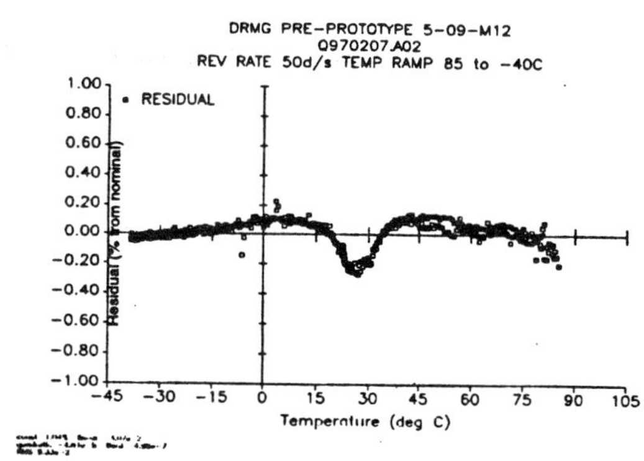

Figure 13. Compensated Scale Factor Performance Across Temperature (85° to -40°C)

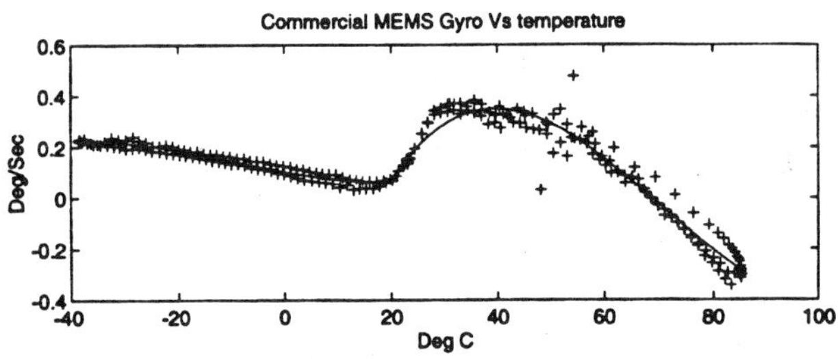

Figure 14. Uncompensated Bias Performance Across Temperature (85° to -40°C)

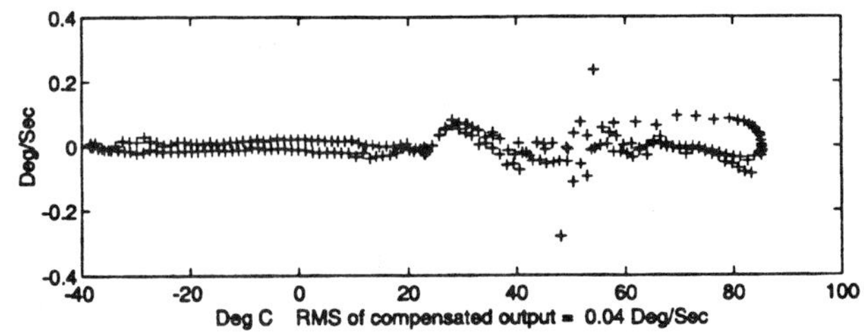

Figure 15. Compensated Bias Performance Across Temperature (85° to -40°C)

Follow on Activities

Follow on product development by both Draper and Boeing is continuing. Higher accuracy sensor designs and integrated units with multiple degrees of freedom (N-DOFs) are being pursued over the next several years by the Draper Boeing inertial sensor development team. This development of improved sensor and electronic designs will produce deices with application in both commercial and defense systems. (reference Figure 16 and 17)

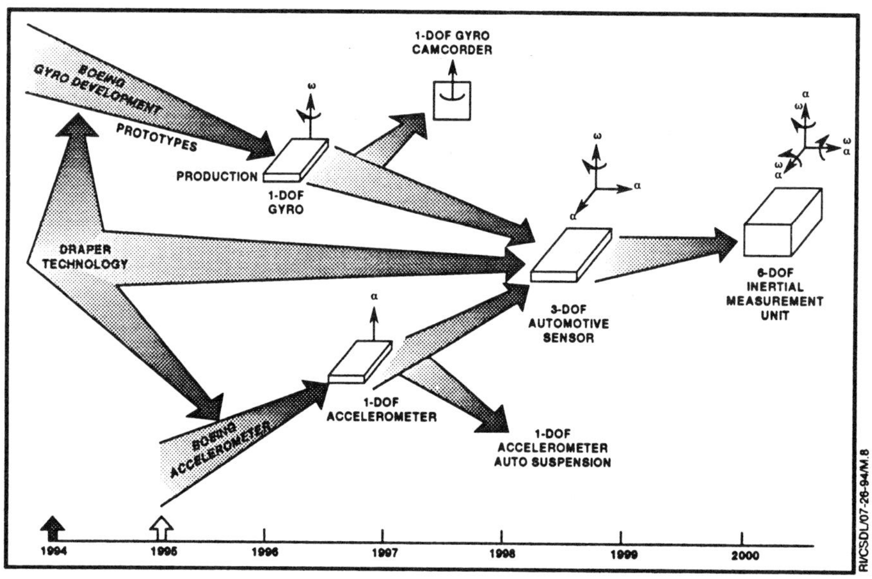

Figure 16. Higher Accuracy Applications

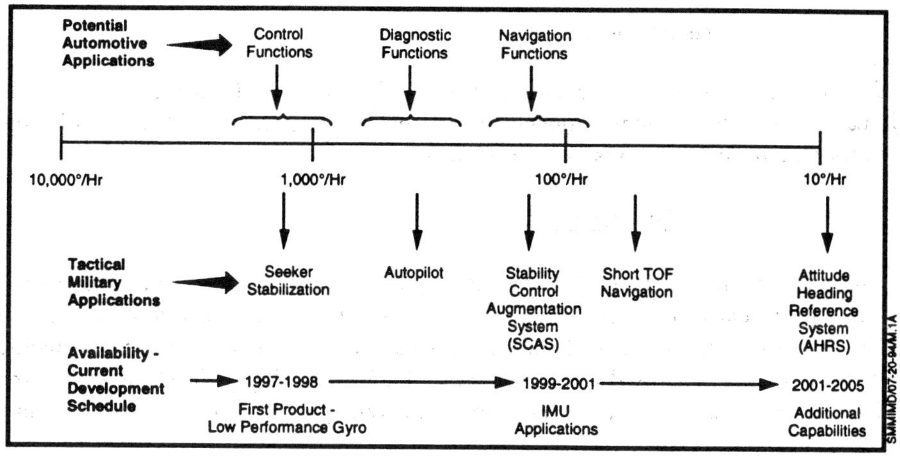

Figure 17. Multiple Degree of Freedom Development

Higher accuracy and multi axis device development will capitalize on the existing processes developed for low-cost commercial applications. Draper and Boeing are both involved in two DARPA funded developments to pursue higher accuracy/N-DOFs designs, and to investigate methods to improve producibility and testability of micromachined devices.

Test Improvements

Several experimental test methods are in development to minimize required test time and to better qualify sensor performance at the earliest possible process step. Any reduction in test time will reduce the ultimate end module cost. Efforts are primarily focused on reducing the test time of the existing test steps in the product line. As resolution and accuracy of the device is improved for future applications, so also must the test processes be improved to maintain the low cost and high volume advantages of the technology.

Within the automotive product presently under development, several non-inertial test methods and non-tactile/non-destructive tests to evaluate device performance and dimensions, are being investigated. Parametric analysis of a sensor's inertial characteristics can presently be performed by reviewing the measured process and probe information. Significant device dimensional analysis can be made by performing non-tactile interfermetric measurements with available standard commercial equipment.

Conclusion

Low-cost rate sensors and large volume production require special test methods to minimize cost and maximize product throughput. Additionally, the understanding of device performance as related to device dimensions and fabrication processes, is necessary to produce high-quality inertial sensors. Low-cost, high-quality, angular rate sensors will open up new arenas for inertial sensors in both military and commercial guidance and control applications. Follow on high-accuracy sensor and multiple axis unit development will lead into applications that have not yet been defined.

References

Micromachined Silicon Inertial Sensors, A Defense Conversion Technology, G. Brand, J. Connelly, 1994 JSDE
Fundamentals of Vehicle Dynamics, T. Gillespie, Society of Automotive Engineers
Car and Driver, "Stability-Enhancement Systems: Trick or Treat?", June 1996
Boeing Development Specification, Automotive Yaw Rate Sensor, Revision 2.0, G. Murray

Acknowledgements

Significant technical review and content suggestions have been received and incorporated from the Boeing Team during the development of this paper. The suggestions supplied have allowed the paper to better describe the micromachined technology and to address the test processes in place at Boeing.

Biographies

Randy Jaffe is the Project Manager and Team Lead of Military and High performance micromachined inertial products for Boeing. His design team is currently developing both single axis and IMU based inertial products. In his 12 years at Boeing, Mr. Jaffe has worked in the areas of inertial sensors and systems development, including developing systems based on micromachined quartz, and ring laser gyros. He holds a degree in physics from the University of California, Irvine and an MBA from the University of Southern California.

Jim Heiertz is the Test Team Leader of the Micromachined Sensors group for Boeing. He leads a multifunctional team, with specific skills in test electronics design, systems engineering, silicon device testing, high-volume/low-cost packaging, and has responsibility for all production development testing, including the assessment of the ongoing process transfer of Draper micromachined rate sensors. His past experience consists of more than 11 years in both the technical development and customer support for the Navy Strategic Systems Programs submarine inertial navigation equipment. He was a member of the Systems Engineering group at Boeing for SINS, ESGM and ESGN equipment for the Poseidon and Trident programs. He also has extensive experience in inertial performance analysis, test, and design. He holds a BS and MS in Electrical Engineering from the California State University, Fullerton, and is a registered Professional Engineer.

Scott Ventresco is the lead engineer for inertial and system level testing of micromachined sensors for Boeing Autonetics & Missile Systems Division, Integrated INS/GPS Engineering, micromachined Sensors group. He is responsible for device/design characterization test, and production test system development. His past experience consists of more than 10 years in both the technical development and customer support for the Navy Strategic Systems Programs submarine inertial navigation equipment. He was a member of the Systems Engineering and Software Engineering groups at Boeing for ESGM and ESGN equipment for the Trident programs. He also has extensive experience in inertial performance analysis, test, and design. He holds a BS in Electrical Engineering from the Ohio Institute of Technology.

AIAA-97-3552

PERFORMANCE OF SMALL, LOW COST RATE SENSORS FOR MILITARY AND COMMERCIAL APPLICATIONS

A. Kourepenis, J. Borenstein, J. Connelly, P. Ward, M. Weinberg

Charles Stark Draper Laboratory
Cambridge, Massachusetts, 02139

Abstract

Draper Laboratory and Boeing North American have formed an alliance to develop small, low cost rate sensors for commercial and military applications. Advanced silicon micromachining techniques produce sensors of high performance, ruggedness, and inherent symmetry. When integrated with Applications Specific Integrated Circuits (ASIC's), the rate sensor will fit in a 3 cm per side flat pack and operate from a single 5 Vdc supply.

Gyroscopes are fabricated using a dissolved wafer process that features single crystal silicon anodically bonded to a glass substrate, resulting in a sensor die size of approximately 1 mm. Uncompensated bias and scale factor performance of 0.5 °/s and 1.0% are nominally demonstrated over the automotive temperature range of -40 °C to +85 °C. Bias stability over smaller temperature ranges of 0.5 °C has surpassed 10 °/hr (0.003 °/s) in overnight (6 hour) tests. Nominal resolution is 150 ° /hr in a 60 Hz bandwidth, yielding an angle random walk of 0.25 °/√h. Best to date resolution and angle random walk performance is 25 °/hr in 60 Hz and 0.04 °/√h respectively. The robustness of the design with regards to environment has been demonstrated in the ability to survive air guns tests in excess of 60,000 g's.

This paper discusses the principle of operation, fabrication techniques, measured and projected performance, with a principal focus on recent tuning fork gyro (TFG) test results. Associated electronics, controls, and applications issues will also be addressed.

Introduction

The Charles Stark Draper Laboratory has invented and developed inertial guidance systems for earth and space applications for over 50 years. Draper has been developing miniature micromachined gyroscopes for over 10 years and was the first to demonstrate rate sensing with micromachined silicon sensor. [1] Micromachining borrows processes from the semiconductor industry to fabricate tiny sensors and actuators on silicon chips. Many new applications are enabled by the reduced size and lower cost inherent to sensors developed with this process.

Draper has made great strides in simplifying the device fabrication process and realizing rate sensors of increasingly higher performance [2,3,4,5]. Initial sensors invented at Draper were fabricated using a 17 step process and had resolution performance (integrated PSD) of 1,000,000 °/hr in a 60 Hz. bandwidth. Combining the tuning fork gyro architecture and dissolved wafer fabrication process, the same performance standard has improved to 150 °/hr nominal with a best to date score of 25 °/hr in the same 60 Hz bandwidth. Sensors are now fabricated with a simpler three step process, increasing yield and lowering unit production costs.

Consistent with Draper's mission of transferring technology to industry, an Alliance with Rockwell Autonetics (now Boeing North American (BNA)) was consummated in 1993 for the purpose of transitioning Draper developed inertial micromechanical technology for production and commercialization. The intended initial application for the sensors is for use in automated skid-control/anti-lock braking systems for automotive applications. Additionally, the higher performance of recent instruments enables insertion into military systems, the first demonstration being low-end, unguided artillery shells and rockets.

Theory of Operation

The TFG (Figure 1) consists of a silicon structure suspended above a glass substrate containing metallization deposited for sensor interfacing. The silicon structure contains two masses suspended by a sequence of beams that are anchored to the substrate at specific points. By applying voltages to the outer motor drives, the two masses are electrostatically forced to generate lateral, in-plane oscillatory motion. An angular rate, Ω, applied about the input axis, perpendicular to the velocity vector of the masses, generates a Coriolis Force that acts to push the masses in and out of the plane of oscillation (labeled F_1 and F_2 in Fig. 2). Because the instantaneous velocity vectors of the relative masses are equal and opposite, antiparallel motion is induced in response to the Coriolis Force. This resultant motion is measured by capacitor plates under each of the two masses, providing a signal proportional to the rate input. Closed loop control is employed to maintain the mass at a null position using similar rebalance electrodes, however, open loop operation is adequate for many identified applications.

Figure 1. Micromachined comb drive tuning fork gyro

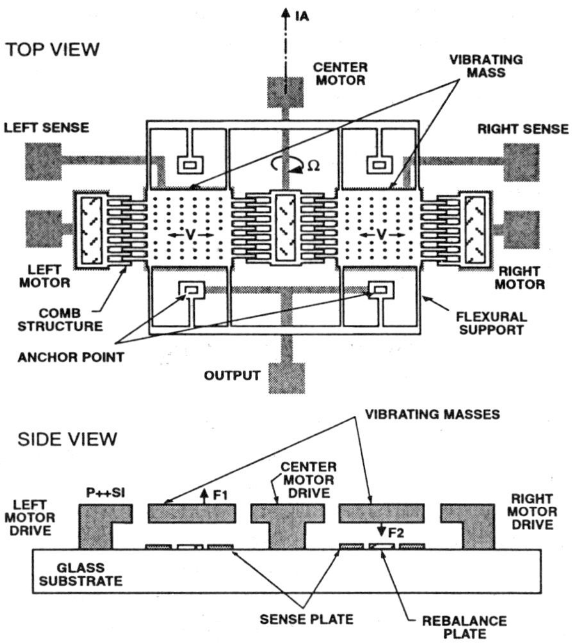

Figure 2. Operational representation of TFG

Sensor Fabrication and Packaging

Sensors are manufactured by the dissolved wafer process (Figure 3). In the Mask 1 step, recesses are etched in a doped silicon wafer defining the height of the silicon above the glass and gap spacing for the capacitive sensing plates. A boron diffusion step follows, which defines the thickness of the structure. The structure's pattern features are defined by Mask 2 and then micromachined using a reactive ion etch (RIE) process. The glass wafer is processed separately. Mask 3 defines the glass recess and metal electrode pattern. The silicon wafer is turned upside down and bonded to the glass wafer. A combination of heat, pressure, and electrostatic field is used to form strong chemical bonds between silicon atoms of the wafer and oxygen atoms in the glass. The final etch dissolves the undoped silicon and leaves the free-standing device. Hundreds of sensors are made on a single wafer.

Gyro sensor devices, requiring a vacuum to obtain the high Q resonance, are sealed in hermetic packages that are designed to maintain a stable environment over a 20-year life. Leadless chip carriers (LCC) with braze-sealed lids are being used to package sensors in pilot production. Sensor chips are installed in packages using compression bonding and wire bond interconnects. The sensor package is then mechanically aligned to a printed circuit board (PCB) containing the electronics and secured by conductive epoxy. This LCC is 0.25 in square by 0.10 in high.

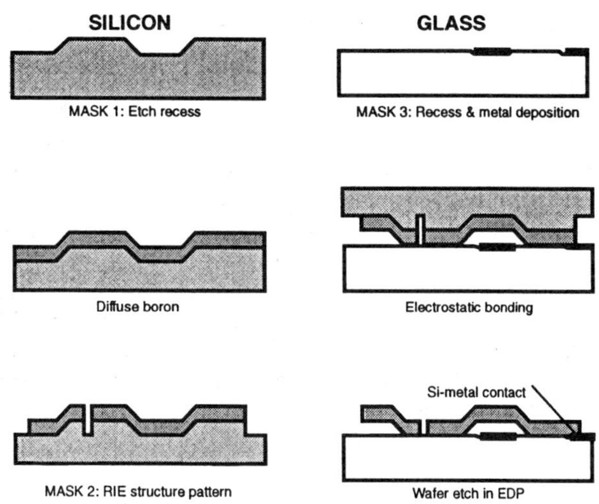

Figure 3 Dissolved wafer, silicon on glass process for TFG

Electronics

Electronics are an integral part of the TFG operation and greatly influence the level of instrument performance achievable for given applications. TFG electronics designs address the principal needs for instrument operability, the motor loop and the sense loop, and the interaction between the two (Figure 4).

The motor loop consists of the a self-drive oscillator that provides voltages to generate electrostatic force and induce proof mass motion. Proof mass position is detected and properly gained and phase shifted to realize and sustain constant drive amplitude. Maintenance of drive amplitude with a high degree of accuracy is necessary for realizing stable performance in sensor scale factor. In addition to the self-sustaining drive amplitude, the reference signal required for demodulation to baseband is precisely derived from proof mass motion by the oscillator loop. Electronics needed to provide phase shift required for closed loop regeneration, also yield the velocity signal reference, which when multiplied by the sense axis signal output, results in a baseband signal proportional to the rate input. Frequency response of the instrument is determined by the baseband filters designed to reject higher harmonic signals generated from the down conversion. Usable bandwidth on typical TFG designs is typically 1.0 to 1.5 KHz.

Detection of Coriolis induced motion of the sensitive axis is accomplished using a preamplifier to detect the charge generated from the change in proof mass displacement. Signal to noise is generally determined by the sensitivity of the sensor and the input voltage noise of the preamplifier. The preamplifier output is amplified and presented to the aforementioned demodulator to generate the baseband representation of the rate input. In open loop applications, this signal provides the output proportional to input rate, while in closed loop applications, this signal is nulled through servo control and the potentials used in mass rebalance present the proportioned output.

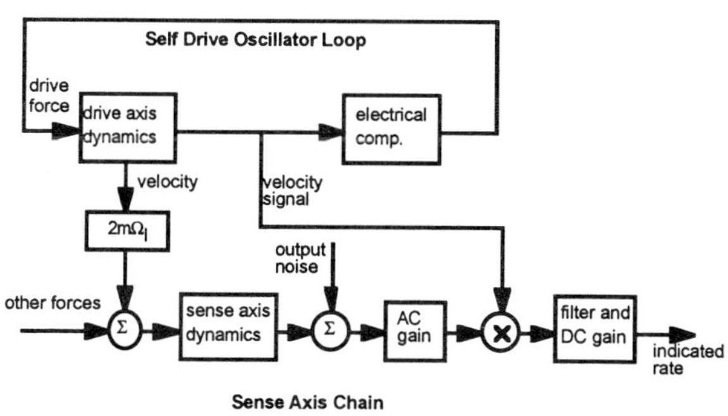

Figure 4. Block diagram representation of TFG electronics

Additional electronic functions for Micro Electro Mechanical Systems (MEMS) include self-test and/or self-calibration, bias and scale-factor correction and/or compensation, and special application interfaces. ASIC electronics using CMOS processes provide excellent performance at low power and cost, while maintaining the small size required by many applications of this technology. Mission needs are addressed by versatile electronic configurations that are adapted for specific requirements.

Test Results

Significant strides made in the sensor design, fabrication, electronics architecture, and packaging have resulted in the excellent data of tested instruments. Since the advent of the TFG, approximately 500 gyros have been packaged and tested at Draper and several thousand at Boeing's companion sites. Qualification tests performed on recent units have resulted in excellent thermal, turn on, noise, and start-up performance. Additionally, specific tests have demonstrated the linearity of the sensor scale factor to high angular rates, lumped drift performance under thermal control, and minimum resolvable signal.

Automotive Applications Specific Results

Initial automotive applications involve using a yaw rate sensor for skid control in anti-lock braking applications. The systems are now being realized in the high cost, luxury models of Mercedes Benz, but low cost MEMS sensors will enable this safety feature to be added to standard automotive platforms at minimal cost. Adding a lateral accelerometer to the gyroscope provides the capability to conduct real time suspension control to provide better handling, stability, and ride optimization. Future applications involve aided navigation, smart cruise control, and detection of roll over conditions.

Tests aimed specifically at performance for the automotive applications have yielded data summarized in Table 1. Data represents tests on an earlier instrument architecture, with sample sizes of 10's of instruments. Data indicate raw and uncompensated performance for the given parameter and to our knowledge, represent the lowest published values reported in any literature of micromachined silicon rate sensors.

Parameter	Units	Mean	Sigma
Scale Factor (-40 °C to +85 °C)	% p-p	2.35	0.86
Bias (-40 °C to +85 °C)	°/s p-p	2.59	1.95
Scale Factor Turn-on	% p-p	1.38	0.56
Bias Turn-on	°/s p-p	0.09	0.07
Noise Density	°/hr/√Hz	21.2	5.52
Angle Random Walk	°/sqrt(hr)	0.25	0.06
Time to Valid Data	seconds	0.73	0.11

Table 1. Summary of automotive yaw rate sensor tests

Maintaining the requisite accuracy across the harsh automotive environment is the principal difficulty involved in the application. From Table 1, scale factor and bias performance across temperature averages approximately 2% and 2.5 °/s respectively, peak to peak error across the entire temperature range. Scale factor and bias accuracy from a cold turn on are approximately 1% and 0.1 °/s peak to peak respectively. Noise density of 20 °/hr/√Hz and angle random walk of 0.25 °/√Hz are typical. Time to valid data is defined as the time period elapsing from application of power to when the output has settled to 2% of the final value. This typically takes about 0.8 seconds for these instruments.

Significant progress in the performance and repeatability of these devices has been accomplished since initial results were published. [3], [4], [5]. Figure 5 and 7, are data recently acquired on later generation sensor and electronics and represent the architecture BNA is currently implementing in the automotive yaw rate sensor pilot production. The present performance indicated by these devices meets or exceeds the needs of the automotive manufacturers without incurring additional cost of further test and compensation, thus maintaining unit production cost goals. The data exhibits very little drift to best fit model coefficients over either temperature or time, indicating additional applications can be served by using production line automotive instruments that are further tested and modeled to implement either family or individual compensation models tailored to the specific needs.

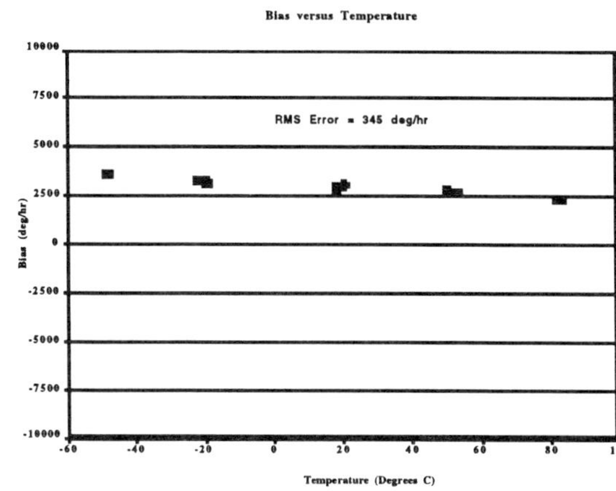

Figure 5 Bias versus Temperature

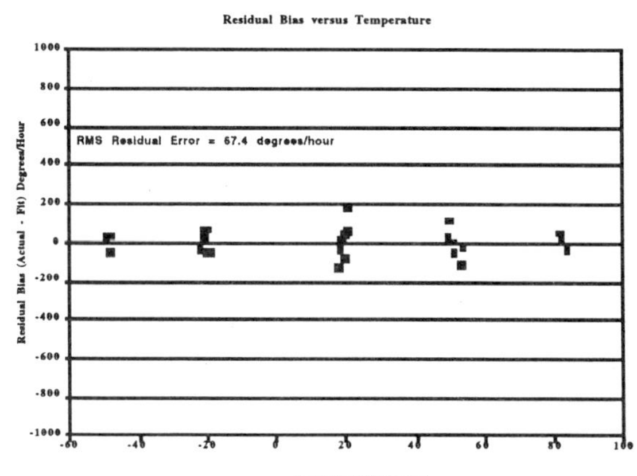

Figure 6. Comp. Bias versus Temperature

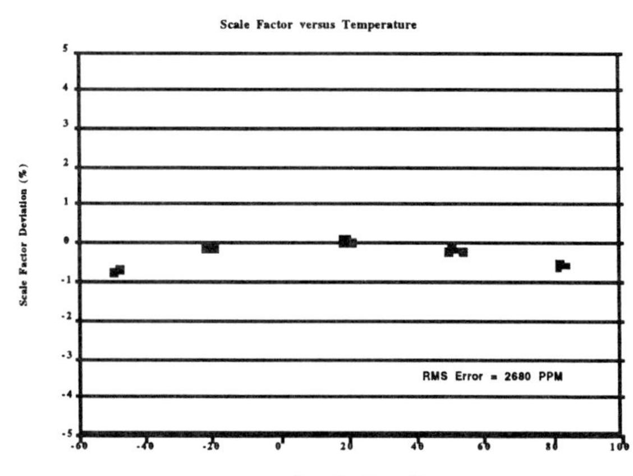

Figure 7. Scale Factor vs Temperature

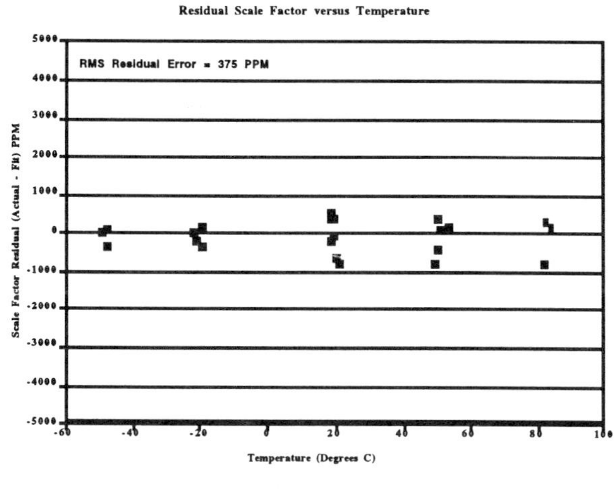

Figure 8. Compensated Scale Factor vs Temperature

Figure 5 and 7 summarize bias and scale factor performance across three thermal cycles from -50 °C to +85 °C conducted over a one week period. Bias error (Figure 5) over the full thermal range is approximately 2200 °/hr peak to peak. As indicated by the data, the instrument is stable over time and offers excellent repeatability over temperature. This type of test reflects a worst case type of scenario in the automotive environment, where the rate sensor is exposed to the maxima and minima thermal extremes. This test would represent a scenario where a car is started at room temperature, 25 °C, driven to the beach and left in the sun, 85 °C, then driven to the mountains for skiing during extremely cold weather, -40 °C, and then returning home after the trip. Scale factor data is presented in Figure 7 where the error is roughly 0.08%. This data far surpasses both the magnitude and stability required for automotive applications demonstrates performance required for guided munitions and other GPS-aided types of applications. This is demonstrated when thermal compensation is applied to the data, as shown in figures 6 and 8. Applying a simple, third order least squares fit reduces the raw bias and scale factor by an order of magnitude to the 67 °/hr and 375 PPM RMS values for bias and scale factor respectively.

Additional Applications Specific Results

Excellent performance, low cost, and small size inherent to instruments fabricated with this technology, are enabling many other applications. Draper teamed with The Naval Surface Warfare Center Dahlgren Division (NSWCDD) to demonstrate the concept of guiding gun launched munitions under NAVSEA funding [6]. Extended Range Guided Munitions (ERGM) for Littoral Warfare Applications is the most immediate need identified for naval surface ships. Draper is also executing a Competent Munitions Advanced Technology Demonstration (CMATD) Program from Office of Naval Research to demonstrate guidance in the correct form factor of the current rounds. Figure 9 shows the CMATD nose replacement concept. All platforms employ GPS-aiding, where the GPS calibrates the inertial measurement unit (IMU) during at least the initial portions of the flight, providing the option of flying the last portion of the mission fully inertial to circumvent jamming or other electronic counter measures and any additional GPS performance or functionality related issues.

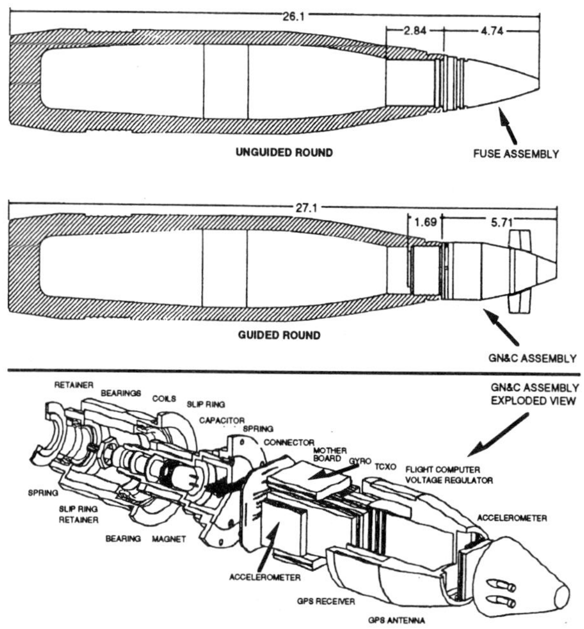

Figure 9 GPS/INS for 5" Guided Munitions Round

Additional applications being pursued involve re-entry vehicles targeting deeply buried targets. These vehicles perform very exacting and stringent maneuvers, requiring high accuracy and performance under stress of the guidance system. GPS-aided systems are the platform being discussed for these applications, where the GPS again calibrates the IMU through the initial portion of the flight and then the vehicle is inertially guided to the target after post plasma blackout of GPS occurs. Preliminary concepts for various applications of surface to surface, surface to air, or air to surface missiles are also being discussed.

Resolution, Noise Density, Angle Random Walk

As stated previously, the dominant noise source of the TFG is the input voltage noise imparted by the preamplifier used to detect charge generated when the mass is forced by Coriolis Acceleration. This is manifested in the instrument output as white noise in angular rate. For micromechanical gyroscope development at Draper, the selected performance benchmark is resolution, defined as the gyro's rate equivalent RMS noise in a bandwidth from 0.1 to 60 Hz. Figure 9 is provided to indicate the performance strides made in Draper gyroscope development. Since the noise is white in angular rate, resolution may be stated as angle random walk in o/√h. Nominal resolution of 100 to 200 °/hr in 60 Hz is achieved, with best to date performance of 25 °/hr, translating to angle random walk of about 0.25 °/√h on average, with a best to date of 0.04 °/√h. Figure 11 is a single sided PSD of the gyro output in a bandwidth from 0.1 to 100 Hz.

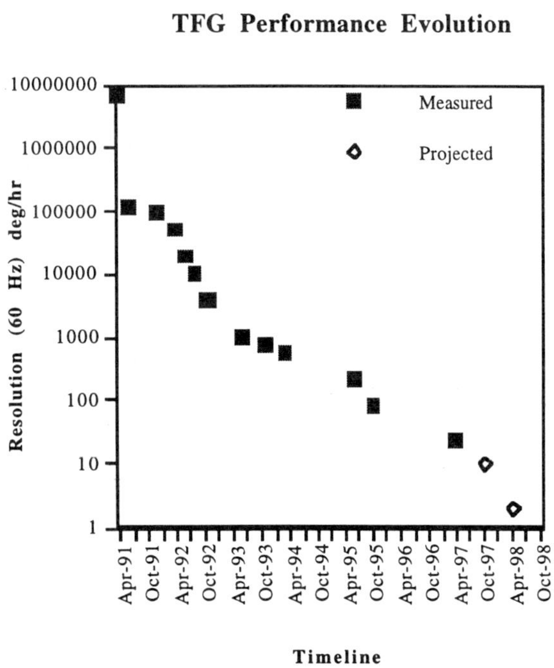

Figure 10. Evolution of microgyro resolution performance at Draper

In 1987, Draper's double gimbal gyro was the first micromechanical gyro to sense angular rate. Resolution was 8 x10^6 °/hr. The double gimbal was bulk micromachined into the silicon substrate and resolution was improved to 4,000 °/hr by 1993.

TFG work began in 1992 and demonstrated 4,000 °/hr resolution in 1993. Draper converted from the double gimbal to the tuning fork because: (1) the tuning fork construction was simpler, requiring many less masks; and (2) the larger drive amplitudes promised better performance. As concurrent development of sensor and electronics progressed, the value of 500 °/hr in 60 Hz was achieved in early 1994. In early 1995, improvements were made to the preamplifier pick-off to lower the input voltage noise, increasing the signal to noise ratio and resulting in the aforementioned nominal levels of resolution and angle random walk. Additional improvements in TFG performance, both sensor and electronics, are being pursued.

The single sided PSD below is used to derive the signal to noise ratio and resolution of the instrument and is measured at the baseband rate output. Rate equivalent units are provided by normalizing the density function to the scale factor of the device as measured at the baseband rate output. From the density, the resolution is determined as a function of bandwidth. Typical densities are 15 to 20 °/hr√Hz, and the figure below provides resolution numbers for 25, 60, and 100 Hz bandwidths.

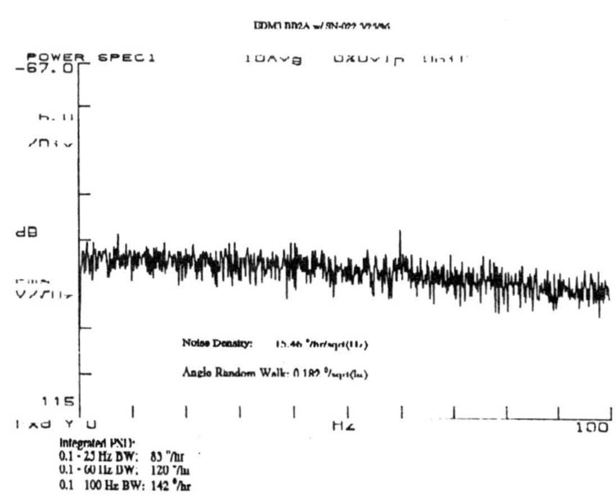

Figure 11. TFG PSD in 0.1 to 100 Hz. Bandwidth

Thermally Controlled Drift Performance

Re-entry types of applications have the most stringent micromechanical systems performance requirements known to date. To optimize performance for this type of platform, temperature control is employed to reduce thermally imparted drift in error model coefficients. Tests on TFG instruments were performed with temperature controlled to roughly 0.5 °C. Figure 11 shows raw, uncompensated data that exhibits a total RMS drift of 4.4 °/hr over the 6 hour period of acquisition. Testing was stationary, i.e. no angular rate reference, meaning scale factor and bias performance errors are lumped into the observed drift.. Data was acquired at 250 samples/second through an anti-aliasing low pass filter with a 0.1 Hz bandwidth and 40 dB/decade roll-off and rectangularly averaged to 1 second samples. For the instrument under test, the noise density was approximately 15 °/hr√Hz, which, when integrated across the 0.1 Hz bandwidth, provides an RMS value of about 4.74 °/hr, approximately the value measured, implying that the observed drift is buried within the white noise in rate. From eyeball estimates of the plot in Figure 11, the low frequency drift observed is less that 10 °/hr peak to peak over the time of the test. This demonstrates the potential of a micromachined TFG configured to meet re-entry mission requirements.

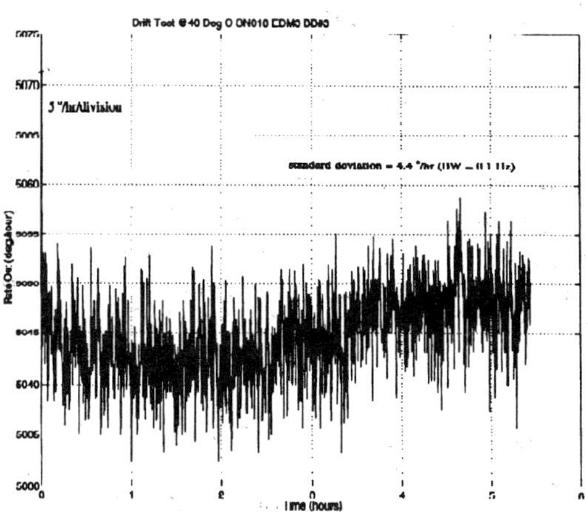

Figure 12. TFG lumped drift performance under thermal control

This same instrument, while maintained under the same thermal conditions as described previously, was monitored aver a 16 day period. Lumped drift over this period was approximately 800 °/hr peak to peak as shown in figure 12. Mechanisms have been identified and are in the process of being corrected. These types of tests indicate the frequency of calibration required for the desired mission application.

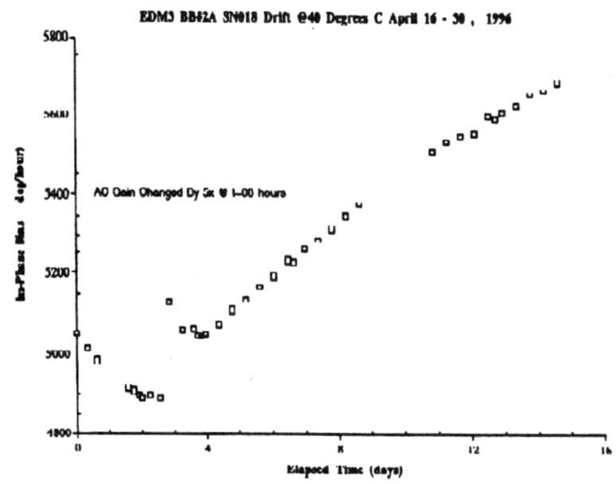

Figure 13 Lumped drift performance over 16 days

High Rate Scale Factor Linearity

Additional importance for various application involves determining the overall ability of the scale factor to maintain accuracy as a function of angular rate. This has implications in the system for dynamic range requirements and also contributes to the computational complexity of compensation models used for guidance and navigation applications requiring solutions for GPS-aided or wholly unaided applications. As the guidance computer requires larger and more complex models to meet the accuracy requirements, power and throughput issues become critical for system level performance and functionality.

Theoretically, the inherent symmetry of the TFG design should provide extremely linear output over large ranges of angular rates. This is verified in the data presented showing the TFG response to input rates varying from 0 to 1000 °/s (Figure 14). Figure 15 is a plot of the mean values calculated at each rate input, plotted as a function of time, and fit with a 2nd order least squares estimate. From the coefficients, one can determine that the instrument output is linear to within 1% at the extrema values of ± 1000 °/s.

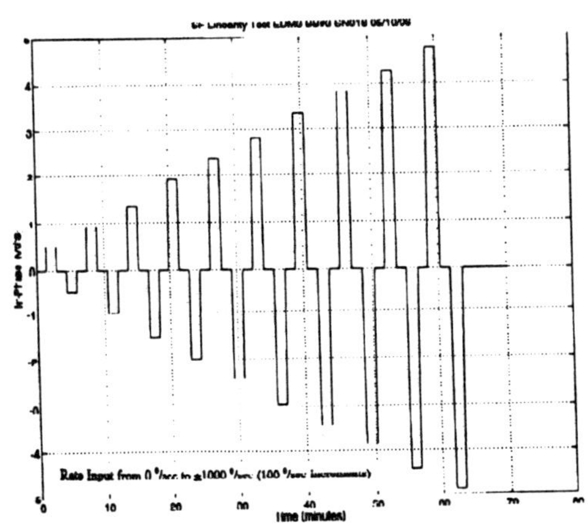

Figure 14 TFG response to angular rate input

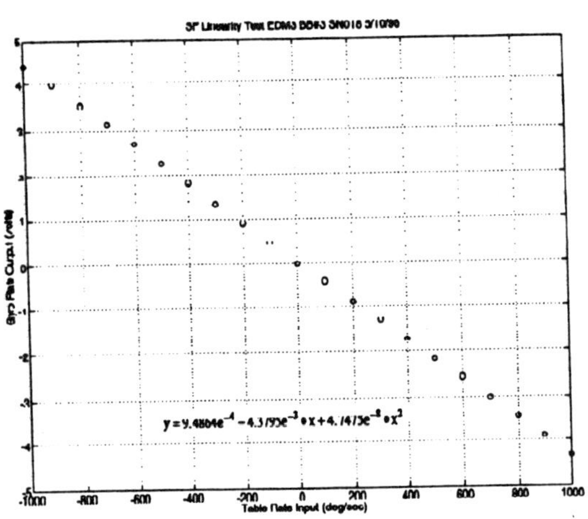

Figure 15. Least squares estimate of TFG output to angular rate input

Conclusions

Applications requiring gyroscopes of high ratios in performance to cost continue to grow in number. Draper and Boeing North American are striving to provide state of the art components to fulfill this market and application need. Micromechanical gyroscope development has continued to make marked and timely progress in achieving higher performance, thus enabling more applications for this unique and innovative technology. The data presented in this paper is indicative of the performance levels of the products eagerly anticipated by numerous customers and applications.

Acknowledgments

The authors gratefully acknowledge continued support from Draper Laboratory and Boeing North American, the dedication of the Alliance technical team of scientists, engineers, and fabrication specialists. Draper personnel specifically acknowledged are Tom Farnkoff, Rich Elliott, James Campbell, Paul Magnussen, Larry Weldon, and Robert Fields for providing test support, Connie Cardoso, James Cousens, Lance Niles, Michelle Lind, Doug Preble and Brenda Hugh for their efforts in device layout and fabrication, Dave Kang and Bernie Antkowiak for mechanical design, Peter Chiacchi Tony Petrovich, Dom Fulginiti, and Frank Petkunas for packaging, Ralph Haley and Tom King for electronics and James Sitomer, Johnathan Bernstein, and Paul Greiff for having provided the necessary vision at various stages of instrument development.

References

1. Greiff, B. Boxenhorn, T. King, and L. Niles, "Silicon Monolithic Gyroscope," *Transducers '91, Digest of Technical Papers*, pp. 966-969, 1991 International Conference on Solid State Sensors and Actuators, IEEE Cat. #91CH2817-5, June 24, 1991.

2. Weinberg, J. Bernstein, S. Cho, A. T. King, A. Kourepenis, and P. Maciel, "A Micromachined Comb-Drive Tuning Fork Rate Gyroscope," *Proceedings of the 49th Annual Meeting "Future Global Navigation and Guidance,"* The Institute of Navigation, June 21-23, 1993.

3. Weinberg, J. Bernstein, S. Cho, A. T. King, A. Kourepenis, and P. Ward, "Micromechanical Tuning Fork Gyroscope Test Results," AIAA GN&C Conference, Scottsdale, AZ, August 1-3, 1994

4. Weinberg, M., J. Bernstein, S. Cho, A.T. King, A. Kourepenis, P. Ward, "Initial Results on Micromachined Comb Drive Gyroscope with EDM2 Electronics," CSDL-P-3398, presented at the 1994 Joint Service Data Exchange, October 31-November 3, 1994, Scottsdale Arizona.

5. N. Barbour, J. Connelly, J. Gilmore, P. Greiff, A. Kourepenis, M. Weinberg, "Micromechanical Instrument and Systems Development at Draper Laboratory," AIAA Guidance, Navigation and Control Conference on Integrated Navigation Systems, San Diego, CA, July 29-31, 1996.

6. J. Dowdle, K Flueckiger, " An Integrated GPS/Micromechanical IMU for Navy 5" Projectiles," The Institute of Navigation Conference, Cambridge, MA, June 18-20, 1996

7. *Microelectromechanical Systems: A DoD Dual Use Technology Industrial Assessment*, Final Report Department of Defense, December 1995.

AIAA-97-3567

Inertial Navigation System for a Micro Unmanned Air Vehicle

Abstract

This paper discusses the design of a Miniature Aided Inertial Measurement Unit (MAIMU) for a micro unmanned air vehicle (MUAV). This work was performed under the SBIR program SB962-081 Micro Unmanned Air Vehicle Critical Technologies. Due to the severe weight and volume constraints imposed on MUAV's an innovative electronic and mechanical design is required. The inertial sensor cluster required for both stabilization and navigation must be small, light weight and low power. These requirements can be met by utilizing micro-machined and new low power FOG sensors. Long term accuracy is also required from the navigation system. The long term accuracy can be provided by a number of devices including small low power GPS.

The challenge is to package the elements of inertial sensors, navigation aids and processor in a small volume requiring extremely low power (1watt) that can become an integral part of the airframe structure. The design activity must focus on what scale of integration is required to meet the objectives (discrete components, multi chip modules, ASICS).

The performance obtainable and the factors affecting the choice of such a navigation system are discussed. The conclusion reached is that although there are many challenges such a system is feasible.

Author

Ian Humphrey, Fibersense Technology Corporation, 755 Dedham St, Canton, MA 02021

Acknowledgments

Fibersense is grateful for the direction, support and encouragement provided by :-
James M. McMichael, MUAV Program Manager, Tactical Technology Office, DARPA.
Jack Suddreth, SETA Support, MAV Program Manager, SRS Technologies.

Introduction

The MUAV program is a program to develop an extremely small air vehicle one such example is shown in figure 1.

A large number of applications required the development of a very small, low power inertial navigation system. The applications include micro-Unmanned Air vehicle, personal navigation, Artillery Shell Guidance, robots etc. Typical goals for such an IMU are as follows :-

- 1 to 10 deg/hr
- <500 ug Accel Bias
- < 10 in^3
- < 1.0 Watt Power drain
- 2 to 4 ozs
- $1,200 per unit in large Quantities.

The goals of the program are ambitious. In particular the performance weight and power consumption numbers are challenging.

Due to the small size and low power requirements of the system the most obvious choice for the inertial sensors is the use of micro-machined sensors. The inertial sensors will be required to provide high bandwidth data for vehicle stabilization and also for longer term navigation data. The mission length is 30 to 60 minutes. The errors produced by a purely inertial system will be very large over this length of mission. Some form of long term aiding will be required for the inertial system. The most obvious choice for long term aiding would be a GPS receiver. This combination of GPS and an inertial system is ideal in many ways. However, due to the extreme power and size constraints other aiding sources must also be considered (e.g. pressure sensors, magnetic heading sensors).

Recent advances in low power Fiber Optic Sensors (FOG's) mean that they should not be ruled out without being evaluated. The FOG alternative is discussed later in the paper.

The current technology makes it possible for the processor and GPS to be contained within a single device if necessary. The RF sections of a GPS receiver are currently very small. A number of venders are also developing digital RF systems. There is no doubt that an extremely small integrated GPS, inertial cluster and processor can be developed. The results of trade studies to determine the optimum level of integration are presented. The size, weight and power of the eventual design must be balanced with the required development cost and schedule.

Figure 1 : Micro Unmanded AirVehicle

Use of automotive class sensors allows for pricing of a ***MAIMU at less than $1,000*** versus $3,000 to $7,000 for other technologies. Traditional IMUs are too large, heavy and expensive to be considered for MUAV applications.

The use of micro-machined inertial sensors provides the high bandwidth, high data rate required for MUAV stabilization and control. The GPS receiver provides the long term navigation accuracy required for the MUAV guidance functions.

The processor within the MGIMU will have sufficient reserve capacity to perform the autopilot and guidance functions required by the MUAV.

1.2 Concept

To meet the navigation performance requirements using small, light weight, low cost sensors, a Kalman filter will be used to estimate MGIMU inertial sensor errors and perform transfer alignment. Each MUAV will have a navigation computer that will compensate inertial sensor outputs and generate a navigation solution. This solution will be compared to the GPS solution using a Kalman filter so that navigation and sensor errors can be estimated. This process should reduce MGIMU errors to 1 to 10 deg/hr and provide accurate free flight navigation.

Typical MUAV applications require gyro drift to be calibrated to 1 to 10 deg/hr this allows them to cope with loss of the GPS signal. The automotive industry requires gyros with drift rates of slightly better than 3600 deg/hr. Analog Devices' engineers expect the typical performance of their gyros to be approximately 360 deg/hr.

Factory calibration and compensation of the MIMU should reduce the long term drift to approximately 40 deg/hr. During pre-release of the MUAV a Kalman filter (a software multi-state filter, with states for each significant inertial sensor and GPS error terms) could potentially be used to compare inertial performance with the GPS data which should further reduce MIMU errors to 1 to 10 deg/hr. The calibration and alignment process would continue during the flight of the MUAV. This would allow the MGIMU to navigate to a relatively high accuracy during short periods of GPS signal loss.

Phase I of the proposed SBIR effort is concentrating on:

- Development of MGIMU specification.
- Processor requirements analysis (navigation, Kalman filter, autopilot/guidance functions)
- Conceptual system design (electrical and mechanical) of brassboard and production MIMU.
- Identifying a UAV airframe partner **

Phase II of the proposed SBIR effort will concentrate on the demonstration of a brassboard MGIMU's capability to satisfy the transfer alignment, navigation and autopilot/guidance requirements. The brassboard will be designed to fit within the size and weight requirements to permit future flight test on the MUAV.

Functional Block Diagram

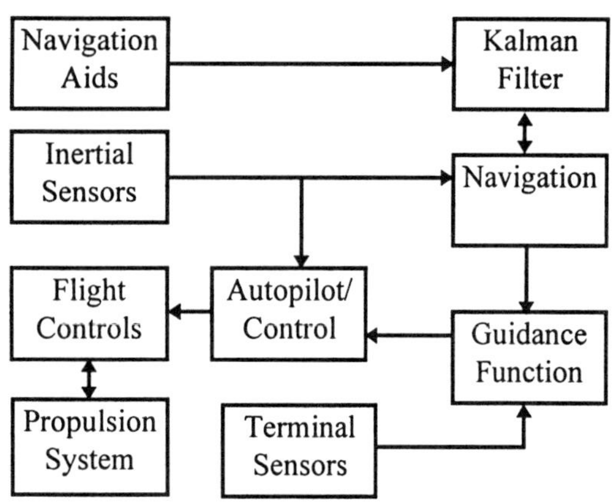

Figure 2 : Functional Block Diagram

Although the system is required to be very small and low power it is also required to be a fully integrated flight guidance and control system. The following section provides a brief overview of the system.

Inertial Sensors

Due to the severe size and weight constraints placed on the air vehicle it is unlikely to be inherently stable. It will require, therefore, active stabilization. This necessitates sensors to provide high bandwidth, acceleration and angular rate information. Micro machined inertial sensors or very low power FOG with micro machined accelerometers are the most likely candidates.

Navigation Aids

The length of the proposed mission is 30 to 60 minutes. This length of time will produce very large errors if only inertial sensors are used. Some form of navigation aid will be required to meet the mission requirements. The most suitable aids would be GPS, Magnetic heading and or pressure sensors. The GPS system could provide accurate long term position and velocity data. The combination of magnetic heading and a pressure sensor would provide long term heading and altitude data. The most suitable navigation aid would depend on the mission.

Terminal Sensors/Seeker

To perform a useful function the system will require addition components such as IR , video or acoustic sensors.

Flight Controls

Innovative flight controls will be required to stabilize and guide the air vehicle.

Propulsion System

The ability to power such a small air vehicle for flights of 30 to 60 minutes will require a very special propulsion system. The most popular candidate appears to be a pulse jet. Other alternatives being considered are autogyros and flapping wings.

Autopilot / Control

The autopilot function that provides the stabilization and control of the air vehicle will need to be extremely robust. The response of the vehicle in likely to be highly non-linear and time varying. The autopilot function will need to be capable of adapting its response to the changing performance of the air vehicle.

Design Trade Studies

A very critical choice in the design of the navigation system for the MUAV is the choice of Inertial sensors. The micro machined accelerometers are proven technology and are in high volume production extensive testing by Fibersense has shown that these can meet the Power performance and size constraints. Duty cycle approaches can reduce the power consumption to acceptable values.

The choice of gyro is more difficult. The most obvious choice due to the small size and low power is a micro machined gyro. The performance of the micro machined gyro may not meet the goals of the MUAV program. Development of new low power FOG's is also a viable option.

The need for a navigation aid is demonstrated later in this paper. The initial response is to consider GPS. Even current low power GPS systems require more power than is available. Lower power GPS systems and other alternative need to be considered. These consist of magnetic compass, pressure sensors (for altitude). The use of a video input to obtain ground speed should also be considered.

Low Power FOG System

Recent advances in FOG technology provide a feasible alternative to the micro machined gyros. Fibersense Technology was part of a consortium consisting of Allied Signal, AT&T and Fibersense. This consortium under a DARPA Technology Reinvestment Program (TRP) MDA972-94-3-0017 exploited developments in the fiber optic communications industry. The program developed a FOG that could utilize components developed for the Fiber To The Home initiative. This resulted in a very low cost , small FOG. This device which contained both the optical bench and the required FOG electronics (IOEM Module).

Fibersense has also been exploring the use of extremely small coils of fiber. An example of small fiber coil is shown in figure 3.

The performance obtainable from the IOEM module containing a low power laser diode source is encouraging.

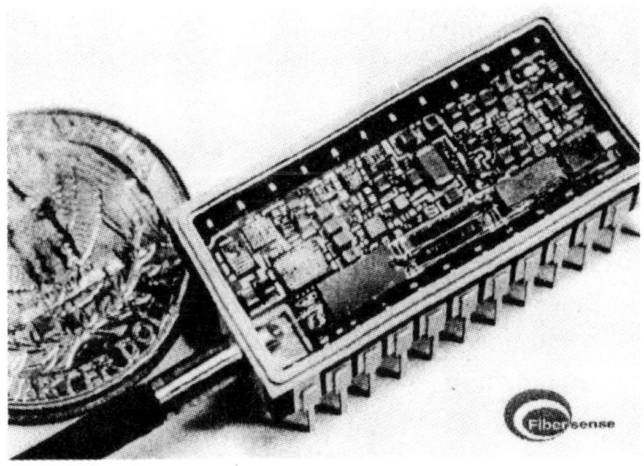

Figure 4 : IOEM Module

The Gyro data varied by only approximately +/- 10 deg/hr over the full temperature range.

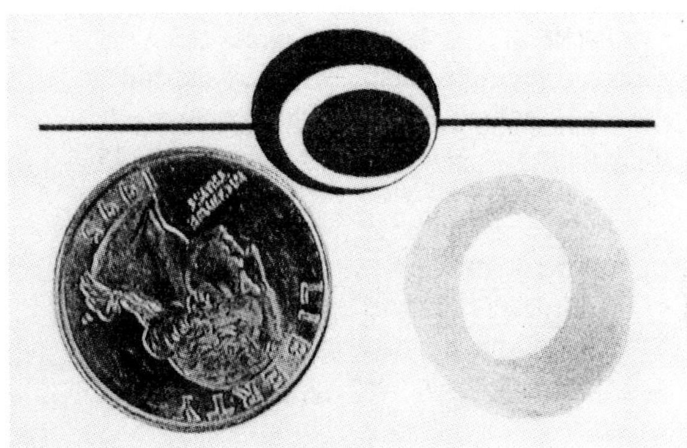

Figure 3: Small Fiber coil

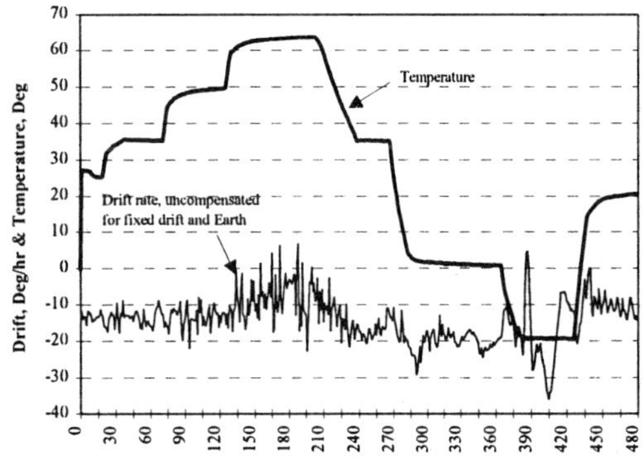

Figure 5 : IOEM Gyro Data

Some development would be required to meet the power constraints but Fibersense is convinced that a power level comparable with a micro machined gyro can be achieved.

Performance Predictions

The performance obtainable from such the inertial system is extremely important. To evaluate the performance achievable from the inertial system a number of covariance simulations have been performed. The covariance simulation models all the major error sources in the (49 Inertial error states).

When considering instruments such as micro machined sensors it is important to consider the errors introduced by environmental effects. The initial reaction when considering the performance of a gyro is to focus on the long term variation in the bias (e.g day-to-day Drift, Deg/Hr) and the short term noise (deg/s/(RootHz)). The error sources are certainly very important. The micro machined gyro is essentially a mechanical device. Even though the size of the moving parts are small it is NOT a solid state device. As such a sensitivity to vibration and g loading must exist. Thermal sensitivities will be present in the device. Some of these sensitivities will be predictable and can be compensated for. Others will be time varying and so cannot be modeled.

If only the bias and noise terms are considered an unrealistically optimistic result can be obtained. The results would indicate that the bias can be estimated. If a long enough integration interval is chosen the affects of even large noise levels can be removed.

A micro machined sensor in development by Analog Devices (ADRS50) is expected to be sold in very large quantities to the automotive industry. The long term drift (15 year) Drift is specified as 1 deg/s. the short term noise is specified as 0.05 deg/sec/Rt.Hz . If these two numbers are considered in isolation a very optimistic performance number can be computed. If an averaging time of 5 minutes is chosen the effective noise of the gyro can be calculated to be approximately 10 deg/hr. This would suggest that the initial drift of 1 deg/s could be calibrated to 10 deg/hr if averaging intervals of 5 minutes could be used. The navigation and Kalman filter algorithms do essentially integrate the gyro data over long intervals. Integration intervals of several minutes are reasonable intervals to consider during an alignment and calibration procedure. In reality other components within the gyro begin to influence the performance at such long integration intervals (such as the noise in the electronics).

The other significant factors to be considered are the environmental effects mentioned previously. The residual time varying errors due to temperature changes, vibration levels and g loading can produce a very different result. Fractions of a degree per second (several hundred degrees/hr) changes in effective drift of the gyro can be introduced by these environmental factors. Working backwards from the predicted instrument sensitivities it would need to be a very benign environment to obtain gyro calibration approaching the 10 deg/hr desired.

Due to the level of development of the current Micro Machined gyros it is very difficult to obtain accurate performance data under realistic environmental conditions. Two levels of performance have been investigated to provide an indication of what level of performance is possible from a representative MUAV inertial system.

Inertial Aiding And Calibration

The mission length of the MUAV is expected to be in 30 to 60 minute range. To maintain a reasonable level of navigational accuracy some form of external aid will be required. For most applications a GPS receiver would be an ideal source of aiding data. A GPS system provide real time absolute position and velocity data. Using this data the inertial system can be aligned and calibrated. This allows the major error sources of the inertial system to be identified and compensation applied. If the GPS signal is then lost (due to external interference or physical shielding) the inertial system can navigate using the compensated inertial data. For relatively short periods the calibrated inertial system can provide accurate navigation.

The performance of the system after the loss of the aiding source depends on a number of factors. These are :- the quality / type of aiding source, the maneuvers performed and the quality of the inertial sensors. The quality and type of aiding source impacts the alignment and calibration accuracy obtained. The greater the noise level the longer it takes and less accurate is the sensor calibration. The maneuvers performed by the vehicle also plays a significant role. The MUAV will not be flying at a high speed. It is also unlikely to be capable of high g turns. This affects how observable the inertial system errors are. An indication of how accurately heading can be estimated is approximated by the ratio of the velocity noise of the GPS and change in velocity performed by the vehicle. The MUAV will not be flying at high velocity (20 m/s say).
The maximum velocity change is, therefore, 40 m/s (a 180 degree turn). If the measurement noise of the GPS in assumed to 0.1 m/s the heading accuracy obtainable would be approximately 2 milli radians (0.1 deg).

An MUAV mission was simulated to give an indication of the accuracy obtainable. Two levels of inertial sensor performance was considered. One represented realistic expectation of current micro machined gyro performance, the other represents a low power FOG (or a greatly improved micro machined gyro.). The initial sensor performance levels are given in table 1.

Gyro Errors	MEM	FOG
Day to Day Drift (deg/hr) Roll, Pitch, Yaw	1,200	10
In run stability (deg/hr)	180	1
Noise	0.05 Deg/s/RtHz	0.15 (deg/RtHr)
Scale Factor (ppm)	1,000	200
Non orthog. (mrad)	5.0	1.0
G-sensitivity (deg/hr/g)	600	0.0
Accel Errors		
Day to Day Bias (mg) X, Y, Z	20.0	20.0
Noise (mg)	0.5	0.5
Scale factor (ppm)	200	200
Non orthog. (mrad)	1.0	1.0

Table 1: Initial Sensor Errors

The MUAV performed a search pattern consisting of a maximum of 90 degree turns coordinated turns at a 5 degree bank angle. For the first 18 minutes the GPS aiding data is available for the remaining 18 minutes the system is navigating using the inertial data only.

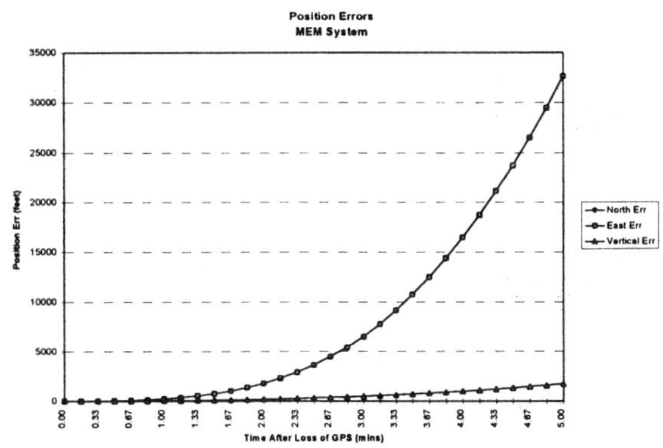

Figure 7 : Position Error MEM

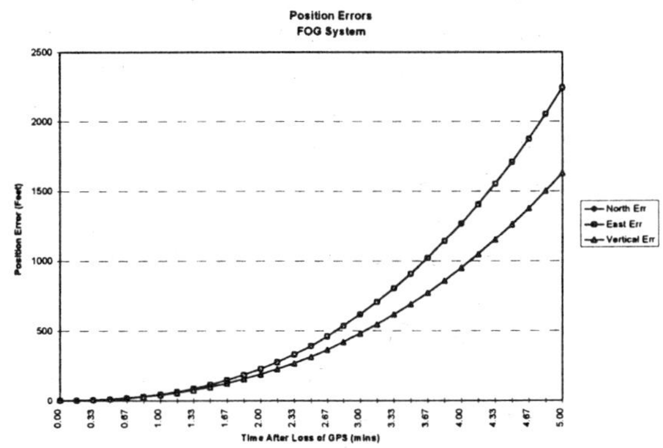

Figure 6 : Position Error FOG

The position errors after up to 5 minutes without GPS data can be summarized as follows.

Time (mins)	MEM err (feet)	FOG err (feet)
1	226	47
2	1,797	228
3	6,531	617
4	16,470	1268
5	32,610	2240

Table 2 : Position Err Vs Time

The predominant error source is the gyro drift. Over this interval the gyro drift propagation into a position error can be approximated as being proportional to the cube of time ($k.t^3$).

The degree to which the two systems were calibrated during the period with GPS is summarized in the following table.

Error Source	MEM	FOG
Roll & Pitch (Deg)	0.3	0.1
Heading (Deg)	3.6	0.4
Gyro Drift (deg/hr)	25*	2
Head. Gyro(deg/hr)	170	3
Accel Bias (mg)	5.2	1.8

Table 3 : Calibration Accuracy

*The g-sensitive gyro drift was essentially uncalibrated at 150 deg/hr/g.

Discussion of Results

The results indicate that some form of navigation aid is essential for use in the MUAV application. Two classes of system were considered one represented the performance expected from micro machined sensors currently in development. The other represented a low power FOG system or a much improved micro machined gyro. Even after calibration using a navigation aid such as GPS the performance obtainable from an all micro machine system is only satisfactory for short intervals of time. Within 2 minutes the velocity error approaches the actual vehicle velocity. The higher performance system can provide satisfactory performance for a much longer time. Which level of performance is acceptable depends on the details of the MUAV mission and the eventual choice of sensors.

The challenge is to meet the performance requirements within the available power constraints. If the navigation aid (e.g. GPS, Magnetic compass etc.) can be relied on for continuous operation during the whole mission an all micro machined solution can be provided. If the system is required to perform through periods of loss of the navigation aid a system using low power FOG's and micro machined gyros would be required. The preliminary studies indicate that although developments are required in several areas the navigation system for the MUAV concept is feasible with the currently emerging technologies.

AIAA-97-3555

3D LOCAL TRAJECTORY PLANNER

Jerzy Z. Sasiadek *
Dept. of Mechanical and Aerospace Eng., Carleton University
1125 Colonel by Drive, Ottawa K1S 5B6, Canada

Ignacy Dulęba
Inst. of Engng Cybernetics, Technical University of Wrocław
Janiszewski St. 11/17, 50-362 Wrocław, Poland

Abstract

In this paper a local trajectory planner is described and applied. This planner works in three dimensional environment populated with static and passive movable obstacles. Dynamics of a moving vehicle and its environment determine the vehicle performance. The performance evaluation is done by minimizing a criterion function. Due to two stage motion planning (decision-trace modes) problem of being trapped by the vehicle in a local minimum of the criterion function is avoided. Environment is perceived at an abstract level therefore the source of environment related data is not very important. The planner, although autonomous, may be adjusted by higher order system (strategic behavior) by changing its parameters. It is not computationally intensive and therefore can be used in real-time applications.

1 Introduction

Controlling a vehicle is not a simple task, because environment as well as its dynamic properties must be taken into account while planning its motion. The vehicle can be a free-flying structure, mobile robot or any autonomous guided vehicle. Usually, the task of motion planning is divided into two stages: a path planning and a trajectory planning. The path planning has an environment description as an input with its start and goal points. It generates a geometrical path the vehicle needs to follow. The geometrical path, in sequence, is time-parameterized by a trajectory planner. The trajectory planner based on dynamic characteristices of the vehicle determines time-dependent characteristics like positions, velocities and accelerations. The trajectory is executed by trajectory tracking system which keeps real values of the positions as closed as possible to their preplanned values. At the path planning level there is a tendency to make assumptions idealizing environment. Obstacles are convex polygons which helps to apply techniques taken from a graph theory, in order to find the shortest path. In real situations the assumptions are hardly ever valid.

The global path planning is performed in off-line mode. Moreover, the situation is even more complex when we consider an environment as an open world, never really known in details. When a global information known in advance is neither perfect nor predictable, there is a tendency to design so-called local path planners[1,6]. Although such planners lead to the loss of a path optimallity yet their actions are focused on a target reaching while avoiding obstacles.

The paper addresses a problem which falls in somewhere between a path planning and a trajectory planning. It covers a situation when environment data acquired before trajectory planning are not perfectly reliable, yet they exist. A rough map of the environment is built using environment data and a path is planned. A task of local trajectory planner is to plan trajectory based on the path, yet the planner must perform tactical decision in order to modify preplanned path when the path, due to lack of reliable data, may collide with obstacles. Please note, that the situation fits neither into global path planning, nor into the trajectory planning. It does not fit into the first group because local modifications of the path are required. Also, it does not fit into the second group because a global path planner is not invoked in case of an unpredictible event.

In this paper three dimensional version of the local trajectory planner, developed for the 2D case

*Professor,
Associate Fellow

in ref. 15, will be presented.

Before giving a detailed description of a three dimensional (3D) local trajectory planner let's properly specify the system amongs other planners known in literature of the subject. Path planners are generally divided into local and global ones and, as a rule, they do not take into account dynamic properities of the vehicle. Former group works in on-line mode while the latter in off-line mode. Known approaches to path planning might be divided into four categories: deterministic, stochastic, learning-based, reflexive (behavioral). Deterministic approaches are used most often in global path planners and they employ various techniques (eg. Voronoi diagrams[13]). This technique applies to the stationary, known obstacles. When movable obstacles are considered and their characteristics are known in advance they may be incorporated in the global path/trajectory planner by adding to spatial variables a time variable and solving the task problem in the such augmented space[8,10].

More often, to represent dynamic environments, i.e. with movable obstacles and unknown or partly unknown characteristics, a stochastic approach is used[3,14]. In stochastic approaches, planners usually work as a local planners and a decision of moving vehicle in a particular direction is made on the basis of minimizing probability of collisions. The machine-learning techniques are applied[5] in unfamiliar environments. These techniques are based on an assumption that although the world is unknown in advance, it may be learned, so there is a regularity in the world and vehicle's navigation amongst obstacles gains knowledge about the environment. This approach may be useful if the environment is regular without any sudden changes. Therefore, vehicle can learn it before being damaged by moving obstacles. All methods mentioned above build a model of the world and planned actions consistent with this model. Also, there are other methods which do not employ any world-model building procedure and they act behaviorally (reflexively).

An important problem in mobile robotics is a data source and collection modules. Currently, the most popular source of data about environment is a sonar-sensor, which gives fast and useful information.

A task for a local trajectory planner is as follows: given data, geometrical path a vehicle (mobile/free-flying robot) needs to follow (environment is populated by static and movable obstacles) as an input find a collision free path and velocity profile along it. In our case vehicle is assumed to be a point and its dynamic behavior is incorporated into maximal allowable velocity and acceleration. When the vehicle occupies some volume it can be reduced to a point by growing obstacles by a half of its diameter. Orientation changes of the vehicle are constrained and the constraints are velocity dependent. The vehicle is equipped with sonar sensors. The system, we introduce, builds at first a histographic map (*physical map*) of an environment which in turn is translated into *a control map*. The histogram techniques has been used previously by Borenstein and Koren[1,2]. It allows to condense all sonar-range information into a sectorial data (distances to the nearest obstacle in the sectors). The data condensed in sectors are virtually sensor independent. In our case the physical map of obstacles is supplemented by a *physical goal*, i.e. the goal the vehicle would direct to from its current position if there is no obstacles around it. The control map, absent in original histogram approach, modifies the physical map as follows: a *real goal* supplements the physical goal and movable obstacles influence the map. The real goal shows direction the vehicle will direct to in obstacle's populated environment. When there is no obstacles both goals are the same. Movable obstacles are either active ones, when they may cause collision with the vehicle or passive ones, when, although visible may not cause any collision. Passive obstacle, when detected, should be filtered out from the control map. In an 3D local trajectory planner only static obstacles and passive movable obstacles will be considered. To deal with active movable obstacles higher order system should be impleneted which having some global environment perception and knowledge may develop a right decision. Nevertheless, some heuristics even at the local trajectory planner can be put forward. Couple of them are listed below:

- the active obstacle is put at the control map at the place(s) the obstacle may collide with the path of the vehicle,
- an extra factor in criterion function (cf. Eq. 11 Section 2) is added to penalize a path for approaching to a movable obstacle,
- The two above heuristics do not advise how velocity should be controlled. Should it be increased to avoid obstacles by runing it out, or decreased to wait for being passed by a movable obstacle.

Obviously, one can find examples when all strategies of avoiding active obstacles may fail, mostly due to the lack of global information. Therefore, active obstacles are not included in our considerations, and were put as a task for higher order system.

A control map is a base to work out a control law determining next velocity and attitude of the vehicle based on the map and current values of the parameters. The most demanding task in the system is a real goal determination. Recently, the most popular approach to determine a real goal is based on a potential field approach[4,7,9,11] although the approach has its own disadvantages[12]. Taking advantage of a preplanned, but not necessarily collision free path, we need not use the potential field approach. A real goal is determined in two different modes. First (*decision mode*), is applied when the vehicle starts its motion. In this mode a criterion function is defined on a set of possible directions of motion. The function should take into account the following factors: distance to the physical goal, difficulty in reorienting the vehicle, obstacle avoidance. The function is minimized to give the best direction for the next step of control. Unfortunatelly, the mode cannot be applied alone due to the fact that the vehicle can be trapped in a local minimum of the criterion function, which means that being far from the goal the vehicle is stopped permanently. To avoid this setback *a trace mode* is added. This mode is switched on when a decision mode can not determine the approaching direction to the goal due to obstacles. In the trace mode the vehicle follows boundary of the obstacle on the way to the physical goal, as long as conditions are not met to switched on to the decision mode again.

The paper is organized as follows: Section 2 describes a local trajectory planner working in 3D environment populated with static and passive movable obstacles. Section 3 presents numerical simulations of the system. Section 4 concludes the paper.

2 A local trajectory planner - modules.

In this section, after introducing nomenclature, the 3D local trajectory planner scheme is given and its components discussed.

2.1 Nomenclature

$|\cdot|$, $||\cdot||$ absolute value and Euclidean norm, respectively.

$<\cdot,\cdot>$, $\cdot\times\cdot$ scalar and vector products, respectively.

$||\cdot,\cdot||_{ang}$ an angle distance between normalized vectors defined as follows:

$$||x,y||_{ang} = arc\cos(<x,y>)\ ,\ ||x|| = ||y|| = 1\ . \tag{1}$$

$Trian$: triangularization of the sphere in R^3 (S^2), is a set of vectors (directions) uniformly distributed on the sphere.

$$Trian(S^2) = \{x \in R^3 \mid \forall x \exists z \quad ||x-z|| < \epsilon\ , \\ ||x|| = 1, ||z|| = 1\} \tag{2}$$

with small positive constant ϵ determining the number of directions constituting the triangulating set. Triangulation is a domain of a criterion function evaluating possible directions of motion. Each of the direction defines a *sector* centered around the vector.

$Strip$: subset of a triangulation set, defined as follows: let be given non collinear unit vectors $x, y,\ z = \frac{x\times y}{||x\times y||}$, and an strip-angle parameter ξ

$$Strip(z,\xi) = \{r \in Trian(S^2)\ , \\ |\ ||r,z||_{ang} - \tfrac{\pi}{2}| < \xi\} \tag{3}$$

The strip groups all directions close to the plane spanned by vectors x, y.

$MAJOR$ is a constant denoting a sonar range, i.e a radius of a ball observed by sensors.

Map is a function which assignes a distance for each direction x from $Trian$ to the nearest obstacle in this direction.

$$dist(x, Obst) > MAJOR \Rightarrow dist(x, Obst) = \infty \tag{4}$$

Sectors with $dist(x, Obst) = \infty$ are called *obstacle free* and form $OFree$ set,
those sectors with $dist(x, Obst) < \infty$ are called *obstacle populated* and they are collected in $OPop$ set.

$dist_F(x, OPop)$ is a distance for $x \in OFree$ sector to the nearest obstacle-populated sector:

$$dist_F(x \in OFree, OPop) = \\ \min_{y\in OPop} ||x,y||_{ang},\ \in (0,\pi) \tag{5}$$

$dist_P(x, OFree)$ is a distance for $x \in OPop$ sector to the nearest obstacle-free sector:

$$dist_P(x \in OPop, OFree) = \\ \min_{y\in OFree} ||x,y||_{ang}\ \in (0,\pi). \tag{6}$$

v_{max} amplitude of a maximal allowable linear velocity,

a_{max} amplitude of a maximal allowable acceleration/deceleration,

$pos(t), v(t), ATT(t)$ current position, velocity, attitude of a vehicle. Let's assume $||ATT(t)|| = 1$. For visualisation purposes the attitude is expressed in spherical coordinates.

$knot_{curr}$ current (sub)goal (knot point),

$MINOR$ is a distance allowing to stop the vehicle moving with maximal velocity, $MINOR = \frac{v_{max}^2}{2 \cdot a_{max}}$

All the vector-based functions have one more argument not shown here. This argument is a position in R^3 when the vectors have their starting point, i.e. the current position of the vehicle.

The task of a local trajectory planner relies on determining a path, velocity and orientation along it for a vehicle following a prescribed geometrical path, possibly populated by static or/and passive movable obstacles. The scheme of a local trajectory planner is depicted in Fig. 1.

Figure 1: Modules of a local trajectory planner.

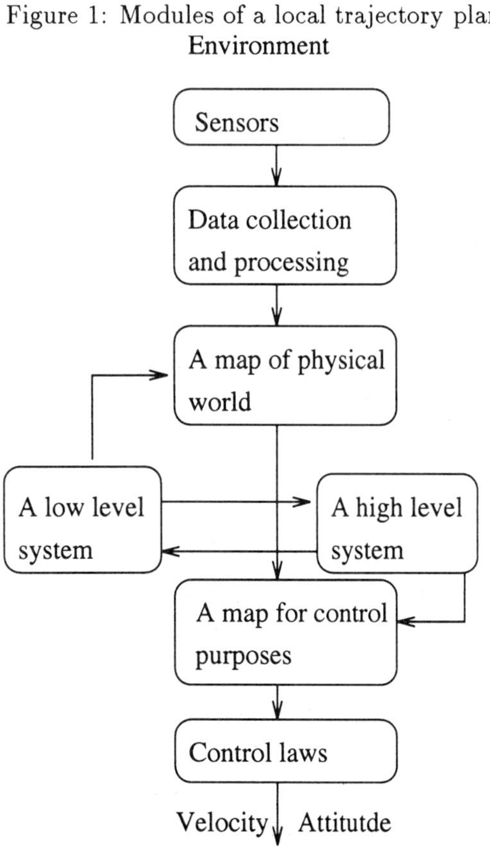

A path to follow is supplied to a local trajectory planner by a high level system. It takes a form of a set of knot-points (*knots*) joined by a straight lines in Cartesian space (*segments*). An *active segment* is the segment on which base the trajectory is being planned. The first active segment has its left boundary at the starting point. Two knots are distinguished. The initial knot where the vehicle begins its motion and the final one which is a goal the vehicle is planning to reach. Knots are encircled by a specific radius. When the vehicle reaches an area around knot point within the radius, the active segment is changed. Segment changes have been introduced to smooth the resulting trajectory. Otherwise, with segment changes at knots, the trajectory would be sharp at knots. Formally, the path can be described as follows:

$$\text{PATH} = \{knot_i \in R^3 \mid i = 0, \ldots, N\} \tag{7}$$

$$\text{RADII} = \{r_i \in R \mid i = 1, \ldots, N-1\}$$

where $N+1$ is the number of knots (k_0, k_N the initial, and the final knot, respectively). r_i a radius of segment change for knot k_i.

Vehicle equipped with sonar sensors obtains information about its environment. After preliminary processing the data, it creates a physical map of the environment. The map is just a sphere with unit radius divided into sectors. In each sector (x) a resultant obstacle is assigned which includes obstacles within the sector. The obstacle is characterized by a distance $map(x)$ from the current position of the vehicle. The vehicle occupies a centre of the physical map and it is characterized by its current velocity and orientation. Vehicle's motion is constrained by parameters reflecting its dynamic properties. Vehicle's velocities, accelerations and reorientation abilities are constrained:

$$0 \leq v(t) \leq v_{max} \quad , \quad a(t) \leq a_{max} \quad , \tag{8}$$

$$\Delta\alpha = f(v(t)) \quad . \tag{9}$$

where $v(t), a(t)$ are amplitudes of velocity and acceleration of the vehicle, respectively; $\Delta\alpha$ is a maximal allowed orientation change in a unit interval of time. It is reasonable to assume that the last quantity is a velocity dependent. Actually, it is computed according to the formula:

$$\Delta\alpha(v(t)) = \alpha_{min} + (\alpha_{max} - \alpha_{min}) \cdot (1 - \frac{v(t)}{v_{max}}) \tag{10}$$

where $\alpha_{min}, \alpha_{max}$ are prescribed values of minimal and maximal orientation changes.

The physical map is supplemented by a physical goal (PG). PG is a unit vector pointing at a current segment point towards which the vehicle would direct itself to in the absence of obstacles.

In Fig. 2 four different cases of determining PG are shown:

a) if a current segment lies within vehicle's sonar range, then PG is one of the two points where sonar range intersects with the segment line. It is always, the nearest point to the given segment end-point,

b) if a goal (final knot) is within the sonar range, then PG coincides with the goal-point;

c) if a current segment is out of vehicle's sonar range, then PG is a point where a bisector of an angle between a line perpendicular to the segment from current vehicle location and a line to the segment's end point crosses the segment;

d) if the vehicle is within an area of knot radii from the knot (segment end-point), then a current segment is changed for the next in sequence of segments, Then one of the a)-c) cases applies.

Figure 2: Three cases of determining PG

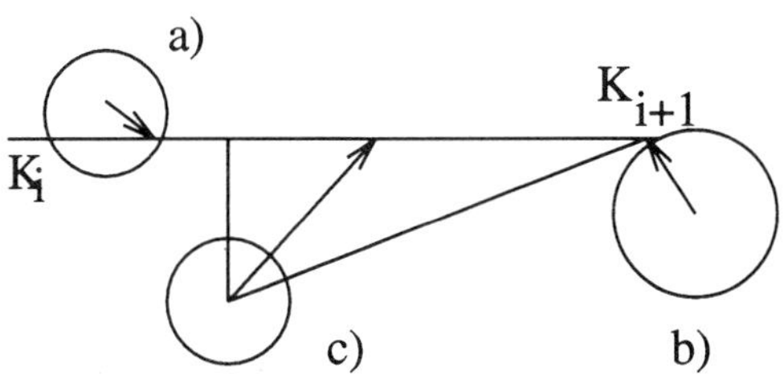

The map of physical world can be summarized as a picture taken from the current position of the vehicle a physical goal added. Real obstacles are lumped together to form a virtual obstacle which populates any given sector. Unfortunately, such a map cannot be used directly for control purposes because the physical goal may be obscured. One of the problems is related to movable obstacles. If obstacles are passive they should be filtered out from the map. If obstacles are active they must be transfered to locations which are the most dangerous to the vehicle and their previous positions (should be) filtered out as well. These modifications are the basis for designing of a control map which is a modified physical map. PG has also its control map counterpart. In general, the vehicle will be directed to the real goal (RG). RG may be just PG if there is no obstacles but it is not true in obstacle crowded environment. RG is determined in two possible modes:

a) Decision mode. The mode in which a vehicle has difficulties in reaching PG. In this mode strategic decisions are made on the basis of three factors: safety, goal reaching and difficulties in reorientating a vehicle.

b) Trace mode. In this mode decisions worked out by the decision mode are executed. the vehicle follows obstacle boundaries when the decision mode is invoked again.

Below details of the two modes are presented.

2.1.1 Decision mode

In the decision mode all obstacle-free ($OFree$) sectors are evaluated according to the formula:

$$J(x) = k_1 \cdot J_1(x) + k_2 \cdot J_2(x) + k_3 \cdot J_3(x),\ ||x|| = 1\ , \tag{11}$$

and the optimal sector becomes the real goal.

$$J(RG) = \min_{x \in OFree} J(x)\ . \tag{12}$$

Components of the criterion function $J(x)$ are following:

$J_1(x)$ represents the angle distance from the physical goal of the sector x

$$J_1(x) = ||x, PG||_{ang}\ ,\ ||PG|| = ||x|| = 1\ , \tag{13}$$

$J_2(x)$ represents safety measures in maneuvering among obstacles.

$$J_2(x) = \pi - dist_F(x, OPop) \tag{14}$$

π term has been introduced to reflect small values of this component for the safest sectors. Note, that in case of the $J_2(x)$ sectors which have smaller distance from PG than the prescribed J_2 assumes 0. That design allows for not to punishing sectors which are fairy safe.

$J_3(x)$ represents difficulties in reorienting the vehicle.

$$J_3(x) = ||x, ATT||_{ang}\ ,\quad ||ATT|| = ||x|| = 1\ , \tag{15}$$

where ATT is current attitude of the vehicle.

Positive k_1, k_2, k_3 were introduced to allow comparison among different quantities and, more important, to choose one or more functions over others. These are the weighting factors.

In the algorithm after selecting the real goal, a distance from the current position to the next knot point is saved as $dist_{PG} = ||knot_{curr} - pos||$. This distance is important in switching from a trace to a decision mode.

2.1.2 Trace mode

This mode has been introduced to prevent the vehicle to stop at a local minimum of the criterion function defined in the decision mode. The real condition to switch from decision to trace mode do not require stopping the vehicle. To switch modes the following condition must be satisfied:

$$dist_P(PG, OFree) \geq dist_{D2T} \ . \tag{16}$$

where $dist_{D2T}$ is a given constant. The switching condition states that the physical goal and its neighborhood is populated with obstacles. The condition (16) prevents the vehicle to get too close to obstacle just to minimize the distance to the goal-point. It invokes a trace mode when the vehicle is relatively far from obstacles on its way to the goal. In the trace mode vehicle traces obstacles boundaries (at some distance). The tracing procedure is as follows:

1. when trace mode is recalled $dist_{PG}$, PG_{mem} and RG_{mem} are recorded. PG_{mem} and RG_{mem} are values of PG and RG in the last moment of the decision mode, respectively. Moreover, a point A is determined on the PG_{mem}, RG_{mem} plane which shows boundary of the obstacle to be traced, (Fig. 3)

2. all directions evaluated in the trace mode to determine a real goal should belong to the strip, $RG \in Strip(\frac{RG_{mem} \times PG_{mem}}{||RG_{mem} \times PG_{mem}||}, \eta)$. The condition imposed on RG states that RG should keep in proximity of the plane PG_{mem}, RG_{mem} within a prescribed angle η.

3. in the succesive iterations, when new maps are built, obstacle populated sectors from the strip are evaluated to determine "succesor" of the boundary point A. The "succesor" is simply a new coordinate of point A in the next iteration. Point A indicates the boundary of the currently traced obstacle (edge of the obstacle). Let's assume β as a safety-range parameter. If the succesor has no obstacles in the range of 2β measured clockwise on the PG_{mem}, RG_{mem} plane, then a real goal is determined according to the formula:

$$RG = Rot(\frac{RG_{mem} \times PG_{mem}}{||RG_{mem} \times PG_{mem}||}, -\beta) \cdot Succ(A) \ . \tag{17}$$

where $Rot(k, \Theta)$ is a rotation transforamtion about given vector $k = (k_1, k_2, k_3)^T$ by an angle Θ,[13]:

$$Rot(k,\Theta) = \begin{bmatrix} k_1^2 vers_\Theta + \cos(\Theta) & k_1k_2 vers_\Theta - k_3\sin(\Theta) & k_1k_3 vers_\Theta + k_2\sin(\Theta) \\ k_1k_2 vers_\Theta + k_3\sin(\Theta) & k_2^2 vers_\Theta + \cos(\Theta) & k_2k_3 vers_\Theta - k_1\sin(\Theta) \\ k_1k_3 vers_\Theta - k_2\sin(\Theta) & k_2k_3 vers_\Theta + k_1\sin(\Theta) & k_3^2 vers_\Theta + \cos(\Theta) \end{bmatrix}$$

where $vers_\Theta = 1 - \cos(\Theta)$. If the succesor does not satisfy the safety condition, next succesor of A is searched clockwise, cf. Fig. 3.

Figure 3: Determining RG in trace mode: a) after switching from decision mode, b) couple of iterations later

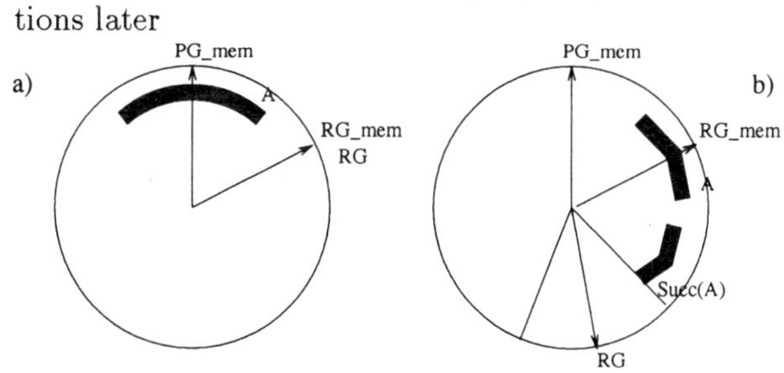

Keeping close to the plane prevents the vehicle from wandering on the obstacle surface.

The decision mode is called back when PG is clearly visible and a distance to the current goal is smaller than the value stored in the memory when the trace mode was originally invoken, , i.e.:

$$\begin{aligned} dist_F(PG, OPop) \geq dist_{T2D} \ , \\ ||knot_{curr} - pos|| \leq dist_{PG} \end{aligned} \tag{18}$$

with constant parameter $dist_{T2D}$. It should be noted that by following boundaries of obstacles the vehicle could even increase the distance to the (sub-)goal, yet, purposefully, it avoids to be trapped in a local optimum of the criterion function.

2.2 Velocity and orientation control laws

All concepts introduced previously were geometric in nature. The control laws module (cf. Fig. 1) which takes into account robot's dynamic charateristics will be discussed next. The module (on the basis of control map and a current vehicle's attitude and velocity) works out the next values of the atti-tutde and velocity. Instead of using rather complicated procedures[6] it is proposed that a very simple mechanism based on the concept of negative feedback should be considered. The feedback works as follows: the vehicle compares its current velocity with a velocity considered safe in the given environment. If the difference is negative, it increases the current velocity, otherwise decreases the velocity. Below, this concept is presented more formally.

The objective of control system laws is to derive $(v(t+\Delta t), ATT(t+\Delta t))$ on a basis of $(v(t), ATT(t))$ and $CONTROL\ MAP(t)$. Lets start with the simpler component: the attitude.

$$ATT(t+\Delta t) = Rot\left(\frac{ATT(t) \times RG(t)}{||ATT(t) \times RG(t)||}, \gamma\right) \tag{19}$$

where

$$\gamma = \begin{cases} \Delta\alpha\,\Delta t & \text{iff} \quad \Delta\alpha\,\Delta t \leq ||ATT(t), RG(t)||_{ang} \\ ||ATT(t), RG(t)||_{ang} & \text{otherwise} \end{cases} \tag{20}$$

$\Delta\alpha$ is computed according to the equation (10).

The velocity control law is as follows:

$$\begin{aligned} \tilde{v} &= v(t) + g \cdot \Delta t \cdot a_{max}, \\ v(t+\Delta t) &= \begin{cases} 0 & \text{iff} \quad \tilde{v} \leq 0 \\ \tilde{v} & \text{iff} \quad 0 < \tilde{v} < v_{max} \\ v_{max} & \text{otherwise} \end{cases} \end{aligned} \tag{21}$$

where $g = g(v(t), CONTROL\ MAP(t)) \in [-1, 1]$ is a function assumed by designer. That should increase the speed if there is no obstacles, and decrease, if obstacles might be dangerous for the vehicle. In order to formalize what may constitute "the dangerous situation" for vehicle the measure of "safe distance" from obstacles has been introduced. This measure takes form of the following function:

$$dist_{safe} = \frac{v^2}{2 \cdot a_{max}} = MINOR \cdot \frac{v^2}{v_{max}^2} \tag{22}$$

That distance is compared with other distances to obstaclea in *an active area*, AA.

$$dist = \min_{x \in AA} map(x), \tag{23}$$

$$AA = \{x \in Trian(S^2)\ ,\ \ |\ ||x, ATT||_{ang} + ||x, RG||_{ang} - ||ATT, RG||_{ang}| < \delta\} \tag{24}$$

where δ is a prescribed constant. The difference $a = dist - dist_{safe}$ is an argument for $g(t) = g(a)$ function. The function is depicted in Figure 4.

Figure 4: $g(a)$ function.

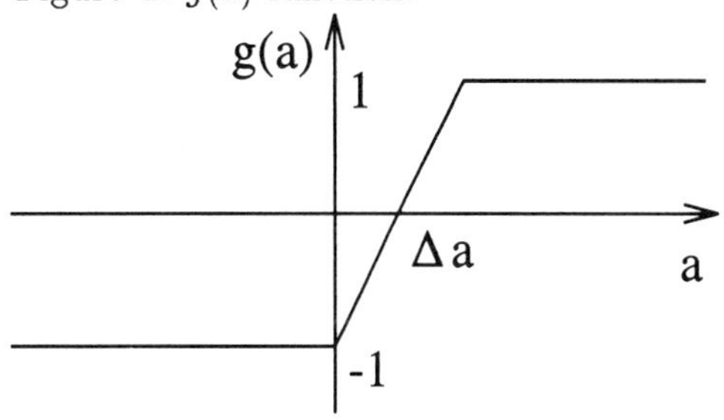

Let's analyze how the negative feedback works in this case. If, there is no obstacles $dist > dist_{safe}$ and the vehicle is sped up until it reaches its maximal allowable velocity v_{max}. If, there are some obstacles, all depends on vehicle's current velocity. When it is small enough that $dist - dist_{safe} > \Delta a$ the vehicle is sped up slightly. When the difference is positive, yet smaller than Δa the vehicle is slowed down, but not drastically. When obstacles are dangerous for the vehicle $dist - dist_{safe} \leq 0$, maximal allowable decceleration is applied. The vehicle reduces its velocity (sometimes even to zero) and by reorienting itself finds sectors for which $dist - dist_{safe}$ has a value which allows to increase velocity.

3 Numerical examples.

The objective of the first test was to verify overall performance of the 3D local planner. The assumptions for the experiment are as follows: there are only static obstacles (paralleloids) with their parameters shown in Table 1. The path to follow is given in Table 1. Data in the tables is dimensionless. The following numbers have been assumed: $ATT(0) = (0, -1, 0)$, $v(0) = 0$, $k_1 = 1, k_2 = 0, k_3 = 1$, $v_{max} = 10$, $\alpha_{min} = 20°/s, \alpha_{max} = 100°/s$. Simulation results are presented in Figures 5, 6, 7, 8.

Some important points on the path (cf. Fig. 5):

A PG becomes obscured by obstacles and the trace mode is invoken.

B switching back to the decision mode,

C movement along a transition segment. Due to high velocity the velocity with delay traces the prescribed path. The effect can be reduced by increasing reorienting abilities of the vehicle (cf. Figure 9 where $\alpha_{max} = 130°/s$ and all other data are the same as in the basic task)

Figure 5: Motion in 3D space

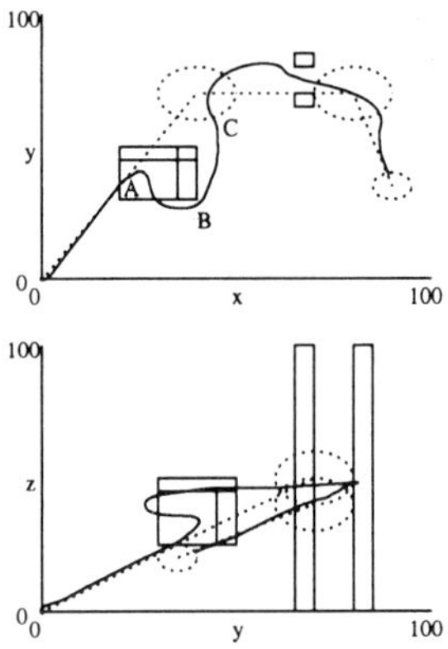

Figure 6: Velocity along the path and the distance to the nearest obstacle

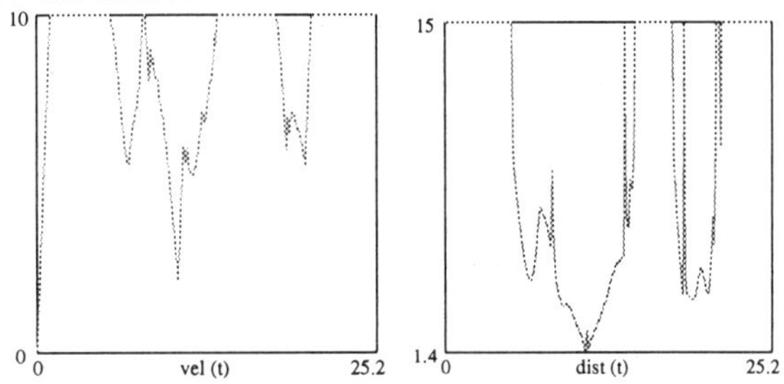

Figure 7: Object orientation - a nutation (alpha) and a precession (beta) angles

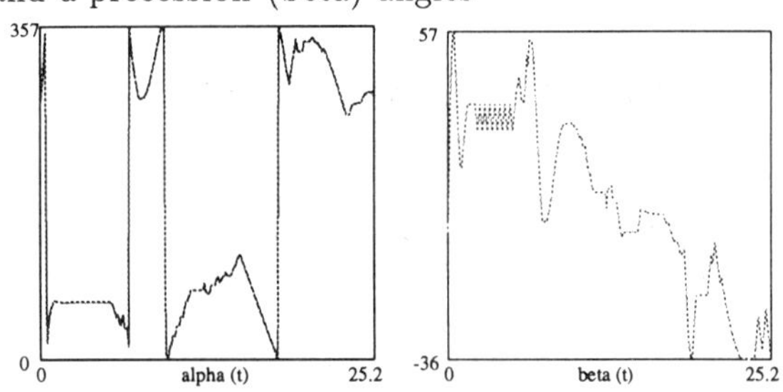

Figure 8: Coordinates of the object along the path

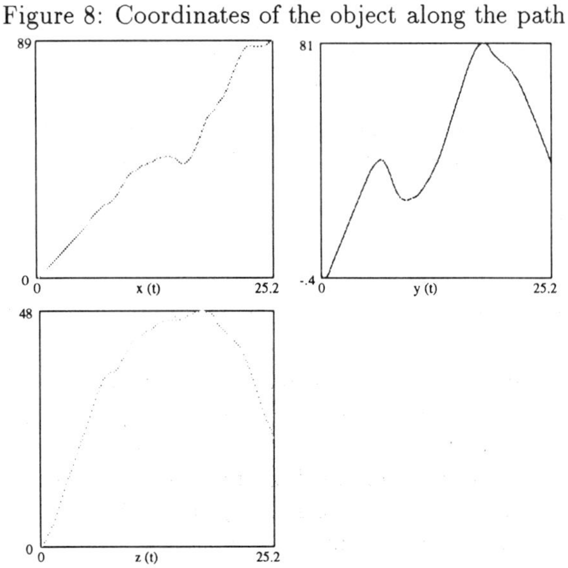

Figure 9: Motion in 3D space

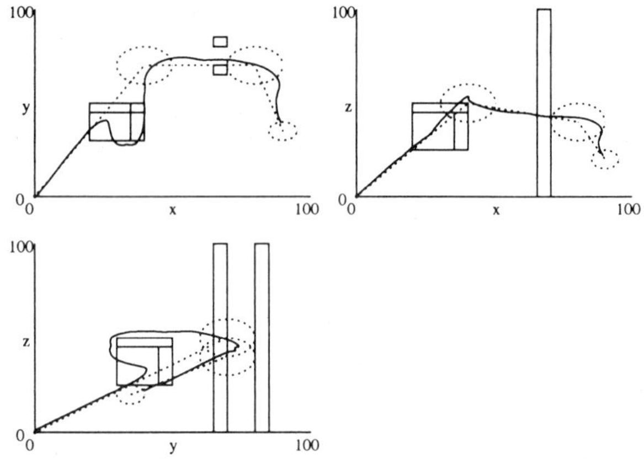

Table 1: Obstacles' parameters and knot points - a path to follow.

Obstacles					
x	y	z	Δx	Δy	Δz
20	30	45	20	20	5
35	30	25	5	20	20
20	45	25	20	5	20
65	65	0	5	5	100
65	80	0	5	5	100

Path			
x	y	z	Radius
0	0	0	0
40	70	50	10
80	70	40	10
90	35	20	5

The second experiment shows the need for a two level map. In the case of passive movable obstacles they are just filtered out from the physical map and on the control map they are not seen. If the filtering has not been done (only the physical map) vehicle may react on moving obstacles which do not interrupt its motion. Assumptions for this experiment are as follows: there is only one obstacle ($x = 0$, $y = 10$, $z = 0$) , ($\Delta x = 20$, $\Delta y = 5$, $\Delta z = 20$) moving along x-axis with different speed. The vehicle should reach the point $(50, 50, 0)$. Other data are as follows: $ATT(0) = (0.707, 0.707, 0)$, $v(0) = 10$. Results are presented in Fig. 10, 11. It could be seen from these figures the vehicle may not choose the best strategy of avoiding movable obstacles (the best strategy would be to pass the obstacle by moving to the left from the goal direction not to the right as in the plots), because local information can be sometimes too poor to work out a right decision.

Figure 10: Path followed by object in xy plane for obstacle's velocity equal to 3/5/7

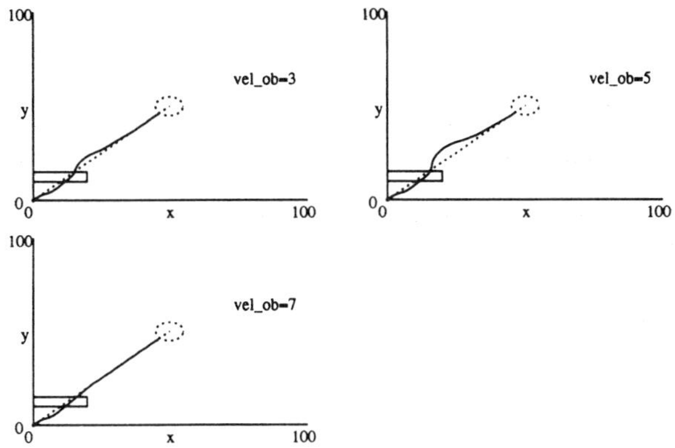

Figure 11: Object's velocity profile for obstacle's velocity equal to 3/5/7.

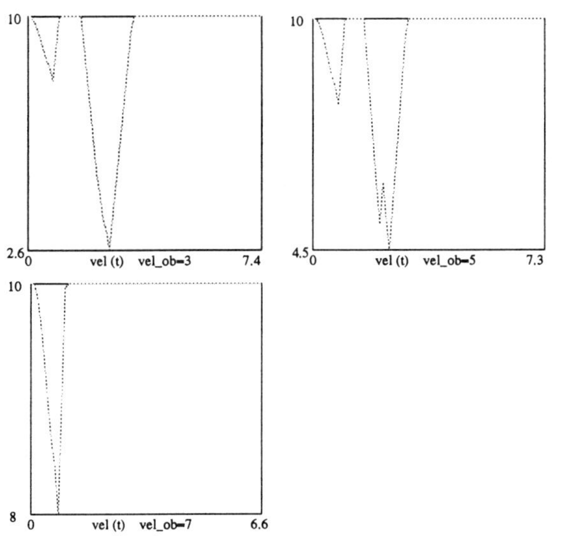

4 Conclusions

In this paper we have considered a local trajectory planner working in 3D environment populated with static and passive mobile obstacle. The task for this planner falls somewhere between a local path planner and a global planner. The novelty of the proposed algorithm is in splitting map of a vehicle's world into two components: the physical and the control map which allows to represent the world effectively for control purpose and avoid a serious drawback potential field based planners of being trapped in local optimum of a criterion function. The second advantage of the system is that a mechanism based on a negative feedback from an environment has been proposed to control velocity of the vehicle. Computational requirements needed to apply the algorithm in real situations are not very heavy (especially that most of the calculations can be parallelized) and therefore it may be applied to the 3D case for free-flying robots as well as for 2D mobile ones in real time mode.

References

[1] Borenstein, J., and Koren Y., "The Vector Field Histogram - Fast Obstacle Avoidance for Mobile Robots," 1991 *IEEE Trans. on Robotics and Autom.* Vol. 7, No. 3, pp. 278-288.

[2] Borenstein, J., and Koren Y., "Histogramic In-motion Mapping for Mobile Robot Obstacle Avoidance", 1991 *IEEE Trans. on Robotics and Autom.* Vol. 7, No. 4, pp. 535-539,

[3] Burlina, P., DeMenthon, D., and Davis, L.S., "Navigation with Uncertainty: Reaching a Goal in a High Collision Risk Region", 1992, *IEEE Int. Conf. on Robotics and Autom.*, Nice, pp. 2440-2445.

[4] Chuang, J.H., and Ahuja N., "Path planning using the Newtonian potentials", 1991, *IEEE Int. Conf. on Robotics and Autom.*, Sacramento, pp. 558-563.

[5] Clark, R.J., Arkin, R.C., and Ram, A., "Learning Momentum: On-line Performance Enhancement for Reactive Systems", 1992 *IEEE Int. Conf. on Robotics and Autom.*, Nice, pp. 111-116.

[6] Elnagar, A., and Base, A., "Heuristics for Local Path Planning", *IEEE Syst. Man and Cybern.*, Vol. 23, No. 2, 1993, pp. 624-634.

[7] Hwang, Y.K., and Ahuja, N., "A Potential Field Approach to Path Planning", *IEEE Trans. on Robotics and Autom.* Vol. 8, No. 1, 1992, pp. 23-32.

[8] Kant, K., and Zucker, S., "Planning Collision-Free Trajectories in Time-Varying Environments: a Two-Level Hierarchy", 1988 *IEEE Int. Conf. on Robotics and Autom.*, Philadelphia, pp. 1644-1649.

[9] Khatib, O., "Real-Time Avoidance for Manipulators and Mobile robots", *Int. Journ. of Rob. Res.* Vol. 5, No. 1, 1986, pp. 90-98.

[10] Kyriakopoulos, K.J., and Saridis, G.N., "An Integrated Collision Prediction and Avoidance Scheme for Mobile Robots in Non-stationary Environments", 1992 *IEEE Int. Conf. on Robotics and Autom.*, Nice, pp. 194-199.

[11] Ratering, S., and Gini, M., "Robot Navigation in a Known Environment with Unknown moving Obstacles", 1993 *IEEE Int. Conf. on Robotics and Autom.*, Atlanta, pp. 25-30.

[12] Slack, M.G., "Navigation Templates: Mediating Qualitative Guidance and Qualitative Control in mobile robots", *IEEE Syst. Man and Cybern.*, Vol. 23, No. 2, 1993, pp. 452-466.

[13] Takahashi, O., and Schilling, R.J., "Motion Planning in a Plane Using Generalized Voronoi Diagrams", 1989 *IEEE Trans. on Robotics and Autom.* Vol. 5, No. 2, pp. 143-150.

[14] Zhu, Q., "Hidden Markov Model for Dynamic Obstacles Aviodance of Mobile Robot Navigation", *IEEE Trans. on Robotics and Autom.* Vol. 7, No. 3, 1991, pp. 390-397.

[15] Sasiadek, J., and Dulęba I., "Local Trajectory Planner", 1995 *AIAA Guidance, Navigation and Control Conf.*, Baltimore, pp. 1474-1483.

[16] Spong, M., and Vidyasagar, M., "Introduction to Robotics", Adisson Wesley, 1989.

AIAA-97-3556

EXPERIMENTS ON SPACE ROBOT PATH PLANNING USING THE SENSORS DATABASE

Toshiaki IWATA*, Kenzo KODAMA, Fumio NUMAJIRI and Hiroshi MURAKAMI
Electrotechnical Laboratory
1-1-4 Umezono, Tsukuba, Ibaraki 305, JAPAN

Abstract

We propose three sensor-based path-planning methods to simultaneously control the orientation of the main body and the arm position of a space robot using only arm motion. In our study, the database of relationships among the current sensor output, the robot motion, the change of sensor output and/or the sensor output after the motion is required. We then use breadth-first search, modified breadth-first search or A-search to determine the arm path. It is theoretically possible to operate the robot and gather data simultaneously as well as to easily modify the database. We confirmed the validity of our methods in experiments using a drop shaft, which provides ten seconds of microgravity during free fall. This experiment with a space robot using a drop shaft was the first trial of its kind in the world.

1. INTRODUCTION

A space robot can simultaneously control the orientation of its main body and arm position using only arm motion due to its nonholonomic characteristics.[1] This means that it is possible to control many parameters (orientation and arm position of the robot) using fewer inputs (joint velocities). There are some control laws based on which systems with many degrees of freedom can be controlled using fewer inputs than the number of degrees of freedom. Nakamura and Mukherjee presented a space robot control method using the Lyapunov function.[1] Yamada demonstrated a method using optimal control.[2] These methods assume a precise robot model and the angular momentum conservation law. However, some robot model parameters, such as the inertia tensor, are difficult to measure precisely, and atmospheric drag and/or gravity-gradient force violates the angular momentum conservation law. Therefore, we propose three sensor-based path-planning methods which are not based on these assumptions. Our methods require a database of the relationships among the current sensor output, the robot motion, the change of sensor output and/or the sensor output after the motion. It is theoretically possible to operate the robot and gather data simultaneously as well as to easily modify the database.

There are several methods of achieving a three-dimensional microgravity field to perform space robot experiments on earth: (1) using parabolic flight of an aircraft,[3] (2) using buoyancy in a water pool,[4] (3) suspension [5,6] or using special mechanisms such as a parallel link [7] and (4) using a drop shaft.[8] In case (1), the microgravity environment is of low quality (10^{-2} G, G=9.8 m/s^2), the experiment time is short (about 20 seconds) and the standard frame (fixed with respect to the aircraft) is not inertial. However, the experimenter can accompany the equipment so that interactive experimentation is possible. In case (2), there is viscous resistance and it is difficult to achieve a balance between gravity and buoyancy, but the experiment time can be long, which makes it suitable for demonstrating the efficiency of a method or a strategy of task sequence. Moreover, there is no limit on space or motion. In case (3), control of the wire tension or mechanism is complicated, and motion of the robot (in particular, free rotation) is limited. However, the experiment time is not restricted. In case (4), time and space are restricted, but the quality of the microgravity environment is good (10^{-4} G), observation in the inertial frame is feasible and it is suitable for dynamic experiments.

We used a drop shaft to demonstrate the validity of path-planning methods based on the database of sensor output and robot motion in microgravity because the experiment requires short time, the comparison with the simulation is easy and we can observe the robot motion in an inertial frame. This experiment with a space robot using a drop shaft was the first trial of its kind in the world.

2. PATH-PLANNING METHODS

In this chapter, we propose three sensor-based path-planning methods, A, B and C. Method A is fundamental and based on a finite-state automaton. Method B is a modified version of Method A, and Method C is an optimized variation of Method B. We consider the space robot situation shown in **Fig. 1**. The x-, y-, and z-axes are fixed on the robot body (L_0). The robot has one arm with a shoulder joint (q_1) and an elbow joint (q_2). Joint q_1 rotates $0 \le q_1 \le 360$ deg and joint q_2 moves $-90 \le q_2 \le +90$ deg. This is the simplest arm configuration which can cause three-dimensional orientation variation. The robot is equipped with a

* Frontier Technology Division, Senior Researcher. Member.

CCD camera which is installed along the x-axis of the body. The CCD camera detects the light source, similarly to a star tracker. The light source is located in front of the robot and its position on the image is indicated as (p_1, p_2). This expresses part of the main body orientation; the robot estimates its orientation based on this position. However, the light source position gives only the direction of the light source relative to the robot. The robot cannot estimate the rotation around the direction vector from the CCD camera to the light source. If the robot moves its arm, then the body also moves, so that the image (p_1, p_2) changes. As a result, this space robot uses two input values (shoulder and elbow joint angle velocities, $\dot{q}_1 = ac_1$ and $\dot{q}_2 = ac_2$) and controls four degrees of freedom (two joint angles q_1, q_2 and light source position (p_1, p_2) on the image, which represents the robot orientation). This is possible because the considering system is nonholonomic.

These path-planning methods would be used for other robots if the sensor output changes while the robot moves.

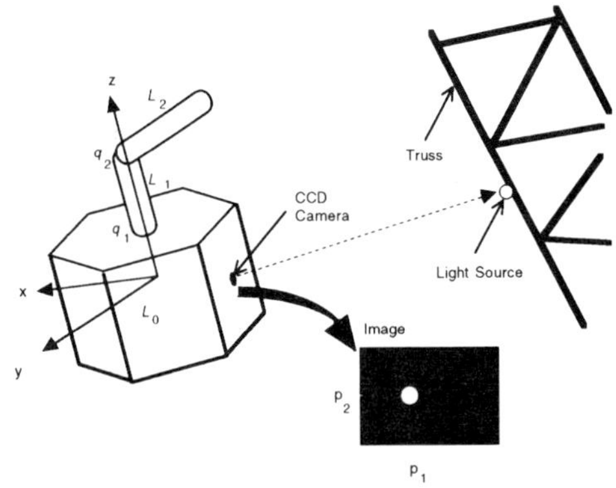

Fig. 1 Situation of Space Robot

Fig. 2 shows the robot model used in the computer simulation. This model, which has the same configuration as that in Fig. 1, was based on the robot used in the drop shaft experiment, as shown in the photograph in the figure.

Link No.	Dimensions (m)			Mass (kg)
	x	y	z	
0	0.19	0.19	0.14	3.25
1	0.04	0.04	0.10	0.45
2	0.04	0.04	0.10	0.47

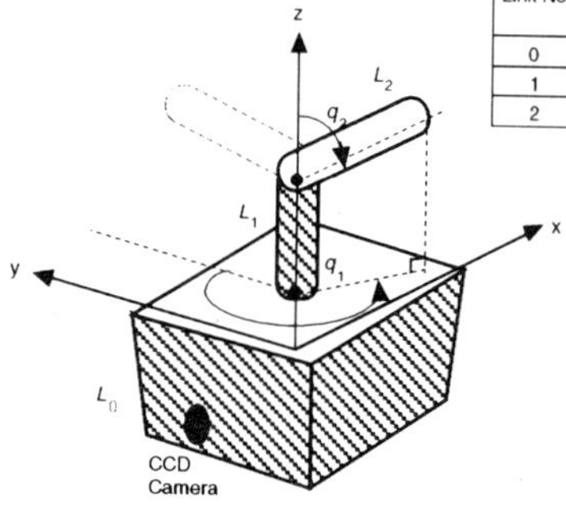

Fig. 2 Robot Model

2.1 Method A -- Finite-State Automaton Approach

Method A is a basic concept of our study. The process of Method A is shown in **Fig. 3**. We consider the discrete robot sensory output (joint angles and the light source positions on the image) to be a node on a graph. The robot motion between discrete joint angles causes a transition between nodes. After the robot collects the data of the current sensory output (q_1, q_2, p_1, p_2), the robot motion (ac_1, ac_2) and the sensory output after the motion $(q_{1n}, q_{2n}, p_{1n}, p_{2n})$, we can search for a path from the initial node to the goal node. This method requires considerable data in spite of the use of coarse discrete samples. For instance, when joints q_1 and q_2 are divided every 30 degrees and 10×10 image positions of the light source are created, the number of nodes becomes $12 \times 6 \times 10 \times 10 = 7{,}200$. Since the number of nodes is finite even though many, we can regard this method as an application of a finite-state automaton. Therefore, we can use some well-known search methods, such as a breadth-first search [9] to plan the path.

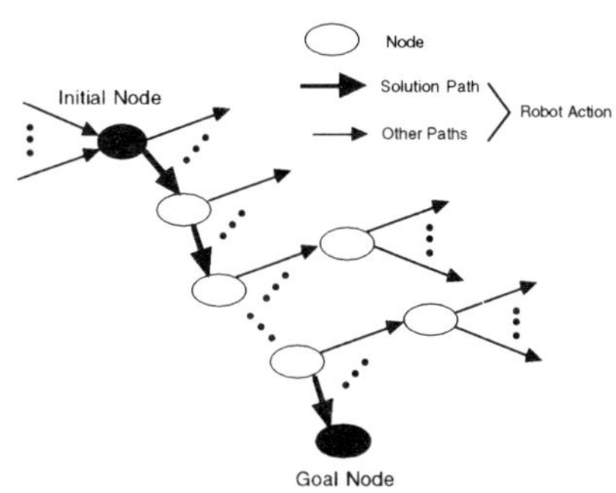

Fig. 3 Process of Method A

It is theoretically possible to compile the entire database through experiments. However, in practice, it is difficult to obtain the database experimentally because of the large size of the required database. For use of this method in space, we construct a database by performing simulations on the ground and then gradually modify the database while operating the robot in space. Part of the actual expressions of the nodes and labels is shown in **Fig. 4**. In this database, the joint angles q_1 and q_2 are discrete every 30 deg; i.e., the shoulder joint is at q_1=15, 45, 75, ..., 345 deg and the elbow joint at q_2=-75, -45, -15, 15, 45 and 75 deg. The light source position on the image is divided into a 10×10 grid (p_1 and p_2=0, 1, 2, 3, ..., 9), where the top-left position is (0, 0) and bottom-right position is (9, 9). The transitions between the nodes (joint motions of q_1 and q_2 are indicated as ac_1 and ac_2, respectively) are limited to ±30 deg/s for one second. The eight transition patterns are expressed as (ac_1, ac_2)= (-30, -30), (0, -30), (30, -30), (-30, 0), (30, 0), (-30, 30),

(0,30) and (30, 30) deg/s. The total size of the database therefore becomes 7,200 × 8 ~ 56,000 (in practice, it is less than 56,000 because some boundary joint angles cannot move in eight ways).

Since the number of nodes searched is finite, a "pruning" (if the node of the expanded path has already been searched, then the node does not need to be investigated) can often be implemented, which makes the search time considerably short.

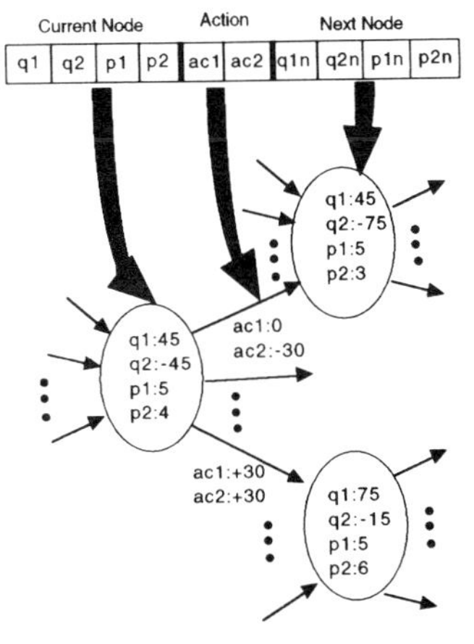

Fig. 4 Actual Node Expression of Method A

Coarse sampling induced the discrepancy between the goal position and the simulation result, causing errors. A sampling error such as (p_1, p_2)=(0, 0)→(0, 1) or (1, 0) or (1, 1), will occur depending on the initial state of (0, 0). A sampling error of this case is shown in **Fig. 5**.

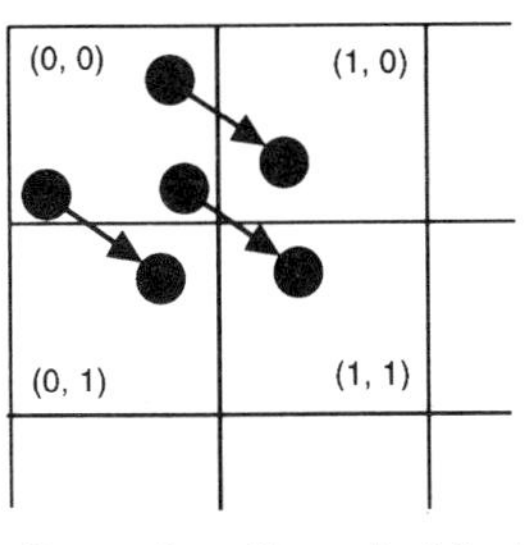

Fig. 5 Sampling Error in Method A

For instance, the path-planning process from the node of (q_1=195, q_2=-75, p_1=5, p_2=4) to the node of (q_1=15, q_2=-75, p_1=5, p_2=4) according to breadth-first search is expressed as follows (see **Fig. 6**):

Step 1 Find the initial node (q_1=195, q_2=-75, p_1=5, p_2=4).

Step 2 Expand the possible paths from the initial node (in this case, five paths because q_2 cannot move -30 deg) to the next nodes.

Step 3 Expand the paths from the leftmost node and find the consecutive nodes and check "pruning".

Step 4 Repeat Step 3 for each node of this generation moving in a direction from left to right.

Step 5 Expand the next-generation nodes as in Steps 3 and 4 until the goal node (q_1=15, q_2=-75, p_1=5, p_2=4) is reached.

Step 6 Repeat Step 5.

Breadth-first search provides the minimum number of steps for robot motion to achieve the goal.

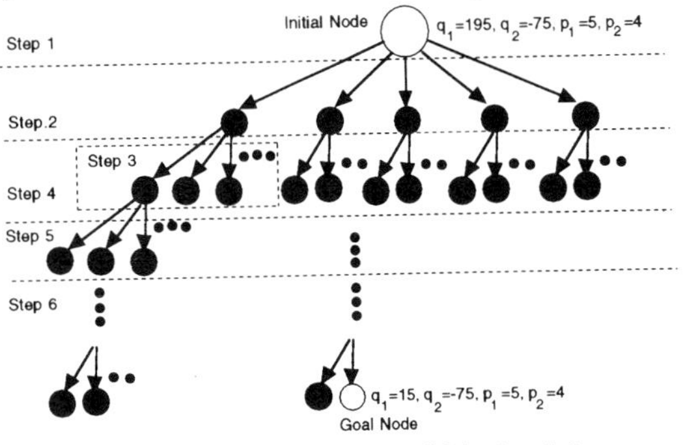

Fig. 6 Search Process of Method A

The light source position on the image expresses some of the orientation of the main body, so a node transition implies the integral of the angular velocity of the main body along a specified path. In the case that both joints q_1 and q_2 move 30 deg, three paths are possible:

(1) the shoulder rotates first, and then the elbow moves ((ac_1, ac_2)=(30, 0)→(0, 30) deg/s);

(2) the shoulder and the elbow move simultaneously ((ac_1, ac_2)= (30, 30) deg/s); and

(3) the elbow moves first, and then the shoulder rotates ((ac_1, ac_2)=(0, 30)→(30, 0) deg/s).

Even if the initial nodes are the same in these three cases, the final nodes are different because the light source positions on the image are eventually different. This corresponds to the nonholonomic property of the system.

The summary of Method A is that (1) a large database is required, (2) the potential for sampling error is large, (3) the search time is not long and (4) the minimum path to the goal is obtained.

2.2 Method B -- Modified Method A

Method B is a modified version of Method A. In this method, the drawback of Method A is resolved by reducing the required database size. The process of Method B is shown in **Fig. 7**. We consider the discrete joint angles to be a node of a graph, as in Method A, but the light source position on the image (p_1, p_2) is a *label* of the node. In method B, (p_1, p_2) indicates an ordinary pixel position, such as 512 × 512, where the top-left position is (1, 1) and the bottom-right one is (512, 512). The change of the light source position on the image (dp_1, dp_2) is included in the node transition. In Method A, the light source position (p_1, p_2) was included in a node domain. The initial label is measured, but the labels of all other nodes are calculated using this data (dp_1, dp_2) when the path is searched. Part of the actual expressions of the nodes and labels are shown in **Fig. 8**. After the robot collects the data of current sensory output (q_1, q_2), the motion (ac_1, ac_2) and the change of the light source position on the image (dp_1, dp_2), it can search for the path from the initial node to the goal node. The database size is $12 \times 6 \times 8 \sim 560$, which is greatly reduced that used in Method A (in our case, 1/100). We may use continuous values as the label (p_1, p_2) and transition (dp_1, dp_2). In this case, the system is no longer a *finite*-state automaton, and we should modify the search methods. When the current node is determined, we use the database to expand the path to the next node. We must distinguish the nodes of the same joint angles if the labels are different. However, if there is a node which has the same label as the node just expanded, we will not expand the path to that node ("pruning"). This means that many nodes and labels must be searched for and evaluated, which is a time-consuming task.

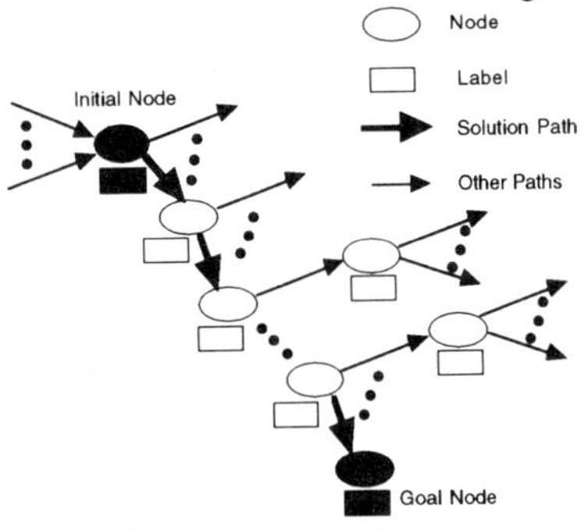

Fig. 7 Process of Method B

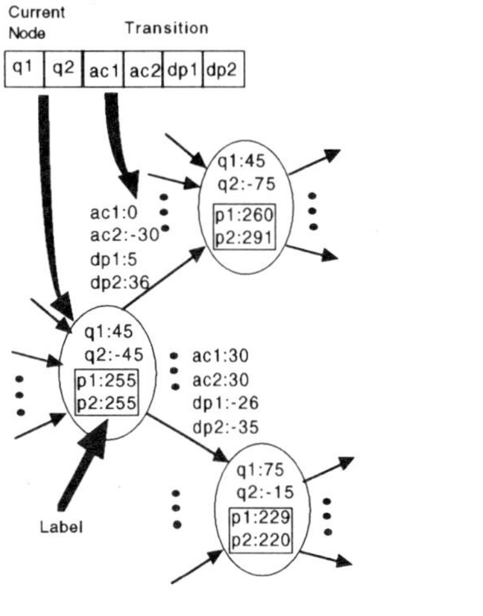

Fig. 8 Actual Node Expression of Method B

The goal node and label should be given as discrete joint angles (the node values) and the permissible range of the CCD camera output as the label values. If the range is narrow, the search might fail within the limit of time or node and label number.

The assumption that the change of the image (dp_1, dp_2) is the same in the whole image plane is debatable. Although this assumption made the database small, if rotation along the viewing direction takes place or the relative position between the light and the camera changes during the motion, this assumption is not valid. In our case, the validity depends on the motion and the position of the center of the robot mass. **Fig. 9** shows the examples of the change of the image in the whole image. In the figure, the initial positions of joint angles are the same (q_1=15 and q_2=-75). However, the motions are different. In (a), ac_1=30 and ac_2=0. The image change remains similar in the entire plane because the rotation of the main body is approximately along the z-axis. In (b), ac_1=0 and ac_2=30, however, the image change is distributed widely because the main body rotates around the x-axis (camera installed direction). In both cases, the database was represented by the change in the motion at the center of the image (circled in the figure) as (dp_1, dp_2).

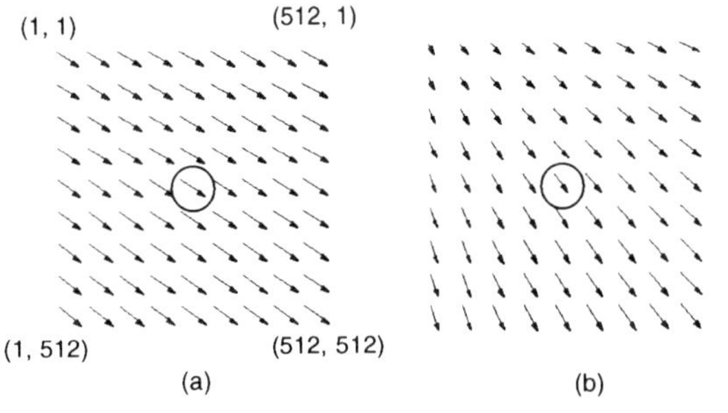

Fig. 9 Image Distribution

For instance, the path-planning process from (q_1=195, q_2=-75, p_1=p_2=255) to (q_1=15, q_2=-75, p_1=p_2=255 with the permissible range of error of 100 pixels) according to modified breadth-first search can be expressed as follows (see **Fig. 10**) :

Step 1 Find the initial node of (q_1=195, q_2= -75, p_1=p_2=255, which are given as the label).

Step 2 Expand the possible paths from the initial node (in this case, five paths), find the next nodes, and calculate the labels.

Step 3 Expand the paths from the leftmost node to find the consecutive nodes and calculate the labels with checking "pruning".

Step 4 Repeat Step 3 for each node of this generation moving in a direction from left to right.

Step 5 Expand the next-generation nodes as in Steps 3 and 4 until the goal node (q_1=15, q_2=-75) is reached and the label (p_1=p_2=255 within the permissible range of error).

Step 6 Repeat Step 5.

Modified breadth-first search provides the minimum number of steps for robot motion to achieve the goal.

The summary of Method B is that (1) the database is small, (2) the changes in the image are not the same in the whole image plane, (3) the search time is long and if the permissible error is small, the solution would not be found within the limited time or searched node and label number and (4) the minimum path to the goal is obtained.

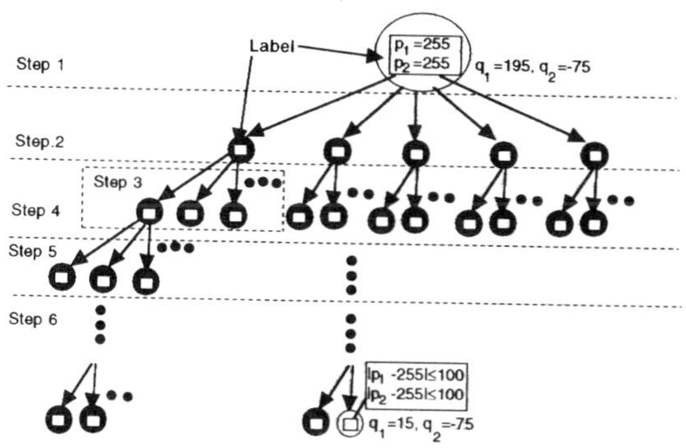

Fig. 10 Search Process of Method B

2.3 Method C --Using A-Search

To overcome the drawback of Method B (long search time), the concept of heuristics and cost was used in Method C. This type of search method is called A-search in the artificial intelligence field, and can drastically shorten the search time. A-search is one of the optimal search patterns and combines the dynamic-programming principle. The database used is the same as Method B. In this method, only one of the expansible paths expands to the next node in each step. To determine the expanding path, we use the heuristics and cost. In our case,

$$c = \sum_{Path=initial}^{current+1} (|ac_1| + |ac_2|) \tag{1}$$

is determined as the cost, and

$$h = 1.0|q_1^{goal} - q_1^{current+1}| + 1.0|q_2^{goal} - q_2^{current+1}| + 0.05(|255 - p_1^{curent+1}|^2 + |255 - p_2^{curent+1}|^2) \tag{2}$$

is determined as heuristics, where "current" indicates the open node which is being investigated ("open node" indicates the node which has been investigated and includes extensible paths), "current+1" indicates the candidate node of the expanding path, and (255, 255) is the central pixel position of the image. Then,

$$f = c + h \tag{3}$$

is the evaluation function. The next node selected is that at which the evaluation function reaches a minimum. In this case, the expanding path is determined by the cost of total joint motion to be as small as possible, the joint angles to be as close to the goal as possible and the light source position to be as motionless as possible. Therefore, the found path is not always the shortest as in breadth-first search. A-search does not always find the optimal path (in order to guarantee the optimum, we must find an evaluation function which satisfies additional conditions, and in such a case, this search method is called A*-search [9]). Since the evaluation function is subjective, what is the optimal path is debatable.

For instance, the path-planning process from (q_1=195, q_2=-75, p_1=p_2=255) to (q_1=15, q_2=-75, p_1=p_2=255 with the permissible range of error 100 pixels) according to A-search can express as follows (see **Fig. 11**) :

Step 1 Find the initial node of start point of (q_1=195, q_2= -75, p_1=p_2=255 as the label) and add this node to the open nodes set.

Step 2 Expand the path of the smallest evaluation function from the initial node to the corresponding node, calculate the label with checking "pruning" and add it to the open nodes set.

Step 3 Expand the path of the smallest evaluation function from the only node in the open nodes set, find the reached node, and calculate the label with checking "pruning". Add the node to the open nodes set. If all paths of the open node are searched, the node is deleted from the open nodes set. If the reached node is the goal node and label is within the permissible range of error, the search is successfully concluded..

Step 4 Repeat Step 3.

The summary of Method C is that (1) the search time is short, (2) the path is not always optimal and (3) the evaluation function is not given *a priori*.

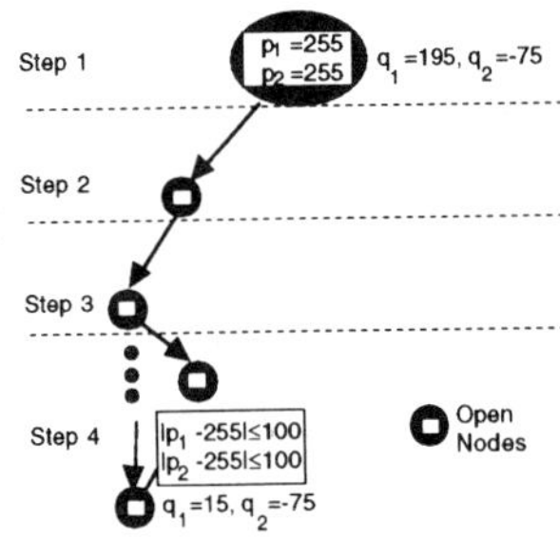

Fig. 11 Search Process of Method C

3. EXPERIMENTS

3.1 Drop Shaft

The drop shaft used in our experiment, formerly a coal mine, is 710 m deep, and the free-fall height is 490 m, which provides a 10-second period of microgravity, as shown in **Fig. 12**. This facility belongs to the Japan Microgravity Center (JAMIC).[10] **Fig. 13** shows the capsule used in the experiment, which consists of an outer capsule and an inner capsule; the experimental apparatus is assembled in the inner capsule. The capsule falls along guide rails. The space between the inner and outer capsules is evacuated (100 Pa) and the inner capsule is levitated in the outer capsule during the free fall. The outer capsule is equipped with thrusters for acceleration and balance.

The typical microgravity condition during the experiment is shown in **Fig. 14**. In the figure, x-, y-, and z-directions are the same as those in Fig. 15 (i.e., the y-direction indicates the direction of gravity). These directions are in agreement with the robot frame in the initial state. The condition data were not adjusted to absolute values, so only the magnitude of change is meaningful.

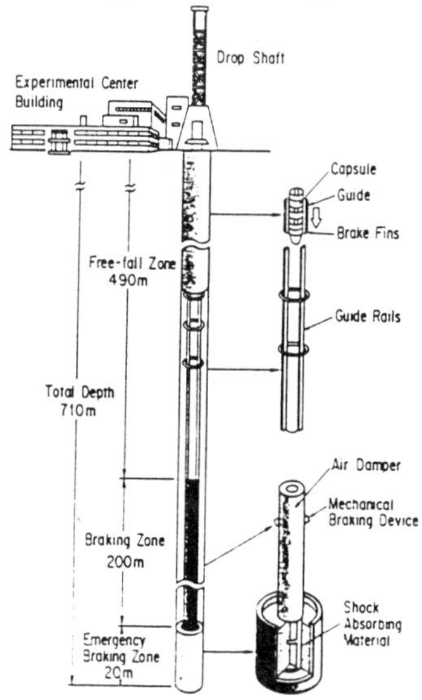

Fig. 12 Drop Shaft

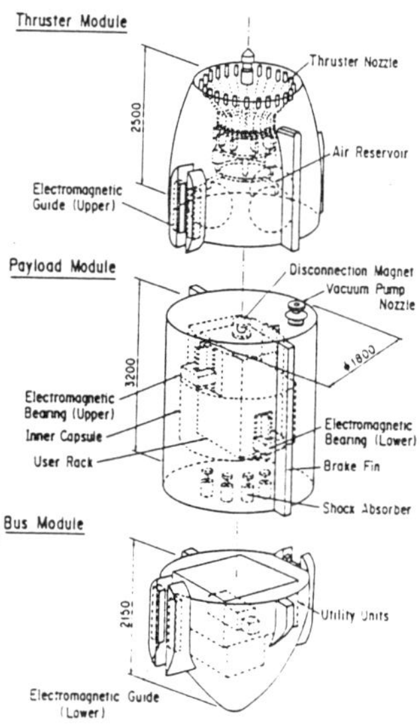

Fig. 13 Capsule

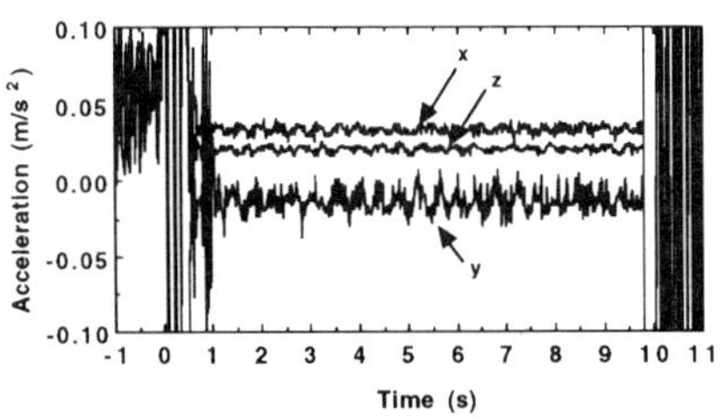

Fig. 14 Microgravity Environment

Fig. 15 shows the rack used in our experiment, which is assembled in the inner capsule. The full size of the rack is 0.870(W) × 0.870(L) × 0.886(H) meters; however, users divide the space into quarters for different experiments. We used a half-rack space of 0.870(W) × 0.870(L) × 0.443(H) meters. The rack is equipped with CCD cameras to observe the robot motion in the inertial frame, an infrared remote controller to drive the robot, a power supply, a command controller to levitate the robot in microgravity, a video deck which receives UHF waves from the CCD on the robot body and silicone rubber (α-gel) to absorb the shock of collision. The levitation method is one of the most difficult techniques in drop shaft experiments. We use a holding mechanism with grippers to hold the robot until the free fall condition. The grippers are opened for 0.5 seconds after achieving the free-fall condition, then closed for 0.5 seconds, then opened again to realize stable levitation. Since the inside of the rack was too dark to observe the robot motion, six small lamps were provided.

In the original setup, we used a light source to estimate the orientation. However, if the robot loses the light source outside the visual field, we cannot evaluate the method. Therefore, we use a sheet of grid graph paper instead of the light source.

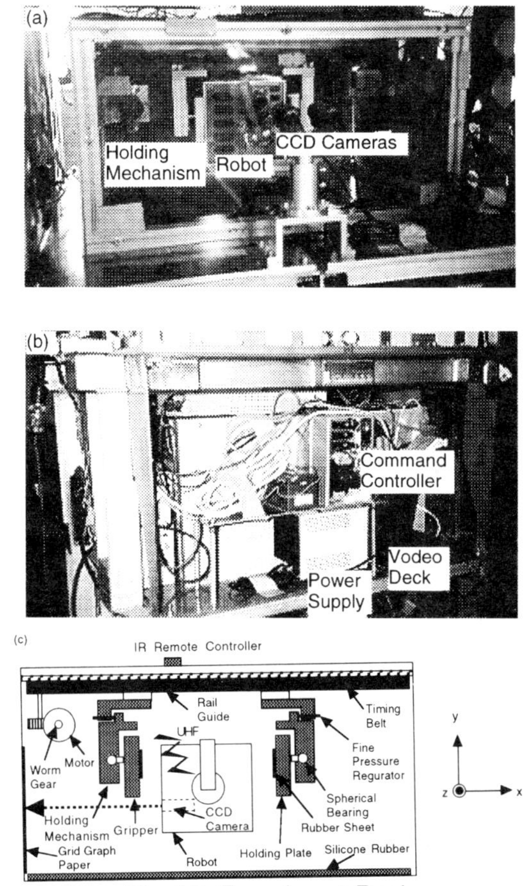

Fig. 15 Experiment Rack

Fig. 16 shows the block diagram of the robot (the specifications and photograph are shown in Fig. 3). The robot was controlled by a V25S CPU; the power was switched on via an infrared remote controller. The robot was equipped with 8-bit D/A and A/D converters to control the motors and measure the joint angles. A CCD camera was installed on the robot and the image was transmitted to the video deck on the UHF band. The viewing angles are 18 degrees for the horizontal direction and 12 degrees for the vertical direction. The number of pixels is 512 × 480 (in the simulation, 512 × 512).

3.2 Experiment for Method B

Method A requires a large database, which makes its use in space difficult. Therefore, the experiment for Method A was omitted in spite of its fundamental concept.

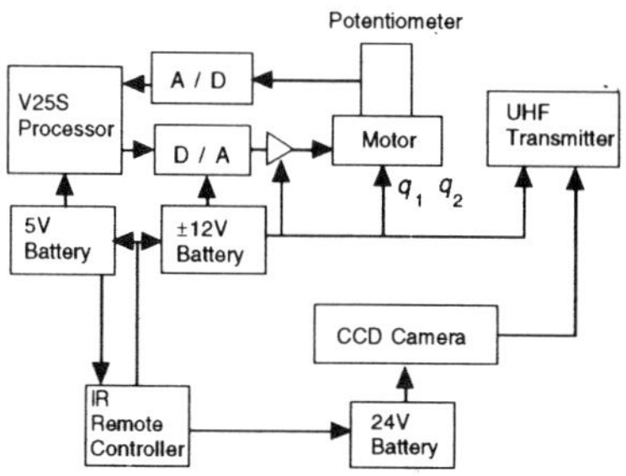

Fig. 16 Block Diagram of Robot for Drop Shaft

In the case of Method B, although the amount of data required for the database is considerably smaller (~560) than that for Method A, it takes too long a time to compile the data using the drop shaft. Therefore, the path-planning database was compiled by computer simulation. We searched for the path from the initial node and label (q_1=195, q_2=-75 and p_1=p_2=255) to the goal node and label (q_1=15, q_2=-75 and p_1=p_2=255 within a pixel error range of 100) using Method B . This indicates that only the shoulder joint rotates 180 degrees, without a change of the camera direction and without movement of the elbow joint. Eventually, a 7-step path was obtained after a 93,452-node search (5,179 seconds of CPU time using Sun Sparc 10) (**Fig. 17**).

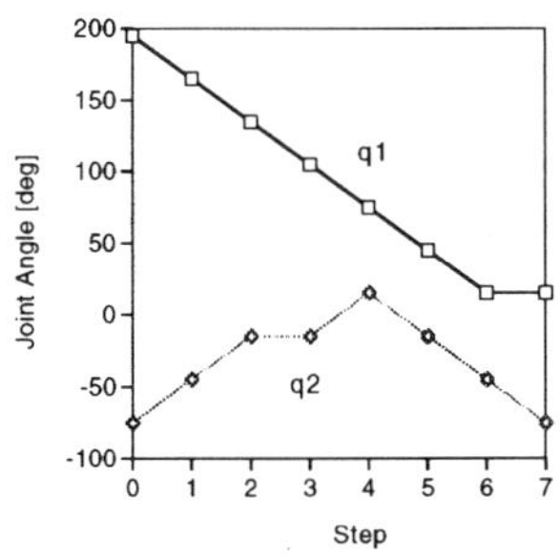

Fig. 17 Results of Path Planning with Method B

The change of the light source position on the image (p_1-c_x, p_2-c_y) (in experiment, c_x=277 and c_y=223, in simulation, c_x=c_y=255) between the initial and the goal position is shown in **Fig. 18**. In the figure, (p1e, p2e), (p1s, p2s) and (p1c, p2c) indicate the traced point positions on the grid graph paper (instead of the original light source position) of experiment, simulation of experiment and calculation from the database, respectively. The difference between the point of (p1s, p2s) and (p1c, p2c) is due to the following: (1) the database was obtained when the light source position was 0.9 meters away from the front of the robot, but the simulation of the experiment had a distance of 0.25 meters; (2) (p1c, p2c) assumes that the light source position is centered in the image, but in the simulation, this assumption is violated (see Fig. 9). The final position determined in the experiment was better than that determined by simulation. The difference between the point of (p1e, p2e) and (p1s, p2s) is due to the incompleteness of the model, the joint motion inaccuracy and the initial condition of levitation (initial momentum and angular momentum). However, the global tendencies of the simulation are in agreement with the results of the experiment.

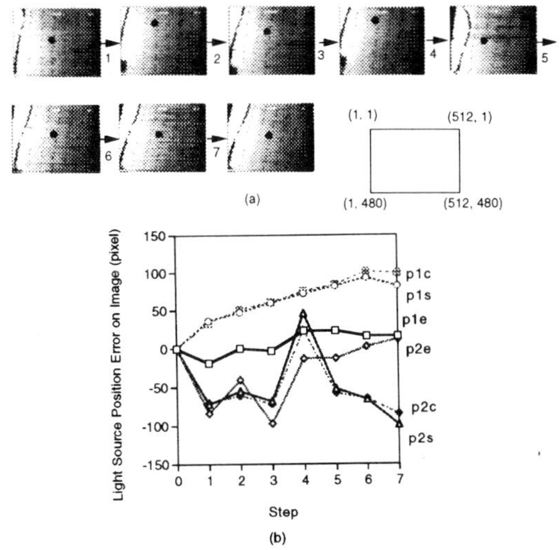

Fig. 18 Traced Point Position on Image
(a) Robot Camera Image
(b)Traced Point Position of Each Step

3.3 Experiment for Method C

In this case, the database used is the same as that in Method B. The searched path is also the same as that in Method B, that is, the path from the initial node and label (q_1=195, q_2=-75 and p_1=p_2=255) to the goal node and label (q_1=15, q_2=-75 and p_1=p_2=255 within a pixel error range of 100). Form (3) of the evaluation function is used. The search time is drastically shorter than that in Method B. An 11-step path was found after a 282-node search (1 second of CPU time). (**Fig. 19**). This result fits our intuitive estimation to rotate the main body as small as possible; first, minimize the arm inertia by straightening the arm (q_2~0), then rotating the arm (moves q_1), and bending joint q_2.

The change of the light source position (p_1-c_x, p_2-c_y) (in experiment, c_x=228 and c_y=226, in simulation, c_x=c_y=255) on the image between the initial position and the goal position is shown in **Fig. 20**.

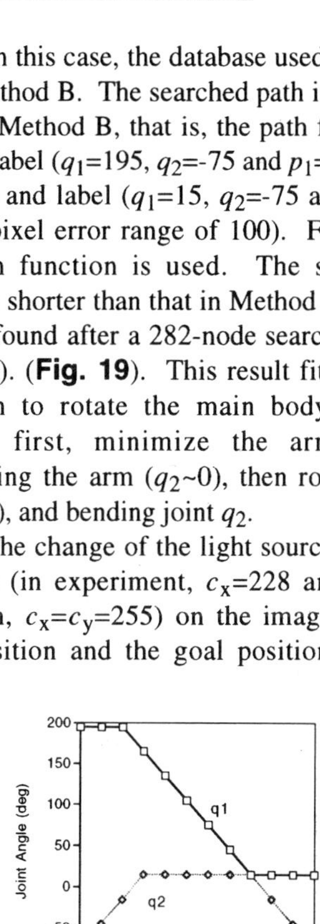

Fig. 19 Results of Path Planning with Method C

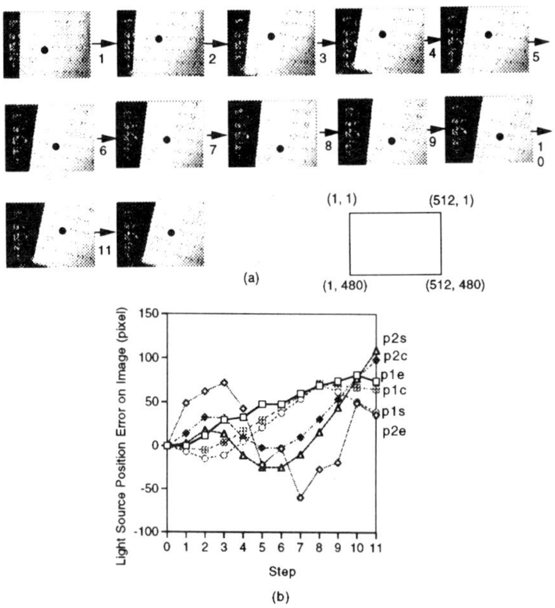

Fig. 20 Traced Point Position on Image
(a) Robot Camera Image
(b)Traced Point Position of Each Step

4. DISCUSSION

4.1 Generality

In this section, we discuss the generality of the proposed methods (i.e., we consider whether or not these methods can solve any path-planning problem). In the case of Method A, the number of searched nodes and paths are finite. Therefore, it is easy to judge the generality. If a database includes some groups of nodes not connected with each other, then paths which do not reach the goal node exist. However, if there is no such group, any path will reach the goal node. In **Fig. 21**, the path from the initial node to goal node 1 does not exist, but the path from the initial node to goal node 2 exists.

In the case of Methods B and C, the problem becomes more complicated. There are no isolated groups of nodes which represent the joint angles. However, the label which represents the image motion has many states. We introduced the permissible range of error for the label in Sec. 2.2. If the permissible range of error is small, it is difficult to find the path within a limited time (in particular, by Method B). Therefore, if the search time is limited, we should make the robot report the paths which satisfy the final joint angles and their error for the label and select the best path.

In the case of Method C, we have more parameters for determining the path, i.e., definition of the evaluation function. In eq. (2), there are many coefficients to be determined (we tentatively put them as 0.05 and 1.0). Moreover, the definition of the evaluation function itself is arbitrary. The optimization of the estimation function affects the results.

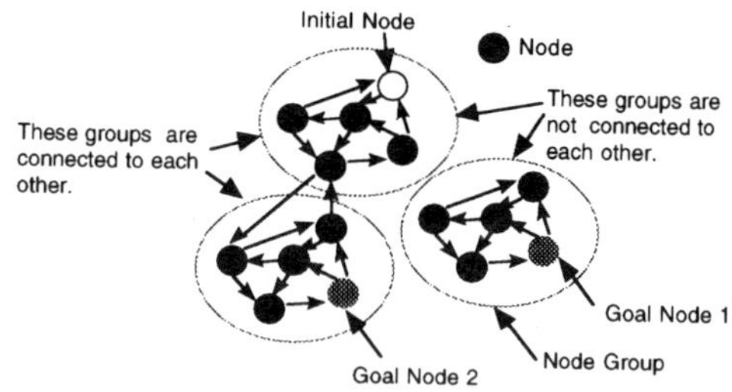

Fig. 21 Existing Path and Nonexisting Path

4.2 Reference Experiment

The simplest path of joint motion from the initial state (q_1=195, q_2=-75) to the final state (q_1=15, q_2=-75) without path planning is one in which joint q_1 rotates uniformly, whereas joint q_2 does not move at all (**Fig. 22**). We experimentally investigated the change of the grid graph paper image. The results are shown in **Fig. 23**. The robot started its motion three seconds after the microgravity state was achieved, and ended its motion at 5.5 seconds. Fig. 23 (a) shows the raw images of initial (3 s) and final state (5.5 s). Fig. 23 (b) shows plots of the image motion from 3 s to 5.5 s. Since point A, which was located at the center ((258, 282) in pixel value) of the image at the initial state, disappeared during the motion, Point B, which was observed throughout the motion, is also shown in the figure. If we assume that points A and B move parallel to each other in the image, the final position of point A on the image is (-36, 261) in pixel value. We estimated 11 deg to be the difference between the initial and final directions, using the CCD images of the grid graph paper.

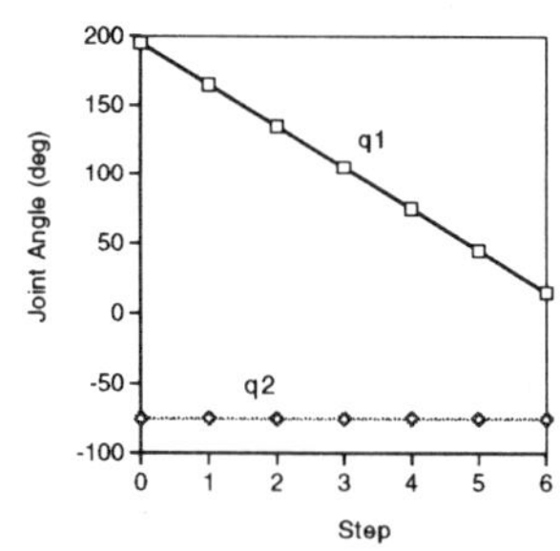

Fig. 22 Simplest Path

5. CONCLUSIONS

We proposed three methods (A, B and C) to plan a path to simultaneously control the orientation and arm position of a space robot, using only arm motion. The database of relationships among the current sensor output, the robot motion, the change of sensor output and/or the sensor output after the motion is required. Method A is fundamental and based on a finite-state automaton, and the search time is short. However, the database and sampling error become large. Method B, which is a modified version of Method A, needs a smaller database, however, it no longer involves a finite-state automaton, so the search time becomes lengthy. The image change distribution also causes error. In order to overcome the drawback of Method B, Method C was proposed. It uses A-search for the same database as used in Method B. The search time becomes very short, but does not always yield the optimal path. We performed experiments with Methods B and C using a drop shaft which provides ten seconds of microgravity during free fall. This was the first test of a space robot using a drop shaft. The results show that these path-planning methods are effective in controlling both orientation and the arm position. **Fig. 24** summarizes motion results of these methods. The robot is parallel to the grid graph paper (see Fig. 9 (c)) in the case of Methods B and C, but is tilted in the case of the simplest path.

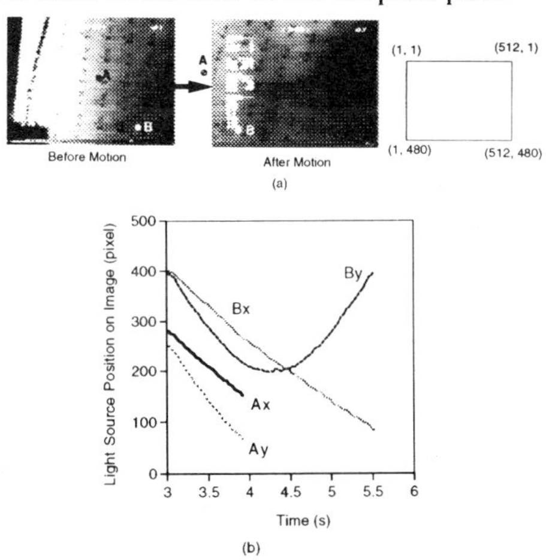

Fig. 23 Result of Simplest Path Determination
(a) Robot Camera Image,
(b) Traced Point Position during Motion

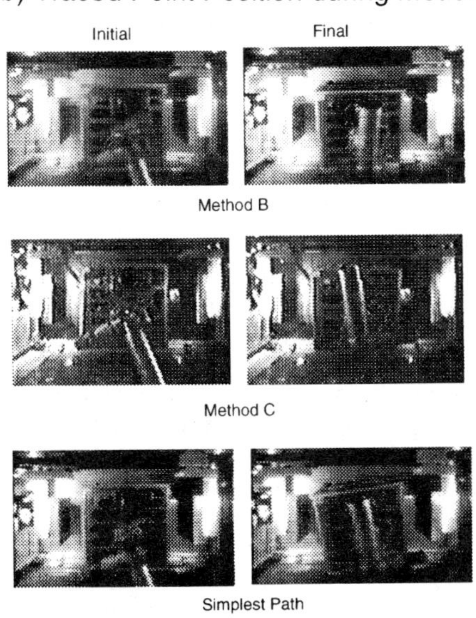

Fig. 24 Motion Results Obtained Using Three Methods

This work is supported by the New Energy and Industrial Technology Development Organization (NEDO) through the Japan Space Utilization Promotion Center (JSUP) in the program of the Ministry of International Trade and Industry (MITI).

References

1) Y. Nakamura and R. Mukherjee: "Nonholonomic Path Planning of Space Robots Via a Bidirectional Approach", IEEE Trans. Robotics and Automation, 7, pp. 500-514, 1991.
2) K. Yamada: "Arm Path Planning for a Space Robot", IEEE/RSJ Intr. Conf. Intelligent Robots and Systems '93, July 26-30, 1993, Yokohama, Japan.
3) R. S. Norsworthy: "Grasping Objects Autonomously in Simulated KC-135 Zerp-G", AIAA Conf. on Intelligent Robots for Factory, Field, Service, and Space, Houston, pp. 757-62, 1994.
4) D. L. Akin and J. R. Spofford: "Redundancy Control of a Free-Flying Telerobot", J. Guidance, Control, and Dynamics, Vol. 13, pp. 515-523, 1990.
5) Y. Sato: "Micro-G Emulation System Using Constant-Tension Suspension for a Space Manipulator", Proc. 1991 IEEE Int. Conf. Robotics and Automation, Sacramento, Cal., pp. 1893-1990, April. 9-11, 1991.
6) P. D. Campbell, P. L. Swaim and C. J. Thompson: "Charlotte Robot Technology for Space and Terrestrial Applications", 25th International Conf. on Environmental Systems, San Diego, CA, July, 10-13, 1995.
7) H. Shimoji, M. Inoue, N. Inaba and Y. Wakabayashi: "Evaluation of Space Robot Behavior Using Berthing Dynamics Simulator", J. Robot Society of Japan (in Japanese), 13, pp. 127-133, 1995.
8) T. Iwata, K. Kodama, F. Numajiri and H. Murakami: "Experiment on Robotic Motion Using Drop Shaft", Strengthen Cooperation in the 21 Century (6th ISCOPS) Vol. 91, Advances in the Astronautical Sciences, Published for American Astronautical Society by Univelt, 1996.
9) P. H. Winston: "Artificial Intelligence" second edition, chap. 4, pp. 87-132, Addison-Wesley, 1984.
10) I. Kudo: "Microgravity Research in Japanese Industry", Adv. Space Res. Vol. 13, No. 7, pp. 13-21, 1993.

AIAA-97-3557

ADAPTIVE CONTROL OF FREE-FLYING SPACE ROBOT WITH POSITION/ATTITUDE CONTROL SYSTEM

Kei SENDA*, Hideyuki NAGAOKA†, and Yoshisada MUROTSU‡

Department of Aerospace Engineering
College of Engineering
Osaka Prefecture University
Sakai, Osaka 593, JAPAN

Abstract

This paper is concerned with an adaptive control of free-flying space robots with uncertain inertial parameters of payloads. It is applicable to control all of the position and attitude of the satellite vehicle as well as that of the manipulator hand if the satellite has position/attitude control devices, e.g., thrusters and control momentum gyros. The proposed adaptive control realizes high control performance by estimating unknown parameters even if a control model has an error to the real plant caused by the payload. Asymptotic stability of the adaptive control is proven by Lyapunov's second method. The effectiveness is examined by numerical simulations for a hardware experimental system.

Introduction

The future space mission will require free-flying space robots which will carry out various extra vehicular activities instead of astronauts, e.g., the construction, repair and maintenance of space structures. The space robot will consist of a satellite vehicle to fly freely and manipulators to handle various objects. The space robot will often manipulate payloads with unknown inertial parameters, e.g., mass, mass center position and inertial tensor. Position and orientation of the satellite vehicle will be interacted by the manipulator dynamics because of the lack of an inertially-fixed base on which to mount the space robot. Hence, control performance may get worse when the manipulators handle a payload with uncertain inertial parameters[1]. Therefore, it is very important to develop control methods for space robots manipulating payloads with unknown inertia parameters.

The kinematics of the space manipulator has been discussed by Vafa and Dubowsky,[2] and Umetani and Yoshida.[3] They have proposed a concept of virtual manipulator and a generalized Jacobian matrix, respectively, and they are taking account of dynamical interactions between the manipulators and the satellite. The virtual manipulator and the generalized Jacobian matrix are computed by using geometric parameters together with inertial parameters of all the links constructing the space robot. This means that the inertial parameters change the kinematics of the space robot. Therefore, the kinematics and dynamics of the whole system cannot be predicted when the manipulator grasps an unknown object. Further, when the objects are not grasped in their nominal position and orientation, the computed virtual manipulator and generalized Jacobian matrix have some errors even if the inertial parameters of the grasped objects are exactly known.

Some of the control schemes for the space robot are proposed by using the generalized Jacobian matrix.[4-7] All the studies have discussed the only case when the manipulator does not handle any objects. Robust stability of those control schemes has never been verified against the kinematical and dynamical variation, e.g., the unknown object. Therefore, it is very important to estimate the inertial parameters accurately for precise control after the manipulator has grasped the objects.

Murotsu, et al.[1] have proposed two parameter identification methods. A two stage procedure is necessary, i.e., a controller must use the estimated parameters after the identification using Murotsu's methods. To avoid this inefficiency, one alternative is an adaptive control which controls the manipulators under estimating the inertia parameters. Slotine et al.[8] proposed an adaptive control for the unknown payload manipulation on the ground, but it cannot directly apply to the space robots. Yamamoto et al.[9] proposed a kinematics-based adaptive control for space robots under the translational and angular momentum conservation. However, it is desirable to control the satellite attitude by using attitude control devices because of the following reasons. Orientation of the satellite vehicle should be fixed when the robot holds a radio communication connection to an external commander during operation. The position and attitude must be controlled for looking closely at an object using vision sensors or for a specific task. Yamamoto's control is not applicable to the space robot with position/attitude control devices, e.g., thrusters and control momentum gyros, because its translational and angular momenta aren't conserved. Moreover, dynamics-based controls, e.g., resolved acceleration control (RAC),[6, 7] are desirable when the robot handles massive payloads or manipulates quickly. Iwata et al.[10] proposed a RAC augmented by a parameter estimator for a space robot with the position/attitude control devices. Its effec-

*Associate Professor, Member AIAA

†Professor, Member AIAA

‡Graduate Student

tiveness was shown by numerical simulations but its stability has not been proven.

This paper proposes a dynamics-based adaptive control for a space robot with a position/attitude control system when the robot manipulates a payload. The proposed control is a unified method which can control all or a part of the position and attitude of the satellite vehicle as well as that of the manipulator hand. Asymptotic stability of the control method is proven by Lyapunov's second method. Numerical simulations show that the adaptive control results in a good performance while the RAC gets worse because of an unknown payload.

Equations of Motion

The equations of motion of the space robot can be derived as:

$$\boldsymbol{M}\dot{\boldsymbol{u}} + \boldsymbol{h} = \boldsymbol{n} \tag{1}$$

where

$$\boldsymbol{u} = \begin{bmatrix} \boldsymbol{v}_0 \\ \boldsymbol{\omega}_0 \\ \dot{\boldsymbol{\theta}} \end{bmatrix}, \quad \boldsymbol{n} = \begin{bmatrix} \boldsymbol{f}_0 \\ \boldsymbol{\tau}_0 \\ \boldsymbol{\tau}_\theta \end{bmatrix}$$

$\boldsymbol{M}$ is an inertia matrix, and $\boldsymbol{h}$ denotes the centrifugal and Coriolis force. The variables $\boldsymbol{v}_0$, $\boldsymbol{\omega}_0$, and $\boldsymbol{\theta}$ are the translational, the angular velocities of the satellite vehicle, and the joint angles of the manipulators, respectively. Their generalized forces $\boldsymbol{f}_0$, $\boldsymbol{\tau}_0$, and $\boldsymbol{\tau}_\theta$ represent the translational, the rotational forces applied to the satellite, and the joint torques, respectively.

If the satellite attitude variable $\boldsymbol{\phi}_0$ satisfies $\boldsymbol{\omega}_0 = \dot{\boldsymbol{\phi}}_0$, Eq. (1) can be rearranged as:

$$\boldsymbol{M}\ddot{\boldsymbol{q}} + \boldsymbol{C}\dot{\boldsymbol{q}} = \boldsymbol{n} \tag{2}$$

where $\dot{\boldsymbol{q}} = \boldsymbol{u}$, $\boldsymbol{C}$ is a matrix of the centrifugal and Coriolis force. A related matrix $(\dot{\boldsymbol{M}} - 2\boldsymbol{C})$ is skew-symmetric.

Parameter Linearization

The unknown parameters to be estimated are inertia parameters of a combined body of the hand and the payload manipulated by the robot. The left side of Eq. (2) can be linearized with respect to the unknown parameters $\boldsymbol{\beta}$ as:

$$\boldsymbol{M}\ddot{\boldsymbol{q}} + \boldsymbol{C}\dot{\boldsymbol{q}} = \boldsymbol{P}(\boldsymbol{q}, \dot{\boldsymbol{q}}, \dot{\boldsymbol{q}}, \ddot{\boldsymbol{q}})\boldsymbol{\beta} + \boldsymbol{Q} \tag{3}$$

where $\boldsymbol{P}(\boldsymbol{q}, \dot{\boldsymbol{q}}, \dot{\boldsymbol{q}}, \ddot{\boldsymbol{q}})$ called regressor is a matrix function of $\boldsymbol{q}$, $\dot{\boldsymbol{q}}$, and $\ddot{\boldsymbol{q}}$. The two $\dot{\boldsymbol{q}}$s in $\boldsymbol{P}$ are distinguished each other. The former is in $\boldsymbol{C}$, and the latter is premultiplied by $\boldsymbol{C}$. The $\boldsymbol{Q}$ is a matrix function of the known parameters and the observable state variables.

The estimation of unknown parameters are denoted by $\hat{\boldsymbol{\beta}}$. The $\hat{\boldsymbol{M}}$ and $\hat{\boldsymbol{C}}$ represent the $\boldsymbol{M}$ and $\boldsymbol{C}$ computed from $\hat{\boldsymbol{\beta}}$. Then, the following equations can be obtained:

$$\hat{\boldsymbol{M}}\ddot{\boldsymbol{q}} + \hat{\boldsymbol{C}}\dot{\boldsymbol{q}} = \boldsymbol{P}(\boldsymbol{q}, \dot{\boldsymbol{q}}, \dot{\boldsymbol{q}}, \ddot{\boldsymbol{q}})\hat{\boldsymbol{\beta}} + \boldsymbol{Q} \tag{4}$$

Subtracting both members of Eq. (4) from Eq. (3) eliminates the known parameter term $\boldsymbol{Q}$ and yields

$$\tilde{\boldsymbol{M}}\ddot{\boldsymbol{q}} + \tilde{\boldsymbol{C}}\dot{\boldsymbol{q}} = \boldsymbol{P}(\boldsymbol{q}, \dot{\boldsymbol{q}}, \dot{\boldsymbol{q}}, \ddot{\boldsymbol{q}})\tilde{\boldsymbol{\beta}} \tag{5}$$

where

$$\begin{aligned} \tilde{\boldsymbol{M}} &= \hat{\boldsymbol{M}} - \boldsymbol{M} \\ \tilde{\boldsymbol{C}} &= \hat{\boldsymbol{C}} - \boldsymbol{C} \\ \tilde{\boldsymbol{\beta}} &= \hat{\boldsymbol{\beta}} - \boldsymbol{\beta} \end{aligned} \tag{6}$$

Adaptive Control

Consider the following Lyapunov function to derive an adaptive controller:

$$V = \frac{1}{2}(\boldsymbol{s}^T\boldsymbol{M}\boldsymbol{s} + \tilde{\boldsymbol{\beta}}^T\boldsymbol{\Gamma}\tilde{\boldsymbol{\beta}}) \tag{7}$$

where $\boldsymbol{s}$ is defined by:

$$\boldsymbol{s} = \dot{\tilde{\boldsymbol{q}}} = \dot{\boldsymbol{q}} - \dot{\boldsymbol{q}}_r \tag{8}$$

for a reference input $\boldsymbol{q}_r$ corresponding to $\boldsymbol{q}$. Time derivative of V yields

$$\begin{aligned} \dot{V} &= \boldsymbol{s}^T\boldsymbol{M}\dot{\boldsymbol{s}} + \frac{1}{2}\boldsymbol{s}^T\dot{\boldsymbol{M}}\boldsymbol{s} + \tilde{\boldsymbol{\beta}}^T\boldsymbol{\Gamma}\dot{\tilde{\boldsymbol{\beta}}} \\ &= \boldsymbol{s}^T[\boldsymbol{M}(\ddot{\boldsymbol{q}} - \ddot{\boldsymbol{q}}_r) + \frac{1}{2}\dot{\boldsymbol{M}}\boldsymbol{s}] + \tilde{\boldsymbol{\beta}}^T\boldsymbol{\Gamma}\dot{\tilde{\boldsymbol{\beta}}} \\ &= \boldsymbol{s}^T[\boldsymbol{n} - \boldsymbol{C}\dot{\boldsymbol{q}} - \boldsymbol{M}\ddot{\boldsymbol{q}}_r + \frac{1}{2}\dot{\boldsymbol{M}}\boldsymbol{s}] + \tilde{\boldsymbol{\beta}}^T\boldsymbol{\Gamma}\dot{\tilde{\boldsymbol{\beta}}} \\ &= \boldsymbol{s}^T[\boldsymbol{n} + \frac{1}{2}(\dot{\boldsymbol{M}} - 2\boldsymbol{C})\boldsymbol{s} - \boldsymbol{C}\dot{\boldsymbol{q}}_r - \boldsymbol{M}\ddot{\boldsymbol{q}}_r] \\ &\quad + \tilde{\boldsymbol{\beta}}^T\boldsymbol{\Gamma}\dot{\tilde{\boldsymbol{\beta}}} \\ &= \boldsymbol{s}^T[\boldsymbol{n} - \boldsymbol{M}\ddot{\boldsymbol{q}}_r - \boldsymbol{C}\dot{\boldsymbol{q}}_r] + \tilde{\boldsymbol{\beta}}^T\boldsymbol{\Gamma}\dot{\tilde{\boldsymbol{\beta}}} \end{aligned} \tag{9}$$

The skew-symmetry of $\dot{\boldsymbol{M}} - 2\boldsymbol{C}$ is used in the rearrangement.

Select a control law as:

$$\boldsymbol{n} = \hat{\boldsymbol{M}}\ddot{\boldsymbol{q}}_r + \hat{\boldsymbol{C}}\dot{\boldsymbol{q}}_r - \boldsymbol{K}_D\boldsymbol{s} \tag{10}$$

Substitute it into Eq. (9), then

$$\dot{V} = \boldsymbol{s}^T[\tilde{\boldsymbol{M}}\ddot{\boldsymbol{q}}_r + \tilde{\boldsymbol{C}}\dot{\boldsymbol{q}}_r - \boldsymbol{K}_D\boldsymbol{s}] + \tilde{\boldsymbol{\beta}}^T\boldsymbol{\Gamma}\dot{\tilde{\boldsymbol{\beta}}} \tag{11}$$

is obtained. Because $\boldsymbol{M}$ and $\boldsymbol{C}$ are linear with respect to the unknown parameters, one obtains

$$\tilde{\boldsymbol{M}}\ddot{\boldsymbol{q}}_r + \tilde{\boldsymbol{C}}\dot{\boldsymbol{q}}_r = \boldsymbol{P}(\boldsymbol{q}, \dot{\boldsymbol{q}}, \dot{\boldsymbol{q}}_r, \ddot{\boldsymbol{q}}_r)\tilde{\boldsymbol{\beta}} \tag{12}$$

Substituting Eq. (12) into Eq. (11) yields

$$\dot{V} = -\boldsymbol{s}^T\boldsymbol{K}_D\boldsymbol{s} + \tilde{\boldsymbol{\beta}}^T[\boldsymbol{\Gamma}\dot{\tilde{\boldsymbol{\beta}}} + \boldsymbol{P}^T\boldsymbol{s}] \tag{13}$$

Because $\dot{\tilde{\boldsymbol{\beta}}} = \dot{\hat{\boldsymbol{\beta}}}$ is true, an adaptation law

$$\dot{\hat{\boldsymbol{\beta}}} = -\boldsymbol{\Gamma}^{-1}\boldsymbol{P}^T(\boldsymbol{q}, \dot{\boldsymbol{q}}, \dot{\boldsymbol{q}}_r, \ddot{\boldsymbol{q}}_r)\boldsymbol{s} \tag{14}$$

leads:

$$\dot{V} = -\boldsymbol{s}^T \boldsymbol{K}_D \boldsymbol{s} \le 0 \tag{15}$$

This inequality shows that $\boldsymbol{s} \to \boldsymbol{0}$ as $t \to \infty$. Therefore, an adaptive controller is constructed by renewing $\hat{\boldsymbol{\beta}}$ from Eq. (14), computing $\hat{\boldsymbol{M}}$ and $\hat{\boldsymbol{C}}$ from the $\hat{\boldsymbol{\beta}}$, and obtaining the control input from Eq. (10).

Though the previous control is a joint variable control, a manipulation variable control is necessary for space robots because the joint variables cannot be predicted for the desired manipulation variables. A control law is derived below in the case that the hand and/or the satellite vehicle are controlled. Hence, the position and attitude of the satellite vehicle can be selected as the manipulation variables as well as that of the hand. But, the following controller is applicable when some of the variables aren't controlled, e.g., only the hand is controlled.

The manipulation variables $\boldsymbol{\psi}$ are related to the variables $\boldsymbol{q}$ by Jacobian matrix $\boldsymbol{J}$ as:

$$\dot{\boldsymbol{\psi}} = \boldsymbol{J}\dot{\boldsymbol{q}} \tag{16}$$

where

$$\dot{\boldsymbol{\psi}} = \begin{bmatrix} \boldsymbol{v}_0 \\ \boldsymbol{\omega}_0 \\ \dot{\boldsymbol{x}} \end{bmatrix}, \quad \boldsymbol{J} = \begin{bmatrix} \boldsymbol{I} & 0 & 0 \\ 0 & \boldsymbol{I} & 0 \\ \boldsymbol{J}_v & \boldsymbol{J}_\omega & \boldsymbol{J}_\theta \end{bmatrix} \tag{17}$$

$\boldsymbol{I}$ is the unit matrix, and $\dot{\boldsymbol{x}}$ is the velocity vector of the hand position and orientation. The $\dot{\boldsymbol{q}}_r$ and $\ddot{\boldsymbol{q}}_r$ are given as follows by using the desired manipulation variables $\boldsymbol{\psi}_d$ and positive definite gain matrix $\boldsymbol{\Lambda}$:

$$\dot{\boldsymbol{q}}_r = \boldsymbol{J}^{-1}\{\dot{\boldsymbol{\psi}}_d - \boldsymbol{\Lambda}(\boldsymbol{\psi} - \boldsymbol{\psi}_d)\} \tag{18}$$

$$\ddot{\boldsymbol{q}}_r = \boldsymbol{J}^{-1}\{\ddot{\boldsymbol{\psi}}_d - \boldsymbol{\Lambda}(\dot{\boldsymbol{\psi}} - \dot{\boldsymbol{\psi}}_d) - \dot{\boldsymbol{J}}\dot{\boldsymbol{q}}\} \tag{19}$$

Then the following relation is obtained because of Eq. (15):

$$\dot{V} = -\{\dot{\tilde{\boldsymbol{\psi}}} + \boldsymbol{\Lambda}\tilde{\boldsymbol{\psi}}\}^T \boldsymbol{J}^{-T}\boldsymbol{K}_D\boldsymbol{J}^{-1}\{\dot{\tilde{\boldsymbol{\psi}}} + \boldsymbol{\Lambda}\tilde{\boldsymbol{\psi}}\} \le 0 \tag{20}$$

where $\tilde{\boldsymbol{\psi}} = \boldsymbol{\psi} - \boldsymbol{\psi}_d$ is the manipulation variable errors. The manipulation variable errors converge to zeros because $\dot{\tilde{\boldsymbol{\psi}}} + \boldsymbol{\Lambda}\tilde{\boldsymbol{\psi}} \to \boldsymbol{0}$ as $t \to \infty$ for a non-singular $\boldsymbol{J}$. Hence the manipulation variables can be controlled by calculating $\dot{\boldsymbol{q}}_r$ and $\ddot{\boldsymbol{q}}_r$ from Eqs. (18) and (19), the control law of Eq. (10), and the adaptation law of Eq. (14)

When any manipulation variables in $\boldsymbol{\psi}$ aren't controlled, their current variables are used for the desired variables. For instance, $\dot{\boldsymbol{v}}_{0d} = \dot{\boldsymbol{v}}_0$, $\boldsymbol{v}_{0d} = \boldsymbol{v}_0$, and $\boldsymbol{r}_{0d} = \boldsymbol{r}_0$ are used for the satellite translational acceleration, the velocity, and the position, respectively, when the satellite position is not controlled. Then the control input $\boldsymbol{f}_0$ corresponding to the satellite position becomes zero. In the same manner, any selection of manipulation variables is available from the positions and orientations of the satellite vehicle and the hand. The asymptotic stability is invariable from this because this procedure only gives the desired manipulation variables.

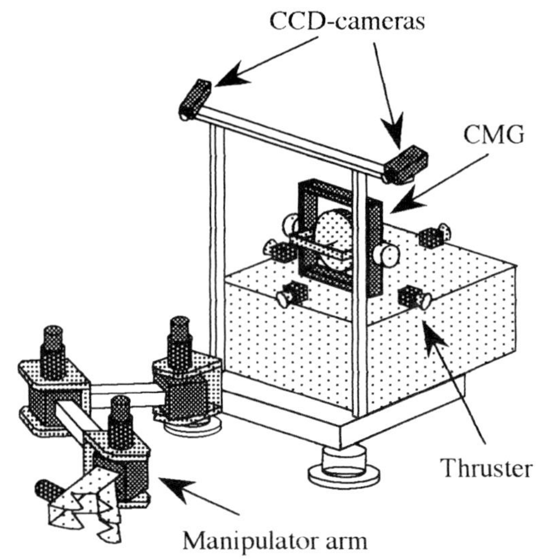

Figure 1. Hardware Space Robot Model

Numerical Simulations

Model and Condition

Numerical simulations are carried out to examine effectiveness of the proposed control method. The plant is a mathematical model of a hardware robot composed of a satellite vehicle and a 3-link manipulator, which is a planar experimental system on the ground simulating a space robot (Figure 1).[11] The robot has a control momentum gyro (CMG) and thrusters for a position/attitude control system. The CMG maintains the attitude of satellite, and the thrusters are for the position control. The parameters of the robot model grasping a payload are listed in Table 1, where m_i, ℓ_i, a_i, and i_i indicate the mass of link i, the length of the link, the distance of mass center from the revolution axis of the joint i, and the mass moment of inertia around its mass center, respectively. The links and joints are numbered from the base to the tip of the manipulator. Link 0 is the satellite vehicle, link 1 is the first link of manipulator, and link 3 is the manipulator hand. The parameters of link 3 are computed for the link 3 with a payload.

In this case, the following inertia parameters of the link 3 and payload are unknown:

$$\boldsymbol{\beta} = \begin{bmatrix} \beta_1 \\ \beta_2 \\ \beta_3 \\ \beta_4 \end{bmatrix} = \begin{bmatrix} m_3 \\ m_3 a_{3x} \\ m_3 a_{3y} \\ I_3 \end{bmatrix} \tag{21}$$

where a_{3x} and a_{3y} are x and y-coordinates of the mass center in link 3 frame. In the following simulations, the inertia parameters of link 3 without payload are used

Table 1. **Specification of Space Robot Model**

Link i	Mass m_i[kg]	Length ℓ_i[m]	Mass center a_i[m]	Inertia i_i[kg·m^2]
Link 0	46.94	0.459	0.000	1.413
Link 1	2.75	0.300	0.206	0.031
Link 2	1.62	0.260	0.138	0.017
Link 3	3.29	0.095	0.047	0.042

Table 2. **Initial Values of $\hat{\beta}$ and Real Values of β**

	β_1	β_2	β_3	β_4
Initial values	1.34	0.060	0.0189	0.0047
Real values	3.29	0.245	0.0189	0.0419

as the initial values of estimated parameters $\hat{\boldsymbol{\beta}}$. The inertia parameters of link 3 together with the payload are the real parameters $\boldsymbol{\beta}$. The initial values of $\hat{\boldsymbol{\beta}}$ and the real values of $\boldsymbol{\beta}$ are listed in Table 2.

Figure 2 shows the initial posture of the space robot. In all the following simulations, the hand is desired to move straightly from the initial state to -0.2 m in x-direction, $+0.2$ m in y-direction, and $+60°$ in orientation. The seventh-order polynomial curves of time are used to interpolate the path of manipulation variables between the initial and the final states.

The conditions of the experimental system, e.g., quantization, time delay and so on, are contained in the numerical simulations. For example, the angular velocity of the satellite is measured by a rate gyro quantized by 0.1 deg/sec and a sampling time of the controller is 1 ms.

The following feedback gains are used for the adaptive control.

$$\begin{aligned} \boldsymbol{K}_D &= \mathrm{diag}[\ 6.0\quad 6.0\quad 6.0\quad 6.0\quad 6.0\quad 6.0\ ] \\ \boldsymbol{\Lambda} &= \mathrm{diag}[\ 160\quad 160\quad 20\quad 160\quad 160\quad 20\ ] \\ \boldsymbol{\Gamma} &= \mathrm{diag}[\ 0.001\quad 0.1\quad 1.0\quad 1.0\ ] \end{aligned}$$

Numerical simulations are carried out for a resolved acceleration control (RAC) to compare control performances with those of the adaptive control. The control law is

$$\boldsymbol{n} = \boldsymbol{M}\dot{\boldsymbol{u}}_d + \boldsymbol{h} \tag{22}$$

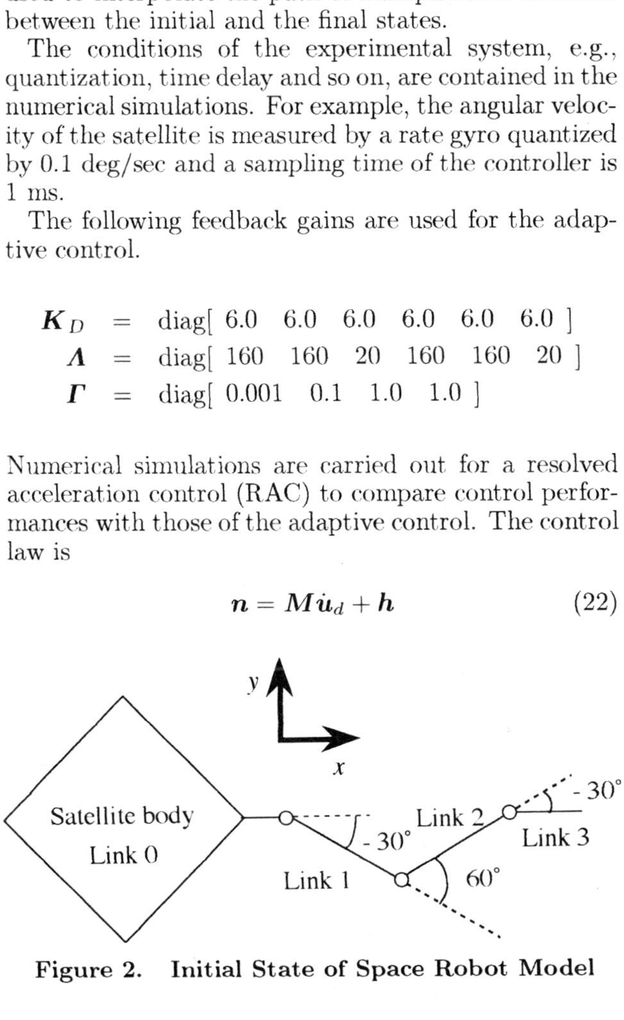

Figure 2. Initial State of Space Robot Model

where

$$\begin{aligned} \dot{\boldsymbol{u}}_d &= \boldsymbol{J}^{-1}\{\ddot{\boldsymbol{\psi}}_d - \boldsymbol{K}_D\dot{\tilde{\boldsymbol{\psi}}} - \boldsymbol{K}_P\tilde{\boldsymbol{\psi}} - \dot{\boldsymbol{J}}\dot{\boldsymbol{q}}\} \\ \tilde{\boldsymbol{\psi}} &= \boldsymbol{\psi} - \boldsymbol{\psi}_d \end{aligned}$$

The derivative and the proportional feedback gains $\boldsymbol{K}_D$ and $\boldsymbol{K}_P$ are

$$\begin{aligned} \boldsymbol{K}_P &= \mathrm{diag}[\ 160\quad 160\quad 120\quad 160\quad 160\quad 120\ ] \\ \boldsymbol{K}_D &= \mathrm{diag}[\ 3.0\quad 3.0\quad 2.0\quad 3.0\quad 3.0\quad 2.0\ ] \end{aligned}$$

When any of manipulation variables in $\boldsymbol{\psi}$ aren't controlled, their desired acceleration should be computed from the condition that their corresponding generalized forces are zeros. The desired acceleration is computed based on Yamada's algorithm.[6]

Hand Path and Satellite Attitude Control (#1)

A teleoperated space robot should fix orientation of a communication antenna installed on the satellite vehicle to hold a radio communication connection to an external commander during operation. The space robot must keep its solar cell panels. It shouldn't swing a view of vision sensor installed on the satellite vehicle. Shown here are the numerical simulations corresponding to those cases when the satellite attitude rests and the hand tracks a desired path. This situation will arise most frequently in real operations.

For the adaptive control, Figures 3, 4, and 5 show the robot motion, the error norm of the hand position, and the estimations of the unknown parameters, respectively. Figure 5 shows the rates ($\hat{\boldsymbol{\beta}}/\boldsymbol{\beta}$) of the estimated values to the real values. The rate being closer to one represents the estimation being more accurate. For the RAC, Figures 6 and 7 show the robot motion and the error norm of hand position, respectively.

Hand Path Control (#2)

Shown here are the numerical simulations when the only hand tracks a desired path to show that the proposed adaptive control is capable of the other situations. Here the only joint torques are the control inputs whereas no position/attitude control device on the satellite vehicle is used. This situation arises when the robot operates autonomously because it doesn't have to hold the radio communication connection, and the satellite position/attitude change is allowed.

For the adaptive control, Figures 8, 9, and 10 show the robot motion, the error norm of the hand position, and the estimation rates ($\hat{\boldsymbol{\beta}}/\boldsymbol{\beta}$) of the unknown parameters, respectively. For the RAC, Figures 11 and 12 show the robot motion and the error norm of hand position, respectively.

Hand Path and Satellite Position/Attitude Control (#3)

The space robot should control the position and attitude of the satellite vehicle as well as that of the hand for looking closely at an object using vision sensors or for a specific task. Shown here are the numerical simulations corresponding to the cases when both the satellite position/attitude and the hand track their desired paths. The desired path of the hand is the same

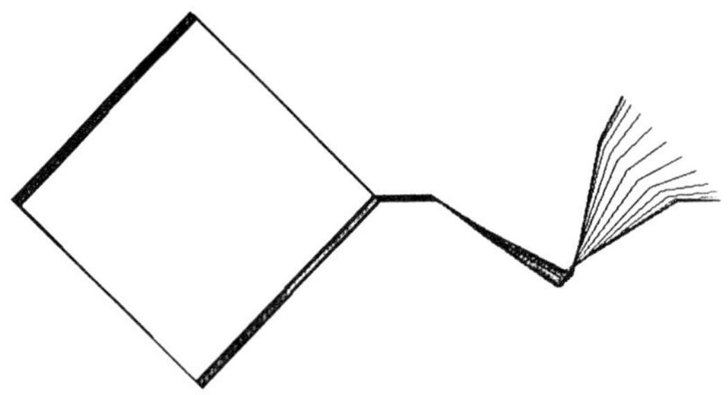

Figure 3. Motion of Robot Controlled by Adaptive Control (#1)

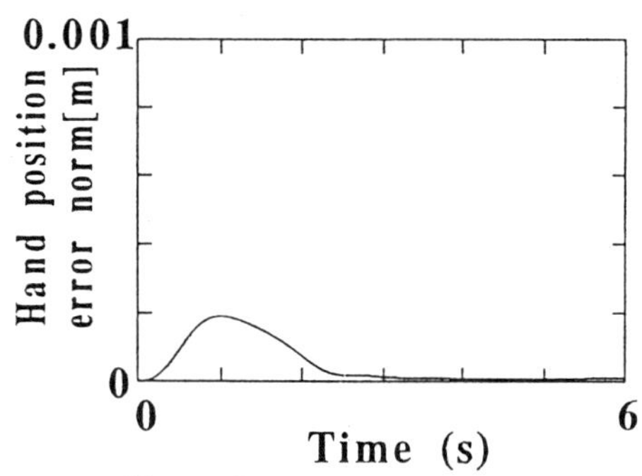

Figure 4. Error Norm Controlled by Adaptive Control (#1)

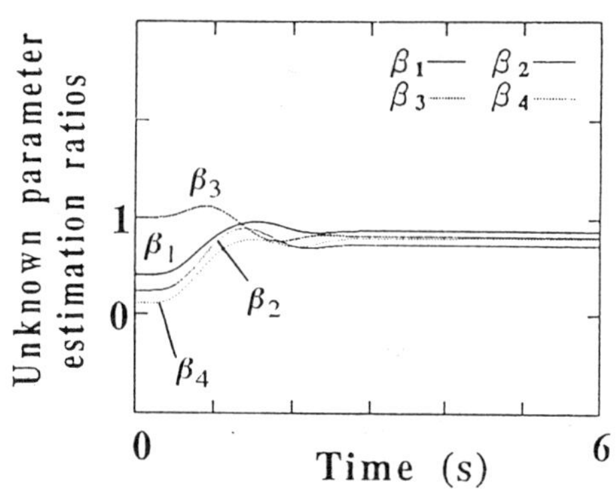

Figure 5. Parameter Estimation (Estimated/ Real) in Adaptive Control (#1)

as the previous cases. As the desired path of the satellite vehicle, the satellite attitude rests, and the position moves straightly from the initial state to +0.1 m in x-direction, -0.2 m in y-direction. The position control devices are supposed to generate any magnitude of control forces to guarantee the stability of closed-loop system. Contrivance, e.g., pulse-width-modulation control, is necessary when thrusters are used because the thrusters are available for only on/off control.

For the adaptive control, Figures 13, 14, and 15 show the robot motion, the error norm of the hand position, and the estimation rates $(\hat{\boldsymbol{\beta}}/\boldsymbol{\beta})$ of the unknown parameters, respectively. For the RAC, Figures 16 and 17 show the robot motion and the error norm of hand position, respectively.

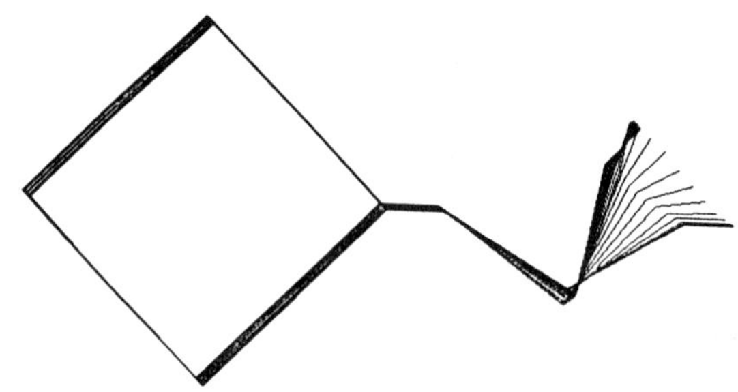

Figure 6. Motion of Robot Controlled by RAC (#1)

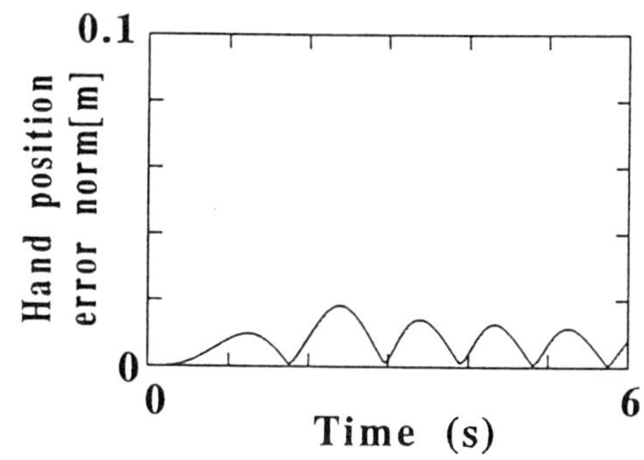

Figure 7. Error Norm Controlled by RAC (#1)

Discussions

In all cases, the figures show that the adaptive control achieves the better control performance than the RAC. Hence, the adaptive control is more effective than the RAC when the space robot manipulates an unknown payload. The estimated parameters tend to converge to the real values during the motion for three seconds as illustrated in Figures 5, 10, and 15, whereas the convergence of the estimation is not guaranteed. In those cases, the parameters converge to the real values when the robot repeats the same motion twice, whereas the results are not shown here. In general, such a continuous motion improves the estimation, and the control performance gets better.

Concluding Remarks

The adaptive control which estimates the unknown inertia parameters of the payload has been proposed for space robots. The asymptotic stability of the closed-loop system is guaranteed unless Jacobian matrix becomes singular because the stability has been proven by Lyapunov's second method. The controller has been able to control any variables of the position and attitude of the satellite vehicle and that of the hand. For instance, the payload manipulation is available with the satellite attitude at rest. Some numerical simulations have examined the availability. All the numerical simulations have shown that the adaptive control can achieve better control performance than the resolved acceleration control for a space robot manipulating an unknown payload.

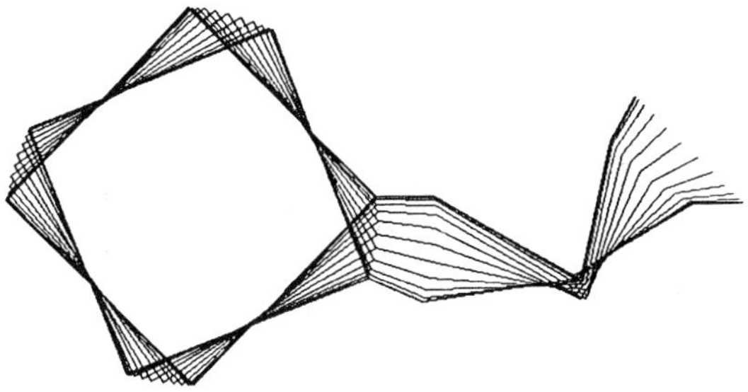

Figure 8. Motion of Robot Controlled by Adaptive Control (#2)

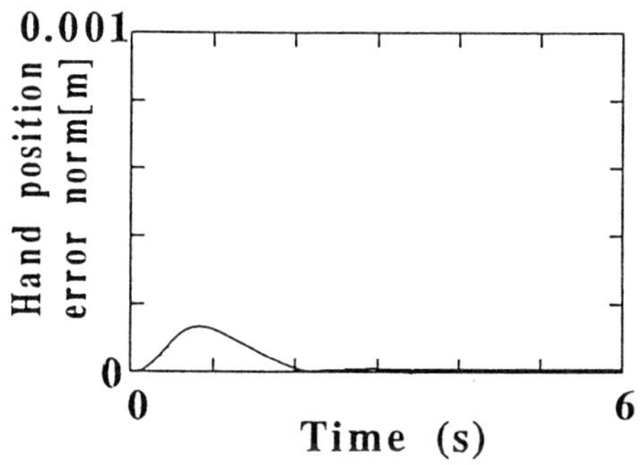

Figure 9. Error Norm Controlled by Adaptive Control (#2)

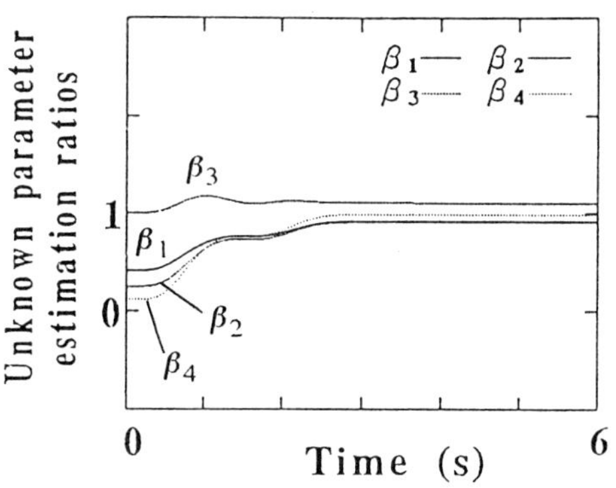

Figure 10. Parameter Estimation (Estimated/ Real) in Adaptive Control (#2)

Acknowledgment

A part of this work was financially supported by a Grant-in-Aid for Scientific Research from Ministry of Education, Science, Sports, and Culture of Japan.

References

[1] Murotsu, Y., Senda, K., Ozaki, M., and Tsujio, S., "Parameter Identification of Unknown Object Handled by Free-Flying Space Robot," *AIAA Journal of Guidance, Control, and Dynamics*, Vol. 17, No. 3, May–June, 1994, pp. 488–494.

[2] Vafa, Z. and Dubowsky, S., "On the Dynamics of Manipulators in Space Using the Virtual Manipulator Approach," *Proceedings of IEEE International Conference on Robotics and Automation*, Raleigh, NC, March–April 1987, pp. 579–585.

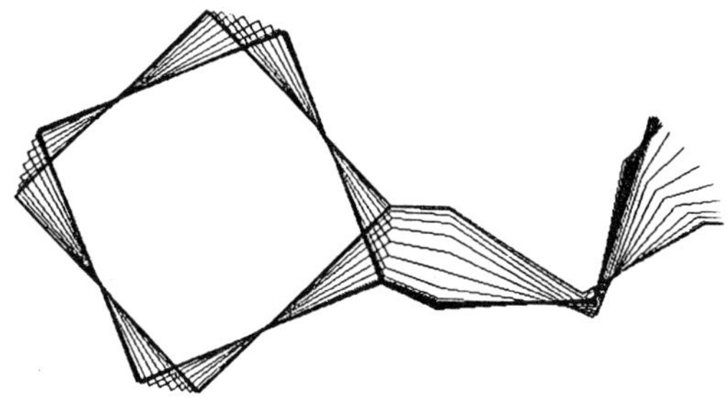

Figure 11. Motion of Robot Controlled by RAC (#2)

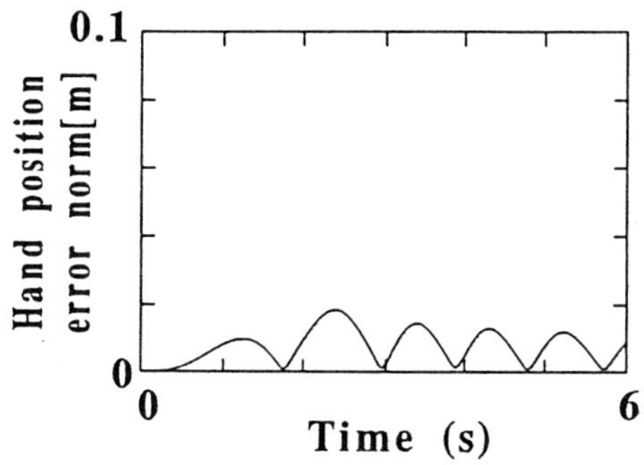

Figure 12. Error Norm Controlled by RAC (#2)

[3] Umetani, Y. and Yoshida, K., "Continuous Path Control of Space Manipulators Mounted on OMV", *Acta Astronautica*, Vol. 15, No. 12, 1987, pp. 981–986.

[4] Masutani, Y., Miyazaki, F., and Arimoto, S., "Sensory Feedback Control for Space Manipulators," *Proceedings of IEEE International Conference on Robotics and Automation*, Scottsdale, AZ, May 1989, pp. 1346–1351.

[5] Umetani, Y. and Yoshida, K., "Resolved Motion Rate Control of Space Manipulators with Generalized Jacobian Matrix", *IEEE Transactions on Robotics and Automation*, Vol. 5, No. 3, 1989, pp. 303–314.

[6] Yamada, K. and Tsuchiya, K., "Efficient Computation Algorithms for Manipulator Control of a Space Robot," *Transactions of Society of Instrument and Control Engineers*, Vol. 26, No. 7, 1990, pp. 765–772. (in Japanese)

[7] Yoshikawa, T., Yokokohji, Y., and Toyoshima, T., "Efficient Computational Algorithms for Trajectory Control of Multi-Arm Space Manipulators," *Proceedings of International Symposium on Artificial Intelligence, Robotics and Automation in Space*, Kobe, Japan, Nov. 1990, pp. 303–306.

[8] Slotine, J-J. E. and Li, W., "On the Adaptive Control of Robot Manipulators," *The International Journal of Robotics Research*, Vol. 6, No. 3, 1987, pp. 49–59.

[9] Yamamoto, T., Kobayashi, J., Ohkawa, F., and Katoh, R., "Digital Adaptive Control of a Manipulator Mounted on Free-Flying Robot in Space," *Proceedings of IFToMM*, pp. 2151–2155, 1995.

[10] Iwata, T., Machida, K., and Toda, Y., "Unified Control Method of Free-Flying Robot in Maneuvering to

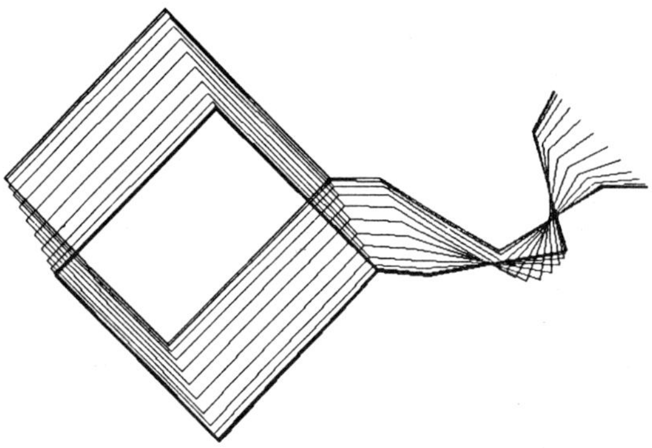

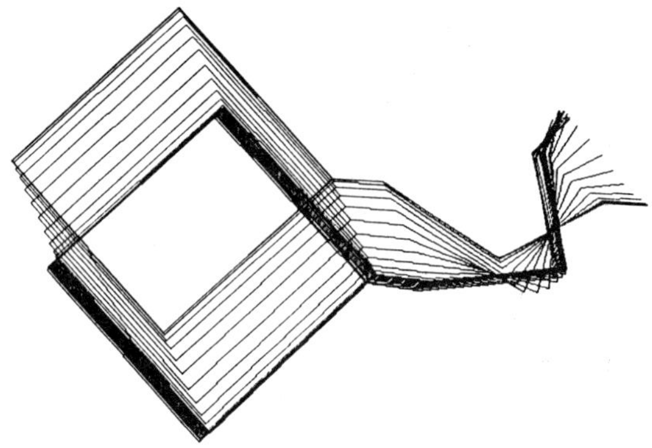

Figure 13. Motion of Robot Controlled by Adaptive Control (#3)

Figure 16. Motion of Robot Controlled by RAC (#3)

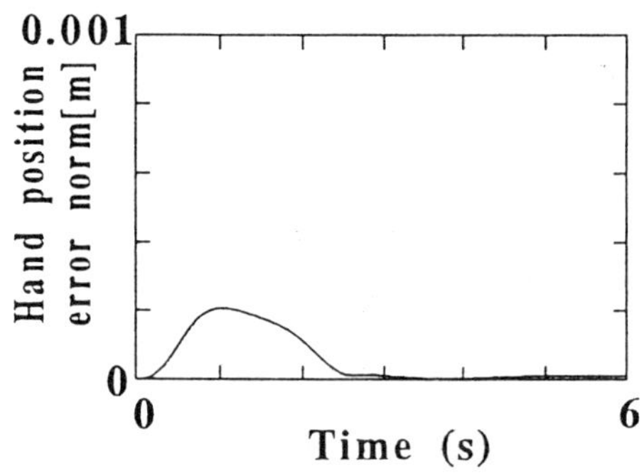

Figure 14. Error Norm Controlled by Adaptive Control (#3)

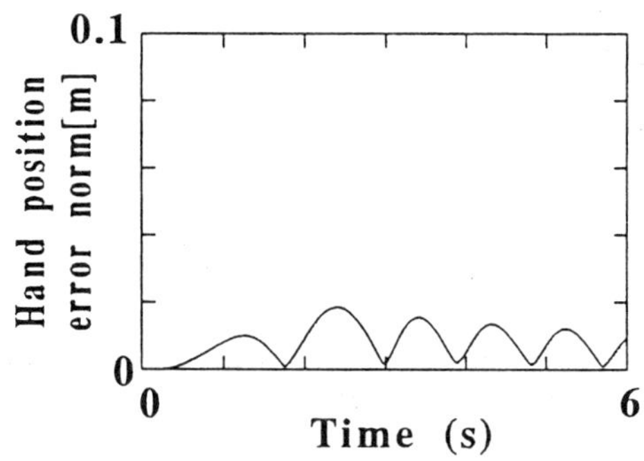

Figure 17. Error Norm Controlled by RAC (#3)

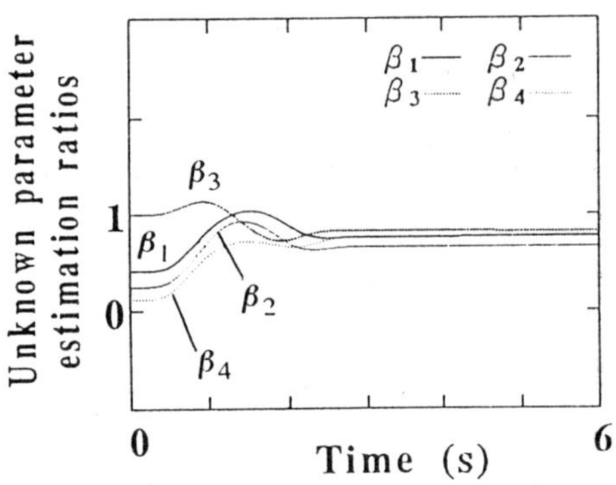

Figure 15. Parameter Estimation (Estimated/ Real) in Adaptive Control (#3)

Capture a Floating Object", *Journal of Japan Society for Aeronautical and Space Sciences*, Vol. 43, No. 502, 1995, pp. 634–640. (in Japanese)

[11] Senda, K., Murotsu, Y., and Adachi, H., "Control Experiment for Space Robot with Visual Information," *AIAA Guidance, Navigation, and Control Conference*, San Diego, CA, July 29-31, 1996, AIAA-96-3900.

AIAA-97-3558

SPACE TELEROBOT EXPERIMENT SYSTEM BASED ON NASDA ETS-VII SATELLITE

Mitsushige Oda*, Megumu Nishida†

National Space Development Agency of Japan (NASDA)
2-4-1 Hamamatsu-cho, Minato-ku, Tokyo, Japan
e-mail: nishida.megumu@nasda.go.jp

Abstract

National Space Development Agency of Japan (NASDA) has developed Engineering Test Satellite-VII(ETS-VII) in order to obtain space robotics technology and Rendezvous Docking technology. ETS-VII will be launched into a 550km altitude orbit by a NASDA H-II rocket in 1997. ETS-VII has a 6-dof robot arm of about 2m long in order to make space telerobot experiments. The telerobot experiment system is designed to reduce operator's work load and ease many robot users who are not special robot operator, and consists of an onboard robot system and an on-ground control system. This paper describes the telerobot experiment system of the ETS-VII satellite .

Key words; space robot, teleoperation, telemanipulation, time delay, telerobot, remote autonomy

1 Introduction

Current space robot such as the space shuttle's remote manipulator is operated by an astronaut on the shuttle to capture an orbiting satellite or to position an astronaut in space who conducts extravehicular activities. Tasks to be performed in space either by astronauts or robots in near future include building of orbiting facilities such as the international space station and exploration of the Lunar, the Mars and other planets. Astronauts alone can not conduct every required task from various reasons such as safety and work load, however, technologies to build a fully autonomous intelligent space robot for these missions are still premature. Therefore, the teleoperational space robots are desired for the near future applications. In realizing the teleoperational or remote controllable space robots, many new technology are still required besides the traditional robot technology for the on-ground applications. Those includes mechanism which can work in the space environment such as vacuum and large temperature change, teleoperation related technology.

NASDA will launch ETS-VII in the year 1997 to develop and verify these technologies in-orbit, and its mission life is 1.5 years.[1-3] The ETS-VII satellite has a robot arm as is shown in Fig. 1. This robot arm will be teleoperated from an on-ground control station using the

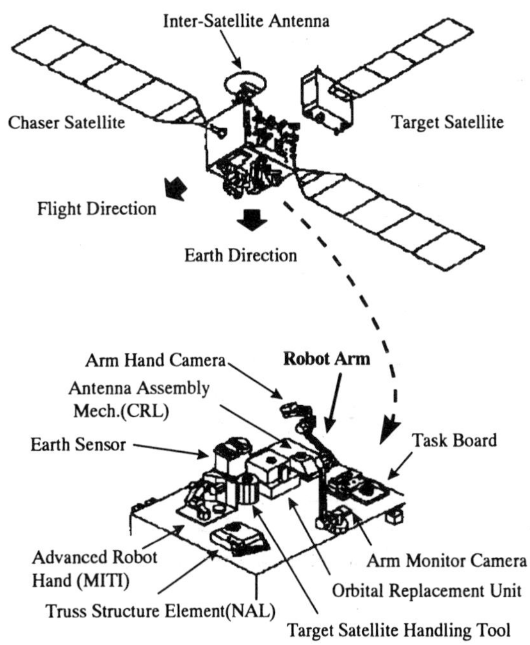

Fig. 1 Engineering Test Satellite -VII (ETS-VII)

---

* Senior engineer, AIAA member

† Associate senior engineer, AIAA member

communication link through a data relay satellite named COMETS in the geo-stational orbit as is shown in Fig. 2. The time delay of the communication link is about 4 to 6 seconds in the round trip. Since ETS-VII is a rare opportunity for the space robot experiments, some Japanese national laboratories join the ETS-VII robot experiments. They are Ministry of International Trade and Industries (MITI), Communication Research Laboratory (CRL), National Aerospace Laboratory (NAL). These laboratories develop their own robot experiment payload and on-ground robot control system in order to make their robot experiments using the NASDA robot arm.

This paper is organized as follows; overview of the teleoperation system of the space robot including ETS-VII is shown in section 2, the onboard robot system of ETS-VII is shown in section 3, the on-ground control system for the robot system is shown in section 4, the conclusion of this paper is section 5.

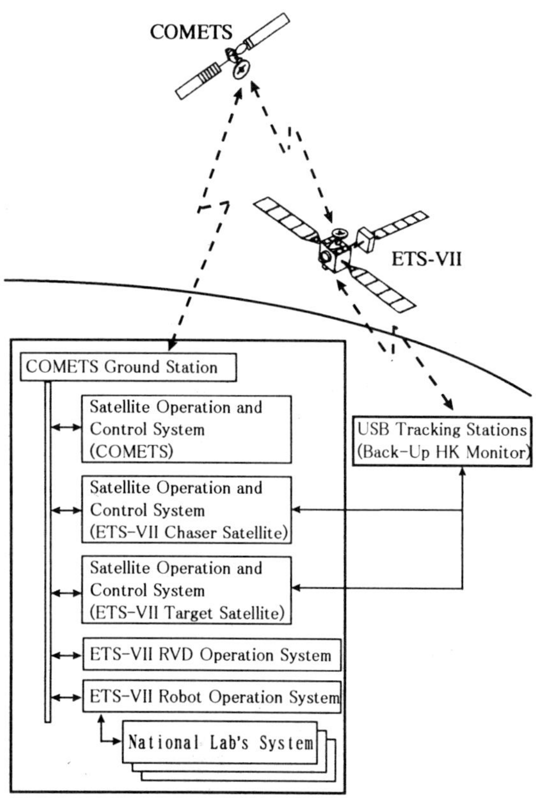

Fig. 2 Communication system of ETS-VII

2 Teleoperation system of the space robots

2.1 Overview of the existing teleoperation system

There are already some space robot systems in the world. The space shuttle remote manipulator system is frequently used since 1981.[4] Operation of the shuttle remote-manipulator system is realized by an astronaut in the space shuttle. One astronaut operates two joysticks to control position and attitude of the manipulator. Another astronaut controls other joysticks to control position and attitude of the shuttle orbitor against the payload such as a satellite. Reaction of the manipulator acts on the shuttle orbitor dynamics. Coordination of the manipulator and the orbitor is realized by the coordination of the two astronauts.

In the year 1993, Germany conducted a space robot experiment (ROTEX) on the space shuttle's pressurized experiment module.[5] The robot arm was 1.4m long and 5-dof. The robot arm was operated by an astronaut in the orbitor and also an operator on the ground. Teleoperation from an on-ground control station in Germany was realized using a data relay satellite and a communication satellite. Time delay in the control loop was about 6 seconds in the round trip. In order to overcome the time delay, the predictive stereo computer graphics image of the onboard robot system was used. In this case, the robot experiments could use fairy good communication capacity to transmit the color video images and telemetry data.

There are also some research that handle teleoperation under the time delayed environment.[6-8] These research consider limited amount of time delay, however, they need not to consider the limitation of the communication capacity. In most cases, the satellite has limited communication capacity for the operation of the satellite itself. Data relay satellites are used to transmit data for the manned spacecraft, earth observation satellite and astronomy satellite. If a robot on a satellite requires large volume of communication capacity, it will jeopardize other users of the space communication link. Therefore, lower communication capacity will be desirable to operate the satellite mounted robot system.

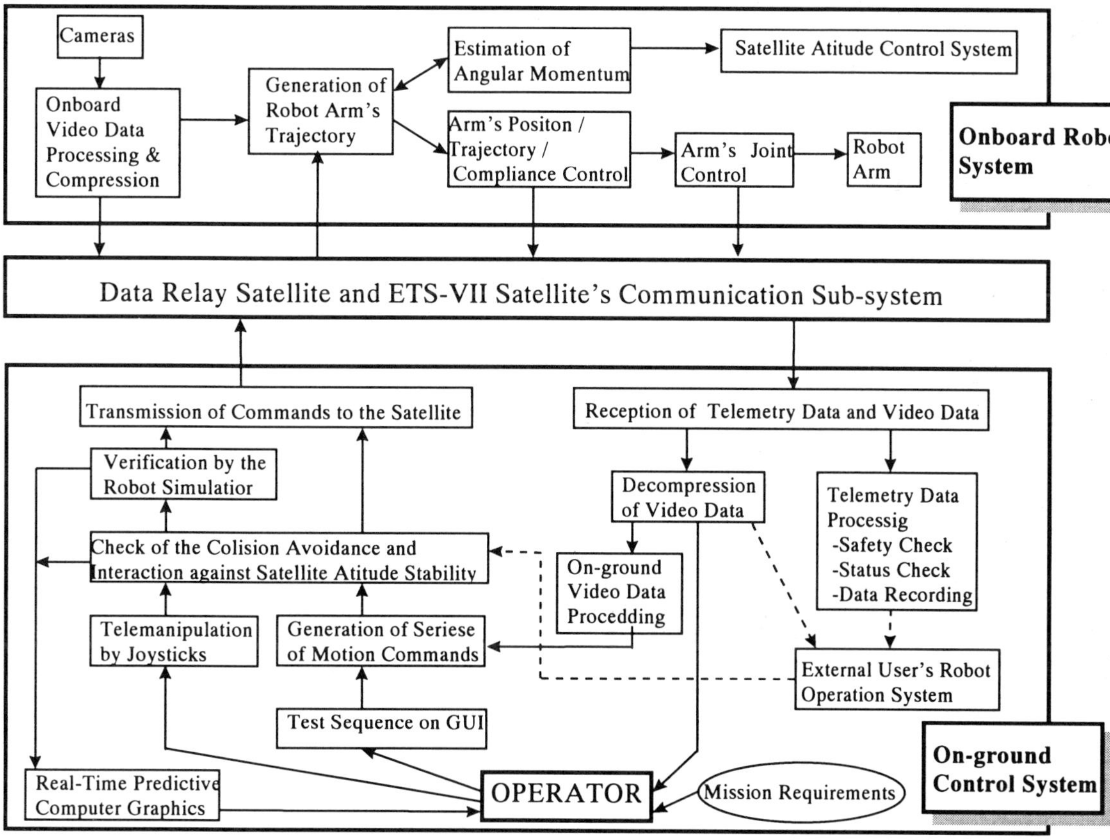

Fig. 3 Block diagram of ETS-VII teleoperation system

2.2 ETS-VII robot teleoperation system

From the previous consideration, the communication capacity for the operation of ETS-VII including robot arm and satellite platform are set as low as 1.5Mbps for the down link and 4kbps for the uplink. Fig. 3 shows the block diagram of ETS-VII teleoperation system. As is shown in Fig. 3, the ETS-VII teleoperation system consists the onboard robot system and the on-ground control system. Design requirements for the teleoperation system are :

(1) to control the onboard robot system by a minimum number of operators; the number of operators required to monitor and operate the satellite mainly determined the cost of satellite operations.

(2) easy to operate it even if the time delay and the limitation of the communication capacity exist; it is also known that the time delay more than a few seconds makes the telemanipulation difficult.

3 Onboard robot system

3.1 ETS-VII satellite

Fig. 4 shows ETS-VII proto-flight model under the system environment test at the NASDA Tsukuba Space Center. ETS-VII is a satellite to conduct the rendezvous docking and space robot technology experiments. For the rendezvous docking experiments, ETS-VII satellite consists of two satellite; a chaser satellite and a target satellite. Mass of the chaser satellite is about 2.5t and that of the target satellite is about 0.4 t. Both satellites will be launched together by NASDA H-II rocket into a low Earth orbit (orbit height: 550km). The target satellite will be released while the rendezvous docking experiments are conducted. During the robot experiments, the target satellite will be connected to the chaser satellite to save the attitude control fuel.

Fig. 4 ETS-VII proto-flight model (faced robot panel)

3.2 Onboard robot system

Table. 1 shows the principal characteristics of the ETS-VII onboard robot system. The onboard robot system consists of following equipment.

3.2.1 Robot arm

Fig. 5 shows the robot arm. The robot arm has 6 joints and its length is about 2m. Each joint is driven by the combination of a DC brushless motor and a harmonic drive gear. Angular velocity of the motor is detected by a resolver. An end effector is attached at the end of the robot arm. The end effector can grasp a grapple fixture which is an attachment to standardize the interface between payloads and the end effector. The end effector can also provide torque and electrical power/signals to the payload using the adequate grapple fixture.[9]

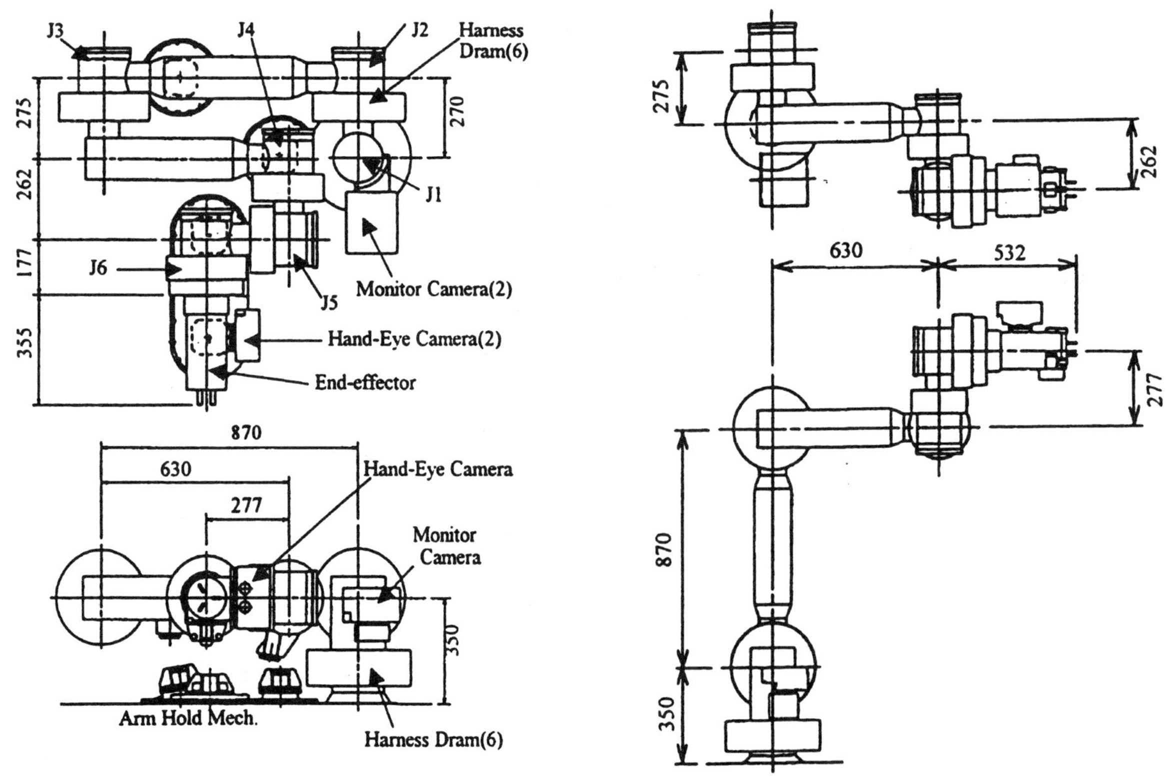

Fig. 5 ETS-VII robot arm (stored position and standby position)

Table 1 Characteristics of the ETS-VII robot system

arm length	~ 2m
arm mass	~ 140kg
tip pose accuracy	10mm, 1deg
tip pose repeatability	2.5mm, 0.13deg
max. tip speed	50mm/s, 5.0deg/s
tip force	> 20N
max. payload	430kg, 700kg m^2

3.2.2 Vision system (hand eye camera and monitor camera on the robot arm)

A hand eye camera is mounted on the robot arm's end effector. A monitor camera is mounted on the first joint of the robot arm. The first joint of the robot arm is used as a pan unit of the monitor camera. They are black and white CCD cameras and each camera has redundancy. The primary and the redundant cameras are built in one box, and can also be used as a stereo camera. These camera image data are sent to the ground control station in the compressed form.

3.2.3 Onboard robot control system

Fig. 6 shows the block diagram of the onboard system of ETS-VII satellite. To maintain the communication link through the data relay satellite, the attitude of the satellite platform is controlled by the onboard satellite attitude control system using the reaction wheels and the gas jet thrusters. When the arm handles the target satellite quickly, the reaction of the robot arm will be too large for the satellite attitude control system to keep the satellite attitude adequately.

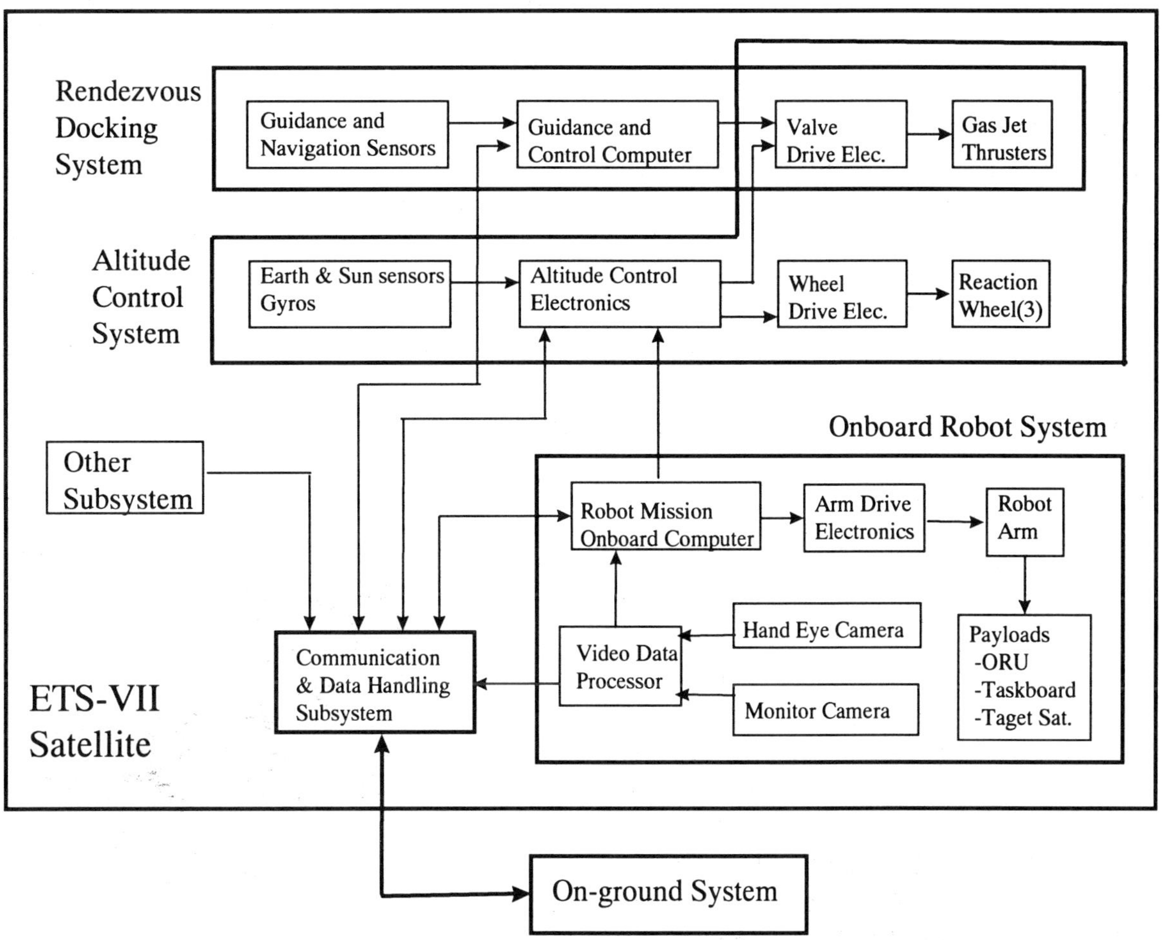

Fig. 6 Block diagram of the onboard system of ETS-VII

Therefore, the coordinated control of the satellite attitude and control and the robot arm is realized through the coordination of the onboard satellite attitude control system, onboard robot control system, and the on-ground robot control system. The onboard robot control system estimates the angular momentum that the planned or commanded robot arm motion will produce. This estimated angular momentum is provided to the onboard satellite attitude control system to conduct the feed-forward angular momentum compensation. The on-ground robot control system also estimates the angular momentum which the instructed robot arm motion will produce. If the estimated angular momentum is too large for the satellite attitude control, then the planned robot arm motion will be modified. The onboard robot control system can also check the robot arm motion by its estimated angular momentum. This function is necessary when the robot arm motion is generated by the onboard control system.[10-12]

The ETS-VII robot arm is controlled by the arm drive electronics (ADE) and the robot mission onboard computer (RMOC). The ADE drives joint motors and controls joint's angular velocity. Angular position of each joints, pose and trajectory of the robot arm, and compliance of the robot arm are managed by RMOC. RMOC interpolates the teleoperation commands and generates the robot arm's trajectory to realize the position loop control of the robot arm. RMOC also performs the onboard video data processing to measure position and attitude of the target marker. Since the location of the target markers are pre-defined, the onboard robot controller can generate a robot arm's path to grasp the payload.

3.2.4 Robot arm's payloads

Payloads which the ETS-VII robot arm will handle are:

(a) Orbital replacement unit (ORU)

ORU is the unit to replace equipment and supply consumer goods for the satellite and space station in orbit. A robot arm will grasp the grapple fixture of ORU and release it from the satellite platform, then remove it from the satellite and replace with the new ORU. On the ETS-VII satellite, only one ORU is used to simulate these tasks. This ORU handling also simulates fuel supply. The ORU has a fuel tank inside the unit and a pair of liquid connector which connects this tank with other tank in the satellite platform. The liquid connector will be mated / de-mated by providing torque from the robot arm's end effector. Fig. 7 shows the ETS-VII ORU proto-flight model.

Fig. 7 ORU proto-flight model

(b) Taskboard

Fig. 8 shows the ETS-VII taskboard proto-flight model. Taskboard includes a peg & hole, a slide handle, a force torque sensor calibration mechanism, a small tethered metal ball to be released for capture and other experimental equipment. The taskboard handling tool will be attached to the robot arm's end effector to handle these small payloads.

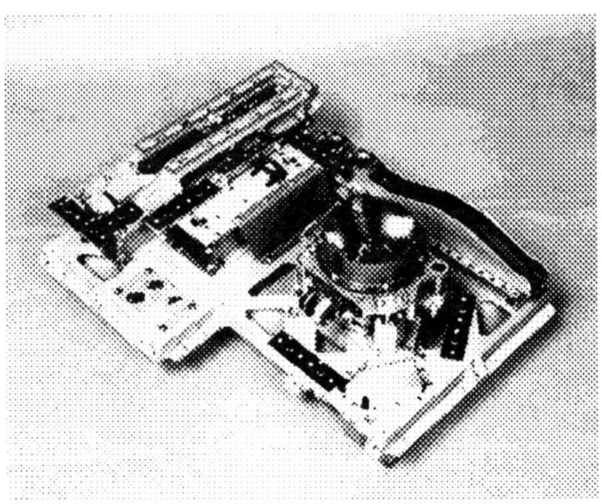

Fig. 8 Taskboard proto-flight model

(c) Target satellite handling tool (TSTL)

Fig. 9 shows TSTL proto-flight model. The ETS-VII robot arm works in the micro-gravity condition and can handle the target satellite whose mass is about 0.4t. Since the capability of the end effector of ETS-VII is not appropriate to grasp / capture the target satellite, TSTL is equipped. The tool will be attached to the robot arm's end effector by grasping the grapple fixture of the tool. This tool exchange system makes the utility of the robot arm widen.

Fig. 9 TSTL proto-flight model

4 On-ground control system

Fig. 10 shows the on-ground control system. The on-ground system has two robot arm control modes, which are the telemanipulation mode and the program control (autonomous remote control) mode. In the telemanipulation mode, a ground operator controls the satellite mounted robot arm by two joysticks. Status of the satellite mounted robot system is provided to the operator by the raw video image from the CCD camera on the robot arm, and the telemetry data from the satellite. Since these information are not enough to teleoperate the onboard robot system under the time delayed environment, an augmented graphics image and supplemental information will be given to the operator.

Fig. 10 On-ground control system

4.1 Telemanipulation mode

In the telemanipulation mode, robot arm control commands are in the form of the robot arm's position and attitude data, and sent at each 250msec to the satellite. Velocity commands to the robot arm from a ground station are not used, since discontinuation of the commands by such as disruption of the communication link will endanger the robot arm and satellite. The telemanipulation commands are generated by the operation of joysticks. Video images from the hand eye camera and monitor camera are provided to the operators. In the communication link the time delay is estimated 4 to 6 seconds. This delay make it difficult to telemanipulate the robot arm by watching the delayed video image and telemetry data. Therefore, the real time computer graphics images of the onboard robot system are provided to the operator. Beside the predictive graphics image of the robot system, some other tools such as a virtual tube to guide the robot arm's tip position to the aimed position, virtual wall to restrict robot arm motion not to collide with other equipment are used.

The specification of the joysticks were determined by the consideration based on the experiments of the on-ground space robot testbed with a 6 dof master-arm.[13] Update cycle of the video images from the satellite is 5 images per second including the hand eye camera and the monitor camera. This update cycle was also determined by the experiment on the testbed.

4.2 Program control mode

Since telemanipulation under the time delayed and limited communication capacity's environment is not an easy task especially for robot users who are not special robot operator, it is necessary to develop an easy way to use the space robot. In the program control mode, robot control commands are in the form of robot motion commands such as "MOVE to A". These commands are generated in the on-ground robot control system based on the operator's higher level instruction. The on-ground operator gives task sequence to the on-ground robot control system using the GUI. The on-ground robot control system generates series of robot control commands from this given task sequence. The on-ground robot control system has a simulator of the onboard robot system which consists of the robot mission onboard computer (RMOC) electrical model and satellite/robot dynamics simulator on an engineering workstation. Series of generated commanded are sent to this simulator and verified. This sequence of commands are saved in the on-ground robot control system and will be sent to the satellite when it is necessary. If necessary, operator can interrupt execution of this command sequence and override it by the manual commanding.

4.3 External user's control mode

Robot control commands of external users (Ministry of International Trade and Industries, Communication Research Laboratory, National Aerospace Laboratory) are sent to the onboard robot system through NASDA on-ground robot control system in order to check the propriety for the robot arm and the safety for the satellite attitude control in real-time. Necessary telemetry data including image data to monitor and control the onboard robot system are provided from NASDA on-ground control system to the users' facilities.

5 Conclusion

This paper introduced the teleoperation system of ETS-VII satellite. ETS-VII will be launched by NASDA H-II rocket in the year 1997, and its mission life is 1.5 years. The concept of ETS-VII teleoperation system is to reduce operator's work load and make it easy to use the space robot for many robot users who are not necessarily robot engineer. The teleoperation system consists an onboard robot system and an on-ground control system. The onboard robot system has a 6-dof robot arm, a vision system, an onboard robot control system and robot arm's payloads. The on-ground control system has telemanipulation mode and program mode. We will report the result of these robot experiments in the year 1998.

References

1 M.Oda, et al.; "ETS-VII, the world first telerobotics satellite (Mission and its design concept)", in Proceedings of the 2nd International Symposium on Artificial Intelligence, Robotics and Automation in Space, Sept.30-Oct.2, 1992, Toulouse, France

2 M.Oda, I.Kawano and F.Yamagata; "ETS-VII (Engineering Test Satellite #VII) - a Rendezvous Docking and Space Robot Technology Experiment Satellite", 46th International Astronautical Congress, Oct.2-6, 1995, Oslo, Norway, AF-95-U.2.01

3 M.Oda, et al. "ETS-VII, Space Robot In-Orbit Experiment Satellite", Proc. of 1996 IEEE Int. Conf. on Robotics and Automation, April 22-28, 1996, Minneapolis, USA, pp.739-744

4 R.Ravindran, K.H.Doetsch, "Design aspect of the shuttle remote manipulator system", Proc. of the AIAA 1982 Guidance and control conference, pp.456-465, AIAA-82-1581, 1982

5 G.Hirzinger, et al., "The sensory and telerobotic aspect of the space robot technology experiment ROTEX", Proc. of the 2nd Int. symp. On Artificial Intelligence, Robotics and Automation in Space, Sept. 1992, Toulouse, France, pp.83-110

6 P.G.Backes, et al., "A Local-Remote Telerobot System for Time-delayed Traded and Shared Control", Proc. of the 1991 Int. Conf. On Advanced Robotics, 1992

7 W.Kim, et al., "Force-reflecting and Shared Compliant Control in Operating Telemanipulators with Time Delay", IEEE Trans. Robotics and Automation, Vol.8, No.2, April 1992, pp.176-185

8 J.Funda, et al., "Teleprogramming: Toward Delayed Invariant Remote Manipulation", Presence, Vol.1, No.1 MIT Press, 1992

9 M.Oda, M.Nishida, et al.; "Development of the EVA End Effector, Grapple Fixtures and Tools for the

Satellite-Mounted Robot System", Proc. of 1996 IEEE/RSJ International Conference on Intelligent Robots and Systems, Nov.1996, Osaka, Japan, pp.1536-1543

10 M.Oda, "Coordinated Control of Spacecraft Attitude and its Manipulator", Proc. of 1996 IEEE Int. Conf. on Robotics and Automation, April 22-28, 1996, Minneapolis, USA, pp.732-738

11 M.Oda, "Motion Control of the Satellite Mounted Robot Arm Which Assures Satellite Attitude Stability", 47th International Astronautical Congress, Oct. 7-11, 1996, Beijin, China, IAF-96-A.7.09

12 M.Oda and Y.Ohkami, "Coordinated Control of Spacecraft Attitude and Space Manipulators", Control Engineering Practice, Vol.5, No.1, pp.11-21, 1997

13 M.Oda, "System Engineering Approach in Designing the Teleoperation System of the ETS-VII Robot Experiment Satellite", Proc. of 1997 IEEE Int. Conf. on Robotics and Automation, April, 1997, Albuquerque, NM, USA, pp.3054-3062

AIAA-97-3559

Dual Mode Inverse Control and Stabilization of Flexible Space Robots

G. de Rivals-Mazères,* W. Yim,†
‡ F. Mora-Camino† and S. N. Singh‡

Abstract

The question of control and stabilization of flexible space robots is considered. Although, this approach is applicable to space robots of other configurations, for definiteness, a flexible planar two-link robot, mounted on a rigid floating platform in space, is considered. The robotic arm has two revolute joints and its links undergo elastic deformation in the plane of rotation. Based on nonlinear inversion technique, a control law is derived for controlling output variables describing the position and orientation of the platform and the joint angles of the robot. Although, the inverse controller accomplishes reference trajectory tracking, it excites the elastic modes of the arm. Using linear quadratic optimal control theory, a composite stabilizer for stabilization of the rigid and flexible modes and a decoupled flexible mode stabilizer are designed for regulating the end point of the robot to the target point and vibration suppression. Stabilization using only elastic mode velocity feedback is also considered. For large maneuvers, first the inverse controller is active, and the stabilizer is switched for regulation when the motion of the robot lies in the neighborhood of the terminal equilibrium state. Simulation results are presented to show that in the closed-loop system including the inverse controller and the stabilizer, trajectory tracking and stabilization of elastic modes are accomplished.

1 Introduction

In future space missions, space robots will play important role in the deployment or retrieval of payloads and will be used for the repair and construction of space structures. These robots have long flexible arms. Space robots have complex dynamical behavior. Lack of an inertially-fixed base on which the manipulators are mounted and control-structure interaction pose many difficulties in control system design. Inclusion of structural deformation in the model gives high-order dynamical systems.

In recent years, researchers have focused attention on the problem of dynamics and control of space robots. Research has been done related to free-flying robots aswell to space robots for docking and intercept maneuvers [1–8]. Based on perturbation approach [9, 10] and Lyapunov stability theory, controllers have been designed in [7, 8, 11, 12] for space robots. Nonlinear inversion technique has been applied in [13, 14] for the space robots, however, control of the spacecraft has not been considered in these. Singular perturbation method has been used to develop continuous path tracking control system for free-flying space robots [15].

In this paper, a dual mode control system is designed for the control of space robots. Although, this approach is applicable to robots of other configurations, for simplicity, control of a two-link planar robotic arm mounted on a rigid platform, which can translate and rotate about one axis, is considered. The long links of the arm experience transverse deformation in the plane of rotation of the arm. Based on the nonlinear inversion technique, a control system is derived for the trajectory control of the rigid modes (position and orientation of spacecraft and joint angles of manipulator) using the joint torquers and the actuators on the rigid platform. The inverse controller accomplished reference trajectory tracking, however, it excites the elastic modes of the flexible links. For regulation and vibration suppression to the target equilibrium point, a linear composite stabilizer and a decoupled flexible mode stabilizer are designed using optimal control theory. A stabilizer using only velocity feedback based on Lyapunov approach is also designed. Simulation results are presented which show that in the closed-loop system including the inverse controller and the stabilizer, trajectory following and regulation are accomplished. By suitable planning of reference trajectory, desired tasks in space can be performed.

*Ecole Nationale de l'Aviation Civile, 7 rue E. Belin, 31055 Toulouse, France

†University of Nevada Las Vegas, College of Engineering, NV 89154, USA

The organization of the paper is as follows. Section 2 presents the problem formulation. An inverse control law is obtained in Section 3. Section 4 presents design of stabilizers, and numerical results are presented in Section 5.

2 Mathematical Model and Control Problem

Figure 2 shows the space robot mounted on the platform. This configuration of the space robot is similar to that of [7], however, for simplicity, the dynamics of the wrist have not been considered here. The Cartesian coordinates of the platform are denoted as x_0, y_0 and the rotational angle is θ_1 with respect to an inertial frame. The joint angles of the robot are θ_2 and θ_3, measured in the inertial frame. The elastic deformation of the links can be expanded in the series:

$$\delta_i(l_i,t) = \sum_{j=1}^{n_e} \Phi_{ij}(l_i)q_{ij}(t) \ , \ i = 1,2$$

where $\Phi_{ij}(l_i)$ are suitable mode shapes, $q_{ij}(t)$ are the generalized coordinates, and l_i is a point on the link i at a distance l_i from the joint.

A complete derivation of equations of motion using Lagrangian approach is given in [7]. For the purpose of derivation, one obtains the kinetic and the potential energy. Then, using Lagrange's equations, one obtains the mathematical model given by:

$$M(z)\ddot{z} + h(z,\dot{z}) + Kq = E(\theta_1)u \tag{1}$$

where the vector of elastic modes is $q = (q_1^T, q_2^T)^T \in \Re^{2n_e}$, $q_i = (q_{i1}, \cdots, q_{in_e})^T \in \Re^{n_e}$, $z = (x_0, y_0, \theta^T, q^T)^T \in \Re^n$, $n \doteq (5+2n_e)$, $\theta = (\theta_1, \theta_2, \theta_3)^T$, the control input is $u = (u_r^T, u_f^T)^T$, $u_r = (F_x, F_y, T_1, T_2, T_3)^T \in \Re^5$ and $u_f \in \Re^s$. Here, $M(z)$ is the positive definite symmetric inertia matrix, $h(z,\dot{z})$ represents the Coriolis and centrifugal forces, $K = (0_{5\times 2n_e}^T, K_1)^T$, K_1 is $(2n_e \times 2n_e)$ positive definite symmetric stiffness matrix, 0 denotes a null matrix of indicated dimensions and E is the control influence matrix. The input u_r consists of the control forces (F_x, F_y) and moment (T_1) acting on the platform and the two joint torques (T_2, T_3) acting on the arm, and u_f denotes the vector of moments applied by actuators located on the flexible links (s is the number of these actuators).

Defining a state vector $x = (z^T, \dot{z}^T)^T \in \Re^{2n}$, the system (1) can be written in a state variable form as:

$$\dot{x} = f(x) + g(x)u \tag{2}$$

where

$$f(x) = \begin{pmatrix} \dot{z} \\ M^{-1}(z)[-h(z,\dot{z}) - Kq] \end{pmatrix}$$
$$g(x) = \begin{pmatrix} 0_{n\times m} \\ M^{-1}(z)E(\theta_1) \end{pmatrix}$$

where $m = 5 + s$.

We associate to system (2) an output vector $y \in \Re^5$ given by:

$$\begin{aligned} y &= (x_0, y_0, \theta^T)^T \\ &\triangleq C_0 z \end{aligned} \tag{3}$$

where $C_0 = (I_5, 0_{5\times 2n_e})$ and I denotes the identity matrix of indicated dimension.

Suppose that a reference trajectory $y_r(t)$ for $y(t)$ is given. We are interested in deriving a control law such that $y(t)$ tracks $y_r(t)$ and the elastic modes are stabilized. We note that by the appropriate selection of reference trajectories y_r, one can maneuver the platform and the robotic arm to perform variety of tasks.

3 Inverse Control

In this section, a control law is obtained by the inversion of the input(u_r)-output(y) map of the systems (2) and (3) for the trajectory tracking. This choice of input-output map allows independent decoupled control of rigid modes using the joint torquers and the actuators on the platform.

For the inversion of the map, we differentiate y successively till the input appears. Differentiating y and using (2) gives:

$$\begin{aligned} \dot{y} &= C_0\dot{z} \\ \ddot{y} &= C_0 M^{-1}(z)[(-h(z,\dot{z}) - Kq) + E_1 u_r + E_2 u_f] \\ &\triangleq a(x, u_f) + D(z)u_r \end{aligned} \tag{4}$$

where $E = (E_1, E_2)$, $a(x,u_f) = C_0 M^{-1}(z)[-h(z,\dot{z}) - Kq + E_2 u_f]$, and $D(z) = C_0 M^{-1}(z)E_1$. Thus the relative degree of $y_i (i = 1,2)$ is two and the input-output map is inversible since, for the space robot, the 5×5 decoupling matrix $D(z)$ is nonsingular at each z [16, 17].

In view of (4), we choose a decoupling control law of the form:

$$u_r = D^{-1}(z)[-a(x,u_f) + v] \tag{5}$$

where the input v is chosen as:

$$\begin{aligned} v &= \ddot{y}_r - P_2\dot{\tilde{y}} - P_1\tilde{y} - P_0 x_s \\ \dot{x}_s &= \tilde{y} \end{aligned} \tag{6}$$

Here, $P_i, i \in \{0,1,2\}$ are diagonal matrices and $\tilde{y}$ is the tracking error $y - y_r$. The control signal v is of PID (proportional, integral and derivative) feedback type. The choice of integral feedback helps in eliminating any constant steady state error in $\tilde{y}$ in the presence of parameter uncertainty.

Substituting control law (5) and (6) in (4) gives:

$$\ddot{\tilde{y}} + P_2\dot{\tilde{y}} + P_1\tilde{y} + P_0 x_s = 0 \tag{7}$$

Differentiating (7) once, one obtains a third order linear differential equation for the tracking error given by:

$$\dddot{\tilde{y}} + P_2\ddot{\tilde{y}} + P_1\dot{\tilde{y}} + P_0\tilde{y} = 0 \tag{8}$$

The matrices P_i are chosen appropriately so that the characteristic polynomial

$$\Pi(\lambda) = det[\lambda^3 I + P_2\lambda^2 + P_1\lambda + P_0]$$

associated with (8) is Hurwitz and thus $\tilde{y}(t) \to 0$ as $t \to \infty$ for any initial error $\tilde{y}(0)$. Moreover, if $\tilde{y}(0) = \dot{\tilde{y}}(0) = x_s(0) = 0$, then one obtains from (8) that $y(t) \equiv y_r(t), t \geq 0$ and in the closed-loop system exact tracking of $y_r(t)$ is accomplished.

Although inverse controller is sufficient for the control of the platform and the joint angles of the robotic arm it causes elastic deformation of the links during maneuver. For point to point control of the end effector, it will be interesting to suppress the vibration of the elastic arm once the end effector has reached the target point.

4 Vibration Suppression

In this section, stabilization of the elastic modes is considered. Suppose that the chosen reference trajectory $y_r(t)$ converges to y^* (a fixed point) as $t \to \infty$. It is interesting to note that using the inverse controller, one obtains $\tilde{y} \approx 0$ (or $y \approx y^*$) after the initial transient period. As y converges to y^*, only elastic mode oscillations persist in the closed-loop system including the inverse controller. Thus, in the terminal phase, the motion of the space robot lies in the vicinity of the equilibrium point $x^{*T} = (z^{*T}, \dot{z}^{*T} = 0)$ where $z^{*T} = (y^{*T}, q^{*T} = 0)$ and therefore, the system can be well approximated by a linear system in the neighborhood of x^*.

For the regulation to the terminal state, we shall consider three kinds stabilizer design in this section.

4.1 Inverse control, and Rigid and elastic mode stabilizer

For the design of a stabilizer, we linearize the system about the equilibrium point x^*. Neglecting the second order terms in velocity $\dot{z}$, one obtains the linearized system from (1) of the form:

$$M(z^*)\Delta\ddot{z} + Kq = E(\theta_1^*)u \tag{9}$$

where $\Delta z = z - z^*$, $\Delta\dot{z} = \dot{z}$, and $\Delta\ddot{z} = \ddot{z}$. In view of (1), at the equilibrium point x^*, one has $u = u^* = 0$.

Defining $\tilde{x} = (\Delta(z)^T, \Delta(\dot{z})^T)^T$, one obtains the linearized system:

$$\dot{\tilde{x}} = A\tilde{x} + Bu \tag{10}$$

where

$$A = \begin{pmatrix} 0_{n\times n} & I_{n\times n} \\ A_{21}(z^*) & 0_{n\times n} \end{pmatrix} B = \begin{pmatrix} 0_{n\times m} \\ M^{-1}(z^*)E(\theta_1^*) \end{pmatrix},$$

$$\text{and } A_{21}(z^*) = (0_{n\times 5}, -M^{-1}(z^*)K)$$

For stabilization of the system (10), one uses the linear quadratic optimal control technique. We choose a performance index of the form:

$$J = \int_0^\infty (\tilde{x}^T Q\tilde{x} + u^T Ru)dt \tag{11}$$

By a suitable choice of the positive definite symmetric weighting matrices Q and R, and by minimizing J, one obtains the optimal control input:

$$u = -R^{-1}B^T S\tilde{x} \triangleq -F\tilde{x} \tag{12}$$

where S is the solution of the Riccati equation [18].

For the maneuver of the space robot, the controller is implemented as follows. First, the inverse control u_r is made active for large executing translational and rotational maneuvers of the space robot. When the output vector $y(t)$ reaches the vicinity of y^*, the inverse controller is switched off and the elastic mode stabilizer is switched on. Thus, in the terminal phase, only the elastic mode stabilizer is active and the terminal regulation of the state x to the equilibrium point x^* is accomplished by the linear feedback law (12).

The closed-loop system is shown in Figure ??.

4.2 Inverse control, and Decoupled flexible mode stabilization

Now we consider decoupled rigid mode control and flexible mode stabilization using only elastic link actuators. Since the inverse controller accomplishes steering of y to y^*, setting $\dot{y} = 0$, $\ddot{y}$ in (9) gives:

$$M_{12}(z^*)\ddot{q} = E_{11}(\theta_1^*)u_r + E_{12}u_f \tag{13}$$

$$M_{22}(z^*)\ddot{q} + K_1 q = E_{21}u_r + E_{22}u_f \tag{14}$$

Substituting the value of u_r from (13) into (14) gives:

$$[M_{22}(z^*) - E_{21}E_{11}^{-1}M_{12}(z^*)]\ddot{q} + K_1 q = (E_{22} - E_{21}E_{11}^{-1}E_{12})u_f \tag{15}$$

Defining $x_f = (q^T, \dot{q}^T)^T$, a state variable representation of (15) is:

$$\dot{x_f} = A_f x_f + B_f u_f \tag{16}$$

where
$\widehat{M}_{22} = (M_{22} - E_{21}E_{11}^{-1}M_{12})$
$E_f = (E_{22} - E_{21}E_{11}^{-1}E_{12})$
$A_f = \begin{pmatrix} 0_{2n_e \times 2n_e} & I_{2n_e \times 2n_e} \\ -\widehat{M}_{22}^{-1}K_1 & 0_{2n_e \times 2n_e} \end{pmatrix}$
$B_f = \begin{pmatrix} 0_{2n_e \times 2n_e} \\ \widehat{M}_{22}^{-1}E_f \end{pmatrix}.$

4.2.1 Optimal decoupled stabilization

For obtaining an optimal stabilizer we choose the performance index

$$J_f = \int_0^\infty (x_f^T Q_f x_f + u_f^T R_f u_f)dt \tag{17}$$

where Q_f and R_f are positive definite. The optimal control is obtained using minimized linear quadratic optimization technique. It is of the form:

$$u_f = -R_f^{-1}B_f S_f x_f \triangleq -F_f \times [q^T, \dot{q}^T]^T \tag{18}$$

where S_f is the solution of the Riccati equation.

4.2.2 Decoupled stabilization using a velocity feedback

It is interesting to examine stabilization of the flexible modes using only the velocity feedback. For the design of stabilizer, we neglect the matrix E_{21} in (14) since it is small. We choose the stabilization signal of the form:

$$u_f = -\lambda E_{22}^T \dot{q} \tag{19}$$

where $\lambda > 0$. For proving stability of the system (14) and (19) with $E_{21} = 0$, we use the Lyapunov stability theory. We choose a positive definite function:

$$W(x_f) = \dot{q}^T M_{22}(z^*)\dot{q} + q^T K_1 q \tag{20}$$

Then the derivative of W along the solution of the system (14) and (19) is given by:

$$\dot{W}(x_f) = -2\lambda \dot{q}^T E_{22}E_{22}^T \dot{q} = -2\lambda \|E_{22}^T \dot{q}\|^2 \leq 0$$

Thus, $\dot{W} \equiv 0$ if $E_{22}^T \dot{q} \equiv 0$. It easily follows that if the matrix pair (C_f, A_f) is observable where $C_f = (0, E_{22})$, then the largest invariant set contained in the set $\Omega = \{x_f \in \Re^{4n_e} : \dot{q} = 0\} = \{0\}$ and the origin $x_f = 0$ of the system (16) and (19) is asymptotically stable [19].

In this section we have obtained three stabilizers (12), (18), and (19).

Using stabilizer (12), one obtains efficient stabilization of the rigid aswell as flexible modes since all the actuators are used for regulation. However, computation of gain matrix F is relatively complex due to the high order of the system. Unlike stabilizer (12), only the actuators on the flexible links are used for stabilization of flexible modes. The inverse controller independently accomplishes reference trajectory tracking. Although the stabilizer (18) or (19) are switched on in the vicinity of the terminal state, it does not affect the tracking ability of the inverse controller.

The stabilizer (19) uses only the velocity feedback and it does not need the computation of the Riccati equation.

5 Simulation Results

In this section, numerical results are presented. The complete closed-loop system (1) with the inverse controller (5), and the stabilizer (11) or (12) is simulated. The parameters of the space robot are:

$a = 0.5m$, $b = 0.5m$
$l_2 = 1m$, $l_3 = 1m$
$m_{platform} = 10kg$, $m_{l_2} = 1kg$, $m_{l_3} = 1kg$, $pm = 200g$
$EI = 120N/m^2$

The feedback gains are chosen as $P_2 = 3\lambda_e I_{5\times5}$, $P_1 = 3\lambda_e^2 I_{5\times5}$, and $P_0 = \lambda_e^3 I_{5\times5}$. With these choices of P_i, the characteristics roots of (8) are at $\lambda = -\lambda_e$. For the simulation, λ_e was equal to 1. Taking the notations of Section 2, we have $n_e = 2$, $n = 9$, $s = 4$: this means we have considered two of the elastic modes of each half-arm and we have added four actuators.

The initial and final conditions are as follow:

	Initial	Final
x_0 (m)	5	0
y_0 (m)	-10	0
θ_1 (°)	60	90
θ_2 (°)	45	90
θ_3 (°)	90	0
Elastic deflection	null	null

We have chosen $\dot{z}(0) = 0$.

Each component y_{rk} of the command trajectory was computed as a "S" curve using the initial and final values of y and the following equations (here $\tau = 5s$):

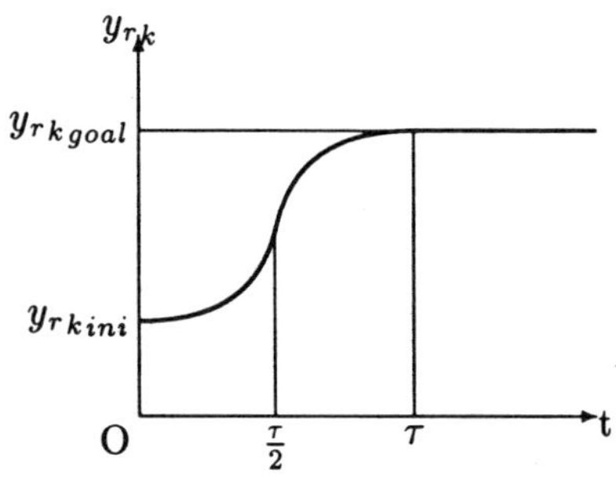

$$t \in [0; \frac{\tau}{2}] \quad y_{rk} = y_{rk_{ini}} + \frac{y_{rk_{goal}} - y_{rk_{ini}}}{2} \frac{4}{\tau^2} t^2$$

$$t \in [\frac{\tau}{2}; \tau] \quad y_{rk} = y_{rk_{goal}} - \frac{y_{rk_{goal}} - y_{rk_{ini}}}{2} \frac{4}{\tau^2} (t - \tau)^2$$

$$t \in [\tau; +\infty[\quad y_{rk} = y_{rk_{goal}}$$

5.1 Trajectory Control: $u_f = 0$

The closed-loop system (1) and (5) was simulated without the stabilizer. Selected responses are shown in Figure 3. In this case, perfect tracking of y_r is accomplished ($y \equiv y_r$), however persistent bounded elastic mode oscillation exists.

5.2 Tracking, and Rigid and Elastic Mode Stabilization

The complete closed-loop system (1), (5) and (12) was simulated. The optimal stabilizer was designed using the weighting matrices Q and R selected as:

$$\begin{aligned} Q &= diag[1; 1; 1; 1; 1; 0.8; 0.8; 0.8; 0.8; \\ &\quad 0.3; 0.3; 0.3; 0.3; 0.3; 0.2; 0.2; 0.2; 0.2] \\ \text{and } R &= I_{9\times 9}. \end{aligned}$$

The eigenvalues of the closed-loop system matrix $A_c = (A - BF)$ are:

$$\begin{cases} -0.200776 \pm 0.198379\jmath \\ -0.203102 \pm 0.200635\jmath \\ -0.487423 \pm 0.455377\jmath \\ -1.09089 \pm 0.820105\jmath \\ -1.8525 \pm 0.882634\jmath \\ -12.8329 \pm 83.6529\jmath \\ -8.92213 \pm 98.3441\jmath \\ -189.774 \pm 423.539\jmath \\ -146.679 \pm 455.667\jmath \end{cases}$$

Initially, only the inverse controller is active and stabilizer is switched at $t = 5s$. Figure 4.a shows the evolution of the coordinates of the platform, its orientation, and the evolution of the angles of the joints. Switching on stabilizer after 5 seconds causes only a small tracking error (the maximum tracking error is $(y - y_r)_{max} = (10mm, 0.6mm, 0.7°, 0.9°, 0.1°)^T$). The control magnitudes are reasonable (see Figure 4.b). We observe smooth regulation of the space robot to the terminal state: the deflections of the end of each half-arm are shown in Figure 4.c. Of course, there exist enough flexibility in the control system, and the parameters P_i, Q, R command trajectory, and the switching instant are properly chosen to obtain desirable responses. One can reduce control magnitude by choosing slower command trajectory. The evolution of the whole robot is shown in Figure 4.d.

To examine the capability of the stabilizer, the system (1) and (12) was simulated, but inverse controller (5) was not used: optimal stabilizer was switched on from the beginning. Here, there is no reference trajectory: the controller aims at minimizing $z - z^*$ and $\dot{z}$ without any other goal. The state vector does converge to the terminal point but very slowly ($40s$ instead of $5s$ previously): see Figure 5.a. On the other hand, the deflections keep low values all the tracking long (see Figure 5.b). This simulation shows that the robot may be controlled with the optimal stabilizer alone even if the start state is quite far from the terminal point.

5.3 Inverse Control and Decoupled Elastic Mode Stabilization

Simulation was done using the closed-loop system (1), (5) and the stabilizers of section 4.2 (stabilization of the flexible modes only: 4 actuators are used instead of 9 in the previous section).

5.3.1 Optimal decoupled stabilization

An optimal stabilizer (18) was designed using the weighting matrices as:

$Q = diag[1; 1; .8; .8; .3; .3; .2; .2]$ and $R = I_{4\times4}$

The eigenvalues of the closed-loop system matrix $A_{cf} = (A_f - B_f F_f)$ are:

$$\begin{cases} -0.645224 \pm 11.1454\jmath \\ -0.76032 \pm 20.2767\jmath \\ -3.79561 \pm 124.082\jmath \\ -4.07113 \pm 141.854\jmath \end{cases}$$

Stabilizer was switched on at $t = 5s$. Selected responses are shown in Figure 6. We observe smooth stabilization of the elastic modes. Unlike Figure 4.a, perfect tracking is accomplished in this case ($\tilde{y} \equiv 0$).

5.3.2 Decoupled Stabilization Using Velocity Feedback

The closed-loop system (1), (5) and stabilizer (19) with $\lambda = 1$ was simulated. This stabilizer uses only the elastic mode velocity $\dot{q}$ feedback. Selected responses are shown in Figure 7. Interestingly, smooth suppression is accomplished only by velocity feedback. As predicted, the tracking error is null: $y - y_r \equiv 0$ because u_f and u_r are independent. Therefore, this is a nice control law: 5 actuators aim at tracking the reference trajectory and 4 actuators aim at stabilizing the robot.

6 Conclusion

A design approach for the control of space robots based on nonlinear inversion technique and optimal control theory was presented. Space vehicle actuators and the joint torquers were used for reference trajectory tracking of rigid modes. Although, the inverse controller accomplishes trajectory control, it excites the elastic modes of the arm. For vibration suppression and regulation to the terminal state, stabilizers were designed using linear quadratic optimal control theory. Decoupled elastic mode stabilization using only elastic mode velocity feedback was also considered. Simulation results showed that in the closed-loop system precise tracking and regulation are accomplished using the inverse controller and the stabilizers.

There are several interesting problems which need to be considered. Design of state estimator using sensor data and design of controller for direct control of the tip trajectory are important. Furthermore, control law simplification and digital implementation of controller are also interesting.

References

[1] Longman, R. W., "The Kinetics and Workspace of a Satellite-Mounted Robot", *Journal of the Astronautical Sciences*, Vol.38, No.4, 1990, pp.423-439.

[2] Lindberg, R.E., Longman, R. W., and Zedd, M. F., "Kinetic and Dynamics Properties of an Elbow Manipulator Mounted on a satellite", *Journal of the Astronautical Sciences*, Vol.38, No.4, 1990, pp.397-421.

[3] Vafa, Z., and Dubowsky, S., "On the Dynamics of Space Manipulators Using the Virtual Manipulator, with Applications to path planning", *Journal of the Astronautical Sciences*, Vol.38, No.4, 1990, pp.441-472.

[4] Alexander, H. L., and Cannon, R. H., "An Extended Operational-Space Control Algorithm for Satellite Manipulators", *Journal of the Astronautical Sciences*, Vol.38, No.4, 1990, pp.473-486.

[5] Nenchev, D., Umetani, Y., and Yoshida, K., "Analysis of a Redundant Free-Flying Spacecraf/Manipulator System", *IEEE Transactions on Robotics and Automation*, Vol.8, No.1, 1992, pp,1-6.

[6] Nakamura, Y., and Mukherjee, R., "Nonholonomic Path Planning via a Bidirectional Approach", *IEEE Transactions on Robotics and Automation*, Vol.7, No.4, 1991, pp.500-514.

[7] Meirovitch, L., and Seungchul, L., "Maneuvering and Control of Flexible Space Robots", *Journal of Guidance, Control, and Dynamics*, Vol.17, No.3, 1994, pp.520-528.

[8] Chen, Y., and Meirovitch, "Control of a Flexible Space Robot Executing a Docking Maneuver", *Journal of Guidance, Control, and Dynamics*, Vol.18, No.4, 1995, pp.756-766.

[9] Meirovitch, L., and Quinn, R. D., "Equations of Motion for Maneuvering Flexible Spacecrafts", *Journal of Guidance, Control, and Dynamics*, Vol.10, No.5, 1987, pp.453-465.

[10] Meirovitch, L., and Kwak, M. K., "Control of flexible Spacecraft with Time-Varying Configuration", *Journal of Guidance, Control, and Dynamics*, Vol.15, No.2, 1992, pp.314-324.

[11] Modi, V. J., and Chang, J. K., "Performance of an Orbiting Flexible Mobile Manipulator", *Transactions of the ASME: Journal of Mechanical Design*, Vol.113, 1991, pp.516-524

[12] Meirovitch, and Chen, Y. "Trajectory and Control Optimization for Flexible Space Robots", *Journal of Guidance, Control, and Dynamics*, Vol.18, No.3, 1995, pp.493-502.

[13] Singh, S. N., and Schy, A. A., "Control of elastic robotic systems by nonlinear inversion and modal damping", *Journal of Dynamic Systems, Measurement and Control*, Vol.108, Sept. 1986, pp.180-189.

[14] Singh, S. N., and Schy, A. A., "Elastic robot control: Nonlinear inversion and Linear stabilization", *IEEE Transactions on Aerospace and Electronic Systems*, Vol.AES-22, No.4, July 1986, pp.340-348.

[15] Murotsu, Y., Senda, K., Mitsuya, A., and Yamane, K., "Theoretical and experimental studies for continuous path control of flexible manipulator mounted on a free-flying space robot", *AIAA Guidance, Navigation and Control Conference*, Montevery, C.A., August 1993, pp.1458-1471.

[16] Singh, S. N., and Rugh, W. J., "Decoupling in a class of nonlinear systems by state variable feedback", *Journal of Dynamic Systems, Measurement and Control*, Vol.94, 1972, pp.323-329.

[17] Isidori, A., "Nonlinear Control Systems", Springer-Verlag, New York, 1985.

[18] Athans, M., and Falb, P. L., "Optimal Control: An Introduction to The Theory and its Applications", McGraw-Hill, New York, 1966

[19] Vidyasagar, M., "Nonlinear Systems Analysis", Prentice Hall, Englewood Cliffs, New Jersey, 1978.

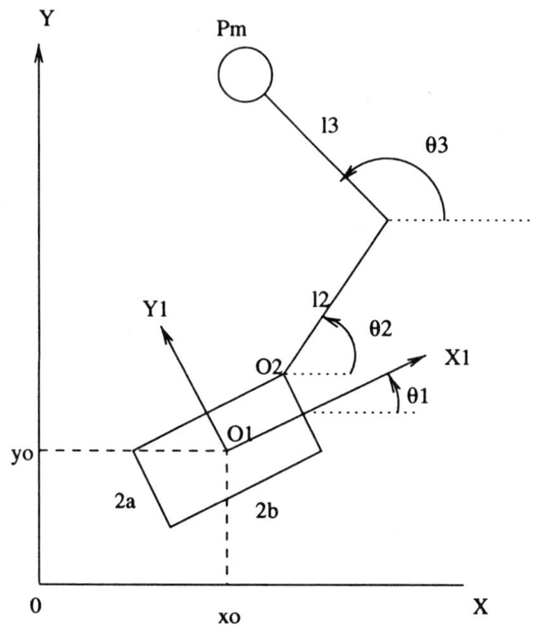

Figure 1: Space robot

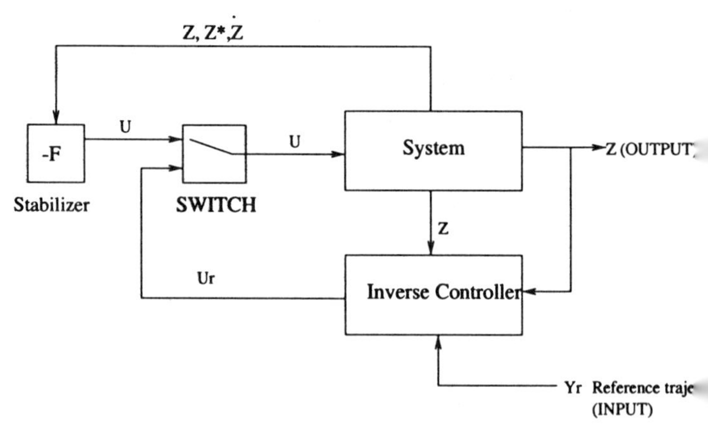

Figure 2: Closed-loop control system

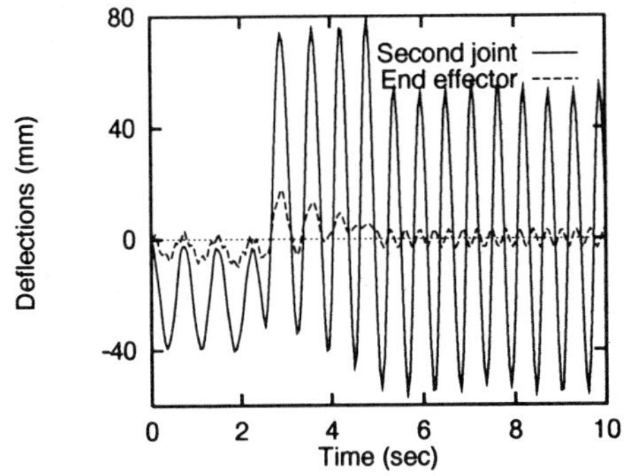

Figure 3: Inverse Controller alone: deflections at the end of each half-arm

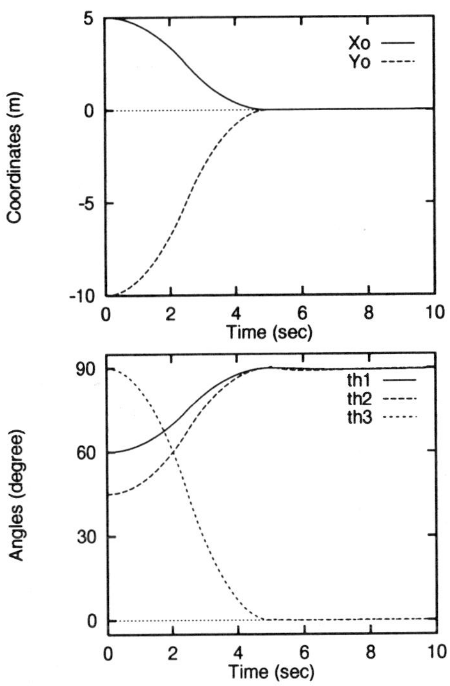

Figure 4.a: Coordinates and Orientation of the platform; angles of the joints

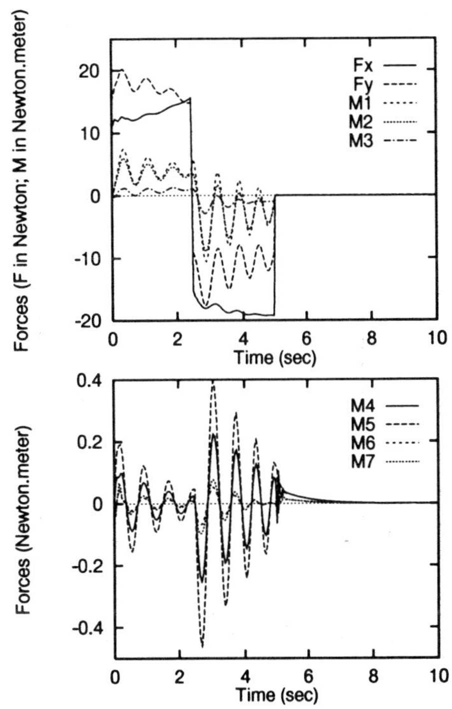

Figure 4.b: Value of the actuators

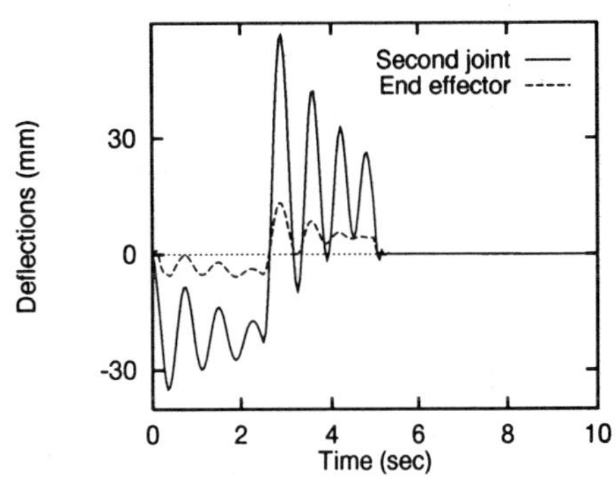

Figure 4.c: Deflections at the end of each half-arm

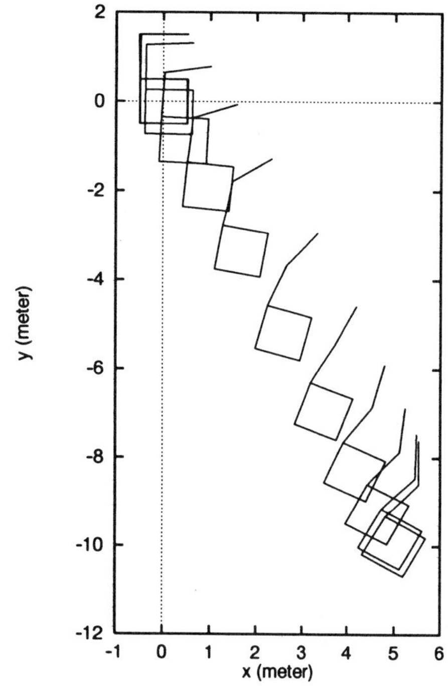

Figure 4.d: Evolution of the robot

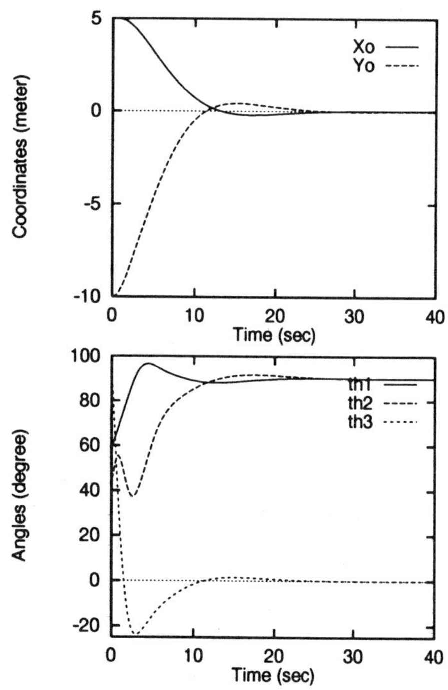

Figure 5.a: Coordinates and Orientation of the platform; angles of the joints

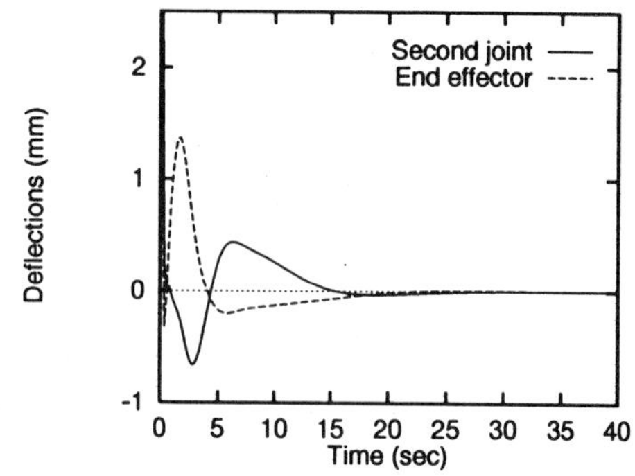

Figure 5.b: Deflections at the end of each half-arm

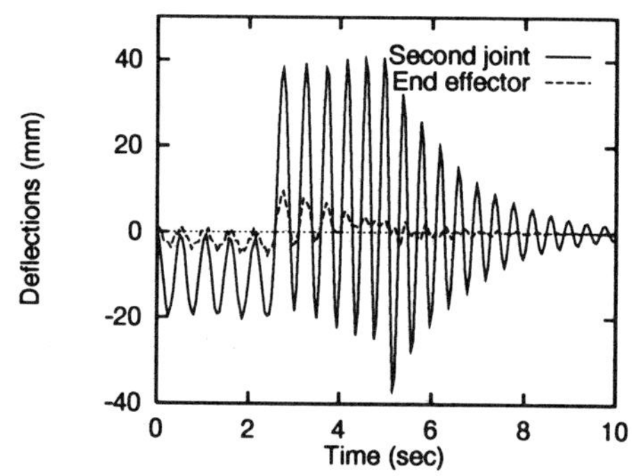

Figure 6: Inverse Control and stabilization of flexible modes with the 4-actuator optimal stabilizer switched on at $t = 5s$: deflections at the end of each half-arm

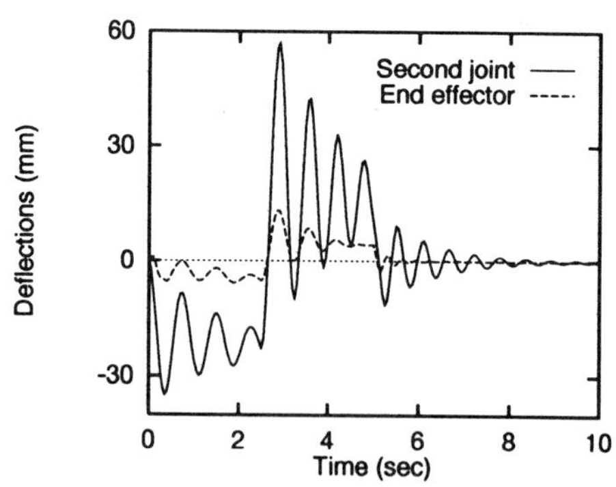

Figure 7: Inverse Control and Velocity feedback stabilizer: deflections at the end of each half-arm

AIAA-97-3560

REAL-TIME SIMULATION OF FLEXIBLE SPACE MANIPULATOR DYNAMICS

Bernd E. Schaefer*, Martin Loesch, Wolfgang Rulka

German Aerospace Research Establishment (DLR)
Oberpfaffenhofen, D-82234 Wessling, Germany

ABSTRACT

Accurate and time-efficient dynamic models for space manipulator systems are required for supporting both the design and the development of the system itself, and furthermore, for incorporating real-time models within an operational environment be it on ground or in space. This paper treats the problem area encountered in both, non-real-time and real-time modelling and simulation of multibody systems with the specific emphasis on space robotics. The most important topics are covered that start from the development of a mathematical model derived from the real-world system, and extend to the special features in finding the adequate multibody formalism, the appropriate numerical integration method, and finally approaches such as the symbolic code formalism to meet real-time requirements. Simulation results are presented for the Experimental Servicing Satellite manipulator system.

1. INTRODUCTION

A real-time simulator for satellite-based robotic tasks is a basic item for the preparation of a telerobotic mission, with conventional use dedicated to the training of human teleoperators on ground, and also for the operational database and flight control procedures validation. Its use can be extended to parallel activities such as failure diagnosis and mission planning. Moreover, already during the entire life cycle of the development of the telerobotic components, the use of appropriate simulators accompanies the different phases guaranteeing for a successful mission. This is valid for system engineering activities such as robotic mission definition, system prototyping, design validation and tele-operations preparations. The reliability of these exercises depends strongly upon the quality and the capability of the simulator.

The purpose of this paper is to report on the simulation capability of the dynamic behaviour of space manipulator systems as part of a real-time system simulation environment. The investigations are focused on analysing the various items impacting real-time performance and on adapting an existing dynamics analysis tool for multibody systems (MBS) to automatically generate real-time dynamics simulation code that can be easily configured and used in a variety of simulation environments, e.g. MATRIX-x or EUROSIM (from ESA). The generated dynamics simulation code is to be suitable for supporting engineering activities during the entire life cycle of a project.[1,2] Hence, it must have a flexible configuration, e.g. being able to run in real-time and also in non-real-time modes as well as in open loop and in closed loop simulations.

Existing multibody dynamics tools generally are targeted to analysis work, typically in non-real-time, and therefore do not provide adequate interfaces to other hardware and software systems and are not optimized for real-time applications. During the definition phase of an advanced future robotized satellite mission for maintenance and repair of geostationary telecommunications satellites, many of these new impacts have been carefully investigated and implemented by the authors in order to achieve real-time dynamics capabilities and performance of multibody system code for manipulator / satellite interaction analysis. This planned service vehicle or satellite is called the 'Experimental Servicing Satellite (ESS)', nationally funded by Germany (Fig. 1). It features challenging telerobotic tasks like rendezvousing, berthing to a non-cooperative target satellite (the 1987 failed TV-Sat1 satellite, still

* Space Operations Division, AIAA Senior Member

orbiting and tumbling in near-GEO) and finally docking and potentially performing repair tasks.

Fig.1: *ESS manipulator during TV-Sat capturing.*

Moreover, prior to a mission like ESS, basic dynamic analysis studies had been performed, regarding manipulators of larger working volume like ESA's European Robotic Arm (ERA) of about 11 m length[3], revealing very delicate dynamic features involved with their inherent elasticity in arms and joints and with their payload mass varying in a large range from some kilograms to about one ton. Here, importantly, adaptive robotic controller designs are required that deal with the changing dynamic parameters during the operational tasks.[4]

The principal goal of most technical simulations is optimization of system dynamic behaviour. This optimization is performed off-line, generally in both time and frequency domain. Moreover, on-line calculation of system dynamics is obtaining increasing importance, e.g. due to optimized robot trajectories with respect to speed and precision in path line, and their advanced and complex control algorithms. In the rising and challenging field of space telerobotics, where a human teleoperator on ground is involved within the overall control loop, the operations task is getting much more complex. Here, inherent time delays in up- and down-link data communications in the order of some seconds for LEO, HEO or GEO (Low, High or Geostationary Earth Orbit) robotized space missions make remote robot control a challenging task. For planetary rover applications, e.g. on Moon or Mars, these delay time numbers very quickly increase to some minutes or some ten minutes.

Additionally, in all these missions, either planned so far (e.g. ESA's Geostationary Servicing Vehicle GSV, Germany's ESS), getting realized soon (US Mars Pathfinder with Sojourner, Japan's ETS-7), or having been realized already (Germany's 1993 Spacelab based ROTEX, Robotics Technology Experiment), in-situ visual observation and supervisory ground control is only restrictedly possible by some few on-board cameras, and, moreover, the image data transfer rate is very limited. To largely overcome both problems, a solution mainly based upon multisensory feedback of the robot's gripper and computer graphics animation of the whole teleoperations scenario has been developped and successfully demonstrated for ROTEX. Similar but more advanced control schemes are now employed for ESS. This kind of teleoperation's philosophy means that the dynamic behaviour of the on-board manipulator or of the in-orbit robotized platform or the planetary rover has to be pre-calculated in real-time and demonstrated in hardware-in-the-loop ground simulators.[5-7]

As in space telerobotics, similar requirements in terrestrial applications are found in real-time automobile driving simulators. Here, the human driver generates steering signals and the car's dynamic reactions are fed back to the drivers cabin motion, calculated on the basis of the underlying mechanical car model. Related applications are found in aircraft training simulators and railway vehicle test rig simulators.[8]

2. General Problem Area Overview

Precise dynamic models for space manipulator systems are necessary, mainly for the two reasons of, firstly, supporting both the design and the development of the system itself, and secondly, of incorporating real-time models within an operational environment be it on ground or in space. For the first application, models of the dynamics and control usually are created and simulation is performed both in non-real-time which allows to increase modelling accuracy as far as possible. However, in real-time applications a partially different approach is required that copes with the specific and inherent features of real-time simulation. The main problems encountered here are the demand of delivering the simulation results at fixed time intervals, say every 10 ms, and hopefully with a modelling accuracy that does not deviate too much from the real hardware system.

The envisaged real-time applications normally focus on areas such as providing complete space manipulator models for a) ground-based test facilities for teleoperator training or operational test purposes, or b) for in-orbit ground-based teleoperations with new problems given by communication time delays and sparsely available vision information, where predictive motion information of the system kinematics and dynamics is mandatory. While for ROTEX a merely kinematically based approach had been sufficient, mainly due to its small length and largely rigid limbs and joints, this may be different for future light-weight robotic space systems (Fig. 2) of considerable length and elasticity.

The large area encountered in both, non-real-time and real-time modelling and simulation of multibody systems with the specific emphasis on robotics therefore has to cover most important topics such as:

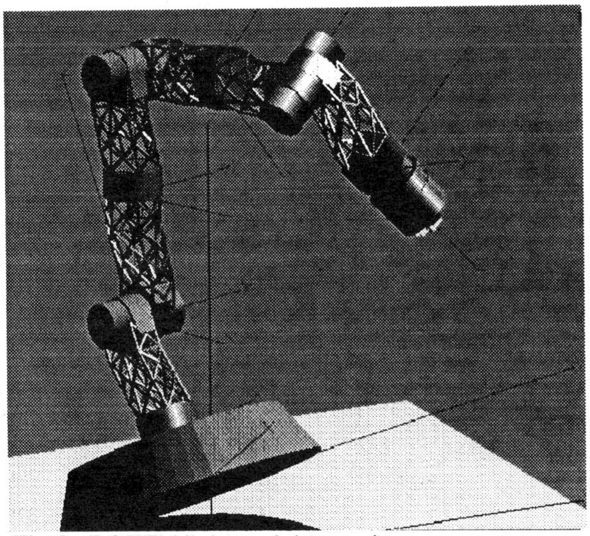

Fig.2: *ROTEX light-weight version.*

1. development of mathematical models derived from the real-world system,
2. extension of these models to the special features in finding the adequate multibody formalism for real-time applications,
3. finding of appropriate numerical integration methods,
4. speed up of computation time by e.g. symbolic code formalism,
5. and the implementation process concerned with the selection of specific computer hardware features (16bit or 32bit calculations, platform-dependent features, etc.).

Furthermore, special attention has to be given to the verification and the validation process, i.e. to the question how well do the simulation results fit to the mathematical model or even to the real-world system. The mathematical modelling process is seen as the most important impact for reaching real-time performances. This process has to address such interesting items as:

1. Stiff and non-stiff differential equations have to be detected, whereas high damping and control action within the models may increase stiffness that, anew, may importantly influence the proper selection of numerical integration methods to cope with these problems (e.g. ODASSL, LSODAR that automatically detect stiff or non-stiff behaviour).
2. Moreover, methods working with constant integration stepsizes are required in real-time simulations, thus potentially deteriorating accuracy.
3. Multibody formalisms have to be treated that may be based upon
4. either relative coordinate approaches (e.g. formalisms based upon methods from d'Alembert, Schiehlen, Kane, etc.)[9]
5. or on the O(n) method
6. or on absolute coordinate approaches (e.g. realised in ADAMS) that again increase system order.
7. For real-time simulation of mathematical models, computer codes have to be optimised (which has been done for the 'symbolic code' approach) in order to efficiently increase computation time.[10,11]
8. Also, the specific kinematic structure (e.g. tree structure in a one-robot application) has to be regarded in order to select the right model approach.
9. For computer hardware implementation, parallelising of codes may be important in cases where two spacecraft perform rendezvous manoeuver and finally dock or where two (or even more) robots work together.

3. On Multibody System Modelling and Simulation Tool

In order to meet the requirements for future robotized missions, advanced real-time simulators need to accommodate more accurate models of all three subsystems: (1) the manipulator itself, (2) its spacecraft being attached to, and (3) the target satellite to be berthed or docked to in case of robotic servicing missions. Moreover, often one is confronted with system models, in which the deformations of the flexible bodies (as in ERA, with two large limbs of about 4.50 m length each) are small but superimposed on a large reference or 'rigid-body' motion. In MBS simulation one exploits this fact to reduce the computational burden for such applications by linearising the equations of motion assuming small deflections. Using relative state variables or coordinates to represent the reference motion and applying O(n)-formalisms, MBS-codes provide an efficient alternative for system analysis via simulation. Following such arguments, flexible multibody formalisms have been used frequently.

MBS codes, that had been applied to spacecraft dynamic analyses and more specifically, to some deployment investigations, were either developed for their very special own purposes at universities etc. or have been implemented and can now be found in large simulation packages such as ADAMS, DADS, DCAP and SIMPACK. Their modelling methodology in generally is different from each other (using Lagrange's formalism, Kane's equations, Newton-Euler approach, O(n) formalism, symbolic code, etc.). Since the efficiency of the various codes with respect to their achieved numerical accuracy (depending upon the various numerical integration methods used in solving the equations) and the computing time obviously is very different, this situation involves a very careful investigation of the best suited algorithm and implemented codes for, e.g. the deployment analyses of a manipulator being located on a satellite platform.

3.1 General Features

A MBS modelling and simulation tool that copes with space robotics has to account for the following specific features:

1. structural elasticity in limbs,
2. model structure changes during the robot's movement which may occur during
   a) contact closures or in short time during contact impacts,
   b) berthing operations, i.e. contact closure by a manipulator in between two approaching spacecraft,
   c) contact closure of two robots on a spacecraft (closure of two kinematic chains),
   d) stick-slip transitions,
3. modelling of differential-algebraic transfer units for controllers and joint drives,
4. capability to use the overall system equations as ordinary differential equations (ODE) reduced to a set of the minimal system coordinates, or more generally, as differential-algebraic equations (DAE).

These features are to be implemented in the MBS tool without having the need to generate the governing equations of motion and the programming code again from the very beginning. This kind of generally implemented equations of motion is often termed as the *numerical code*. Having verified the system behaviour by off-line simulations, the MBS tool then should be able to automatically generate the so-called *symbolic code*. This code can be a stand-alone program (in Fortran or C) that, for the given simulation model, is highly optimized in order to speed up computation time, and therefore it is extremely well suited for real-time simulation applications. Compared to the numerical code, the number of freely variable parameters is here, in general, strongly limited.

Furthermore, for time integration of MBS models various integration methods are required that can handle

1. stiff and non-stiff differential equations,
2. system models composed by ODEs as well as DAEs whereas the integrators for DAEs should be able to cope with the stabilization of robotic joint constraints,
3. the use of integrators with variable stepsize and order for high precision computations,
4. and the use of integrators with constant stepsize for real-time simulations.

This manifold of modelling and simulation features is important to judge and evaluate the derived real-time models and their underlying simulation environment.

3.2 System Coordinates

The proper choice of the system coordinates, i.e. the state variables that describe the mechanical system is important to guarantee both increased generality of the MBS tool and time-efficiency of the derived simulation code. Here, *relative coordinates* as state variables, i.e. the degrees of freedom (DOF) of the robotic joints that connect adjacent bodies, are favourized mainly by the reason that this set of coordinates includes the smaller set of *absolute coordinates* (Fig. 3). For the specific space robotic applications the equations of motion for relative coordinates are then ODEs. Furthermore, the use of suitable O(n)-formalisms[12] produce very time-efficient codes that can be improved again by the use of the so-called *residuum algorithm.*[13]

The force and control elements of the MBS model usually require kinematic relative measurements as input, mostly between the coupling points of these elements (robotic joints). For small relative rotations of two adjacent bodies these measurements can be easily calculated approximately by linearization according to the skew-symmetric rotation matrix

$$A_{ij} = \begin{pmatrix} 1 & -\gamma & \beta \\ \gamma & 1 & -\alpha \\ -\beta & \alpha & 1 \end{pmatrix}$$

between the adjacent coordinate systems i and j with rotation angles α, β and γ. This is valid for force elements like torsional or spherical spring and damping elements, but is not applicable to elements that exhibit large relative input angles like many drive motors and control elements like sensors and moveable cameras. In all cases where the body movements have to be described in absolute coordinates, the input angles are to be calculated from the relative rotation matrix of the force element coupling points:

$$A_{M_i,M_j} = A^T_{I,M_i}(x_i) \cdot A^T_{I,M_j}(x_j)$$

In many cases structural elasticity modelling has proven to be of minor influence on dynamic behaviour,[3] mainly due to small arm length and strong materials with high modulus of elasticity. While this influence mostly is neglected, proper modelling of joint elasticities, also for real-time applications, is getting increased importance, depending upon the chosen drive gear system (harmonic drive e.g. is a candidate for the ESS manipulator). Here, the relative coordinate

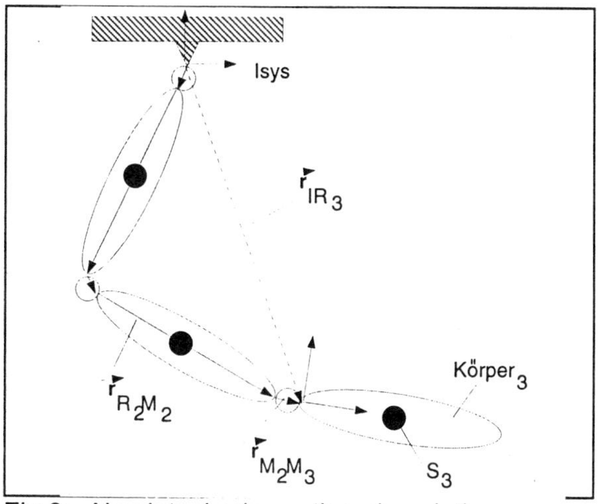

Fig.3: *Absolute body motion description generated by combination of relative motion description.*

choice is in favour over the absolute coordinates, since the determination of the initial conditions to start the simulation is no simple task at all if absolute state variables had been chosen (Fig. 4).

3.3 Time Integration Methods

Choice and performance of the integrators are predominantly determined by stability and accuracy demands of the simulation. While accuracy is given by the respective simulation task, stability characteristics depend upon the chosen integration method. For real-time applications explicit time integration methods with constant stepsize h are preferred. Here, stability of these 'non-stiff' methods is given by the ratio of the eigenvalues λ concerning their maximum $|\lambda_{max}|$ and minimum $|\lambda_{min}|$ absolute value of the linearized equations of motion. Non-stiff methods, like Euler, Runge-Kutta, Adams methods etc. perform stable if

$$h \cdot |\lambda_{max}| < F$$

where F is a parameter marking the selected integration method; for the unfavourable Euler method, $F = 2$, which gives a maximum stepsize of $h_{max} = 1.14 \cdot 10^{-3}\text{sec}$ with $|\lambda_{max}| = 1400$ (Fig. 5). For the very slow movements usually encountered in telerobotics operations this of course is a very small number, requiring very large computation times which is not compatible with the demand on real-time performance. On the other side, integration methods for numerically stiff systems do not exhibit these problems. According to the stability regions of the BDF methods, integrators can be selected large enough for variable orders k and stepsize h to allow stable solutions throughout the simulation. Anyhow,

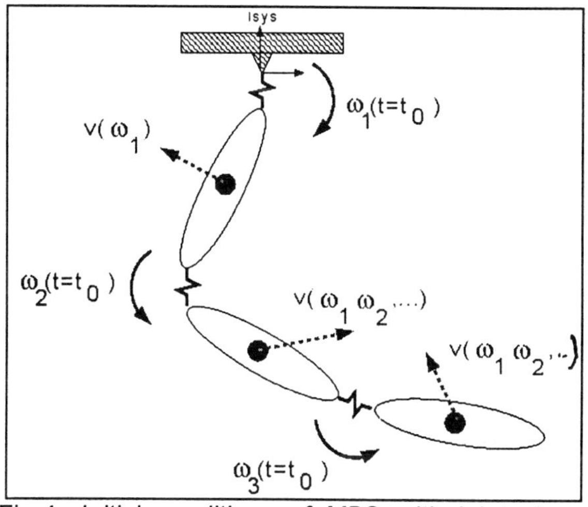

Fig.4: *Initial conditions of MBS with joint elastokinematics.*

BDF methods for orders $k = 1$ and $k = 2$ produce numerically induced damping that may damp out mechanical high-frequent oscillations of low amplitudes. For the ESS robotic system as a typical simulation example a mean stepsize of $h = 2 \cdot 10^{-2}\,\text{sec}$ has proven sufficient with a tolerance factor $tol = 3 \cdot 10^{-4}$ for the accuracy quality.

In SIMPACK the integration methods can be controlled such that they maintain a discrete time step, which is needed in case discrete controller elements are part of the system model, or in case real-time environments have to be faced with. Furthermore, the stepsize then remains within reasonable bounds if the control of accuracy is skipped. The method then works with pseudo-constant stepsize and the control of the accuracy is left to the user by selecting the appropriate discrete time step. By acquiring pseudo-constant stepsize for BDF methods, the overall computation time will be increased: Compared to explicit methods, the BDF methods require the calculation of the Jacobian,

$$J = \frac{\partial f(x)}{\partial x},$$

of the complete system model $\dot{x} = f(x,t)$, where x is the state vector. Usually, this matrix is determined by the integration methods chosen by means of applying numerically the difference quotient

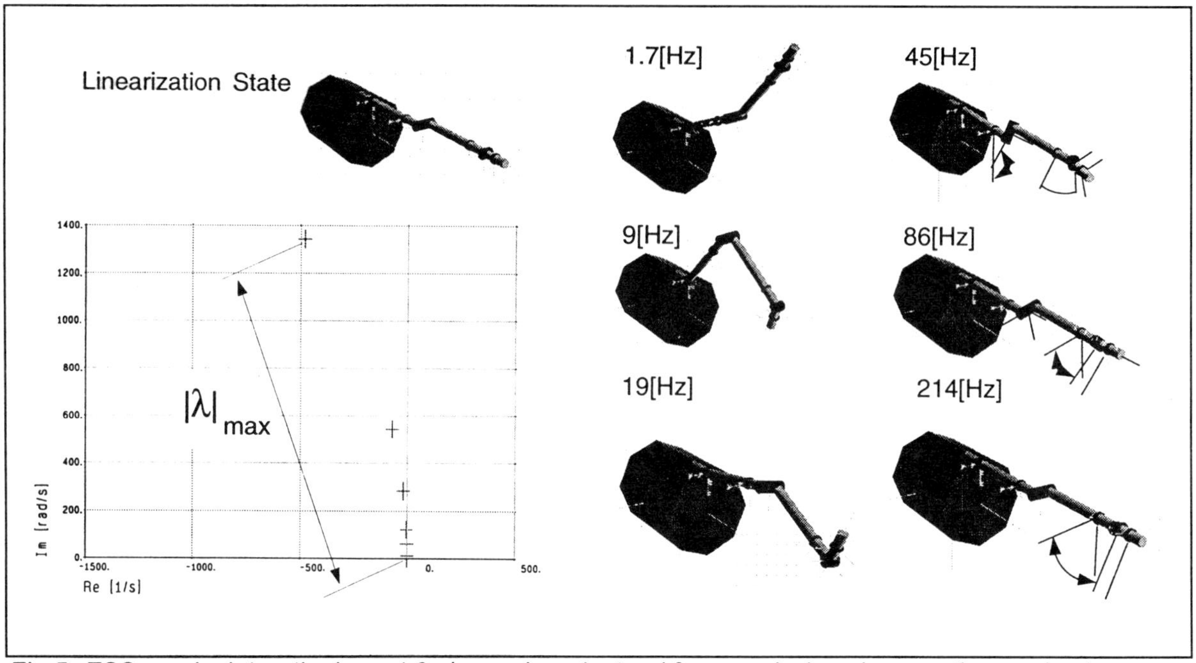

Fig.5: *ESS manipulator, the lowest 6 eigenvalues (natural frequencies) and respective mode shapes.*

$$J_{*,j} = \frac{\partial f(x)}{\partial x} \cong \frac{f(x, x_j + \Delta x_j) - f(x)}{\Delta x_j},$$

which requires N_x calculations of $\dot{x} = f(x,t)$, and finally influences CPU-time very strongly.

Often used is Runge-Kutta RK3 or RK4 with fixed stepsize for the integration of the equations of motion. The reason is found in its robustness of the large range of functions which can be present on the right hand side (RHS) of the equations of motion. RK3 and RK4 with fixed time step provides integration with very good robustness to various dynamical conditions. Numerical stability mainly depends upon time stepsize value. Values of at least ten times smaller than the period corresponding to the highest frequency of interest is usually recommended[2].

4. Approaches to Meet Real-Time Simulation Requirements

Here, real-time means that within a given sampling interval or cycle time Δt, typically in the range between 10 and 100 ms, the system state at time $t_{i+1} = t_i + \Delta t$ is pre-calculated beginning with the state $x(t_i)$ at time t_i. For strongly nonlinear systems as in telerobotics this is performed by time integration of the governing differential equation $\dot{x} = f(x,t)$. Due to the small sampling interval and due to the limited processor performance (the workstation class is being reasonably used for medium budget projects), this forces to optimize the calculation process by several means such as

a) utilizing parallel computations wherever possible (e.g. in case of a two-robot scenario),
b) reducing model complexity as far as possible to account, at least, for the very dominating dynamic effects,
c) adapting the integration methods to the given simulation problem by proper choice of the accuracy of the results obtained,
d) making use of potential robustness of the designed controllers in order to account for deviations between simulated and real world by appropriately equalizing at certain wait intervals (e.g. every 5 minutes, depending upon the chosen cycle time), if admitted by the teleoperations scenario.

In telerobotics, if no experimental results to be compared with are available, one is often confronted with the problem of judging whether the model and the simulation are sufficiently accurate. 'Simulating the simulation' is then required which means to compare the results of a simplified model without controlling its accuracy, with those of a much more complex or full model. This refers also to the selected integration methods. The full model, naturally being of high modelling quality, obviously is calculated off-line, and moreover, critical sub-systems such as complex joint models or structural flexibility in limbs, can be calculated and verified separately. Numerical integration is then performed using methods of variable time step and variable order of the underlying integrators. That also allows to control the quality of the simulated results and one can react immediately to unsteady system excitations like impacts / shocks during contact, and stick-slip transitions.

4.1 Tuning Possibilities

Off-line calculations and analyses of the linearized system give evidence and feeling for the behaviour of the simulated system model with respect to robustness of the results and integration methods. To approach real-time performance much more closer, some kind of tuning of the simulation may be performed by the following methods (as having been realized within SIMPACK):

1. Reduction of the number of required calculations of the Jacobian:
   Implicit methods calculate the matrix
   $$F = \frac{\partial \Delta(\dot{x},x)}{\partial \dot{x}} \cdot h + J \quad \text{with} \quad J = \frac{\partial \Delta(\dot{x},x)}{\partial x}$$
   which usually is calculated only in cases when the stepsize is varied. However, F is obtained merely by simple algebraic additions,
   $$J_{save} = F_{old}(h,x) - E \cdot h_{old} \text{ giving}$$
   $$F_{new} = E \cdot h_{new} + J_{save} \,,$$
   if the integrator has been given additional informations, namely that the equations of motion derived from the O(n) formalism are the explicit representation of the DAE system,
   $$\Delta(\dot{x},x) = \dot{x} - f(x) = 0 \,.$$
   By this simple measure, depending obviously upon the specific model, computation time is strongly reduced.
2. Reduction of CPU-time by appropriate calculation of the equations of motion:
   a) making use of the symbolic code formalism,
   b) automatically combining, to only one single body, all those bodies in the model that exhibit no relative motion to each other; this virtual joint then is called a 0-DOF joint,
   c) making use of the residuum algorithm instead of the explicit DAE representation,[13]
   d) forcing model deterioration but with acceptable deviations in accuracy:
   performing complete linearization of the kinematics, but accounting for non-linear force elements $f(x)$ which simplifies the equations of motion to
   $$\dot{x} = Ax + Bu + B_f f(x)$$
   which may only be used in cases where the interest is directed towards fine manipulation without performing gross robot motions,
   e) neglecting all those elements of the inertia tensor, that reveal only small influence on the motion,
   f) neglecting terms of order $\tilde{\omega} I \omega$.

4.2 Structural Flexibility

The data characterising spacecraft structural flexibilities usually are obtained from FE models using standard codes such as NASTRAN, ASKA or PERMAS. The resulting flexible modes comprise the modal frequency, the modal mass and inertia, the modal damping and the mode shape. Based upon this approach, quite different but more handsome representations can be derived to model the structural flexible parts[14] by using

a) spring-mass systems
b) standard (i.e. simpler) flexible elements such as rods, beams, plates, etc.
c) assumed modes or Rayleigh-Ritz method.

Obviously, the assumed modes method now is the most common implemented in many multibody dynamics software codes allowing an easy transition from finite element to multibody system modelling. Regarding real-time applications, the assumed modes method is certainly the most efficient in terms of CPU time, since the number of critical flexible modes can be selected at a minimum, depending upon the degree of accuracy.

4.3 Symbolic Code Pre-Processing

Symbolic manipulation pre-processing of the equations of motion very substantially increases simulation efficiency. This increase in efficiency, usually by a factor of 10 in CPU-time, is achieved through simplifications that are possible because of special configuration characteristics, as well as arithmetic and algebraic simplifications. This important feature has been realised in computer codes SIMPACK. Code optimization is model-dependant. Generally, in O(n)-formalism and relative coordinate choice, the system matrices are sparsely occupied: most matrix elements are 0 or 1, a fact that favouraby is utilized by the pre-processing algorithms to generate the so-called symbolic code in order to save many computations for matrix manipulation.

4.4 MBS Simulation Tool Used

Most of the important demands for reaching efficient real-time simulation performance have been incorporated in SIMPACK *(Simulation Package,* in-house developed and now commercially available), which is the tool we used throughout the simulations presented in chapter 5. Moreover, this tool deals with the important fact of consistency during the whole life cycle of the design and simulation process of space robotics. It allows to attend the space robot's 'dynamic life' beginning with the design phase, the availability of a high quality simulation environment and finally the generation of efficient symbolic code for hardware-in-the-loop simulations. This consistency in the various project phases is extremely important, if time-consuming system model changes are to be avoided. This refers to design changes that might be required at the one end of the project phase, and that do influence the real-time simulations at the other end.

5. Simulation Results

The main goal of the simulations performed was to identify the influence and hence the efficiency of the different measures described in order to speed up the CPU-time. These results then were the basis for the decision what type of simulation code can be chosen for the real-time applications. The way and the steps how to meet real-time performance are described for a typical example. The basic scenario of investigation was the deployment procedure of the ESS manipulator system, beginning from the folded position to the fully extended arm. An optimization procedure had been applied before in order to determine the respective angular velocities of each of the 6 joints (whereas the joints of number 2 and 6 were not used for the deployment phase), and the complete simulation time required 60 sec (Fig. 6). The total length of the manipulator (neglecting its fixed base point mechanical parts) is 4.60 m with lower and upper limb length of 1.50 m each, the total mass is 18.92 kg. The inertia tensor (only the main diagonal terms) for limb structural masses, motor and drive mechanics has been considered, too. Each joint is modelled regarding motor inertia, rotational stiffness and damping (harmonic drives) and its transmission ratio.

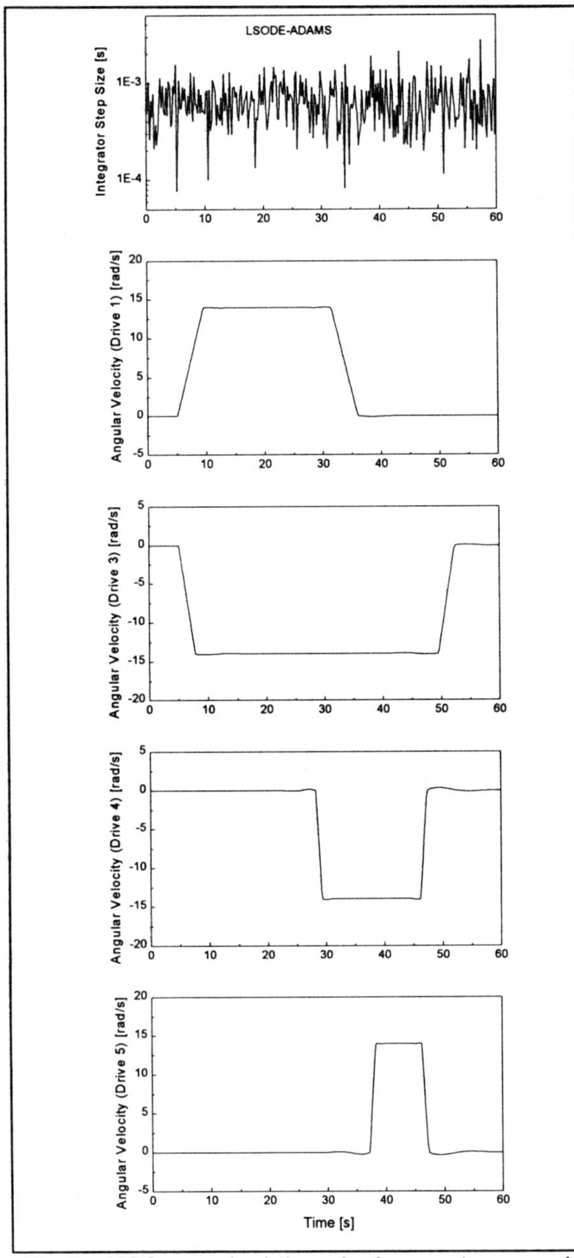

Fig.6: *ESS manipulator deployment scenario, stepsize used by integrator (upper graph, LSODE-Adams) and joint motor angular velocities.*

Fig. 6 gives the angular velocities for the respective joints and the variable stepsize of the integrator used. The influence of the different integration methods (explicit and implicit) is shown in Fig. 7 where the reduction ratio of numerical to symbolic code varies by a factor between 0.12 and 0.18. For a fixed stepsize (RK3 integrator was chosen), the influence of model complexity, i.e. model deterioration by reducing the number of DOF is shown in Fig. 8. Here, the 6-joint model gives 157,2 sec CPU-time at stepsize of 1 msec which is not compatible with the simulation time of 60 sec. Moreover, for larger stepsizes this method gets unstable. Reducing the DOF to a 5-joint or even a 4-joint model, finally reveals acceptable values around the given simulation time. A further approach was taken to save the original number of DOF, but to reduce model complexity in 1 or 2 joints, i.e. to neglect joint elasticity and to guide the joints rheonomically: Fig. 8 shows the results, the respective graphs are denominated with 6-joint (5 elastic, 4 elastic); both graphs are almost identical, and they exhibit practicable results for real-time applications. Table 1 gives the total CPU-time and the maximum stepsize that can be reached before the time integration gets unstable (cf. Fig.8).

Fig. 9 shows stepsize and order of the integrator (ODASSL, implicit, for stiff ODEs): in the upper graph stepsize and order have been automatically adapted by the integrator, while in the lower figure an upper limit of $h = 1 \cdot 10^{-2}\,\mathrm{sec}$ has been chosen but allowing for variable order. Since the stepsize in the first case varies strongly with simulation time, the change in order of the integrator is not as strongly exhibited as in the second case of pseudo-constant stepsize.

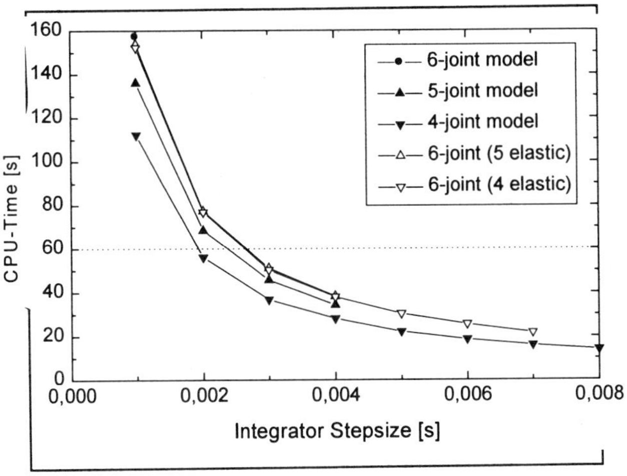

Fig. 7: *CPU-time ratio between symbolic and numerical code.*

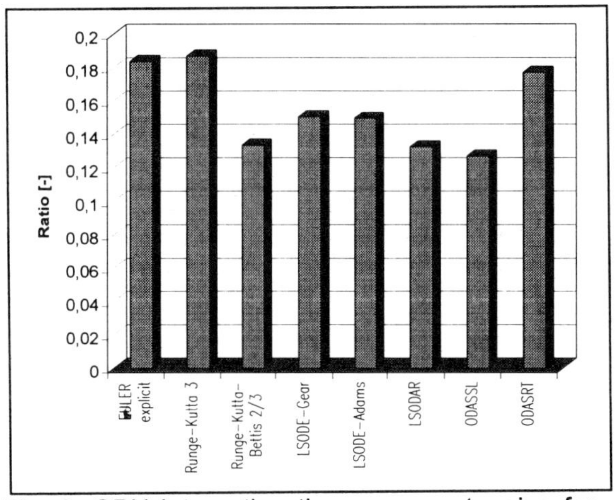

Fig. 8: CPU integration time versus stepsize for different model complexity (cf. Table 1).

system model	max. step-size (sec)	CPU-time (sec)
6-joint	0.001	157.20
5-joint	0.004	34.25
4-joint	0.016	7.17
6-joint (5 elastic)	0.004	38.26
6-joint (4 elastic)	0.007	21.69

Table 1: *CPU-times for different complexity (cf. Fig.8)*

Table 2 gives an overview of the CPU-times achieved for the different measures taken to speed up the simulation: i.e. numerical versus symbolic code, explicit versus implicit integration methods, combination of adjacent bodies to a 0-DOF joint wherever feasible, neglection of small I-tensor elements, linearization of state vector x, use of the residuum algorithm, and finally the full linearization of the equations of motion. The CPU-time needed in case of free variable stepsize of the integrator was 43.5 sec for case (b) and 6.1 sec for case (d), and in case of pseudo-constant stepsize the respective numbers are 88.6 sec (case (b)) and 12.4 sec (case (d)), which gives a reduction in CPU-time of about 0.14 for both integration methods. These calculations have been performed on a HP-735 and a SGI Crimson (150 MHz CPU) platform, which behaved similarly in performance.

6. Concluding Remarks and Further Activities

The different features impacting the performance of real-time simulation have been clearly pointed out. An approach based on a multibody system code has been given in order to realize an appropriate simulation environment that is compatible with the real-time requirements. Increased effort still has to be directed towards optimization of numerical code implementation features. Further activities are dedicated to model reduction items including structural flexibilities without essentially deteriorating the accuracy of the performance behaviour.

Based upon the available workstation hardware, the simulation results have shown that modelling improvements are required for more complex models in order to meet real-time performance. Off-line model reduction approaches are obviously one solution. Integrator type, stepsize and accuracy tolerances have to be carefully studied and adapted prior to apply the simulations to real-time scenarios.

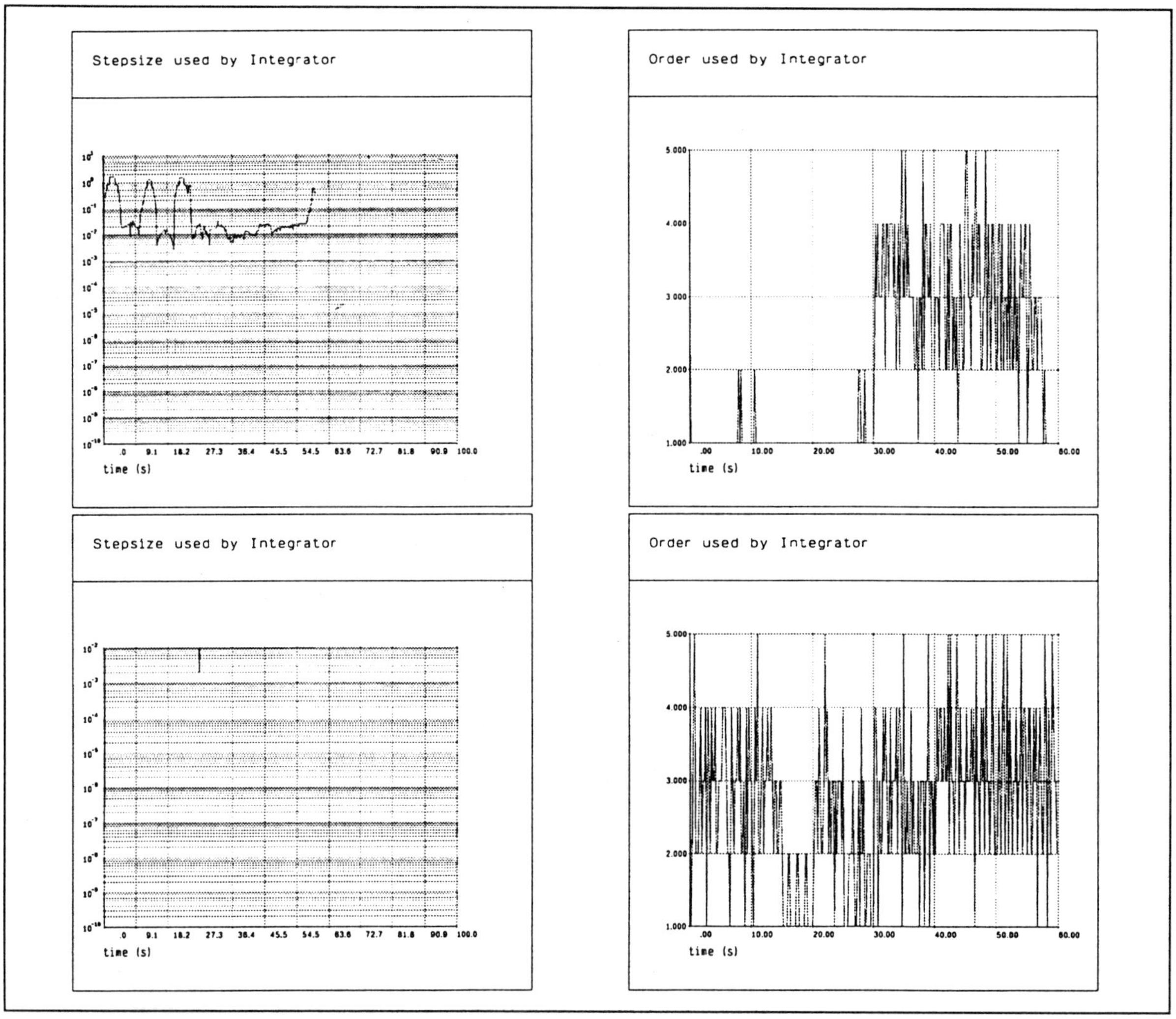

Fig.9: *Freely variable (upper graphs) and pseudo-constant (lower graphs) integrator stepsize.*

type of code		CPU-time normalized
numerical code	(a) explicit, original MBS version	2.05
	(b) explicit, combining bodies to 0-DOF	1.0
	(c)implicit, combining bodies to 0-DOF	0.60
symbolic code	(d) explicit ODE, combining bodies to 0-DOF	0.14
	(e) explicit ODE, $\boldsymbol{I}$-tensor elements given only for axis of rotation	0.135
	(f) explicit ODE, $\boldsymbol{x}$ linearized, $\boldsymbol{u}$-excitation large	
	(g) implicit ODE, residuum algorithm	0.13
		0.10
linearized ODE	(h) $\dot{\boldsymbol{x}} = \boldsymbol{A}\boldsymbol{x} + \boldsymbol{B}\boldsymbol{u}$	0.047

Table 2. *CPU-times achieved for the different measures taken to speed up the simulation.*

REFERENCES

1 Duc Minh Tran; *Equations of Motion of Multibody Systems in the ESA-MIDAS Software,* Proc. Spacecraft Structures and Mechanical Testing, ESA SP-321, Vol.2, Noordwijk, NL, 24-26 April 1991.

2 *Spacecraft Modelling Methods in Support of Simulation of Future Missions,* Final Report, SMM/TN/007, ESA-CR(P)-3947, Noordwijk, The Netherlands, Feb. 1995.

3 Schlenger, B. Schaefer, W. Rulka, R. Fonseca; *Real-Time Simulation of the Dynamics of Elastic Space Manipulators by MBS Software,* 3rd eioots European In-Orbit Operations Technology Symposium, ESTEC, Noordwijk, NL, 22-24 June 1993.

4 Fonseca, R.V. de Moraes, B.E. Schaefer; *Adaptive Control and Simulation of an Elastic Space Manipulator Replica,* Paper No. IFAC-2390, 13th IFAC World Congress, San Francisco, USA, 30 June - 5 July 1996.

5 Loesch, B. Schaefer, R. Krenn, R. Rehm, D. Garton; *Integration of Simulation Tools for Efficient Telerobotic Operations,* 4th Workshop on 'Simulators for European Space Programmes', ESA/ESTEC, Noordwijk, NL, 22-24 Oct 1996.

6 Loesch, B. Schaefer, R. Rehm, D. Garton; *Advanced Ground-Based Telerobotic Operations and Simulation Environment,* 'Space Ops 96' 4th Int. Symposium on Space Mission Operations and Ground Data Systems, 16-20 Sept 1996, Munich, Germany.

7 Garton, R. Rehm, B. Schaefer; *A Man-Machine Interface for the Real-Time Control of Robots and the Simulation of Satellite Teleoperations,* 14th European Annual Conference on Human Decision Making and Manual Control, Delft, NL, 14-16 June 1995.

8 *Status Report 1997 for the Period 1993-1996,* Institute of Robotics and System Dynamics, DLR Oberpfaffenhofen, Germany, Feb. 1997.

9 Haug; *Computer Aided Kinematics and Dynamics of Mechanical Systems,* Allyn & Bacon, 1989.

10 Schwertassek, W. Rulka; *Aspects of Efficient and Reliable Multibody System Simulation,* in *Real-Time Integration Methods for Mechanical System Simulation,* (E. J. Haug, R. C. Deyo, ed.), Springer-Verlag, Berlin, 1991, pp. 55-96.

11 Wallrapp, R. Schwertassek; *Representation of Geometric Stiffening in Multibody System Simulation,* Int. Journal for Numerical Methods in Engineering, 32, 1991, pp. 1833-1850.

12 Rulka; *SIMPACK, a Computer Program for Simulations of Large-Motion Multibody Systems,* in Multibody System Handbook, W. Schiehlen (ed.), VDI-Verlag Düsseldorf, Germany, 1993.

13 Eichberger; *Simulation von Mehrkörpersystemen auf parallelen Rechnerarchitekturen,* Fortschritt-Berichte VDI, Reihe 8, Nr. 332, VDI Verlag Düsseldorf, Germany, 1993.

14 Roberson, R. Schwertassek; *Dynamics of multibody systems,* Springer Verlag, Berlin, 1988.

AIAA-97-3562

TERMINAL AREA NAVIGATION USING RELATIVE GPS BIAS STATES WITH A TERRAIN SCENE

Jerome J. Krempasky

Hughes Missile Systems Company

P.O. Box 11337, Tucson, Az 85734

ABSTRACT

The effects of large GPS pseudorange bias errors and large absolute target location errors can be removed almost entirely from a guided weapon incorporating GPS updates in its Kalman filter by utilizing GPS bias states following a high quality terrain scene. Specifically, target relative bias states are added to the Kalman filter, and these states provide an accurate estimate of the total GPS bias after the terrain scene update is accepted by the Kalman filter. The performance of two different bias state configurations, one having three states and the other seven, is analyzed by means of a covariance simulation. The scenarios investigated include short and long terrain scene to target separations, both with and without the presence of high-powered GPS jammers and a software-implemented satellite constellation freeze. It is found that because of its virtual elimination of the pseudorange and absolute target location errors, and its affordability when compared to a terminal guidance seeker, the relative GPS bias state implementation is a viable precision strike candidate.

INTRODUCTION

Differential or relative GPS schemes, used for military purposes, provide affordable, all-weather precision strike capability for a guided weapon without the need to employ a terminal guidance seeker. The differential or relative GPS schemes proposed to date utilize two GPS receivers, one on the guided weapon and a second "cooperative receiver", with communication established between the two receivers[1]. The location of the cooperative receiver can be either ground-based or on-board a mobile targeting system such as an aircraft. Typical differential GPS schemes employ a cooperative GPS receiver situated at a precisely known position, referred to as the base station[2]. The purpose of this GPS receiver is to provide an estimate of the pseudorange bias-like errors that are present in the pseudorange measurements to each of the visible satellites. The pseudorange bias estimates are then to be transmitted to any guided weapons equipped with GPS receivers that are operating in the neighborhood of the cooperative receiver. Corrections received by a given weapon's GPS antenna are applied to the pseudorange measurements performed in the weapon's GPS processor before it computes its position solution. Depending upon the degree of spatial and temporal correlation that the bias errors exhibit between the satellite signals received at the cooperative and the weapon GPS receivers, the pseudorange errors are removed to a certain extent from the weapon's calculated position solution. The drawback to this scheme is that these correlations degrade substantially over the times and distances typical of most tactical engagement scenarios, thus rendering the pseudorange bias cancellation incomplete[3].

If, instead, the cooperative GPS receiver is located on-board a mobile targeting platform such as an AWACS, then the precision strike scheme is more properly referred to as relative GPS. In this mechanization, the location of the target with respect to the targeting platform is known very accurately, e.g., through the use of the AWAC's synthetic aperature radar (SAR). By contrast, it is not necessary to know the absolute location of either the target or the cooperative GPS receiver in the WGS-84 earth-centered, earth-fixed reference frame. Typically, these absolute location errors are quite large in comparison to relative target location errors. With the relative GPS mechanization, the accuracy achieved at the target can be significantly better than the standard GPS PPS position solution which is fully corrupted by the pseudorange random biases. Such accuracy is achievable only if the precision strike weapon's GPS receiver can maintain a high degree of correlation of its GPS position solution bias errors with those of the targeting platform. The problems involved in maintaining this correlation have been discussed in the literature[3], and certain schemes have been proposed to enhance the probability of maintaining a strong correlation.

In the literature[4], a high quality terrain scene is used in place of the cooperative GPS receiver to achieve precision strike capability. In this mechanization, referred to as the relative GPS correction vector scheme, the location of the terrain scene relative to the target is known very precisely, a fact guaranteed by the appearance of both on the same satellite photograph (geo-cell). At the time of overflying the terrain scene, the guided weapon forms a three-dimensional vector in its navigational software. This correction vector, which represents the difference between the GPS derived position solution of the weapon and the center of the terrain scene, is then applied outside of the system Kalman filter to correct all computed GPS position solutions following passage of the terrain scene. Application of the correction vector has the effect of essentially eliminating the large pseudorange bias-like errors since GPS updates following the terrain scene are highly correlated with that at the terrain scene given scene/target separations that are not excessive. In addition, application of the correction vector forces the weapon to navigate in the terrain scene/target relative coordinate system where the relative location error, not the absolute errors, are significant. A disadvantage inherent in this mechanization is that the GPS updates following the terrain scene become less correlated with the GPS update at the terrain scene as the separation between the scene and target becomes substantial. This loss of correlation implies a corresponding decrease in the accuracy of the weapon position relative to the target.

In the relative GPS scheme considered in the present paper, a high quality terrain scene is still utilized, but the establishment of a correction vector is replaced by the addition of target relative bias states to the system Kalman filter. The function of these extra bias states is to provide an estimate of the total bias present in the GPS pseudorange updates following passage of the terrain scene. One component of the total bias is the mission planning error incurred in establishing the location, in ECEF coordinates, of the terrain scene and the target (both of which are assumed to lie on the same satellite photograph). The other component of the total bias is, of course, the pseudorange bias present in the range measurement to each of the tracked satellites. The estimation of the total GPS bias is accomplished by accepting the terrain scene update directly into the system Kalman filter. This is to be contrasted with the implementation of the correction vector scheme, wherein the correction vector is established outside the Kalman filter and, as such, the terrain scene updating is transparent to the Kalman filter. Two bias state configurations are analyzed herein, one having three target relative states, and the other having seven states altogether, namely, three target relative states plus four pseudorange bias states. In the first configuration, three bias states are employed to estimate the total bias since presumably this number is sufficient to span the entire three-dimensional space in which the total vector GPS bias exists. In recognition that this might be inadequate due to the fact that the two components of GPS bias are quite distinct in nature, the second bias state configuration is also proposed as a plausible candidate.

In the main body of this report, a more detailed discussion of the nature of the two relative GPS bias state configurations is found, along with a presentation of the analytical forms of the observation matrices associated with them. Expressions for the accuracy of the calculated weapon position relative to the target are developed, in order that a comparison of the benefits of both configurations relative to each other and to the relative GPS correction vector scheme can be effected. A discussion of the pertinent features of a covariance simulation that is employed to accomplish these accuracy comparisons then follows. Since the covariance simulation is run several times for a given scenario, each time using a different, randomly selected GPS starting time, a feasible analytical approach is presented for handling the "Monte-Carlo" procedure applied to GPS starting time. Finally, CEP results are shown for the two GPS bias state configurations and the correction vector scheme, both for short and long separations of the terrain scene and target. In addition, CEP results are deduced for the case where three high powered GPS jammers are positioned along the route to the target, two located at midcourse and one positioned directly over the target. For completeness, all scenarios examined are analyzed both with and without the presence of a GPS satellite constellation freeze established in the GPS processor software.

RELATIVE GPS BIAS STATE CONFIGURATIONS

The dynamics occurring in the vicinity of the terminal area are shown in Figure 1 to illustrate the performance of the relative GPS implementation schemes involving a terrain scene. From this figure, it is seen that the weapon navigates a trajectory that deviates from the planned trajectory, as determined by mission planning, by the navigation position error. The pseudorange random bias errors associated with the pseudorange measurements to each of the four satellites being tracked form the primary components of the navigation position error. If no navigation error were present, the planned trajectory would take the guided weapon to the planned WGS-84 coordinates of the terrain scene, which differ from the actual scene coordinates by the large absolute scene location error. The planned WGS-84 target coordinates also differ from the actual target coordinates by the large absolute target location error, which is nearly equal to the absolute scene location error assuming that the terrain scene and target are on the same geo-cell. The difference between the two absolute errors is the small scene/target relative error. Shortly after the acceptance of the terrain scene update, the relative GPS bias state implementation estimates the total GPS bias present in the GPS updates occurring after the terrain scene. Application of appropriate corrections based upon these esimates then restores the navigated trajectory to within a small distance of the actual scene center, i.e., the scene center error. This error arises from the inability to precisely locate the scene center, and is related to the size of the digitized terrain scene cells. It is noted that if the correction vector scheme is instead employed, the correction vector applied to GPS updates after the terrain scene will restore the navigated trajectory to the indicated path. Flying a trajectory parallel to the planned heading finally results in the weapon arriving at the indicated impact point. This point differs from the actual target location by the sum of the scene center and scene/target relative errors, which is significantly less than either the navigation or absolute location errors.

In the first relative GPS bias state configuration to be considered, three target relative bias states are added to the existing states of the system Kalman filter. As indicated previously, these bias states attempt to estimate the total GPS bias, one component of which is the mission planning error in establishing the absolute terrain scene location, and the other being the pseudorange bias error present in each weapon to satellite range measurement. As implemented, the three target relative bias states are intended to estimate the east, north, and up components of the total GPS bias present in the GPS pseudorange updates. These states, which are turned on at the location of the terrain scene, are regarded as random constants and thus each satisfies the error state differential equation, $\delta\dot{x} = 0$. When (and if) a satellite constellation change occurs after acceptance of the terrain scene update, the covariances of all three target relative states are reset to their initial values in the system Kalman filter. In addition, all off-diagonal elements in the rows and columns containing the three target relative states are set to zero.

The three target relative bias states of the first configuration are coupled into the other error states of the system Kalman filter via their appearance in the observation matrix, $\mathbf{H}$. For the GPS pseudorange updates following the terrain scene update, the system Kalman filter observation matrix corresponding to the ith tracked satellite has the following form:

$$\begin{aligned}\mathbf{H}_i = \big[&-RU_{iy} \;\; RU_{ix} \;\; G_{iu} \;\; 0 \;\; \ldots \;\; 0 \;\; c \;\; 0 \\ &\ldots \;\; 0 \;\; -G_{ie} \;\; -G_{in} \;\; -G_{iu}\big]\end{aligned} \tag{1}$$

Here R denotes the earth radius plus the guided weapon altitude, c the speed of light, and $-G_{ij}$ the jth component of the unit vector, in the ENU (East, North, Up) system, pointing from the ith satellite to the weapon ($j = e, n, u$). The quantities U_{ix} and U_{iy} are defined as

$$\begin{aligned}U_{ix} &= G_{ie}\cos\alpha + G_{in}\sin\alpha, \\ U_{iy} &= -G_{ie}\sin\alpha + G_{in}\cos\alpha\end{aligned} \tag{2}$$

where α is the wander azimuth angle associated with the wander azimuth coordinate frame (with vertical axis up), which serves as the local level system for the weapon. The first three entries in the row matrix of Equation (1) correspond, respectively, to the weapon angular position error states, $\delta\theta_x$ and $\delta\theta_y$, and the weapon altitude error, δh, in the Kalman filter error state vector, $\delta\mathbf{x}$. In addition, the parameter c corresponds to the clock error state, Δt, and G_{ij} to the jth component of the target relative bias error state. It is to be noted that the terrain scene update utilizes an observation matrix which couples directly into the guided weapon's position error states. Given the fact that the observation matrix of Equation (1) also directly couples into the weapon's position error states, besides coupling into the target relative bias states, the usage of this scheme therefore achieves estimation of the target relative bias states following the terrain scene update.

In the second bias state configuration under investigation, four pseudorange bias states are added to the system Kalman filter which encompasses the first bias state configuration. This is done in an attempt to break up the estimation of the total GPS bias into three target relative states which accommodate the three components of the mission planning error, and four pseudorange bias states which accommodate the four independent pseudorange biases. In contrast to the three target relative bias states, the four pseudorange bias states are active during the entire mission, i.e., before and after overflying the terrain scene. Immediately before the terrain scene update, the four pseudorange bias state covariances are "bumped up" in the system Kalman filter. This action is consistent with the usual increasing of the position error state covariances just prior to the terrain scene update to force the Kalman filter to accept virtually the entire terrain scene update as the truth, and thus transition over to a target relative coordinate system.

The observation matrix for the second bias state configuration, which is characterized by seven bias states added to the system Kalman filter, has the following form for a GPS pseudorange update associated with the ith tracked satellite:

$$\begin{aligned}\mathbf{H}_i = \big[&-RU_{iy} \;\; RU_{ix} \;\; G_{iu} \;\; 0 \;\; \ldots \;\; 0 \;\; c \;\; 0 \;\; \ldots \;\; 0 \\ &-G_{ie} \;\; -G_{in} \;\; -G_{iu} \;\; -1 \;\; 0 \;\; 0 \;\; 0\big]\end{aligned} \tag{3}$$

In terms of the system error state vector, $\delta\mathbf{x}$, the -1 appearing here corresponds to the pseudorange bias error state associated with the first tracked satellite. By moving the -1 to any one of the positions occupied by the three zeroes that follow, it will then correspond to the pseudorange bias error state associated with the ith tracked satellite, where $i = 2,3,4$. As far as resetting the Kalman filter is concerned, only the pseudorange bias error state corresponding to a switched satellite is reset to its initial value whenever a constellation change occurs. Simultaneously, the off-diagonal elements relating to the affected pseudorange bias error state are zeroed in the Kalman filter covariance matrix. Once this is accomplished, it is not necessary to reset the covariances corresponding to the three target relative states, as is done in the first bias state configuration discussed above.

MATHEMATICAL PREDICTION OF ACCURACY

To assess the accuracy of this approach, a method for determining the uncertainty in the position of the precision strike weapon relative to the target after the terrain scene update needs to be established. The weapon position relative to the target, $\mathbf{x}_{tw}$, is the difference between the weapon position, $\mathbf{x}_w$, and the target position, $\mathbf{x}_t$, both referenced to the ECEF coordinate system, i.e., $\mathbf{x}_{tw} = \mathbf{x}_w - \mathbf{x}_t$. The weapon position, in turn, can be written as $\mathbf{x}_w = \mathbf{x}_{est} + \delta\mathbf{x}_s$, where $\mathbf{x}_{est}$ is the Kalman filter estimated weapon position (which is sent to the weapon's autopilot), and $\delta\mathbf{x}_s$ is the steering error, which represents the ability of the guided weapon to steer to the commanded position. Taken all together, the weapon position relative to the target at time k is given by

$$\mathbf{x}_{tw}(k) = \mathbf{x}_{est}(k) - \mathbf{x}_{ter} + \delta\mathbf{x}_s(k) + \mathbf{x}_{ter} - \mathbf{x}_t \qquad (4)$$

Here $\mathbf{x}_{ter}$, the position vector locating the terrain scene center relative to ECEF coordinates, has been added and subtracted on the right hand side. The uncertainty in the target relative weapon position can now be found by taking the covariance of this expression.

Since $\mathbf{x}_{est}(k) - \mathbf{x}_{ter}$, $\delta\mathbf{x}_s(k)$, and $\mathbf{x}_{ter} - \mathbf{x}_t$ are all uncorrelated, taking the covariance of Equation (4) yields

$$\begin{aligned} E(\mathbf{x}_{tw}(k)\mathbf{x}_{tw}^T(k)) &= E(\delta\mathbf{x}_{est}(k)\delta\mathbf{x}_{est}^{\mathrm{T}}(k)) \\ &+ E(\delta\mathbf{x}_s(k)\delta\mathbf{x}_s^{\mathrm{T}}(k)) + E((\mathbf{x}_{ter} - \mathbf{x}_t)(\mathbf{x}_{ter} - \mathbf{x}_t)^T) \end{aligned} \qquad (5)$$

where $\delta\mathbf{x}_{est}(k)$ denotes $\mathbf{x}_{est}(k) - \mathbf{x}_{ter}$. The errors in $\mathbf{x}_{ter}$ and $\mathbf{x}_t$ can be written as

$$\delta\mathbf{x}_{ter} = \delta\mathbf{x}_{ec}, \qquad \delta\mathbf{x}_t = \delta\mathbf{x}_{ec} + \delta\mathbf{x}_{ct} \qquad (6)$$

where $\delta\mathbf{x}_{ec}$ is the error in locating the true terrain scene center relative to ECEF and $\delta\mathbf{x}_{ct}$ is the target to terrain scene relative error. If, as is the case assumed here, the terrain scene and target are on the same satellite geo-cell, then $\delta\mathbf{x}_{ct}$ is much smaller than $\delta\mathbf{x}_{ec}$. Upon combining Equations (5) and (6), the desired accuracy expression is found to be

$$\begin{aligned} E((\mathbf{x}_{tw}(k)\mathbf{x}_{tw}^{\mathrm{T}}(k)) &= E(\delta\mathbf{x}_{est}(k)\delta\mathbf{x}_{est}^T(k)) \\ &+ E(\delta\mathbf{x}_s(k)\delta\mathbf{x}_s^T(k)) + E(\delta\mathbf{x}_{ct}\delta\mathbf{x}_{ct}^T) \end{aligned} \qquad (7)$$

The first error term on the right hand side is the navigational error that is associated with the Kalman filter estimated weapon position in the target relative coordinate frame. It is driven mainly by the uncertainties arising from the white measurement noise corresponding to the GPS pseudorange updates after the terrain scene, and from the measurement noise corresponding to the terrain scene update itself. The second error term is that associated with steering the guided weapon to the target in response to the weapon position estimates, and is typically the smallest of the three error terms. This error includes environmental effects such as wind gusts. The third, and final, source of error inherent in the implementation of the relative GPS bias state approach is the relative error existing between the terrain scene and target. This error is the largest of the three, and grows in size as scenarios with larger terrain scene to target separations are considered. Nevertheless, due to the high accuracy of present day satellite imagery which constrains the magnitude of the third source of error, the total of all three errors (in the proper rss sense) is significantly less than that achieved with GPS only navigation.

It is interesting at this point to compare Equation (7) with the corresponding expression for the target relative weapon position error appropriate for the relative GPS correction vector scheme. This expression is[4]

$$\begin{aligned} E((\mathbf{x}_{tw}(k)\mathbf{x}_{tw}^{\mathrm{T}}(k)) &= E(\Delta\mathbf{x}_{est}(k)\Delta\mathbf{x}_{est}^T(k)) \\ &+ E(\delta\mathbf{x}_s(k)\delta\mathbf{x}_s^T(k)) + E(\delta\mathbf{x}_{cd}\delta\mathbf{x}_{cd}^T) + E(\delta\mathbf{x}_{ct}\delta\mathbf{x}_{ct}^T) \end{aligned} \qquad (8)$$

where $\Delta\mathbf{x}_{est}(k)$ denotes $\mathbf{x}_{est}(k) - \mathbf{x}_{est}(0)$ and $\mathbf{x}_{est}(0)$ denotes the estimated weapon position at the terrain scene. The term $\delta\mathbf{x}_{cd}$ is the error in locating the true center of the terrain scene incurred in the process of determining the correction vector. An examination of this equation reveals that of the four error terms present, two are identical to those appearing in Equation (7), namely, the steering error and the scene/target relative error. The first error term appearing on the right hand side of this equation is similar in nature to the first error source in Equation (7) in the sense that they are both navigational errors. The first term on the right hand side of Equation (8), however, arises from the manner in which the correction vector is established, and is driven mainly by the decorrelations existing between the GPS updates occurring after the terrain scene and the GPS update occurring at the terrain scene. These decorrelations grow slowly in time and thus this error term can, when evaluated at the location of the target, become somewhat large for significant scene/target separations. The effect of the uncertainty in locating the precise center of the terrain scene, which is a function of the digitized scene cell size, does not appear in this error term. Instead, it appears explicitly in the fourth term on the right hand side of Equation (8). This is to be contrasted to the relative GPS bias state implementation under investigation, wherein the effects of the terrain scene measurement noise are filtered within the Kalman filter and appear in the navigational error term. It is noted in passing that to evaluate the navigational error term for the correction vector scheme, it is necessary to evaluate the matrix autocorrelation function which relates the Kalman filter states at the terrain scene to those at a later time.

COVARIANCE SIMULATION DESCRIPTION

To deduce specific values of the accuracy associated with each of the relative GPS bias state configurations introduced in this document, a full covariance simulation has been developed. The Kalman filter modeled in the simulation has 21 basic states, namely, three position, three velocity, three atitude, three gyro bias, three accelerometer bias, three accelerometer scale factor, one barometric altimeter bias, and two GPS clock bias error states. A much larger true world filter, from which the actual position accuracy is deduced, contains these same error states along with an additional 60 error states. Included in these extra states are three error states modeling the effects of gravitational anomaly and deflections, and eight GPS states representing the four independent pseudorange biases and the four independent delta-range biases associated with the tracked GPS satellites. Three states modeling the three components of the mission planning error also appear among the true world filter states. The remainder of the true world states model INS instrument errors that are relatively small in magnitude. With regard to the INS itself, it is assumed to be a generic, high quality strapdown system with the instrument error values listed in Table 1.

Incorporated in the covariance simulation is the capability to accept a wide variety of measurement inputs to the system Kalman filter. Among these are the velocities originating from the launch platform during transfer alignment of the guided weapon, GPS pseudorange and delta-range measurements, barometric altimeter measurements, and terrain scene position fixes. The GPS receiver, which is tightly coupled to the INS, is assumed to be a single channel, sequential device, augmented by a robust CRPA which ensures a high degree of GPS jammer immunity. To exploit this anti-jam capabilty, different kinds of GPS jammers of arbitrary strength and location can be modeled in the simulation. The operations of acquiring and tracking GPS satellites, along with the operations of reacquiring or replacing satellites lost due to jamming or geometric masking are modeled in the covariance simulation to afford a realistic representation of a guided weapon navigating by GPS. The simulation also has the capability of readily incorporating additional states in its system Kalman filter, a feature which renders the inclusion of the target relative bias states and pseudorange bias states a straightforward matter. Arbitrary launch platform profiles and guided weapon flight trajectories can be read into the simulation, thus affording the user the ability to study any desired weapon scenario.

INSTRUMENT ERROR	ONE SIGMA VALUE
Gyro Bias	0.01 deg/hr
Gyro Scale Factor	10 ppm
Gyro Misalignment	6 arcsec
Gyro Scale Factor Asym.	1 ppm
Gyro G-sensitive Bias	0.0005 deg/hr/g
Gyro Random Walk	0.0075 deg/$\sqrt{hr}$
Gyro Compliance	1 arcsec/g
Accel Bias	150 μg
Accel Scale Factor	200 ppm
Accel Misalignment	15 arcsec
Accel Scale Factor Asym.	40 ppm
Accel Nonlinearity	20 $\mu g / g^2$
Accel Random Walk	0.05 ft/sec $\sqrt{hr}$

TABLE 1. Assumed INS Instrument Errors

GPS STARTING TIME CONSIDERATIONS

GPS starting time has a significant effect upon the covariance simulation generated CEP predictions since it affects the entire time-dependent geometry of the satellites and, consequently, their interplay with the GPS receiver on-board the guided weapon. In view of this consideration, CEP results for a given scenario have been generated by running the covariance simulation 50 times, with each run characterized by a randomly selected GPS starting time. Specifically, random draws on GPS starting time from an almanac representing the calender year 1994 have been performed for the present analysis to allow the simulation to be exercised in this "Monte-Carlo" fashion. It should be noted that the number of simulation runs for a given scenario has been limited to 50 for computer run-time considerations The question now naturally arises as to how to treat the 50 different CEP's, each corresponding to a different GPS starting time. In particular, how should the data be processed to obtain a single figure of merit that closely matches the single CEP that would be obtained from a comparable, true Monte Carlo simulation where GPS starting time, among many other parameters, is randomized (Monte-Carloed) on each pass.

To begin, it is observed that the "usual" form of a closed form analytical expression for CEP is

$$CEP_o = K\sqrt{E\{\mathbf{y}^2\}}, \qquad K = \sqrt{2\ln 2} = 1.17741 \tag{9}$$

Here $\mathbf{y}$ is a random variable denoting the crosstrack position error. From each of the 50 covariance simulation runs for a given scenario, a CEP is obtained which can be represented as

$$CEP = K\sqrt{E\{\mathbf{y}^2 \mid \mathbf{x} = x\}} = K\sqrt{E\{\mathbf{y}^2 \mid x\}} \tag{10}$$

where x signifies the particular GPS starting time. This CEP is a function of x and can thus be regarded as a random variable if x is replaced by $\mathbf{X}$. The result is

$$\mathbf{CEP} = K\sqrt{E\{\mathbf{y}^2 \mid \mathbf{x}\}} \quad \text{or} \quad \mathbf{CEP}^2 = K^2 E\{\mathbf{y}^2 \mid \mathbf{x}\} \tag{11}$$

Extracting the expected value of this random variable yields

$$E\{\mathbf{CEP}^2\} = K^2 E\{E\{\mathbf{y}^2 \mid \mathbf{x}\}\} \tag{12}$$

The right hand side of this expression, which contains a repeated occurrence of the expected value operator, must now be evaluated.

From the theory of probability, the expected value of the random variable $E\{\mathbf{y}^2 \mid \mathbf{x}\}$ is just $E\{\mathbf{y}^2\}$; hence

$$E\{\mathbf{CEP}^2\} = K^2 E\{\mathbf{y}^2\} = CEP_o^2 \tag{13}$$

This relationship shows that a good fit to the desired quantity CEP_o can be deduced from the various CEP's obtained from the 50 runs of the covariance simulation by taking the square root of the average of the squares of the 50 CEP's, i.e.,

$$CEP_o = \sqrt{E\{\mathbf{CEP}^2\}} \cong \sqrt{\frac{CEP_1^2 + CEP_2^2 + \ldots + CEP_{50}^2}{50}} \tag{14}$$

Since this seems to be a good match to the CEP that one would obtain from a true Monte Carlo simulation, it is therefore used as the figure of merit to evaluate the performance of the various cases to be studied in the next section. Accordingly, all "CEP" values listed in Tables 2 through 5 are obtained by means of the averaging operation indicated on the right hand side of Equation (14).

COVARIANCE SIMULATION RESULTS

An air-launched cruise missile is selected as the GPS receiver-equipped, guided weapon which will serve as a test platform for the relative GPS bias state implementation under investigation. Following a four minute in-air transfer alignment in which the weapon accepts velocity measurement inputs from the host aircraft's INS, the guided weapon flies, at a subsonic velocity and low altitude, the trajectory whose groundtrack is depicted in Figure 2. In the immediate vicinity of the target, the weapon executes a 15 degree terminal dive to impact the target. The four turn trajectory that the weapon undergoes between the terrain scene and the target constitutes the long terrain scene to target separation, and is referred to herein as trajectory 2. The short terrain scene to target separation, designated as trajectory 1, is simply a straight line (constant heading) path between the terrain scene and target and, as such, is not depicted here. The straight line distance between the terrain scene and target for trajectory 2 is about four times that associated with trajectory 1. On the other hand, the distance along the flight path between the terrain scene and target for trajectory 2 is about seven times that associated with trajectory 1. Depicted also in Figure 2 are the location of the high powered broadband jammers whose presence will be considered in half of the scenarios that are run. As seen, two of these jammers straddle the flight path at a midcourse location, while the third jammer lies directly above the target.

In the results to be presented in Tables 2 through 5, values are listed for CEP at the target, calculated from the covariance simulation in the manner discussed in the previous section. The plane in which CEP is calculated is that which is perpendicular to the velocity vector, which seems to make sense from miss distance considerations. Given this observation and the fact that the weapon executes a 15 degree terminal dive, the CEP values presented will be conservative since they incorporate a substantial portion of the vertical component of vehicle position error, which tends to be approximately twice as large as either of the horizontal components. Certainly, the CEP values listed in Tables 2 through 5 are significantly larger than if CEP were calculated in a horizontal plane, which is often done.

Shown in Table 2 are the covariance simulation predictions for CEP at the target in the case of trajectory 1 with no GPS jamming present. Besides the relative GPS bias state configurations under consideration, it is seen that the table, for the sake of comparison, also lists results for the relative GPS correction vector scheme[4]. It is noted that the CEP values are normalized such that the 7 bias state configuration has a value of unity for the "constellation freeze" case. The meaning of the term "constellation freeze" is that the GPS receiver software is not permitted to accept a change in the GPS satellite constellation after the terrain scene update has occurred, even if a tracked satellite is lost. The receiver software is permitted, of course, to allow reacquisition of a lost satellite, provided that its carrier signal-to-noise ratio is acceptable for tracking. Normally, a change in satellite constellation takes place if, after making a failed attempt to reacquire a satellite lost due to jamming or geometric masking, the receiver software determines that another visible satellite is acceptable for tracking. Also, it is normal for a constellation change to occur due to the so-called GDOP refinement algorithm, if it is determined that a visible GPS constellation with an improved GDOP exists. The nominal period for the implementation of this algorithm, which forces a constellation change only if the GDOP improvement is 10%, is every two minutes. Note that no result is listed in Table 2 for the correction vector scheme under the "no constellation freeze" column, because the establishment and usage of the correction vector is predicated on freezing the GPS constellation.

RELATIVE GPS SCHEME	CEP * - NO CONST. FREEZE	CEP * - CONST. FREEZE
3 Bias State	1.12	1.11
7 Bias State	1.00	1.00
Correction Vector	N/A	1.05

* Normalized

TABLE 2. CEP Results for Trajectory 1 with no GPS Jamming

An examination of Table 2 reveals that there is little difference in the covariance simulation generated values for CEP at the target amongst the three relative GPS implementations, both with and without a constellation freeze. The reason is that after the terrain scene update, only very few constellation changes occur - due to the refinement algorithm - out of the 50 runs (each corresponding to a randomly selected GPS starting time) for the case of no constellation freeze. Furthermore, no satellites are observed to be lost due to geometric masking (none are lost due to jamming since the scenario is jammer free). With regard to the CEP values themselves, it can be stated that although they are presented in normalized form, the actual numbers are indeed in the precision strike category. It is expected that larger differences will be observed amongst the three relative GPS schemes when GPS jamming is included in the scenario, since this will induce satellite loss between the terrain scene and the target. This effect should also be enhanced through the use of the long terrain scene to target separation, i.e., trajectory 2, since significantly more time is available for losing satellites between the terrain scene and the target.

If GPS jamming is now turned on for trajectory 1, the CEP results do indeed show more of a separation as seen from Table 3. The ordering is such that the 7 bias state configuration is superior to the 3 bias state configuration, which in turn is superior to the correction vector scheme. As suggested earlier, the performance of the 7 bias state configuration exceeds that of the 3 bias state configuration since it can do a better job in separating out the mission planning errors and the pseudorange bias errors, which are different in character, in the presence of an environment in which satellites are lost. The performance of the correction vector scheme is degraded relative to the other two implementations due to the fact that the purely navigational error in this scheme grows in time as the GPS updates following the terrain scene become more uncorrelated with the GPS update at the terrain scene, an effect which is enhanced by satellite loss.

RELATIVE GPS SCHEME	CEP * - NO CONST. FREEZE	CEP * - CONST. FREEZE
3 Bias State	1.18	1.15
7 Bias State	1.03	1.03
Correction Vector	N/A	1.20

* Normalized

TABLE 3. CEP Results for Trajectory 1 with GPS Jamming

Depicted in Tables 4 and 5 are the predictions for CEP that result from running the covariance simulation for trajectory 2, both with and without the presence of GPS jamming. As indicated above, employment of the longer terrain scene to target separation scenario does accentuate, especially in the GPS jamming environment, the differences between the three relative GPS implementations. The ordering of the performances is maintained as before, for the reasons that are discussed. To understand the apparent advantage afforded by the "constellation freeze" option in the case of jamming, it is observed that in the case of no constellation freeze, nearly two new constellations occur on the average (over the 50 member ensemble) as a result of a satellite lost due to jamming being replaced by a visible satellite in the opposite part of the sky. Furthermore, it is observed that most of the time when a new constellation is formed as a result of jamming, the refinement algorithm subsequently selects another new constellation because the GDOP of the new constellation formed as a result of jamming is degraded. Also, detailed examination of the constellation freeze runs indicates that a satellite lost due to jamming is almost always reacquired soon after being lost if the GPS receiver software is commanded not to give up, that is, keep trying to reacquire the lost satellite after failing to reacquire it on the first attempt, rather than trying to secure a replacement. The problem with the multiple constellation changes observed under the "no constellation freeze" option is that the biases are reset in the Kalman filter at each occurrence of a constellation change, and this action leads to a degradation in position error. Indeed, the multiple resetting of the biases, in the case of no constellation freeze, causes more of a degradation in CEP than is experienced in the case of a constellation freeze, which suffers only due to the degraded mode tracking of three satellites over a relatively short period of time. The high anti-jam immunity of the assumed GPS receiver is undoubtedly responsible for the relatively quick reacquisition of lost satellites when a constellation freeze is implemented.

RELATIVE GPS SCHEME	CEP * - NO CONST. FREEZE	CEP * - CONST. FREEZE
3 Bias State	2.27	1.97
7 Bias State	1.98	1.94
Correction Vector	N/A	2.61

* Normalized

TABLE 4. CEP Results for Trajectory 2 with no GPS Jamming

RELATIVE GPS SCHEME	CEP * - NO CONST. FREEZE	CEP * - CONST. FREEZE
3 Bias State	3.67	2.75
7 Bias State	2.42	2.36
Correction Vector	N/A	6.64

* Normalized

TABLE 5. CEP Results for Trajectory 2 with GPS Jamming

SUMMARY

CEP results generated from a covariance simulation show that the "internal" implementation of a relative GPS scheme appears to perform better than the "external" implementation for a guided weapon employing a high quality terrain scene update. In the internal implementation, bias states added to the system Kalman filter attempt to estimate the total GPS bias once the terrain scene update is accepted into the Kalman filter, whereas in the external implementation, a correction vector is established outside the Kalman filter and used for all GPS updates after the terrain scene. The external implementation suffers somewhat, especially in the event of long scene to target separations and/or high levels of jamming, due to the fact that the GPS updates following the terrain scene become increasingly uncorrelated with the GPS update at the terrain scene, thus leading to position error degradation. With regard to the two internal implementations, the configuration having 7 bias states appears to out-perform the configuration having only 3 bias states. The reason is that the 7 bias state configuration is naturally set up to more readily estimate the three components of mission planning error and the four pseudorange biases associated with each of the tracked GPS satellites than is the 3 bias state configuration.

In the presence of strong levels of GPS jamming, the results indicate a preference for the adoption of the "constellation freeze" option. This is because any disadvantage incurred by the constellation freeze in tracking less than four satellites while waiting for reacquisition of the lost satellites is less than the disadvantages associated with not freezing the constellation after the terrain scene update. Specifically, these disadvantages arise from the CEP degrading effects of the biases being reset and the GDOP being raised at each occurrence of a constellation change resulting from jamming. Also, the biases are reset at each constellation change resulting from the refinement algorithm, an action which usually follows a constellation change due to jamming. In conclusion, it can be stated that the relative GPS bias state implementation, in particular the 7 state configuration, is a viable option for obtaining precision strike accuracy, especially in view of its affordability in comparison to the other main alternative, i.e., weapons equipped with terminal guidance seekers.

REFERENCES

1. Schmidt, G., and R. Setterlund, "Precision Strike Concepts Exploiting Relative GPS Techniques", *Navigation, Jour. of the Inst. of Navigation,* Vol. 41, No.1, Spring 1994.

2. Brown, R. G., and P. Y. Hwang, *Introduction to Random Signals and Applied Kalman Filtering*, New York: Wiley, 1992.

3. Youhanaie, M., et al, "Relative GPS Guidance for Precision Weapon Delivery", *Institute of Navigation, GPS-95 Technical Meeting*, September 1995.

4. Krempasky, J. J., "Terminal Area Navigation Using a Relative GPS Correction Vector Scheme", *AIAA Guidance, Navigation and Control Conference*, July 1996.

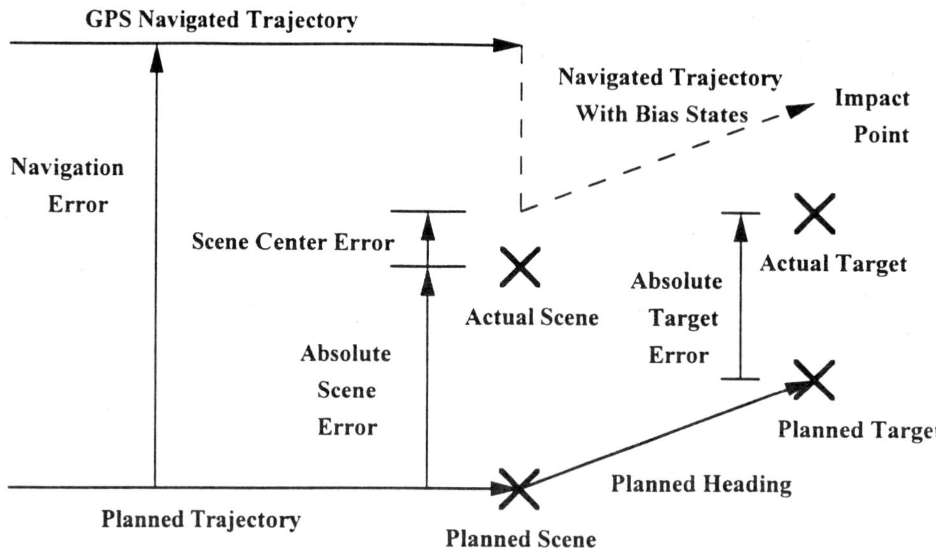

FIGURE 1. Application of Relative GPS Bias State Scheme

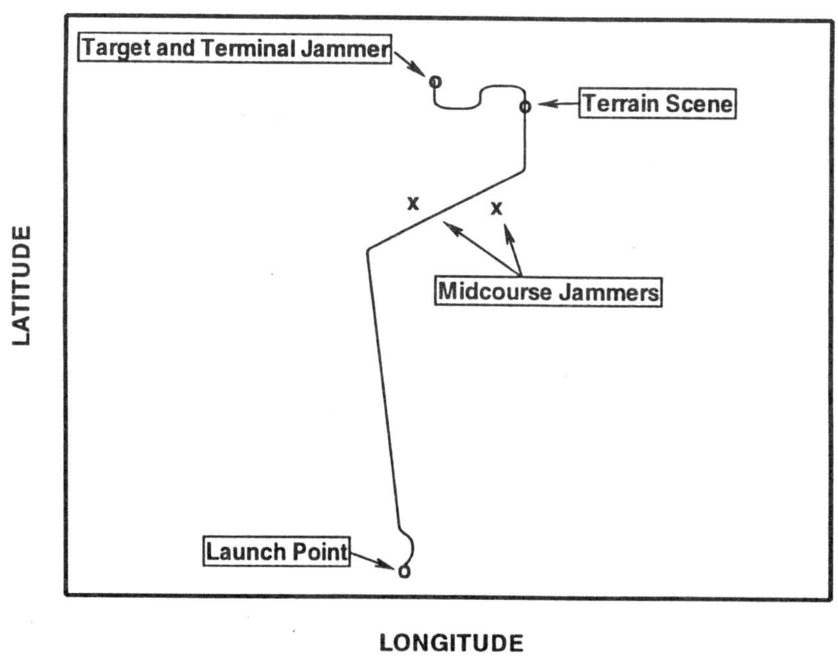

FIGURE 2. Groundtrack for Guided Weapon

AIAA-97-3563

Equivalent Nonlinear Error Models of Strapdown Inertial Navigation System

Myeong-Jong Yu[*], Heung-Won Park[†] and Chang-Bae Jeon[‡]
Agency for Defense Development
Yuseong P.O. Box 35-4, Taejon, 305-600, Korea

Abstract

The conventional linearized error models of strapdown inertial navigation system(SDINS) are found not to be effective to represent the nonlinear characteristics of navigation error propagation in case of the system which causes the large navigation errors. To represent the large attitude error of the SDINS, the rotation vector error is introduced and the relationship between the rotation vector error and the quaternion error is derived. Using this relationship the nonlinear error models of the SDINS are presented with the rotation vector error and the quaternion error, respectively. And it is shown that the nonlinear SDINS error models presented here are equivalent to each other. In addition it is shown that in the case of the system with the small enough attitude error, the nonlinear error models can be reduced to the conventional linearized error models.

1 Introduction

SDINS provides the navigation information such as the velocity, the position, and the attitude of a vehicle using output signals from the inertial sensors rigidly attached to the body of the vehicle. Compared with the GINS(Gimbaled Inertial Navigation System), the SDINS has lots of advantages, such as small size, light weight, low cost, easy maintenance, etc.. According to the development of high accuracy strapdown gyros and powerful micro computers, the GINS has been replaced by the SDINS recently.

The navigation equations of SDINS consist of the nonlinear equations. The SDINS error model plays an important role in analyzing the characteristics of the navigation error propagation or in implementing the Kalman filter for an aided navigation system. Assuming that the characteristics of the navigation error propagation is linear, the linearized error model is generally used. According to the used attitude error state, the linearized SDINS error models can be classified by the Psi-angle, the Phi-angle, or the quaternion error model and so on [1-7]. It has been recently shown that those linearized SDINS error models are equivalent to each other [5-7] . But the linearized SDINS error models presented, so far, are not effective to exactly represent the nonlinear characteristics of the system where the large navigation errors are generated such as in-flight alignment.

Therefore, in this paper nonlinear SDINS error models, which are applicable to the system whose navigation errors are large, are derived by using the rotation vector error and the quaternion error, respectively. And the exact relationship between the rotation vector error and the quaternion error is derived. Using the relationship it is shown that nonlinear SDINS error models are equivalent each other. Furthermore, it is proved that the conventional linearized quaternion error model is equal to the nonlinear error model using the rotation vector error presented in this paper. And it is also shown that in the case of the small attitude error, the nonlinear error model derived using the rotation vector error can be reduced to the Phi-angle model expressed by the equivalent tilt angle.

In the next section, the relationship between the rotation vector error and the quaternion error is derived. The nonlinear attitude error model with the rotation vector error is obtained from the conventional linearized quaternion error model in section 3. In section 4, the nonlinear velocity error models are derived using the rotation vector error and the quaternion error, respectively. Conclusions are described in section 5. Finally, it is shown that the new approach method to derive the nonlinear error mod-

---

[*] Senior Researcher
[†] Senior Researcher
[‡] Principal Researcher

els for SDINS can be similarly applied to another rotation vector error in Appendix A.

2 Relationship between Attitude Errors

The attitude error has been represented using the equivalent tilt angle or the quaternion error[1-7]. However, when a system has a large attitude error, it is difficult to represent the exact relationship between attitude errors using the above error parameters. Thus, to solve this problem, the rotation vector error parameter is introduced as a new variable. That is, $\Delta\Phi^n$ is defined as the rotation vector error between a true navigation frame and an indicated navigation frame in the local-level frame. And $\delta Q(\equiv Q_b^n - Q_b^{n'})$ is defined as the quaternion error which corresponds to $\Delta\Phi^n$. Then the new relationship between the rotation vector error and the quaternion error can be derived.

We suppose that $\tilde{Q}_b^n$ is the indicated quaternion, such as $\tilde{Q}_b^n(\equiv [\tilde{q}_0, \tilde{q}_1, \tilde{q}_2, \tilde{q}_3]^T)$, it can be decomposable as (1).

$$\tilde{Q}_b^n \equiv Q_b^{n'} = Q_n^{n'} Q_b^n, \qquad (1)$$

where the subscripts, n and b, denote a navigation and a body frame, respectively. n' represents an indicated navigation frame with the error. Q_b^n denotes the true quaternion, such as $Q_b^n(\equiv [q_0, q_1, q_2, q_3]^T)$. And $Q_n^{n'}$ is the error quaternion given in terms of $\Delta\Phi^n(\equiv [\Delta\phi_N, \Delta\phi_E, \Delta\phi_D]^T)$.

Without loss of generality, (1) can be converted into (2).

$$Q_b^n = Q_{n'}^n Q_b^{n'} \qquad (2)$$

where using the conventional relationship between the rotation vector and the quaternion, the four elements of the error quaternion , $Q_{n'}^n$, are defined as

$$\begin{aligned} Q_{n'}^n &= [q_{0n}, q_{1n}, q_{2n}, q_{3n}]^T \\ &= [1 + c_n, s_n\Delta\phi_N, s_n\Delta\phi_E, s_n\Delta\phi_D]^T, \end{aligned} \qquad (3)$$

$$\begin{aligned} c_n &= cos(\frac{\Phi_{0n}}{2}) - 1, s_n = \frac{1}{\Phi_{0n}} sin(\frac{\Phi_{0n}}{2}), \\ \Phi_{0n} &= \sqrt{(\Delta\phi_N)^2 + (\Delta\phi_E)^2 + (\Delta\phi_D)^2}. \end{aligned} \qquad (4)$$

And (2) can be written as (5) based on the quaternion multiplication in [3].

$$Q_b^n = \begin{bmatrix} q_{0n} & -q_{1n} & -q_{2n} & -q_{3n} \\ q_{1n} & q_{0n} & -q_{3n} & q_{2n} \\ q_{2n} & q_{3n} & q_{0n} & -q_{1n} \\ q_{3n} & -q_{2n} & q_{1n} & q_{0n} \end{bmatrix} \begin{bmatrix} \tilde{q}_0 \\ \tilde{q}_1 \\ \tilde{q}_2 \\ \tilde{q}_3 \end{bmatrix} \qquad (5)$$

From (5) the relationship between the quaternion error and the rotation vector error can be obtained as (6).

$$\delta Q = c_n Q_b^{n'} + s_n Y(Q_b^{n'})\Delta\Phi^n, \qquad (6)$$

where

$$Y(Q_b^{n'}) \equiv Y = \begin{bmatrix} -\tilde{q}_1 & -\tilde{q}_2 & -\tilde{q}_3 \\ \tilde{q}_0 & \tilde{q}_3 & -\tilde{q}_2 \\ -\tilde{q}_3 & \tilde{q}_0 & \tilde{q}_1 \\ \tilde{q}_2 & -\tilde{q}_1 & \tilde{q}_0 \end{bmatrix}. \qquad (7)$$

Multiplying both sides of (6) by the transpose of the matrix Y, Y^T, $\Delta\Phi^n$ is given by

$$\Delta\Phi^n = \frac{1}{s_n} Y^T \delta Q. \qquad (8)$$

Then, multiplying both sides of (8) by Y, another δQ can be derived as follows:

$$\delta Q = s_n Y \Delta\Phi^n + \tilde{Q}_b^n \tilde{Q}_b^{nT} \delta Q. \qquad (9)$$

If the attitude error is small enough, (8) and (9) are simplified as (10) and (11), respectively.

$$\Delta\Phi^n = 2Y^T \delta Q \qquad (10)$$

$$\delta Q = \frac{1}{2} Y \Delta\Phi^n + \tilde{Q}_b^n \tilde{Q}_b^{nT} \delta Q \qquad (11)$$

(10) and (11) have the same form as the results presented previously in [7]. But if the system has a large attitude error, they are not useful because they do not express the exact relationship between the rotation vector error and the quaternion error. In addition, it is difficult to get the quaternion error based on the known rotation vector error in (11). Compared with the previous results, the newly proposed relation equations, (6) and (8), can completely describe their relationship and the quaternion error is easily obtained from the rotation vector error.

From (3), (4), (6), and (9), We can obtain c_n as

$$\begin{aligned} c_n &= cos(\frac{\Phi_{0n}}{2}) - 1 = \tilde{Q}_b^{nT} \delta Q \\ &= \tilde{q}_0\delta q_0 + \tilde{q}_1\delta q_1 + \tilde{q}_2\delta q_2 + \tilde{q}_3\delta q_3. \end{aligned} \qquad (12)$$

According to the several papers [5-7], it is known that this term, c_n, can be neglected by normalizing the quaternion. But the above presented equations, (4) and (12), show that it can not be neglected in spite of the normalization procedure for the system in the presence of the large attitude error. Therefore, it must be considered as an important factor in analyzing and deriving the new error models.

3 Attitude Error Model

In this section a nonlinear attitude error model for SDINS is derived using the rotation vector error from the conventional linearized quaternion error model and the new nonlinear error model is compared with the conventional equivalent tilt angle model presented in the recent papers [1-7].

The nonlinear attitude error model for $\Delta\Phi^n$ can be obtained by using the relation equation between the rotation vector error and the quaternion error.

First differentiating both sides of (8), we get

$$\Delta\dot{\Phi}^n = \frac{1}{s_n}Y^T\delta\dot{Q} + \frac{1}{s_n}\dot{Y}^T\delta Q + \frac{d}{dt}(\frac{1}{s_n})Y^T\delta Q \quad (13)$$

And it is known that the conventional quaternion error model, $\delta\dot{Q}$, is equal to (14) in [3], [6], and [8].

$$\delta\dot{Q} = \frac{1}{2}M\delta Q + \frac{1}{2}(U\delta\omega_{ib}^b - Y\delta\omega_{in}^n) \quad (14)$$

where $\delta\omega_{ib}^b$ denotes the gyro error model and M, U are represented by

$$M \equiv \frac{1}{2}\begin{bmatrix} 0 & -\omega_X & -\omega_Y & -\omega_Z \\ \omega_X & 0 & \omega_Z & -\omega_Y \\ \omega_Y & -\omega_Z & 0 & -\omega_X \\ \omega_Z & \omega_Y & -\omega_X & 0 \end{bmatrix} - \frac{1}{2}\begin{bmatrix} 0 & -\omega_N & -\omega_E & -\omega_D \\ \omega_N & 0 & -\omega_D & \omega_E \\ \omega_E & \omega_D & 0 & -\omega_N \\ \omega_D & -\omega_E & \omega_N & 0 \end{bmatrix}, \quad (15)$$

$$U(Q_b^{n'}) \equiv U = \begin{bmatrix} -\tilde{q}_1 & -\tilde{q}_2 & -\tilde{q}_3 \\ \tilde{q}_0 & -\tilde{q}_3 & \tilde{q}_2 \\ \tilde{q}_3 & \tilde{q}_0 & -\tilde{q}_1 \\ -\tilde{q}_2 & \tilde{q}_1 & \tilde{q}_0 \end{bmatrix}. \quad (16)$$

Inserting (6) and (14) into (13) results in

$$\begin{aligned} \Delta\dot{\Phi}^n &= \frac{1}{s_n}Y^T\delta\dot{Q} + \frac{1}{s_n}\dot{Y}^T\delta Q + \frac{d}{dt}(\frac{1}{s_n})Y^T\delta Q \\ &= [Y^TMY + \dot{Y}^TY + \frac{d}{dt}(\frac{1}{s_n})s_nY^TY]\Delta\Phi_n \\ &+ \frac{c_n}{s_n}[Y^TM\tilde{Q}_b^n + \dot{Y}^T\tilde{Q}_b^n + \frac{d}{dt}(\frac{1}{s_n})s_nY^T\tilde{Q}_b^n] \\ &+ \frac{1}{2s_n}Y^T[U\delta\omega_{ib}^b - Y\delta\omega_{in}^n]. \end{aligned} \quad (17)$$

Substituting the elements of the quaternion differential equation in [3] and [6], $\dot{\tilde{Q}}_b^n = \frac{1}{2}[\tilde{Q}_b^n]\omega_{nb}^b$,into the elements of $\dot{Y}^T$, and then using the matrices Y and U, the following relationships can be obtained.

$$Y^TY = I, \quad (18)$$

$$Y^TU = \tilde{C}_b^n, \quad (19)$$

$$\frac{d}{dt}(\frac{1}{s_n})s_n\Delta\Phi^n = c\Delta\Phi^n(\Delta\Phi^n)^T\Delta\dot{\Phi}^n, \quad (20)$$

$$c = \frac{1}{\Phi_{0n}^2}(1 - \frac{\Phi_{0n}}{2}cot(\frac{\Phi_{0n}}{2})), \quad (21)$$

$$Y^TMY + \dot{Y}^TY = -\omega_{in}^nX, \quad (22)$$

$$Y^TM\tilde{Q}_b^n + \dot{Y}^T\tilde{Q}_b^n = 0, \quad (23)$$

$$Y^T\tilde{Q}_b^n = 0, \quad (24)$$

where $\tilde{C}_b^n$ is an indicated transformation matrix from the body frame to the navigation frame and ω_{in}^nX denotes a skew-symmetric matrix of the ω_{in}^n.

Inserting (18) - (24) into (17) results in the following nonlinear attitude error model.

$$\begin{aligned} \Delta\dot{\Phi}^n &= [I - c\Delta\Phi^n(\Delta\Phi^n)^T]^{-1}[-\omega_{in}^nX\Delta\Phi^n \\ &+ \frac{1}{2s_n}[\tilde{C}_b^n\omega_{ib}^b - \delta\omega_{in}^n]]. \end{aligned} \quad (25)$$

Using $Q_b^n = Q_b^{n'} + \delta Q$ instead of $Q_b^n = Q_b^{n'} - \delta Q$ in the definition of the quaternion error, another model is given by

$$\begin{aligned} \Delta\dot{\Phi}^n &= [I - c\Delta\Phi^n(\Delta\Phi^n)^T]^{-1}[-\omega_{in}^nX\Delta\Phi^n \\ &- \frac{1}{2s_n}[\tilde{C}_b^n\omega_{ib}^b - \delta\omega_{in}^n]]. \end{aligned} \quad (26)$$

The sign of the forcing term in (26) differs from that of (25). It is shown that the sign of the forcing term can be changed according to the definition of the quaternion error.

From (14), (25), and (26), it is obvious that the conventional quaternion error model is exactly equal to the nonlinear attitude error model rather than the conventional perturbation error model presented in [5-7]. Compared with (25) or (26), the conventional quaternion error model is very helpful in the sense that it provides the linear error model instead of the nonlinear error model. For this reason, with the quaternion error model it is easier to implement a linear Kalman filter for the system in the presence of the large attitude error.

Inserting the first order term of $[I - c\Delta\Phi^n(\Delta\Phi^n)^T]^{-1}$ into (25) based on a Taylor series, (25) can be readily simplified as (27) because $(\Delta\Phi^n)^T\omega_{in}^nX\Delta\Phi^n$ is zero.

$$\begin{aligned} \Delta\dot{\Phi}^n &= -\omega_{in}^nX\Delta\Phi^n + \frac{1}{2s_n}[I \\ &+ c\Delta\Phi^n(\Delta\Phi^n)^T][\tilde{C}_b^n\omega_{ib}^b - \delta\omega_{in}^n]. \end{aligned} \quad (27)$$

It is easy to obtain the linear attitude error model from the proposed nonlinear model. Supposing that the attitude error is very small and neglecting the high order terms of an error product, (25) is simplified as follows:

$$\Delta\dot{\Phi}^n = -\omega_{in}^n X \Delta\Phi^n + [\tilde{C}_b^n \omega_{ib}^b - \delta\omega_{in}^n] \tag{28}$$

which shows that the nonlinear rotation vector error model of (25) can be reduced to the the conventional equivalent tilt angle model presented in [7]. The sign of the forcing term of the equivalent tilt angle model in [7] is opposite to that of (28), because the sign of the navigation errors defined in this paper is opposite to that of the navigation errors defined as the differences between the indicated and the true quantites in [7].

4 Velocity Error Model

The conventional velocity equation of SDINS is represented by

$$\dot{V}^n = C_b^n f^b - (2\omega_{ie}^n + \omega_{en}^n) X V^n + g^n \tag{29}$$

and from (29) the linear velocity error model (30) can be obtained by perturbation as follows. But the velocity error model derived based on the large navigation errors is equal to (31).

$$\begin{aligned} \delta\dot{V}^n &= \Delta C_b^n \tilde{f}^b - (2\tilde{\omega}_{ie}^n + \tilde{\omega}_{en}^n) X \delta\tilde{V}^n + \tilde{C}_b^n (\delta\tilde{f}^b \\ &+ \eta_a) + \tilde{V}^n X (2\delta\tilde{\omega}_{ie}^n + \delta\tilde{\omega}_{en}^n) + \delta g^n. \end{aligned} \tag{30}$$

$$\begin{aligned} \delta\dot{V}^n &= \Delta C_b^n \tilde{f}^b - (2\tilde{\omega}_{ie}^n + \tilde{\omega}_{en}^n) X \delta\tilde{V}^n + \tilde{C}_b^n (\delta\tilde{f}^b \\ &+ \eta_a) + \tilde{V}^n X (2\delta\tilde{\omega}_{ie}^n + \delta\tilde{\omega}_{en}^n) \\ &+ \delta\tilde{V}^n X (2\delta\tilde{\omega}_{ie}^n + \delta\tilde{\omega}_{en}^n) + \delta g^n. \end{aligned} \tag{31}$$

In the velocity error model, $\Delta C_b^n \tilde{f}^b$ is strongly related to the attitude error. It is important to compute this term exactly to describe the correlated relationship between the attitude error and the velocity error. Therefore, we are concerned with computation of the transformation matrix error, ΔC_b^n , which is defined as the difference between a true transformation matrix, C_b^n and an indicated transformation matrix, $\tilde{C}_b^n$.

First, we compute this term with the quaternion error. Then the equivalent term with the rotation vector error is derived. Substituting the quaternion error, $\delta Q (\equiv Q_b^n - Q_b^{n'})$, into elements of true transformation matrix given by (32), the transformation matrix error given in terms of δQ and the indicated quaternion can be obtained as (33).

$$C_b^n = \tilde{C}_b^n + \Delta C_b^n =$$

$$\begin{bmatrix} q_0^2 + q_1^2 - q_2^2 - q_3^2 & 2(q_1 q_2 - q_0 q_3) & 2(q_1 q_3 + q_0 q_2) \\ 2(q_0 q_3 + q_1 q_2) & q_0^2 - q_1^2 + q_2^2 - q_3^2 & 2(q_2 q_3 - q_0 q_1) \\ 2(q_1 q_3 - q_0 q_2) & 2(q_0 q_1 + q_2 q_3) & q_0^2 - q_1^2 - q_2^2 + q_3^2 \end{bmatrix} \tag{32}$$

$$\begin{aligned} \Delta C_{11} &= 2(\tilde{q}_0\delta q_0 + \tilde{q}_1\delta q_1 - \tilde{q}_2\delta q_2 - \tilde{q}_3\delta q_3) \\ &\quad +\delta q_0^2 + \delta q_1^2 - \delta q_2^2 - \delta q_3^2 \\ \Delta C_{12} &= 2(\tilde{q}_1\delta q_2 + \tilde{q}_2\delta q_1 - \tilde{q}_0\delta q_3 - \tilde{q}_3\delta q_0 \\ &\quad +\delta q_1\delta q_2 - \delta q_0\delta q_3) \\ \Delta C_{13} &= 2(\tilde{q}_1\delta q_3 + \tilde{q}_3\delta q_1 + \tilde{q}_0\delta q_2 + \tilde{q}_2\delta q_0 \\ &\quad +\delta q_1\delta q_3 + \delta q_0\delta q_2) \\ \Delta C_{21} &= 2(\tilde{q}_1\delta q_2 + \tilde{q}_2\delta q_1 + \tilde{q}_0\delta q_3 + \tilde{q}_3\delta q_0 \\ &\quad +\delta q_1\delta q_2 + \delta q_0\delta q_3) \\ \Delta C_{22} &= 2(\tilde{q}_0\delta q_0 - \tilde{q}_1\delta q_1 + \tilde{q}_2\delta q_2 - \tilde{q}_3\delta q_3) \\ &\quad +\delta q_0^2 - \delta q_1^2 + \delta q_2^2 - \delta q_3^2 \\ \Delta C_{23} &= 2(\tilde{q}_2\delta q_3 + \tilde{q}_3\delta q_2 - \tilde{q}_0\delta q_1 - \tilde{q}_1\delta q_0 \\ &\quad +\delta q_2\delta q_3 - \delta q_0\delta q_1) \\ \Delta C_{31} &= 2(\tilde{q}_1\delta q_3 + \tilde{q}_3\delta q_1 - \tilde{q}_0\delta q_2 - \tilde{q}_2\delta q_0 \\ &\quad +\delta q_1\delta q_3 - \delta q_0\delta q_2) \\ \Delta C_{32} &= 2(\tilde{q}_0\delta q_1 + \tilde{q}_1\delta q_0 + \tilde{q}_2\delta q_3 + \tilde{q}_3\delta q_2 \\ &\quad +\delta q_0\delta q_1 + \delta q_2\delta q_3) \\ \Delta C_{33} &= 2(\tilde{q}_0\delta q_0 - \tilde{q}_1\delta q_1 - \tilde{q}_2\delta q_2 + \tilde{q}_3\delta q_3) \\ &\quad +\delta q_0^2 - \delta q_1^2 - \delta q_2^2 + \delta q_3^2 \end{aligned} \tag{33}$$

where ΔC_{ij} is the component of ΔC_b^n.

It is possible to obtain the equivalent matrix, ΔC_b^n, which is expressed in terms of $\Delta\Phi^n$, by inserting (6) into (33). we get

$$\Delta C_b^n = 2 \begin{bmatrix} -s_n^2(\Delta\phi_E^2 + \Delta\phi_D^2) & s_n(1+c_n)\Delta\phi_D + s_n^2\Delta\phi_N\Delta\phi_E & -s_n(1+c_n)\Delta\phi_E + s_n^2\Delta\phi_N\Delta\phi_D \\ -s_n(1+c_n)\Delta\phi_D + s_n^2\Delta\phi_N\Delta\phi_E & -s_n^2(\Delta\phi_N^2 + \Delta\phi_D^2) & s_n(1+c_n)\Delta\phi_N + s_n^2\Delta\phi_E\Delta\phi_D \\ s_n(1+c_n)\Delta\phi_E + s_n^2\Delta\phi_N\Delta\phi_D & -s_n(1+c_n)\Delta\phi_N + s_n^2\Delta\phi_E\Delta\phi_D & -s_n^2(\Delta\phi_N^2 + \Delta\phi_E^2) \end{bmatrix} \tilde{C}_b^n. \tag{34}$$

The derived (33) and (34) are the exact trasformation matrix errors for the system in the presence of the large attitude error. The nonlinear models of the velocity error can be obtained by substituting (33) or (34) into (31). Compared with (34), (33) has some advantages in implementing an optimal filter because it has less the high order terms of the error product than (34). On the other hand, (34) has some advantages in analyzing error models.

For the case where the system has the sufficiently small attitude error, (34) is simplified as (35) which is similar to the equivalent tilt angle model presented in [7].

$$\Delta C_b^n = \begin{bmatrix} 0 & -\Delta\phi_D & \Delta\phi_E \\ \Delta\phi_D & 0 & -\Delta\phi_N \\ -\Delta\phi_E & \Delta\phi_N & 0 \end{bmatrix} \tilde{C}_b^n \quad (35)$$

To derive the linear model of the velocity error, $\Delta C_b^n \tilde{f}^b$ which is expressed with the quaternion error can be obtained by using (35) and(10) as follows:

$$\Delta C_b^n \tilde{f}^b = -[\tilde{C}_b^n \tilde{f}^b] X \Delta\Phi^n = -2[\tilde{C}_b^n \tilde{f}^b] X Y^T \delta Q. \quad (36)$$

For the special case of having the large $\Delta\phi_D$, the very small $\Delta\phi_N$, and the very small $\Delta\phi_E$, (34) is approximately simplified as (37) by using (4).

$$\Delta C_b^n = 2 \begin{bmatrix} 2c_n & -s_n\Delta\phi_D & s_n\Delta\phi_E \\ s_n\Delta\phi_D & 2c_n & -s_n\Delta\phi_N \\ -s_n\Delta\phi_E & s_n\Delta\phi_N & 0 \end{bmatrix} \tilde{C}_b^n \quad (37)$$

In this case, $\Delta C_b^n \tilde{f}^b$ is derived by using (8), (12), and (37) as follows:

$$\Delta C_b^n \tilde{f}^b = -2[\tilde{C}_b^n \tilde{f}^b] X Y^T \delta Q + 4 \begin{bmatrix} 1 & 0 & 0 \\ 0 & 1 & 0 \\ 0 & 0 & 0 \end{bmatrix} \tilde{C}_b^n \tilde{f}^b Q^T \delta Q. \quad (38)$$

(38) is the linear error model with the quaternion error, while (37) is the nonlinear error model with the rotation vector error. These results indicate that it is possible to transform the nonlinear error model into the linear error model by using the proposed models and relation equations.

In a similar manner, we can also consider $\Delta\Phi^b$, which is defined as another rotation vector error between an indicated body frame and a true body frame in the local-level frame. The relation equations and the nonlinear error models with respect to $\Delta\Phi^b$ are derived in Appendix A.

5 Conclusions

Nonlinear SDINS error models using the rotation vector error and the quaternion error, respectively, have been presented to represent the nonlinear characteristics of navigation error propagation for the system in the presence of the large navigation errors. To represent the large attitude error of the system, the rotation vector error is introduced and the relationship between the rotation vector error and the quaternion error is derived. The nonlinear attitude error model with the rotation vector error is derived from the conventional linearized quaternion error mode by using the obtained relationship. And nonlinear velocity error models are also derived using the rotation vector error and the quaternion error, respectively. In addition, it is shown that the derived SDINS error models are equivalent to each other. The proposed results will contribute to the analysis of the nonlinear error dynamics and be effective for the implementation of nonlinear filters for the SDINS. And the new approach method presented in this paper can be also extended to the GINS error model.

References

1. A. Weinreb and I. Y. Bar-Itzhack, "The psi-angle error equation in strapdown inertial navigation systems", IEEE Trans. Aerosp. Electron. Syst., AES-14. NO. 3, May 1978, pp.539-542.

2. Britting K. R., *Inertial Navigation Systems Analysis*, Wiley Interscience, New York, 1971.

3. G. M. Siouris, *Aerospace Avionics Systems: A Modern Synthesis*, Academic Press, 1993.

4. D. Goshen-Meskin and I. Y. Bar-Itzhack, "Unified Approach to Inertial Navigation System Error Modeling", J. Guidance, Control, and Dynamics VOL. 15, NO. 3, May-June 1992, pp.648-653.

5. Bernard Friedland, "Analysis Strapdown Navigation Using Quaternions", IEEE Trans. Aerosp. Electron. Syst., AES-14, NO. 5, Sep. 1978, pp.764-768.

6. Minoru Shibata, "Error Analysis Strapdown Inertial Navigation Using Quaternions", J. Guidance, VOL. 9, NO. 3, May-June 1986, pp.379-381.

7. D. Chung, J. G. Lee, C. G. Park, and H. W. Park "Strapdown INS Error Model for Multiposition Alignment", IEEE Trans. Aerosp. Electron. Syst., AES-32, NO. 4, Oct. 1996, pp.1362-1366.

8. S. Vathsal, "Derivation of the Relative Quaternion Differential Equation", J. Guidance, VOL. 14, NO. 5, Sept.-Oct. 1991, pp.1061-1064

Appendix A

The above suggested results are similarly applied to obtain the relation equations and the nonlinear error models associated with $\Delta\Phi^b$. Let $\Delta Q(\equiv Q_b^n - Q_{b'}^n)$ be the quaternion error with respect to $\Delta\Phi^b$. We suppose that the indicated quaternion, $\tilde{Q}_b^n$, can be decomposable as (39).

$$\tilde{Q}_b^n \equiv Q_{b'}^n = Q_b^n Q_{b'}^b \tag{39}$$

where the subscript, b', represents the indicated body frame with the error, $Q_{b'}^b$ denotes the error quaternion involved $\Delta\Phi_b(\equiv [\Delta\phi_X, \Delta\phi_Y, \Delta\phi_Z]^T)$.

Without loss of generality, (39) is converted into (40). And (40) can also be written as (41) based on the quaternion multiplication.

$$Q_b^n = Q_{b'}^n Q_b^{b'} \tag{40}$$

$$Q_b^n = \begin{bmatrix} \tilde{q}_0 & -\tilde{q}_1 & -\tilde{q}_2 & -\tilde{q}_3 \\ \tilde{q}_1 & \tilde{q}_0 & -\tilde{q}_3 & \tilde{q}_2 \\ \tilde{q}_2 & \tilde{q}_3 & \tilde{q}_0 & -\tilde{q}_1 \\ \tilde{q}_3 & -\tilde{q}_2 & \tilde{q}_1 & \tilde{q}_0 \end{bmatrix} \begin{bmatrix} q_{0b} \\ q_{1b} \\ q_{2b} \\ q_{3b} \end{bmatrix} \tag{41}$$

Where

$$\begin{aligned} Q_b^{b'} &= [q_{0b}, q_{1b}, q_{2b}, q_{3b}]^T \\ &= [1 + c_b, s_b\Delta\phi_X, s_b\Delta\phi_Y, s_b\Delta\phi_Z]^T, \end{aligned} \tag{42}$$

$$\begin{aligned} c_b &= cos(\frac{\Phi_{0b}}{2}) - 1, s_n = \frac{1}{\Phi_{0b}} sin(\frac{\Phi_{0b}}{2}), \\ \Phi_{0b} &= \sqrt{(\Delta\phi_X)^2 + (\Delta\phi_Y)^2 + (\Delta\phi_Z)^2}. \end{aligned} \tag{43}$$

In a similar manner, the relationship between ΔQ and $\Delta\Phi^b$ can be obtained as follows:

$$\Delta Q = c_b Q_{b'}^n + s_b U(Q_{b'}^n)\Delta\Phi^b \tag{44}$$

$$\Delta\Phi^b = \frac{1}{s_b} U^T \Delta Q \tag{45}$$

$$\Delta Q = s_b U \Delta\Phi^b + \tilde{Q}_b^n \tilde{Q}_b^{nT} \Delta Q \tag{46}$$

where

$$\begin{aligned} c_b &= cos(\frac{\Phi_{0b}}{2}) - 1 = \tilde{Q}_b^{nT} \Delta Q \\ &= \tilde{q}_0\Delta q_0 + \tilde{q}_1\Delta q_1 + \tilde{q}_2\Delta q_2 + \tilde{q}_3\Delta q_3 \end{aligned} \tag{47}$$

For a sufficiently small attitude error, (45), and (46) are simplified as (48) and (49), respectively. The results are equal to these presented in [5] and [6].

$$\Delta\Phi^b = 2U^T \Delta Q \tag{48}$$

$$\Delta Q = \frac{1}{2} U \Delta\Phi^b + \tilde{Q}_b^n \tilde{Q}_b^{nT} \Delta Q \tag{49}$$

From the obtained results a dual relation between $\Delta\Phi^n$ and $\Delta\Phi^b$ is given by

$$\Delta\Phi^n = \tilde{C}_b^n \Delta\Phi^b. \tag{50}$$

Using (50), the equivalent error models can be easily obtained from the known error models.

In a similar manner, the quaternion error model with respect to $\Delta\Phi^b$ is obtained by replacing δQ by ΔQ in (14). The nonlinear attitude error model for $\Delta\Phi^b$ can be also derived as follows:

Differentiating both sides of (45) and substituting (14) into $\Delta\dot{Q}$, (51) can be obtained.

$$\begin{aligned} \Delta\dot{\Phi}^b &= \frac{1}{s_b} U^T \Delta\dot{Q} + \frac{1}{s_b}\dot{U}^T \Delta Q + \frac{d}{dt}(\frac{1}{s_b}) U^T \Delta Q \\ &= [U^T M U + \dot{U}^T U + \frac{d}{dt}(\frac{1}{s_b}) s_b U^T U]\Delta\Phi_b \\ &+ \frac{c_b}{s_b}[U^T M \tilde{Q}_b^n + \dot{U}^T \tilde{Q}_b^n + \frac{d}{dt}(\frac{1}{s_b}) s_b U^T \tilde{Q}_b^n] \\ &+ \frac{1}{2s_b} U^T [U\delta\omega_{ib}^b - Y\delta\omega_{in}^n]. \end{aligned} \tag{51}$$

And the following relationships can also be derived.

$$U^T U = I, \tag{52}$$

$$U^T Y = \tilde{C}_n^b, \tag{53}$$

$$\frac{d}{dt}(\frac{1}{s_b}) s_b \Delta\Phi^b = d\Delta\Phi^b (\Delta\Phi^b)^T \Delta\dot{\Phi}^b, \tag{54}$$

$$d = \frac{1}{\Phi_{0b}^2}(1 - \frac{\Phi_{0b}}{2} cot(\frac{\Phi_{0b}}{2})), \tag{55}$$

$$U^T M U + \dot{U}^T U = -\omega_{ib}^b X, \tag{56}$$

$$U^T M \tilde{Q}_b^n + \dot{U}^T \tilde{Q}_b^n = 0, \tag{57}$$

$$U^T \tilde{Q}_b^n = 0 \tag{58}$$

Rearranging each term on the right hand side of (51) using the above results, the nonlinear attitude error model for $\Delta\Phi^b$ is given by

$$\begin{aligned} \Delta\dot{\Phi}^b &= [I - d\Delta\Phi^b(\Delta\Phi^b)^T]^{-1}[-\omega_{ib}^b X \Delta\Phi^b \\ &- \frac{1}{2s_b}[\delta\omega_{ib}^b - \tilde{C}_n^b \delta\omega_{in}^n]]. \end{aligned} \tag{59}$$

From a Taylor series to $[I - d\Delta\Phi^b(\Delta\Phi^b)^T]^{-1}$ (59) becomes

$$\begin{aligned} \Delta\dot{\Phi}^b &= -\omega_{ib}^b X \Delta\Phi^b + \frac{1}{2s_b}[I \\ &+ d\Delta\Phi^b(\Delta\Phi^b)^T][\delta\omega_{ib}^b - \tilde{C}_n^b \delta\omega_{in}^n]. \end{aligned} \tag{60}$$

Supposing that the attitude error is very small and neglecting the second order terms of the error product, (60) can be simplified as

$$\Delta\dot{\Phi}^b = -\omega_{ib}^b X \Delta\Phi^b + [\delta\omega_{ib}^b - \tilde{C}_n^b \delta\omega_{in}^n]. \tag{61}$$

(61) is identical to the equivalent tilt angle model presented by Shibata in [6]. But like the case discussed previously, their equivalent relationship is

only available for the case where the attitude error is small enough. And the sign of the forcing term in (61) differs from that of the model given in [6], because the direction of the rotation vector error is oppositely defined each other. As mentioned above, because the sign of the forcing term of the attitude error models in SDINS can be changed according to the definition of the navigation errors, the rotation vector error or both, its sign becomes unimportant.

The transformation matrix error with ΔQ is also easily obtained by substituting these of ΔQ into the elements of δQ in (33). And the transformation matrix error can be also derived in terms of $\Delta\Phi^b$ using $\Delta Q (\equiv Q_b^n - Q_{b'}^n)$, (33), and (44), as follows:

$$\Delta C_b^n = 2\tilde{C}_b^n \begin{bmatrix} -s_b^2(\Delta\phi_Y^2 + \Delta\phi_Z^2) & s_b(1+c_b)\Delta\phi_Z + s_b^2\Delta\phi_X\Delta\phi_Y & -s_b(1+c_b)\Delta\phi_Y + s_b^2\Delta\phi_X\Delta\phi_Z \\ -s_b(1+c_b)\Delta\phi_Z + s_b^2\Delta\phi_X\Delta\phi_Y & -s_b^2(\Delta\phi_X^2 + \Delta\phi_Z^2) & s_b(1+c_b)\Delta\phi_X + s_b^2\Delta\phi_Y\Delta\phi_Z \\ s_b(1+c_b)\Delta\phi_Y + s_b^2\Delta\phi_X\Delta\phi_Z & -s_b(1+c_b)\Delta\phi_X + s_b^2\Delta\phi_Y\Delta\phi_Z & -s_b^2(\Delta\phi_X^2 + \Delta\phi_Y^2) \end{bmatrix}. \tag{62}$$

The nonlinear models of the velocity error can be obtained by combining the above results with (31). For a sufficiently small attitude error, (62) is simplified as

$$\Delta C_b^n = \tilde{C}_b^n \begin{bmatrix} 0 & -\Delta\phi_Z & \Delta\phi_Y \\ \Delta\phi_Z & 0 & -\Delta\phi_X \\ -\Delta\phi_Y & \Delta\phi_X & 0 \end{bmatrix}. \tag{63}$$

This is equal to the equivalent tilt model presented in [6] as well as the conventional model. The linear velocity error model given in terms of the quaternion error, ΔQ, can be also derived using (30), (48), and (63). In this case, $\Delta C_b^n \tilde{f}^b$ is obtained as follows:

$$\Delta C_b^n \tilde{f}^b = -\tilde{C}_b^n [\tilde{f}^b] \times \Delta\Phi^b = -2\tilde{C}_b^n [\tilde{f}^b] \times U^T \Delta Q. \tag{64}$$

Even though the results presented above have derived for SDINS in the local-level frame, the new approach method used to derive the relationship between the attitude errors and the nonlinear error models can be similarly applied to the error model developed by Friedland in the inertial frame in [5].

Optimal Normalization Scheme for Computed Quaternion Using Optimal Regulator Theory

Der-Ren Taur* and Jeng-Shing Chern**
Chung Shan Institute of Science and Technology
P.O. BOX 90008-6-7
Lungtan, Taiwan 32526,ROC

I . Introduction

The quaternion are widely used in aerospace navigation applications[1].The constraint of its norm can be satisfied by adjoining the gradient of the square of the constraint directly into the governing equations[2,3] without invoking renormalization scheme [4,5]. In Ref.6, a finite-time optimal terminal controller has been employed to obtain the normalization scheme of a computed quaternion. However, the implementation is not very practical for application since a matrix Riccati equation has to be solved on-line.

In this study, an infinite-time optimal regulator problem has been solved to obtain an optimal normalization scheme for a computed quaternion and to satisfy its norm constraint. This new scheme is easier to implement and is much more practical for real-time applications.

II . Governing Equations

Assuming that the true quaternion vector $\bar{p}_T$ propagate according to the following equations of motion.

$$\dot{\bar{p}}_T = 1/2L(\bar{p}_T)\begin{bmatrix}0 & W_1 & W_2 & W_3\end{bmatrix}^T \quad (1)$$

where W_1 , W_2 and W_3 are the projections of the body rate $\vec{W}$ on the body reference axes. The quaternion vector $\bar{p}$ and the orthonormal matrix L are defined as:

$$\bar{p} = \begin{pmatrix}e_0 & e_1 & e_2 & e_3\end{pmatrix}^T \quad (2)$$

$$L(\bar{p}) = \begin{bmatrix} e_0 & -e_1 & -e_2 & -e_3 \\ e_1 & e_0 & -e_3 & e_2 \\ e_2 & e_3 & e_0 & -e_1 \\ e_3 & -e_2 & e_1 & e_0 \end{bmatrix} \quad (3)$$

Besides, the true quaternion vector $\bar{p}_T$ satisfies the following norm constraint

$$N^2(\bar{p}_T) = 1 \quad (4)$$

On the other hand, the computed quaternion vector $\bar{p}_c$ is governed by the following differential system.

$$\dot{\bar{p}}_c = 1/2L(\bar{p}_c)\begin{bmatrix}0 & W_1 & W_2 & W_3\end{bmatrix}^T \quad (5)$$

However, due to the propagation errors it is desired to add a certain control vector $\bar{u}$ into Eq.(5) to enforce the error quaternions $\delta\bar{p} = \bar{p}_T - \bar{p}_c$ to zero[7]. Thus the governing equations of the computed quaternions can be rewritten as:

$$\dot{\bar{p}}_c = 1/2L(\bar{p}_c)\begin{bmatrix}0 & W_1 & W_2 & W_3\end{bmatrix}^T + \bar{u} \quad (6)$$

---

*Associate Scientist, Member AASRC
**Senior Scientist, Associate Fellow AIAA, permanent Member AASRC, Member AIAA Committee on Standards

Accordingly, the equations of motion for the error quaternions $\delta\bar{p}$ are given by

$$\delta\dot{\bar{p}} = 1/2A(\bar{W})\delta\bar{p} + B\bar{u} \quad (7)$$

where the plant matrices A and B are defined as:

$$A(\bar{W}) = \begin{bmatrix} 0 & -W_1 & -W_2 & -W_3 \\ W_1 & 0 & W_3 & -W_2 \\ W_2 & -W_3 & 0 & W_1 \\ W_3 & W_2 & -W_1 & 0 \end{bmatrix}$$

$$B = \begin{bmatrix} -1 & 0 & 0 & 0 \\ 0 & -1 & 0 & 0 \\ 0 & 0 & -1 & 0 \\ 0 & 0 & 0 & -1 \end{bmatrix} \quad (8)$$

In fact, Eq.(7) is a linear time-varying system since the body rate is a function of time.

III . Infinite-Time Optimal Regulator Problem

The control $\bar{u}$ can be designed optimally by minimizing a quadratic performance index of the type given below[8]:

$$J(\bar{u}) = 1/2\int_0^\infty [\delta\bar{p}^T Q\delta\bar{p} + \bar{u}R\bar{u}]dt \quad (9)$$

where the weighting matrices Q and R are positive definite and are defined as

$$Q = CI_{4\times4}$$
$$R = I_{4\times4} \quad (10)$$

and C is a positive constant.

Assuming that the optimal control $\bar{u}^*$ can be expressed as:

$$\bar{u}^* = -K\delta\bar{p} \quad (11)$$

where K is a constant gain matrix and can be obtained by solving an associated parameter optimization problem[9].

By substituting Eq.(11) into Eq.(7), we have the closed loop state equation

$$\delta\dot{\bar{p}} = (1/2A - BK)\delta\bar{p} \quad (12)$$

Since $[A,B]$is uniformly controllable and $[A,Q^{1/2}]$is uniformly observable, the closed loop system in Eq.(12) is uniformly asymptotically stable[8,10-12].

By substituting Eq.(11) into Eq.(9), we obtain

$$J = 1/2\int_0^\infty (\delta\bar{p}^T Q\delta\bar{p} + \delta\bar{p}^T K^T RK\delta\bar{p})dt$$
$$= 1/2\int_0^\infty \delta\bar{p}^T (Q + K^T RK)\delta\bar{p}dt \quad (13)$$

For a stable system a Liapunov function exists. Let us define $\delta\bar{p}^T S\delta\bar{p}$ be a Liapunov function and the matrix S be a constant non-negative definite symmetric matrix [11,12]. Then we may set

$$\delta\bar{p}(Q + K^T RK)\delta\bar{p} = -d/dt(\delta\bar{p}^T S\delta\bar{p}) \quad (14)$$

From Eqs.(12) and (14), it results that

$$\delta\bar{p}^T (Q + K^T RK)\delta\bar{p} =$$
$$-\delta\bar{p}^T[(1/2A - BK)^T S + S(1/2A - BK)]\delta\bar{p} \quad (15)$$

Since Eq.(15) must hold true for arbitrary $\delta\bar{p}$,we have

$$(1/2A - BK)^T S + S(1/2A - BK)$$
$$= -(Q + K^T RK) \quad (16)$$

We may decompose the positive definite matrix R into the form:

$$R = D^T D \quad (17)$$

where D is a nonsingular matrix. Then, Eq.(16) can be written as:

$$1/2\,A^T S + 1/2\,SA - SBR^{-1}B^T S + Q + [DK-(D^T)^{-1}B^T S]^T[DK-(D^T)^{-1}B^T S] = 0 \tag{18}$$

The minimization of index J in Eq.(13) with respect to the parameter K requires the minimization of the left-hand side of Eq.(18) with respect to K. The solution is

$$K = R^{-1}B^T S \tag{19}$$

Hence, the optimal control law is given by

$$\vec{u}^{*} = -K\delta\vec{P} = -R^{-1}B^T S\delta\vec{P} \tag{20}$$

The substitution of Eq.(19) into Eq.(18) gives the following algebraic Riccati equation (ARE)[11,12]

$$1/2\,A^T S + 1/2\,SA - SBR^{-1}B^T S + Q = 0 \tag{21}$$

We can use Eq.(21) to solve for the non-negative definite matrix S.

Using Eqs.(8) and (10) into Eq.(21), the ARE can be rewritten as

$$1/2\,A^T S + 1/2\,SA - S^2 + CI = 0 \tag{22}$$

where the symmetric matrix S is defined as

$$S = \begin{bmatrix} S_{11} & S_{12} & S_{13} & S_{14} \\ S_{12} & S_{22} & S_{23} & S_{24} \\ S_{13} & S_{23} & S_{33} & S_{34} \\ S_{14} & S_{24} & S_{34} & S_{44} \end{bmatrix} \tag{23}$$

By substituting Eqs.(8) and (23) into Eq.(22), the following algebraic relations can be obtained

$$(W_1S_{12} + W_2S_{13} + W_3S_{14})$$

$$-(S_{11}{}^2 + S_{12}{}^2 + S_{13}{}^2 + S_{14}{}^2) + C = 0$$

$$1/2\,W_1(S_{22} - S_{11}) + 1/2\,W_2(S_{14} + S_{23}) + 1/2\,W_3(S_{24} - S_{13})$$

$$-(S_{11}S_{12} + S_{22}S_{12} + S_{23}S_{13} + S_{14}S_{24}) = 0$$

$$1/2\,W_1(S_{23} - S_{14}) + 1/2\,W_2(S_{33} - S_{11})$$

$$+1/2\,W_3(S_{34} + S_{12})$$

$$-(S_{13}S_{11} + S_{23}S_{12} + S_{33}S_{13} + S_{34}S_{14}) = 0$$

$$1/2\,W_1(S_{24} + S_{13}) + 1/2\,W_2(S_{34} - S_{12})$$

$$+1/2\,W_3(S_{44} - S_{11})$$

$$-(S_{14}S_{11} + S_{24}S_{12} + S_{34}S_{13} + S_{44}S_{14}) = 0$$

$$(-W_1S_{12} + W_2S_{24} - W_3S_{23}) -$$

$$(S_{12}^2 + S_{22}^2 + S_{23}^2 + S_{24}^2) + C = 0$$

$$-1/2\,W_1(S_{13} + S_{24}) + 1/2\,W_2(S_{34} - S_{12})$$

$$+1/2\,W_3(S_{22} - S_{33})$$

$$-(S_{13}S_{12} + S_{23}S_{22} + S_{33}S_{23} + S_{34}S_{24}) = 0$$

$$(-W_1S_{34} - W_2S_{13} + W_3S_{23})$$

$$-(S_{13}^2 + S_{23}^2 + S_{33}^2 + S_{34}^2) + C = 0$$

$$1/2\,W_1(S_{23} - S_{14}) + 1/2\,W_2(S_{44} - S_{22})$$

$$-1/2\,W_3(S_{12} + S_{34})$$

$$-(S_{14}S_{12} + S_{24}S_{22} + S_{34}S_{23} + S_{44}S_{24}) = 0$$

$$1/2\,W_1(S_{33} - S_{44}) - 1/2\,W_2(S_{23} + S_{14})$$

$$+1/2\,W_3(S_{24} - S_{13})$$

$$-(S_{14}S_{13} + S_{24}S_{23} + S_{34}S_{33} + S_{44}S_{34}) = 0$$

$$(W_1S_{34} - W_2S_{24} - W_3S_{14}) -$$

$$(S_{14}^2 + S_{24}^2 + S_{34}^2 + S_{44}^2) + C = 0 \tag{24}$$

Since W_1, W_2 and W_3 in Eq.(24) are arbitrary, their coefficients must be equal to zero. Thus we have the following equality relations

$$S_{12} = S_{13} = S_{14} = 0$$

$$S_{23} = S_{24} = S_{34} = 0 \tag{25}$$

$$S_{11} = S_{22} = S_{33} = S_{44} = C^{\frac{1}{2}}$$

In other words, the unique solution to the ARE is given by

$$S = \begin{bmatrix} 1 & 0 & 0 & 0 \\ 0 & 1 & 0 & 0 \\ 0 & 0 & 1 & 0 \\ 0 & 0 & 0 & 1 \end{bmatrix} = C^{\frac{1}{2}} I_{4\times4} \tag{26}$$

IV . Optimum Normalization Scheme

By substituting Eqs.(20) and (26) into Eq.(6), we obtain the new equations of motion for the computed quaternions :

$$\dot{\bar{p}}_c = 1/2L(\bar{p}_c)\begin{bmatrix} 0 & W_1 & W_2 & W_3 \end{bmatrix}^T$$
$$+ C^{\frac{1}{2}}(\bar{P}_T - \bar{P}_C) \tag{27}$$

Now we can define the true quaternion vector $\bar{p}_T$ as follows [4,5]

$$\bar{p}_T = \bar{p}_c / N(\bar{p}_c) \tag{28}$$

Then Eq.(27) can be rewritten as

$$\dot{\bar{p}}_c = 1/2L(\bar{p}_c)\begin{bmatrix} 0 & W_1 & W_2 & W_3 \end{bmatrix}^T$$
$$+ C^{\frac{1}{2}}\left(\frac{1 - N(\bar{p}_c)}{N(\bar{p}_c)}\right)(\bar{p}_c) \tag{29}$$

Or, the optimal normalization scheme for a computed quaternion to satisfy its norm constraint is given by

$$\dot{\bar{p}}_c = \left[1/2A(\vec{W}) + C^{\frac{1}{2}}\left(\frac{1 - N(\bar{p}_c)}{N(\bar{p}_c)}\right)I_{4\times4}\right]\bar{p}_c \tag{30}$$

The above algorithm can be easily implemented for real time on-line computation.

V. Simulated Results

The algorithm in Eq.(30) can be rewritten in component form as follows

$$\dot{e}_0 = -\frac{1}{2}(e_1W_1 + e_2W_2 + e_3W_3) + C_1e_0$$

$$\dot{e}_1 = \frac{1}{2}(e_0W_1 + e_2W_3 - e_3W_2) + C_1e_1$$

$$\dot{e}_2 = \frac{1}{2}(e_0W_2 - e_1W_3 + e_3W_1) + C_1e_2 \tag{31}$$

$$\dot{e}_3 = \frac{1}{2}(e_0W_3 + e_1W_2 - e_2W_1) + C_1e_3$$

where

$$C_1 = C^{\frac{1}{2}}(1 - N(\bar{p}_c))/N(\bar{p}_c)$$

$$N(\bar{p}_c) = (e_0^2 + e_1^2 + e_2^2 + e_3^2)^{\frac{1}{2}} \tag{32}$$

To evaluate the performance of this new algorithm, Eqs.(31) and (32) have been simulated assuming different $\delta\bar{p}(t_0)$ and using sinusoidal body rates expressed in all three axes of an Earth-pointing satellite. The body rates can be presented as follows[6]:

$$W_1 = \frac{1}{120}\frac{\pi}{180}\sin(10t)$$

$$W_2 = \frac{1}{15}\frac{\pi}{180} + \frac{1}{120}\frac{\pi}{180}\sin(10t + \frac{2\pi}{3})$$

$$W_3 = \frac{1}{120}\frac{\pi}{180}\sin(10t + \frac{4\pi}{3}) \qquad (33)$$

In all the cases simulated, the quantity $\varepsilon = 1 - N^2(\bar{p}_c)$ has been found to converge to a very small quantity of the order of 10^{-13} from an initially large value of the order of $|\varepsilon|$=0.3. The converging speed is a function of the positive constant gain factor C in Eq.(32), the larger the gain factor is the faster the converging rate. For example, when C=16 the converging time is less than 6.7 sec, but when C=64 the converging time reduces to 3.4 sec. Actually, these numerical results are quite close to the previously proposed normalization scheme with derivative constraint.

VI. Conclusions

A new optimal normalization scheme for the computed quaternions has been obtained by using the infinite-time optimal regulator theory. The infinite-time regulator problem results in a steady state optimal solution to alleviate the necessity of solving a matrix Riccati equation on-line for the computation of the optimal gain matrix. This new scheme is easy to implement and practical for application .It also provides a rigorous theoretical foundation for the derivative constraint in the previously proposed normalization scheme for the computed quaternions.

References

1. Lefferts, E.J., Markley,F.L., and Shuster, M.D.," Kalman Filtering for Spacecraft Attitude Estimation," *Journal of Guidance, Control, and Dynamics*,Vol.5, Sept-Oct, 1982, PP. 417-429
2. Robinson, A.C., On the Use of Quaternions in Simulation of Rigid Body Motion, WADC TR 58-17, Decem 1958.
3. Mitchel, E.E., and Rogers, A.E., "Quaternion parameters in the Simulation of Spinning Body," *Simulation*, Vol.4, No.6, June 1965, PP.390-396
4. Bar-Itzhack, I.Y., "Optimum Normalization of a Computed Quaternion of Rotation," *IEEE Transaction on Aerospace and Electronic systems*, March 1976 PP. 401-402.
5. Giardina, C.R., Bronson, R., and Wallen, L., "An Optimal Normalization scheme," *IEEE Transactions on Aerospace and Electronic Systems*, "Vol. AES-11, July 1975, PP. 443-446.
6. Vathsal, S., "Optimal Control of Quaternion Propagation Errors in Spacecraft Navigation," *Journal of Guidance, Control, and Dynamics*, Vol. 9, No.3, May 1986, PP. 382-384.
7. Friedland, B., "Analysis Strapdown Navigation Using Quaternions," *IEEE Transactions on Aerospace and Electronic systems*, Vol. AES-14, Sept. 1978, PP. 764-768.
8. Anderson, B.D.O., and More, J.B., Optimal Control, Prentice Hall, 1989 , pp.35-67.
9. Schultz, D.G., and Melsa, J.L., State Functions and Lineal Control System, Mcrraw-hill, 1967, pp.275-280.
10. Silverman, L.M., and Anderson, B.D.O., "Controllability, Observability and Stability of Linear System," *SIAM J. control*, Vol.6.

No.1, 1968, PP. 121-130.

11. Kalman, R.E., "Contributions to the theory of Optimal Control. "*Bol. Soc. Mat.* Mexicana (2), 5 (1960), PP. 102-119.

12. Kwakervaak, H., and Sivan, R., Linear Optimal Control system, Wiley-Interscience, 1972, pp.220-236.

Estimation of Target Lateral Acceleration for Augmented Proportional Navigation.

B. Boberg
J. Lundberg
J. Wiberg
S-L Wirkander
Defence Research Establishment (FOA),
172 90 Stockholm, Sweden.
Email: boberg@sto.foa.se wirkander@sto.foa.se

Extending the ordinary Proportional Navigation control law (PN) for a homing missile to the so called Augmented Proportional law (APN) can reduce miss distances, e.g. if there are limitations in realizable lateral acceleration for the missile. However, APN requires knowledge about the target's lateral acceleration.

This paper is an investigation and comparison of four different algorithms for estimation of the lateral acceleration of the target for a homing missile. These algorithms are:

Kalman estimator **An ordinary linear Kalman filter with states describing relative position, velocity, and target acceleration.**

Two Stage Kalman estimator **A linear Kalman filter which basically contains states only for relative position and velocity, whose innovations are used to separately estimate the target acceleration.**

Interactive Multiple Model estimator, IMM **Two Kalman filters, each one working on its own system model. One of the models contains only position and velocity states, the other one also a target acceleration state. Probabilities for each model to be true are continuously maintained, and the target acceleration is estimated as the second model's acceleration estimate, weighted with the corresponding probability.**

Interactive Acceleration Compensation estimator, IAC **A combination of the Two-Stage and the IMM algorithm. The purpose is to get an algorithm with smaller computational requirements than IMM.**

Introduction

Proportional Navigation (PN) has been used for more than four decades, in radar as well as in infrared homing missiles, because of its simplicity and efficiency. During the sixties, control theory was used for deriving guidance laws in closed form for homing missiles. It then turned out that PN is the optimal solution to the problem of minimizing the quadratic integral of the missile's lateral acceleration when intercepting a non-manœuvring target, under the assumption of no dynamics in the missile.

Extending PN to APN

The simplicity of PN lies in the fact that only measurements of the rate of change in line-of-sight angle from the missile to the target relative to a fixed reference are needed, $\dot{\sigma}$, Figure 1. The commanded missile lateral acceleration is then expressed as

$$n_C = NV_C\dot{\sigma}, \quad (1)$$

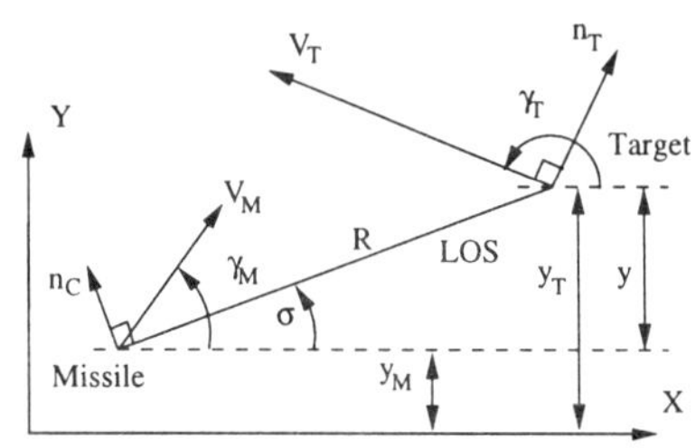

Figure 1 ***Geometry of the planar tracking problem***

where
σ = line-of-sight angle,
$V_C = -\dot{R}$ = closing velocity,
N = navigation constant, $3 \le N \le 5$.

In order to motivate the two guidance laws, we observe that for small σ we have approximately $\sigma = y/R$, Figure 1, which gives, after differentiation and inserting into the PN guidance law:

$$n_C t_{go}^2/2 = N(y + \dot{y}\, t_{go})/2, \quad (2)$$

where
$t_{go} = \frac{R}{V_C}$ = time to go, and
y = target-missile distance in the Y-direction.

The left hand side of (2) is the vertical displacement during the remaining flight time, t_{go}, as a result of the commanded acceleration n_C. The expression inside the parenthesis of the right hand side is called the "zero effort miss distance", which is a linear prediction of what the miss distance in the y-direction would become if the missile kept its lateral acceleration zero, $n_C = 0$. If $N = 2$, the PN control law would try to catch up with the target exactly during the time remaining, t_{go}. When $N > 2$ the missile gets a certain lead angle when using PN. A natural extension of the PN law would be to add a second order term in the zero effort miss distance expression, taking into account that the target manœuvres with constant lateral acceleration n_T. The new second order expression for the zero effort miss distance would then be $y + \dot{y}\, t_{go} + n_T \frac{t_{go}^2}{2}$, and this gives rise to the APN guidance law, which, as is seen by combining (1) and (2), then becomes

$$n_C = N V_C \dot{\sigma} + n_T N/2. \tag{3}$$

Effects of Dynamics and Limitations

We know so far[1] that the two guidance laws described above (PN and APN) both will hit a constantly manœuvring target if

1. the missile has no dynamics, i.e. the commanded lateral acceleration is effectuated instantly, and

2. there are no limitations in the realizable acceleration.

Furthermore, APN is optimal if these conditions are satisfied. If they are not, we will in general get non-zero miss distance for both methods. However, it turns out that in a case with both missile dynamics and acceleration limitations, APN will give better performance in terms of miss distance. In Figure 2 the miss distances as functions of total flight time (t_F) are shown for a 1:st order missile dynamics (time constant, $\tau = 1$ *sec*). The different plots correspond to different degrees of limitation in n_C. In Figure 2 a) there is no limitation at all, in Figure 2 b) and 2 c) the limitations are 25g and 10g, respectively. From these plots we observe three things:

1. APN gives smaller miss distances than PN for short flight times, in these cases at least for $0 < t_F < 2$ s.

2. The flight time interval in which APN gives smaller miss distances than PN increases with decreasing limits in n_C.

3. The relative (and absolute) differences in miss distances between the two methods also increase with decreasing limits in n_C (of course the absolute miss distances are increasing with decreasing limits also for APN).

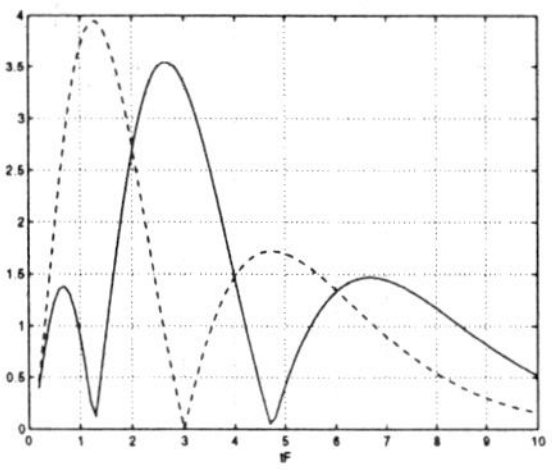

a) Unlimited lateral acceleration n_C

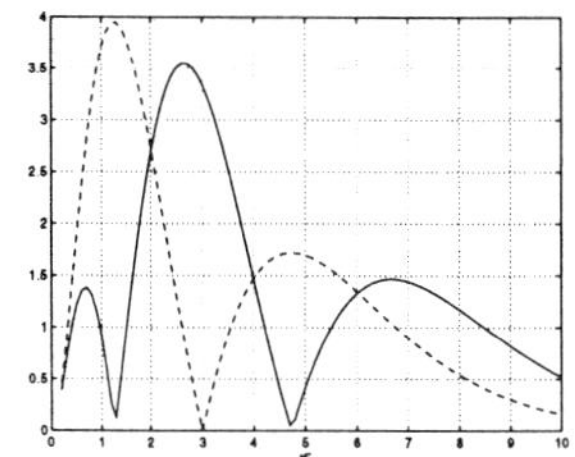

b) $n_{C,max} = 25g$

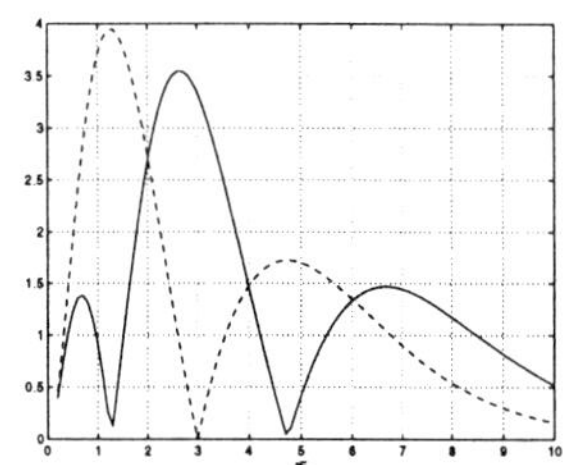

c) $n_{C,max} = 10g$

Figure 2 Miss distance for PN (dashed line) and APN (solid line) plotted against flight time using a 1:st order missile dynamics.

The explanation to these differences is, loosely speaking, that APN needs less control lateral acceleration (n_C) than PN and is thus less prone to hit the limit.

Estimation of Target Acceleration

In order to use the Augmented Proportional Navigation guidance law, a sufficiently accurate estimate of target lateral accelerations is needed. The most straightforward way to achieve this is to construct a model of the total missile-target system, including the target acceleration, and apply a Kalman estimator to it with states describing relative positions and velocities as well as the target accelerations.

If we know that the target does not accelerate, this estimator will, if it is working well, give an acceleration estimate which is near zero. In this case, which is a very common one, it would however be nice to be able to reduce the dimension of the filter by excluding the target acceleration state from it, only keeping the states corresponding to the position and velocity. This is because it is always a waste of information to estimate states whose values are known, the estimation quality of the remaining states being unnecessarily reduced. For this purpose the so called *Two Stage Kalman estimator*[2,3] can be used, where the position/velocity estimates are first calculated as if there were no accelerations. It requires the ability to independently detect presence and absence of target accelerations, e.g. by analyzing the innovations of the estimator without accelerations. If the accelerations are detected in this way, they can be estimated afterwards, and can also be used to correct the position/velocity estimates.

Another way to avoid estimating target accelerations which are zero is to construct two models, one low-order, corresponding to the positions and velocities only, and one high-order, which includes also the accelerations. The *Interactive Multiple Model algorithm* (IMM)[4,5] keeps Kalman filters running on both these models simultaneously, using the innovations in order to continuously compute the probability for each model to be in effect. These probabilities replace the decision mechanism used in the Two Stage estimator, and optimal estimates of all states, including the accelerations, are calculated as the sums of the estimates from the models, weighted with the respective probabilities.

Finally, the *Interactive Acceleration Compensation algorithm* (IAC)[6] is a combination of IMM and the Two Stage estimator, in that it estimates the position and velocity states as if there were no acceleration (like the Two Stage estimator) and calculates continuously the probability of acceleration (like the IMM estimator). Then it finally calculates the acceleration estimates and corrects the position/velocity estimates in proportion to the probability for acceleration. This eliminates the need for an explicit acceleration detector (like in the Two Stage estimator), but does not require two full models running in parallel.

These four algorithms will now be tested in the context of a homing missile using Augmented Proportional Navigation towards a target that switches between manœuvring and non-manœuvring mode. In section the algorithms will be described in general, and in section they will be applied to a particular missile-target system and simulation results will be shown. Finally, section contains conclusions and suggestions for future research.

Description Of The Estimation Algorithms

System Description

Consider a linear missile-target system whose dynamics is modeled by

$$\begin{aligned} X_{k+1} &= A_k X_k + B_k b_k + w_k^X, \\ b_{k+1} &= b_k + w_k^b, \end{aligned} \tag{4}$$

where X_k is the system state, containing position and velocity information, b_k is a bias vector with accelerations, and w_k^X and w_k^b are white Gaussian sequences with zero means and variances given by

$$\begin{aligned} E\left[w_k^X (w_l^X)^T\right] &= Q_k^X \delta_{kl}, \\ E\left[w_k^X (w_l^b)^T\right] &= Q_k^{Xb} \delta_{kl}, \\ E\left[w_k^b (w_l^b)^T\right] &= Q_k^b \delta_{kl}. \end{aligned} \tag{5}$$

The measurement vector at time k consists of position and velocity information according to

$$z_k = H_k X_k + n_k, \tag{6}$$

where the measurement noise process n_k is a white Gaussian sequence with zero mean and variance

$$E\left[n_k n_l^T\right] = R_k \delta_{kl}. \tag{7}$$

Kalman estimator

Using (4) - (7), an augmented state Kalman filter can be constructed to estimate the state and bias in an optimal way. We present the estimator in a partitioned form where the state vector is divided into an "X" part with states defining the positions and velocities, and a "b" part with accelerations, consistent with the system description. The algorithm corresponding to the "X" part will then be

$$\begin{aligned} X_{k|k-1} &= A_{k-1} X_{k-1|k-1} + B_{k-1} b_{k-1|k-1}, \\ P^X_{k|k-1} &= A_{k-1} P^X_{k-1|k-1} A^T_{k-1} \\ &\quad + B_{k-1} P^b_{k-1|k-1} B^T_{k-1} \\ &\quad + B_{k-1} (P^{Xb}_{k-1|k-1})^T A^T_{k-1} \\ &\quad + A_{k-1} P^{Xb}_{k-1|k-1} B^T_{k-1} + Q^X_{k-1}, \\ K^X_k &= P^X_{k|k-1} H^T_k [H_k P^X_{k|k-1} H^T_k + R_k]^{-1}, \\ \nu_k &= z_k - H_k X_{k|k-1}, \\ X_{k|k} &= X_{k|k-1} + K^X_k \nu_k, \\ P^X_{k|k} &= (I - K^X_k H_k) P^X_{k|k-1}, \end{aligned} \tag{8}$$

and for the "b" we have

$$\begin{aligned} b_{k|k-1} &= b_{k-1|k-1}, \\ P^b_{k|k-1} &= P^b_{k-1|k-1} + Q^b_{k-1}, \\ P^{Xb}_{k|k-1} &= A_{k-1} P^{Xb}_{k-1|k-1} + B_{k-1} P^b_{k-1|k-1} + Q^{Xb}_{k-1}, \\ K^b_k &= P^{Xb}_{k|k-1} H^T_k [H_k P^X_{k|k-1} H^T_k + R_k]^{-1}, \\ b_{k|k} &= b_{k|k-1} + K^b_k \nu_k, \\ P^b_{k|k} &= -K^b_k H_k P^{Xb}_{k|k-1} + P^b_{k|k-1}, \\ P^{Xb}_{k|k} &= (I - K^X_k H_k) P^{Xb}_{k|k-1}. \end{aligned} \tag{9}$$

The Two Stage Kalman Estimator

The estimation of the missile-target motion can be separated into two stages, as introduced in.[3] The first stage, considering the system (4), consists of estimating positions and velocities while ignoring the bias vector, i.e. assuming that the velocity is constant. In the second stage the bias, i.e. the target acceleration, is estimated using the innovations from the first stage. Given a certain algebraic constraint, the optimal estimate of the system state can be obtained as a linear combination of the outputs from the first and the second stage. Note that this combination needs to be done only when a significant acceleration has been detected. The detector is preferrably based on the squared and normalized innovations from the constant velocity filter, $(\bar{\nu}_k^X)^T [Cov(\bar{\nu}_k^X)]^{-1} \bar{\nu}_k^X$. If $Q_k^b = 0$, i.e. the bias is constant (but unknown), then the Two Stage estimator is equivalent to the Kalman estimator, as shown in.[2] In[7,3] the bias is assumed to be corrupted by a white noise process which may be correlated with the state input process. This gives an optimal two-stage estimator only if a certain algebraic relation holds between the cross covariance of state noise and bias noise and the bias noise itself. As this condition is practically never satisfied, the Two Stage estimator is a suboptimal (i.e. non-optimal) filter.

The system defined by (4) - (7) will now be used in the presentation of the Two Stage estimator. Beginning with the first stage, the bias b_k is set equal to zero, in order to obtain the constant velocity or "bias free" estimator:

$$\begin{aligned} \bar{X}_{k|k-1} &= A_{k-1}\bar{X}_{k-1|k-1}, \\ U_k &= A_{k-1}V_{k-1} + B_{k-1}, \\ \bar{Q}^X_{k-1} &= Q^X_{k-1} - U_k Q^b_{k-1} U_k^T, \\ \bar{P}^X_{k|k-1} &= A_{k-1}\bar{P}^X_{k-1|k-1}A^T_{k-1} + \bar{Q}^X_{k-1}, \\ \bar{K}^X_k &= \bar{P}^X_{k|k-1}H_k^T[H_k\bar{P}^X_{k|k-1}H_k^T + R_k]^{-1}, \\ \bar{\nu}^X_k &= z_k - H_k\bar{X}_{k|k-1}, \\ \bar{X}_{k|k} &= \bar{X}_{k|k-1} + \bar{K}^X_k\bar{\nu}^X_k, \\ \bar{P}^X_{k|k} &= (I - \bar{K}^X_k H_k)\bar{P}^X_{k|k-1}. \end{aligned} \tag{10}$$

The next step in the design is to build a bias estimator based upon the innovations $\bar{\nu}^X_k$ from the first stage filter.

$$\begin{aligned} \bar{b}_{k|k-1} &= \bar{b}_{k-1|k-1}, \\ S_k &= H_k U_k, \\ \bar{P}^b_{k|k-1} &= \bar{P}^b_{k-1|k-1} + Q^b_k, \\ \bar{K}^b_k &= \bar{P}^b_{k|k-1}S_k^T \times \\ &\quad [S_k\bar{P}^b_{k|k-1}S_k^T + H_k\bar{P}^X_{k|k-1}H_k^T + R_k]^{-1}, \\ \bar{\nu}^b_k &= \bar{\nu}^X_k - S_k\bar{b}_{k|k-1}, \\ \bar{b}_{k|k} &= \bar{b}_{k|k-1} + \bar{K}^b_k\bar{\nu}^b_k, \\ \bar{P}^b_{k|k} &= (I - \bar{K}^b_k S_k)\bar{P}^b_{k|k-1}, \\ V_k &= U_k - \bar{K}^X_k S_k. \end{aligned} \tag{11}$$

If we define the acceleration detector switch variable Δ_k according to

$$\Delta_k = \begin{cases} 1 & \text{if switch is on,} \\ 0 & \text{if switch is off} \end{cases} \tag{12}$$

at time k, then the state estimate after compensation can be written as

$$\begin{aligned} X^{2Stage}_{k|k-1} &= \bar{X}_{k|k-1} + \Delta_k U_k \bar{b}_{k|k-1}, \\ X^{2Stage}_{k|k} &= \bar{X}_{k|k} + \Delta_k V_k \bar{b}_{k|k}, \end{aligned} \tag{13}$$

where $X^{2Stage}_{k|k-1}$ is the estimate based on the measurements Z^{k-1} and $X^{2Stage}_{k|k}$ is the estimate based on Z^k. The corresponding error covariances are

$$\begin{aligned} P^{2Stage,X}_{k|k} &= \bar{P}^X_{k|k} + \Delta_k V_k \bar{P}^b_{k|k} V_k^T, \\ P^{2Stage,X}_{k|k-1} &= \bar{P}^X_{k|k-1} + \Delta_k U_k \bar{P}^b_{k|k-1} U_k^T, \\ P^{2Stage,Xb}_{k|k} &= V_k \bar{P}^b_{k|k}, \\ P^{2Stage,Xb}_{k|k-1} &= U_k \bar{P}^b_{k|k-1}, \\ P^{2Stage,b}_{k|k} &= \bar{P}^b_{k|k}, \\ P^{2Stage,b}_{k|k-1} &= \bar{P}^b_{k|k-1}. \end{aligned} \tag{14}$$

This algorithm is equivalent to the Kalman estimator algorithm shown in section if and only if the condition

$$Q^{Xb}_k = U_k Q^b_{k+1} \tag{15}$$

is satisfied and the modified error covariance of the process noise used in the constant velocity filter given by

$$\bar{Q}^X_{k-1} = Q^X_{k-1} - U_k Q^b_{k-1} U_k^T \tag{16}$$

is positive semidefinite. A proof of this is given in.[3]

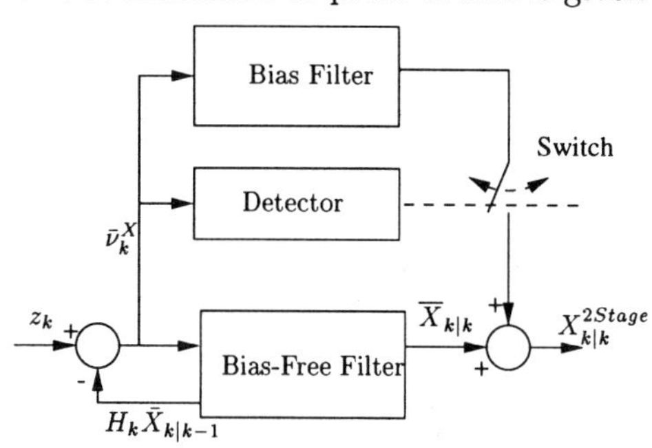

Figure 3 ***The block scheme for the Two Stage estimator***

The Interacting Multiple Model Algorithm

A system with Markovian switching coefficients can be represented as

$$\begin{aligned} X_k &= F_k[M_k]X_{k-1} + G_k[M_k]w_k, \\ z_k &= H_k[M_k]X_k + n_k, \\ E[w_k w_l^T] &= Q_k\delta_{kl}, \\ E[n_k n_l^T] &= R_k\delta_{kl}, \end{aligned} \tag{17}$$

where M_k denotes the model used at time k and in effect during the sampling period ending at time k. We assume that M_k is picked from a set of r models $\{M^i\}_{i=1}^r$. In the case treated in this work there are two models: M^1, which is a low order constant velocity model, and M^2, which is a high order constant acceleration model. (Alternatively, both models can be high order, if the acceleration states of M^1 is set identically to zero.) The details concerning the treated planar tracking problem will be discussed later. It is assumed that the jumping between the models is a Markov process with known mode transition probabilities p^{ij} according to

$$P\{M_k = M^j | M_{k-1} = M^i\} \hat{=} P\{M_k^j | M_{k-1}^i\} \hat{=} p^{ij}, \tag{18}$$

i.e. p^{ij} is the probability that the system will be in mode M^j during the time interval ending at k, given that it was in mode M^i during the interval ending at k-1. These probabilities are assumed to be constant and independent of the system state.

The probability density function (pdf) for X_k, given the measurements up to and including time k, Z^k, can, if we expand it over all models, be written

$$p[X_k|Z^k] = \sum_{j=1}^{r} p[X_k|M_k^j, Z^k] P\{M_k^j|Z^k\}. \tag{19}$$

The last factor in this expression is the probability that model M^j is in effect at time k, and applying Bayest́ rule, it can be written

$$\begin{aligned} \mu_k^j &\hat{=} P\{M_k^j|Z^k\} \\ &= \frac{1}{c} p[z_k|M_k^j, Z^{k-1}] P\{M_k^j|Z^{k-1}\} \\ &= \frac{1}{c} p[z_k|M_k^j, Z^{k-1}] \times \\ &\sum_{i=1}^{r} P\{M_k^j|M_{k-1}^i, Z^{k-1}\} P\{M_{k-1}^i|Z^{k-1}\}, \end{aligned} \tag{20}$$

where c is a normalizing constant such that $\sum_{j=1}^{r} \mu_k^j = 1$. The first factor in the nominator, $p[z_k|M_k^j, Z^{k-1}]$, is called Λ_k^j, and is interpreted as the likelihood function for the measurements at time k, given previous (i e up to k-1, inclusive) measurements, and that model M^j is in effect during the time interval from k-1 to k. Inserting the Markov transition probabilities according to

$$P\{M_k^j|M_{k-1}^j, Z^{k-1}\} = P\{M_k^j|M_{k-1}^j\} = p^{ij}, \tag{21}$$

where the first equality is justified by the fact that the model transitions are independent of the observations, we can write (20) as a recursive updating scheme for the model probabilities:

$$\mu_k^j = \frac{1}{c} \Lambda_k^j \sum_{i=1}^{r} p^{ij} \mu_{k-1}^i. \tag{22}$$

Bayes' rule applied to the model-conditioned pdf for X_k, i.e. $p[X_k|M_k^j, Z^k]$, in (19) gives

$$p[X_k|M_k^j, Z^k] = \frac{p[z_k|X_k, M_k^j, Z^{k-1}] p[X_k|M_k^j, Z^{k-1}]}{p[z_k|M_k^j, Z^{k-1}]}. \tag{23}$$

All factors in this expression are density functions for random variables conditioned on M^j being the model in effect. As the measurement noise n_k in (17) is independent of previous measurements Z^{k-1}, the first factor in the nominator is

$$p[z_k|X_k, M_k^j] = \mathcal{N}(z_k, H_k[M^j]X_k, R_k), \tag{24}$$

where $\mathcal{N}$ is the Gaussian pdf function, i.e.

$$\mathcal{N}(x, m, R) \hat{=} \frac{1}{\sqrt{det(2\pi R)}} e^{-\frac{1}{2}(x-m)^T R^{-1}(x-m)}. \tag{25}$$

The second factor in the nominator of equation (23) is the prior density function for the state at time k. It will now be approximated by a Gaussian in the following way:
First, we can write

$$\begin{aligned} & p[X_k|M_k^j, Z^{k-1}] \\ &= \sum_{i=1}^{r} p[X_k|M_k^j, M_{k-1}^i, Z^{k-1}] \mu_{k-1|k-1}^{i|j}, \end{aligned} \tag{26}$$

where

$$\mu_{k-1|k-1}^{i|j} \hat{=} P\{M_{k-1}^i|M_k^j, Z^{k-1}\}, \tag{27}$$

i.e. the probability that model M^i was in effect at time k-1, given that M^j is in effect at time k, conditioned on old data (Z^{k-1}). Applying Bayes rule again, $\mu_{k-1|k-1}^{i|j}$ can be calculated from old model probabilities and the Markov transition probabilities according to

$$\begin{aligned} & \mu_{k-1|k-1}^{i|j} \\ &= \frac{1}{c_j} P\{M_k^j|M_{k-1}^i, Z^{k-1}\} P\{M_{k-1}^i|Z^{k-1}\} \\ &= \frac{1}{c_j} p^{ij} \mu_{k-1}^i, \end{aligned} \tag{28}$$

with the normalizing factor

$$c_j = \sum_{i=1}^{r} p^{ij} \mu_{k-1}^i. \tag{29}$$

Second, the first factors under the summation sign in (26) can be approximated in the following way:

$$\begin{aligned} & p[X_k|M_k^j, M_{k-1}^i, Z^{k-1}] \\ &\approx p[X_k|M_k^j, \hat{X}_{k-1|k-1}^i, P_{k-1,k-1}^i], \end{aligned} \tag{30}$$

where the conditioning on M_{k-1}^i and Z^{k-1} is replaced by the approximate fact that the state at k-1, X_{k-1}, was Gaussian with mean $\hat{X}_{k-1|k-1}^i$ and covariance $P_{k-1|k-1}^i$. Because of the linearity of the system, X_k will, on this condition, also be Gaussian so that

$$\begin{aligned} & p[X_k|M_k^j, \hat{X}_{k-1|k-1}^i, P_{k-1,k-1}^i] \\ &= \mathcal{N}(X_k, F_k^j \hat{X}_{k-1|k-1}^i, T_k^j), \end{aligned} \tag{31}$$

where $F_k^j \hat{=} F_k[M^j]$, $T_k^j \hat{=} F_k^j P_{k-1|k-1}^i (F_k^j)^T + G_k^j Q_k (G_k^j)^T$, and $G_k^j \hat{=} G_k[M^j]$. We now have a Gaussian sum approximating the X_k density function.

It can, however, be further approximated. Equations (26), (30), and (31) give

$$p[X_k|M_k^j, Z^{k-1}] \approx \sum_{i=1}^{r} \mathcal{N}(X_k, F_k^j \hat{X}_{k-1|k-1}^i, T_k^j) \mu_{k-1|k-1}^{i|j} \approx \mathcal{N}(X_k, F_k^j \sum_{i=1}^{r} \mu_{k-1|k-1}^{i|j} \hat{X}_{k-1|k-1}^i, C_j), \quad (32)$$

where C_j is determined by means of moment matching. The expression for the mean value in this Gaussian distribution shows that $F_k^j \sum_{i=1}^r \mu_{k-1|k-1}^{i|j} \hat{X}_{k-1|k-1}^i$ is the predicted estimate of X_k, given that model M^j is in effect at time k. Let us make the definition

$$E[X_{k-1}|M_k^j, Z^{k-1}] \approx \sum_{i=1}^{r} \mu_{k-1|k-1}^{i|j} \hat{X}_{k-1|k-1}^i \hat{=} \hat{X}_{k-1|k-1}^{0j}. \quad (33)$$

Now, for a linear system with Gaussian noises, where model M^j is in effect in the time interval ending at k, (23) is equivalent to a Kalman filter with the start estimate $E[X_{k-1}|M_k^j, Z^{k-1}] \approx \hat{X}_{k-1|k-1}^{0j}$! The corresponding covariance matrix (not derived here) is

$$P_{k-1|k-1}^{0j} = \sum_{i=1}^{r} \mu_{k-1|k-1}^{i|j} \{P_{k-1|k-1}^i + [\hat{X}_{k-1|k-1}^i - \hat{X}_{k-1|k-1}^{0j}][\hat{X}_{k-1|k-1}^i - \hat{X}_{k-1|k-1}^{0j}]^T\}, \quad (34)$$

where the matrices $P_{k-1|k-1}^i, i = 1, ..., r$, are the covariances from time k-1.

A summary of the IMM algorithm will now be given. Suppose that all model conditioned estimates and covariances, and all the model probabilities at the previous time k-1 are known, i.e. $\hat{X}_{k-1|k-1}^j$, $P_{k-1|k-1}^j$, and μ_{k-1}^j for $j = 1, ...r$. Then the algorithm consists of the following steps:

1. Calculate the mixing probabilities $\mu_{k-1|k-1}^{i|j}$ for $i, j = 1, ..., r$ from equations (28) and (29).
2. Calculate the mixed initial condition $\hat{X}_{k-1|k-1}^{0j}$, $P_{k-1|k-1}^{0j}$ for the Kalman filter matched to model M^j, $j = 1, ..., r$ from equations (33) and (34), respectively.
3. Update model conditioned estimate $\hat{X}_{k|k}^j$ and covariance $P_{k|k}^j$ from the Kalman estimator corresponding to model M^j, $j = 1, ..., r$.
4. Calculate the likelihood function $\Lambda_k^j = \mathcal{N}(z_k, H_k^j F_k^j \hat{X}_{k-1|k-1}^{0j}, H_k^j P_{k|k-1}^j (H_k^j)^T + R_k)$, where $H_k^j \hat{=} H_k[M^j]$, $j = 1, ..., r$.
5. Update the model probability μ_k^j from equation (22), $j = 1, ..., r$.
6. Calculate the optimal estimate $X_{k|k}^{IMM}$ of X_k by summing all the model conditioned estimates, weighted with their respective model probabilities, i.e. $X_{k|k}^{IMM} = \sum_{j=1}^r \mu_k^j \hat{X}_{k|k}^j$.
   Finally, calculate the corresponding covariances $P_{k|k}^{IMM}$ for $j = 1, ...r$ for the total estimation error according to $P_{k|k}^{IMM} = \sum_{j=1}^r \mu_k^j \{P_{k|k}^j + [\hat{X}_{k|k}^j - X_{k|k}^{IMM}][\hat{X}_{k|k}^j - X_{k|k}^{IMM}]^T\}$

The Interacting Acceleration Compensation Algorithm

The IAC algorithm is an application of the IMM algorithm, where the estimates of the two models are calculated using the technique of the Two Stage Kalman estimator (subsection).

There are two models, M^1 and M^2, where M^1 has a state vector that contains only position and velocity information, thus implicitly considering the accelerations equal to zero. It is described by the first of the equations in (4) with $b_k \equiv 0$, together with the first equation in (5), (6), and (7). One step of the Kalman estimator for M^1 is therefore given by the so called "bias free" part of the Two Stage estimator, i.e. the equations (10) under the assumption that $\bar{X}_{k-1|k-1}$ is an optimal estimate at time k-1.

Model M^2 has a state vector that contains only the accelerations and is described by the second equation in (4) and the last equation of (5). The measurements from M^2 are the innovations of M^1, which are defined in the 6:th equation of (10). Its error covariance is $[H_k \bar{P}_{k|k-1}^X H_k^T + R_k]$.

Although no rigorous derivation of the IAC algorithm will be given here (see[6]), its calculation steps are now listed:

Assume that we have the M^1 estimate $\bar{X}_{k-1|k-1}$ and covariance $\bar{P}_{k-1|k-1}^X$ at time k-1, together with the probability μ_{k-1}^1 for this model to be true, i.e. the probability that there is no acceleration. Assume also that we know the M^2 estimate $\bar{b}_{k-1|k-1}$ and covariance $\bar{P}_{k-1|k-1}^b$ at time k-1, together with the probability μ_{k-1}^2 that there is acceleration. Finally, assume that we know V_{k-1}.

Then the IAC algorithm can be written

1. Calculate the mixing probabilities $\mu_{k-1|k-1}^{i|j}$ for $i, j = 1, 2$ from equations (28) and (29).

2. Calculate mixed initial conditions as

$$\bar{X}^0_{k-1|k-1} = \bar{X}_{k-1|k-1}$$
$$\bar{P}^{0X}_{k-1|k-1} = \bar{P}^X_{k-1|k-1} + \{\sum_{i=1}^{2} \mu^i_{k-1}\mu^{1|i}_{k-1|k-1}\mu^{2|i}_{k-1|k-1}\} \times (V_{k-1}\bar{b}_{k-1|k-1})(V_{k-1}\bar{b}_{k-1|k-1})^T$$

to be used in equation (10) for updating of $\bar{X}_{k|k}$ and $\bar{P}^X_{k|k}$, and

$$\bar{b}^0_{k-1|k-1} = \mu^{2|2}_{k-1|k-1}\bar{b}_{k-1|k-1}$$
$$\bar{P}^{0b}_{k-1|k-1} = \mu^{2|2}_{k-1|k-1} \times \{\bar{P}^b_{k-1|k-1} + \mu^{1|2}_{k-1|k-1}\bar{b}_{k-1|k-1}\bar{b}^T_{k-1|k-1}\}$$

to be used in equation (11) for updating of $\bar{b}_{k|k}$ and $\bar{P}^b_{k|k}$.

3. Using the mixed initial conditions calculated in step 2 as starting values, update model conditioned estimates $\bar{X}_{k|k}$ and $\bar{b}_{k|k}$ and the corresponding covariances $\bar{P}^X_{k|k}$ and $\bar{P}^b_{k|k}$ from equations (10) and (11), respectively.

4. Calculate the likelihood functions

$$\Lambda^1_k = \mathcal{N}(z_k, H_k\bar{X}_{k|k-1}, H_k\bar{P}^X_{k|k-1}H^T_k + R_k)$$
$$\Lambda^2_k = \mathcal{N}(\bar{\nu}^X_k, S_k\bar{b}_{k|k-1}, S_k\bar{P}^b_{k|k-1}S^T_k + H_k\bar{P}^X_{k|k-1}H^T_k + R_k).$$

5. Update the model probabilities μ^1_k and μ^2_k from equation (22):

6. Calculate the optimal estimate $b^{IAC}_{k|k}$ of b_k and $X^{IAC}_{k|k}$ of X_k according to

$$b^{IAC}_{k|k} = \mu^2_k\bar{b}_{k|k}$$
$$X^{IAC}_{k|k} = \bar{X}_{k|k} + V_k b^{IAC}_{k|k}.$$

Finally, calculate the corresponding covariances

$$P^{b,IAC}_{k|k} = \mu^2_k\{\bar{P}^b_{k|k} + \mu^1_k\bar{b}_{k|k}(\bar{b}_{k|k})^T\}$$
$$P^{X,IAC}_{k|k} = \bar{P}^X_{k|k} + V_k P^{b,IAC}_{k|k}V^T_k.$$

Application of the algorithms

The different algorithms described above will now be applied to a system consisting of a missile with no dynamics and a target.

Description of missile-target system

The combined missile-target system, upon which the different estimation algorithms now will be tested, is a simple linearized 3:rd order model, consisting of the line-of-sight angle, its rate of change, and the target's lateral acceleration (for a geometric description of the variables, see Figure 1). In these simulations, the rate of change of the line-of-sight angle ($\dot{\sigma}$) is used as measurement, as we normally have access to this variable when using ordinary Proportional Navigation guidance. Also, we presuppose knowledge about the missile-target distance (R) and closing velocity (V_c). The target is modeled as integrated white noise in its lateral acceleration (n_T). The state-space model for the system can thus be written

$$\ddot{\sigma} = \dot{\sigma}2V_c/R + (n_T - n_C)/R, \quad \dot{n_T} = w, \quad z = \dot{\sigma} + v. \tag{35}$$

The measurement error is white noise with variance $E[v^2] = 4 * 10^{-6}rad^2/s^2$. The acceleration noise covariance $q \hat{=} E[w^2]$ is used as a parameter to be optimized for the different algorithms.

q optimization

The value of q is chosen to minimize the RMS error of the n_T estimate at a number of runs, where in each run the target acceleration is changed according to a fixed sequence 0g, 6g, -6g, 0g,..., but at random points in time, the mean interval length being 3-4 seconds. For each algorithm this is done for different q values, and for the given algorithm that q value is chosen that gives the lowest RMS error value. The resulting optimal values are: $q^{Kalman} = 900m^2/s^4, q^{2Stage} = 63300m^2/s^4$, $q^{IMM} = 1250m^2/s^4, q^{IAC} = 5300m^2/s^4$

Estimator comparisons

The estimation of target acceleration is the main object of investigation in this paper. Therefore, typical estimates from simulations of 4 different scenarios will be presented below. These scenarios are characterized in Figure 4, showing the target trajectories, and Table 1, which contains missile and target velocities, the values of the constant target accelerations, the initial relative distance, and the time after which the target manœuvre is started (t_0).

As an example, the simulation plots for scenario 4 are shown in Figure 5 for the different algorithms. The solid lines show true values and the dashed lines estimated ones. Note that the estimators calculate the vertical component of the lateral target acceleration, because of the linearization of the model (35). When the target trajectory becomes less horizontal, the vertical part of its acceleration therefore

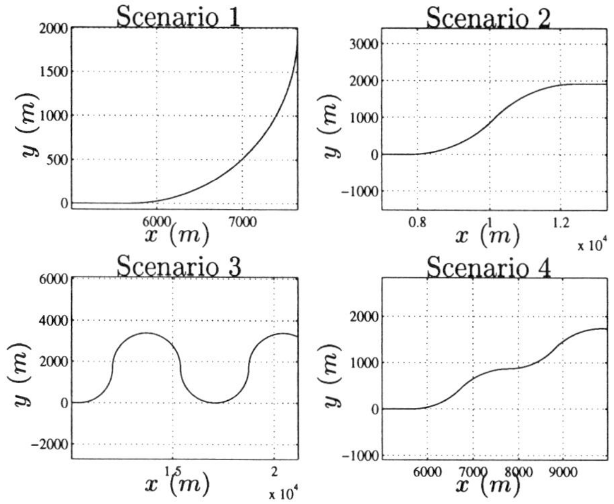

Figure 4 ***The four scenarios of a target's evasive flight path***

SCE-NA-RIO	Target Velocity (m/s)	Missile Vel. (m/s)	Init. Dist. (km)	n_T (g)	t_0 (s)
1	340	800	5	6	2
2	400	770	7	±5	1
3	340	600	10	±7	1
4	340	700	5	±8	1

Table 1 Scenarios characteristics

decreases. It is these time varying vertical components that are shown in the plots, in order to give as meaningful comparisons between true and estimated accelerations as possible.

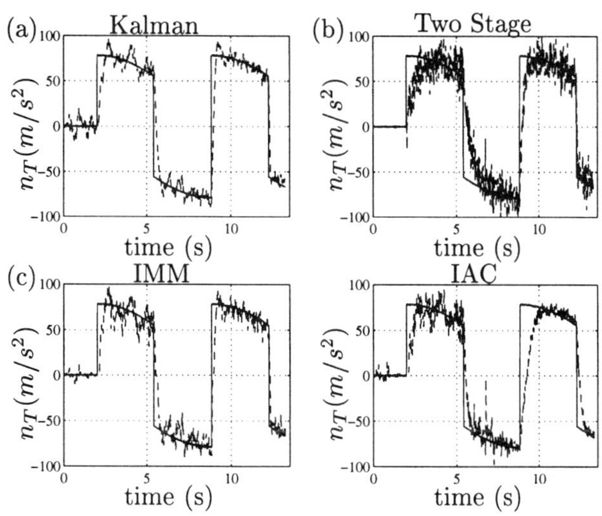

Figure 5 Scenario 4: Target acceleration in y direction.

The main differences between the algorithms are in the estimates' noise frequency contents, that are quite in accordance with the choices of q values (subsection). The Two Stage algorithm, that has more than 10 times higher q value than the others, gives the by far fastest varying estimates, as could be expected, and the slowest varying estimates are given by the Kalman estimator, which has the smallest q value. Other characterstics, such as rise times and overshoot, seem to be approximately the same for all the estimators. It seems, though, that IAC responds significantly slower to changes than the other algorithms. Note, however, that the q value q^{IAC} for IAC still is the optimal value for the estimation error of n_T. If we chose a lower q value, the estimator's response to changes in n_T would be still slower, which would degrade the quality even more. Figure 6 shows accumulated quadratic errors as functions of time for the different scenarios. We see that the ordinary Kalman estimator (solid lines) does very well in all the scenarios. In scenario 4, however, IMM (dash-dotted) shows its strength in following fast changes. Note also that only the Kalman estimator has a constant noise level for all the three acceleration levels. This is explained by the fact that this estimator is the only one that has the same model order all the time. When $n_T = 0$, model M^1 of IMM and IAC gets a high probability value, thus giving an $\hat{n}_T$ value near 0, practically independently of the noisy estimate in model M^2.

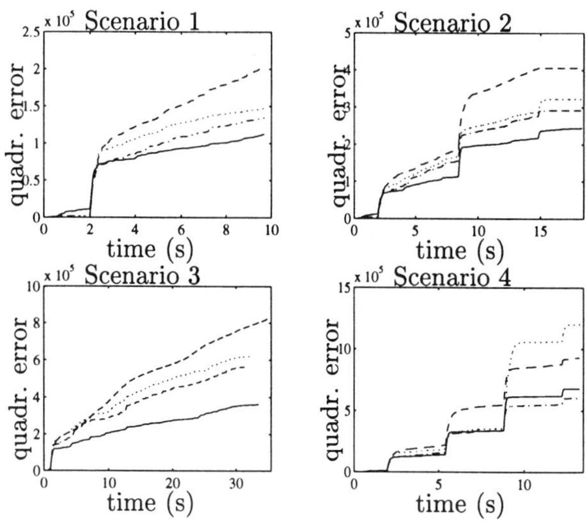

Figure 6 Accumulated quadratic estimation errors as functions of time for the different scenarios and estimation algorithms. Solid line: Kalman, dashed: 2-stage, dash-dotted: IMM, dotted: IAC

Monte Carlo simulations

In order to get quantitative results by which the 4 estimation algorithms can be compared, two series of Monte Carlo simulations were performed, one with a tail chase, with a start distance of 6km, and the other with a head on case, start distance 13km. In both cases, the missile and the target velocities were 800 m/s and 300 m/s, respectively. The target lateral acceleration (n_T) was cyclically changing between 0g, 6g, and -6g during a simulation, starting with 0g. The length of the time interval during which n_T was constant was random with an expected value of 4s.

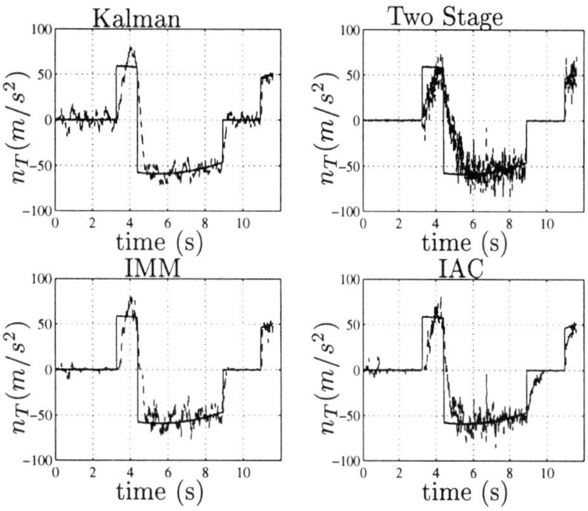

Figure 7

500 runs per Monte Carlo simulation were done. The RMSE values of the estimation errors in n_T and $\dot{\sigma}$ were registered, as well as miss distances. Also, the so called Time Ratio, i.e. the total simulation time divided by the flight time for the missile, together with the number of flops per estimation cycle were calculated as measures of the amount of CPU work that was required from each algorithm. The mean values from all the 500 Monte Carlo simulations in the tail chase case are shown in Table 2. As can be seen from both estimate columns in both Tables, IMM gives sligthly better estimates than the Kalman estimator to the prize of about 3 times higher CPU load. The smaller load of IAC, compared to IMM, does not motivate the use of IAC, since Kalman still does the job better and moreover uses less CPU. The result of this Monte Carlo simulation thus shows that we should prefer the Kalman estimator if we care about the computer load, otherwise the IMM.

Algo-rithm	*RMSE* n_T (m/s^2)	*RMSE* $\dot{\sigma}$ $(10^{-5}$ $rad/s)$	*Miss Dist-ance* (m)	*Time Ratio*	*Flops per Cycle*
Kalman	0.393	2.07	1.41	1.16	680
2-Stage	0.453	3.83	1.33	1.22	629
IMM	0.383	1.89	1.50	2.97	2005
IAC	0.415	2.55	1.50	1.61	722

Table 2 Monte Carlo mean values, tail chase case.

Summary and conclusions

A quality comparison by means of computer simulations has been made between four different algorithms for estimation of target lateral acceleration when applying the Augmented Proportional Navigation guidance law to a homing missile. The algorithms are a Kalman estimator, a Two Stage estimator, an Interactive Multipel Model (IMM) estimator, and an Interacting Acceleration Compensation (IAC) estimator.

A simple two dimensional model was used for the tests. Inspection of four different scenarios with piecewise fixed target acceleration (n_T) shows that the Kalman estimator has the lowest RMS error if n_T does not change too often, but with more frequent changes in n_T, IMM performs slightly better. It is worth noting that IMM requires about 3 times as much CPU time as the Kalman estimator.

Because of the simplicity of the model these results probably require confirmation from future studies with the following characteristics:

- Nonlinear missile-target model in order to estimate the true target lateral acceleration.
- Extended Kalman filters (or equivalent) in the estimators.
- Increased number of models for IMM, e.g. different models for positive and negative target accelerations.

References

[1]Kerakos, R., "Guidance Laws For Homing Missiles Against Maneuvering Target - Study Of Robustness (in Swedish)," FOA-R-96-00201-2.1-SE ISSN 1104-9154, National Defence Research Establishment, Department of Guidance & Control, Materials, and Underwater Sensors, S-172 90 Stockholm, Sweden, Jan 1996.

[2]Friedland, B., "Treatment of bias in recursive filtering." *IEEE Transactions on Automatic Control*, Vol. 14, Aug 1969.

[3]Alouani, A. T., Xia, P., and Blair, W., "A two-stage Kalman estimator for state estimation in the presence of random bias and tracking maneuvering targets." *Proceedings of 30th IEEE Conferende on Decision and Control*, No. 30, IEEE, Dec 1991, pp. 2059 – 2062.

[4]Bar-Shalom, Y., Chang, K., and Blom, H., "Tracking a Maneuvering Target Using Input Estimation Versus the Interacting Multiple Model Algorithm." *IEEE Transactions on Aerospace and Electronic Systems*, Vol. 25, Mar 1989, pp. 296–300.

[5]Blom, H. and Bar-Shalom, Y., "The Interacting Multiple Model Algorithm for Systems with Markovian Switching Coefficients." *IEEE Transactions on Automatic Control*, Vol. 23, Aug 1988, pp. 780–783.

[6]Watson, G. and Blair, W., "Interacting Acceleration Compensation Algorithm For Tracking Maneuvering Targets," *IEEE Transactions on Aerospace and Electronic Systems*, Vol. 31, No. 3, Jul 1995, pp. 1152–1159.

[7]Ignagni, M. B., "An Alternate Derivation and Extension of Friedland's Two-Stage Kalman Estimator," *IEEE Trans. Auto Cont.*, Vol. 26, No. 3, June 1981.

AIAA-97-3566

ADJOINTS REVISITED: A SOFTWARE TOOL TO FACILITATE THEIR APPLICATION

Domenic Bucco*
Defence Science and Technology Organisation,
Salisbury, South Australia, 5108

Abstract

In the area of weapons analysis, the method of adjoints has been used quite cleverly as an efficient alternative to Monte Carlo simulations when the system under study is described by ordinary differential equations with time varying coefficients. Traditionally, when using this approach, an adjoint system is constructed using a set of rules and the solution of the original system is then sought by invoking the impulse response of the new system. However, the construction of the adjoint system has relied largely on the manual conversion of the original system block diagram in accordance with the given set of rules. This not only becomes tedious for a complex system having many feedback paths and time varying coefficients but is prone to human error. In this paper, we describe a MATRIXx toolbox which automates the construction process for any given linear time varying system described in SystemBuild and aids in the rapid adjoint analysis of the system. Examples drawn from the literature on the adjoint analysis of missile guidance systems are included as demonstrations of the toolbox.

1. INTRODUCTION

In the performance analysis of RF guided missiles, the miss distance is one of the key indicators of success or failure of the interceptor's mission. Among the many techniques generally employed to predict miss distance performance is the adjoint simulation method. This technique has been very popular during the preliminary stages of missile design and is valuable in performance projection when the system under study is subject to various noise and error sources. Excellent references on the technique include the book by Zarchan[1] and the report by Bibel[2].

The adjoint technique is based on the system impulse response and can be used to exactly analyse linear, time-varying noise driven systems in one computer run. Therefore, if the missile system to be studied is approximately linear, then the adjoint approach offers considerable reductions in both time and cost over other methods such as the Monte Carlo approach. A particularly attractive feature of the adjoint method is that, in addition to providing the RMS value of the miss distance as a function of the time to go, it also computes an error budget in the same run.

The popularity of the adjoint technique for missile miss distance calculations led to a misconception that the technique is a miss distance computation scheme[3]. Actually, the technique allows the computation of the mean square value of any state of a linear time varying system at any desired observation time in one computer run. As exemplified in references[1,2], in addition to miss distance performance, one can use the adjoint technique to assess the performance of the system with respect to other parameters such as missile acceleration or fin deflection. An extension of the adjoint simulation method to accommodate certain system nonlinearities was proposed by Zarchan[4].

In all these cases, the success of the method lies in the proper construction of the adjoint system. In the past, this system has generally relied on manual manipulation of the original system block diagram followed by computer implementation of the newly formed adjoint block diagram. Guidelines for the block diagram manipulations were originally enumerated by Laning and Battin[5]. The purpose of this paper is to present a new software toolbox, that runs under the MATRIXx® SystemBuild™ environment, for the automatic construction of an adjoint system from a given linear time varying system that has been modelled using SystemBuild. Thus, with this toolbox, the process of adjoint construction is not prone to human error and simulation of the adjoint system can then proceed immediately using the standard SystemBuild simulation tools.

---

*Senior Research Scientist, Member AIAA
MATRIXx is a Registered trademark of Integrated Systems, Inc.
SystemBuild is a trademark of Integrated Systems, Inc.

2. REVIEW OF ADJOINT MATHEMATICS

In this section, we briefly review features of linear systems theory necessary to the development and understanding of the adjoint simulation method.

2.1 Deterministic Inputs

Consider a system described by a linear differential equation of the form

$$\frac{d^n y}{dt^n} + a_{n-1}(t)\frac{d^{n-1} y}{dt^{n-1}} + \cdots a_1(t)\frac{dy}{dt} + a_0(t)y = u \tag{1}$$

where y denotes the system output, u is its deterministic input and the coefficients are functions of time. According to linear systems theory[2], the solution to the above differential equation is given by

$$y(t) = \int_{t_i}^{t} u(\tau)h(t,\tau)d\tau \tag{2}$$

where t_i is the time at which the input is applied while $h(t,\tau)$ denotes the system weighting function. Physically, the function $h(t,\tau)$, represents the response of the system at time t to a unit impulse applied at the input at time τ. With the aid of this weighting function, the system described by equation (1) is conveniently represented in block diagram form as shown in figure 1.

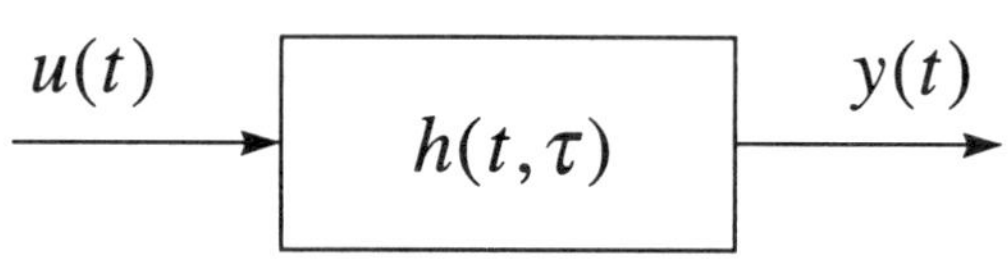

Figure 1. Linear time varying system

While equation (2) gives the formal solution to the problem posed in figure 1, its practical usefulness is limited. The problem here lies in the fact that the variable of integration corresponds to the impulse application time. Thus, in order to generate $h(t,\tau)$ required for the above solution, it becomes necessary to simulate the system response for each impulse application time τ. Once the weighting function is obtained in this manner, it is then necessary to carry out the integration in equation (2) numerically.

Since every linear time-varying system has a corresponding adjoint system, a better and more efficient alternative is to reformulate equation (2) in terms of the weighting function of an associated adjoint system. To this end, utilising the so-called modified adjoint technique[6] has the effect of requiring only one computer simulation run to yield the desired solution.

By definition, the differential equation describing the modified adjoint system corresponding to the above homogeneous system (that is, with $u = 0$) is given by

$$\frac{d^n z}{dt^n} + \frac{d^{n-1}[a_{n-1}(t^*)z]}{dt^{n-1}} + \cdots + \frac{d[a_1(t^*)z]}{dt} + a_0(t^*)z = 0 \tag{3}$$

where z is a function of adjoint time $t^* = t_f - t$ and t_f denotes final time.

Now, if the adjoint system, described by the above differential equation, is subjected to an impulsive input at time τ^*, then the resulting system response, known as the adjoint weighting function, will be denoted by $h^*(t^*,\tau^*)$.

A fundamental relationship exists between a linear time varying system and its corresponding adjoint system[1,2], namely,

$$h^*(t_f - t_1, t_f - t_2) = h(t_2, t_1) \tag{4}$$

where t_1 and t_2 are the impulse application and observation times, respectively, of the original system. Thus, on substitution of equation (4) into equation (2) and, after a change of variable, we get

$$y(t) = \int_{t_f - t}^{t_f - t_i} u(t_f - t^*)h^*(t^*, t_f - t)dt^* \tag{5}$$

It is clear now that the variable of integration appears as the first argument in the adjoint weighting function and thus the solution for the output can be computed with just one computer simulation run. For example, if the input to the system is a step of magnitude K which is applied at time zero, and we seek the solution at final time, then equation (5) becomes

$$y(t_f) = K\int_0^{t_f} h^*(t^*,0)\, dt^* \tag{6}$$

Consequently, the original system response at final time can be readily computed in one simulation run by integrating the weighting function of the corresponding adjoint system. This is represented in block diagram form in figure 2.

This procedure can be extended to include many step input disturbances as well as other types of deterministic inputs such as ramp or sinusoidal inputs.

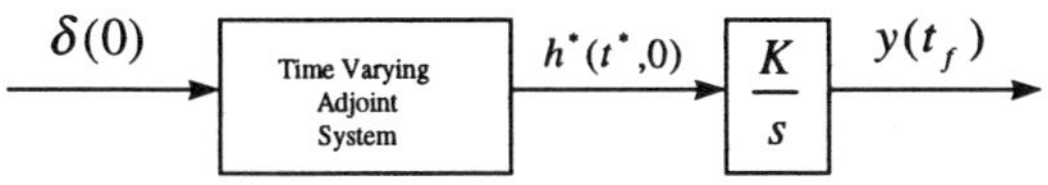

Figure 2. Adjoint solution for step input

In this case, one adjoint computer simulation yields the system response for all final times, along with a detailed error budget showing how each disturbance influenced total system performance.

2.2 Stochastic Inputs

Suppose the linear system is driven by some noise process $n(t)$. The system output is now random in nature and given by

$$y(t) = \int_{t_i}^{t} n(\tau)h(t,\tau)d\tau \tag{7}$$

Squaring both sides of equation (7) and taking expectations yields

$$E\left[y^2(t)\right] = \int_{t_i}^{t} \Phi_0(\tau)h^2(t,\tau)\, d\tau \tag{8}$$

where we have assumed the noise input to be non-stationary white noise with autocorrelation function $R(t,\tau) = \Phi_0(t)\delta(t,\tau)$. For stationary white noise input, the output mean square value reduces to

$$E\left[y^2(t)\right] = \Phi_0 \int_{t_i}^{t} h^2(t,\tau)\, d\tau \tag{9}$$

where the white noise power spectral density Φ_0 is double sided.

Again, the problem with computation of the weighting function can be overcome by considering the corresponding adjoint system. In this case, we have

$$E\left[y^2(t)\right] = \Phi_0 \int_{t_f-t}^{t_f-t_i} [h^*(t^*,t_f-t)]^2\, dt^* \tag{10}$$

If the noise input begins at $t_i = 0$ and the observation time is taken as final time, the mean square response simplifies to

$$E\left[y^2(t_f)\right] \;=\; \Phi_0 \int_0^{t_f} [h^*(t^*,0)]^2\, dt^* \tag{11}$$

Thus the mean square response of the original system due to a white noise input can be determined by computing the weighting function of the adjoint system in one computer simulation run. This is conveniently depicted in block diagram form in figure 3.

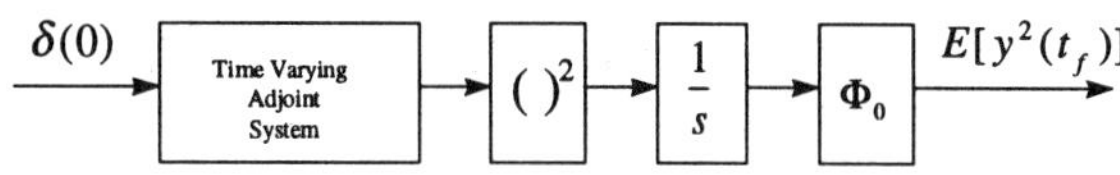

Figure 3. Adjoint solution for white noise input

Because of the superposition principle, the above procedure can be extended to the case of many white noise inputs. In this case, one adjoint simulation run yields an exact statistical analysis of the noise-driven system plus a statistical error budget showing how each white noise error source contributed to the total system response. Coloured noise inputs can be treated in a similar fashion by the use of appropriate shaping filters.

3. ADJOINT CONSTRUCTION RULES

From the above theory, it is clear that, for linear time varying systems, the adjoint simulation method offers tremendous efficiencies over other methods as a performance projection tool. This fact had been realised early on by many weapons analysts and cleverly utilised in the preliminary design stages of many current missile systems.

One of the difficulties with this approach is the construction of the adjoint system from which the desired weighting function must be computed. Although three methods for the construction process have appeared in the literature, by far the most popular method has been that proposed by Laning and Battin[5].

The three methods are:

(1) State space method, which follows directly from the definition of the adjoint system when the system equations are given in state space form[6].

(2) Semi-analytical method, which uses the state space approach to determine the adjoint of the system matrix but then converts this matrix into a system block diagram for computer simulation purposes[3].

(3) Adjoint construction rules method proposed by Laning and Battin[5] to preserve the block diagram configuration of the original system and to highlight the physical relationship between the

adjoint and the original linear time varying system.

MAAT implements the adjoint construction rules method when generating the desired adjoint system from a given SystemBuild block diagram. The adjoint construction rules are:

(a) Convert all system inputs into equivalent systems driven by impulses or white noise processes.
(b) Reverse all signal flow directions.
(c) Switch all system inputs to adjoint outputs and all system outputs to adjoint inputs.
(d) Change all summing points to branch points and all branch points to summing points.
(e) Replace time t in all time varying elements with adjoint time t^* where $t^* = t_f - t$.
(f) Change the output of the original system to an input summing point and apply an impulse signal.
(g) At all the adjoint system stochastic output points, add the adjoint solution sequence as illustrated in figure 3 to yield the mean square value of the system response.

Traditionally, the construction of an adjoint system has been carried out manually by following the above set of rules. However, for very complex systems with many feedback paths, this can be extremely tedious, time consuming and error prone. Furthermore, it is difficult to establish whether any errors have been made in the required manipulations.

Using MAAT as an interface to SystemBuild alleviates the above shortcomings by automating the entire adjoint construction process. Standard SystemBuild tools are then utilised to carry out the desired simulations.

4. THE MAAT SOFTWARE

The MATRIXx Adjoint Analysis Toolbox (MAAT) software has been developed using SystemBuild Access (SBA). The SBA feature of MATRIXx provides an Xmath interface into the SystemBuild catalogue database. This interface is a collection of commands and functions that enable the user to create, modify, query and delete objects in SystemBuild.

Consequently, this feature of MATRIXx is highly suited to the implemention of the adjoint construction rules listed in the last section.

The present version of MAAT accommodates three types of system inputs, namely, step, sinusoidal and stationary white noise inputs. The software allows any of the original system outputs to be selected as the adjoint input and, prior to adjoint construction, performs a check to ensure system linearity. Further details on the functional and user aspects of MAAT can be found in reference[7].

5. ILLUSTRATIVE EXAMPLES

Two missile related examples taken from the literature will be used as illustrations of the software.

5.1 Deterministic Example

Consider the single lag proportional navigation (PN) homing loop example taken from Zarchan[1] and depicted in figure 4. However, instead of a step in target manoeuvre considered by Zarchan, we have chosen here a sinusoidal input. The other deterministic input is an initial angular deviation of the missile from the PN collision triangle known as a heading error (HE). This is modelled as an initial condition on the relative velocity.

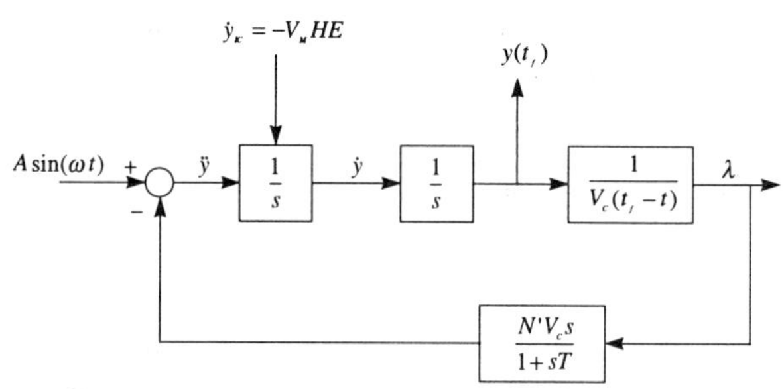

Figure 4. System with two deterministic inputs

The MATRIXx implementation of figure 4 is shown in figure 5. On invoking MAAT from the SystemBuild main menu, and following the prompts for input information concerning the adjoint system, as depicted in figure 6, we obtain the automatically constructed adjoint system block diagram shown in figure 7.

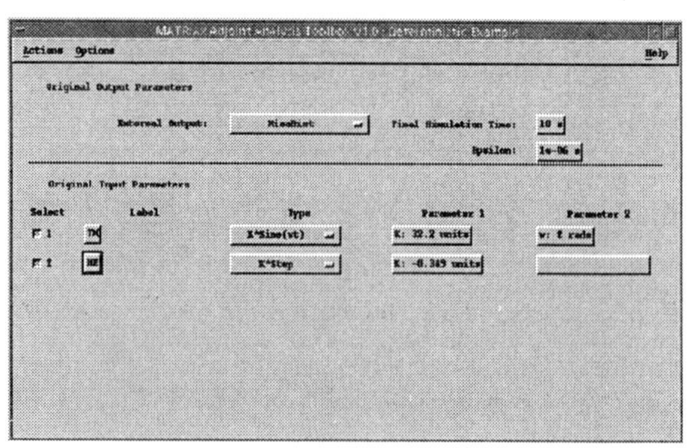

Figure 6. Snapshot of MAAT window for the deterministic example

Once the adjoint system is available in SystemBuild, it is an easy process to determine its weighting function and hence the required forward loop solution according to the theory outlined in section 2 above. In this case, since we are interested in the value of $y(t_f)$ due to each input, we apply an impulse at time zero to the adjoint system at the point where $y(t_f)$ is output in the forward loop system, as shown in figure 4.

In general, there are three methods for simulating the impulse delta function using MATRIXx and SystemBuild. One method, which shall be referred to here as the "brute force" approach, approximates the impulse delta function at time zero as

$$\delta(t) = \begin{cases} \dfrac{2}{\Delta t}, & t = 0 \\ 0, & t \neq 0 \end{cases} \tag{12}$$

where Δt denotes the update interval for the numerical differential equation solver. Note that the area under the impulse function, as defined above, is one. Also, a fixed step integration routine should be used. This approach has been adopted here as it requires no modification of the MAAT generated system.

Another method is to apply a unit initial condition on the first integrator downstream of the impulse entry point[1, 2] and to impose a zero input function in SystemBuild.

Finally, a third method is to impose a unit step input at a point past the first integrator downstream of the impulse entry point. This follows since the integration of an impulse function is a step function. However, this approach requires block diagram manipulation following the construction of the complete adjoint system by MAAT.

Having determined the input function to the adjoint system as outlined above, standard MATRIXx simulation tools are used to generate the desired error budget plots. These plots are displayed in figure 8.

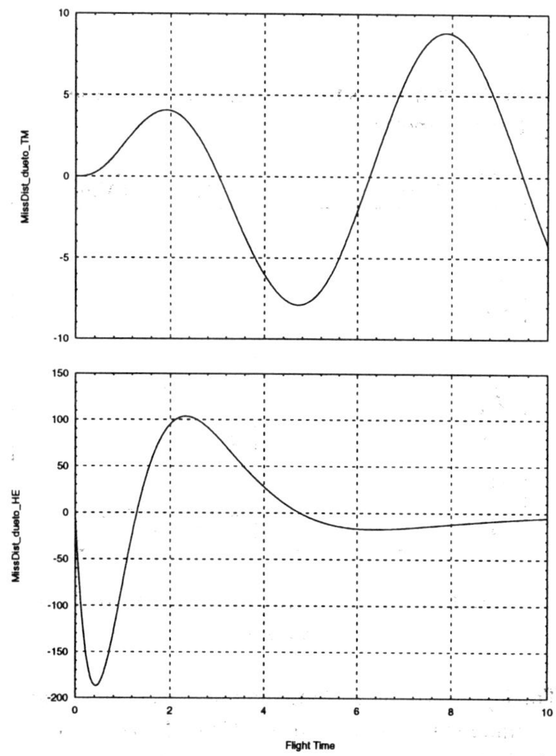

Figure 8. Error budget plot for the deterministic example

5.2 Stochastic Example

This example is taken from Bibel[2] and is depicted in figure 9. The missile is modelled by a derivative/lag seeker, a first order lag noise filter, PN guidance and a third order binomial lag autopilot/airframe. The target manoeuvre is characterised by a 1-g step in target acceleration. In this example, the line of sight angle measurement is corrupted by glint, fading and receiver noises. A semiactive seeker has been assumed and the receiver noise model is as described by Bibel[2]. Thus the system has four inputs; one deterministic input and three stochastic inputs. The stochastic inputs are modelled as white noise sources with given power spectral density.

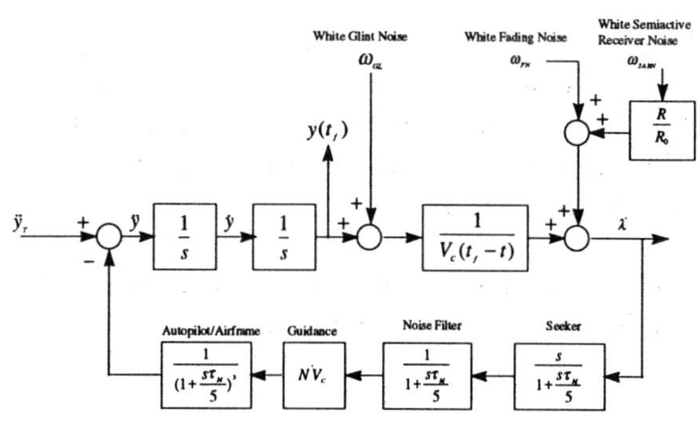

Figure 9. System with three white noise inputs

Table 1 summarises the nominal values for the system parameters.

Parameter	Nominal Value
t_f	10 s
V_C	4000 ft/s
N′	3
τ_M	1 s
R_0	10000 ft
Φ_{GL}	20 ft^2/Hz
Φ_{FN}	$2x10^{-8}$ rad^2/Hz
Φ_{SARN}	$2x10^{-6}$ rad^2/Hz

Table 1. Nominal values used for the stochastic example

The SystemBuild implementation of the stochastic example is given in figure 10. Using MAAT, we obtain the constructed adjoint system as shown in figure 11.

Figure 12 depicts a snapshot of the MAAT window prior to construction of the adjoint system for this example.

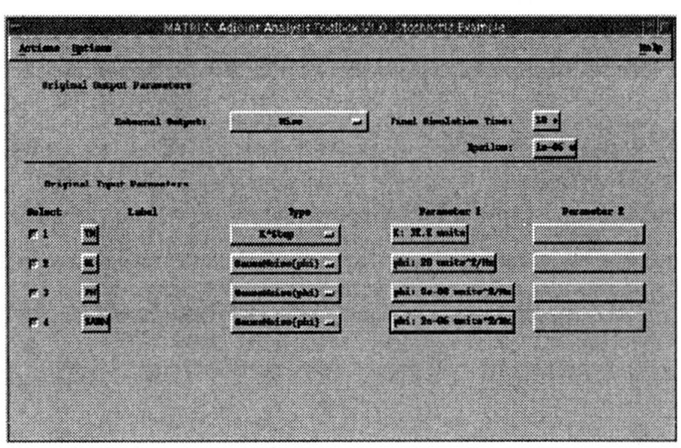

Figure 12. Snapshot of MAAT window for the stochastic example

Following the procedure outlined previously for the numerical integration of the adjoint system, and the theory in section 2 above for the calculation of the RMS response, we obtain the error budget plot reproduced as figure 13.

Computer demonstrations of the above two examples will be presented at the conference.

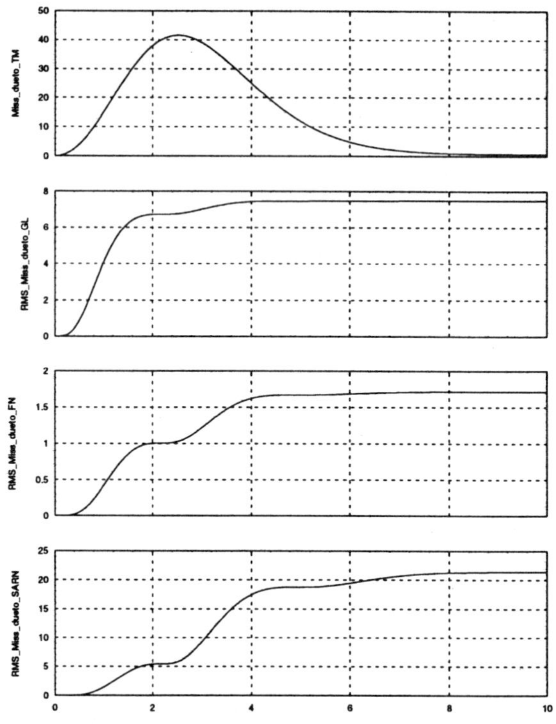

Figure 13. Error budget plot for the stochastic example

6. CONCLUSION

In this paper, we have presented a brief description of a new MATRIXx toolbox for the rapid construction and subsequent analysis of adjoint systems. Two missile examples selected from the literature have been used as illustrations of the toolbox. Currently, the MATRIXx Adjoint Analysis Toolbox (MAAT) software only handles continuous systems. Future enhancements to the toolbox will include a capability to construct and analyse the adjoint block diagram of mixed continuous/discrete systems. It is envisaged that the toolbox can also be extended to accommodate the analysis of nonlinear systems using statistical linearisation techniques in a similar fashion to the approach adopted for SLAM[4].

Continuous SuperBlock	Inputs	Outputs
Deterministic Example	2	1

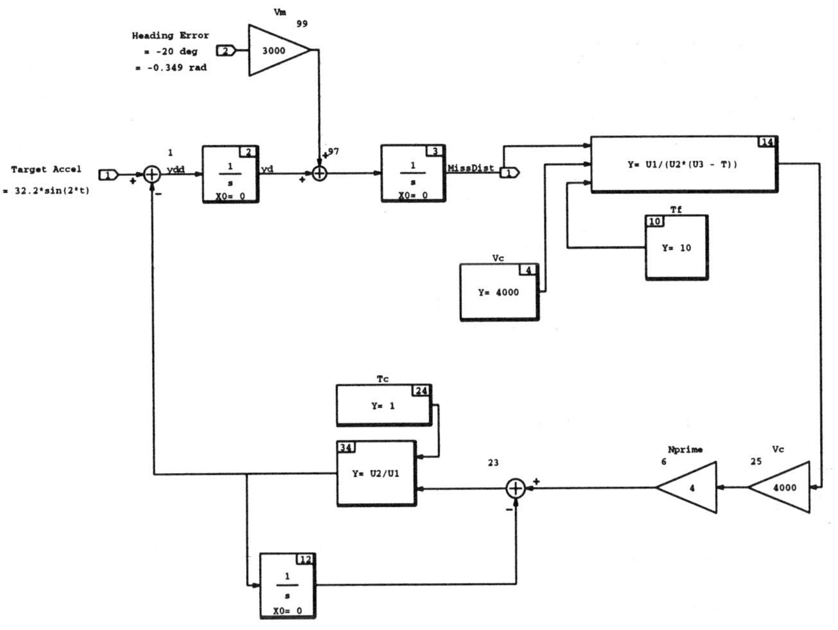

Figure 5. SystemBuild block diagram of deterministic example

Continuous SuperBlock	Inputs	Outputs
MissDist_Adjoint	1	2

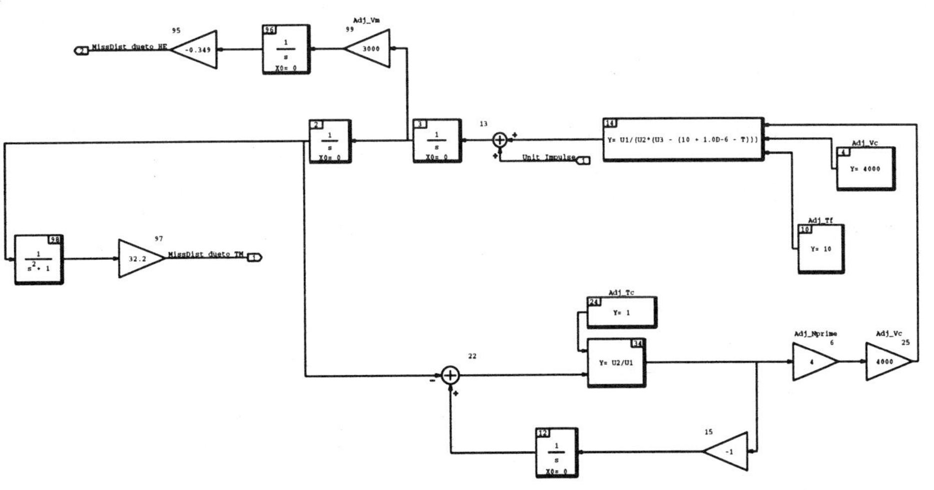

Figure 7. Adjoint block diagram of deterministic example

Continuous SuperBlock	Inputs	Outputs
Stochastic Example	4	1

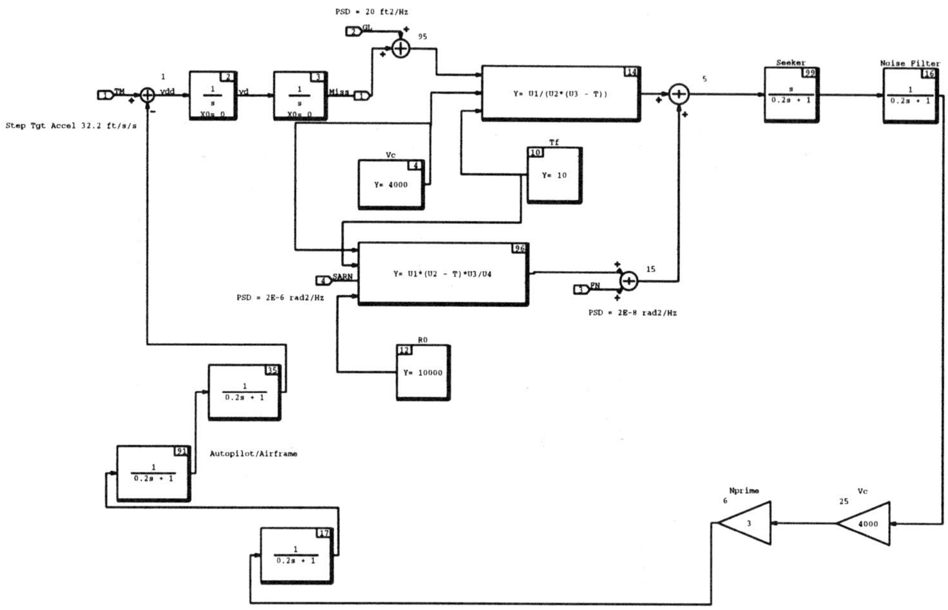

Figure 10. SystemBuild block diagram of stochastic example

Continuous SuperBlock	Inputs	Outputs
Miss_Adjoint	1	4

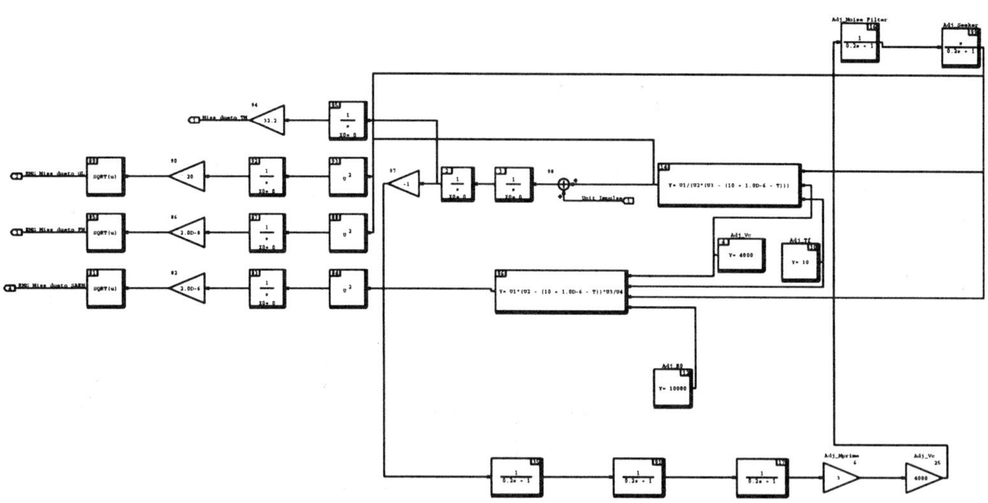

Figure 11. Adjoint block diagram of stochastic example

7. ACKNOWLEDGEMENTS

The author gratefully acknowledges the many fruitful discussions with Dr Paul Zarchan of C. S. Draper Laboratories, Inc. on the subject of adjoint techniques. Discussions with a colleague, Dr David Brown of WSD, DSTO, helped to clarify some of the MATRIXx implementations used in MAAT.

8. REFERENCES

1. P.Zarchan, "Tactical and strategic missile guidance," 2nd ed., AIAA, Inc. 1994.

2. J.E. Bibel, "Application of the adjoint method in homing missile performance analysis," NSWC TR 88-311, May 1990.

3. G.C. Willems, "An improved approach to the adjoint method applied to the statistical analysis of a homing missile," Report No. RE-TR-67-8, August 1967.

4. P. Zarchan, "Complete statistical analysis of nonlinear missile guidance systems - SLAM," AIAA Journal of Guidance and Control, Vol. 2, No. 1, January-February 1979.

5. J.H. Laning & R.H. Battin, "Random processes in automatic control," McGraw-Hill Book Company, New York, NY 1956.

6. P.M. Derusso, R.J. Roy & C.M. Close, "State variables for engineers," John Wiley and Sons, Inc., New York, NY 1967.

7. B. Ramsay & M. Kew, "MATRIXx Adjoint Analysis Toolbox User's Manual," Sydac Software, Pty Ltd, Internal Document No. PE-6004-UM-01, July 1996.

AUTHOR INDEX

AUTHOR INDEX